MW01201343

Biology and Management of White-tailed Deer

Biology and Management of White-tailed Deer

Edited by David G. Hewitt

CRC Press is an imprint of the
Taylor & Francis Group, an **informa** business

CRC Press
Taylor & Francis Group
6000 Broken Sound Parkway NW, Suite 300
Boca Raton, FL 33487-2742

Printed in the United States of America on acid-free paper
Version Date: 20110511

International Standard Book Number: 978-1-4398-0651-7 (Hardback)

Visit the Taylor & Francis Web site at
http://www.taylorandfrancis.com

and the CRC Press Web site at
http://www.crcpress.com

Contents

Part III Management

Part IV The Future

Preface

The white-tailed deer is an American original, having evolved in North America. The white-tailed deer is a survivor. During its existence in the Americas, innumerable other species have come, through migration and speciation, and gone, through extinction. The whitetail is a survivor because it is highly adaptable. Whitetails are found from the boreal forests of central Canada to the equatorial savannahs and forests of Peru, from the Rocky Mountains and Andes to the llanos of Venezuela and swamps of the United States Gulf Coast. The species' adaptability over this broad range is shown not only in the habitats it uses, but in the diets it selects, a twofold difference in body size, variable pelage, and reproduction that varies from highly seasonal to year-round.

Interactions between humans and white-tailed deer began as soon as people migrated to the Americas. Hunting and eating deer were surely important to early humans. Deer also became prominent in the art and religion of the first Americans, and once agriculture arose, the concept of deer causing damage also developed.

Across the species' vast range and during the past 200 years, management of white-tailed deer has encompassed all aspects of wildlife biology. In some places at times, white-tailed deer have become endangered and even extirpated. In other places at times, white-tailed deer have been managed sustainably for decades. And in still other places at times, especially during the past 25 years, white-tailed deer have become too numerous relative to society's desires and even have negatively impacted diverse and naturally functioning ecosystems. In line with other wildlife conservation issues of the day, biologists are beginning to consider the effects of changing climate on deer distribution, particularly in the southwestern United States and along northern edges of the species' range.

To complicate this already complex management milieu, white-tailed deer have both positive and negative values to society and have constituencies promoting these diverse values. Many people have positive feelings toward deer because whitetails are large, graceful, and beautiful animals. Other people value deer as a challenging species to hunt. Some constituencies advocate much smaller deer populations because of damage to vehicles, landscaping, crops, forest products, and ecosystems. Balancing these competing desires of society is the challenge of wildlife biologists and agency administrators.

This book seeks to compile current understanding of white-tailed deer biology and management. This goal was last attempted over 25 years ago by Halls (1984). During the past quarter century, deer populations have flourished, particularly in urban areas. Hunting, the traditional tool used to manage deer population size, is difficult to apply in some instances and is becoming less effective in others. Thus, managing conflicts between people and whitetails is increasingly difficult and complex. Another prominent change in deer management concerns the time and resources landowners and hunters invest in deer management. A growing realization that a quality hunting experience is based on a quality deer herd has changed deer management paradigms from buck-only harvest and high deer densities toward more natural age structures and balancing deer density with the habitat. The intensity of deer management has reached its zenith in the captive deer facilities that occur where allowed by state law.

The prominence of and interest in white-tailed deer has spurred a great deal of research on the species' ecology and management. An exceedingly conservative estimate of research involving white-tailed deer can be obtained by a literature search of JSTOR, which, in June 2010, found 631 papers published before 1985 and 834 papers published since January 1, 1985 that contained the words "white-tailed deer" in the title or abstract. Summarizing this body of research and the vast amount of gray literature on white-tailed deer for this book was accomplished by 35 authors from throughout the white-tail's range, including personnel from state and federal agencies, nongovernmental organizations, private consulting companies, and universities. These experts found time in their busy schedules to review the literature for their respective chapters, decide which issues were essential to cover and which would need to be omitted because of space limitations, and rendered the stacks of research papers and knowledge into the chapters of this book. These chapters can serve as the gateway for white-tailed deer enthusiasts to a greater understanding of this fascinating species.

Also to enhance reader understanding, this book includes a companion CD-ROM disc containing full-color versions of all figures from the book.

As editor of this volume, I would like to thank Fred Bryant, Director of the Caesar Kleberg Wildlife Research Institute (CKWRI), for encouraging such a project and providing the freedom and resources to complete it. I appreciate the support and confidence of Stuart Stedman, a generous benefactor of the Deer Research Program at the CKWRI and the professors' professor when it comes to insight in producing giant white-tailed deer in southern Texas. The invigorating work environment and interactions provided by my colleagues and graduate students at Texas A&M University–Kingsville have been invaluable. I am thankful to Judy Hartke who graciously provided beautiful and original artwork for the book's cover and section breaks. John Sulzycki and Jill Jurgensen with CRC Press, Taylor & Francis Group, were a pleasure to work with; I appreciate greatly their support and productive input as this project developed. My grandfather, W. L. Robinette, through his pioneering research on mule deer and his spending time outdoors with his grandchildren, showed me that it is possible to make a living while pursuing your passion. And finally, my parents, Glenn and Lee Hewitt, my wife Liisa, and my kids Nicole and Matt deserve all the accolades I can bestow for their support in this project and throughout my career.

David G. Hewitt
Professor and Stuart Stedman Chair for White-tailed Deer Research

Editor

David Glenn Hewitt is the Stuart Stedman Chair for White-tailed Deer Research at the Caesar Kleberg Wildlife Research Institute at Texas A&M University–Kingsville. He graduated with highest distinction and honors from Colorado State University in 1987 with a bachelor of science degree in wildlife biology. He earned a master's degree in wildlife biology from Washington State University and then worked for a year as a research associate at the Texas Agriculture Experiment Station in Uvalde, Texas. In 1994, David completed a PhD in wildlife biology at Virginia Tech. David taught wildlife courses at Humboldt State University during the 1994–1995 academic year and then spent a year as a postdoctoral scientist at the Jack Berryman Institute at Utah State University. In 1996, he was hired as an assistant professor at Texas A&M University–Kingsville. His primary research interests are in wildlife nutrition and white-tailed deer ecology and management. He has authored or coauthored more than 60 peer-reviewed scientific publications and three book chapters and has co-edited the book *Wildlife Science: Linking Ecological Theory and Management Applications.*

David served as associate editor of the *Journal of Wildlife Management* during 1997–1998 and *Rangeland Ecology and Management* during 2004–2006. He was recognized as the Outstanding Young Alumnus from the College of Natural Resources at Virginia Polytechnic Institute and State University in 1999 and received the Javelina Alumni Award for Research Excellence in 2004, the Presidential Award for Excellence in Research and Scholarship from the College of Agriculture and Human Sciences, Texas A&M University–Kingsville in 2004, and the Educator of the Year Award from the Texas Chapter of The Wildlife Society in 2010.

Contributors

Kip P. Adams
Quality Deer Management Association
Knoxville, Pennsylvania

Warren Ballard
Department of Natural Resources Management
Texas Tech University
Lubbock, Texas

Jacob L. Bowman
Department of Entomology and Wildlife Ecology
University of Delaware
Newark, Delaware

R. Terry Bowyer
Biological Sciences
Idaho State University
Pocatello, Idaho

Tyler A. Campbell
National Wildlife Research Center Texas
Field Station
Texas A&M University–Kingsville
Kingsville, Texas

Michael R. Conover
Wildland Resources Department
Utah State University
Logan, Utah

Martín Correa-Viana
Universidad Ezequiel Zamora (UNELLEZ)
Guanare, Portuguesa, Venezuela

Steeve D. Côté
Département de Biologie
Université Laval
Québec City, Canada

Steve Demarais
Department of Wildlife, Fisheries, and Aquaculture
Mississippi State University
Starkville, Mississippi

Charles A. DeYoung
Caesar Kleberg Wildlife Research Institute
Texas A&M University–Kingsville
Kingsville, Texas

Randy W. DeYoung
Caesar Kleberg Wildlife Research Institute
Texas A&M University–Kingsville
Kingsville, Texas

Duane R. Diefenbach
Pennsylvania Cooperative Fish and Wildlife
Research Unit
The Pennsylvania State University
University Park, Pennsylvania

Stephen S. Ditchkoff
School of Forestry and Wildlife Sciences
Auburn University
Auburn, Alabama

Timothy E. Fulbright
Caesar Kleberg Wildlife Research Institute
Texas A&M University–Kingsville
Kingsville, Texas

R. Joseph Hamilton
Quality Deer Management Association
Walterboro, South Carolina

Lonnie Hansen
Missouri Department of Conservation
Columbia, Missouri

James R. Heffelfinger
Arizona Game and Fish Department
Tucson, Arizona

David G. Hewitt
Caesar Kleberg Wildlife Research Institute
Texas A&M University–Kingsville
Kingsville, Texas

Scott E. Hygnstrom
School of Natural Resources
University of Nebraska
Lincoln, Nebraska

Harry A. Jacobson
Department of Wildlife, Fisheries and Aquaculture
Mississippi State University
Starkville, Mississippi

Jonathan A. Jenks
Department of Wildlife and Fisheries Sciences
South Dakota State University
Brookings, South Dakota

David M. Leslie, Jr.
Department of Natural Resource Ecology and Management
Oklahoma State University
Stillwater, Oklahoma

Hugo López-Arevalo
Instituto de Ciencias Naturales
Universidad Nacional de Colombia
Bogotá, Colombia

Salvador Mandujano
Red de Biología y Conservación de Vertebrados
Instituto de Ecología, A.C.
Xalapa, Veracruz, Mexico

Ma. Isabel Di Mare
Instituto Internacional de Conservación y Manejo de Vida Silvestre
Heredia, Costa Rica

Karl V. Miller
Warnell School of Forestry and Natural Resources
University of Georgia
Athens, Georgia

Misael Molina
Universidad Nacional Experimental Sur del Lago
Campo Universitario
Santa Bárbara de Zulia, Venezuela

Brian P. Murphy
Quality Deer Management Association
Bogart, Georgia

J. Alfonso Ortega-S.
Caesar Kleberg Wildlife Research Institute
Texas A&M University–Kingsville
Kingsville, Texas

Stephen M. Shea
The St. Joe Timberland Company
Panama City Beach, Florida

Kelley M. Stewart
Department Natural Resources and Environmental Sciences
University of Nevada
Reno, Nevada

Bronson K. Strickland
Department of Wildlife, Fisheries, and Aquaculture
Mississippi State University
Starkville, Mississippi

Kurt C. VerCauteren
USDA APHIS—Wildlife Services
National Wildlife Research Center
Fort Collins, Colorado

Jorge G. Villarreal-González
Consejo Estatal de Flora y Fauna del Estado de Nuevo Leon
Monterrey, Nuevo Leon, Mexico

Peter J. Weisberg
Department Natural Resources and Environmental Sciences
University of Nevada
Reno, Nevada

Part I

The Past

1

Taxonomy, Evolutionary History, and Distribution

James R. Heffelfinger

CONTENTS

Taxonomy

Taxonomy is the process of naming, describing, and organizing plants and animals into categories based on similarities and differences. These categories indicate evolutionary relationships because similar animals generally have common ancestors. This structured system of classification was originally based on morphology, but as molecular analyses became refined, genetic data became very useful in elucidating relationships that remained unresolved. Both morphological and genetic evaluations have shortcomings, and so it is important to use all available information when inferring taxonomic relationships.

In 1758, Swedish physician and botanist Carl von Linnaeus finalized a system for naming plants and animals in a classification scheme he called *Systema Naturae* (Linnaeus, 1758). Linnaeus' naming convention consisted of a hierarchy of seven classifications that grow progressively more specific (Table 1.1). This system is still called *binominal nomenclature* because it uses two names for each species; the first name is the *genus* and the second is the *species*.

TABLE 1.1
Classification of White-tailed Deer within Linnaeus' *Systema Naturae* (Linnaeus, 1758)

Kingdom	Animalia (Animal Kingdom)
Phylum	Chordata (animals with a backbone)
Class	Mammalia (mammals)
Order	Artiodactyla (even-toed hoofed mammals)
Family	Cervidae (the deer family)
Genus	*Odocoileus* (medium-sized North American deer)
Species	*virginianus* (white-tailed deer)

With this naming system each plant and animal in the world has a unique scientific name (*Genus species*) used by scientists in all countries regardless of their primary language. Scientific names are sometimes in Greek, but usually Latin. Because Latin is a "dead" language and not subject to continual change, it is the international language of science. The *subspecies* category was not part of the original classification system, but was added later in an attempt to describe variations (sometimes called *races* or *ecotypes*) within the same species. The subspecies name is added to the end of the two-word scientific name (*Genus species subspecies*). In the field of taxonomy, there are some biologists who are considered "lumpers" and others who are "splitters." Lumpers prefer to focus on the similarities among animals and group several similar forms into one category. Splitters, on the other hand, prefer to separate even slightly different forms into different taxonomic categories.

Taxonomy was a full-time job for many early naturalists. Historically, especially in the eighteenth and nineteenth centuries, taxonomic splitting was very common. There were very few specimens available and the exploration of new lands resulted in new specimens that seemingly had unique characteristics. Many new categories were established based on only a few specimens. In some cases, a small and barely discernible difference resulted in the naming of a new species. Merriam (1918) examined grizzly bear skulls and declared that there were 86 species of grizzlies in North America, with 27 species in Alaska alone.

Many of these early "species" were later reduced to subspecies status or dissolved completely resulting in a series of synonyms for many subspecies. These early efforts at categorizing animals introduced much confusion and bad science into the taxonomic realm when further analysis of many more samples showed characters not to be diagnostic of anything meaningful. Unfortunately, one only needs a single specimen and a short mention in print to establish a scientific name, leaving the scientific community the burden of conducting a comprehensive morphologic, genetic, and ecological study throughout the animal's entire range to properly evaluate its validity.

Deer and Other Ungulates

Deers are members of the Class Mammalia; which contains all warm-blooded animals that produce milk for their young, usually have fur, and possess seven neck vertebrae. Within the Class Mammalia are 26 Orders; two of these are groups of animals that walk on thick, modified toenails called hooves (Wilson and Reeder, 2005). These animals are called ungulates from the Latin word *unguis* meaning "claw" or "toenail" (Gotch, 1995).

Ungulates with an odd number of toes (one or three) on each hoof belong to the Order Perissodactyla (horses, rhinos, tapirs), while the Order Artiodactyla ("artios"= even, "daktulos"= toes) contains all even-toed ungulates like cattle, deer, goats, antelope, and pigs. Within Artiodactyls, there are many different taxonomic families that have been traditionally recognized, but only four occur naturally in North America: Bovidae (sheep, cattle, goats, bison), Antilocapridae (pronghorn antelope), Tayassuidae (collared peccary), and Cervidae (deer, elk, moose) (Nowak, 1999).

The most remarkable taxonomic discovery in recent years is the well-supported placement of whales and dolphins (Cetacea) deep within the Order Artiodactyla. Genetic and fossil evidence (astragalus bones) confirms that Cetacea evolved from an artiodactyl ancestor, similar to hippopotamus (Geisler et al., 2007). Some are now calling the order "Cetartiodactyla."

White-tailed Deer and Other Cervids

The deer family (Cervidae) is comprised of all animals that shed antlers annually, including moose, elk/red deer, caribou/reindeer, white-tailed and mule deer, as well as several Asian, European, and South American species. Only males have antlers, except for caribou/reindeer in which females bear a smaller version of the males'. Cervids, as members of the family are called, walk on the hooves (toenails) of the third and fourth toes, but no longer have the first digit (thumb or big toe). The second and fifth toes have been reduced and assume a nonfunctioning role in what are called dew claws. True cervids have a four-chambered stomach like other ruminants, but lack a gall bladder.

Worldwide, there are 18 genera in the deer family (Groves, 2007) containing about 51 species (Wilson and Reeder, 2005). There is still some question about the distinctness of some of these species and some disagreement about what constitutes species versus subspecies differences. Taxonomic revision is a continual process as additional morphometric and especially genetic information become available. Regardless of the number of species, there is little doubt about the worldwide success of the deer family, which is native to all continents except for Australia and Antarctica. The family ranges from the 4-kg pudu of South America to the 725-kg Alaskan moose and occupies habitats from arctic tundra to tropical forest.

Chinese water deer are included in Cervidae even though this species lacks antlers. Rather than antlers, male Chinese water deer have the large protruding upper canines reminiscent of several extinct cervids. This species has been used as an example of a cervid that retained its primitive form, but there are indications that it may have had antlers in the past and reverted to an antlerless and tusked condition secondarily (Groves, 2007).

Two deer-like ruminants have been associated with the deer family at times, but are not true cervids. The first are the diminutive (2.3 kg) mouse deer and chevrotains. The chevrotain is a small antlerless animal that lives in the tropical forests of Africa and Southeast Asia. These solitary animals have upper canines, no antlers, and represent a very primitive form of ruminant. The musk deer is a 7–15-kg animal resembling the Chinese water deer with enlarged saber-like canines. This too was originally considered a cervid, but some morphologic differences, such as the presence of a gall bladder and abdominal musk gland, have always been enigmatic. Increasingly sophisticated genetic work has recently shown it to be more closely related to the cattle family, Bovidae (Groves, 2007). Separate taxonomic families are now used for both the chevrotain (Tragulidae) and musk deer (Moschidae).

The genus *Odocoileus* includes two species of medium-sized deer whose distribution is centered on North America: the mule deer and the white-tailed deer. This genus name was given by Constantine Rafinesque based on a few teeth given to him by a colleague who had broken them out of a jaw protruding from the wall of a limestone cave in Pennsylvania. Rafinesque returned to the cave in hopes of finding more material to examine, but found none. Although he described the teeth in detail, he was unsuccessful in matching them to any living animal. Despite naming not only the genus *Odocoileus*, but also giving mule deer their scientific name, it is apparent that Rafinesque had not yet seen deer teeth for comparison. He named his mysterious cave teeth with the genus "*Odocoileus*," meaning "*teeth well hollowed*" because of the crescent-shaped infundibula in the chewing surface of the teeth (Rafinesque, 1832). Later taxonomists realized these teeth were from white-tailed deer and designated this publication as the first to assign a valid genus to these North American deer.

The name *Odocoileus* was used widely without question, until Hershkovitz (1948) pointed out that it was wrong. The genus name *Dama* was used for white-tailed deer by Zimmerman (1780), 52 years before Rafinesque pondered the origin of his cave teeth. Under taxonomic rules this makes *Dama* the correct genus name for white-tailed and mule deer in North America. Shortly thereafter, some sources began to use "*Dama virginianus*" for white-tailed deer (Hall and Kelson, 1959). This caused considerable confusion in the taxonomic community since *Dama* was already being used for European fallow deer (*Dama dama*). These New World and Old World deer are not closely related, which necessitated finding a new genus name for fallow deer. The cascading confusion of changing at least two well-established genera was deemed unacceptable and so the International Commission on Zoological Nomenclature used its plenary powers to issue Opinion 581. This decision acknowledged that although *Dama* is technically correct, *Odocoileus* would be recognized as the official genus name for white-tailed and mule deer (International Code of Zoological Nomenclature, 1960).

White-tailed deer were first described in notes made by Thomas Hariot, who was part of Sir Walter Raleigh's attempt to establish a settlement on the North Carolina coast in 1584 (referred to generally as "Virginia" at the time). At the age of 25, Hariot produced the first detailed account of the New World published in English. His *A Briefe and True Report of the New Found Land of Virginia* included a notation that deer were common and that the native inhabitants traded thousands of deer hides annually for firearms (Hariot, 1588). He also offered that deer were of "ordinary" size near the sea coast and larger inland where the habitat was better. Having only red deer and roe deer as reference, Hariot (1588) described this new deer species by writing that "... they differ from ours onely in this, their tailes are longer and the fnags [snags] of their hornes looke backward." The species name *virginianus* reflects the location where it was first described by Hariot (1588) and the species is still commonly referred to as "Virginia Deer." Ironically, Thomas Hariot was describing deer in what is now North Carolina.

Evolutionary History

The earliest hoofed animals with an even number of toes (Artiodactyls) appeared during the early Eocene Epoch, 56–34 million years ago. Rabbit-sized ungulate ancestors, such as *Diacodexis* and other similar forms, were distributed throughout North American and Eurasia (Theodor et al., 2007). *Diacodexis* possessed a unique ankle bone, called the astragalus, which acts as a double pulley providing great flexibility in the hind foot. This bone marks this animal unmistakably as the first known artiodactyl; all even-toed ungulates have this bone. Like the artiodactylids that followed, these animals possessed long limbs for running. Although they walked on all four (rear) or five (front) hoofed toes, they supported most of their weight on the two central toes on each foot. Thus, even at this early stage, one can see the development toward the two-toed ungulates of today and their unused lateral dew claws.

Primitive artiodactyls diversified and increased in abundance throughout the Eocene as the climate became dryer and possibly cooler, allowing ruminants to flourish (Metais and Vislobokova, 2007). By the close of the Eocene there were several groups of primitive ruminants that were precursors to the cattle, pronghorn, camel, and deer families. A small ruminant like the *Archaeomeryx* in Asia gave rise to a subsequent diversification and radiation into forms exemplified by *Eumeryx* in Eurasia. *Eumeryx* already possessed many characteristics seen in today's deer and bovids, such as no upper incisors, incisor-like lower canines, low-crowned molars, and much reduced first lower premolars (Figure 1.1) (Stirton, 1944).

Evolutionary development of these ruminants continued through the Oligocene Epoch (34–24 million years ago) with the appearance of increasingly complex forms such as the Moschidae family.

FIGURE 1.1 The Eurasian *Eumeryx* represents the transition between very primitive artiodactyls and the more graceful forms that eventually evolved into deer. (Illustration by R. Babb from Heffelfinger, J. R. 2006. Deer of Southwest. College Station: Texas A&M University Press.)

Moschids, like the North American *Blastomeryx* and the Eurasian *Dremotherium*, are primitive deer-like mammals with no antlers, but exaggerated tusk-like canines (Prothero, 2007). A Eurasian form of these sabre-toothed deer, such as *Dremotherium*, is the most probable ancestor to all cervids. Moschids disappeared by the end of the Miocene, with the exception of one genus, the present-day musk deer. Musk deer of eastern Asia are not actually cervids, but represent direct descendants of these primitive Moschid forms. Canine tusks are not normally associated with Cervidae today, but the Chinese water deer provides an example of a true deer that lacks antlers and possesses large canines remarkably similar to fossil deer.

Antlered Deer Appear

Despite the abundance and diversity of ruminants in North America, none of these forms gave rise to North American deer. Eurasian deer-like animals such as the tusked and antlerless *Dremotherium* are recognized as the types of primitive ruminants that eventually gave rise to all cervids. Later Miocene forms in the Family Lagomerycidae offer important clues, and a probable "missing link," to the early development of deer (Gentry, 1994). Many of the Lagomerycids, such as *Procervulus*, not only possessed large canine tusks, but also forked or palmated antlers that were shed, but probably not every year (Figure 1.2). Thus, with the occasional casting of antlers, *Procervulus* and related forms were positioned precisely at the genesis of the deer family.

The earliest true deer (Cervidae) appeared in Eurasia in the middle of the Miocene (Scott and Janis, 1987). One of these ancestral deer had small antlers that normally formed a single fork (Figure 1.3) (*Dicrocerus*). Another Miocene deer, *Stephanocemas*, had tusk-like canines and antlers that formed a bowl-shaped palm (Figure 1.3) (Gentry, 1994). The antlers of these early deer were shed annually from long antler bases much like the present-day muntjac of Asia (Figure 1.4).

FIGURE 1.2 The increasingly deer-like forms such as *Dremotherium* (left) and *Procervulus* (right) appear in the Eurasian fossil record during the Miocene. *Dremotherium* had no antlers, but the males had large saber-like canines. The horns/antlers of *Procervulus* may have been shed at irregular intervals. (Illustration by R. Babb from Heffelfinger, J. R. 2006. Deer of Southwest. College Station: Texas A&M University Press.)

FIGURE 1.3 *Stephanocemas* (left) and *Dicrocerus* (right) are the first animals to shed their antlers on a regular and recurrent basis. All of today's true (antlered) deer arose from early deer such as these. (Illustration by R. Babb from Heffelfinger, J. R. 2006. Deer of Southwest. College Station: Texas A&M University Press.)

FIGURE 1.4 The musk deer (left), muntjac (center), and roe deer (right) provide an illustrative example of the evolutionary progression from oral to cranial weaponry in Cervidae. (Photo by P. Myers. With permission.)

With the evolutionary development of increasingly elaborate antlers, the occurrence of tusk-like canines was much reduced in the deer family (Figure 1.4) (Eisenberg, 1987). The antlerless Chinese water deer have prominent canines, while other antlered deer have lost their canines entirely or they are very much reduced (as in elk). The muntjac and tufted deer of Asia occupy an intermediate position with small antlers and small canines. It has been hypothesized that the reduction of large canines occurred because the development of elaborate antlers supplanted the need for these teeth as weapons or sexual display organs. An alternate theory surmises that the reduction of these enormous canines was caused simply by a change to a browse-dominated diet and the need to grind food with side-to-side jaw movements.

The First North American Deer

There is no record of true (antlered) deer in North America until the close of the Miocene (Webb, 2000), when an ancestral stock immigrated from Eurasia by way of the Bering Land Bridge seven to five million years ago. These new immigrants found the North American continent in ecological turmoil and experiencing a high level of extinction in many of the large assemblages of native ruminants. The earliest fossils of deer in North America are represented by *Eocoileus gentryorum* found in five million-year-old deposits in Florida (Figure 1.5) (Webb, 2000). The antlers of *Eocoileus* rose straight up from the frontal bones and were very similar to present-day roe deer (Figure 1.6). This North America form was probably not far removed from the ancestor of all roe deer, Chinese water deer, and deer of the New World (Webb, 2000; Pitra et al., 2004).

Another very early North American cervid is found in similarly aged deposits (4.8–3.4 million years ago) in Nebraska. *Bretizia pseudalces* was very similar to *Odocoileus*, but differed in that the antlers were strongly palmated and spread laterally much like a moose (pseudalces = "fake moose"). This deer was found on the west coast and Great Plains, but disappears from the fossil record at the end of the Pleistocene (Fry and Gustafson, 1974; Gustafson, 1985; Gunnell and Floral, 1994).

Navahoceros was a short-legged, stocky deer found in the Rocky Mountains starting about three million years ago. This animal was built for life in the mountains with a body form similar to other mountain ungulates (Kurten and Anderson, 1980). This deer sported three-tined antlers not unlike white-tailed deer, but had cranial characteristics similar to *Rangifer* (Webb, 1992). By 11,500 years before present, *Navahoceros* falls out of the fossil record (Kurten and Anderson, 1980), although it may live on today in South American descendants (Webb, 2000).

FIGURE 1.5 Sometime prior to five million years ago, an early deer ancestor crossed the Bering Land Bridge and thus true deer were introduced into North America. *Eocoileus gentryrorum* fossils from that time unearthed in Florida represents the earliest known cervid in North America. (Illustration by R. Babb.)

FIGURE 1.6 The fossil antlers of *Eocoileus* (UF90400, left) bear a remarkable resemblance to extant European roe deer (right), illustrating the close relationship between the latter and all New World deer. (Photo by J. Heffelfinger.)

The oldest fossils identified as *Odocoileus* are found in Kansas and dated to the middle Pliocene (about four million years ago; Oelrich, 1953). These fossils consisted of large molars that were unmistakably *Odocoileus*, but appeared shorter and wider than the teeth of present day white-tailed and mule deer. Oelrich (1953) named this deer *Odocoileus brachyodontus*, but more recent evaluation of material attributed to this name has shown there are no characteristics to reliably differentiate these molars from those of living *Odocoileus* (Wheatley and Ruez, 2006).

New World Deer Find a New Continent

In the late Pliocene Epoch (2.7 million years ago), a shifting of tectonic plates formed the Isthmus of Panama, which joined South America to North America and facilitated the Great American Biotic Interchange (Webb, 2006). Deer did not occur in South America prior to establishment of this land bridge, but immigrating cervids found abundant resources to exploit and thus began a remarkable evolutionary radiation. From this immigration event, there was an explosive diversification of deer resulting in six genera and at least 13 species of deer currently occupying South America (Webb, 2000; Gilbert et al., 2006).

The extant forms such as pudu, pampas deer, marsh deer, brocket deer, and taruka/huemul represent an amazing diversity of morphology and ecological adaptation resulting from this radiation. White-tailed deer are sometimes cited as the source of all South American deer, but it is more likely they all derived from one or more North American ancestors. The current view is that one or more ancestral deer may have entered South America in the Pliocene.

Morphological similarities and recent genetic analysis indicate that at least two ancestral forms crossed the isthmus into South America. The first clade consists of the diminutive pudu, taruka, and huemul occupying the Andes. Taruka and huemul share morphological affinities to *Navahoceros* in that they have keeled basioccipital bone on the underside of the skull and relatively short legs indicative of mountain dwellers. The fossil record is nearly silent for pudu, but morphology and genetic analysis of both nuclear DNA (nDNA) and mitochondrial DNA (mtDNA) show they are closely aligned with the taruka and huemul meaning they probably share a common immigrating ancestor (Webb, 2000; Gilbert et al., 2006).

Another immigrant closely resembling a primitive *Odocoileus* probably represents a separate cervid to enter South America in the mid-Pliocene. This source stock eventually diversified into marsh deer, pampas deer, and red brocket deer (Webb, 2000). The rest of the brocket species (gray brockets) are genetically very different (Gilbert et al., 2006; Duarte et al., 2008), which may indicate a Central American origin and several immigrations. The fact that white-tailed deer presently occur in South America indicates they arrived in their present form after the great cervid diversification, or that they immigrated with a more *Eocoileus*-like ancestor that evolved into marsh and pampas deer.

Duarte et al. (2008) proposed a scenario with eight different forms of cervid immigrants in the Pliocene. This analysis was based on mtDNA and sometimes mtDNA lineages do not correspond to true evolutionary lineages (Cronin, 1993). Rapid radiation and diversification in the newly colonized South American continent seems a more plausible scenario given the near lack of large ungulates and associated predators at the time. Additional analysis with nDNA markers will help tell the story of the origin of New World deer.

The Rise of White-tailed Deer

During the late Pliocene, there were at least three recognizable types of deer in North America: *Navahoceros* in the mountains, *Bretzia* in the West, and the widespread, increasingly common *Odocoileus*. Only *Odocoileus* made it out of the Pleistocene alive and emerged as the only medium-sized cervid in North America. Remarkably, fossils indistinguishable from living *Odocoileus* have been found dating back nearly four million years and the subsequent fossil record includes locations throughout most of this species' current geographic range in North America (Kurten and Anderson, 1980). *Odocoileus* had clearly found the environmentally tumultuous Pleistocene and Holocene environment (last two million years) conducive to its incredible success as a species.

White-tailed deer and mule deer are closely related and the product of the same Pliocene *Odocoileus* stock. These two forms probably started to differentiate during the early to mid-Pleistocene. Several recurring glaciation events occurred during the last two million years and produced a complicated and poorly understood pattern of geographic barriers in the northern latitudes of North America. The most likely cause for speciation of mule deer and white-tailed deer is physical isolation due to climate-induced habitat changes. These cyclical changes in the distribution of forests, shrublands, and grasslands occurred through the many glacial/interglacial changes throughout the Pleistocene. Any one of these glacial cycles lasting 10,000–100,000 years could be enough to differentiate the two species. The last Pleistocene glaciation (11,000–20,000 years before present) may have facilitated the subspecific division between mule deer and black-tailed deer, as the latter was isolated in coastal refugia of the Pacific Northwest (Latch et al., 2009).

The current geographic distribution of white-tailed deer overlaps that of mule deer in many places. This overlap represents a secondary contact between the two species after their post-Pleistocene range expansion. Where they are sympatric, hybridization has been documented (Heffelfinger, 2000), but occurs at a low rate and does not represent an ecological problem.

Unravelling the complete story of deer evolution throughout the late Pliocene/early Pleistocene (4 million years to 600,000 years ago), has been hampered by repeated glaciations that scoured the landscape for thousands of years, destroying most evidence of early North American deer evolution (Geist, 1998). Additionally, white-tailed deer and mule deer are difficult to distinguish without lacrymal fossa, certain leg bones (Jacobson, 2004), or antlers from mature males. Most materials designated as one or the other species have been done so based solely on geography or size. Because Pliocene and Pleistocene geographic ranges are not clearly known and body size is variable, most of these species assignments are suspect. A reevaluation is needed of New World deer fossils, particularly *Odocoileus* using all the information currently available (Jacobson, 2004).

With the lack of a strong fossil record, science has turned to genetic analysis to investigate the evolutionary relationships of white-tailed and mule deer. By making assumptions about the rate that a genome accumulates mutations, geneticists can estimate the time since two organisms diverged from a common ancestor (Avise, 1994). These molecular clocks are notoriously sensitive to the assumptions that are used and should be viewed with healthy skepticism. However, molecular clocks provide another way to estimate the evolutionary history of an animal. Various attempts to estimate the time of divergence between white-tailed and mule deer have resulted in a range of 750,000–3.7 million years (Baccus et al., 1983; Carr and Hughes, 1993; Douzery and Randi, 1997).

Distribution

Geographic Variation in White-tailed Deer

White-tailed deer have emerged as the most abundant and widespread of all the New World deer species. With this success and vast geographic distribution, we see phenotypic and genotypic variations throughout their range. Some of these differences are due to genetic changes brought about by isolation of some populations and others are simply examples of the phenotypic plasticity as populations in local areas adapt to habitat, forage, or climatic conditions (Strickland and Demarais, 2000, 2008).

Phenotypic variation of white-tailed deer has been expressed in discrete taxonomic subspecies, complete with multicolored maps showing well-defined distributions. Currently there are 38 subspecies of white-tailed deer commonly recognized, but many more have been described (Table 1.2). Most of these descriptions were based on only a few specimens and have not been evaluated sufficiently to determine whether they are valid. Overlap in characteristics among most deer subspecies is so great that no list of differences can be written to allow biologists to differentiate subspecies. Most authorities simply keep using these names because there is no information available to support or reject their subspecies designation. In the early years of the field of natural history, it took only a single specimen and a polygon on a map to create a subspecies. Science is a process of disproving theories and so we are stuck with our multicolored maps and nebulous descriptions until someone is able to conduct a range-wide comprehensive evaluation of subspecies using genetic, morphological, and perhaps ecological characteristics.

The current geographic distribution and genetic integrity of white-tailed deer has not escaped human influence. In addition to internal and external natural forces shaping the whitetail phenotype, humans have influenced local phenotypes by moving tens of thousands of deer back and forth across the United States. Reviewing the history of white-tailed deer translocation illustrates the folly of our current topological view of subspecies. The state of Virginia was restocked with deer from 11 states (Marchington et al., 1995). Florida, Georgia, Arkansas, Kentucky, Louisiana, Mississippi, North Carolina, Tennessee, Virginia, and West Virginia each received hundreds of deer from Wisconsin. Mississippi received at least 72 deer from Mexico (Handley, 1952). In addition to Wisconsin deer, deer from Texas were released in Florida (437), Louisiana (>167), and Georgia (1058) to restock deer habitat (Marchington et al., 1995). White-tailed deer restoration programs throughout the continent are rightly hailed as a conservation success story, but they have also gone a long way to befuddle an already poor characterization of geographic differentiation (Leberg and Ellsworth, 1999).

The whole concept of subspecies has been under attack for some time (Wilson and Brown, 1953). Subspecies' boundaries, when taken literally (as they usually are), frequently create a nonsensical pattern of geographic differentiation. For example, it is doubtful if Minnesotans find value in differentiating between the three subspecies of white-tailed deer in their state. However, in some cases, recognizing these animals as a different "race" or "ecotype" may be helpful in addressing unique conservation problems facing animals in that area. Recognizing ecotypes can aid the management of those populations by encouraging management actions that are critical, but may not be needed elsewhere. However, when the difference between geographic variants is not well documented and they are given the official status of a scientific name (subspecies), disproportionate legal repercussions may occur (O'Brien and Mayr, 1991; Geist, 1992; Cronin, 1997). For example, the Guatemalan white-tailed deer (*O. v. mayensis*) was placed on Appendix III of the Convention on International Trade in Endangered Species (CITES) in 1981. The CITES "Information Sheet" containing the reasons for listing, distribution, and description of the taxon has never been submitted. This subspecies has enjoyed international legal protection for nearly 30 years and yet it has never been described and appears nowhere in the scientific literature. This CITES designation on a nonexistant subspecies could have serious legal repercussions for anyone hunting or exporting white-tailed deer from Central America.

Increasingly, sophisticated genetic analyses are being employed to identify genetic differences among populations of a species throughout its range, and have yielded useful taxonomic guidance. In most genetic studies of geographical variation, the genetic patterns do not match previously defined subspecies (DeYoung et al., 2003; Latch et al., 2009).

TABLE 1.2
Subspecies Described for White-tailed Deer (*Odocoileus virginianus*)

Subspecies	Common Name of Subspecies	Subspecific Synonyms	Distribution	Source of Designation
O. v. acapulcensis	Acapulco		Pacific coast of southwest Mexico from states of Colima to Oaxaca	Caton (1877), Méndez (1984)
O. v. borealis	Northern woodland		Northeastern United States and southeastern Canada	Miller (1900)
O. v. cariacou	Brazilian	*sylvaticus, campestris, spinosus, suacuapara*	French Guiana and northeastern Brazil, South America	Boddaert (1784), Brokx (1972, 1984)
O. v. carminis	Carmen Mountains		Sierra del Carmen, northern Coahuila, Mexico; Chisos and surrounding mountains in West Texas	Goldman and Kellogg (1940)
O. v. chiriquensis	Chiriquí		Southern tip of Costa Rica and Panama, Central America	J. A. Allen (1910), Méndez (1984)
O. v. clavium	Florida Key		Florida Keys, Florida	Barbour and Allen (1922)
O. v. couesi	Coues	*battyi, baileyi*	Arizona, New Mexico, Sonora, Chihuahua, Durango	Coues and Yarrow (1875), Smith (1991)
O. v. curassavicus	Curaçao		Curaçao Island, Venezuelan Coast, South America	Hummelinck (1940), Brokx (1972, 1984)
O. v. dacotensis	Dakota		North Central United States and South Central Canada	Goldman and Kellogg (1940)
O. v. goudotii	Northern Andes	*columbicus, lasiotis*	Andes Mountains in Colombia and Mérida Andean highlands in western Venezuela, South America	Gay and Gervais (1846), Brokx (1972, 1984)
O. v. gymnotis	Savannah	*savannarum, wiegmanni, tumatumari*	Savannahs of Venezuela, Guyana and Surinam, eastern llanos of Colombia, South America	Wiegmann (1833), Brokx (1972, 1984)
O. v. hiltonensis	Hilton Head Island		Hilton Head Island, South Carolina Coast	Goldman and Kellogg (1940)
O. v. leucurus	Columbia River		Mouth of the Columbia River, Washington and Oregon	Douglas (1829)
O. v. macrourus	Kansas	*louisianae*	Iowa south to Louisiana	Rafinesque (1817)
O. v. margaritae	Margarita Island		Isla de Margarita, Venezuelan Coast, South America	Osgood (1910), Brokx (1972, 1984)
O. v. mcilhennyi	Avery Island		Coastal areas of Louisiana and East Texas	Miller (1928)
O. v. mexicanus	Mexican Tableland	*lichtensteini*	Southcentral Mexico, encompassing Queretaro, Hidalgo, México, Morelos, Tlaxcala, Puebla and portions of surrounding states	Gmelin (1788), Méndez (1984)
O. v. miquihuanensis	Miquihuana		Sierra Madre Oriental in Coahuila, Nuevo Leon, San Luis Potosí and portions of surrounding Mexican states	Goldman and Kellogg (1940), Méndez (1984)
O. v. nelsoni	Chiapas		Highlands from Chiapas, Mexico to Nicaragua, Central America	Merriam (1898), Méndez (1984)
O. v. nemoralis	Nicaragua	*truei, costaricensis, mayensis*	Lowland areas throughout Central America in Belize, Guatemala, Honduras, San Salvador, Nicaragua and Costa Rica	Hamilton-Smith (1827), Méndez (1984)

continued

TABLE 1.2 (continued)
Subspecies Described for White-tailed Deer (*Odocoileus virginianus*)

Subspecies	Common Name of Subspecies	Subspecific Synonyms	Distribution	Source of Designation
O. v. nigribarbis	Blackbeard Island		Blackbeard Island, Georgia Coast	Goldman and Kellogg (1940)
O. v. oaxacensis	Oaxaca		Central Oaxaca, Mexico	Goldman and Kellogg (1940), Méndez (1984)
O. v. ochrourus	Northwest		Southwestern Canada and northwestern United States	Bailey (1932)
O. v. osceola	Florida Coastal	*fraterculus*	Northwest Florida	Bangs (1896)
O. v. peruvianus	Peruvian	*brachyceros, peruviana, philippii, peruanus*	Slopes of the Andes in Peru and possibly Bolivia, South America	Gray (1874), Brokx (1972, 1984)
O. v. rothschildi	Coiba Island		Coiba Island, Panama Coast, Central America	Thomas (1902), Méndez (1984)
O. v. seminolus	Florida		Florida	Goldman and Kellogg (1940)
O. v. sinaloae	Sinaloa		Mexican states of Sinaloa, Nayarit, Jalisco, parts of Michoacan and Guanajuato	Allen (1903), Méndez (1984)
O. v. taurinsulae	Bull Island		Bull Island, South Carolina Coast	Goldman and Kellogg (1940)
O. v. texanus	Texas		Texas north through the Central Plains	Mearns (1898)
O. v. thomasi	Mexican Lowland		Eastern Veracruz and Oaxaca, Villahermosa, portions of Chiapas and Campeche, Mexico	Merriam (1898), Méndez (1984)
O. v. toltecus	Rain Forest		Northern Oaxaca and southern Veracruz, Mexico	Saussure (1860)
O. v. tropicalis	Tumbesian	*columbicus, punensis*	West of the Andes along the Pacific coast of Ecuador and northern Peru, South America	Cabrera (1918), Brokx (1972, 1984)
O. v. ustus	Ecuador	*consul, abeli, gracilis antonii, aequatorialis*	The Andes of Ecuador and maybe southern Colombia, South America	Trouessart (1910), Brokx (1972, 1984)
O. v. venatorius	Hunting Island		Hunting Island, South Carolina Coast	Goldman and Kellogg (1940)
O. v. veraecrucis	Northern Veracruz		Southeastern Tamaulipas, Mexico and most of Veracruz, Central America	Goldman and Kellogg (1940), Méndez (1984)
O. v. virginianus	Virginia	*wisconsinensis*	Southeastern United States	Zimmerman (1780)
O. v. yucatanensis	Yucatan		Throughout Yucatan and the northern half of Campeche and Quintana Roo, Mexico	Hays (1874), Méndez (1984)

Regardless of whether one uses genetic or morphological characteristics to differentiate subspecies, the difficulty lies in delineating categories out of what is usually a continuum of differences. Animals at each end of the continuum may look different, but there is no place in between that offers a clear division. This is the crux of the subspecies dilemma. In some cases, a characteristic will change more abruptly at a point along the continuum, but the question remains, "How different is different enough?" to consider using two names (or 38!) to designate this difference. Unfortunately, there is no answer that is universally applied.

Heffelfinger (2005) and Villarreal et al. (2009) argued that creating categories for trophy record-keeping purposes could improve deer conservation. In some parts of white-tailed deer distribution, deer almost never grow antlers large enough to make the record books. Separating record book categories geographically could encourage conservation interest and funds for deer in otherwise neglected areas of their range. Especially in Central and South America and Mexico, this interest could lead to an economic benefit to local communities, who would then see local deer populations as a resource with high economic benefit to be protected and promoted. As beneficial as these categories may be for conservation, they must not be thought of as taxonomic entities (i.e., subspecies) unless defensible with morphological and genetic data.

North America

In general, the northern extent of the geographic range of white-tailed deer is limited by harsh winter conditions (i.e., deep snow), exacerbated by short growing seasons, and boreal forest that lacks large areas in early successional stages that are nutritionally important to deer (Figure 1.7a). One or two harsh winters can reduce local deer populations by half at the northern periphery of whitetail range. During these events, the remaining deer survive in pockets of the best habitat or agricultural areas and then repopulate

(a)

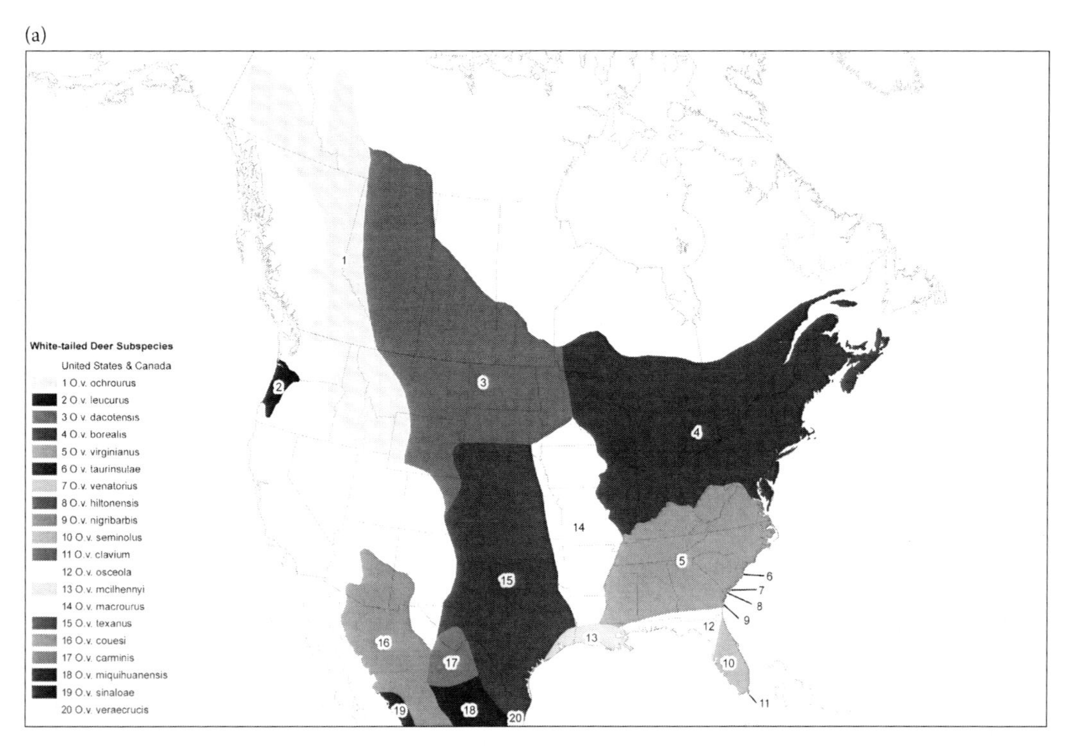

FIGURE 1.7 Distribution of white-tailed deer in the United States and Canada (a), Mexico (b), and Central and South America (c) with general areas representing the subspecies that have been described in the literature. (Cartography by C. Query. With permission.)

(b)

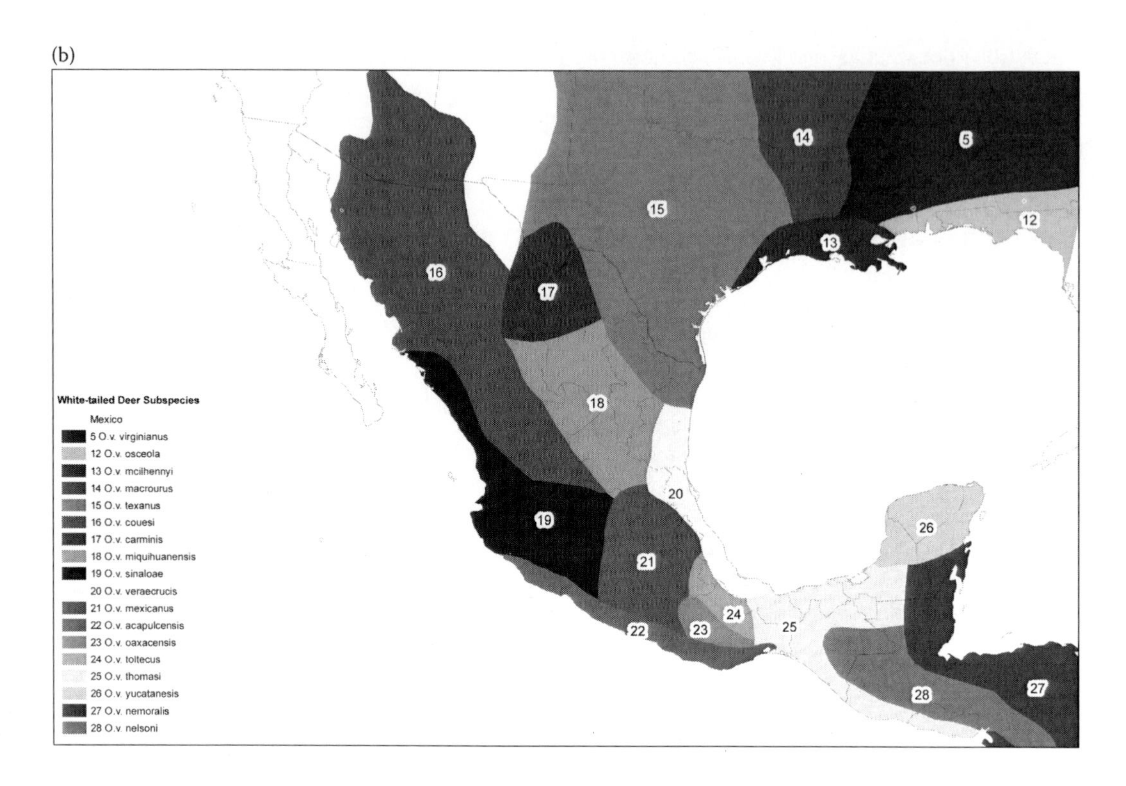

(c)

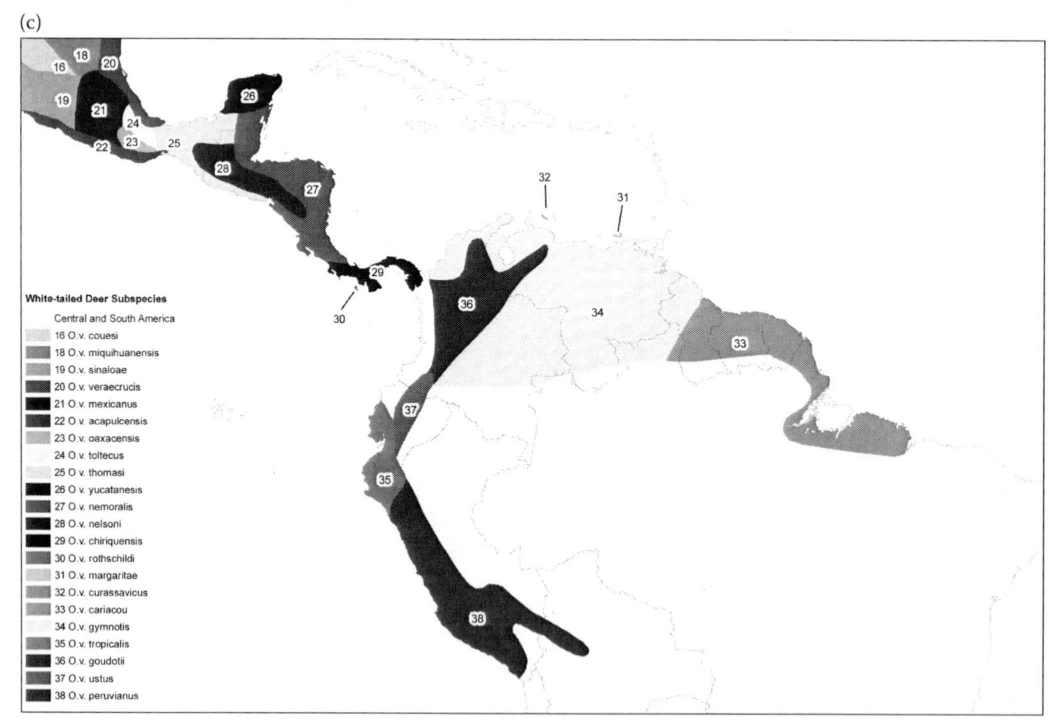

FIGURE 1.7 *continued*

FIGURE 1.8 White-tailed deer in the northern extent of their geographic range are much larger than their southern counterparts and their distribution is limited primarily by harsh winters. (Photo by T. Daniel. With permission.)

surrounding habitat during mild winters (A. Schmidt, Saskatchewan Ministry of Environment, personal communication). Long, severe winters in the northern latitudes are harder for deer to survive because the summer growing season is relatively short and there is less opportunity to obtain optimum nutrition for reproduction and store fat reserves to survive the winter (Figure 1.8). This combination of short summers and relatively long winters creates an ebb and flow at the northern edge of whitetail distribution and that balance can be influenced by other natural and anthropic factors.

White-tailed deer are expanding northward along their northern boundary, in part as a continuation of the post-Pleistocene range expansion of many plants and animals. However, there are many other potential factors that could be accelerating this increase in distribution in recent decades. Human alteration of forests may increase white-tailed deer survival and reproduction. Thus, timber harvest in the Boreal Forest near the edge of whitetail range could enable deer populations to exist in areas that would otherwise be unsuitable. Also, the establishment of long linear seismic lines through the forest has been implicated as a factor facilitating expansion of white-tailed deer populations (Gainer, 1995). These seismic lines are kept clear of trees and revegetate with shrubs and forbs that favor deer use (K. Morton, Alberta Fish and Wildlife Division, personal communication). Still, the effects of these forest alterations are not consistent across the northern periphery. For example, when clearcuts are replanted with trees, herbicides may be used that eliminate important hardwood browse, making the area unsuitable for deer. Fires can also have a beneficial effect on deer habitat by opening patches of closed forest and encouraging aspen and other valuable early successional plants.

Agricultural production is known to attract and hold white-tailed deer in new areas and is generally considered an important factor promoting range expansion (K. Dawe, University of Alberta, personal communication). Agriculture brings high-energy and high-protein foods that compensate for the shorter growing season. These high-quality, abundant foods allow deer to obtain the nutrients necessary not only to survive winter, but also to have the reproductive capacity for population growth when conditions are favorable. In some cases, open windswept agricultural fields offer areas of limited snow cover where deer can access food during winter and that are free of snow weeks earlier than nearby forest during spring (T. Nette, Nova Scotia Department of Natural Resources, personal communication).

A series of mild winters certainly allow for range expansion to the north because of increased survival in peripheral populations. This leads to speculation of how much of the recent expansion may be due to changes in the global climate. In the Boreal Forest it is not entirely clear whether winter severity has changed significantly during this recent expansion of white-tailed deer. Evaluation of winter severity by decade in British Columbia failed to detect any obvious patterns (Baccante and Woods, 2008). If the overall climate continues to warm significantly for the next 50–100 years, northward expansion of the white-tailed deer's range should continue (Veitch, 2001). In the southwestern United States and northern Mexico, continued warming may cause a reduction in distribution as marginal arid habitat becomes unsuitable. There is probably little value in speculating about future distribution too much farther into the future than the next update of this book. As is frequently the case, a combination of weather and habitat changes will dictate where we find whitetails in the coming decades.

Whitetails were not a part of the native fauna in the Yukon Territory, but first appeared along the British Columbia border in 1975 and had reached Moose Creek near Stewart Crossing by 1998 (Hoefs, 2001). They remain rare, but stable, and scattered in small pockets rather than a continuous distribution as the map implies (Figure 1.7a). On the east side of the MacKenzie Mountains, in the Northwest Territories, whitetails are expanding northward. White-tailed deer were also not native to the Northwest Territories, but were reported in the early 1900s and documented more frequently beginning in the early 1960s. In 1996, a hunter harvested a healthy doe along the MacKenzie River north of Norman Wells, which is only about 100 km south of the Arctic Circle (Veitch, 2001). This represents the northernmost whitetail record in North America to date. Although still existing in scattered and small populations, whitetails are common in some areas and overall seem to be increasing in abundance (R. Gau, Northwest Territories Department of Environment and Natural Resources, personal communication).

White-tailed deer are native to southern Alberta, but became established in the Peace River Parklands in the northwestern part of the province in the 1980s (Wishart, 1984; Gainer, 1995), and now extend north farther than mule deer (K. Morton, Alberta Fish and Wildlife Division, personal communication). Whitetails do not extend as far to the north in neighboring Saskatchewan, being limited to the southern half of the province. In both Alberta and Saskatchewan, whitetails are not as common in the Boreal Forest, but have expanded in the last 10–15 years to occupy areas with hardwood forest stands or mixed forest with a significant aspen component (A. Schmidt, Saskatchewan Ministry of Environment, personal communication). Whitetails expanded northward with agricultural development, but there are also areas where agriculture stops at the Boreal Forest and whitetail distribution continues about 160 km farther north. This is marginal habitat for whitetails, where arboreal lichen has become an important source of nutrition (Latham, 2009).

To the west in British Columbia, white-tailed deer have expanded northward and westward almost to the Pacific Ocean, with a population near the town of Hope in the extreme southwestern corner of the province (G. Kuzyk, British Columbia Ministry of Environment, personal communication). Most white-tailed deer in British Columbia occur in the southeast corner with highest densities along river corridors (Mowat and Kuzyk, 2009). Extra-limital occurrences have been documented throughout much of the southcentral part of the province connecting with low-density populations in the northeast.

Whitetails are also found across the international boundary in eastern Washington and in northeast Oregon. Along the Pacific coastal area there are two main subpopulations of whitetails that are isolated in distribution from each other and from those to the east. One subpopulation occupies the mouth of the Columbia River at the Washington–Oregon border and the other is about 320 km south near Roseburg in southern Oregon. On February 19, 1806, Lewis and Clark wrote about the deer they encountered at the mouth of the Columbia River by saying "These do not appear to differ essentially from those of our country being about the same size, shape, and appearance in every respect except their length of tail which is more than half as long again as our deer" (Lewis, 1806). These Columbia River white-tailed deer (*O. v. leucurus*) were designated as a different subspecies by Douglas (1829) based on one specimen shot, but not preserved. Because everyone considered these deer to be unique and their range very limited, they were designated as Endangered Species in 1967 (Figure 1.9). Eleven years later the Roseburg population was designated to be the same subspecies as Columbia River whitetails and also became endangered by association (Smith et al., 2003).

Despite the legal designations, no one had evaluated the uniqueness of the Columbia River whitetail until Gavin and May (1988) used early genetic analysis techniques which led them to question the uniqueness of this whitetail subspecies and also identified a low level of hybridization with black-tailed deer. More recent morphological evaluation of cranial variation in these deer showed that these two endangered deer subpopulations differed from one another as much as they differed from the nearby nonendangered whitetail subspecies (Smith et al., 2003). After a robust population recovery to more than 6000 deer and acquisition of more habitat, the Roseburg population was delisted in 2003 and Oregon authorized a hunt in that population beginning in 2005. Piaggio and Hopken (2009) completed a relatively

FIGURE 1.9 Columbia River white-tailed deer have been designated as Endangered Species since 1967, but recent genetic work questions the uniqueness of this population. (Photo by J. David. With permission.)

comprehensive genetic analysis using both mtDNA and nDNA to evaluate the taxonomic status of this endangered subspecies. Their analysis of samples taken from the Roseburg area, mouth of the Columbia River, and northeastern Oregon led them to conclude that the subspecific designation of the Columbia River white-tailed deer is not warranted.

Large, cold-adapted whitetails are found in western Canada, across the prairie-dominated areas of southern Canada and the northcentral United States, into the midwestern farmland, and up through the northeastern forests. The upper Midwest is one of the most fertile areas in North America and likewise produces the most record book bucks. The four leading states for the production of bucks qualifying for the Boone and Crockett record book are: Iowa, Minnesota, Wisconsin, and Illinois, with those four states representing 39% of the top 100 entries (Boone and Crockett Club, 2005). This geographic pattern of the largest number of trophy bucks from the most fertile soils is not a coincidence.

In the northeast, white-tailed deer are distributed throughout southern Ontario, Quebec, and the northeastern United States. Whitetail distribution in parts of the northeast has shifted northward in recent years, but is probably stable at present (C. Curley, Ontario Ministry of Natural Resources, personal communication). Deer are found throughout Nova Scotia (including Cape Breton Island), and New Brunswick, but do not occur on Isles de la Madeleine or Newfoundland. Whitetails were absent on Prince Edward Island until very recently when a few whitetails appeared unexpectedly. It is not clear how they navigated the 16 km of icy waters and strong currents of the Northumberland Strait between the mainland and the island. It is not likely they walked the 12-km Confederation Bridge on to the island, but possible some were transported to the island by well-meaning people as fawns. Dispersal atop floating ice would be unusual, but this has been observed in the area (Figure 1.10). Regardless of their mode of arrival, deep snows and very little remaining forest will limit the permanent establishment of a large deer population on Prince Edward Island (T. Nette, Nova Scotia Department of Natural Resources, personal communication). Whitetails occur on Anticosti Island (Quebec), but were not native there and they did not arrive through natural dispersal. Henri Menier released about 200 deer onto the island in 1896–1897 and their population has grown to exceed 120,000 deer (Plante et al., 2004).

From a taxonomic standpoint, there is probably little physical difference among white-tailed deer across the northern portion of their range. Each population adapts to local conditions and nutrition levels, but in a broad sense, are exposed to similar environmental stresses. Minnesotans do not spend much time thinking about the three different whitetail subspecies designated in their state, nor should they.

White-tailed deer become somewhat smaller as one moves into the southeastern United States. However, Barbour and Allen (1922) evaluated skull size and pelage variation and had difficulty separating specimens of northeastern deer from others collected along the eastern seaboard as far south as the Florida peninsula. They concluded the characters used to describe these subspecies were "hardly diagnostic."

FIGURE 1.10 White-tailed deer are occasionally seen on ice floes off the coast of New Brunswick, Canada which may play a role in dispersal to islands. (Photo by J. Mundle. With permission.)

Several studies in the southeastern United States have evaluated patterns of genetic diversity. Some early work was contradictory (Ellsworth et al., 1994; Leberg et al., 1994), but a combined and more comprehensive effort showed that deer translocations had substantial and long-lasting effects on the genetic composition of populations receiving deer (Leberg and Ellsworth, 1999). Other nearby populations were not significantly affected due to limited dispersal of translocated deer and their offspring. Early genetic work with allozymes at the Savannah River Ecology Laboratory found no significant genetic differentiation among six subspecies covering the northeast, Blackbeard Island, Florida, Texas, and Virginia (Smith et al., 1984). Other studies have found some regional differentiation among whitetails in the southeastern United States, but the genetic divisions do not match described subspecies ranges (Ellsworth et al., 1994; Leberg and Ellsworth, 1999; DeYoung et al., 2003).

On the coast of Georgia and South Carolina there are four "island" populations that have been designated as unique subspecies (Blackbeard/Sapelo, Hilton Head, Hunting, and Bull islands). Most authors have considered these valid subspecies as they are island forms and assumed to be isolated from other deer populations. However, upon closer inspection one sees that these are not discrete and completely isolated islands, but simply coastal areas separated from the mainland by a river or marshy area. These subspecies have not been evaluated comprehensively using modern genetic analyses. Leberg and Ellsworth (1999) analyzed samples from one of these four insular subspecies (*O. v. nigribarbis*) in a comparison of mtDNA among whitetails from the southeastern United States, including other islands where populations were not designated as unique. They concluded that coastal island whitetail populations retained the ancestral genetic variation lost from many southeastern deer populations because of translocations and genetic bottlenecks during times of low deer abundance. Deer on these islands are smaller than deer on the mainland (Klimstra et al., 1991), but Leberg and Ellsworth (1999) found that the islands were not genetically isolated from the mainland, which is consistent with reports that deer swim between the mainland and some of these barrier islands. These four coastal populations are worthy of conservation, but if they represent valid taxonomic subspecies, then there are potentially hundreds more undescribed subspecies up and down the Atlantic Coast.

The subspecific status of white-tailed deer inhabiting the Florida Keys (*O. v. clavium*) is unquestionable, being geographically, phenotypically, and genetically differentiated. In the 1940s, it was thought there were fewer than 50 Key deer remaining (Dickson, 1955), but their population has increased in recent decades. About 600 Key deer now occupy 20–25 islands in a 40-km stretch between Johnson and Saddlebunch Keys (Figure 1.11). Historic reports indicate they may have been distributed as far as Key West (Dickson, 1955), but most deer are now found on Big Pine Key ($n = 400$), No Name Key ($n = 100$), Big Torch, Little Torch, Cudjo, and Sugarloaf ($n = 100$) (P. Hughes, National Key Deer Refuge, personal

FIGURE 1.11 The diminutive Key deer represents a classic example of an island phenotype that has become smaller to cope with reduced forage availability inherent in an isolated existence. (Photo by M. Averette. With permission.)

communication). The northernmost Key deer are separated from the mainland by at least 48 km of open water and an even longer string of uninhabited (by deer) islands. Key deer were placed on the endangered species list in 1967 and restoration efforts are ongoing (U.S. Fish and Wildlife Service, 2008). As part of the recovery plan, a few dozen deer have been translocated from Big Pine Key to Upper Sugarloaf and Cudjo keys in an effort to assist natural dispersal.

Early studies with relatively small sample sizes documented that Key deer were smaller than other whitetails (Barbour and Allen, 1922; Dickson, 1955). They typically stand about 61 cm at the shoulder with an average weight of 30 and 40 kg for does and bucks, respectively. A more recent study by Maffei et al. (1988) using 20 measurements of about 400 Key deer skulls and mandibles showed that this population was clearly smaller from those on the Florida mainland. Analysis by age class of antlers from 501 male Key deer and 601 of their mainland counterparts in the Everglades (*O. v. seminolus*) showed Key deer antlers to be significantly smaller (Folk and Klimstra, 1991). It is not surprising that these island deer would be smaller because insular mammals, following a well-documented ecological pattern, typically adapt to a more efficient body size in response to chronically inadequate resources (Case, 1978; Brisbin and Lenarz, 1984). Few genetic studies have been done on Key deer. Small insular gene pools would be expected to have low genetic diversity and change rapidly due to their isolation, founder effect during colonization, periodic genetic bottlenecks, and genetic drift. When Ellsworth et al. (1994) analyzed mtDNA variation in 142 deer from the southeastern United States; they found that all 15 Key deer in their sample had the same haplotype that was not shared with any deer sampled from the mainland.

Deer throughout the Florida peninsula and westward along the Gulf coast to Texas have been described as three different subspecies, but none have characteristics that would clearly distinguish them as unique from other deer in the southeastern United States. One of these subspecies was first collected near Avery Island, Louisiana (*O. v. mcilhennyi*) and named after the family (McIlhenny) that produced Tabasco hot sauce there (Miller, 1928).

Three forms of white-tailed deer have geographic ranges straddling the United States–Mexico border. The Coues white-tailed deer (*O. v. couesi*; pronounced "cows") are found in scattered populations throughout most of the southwestern United States in central and southeastern Arizona and southwestern New Mexico, and also in the Mexican states of Sonora, Chihuahua, Durango, and Zacatecas (Heffelfinger, 2006). Coues whitetails prefer oak woodland habitat between 1220 and 2440 m in elevation (Figure 1.12). This is the only form of whitetail that is recognized with a separate category in the Boone and Crockett Club record book because it is morphologically different and geographically isolated from other whitetail subspecies, except to the south where it blends into other recognized forms in Sinaloa and Central Mexico. Mature bucks of this diminutive race commonly have field dressed weights around 45 kg.

FIGURE 1.12 Coues white-tailed deer of the southwestern oak woodlands are the only whitetail subspecies recognized as a separate category in the Boone and Crockett scoring system. (Photo by G. Andrejko. With permission.)

Preliminary investigation into the genetic uniqueness of Coues whitetails has begun using a suite of microsatellite markers and early results are promising (D. Paetkau, Wildlife Genetics International, unpublished data). Because of partial geographic isolation, further analyses may reveal molecular markers that differentiate this small southwestern race from other North American subspecies.

The Texas white-tailed deer (*O. v. texanus*) occur throughout that state, but has become famous for the large-antlered bucks produced in South Texas under conservative harvest and sometimes intensive management. This deer is smaller than northern whitetails, but larger than other southwestern races with body size increasing to the north in the southern Great Plains. Texas white-tailed deer in the Great Plains have no recognizable physical differences from other adjacent mid-continent subspecies. In Mexico, the Texas whitetail occupies northeastern Coahuila, Nuevo Leon, and Tamaulipas before fading off to the West into the scattered mountain ranges occupied by the Carmen Mountains whitetail or merging into other ill-defined subspecies to the south.

The Carmen Mountains white-tailed deer (*O. v. carminis*) was first described in 1940 as being different from Texas whitetails because they were smaller and had antlers with shorter tines (Goldman and Kellogg, 1940). The original description placed them in the Sierra del Carmen in northern Coahuila and the Chisos Mountains in the Big Bend Region of Texas (Figure 1.13). Later authors and local biologists have noted that smaller whitetails also occupy other isolated mountain ranges on both sides of the international boundary (Baker, 1956; Krausman and Ables, 1981). The smaller Carmen Mountains phenotype is believed to inhabit the Chisos, Sierra Quemada, Sierra del Caballo Muerto, Chinati, and Sierra Vieja in West Texas (B. Tarrant, Texas Parks and Wildlife Department, personal communication). They have occurred in the Christmas and Rosillos mountains in the past, but apparently not in recent years. The Del Norte and Glass mountains are also reported to have Carmen Mountains whitetails at higher elevations, but larger deer resembling the Texas whitetail in the lower surrounding desert scrub. Whitetails

FIGURE 1.13 Carmen Mountains white-tailed deer occupying a cluster of small, isolated mountains in northern Coahuila, Mexico, and West Texas appear to be an intermediate form between the diminutive Coues whitetail and those in Texas. (Photo by T. Fulbright. With permission.)

are found throughout the Davis Mountains, but many of them are more similar to Texas whitetail in size (Krausman et al., 1978).

South of the border, the Carmen Mountains deer are said to occur in the Serranias del Burro, Sierra del Carmen, Sierra Encantada, Hechiceros, Sierra Santa Rosa, and Sierra Santa Fe del Pino in northern Coahuila and Chihuahua (Krausman and Ables, 1981; B. McKinney, CEMEX, personal communication). Whitetails resembling the Carmen Mountains phenotype occur in other scattered mountains to the south as far as Jaral, Coahuila (Baker, 1956), but very little work has been done to evaluate phenotypic or genetic variation in that portion of their range.

Krausman et al. (1978) recorded 15 measurements from 167 skulls and antlers representing Texas, Carmen Mountains, and Coues whitetails. Their measurements clearly showed a clinal change in skull morphology from small Coues to larger deer in the Sierra del Carmen and then a gradual increase in size northward and eastward through West Texas to the range of the Texas whitetail. Carmen and Coues whitetails were more similar to each other than to the Texas whitetail, but intergradation of body size through the region was evident. Carmen Mountains whitetails give way to the larger Texas whitetail north and east of Alpine, Texas and in Coahuila on the east side of the Serranias del Burro (Baker, 1956) and the foothills of the Sierra de Santa Rosa (C. Sellers, Rancho la Escondida, personal communication). This body size cline was recognized by Goldman and Kellogg (1940, p. 82) in the original description of the subspecies when they wrote "that complete intergradation of the two must occur along the basal slopes of the mountains."

Most of the current geographic range of white-tailed deer in the southwestern United States and northern Mexico was pine–juniper–oak woodland until about 8000–9000 years ago when a change in climate brought in desert scrub communities and pushed the mesic woodlands to remnant "islands" in higher elevations (Van Devender, 1977). This shift occurred throughout the present ranges of Coues and Carmen Mountains whitetails, thereby isolating them across a fragmented landscape and allowing selective processes to operate on these populations independently. Today the Coues deer have a fairly continuous distribution in the Sierra Madres and the Texas whitetail has a continuous distribution in northeastern Mexico. Between these distributions is an area where an intermediate white-tailed deer occurs sparsely in a few scattered, isolated mountain ranges in eastern Chihuahua and western Coahuila. Considering geographic distribution and morphology of the deer referred to as Carmen Mountains whitetails, it seems that this is simply a series of isolated populations from a formerly continuous cline between the smaller Coues whitetail to the west and the larger Texas whitetail to the east.

In addition to the three aforementioned deer, Mexico has had no less than 11 designated subspecies, many with relatively small geographic ranges and without characteristics that are unique to deer in that area (Figure 1.7b). Mexican white-tailed deer inhabit an incredible variety of habitat conditions throughout the country from dry deserts to mixed conifer forest to tropical rainforest with annual precipitation ranging from 25 to 280 cm/year (Mendez, 1984). Because of the great diversity of elevations and vegetation associations, white-tailed deer vary in a myriad of ways throughout the country, tending to be larger in more mesic forested conditions and smaller where nutrition is limited and a more efficient body size is advantageous. In general, deer continue the clinal body size reduction as one moves south into central and southern Mexico (Goldman and Kellogg, 1940).

Unfortunately, there have been few studies of white-tailed deer in central and southern Mexico and almost no work to clarify subspecific taxonomy (Mandujano, 2004). Mandujano et al. (2008) presented a sensible approach to recognizing geographical differences in white-tailed deer in Mexico by grouping regions of the country together into ecotypes based on vegetation associations. This approach has been used successfully in mule deer (Heffelfinger et al., 2003) and represents a logical way to improve management and conservation without relying on weakly supported subspecies taxonomy.

Central and South America

White-tailed deer distribution continues southeasterly through the Central American countries of Guatemala, Belize, El Salvador, Honduras, Nicaragua, Costa Rica, and Panama. These countries provide diverse vegetation communities from tropical forest to open savanna, but whitetails reach their highest densities in the thorn scrub and forest–savanna ecotones (Mendez, 1984). The unbroken tropical forests such as those on the Caribbean side and eastern Panama are not considered ideal for white-tailed deer and in these areas brocket deer are the more common cervid (J. Barrio, Centro de Ornitología y Biodiversidad–Peru, personal communication).

Almost no taxonomic work has been done in Central America (Gallina et al., 2010). These deer are very much like those occupying similar habitats in southern Mexico, although a higher percentage of specimens lack a metatarsal gland (Lydekker, 1898). Indeed when describing the white-tailed deer in Honduras, Hershkovitz (1951, p. 568) wrote: "Whatever the characters, the British Honduran Virginia Deer could be assigned indifferently to any one of a half dozen forms described from southern Mexico." One general trend that is evident in whitetails throughout their range is a trend toward more compressed breeding seasons in the northern extent of their range due to high seasonality. In Central America and locations further south, this breeding synchrony weakens because seasonal fluctuations are less evident (Klein, 1982; Branen and Marchington, 1987).

Whitetails are native to the 673-km^2 island of Coiba situated 24 km off the Pacific Coast of Panama. Now a national park, the island has been separated from the mainland for about 15,000 years. This island form of whitetail is considered a separate subspecies (*O. v. rothschildi*) and is demonstrably smaller than other Central American whitetails, rivaling the Key deer in south Florida (Halls, 1978). Whitetails are also abundant on Contadora Island, 48 km off Panama's Pacific Coast; however, this island population is the result of escapes (or liberations) from a captive herd brought from the mainland.

White-tailed deer in South America are present throughout the northern and northwestern portions of the continent (Figure 1.7c) in diverse environmental conditions ranging from islands and marshy savannas near sea level to 4200 m in the Andes Mountains. Body size of these deer is generally a little larger than those of Central America and southern Mexico (Brokx, 1984). A higher incidence of maxillary canine teeth, proportionately larger molars, and shorter tails are also characteristics that have been associated with whitetails in South America (Gallina et al., 2010). Whitetails in South America also universally lack a metatarsal gland although some may have a remnant tuft of hair in a slightly different location than North American white-tailed deer (Husson, 1960; Brokx, 1972).

Several authors have noted that all the different whitetail subspecies in South America seem to group into two ecotypes based mostly on pelage similarities shaped by environmental pressures (Hershkovitz, 1958; Brokx, 1984). One ecotype consists of the deer found in the High Andes or temperate zone and the other inhabits the lowlands. The difference between the two forms is mostly in appearance of the pelage. The temperate ecotype is sometimes referred to by the locals as Venado Gris (Gray Deer) because it has a gray coat year-round (Brokx, 1984). The lowland whitetail ecotype appears reddish-brown throughout the year.

Of the lowland ecotype, there are two island deer populations that are isolated from the mainland and have their own unique subspecies designations. Deer on Isla de Margarita (*O. v. margaritae*) off the Venezuelan Caribbean coast are the smallest whitetails in South America and have support as a valid subspecies because of geographic isolation and demonstrated genetic (Moscarella et al., 2003) and morphological (Molina and Molinari, 1999) differentiation. Curiously, one sample from the mainland adjacent to Isla de Margarita included in the analysis by Moscarella et al. (2003) clustered with all other Margarita Island deer and not with mainland deer.

The other island population is not as clearly defined as a subspecies. White-tailed deer on the island of Curacao, 60 km off Venezuela's northwest coast, have not been evaluated taxonomically (*O. v. curassavicus*). Hummelinck (1940) described this subspecies as also occurring on the mainland on the Guajira Peninsula in Colombia. White-tailed deer remains are not present in the island's archaeological sites from Archaic Age inhabitants (prior to 500 AD), but only appear after the arrival of Caquetio immigrants sometime after 500 AD (Hooijer, 1960; Havier, 1987). By the time the Spaniards arrived on the island (1499 AD), white-tailed deer were common. Hernandez de Alba (1963) notes that Caquetios traded venison and live deer between Venezuela and the islands prior to Spanish contact. This strongly suggests white-tailed deer were not native to the island, but were introduced by this immigrating culture from the South American mainland (Husson, 1960). Caquetios transporting live deer in wooden canoes across 60 km of ocean represents the earliest known deer translocation. The population now numbers a few hundred deer and has been protected since 1931. Most of the deer are currently found on the northwest end of the island in protected areas, such as Christoffel Park. Their habitat is currently being dramatically altered by rapid housing development to support the tourism industry (J. de Freitas, CARMABI, personal communication).

The lowland ecotypes on the mainland extend from the dry deciduous forests on the northwest (coastal Peru and Ecuador) around to the northeast (Colombia through Brazil) in the more extensive open marshes and grassland savannas (llanos). These lowland deer have a thin reddish-brown summer coat similar to North American white-tailed deer, but their "winter" pelage has almost no insulative underfur. Interestingly, Brokx (1972) reported that after moving a lowland deer to colder elevations, it grew thick wooly underfur which provides some insight into the problems of using pelage as a taxonomic character.

Brokx (1972, 1984) advocated a new subspecies from the llanos of eastern Colombia and the Apure Region in western Venezuela (*O. v. apurensis*) (Figure 1.14). Deer in this region appear to have several phenotypic characteristics that differ from other lowland whitetails in northern Venezuela such as a yellow-tan pelage (due to a different banding pattern on the hairs), smaller body size, and skull characteristics (Brokx, 1972; Molinari, 2007). Interestingly, recent genetic analysis included only two samples from and adjacent to the Apure Region, but both shared a unique haplotype that was not found in any of the other samples (Moscarella et al., 2003). Even though whitetails in this area show some differentiation, it makes little sense to establish a 39th white-tailed deer subspecies without a more comprehensive taxonomic evaluation of all South American forms.

FIGURE 1.14 White-tailed deer in the Apure Region of the Venezuelan llanos face dramatically different conditions and yet are unmistakably whitetails. (Photo by W. Atkinson. With permission.)

FIGURE 1.15 The dry deciduous Tumbesian Forest on the west side of the Andes Mountains are home to a race of white-tailed deer that was originally described as having a short tail and a short brown coat year-round. (Photo by J. Barrio. With permission.)

Along the Pacific Coast, west of the Andes in Ecuador and northern Peru exists the dry, deciduous Tumbesian Forest. Another lowland form of whitetail occurs here and is characterized by short brown pelage year round and a short tail (*O. v. tropicalis*) (Figure 1.15). These whitetails are reportedly smaller than those of the llanos in Colombia and Venezuela (Brokx, 1984).

The temperate ecotype inhabits extremely high-elevation habitat in the Andes Mountains up to at least 4200 m in Colombia, Ecuador, Peru, and Bolivia (J. Barrio, Centro de Ornitología y Biodiversidad–Peru, personal communication). Separate subspecies have been designated for Colombia (*O. v. goudotii*), Ecuador (*O. v. ustus*), and Peru (*O. v. peruvianus*), but no genetic or phenotypic characteristics differentiate them. No compelling data suggest there is more than one form of white-tailed deer throughout the Andes Mountains in these countries (Brokx, 1972). White-tailed deer in the Colombian Andes do not come into contact with those in the lowlands to the east and are thus isolated (Moscarella et al., 2003). Deer in these high-elevation ranges have a more synchronous breeding season, an obvious thick gray winter coat with underfur on their body, and thicker pelage on the ears. They have a shorter summer coat, but even that appears gray. High-elevation deer weigh 52–57 kg (Molinari, 2007), which is slightly larger than lowland deer.

Distribution of white-tailed deer in the Andes is not limited by elevation per se, but by steep and arid habitat above and thick rainforest on mountain slopes below (Brokx, 1984). The extreme southern distribution of white-tailed deer is represented by an extension from southern Peru into Bolivia along a portion of the Andes. In Bolivia, white-tailed deer are found at least as far south as the village of Pelechuco, and maybe somewhat farther based on habitat (Jungius, 1974). At elevations around and above 4000 m in Peru and Bolivia, whitetails share the range with taruka (Jungius, 1974; Tarifa et al., 2001; Barrio, 2006).

Molina and Molinari (1999) compared skull and mandible characteristics of 140 Venezuelan white-tailed deer to similar published information from North America (Rees, 1969). The analysis found substantial diversity in skull and mandible characters within Venezuela, with the Isla de Margarita and the Andes subspecies differentiated from each other and from the lowland ecotype. This is probably not surprising for a sample of skulls ranging from the small insular race to the temperate ecotype at 3650 m elevation. They also found differences between white-tailed deer in Venezuela and North America which led them to suggest that North and South American whitetails were different species. The suggestion to split whitetails at the species level was based on the presence/absence of a metatarsal gland, 13 discrete cranial–mandibular characters, and principle component analysis of mandible measurements. The metatarsal gland has been long recognized as being of little value as a taxonomic character (Lydekker, 1898; Brokx, 1972) as it is sometimes absent in Central American whitetails and has even been documented missing from Coues white-tailed deer in Arizona (Arizona Game and Fish Department, unpublished data). While the cranial–mandibular analysis illustrated a Venezuela–North America difference, data were not complete nor compelling enough for establishing three new *Odocoileus* species in South America.

Moscarella et al. (2003) compared mtDNA sequences from 26 samples representing three subspecies from Venezuela to help clarify taxonomic discussions. They confirmed a remarkable divergence among haplotypes in Venezuela and North America and genetic support for differentiation of deer from Isla de Margarita and the temperate ecotype in the Andes. However, Moscarella et al. (2003) argue that genetic differentiation was within the range of other subspecies and splitting whitetails in Venezuela into three species is not supported by the genetic data. The original analysis of Molina and Molinari (1999) was revisited by Molinari (2007) and expanded to incorporate the genetic information from Moscarella et al. (2003). However, the expanded discussion did not add substantively new information that would make a compelling case for three new species of Venezuelan white-tailed deer.

Although others have reported differences between the continents (Smith et al., 1986), a separation of *Odocoileus virginianus* at the species level with the information currently on hand would be incongruent with the level of differentiation seen in other ungulate species. Only a comprehensive analysis that includes specimens representing all recognized subspecies (not just those in Venezuela) will help answer this question. As phylogenies inferred with mtDNA can be misleading (Cronin, 1991, 1993), especially with the *Odocoileus* complex, the analysis will have to include nDNA and possibly a Y chromosome marker to be able to tell the whole story (Gallina et al., 2010). Much of the good work of Moscarella et al. (2003) and Molinari (2007) will be helpful in coming to a final resolution once additional data are available.

Extra Limital Distribution

New Zealand

Three species of bats were the only mammals native to New Zealand before human intervention. Since the arrival of Captain Cook in 1769, at least 50 mammal species have been introduced into New Zealand and of these eight nonnative deer species have become established (Harris, 1984). White-tailed deer were first introduced with two bucks and two does from Kansas in the Takaka Valley in 1901 (Whitehead, 1972). Four years later, 19 more from New Hampshire were released with two bucks and seven does going to Cook's Arm on Stewart Island, and nine more (three bucks, six does) released on the north end of Lake Wakatipu on South Island (Harris, 1984). One buck was also released in the Takaka Valley in 1905 in an unsuccessful attempt to maintain the original release.

From these releases both the Lake Wakatipu and the Stewart Island population have become established. Under protection from 1905 through 1919, they increased steadily before a hunting season was established. By 1926 it was apparent that hunting was not going to limit the population and all restrictions on harvest were lifted. The Lake Wakatipu population occupies only about 350 km^2 north of the lake and has been plagued by diseases and low productivity (King, 2005). It is hunted on a small portion of Aspiring National Park and on a few large properties in the area (S. Laing, New Zealand Hunting Info, personal communication).

The release on the 1720-km^2 Stewart Island was much more successful and deer quickly spread throughout the island. The island consists of steep slopes and ridges covered by thick forest and scrub

FIGURE 1.16 New Zealand white-tailed deer were not native to that country, but have adapted to the new environment as illustrated by these deer eating kelp on Stewart Island. (Photo by P. Peychers. With permission.)

habitat. This, and a lack of predators, provided conditions for rapid population growth and by 1926 not only were protections lifted, but a bounty was paid for every deer tail brought in. Because whitetails were not distributed completely throughout the island until the 1950s, there is some question how much damage by red deer was being wrongly attributed to whitetails. Throughout the 1930s and 1940s, government shooters killed thousands of white-tailed deer (Harris, 1984). Today Stewart Island is a popular whitetail hunting destination with 50 hunting blocks around the perimeter of the island and unlimited hunting in the central area where deer are much less common. White-tailed deer are strong swimmers and occasionally swim to some of the nearby islands such as Earnest, Pearl, Bravo, Owen, and Noble islands (King, 2005; Harper, 2006). Two other populations were established briefly in the 1990s in Canterbury, but none persist as wild populations (King, 2005; J. DeLury, Stewart Island Whitetail Research Group, personal communication).

New Zealand white-tailed deer are smaller than their source stock in New Hampshire with weights averaging 54 kg for bucks and 40 kg for does (Davidson and Challies, 1990). Interestingly, because these northern hemisphere deer are living in the southern hemisphere, their annual cycles have adjusted to be exactly the opposite of their homeland. The peak of rut is April–May and fawns are born during December–January, with twin births rare (Harris, 1984). The ever-adaptable whitetail in New Zealand has found seaweed such as kelp to be an important source of nutrition (Figure 1.16).

Finland

Wanting to present a gift to their motherland, Minnesotans of Finnish descent captured and donated three male and four female fawns to the Public Finnish Hunters Association. These seven "Virginia deer" were sent by train from (ironically enough) Virginia, Minnesota to the east coast where they were loaded on a ship during August 1934 (Kairikko and Ruola, 2005). Only one male and four female fawns survived the two-week voyage to Helsinki, Finland and the sole buck fawn was near death upon arrival. After some nurturing he recovered and was released with the does into a 3-hectare enclosure on the Laukko Estate, 170 km north of Helsinki. One of the females was blind, never produced a fawn, and was killed by a golden eagle in 1937. The other does successfully reproduced and so began the Finnish white-tailed deer herd.

Four additional deer (two males, two females) were sent from New York State in 1937 and upon arriving in autumn were placed in the Korkeasaari Zoo to be released into the wild the next spring (Kairikko and Ruola, 2005). Records are sketchy, but it does not appear any of these deer were ever released into the wild. All deer from the original translocation escaped from the Laukko enclosure in March of 1938.

FIGURE 1.17 White-tailed deer translocated to Finland in the 1930s have been tremendously successful despite originating from less than 10 founding members. Finnish white-tailed deer have distinctively dark, long noses and golden foreheads, perhaps due to this genetic bottleneck. (Photo by P. Peychers. With permission.)

The does were enticed back in, but the buck stayed out and roamed close by until they were all released in May of that year (one buck, three does, and two male fawns). The population grew steadily in the Laukko Estate area, but authorities were concerned that all deer born in Finland were sired by the same male with only three maternal lines.

Late in 1948, six white-tailed deer (three male, three female) were flown from Minnesota to Helsinki and placed in the Laukko enclosure where two males quickly died (Kairikko and Ruola, 2005). The next spring, one buck and three does were released to join the growing population in the vicinity (about 100 deer). The population continued to grow and expand each year until the first hunting season was established in 1960 when nine bucks were harvested. The current license allows the harvest of one adult and two fawns (the latter for population control). By 2008, more than 25,000 deer were harvested annually in Finland (S. Laaksonen, Finnish Fish and Wildlife Health Research Unit, personal communication). White-tailed deer are now abundant in the southern half of Finland (Figure 1.17). To the northwest, they have reached the Swedish border, but have not become established in this area. Some white-tailed deer are even being reported just over the Russian border on the Karelian Isthmus and north near Ladoga Lake (K. Nygren, Game and Fisheries Research Institute—Finland, personal communication).

In the late 1960s and early 1970s, nearly 100 deer were translocated over several years to other areas in Finland in an attempt to expand the population. Some of these were successful, including those near the eastern border with Russian (Kairikko and Ruola, 2005). During this effort, 10 whitetails were also translocated to Russia and split between the Moscow Zoo and other locations. Some of these may have gone to the Zavidovo wildlife management area 100 km northwest of Moscow (K. Nygren, Game and Fisheries Research Institute—Finland, personal communication).

British Isles

One vague reference to white-tailed deer being released on the Scottish Isle of Arran in 1832 marks the earliest release of this species into a wild setting in Europe (Fitter, 1959). These deer are said to have thrived for a time before dying out sometime after 1872.

Woburn Park in northern England has a long history of captive deer and deer conservation. It was the 11th Duke of Bedford who saved the Peré David's deer from extinction in the 1880s by holding and propagating them in his park. Meticulous inventory records were kept of the deer at Woburn since the late 1800s. Several scientific names were used for white-tailed deer throughout the period of record-keeping, but they were universally referred to as "Virginia Deer" in the records. An entry in *A record of the collection of animals kept in Woburn Park, the property of the Duke of Bedford 1892–1905*, states "Virginian deer Mazama Americana [sic] imported=140, born=44, died=145, killed=1. Present total [in 1905] 38. Date of importation=1894" (A. Mitchell, Woburn Enterprises Limited, personal communication).

For some reason the highly adaptable white-tailed deer did not fare well in Woburn Park, but when released out of the enclosure sometime before 1905 and allowed to roam free in the nearby forest, the small nucleus of animals became more numerous. In 1913, personal records kept by the Duchess of Bedford recorded only nine whitetails at Woburn Park, but noted that those outside the fence could not be counted (A. Mitchell, Woburn Enterprises Limited, personal communication). Whitehead (1950, p. 121) writes that the last white-tailed deer were seen at the beginning of World War II and since they were free-ranging in the nearby woods, "the troops occupying those areas soon exterminated them." No records of deer inventory survived the period from 1914 to 1946, but records from 1947 to the present make no mention of the white-tailed "Virginia" deer.

Since very early times there have been hundreds of deer parks and zoological gardens scattered across the United Kingdom that have exchanged deer. Many of these parks did not keep records that survived the ages so a complete reconstruction of all white-tailed deer held in captivity is not possible. An inventory of zoological gardens in 1949 recorded two male white-tailed deer in Whipsnade Park and four does (one from South America) in Regent's Park (Whitehead, 1950). As of the end of 2008, the three largest zoological parks in Great Britian: Whipsnade, Woburn Park, and the London Zoo held no white-tailed deer. There are currently no free-ranging whitetails in the British Isles.

Austria

In 1870, some white-tailed deer were translocated from the United States to the Grafenegg Castle in lower Austria. Five years later, some of those deer were then moved to an enclosure near Weidlingau and then to Vienna in 1910 (Bojović and Halls, 1984, Kairikko and Ruola, 2005). There are few details of these early translocations, but all deer seemed to disappear during the turmoil of World War I.

Czech Republic and Slovakia

Some records place the earliest releases of white-tailed deer in the Czech Republic around 1840, but no details are available (Bartoš, 1994). In 1853, seven whitetails were released in then Czechoslovakia, but it was probably not until 15 more deer, mostly from Canada, were released in 1892 and 1893 when the population really became established (Bojović and Halls, 1984; Bartoš, 1994). These deer were released into an enclosure within the Dobris forest about 30 km southwest of Prague (Czech Republic) between the Vltava and Berounka rivers. Sixteen more deer were added to this population in 1906 to bolster numbers. During the turmoil of World War I, the Dobris forest enclosure was destroyed and the deer escaped their confinement, with most remaining in the area.

To help spread white-tailed deer to other parts of the country, eight deer were translocated to a 17-hectare enclosure in Holovous (northern Czech Republic) and two others to a small pen near Košice in Slovakia. The deer at Holovous were accidentally released into the wild in 1965 (Bartoš, 1994). The main population still localized in the Dobris forest is considered well established and stable at about 700 deer (L. Bartoš, Research Institute of Animal Production, personal communication).

Poor reproduction has been typical of this population from the beginning (Bartoš et al., 2002), which has limited its ability to increase and expand as white-tailed deer have in Finland. Unfortunately, introduction of white-tailed deer inadvertently also introduced the large liver fluke to Europe. These parasites have now infected other native cervids such as red deer and are considered a major management problem (Bojović and Halls, 1984).

Serbia and Croatia

Establishment of white-tailed deer in the former Yugoslavia is a more recent accomplishment. Twenty-one deer from eastern and southern United States formed the nucleus of this effort. Seven deer were translocated from Virginia and Maryland between 1970 and 1971 and released into a 5-hectare enclosure about 15 km from Belgrade, Serbia. In 1973, two shipments from Pennsylvania and Louisiana totalling 14 white-tailed deer arrived and were placed in an enclosure near Karadjordjeve along the Danube River in northwestern Serbia, about 120 km west of Belgrade (Bojović and Halls, 1984). Several deer escaped the enclosure and were seen in the area for many years, but no wild population persisted there (Paunovic et al., 2010).

In spring 1975, the captive populations were doing well and so they were used as source stock to increase their distribution in the region. One buck and two does were moved to the Deliblato forest east of Belgrade near the Serbian–Romanian border. Another two bucks and three does were moved to Brać, a large island on the Croatian coast of the Adriatic Sea (Bojović and Halls, 1984). In both of these new areas the deer were confined to small enclosures for at least three years and then released. A hunting season was established in 1980, an indication of the deer's success. The population at Karadjordjeve increased to at least 150 deer by 1983, but only about 40 remained in 2007. This fenced population is hunted, but annual legal harvest is only about two deer with another 10 poached annually (Paunovic et al., 2010). The current hunting season for white-tailed deer opens on September 16 for bucks, and on October 1 for does. The season closes for both sexes January 31 (Paunovic et al., 2010). Persistence of deer at the other release sites in Serbia and Croatia is unknown, but it is thought that Karadjordjeve holds the only white-tailed deer in the country. War and economic turmoil with the break-up of Yugoslavia in the 1990s diverted attention away from wildlife management.

Bulgaria

The Finnish white-tailed deer population was so productive, they sold six does and four bucks to Bulgaria in 1977. These deer were placed in the Kozy Rog area near the Greek border. After growing to 20–30 animals that population appears to be in decline or already extirpated (Whitehead, 1993).

Caribbean Islands

Cuba

White-tailed deer were introduced to Cuba around 1850. The source of these animals is not clear, but originally thought to be Mexico or the southeastern United States (DeVos et al., 1956). Emerging genetic information indicates Cuban whitetails are not from the nearby Florida Keys nor the Southeastern United States (D. Reed, personal communication). Whitetails occupy many of the forested and mountainous areas throughout Cuba (Borroto-Páez, 2009), but are more common in the eastern and western portions and a few locations in the center such as Cienaga de Zapata, Sierra Najasa, and the Escambray Mountains. There is no open hunting season for whitetails, but poaching and widespread forest clearing have caused populations to decline (R. Borroto-Páez, Instituto de Ecologia y Sistematica, personal communication). White-tailed deer are strong swimmers and have populated peripheral islands such as Cayo Sabinal, Cayo Romano, Isla de la Juventud, and the Camaguey Archipelago (Borroto-Páez, 2009) (Figure 1.18).

Jamaica

When Christopher Columbus landed on Jamaica in 1494, he did not find white-tailed deer among the native fauna. Whitetails became established by accident on this 11,396-km^2 island. Hurricanes Allen in 1980 and Gilbert in 1988 damaged a captive facility near Sommerset Falls on the island's northeastern coast and freed the captive deer. It is not known how many escaped in 1980, but it is thought that three bucks and three does were liberated in 1988 (Chai, 2003).

Jamaica is dominated by mountains rising to more than 2134 m and covered with lush forest full of endemic species so there is great concern that white-tailed deer will overpopulate the island and cause damage to native flora and small independent farms. Some local farmers are organizing deer hunts to reduce crop damage and one community leader reports that more than 300 whitetails were killed during a three- to four-year period (Chai, 2003).

United States Virgin Islands

The U.S. Virgin Islands consist of four main islands (St. Thomas, St. John, St. Croix, Water Island) and dozens of smaller islands that are considered part of a group called the Leeward Islands. Christopher

FIGURE 1.18 White-tailed deer are strong swimmers and can swim from the Cuban mainland to nearby islands and back again. (Photo by Christopher Creighton, U.S. Naval Station, Guantanamo Bay, Cuba. With permission.)

Columbus documented and named the Virgin Islands and brought them into the written record of history in 1493. Through time, the islands were claimed by many nations, but eventually became a Dutch colony in the mid-1700s. It is through an early colonial document that we first learn about white-tailed deer in this region. There is a Danish record by a ship captain that mentions five white-tailed deer being released on St. Croix during or before 1790. The nucleus of this population may have included more than the five deer mentioned, but we know that they proliferated on the island and in 1840 were said to "inhabit the mountainous parts of the island" (Seaman, 1966).

By the time the United States purchased the Virgin Islands from the Danish in 1917, there was an estimated 3000 deer well distributed throughout the 212 km^2 St. Croix Island (Seaman, 1966). This was probably the peak in that population because shortly after that time, commercial venison hunters started to use spotlights and buckshot at night to shoot whitetails indiscriminately. To stop the slaughter and conserve deer on the island, Governor Paul Pearson established a restrictive open season and made it illegal to kill deer outside of that season. This era of conservation was short-lived, however, with the 1938 initiation of a cattle-fever tick eradication program. In support of this program, the Governor approved a bill to eradicate all deer from the island of St. Croix (Seaman, 1966). This program only lasted until 1941 and hundreds of deer remained and persist on the island today.

In 1854, some of the St. Croix deer were moved across 56 km of open sea to St. Thomas Island. Later a few of those deer swam 6.4 km to populate St. John Island. More deer were reportedly brought from Texas and the Carolinas until there were 1000 deer estimated to be living on St. Croix and 600 on St. Thomas in 1979 (Baker, 1984).

White-tailed deer on St. John (52 km^2) are found over most of the island and increasing in abundance as evidenced by recent trends in vehicle collisions and anecdotal sightings (C. Stengel, Virgin Islands NP, personal communication) (Figure 1.19). There are currently no deer on Water Island, the fourth largest of the Virgin Island complex.

St. Thomas (83 km^2) is also home to white-tailed deer where they are found in pockets where there is little development. They occur in highest densities in undisturbed areas on the west end, north side, and some residential areas on the east end of the island. They swim to nearby islands and have been seen recently on St. James, Thatch, Congo, and other small islands (R. Platenberg, Virgin Islands NP, personal communication).

True to its long-term island existence, whitetails on the island weigh less than their assumed parent population in the southeastern United States. Seaman (1966) reports St. Croix bucks weighing 41–50 kg and does ranging 32–41 kg. White-tailed deer are not established on any of the British Virgin Islands (C. Petrovic, Econcerns, personal communication), but individuals may periodically swim there from St. John.

FIGURE 1.19 Despite more than 220 years of island isolation, white-tailed deer in the U.S. Virgin Islands could be mistaken for their conspecifics in the southeastern United States. (Photo by C. Stengel. With permission.)

Puerto Rico

White-tailed deer were introduced to the Puerto Rican island of Culebra in 1966. This island is only 20 km west of St. Thomas Island (U.S. Virgin Islands) and 27 km east of Puerto Rico (Philibosian and Yntema, 1977). Whitetails are still present on Culebra (mostly on the east end of the island) and on the small islands of Luis Peña and Cayo Norte (Long, 2003).

Other Islands

There are other reported translocations of white-tailed deer to various islands of the West Indies, but very few details are available. Records indicate whitetails and fallow deer were released on the islands of Antiqua and Barbuda (Leeward Islands) in the seventeenth century, but today only fallow deer remain and those are on Barbuda and the small island of Guiana north of Antiqua (Lever, 1985; Long, 2003). Other vague reports mention Dominica, Grenada, and the Dominican Republic as receiving white-tailed deer at some point. There is no evidence of free-ranging whitetail herds at any of these locations (Whitehead, 1993).

Acknowledgments

George Andrejko (Arizona Game and Fish Department), Wayne Atkinson, Mark Averette, Randy Babb (Arizona Game and Fish Department), Javier Barrio (Centro de Ornitología y Biodiversidad-Peru), Luděk Bartoš (Research Institute of Animal Production), David Bergman (USDA-APHIS-WS), Rafael Borroto-Páez (Instituto de Ecologia y Sistematica), Rafe Boulon (Virgin Island NP), Clay Brewer (Texas Park and Wildlife Department), David E. Brown (Arizona State University), George Bubenik (University of Guelph), Neil B. Carmony, Rod Cumberland (New Brunswick Department of Natural Resources), Christie Curley (Ontario Ministry of Natural Resources), Tim Daniel (Ohio Department of Natural Resources), Joel David (Julia Butler Hansen National Wildlife Refuge), Kim Dawe (University of Alberta), Herman de Bour, John de Freitas (CARMABI), Jim Dell (National Key Deer Refuge), John DeLury, Rene Dube, Rob Florkiewicz (Yukon Department of Environment), Sonia Gallina (Tessaro, Instituto de Ecologia), Rob Gau (Department of Environment and Natural Resources, NW territories), Louis Harveson (Sul Ross State University), Phillip Hughes (National Key Deer Refuge), Thomas Jung (Yukon Department of Environment), Dean Konjevic (University of Zagreb, Croatia), Gerry Kuzyk (British Colombia Ministry of Environment), Sauli Laaksonen (Finnish Fish and Wildlife Health Research Unit), Steuart Laing (New Zealand Hunting Info), John Linnell (Norwegian Institute for Nature Research), Walburga Lutz (Landesbetrieb Wald und Holz), Salvador Mandujano, Bonnie McKinney (CEMEX), Heikki Mikkola,

Aaron Miller (Arizona Game and Fish Department), Ann Mitchell (Woburn Enterprises Limited), Kim Morton (Alberta Fish and Wildlife Division), James Mundle (New Brunswick Department of Natural Resources), Phil Myer (University of Michigan), Tony Nette (Nova Scotia Department of Natural Resources), Kaarlo Nygren (Game and Fisheries Research Institute, Finland), Milan Paunovic, Dave Person (Alaska Department of Fish and Game), Clive Petrovic (British Virgin Islands), Paul Peychers, Renate Platenberg (Virgin Islands NP), Rory Putnam, Chris Query (SWCA Environmental Consultants), David L. Reed (Florida Museum of Natural History), Adam Schmidt (Saskatchewan Ministry of Environment), Raymond Skiles (Big Bend National Park), Carrie Stengel (Virgin Islands NP), Billy Tarrant (Texas Parks and Wildlife Department), Don Whitaker (Oregon Department of Fish and Wildlife), Gary Witmer (USDA-APHIS).

REFERENCES

Allen, J. A. 1903. A new deer and a new lynx from the state of Sinaloa, Mexico. *Bulletin of the American Museum of Natural History* 19:613–615.

Allen, J. A. 1910. Additional mammals from Nicaragua. *Bulletin of the American Museum of Natural History* 28:87–115.

Avise, J. C. 1994. *Molecular Markers Natural History and Evolution*. New York, NY: Chapman & Hall.

Baccante, N. and R. B. Woods. 2008. Relationship between weather factors and survival of mule deer fawns in the Peace Region of British Columbia. Peace Region Technical Report, Wildlife and Fish Section, Ministry of Environment. Fort Saint John, British Columbia.

Baccus, R., N. Ryman, M. H. Smith, C. Reuterwall, and D. Cameron. 1983. Genetic variability and differentiation of large grazing mammals. *Journal of Mammalogy* 64:109–120.

Bailey, V. 1932. The northwestern white-tail deer. *Proceedings of the Biological Society of Washington* 45:43–44.

Baker, R. H. 1956. Mammals of Coahuila, Mexico. *University of Kansas Publications, Museum of Natural History* 9:125–335.

Baker, R. H. 1984. Origin, classification and distribution of the white-tailed deer. In *White-tailed Deer: Ecology and Management*, ed. L. K. Halls, 1–18. Harrisburg, PA: Stackpole Books.

Bangs, O. 1896. The Florida deer. *Proceedings of the Biological Society of Washington* 10:25–28.

Barbour, T. and G. M. Allen. 1922. The white-tailed deer of eastern United States. *Journal of Mammalogy* 3:65–78.

Barrio, J. 2006. Manejo no intencional de dos especies de cérvidos por exclusión de ganado en la parte alta del Parque Nacional Río Abiseo, Perú. *Manejo de Fauna Silvestre en Latinoamérica* 1:1–10.

Bartoš, L. 1994. Ecology of white-tailed deer in Czechia: Investigation of an unsuccessful stocking of an exotic ungulate. Unpublished project proposal. Prague, Czech Republic.

Bartoš, L., D. Vankova, K. V. Miller, and J. Siler. 2002. Interspecific competition between white-tailed, fallow, red, and roe deer. *Journal of Wildlife Management* 66:522–527.

Boddaert, P. 1784. *Elenchus Animalium*. Rotterdam, the Netherlands: C. R. Hake.

Bojović, D. and L. K. Halls. 1984. Central Europe. In *White-tailed Deer: Ecology and Management*, ed. L. K. Halls, 557–560. Harrisburg, PA: Stackpole Books.

Boone and Crockett Club. 2005. *Records of North American Big Game*. Missoula: Boone and Crockett Club.

Borroto-Páez, R. 2009. Invasive mammals in Cuba: An overview. *Biological Invasions*. Published online: DOI 10.1007/s10530–008–9414-z. Accessed September 1, 2009.

Branan, W. V. and R. L. Marchington. 1987. Reproductive ecology of white-tailed and red brocket deer in Suriname. In *Biology and Management of the Cervidae*, ed. C. M. Wemmer, 972–976. Washington, DC: Smithsonian Institution Press.

Brisbin, I. L. and M. S. Lenarz. 1984. Morphological comparisons of insular and mainland populations of southeastern white-tailed deer. *Journal of Mammalogy* 65:44–50.

Brokx, P. A. 1972. A study of the biology of Venezuelan white-tailed deer (*Odocoileus virginianus gymnotis* Wiegman, 1833), with a hypothesis on the origin of South American cervids. PhD dissertation, University of Waterloo, Ontario.

Brokx, P. A. 1984. White-tailed deer populations and habitats of South America. In *White-tailed Deer: Ecology and Management*, ed. L. K. Halls, 525–546. Harrisburg, PA: Stackpole Books.

Cabrera, A. 1918. Sobre los *Odocoileus* de Colombia. *Boletin de la Real Sociedad Española de Historia Natura* 18:300–307.
Carr, S. M. and G. A. Hughes. 1993. Direction of introgressive hybridization between species of North American deer (*Odocoileus*) as inferred from mitochondrial-cytochrome-b sequences. *Journal of Mammalogy* 74:331–342.
Case, T. J. 1978. A general explanation for insular body size trends in terrestrial vertebrates. *Ecology* 59:1–18.
Caton, J. D. 1877. *The Antelope and Deer of America*. New York, NY: Hurd and Houghton.
Chai, S. C. 2003. An assessment of the invasive white-tailed deer in Portland—Distribution, numbers and socioeconomic impacts. MS thesis, The University of the West Indies.
Coues, E. and H. C. Yarrow. 1875. Report upon the collection of mammals made in portions of Nevada, Utah, California, Colorado, New Mexico, and Arizona during the years 1871, 1872, 1873, and 1874. In *Report upon Geographical and Geological Explorations and Surveys West of the One Hundredth Meridian*, ed. G. M. Wheeler, 35–129. Washington, DC: United States Army.
Cronin, M. A. 1991. Mitochondrial-DNA phylogeny of deer (Cervidae). *Journal of Mammalogy* 72:533–566.
Cronin, M. A. 1993. Mitochondrial DNA in wildlife taxonomy and conservation biology: Cautionary notes. *Wildlife Society Bulletin* 21:339–348.
Cronin, M. A. 1997. Systematics, taxonomy, and the endangered species act: The example of the California gnatcatcher. *Wildlife Society Bulletin* 25:661–666.
Davidson, M. M. and C. N. Challies. 1990. White-tailed deer. In *The Handbook of New Zealand Mammals,* ed. C. M. King, 507–514. Auckland, New Zealand: Oxford University Press.
DeVos, A., R. H. Manville, and R. G. Van Gelder. 1956. Introduced mammals and their influence on native biota. *New York Zoological Society* 41:163–194.
DeYoung, R. W., S. Demarais, R. L. Honeycutt, A. P. Rooney, R. A. Gonzales, K. L. Gee. 2003. Genetic consequences of white-tailed deer (*Odocoileus virginianus*) restoration in Mississippi. *Molecular Ecology* 12:3237–3252.
Dickson, J. D., III. 1955. An ecological study of the Key deer. Florida Game and Freshwater Fish Commission Tech. Bull. Number 3. Tallahassee, Florida.
Douglas, D. 1829. Observations on two undescribed species of North American mammals (*Cervus leucurus* et *Ovis californicus*). *Zoological Journal* 4:330–332.
Douzery, E. and E. Randi. 1997. The mitochondrial control region of cervidae: Evolutionary patterns and phylogenetic content. *Molecular Biology and Evolution* 14:1154–1166.
Duarte, J. M., S. Gonzalezm, and J. Maldonado. 2008. The surprising evolutionary history of South American deer. *Molecular Phylogenetics and Evolution* 49:17–22.
Eisenberg, J. F. 1987. The evolutionary history of the Cervidae with special reference to the South American Radiation. In *Biology and Management of the Cervidae*, ed. C. M. Wemmer, 61–64. Washington, DC: Smithsonian Institution Press.
Ellsworth, D. L., R. L. Honeycutt, N. J. Silvy, J. W. Bickham, and W. D. Klimstra. 1994. Historical biogeography and contemporary patterns of mitochondrial DNA variation in white-tailed deer from the southeastern United States. *Evolution* 48:122–136.
Fitter, R. S. R. 1959. *The Ark in our Midst*. London: Collins.
Folk, M. J. and W. D. Klimstra. 1991. Antlers of white-tailed deer (*Odocoileus virginianus*) from insular and mainland Florida. *Florida Field Naturalist* 19:97–132.
Fry, W. E. and E. P. Gustafson. 1974. Cervids from the Pliocene and Pleistocene of central Washington. *Journal of Paleontology* 48:375–386.
Gainer, R. S. 1995. Range extension of white-tailed deer. *Alberta Naturalist* 25:34–36.
Gallina, S., S. Mandujano, J. Bello, H. F. Lopez-Arevalo, and M. Weber. 2010. White-tailed deer (*Odocoileus virginianus*, Zimmerman 1780). In *Neotropical Cervidology*, eds. J. M. B. Duarte and S. Gonzalez, 101–118. Jaboticabal, Brazil: Funep.
Gavin, T. A. and B. May. 1988. Taxonomic status and genetic purity of Columbian white-tailed deer, an endangered species. *The Journal of Wildlife Management* 52:1–10.
Gay, M. M. and P. Gervais. 1846. Remarques sur le Capra pudu et L'Equus bisculus de Molina. *Annales des Sciences Naturelles pour la Zoologia* 5:87–94.
Geisler, J. H., J. M. Theodor, M. D. Uhen, and S. E. Foss. 2007. Phylogenetic relationships of cetaceans to terrestrial artiodactyls. In *The Evolution of Artiodactyls*, eds. D. R. Prothero and S. E. Foss, 19–31. Baltimore, MD: Johns Hopkins University Press.

Geist, V. 1992. Endangered species and the law. *Nature* 357:274–276.

Geist, V. 1998. *Deer of the World: Their Evolution, Behaviour and Ecology*. Mechanicsburg, PA: Stackpole.

Gentry, A. 1994. The Miocene differentiation of Old World Pecora (Mammalia). *Historical Biology* 7:115–158.

Gilbert, F., A. Ropiquet, and A. Hassanin. 2006. Mitochondrial and nuclear phylogenies of Cervidae (Mammalia, Ruminantia): Systematics, morphology, and biogeography. *Molecular Phylogenetics, and Evolution* 40:101–117.

Gmelin, J. F. 1788. *Systema Naturae*. Impensis Georg Emmanuel Beer, Lipsiae, 1:1–500.

Goldman, E. A. and R. Kellogg. 1940. Ten new white-tailed deer from North and Middle America. *Proceedings of the Biological Society of Washington* 53:81–90.

Gotch, A. F. 1995. *Latin Names Explained*. London: Cassel Publishing.

Gray, J. E. 1874. On *Xenelaphus*, *Furcifer* and *Coassus peruvianus* of Peruvian Alps. *Annals and Magazine of Natural History* 13:331–332.

Groves, C. P. 2007. Family Cervidae. In *The Evolution of Artiodactyls*, eds. D. R. Prothero and S. E. Foss, 249–256. Baltimore, MD: Johns Hopkins University Press.

Gunnell, G. F. and A. Foral. 1994. New species of *Bretzia* (Cervidae; Artiodactyla) from the latest Pleistocene or earliest Holocene of Nebraska and South Dakota. *Journal of Mammalogy* 75:378–381.

Gustafson, E. P. 1985. Antlers of *Bretzia* and *Odocoileus* (Mammalia, Cervidae) and the evolution of New World deer. *Transactions of Nebraska Academy of Sciences* 13:83–92.

Hall, E. R. and K. R. Kelson. 1959. *The Mammals of North America*. New York, NY: Ronald Press Company.

Halls, L. K. 1978. White-tailed deer. In *Big Game of North America: Ecology and Management*, eds. J. L. Schmidt and D. L. Gilbert, 43–65. Harrisburg, PA: Stackpole Books.

Hamilton-Smith, C. 1827. Ruminantia. In *The Animal Kingdom Arranged in Conformity with its Organization*, ed. E. Griffith, 296–376. London: G. B. Whittaker.

Handley, R. B. 1952. Deer management in Mississippi. *Mississippi Game and Fish* 16:3–10.

Hariot, T. 1588. *A Briefe and True Report of the New Found Land of Virginia*. New York, NY: Dodd, Mead and Company.

Harper, G. A. 2006. Habitat use by three rat species (*Rattus* spp.) on an island without other mammalian predators. *New Zealand Journal of Ecology* 30:321–333.

Harris, L. H. 1984. New Zealand. In *White-tailed Deer: Ecology and Management*, ed. L. K. Halls, 547–556. Harrisburg, PA: Stackpole Books.

Havier, J. B., Jr. 1987. *Amerindian Cultural Geography on Curaçao*. PhD dissertation, Leiden University, Amsterdam.

Hays, W. J. 1874. Description of a species of *Cervus*. *Annals of the Lyceum of Natural History of New York* 10:218–219.

Heffelfinger, J. R. 2000. Hybridization in large mammals. In *Ecology and Management of Large Mammals in North America,* eds. P. R. Krausman and S. Demarais, 27–37. Upper Saddle River, NJ: Prentice-Hall.

Heffelfinger, J. R. 2005. Mule deer subspecies: Nonsensical names or practical units of conservation and record-keeping? *Fair Chase* 24(1):48–52.

Heffelfinger, J. R. 2006. *Deer of Southwest*. College Station: Texas A&M University Press.

Heffelfinger, J. R., L. H. Carpenter, L. C. Bender, et al. 2003. Ecoregional differences in population dynamics. In *Mule Deer Conservation: Issues and Management Strategies*, eds. J. C. DeVos, M. R. Conover, and N. E. Headrick, 63–92. Logan, UT: Jack H. Berryman Institute.

Hernandez de Alba, G. 1963. The Highland Tribes of Southern Columbia. In *Handbook of South American Indians*, ed. J. Steward, 915–960. New York, NY: Cooper Square.

Hershkovitz, P. 1948. The technical name of the Virginia deer with a list of the South American forms. *Proceedings of the Biological Society of Washington* 61:41–48.

Hershkovitz, P. 1951. Mammals from British Honduras, Mexico, Jamaica, and Haiti. *Fieldiana Zoology* 31:547–569.

Hershkovitz, P. 1958. Technical names of the South American marsh deer and pampas deer. *Proceedings of the Biological Society of Washington* 71:13–16.

Hoefs, M. 2001. Mule, *Odocoileus hemionus*, and white-tailed, *O. virginianus*, deer in the Yukon. *Canadian Field Naturalist* 115:296–300.

Hooijer, D. A. 1960. Mammalian remains from Indian sites on Aruba. *Studies on the Fauna of Curaçao and other Caribbean Islands* 49:154–157.

Hummelinck, P. W. 1940. Studies on the fauna of Curacao, Aruba, Bonaire and the Venezuelan Islands. *Martinus Nijhoff, The Hague* 1:1–130.

Husson, A. M. 1960. De zoogdieren van de Nederlandse Antillen. *Fauna Nederlandse Antillen* 2:1–70.

International Code of Zoological Nomenclature. 1960. Opinion 581: Determination of the generic names for the fallow deer of Europe and the Virginia deer of America (class Mammalia). *The Bulletin of Zoological Nomenclature* 17:267–275.

Jacobson, J. A. 2004. Determining human ecology on the plains through the identification of mule deer (*Odocoileus hemionus*) and white-tailed deer (*Odocoileus virginianus*) postcranial material. PhD dissertation, University of Tennessee.

Jungius, V. H. 1974. Beobachtungen am weißwedelhirsch und an anderen Cerviden. *Säugetierkunde* 39:373–383.

Kairikko, J. K. and J. Ruola. 2005. *White-tailed Deer in Finland*. Jyväskylä, Finland: The Finnish Hunters Association.

King, C. M. 2005. *The Handbook of New Zealand Mammals*. Victoria, Australia: Oxford University Press.

Klein, E. H. 1982. Phenology of breeding and antler growth in white-tailed deer in Honduras. *Journal of Wildlife Management* 46:826–829.

Klimstra, W. D., M. J. Folk, and R. W. Ellis. 1991. Skull size of two insular and one mainland subspecies of *Odocoileus virginianus* from the southeast. *Transactions of the Illinois State Academy of Science* 84:185–191.

Krausman, P. R., and E. D. Ables. 1981. *Ecology of the Carmen Mountains White-tailed Deer.* Scientific Monograph Series Number 15. U. S. Department of the Interior, Washington, DC.

Krausman, P. R., D. J. Schmidly, and E. D. Ables. 1978. Comments on the taxonomic status, distribution, and habitat of the Carmen Mountains white-tailed deer (*Odocoileus virginianus carminis*) in Trans-Pecos Texas. *Southwestern Naturalist* 23:577–590.

Kurten, B. and E. Anderson. 1980. *Pleistocene Mammals of North America*. New York, NY: Columbia University Press.

Latch, E. K., J. R. Heffelfinger, J. A. Fike, and O. E. Rhodes, Jr. 2009. Species-wide phylogeography of North American mule deer: Cryptic glacial refugia and post-glacial recolonization. *Molecular Ecology* 18:1730–1745.

Latham, A. D. M. 2009. Wolf ecology and caribou–primary prey–wolf spatial relationships in low productivity peatland complexes in northeastern Alberta. PhD dissertation, University of Alberta.

Leberg, P. L. and D. L. Ellsworth. 1999. Reevaluation of the genetic consequences of translocations on deer populations. *Journal of Wildlife Management* 63:327–334.

Leberg, P. L., P. W. Stangel, H. O. Hillestad, R. L. Marchington, and M. H. Smith. 1994. Genetic structure of reintroduced wild turkeys and white-tailed deer populations. *Journal of Wildlife Management* 58:698–711.

Lever, C. 1985. *Naturalized Mammals of the World.* London: Longman.

Lewis, W. 1806. The journals of Lewis and Clark. http://www.gutenberg.org/etext/8419. Accessed October 1, 2009.

Linnaeus, C. 1758. *Tomus I. Systema naturae per regna tria naturae, secundum classes, ordines, genera, species, cum characteribus, differentiis, synonymis, locis*. Tenth Edition. Holmiae. (Laurentii Salvii): [1–4], 1–824. http://www.biodiversitylibrary.org/item/10277#266. Accessed November 5, 2009.

Long, J. L. 2003. *Introduced Mammals of the World: Their History, Distribution and Influence*. Australia: CSIRO.

Lydekker, R. 1898. *The Deer of All Lands: A History of the Family Cervidae Living and Extinct*. London: Rowland Ward, Ltd.

Maffei, M. D., W. D. Klimstra, and T. J. Wilmers. 1988. Cranial and mandibular characteristics of the Key deer (*Odocoileus virginianus clavium*). *Journal of Mammalogy* 69:403–407.

Mandujano, S. 2004. Analysis bibliografico de los estudios de venados en Mexico. *Acta Zoologica Mexicana* 20:211–251.

Mandujano, S., C. A. Delfin-Alfonso, and S. Gallina. 2008. Analysis biogeografico de las subespecies del venado cola blanca *Odocoileus virginianus* in Mexico. *Simposio de Venados en Mexico* 11:1–15.

Marchington, R. L., K. V. Miller, and J. S. McDonald. 1995. Genetics. In *Quality Whitetails*, eds. K. V. Miller and L. K. Marchington, 169–189. Mechanicsburg, PA: Stackpole.

Mendez, E. 1984. Mexico and Central America. In *White-tailed Deer: Ecology and Management*, ed. L. K. Halls, 513–524. Harrisburg, PA: Stackpole Books.

Merriam, C. H. 1898. The earliest generic name for the North American deer, with descriptions of five new species and subspecies. *Proceedings of the Biological Society of Washington* 12:99–104.

Merriam, C. H. 1918. Review of the grizzly and big brown bears of North America. *North American Fauna* Number 41. U.S. Department of Agriculture, Bureau of Biological Survey, Washington, DC.

Metais, G. and I. Vislobokova. 2007. Basal ruminants. In *The Evolution of Artiodactyls*, eds. D. R. Prothero and S. E. Foss, 189–212. Baltimore, MD: Johns Hopkins University Press.

Miller, F. W. 1928. A new white-tailed deer from Louisiana. *Journal of Mammalogy* 9:57–59.

Miller, G. S. 1900. Key to the land mammals of northeastern North America. *Bulletin of the New York State Museum of Natural History* 8:61–160.

Molina, M. and J. Molinari. 1999. Taxonomy of Venezuelan white-tailed deer (Mammalia, Cervidae, Odocoileus) based on cranial and mandibular traits. *Canadian Journal of Zoology* 77:632–645.

Molinari, J. 2007. Variación geográfica en los venados de cola blanca (Cervidae, *Odocoileus*) de Venezueala, con enfasis en *O. margaritae*, la especie de la Isla de Margarita. *Memoria de la Fundacion La Salle de Ciencias Naturales* 167:29–72.

Moscarella, R. A., M. Aguilera, and A. A. Escalante. 2003. Phylogeography, population structure, and implications for conservation of white-tailed deer (*Odocoileus virginianus*) in Venezuela. *Journal of Mammalogy* 84:1300–1315.

Mowat, G. and G. Kuzyk. 2009. Mule deer and white-tailed deer population review for the Kootenay Region of British Columbia. Unpublished report. British Columbia Ministry of Environment, Nelson, British Columbia.

Nowak, R. M. 1999. *Walker's Mammals of the World* (6th edition). Baltimore, MD: Johns Hopkins University Press.

O'Brien, S. J. and E. Mayr. 1991. Bureaucratic mischief: Recognizing endangered species and subspecies. *Science* 251:1187–1188.

Oelrich, T. M. 1953. Additional mammals from the Rexroad fauna. *Journal of Mammalogy* 34:373–378.

Osgood, W. H. 1910. Mammals from the coast and islands of northern South America. *Field Museum of Natural History, Zoological Series, Publication 149*, 10:23–32.

Paunovic, M., D. Cirovic, and J. D. C. Linnell. 2010. Ungulates and their management in Serbia. In *European Ungulates and their Management in the 21st Century*, eds. M. Apollonio, R. Andersen, and R. J. Putnam, 563–571. Cambridge: Cambridge University Press.

Philibosian, R. and J. A. Yntema. 1977. Annotated checklist of the birds, mammals, reptiles, and amphibians of the Virgin Islands and Puerto Rico. Frederiksted, St. Croix: Information Services.

Piaggio, A. J. and M. W. Hopken. 2009. Evolutionary relationships and population genetic assessment of Oregon white-tailed deer. U.S. Department of Agriculture/APHIS/WS/National Wildlife Research Center Report. Fort Collins, CO.

Pitra, C., J. Fickel, E. Meijaard, and P. C. Groves. 2004. Evolution and phylogeny of old world deer. *Molecular Phylogenetics and Evolution* 33:880–895.

Plante, M., K. Lowell, F. Potvin, B. Boots, and M. Fortin. 2004. Studying deer habitat on Anticosti Island, Quebec: Relating animal occurrences and forest map information. *Ecological Modelling* 174:387–399.

Prothero, D. R. 2007. Family Moschidae. In *The Evolution of Artiodactyls*, eds. D. R. Prothero and S. E. Foss, 221–226. Baltimore, MD: Johns Hopkins University Press.

Rafinesque, C. S. 1817. Museum of Natural Sciences. *The American Monthly Magazine and Critical Review* 1:431–442.

Rafinesque, C. S. 1832. Description of some of the fossil teeth in a cave in Pennsylvania. *Atlantic Journal* 1:109–110.

Rees, J. W. 1969. Morphologic variation in the mandible of the white-tailed deer (*Odocoileus virginianus*): A study of populational skeletal variation by principal component analyses. *Journal of Morphology* 128:113–130.

Saussure, H. D. 1860. Note sur quelques mammiferes du Mexique. *Revue et Magasin de Zoologie Pure et Applique* 12:1–494.

Scott, K. M. and C. M. Janis. 1987. Phylogenetic relationships of the Cervidae, and the case for a superfamily "Cervoidea." In *Biology and Management of the Cervidae,* ed. C. M. Wemmer, 3–20. Washington, DC: Smithsonian Institution Press.

Seaman, G. A. 1966. A short history of the deer of St. Croix. *Caribbean Journal of Science* 6:33–41.

Smith, M. H., W. V. Branan, R. L. Marchington, P. E. Johns, and M. C. Wooten. 1986. Genetic and morphological comparisons of brocket deer, brown deer and white-tailed deer. *Journal of Mammalogy* 67:103–111.

Smith, M. H., H. O. Hillestad, R. Baccus, and M. N. Manlove. 1984. Population genetics of the white-tailed deer. In *Ecology and Management of White-tailed Deer*, ed. L. K. Halls, 119–128. Harrisburg, PA: Stackpole Books.

Smith, P. 1991. *Odocoileus virginianus. Mammalian Species* 388:1–13.

Smith, W. P., L. N. Carraway, and T. A. Gavin. 2003. Cranial variation in Columbian white-tailed deer populations: Implications for taxonomy and restoration. *Proceedings of the Biological Society of Washington* 116:1–15.

Stirton, R. A. 1944. Comments on the relationships of the cervoid family Palaeolomerycidae. *American Journal of Science* 242:633–655.

Strickland, B. K. and S. Demarais. 2000. Age and regional differences in antlers and mass of white-tailed deer. *Journal of Wildlife Management* 64:903–911.

Strickland, B. K. and S. Demarais. 2008. Influence of landscape composition and structure on antler size of white-tailed deer. *Journal of Wildlife Management* 72:1101–1108.

Tarifa, T., J. Rechberger, R. B. Wallace, and A. Nunez. 2001. Confirmation de la presencia de *Odocoileus virginianus* (Artiodactyla, Cervidae) en Bolivia, y datos preliminares sobre su ecologia y su simpatria con *Hippocamelus antisensis. Ecologia en Bolivia* 35:41–49.

Theodor, J. M., J. Erfurt, and G. Metais. 2007. The earliest artiodactylas: Diacodexeidae, Dichobunidae, Homacodontidae, Leptochoeridae, and Raoellidae. In *The Evolution of Artiodactyls*, eds. D. R. Prothero and S. E. Foss, 32–58. Baltimore, MD: Johns Hopkins University Press.

Thomas, O. 1902. On the generic names *Notophorus*, *Alces*, *Dama*, and *Cephalotes*, with remarks on the "one-letter rule" in nomenclature. *Proceedings of the Biological Society of Washington* 15:197–198.

Trouessart, E. L. 1910. Mammiferes de la Mission de l'Equateur, d'apres les collections formees par Rivet. *Mission du Service Geographique de l'Armee pour la Mesure d'un Arc Meridian Equatorial en Amerique du Sud* 9:1–31.

U.S. Fish and Wildlife Service. 2008. Technical/agency draft, Key deer recovery plan (*Odocoileus virginianus clavium*), third revision. U.S. Fish and Wildlife Service. Atlanta, Georgia.

Van Devender, T. R. 1977. Holocene woodlands in the southwest deserts. *Science* 198:189–192.

Veitch, A. M. 2001. An unusual record of a white-tailed deer, *Odocoileus virginianus*, in the Northwest Territories. *Canadian Field-Naturalist* 115:172–175.

Villarreal, O. A., G. M. Martinez, J. E. Hernandez, F. J. Franco, and J. C. Camacho. 2009. Conformation del libro de record de venado cola blanca para la subespecie Mexicanus. In *Proceedings of the 4th Simposio Sobre Fauna Cinegetica de Mexico*, ed. O. A. Villarreal, 154–164. Puebla, Mexico: Benemérita Universidad Autónoma de Puebla.

Webb, S. D. 1992. A cranium of Navahoceros and its phylogenetic place among new world cervidae. *Annales Zoologici Fennici* 28:401–410.

Webb, S. D. 2000. Evolutionary history of new world Cervidae. In *Antelopes, Deer, and Relatives: Fossil Record, Behavioral Ecology, Systematics, and Conservation*, eds. E. S. Vrba and G. B. Schaller, 38–64. New Haven, CT: Yale University Press.

Webb, S. D. 2006. The great American biotic interchange: Patterns and processes. *Annuals of the Missouri Botanical Gardens* 93:245–257.

Wheatley, P. V., and D. R. Ruez. 2006. Pliocene O*docoileus* from Hagerman Fossil Beds National Monument, Idaho, and comments on the taxonomic status of *Odocoileus brachyodontus. Journal of Vertebrate Paleontology* 26:462–465.

Whitehead, G. K. 1950. *Deer and Their Management in the Deer Parks of Great Britain and Ireland.* London: Country Life Limited.

Whitehead, G. K. 1972. *Deer of the World.* London: Constable.

Whitehead, G. K. 1993. *The Whitehead Encyclopedia of Deer.* London: Quiller.

Wiegmann, A. F. A. 1833. Allgemeine Betrachtungen fiber die Hirscharten. *Isis von Oken, Brockhaus zu Leipzig, Zurich* 10:950–980.

Wilson, E. O. and W. L. Brown, Jr. 1953. The subspecies concept and its taxonomic application. *Systematic Zoology* 2:97–111.

Wilson, D. E. and D. M. Reeder. 2005. *Mammal Species of the World: A Taxonomic and Geographic Reference.* Baltimore, MD: Johns Hopkins University Press.

Wishart, W. D. 1984. Western Canada. In *White-tailed Deer: Ecology and Management*, ed. L. K. Halls, 475–486. Harrisburg, PA: Stackpole.

Zimmerman, E. A. W. 1780. Geographische Geschichte des Menschen und der vierfüßigen Thiere. In der *Weygandschen Buchhandlung, Leipzig* 2:1–432.

Part II

Biology

2

Anatomy and Physiology

Stephen S. Ditchkoff

CONTENTS

The white-tailed deer exhibits considerable variation in anatomy and physiology across its range. Factors such as latitude, climate, and habitat influence morphological characteristics, as well as physiological adaptations that allow white-tailed deer to thrive across North and South America. In all cases, these adaptations have evolved over thousands of years in response to local selective pressures to enhance survival and productivity of the species. Detailed knowledge and understanding of anatomy and physiology is critical to the management and research of white-tailed deer. Most behavioral aspects (e.g., reproduction, foraging, predator avoidance, social interactions) of white-tailed deer are physiologically driven, and a thorough understanding of these processes requires intimate knowledge of their physiological basis. Anatomical attributes of many internal and external structures in white-tailed deer strongly influence function (e.g., digestion, vision), and so it is also critical that managers and researchers understand anatomy as well. Finally, hunters and others who interact with white-tailed deer from a recreational perspective could increase their enjoyment of this renewable resource by understanding anatomy and physiology.

Anatomy and physiology generally correspond to local conditions and are highly predictable. However, wide-scale translocations and reintroductions have resulted in a mosaic of genetic strains across North America. For example, in the early and mid-1900s, many states across the Southeast began restocking programs to revitalize populations that had been extirpated or driven to rarity. Sources of deer used in these restocking programs were highly variable, and often translocated deer had substantially different physical or physiological characteristics than local deer. For example, from 1926 to 1998 there were 44 documented translocations of white-tailed deer in the state of Alabama (McDonald and Miller, 2004). While about 85% of these deer were relocated from populations within the state, 579 deer were moved to Alabama from Arkansas, Georgia, Michigan, Ohio, North Carolina, Texas, and Wisconsin. In all, these restocking programs involved deer from six subspecies. Deer from northern regions were larger than those translocated from instate populations. Additionally, the Alabama stock used during these restocking efforts had a traditionally late breeding season (January), which has resulted in a mosaic of physiologically driven breeding dates across the state (Causey, 1990; Gray et al., 2002). Similar patterns of varying physiology and physical characteristics are found in other states where white-tailed deer populations have become a mosaic of highly variable genetic strains. As a result, white-tailed deer anatomy and physiology may not always be predictable based solely on geography.

Physical Characteristics

Pelage

The pelage of an adult white-tailed deer is normally a uniform reddish-brown to gray on the head, back, sides, and legs. During summer, the coat thickness is light with little underfur, and typically the coloration is red or rust. However, as cold weather sets in, white-tailed deer grow their winter pelage which is brown intermixed with gray, is generally thicker than the summer coat, and has a well-developed underfur layer. White fur is found on the abdomen and chest, the inside of the legs, around the chin, inside the ears, and on the underside and edges of the tail. Some adult deer have a defined white throat patch that may be continuous with white fur under the chin. Markings on the neck and face may vary considerably among deer. Most deer have some black hair immediately posterior to the nose pad, and then a patch of white hair immediately posterior to the black. Both the black and white hair posterior to the nose pad may continue onto the lower jaw. There is normally some white hair around the eyes, with the remainder of the hair on the head and neck being brown. Variation in facial markings can often be used to uniquely identify individual deer. Other areas on the body that may exhibit unique color patterns are the back of the ears and tail. Some deer will carry the prototypical red-brown color on the back of the ears and tail, while others will have black.

Fawns are born with a cryptic, camouflage coloration. The base color is red or brown with white spots 1–2 cm in diameter along the back and sides. There is normally a line of white spots from the neck to the base of the tail on each side of the spine. This coloration enables the bedded fawn to blend into its surroundings because spots break up the fawn's outline and mimic patches of sunlight filtering through vegetation. The spots begin to fade after about two months of age and normally are not visible after four months. In many cases, fawns will retain a reddish hue during winter, as opposed to adults that replace their reddish summer pelage with brown or gray fur. Observant sportsmen can sometimes use this difference in coloration to help classify deer during the hunting season.

While most white-tailed deer exhibit the pelage coloration pattern described above, occasionally individual deer are found to be all, or partially, white (Newsom, 1937; Taylor, 1956; Ryel, 1963; Hesselton, 1969; Martin and Rasmussen, 1981). Albinism is a rare recessive trait in which individuals exhibit white pelage and pink eyes because of a lack of pigmentation in the eyes, skin, and hair. Although very rare

FIGURE 2.1 A white-tailed deer buck with melanistic pelage. Notice the lack of white hairs around the head, neck, chest, and legs. Compare the coloration to that of a normal white-tailed deer on the right. (Photo by J. T. Baccus. With permission.)

in natural populations, the frequency of albinism in a population can increase due to inbreeding (Smith et al., 1984) or laws that protect white animals from harvest (Martin and Rasmussen, 1981). More common than albinos are partially white or piebald deer. These deer have variable amounts of white pelage across the body but lack the pink eyes that are a defining characteristic of true albinos. Smith et al. (1984) reported that less than 1% of deer in hunted populations are piebald. The low frequency of white, or partially white, deer in natural populations has been attributed to increased susceptibility to predators and to such deer having a greater incidence of physical deformities such as dorsal bowing of the nose, short legs, scoliosis, or short mandibles that increase the likelihood of death at an early age (Davidson and Nettles, 1997).

Even rarer than albinos are melanistic white-tailed deer (Rue, 1978). Melanistic animals lack distinctive variations in color such as brown or white pelage and are dark or even black across the majority of the body (Figure 2.1). In white-tailed deer, melanistic individuals normally are black across the entire body, with the exception of white hairs on the ventral surface of the tail extending onto the anal region and on the tarsal and metatarsal glands (Baccus and Posey, 1999). Melanistic deer may have a distinctive darker mid-dorsal stripe that extends from the head to the tail. Baccus and Posey (1999) described semimelanistic deer as those who retain the typical white pelage patterns (e.g., face, neck, tail, and ventral surface) of nonmelanistic deer, but black hairs have replaced the reddish-brown to gray hues that normally cover the rest of the body. Melanistic deer have been documented in Texas (Smith et al., 1984; Baccus and Posey, 1999), Wisconsin (Wozencraft, 1979), South Carolina, Michigan (Rue, 1978), Pennsylvania (D'Angelo and Baccus, 2007), Idaho (Severinghaus and Cheatum, 1956), and New York (Townsend and Smith, 1933). Melanism is a genetic morphism that coexists with typical color morphs in temporary or permanent balance such that this condition may occur in the same locale on a semiregular basis (Ford, 1945). Although melanism is extremely rare, Baccus and Posey (1999) reported an incidence of 8.5% of melanistic white-tailed deer in eight counties in central Texas.

Fetal Development

Gestation in white-tailed deer is approximately 200 days, although gestation lengths have been reported as short as 187 days (Haugen, 1959) and as long as 213 days (Verme, 1969). By about 37 days of gestation, the period of tissue differentiation and organ development is complete, and what was once considered an embryo is now a fetus (Armstrong, 1950). External pinnae become apparent between 58 and 65 days of gestation and pigmentation of the skin first becomes apparent between 79 and 95 days. Spots appear on the fetus at about 145 days, pigmentation and hair patterns are complete by 160 days, and by 180 days the fetus has all the appearances of a neonate (Short, 1970). Fetal development is curvilinear, such that 75–80% of fetal growth occurs during the last trimester (Armstrong, 1950). Peak energetic costs of gestation during the third trimester are 84% greater than fasting metabolic rate (Pekins et al., 1998). This pattern of fetal growth is referred to as delayed development, which ensures that the greatest nutrient demands of gestation generally occur when forage availability is adequate, rather than during winter (Robbins et al., 1975). From about 40 days after conception until parturition, fetal growth is highly predictable (Soprovich, 1992), and measures of fetal development such as forehead-rump measurements can be used to accurately estimate age, and consequently conception dates (Hamilton et al., 1985).

Neonates and Fawns

At birth, white-tailed deer weigh from 1.8 to 4.1 kg (Trodd, 1962; Verme, 1989) depending on litter size, geographic region, and maternal nutrition. Small neonates have a difficult time suckling due to less physical strength and endurance, and many studies have documented positive relationships between birth mass and survival in ungulates (Clutton-Brock et al., 1982; Fairbanks, 1993; Sams et al., 1996).

Litter size in white-tailed deer is normally one or two, and the average litter size of adult deer in a healthy population ranges from 1.6 to 1.8 (Roseberry and Klimstra, 1970; Wilson and Sealander, 1971; Haugen, 1975; Johns et al., 1977; Jacobson et al., 1979; Kie and White, 1985; Rhodes et al., 1985; Ozoga, 1987; Verme, 1989). Triplets are not uncommon in white-tailed deer and litter sizes of four and even five have been reported (Trodd, 1962; Van Deelen et al., 2007). Litter size is associated positively with female age (Table 2.1). Younger does tend to have smaller litter sizes because of the competing demands of growth and reproduction, and normally, fawns do not reproduce unless they are in exceptional nutritional condition. The high nutritional demands of producing twin fawns normally results in lower birth mass per fawn compared to singleton litters (Verme, 1963). For example, male and female singleton fawns in Michigan averaged 4.1 and 3.9 kg, respectively, while twin males and twin females averaged 3.7 and 3.5 kg each, respectively (Verme, 1989). Male and female fawns from mixed-sex litters averaged 3.6 and 3.4 kg, respectively. In most ungulate species females differentially invest in male fetuses, which results in greater birth weight of male offspring (Clutton-Brock et al., 1982; San José et al., 1999; Adams, 2005). Clutton-Brock et al. (1981) speculated that females in polygynous mammals invest greater resources in sons than daughters during gestation because of the greater potential reproductive success of sons relative to daughters, and the influence that birth mass has on lifetime reproductive success in sons.

Gestation is a productive process that requires nutritional resources beyond those needed for maintenance of the body. Therefore, females in better condition or on a higher nutritional plane will have greater resources available for investment in productive processes, and the result will be larger fawns at birth. Verme (1963) found during experimental trials that does with adequate nutrition and in good condition produced fawns that were 11% greater in mass on average than does that had experienced

TABLE 2.1

Litter Size of Female White-tailed Deer in Relation to Age at the Time of Breeding

	Doe Age at the Time of Breeding				
State	**0.5**	**1.5**	**2.5**	**≥3.5**	**References**
Illinois	1.00	1.76	1.90	1.93	Roseberry and Klimstra (1970)
Manitoba	1.14	1.34	1.91[a]		Ransom (1967)
Minnesota		1.30	1.80[a]		DelGiudice et al. (2007)
Mississippi	1.00	1.40	1.66[a]		Jacobson et al. (1979)
New York	1.26	1.58	1.84[a]		Hesselton and Jackson (1974)
South Carolina	1.06	1.56	1.73	1.76	Rhodes et al. (1985)

[a] Mean number of fetuses is for does that are ≥2.5 years of age.

nutritional restriction during gestation. A variety of factors including climate, deer density, habitat quality, doe age, and prior reproductive output can influence the nutritional status of gravid does. Climatic factors such as extreme winter can negatively influence resources available to does and result in lower birth weights (Verme, 1965). Mech et al. (1987) speculated that climatic influences on maternal condition can be cumulative over a number of years, and periods of nutritional restriction that occurred several years previously may still negatively influence reproduction. Additionally, Mech et al. (1991) surmised that climatic influences can be multigenerational and reported that fawns born to does whose mothers experienced harsh winters during gestation had greater probability of mortality due to wolf predation. Low birth weights of mothers that had experienced reduced prenatal growth due to harsh climatic conditions during gestation negatively influenced prenatal and/or postnatal growth of their offspring.

Population density (Kie and White, 1985) and habitat quality (Rhodes et al., 1985) strongly influence food availability and reproduction in white-tailed deer. Young females (fawns or yearlings) normally produce fewer fawns or fawns with lower birth mass than prime-aged does (Roseberry and Klimstra, 1970; Jacobson et al., 1979; Rhodes et al., 1985; Verme, 1989). Younger deer must balance the competing nutritional demands of reproduction and growth, and cannot invest the same resources in reproduction as older does (Rhodes et al., 1985). Reproductive expenditures (e.g., gestation, lactation) during the previous year can also negatively influence the condition of a doe and result in reduced productivity (Cheatum and Severinghaus, 1950; Verme, 1967).

Postnatal growth of fawns is a function of maternal nutrition and milk production. Lactation is probably the most costly activity among any mammalian species; the nutritional demands of lactation far exceed those of gestation (Moen, 1973; Robbins, 1993). Postnatal growth of fawns will be influenced by litter size (e.g., competition with siblings for milk), milk production of the dam, and, like all other aspects of reproduction, by population density, habitat quality, and climatic factors. Verme (1989) reported that by three months of age, singleton fawns could weigh up to 5 kg more than twin fawns. This trend was particularly apparent among fawns born to 2-year-old does; older does seemed better able to obtain nutrients required to sustain lactation for twin fawns. These data suggest, like birth weights, that nutrient demands of growth in young does compete with lactation and thereby negatively influence fawn growth. In cases where poor maternal condition negatively influences birth weight and one fawn of a pair dies shortly after parturition, the remaining fawn may exhibit a growth rate greater than twin fawns of does in better condition because of greater availability of milk (Verme, 1963). Male fawns exhibit greater growth rates than female fawns (Verme, 1989), which is partly a function of greater milk consumption because of more and longer suckling bouts relative to females (Clutton-Brock et al., 1981).

Subadults and Adults

Body mass and size vary considerably in both subadults and adults across the range of white-tailed deer. Body size in white-tailed deer tends to be associated positively with latitude. This general tendency holds true for adult males and females, yearlings, and even fawns. For example, body mass of yearling bucks ranges from less than 30 kg in parts of Texas to over 50 kg in many northern states (Table 2.2). Adult body mass follows similar patterns. Mature, male white-tailed deer in northern regions may exceed 180 kg, but body mass commonly ranges from 60 kg to over 100 kg across most of their range (Sauer, 1984). Subspecies, habitat quality, age, season, genetic quality, and other factors may also influence body size in adult deer. The smallest subspecies of white-tailed deer is the Key deer, found only on a group of small islands off the southern coast of Florida. At one year of age, bucks weigh only 19.1 kg and does weigh only 16.7 kg (Hardin et al., 1984). Geographic isolation to islands and the resulting ecological and evolutionary pressures associated with island populations have likely influenced the body size of this subspecies.

The body size of adult deer is generally attributed to effects of genetics, nutrition, and other factors such as disease and injury. However, a recent study has suggested that other less obvious factors may influence adult body size. Mech et al. (1991) originally proposed that nutritional effects can be multigenerational in white-tailed deer, and found that regardless of nutrition through life, the nutritional conditions experienced by the grandmother of a deer may influence life-history traits throughout that individual's life. Recently, Monteith et al. (2009) reported convincing evidence that multigenerational

TABLE 2.2

Eviscerated Carcass Mass (kg) of Yearling Male and Female White-tailed Deer in North America

	Males		Females		
Location	**n**	**Mass**	**n**	**Mass**	**References**
Texas					
Edwards Plateau	101	25.7	36	23.8	Teer et al. (1965)
Llano Basin	43	24.4	146	23.0	
Florida	253	41.5[a]			Shea et al. (1992)
Georgia	18	31.4	8	27.1	Wentworth et al. (1992)
Mississippi					
Coastal Flatwoods	138	34.9	199	28.5	Strickland et al. (2008)
MS River Delta	4164	47.5	6104	37.5	
South Carolina	131	43.9[a]	372	37.3[a]	Ditchkoff et al., unpublished data
Oklahoma	250	36.8	94	32.0	Ditchkoff et al. (1997)
Tennessee	NA	42.6[c]	NA	35.6[c]	Jenks et al. (2002)
Kentucky					
Central	NA	50.5	NA	40.3	Dechert (1967)
Western	235	43.6	60	37.0	Feldhammer et al. (1989)
Missouri					
Northeast			214	44.3	Stoll and Parker (1986)
Southern			34	34.1	
Illinois	88	44.8[b]	114	36.6[b]	Roseberry and Klimstra (1975)
West Virginia					
Eastern	451	35.9	295	32.7	Gill (1956)
Western	436	47.7	417	41.3	
Ohio					
Northwest	531	56.1	244	47.1	Tonkovich et al. (2004)
Southeast	824	48.3	468	42.2	
Michigan	42	61.2[a]	50	54.3[a]	Ozoga and Verme (1982)
New York	166	37.9	144	35.9	Severinghaus (1955)
			71	43.5	Hesselton and Sauer (1973)
Minnesota	278	50.2	55	44.0	Fuller et al. (1989)

[a] Whole-body mass.
[b] Eviscerated carcass mass was calculated from whole-body mass using regression equations from Roseberry and Klimstra (1975).
[c] Reported figures are an estimate of data described in graphical format in the literature.

nutritional effects influence the body size of white-tailed deer. In the wild, deer from southwestern South Dakota are about 30% smaller than their counterparts in agricultural areas of eastern South Dakota: antler size is also smaller in these animals. In captivity, first-generation animals from these same genetic stocks showed these same patterns in body size, even though they were all raised on high-quality food in similar conditions. However, second-generation males from southwestern South Dakota attained body mass and antler size approaching that of second-generation captive males from the eastern part of the state (Figure 6.7). These data suggest that nutritional restriction takes several generations to overcome, and body size of individual deer may be strongly influenced by nutritional conditions during that deer's life time and the nutritional plane of its mother and perhaps grandmother.

Adult male and female white-tailed deer display different patterns of growth into adulthood. By two to four years of age, females normally attain maximum body size, while males increase in body size well past this age (Figure 2.2). This general pattern is apparent across the white-tailed deer's range. Growth patterns differ between males and females because of differences in reproductive strategies

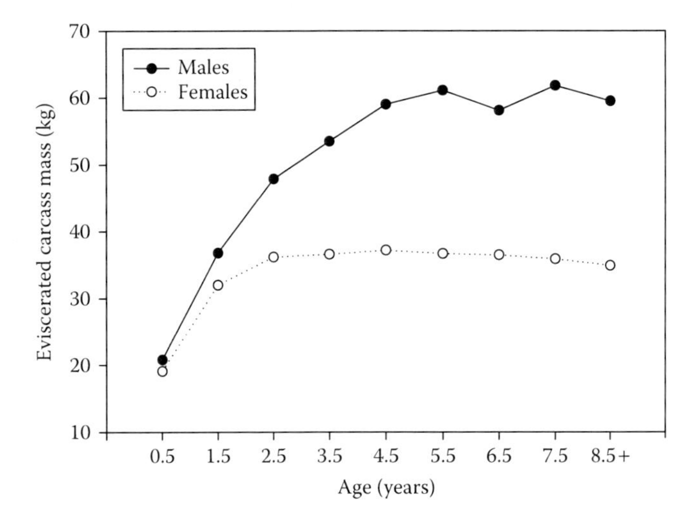

FIGURE 2.2 Eviscerated carcass mass of adult male and female white-tailed deer harvested at the McAlester Army Ammunition Plant in southeast Oklahoma during a period of quality management from 1989 to 1996. (Data from deer in Ditchkoff, S. S. et al. 1997. *Proceedings of the Annual Conference of the Southeastern Association of Fish and Wildlife Agencies* 51:389–399.)

and patterns of reproductive success between the sexes. In white-tailed deer and other polygynous species, female body size does not influence breeding opportunity: small females will be bred just as readily as large females. Thus, there is little advantage, from a reproductive perspective, for a female to invest resources to increase body size. Instead, reproductive success of female deer and survival and reproductive success of their offspring benefit from resources invested in offspring during gestation and lactation. Because fawn birth mass is associated positively with survival (Sams et al., 1996), resources diverted by a doe from adult growth to fetal development improve the chance her fawns will survive. The energetic costs of gestation are 16.4% greater than the energetic requirements for non-pregnant does (Pekins et al., 1998), and thus pregnant females cannot afford to put resources toward adult growth if reproduction is a priority. Reproductively active females also need to cope with the cost of lactation. In terms of energy and nutrients, lactation is two to three times more costly than gestation (Robbins, 1993). Moen (1973) reported that the energetic cost to a female with twins at the end of gestation was 1.64 times the basal metabolic rate (BMR), and at the peak of lactation was 2.3 times the BMR. Reproduction is a long-term investment for females, and lifetime reproductive success will ultimately be a function of resource investment. Because a doe's ability to acquire nutrients and support reproduction is partially a function of body size, young does are forced to balance the costs of growth with those of reproduction, and litter sizes are reduced in younger does. When nutrient availability is limited, females will adopt a strategy that favors their own survival, reproductive potential, and growth over that of their offspring (Therrien et al., 2007). In these cases, fawn growth rates and survival will be less than normal.

In contrast to females, males continue to increase in body size until five years of age or older. Reproductive success in males is a function of their ability to acquire breeding opportunities. Male breeding success will be a function of dominance, and dominance is at least partially driven by body size. In short, larger males will acquire more breeding opportunities than smaller-bodied males, and hence have greater reproductive success. In contrast to females, males have the luxury of being able to divert resources to growth without compromising reproduction. In fact, their reproductive success is dependent upon this allocation of resources. Those males who are more efficient at or are better able to allocate resources to growth will achieve greater lifetime reproductive success than their counterparts. Additionally, larger males may benefit by being able to store greater energy reserves, enabling them to spend more time searching for and tending mates and less time searching for food during the breeding season. Because large males already possess a body size that allows them to effectively compete for breeding opportunities, they have the luxury of diverting resources to energy stores. In contrast, younger, smaller males must continue to invest in body growth at the expense of fat deposition because they do not yet possess a body size that enables them to effectively compete for potential mates. As a result, their effort during the breeding season will be limited by competing demands of feeding and searching for mates.

Senescence

Declines in productivity, body mass, antler size, or physical and physiological condition with advancing age are evidence of senescence, and it is generally believed that white-tailed deer begin to senesce by 10 years of age. Although understanding senescence is important for predicting population dynamics in herds with low adult mortality, few studies have examined senescence in white-tailed deer. DelGiudice et al. (2007) examined fertility in free-ranging female white-tailed deer through 15 years of age and found no measurable reduction in the number of young produced per female. Similarly, Masters and Mathews (1990) found that does greater than 10 years of age exhibited little evidence of reproductive senescence. Mech and McRoberts (1990) reported no evidence of declines in body mass of female deer up to 12 years of age. There are few data available for male body mass at older ages. The probability of mortality increased after six years of age in female deer (DelGiudice et al., 2002, 2006) and at a similar age in male deer (Ditchkoff et al., 2001c; Webb et al., 2007). Increased mortality at older ages is likely a function of nutritional decline due to dentition wear and the subsequent increase in susceptibility to other mortality factors. Misrepresentation of productivity or the presence of extremely old deer in population models likely has little measurable impact for most populations because very few animals live past 10 years of age. However, in populations that have older age structures because hunting is either tightly controlled or absent, old individuals could influence population productivity (Masters and Mathews, 1990).

Exocrine Glands

Seven glands or regions of enhanced glandular activity have been identified in white-tailed deer (Figure 10.4). Glands in white-tailed deer are important during olfactory communication, and may communicate information such as sex, social status, reproductive status, individual identity, genetic characteristics, and condition (see Chapter 10). Tarsal glands are located on the medial surface of the hind legs at the tarsal joint. Metatarsal glands are located on the outside of each hind leg 10–15 cm above the hoof. Quay (1971) noted regional variation in the development of metatarsal glands. North of the Mexican border, almost all deer exhibit fully developed metatarsal glands, whereas the frequency of deer exhibiting metatarsal gland development in populations south of the United States ranges from 0% to 94%. Interdigital glands are located on both the front and rear legs between the hooves. Substances secreted by the interdigital, tarsal, and metatarsal glands are believed to serve as kairomones to some tick species (Carroll et al., 1998; Carroll, 2001). The presence of these secretions on the ground and vegetation may assist ticks in identifying ambush sites and locating deer hosts. Additionally, Wood et al. (1995) found that interdigital gland secretions in mule deer have antimicrobial properties and speculated that these secretions may serve as a defense mechanism against microorganisms.

Preorbital glands are found on the lower-front portion of the eye (Sauer, 1984). The forehead region of deer contains large numbers of apocrine glands. Activity of and secretions by these glands appear to be greater in males than females, and greater in dominant than subordinate males (Atkeson and Marchinton, 1982). Preputial glands are located on the ventral surface of the prepuce, and have been described by Odend'hal et al. (1992) as enlarged sebaceous glands that are normally associated with a hair follicle. Nasal glands are located within the haired skin of the lateral wall of the nostrils (Atkeson et al., 1988). Nasal glands, unlike the other glands described here, do not appear to function in chemical communication. The chemical composition of some deer glands and the association of these compounds with age and sex have been reported (Gassett et al., 1996, 1997).

Reproductive Physiology

Females

Endocrinology

White-tailed deer are seasonal polyestrous breeders, with a breeding season during autumn to early winter across most of their range. Timing of the breeding season is linked to photoperiod, and as such, there is a general continuum in breeding season timing associated with latitude. Deer in more northern regions tend to breed in November, whereas the breeding season in southern regions may be as late as January or February (Verme and Ullrey, 1984) (Figure 16.1). Melatonin serves as a physiological calendar for white-tailed deer and other mammals. Melatonin is produced during periods of darkness by the pineal gland, and as the day length decreases from summer to autumn, production and circulating levels of melatonin gradually increase. When melatonin levels reach a critical concentration, a series of hormonal events ensue.

Hormonal control of estrus is governed by the hypothalamus, the anterior lobe of the pituitary, and ovaries (Figure 2.3). A few days prior to ovulation, melatonin stimulates the hypothalamus to release gonadotropin-releasing hormone (GnRH). This surge of GnRH causes a sharp increase in the production of luteinizing hormone (LH) by the anterior pituitary, which is responsible for stimulating ovulation. Plotka et al. (1980) noted that serum LH was no more than 1.0 ng/mL, except for the day of ovulation when the mean concentration was 26.4 ng/mL. Knox et al. (1992) found that LH levels ranged from 35.0 to 60.1 ng/mL on the day of ovulation, but also noted that LH concentrations were slightly elevated in the days preceding ovulation and, in some study animals, were elevated prior to entering estrus. They described these LH surges as pre-estrus peaks associated with the termination of seasonal anestrous. Following ovulation, corpora lutea develop and produce progesterone, which is essential for the development and maintenance of the uterine environment during pregnancy. Prior to ovulation, serum progesterone is normally less than 2 ng/mL, but increases to more than 5 ng/mL following ovulation (Plotka et al., 1977).

Estrogen is produced by the ovaries, and its role in reproduction is multifaceted. The rise in estrogen during the days prior to estrus serves to stimulate breeding behavior and development of the uterine environment for receipt of a fertilized embryo. Estrogen levels are normally 5–30 pg/mL during the month preceding estrus (Plotka et al., 1980; Knox et al., 1992), and peak at or near the day of estrus. Following estrus, estrogen concentrations decline, but slowly rise throughout gestation and peak at parturition (Plotka et al., 1977), serving as a signal to the female body that parturition is approaching.

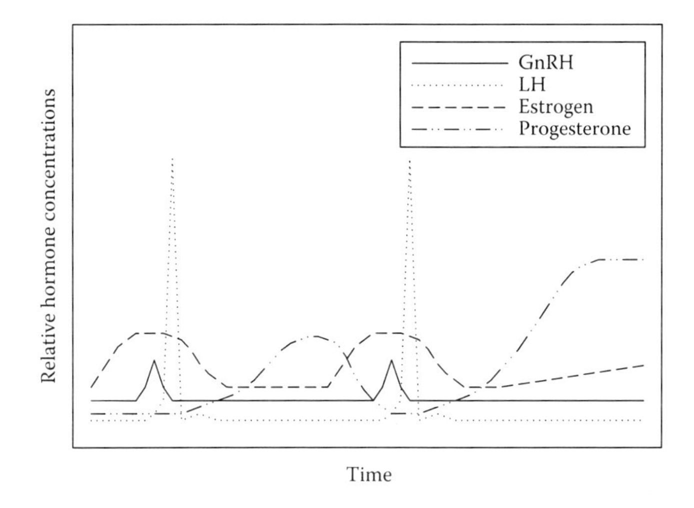

FIGURE 2.3 Timing and pattern of hormone production in adult, female white-tailed deer during the estrous cycle. Peaks in LH concentration occur at ovulation, and sustained elevated concentrations of progesterone following the second ovulation are representative of pregnancy.

If a female fails to become pregnant during an estrous cycle, she will enter estrus again 21–30 days later. This cycle of recurrent estrus in deer may repeat up to seven times before the female stops cycling and enters anestrus (Knox et al., 1988). If pregnancy does not occur, the corpora lutea will decrease in size, and progesterone concentrations will decline about seven days prior to the next ovulation (Plotka et al., 1980). When the ratio of progesterone to estrogen reaches a critical level, GnRH will once again trigger a surge in LH, which will lead to ovulation.

Fertility control in white-tailed deer is based largely on understanding the endocrine control of the reproductive cycle. The three primary approaches to reducing fertility in deer include hormonal implants, contragestational agents, and contraceptive vaccines. Norgestomet and levonorgestrel, synthetic derivatives of progesterone, have both been studied for use in white-tailed deer. These compounds simulate hormonal conditions of pregnancy and prevent ovulation. DeNicola et al. (1997a) reported that norgestomet could be effective at reducing fertility rates in white-tailed deer, while White et al. (1994) had limited success in reducing the fertility of deer with levonorgestrel implants. Prostaglandin $F_{2\alpha}$, a contragestational agent that acts on the corpus luteum to cause luteolysis, can also reduce pregnancy rates in deer (DeNicola et al., 1997b; Waddell et al., 2001). Another approach to reduce fertility is by blocking the GnRH pathway. GnRH agonists prevent GnRH from binding to receptors in the anterior pituitary and subsequently stop production of LH and other hormones essential for ovulation. GnRH agonists are effective at reducing pregnancy rates in deer (Miller et al., 2000; Baker et al., 2004), but may also cause behavioral changes because GnRH is the initial hormone responsible for stimulation of the behavior and physiology of reproduction.

Porcine zona pellucida (PZP) has received the most attention as an immunocontraceptive in recent years (Turner et al., 1996; Miller et al., 2001; Walter et al., 2002; Locke et al., 2007). The zona pellucida is a glycoprotein coating that surrounds the cell membrane of an oocyte and is involved in sperm binding. When PZP is injected into a female, her immune system mounts an immunological defense against the foreign material and, concurrently, against her own zona pellucida, thereby preventing pregnancy. The greatest advantage of immunocontraceptives compared to other approaches to fertility control is that they do not alter hormonal balance, and thus cause associated negative side effects. While many feel that these forms of fertility control show promise as tools to reduce overpopulated herds of white-tailed deer, inherent limitations associated with remotely delivering these compounds to enough deer to have a measurable impact on herd productivity preclude use in most areas (Warren, 1995; Muller et al., 1997). The greatest limitation of these compounds is that they are not effective if delivered orally, and thus require much time and money to administer.

Lactation

Near the end of gestation, the anterior pituitary produces prolactin, which signals the mammary tissue to begin development. This stage in the reproductive cycle is a critical, yet often overlooked aspect of deer population productivity. Production of milk is a defining characteristic of a mammal and a critical stage in the development of young. Neonatal development and survival is largely a function of milk consumption, and fawns that consume inadequate quantities of milk have reduced growth rates and survival. The first few days of a neonate's life are critical with regard to milk consumption. Additionally, immune system health is dependent upon adequate milk consumption during the first 24 hours because colostrum is present in the milk during this period. Colostrum is comprised of large numbers of immunoglobulins representing antibodies present in the maternal blood at the time of birth. This passive transfer of immunity from doe to fawn is the fawn's first line of immunological defense. Immunity to additional pathogens will be acquired as the fawn is exposed to novel immune challenges. When colostrum transfer is compromised, gamma globulin levels in neonate serum are generally lower, and the probability of mortality is elevated during the first 21 days of life (Sams et al., 1996). When the immune system is compromised in neonates, they are more susceptible to parasitic infection, and levels of tumor necrosis factor-α (TNF-α) may be elevated (Ditchkoff et al., 2001a). Elevated TNF-α stimulates the mobilization of peripheral energy reserves in support of metabolic demands associated with an inflammatory response. Thus, inefficient nursing during the time of colostrum transfer in fawns with low birth mass may lead to immune system deficiencies (Sams et al., 1996; Ditchkoff et al., 2001a).

Milk production is often limited by the high nutritional demands placed on the lactating female. As a result, there is considerable variation in the quantity of milk produced as a function of a female's condition and ability to acquire the necessary nutritional resources. White-tailed deer milk ranges in water content from 66.5% to 77.8%, in protein from 10.1% to 11.5%, in sugar from 2.2% to 3.0%, and in ash from 1.6% to 1.8% (Silver, 1961). Fat content varies from 7.5% to 18.0% and is greatest near weaning. Undernutrition of the female during lactation has a greater impact on milk production than milk composition. In most cases, milk composition is not affected.

Secondary Sex Ratios

At birth, the sex ratio of fawns is approximately 1:1, and yet several theories that describe differential investment of mothers in sons and daughters have been put forth. Trivers and Willard (1973) suggested that females in good condition should invest in sons, whereas females in below average condition should produce daughters. The differential cost of producing sons and daughters, and variation in potential lifetime reproductive success of well-nourished and undernourished sons and daughters is the basis of this theory. Verme (1983) reported that the opposite holds true in white-tailed deer. More male offspring tend to be produced by younger females, females in poor condition, and females in poor-quality habitat, the opposite of what would be expected according to Trivers and Willard (1973). Observations of females in good condition producing more daughters than sons have been explained by the local resource competition hypothesis (Clark, 1978), the advantaged daughter hypothesis (Hiraiwa-Hasegawa, 1993), and the advantaged matriline hypothesis (Leimar, 1996). The local resource competition hypothesis suggests that a female should produce the dispersing sex when in poor condition to reduce competition for resources. The advantaged daughter and advantaged matriline hypotheses suggest that mothers are better able to influence reproductive success in daughters than sons, and thus should invest in daughters when in good condition. Date of conception may also influence the sex of offspring and further complicate interpretations of fawn sex ratio data. For example, the ratio of male versus female offspring varied throughout the birthing period in Michigan (Saalfeld et al., 2007) (Figure 2.4), and females in Alabama produced 54–55% male offspring during the first half of the birthing season and only 47% males after the peak of birth (Ditchkoff et al., 2009; Figure 2.5). Early-born males have a developmental advantage over late-born males that could translate into variation in lifetime reproductive success; the effect of this developmental advantage on lifetime reproductive success of females would not be as dramatic.

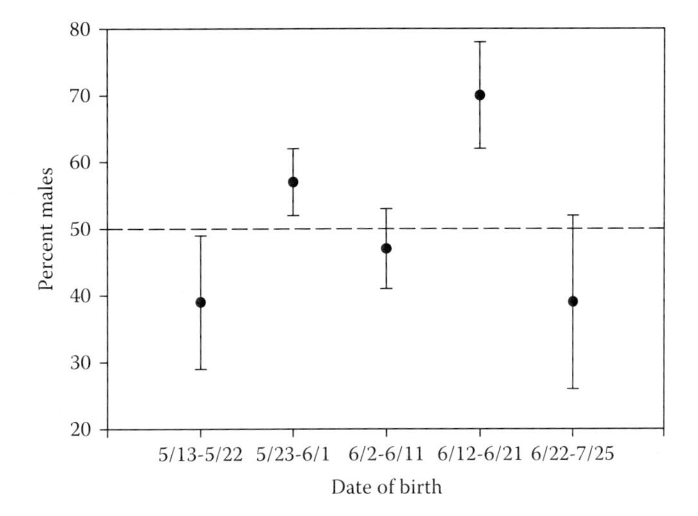

FIGURE 2.4 Fetal sex ratios of white-tailed deer from the Cusino enclosure, Michigan from 1973 to 1984. (Data from Saalfeld, S. T. et al. 2007. *Canadian Field-Naturalist* 121:412–419.)

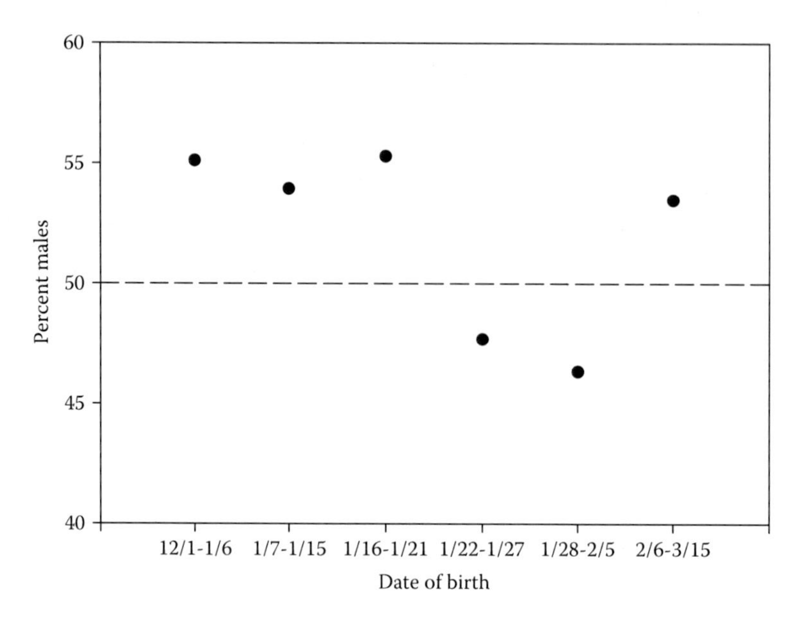

FIGURE 2.5 Fetal sex ratios of white-tailed deer in Alabama from 1995 to 2002. (Data from Ditchkoff, S. S. et al. 2009. *Journal of the Alabama Academy of Science* 80:27–34.)

Males

Endocrinology

As in females, the reproductive cycle in male white-tailed deer is initiated by surges of GnRH from the hypothalamus. The anterior pituitary responds to the GnRH surge by producing LH. This initial surge in LH occurs during July in northern regions (Mirarchi et al., 1978; Bubenik et al., 1982) and possibly later in more southern regions (Bubenik et al., 1990). Concentrations of LH during this initial surge normally range from 2.0 to 3.0 ng/mL (Mirarchi et al., 1978; Bubenik et al., 1982). Elevated levels of circulating LH cause Leydig cells in the testes to increase the production of testosterone. Testosterone concentrations begin to rise in August and continue to rise until the peak of the breeding season, after which concentrations decline rapidly (Mirarchi et al., 1978; Bubenik et al., 1982, 2005). During the peak breeding season, circulating testosterone may reach concentrations of 20 ng/mL (Mirarchi et al., 1978). This steady increase in testosterone is important for the initiation of spermatogenesis and the development of secondary sex characteristics (e.g., neck swelling, aggressive behavior). Elevated testosterone also serves as a negative feedback, along with other hormones, to LH production. Seasonal peaks in testosterone production increase annually until five to seven years of age (Bubenik and Schams, 1986; Ditchkoff et al., 2001b), after which concentrations decline. Younger males tend to have lower testosterone concentrations than older males because of the negative effects of testosterone on survival. Testosterone is an immune suppressant and behavioral changes associated with elevated testosterone can lead to increased probability of natural mortality (Ditchkoff et al., 2001c). Because young males are relatively poor competitors for mating opportunities, benefits derived from elevated testosterone normally will not outweigh the costs. The role of hormones in antler growth is discussed in Chapter 4.

Reproductive Organs and Spermatogenesis

In response to increasing testosterone concentrations, reproductive organs of male deer begin to develop. In July, testis and epididymal weights begin to increase, with peak weights occurring during the breeding season (Mirarchi et al., 1977). Spermatozoa numbers follow a similar pattern in which greatest numbers are found in both the epididymides and testes during the peak breeding season. Three months following the breeding season, both epididymides and testes decline to prereproductive mass. Although reproductive organs of males atrophy outside of the breeding season, the epididymides and testes still contain spermatozoa, and so the potential exists for males to successfully breed receptive does well into spring or even summer. Epidydimal and testis weights are associated positively with age (Lambiase et al., 1972), and this pattern is also generally found with spermatozoan production (Mirarchi et al., 1977; Peles et al., 2003). The ratio of X- and Y-sperm produced by male deer does not differ from 1:1 (DeYoung et al., 2004), as has been described in some other species.

Digestion

Anatomy

The digestive anatomy of white-tailed deer consists of structures and organs designed to physically and chemically reduce forage particle size so that nutrients can be absorbed. The first structures involved in the digestive process are the lips and tongue, which aid in acquisition of food, and help position food in the mouth for mastication. White-tailed deer possess six incisors and two incisiform canines on the bottom jaw, and only a dental pad on the top jaw (Figure 2.6). Because deer do not have upper incisors, and consequently the ability to "scissor" vegetation when biting, they grasp vegetation between the lower incisors and upper palate and tear it away from the plant. Immediately posterior to the incisors is the diastema, a gap in the tooth row 7–8 cm in length. Posterior to the diastema are three premolars and three molars which are used to chew and grind forage. Salivary glands secrete saliva into the oral cavity. Saliva serves five general purposes in white-tailed deer and other ruminants: (1) lubricate food and aid in passage through the esophagus to the stomach, (2) begin the process of enzymatic digestion, (3) help protect against pathogens (bacteria, viruses, and fungi), (4) provide a line of protection against plant-defensive compounds (e.g., tannins), and (5) help maintain rumen pH (Fickel et al., 1998).

Swallowed food passes through the esophagus to the rumen, the first of four chambers, along with the reticulum, omasum, and abomasum that make up the stomach complex in white-tailed deer and other ruminant animals (Figure 2.7). The rumen and reticulum form a chamber in which forage is fermented by microbes. The rumen has papilla up to 1.5 cm in length, which increase the rumen's surface area and promote nutrient absorption (Figure 2.8). The reticulum can be differentiated from the rumen by the reticulated pattern of folds on its internal surface. The omasum is responsible for ensuring that only sufficiently small forage particles are able to exit the rumenoreticular complex. The omasal orifice and folds of omasal

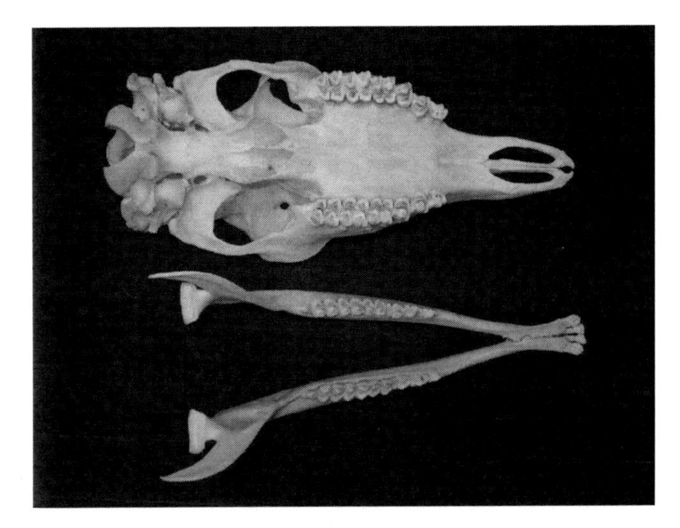

FIGURE 2.6 White-tailed deer lack upper incisors and have a gap, or diastema, between the lower incisors and premolars on the lower jaw. (Photo by S. S. Ditchkoff.)

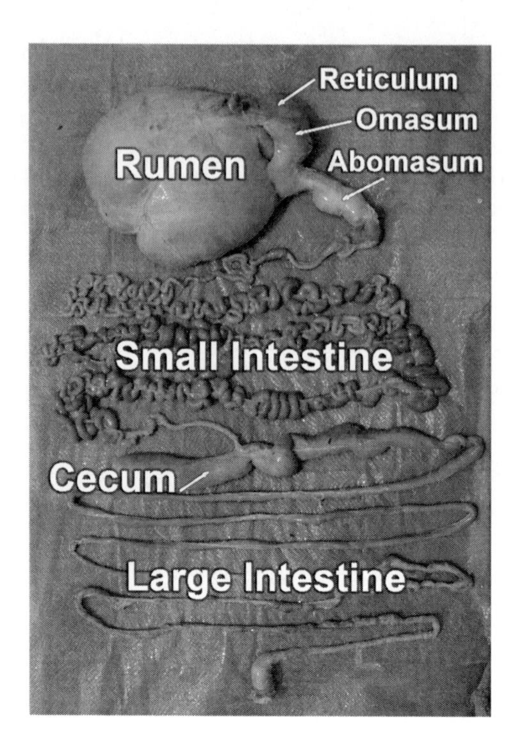

FIGURE 2.7 The gastrointestinal tract of the white-tailed deer consists of the rumenoreticular complex, omasum, and abomasum, small intestine, cecum, and large intestine. (Photo by D. G. Hewitt.)

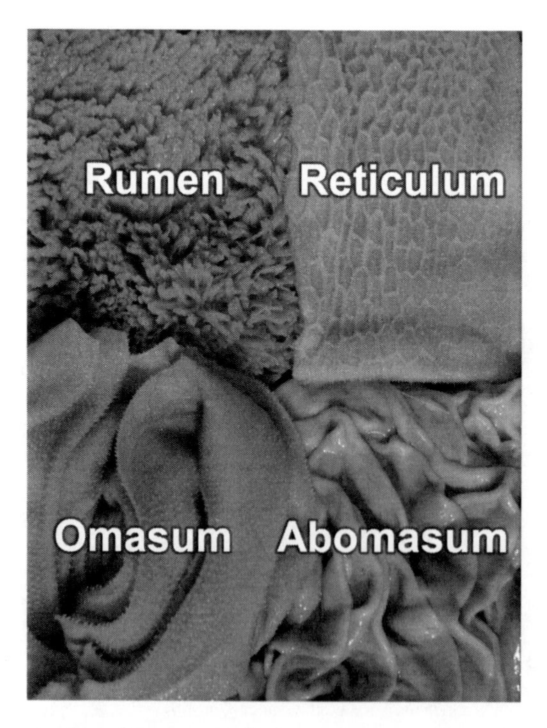

FIGURE 2.8 Internal surface structure of the rumen, reticulum, omasum, and abomasum. (Photo by D. G. Hewitt.)

tissue act as barriers to large particles, ensuring that they remain in the rumenoreticular complex until they have been adequately masticated and fermented. The omasum also absorbs excess fluid and separates the basic environment of the rumenoreticular complex from the acidic conditions of the abomasum (Robbins, 1993). The abomasum appears and functions similar to the stomach of a monogastric animal, with smooth, slightly folded tissue where gastric digestion occurs. After leaving the abomasum, digesta enters the small intestine where dietary and microbial proteins and lipids are enzymatically digested and absorbed. At the junction of the small and large intestine is the cecum, a blind pouch up to 20 cm long that is important in digestion for some herbivores, but probably contributes little to digestion in white-tailed deer. The primary function of the large intestine in white-tailed deer is for water and electrolyte absorption. The combined length of the small and large intestines may exceed 20 m (Jenks et al., 1994).

Digestive Physiology

The main components of white-tailed deer diets are structural and nonstructural carbohydrates. Structural carbohydrates comprise the majority of the diet, consist of cellulose and hemicellulose, and are found in plant cell walls. Common nonstructural carbohydrates include cellular contents, starch, and simple sugars associated with fruits and other mast. Lignin, an indigestible organic polymer, can also be found in large quantities in plant cell walls, and may be chemically linked to structural carbohydrates, thereby reducing their digestibility. No vertebrate species produce enzymes capable of digesting structural carbohydrates in plant cell walls. Thus, herbivorous mammals dependent on plant material as a source of energy have a problem. Ruminant animals (e.g., cervids, bovids, giraffes, tragulids, camels, pronghorn antelope) have developed symbiotic relationships with microorganisms to digest these otherwise indigestible structural carbohydrates. The microflora ferment ingested material in the rumen and reticulum, also referred to as the rumenoreticular complex. Rumen fluid contains approximately 40.0×10^9 bacteria (Church, 1979) and 45.4×10^4 protozoa (Dehority et al., 1999) per milliliter, although numbers vary depending on diet. Church (1979) described 10 different types of bacteria based on the type of substrate used as their primary energy source: cellulolytic (cellulose-digesting), hemicellulolytic (hemicelluloses-digesting), amylolytic (starch-digesting), sugar-utilizing, acid-utilizing, proteolytic (protein-digesting), ammonia-producing, methanogenic (methane-producing), lipolytic (lipid-digesting), and vitamin-synthesizing.

Fermentation requires a basic and anaerobic environment. Ruminants produce copious amounts of saliva that aid in mastication and swallowing. The saliva contains high concentrations of salts that serve as buffers to help maintain an appropriate pH in the rumenoreticular cavity. To further aid in the physical reduction of plant material, deer ruminate or "chew their cud." Deer ruminate during periods of inactivity by regurgitating and remasticating previously ingested plant material. The bolus is then reswallowed. Rumination is an extremely important component of the ruminant digestive strategy. The process of fermenting structural carbohydrates is slow relative to enzymatic digestion of simple carbohydrates. Remasticating forages reduces particle size, which increases surface area and fermentation rates. As particle size decreases, passage rate increases, and limitations on consumption due to gut fill are decreased. Rumination may have evolved as an antipredatory behavior that enabled ruminant herbivores to reduce feeding time and subsequent vulnerability to predation (Phillips, 1993; Gregorini et al., 2006). When feeding, animals tend to be less vigilant and more susceptible to predation (Winnie and Creel, 2007). This hypothesis contends that to reduce feeding time and susceptibility to predation, ruminants do not fully masticate their food. Later, in a less dangerous environment, they can masticate the ingested material at their leisure.

Fermentation of structural carbohydrates is a slow digestive process. Passage rate (time from consumption to defecation) in white-tailed deer of 80% of ingested material is about 48 hours (Mautz and Petrides, 1971), and the majority of this time is for fermentation in the rumenoreticular complex. Because ingested forage must remain in the rumenoreticular chamber for such a long period, the rumen and reticulum must be large organs. Short et al. (1969) estimated the contents of the rumen and reticulum to be 7–10% of the adult body mass.

The major source of energy for white-tailed deer comes from fermentation of carbohydrates. Rumen microbes ferment ingested carbohydrates as a source of energy, and the by-products of fermentation are then absorbed by the deer and utilized. Oxidation of one mole of glucose, which is the basic building block of cellulose, generates 38 moles of ATP. During fermentation of glucose, rumen microflora use

2–6 moles of ATP, leaving 32–36 moles of ATP for absorption by the host in the form of volatile fatty acids (VFA) (Robbins, 1993). These VFAs are absorbed either directly through the rumen wall or from contents in the small intestine.

The differential energy yield from fermentation compared to enzymatic digestion poses some hypothetical trade-offs associated with digestion. Structural carbohydrates can only be digested through fermentation, and so must be digested in the rumen. However, if nonstructural carbohydrates could somehow bypass the rumenoreticular complex so as not to be fermented, they could be digested enzymatically without the 5–15% loss of energy associated with fermentation. Nursing fawns are able to accomplish this feat. Milk is comprised of water, protein, fats, simple carbohydrates, and minerals (Silver, 1961), which are all digestible or absorbable without fermentation. To avoid the energetic losses associated with fermentation, neonatal ruminants are able to shunt milk past the rumenoreticular complex via the reticular groove. The reticular groove is a section of the reticulum that is able to fold over on itself such that it forms a tube from the distal end of the esophagus to the omasal opening. The act of suckling stimulates the reticular groove to form, thereby allowing milk to bypass the rumenoreticulum and avoid fermentation. It has been hypothesized that some small ruminant species (possibly deer) that are selective feeders and consume greater concentrations of soluble carbohydrates than grazing species, may have the ability to selectively allow soluble carbohydrates to bypass the rumenoreticular complex via this mechanism (Hofmann, 1989). However, it has generally been believed that functionality of the reticular groove disappears in adulthood. Ørskov et al. (1970) demonstrated that domestic ruminants could be trained to maintain functionality of the reticular groove into adulthood, although this selective bypass mechanism in adult animals has yet to be demonstrated in free-ranging species (Ditchkoff, 2000).

Bacteria and other microorganisms in the rumenoreticulum also aid in protein digestion. In contrast to monogastric species, white-tailed deer and other ruminants do not rely entirely upon gastric digestion of proteins. Rather, rumen microflora metabolize 65–85% of ingested protein and reformulate it into amino acids that meet their needs. Then, when the rumen microflora are passed out of the rumenoreticular complex, these reformulated proteins are digested in the acidic environment of the abomasum and absorbed in the small intestine. Concurrent with protein digestion in the gastrointestinal tract, excess ammonia in the rumen is absorbed across the rumen wall and transported to the liver. This ammonia can then be converted into urea and transported in the blood to the salivary glands where it can be recycled and sent back to the rumen. Because rumen microflora are capable of using inorganic nitrogen, this recycled nitrogen can be incorporated into microbial protein and later absorbed by the animal. The abomasum is the gastric stomach of the ruminant animal, and the point where protein digestion is initiated. Microbial protein and dietary protein that avoided microbial uptake are exposed to a highly acidic environment that denatures proteins. Once proteins are broken down into amino acids, they can then be absorbed across the wall of the small intestine.

White-tailed deer have the ability to modify the morphology of the gastrointestinal tract based on body condition, reproductive status, forage quality, and nutrient requirements. Intestinal length tends to be shortest during autumn and generally increases from autumn to summer (Weckerly, 1989). These changes in gut length are likely due to variation in seasonal nutrient requirements and food quality. There is evidence that lactating females have longer gastrointestinal tracts than males, perhaps because of the high nutrient demands of lactation (Jenks et al., 1994; Zimmerman et al., 2006). Although gut tissues are metabolically expensive, the increase in gut tissue is likely offset by increased digestive efficiency. In addition to changes in intestinal length, gastrointestinal morphology (e.g., length, number, and size of papillae) can change in as short as two weeks in response to changes in diet quality (Zimmerman et al., 2006). These data suggest that white-tailed deer are able to rapidly alter the structure, and possibly function, of their digestive system to maximize digestive efficiency relative to nutrient requirements and availability.

Winter Undernutrition

White-tailed deer in northern regions exhibit marked changes in their energetic strategy during winter compared to other seasons. Like hibernating animals, deer accumulate fat reserves prior to winter and then use those resources during periods of nutritional restriction. As a result, deer may lose 20–30% of their body mass during winter (Mautz, 1978). The strategy employed by deer during winter is to reduce energetic costs at the expense of reducing energy intake. White-tailed deer voluntarily restrict feed intake and activity during winter (Thompson et al., 1973) for several reasons. The cost of searching for food is normally greater than the energy obtained from food during this period because of the combination of low food availability and low food quality in most deer wintering areas (Ditchkoff and Servello, 1998). Furthermore, extremely high energetic costs of locomotion through snow limit deer's ability to move during winter in northern portions of the species' range (Parker et al., 1984) (Figure 3.2). As a result, white-tailed deer have evolved a strategy whereby they rely on fat reserves to meet a substantial portion of their energetic needs. Deer that search extensively for food have a greater probability of dying due to winter starvation than deer that forage opportunistically. While some studies have suggested that metabolic rates are less during winter in white-tailed deer (Silver et al., 1969; Thompson et al., 1973), more recent studies have suggested that this is probably not the case (Mautz et al., 1992).

During winter, white-tailed deer catabolize both fat and protein reserves to meet energetic requirements (Torbit et al., 1985). The process by which deer catabolize these tissues is similar physiologically to starvation, and could best be described as controlled starvation. During phase I of starvation, deer rely primarily upon catabolism of proteins to meet their energetic needs. This phase is characterized by a high rate of weight loss because of the lower energetic content of protein relative to fat tissue (Robbins, 1993). Phase II of starvation is more prolonged and characterized by a lower rate of weight loss due to catabolism primarily of fat tissue (Robbins, 1993). Because catabolism of fat does not generate glucose, and the brain and red blood cells require glucose as an energy source (Torbit et al., 1985), there is continued catabolism of some protein during this phase because of its suitability for gluconeogenesis. When most fat reserves have been mobilized, white-tailed deer enter phase III of starvation. Protein catabolism once again becomes the major source of energy, and the rate of weight loss accelerates. Deer are able to recover from early phase III starvation, but because this is a short phase characterized by rapid weight loss, the negative energy balance must be reversed quickly (Robbins, 1993). The length of time that an animal is in any phase of starvation is ultimately a function of body size, energetic reserves, energetic demands, and nutrient availability. As a result, there is high interanimal and regional variability in the duration of these different phases of starvation.

Mautz (1978) described the manner in which deer attempt to slow progression through the phases of starvation and how external factors may influence starvation. When food is available, deer use ingested forage as a source of energy and thus slow catabolism of body tissues. Conserving endogenous reserves delays the onset of phase III. Therefore, if deer are able to acquire sufficient fat reserves during autumn and the wintering period is not unusually long, rates of overwinter starvation are typically low. However, when winter conditions arrive early or break late and the length of winter is above average, high rates of starvation are possible.

Hematology and Serum Chemistry

Effect of Age

Hematology and serum chemistry are highly dynamic in neonatal white-tailed deer because of their rapid growth and dramatic changes in digestion and physiology (Seal et al., 1981). During the first few weeks following birth, fawns are developing a fully functional physiological system that previously had operated only in unison with that of their dam. Additionally, fawns are shifting from a primarily monogastric digestive system to a system of foregut fermentation based in the rumenoreticular complex. Red blood cell, hemoglobin, total protein, mean corpuscular hemoglobin concentration (MCHC), and packed cell volume (PCV) (hematocrit) in the first week of life may be 40–60% of values at three–four months of age (Figure 2.9; Johnson et al., 1968). Conversely, mean corpuscular volume (MCV), mean corpuscular hemoglobin (MCH), and glucose levels decline with age up to four months of age. By about three months of age, hematology values approach those of adults (Tumbleson et al., 1970; Seal et al., 1981; Rawson et al., 1992). Increases in erythrocytes, hemoglobin, and MCHC with age suggest an improved ability to utilize and carry oxygen that will be needed for increasing metabolic demands associated with

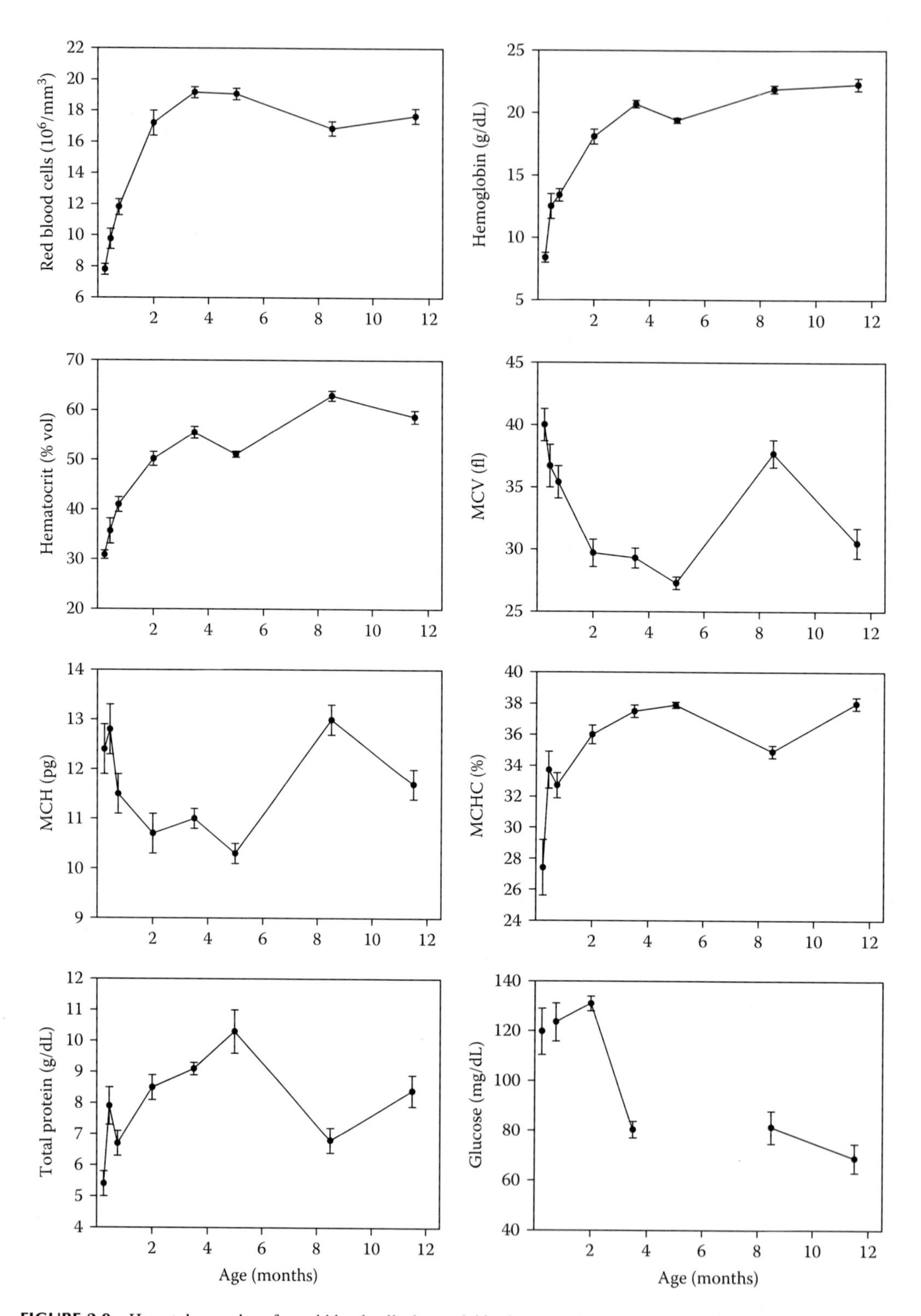

FIGURE 2.9 Hematology values for red blood cells, hemoglobin, hematocrit, mean corpuscular volume (MCV), mean corpuscular hemoglobin (MCH), mean corpuscular hemoglobin concentration (MCHC), total protein, and glucose of juvenile white-tailed deer in Michigan. (Reprinted from Johnson, H. E. et al. 1968. *Journal of Mammalogy* 49:749–754. With permission of Allen Press.)

growth (Rawson et al., 1992). Decreases in blood glucose are likely a function of the shift from monogastric digestion to fermentation.

After one year, hematology and serum chemistry do not exhibit marked changes with age. Rather, season and body condition are the primary influences on blood chemistry. This predictability allows serum chemistry and hematological data to be used effectively to examine questions concerning condition, reproduction, and other physiological parameters. Unlike other serum chemistry data, gamma globulins tend to increase with age to two or three years before stabilizing (Seal et al., 1981). Table 2.3 provides reference values for many hematological and blood chemistry parameters.

TABLE 2.3

Normal Range of Hematologic, Hormonal, and Serum Chemistry Values for White-tailed Deer

Blood Parameter	**Units**	**Range[a]**
Hematology		
Hematocrit	% Vol.	39–58
Hemoglobin	g/dL	14.5–22.5
Mean corpuscular hemoglobin (MCH)	Pg	7–15
Mean corpuscular hemoglobin concentration (MCHC)	g/dL	34–40
Mean corpuscular volume (MCV)	Fl	28–38
Red blood cells	10^6/µL	10.0–16.7
White blood cells	/µL	1000–4200
Basophils	/µL	<10
Eosinophils	/µL	<100
Lymphocytes	/µL	600–1800
Monocytes	/µL	<10
Neutrophils	/µL	600–2800
Proteins		
Albumin	g/dL	2.0–4.2[b]
Fibrinogen	mg/dL	170–300
Gamma globulins	g/dL	3.0–4.5[b]
Haptoglobin	mg/dL	<10
Total protein	g/dL	5.0–7.8
Transferrin	mg/dL	250–450
Vitamins and minerals		
Calcium	mg/dL	6.6–10.8[c]
Chloride	mEq/L	100–110
Iron	µg/dL	70–225
Phosphorus	mg/dL	1.0–8.5[c]
Potassium	mEq/L	3.4–11.5[d]
Sodium	mEq/L	132–156
Chemistry		
Bicarbonate	mEq/L	12–24
Bilirubin	mg/dL	0.1–1.0
Cholesterol	mg/dL	30–100
Creatinine	mg/dL	0.4–2.0
Glucose	mg/dL	60–320
Nonesterified fatty acids	µEq/L	150–600
Triglycerides	mg/dL	3–50[c]
Urea nitrogen	mg/dL	8–45[b]
Hormones		
Cortisol	µg/dL	2–8
Estradiol (females)	pg/mL	5–35

continued

TABLE 2.3 (continued)

Normal Range of Hematologic, Hormonal, and Serum Chemistry Values for White-tailed Deer

Blood Parameter	Units	Range[a]
Glucagon	pg/mL	200–600
Growth hormone	ng/mL	3–16
Insulin	μU/mL	18–50
Insulin-like growth factor I	ng/mL	55–80[e]
Progesterone (females)	ng/mL	0.2–14.0
Prolactin	ng/mL	50–5000
Testosterone (males)	ng/dL	50–2000
Thyroxine (T_4)	μg/dL	7–30[c]
Triiodothyronine (T_3)	ng/dL	80–150[d]
Enzymes		
Alkaline phosphatase	IU/L	10–200[b]
Creatinine phosphokinase	IU/L	20–400
Lactic dehydrogenase	IU/L	300–700[b]
Serum glutamic oxaloacetic transaminase	IU/L	40–150
Serum glutamic pyruvic transaminase	IU/L	20–150
Gamma glutamyl transpeptidase	IU/L	40–100

[a] Data ranges were reported by Seal et al. (1981) unless otherwise indicated.
[b] Data from Waid and Warren (1984).
[c] Data reported from DelGiudice et al. (1992).
[d] Data reported from DeLiberto et al. (1989).
[e] Data from Ditchkoff et al. (2001b).

Effect of Nutrition and Season

Many blood parameters are strongly linked to changes in body condition or nutritional plane, and because of the close link between body condition and season in white-tailed deer, seasonal changes in blood parameters are often predictable (Seal et al., 1981). PCV, hemoglobin, and red blood cell count are often elevated during winter in northern regions. Declining body condition and nutritional deprivation lead to dehydration and subsequent elevations in these blood parameters (DelGiudice et al., 1992). DeLiberto et al. (1989) found low PCV during summer in Oklahoma. They suggested that in warmer regions where winter conditions are not as harsh, nutritional deficits may be greatest during summer. In Oklahoma and surrounding areas, food availability and quality often decline during summer because of semiarid conditions, and the nutritional demands of lactation can lead to some blood parameters being similar to those of deer in northern regions during winter. Body weight and fat reserves are at their lowest levels in August in Texas, and serum chemistry parameters such as albumin and alkaline phosphatase may be lower (Waid and Warren, 1984). Glucose, potassium, chloride, calcium, serum glutamic oxaloacetic transaminase, lactic dehydrogenase (Waid and Warren, 1984), white blood cells, MCV, thyroxine, sodium, and triglycerides (DelGiudice et al., 1992) tend to remain stable seasonally. A more thorough discussion on associations between blood chemistry and nutrition can be found in Chapter 3.

Nervous System

Vision

The physical structure and location of deer's eyes have evolved to maximize field of view. White-tailed deer, like most prey species, have eyes set on the side of the head which provides a wider field of view than most predatory species whose eyes are set in the front of the head. In addition, both the cornea and pupil are horizontally oval (Figure 2.10), corresponding to a linear concentration of

FIGURE 2.10 The eyes of white-tailed deer are on the sides of the head, providing a wide field of view. Notice the horizontal orientation of the pupil, which corresponds to a linear concentration of receptor cells on the retina. (Photo by D. G. Hewitt.)

receptor cells on the retina called the visual streak. Light traveling through the pupil is focused on this linear band of heightened visual acuity (Müller-Schwarze, 1994). These adaptations allow deer to distinguish distant objects across a 310 degree field of view without moving their head (Müller-Schwarze, 1994; VerCauteren and Pipas, 2003). White-tailed deer, like some other nocturnal mammals, possess a layer of dense fibrous tissue on the retina called the tapetum lucidum. This layer serves to reflect light back through the retina so that rods and cones have two opportunities to be stimulated, thus increasing apparent differences in brightness (VerCauteren and Pipas, 2003). Light reflected from the tapetum lucidum causes the eyes of deer to shine when exposed to a bright light at night (Müller-Schwarze, 1994).

Early reports on deer vision suggested that white-tailed deer had no ability to differentiate colors (Dalrymple, 1975). However, Witzel et al. (1978) reported cones (photoreceptors capable of detecting color) on the white-tailed deer retina at a density of about $10{,}000/mm^2$. Further investigations revealed that deer can distinguish among different colors (Zacks and Budde, 1983; Zacks, 1985; Smith et al., 1989). In addition to rods with heightened sensitivity around 497 nm, deer possess short-wavelength-sensitive cones with maximum sensitivity between 450 and 460 nm, and medium-wavelength-sensitive cones with maximum sensitivity around 537 nm (Jacobs et al., 1994). Medium-wavelength-sensitive cones have greater density in the visual streak, while short-wavelength-sensitive cones are distributed evenly across the retina (D'Angelo et al., 2008). The presence of two classes of cone pigments indicates that white-tailed deer have the basis for dichromatic color vision. Deer may be able to discriminate among colors using signals from both cone and rod receptors (Jacobs et al., 1994). Under low-light conditions, rods are more sensitive to light than cones, but become saturated at moderate light intensities. Thus, during low-light conditions, deer are likely most sensitive to the blue to blue-green portion of the spectrum because of the contribution of rods to color differentiation (VerCauteren and Pipas, 2003). Birgersson et al. (2001) speculated that color vision may allow deer to discriminate among plants or plant parts based on concentrations of nutrients or plant-defensive compounds. However, Smith et al. (1989) surmised that color vision is not a dominant cue for white-tailed deer, although they have the ability to discriminate based on color.

Hearing

Hearing in white-tailed deer is an enhanced sense that provides real-time data regarding the presence of other animals. This sense relies on two large-cupped external pinnae that have the ability to rotate independently. White-tailed deer's greatest auditory sensitivity is from 4 to 8 kHz, but they have the ability to hear sounds within the range of 0.25–30.0 kHz (D'Angelo et al., 2007). Humans cannot detect sounds above 20 kHz (Durrant and Lovrinic, 1995), suggesting that ultrasonic auditory stimuli could be used to reduce deer–human conflicts such as deer–vehicle collisions (D'Angelo et al., 2007). However, the physical properties of sound waves, relationships between sound intensity and distance, and concerns for human safety (e.g., sounds too loud) may preclude their effective use on moving vehicles (D'Angelo et al., 2007).

Chemical Senses

As in humans, white-tailed deer have the ability to sense chemicals through both taste and smell. Deer have taste buds located on the tongue and arranged in groups of mushroom-shaped and circumvallate papillae. Circumvallate papillae are arranged in two rows along either side of the tongue's midline, while mushroom-shaped papillae are clustered behind the highest point on the tongue's upper surface (Müller-Schwarze, 1994). Taste may assist deer in testing palatability of forages. For example, many plant-defensive compounds taste bitter. Olfaction is an extremely important sense in deer as it aids in the location of preferred forages, the detection of potential predators, and social interactions among deer. The long rostrum of deer suggests their sense of smell is acute. Considering how important olfaction is to deer, it is surprising that almost no research has been conducted in this area. While there have been investigations of the response of deer to olfactory repellents and other odors, little or no information is available regarding morphology or sensitivity of their olfactory system.

A third form of chemical detection in white-tailed deer is vomolfaction. Located in the upper palate of the mouth are two pores called the incisive foramina. These pores lead to the vomeronasal organ, which is responsible for the detection of chemicals that signal a female is in estrus. Vomolfaction occurs when a male licks the urine of a potential mate. The behavior referred to as flehmen (see Chapter 10), when a buck curls his upper lip, is actually a physical process whereby the male transfers the chemical compounds of interest from the tongue to the incisive foramina, where they can be sensed by the vomeronasal organ (Müller-Schwarze, 1994).

Stress

Glucocorticoids

Stress is an often overlooked phenomenon that can influence the survival and productivity of white-tailed deer. Stress can manifest itself in two forms: acute and chronic. When an animal perceives a threat, the anterior pituitary secretes adrenocorticotrophic hormone (ACTH), which causes the adrenal cortex to secrete glucocorticoids (Asterita, 1985). This stress response enhances the ability of an individual to cope with the stressor by preparing energy necessary for the response and minimizing energy expenditures in tissues that are not necessary for immediate survival. The stress response is normally an acute response and lasts only minutes to hours. However, prolonged exposure to stress can alter behavior and reduce growth, survival, reproductive success, and resistance to disease in many species (Blanc and Thériez, 1998; Stefanski, 2001b). In fact, stress has the potential to influence all aspects of life history because it affects hormonal pathways which are at the foundation of physiological health. Factors that can induce stress in white-tailed deer include, but are not limited to, undernutrition, pathogens, climatic extremes, and social interactions. Social stress is a psychological stress experienced by organisms, and may be caused by high population density (Blanc and Thériez, 1998), disruptions in social structure (Foley et al., 2001), or exposure to high levels of agonistic behavior (Creel, 2001). While social stress is often considered a psychological disorder of little concern, in actuality it can have serious impacts on individuals and populations (Creel, 2001; Foley et al., 2001; Stefanski, 2001a).

Until recently, measures of physiologic stress have been limited to blood parameters collected from live animals (Franzmann et al., 1975; DeNicola and Swihart, 1997), but a new, noninvasive technique for assessing stress in free-ranging wildlife has recently been developed (Miller et al., 1991; Creel et al., 1997; Wasser et al., 2000) and validated for use with white-tailed deer (Millspaugh et al., 2002). This technique involves measuring glucocorticoids secreted during times of stress and voided in the feces. Fecal glucocorticoids are associated positively with stress in many species (Harper and Austad, 2000; Wasser et al., 2000), including white-tailed deer (Millspaugh et al., 2002), and provide an estimate of all glucocorticoid secretions during the previous 24–48 hours (Harper and Austad, 2000). Measurement of fecal glucocorticoids is superior to techniques requiring blood collection and physical handling of animals because stress hormones can rise substantially in the blood only minutes after experiencing stress (Creel, 2001). Voided samples generally retain glucocorticoid levels for up to seven days if not exposed to precipitation (Washburn and Millspaugh, 2002). Entire fecal groups, rather than individual pellets, should be collected and homogenized to ensure accurate analysis (Millspaugh and Washburn, 2003). Other metabolites that have been used to measure stress in white-tailed deer and other cervids include urinary cortisol concentrations (Saltz and White, 1991) and TNF-α (Ditchkoff et al., 2001a).

Capture Myopathy

Capture of white-tailed deer, regardless of technique, can impose significant stress. This stress may not be outwardly apparent, but can manifest itself days later and cause mortality. Physical restraint and associated stresses often result in extreme exertion in white-tailed deer. Such exertion can ultimately lead to a complex degenerative disease of skeletal muscle known as capture myopathy (Davidson and Nettles, 1997). In severe cases, metabolic acidosis and cardiac and circulatory problems may arise and result in death in only a few hours. In less severe cases, animals may have damage to skeletal muscle and internal organs, and may finally succumb to the disease days or weeks later due to kidney failure (Davidson and Nettles, 1997). This condition is characterized by marked elevations of glutamic oxaloacetic transaminase, glutamic pyruvic transaminase, lactate dehydrogenase, and creatinine phosphokinase in the serum. Additionally, serum potassium, urea nitrogen, creatinine, and lactate may also increase because of damage to the kidneys and muscle degeneration (Wobeser, 1981). Animals that die due to capture myopathy typically have pale muscles and dark urine. Davidson and Nettles (1997) suggested trapping rather than chasing animals, blindfolding, minimizing noise and activity, and limiting capture to periods with reduced temperatures to reduce stress on animals and incidence of capture myopathy. Additionally, they indicated that providing sodium bicarbonate, selenium, or vitamin E supplements may reduce complications due to capture myopathy.

Stress associated with capture can influence blood parameters of interest. DelGiudice et al. (1990) reported that hemoglobin, PCV, and possibly creatine phosphokinase were greater in deer caught with Clover traps or cannon nets than in deer tranquilized and captured with use of a capture collar. Collars containing injectable chemical immobilizers have less impact on blood parameters than more traditional capture methods, possibly because of the lower stress experienced by the animals (DelGiudice et al., 1990). Sams et al. (1993) found greater mean values for PCV, MCV, white blood cells, lymphocytes, neutrophils, creatinine, cholesterol, total protein, albumin, globulins, sodium, lactic dehydrogenase, and gamma glutamyl transferase in deer captured with drop nets than hunter-harvested deer. They describe the physiological mechanisms behind these stress-driven changes in clinical blood profiles.

Despite suggestions that chemical immobilization or tranquilization after physical capture may reduce chances of capture myopathy (Henderson et al., 2000), there is little evidence at this time to suggest that this is the case (Peterson et al., 2003). The extent of stress experienced by chemically immobilized deer is not completely understood, although a few studies have attempted to quantify stress in captured deer. Mean serum cortisol concentrations were two to three times greater for deer captured by darting and 15 times greater in deer captured with drop net than baseline values established from deer euthanized by a shot to the head (DeNicola and Swihart, 1997). Serum cortisol concentrations were about eight times greater in deer captured with Clover traps and cannon nets than deer captured with collars containing injectable chemical immobolizers (DelGiudice et al., 1990). Peterson et al. (2003) suggested that because chemical immobilization often requires lengthy induction and recovery times, deer that are efficiently handled and released in a short time period using other capture techniques may experience less stress than those that are chemically immobilized after capture. Webb et al. (2008) reported that of 100 bucks captured via helicopter and net gun and fitted with radiocollars, only one deer died from what might have been capture myopathy. They limited helicopter chases to eight minutes, and kept handling time below 10 minutes: deer were not chemically immobilized. Ditchkoff et al. (2001c) captured and radiocollared 80 bucks using drop nets, and physically restrained them without the aid of chemical immobilizers: handling time generally ranged from 20 to 40 minutes. Of these 80 captures, one deer died within minutes of being trapped, and one buck died approximately a week after capture from what may have been capture myopathy.

Capture techniques that reduce handling time and stress tend to have the greatest probability of success. Beringer et al. (1996) found that as the number of deer captured together with rocket nets increased, the probability of mortality due to capture myopathy increased. In contrast, they reported no postrelease mortalities in deer captured using Clover traps. They hypothesized that the increased stress associated with lengthy handling times in groups of deer captured with rocket nets led to greater incidence of capture myopathy. Conversely, Haulton et al. (2001) found that Clover traps were more likely to lead to capture myopathy than box traps, rocket nets, or darting, highlighting the contradictory information available and the difficulty in identifying factors that cause capture myopathy. It can be extremely difficult, even with thorough physiological postmortem examinations, to distinguish between natural mortality and capture myopathy as the number of days following capture increases. Predation or scavenging may destroy evidence or samples, and statistical analyses may have difficulty distinguishing between natural and capture-related mortality factors because of low sample sizes (DelGiudice et al., 2005).

REFERENCES

Adams, L. G. 2005. Effects of maternal characteristics and climatic variation on birth masses of Alaskan caribou. *Journal of Mammalogy* 86:506–513.

Armstrong, R. A. 1950. Fetal development of the northern white-tailed deer (*Odocoileus virginianus borealis* Miller). *American Midland Naturalist* 43:650–667.

Asterita, M. F. 1985. *The Physiology of Stress.* New York, NY: Human Sciences Press.

Atkeson, T. D. and R. L. Marchinton. 1982. Forehead glands in white-tailed deer. *Journal of Mammalogy* 63:613–617.

Atkeson, T. D., V. F. Nettles, R. L. Marchinton, and W. V. Branan. 1988. Nasal glands in the Cervidae. *Journal of Mammalogy* 69:153–156.

Baccus, J. T. and J. C. Posey. 1999. Melanism in white-tailed deer in central Texas. *Southwestern Naturalist* 44:184–192.

Baker, D. L., M. A. Wild, M. M. Connor, H. B. Ravivarapu, R. L. Dunn, and T. M. Nett. 2004. Gonadotropin-releasing hormone agonist: A new approach to reversible contraception in female deer. *Journal of Wildlife Diseases* 40:713–724.

Beringer, J., L. P. Hansen, W. Wilding, J. Fischer, and S. L. Sheriff. 1996. Factors affecting capture myopathy in white-tailed deer. *Journal of Wildlife Management* 60:373–380.

Birgersson, J., U. Alm, and B. Forkman. 2001. Colour vision in fallow deer: A behavioural study. *Animal Behaviour* 61:367–371.

Blanc, F. and M. Thériez. 1998. Effects of stocking density on the behaviour and growth of farmed red deer hinds. *Applied Animal Behaviour Science* 56:297–307.

Bubenik, G. A., R. D. Brown, and D. Schams. 1990. The effect of latitude on the seasonal pattern of reproductive hormones in the male white-tailed deer. *Comparative Biochemistry and Physiology Part A* 97:253–257.

Bubenik, G. A., K. V. Miller, A. L. Lister, D. A. Osborn, L. Bartos, and G. J. Ven Der Kraak. 2005. Testosterone and estradiol concentrations in serum, velvet skin, and growing antler bone of male white-tailed deer. *Journal of Experimental Zoology* 303A:186–192.

Bubenik, G. A., J. M. Morris, D. Schams, and A. Claus. 1982. Photoperiodicity and circannual levels of LH, FSH, and testosterone in normal and castrated male, white-tailed deer. *Canadian Journal of Physiology and Pharmacology* 60:788–793.

Bubenik, G. A. and D. Schams. 1986. Relationship of age to seasonal levels of LH, FSH, prolactin and testosterone in male, white-tailed deer. *Comparative Biochemistry and Physiology Part A* 83:179–183.
Carroll, J. F. 2001. Interdigital gland substances of white-tailed deer and the response of host-seeking ticks (Acari: Ixodidae). *Journal of Medical Entomology* 38:114–117.
Carroll, J. F., G. D. Mills, Jr., and E. T. Schmidtmann. 1998. Patterns of activity in host-seeking adult *Ixodes scapularis* (Acari: Ixodidae) and host-produced kairomones. *Journal of Medical Entomology* 35:11–15.
Causey, M. K. 1990. Fawning date and growth of male Alabama white-tailed deer. *Proceedings of the Annual Conference of the Southeastern Association of Fish and Wildlife Agencies* 44:337–341.
Cheatum, E. L. and C. W. Severinghaus. 1950. Variations in fertility of white-tailed deer related to range conditions. *Transactions of the North American Wildlife Conference* 15:170–190.
Church, D. C. 1979. *Digestive Physiology and Nutrition of Ruminants. Volume I—Digestive Physiology* (2nd edition). Portland, Oregon: O&B Books.
Clark, A. B. 1978. Sex ratio and local resource competition in a prosimian primate. *Science* 201:163–165.
Clutton-Brock, T. H., S. D. Albon, and F. E. Guinness. 1981. Parental investment in male and female offspring in polygynous mammals. *Nature* 289:487–489.
Clutton-Brock, T. H., F. E. Guinness, and S. D. Albon. 1982. *Red Deer: Behavior and Ecology of Two Sexes.* Chicago, IL: University of Chicago Press.
Creel, S. 2001. Social dominance and stress hormones. *Trends in Ecology and Evolution* 16:491–497.
Creel, S., N. M. Creel, and S. L. Monfort. 1997. Radiocollaring and stress hormones in African wild dogs. *Conservation Biology* 11:544–548.
D'Angelo, G. J., and J. T. Baccus. 2007. First record of melanistic white-tailed deer in Pennsylvania. *American Midland Naturalist* 157:401–403.
D'Angelo, G. J., A. R. De Chicchis, D. A. Osborn, G. R. Gallagher, R. J. Warren, and K. V. Miller. 2007. Hearing range of white-tailed deer as determined by auditory brainstem response. *Journal of Wildlife Management* 71:1238–1242.
D'Angelo, G. J., A. Glasser, M. Wendt et al. 2008. Visual specialization of an herbivore prey species, the white-tailed deer. *Canadian Journal of Zoology* 86:735–743.
Dalrymple, B. W. 1975. When can your quarry see you? *Outdoor Life* 156:61–65, 143–144.
Davidson, W. R. and V. F. Nettles. 1997. *Field Manual of Wildlife Diseases in the Southeastern United States* (2nd edition). Athens, GA: Southeastern Cooperative Wildlife Disease Study.
Dechert, J. A. 1967. The effects of overpopulation and hunting on the Fort Knox deer herd. *Proceedings of the Annual Conference of the Southeastern Association of Game and Fish Commissioners* 21:15–23.
Dehority, B. A., S. Demarais, and D. A. Osborn. 1999. Rumen ciliates of white-tailed deer (*Odocoileus virginianus*), axis deer (*Axis axis*), sika deer (*Cervus nippon*), and fallow deer (*Dama dama*) from Texas. *Journal of Eukaryotic Microbiology* 46:125–131.
DelGiudice, G. D., J. Fieberg, M. R. Riggs, M. Carstensen Powell, and W. Pan. 2006. A long-term age-specific survival analysis of female white-tailed deer. *Journal of Wildlife Management* 70:1556–1568.
DelGiudice, G. D., K. E. Kunkel, L. D. Mech, and U. S. Seal. 1990. Minimizing capture-related stress on white-tailed deer with a capture collar. *Journal of Wildlife Management* 54:299–303.
DelGiudice, G. D., M. S. Lenarz, and M. Carstensen Powell. 2007. Age-specific fertility and fecundity in northern free-ranging white-tailed deer: Evidence for reproductive senescence? *Journal of Mammalogy* 88:427–435.
DelGiudice, G. D., L. D. Mech, K. E. Kunkel, and E. M. Gese. 1992. Seasonal patterns of weight, hematology, and serum characteristics of free-ranging female white-tailed deer in Minnesota. *Canadian Journal of Zoology* 70:974–983.
DelGiudice, G. D., M. R. Riggs, P. Joly, and W. Pan. 2002. Winter severity, survival, and cause-specific mortality of female white-tailed deer in north-central Minnesota. *Journal of Wildlife Management* 66:698–717.
DelGiudice, G. D., B. A. Sampson, D. W. Kuehn, M. Carstensen Powell, and J. Fieberg. 2005. Understanding margins of safe capture, chemical immobilization, and handling of free-ranging white-tailed deer. *Wildlife Society Bulletin* 33:677–687.
DeLiberto, T. J., J. A. Pfister, S. Demarais, and G. Ven Vreede. 1989. Seasonal changes in physiological parameters of white-tailed deer in Oklahoma. *Journal of Wildlife Management* 53:533–539.
DeNicola, A. J., D. J. Kesler, and R. K. Swihart. 1997a. Dose determination and efficacy of remotely delivered norgestomet implants on contraception of white-tailed deer. *Zoo Biology* 16:31–37.

DeNicola, A. J., D. J. Kesler, and R. K. Swihart. 1997b. Remotely delivered prostaglandin $F_{2\alpha}$ implants terminate pregnancy in white-tailed deer. *Wildlife Society Bulletin* 25:527–531.

DeNicola, A. J. and R. K. Swihart. 1997. Capture-induced stress in white-tailed deer. *Wildlife Society Bulletin* 25:500–503.

DeYoung, R. W., L. I. Muller, S. Demarais et al. 2004. Do *Odocoileus virginianus* males produce Y-chromosome-biased ejaculates? Implications for adaptive sex ratio theories. *Journal of Mammalogy* 85:768–773.

Ditchkoff, S. S. 2000. A decade since "diversification of ruminants": Has our knowledge improved? *Oecologia* 125:82–84.

Ditchkoff, S. S., M. S. Mitchell, W. N. Gray, II, and C. W. Cook. 2009. Temporal variation in sex allocation of white-tailed deer. *Journal of the Alabama Academy of Science* 80:27–34.

Ditchkoff, S. S., M. G. Sams, R. L. Lochmiller, and D. M. Leslie, Jr. 2001a. Utility of tumor necrosis factor-α and interleukin-6 as predictors of neonatal mortality in white-tailed deer. *Journal of Mammalogy* 82:239–245.

Ditchkoff, S. S. and F. A. Servello. 1998. Litterfall: An overlooked food source for wintering white-tailed deer. *Journal of Wildlife Management* 62:250–255.

Ditchkoff, S. S., L. J. Spicer, R. E. Masters, and R. L. Lochmiller. 2001b. Concentrations of insulin-like growth factor-I in adult male white-tailed deer (*Odocoileus virginianus*): Associations with serum testosterone, morphometrics and age during and after the breeding season. *Comparative Biochemistry and Physiology Part A* 129:887–895.

Ditchkoff, S. S., E. R. Welch, Jr., R. L. Lochmiller, R. E. Masters, and W. R. Starry. 2001c. Age-specific causes of mortality among male white-tailed deer support mate-competition theory. *Journal of Wildlife Management* 65:552–559.

Ditchkoff, S. S., E. R. Welch, Jr., W. R. Starry, W. C. Dinkines, R. E. Masters, and R. L. Lochmiller. 1997. Quality deer management at the McAlester Army Ammunition Plant: A unique approach. *Proceedings of the Annual Conference of the Southeastern Association of Fish and Wildlife Agencies* 51:389–399.

Durrant, J. D. and J. H. Lovrinic. 1995. *Bases of Hearing Science.* Baltimore, MD: Williams & Wilkins.

Fairbanks, W. S. 1993. Birthdate, birthweight, and survival in pronghorn fawns. *Journal of Mammalogy* 74:129–135.

Feldhammer, G. A., T. P. Kilbane, and D. W. Sharp. 1989. Cumulative effect of winter on acorn yield and deer body weight. *Journal of Wildlife Management* 53:292–295.

Fickel, J., F. Göritz, B. A. Joest, T. Hildebrandt, R. R. Hofmann, and G. Breves. 1998. Analysis of parotid and mixed saliva in roe deer (*Capreolus capreolus* L.). *Journal of Comparative Physiology B* 168:257–264.

Foley, C. A. H., S. Papageorge, and S. K. Wasser. 2001. Noninvasive stress and reproductive measures of social and ecological pressures in free-ranging African elephants. *Conservation Biology* 15:1134–1142.

Ford, E. B. 1945. Polymorphism. *Biological Review* 20:73.

Franzmann, A. W., A. Flynn, and P. D. Arneson. 1975. Serum corticoid levels relative to handling stress in Alaskan moose. *Canadian Journal of Zoology* 53:1424–1426.

Fuller, T. K., R. M. Pace, III, J. A. Markl, and P. L. Coy. 1989. Morphometrics of white-tailed deer in north-central Minnesota. *Journal of Mammalogy* 70:184–188.

Gassett, J. W., D. P. Wiesler, A. G. Baker et al. 1996. Volatile compounds from interdigital gland of male white-tailed deer (*Odocoileus virginianus*). *Journal of Chemical Ecology* 22:1689–1696.

Gassett, J. W., D. P. Wiesler, A. G. Baker et al. 1997. Volatile compounds from the forehead region of male white-tailed deer (*Odocoileus virginianus*). *Journal of Chemical Ecology* 23:569–578.

Gill, J. 1956. Regional differences in size and productivity of deer in West Virginia. *Journal of Wildlife Management* 20:286–292.

Gray, W. N., II, S. S. Ditchkoff, K. Causey, and C. W. Cook. 2002. The yearling disadvantage in Alabama deer: Effect of birth date on development. *Proceedings of the Annual Conference of the Southeastern Association of Fish and Wildlife Agencies* 56:255–264.

Gregorini, P., S. Tamminga, and S. A. Gunter. 2006. Review: Behavior and daily grazing patterns of cattle. *Professional Animal Scientist* 22:201–209.

Hamilton, R. J., M. L. Tobin, and W. G. Moore. 1985. Aging fetal white-tailed deer. *Proceedings of the Annual Conference of the Southeastern Association of Fish and Wildlife Agencies* 39:389–395.

Hardin, J. W., W. D. Klimstra, and N. J. Silvy. 1984. Florida Keys. In *White-tailed Deer: Ecology and Management*, ed. L. K. Halls, 381–390. Harrisburg, PA: Stackpole Books.

Harper, J. M. and S. N. Austad. 2000. Fecal glucocorticoids: A noninvasive method of measuring adrenal activity in wild and captive rodents. *Physiological and Biochemical Zoology* 73:12–22.
Haugen, A. O. 1959. Breeding records of captive white-tailed deer in Alabama. *Journal of Mammalogy* 40:108–113.
Haugen, A. O. 1975. Reproductive performance of white-tailed deer in Iowa. *Journal of Mammalogy* 56:151–159.
Haulton, S. M., W. F. Porter, and B. A. Rudolph. 2001. Evaluating 4 methods to capture white-tailed deer. *Wildlife Society Bulletin* 29:255–264.
Henderson, D. W., R. J. Warren, J. A. Cromwell, and R. J. Hamilton. 2000. Responses of urban deer to a 50% reduction in local herd density. *Wildlife Society Bulletin* 28:902–910.
Hesselton, W. T. 1969. The incredible white deer herd. *New York State Conservationist* 24:18–19.
Hesselton, W. T. and L. W. Jackson. 1974. Reproductive rates of white-tailed deer in New York state. *New York Fish and Game Journal* 21:135–152.
Hesselton, W. T. and P. R. Sauer. 1973. Comparative physical condition of four deer herds in New York according to several indices. *New York Fish and Game Journal* 20:77–107.
Hiraiwa-Hasegawa, M. 1993. Skewed birth sex ratios in primates: Should high ranking mothers have daughters or sons? *Trends in Ecology and Evolution* 8:395–400.
Hofmann, R. R. 1989. Evolutionary steps of ecophysiological adaptation and diversification of ruminants: A comparative review of their digestive system. *Oecologia* 78:443–457.
Jacobs, G. H., J. F. Deegan, II, J. Neitz, B. P. Murphy, K. V. Miller, and R. L. Marchinton. 1994. Electrophysiological measurements of spectral mechanisms in the retinas of two cervids: White-tailed deer (*Odocoileus virginianus*) and fallow deer (*Dama dama*). *Journal of Comparative Physiology A* 174:551–557.
Jacobson, H. A., D. C. Guynn, Jr., R. N. Griffin, and D. Lewis. 1979. Fecundity of white-tailed deer in Mississippi and periodicity of corpora lutea and lactation. *Proceedings of the Annual Conference of the Southeastern Association of Fish and Wildlife Agencies* 33:30–35.
Jenks, J. A., D. M. Leslie, Jr., R. L. Lochmiller, and M. A. Melchiors. 1994. Variation in gastrointestinal characteristics of male and female white-tailed deer: Implications for resource partitioning. *Journal of Mammalogy* 75:1045–1053.
Jenks, J. A., W. P. Smith, and C. S. DePerno. 2002. Maximum sustained yield harvest versus trophy management. *Journal of Wildlife Management* 66:528–535.
Johns, P. E., R. Baccus, M. N. Manlove, J. E. Pinder, III, and M. H. Smith. 1977. Reproductive patterns, productivity and genetic variability in adjacent white-tailed deer populations. *Proceedings of the Annual Conference of the Southeastern Association of Fish and Wildlife Agencies* 31:167–172.
Johnson, H. E., W. G. Youatt, L. D. Fay, H. D. Harte, and D. E. Ullrey. 1968. Hematological values of Michigan white-tailed deer. *Journal of Mammalogy* 49:749–754.
Kie, J. G. and M. White. 1985. Population dynamics of white-tailed deer (*Odocoileus virginianus*) on the Welder Wildlife Refuge, Texas. *Southwestern Naturalist* 30:105–118.
Knox, W. M., K. V. Miller, D. C. Collins, P. B. Bush, T. E. Kiser, and R. L. Marchinton. 1992. Serum and urinary levels of reproductive hormones associated with the estrous cycle in white-tailed deer (*Odocoileus virginianus*). *Zoo Biology* 11:121–131.
Knox, W. M., K. V. Miller, and R. L. Marchinton. 1988. Recurrent estrous cycles in white-tailed deer. *Journal of Mammalogy* 69:384–386.
Lambiase, J. T., Jr., R. P. Amann, and J. S. Lindzey. 1972. Aspects of reproductive physiology of male white-tailed deer. *Journal of Wildlife Management* 36:868–875.
Leimar, O. 1996. Life-history analysis of the Trivers and Willard sex-ratio problem. *Behavioral Ecology* 7:316–325.
Locke, S. L., M. W. Cook, L. A. Harveson et al. 2007. Effectiveness of Spayvac® for reducing white-tailed deer fertility. *Journal of Wildlife Diseases* 43:726–730.
Martin, P. P. and G. P. Rasmussen. 1981. *An Investigation into the Mode of Inheritance of White Coat Color in White-tailed Deer*. New York State Department of Environmental Conservation, Albany, New York.
Masters, R. D. and N. E. Mathews. 1990. Notes on reproduction of old (>9 years) free-ranging white-tailed deer, *Odocoileus virginianus*, in the Adirondacks, New York. *Canadian Field-Naturalist* 105:286–287.
Mautz, W. W. 1978. Sledding on a bushy hillside: The fat cycle in deer. *Wildlife Society Bulletin* 6:88–90.
Mautz, W. W., J. Kanter, and P. J. Pekins. 1992. Seasonal metabolic rhythms of captive female white-tailed deer: A reexamination. *Journal of Wildlife Management* 56:656–661.

Mautz, W. W. and G. A. Petrides. 1971. Food passage rate in the white-tailed deer. *Journal of Wildlife Management* 35:723–731.

McDonald, J. S. and K. V. Miller. 2004. *A History of White-tailed Deer Restocking in the United States. 1878–2004.* Watkinsville, GA: Quality Deer Management Association.

Mech, L. D. and R. E. McRoberts. 1990. Relationship between age and mass among female white-tailed deer during winter and spring. *Journal of Mammalogy* 71:686–689.

Mech, L. D., R. E. McRoberts, R. O. Peterson, and R. E. Page. 1987. Relationship of deer and moose populations to previous winter's snow. *Journal of Animal Ecology* 56:615–627.

Mech, L. D., M. E. Nelson, and R. E. McRoberts. 1991. Effects of maternal and grandmaternal nutrition on deer mass and vulnerability to wolf predation. *Journal of Mammalogy* 72:146–151.

Miller, L. A., K. Crane, S. Gaddis, and G. J. Killian. 2001. Porcine zona pellucida immunocontraception: Long-term health effects on white-tailed deer. *Journal of Wildlife Management* 65:941–945.

Miller, L. A., B. E. Johns, and G. J. Killian. 2000. Immunocontraception of white-tailed deer with GnRH vaccine. *American Journal of Reproductive Immunology* 44:266–274.

Miller, M. W., N. T. Hobbs, and M. C. Sousa. 1991. Detecting stress responses in Rocky Mountain bighorn sheep (*Ovis canadensis canadensis*): Reliability of cortisol concentrations in urine and feces. *Canadian Journal of Zoology* 69:15–24.

Millspaugh, J. J. and B. E. Washburn. 2003. Within-sample variation of fecal glucocorticoid measurements. *General and Comparative Endocrinology* 132:21–26.

Millspaugh, J. J., B. E. Washburn, M. A. Milanick, J. Beringer, L. P. Hansen, and T. M. Meyer. 2002. Non-invasive techniques for stress assessment in white-tailed deer. *Wildlife Society Bulletin* 30:899–907.

Mirarchi, R. E., B. E. Howland, P. F. Scanlon, and R. L. Kirkpatrick. 1978. Seasonal variation in plasma LH, FSH, prolactin, and testosterone concentrations in adult male white-tailed deer. *Canadian Journal of Zoology* 56:121–127.

Mirarchi, R. E., P. F. Scanlon, and R. L. Kirkpatrick. 1977. Annual changes in spermatozoan production and associated organs of white-tailed deer. *Journal of Wildlife Management* 41:92–99.

Moen, A. N. 1973. *Wildlife Ecology: An Analytical Approach.* San Francisco: W. H. Freeman and Co.

Monteith, K. L., L. E. Schmitz, J. A. Jenks, J. A. Delger, and R. T. Bowyer. 2009. Growth of male white-tailed deer: Consequences of maternal effects. *Journal of Mammalogy* 90:651–660.

Muller, L. I., R. J. Warren, and D. L. Evans. 1997. Theory and practice of immunocontraception in wild mammals. *Wildlife Society Bulletin* 25:504–514.

Müller-Schwarze, D. 1994. The senses of deer. In *The Wildlife Series: Deer*, eds. D. Gerlach, S. Atwater, and J. Schnell, 58–65. Mechanicsburg, PA: Stackpole Books.

Newsom, W. M. 1937. Mammals of Anticosti Island. *Journal of Mammalogy* 18:435–442.

Odend'hal, S., K. V. Miller, and D. M. Hoffmann. 1992. Preputial glands in the white-tailed deer (*Odocoileus virginianus*). *Journal of Mammalogy* 73:299–302.

Ørskov, E. R., D. Benzie, and R. N. B. Kay. 1970. The effects of feeding procedure on closure of the oesophageal groove in young sheep. *British Journal of Nutrition* 24:785–795.

Ozoga, J. J. 1987. Maximum fecundity in supplementally-fed northern Michigan white-tailed deer. *Journal of Mammalogy* 68:878–879.

Ozoga, J. J. and L. J. Verme. 1982. Physical and reproductive characteristics of a supplementally-fed white-tailed deer herd. *Journal of Wildlife Management* 46:281–301.

Parker, K. L., C. T. Robbins, and T. A. Hanley. 1984. Energy expenditures for locomotion by mule deer and elk. *Journal of Wildlife Management* 48:474–488.

Pekins, P. J., K. S. Smith, and W. W. Mautz. 1998. The energy cost of gestation in white-tailed deer. *Canadian Journal of Zoology* 76:1091–1097.

Peles, J. D., O. E. Rhodes, Jr., and M. H. Smith. 2003. Spermatozoan numbers and characteristics of associated organs in male white-tailed deer during the breeding season. *Acta Theriologica* 2003:123–130.

Peterson, M. N., R. R. Lopez, P. A. Frank, M. J. Peterson, and N. J. Silvy. 2003. Evaluating capture methods for urban white-tailed deer. *Wildlife Society Bulletin* 31:1176–1187.

Phillips, C. J. C. 1993. *Cattle Behaviour.* Ipswich, UK: Farming Press Books.

Plotka, E. D., U. S. Seal, L. J. Verme, and J. J. Ozoga. 1977. Reproductive steroids in the white-tailed deer (*Odocoileus virginianus borealis*). II. Progesterone and estrogen levels in peripheral plasma during pregnancy. *Biology of Reproduction* 17:78–83.

Plotka, E. D., U. S. Seal, L. J. Verme, and J. J. Ozoga. 1980. Reproductive steroids in deer. III. Estradiol and progesterone around estrous. *Biology of Reproduction* 22:576–581.

Quay, W. B. 1971. Geographic variation in the metatarsal "gland" of the white-tailed deer (*Odocoileus virginianus*). *Journal of Mammalogy* 52:1–11.

Ransom, A. B. 1967. Reproductive biology of white-tailed deer in Manitoba. *Journal of Wildlife Management* 31:114–123.

Rawson, R. E., G. D. DelGiudice, H. E. Dziuk, and L. D. Mech. 1992. Energy metabolism and hematology of white-tailed deer fawns. *Journal of Wildlife Diseases* 28:91–94.

Rhodes, O. E., Jr., K. T. Scribner, M. H. Smith, and P. E. Johns. 1985. Factors affecting the number of fetuses in a white-tailed deer herd. *Proceedings of the Annual Conference of the Southeastern Association of Fish and Wildlife Agencies* 39:380–388.

Robbins, C. T. 1993. *Wildlife Feeding and Nutrition* (2nd edition). San Diego, CA: Academic Press.

Robbins, C. T., A. N. Moen, and J. T. Reid. 1975. Body composition of white-tailed deer. *Journal of Animal Science* 38:871–876.

Roseberry, J. L. and W. D. Klimstra. 1970. Productivity of white-tailed deer on Crab Orchard National Wildlife Refuge. *Journal of Wildlife Management* 34:23–28.

Roseberry, J. L. and W. D. Klimstra. 1975. Some morphological characteristics of the Crab Orchard Deer Herd. *Journal of Wildlife Management* 39:48–58.

Rue, L. L., III. 1978. *The Deer of North America.* New York, NY: Crown Publishers, Inc.

Ryel, L. A. 1963. The occurrence of certain anomalies in Michigan white-tailed deer. *Journal of Mammalogy* 44:79–98.

Saalfeld, S. T., S. S. Ditchkoff, J. J. Ozoga, and M. S. Mitchell. 2007. Seasonal variation in sex ratios provides developmental advantages in white-tailed deer, *Odocoileus virginianus*. *Canadian Field-Naturalist* 121:412–419.

Saltz, D. and G. C. White. 1991. Urinary cortisol and urea nitrogen responses to winter stress in mule deer. *Journal of Wildlife Management* 55:1–16.

Sams, M. G., R. L. Lochmiller, C. W. Qualls, Jr., and D. M. Leslie, Jr. 1993. Clinical blood profiles of stressed white-tailed deer: Drop-net versus harvest. *Proceedings of the Annual Conference of the Southeastern Association of Fish and Wildlife Agencies* 47:198–210.

Sams, M. G., R. L. Lochmiller, C. W. Qualls, Jr., D. M. Leslie, Jr., and M. E. Payton. 1996. Physiological correlates of neonatal mortality in an overpopulated herd of white-tailed deer. *Journal of Mammalogy* 77:179–190.

San José, C., F. Braza, and S. Aragón. 1999. The effect of age and experience on the reproductive performance and prenatal expenditure of resources in female fallow deer (*Dama dama*). *Canadian Journal of Zoology* 77:1717–1722.

Sauer, P. R. 1984. Physical characteristics. In *White-tailed Deer Ecology and Management*, ed. L. K. Halls, 73–90. Harrisburg, PA: Stackpole Books.

Seal, U. S., L. J. Verme, and J. J. Ozoga. 1981. Physiologic values. In *Diseases and Parasites of White-tailed Deer*, eds. W. R. Davidson, F. A. Hayes, V. F. Nettles, and F. E. Kellogg, 17–34. Athens, GA: Southeastern Cooperative Wildlife Disease Study.

Severinghaus, C. W. 1955. Deer weights as an index of range conditions on two wilderness areas in the Adirondack region. *New York Fish and Game Journal* 2:154–160.

Severinghaus, C. W. and E. L. Cheatum. 1956. The life and times of white-tailed deer. In *The Deer of North America*, ed. W. P. Taylor, 57–186. Harrisburg, PA: Stackpole Books.

Shea, S. M., T. A. Breault, and M. L. Richardson. 1992. Herd density and physical condition of white-tailed deer in Florida flatwoods. *Journal of Wildlife Management* 56:262–267.

Short, C. 1970. Morphological development and aging of mule and white-tailed deer fetuses. *Journal of Wildlife Management* 34:383–388.

Short, H. L., E. E. Remmenga, and C. E. Boyd. 1969. Variations in ruminoreticular contents of white-tailed deer. *Journal of Wildlife Management* 33:187–191.

Silver, H. 1961. Deer milk compared with substitute milk for fawns. *Journal of Wildlife Management* 25:66–70.

Silver, H., N. F. Colovos, J. B. Holter, and H. H. Hayes. 1969. Fasting metabolism of white-tailed deer. *Journal of Wildlife Management* 33:490–498.

Smith, B. L., D. J. Skotko, W. Owen, and R. J. McDaniel. 1989. Color vision in the white-tailed deer. *Psychological Record* 39:195–201.

Smith, M. H., H. O. Hillestad, R. Baccus, and M. N. Manlove. 1984. Population genetics. In *White-tailed Deer Ecology and Management*, ed. L. K. Halls, 119–128. Harrisburg, PA: Stackpole Books.

Soprovich, D. W. 1992. Fetal growth rate estimation from length and date of death. In *The Biology of Deer*, ed. R. D. Brown, 63–68. New York, NY: Springer-Verlag.

Stefanski, V. 2001a. Effects of psychosocial stress or food restriction on body mass and blood cellular immunity in laboratory rats. *Stress and Health* 17:133–140.

Stefanski, V. 2001b. Social stress in laboratory rats: Behavior, immune function, and tumor metastasis. *Physiology and Behavior* 73:385–391.

Stoll, R. J., Jr. and W. P. Parker. 1986. Reproductive performance and condition of white-tailed deer in Ohio. *Ohio Journal of Science* 86:164–168.

Strickland, B. K., S. Demarais, and P. D. Gerard. 2008. Variation in mass and lactation among cohorts of white-tailed deer *Odocoileus virginianus*. *Wildlife Biology* 14:263–271.

Taylor, W. P. 1956. *The Deer of North America: Their History and Management*. Harrisburg, PA: Stackpole Books.

Teer, J. G., J. W. Thomas, and E. A. Walker. 1965. Ecology and management of white-tailed deer in the Llano Basin of Texas. *Wildlife Monographs* 15:1–62.

Therrien, J.-F., S. D. Côté, M. Festa-Bianchet, and J.-F. Ouellet. 2007. Conservative maternal care in an iteroparous mammal: A resource allocation experiment. *Behavioral Ecology and Sociobiology* 62:193–199.

Thompson, C. B., J. B. Holter, H. H. Hayes, H. Silver, and W. E. Urban, Jr. 1973. Nutrition of white-tailed deer. I. Energy requirement of fawns. *Journal of Wildlife Management* 37:301–311.

Tonkovich, M. J., M. C. Reynolds, W. L. Culbertson, and R. J. Stoll, Jr. 2004. Trends in reproductive performance and condition of white-tailed deer in Ohio. *Ohio Journal of Science* 104:112–122.

Torbit, S. C., L. H. Carpenter, D. M. Swift, and A. W. Alldredge. 1985. Differential loss of fat and protein by mule deer during winter. *Journal of Wildlife Management* 49:80–85.

Townsend, M. T. and M. W. Smith. 1933. The white-tailed deer of the Adirondacks. *Roosevelt Wildlife Bulletin* 6:161–325.

Trivers, R. L. and D. E. Willard. 1973. Natural selection of parental ability to vary the sex ratio of offspring. *Science* 179:90–92.

Trodd, L. L. 1962. Quadruplet fetuses in a white-tailed deer from Espanola, Ontario. *Journal of Mammalogy* 43:414.

Tumbleson, M. E., J. D. Cuneio, and D. A. Murphy. 1970. Serum biochemical and hematological parameters of captive white-tailed fawns. *Canadian Journal of Comparative Medicine* 34:66–71.

Turner, J. W., Jr., J. F. Kirkpatrick, and I. K. M. Liu. 1996. Effectiveness, reversibility, and serum antibody titers associated with immunocontraception in captive white-tailed deer. *Journal of Wildlife Management* 60:45–51.

Van Deelen, T. R., P. Kaiser, M. A. Watt, and S. R. Craven. 2007. Quintuplet fetuses from a white-tailed deer (*Odocoileus virginianus*) in Wisconsin. *American Midland Naturalist* 157:398–400.

VerCauteren, K. C. and M. J. Pipas. 2003. A review of color vision in white-tailed deer. *Wildlife Society Bulletin* 31:684–691.

Verme, L. J. 1963. Effect of nutrition on growth of white-tailed deer fawns. *Transactions of the North American Wildlife Conference* 28:431–443.

Verme, L. J. 1965. Reproduction studies on penned white-tailed deer. *Journal of Wildlife Management* 29:74–79.

Verme, L. J. 1967. Influence of experimental diets on white-tailed deer reproduction. *Transactions of the North American Wildlife Conference* 32:405–420.

Verme, L. J. 1969. Reproductive patterns of white-tailed deer related to nutritional plane. *Journal of Wildlife Management* 33:881–887.

Verme, L. J. 1983. Sex ratio variation in *Odocoileus*: A critical review. *Journal of Wildlife Management* 47:573–582.

Verme, L. J. 1989. Maternal investment in white-tailed deer. *Journal of Mammalogy* 70:438–442.

Verme, L. J. and D. E. Ullrey. 1984. Physiology and nutrition. In *White-tailed Deer Ecology and Management*, ed. L. K. Halls, 91–118. Harrisburg, PA: Stackpole Books.

Waddell, R. B., D. A. Osborn, R. J. Warren, R. N. Griffin, and D. J. Kesler. 2001. Prostaglandin $F_{2\alpha}$-mediated fertility control in captive white-tailed deer. *Wildlife Society Bulletin* 29:1067–1074.

Waid, D. D. and R. J. Warren. 1984. Seasonal variations in physiological indices of adult female white-tailed deer in Texas. *Journal of Wildlife Diseases* 20:212–219.

Walter, W. D., P. J. Perkins, A. T. Rutberg, and H. J. Kilpatrick. 2002. Evaluation of immunocontraception in a free-ranging suburban white-tailed deer herd. *Wildlife Society Bulletin* 30:186–192.

Warren, R. J. 1995. Should wildlife biologists be involved in wildlife contraception research and management? *Wildlife Society Bulletin* 23:441–444.

Washburn, B. E. and J. J. Millspaugh. 2002. Effects of simulated environmental conditions on glucocorticoid metabolite measurements in white-tailed deer feces. *General and Comparative Endocrinology* 127:217–222.

Wasser, S. K., K. E. Hunt, J. L. Brown et al. 2000. A generalized fecal glucocorticoid assay for use in a diverse array of nondomestic mammalian and avian species. *General and Comparative Endocrinology* 120.

Webb, S. L., D. G. Hewitt, and M. W. Hellickson. 2007. Survival and cause-specific mortality of mature male white-tailed deer. *Journal of Wildlife Management* 71:555–558.

Webb, S. L., J. S. Lewis, D. G. Hewitt, M. W. Hellickson, and F. C. Bryant. 2008. Assessing the helicopter and net gun as a capture technique for white-tailed deer. *Journal of Wildlife Management* 72:310–314.

Weckerly, F. W. 1989. Plasticity in length of hindgut segments of white-tailed deer (*Odocoileus virginianus*). *Canadian Journal of Zoology* 67:189–193.

Wentworth, J. M., A. S. Johnson, P. E. Hale, and K. E. Kammermeyer. 1992. Relationship of acorn abundance and deer herd characteristics in the southern Appalachians. *Southern Journal of Applied Forestry* 16:5–8.

White, L. M., R. J. Warren, and R. A. Fayrer-Hosken. 1994. Levonorgestrel implants as a contraceptive in captive white-tailed deer. *Journal of Wildlife Diseases* 30:241–246.

Wilson, S. N. and J. A. Sealander. 1971. Some characteristics of white-tailed deer reproduction in Arkansas. *Proceedings of the Annual Conference of the Southeastern Association of Game and Fish Commissioners* 25:53–65.

Winnie, J., Jr. and S. Creel. 2007. Sex-specific behavioural responses of elk to spatial and temporal variation in the threat of wolf predation. *Animal Behaviour* 73:215–225.

Witzel, D. A., M. D. Springer, and H. H. Mollenhauer. 1978. Cone and rod photoreceptors in the white-tailed deer *Odocoileus virginianus*. *American Journal of Veterinary Research* 39:699–701.

Wobeser, G. A. 1981. Trauma. In *Diseases and Parasites of White-tailed Deer*, eds. W. R. Davidson, F. A. Hayes, V. F. Nettles, and F. E. Kellogg, 35–42. Athens, GA: Southeastern Cooperative Wildlife Disease Study.

Wood, W. F., T. B. Shaffer, and A. Kubo. 1995. (E)-3-tridecen-2-one, an antibiotic from the interdigital glands of black-tailed deer *Odocoileus hemionus columbianus*. *Experientia* 51:368–369.

Wozencraft, W. C. 1979. Melanistic deer in southern Wisconsin. *Journal of Mammalogy* 60:437.

Zacks, J. L. 1985. Photopic spectral sensitivity of the white-tailed deer, *Odocoileus virginianus*. *Investigative Opthalmology and Visual Science Supplement* 26:184.

Zacks, J. L. and W. Budde. 1983. Behavioral investigations of color vision in the white-tailed deer, *Odocoileus virginianus*. *Investigative Opthalmology and Visual Science Supplement* 24:185.

Zimmerman, T. J., J. A. Jenks, and D. M. Leslie, Jr. 2006. Gastrointestinal morphology of female white-tailed deer and mule deer: Effects of fire, reproduction, and feeding type. *Journal of Mammalogy* 87:598–605.

3

Nutrition

David G. Hewitt

CONTENTS

White-tailed deer are a product of their environment. The North American environment has shaped the species on an evolutionary timescale while individual deer are influenced by their surroundings during their lifetime. One of the fundamental interactions between a deer and its environment is through nutrition, the process by which a deer consumes portions of its environment and uses the ingested chemicals and energy for its own maintenance, growth, and reproduction. Because nutrition affects survival, growth, and reproduction, research has sought to understand, and management to influence, deer nutrition.

Nutrient Requirements

A nutrient is a chemical or class of chemicals ingested by animals that are necessary for normal metabolism and which promote health and productivity. Protein, vitamins, minerals, and water are primary

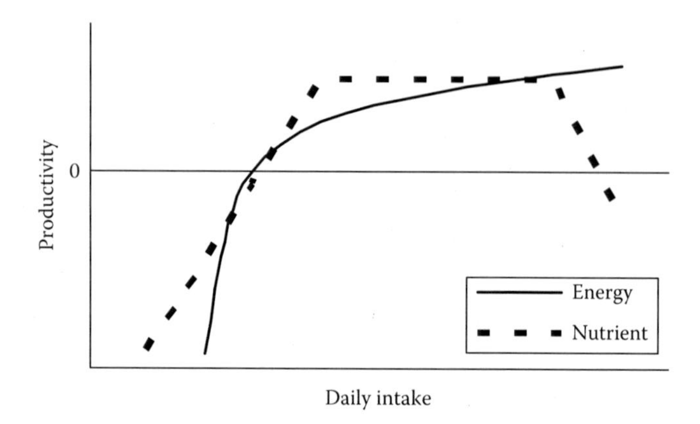

FIGURE 3.1 Deer productivity is related to intake of energy and nutrients. Efficiency of energy use for production varies with energy intake relative to maintenance whereas there is a range of nutrient intake over which productivity requirements are met and beyond which some nutrients become toxic.

classes of nutrients. Although energy is not a chemical, but rather a property of many different chemicals, it behaves as a nutrient in that too little will cause sickness or death, increasing amounts promote health, growth, and reproduction, and too much can be detrimental (Figure 3.1). Thus, animals have nutrient requirements and those requirements vary with many factors. For example, nutrient requirements of nonreproductive adult animals are often less than those for growing or reproductive animals. Nutrient requirements for temperate animals, like white-tailed deer, vary seasonally.

A variety of approaches can be used to estimate nutrient requirements, depending on the objectives (Barboza et al., 2009). Maintenance requirements may be defined by the amount of a nutrient necessary to maintain body mass (Ullrey et al., 1969; Strickland et al., 2005) or to replace obligatory losses of a nutrient in feces and urine (Grasman and Hellgren, 1993; Asleson et al., 1996). Another approach to predicting nutrient requirements is to sum the costs of various components of the requirement (National Research Council, 2007). For example to estimate the maintenance energy requirement of a deer, energy necessary for basal metabolism, activity, and thermoregulation can be summed (Robbins, 1993; Parker et al., 1999). Nutrient requirements for productive processes may be estimated by measuring the rate or outcome of the productive process, such as body mass gain, milk production, or protein or energy balance, under different daily intakes of the nutrient (Ullrey et al., 1967; Holter et al., 1979b). Knowledge of nutrient requirements for maintenance and production enable ecologists and deer managers to assess the ability of habitat to support deer populations, understand factors limiting a deer population, and design projects to meet management objectives.

Energy

Energy is the ability to do work. This standard definition is vague because energy is involved in every biochemical reaction in a deer. Energy is obviously necessary for activity, such as walking, and to stay warm in the cold. Less obvious is the energy used by an animal at rest. Nearly half the energy expenditure of a resting deer is used by body organs for basic functions such as circulation, respiration, nerve impulses, and waste removal. Most of the remaining energy expended by a resting animal is for cellular maintenance processes such as protein turnover and ion transport (Robbins, 1993).

Maintenance

Energy expenditure of a fasting, nonreproductive adult animal that is inactive, but not sleeping, in a thermal neutral environment is referred to as basal metabolism. Basal metabolism is not the lowest metabolic rate of a deer, because metabolic rate when sleeping may be 16% lower than when awake (Heller, 1988). The fed state of an animal is important to consider in calculating metabolic rate, which may be 28–34%

lower when fasting than when fed (Jensen et al., 1999). Large animals have a lower metabolic rate than small animals for each unit of body mass. Expressing metabolic rate relative to the animal's body mass raised to the 0.75 power enables comparisons between animals of different body sizes. Averaging across many species, basal metabolic rate (BMR) of mammals is 70 kcal/kg$^{0.75}$/day (Kleiber, 1961), although some species and species groups may differ from this broad average (Barboza et al., 2009). BMRs of white-tailed deer during winter are above this interspecific average (81–90 kcal/kg$^{0.75}$/day) (Silver et al., 1969, 1971; Mautz et al., 1992) and during summer, some studies have found BMR of white-tailed deer to be 42–48% greater than winter values (Silver et al., 1969; Holter et al., 1976). Other studies have documented only a 2–5% increase in BMR from winter to summer (Silver et al., 1959; Mautz et al., 1992). Although some cervids appear to reduce metabolic rate during winter as an energy conservation strategy (Arnold et al., 2004), Mautz et al. (1992) suggested increased BMR during summer may be an artifact resulting from measurements on deer not at rest and outside their thermal neutral zone. Differences in food intake between the seasons may influence energy use by digestive organs and supporting viscera (National Research Council, 2007), further explaining seasonal changes in BMR of white-tailed deer and other cervids (Weiner, 1977; Ringberg, 1979; Renecker and Hudson, 1986).

Activity is a second component of energy expenditure for maintenance and is necessary for obtaining resources, avoiding predation, and interacting socially. Energy cost of activity varies with the type of activity and the proportion of time a deer spends on each behavior. Alert behavior, grooming, ruminating, and other activities of bedded deer can increase energy expenditure 2–25% above BMR (Table 3.1). Muscle contractions necessary to support a standing animal and maintain balance increase energy

TABLE 3.1

Energetic Cost of Activity and Production in Deer as a Multiple of Interspecific Basal Metabolic Rate (BMR) (70 kcal/kg$^{0.75}$/day)

Activity or Productive State	Multiple of BMR	Species	Source
Lying—inactive in winter	1.11–1.28	White-tailed deer	1, 10, 11, 12
Lying—active	1.27	White-tailed deer	1
Lying—fawn fasted in October	1.60	White-tailed deer	8
Lying—fawn fasted in winter	1.10–1.28	White-tailed deer	7, 8
Lying—fawn fed in winter	1.67–1.96	White-tailed deer	7
Standing	1.40	Elk	3
Standing and less than 50% walking	1.90	White-tailed deer	1
Foraging	1.88	Elk	4
Foraging—assuming 1.4*BMR for standing	1.46–1.77	Moose and elk	2
Walking	2.06	White-tailed deer	1
Walking uphill (5 m vertical/min)	3.77	Deer	1, 3
Walking—brisket deep low density snow	6.50	Elk	3
Running/bounding	3.08–4.75	White-tailed deer	1
Gestation—early (twin fawns metabolism only)	1.15	White-tailed deer	5
Gestation—late (twin fawns metabolism only)	1.79	White-tailed deer	5
Gestation—late (single fawn metabolism & tissue)	1.98	White-tailed deer	6
Gestation—late (twin fawns metabolism & tissue)	2.84	White-tailed deer	6
Lactation—peak (single fawn metabolism & milk)	3.49	White-tailed deer	6
Lactation—peak (twin fawns metabolism & milk)	4.73	White-tailed deer	6
Body growth—fawn maximum in autumn	4.40	White-tailed deer	9
Antler growth	1.49	Cervids	6
Hair growth	1.36	Cervids	6

All energy estimates for productive processes assume that the deer is lying inactive (1.19*BMR), so that heat increment and costs of activity and thermoregulation would need to be added to estimate actual daily energy expenditure.

1, Mautz and Fair (1980); 2, Fancy and White (1985); 3, Parker et al. (1984); 4, Wickstrom et al. (1984); 5, Pekins et al. (1998); 6, National Research Council (2007); 7, Jensen et al. (1999); 8, Thompson et al. (1973); 9, Holter et al. (1979b); 10, Mautz et al. (1992); 11, Silver et al. (1969); 12, Silver et al. (1971).

expenditure about 20% above that of lying (Robbins, 1993) and 40% above BMR (Table 3.1). Walking may increase energy expenditure two times and trotting or running three to four times over BMR.

White-tailed deer living in mountainous and snowy environments have additional energy costs for locomotion. Deer expend 5.99 kcal to lift 1 kg of body mass 1 km vertical, independent of the steepness of the slope (Parker et al., 1984; Robbins, 1993). Traveling uphill at 5 m vertical/min will result in a net increase of 0.7 times BMR (Table 3.1). The effect of snow on the energy cost of movement depends on snow depth, density, and hardness. The most relevant measure of snow depth is relative to brisket height, which is 55–60 cm for deer weighing 50–80 kg (Parker et al., 1984). The effect of snow on the energy cost of locomotion increases dramatically as the depth to which deer sink in snow increases above 50% of their brisket height (Figure 3.2). The cost is also greater for heavy, wet snow than light, fluffy snow.

Because deer are homeotherms, they expend energy to maintain body temperature under extremes of heat and cold. Panting increases energy use in hot conditions and pileoerection and shivering use energy in cold conditions (Parker and Robbins, 1984). Although good relationships can be obtained between an animal's metabolic rate and temperature in controlled environments (e.g., Parker and Robbins, 1984; Jensen et al., 1999) predicting such relationships for free-ranging deer is exceedingly complex (Parker and Gillingham, 1990; National Research Council, 2007). Upper and lower critical temperatures, that is, temperatures above and below which deer increase metabolic rate, are influenced by a great number of external factors, including thermal radiation, wind speed, precipitation, and vegetative cover as well as characteristics of deer, including age, body size, productive state, pelage characteristics, and forage intake (Barboza et al., 2009). For example, the lower critical temperature of a 60 kg temperate ungulate with a winter coat is 25°C lower than with a summer coat (−20°C vs 5°C) (Robbins, 1993). The lower critical temperature of a fasting white-tailed deer fawn is −0.8°C compared to −11.2°C for a fawn that had recently eaten (Jensen et al., 1999), a sufficiently large difference that with adequate dry matter intake, thermoregulation is probably a minor component of a fawn's energy budget during winter.

Maintenance energy requirements including basal metabolism, the cost of digesting and metabolizing food, thermoregulation, and activity have been estimated by either measuring energy intake sufficient to maintain body mass, summing estimates of component costs, or using doubly labeled water which estimates CO_2 production through differential loss of oxygen and hydrogen isotopes from the deer's body. Estimates of digestible energy (DE) sufficient to maintain body mass of captive deer in northern portions of the species' range vary from 144 to 160 kcal/kg$^{0.75}$/day during winter (Ullrey et al., 1969, 1970; Thompson et al., 1973; Holter et al., 1979b). Maintenance requirements of deer from southern portions of the species' range (85–107 kcal/kg$^{0.75}$/day) (Strickland et al., 2005) may be lower than those from northern portions of the range although differences in study techniques make comparisons difficult. Digestible energy for maintenance of fawns during autumn (178 kcal/kg$^{0.75}$/day) and yearling deer during summer (206–219 kcal/kg$^{0.75}$/day) are greater than maintenance of adults during winter (Holter et al., 1979b).

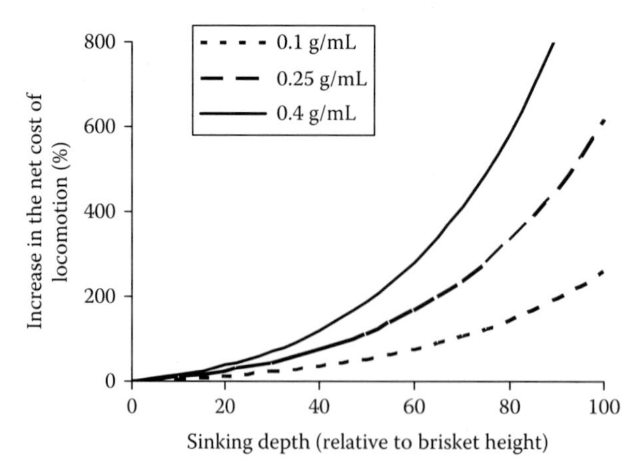

FIGURE 3.2 Net increase in energy expenditure, relative to travel in the absence of snow, for a deer traveling through snow of three densities (g/ml) and as a function of snow depth relative to the height of the deer's brisket. (Based on equation presented by Parker, K. L., C. T. Robbins, and T. A. Hanley. 1984. *Journal of Wildlife Management* 48:474–488.)

Free ranging deer may have energy expenditure 64% greater than that measured in captive deer (National Research Council, 2007). Incorporating seasonal changes in metabolic rates and activity, Moen (1978) estimated energy expenditure of a nonreproductive female white-tailed deer as 123 kcal/kg$^{0.75}$/day during winter and 262 kcal/kg$^{0.75}$/day during summer. Measurements of black-tailed deer energy expenditure using doubly labeled water were 154 kcal/kg$^{0.75}$/day in mild winter conditions and 191 kcal/kg$^{0.75}$/day in more severe winter conditions (DE = 178–219 kcal/kg$^{0.75}$/day) (Parker et al., 1999).

Production

The energy cost of production in deer includes reproduction, body growth in young deer, seasonal fattening, and pelage and antler growth. The cost of reproduction in males begins with the rut which can be quantified by tracking changes in body reserves or accounting for energy expended in activity associated with breeding. Bucks in southern Texas increase movements 5.9 km/day from prerut to the rut (A. Foley, Texas A&M University-Kingsville, unpublished data). Assuming that the net cost of movement is 2.57 kcal/kg $^{0.316}$/km (Robbins, 1993), an 80-kg buck would expend an additional 304 kcal/day, a 7% increase in metabolic rate relative to maintenance. However, increased energy expenditure is only part of the cost of reproduction in bucks. Males that spend too much time eating during the rut are not likely to be as successful breeders as males that spend most of their time searching for, courting, and tending females. Thus, the largest cost of reproduction in males is time not foraging. Fat stores acquired during autumn are used during the rut (Johns et al., 1984), causing bucks to lose up to 30% of body mass. Even a modest body mass loss of 20% in a 90 kg buck would result in release of 51,660 kcal of net energy, or 1260 kcal/day assuming a six-week rut, 3.5 kcal/g of body mass, and net energy efficiency of 0.82 (Robbins, 1993; Barboza et al., 2009).

Although females may show unusual movements around the time they are bred (Ivey and Causey, 1981), the energy cost of reproduction for females largely begins with gestation. The costs of gestation are the sum of increased metabolism and energy deposited in fetal and uterine tissue. The efficiency of energy deposition in the gravid uterus is 12–16% (National Research Council, 2007), making increased metabolism a majority of the cost of pregnancy. Energy cost of maintaining fetal and uterine tissue is reduced by delaying fetal growth until late in pregnancy, explaining why 82% of the energy in the gravid uterus is deposited and 92% of metabolic energy is expended during the final trimester of gestation (Robbins and Moen, 1975; Pekins et al., 1998). Thus, the cost of gestation is minimal during winter and does not increase until after spring green up through most of the white-tailed deer's range.

Gestation is only about 20% of the energy necessary for a doe to complete a reproductive effort (Oftedal, 1985). Lactation accounts for the remaining 80% of the energy used in raising fawns. As in gestation, the cost of lactation changes over time. Peak milk yield occurs 10–37 days after birth and declines exponentially to low amounts 120 days after birth (Sadleir, 1980; National Research Council, 2007). Nursing may occur past this time, but is likely to have greater social than nutritional benefits (Robbins, 1993). Including all maintenance requirements of a free-ranging doe, peak lactation may cost 470 kcal/kg$^{0.75}$/day, or 6.7 times BMR (National Research Council, 2007). This rate of energy expenditure cannot be met by intake alone, causing lactating does to lose body condition as they rely on endogenous reserves.

Energy cost of growth includes increased cellular metabolism and energy in the tissue deposited. Energy in tissue can hypothetically vary from zero if only water or ash is deposited to 9 kcal/g if only fat is deposited, although tissues are generally mixtures of such components. In white-tailed deer, gross energy content of gain varies from 1.1 kcal/g in fawns to 3.5 kcal/g in adult deer because the proportion of water in the gain decreases and the proportion of fat increases (Robbins et al., 1974a). The rate of gain also influences the cost. Thus, a fawn growing 200 g/day would require 220 kcal/day in energy deposited into tissue, whereas a mature buck increasing body mass 120 g/day (growth of 14.4 kg from July to October) would require an additional 420 kcal/day just for energy deposited into tissue. Metabolizable energy requirement for a 10 kg fawn growing at 200 g/day is 373 kcal/kg$^{0.75}$/day and for a mature buck gaining 120 g/day during summer is 239 kcal/kg$^{0.75}$/day (National Research Council, 2007).

Protein

Proteins are diverse molecules that serve contractile, structural, protective, transport, regulatory, immune, and metabolic roles in a deer (Barboza et al., 2009). Despite these varied roles, all proteins are similar in that they are composed of building blocks called amino acids. Animal proteins are composed of 20–25 different amino acids, all of which contain nitrogen (N) in an amine group and thus N metabolism in an animal is closely related to protein metabolism. Amino acids are absorbed from the diet or manufactured by microbes in the deer's digestive tract and then absorbed by the deer. Deer can make some amino acids given the proper precursors. Amino acids that an animal cannot make, or cannot make in sufficient amounts, must be supplied by the diet and are called essential amino acids.

Ruminants, including deer, are different from many other animals in that symbiotic microbes in the rumen manufacture amino acids from dietary carbon and N to meet microbial requirements. The amino acid composition of microbial protein matches closely with the requirements of the host animal. For this reason, deer do not have requirements for essential amino acids (National Research Council, 2007). Deer are also different from nonruminant species in that waste N from amino acid metabolism can be recycled in the rumen where microbes use the N to manufacture amino acids (National Research Council, 2007). The proportion of urea N recycled increases when deer eat diets with low protein concentrations (Robbins et al., 1974b), thus enabling deer to maintain N balance on low protein diets.

Measuring protein in forages is exceedingly difficult, but N is easy to measure and there is a good relationship between N and protein in forages. For these reasons, protein in forage is usually estimated by determining percent N and multiplying by 6.25. This conversion factor is derived from the fact that a wide variety of proteins contain 16% N. The result is termed crude protein because it represents an estimate of protein derived from the N content. Crude protein may differ from true protein in forages with large amounts of nonprotein N or in forages in which proteins do not contain 16% N (Milton and Dintzis, 1981; Izhaki, 1993).

Maintenance

Maintenance protein requirements are the sum of obligatory N losses in the feces and urine (Robbins, 1993). Fecal losses of endogenous N are from digestive secretions, sloughed intestinal cells, and microbial products. Because this metabolic fecal N is largely a function of dry matter intake, it represents a cost of eating. This cost ranges from 2.6 to 7.8 g N/kg dry matter intake (Robbins et al., 1975; Smith et al., 1975; Holter et al., 1977, 1979a; Robbins et al., 1987a; Asleson et al., 1996). Lower values of metabolic fecal N are generally from pelleted diets which may result in less abrasion of the digestive tract and therefore less metabolic fecal N (Smith et al., 1975, although see Robbins et al., 1987a). Obligatory loss of N in urine results from oxidation of amino acids and inefficiencies in protein turnover that occur even when protein intake is low. Excretion of this endogenous urinary N is related to metabolic rate and thus often expressed as a function of metabolic body size. Estimates of endogenous urinary N range from 0.115 to 0.136 g/kg$^{0.75}$/day (Robbins et al., 1974b; Asleson et al., 1996).

Protein intake sufficient to offset metabolic fecal and endogenous urinary N losses range from 0.37 to 0.61 g N/kg$^{0.75}$/day in adult bucks to 0.77 g N/kg$^{0.75}$/day in yearling deer during summer (Holter et al., 1979a; Asleson et al., 1996). At levels of feed intake in these studies, 4.1–5.8% dietary protein is sufficient to meet protein requirements for maintenance. Deer also continually lose N from skin shedding which amounts to 0.04 g N/kg$^{0.75}$/day (National Research Council, 2007). Incorporating these N losses and accounting for digestive and metabolic efficiencies, maintenance requirements for white-tailed deer are 8–9% crude protein during winter and 10% during summer (National Research Council, 2007).

Production

Crude protein deposited into the gravid uterus increases exponentially from 2.1 g/day early in gestation to 39.3 g/day at term (Robbins and Moen, 1975) so that protein requirements for gestation do not become large until the third trimester. Accounting for the efficiency of protein deposition and digestion

(National Research Council, 2007), a doe requires 128 g/day of additional crude protein from her diet, or an increase of 8 g/100 g feed above that required for maintenance, assuming that protein for pregnancy is of dietary origin (but see Parker et al., 2005). At peak lactation, a doe supporting two fawns requires 203 g/day additional crude protein (National Research Council, 2007) above maintenance, which if entirely from dietary sources would increase dietary crude protein content 13 g/100 g feed above maintenance. Estimates of dietary crude protein for late pregnancy and peak lactation are 14–16% (National Research Council, 2007), requirements which assume endogenous protein is used to support lactation.

White-tailed deer fawns may grow 150–210 g/day after weaning (Ullrey et al., 1967; Thompson et al., 1973; Smith et al., 1975), requiring diets with 17–19% crude protein (National Research Council, 2007). Protein requirement of 25% was estimated from growth rates of fawns consuming diets differing in crude protein content (Smith et al., 1975). Male fawns may have higher growth rates and protein requirements than female fawns (Ullrey et al., 1967). Crude protein requirement of mature male deer gaining body mass and growing antlers during summer is 9–10% (Asleson et al., 1996; National Research Council, 2007). Dry antler is 45% protein (Robbins, 1993), but that protein is deposited over at least 130 days so that on a daily basis less than 7% of protein intake is deposited into antlers (Asleson et al., 1996).

Vitamins

Vitamins are complex organic molecules required in small amounts for normal metabolism. Many vitamins are part of enzyme systems so that deficiencies produce specific symptoms. Subclinical deficiencies can affect deer health or performance, but are difficult to diagnose. Vitamins A, D, E, and K are fat-soluble vitamins and are required by deer. Deer can generally synthesize sufficient vitamin D through exposure to UV light in sunlight. Adequate vitamin K is synthesized by microbes in the digestive tract to meet deer requirements. Vitamins A and E can be obtained in sufficient amounts in green forage so that the only time a deer may experience deficiencies is during extended periods of poor plant growth, such as during drought or winter. Deer can store both of these vitamins in their liver, which reduces the likelihood of deficiencies. Vitamin A toxicity is possible in ruminants with excessive supplementation of active forms of the vitamin (National Research Council, 2007). Toxicity does not occur when deer eat forage because conversion of carotene, the primary chemical with vitamin A activity in plants, to an active form of vitamin A declines with increasing carotene consumption (Robbins, 1993). Vitamin E promotes stability of cell membranes and can protect animals from white muscle disease and capture-induced myopathy. White-tailed deer fawns from does supplemented with vitamin E had half the mortality rate from white muscle disease than fawns from does fed a diet deficient in vitamin E (Brady et al., 1978). Surveys of wild deer suggest that deficiencies of vitamins A and E are possible but rare (Youatt et al., 1976; Dierenfeld and Jessup, 1990).

All remaining vitamins, including B vitamins, vitamin C, folic acid, and niacin are water soluble. As such, they are readily excreted in the urine and therefore cannot be stored and are not toxic. Ruminants can meet their requirements for water-soluble vitamins from dietary sources, microbe synthesis in the rumen, and endogenous production (National Research Council, 2007). Deficiencies of water-soluble vitamins in fawns, before they become functional ruminants, are conceivable but have not been explored.

Minerals

There are 14 elements, other than carbon, oxygen, nitrogen, and hydrogen, necessary for normal metabolism in deer. These elements are considered either macro or microminerals depending on the amount required in the diet. Although minerals compose less than 5% of a deer's body (Robbins et al., 1974a), they serve many purposes including structural in bone and antler, transport in hemoglobin, and metabolic as cofactors in a wide variety of enzymes. Macrominerals such as calcium (Ca) and phosphorus (P) are 22–30 mg/g of a deer's body, while microminerals such as iron, zinc, manganese, copper, and molybdenum are 3–165 μg/g (Robbins, 1993). Ca and P nutrition have received moderate attention from researchers in part because they are major components of antlers.

Macrominerals

Ninety-eight percent of Ca and 80% of P in a deer's body are in bones. Both minerals are also components in enzyme systems and P is part of nucleic acids, cell membranes, and many metabolically active compounds. Ratios of dietary Ca:P that deviate dramatically from 2:1 (e.g., greater than 7:1 or less than 1:1) can reduce absorption of the less abundant mineral or cause metabolic changes resulting in deficiency symptoms (Robbins, 1993; National Research Council, 2007). Because forages often have higher concentrations of Ca than P, a deer's Ca requirements can often be met from forages but P maybe deficient. Despite P being considered a limiting nutrient for many herbivores, male white-tailed deer appear to be able to meet requirements for body and antler growth in arid ranges of the southern United States, even when forbs are scarce (Grasman and Hellgren, 1993). Daily requirements of Ca and P may increase over maintenance up to fourfold to meet demands of reproduction and of body and antler growth, but increases in dry matter intake often help meet this increased demand, so that large changes in Ca and P concentrations in the diet may not be necessary (Table 3.2). Furthermore, high Ca and P requirements during late gestation, lactation, and antler growth can be offset through increases in absorption efficiency, decreased endogenous losses, and resorption from bone (Stephenson and Brown, 1984; National Research Council, 2007). By relying on large stores of Ca and P in bones, a deer may experience seasonal osteoporosis but can lengthen the period over which it consumes mineral needed for lactation and antler growth (Brown, 1990). Finally, deer may seek food items with high concentrations of Ca and P, such as bones, antlers, and gastropod shells when mineral requirements are high (Krausman and Bissonette, 1977; G. Timmons, R. Darr, K. Williamson, and L. Garver, unpublished observations), although the contribution of these sources to mineral budgets is unknown.

Deer, like many other mammalian herbivores, have a strong sodium (Na) appetite particularly during spring and summer when vegetation has high water and potassium content. High water content increases sodium loss because of high rates of water excretion and high potassium concentration reduces efficiency of Na absorption from the digestive tract (Weeks and Kirkpatrick, 1976; Barboza et al., 2009). Studies with captive white-tailed deer suggest Na maintenance requirements of 3.27 mg/kg body mass/day, or about 109 mg/kg dry matter intake (Hellgren and Pitts, 1997), with no seasonal differences detected. Calculated Na budgets of white-tailed deer in New Hampshire showed that terrestrial forages could not meet Na losses and production requirements (Pletscher, 1987). Sodium in aquatic vegetation and from salt spread on highways during winter appeared necessary to balance Na budgets. Female deer may have annual Na requirements twice those of males, due in large part to costs of gestation and lactation (Pletscher, 1987).

Mineral licks, areas where deer consume soil or drink highly mineralized water, are frequented by white-tailed deer during spring and summer, primarily to obtain sodium (Weeks, 1978; Kennedy et al., 1995; Campbell et al., 2004); winter use is negligible. In many instances, female deer use licks more than

TABLE 3.2

Calcium and Phosphorus Requirement of a 50-kg White-tailed Deer Doe for Maintenance and Reproduction and for an 80-kg Buck for Body and Antler Growth during Summer

	Calcium		Phosphorus	
	g/day	Percent	g/day	Percent
Maintenance	2.3	0.21	1.9	0.17
Late gestation (twin fetuses)	8.3	0.55	8.2	0.34
Lactation (twin fawns)	8.2	0.34	7.5	0.50
Antler and body growth (60 g/day)	3.4	0.14	2.9	0.12
Antler and body growth (120 g/day)	4.5	0.17	3.7	0.14

Source: Adapted from National Research Council. 2007. *Nutrient Requirements of Small Ruminants Sheep, Goats, Cervids, and New World Camelids.* Washington, DC: National Academies Press.

males (Wiles and Weeks, 1986; Kennedy et al., 1995), but use and even dominance by males may occur (Weeks, 1978). Deer generally use licks within their home range, but may travel up to 5.5 km to use a specific lick (Wiles and Weeks, 1986; Campbell et al., 2004).

Microminerals

There are no well-established requirements for microminerals in deer (National Research Council, 2007). Livestock requirements are often used but maybe incorrect because livestock have not necessarily been selected for efficient use of minerals, leading to overestimates of mineral requirements for deer (Robbins, 1993).

Iodine (I) is deficient in soils and forage in parts of the white-tailed deer's range, such as the Great Lakes region of the United States. No effect on reproductive success or thyroid function was found when a basal diet with 0.26 mg I/kg dry matter was supplemented with 0.2 and 0.7 mg I/kg dry matter (Watkins et al., 1983). Thus, 0.26 mg I/kg dry matter appears sufficient to meet reproduction in white-tailed deer, which in conjunction with forage analysis suggested that I was not limiting in Michigan (Watkins and Ullrey, 1983).

Copper (Cu) and zinc (Zn) are necessary for growth and health of deer because they are components in multiple enzyme systems and required for proper immune system function. Anecdotal reports of high concentrations of dietary Cu and Zn increasing antler size in deer were tested by Bartowkewitz et al. (2007). Diets with up to 200 mg/kg Cu and 1000 mg/kg Zn did not affect antler size in white-tailed deer, but did improve immune response *in vitro*. Despite high concentrations of Cu and Zn, no toxicosis was observed. Toxicosis was a concern because diets in which Cu exceeds 15 mg/kg and Zn exceeds 300 mg/kg maybe toxic to sheep (National Research Council, 2007).

Selenium (Se) is a component of glutathione peroxidase, an enzyme system that protects cell membranes against oxidative damage (Barboza et al., 2009). Deer deficient in Se are susceptible to white muscle disease and myopathy, especially after extreme exertion such as occurs when they are captured. Ninety percent of white-tailed deer in Michigan had Se concentrations in their muscles suggesting possible Se deficiency (Ullrey et al., 1981). Supplemental Se increased glutathione peroxidase of captive deer but had no effect on red blood cell hemolysis or fawn survival (Brady et al., 1978). A manipulative experiment involving free-ranging black-tailed deer in California demonstrated that reproduction was limited by Se (Flueck, 1994).

Water

Water is an essential nutrient that is continually lost from a deer's body through excretion of wastes and evaporation from the respiratory tract and skin. Additional water is lost when a doe gives birth or is lactating. Water is lost from the body at such a rate that a deer's health maybe jeopardized if water is not replaced in 24–48 hours. Water restriction can reduce feed intake up to 63%, potentially influencing body condition and productivity (Lautier et al., 1988). Water is replaced from three sources. The first is free water which comes from drinking, eating snow, and consuming dew. A deer also obtains water from its food, termed preformed water. Water content of food may vary from less than 5% in extremely dry forage to greater than 70% in fruits and succulent plants. Finally metabolic water is produced from oxidation of fats (1.07 g water/g fat), carbohydrates (0.56 g water/g carbohydrate), and protein (0.40 g water/g protein; Robbins, 1993). Water requirements can be difficult to predict because of variability arising from water vapor pressure, air temperature, solar and thermal radiation, energy expenditure, water content of feed and feed intake, growth and reproductive state, and types and timing of behavior. Diet may also influence water requirements, such as when white-tailed deer eating dried alfalfa drank 52% more water than deer eating a diet of 25% dried alfalfa and 75% dried guajillo, a browse species with high concentrations of secondary plant chemicals and common in deer diets in southern Texas and northeastern Mexico (Campbell and Hewitt, 2005). Water intake may have been greater in the alfalfa diet because of a twofold increase in digestible N intake and urinary N excretion.

Estimates of water flux, the amount of water entering and leaving the body at homeostasis, of a 50 kg free-ranging ruminant are 3.8–4.4 L/day (Nagy and Peterson, 1988). This amount of water includes preformed and metabolic water, and so a deer would not need to drink that amount. Averaging across several estimates of free and preformed water intake, domestic goats require 2.2 L/day (range = 1.56–3.36 L/day) (National Research Council, 2007). A deer's reliance on permanent sources of free water is inversely related to precipitation, probably due to greater intake of preformed water and use of ephemeral water sources (Webb et al., 2006; Wallach et al., 2007).

Digestion

For a deer to meet its nutrient requirements, it must ingest forage and other portions of its environment and then reduce the ingested material to a size that can be absorbed or transported across membranes in the digestive tract into the body proper. Digestion is the process of physically and chemically reducing food from bite size to molecular size and then absorbing it into the body. Digestion begins in the mouth where particle size of food is reduced by chewing, either immediately after ingestion or later during rumination. Food is mixed with saliva and swallowed where it enters the reticulum-rumen. A diverse community of symbiotic microbes in the rumen ferments the ingested food, producing volatile fatty acids which are absorbed across the rumen wall and used by the deer for energy (Short, 1963; Pearson, 1965) (Chapter 2). This symbiotic relationship is essential because deer, like all vertebrate animals, produce no enzyme to digest cellulose and hemicellulose, predominant structural molecules in plants. Given the proper temperature, pH, and anaerobic conditions found in the rumen, these symbiotic microbes are able to ferment cellulose. A cost of this relationship is that the microbes may ferment up to 80% of sugars and starch, chemicals the deer could digest without microbial assistance. Ruminal microbes use 13–18% of gross energy ingested for their own metabolism (National Research Council, 2007).

Once particle size has been reduced sufficiently by rumination and microbial action, ingested food passes through an orifice between the reticulum and omasum to enter the omasum where water and minerals are absorbed. The ingesta, along with rumen microbes and many of their by-products, then enter the abomasum where hydrochloric acid and pepsin are secreted to initiate enzymatic digestion. Muscular contractions move the digestive slurry into the small intestine where secretions from the pancreas, liver, and intestinal cells promote digestion and absorption of lipid, protein, and nonstructural carbohydrates. The resulting amino acids and peptides, fatty acids, and sugars are absorbed across the wall of the small intestine where they enter the circulatory system for distribution throughout the body. Any material not absorbed from the small intestine enters the large intestine where a minor amount of microbial fermentation occurs (Allo et al., 1973). The primary digestive function of the large intestine, however, is to remove water, minerals, and vitamins from the intestinal contents before they are excreted from the body.

Digestibility of forage can be predicted from its chemical content. True digestibility of protein (percent of dietary protein absorbed from the digestive tract) in forages without tannin is 92% (Robbins et al., 1987a; Robbins, 1993). Because apparent protein digestibility (difference between ingested and fecal protein expressed as a percent of ingested protein) is affected by crude protein of endogenous origin in the feces, apparent digestible protein can vary from less than 0% to nearly 90%, depending on percent protein in the diet (Mould and Robbins, 1981). Tannins are plant chemicals that can bind protein and lower its digestibility by up to 7 g protein/100 g feed (Robbins et al., 1987a). Digestible protein (DP; g/100 g of feed) can be predicted for many forages from the protein content (CP; % crude protein) and tannin precipitation (TAN; mg/mg of forage dry matter) (see Robbins et al., 1987a) as $DP = 3.87 + 0.9283CP - 11.82TAN$. Some browse may have DP lower than that predicted by this equation, probably due to fiber-bound protein (Robbins et al., 1987a; Barnes et al., 1991; Campbell and Hewitt, 2005).

Protein is part of the cell-soluble fraction of forage, a heterogeneous mixture of chemicals in a plant's cytoplasm. Cell solubles are also called neutral detergent solubles (NDS) because they are quantified by boiling the forage in neutral detergent solution (Goering and Van Soest, 1970). Although true digestibility of NDS is essentially 100%, metabolic fecal losses of protein reduce apparent digestion of NDS in proportion to the concentration of NDS in the diet (Robbins et al., 1987b). Reductions in protein digestion from tannins also decrease NDS digestibility. Incorporating these effects, NDS digestibility (g NDS/100 g feed) $= (-16.03 + 1.02NDS) - 2.8P$, where NDS is percent NDS in the forage and P is the reduction in protein digestion, in percent (Robbins et al., 1987b).

TABLE 3.3

Chemical Composition, Energy Content, and Digestibility of Foods by White-tailed Deer

Forage	Lipid (%)	Ash (%)	Gross Energy (kcal/g)	DDM Coefficient	Digestible Energy	
					Coefficient	kcal/g
Alfalfa	6.8	3.8	5.2	0.55	0.56	2.4
Rye/lambsquarter	4.1	16.9	4.9	0.71	0.67	3.2
Acorn/pellet[a]	9.0	4.0	4.4	0.56	0.60	2.5
Cottonseed	17.8	4.0	5.4	0.50	0.84	4.6

Note that digestible dry matter (DDM) and DE coefficients are similar for foods with low concentrations of lipid or ash, but that the DE coefficient is greater than the DDM coefficient for foods with high concentrations of lipids (i.e., acorn/pellet and cotton seed diets) and is lower for foods with high concentrations of ash (i.e., rye/lambsquarter diet; data from Ullrey et al., 1987; Elston, 2003; National Research Council, 2007; Bullock, 2009).

[a] Diet composed of 30% Shumard oak acorns and 70% pelleted diet.

Digestibility of the fiber fraction of forage by deer is lower than that of larger cervids (Mould and Robbins, 1982; Baker and Hansen, 1985; Baker and Hobbs, 1987) and can vary from 20% to 80% depending on the composition of the fiber (Robbins et al., 1987b). Lignin, cutin, and silica are not only indigestible but also reduce digestion of hemicellulose and cellulose. Digestion of neutral detergent fiber (NDF) is a curvilinear function of the lignin–cutin content of the NDF and a linear function of silica as described by the equation:

$$\text{NDF digestion}(\%) = (0.9231e^{-0.0451A} - 0.03B) \cdot (\text{NDF}),$$

where A = lignin and cutin content as a percent of NDF, B = biogenic silica content (%; in grasses only), and NDF = neutral detergent fiber (%) (Robbins et al., 1987b). The equations to estimate NDS and NDF digestibility can be combined to estimate percent dry matter digestibility of deer forages (Robbins et al., 1987b; Hanley et al., 1992).

Digestibility of gross energy in forages with little crude fat or ash is nearly identical to digestibility of dry matter (Robbins, 1993). Forages with appreciable amounts of lipids will have a higher DE coefficient than predicted from DDM and forages with high ash content may have a lower DE coefficient (Table 3.3).

Forage Intake

A deer's health and productivity are influenced by a complex relationship between nutrient requirements, forage quality, and forage intake. Forage intake is affected by many processes operating on short-term (g forage/min), daily, and seasonal timescales, knowledge of which can provide important insight into white-tailed deer nutritional ecology. Surprisingly, forage biomass is not a good predictor of short-term intake rate in browsing herbivores such as white-tailed deer (Spalinger et al., 1988; Spalinger and Hobbs, 1992). Instead, short-term intake rate is influenced by a complex interaction of bite size, time required to harvest and process a bite, bite density, and the degree to which forage is apparent (Spalinger et al., 1988; Spalinger and Hobbs, 1992; Hobbs et al., 2003). Although deer compensate for small bite sizes by increasing bite rates, high bite rates interfere with chewing and swallowing, thereby causing intake rate to decline (Spalinger et al., 1988). There is a threshold density of plants above which bite size controls intake (Hobbs et al., 2003). Below this threshold density, the distance between plants controls intake. The threshold density in turn depends on bite size such that with large bites, the distance between plants needs to exceed 9 m before spacing affects intake. At very small bites, spacing may be important at distances as small as 1 m (Hobbs et al., 2003).

Search time becomes important when desired forages are intermixed with other vegetation, as for forbs in a grassy meadow or acorns in leaf litter. In such cases, intake rate is hypothesized to be influenced by the density of bites, width of the search path, and the rate at which deer travel while searching (Spalinger and Hobbs, 1992). Spines and thorns may reduce intake rate by increasing harvest time (Cooper and Owen-Smith, 1986, but see Cash and Fulbright, 2005). Forage fiber concentration acts to reduce intake rate by increasing processing time (Spalinger et al., 1988).

These relationships suggest that adding more browse biomass, without changing bite size, may not influence the rate at which deer can harvest forage, at least in the short term. Furthermore, a forage species from which deer can harvest large bites could support a higher intake rate by deer than another forage species with greater biomass, but from which deer can only obtain small bites. Under optimal conditions, white-tailed deer can consume up to 12.8 g dry matter/min of alfalfa (Robbins, 1993), 18.0 g/min of live oak acorns, 6.3 g/min of mesquite pods (Elston and Hewitt, 2010), 2.6–16.6 g/min of foliage from nine browse species (Koerth and Stuth, 1991), and 2.5–6.5 g/min for six browse species in southern Texas (Spalinger et al., 1997). Free-ranging white-tailed deer consumed 2.2 g/min of acorns, averaged across years of low- and high-acorn crops (McShea and Schwede, 1993).

The amount of forage eaten daily could be influenced by short-term intake rate if the short-term intake rate is sufficiently low that deer do not have time in a day to forage, ruminate, and complete other necessary activities. Assuming that short-term intake rate is not limiting, daily intake is a curvilinear function of DE content (Figure 3.3) (Ammann et al., 1973; Robbins, 1993). Below about 2.2 kcal DE/g, daily intake is constrained by rumen capacity and passage rates because higher concentrations of fiber prolong the time necessary for food to pass from the rumen. Forages vary in the rate at which they pass from the rumen, with much of the variation explained by cell wall thickness (Spalinger et al., 1986). Early findings suggested lignin may make cell walls more fragile, promoting quick reductions in size with rumination and thus faster rates of passage and higher intake (Milchunas et al., 1978; Van Soest, 1982). Later studies testing this hypothesis found that lignin reduced particle breakdown and passage rates (Spalinger et al., 1986; Baker and Hobbs, 1987). Deer appear able to consume browse diets high in lignin because, relative to other ungulates, deer have lower mean retention times and are able to increase rumen fill (Baker and Hobbs, 1987).

For DE concentrations greater than 2.2 kcal/g (but see Gray and Servello, 1995 for evidence of a higher threshold), daily intake is limited by physiological responses related to the deer's energy status (Robbins, 1993; National Research Council, 2007). These responses involve complex interactions of hormones, neurotransmitters, dietary nutrients, and body composition, such that deer will consume sufficient forage to meet energy requirements (Rhind et al., 2002; Barboza et al., 2009). Daily dry matter consumption of diets in which intake would be regulated physiologically range from 1.8% to 2.3% of body weight for maintenance to 3.1% of body weight for late gestation and 4.8% during peak lactation (National Research Council, 2007). Bucks during late summer require intake of 3–3.5% body weight, depending on their body size and rate of gain (National Research Council, 2007).

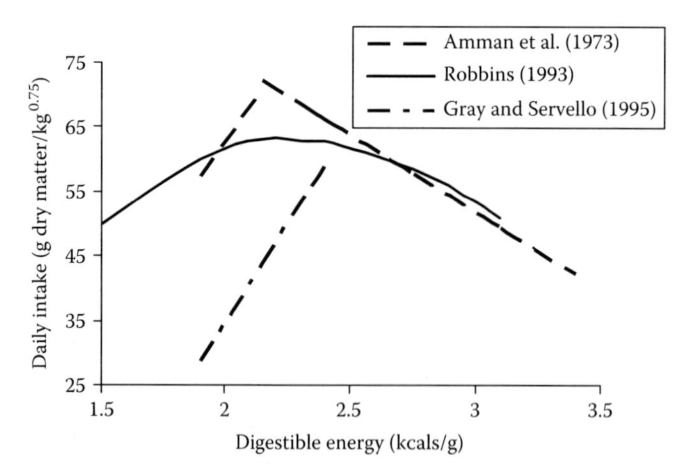

FIGURE 3.3 Dry matter intake of white-tailed deer and mule deer as a function of DE content of forage. (Based on data from Ammann et al., 1973; Robbins, 1993; Gray and Servello, 1995.)

Daily intake of forages containing secondary plant chemicals may be limited by detoxification rates (Campbell and Hewitt, 2005). Deer may be less affected by tannins in forage compared to other ungulates, particularly grazers, because deer produce salivary proteins that have a high affinity for tannin (Austin et al., 1989). By binding a large amount of tannin per unit of salivary protein, the tannin cannot bind dietary protein, essential amino acids may remain available for absorption, and tannin that could be toxic if absorbed is excreted in the feces (Robbins et al., 1991; Makkar, 2003).

White-tailed deer show large seasonal changes in daily forage intake, even when discounting changes in production status, such as high food intake of lactating deer (Rhind et al., 2002). Food intake of bucks during the rut may decline more than 50% (Bartoskewitz et al., 2007) when hormonal changes increase activity and interest in reproduction and decrease appetite. Intake is often high during autumn and then declines 20–50% during winter, even in southern latitudes of the United States where winter is not extreme and in captivity where food is available *ad libitum* (Ozoga and Verme, 1970; Thompson et al., 1973; Holter et al., 1977; Wheaton and Brown, 1983). Seasonal changes in food intake are part of the white-tailed deer's survival strategy in which fat stores obtained during autumn are necessary for winter survival (Mautz, 1978) and their use appears to be obligatory.

Deer Forages

Deer are classified as concentrate selectors (Hofmann, 1989) (Figure 9.1) and as such their diets are dominated by leaves and stems of woody plants, forbs, and hard (nuts and pods of woody plants) and soft mast (fruits and berries; Table 3.4). Although selective in what they eat, deer in an area may consume annually over 100 plant species and several morphological parts from many of those species (Chamrad and Box, 1968; Nixon et al., 1970; Korschgen et al., 1980; Gee et al., 1991). Vegetation that deer eat is often classified as browse, forb, grass, and mast. Despite high variation in the nutritional value of foods within each of these forage categories, broad patterns exist in their availability and quality (Figure 3.4) (Snider and Asplund, 1974). Forbs are generally available during the growing season and have greater digestible protein and DE than browse, which is available in some form in every season. Mast is often ephemeral and is typically low in digestible protein and high in DE, although pods of some legume species may be high in crude protein (Snider and Asplund, 1974; Everitt and Alaniz, 1981; Everitt, 1986; Pekins and Mautz, 1988). Grass consumed by deer is typically young and succulent and therefore consumed either early in the growing season or during the dormant season when cool season grasses remain green. Such grass has high dry matter digestibility (54–73%) and at least 14% crude protein (Robbins et al., 1975, 1987b; Ullrey et al., 1987). Lichens have high DE but exceedingly low protein (<4%) (Robbins, 1987).

Although there has been a vast amount of effort dedicated to white-tailed deer food habits, much less effort has been dedicated to understanding diet quality, a much more important parameter. In most instances, dietary protein decreases from spring through winter (Figure 3.5), although high mast consumption can result in low dietary protein during summer (Timmons et al., 2010). Digestible energy tends to decrease from spring through summer, at least in deer populations in the southern United States (Figure 3.5), with DE less than 2.2 kcal/g during summer suggesting energy may limit reproduction. Digestible energy may increase in autumn if mast is available and remain moderate during winter. In northern portions of the deer's range, dietary protein and DE decline during winter, perhaps equaling 2.03 kcal/g (calculated from Mautz et al., 1976).

Secondary Plant Chemicals

Realized deer diets are not only influenced by the availability and nutrient content of forages, but also by the concentration and type of secondary plant chemicals in potential forages. Secondary plant chemicals are a diverse group of compounds that are not part of a plant's primary metabolic pathways. These chemicals have many functions in plants, including reducing herbivory and microbial infection. Secondary plant chemicals in forages consumed by white-tailed deer may dilute nutrients, reduce digestive efficiency, or cause toxicity (Robbins et al., 1987a,b).

TABLE 3.4

Average (Minimum, Maximum) Seasonal Percent Composition of White-tailed Deer Diets Derived from Published Studies in Midwestern (N = 5–9 Studies, Depending on the Season), Northeastern (N = 4–6), Northwestern (N = 8–12), Southeastern (N = 9–13), and Southwestern (N = 24–30) North America

		Season			
Forage Class	**Region**[a]	**Spring**	**Summer**	**Autumn**	**Winter**
Browse	Midwest	29 (4, 45)	41 (4, 71)	18 (1, 44)	45 (1, 90)
	Northeast	44 (14, 93)	45 (21, 58)	38 (15, 80)	91 (79, 100)
	Northwest	36 (1, 81)	45 (11, 86)	58 (27, 95)	74 (35, 94)
	Southeast	53 (26, 85)	47 (27, 65)	40 (9, 65)	56 (38, 74)
	Southwest	41 (2, 94)	41 (0, 96)	42 (1, 94)	46 (1, 95)
Forb	Midwest	17 (2, 27)	25 (8, 51)	6 (1, 13)	8 (0, 35)
	Northeast	36 (0, 78)	21 (6, 42)	10 (0, 23)	2 (0, 9)
	Northwest	31 (5, 70)	34 (13, 68)	21 (4, 51)	9 (0, 29)
	Southeast	36 (12, 49)	27 (6, 42)	11 (1, 45)	12 (2, 34)
	Southwest	41 (4, 90)	34 (2, 97)	25 (2, 85)	27 (1, 65)
Mast	Midwest	4 (0, 11)	14 (0, 49)	53 (40, 65)	21 (0, 57)
	Northeast	0 (0, 0)	22 (7, 65)	24 (0, 60)	0 (0, 0)
	Northwest	1 (0, 5)	0 (0, 1)	1 (0, 7)	0 (0, 2)
	Southeast	2 (0, 7)	10 (2, 20)	34 (3, 83)	14 (0, 37)
	Southwest	5 (0, 44)	11 (0, 71)	17 (0, 72)	4 (0, 27)
Crops	Midwest	30 (0, 84)	8 (0, 48)	15 (0, 47)	13 (0, 60)
	Northeast	0 (0, 0)	2 (0, 10)	1 (0, 3)	0 (0, 0)
	Northwest	11 (0, 46)	13 (0, 43)	9 (0, 47)	6 (0, 39)
	Southeast	0 (0, 0)	4 (0, 39)	3 (0, 23)	4 (0, 25)
	Southwest	0 (0, 7)	0 (0, 3)	1 (0, 20)	2 (0, 36)
Grass	Midwest	15 (2, 37)	2 (0, 5)	3 (0, 14)	10 (1, 28)
	Northeast	12 (0, 35)	5 (0, 9)	4 (0, 15)	1 (0, 4)
	Northwest	21 (5, 39)	5 (1, 21)	9 (1, 31)	9 (0, 37)
	Southeast	8 (1, 30)	3 (0, 8)	4 (0, 10)	11 (1, 32)
	Southwest	5 (0, 19)	6 (0, 33)	7 (0, 25)	14 (0, 50)
Lichen/fungus	Midwest	1 (0, 3)	3 (0, 7)	2 (0, 5)	1 (0, 4)
	Northeast	8 (0, 23)	3 (0, 7)	19 (1, 60)	5 (0, 21)
	Northwest	0 (0, 1)	1 (0, 4)	1 (0, 5)	1 (0, 3)
	Southeast	1 (0, 2)	6 (0, 20)	7 (0, 21)	3 (0, 9)
	Southwest	1 (0, 15)	1 (0, 34)	2 (0, 32)	0 (0, 7)

[a] States/Provinces in each region: Midwest = IL, IN, MI, MN, MO, OH; Northeast = ME, NH, PA, New Brunswick, Ontario, Quebec; Northwest = CO, ID, MT, ND, SD; Southeast = AR, FL, GA, LA, MA, NC, TN, VA, WV; Southwest = AZ, OK, TX, and northeast Mexico.

Browses typically have higher concentrations of secondary plant chemicals than forbs and grasses. Because white-tailed deer consume browse throughout their range, deer have a long evolutionary history with these chemicals and have evolved coping mechanisms (McArthur et al., 1991). A deer's first defense against secondary plant chemicals is to select forages with low concentrations of these chemicals or with chemicals of low toxicity (Cooper and Owen-Smith, 1985). This strategy influences forage palatability and results in avoidance of some forages. A second defense is to neutralize chemicals before they have substantial impacts. Deer have salivary proteins that bind tannins, thereby reducing toxicity and effects on protein digestion (Austin et al., 1989). Rumen microbes metabolize many secondary plant chemicals, making ruminants less susceptible than nonruminants to toxic effects of these chemicals (McArthur

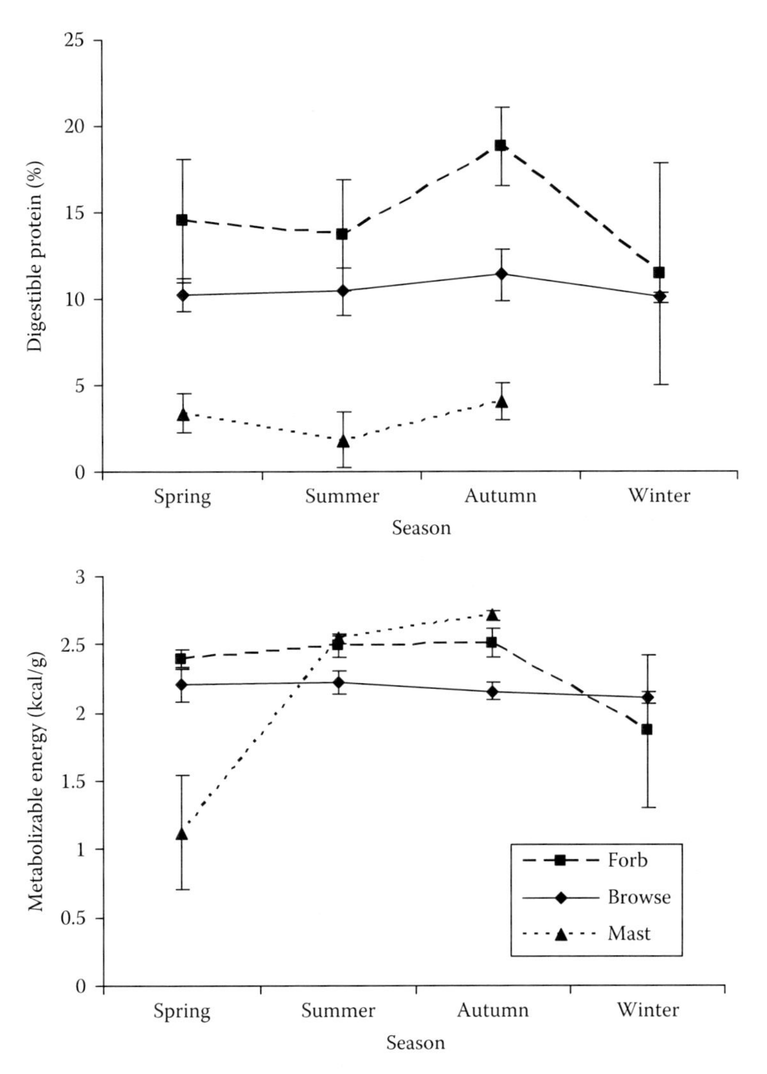

FIGURE 3.4 Changes in digestible protein and metabolizable energy of forbs, browse, and mast eaten seasonally by white-tailed deer in southern Texas. (From Timmons, G. R. et al. 2010. *Journal of Wildlife Management* 74:995–1002. With permission of The Wildlife Society, Bethesda, MD.)

et al., 1991). Defense against secondary plant chemicals that have been absorbed is through enzymatic reactions, primarily in the liver, to detoxify and excrete the chemicals (McArthur et al., 1991). Absorbed chemicals are oxidized and excreted or conjugated to a glucose, amino acid, or sulfate molecule to increase polarity for excretion (Klaassen, 1996; Servello and Schneider, 2000; Campbell and Hewitt, 2005). Because maintenance of enzyme systems and production of conjugates is expensive in energy and nutrients, consumption of secondary plant chemicals can be costly. Incorporating effects of tannin on protein digestion and presence of nonprotein N can reduce nutritional estimates of deer carrying capacity by up to 50% (Windels and Hewitt, in press).

Secondary plant chemicals may have positive effects on deer nutrition and health. Low tannin concentrations may enable passage of protein through the rumen, potentially increasing the biological value of the protein. Some secondary plant compounds have anthelminthic and antioxidant properties so that there could be health and survival benefits of consuming plants with such chemicals (Iason, 2005; Villalba and Provenza, 2007). Despite the prevalence of plant secondary compounds in deer diets, current understanding of effects and implications of these chemicals is rudimentary.

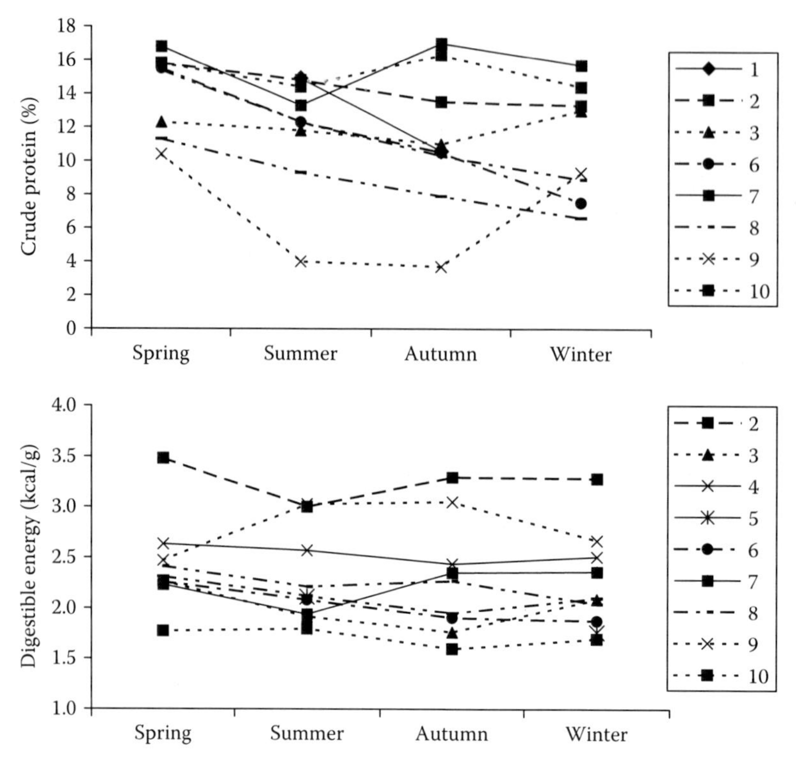

FIGURE 3.5 Seasonal estimates of white-tailed deer diet quality from Tennessee (reference 1), Illinois (2), Texas (3, 6, 7, 8, 9), Louisiana (4), Missouri (5), and Mexico (10). Digestible dry matter was converted into DE by multiplying the decimal percent DDM by 4.4 kcal/g. Reference 9 shows digestible protein instead of crude protein. 1, Weckerly and Nelson (1990); 2, Nixon et al. (1991); 3, Ortega et al. (1997); 4, Thill et al. (1987); 5, Snider and Asplund (1974); 6, Bryant et al. (1980); 7, Meyer et al. (1984); 8, Bryant et al. (1981); 9, Timmons et al. (2010); 10, Ramírez et al. (1996).

Food Habits

Diets selected by white-tailed deer are the outcome of complex processes integrating distributions of forage quality and abundance, search and handling times, perceived risk, and deer nutrient requirements (Hanley, 1997; Barboza et al., 2009). Forage selection is complicated by temporally varying deer nutrient requirements and temporally and spatially varying forage availability and quality. Deer are successful in the seemingly difficult task of selecting an appropriate diet because natural selection has instilled deer with innate forage preferences and deer, like other herbivores, can learn from their foraging experiences (Provenza, 1995; Spalinger et al., 1997).

White-tailed deer diets typically contain dozens of species seasonally and more than 100 species annually. Although diets are diverse when considering presence of species, typically a small number of species compose a majority of the diet. This pattern probably occurs as deer consume small amounts of rare but high-quality forages, large amounts of common but moderate quality forage, and small amounts of many other species. Such diverse diets may enable deer to monitor temporal and spatial changes in forage quality, mix diets to mitigate toxins, and allow deer to meet nutrient requirements, particularly for mineral and vitamins that maybe needed in small amounts (Provenza, 1996; Provenza et al., 2003).

Quantifying food habits without bias can be exceedingly difficult. Sampling rumen contents and feces are the most common techniques of estimating white-tailed deer diets (57% and 27% of studies, respectively; see below). Such postingestive material may have a composition different from the diet because some foods, such as fungi, high-quality forbs, and fruits, remain in the rumen for a short period of time and may be more digestible, causing them to be underrepresented in food habit estimates (Gill et al., 1983; Spalinger et al., 1986). Observation of tame or habituated animals is a technique that avoids problems of differential digestion but may be biased if tame animals forage differently than wild deer or bites of some plants are more difficult to quantify than others (Spalinger et al., 1997). A final technique used to quantify deer food habits has been to count the number of bites on different forage plants at sites where deer have fed. This technique could be biased by the selection of observation areas and by failure to include plants that are entirely consumed.

To assess spatial and temporal patterns in white-tailed deer food habits, I compiled published reports of deer diets, by season. Seasons were spring (Mar–May), summer (Jun–Aug), autumn (Sep–Nov), and winter (Dec–Feb) if not defined by the authors. Forage categories were browse (leaves and twigs of woody plants), forbs (leaves and stems of nonwoody dicot plants), grass (leaves and stems of monocot plants), mast (fruits), crops (grain and hay crops), fungus and lichens, cactus, and other. Forage categories were derived from the authors' classification or, when not reported, from designations in the USDA Plant Database (http://plants.usda.gov/index.html). I considered studies that measured deer diets using rumen content analysis, fecal analysis, and observation of tame deer.

Averaging across regions and seasons, white-tailed deer diets consist of 46% browse, 24% forbs, 11% mast, 8% grass, 4% crops, 2% cactus, 2% fungus, and 3% other. This average diet, however, masks interesting and important geographic and temporal variation (Table 3.4). Average diet varies least among regions during summer, with the exception of mast and crops in the northwest. Spring variation is greater, with deer diets in the Midwest and northwestern regions containing less browse and forbs and more crops and grass than other regions. Autumn diets vary greatly among regions, with mast particularly important in the Midwest and southeast, browse, crops, and grass relatively important in the northwest, lichens and fungi unusually important in the northeast, and browse and forbs composing the majority of the diet in the southwest. There is a strong latitudinal gradient in browse use during winter. Browse averages 74–91% of deer diets in northern regions, whereas forbs remain available and are used during winter in the southwest. Mast is often part of diets in the Midwest and southeast, and grass is consumed in all regions outside the northeast (Table 3.4).

Forage availability greatly influences deer food habits. High-quality forages, such as crops and mast, compose large portions of deer diets when they are available. Crops were exceedingly important to deer during summer and autumn in the Midwestern United States (Korschgen, 1962; Nixon et al., 1970, 1991) and along riparian areas in northwestern portions of the species range (Dusek et al., 1989). When high-quality, preferred forages are unavailable, such as when deer densities are too high or snowfall limits available forage, low-palatability foods may constitute a large portion of the diet (Rose and Harder, 1985; Stromayer et al., 1998; Tremblay et al., 2005). Changes in availability can result in dramatically different food habits in small areas. For example, near the Gulf coast of southern Texas, forbs comprise 50–98% of seasonal diets (Chamrad and Box, 1968; Kie et al., 1980; Kie and Bowyer, 1999), whereas 150 km inland, browse and cactus compose 50–75% of seasonal diets (Everitt and Drawe, 1974; Arnold and Drawe, 1979; Everitt and Gonzalez, 1979). A moisture gradient explains this pattern such that forbs are more available in mesic rangelands near the coast.

Deer are opportunistic and will use forages not widely recognized as deer food. Cactus may compose 20–61% of deer diets in southern (Arnold and Drawe, 1979; Everitt and Gonzalez, 1979) and western Texas (Krausman, 1978). Fungus and lichens may compose over 20% of deer diets in Virginia (Harlow et al., 1975), Texas (Short, 1971; Fulbright and Garza, 1991), Montana (Ward and Marcum, 2005), eastern Canada (Skinner and Telfer, 1974; Lefort et al., 2007), and Maine (Crawford, 1982). Leaf litter and senescent forbs are increasingly recognized as important forages for deer during the dormant season, particularly in northern environments (Ditchkoff and Servello, 1998; Tremblay et al., 2005; Windels and Jordan, 2008).

Assessing Nutritional Status

Estimating composition and quality of deer diets is one approach to understanding the nutritional status of a deer population. However, accurate measures of diets are difficult to obtain and are most meaningful if considered in the context of a deer's nutrient requirements, which are also difficult to estimate. An alternative approach is to evaluate the deer's nutritional status and thereby allow the deer to integrate their nutrient requirements with forage availability and quality. This approach may not provide insight into why deer are performing well or poorly, but can demonstrate when deer are nutritionally deficient and suggest areas of further investigation to understand why. There are many techniques to assess nutritional condition of white-tailed deer (Servello et al., 2005). These can be classified as (1) body mass and growth rates, (2) demographic parameters, (3) body condition, (4) concentrations of metabolites in blood or urine, and (5) indirect measures of diet quality through analysis of rumen contents or feces.

Because of the high nutrient requirements for growth, body mass or growth rates of deer less than 24 months of age are particularly good indicators of the nutritional environment (Verme and Ozoga, 1980, but see Taillon et al., 2006). Body mass of adult deer can change with nutritional conditions, but adult deer can meet requirements over a much broader range of conditions and can more readily alter activity and investment in reproduction to maintain a target body mass. Interpreting adult body mass may require consideration of an animal's reproductive history because does that have raised fawns or bucks at the end of the rut may be underweight (Warren et al., 1981), not because of poor habitat conditions, but because of successful reproduction or annual physiological rhythms. Structural characteristics of a deer's body may be affected by the nutritional environment, such as yearling antler size (Servello et al., 2005), the ratio of femur length to hind foot length (Klein, 1964), and perhaps tooth replacement patterns (Loe et al., 2004).

Demographic rates can be sensitive to nutritional conditions, but reproduction and survival are also affected by other processes such as predation and disease. Thus, corroborating data may be necessary before using demographic data to make inferences about a population's nutritional environment. For example, low survival of fawns or adult deer would not be evidence of poor nutrition unless data on cause of death or body condition were also available. Cause of death data may be difficult to interpret because animals in poor nutritional condition are likely to be more susceptible to predation (Mech, 2007). Pregnancy is a stage of the reproductive cycle influenced by nutrition and not predation. Pregnancy rates can be assessed with hormonal assays or through presence of a corpus luteum or fetus (Figures 3.6 and 8.1). However, pregnancy rates of adult deer are relatively robust to nutritional conditions and over generations, adult deer can maintain reproductive rates in spite of declining forage quality by adjusting other life-history traits, such as growth rate or body size (Simard et al., 2008). In contrast, fawn and yearling pregnancy rates are sensitive to nutritional conditions, especially energy intake during autumn (Verme, 1969; Abler et al., 1976). Eighty percent of fawns receiving a high-energy diet exhibited estrous while no fawns eating a low-energy diet cycled (Abler et al., 1976).

When assessing body condition, a biologist is determining the status of fat and protein components of the deer's body. Fat is particularly important to assess because it represents energy intake above that necessary for maintenance and production and it serves as an energy reserve. Muscle may also be used to provide energy and amino acids when intake is insufficient to meet the deer's requirements. Because muscle is 72% water and protein has half the energy density of fat (Robbins, 1993), nine times more energy is released when a unit of fat is metabolized compared to a unit of muscle. For this reason, when a deer has metabolized most of its fat, it will lose weight rapidly.

The definitive measure of body condition is to homogenize the deer's body and measure fat, protein, ash, and water composition in the lab, but the process is laborious (Barboza et al., 2009). The next best measure of body condition is to measure body water content using dilution of water containing heavy or radioactive isotopes of hydrogen which then allows calculation of the proportions of protein, fat, and ash in the body (Torbit et al., 1985; Barboza et al., 2009). However, this technique also requires substantial time and may be difficult to apply in field studies. Recent research in several cervid species has shown that thickness of subcutaneous rump fat is closely related to whole-body fat, at least above 7% body fat (Stephenson et al., 2002; Cook et al., 2007). Rump fat thickness can be measured quickly with ultrasound (Figure 3.7). When this measurement is combined with a body condition score, percent fat in a deer's body can be measured over a range from 2% to 30% (Cook et al., 2007, 2010). Fat around various organs, particularly the kidney, and in bone marrow is also used as an index to body condition (Riney, 1955; Ransom, 1965; Kistner et al., 1980; Finger et al., 1981). However, kidney and marrow fat are not linearly related to total-body fat, making their resolution poor over some range of body conditions (Watkins et al., 1991) and requiring careful interpretation (Mech and Delgiudice, 1985).

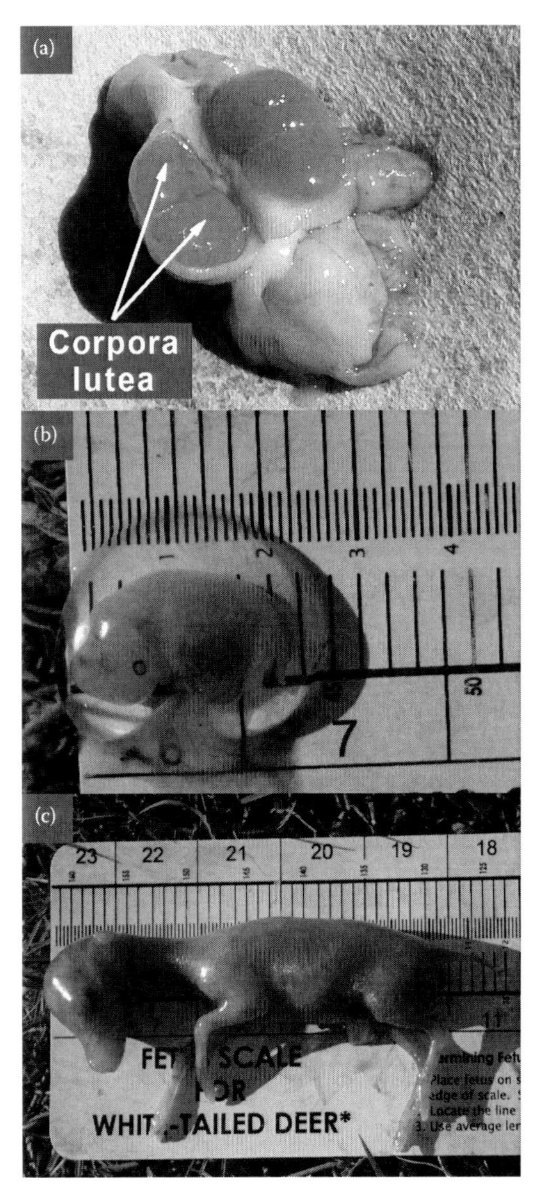

FIGURE 3.6 Corpora lutea (a), ephemeral endocrine structures arising in the ovary at sites where the ovum had been released, and fetuses [44 (b) and 70 (c) days since conception] are evidence of pregnancy and may serve as an index to the nutritional environment, particularly in young deer. (Photos by T. E. Fulbright and D. G. Hewitt.)

Metabolites and hormones in blood and urine can be influenced by a deer's nutrient intake and body condition and therefore are possible nutritional indices. However, relationships between blood and urine chemistry and a deer's nutritional status are often complex, may have daily and seasonal rhythms, and may be influenced by interactions among nutrients, such that only a small number of physiological indices are reliable indicators of nutrient status (Harder and Kirkpatrick, 1994; Brown et al., 1995).

Urea in blood and urine has been investigated more than any other physiological index. Urea concentrations in blood and urine are positively related to protein intake (Warren et al., 1982; Brown et al., 1995), but also increase in later stages of starvation as deer metabolize muscle tissue for energy (DelGiudice et al., 1994; DelGiudice, 1995). Thus, blood or urinary urea could be a useful index to deer protein intake or nutritional condition if high protein intake can be differentiated from high protein catabolism. Ratios

FIGURE 3.7 Body condition of deer can be measured using a combination of body condition score and thickness of subcutaneous fat on the rump, measured using a portable ultrasound machine. (Photo by R. W. DeYoung.)

of N isotopes in urea could indicate whether N in urea was of dietary or endogenous origin, although this approach has not been investigated in white-tailed deer (Barboza and Parker, 2006). Urinary urea has been proposed as an index to nutritional restriction in northern deer during winter because winter is nutritionally limiting, diets of northern deer during winter typically contain low concentrations of protein (Mautz et al., 1976), and urine samples can be readily collected from snow (DelGiudice, 1995; Ditchkoff and Servello, 1999). Metabolites in urine samples collected from snow are expressed as a ratio of creatinine to account for dilution in snow and differences in hydration among deer (DelGiudice, 1995). Urea:creatinine ratios maybe most useful in identifying the proportion of a deer population that has exhausted its fat reserves and is still in negative energy balance (Ditchkoff and Servello, 1999). Using urea from snow-urine samples has been criticized because the index is not strictly related to body condition (Parker et al., 1993; Saltz et al., 1995; Cabanac et al., 2005), creatinine production may vary seasonally (Parker et al., 2005), and there are difficulties in interpreting the results at a population level (Saltz et al., 1995; White et al., 1995). Glucuronic acid:creatinine ratios are related to consumption of browse species with high concentrations of secondary plant chemicals and thus may indicate the proportion of the diet composed of low-palatability forages (Servello and Schneider, 2000; Sauvé and Côté, 2006).

Dietary protein intake is related to N concentration in the feces (Leslie and Starkey, 1985; Osborn and Ginnett, 2001) and because feces can be readily collected, fecal N has been used as an index to dietary protein or diet quality (Leslie et al., 2008). However, fecal N is also influenced by daily forage intake and dry matter digestibility of forage, which could complicate interpretation (Barboza et al., 2009). Furthermore, fecal N may be elevated when deer consume forages containing tannins because tannins can bind dietary protein and reduce its digestibility (Robbins et al., 1987a). White-tailed deer commonly consume woody species with tannins, and therefore fecal N may not be an appropriate measure of diet quality for white-tailed deer in all situations (Robbins et al., 1987a; Osborn and Ginnett, 2001). Another fecal index to diet quality is 2,6-diaminopimelic acid (DAPA), an undigested component of rumen microbes. Fecal DAPA is related to digestibility of the diet and therefore to DE (Brown et al., 1995; Wehausen, 1995).

Managing Nutrition

Nutrition can influence physical and behavioral characteristics of individual deer and many aspects of white-tailed deer demography. Because forage may not be of sufficient quality for deer to meet their genetic or biological potential in some places and times, nutrition is often the target of management. If deer density is high, removing deer will improve the nutritional plane of remaining deer (Kie et al., 1980; Swihart et al., 1998), at least in many portions of the white-tailed deer's range (but see Chapter 5). Managers may also seek to improve deer nutrition through habitat management, food plots, or supplemental feed.

Habitat Manipulation

Habitat management techniques vary widely depending on vegetation type, productivity of soils, precipitation, and resources available. In closed canopy forests, removing trees allows more sunlight to reach the understory, stimulating growth of herbaceous vegetation and shrubs (Hughes and Fahey, 1991; Edwards et al., 2004). Such growth is often high-quality forage and is accessible to deer, unlike vegetation in the forest canopy. Similar responses may occur from manipulation of shrubs in rangelands (Stewart et al., 2000; Fulbright and Ortega-S, 2006). Habitat management to improve deer nutrition should be planned carefully to ensure that improved nutrition in one season is not obtained at the expense of nutrition in another season. Removing shrubs or overstory plants may provide more herbaceous forage during the growing season, but less browse or mast during other seasons (Johnson et al., 1995).

Extensive grass swards may not provide high-quality deer forage. Fire, grazing, and mechanical manipulations may reduce grass cover and promote forb growth (Jenks et al., 1996; Payne and Bryant, 1998), but effects of these treatments are variable depending on timing of application, conditions during and after application, and site productivity (Fulbright and Ortega-S, 2006; Figure 18.4). Because high stocking rates of domestic livestock and poor range condition are detrimental to deer nutritional status (Bryant et al., 1980, 1981; Jenks and Leslie, 2003), conservative grazing pressure can also improve white-tailed deer nutrition.

Supplemental Feeding

When nutritional quality of deer diets is insufficient to support desired survival, growth, or reproduction, deer managers may choose to supplement deer (Kammermeyer and Thackston, 1995; Fulbright and Ortega-S., 2006). Supplementation may occur through food plots or provision of feed or minerals brought to the site. Deer behavior can be modified by providing high-quality foods (Henke, 1997), a tactic used by hunters and photographers to bring deer into the open where they can be more easily seen and harvested.

Food Plots

Food plots are a common management practice in the southeastern United States because habitat quality is poor in many areas (Shea and Osborn, 1995), rainfall is sufficient, and winter conditions are relatively mild. Composition of food plots varies depending on season, management objectives, soil, temperature, and moisture conditions (Payne and Bryant, 1998; Fulbright and Ortega-S., 2006). There is a bewildering array of forages that have been planted in food plots (at least 60 species; Kammermeyer and Thackston, 1995). Cereal grains, such as oats, rye, and wheat, are commonly used in cool season food plots and are palatable and nutritious through much of the winter in the southern United States. Legumes, such as alfalfa and various clovers are also included in winter food plots because they are highly nutritious, provide variety, and may extend the time during which food plots contain palatable forage. Warm season food plots often contain legumes, such as cowpeas, vetch, soybeans, and lablab. Up to 70% of deer diets may be composed of food plot forages (Keegan et al., 1989; Hehman and Fulbright, 1997) and up to 90% of fecal pellet groups may contain food plot forages (Johnson et al., 1987; Keegan et al., 1989).

To be of benefit, forages in food plots must be of higher quality than native forages. For example, a failure of food plots in Louisiana to increase body mass or antler size of yearling deer was attributed to low deer densities and adequate native forage (Johnson and Dancak, 1993). Although research is scarce, under the right conditions food plots may increase diet quality, deer body mass and antler size, and deer harvest (Johnson et al., 1987; Keegan et al., 1989; Vanderhoof and Jacobson, 1989; Kammermeyer and Moser, 1990). Food plots may concentrate deer, but do not appear to increase gastrointestinal parasites in deer (Schultz et al., 1994).

Supplemental Feed

An increasingly common approach to influencing deer nutrition is to provide a nutritional supplement. Supplements have been used to reduce mortality during severe winter conditions in northern portions of the species' range (Sage and Gustafson, 1991; Lewis and Rongstad, 1998; Page and Underwood, 2006) and to increase productivity and survival of deer in areas with poor habitat in the southern United States (Zaiglin and DeYoung, 1989; Bartoskewitz et al., 2003). Pelleted feeds are commonly used for supplementation but various grains, legumes, and legume hays have also been used. Whole cottonseed has been used as a supplement because of its high protein and DE concentrations (Bullock, 2009), resistance to moisture degradation, and lower consumption relative to pelleted feeds by nontarget animals. Although cottonseed contains gossypol, a naturally occurring chemical that can be toxic and reduce antler size and fertility in male cervids (Brown, 2001; Gizejewski et al., 2008), white-tailed deer do not appear particularly sensitive to its effects (Bullock, 2009). Grass hays, even if pelleted, are poor supplements (Ouellet et al., 2001).

Baiting is a form of supplemental feeding whose objective is to bring deer to a specific place so they can be trapped, photographed, censused, or harvested. Foods high in energy, such as corn or apple pumice, are most often used as bait, but salt and alfalfa hay have also been used (Naugle et al., 1995; Koerth and Kroll, 2000; Haulton et al., 2001). Baiting can change deer distribution, primarily through changes in use of areas within the deer's home range although long distances between bait sites and deer harvest sites have been documented (Van Brackle et al., 1995; Kilpatrick and Stober, 2002; Campbell et al., 2006). The effect of baiting on harvest rates may vary between archery and firearm hunters and total harvest may not increase if hunters become more selective or hunt fewer days as a result of using bait (Rudolph et al., 2006; Van Deelen et al., 2006; Kilpatrick et al., 2010).

Supplemental feed can improve the nutritional status of deer (Tarr and Pekins, 2002; Page and Underwood, 2006; Timmons et al., 2010), and increase survival, body mass, antler size, and reproductive rates (Ozoga and Verme, 1982; Zaiglin and DeYoung, 1989; Lewis and Rongstad, 1998; Bartoskewitz et al., 2003). Supplementing deer may also have unintended effects. Close contact among deer at feed sites may help maintain or spread disease, as appears to have occurred with bovine tuberculosis in Michigan (Hickling, 2002; Palmer et al., 2004). Supplemental feed may impact vegetation communities by changing the location of deer foraging (Doenier et al., 1997; Cooper et al., 2006), such that vegetation near feed sites may be overbrowsed, and altering forage selection (Murden and Risenhoover, 1993; Timmons et al., 2010). Impacts on vegetation could be magnified if supplemental feed enables higher deer densities to exist than could be supported by forage alone. Finally, supplemental feed intended for deer may be consumed by other species (Lambert and Demarais, 2001), which may have additional ecosystem effects, such as increased predation on ground-nesting birds (Cooper and Ginnett, 2000).

Mineral sources may be provided either as a supplement or bait in some areas of the white-tailed deer's range. Although deer use supplemental mineral sites, particularly during spring and summer (Schultz and Johnson, 1992a,b; Campbell et al., 2004), their affect has not been well documented. Commercial mineral supplements did not affect deer body mass or antler characteristics in Louisiana (Schultz and Johnson, 1992a).

Acknowledgments

P. S. Barboza provided valuable input on an earlier draft of this chapter. C. Lawson assisted with figure preparation. The financial and administrative support of the Caesar Kleberg Wildlife Research Institute and F. Bryant during preparation of this chapter is appreciated.

REFERENCES

Abler, W. A., D. E. Buckland, R. L. Kirkpatrick, and P. F. Scanlon. 1976. Plasma progestins and puberty in fawns as influenced by energy and protein. *Journal of Wildlife Management* 40:442–446.

Allo, A. A., J. H. Oh, W. M. Longhurst, and G. E. Connolly. 1973. VFA production in the digestive systems of deer and sheep. *Journal of Wildlife Management* 37:202–211.

Ammann, A. P., R. L. Cowan, C. L. Mothershead, and B. R. Baumgardt. 1973. Dry matter and energy intake in relation to digestibility in white-tailed deer. *Journal of Wildlife Management* 37:195–201.

Arnold, L. A. and D. L. Drawe. 1979. Seasonal food habits of white-tailed deer in the south Texas plains. *Journal of Range Management* 32:175–178.

Arnold, W., T. Ruf, S. Reimoser, et al. 2004. Nocturnal hypometabolism as an overwintering strategy of red deer (*Cervus elaphus*). *American Journal of Physiology Regulatory, Integrative and Comparative Physiology* 286:R174–R181.

Asleson, M. A., E. C. Hellgren, and L. W. Varner. 1996. Nitrogen requirements for antler growth and maintenance in white-tailed deer. *Journal of Wildlife Management* 60:744–752.

Austin, P. J., L. A. Suchar, C. T. Robbins, and A. E. Hagerman. 1989. Tannin-binding proteins in saliva of deer and their absence in saliva of sheep and cattle. *Journal of Chemical Ecology* 15:1335–1347.

Baker, D. L. and D. R. Hansen. 1985. Comparative digestion of grass in mule deer and elk. *Journal of Wildlife Management* 49:77–79.

Baker, D. L. and N. T. Hobbs. 1987. Strategies of digestion: Digestive efficiency and retention time of forage diets in montane ungulates. *Canadian Journal of Zoology* 65:1978–1984.

Barboza, P. S. and K. L. Parker. 2006. Body protein stores and isotopic indicators of N balance in female reindeer (*Rangifer tarandus*) during winter. *Physiological and Biochemical Zoology* 79:628–644.

Barboza, P. S., K. L. Parker, and I. D. Hume. 2009. *Integrative Wildlife Nutrition*. Berlin: Springer.

Barnes, T. G., L. H. Blankenship, L. W. Varner, and J. F. Gallagher. 1991. Digestibility of guajillo for white-tailed deer. *Journal of Range Management* 44:606–610.

Bartoskewitz, M. L., D. G. Hewitt, J. C. Laurenz, J. S. Pitts, and F. C. Bryant. 2007. Effect of dietary concentrations of copper and zinc on white-tailed deer antler growth, body size, and immune system function. *Small Ruminant Research* 73:87–94.

Bartoskewitz, M. L., D. G. Hewitt, J. S. Pitts, and F. C. Bryant. 2003. Supplemental feed use by free-ranging white-tailed deer in southern Texas. *Wildlife Society Bulletin* 31:1218–1227.

Brady, P. S., L. J. Brady, P. A. Whetter, D. E. Ullrey, and L. D. Fay. 1978. The effect of dietary selenium and vitamin E on biochemical parameters and survival of young among white-tailed deer (*Odocoileus virginianus*). *Journal of Nutrition* 108:1439–1448.

Brown, C. G. 2001. Evaluation of whole cottonseed consumption on growth and reproductive function in male cervids. MS thesis. Texas A&M University.

Brown, R. D. 1990. Nutrition and antler development. In *Horns, Pronghorns, and Antlers*, eds. G. A. Bubenik and A. B. Bubenik, 426–441. New York, NY: Springer-Verlag.

Brown, R. D., E. C. Hellgren, M. Abbott, D. C. Ruthven, and R. L. Bingham. 1995. Effects of dietary energy and protein restriction on nutritional indices of female white-tailed deer. *Journal of Wildlife Management* 59:595–609.

Bryant, F. C., M. M. Kothmann, and L. B. Merrill. 1980. Nutritive content of sheep, goat, and white-tailed deer diets on excellent condition rangeland in Texas. *Journal of Range Management* 33:410–414.

Bryant, F. C., C. A. Taylor, and L. B. Merrill. 1981. White-tailed deer diets from pastures in excellent and poor range condition. *Journal of Range Management* 34:193–200.

Bullock, S. L. 2009. Nutrition and physiology of white-tailed deer consuming whole cottonseed. MS thesis, Texas A&M University–Kingsville.

Cabanac, A. J., J.-P. Ouellet, M. Crete, and P. Rioux. 2005. Urinary metabolites as an index of body condition in wintering white-tailed deer *Odocoileus virginianus*. *Wildlife Biology* 11:59–66.

Campbell, T. A. and D. G. Hewitt. 2005. Nutritional value of guajillo as a component of male white-tailed deer diets. *Rangeland Ecology and Management* 58:58–64.

Campbell, T. A., C. A. Langdon, B. R. Laseter, et al. 2006. Movements of female white-tailed deer to bait sites in West Virginia, USA. *Wildlife Research* 33:1–4.

Campbell, T. A., B. R. Laseter, W. M. Ford, and K. V. Miller. 2004. Unusual white-tailed deer movements to a gas well in the central Appalachians. *Wildlife Society Bulletin* 32:983–986.

Cash, V. W. and T. E. Fulbright. 2005. Nutrient enrichment, tannins, and thorns: Effects on browsing of shrub seedlings. *Journal of Wildlife Management* 69:782–793.

Chamrad, A. D. and T. W. Box. 1968. Food habits of white-tailed deer in south Texas. *Journal of Range Management* 21:158–164.

Cook, R. C., J. G. Cook, T. R. Stephenson et al. 2010. Revisions of rump fat and body scoring indices for deer, elk, and moose. *Journal of Wildlife Management* 74:880–896.

Cook, R. C., T. R. Stephenson, W. L. Myers, J. G. Cook, and L. A. Shipley. 2007. Validating predictive models of nutritional condition for mule deer. *Journal of Wildlife Management* 71:1934–1943.

Cooper, S. M. and T. F. Ginnett. 2000. Potential effects of supplemental feeding of deer on nest predation. *Wildlife Society Bulletin* 28:660–666.

Cooper, S. M. and N. Owen-Smith. 1985. Condensed tannins deter feeding by browsing ruminants in a South African savanna. *Oecologia* 67:142–146.

Cooper, S. M. and N. Owen-Smith. 1986. Effects of plant spinescence on large mammalian herbivores. *Oecologia* 68:446–455.

Cooper, S. M., M. K. Owens, R. M. Cooper, and T. F. Ginnett. 2006. Effect of supplemental feeding on spatial distribution and browse utilization by white-tailed deer in semi-arid rangeland. *Journal of Arid Environments* 66:716–726.

Crawford, H. S. 1982. Seasonal food selection and digestibility by tame white-tailed deer in central Maine. *Journal of Wildlife Management* 46:974–982.

DelGiudice, G. D. 1995. Assessing winter nutritional restriction of northern deer with urine in snow: Considerations, potential, and limitations. *Wildlife Society Bulletin* 23:687–693.

DelGiudice, G. D., L. D. Mech, and U. S. Seal. 1994. Undernutrition and serum and urinary urea nitrogen of white-tailed deer during winter. *Journal of Wildlife Management* 58:430–436.

Dierenfeld, E. S. and D. A. Jessup. 1990. Variation in serum α-tocopherol, retinol, cholesterol, and selenium of free-ranging mule deer (*Odocoileus hemionus*). *Journal of Zoo and Wildlife Medicine* 21:425–432.

Ditchkoff, S. S. and F. A. Servello. 1998. Litterfall: An overlooked food source for wintering white-tailed deer. *Journal of Wildlife Management* 62:250–255.

Ditchkoff, S. S. and F. A. Servello. 1999. Sampling recommendations to assess nutritional restriction in deer. *Wildlife Society Bulletin* 27:1004–1009.

Doenier, P. B., G. D. DelGiudice, and M. R. Riggs. 1997. Effects of winter supplemental feeding on browse consumption by white-tailed deer. *Wildlife Society Bulletin* 25:235–243.

Dusek, G. L., R. J. MacKie, J. D. Herriges, Jr., and B. B. Compton. 1989. Population ecology of white-tailed deer along the lower Yellowstone River. *Wildlife Monographs* 104:1–68.

Edwards, S. L., S. Demarais, B. Watkins, and B. K. Strickland. 2004. White-tailed deer forage production in managed and unmanaged pine stands and summer food plots in Mississippi. *Wildlife Society Bulletin* 32:739–745.

Elston, J. J. 2003. Comparative nutritional aspects of mast consuming wildlife in south Texas. PhD dissertation, Texas A&M University–Kingsville.

Elston, J. J. and D. G. Hewitt. 2010. Intake of mast by wildlife in Texas and the potential for competition with wild boars. *Southwestern Naturalist* 55:57–66.

Everitt, J. H. 1986. Nutritive value of fruits or seeds of 14 shrub and herb species from south Texas. *Southwestern Naturalist* 31:101–104.

Everitt, J. H. and M. A. Alaniz. 1981. Nutrient content of cactus and woody plant fruits eaten by birds and mammals in south Texas. *Southwestern Naturalist* 26:301–305.

Everitt, J. H. and D. L. Drawe. 1974. Spring food habits of white-tailed deer in the south Texas plains. *Journal of Range Management* 27:15–20.

Everitt, J. H. and C. L. Gonzalez. 1979. Botanical composition and nutrient content of fall and early winter diets of white-tailed deer in south Texas. *Southwestern Naturalist* 24:297–310.

Fancy, S. G. and R. G. White. 1985. Incremental cost of activity. In *Bioenergetics of Wild Herbivores*, eds. R. J. Hudson and R. G. White, 143–159. Boca Raton, FL: CRC Press, Inc.

Finger, S. E., I. L. Brisbin, Jr., M. H. Smith, and D. F. Urbston. 1981. Kidney fat as a predictor of body condition in white-tailed deer. *Journal of Wildlife Management* 45:964–968.

Flueck, W. T. 1994. Effect of trace elements on population dynamics: Selenium deficiency in free-ranging black-tailed deer. *Ecology* 75:807–812.

Fulbright, T. E. and A. Garza, Jr. 1991. Forage yield and white-tailed deer diets following live oak control. *Journal of Range Management* 44:451–455.

Fulbright, T. E. and J. A. Ortega-S. 2006. *White-tailed Deer Habitat: Ecology and Management on Rangelands*. College Station, TX: Texas A&M University Press.
Gee, K. L., M. D. Porter, S. Demarais, F. C. Bryant, and G. Van Vreede. 1991. *White-tailed Deer: Their Food and Management in the Cross Timbers*. Ardmore, OK; Lubbock, TX: Samuel Roberts Noble Foundation; Texas Tech University.
Gill, R. B., L. H. Carpenter, R. M. Bartmann, D. L. Baker, and G. G. Schoonveld. 1983. Fecal analysis to estimate mule deer diets. *Journal of Wildlife Management* 47:902–915.
Gizejewski, Z., B. Szafranska, Z. Steplewski et al. 2008. Cottonseed feeding delivers sufficient quantities of gossypol as a male deer contraceptive. *European Journal of Wildlife Research* 54:469–477.
Goering, H. K. and P. J. Van Soest. 1970. *Forage Fiber Analysis*. Washington, DC: United States Department of Agriculture, Agriculture Handbook 379.
Grasman, B. T. and E. C. Hellgren. 1993. Phosphorus nutrition in white-tailed deer: Nutrient balance, physiological responses, and antler growth. *Ecology* 74:2279–2296.
Gray, B. P. and F. A. Servello. 1995. Energy intake relationships for white-tailed deer on winter browse diets. *Journal of Wildlife Management* 59:147–152.
Hanley, T. A. 1997. A nutritional view of understanding and complexity in the problem of diet selection by deer (Cervidae). *Oikos* 79:209–218.
Hanley, T. A., C. T. Robbins, A. E. Hagerman, and C. McArthur. 1992. Predicting digestible protein and digestible dry matter in tannin-containing forages consumed by ruminants. *Ecology* 73:537–541.
Harder, J. D. and R. L. Kirkpatrick. 1994. Physiological indices in wildlife research. In *Research and Management Techniques for Wildlife and Habitats*, ed. T. A. Bookout, 275–306. Bethesda, MD: The Wildlife Society.
Harlow, R. F., J. B. Whelan, H. S. Crawford, and J. E. Skeen. 1975. Deer foods during years of oak mast abundance and scarcity. *Journal of Wildlife Management* 39:330–336.
Haulton, S. M., W. F. Porter, and B. A. Rudolph. 2001. Evaluating 4 methods to capture white-tailed deer. *Wildlife Society Bulletin* 29:255–264.
Hehman, M. W. and T. E. Fulbright. 1997. Use of warm-season food plots by white-tailed deer. *Journal of Wildlife Management* 61:1108–1115.
Heller, H. C. 1988. Sleep and hypometabolism. *Canadian Journal of Zoology* 66:61–69.
Hellgren, E. C. and W. J. Pitts. 1997. Sodium economy in white-tailed deer (*Odocoileus virginianus*). *Physiological Zoology* 70:547–555.
Henke, S. E. 1997. Do white-tailed deer react to the dinner bell? An experiment in classical conditioning. *Wildlife Society Bulletin* 25:291–295.
Hickling, G. J. 2002. *Dynamics of Bovine Tuberculosis in Wild White-tailed Deer in Michigan*. Michigan Department of Natural Resources Report 3363, Lansing, Michigan.
Hobbs, N. T., J. E. Gross, L. A. Shipley, D. E. Spalinger, and B. A. Wunder. 2003. Herbivore functional response in heterogeneous environments: A contest among models. *Ecology* 84:666–681.
Hofmann, R. R. 1989. Evolutionary steps of ecophysiological adaptation and diversification of ruminants: A comparative review of their digestive system. *Oecologia* 78:443–457.
Holter, J. B., H. H. Hayes, and S. H. Smith. 1979a. Protein requirement of yearling white-tailed deer. *Journal of Wildlife Management* 43:872–879.
Holter, J. B., W. E. Urban, Jr., and H. H. Hayes. 1977. Nutrition of northern white-tailed deer throughout the year. *Journal of Animal Science* 45:365–376.
Holter, J. B., W. E. Urban, Jr., and H. H. Hayes. 1979b. Predicting energy and nitrogen retention in young white-tailed deer. *Journal of Wildlife Management* 43:880–888.
Holter, J. B., W. E. Urban, Jr., H. H. Hayes, and H. Silver. 1976. Predicting metabolic rate from telemetered heart rate in white-tailed deer. *Journal of Wildlife Management* 40:626–629.
Hughes, J. W. and T. J. Fahey. 1991. Availability, quality, and selection of browse by white-tailed deer after clearcutting. *Journal of Forestry* 89:31–36.
Iason, G. 2005. The role of plant secondary metabolites in mammalian herbivory: Ecological perspectives. *Proceedings of the Nutrition Society* 64:123–131.
Ivey, T. L. and M. K. Causey. 1981. Movements and activity patterns of female white-tailed deer during rut. *Proceedings of the Southeastern Association of Fish and Wildlife Agencies* 35:149–166.
Izhaki, I. 1993. Influence of nonprotein nitrogen on estimation of protein from total nitrogen in fleshy fruits. *Journal of Chemical Ecology* 19:2605–2615.

Jenks, J. A. and D. M. Leslie, Jr. 2003. Effect of domestic cattle on the condition of female white-tailed deer in southern pine-bluestem forests. *Acta Theriologica* 48:131–144.

Jenks, J. A., D. M. Leslie, Jr., R. L. Lochmiller, M. A. Melchiors, and F. T. I. McCollum. 1996. Competition in sympatric white-tailed deer and cattle populations in southern pine forests of Oklahoma and Arkansas, USA. *Acta Theriologica* 41:287–306.

Jensen, P. G., P. J. Pekins, and J. B. Holter. 1999. Compensatory effect of the heat increment of feeding on thermoregulation costs of white-tailed deer fawns in winter. *Canadian Journal of Zoology* 77:1474–1485.

Johns, P. E., M. H. Smith, and R. K. Chesser. 1984. Annual cycles of the kidney fat index in a southeastern white-tailed deer herd. *Journal of Wildlife Management* 48:969–973.

Johnson, A. S., P. E. Hale, W. M. Ford et al. 1995. White-tailed deer foraging in relation to successional stage, overstory type and management of southern Appalachian forests. *American Midland Naturalist* 133:18–35.

Johnson, M. K. and K. D. Dancak. 1993. Effects of food plots on white-tailed deer in Kisatchie National Forest. *Journal of Range Management* 46:110–114.

Johnson, M. K., B. W. Delany, S. P. Lynch et al. 1987. Effects of cool-season agronomic forages on white-tailed deer. *Wildlife Society Bulletin* 15:330–339.

Kammermeyer, K. E., and E. B. Moser. 1990. The effect of food plots, roads, and other variables on deer harvest in northeastern Georgia. *Proceedings of the Southeastern Association of Fish and Wildlife Agencies* 44:364–373.

Kammermeyer, K. E. and R. Thackston. 1995. Habitat management and supplemental feeding. In *Quality Whitetails: The Why and How of Quality Deer Management*, eds. K. V. Miller and R. L. Marchinton, 129–154. Mechanicsburg, PA: Stackpole Books.

Keegan, T. W., M. K. Johnson, and B. D. Nelson. 1989. American jointvetch improves summer range for white-tailed deer. *Journal of Range Management* 42:128–134.

Kennedy, J. F., J. A. Jenks, R. L. Jones, and K. J. Jenkins. 1995. Characteristics of mineral licks used by white-tailed deer (*Odocoileus virginianus*). *American Midland Naturalist* 134:324–331.

Kie, J. G. and R. T. Bowyer. 1999. Sexual segregation in white-tailed deer: Density-dependent changes in use of space, habitat selection, and dietary niche. *Journal of Mammalogy* 80:1004–1020.

Kie, J. G., D. L. Drawe, and G. Scott. 1980. Changes in diet and nutrition with increased herd size in Texas white-tailed deer. *Journal of Range Management* 33:28–34.

Kilpatrick, H. J., A. M. LaBonte, and J. S. Barclay. 2010. Use of bait to increase archery deer harvest in an urban -suburban landscape. *Journal of Wildlife Management* 74:714–718.

Kilpatrick, H. J. and W. A. Stober. 2002. Effects of temporary bait sites on movements of suburban white-tailed deer. *Wildlife Society Bulletin* 30:760–766.

Kistner, T. P., C. E. Trainer, and N. A. Hartmann. 1980. A field technique for evaluating physical condition of deer. *Wildlife Society Bulletin* 8:11–17.

Klaassen, C. D. 1996. *Casarett and Doull's Toxicology: The Basic Science of Poisons* (5th edition). New York, NY: Pergamon Press.

Kleiber, M. 1961. *The Fire of Life: An Introduction to Animal Energetics*. New York, NY: John Wiley and Sons.

Klein, D. R. 1964. Range-related differences in growth of deer reflected in skeletal ratios. *Journal of Mammalogy* 45:226–235.

Koerth, B. H. and J. C. Kroll. 2000. Bait type and timing for deer counts using cameras triggered by infrared monitors. *Wildlife Society Bulletin* 28:630–635.

Koerth, B. H. and J. W. Stuth. 1991. Instantaneous intake rates of 9 browse species by white-tailed deer. *Journal of Range Management* 44:614–618.

Korschgen, L. J. 1962. Foods of Missouri deer, with some management implications. *Journal of Wildlife Management* 26:164–172.

Korschgen, L. J., W. R. Porath, and O. Torgerson. 1980. Spring and summer foods of deer in the Missouri Ozarks. *Journal of Wildlife Management* 44:89–97.

Krausman, P. R. 1978. Forage relationships between two deer species in Big Bend National Park, Texas. *Journal of Wildlife Management* 42:101–107.

Krausman, P. R. and J. A. Bissonette. 1977. Bone-chewing behavior of desert mule deer. *Southwestern Naturalist* 22:149–150.

Lambert, B. C., Jr. and S. Demarais. 2001. Use of supplemental feed for ungulates by non-target species. *Southwestern Naturalist* 46:118–121.

Lautier, J. K., T. V. Dailey, and R. D. Brown. 1988. Effect of water restriction on feed intake of white-tailed deer. *Journal of Wildlife Management* 52:602–606.
Lefort, S., J.-P. Trembley, F. Fournier, F. Potvin, and J. Huot. 2007. Importance of balsam fir as winter forage for white-tailed deer at the northeastern limit of their distribution range. *Écoscience* 14:109–116.
Leslie, D. M., R. T. Bowyer, and J. A. Jenks. 2008. Facts from feces: Nitrogen still measures up as a nutritional index for mammalian herbivores. *Journal of Wildlife Management* 72:1420–1433.
Leslie, D. M., Jr. and E. E. Starkey. 1985. Fecal indices to dietary quality of cervids in old-growth forests. *Journal of Wildlife Management* 49:142–146.
Lewis, T. L. and O. J. Rongstad. 1998. Effects of supplemental feeding on white-tailed deer, *Odocoileus virginianus*, migration and survival in northern Wisconsin. *Canadian Field Naturalist* 112:75–81.
Loe, L. E., E. L. Meisingset, A. Mysterud, R. Langvatn, and N. C. Stenseth. 2004. Phenotypic and enviromnetal correlates of tooth eruption in red deer (*Cervus elaphus*). *Journal of Zoology, London* 262:83–89.
Makkar, H. P. S. 2003. Effects and fate of tannins in ruminant animals, adaptation to tannins, and strategies to overcome detrimental effects of feeding tannin-rich feeds. *Small Ruminant Research* 49:241–256.
Mautz, W. W. 1978. Sledding on a bushy hillside: The fat cycle in deer. *Wildlife Society Bulletin* 6:88–90.
Mautz, W. W., and J. Fair. 1980. Energy expenditure and heart rate for activities of white-tailed deer. *Journal of Wildlife Management* 44:333–342.
Mautz, W. W., J. Kanter, and P. J. Pekins. 1992. Seasonal metabolic rhythms of captive female white-tailed deer: A reexamination. *Journal of Wildlife Management* 56:656–661.
Mautz, W. W., H. Silver, J. B. Holter, H. H. Hayes, and W. E. Urban, Jr. 1976. Digestibility and related nutritional data for seven northern deer browse species. *Journal of Wildlife Management* 40:630–638.
McArthur, C., A. E. Hagerman, and C. T. Robbins. 1991. Physiological strategies of mammalian herbivores against plant defenses. In *Plant Defenses against Mammalian Herbivory*, eds. R. T. Palo and C. T. Robbins, 103–114. Boca Raton, FL: CRC Press.
McShea, W. J. and G. Schwede. 1993. Variable acorn crops: Responses of white-tailed deer and other mast consumers. *Journal of Mammalogy* 74:999–1006.
Mech, L. D. 2007. Femur-marrow fat of white-tailed deer fawns killed by wolves. *Journal of Wildlife Management* 71:920–923.
Mech, L. D. and G. D. Delgiudice. 1985. Limitations of the marrow-fat technique as an indicator of body condition. *Wildlife Society Bulletin* 13:204–206.
Meyer, M. W., R. D. Brown, and M. W. Graham. 1984. Protein and energy content of white-tailed deer diets in the Texas coastal bend. *Journal of Wildlife Management* 48:527–534.
Milchunas, D. G., M. I. Dyer, O. C. Wallmo, and D. E. Johnson. 1978. *In-vivo/in-vitro Relationships of Colorado Mule Deer Forages*. Colorado Division of Wildlife Special Report 43. Denver, Colorado.
Milton, K. and F. R. Dintzis. 1981. Nitrogen-to-protein conversion factors for tropical plant samples. *Biotropica* 13:177–181.
Moen, A. N. 1978. Seasonal changes in heart rates, activity, metabolism, and forage intake of white-tailed deer. *Journal of Wildlife Management* 42:715–738.
Mould, E. D. and C. T. Robbins. 1981. Nitrogen metabolism in elk. *Journal of Wildlife Management* 45:323–334.
Mould, E. D. and C. T. Robbins. 1982. Digestive capabilities in elk compared to white-tailed deer. *Journal of Wildlife Management* 46:22–29.
Murden, S. B. and K. L. Risenhoover. 1993. Effects of habitat enrichment on patterns of diet selection. *Ecological Applications* 3:497–505.
Nagy, K. A. and C. C. Peterson. 1988. Scaling of water flux rate in animals. *University California Publications in Zoology* 120:1–172.
National Research Council. 2007. *Nutrient Requirements of Small Ruminants: Sheep, Goats, Cervids, and New World Camelids*. Washington, DC: National Academies Press.
Naugle, D. E., B. J. Kernohan, and J. A. Jenks. 1995. Seasonal capture success and bait use of white-tailed deer in an agricultural-wetland complex. *Wildlife Society Bulletin* 23:198–200.
Nixon, C. M., L. P. Hansen, P. A. Brewer, and J. E. Chelsvig. 1991. Ecology of white-tailed deer in an intensively farmed region of Illinois. *Wildlife Monographs* 118:1–77.
Nixon, C. M., M. W. McClain, and K. R. Russell. 1970. Deer food habits and range characteristics in Ohio. *Journal of Wildlife Management* 34:870–886.
Oftedal, O. T. 1985. Pregnancy and lactation. In *Bioenergetics of Wild Herbivores*, eds. R. J. Hudson and R. G. White, 215–238. Boca Raton, FL: CRC Press, Inc.

Ortega, I. M., S. Soltero-Gardea, D. L. Drawe, and F. C. Bryant. 1997. Evaluating grazing strategies for cattle: Nutrition of cattle and deer. *Journal of Range Management* 50:631–637.

Osborn, R. G. and T. F. Ginnett. 2001. Fecal nitrogen and 2,6-diaminopimelic acid as indices to dietary nitrogen in white-tailed deer. *Wildlife Society Bulletin* 29:1131–1139.

Ouellet, J.-P., M. Crête, J. Maltais, C. Pelletier, and J. Huot. 2001. Emergency feeding of white-tailed deer: Test of three feeds. *Journal of Wildlife Management* 65:129–136.

Ozoga, J. J. and L. J. Verme. 1970. Winter feeding patterns of penned white-tailed deer. *Journal of Wildlife Management* 34:431–439.

Ozoga, J. J. and L. J. Verme. 1982. Physical and reproductive characteristics of a supplementally-fed white-tailed deer herd. *Journal of Wildlife Management* 46:281–301.

Page, B. D. and H. B. Underwood. 2006. Comparing protein and energy status of winter-fed white-tailed deer. *Wildlife Society Bulletin* 34:716–724.

Palmer, M. V., W. R. Waters, and D. L. Whipple. 2004. Shared feed as a means of deer-to-deer transmission of *Mycobacterium bovis*. *Journal Wildlife Disease* 40:87–91.

Parker, K. L., P. S. Barboza, and T. R. Stephenson. 2005. Protein conservation in female caribou (*Rangifer tarandus*): Effects of decreasing diet quality during winter. *Journal of Mammalogy* 86:610–622.

Parker, K. L., G. D. DelGiudice, and M. P. Gillingham. 1993. Do urinary urea nitrogen and cortisol ratios of creatinine reflect body-fat reserves in black-tailed deer? *Canadian Journal of Zoology* 71:1841–1848.

Parker, K. L. and M. P. Gillingham. 1990. Estimates of critical thermal environments for mule deer. *Journal of Range Management* 43:73–81.

Parker, K. L., M. P. Gillingham, T. A. Hanley, and C. T. Robbins. 1999. Energy and protein balance of free-ranging black-tailed deer in a natural forest environment. *Wildlife Monographs* 143:1–48.

Parker, K. L. and C. T. Robbins. 1984. Thermoregulation in mule deer and elk. *Canadian Journal of Zoology* 62:1409–1422.

Parker, K. L., C. T. Robbins, and T. A. Hanley. 1984. Energy expenditures for locomotion by mule deer and elk. *Journal of Wildlife Management* 48:474–488.

Payne, N. F. and F. C. Bryant. 1998. *Wildlife Habitat Management of Forestlands, Rangelands, and Farmlands*. Malabar, FL: Krieger Publishing Company.

Pearson, H. A. 1965. Rumen organisms in white-tailed deer from south Texas. *Journal of Wildlife Management* 29:493–496.

Pekins, P. J. and W. W. Mautz. 1988. Digestibility and nutritional value of autumn diets of deer. *Journal of Wildlife Management* 52:328–332.

Pekins, P. J., K. S. Smith, and W. W. Mautz. 1998. The energy cost of gestation in white-tailed deer. *Canadian Journal of Zoology* 76:1091–1097.

Pletscher, D. H. 1987. Nutrient budgets for white-tailed deer in New England with special reference to sodium. *Journal of Mammalogy* 68:330–336.

Provenza, F. D. 1995. Postingestive feedback as an elementary determinant of food preference and intake in ruminants. *Journal of Range Management* 48:2–17.

Provenza, F. D. 1996. Acquired aversions as the basis for varied diets of ruminants foraging on rangelands. *Journal of Animal Science* 74:2010–2020.

Provenza, F. D., J. J. Villalba, L. E. Dziba, S. B. Atwood, and R. E. Banner. 2003. Linking herbivore experience, varied diets, and plant biochemical diversity. *Small Ruminant Research* 49:257–274.

Ramírez, R. G., G. F. W. Haenlein, A. Treviño, and J. Reyna. 1996. Nutrient and mineral profile of white-tailed deer (*Odocoileus virginianus, texanus*) diets in northeastern Mexico. *Small Ruminant Research* 23:7–16.

Ransom, A. B. 1965. Kidney and marrow fat as indicators of white-tailed deer condition. *Journal of Wildlife Management* 29:397–398.

Renecker, L. A. and R. J. Hudson. 1986. Seasonal energy expenditures and thermoregulatory responses of moose. *Canadian Journal of Zoology* 64:322–327.

Rhind, S. M., Z. A. Archer, and C. L. Adam. 2002. Seasonality of food intake in ruminants: Recent developments in understanding. *Nutrition Research Reviews* 15:43–65.

Riney, T. 1955. Evaluating condition of free ranging red deer (*Cervus elephus*) with special reference to New Zealand. *New Zealand Journal of Science and Technology* 36B:429–463.

Ringberg, T. 1979. The Spitzbergen reindeer—a winter-dormant ungulate? *Acta Physiologica Scandinavica* 105:268–273.

Robbins, C. T. 1987. Digestibility of an arboreal lichen by mule deer. *Journal of Range Management* 40:491–492.
Robbins, C. T. 1993. *Wildlife Feeding and Nutrition* (2nd edition). San Diego, CA: Academic Press.
Robbins, C. T., A. E. Hagerman, P. J. Austin, C. McArthur, and T. A. Hanley. 1991. Variation in mammalian physiological responses to a condensed tannin and its ecological implications. *Journal of Mammalogy* 72:480–486.
Robbins, C. T., T. A. Hanley, A. E. Hagerman et al. 1987a. Role of tannins in defending plants against ruminants: Reduction in protein availability. *Ecology* 68:98–107.
Robbins, C. T. and A. N. Moen. 1975. Uterine composition and growth in pregnant white-tailed deer. *Journal of Wildlife Management* 39:684–691.
Robbins, C. T., A. N. Moen, and J. T. Reid. 1974a. Body composition of white-tailed deer. *Journal of Animal Science* 38:871–876.
Robbins, C. T., S. Mole, A. E. Hagerman, and T. A. Hanley. 1987b. Role of tannins in defending plants against ruminants: Reduction in dry matter digestion? *Ecology* 68:1606–1615.
Robbins, C. T., R. L. Prior, A. N. Moen, and W. J. Visek. 1974b. Nitrogen metabolism of white-tailed deer. *Journal of Animal Science* 38:186–191.
Robbins, C. T., P. J. V. Soest, W. W. Mautz, and A. N. Moen. 1975. Feed analyses and digestion with reference to white-tailed deer. *Journal of Wildlife Management* 39:67–79.
Rose, J. and J. D. Harder. 1985. Seasonal feeding habits of an enclosed high density white-tailed deer herd in northern Ohio. *Ohio Journal of Science* 85:184–190.
Rudolph, B. A., S. J. Riley, G. J. Hickling, et al. 2006. Regulating hunter baiting for white-tailed deer in Michigan: Biological and social considerations. *Wildlife Society Bulletin* 34:314–321.
Sadleir, R. M. F. S. 1980. Milk yield of black-tailed deer. *Journal of Wildlife Management* 44:472–478.
Sage, R. W. and K. A. Gustafson. 1991. *Feeding Adirondack Deer in Winter: Let's Understand What We're Doing*. Syracuse, New York, NY: State University of New York College of Environmental Science and Forestry.
Saltz, D., G. C. White, R. M. Bartmann et al. 1995. Assessing animal condition, nutrition, and stress from urine in snow. *Wildlife Society Bulletin* 23:694–704.
Sauvé, D. G. and S. D. Côté. 2006. Is winter diet quality related to body condition of white-tailed deer (*Odocoileus virginianus*)? An experiment using urine profiles. *Canadian Journal of Zoology* 84:1003–1010.
Schultz, S. R., R. X. Barry, M. K. Johnson, J. E. Miller, and W. A. Forbes. 1994. Effects of feed plots on fecal egg counts of white-tailed deer. *Small Ruminant Research* 13:93–97.
Schultz, S. R. and M. K. Johnson. 1992a. Effects of supplemental mineral licks on white-tailed deer. *Wildlife Society Bulletin* 20:303–308.
Schultz, S. R. and M. K. Johnson. 1992b. Use of artificial mineral licks by white-tailed deer in Louisiana. *Journal of Range Management* 45:546–548.
Servello, F. A., E. C. Hellgren, and S. R. McWilliams. 2005. Techniques for wildlife nutritional ecology. In *Techniques for Wildlife Investigations and Management*, ed. C. E. Braun, 554–590. Bethesda, MD: The Wildlife Society.
Servello, F. A. and J. W. Schneider. 2000. Evaluation of urinary indices of nutritional status for white-tailed deer: Tests with winter browse diets. *Journal of Wildlife Management* 64:137–145.
Shea, S. M. and J. S. Osborn. 1995. Poor quality habitats. In *Quality Whitetails: The Why and How of Quality Deer Management*, eds. K. V. Miller and R. L. Marchinton, 193–209. Mechanicsburg, PA: Stackpole Books.
Short, H. L. 1963. Rumen fermentations and energy relationships in white-tailed deer. *Journal of Wildlife Management* 27:184–195.
Short, H. L. 1971. Forage digestibility and diet of deer on southern upland range. *Journal of Wildlife Management* 35:698–706.
Silver, H., N. F. Colovos, and H. H. Hayes. 1959. Basal metabolism of white-tailed deer: A pilot study. *Journal of Wildlife Management* 23:434–438.
Silver, H., N. F. Colovos, J. B. Holter, and H. H. Hayes. 1969. Fasting metabolism of white-tailed deer. *Journal of Wildlife Management* 33:490–498.
Silver, H., J. B. Holter, N. F. Colovos, and H. H. Hayes. 1971. Effect of falling temperature on heat production in fasting white-tailed deer. *Journal of Wildlife Management* 35:37–46.
Simard, M. A., S. D. Côté, R. B. Weladji, and J. Huot. 2008. Feedback effects of chronic browsing on life-history traits of a large herbivore. *Journal of Animal Ecology* 77:678–686.
Skinner, W. R. and E. S. Telfer. 1974. Spring, summer, and fall foods of deer in New Brunswick. *Journal of Wildlife Management* 38:210–214.

Smith, S. H., J. B. Holter, H. H. Hayes, and H. Silver. 1975. Protein requirement of white-tailed deer fawns. *Journal of Wildlife Management* 39:582–589.

Snider, C. C. and J. M. Asplund. 1974. *In vitro* digestibility of deer foods from the Missouri Ozarks. *Journal of Wildlife Management* 38:20–31.

Spalinger, D. E., S. M. Cooper, D. J. Martin, and L. A. Shipley. 1997. Is social learning an important influence on foraging behavior in white-tailed deer? *Journal of Wildlife Management* 61:611–621.

Spalinger, D. E., T. A. Hanley, and C. T. Robbins. 1988. Analysis of the functional response in foraging in the Sitka black-tailed deer. *Ecology* 69:1166–1175.

Spalinger, D. E. and N. T. Hobbs. 1992. Mechanisms of foraging in mammalian herbivores: New models of functional response. *American Naturalist* 140:325–348.

Spalinger, D. E., C. T. Robbins, and T. A. Hanley. 1986. The assessment of handling time in ruminants: The effect of plant chemical and physical structure on the rate of breakdown of plant particles in the rumen of mule deer and elk. *Canadian Journal of Zoology* 64:312–321.

Stephenson, D. C. and R. D. Brown. 1984. Calcium kinetics in male white-tailed deer. *Journal of Nutrition* 114:1014–1024.

Stephenson, T. R., V. C. Bleich, B. M. Pierce, and G. P. Mulcahy. 2002. Validation of mule deer body composition using *in vivo* and post-mortem indices of nutritional condition. *Wildlife Society Bulletin* 30:557–564.

Stewart, K. M., T. E. Fulbright, and D. L. Drawe. 2000. White-tailed deer use of clearings relative to forage availability. *Journal of Wildlife Management* 64:733–741.

Strickland, B. K., D. G. Hewitt, C. A. DeYoung, and R. L. Bingham. 2005. Digestible energy requirements for maintenance of body mass of white-tailed deer in southern Texas. *Journal of Mammalogy* 86:56–60.

Stromayer, K. A. K., R. J. Warren, A. S. Johnson et al. 1998. Chinese privet and the feeding ecology of white-tailed deer: The role of an exotic plant. *Journal of Wildlife Management* 62:1321–1329.

Swihart, R. K., J. Harmon, P. Weeks, A. L. Easter-Pilcher, and A. J. DeNicola. 1998. Nutritional condition and fertility of white-tailed deer (*Odocoileus virginianus*) from areas with contrasting histories of hunting. *Canadian Journal of Zoology* 76:1932–1941.

Taillon, J., D. G. Sauvé, and S. D. Côté. 2006. The effects of decreasing winter diet quality on foraging behavior and life-history traits of white-tailed deer fawns. *Journal of Wildlife Management* 70:1445–1454.

Tarr, M. D. and P. J. Pekins. 2002. Influences of winter supplemental feeding on the energy balance of white-tailed deer fawns in New Hampshire, U.S.A. *Canadian Journal of Zoology* 80:6–15.

Thill, R. E., A. Martin, Jr., H. F. Morris, Jr., and E. D. McCune. 1987. Grazing and burning impacts on deer diets on Louisiana pine-bluestem range. *Journal of Wildlife Management* 51:873–880.

Thompson, C. B., J. B. Holter, H. H. Hayes, H. Silver, and W. E. Urban, Jr. 1973. Nutrition of white-tailed deer. I. Energy requirements of fawns. *Journal of Wildlife Management* 37:301–311.

Timmons, G. R., D. G. Hewitt, C. A. DeYoung, T. E. Fulbright, and D. A. Draeger. 2010. Does supplemental feed increase selective foraging in a browsing ungulate? *Journal of Wildlife Management* 74:995–1002.

Torbit, S. C., L. H. Carpenter, A. W. Alldredge, and D. M. Swift. 1985. Mule deer body composition: A comparison of methods. *Journal of Wildlife Management* 49:86–91.

Tremblay, J.-P., I. Thibault, C. Dussault, J. Huot, and S. D. Côté. 2005. Long-term decline in white-tailed deer browse supply: Can lichens and litterfall act as alternative food sources that preclude density-dependent feedbacks. *Canadian Journal of Zoology* 83:1087–1096.

Ullrey, D. E., J. T. Nellist, J. P. Duvendeck, P. A. Whetter, and L. D. Fay. 1987. Digestibility of vegetative rye for white-tailed deer. *Journal of Wildlife Management* 51:51–53.

Ullrey, D. E., W. G. Youatt, H. E. Johnson, L. D. Fay, and B. L. Bradley. 1967. Protein requirement of white-tailed deer fawns. *Journal of Wildlife Management* 31:679–685.

Ullrey, D. E., W. G. Youatt, H. E. Johnson et al. 1969. Digestible energy requirements for winter maintenance of Michigan white-tailed does. *Journal of Wildlife Management* 33:482–490.

Ullrey, D. E., W. G. Youatt, H. E. Johnson et al. 1970. Digestible and metabolizable energy requirements for winter maintenance of Michigan white-tailed does. *Journal of Wildlife Management* 34:863–869.

Ullrey, D. E., W. G. Youatt, and P. A. Whetter. 1981. Muscle selenium concentrations in Michigan deer. *Journal of Wildlife Management* 45:534–536.

Van Brackle, M. D., R. L. Marchinton, G. O. Ware et al. 1995. Oral biomarking of a supplementally-fed herd of free-ranging white-tailed deer. *Proceedings of the Southeastern Association of Fish and Wildlife Agencies* 49:372–382.

Van Deelen, T. R., B. Dhuey, K. R. McCaffery, and R. E. Rolley. 2006. Relative effects of baiting and supplemental antlerless seasons on Wisconsin's 2003 deer harvest. *Wildlife Society Bulletin* 34:322–328.

Van Soest, P. J. 1982. *Nutritional Ecology of the Ruminant*. Corvallis, OR: O & B Books.

Vanderhoof, R. E. and H. A. Jacobson. 1989. Effects of agronomic plantings on white-tailed deer antler characteristics. Presentation at the *12th Annual Meeting of the Southeastern Deer Study Group*. Oklahoma City, Oklahoma.

Verme, L. J. 1969. Reproductive patterns of white-tailed deer related to nutritional plane. *Journal of Wildlife Management* 33:881–887.

Verme, L. J. and J. J. Ozoga. 1980. Influence of protein-energy intake on deer fawns in autumn. *Journal of Wildlife Management* 44:305–314.

Villalba, J. J. and F. D. Provenza. 2007. Self-medication and homeostatic behaviour in herbivores: Learning about the benefits of nature's pharmacy. *Animal* 1:1360–1370.

Wallach, A. D., M. Inbar, M. Scantlebury, J. R. Speakman, and U. Shanas. 2007. Water requirements as a bottleneck in the reintroduction of European roe deer to the southern edge of its range. *Canadian Journal of Zoology* 85:1182–1192.

Ward, R. L. and C. L. Marcum. 2005. Lichen litterfall consumption by wintering deer and elk in western Montana. *Journal of Wildlife Management* 69:1081–1089.

Warren, R. J., R. L. Kirkpatrick, A. Oelschlaeger, P. F. Scanlon, and F. C. Gwazdauskas. 1981. Dietary and seasonal influences on nutritional indices of adult male white-tailed deer. *Journal of Wildlife Management* 45:926–936.

Warren, R. J., R. L. Kirkpatrick, A. Oelschlaeger et al. 1982. Energy, protein, and seasonal influences on white-tailed deer fawn nutritional indices. *Journal of Wildlife Management* 46:302–312.

Watkins, B. E. and D. E. Ullrey. 1983. Iodine concentration in plants used by white-tailed deer in Michigan. *Journal of Wildlife Management* 47:1220–1226.

Watkins, B. E., D. E. Ullrey, R. F. Nachreiner, and S. M. Schmitt. 1983. Effects of supplemental iodine and season on thyroid activity of white-tailed deer. *Journal of Wildlife Management* 47:45–58.

Watkins, B. E., J. H. Witham, D. E. Ullrey, D. J. Watkins, and J. M. Jones. 1991. Body composition and condition evaluation of white-tailed deer fawns. *Journal of Wildlife Management* 55:39–51.

Webb, S. L., C. J. Zabransky, R. S. Lyons, and D. G. Hewitt. 2006. Water quality and summer use of sources of water in Texas. *Southwestern Naturalist* 51:368–375.

Weckerly, F. W. and J. P. Nelson, Jr. 1990. Age and sex differences of white-tailed deer diet composition, quality, and calcium. *Journal of Wildlife Management* 54:532–538.

Weeks, H. P. 1978. Characteristics of mineral licks and behavior of visiting white-tailed deer in southern Indiana. *American Midland Naturalist* 100:384–395.

Weeks, H. P. and C. M. Kirkpatrick. 1976. Adaptations of white-tailed deer to naturally occurring sodium deficiencies. *Journal of Wildlife Management* 40:610–625.

Wehausen, J. D. 1995. Fecal measures of diet quality in wild and domestic ruminants. *Journal of Wildlife Management* 59:816–823.

Weiner, J. 1977. Energy metabolism of the roe deer. *Acta Theriologica* 22:3–24.

Wheaton, C. and R. D. Brown. 1983. Feed intake and digestive efficiency of south Texas white-tailed deer. *Journal of Wildlife Management* 47:442–450.

White, P. J., R. A. Garrott, C. A. V. White, and G. A. Sargeant. 1995. Interpreting mean chemical ratios from simple random collections of snow-urine samples. *Wildlife Society Bulletin* 23:705–710.

Wickstrom, M. L., C. T. Robbins, T. A. Hanley, D. E. Spalinger, and S. M. Parish. 1984. Food intake and foraging energetics of elk and mule deer. *Journal of Wildlife Management* 48:1285–1301.

Wiles, G. J. and H. P. Weeks, Jr. 1986. Movements and use patterns of white-tailed deer visiting natural licks. *Journal of Wildlife Management* 50:487–496.

Windels, S. K. and D. G. Hewitt. 2011. Effects of plant secondary compounds on nutritional carrying capacity estimates of a browsing ungulate. *Rangeland Ecology and Management*.

Windels, S. K. and P. A. Jordan. 2008. Winter use of senescent herbaceous plants by white-tailed deer in Minnesota. *American Midland Naturalist* 160:253–258.

Youatt, W. G., D. E. Ullrey, and W. T. Magee. 1976. Vitamin A concentration in livers of white-tailed deer. *Journal of Wildlife Management* 40:172–173.

Zaiglin, R. E. and C. A. DeYoung. 1989. Supplemental feeding of free-ranging deer in south Texas. *Texas Journal of Agriculture and Natural Resources* 3:39–41.

4

Antlers

Steve Demarais and Bronson K. Strickland

CONTENTS

Introduction

For thousands of years man has pursued deer and their deciduous ornaments. Native Americans used the antlers of white-tailed deer as tools, for medicinal purposes, and as ceremonial symbols. In fact, deer antlers have been associated with humans for medicinal reasons as far back as 200 AD (Putman, 1988). Today, modern man continues to be fascinated by these intriguing structures. Researchers have debated their purpose (Darwin, 1874; Geist, 1966; Clutton-Brock, 1982) for over a hundred years, while naturalists and hunters alike strive to glimpse a mature male with immense, complex antlers.

The current fascination with antlers may represent vestigial behavior from our co-evolution with deer, and remind us of hunts long past. Researchers have promoted the theory that antlers serve as indicators of quality and dominance to conspecifics (Geist, 1966). Perhaps antlers serve as a symbol of quality to modern man as well. For some, antlers may appeal to their naturalistic instincts when they interact with the outdoors. To others, antlers may signify a superior and challenging specimen and appeal to their dominionistic tendencies during chase (Kellert and Smith, 2000). Nonetheless, researchers, hunters, and nonhunters alike appreciate antlers of white-tailed deer. In this chapter we will review the why and how of antlers and ways this information can be used to manage one of North America's most valuable natural resources, the white-tailed deer.

What Are Antlers?

Antlers are an amazing natural phenomenon of the Cervidae, especially when one considers their anatomy and physiology. Rapid growth rate of these "boney protuberances" makes antler tissue a potential biological model for bone cancer research, and annual loss and regeneration make antler tissue of interest to researchers working on organ and limb regeneration. Antler characteristics are controlled by the genetic code but with influence from external environmental variables including photoperiod and nutrition. Modifying external sources of variation is a target of management throughout the whitetail's range. Understanding the anatomy and physiology of antlers is desired to effectively manage the pronounced morphological variation present in this most managed of wildlife species.

Antler form or conformation is highly variable and depends on age, genetics, and nutrition, as discussed below. Points or tines are classified as typical (points generally symmetric and arising from the top of the main beam) and abnormal or nontypical (points arising from other than the top of the main beam or asymmetrical with the other side) (Figure 4.1). Abnormalities include kicker points projecting laterally from a tine or beam, drop tines projecting downward from a main beam, extra main beams, palmation of the main beam or between tines, and clustered tines (Figure 4.1b–f). The frequency of abnormal points increases with age. White-tailed deer antlers usually have more prominent brow tines and lack the bifurcated branching present in mule deer. However, abnormal points arising from the second tine, similar to the bifurcated branching of mule deer, are relatively common in mature whitetails, perhaps indicative of the close taxonomic relationship of the two species. A similar situation occurs in the other direction, in that Sitka black-tailed deer antlers resemble those of white-tailed deer (Geist, 1994).

The annual casting and growth of new antlers is an important part of the whitetail's mystique (Figure 4.2). Antlers grow rapidly from their pedicle (base) while in velvet during the spring and summer, 0.3 cm/day for yearlings and 0.6 cm/day for adults (Jacobson and Griffin, 1983). Growth rate slows dramatically during late summer as mineralization of the antler is completed. Restriction of blood supply around the autumn equinox causes the velvet's death; when dry enough, velvet is removed in as little as 24 hours by thrashing on vegetation.

Well after the breeding season, osteoclasts become active along the distal edge of the pedicle and inward over a two-week period (Goss et al., 1992). These osteoclasts de-mineralize the bone along an abscission line where the pedicle meets the antler; the weakened attachment combined with the weight of the antler causes it to drop off or be cast. Although described as an abscission "line," the surface of the detached antler base is rough in texture due to attenuated spicules of bone (Goss, 1983). The abscission line may be either flush with or extend beyond the burr or coronet (Figure 4.3). Brain abscesses, resulting from secondary infection following injury to the frontal bone or pedicle, may result in a jagged abscission line projecting deep into the pedicle and frontal bone (Figure 4.3d).

Once the antler is cast, the top of the pedicle can be considered an open wound. This area reacts like any wound, bleeding for a short period and developing a scab-like covering. Tissue from the outer edge of the pedicle migrates across the exposed surface and creates a covering called wound epithelium within as few as 10 days (Price et al., 2005; Kierdorf et al., 2009) (Figure 4.2). Beneath the wound epithelium are cells which will create the antler growth zones described in detail below. Antler growth resumes shortly after completion of the wound epithelium in white-tailed deer and most temperate cervids; in contrast, moose antlers do not begin growth until two to three months after antler casting.

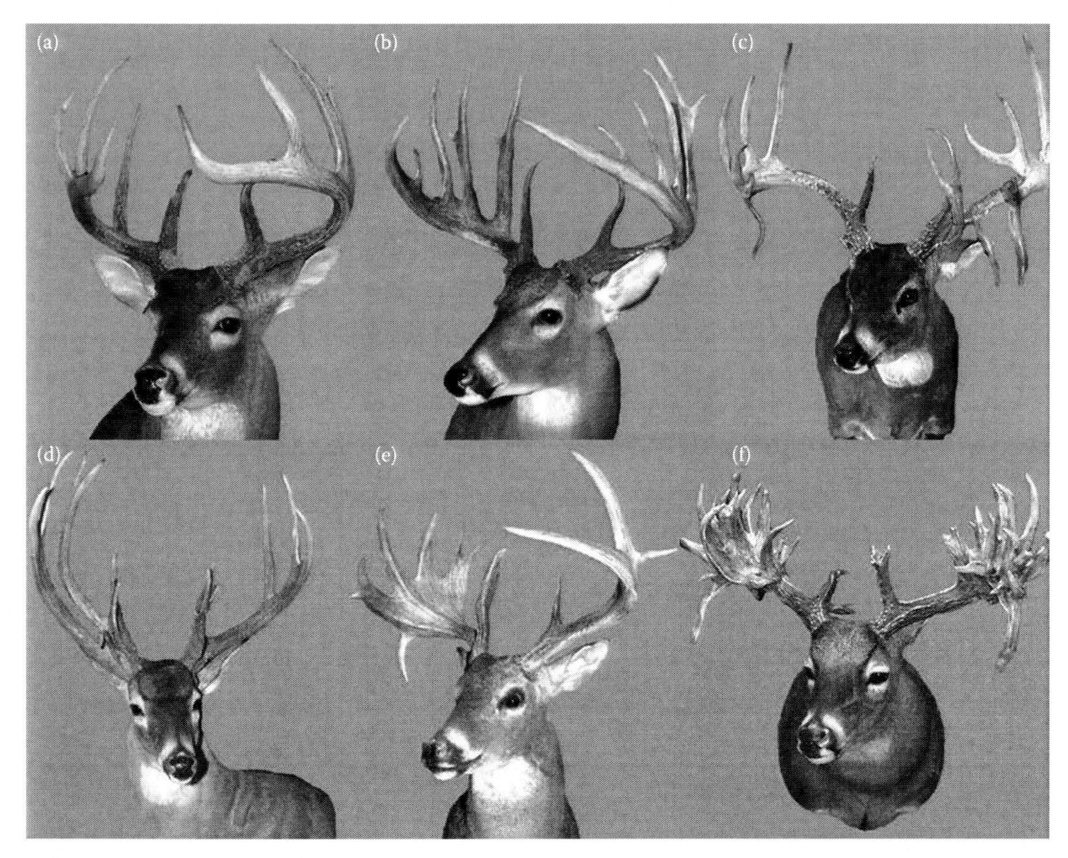

FIGURE 4.1 Examples of antler conformations with Boone and Crockett scores. (a) Typical: gross 173 7/8, net 172 0/8 typical, (b) typical with sticker points: gross 192 3/8, net 171 4/8 typical, (c) drop tines: gross 214 1/8, net 201 3/8 nontypical, (d) double main beam: gross 192 4/8, net 178 7/8 nontypical, (e) palmated: gross 190 4/8, net 175 5/8 nontypical, and (f) cluster points: gross 312 4/8, net 295 6/8 nontypical. All deer were harvested in Mississippi. (Photos by Magnolia Records Program, Mississippi Department of Wildlife, Fisheries and Parks and Mississippi Wildlife Federation. With permission.)

The hardened antler is often considered dead tissue, and certainly it is when cast. However, a well-developed vascular system has been described connecting the pedicle to the spongy core and compact edge of hardened fallow deer antlers as late as three weeks prior to antler casting. It is theorized that this vascular connection keeps the antler moist during the breeding season and thus more resistant to breakage during behavioral interactions (Rolf and Enderle, 1999).

One might think that a "bone" would have similar composition throughout its entirety and that one antler would be like another. However, there are many sources of variation in the composition of antlers. Stage of growth is related to this variation; ash composition of antler varied from 20% in the velvet growth stage to 65% in hardened antler (Ullrey, 1983; Landete-Castillejos et al., 2007a). Chemical composition differs among parts of the antler (Miller et al., 1985; McDonald et al., 2005) due to the differing mechanical roles played by each location and/or to the varying nutritional conditions at the time each portion was grown (Landete-Castillejos et al., 2007a,b). Similarly, rainfall during the period of antler growth was positively correlated with aluminum and manganese concentrations (McDonald et al., 2005). Mineral content of antlers also varies geographically, perhaps due to soil characteristics; such as between Texas and Georgia and among locations within Louisiana (Miller et al., 1985; Schultz et al., 1994; McDonald et al., 2005). Temporal variation existed among years in Texas, probably due to varying rainfall (McDonald et al., 2005). Lower dietary calcium reduced specific gravity of cast antlers (Ullrey et al., 1973). White-tailed deer antlers are composed of 22–24% calcium, 10–11% phosphorus, 0.4–0.5% magnesium, 0.8–0.9% sodium, 0.2–0.3% sulfur, 10–35 ppm aluminum, 390–500 ppm

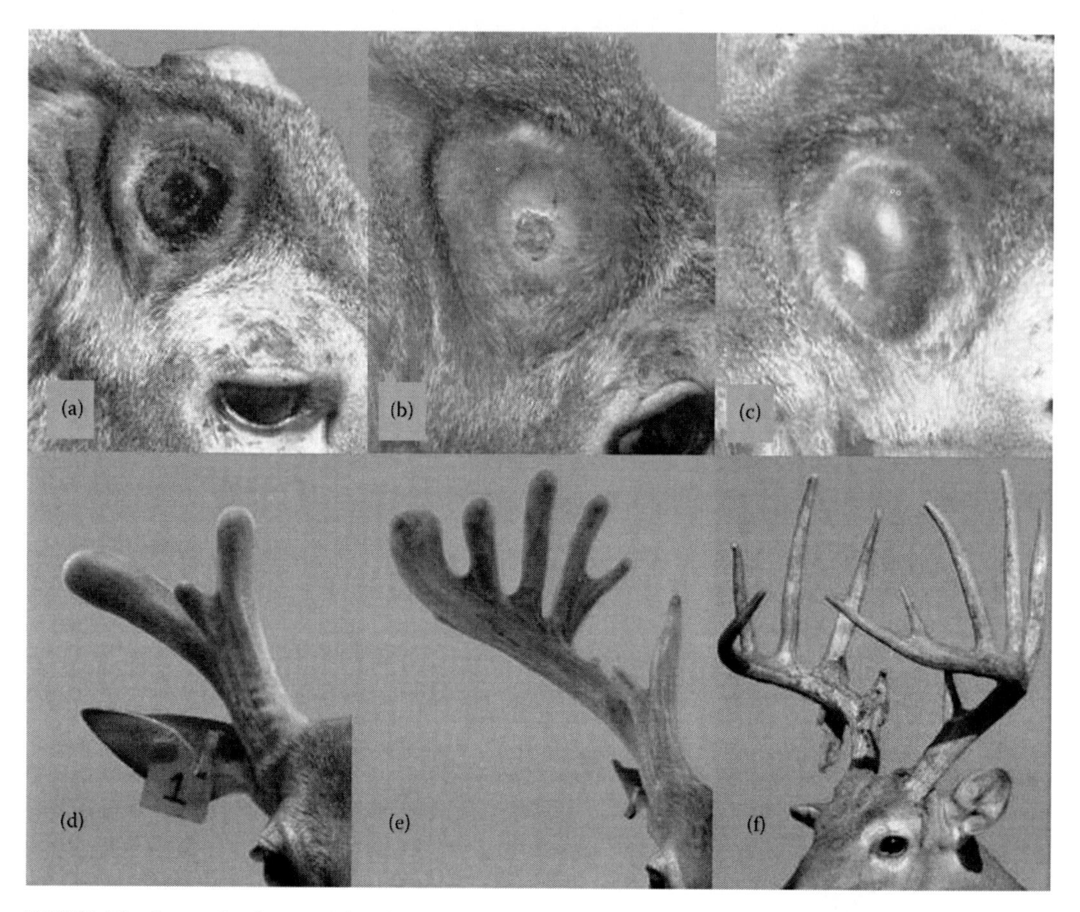

FIGURE 4.2 Stages of antler growth in a 6-year-old white-tailed deer. (a) One day after antler shed, (b) 15 days after shed, scab still attached, (c) 30 days after shed, scab is shed, (d) 90 days after shed, (e) 150 days after shed, antler growth will be completed within 10–20 days, with one additional month used to complete hardening and drying of velvet, and (f) hardened antler with shreds of dried velvet on a second animal. (Photos (a)–(e) by S. Demarais; photo (f) by D. Hewitt. With permission.)

potassium, 1.8–6.4 ppm copper, 28–37 ppm iron, 1–30 ppm manganese, and 58–71 ppm zinc (Schultz et al., 1994; McDonald et al., 2005).

Antlers differ profoundly with regard to composition and growth compared to horns produced by the Bovidae. Horns are not dropped and regenerated annually. There is a boney core within the horn, but the horn itself is composed primarily of keratin, a protein that is also the primary constituent of hair, nails, and skin. Growth originates on the surface of the boney core from a layer of epidermis that produces layers of keratinized tissue. New keratinized epidermis pushes the previous years' growth distally from that surface. Thus, variation in shape is generated from growth rates that vary spatially across the boney core (Goss, 1983). Horns are often said to "grow from the base," but it should be understood that the base refers to the entire boney core, not just the proximal base of the core. Horns expand in size annually from addition of new material and are not shed.

Anatomy of Antler Growth

The first step toward future antler growth is production of the pedicle, an outgrowth of the frontal bone. Pedicle location is established in the embryo as a thickened, fibrous vascular membrane called antlerogenic periosteum (Goss, 1983; Price et al., 2005). Antlerogenic periosteum is capable of autonomous differentiation; it produces an ectopic pedicle and antler even when grafted to other regions of the body (Kierdorf and Kierdorf, 2002).

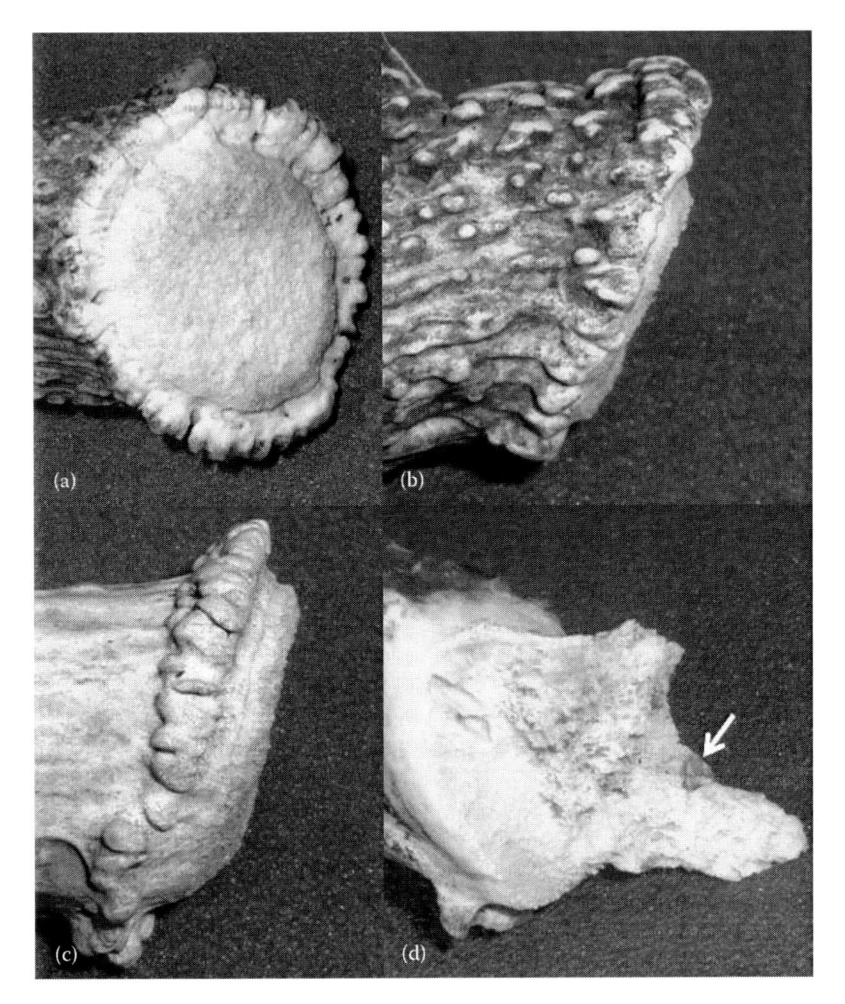

FIGURE 4.3 Abscission line at the antler–pedicle interface. (a) Showing the rough texture due to attenuated spicules of bone, (b) showing a line flush with the antler base, (c) showing a line that would have extended from the base into the pedicle, and (d) showing the likely effect of a brain abscess or bacterial infection (note dried puss at arrow). (Photos by S. Demarais. With permission.)

The pedicle gradually pushes up from its embryonic base on the frontal bone, becoming visible externally at four–five months of age and lengthening over the next two–three months (Figure 4.4). Production of the bone of the pedicle can be thought of as a four-step process including cellular proliferation from the antlerogenic periosteum, production of cartilaginous tissue, formation of osseocartilaginous tissue, and finally hardening into boney tissue (Li and Suttie, 1994). The final pedicle is composed of richly vascularized, spongy (cancellous) bone covered by the pedicle periosteum (Goss, 1983). The extensive vascular supply allows it to be a conduit for important nutrients and growth-regulating hormones into the developing antler.

The periosteum on the side of the pedicle remains an important contributor to future antler growth. From this material additional bone is laid down in subsequent years to broaden the pedicle base to support increasingly larger antlers (Figure 4.5). This material is also the source of antlerogenic periosteum cells which produce the regenerating antler. Early workers suggested that cells migrate from the dermis of the pedicle (Wislocki, 1943). More recently it has been theorized that a form of stem cell that can differentiate into chondrocytes (build cartilage) and osteoblasts (build bone) migrates from the distal pedicle periosteum to form the regenerating antlerogenic periosteum under the wound epithelium (Li et al., 2005; Kierdorf et al., 2007).

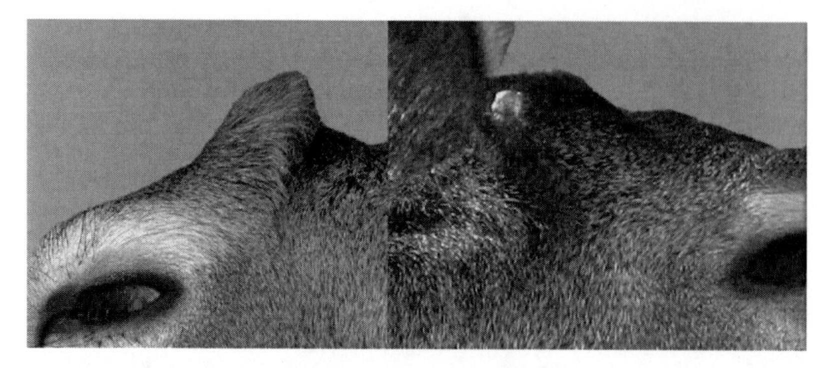

FIGURE 4.4 Pedicle development occurs on fawns at four to seven months of age (left). Some fawns reach sexual maturity during their first winter and produce a hardened, antler button (right). (Photo on right by W. McKinley; photo on left by S. Demarais. With permission.)

The change from pedicle growth into antler growth is subtle but noticeable. The covering of the growing pedicle resembles that of typical skin and fur. The covering of the growing antler takes on the appearance of shiny velvet because of sparser hairs growing straight out and oily secretions from sebaceous glands associated with each hair follicle (Goss, 1983).

The first readily seen antler grows during a buck's first full calendar year of life, but larger fawns may produce small, hardened "buttons" their first winter (Figure 4.4). Prevalence of antler buttons in enclosed populations with supplemental feed has varied from 20% in Mississippi to 84% in Michigan (Ozoga, 1988; Jacobson, 1995).

Beneath the velvet epidermis and dermis of the growing antler is a thick, fibrous protective membrane. This membrane is called perichondrium when it covers the cartilagenous growth stage and periosteum when it covers the boney stage of the growing antler (Figure 4.6).

Antler growth takes place at the distal ends of the main beams and tines (Figure 4.6) (Price et al., 2005). Beneath the velvet and perichondrium lies the mesenchyme or mesenchymal growth zone, an area of intense mitotic activity and thus, rapid cell generation and growth (Kierdorf et al., 2007, 2009). In the chondroprogenitor region these young cells begin to differentiate into chondrocytes and to form the columnar structure characteristic of cartilage and bone. As antler cartilage forms it takes on a characteristic that makes it unique among other vertebrate cartilage: an abundant blood supply. This blood

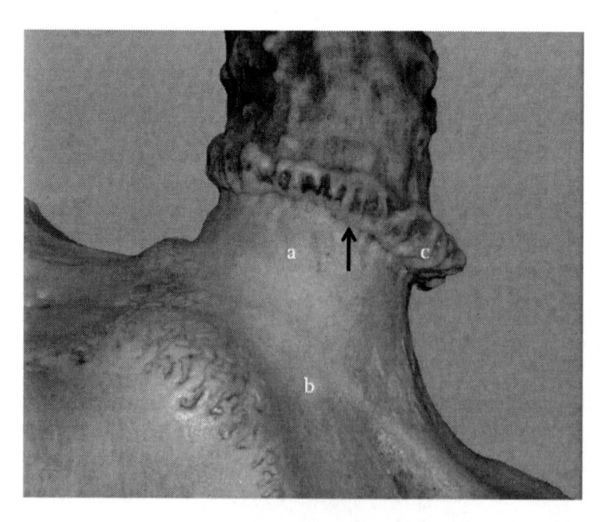

FIGURE 4.5 The pedicle (a) arises from the frontal bone (b) and meets the antler at the abscission line (arrow) proximal to the antler burr or coronet (c). Note the interlacing suture lines where the various skull bones join. (Photo by S. Demarais. With permission.)

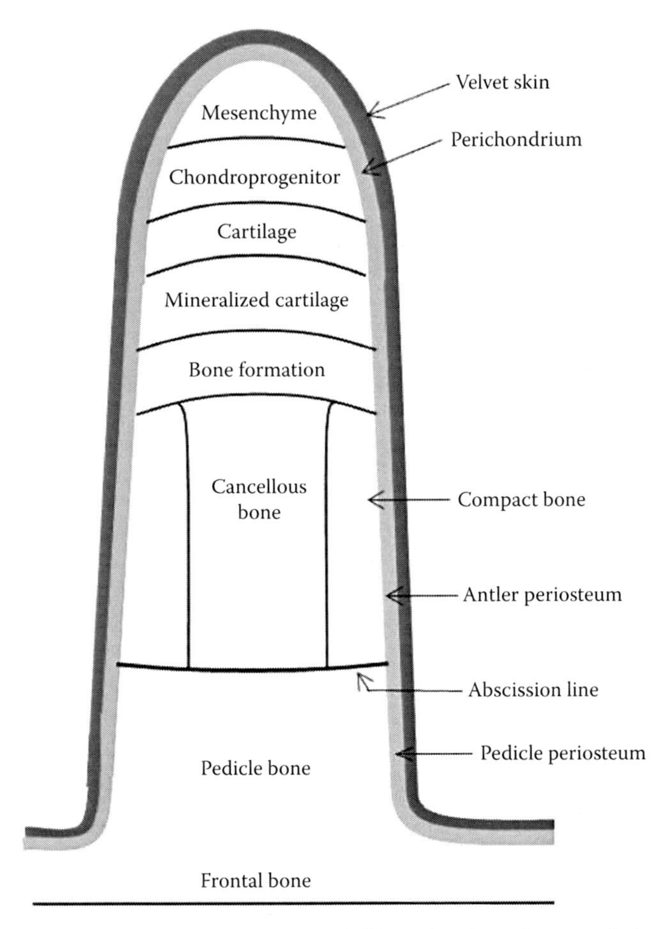

FIGURE 4.6 Longitudinal schematic of a growing antler tip illustrating the main anatomical regions. (Adapted from Price, J. S. et al. 2005. *Journal of Anatomy* 207:603–618.)

supply is needed to support the high metabolic demands of rapid tissue regeneration (Price et al., 2005). The enlarged and columnar chondrocytes then begin the process of mineralization. Once mineralized, chondroclasts resorb the cartilage, and bone is laid down on the remaining "scaffold" (Kierdorf et al., 2009). Osteoblasts continue the mineralization process until the mineralized cartilage is replaced completely with bone.

There are two types of bone within an antler, spongy bone and compact (cortical) bone. Spongy bone makes up the inner portion and is less dense, softer, and weaker. Spongy bone is highly vascularized during growth, which allows the transport of nutrients and growth-regulating hormones. Compact bone forms the outer shell of the antler, its greater density and stiffness providing support and strength needed for behavioral interactions. Spongy bone makes up 48–61% of the diameter of an average antler (McDonald et al., 2005). Although there is considerable variation among antlers, the relative amounts of spongy and compact bone are consistent throughout a given antler (Figure 4.7).

Physiological Regulation of Antler Growth

Understanding antler growth has long been of interest, exhibited as early as 350 BC in Aristotle's *The History of Animals*. More recently there have been abundant efforts to enumerate the physiological mechanisms that regulate antler growth. Perhaps not surprisingly, after more than 60 years of modern study, controversy still exists over some aspects of the physiological regulation of this remarkable process.

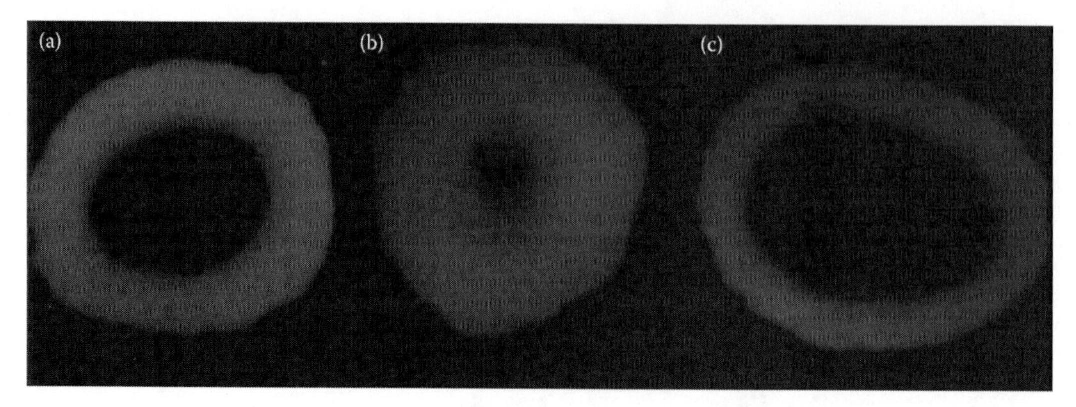

FIGURE 4.7 X-ray of cross sections at antler base showing the range of variation in relative amount of spongy (dark center) and compact bone. (a) Average composition, (b) minimal spongy bone, and (c) maximal spongy bone. (Photo by S. Demarais. With permission.)

Most research on physiological regulation of antler growth has been conducted on cervids other than white-tailed deer, especially the genera *Cervus* and *Dama*. Although most temperate cervids appear to have similar processes and regulatory mechanisms, there are variations in levels of sensitivity. For example, the antler cycle of white-tailed deer is very sensitive to testosterone level variation, and a significant decrease will cause a hardened antler to be cast within two weeks. In contrast, artificial reduction of testosterone in fallow deer and reindeer has little effect on the timing of antler casting (Bubenik et al., 2002). Such variations require caution when extrapolating results among cervid species.

Physiological regulation can be exercised in various ways. Antler growth can be initiated when concentrations of a "permissive" factor reach a minimal threshold or when concentrations of a "repressor" factor fall below a threshold (Price et al., 2005). Hormones produced in one part of the body may be carried in the blood to influence processes within another organ. Testosterone is a good example; produced in the testes and adrenal glands, this hormone stimulates the antlerogenic periosteum to produce the pedicle during a buck's first autumn (Goss, 1983). Additionally, there are localized hormones and growth factors synthesized in and released from endocrine cells that affect nearby cells. An example of such a locally acting chemical would be retinoic acids (derivatives of vitamin A), that may help regulate cellular differentiation in growing antlers (Allen et al., 2002; Price and Allen, 2004). Testosterone and retinoic acids have direct impacts on antler growth. There are other hormones with indirect effects, such as luteinizing hormone, which stimulates testosterone production by Leydig cells in the testes. Lastly, there are a myriad of hormones without direct effects but that support antler growth, such as thyroid hormones that maintain metabolic rates and calcitonin and parathyroid hormone which regulate calcium homeostasis (Bubenik, 1990).

Seasonal Timing

Antler growth must coincide with breeding season because antlers play a major role in reproductive success. Development of the pedicle and subsequent first set of antlers are internally programmed to occur in a defined sequence and interval (Goss, 1983). Thereafter, growth, velvet shedding, hard antlers, and casting take place on approximately an annual cycle. The specific timing of these events is entrained by variation in photoperiod, the same mechanism that entrains the female breeding cycle. The existence of an endogenous circannual rhythm was evidenced by a white-tailed deer born without eyes. Although he never had eyes to sense variation in photoperiod, he exhibited a 373–378-day annual cycle of antler growth as an adult (Jacobson and Waldhalm, 1992).

The regulatory effects of photoperiod on the antler cycle have been reviewed extensively by Goss (1983). When cervids are transported from the northern to the southern hemisphere, their antler cycle shifts by six months and is entrained to the new seasonal pattern. Deer also adapt to changes in frequency of the normal 12-month seasonal photoperiod cycle. Sika deer held under light cycles of six-month duration grew two sets of antlers within a year. When the light cycle equaled four months, deer grew three sets of antlers. Their limit of adaptability to shorter periods was approached when the light cycle equaled three months. Deer did not adjust to a two-month light cycle and maintained a normal 12-month cycle of growth and casting. Extension of the photoperiod cycle also shifted the antler growth cycle. Yearling sika exposed to an extended, 24-month cycle grew one set of antlers every two years. In contrast, adult sika that grew up with a 12-month cycle maintained that cycle when exposed to the 24-month cycle. Clearly, animals respond to photoperiod, which maintains their internal rhythm on a precise cycle that ensures a survival advantage where needed.

The extensive latitudinal distribution of white-tailed and mule deer provided Goss (1983) an opportunity to review how latitude affects variation in the timing of antler casting (Figure 4.8). Variation is greatly reduced in northern latitudes, likely due to the distinct adaptive pressure applied to individuals deviating from that which is most conducive to survival of offspring. Seasonal variation increases at southerly latitudes where seasonal extremes are less harsh, transitioning to year-round breeding at and below 15° North (Goss, 1983).

The pineal gland plays an important role in translating seasonal variation in photoperiod into physiological responses that control the antler cycle (Goss, 1983). The mechanism for this effect is likely variation in production of the hormone melatonin, which affects the hypothalamus and pituitary gland (Bubenik, 1990). Seasonal variation in melatonin production likely fine tunes the inherent internal cycle. Removal of the pineal from white-tailed deer delayed the beginning of antler growth, velvet shedding, and casting compared to controls; the authors concluded that the pineal gland does not originate the seasonal cycle but synchronizes it within individual deer (Snyder et al., 1983).

Growth

Wislocki (1943) proposed the existence of an antler growth stimulus, and attempts to identify it with certainty continue to this day. Possibilities have included growth hormone (Wislocki, 1943) and testosterone (Bubenik, 1983b). The leading candidate for the last 20 years has been insulin-like growth factor I (IGF-I) (Suttie et al., 1985; Price et al., 2005). But even IGF-I is receiving reevaluation, with a suggestion that we need to reconsider testosterone as the antler growth stimulus (Bartoš et al., 2009), and so the final answer may be awaited.

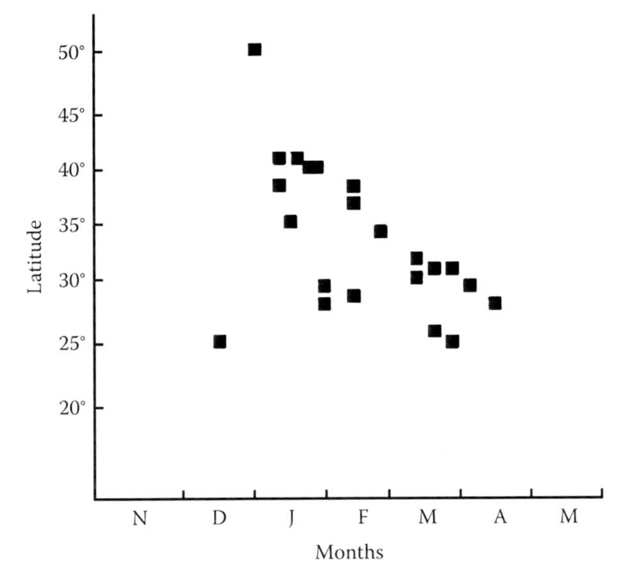

FIGURE 4.8 Scattergram of average dates when white-tailed deer and mule deer cast their antlers as a function of latitude. (Adapted from Goss, R. J. 1983. *Deer Antlers: Regeneration, Function, and Evolution*. New York, NY: Academic Press.)

Testosterone certainly plays a major role in the antler cycle. Short-term production of testosterone in fetal testes may cause temporary development of primordial pedicles, which disappear prior to birth (Lincoln, 1973). Pedicle growth by fawns has been linked to increasing testosterone levels their first autumn; and castration prior to pedicle growth precludes subsequent pedicle and antler growth (Bubenik, 1990; Li et al., 2003). Production of an antler button by some fawns their first winter may depend on adequate testosterone production (Brown et al., 1983). Although female deer have the potential to develop pedicles and antlers, their growth is limited by inadequate testosterone stimulation (Li et al., 2003). A minimal threshold level of male sex hormones may stimulate initial antler growth (Bubenik, 1990; Bartoš et al., 2000); this theory is controversial because testosterone has no effect on proliferation of mesenchymal or cartilage cells *in vitro* (Suttie and Fennessy, 1992; Price and Allen, 2004). Greatly increasing levels of testosterone are associated with completion of antler growth, final mineralization, and velvet shedding. Castration or chemical elimination of testosterone during antler growth precludes hardening and antler growth will continue. Antler casting is due to declining testosterone levels following the breeding season (Goss, 1983).

Although much emphasis is properly placed on the role of testosterone in timing of the antler cycle, injected estrogen has a more potent effect on growing antlers than testosterone (Goss, 1968). Injection of 17-β estradiol during early antler growth in red deer stopped growth and produce hardened antlers in 14 days (Price et al., 2002 cited in Price and Allen, 2004). Testosterone appears to be converted into estrogen locally within the antlers (Bubenik et al., 2005), so estrogen does not show up in normal blood profiles. High-affinity estrogen receptors occur in periosteum and calcified cartilage and within the perichondrium at the growing tips of antlers (Lewis and Barrell, 1994; Barrell et al., 1999). Blocking estrogen receptors during antler growth inhibited antler bone formation and delayed velvet shedding (Bubenik and Bubenik, 1978 cited in Bubenik et al., 2005). Estrogen may cause progenitor cells in the perichondrium to become osteoblasts instead of chondrocytes, thus causing bone formation instead of cartilage formation (Price and Allen, 2004).

Close correlation between blood levels of IGF-I and the antler growth period makes IGF-I a natural choice for the antler growth stimulus. Insulin-like growth factor I is the mechanism by which the more renowned growth hormone has its positive impact on deposition of protein and increased growth in most tissues, especially the skeletal frame (Guyton and Hall, 2000). The presence of IGF receptors in the growing antler tip and the *in vitro* stimulation of cells from this region by IGF-I supports the conclusion that this hormone is important to antler growth (Price et al., 1994, 2005; Sadighi et al., 1994).

Much of the historical antler growth research has focused on circulating hormone physiology, but recently molecular mechanisms at the local level of the antler have shown promise (Price and Allen, 2004). Parathyroid hormone-related peptide was first identified associated with hypercalcemia of various cancers (Suva et al., 1987). This peptide helps regulating chondrocyte cell differentiation and formation of osteoclasts in the growing antler tip, and so it likely affects both cartilage and bone formation (Price and Allen, 2004). Vitamin A (retinol) and its derivatives play important roles in embryonic skeletal development (Allen et al., 2002). Significant amounts of vitamin A are present in antler tissue at all stages of differentiation (Price and Allen, 2004). Retinoic acid, a derivative of vitamin A, accelerated early antler growth and likely plays a functional role in regulating cellular differentiation in growing antlers (Kierdorf and Bartoš, 1999; Allen et al., 2002).

Much has been learned about the physiological control of antler growth during the modern research era. However, there are many more secrets to unlock and more interrelationships to document. This desired knowledge is not just to fulfill mere academic interests. With every increasing effort to maximize antler production, there will be more interest in finding the missing detail that will allow managers to more completely put together the pieces of this fascinating puzzle. Much more will be learned before we have a full understanding of the regulation of the complex process of antler growth.

Why Antlers?

So why do male white-tailed deer have antlers? Antlers are costly for males to produce (Ullrey, 1983), so the cost of possessing antlers should be offset by increased fitness, but how? Antlers are described as a secondary sexual characteristic, which means antlers are structures not directly related to male sexual organs (penis, testicles, etc.), but found only on males (with rare exceptions). Researchers have speculated the purpose of antlers in cervids and the most common explanations reviewed by Clutton-Brock (1982) include (1) antlers are used for defense against predators, (2) antlers display dominance to other males, (3) antlers are ornaments used by females to assess the genetic quality of males, and (4) antlers are used as weapons for intraspecific combat. Other unsupported theories include being a seasonal mineral reservoir and being used for heat dissipation (Clutton-Brock, 1982; Goss, 1983).

Antlers as a Defense against Predators

Undoubtedly, antlers can inflict severe tissue damage or death to an attacking predator, but why would they be found on males only? If the primary adaptive function of antlers was protection from predators, then females should have evolved an antler-like structure to combat predators as well. Instead, the most common defense against predators is running away (Mech, 1984). It stands to reason that the reluctance of males to use their antlers except when cornered, as well as their absence in females, discounts this theory as the primary reason for the evolution of antlers in white-tailed deer (Clutton-Brock, 1982).

Antlers as a Display of Dominance to Other Males

Antlers may serve as a signal to convey general age-related dominance to other males. In this scenario, antlers may be an adaptive display structure to minimize energy expenditure, wounding, and potentially death from intraspecific combat for dominance (Geist, 1966). To augment the visual cue to competitors, Lincoln (1994) noted that large deer species will keep accumulated vegetation in their antlers and postulated that some tines in red deer, elk, and reindeer antlers do not aid in combat, and likely have only a visual function. Experimental reduction of antlers in high-ranking red deer males led to loss of social ranking (Bubenik, 1983a) and their harem; their social ranking was regained during following years with growth of new antlers (Lincoln, 1972, 1994). Bartoš and Hyánek (1983) reported that within age classes, antler size was associated with the most dominant red deer stags, but Clutton-Brock (1982) did not detect similar relationships in a red deer population after accounting for age and body size.

White-tailed deer behavior indicates antlers may have a role in male–male interactions for social status. Prior to and during the breeding season, males commonly use their antlers with threat posturing and sparring to establish dominance hierarchies (Marchinton and Hirth, 1984; Miller et al., 2003). Posturing alone may be enough to discourage the challenge of younger, smaller males. Because antler size is generally related to body size (Figure 4.9), it is difficult to disentangle the proximate (antlers) and ultimate (age, body size) visual cues to competitors. Currently, there is no evidence that relatively large antlers are associated with dominance in white-tailed deer; instead, strength and fighting ability are likely most correlated with age-specific dominance (Townsend and Bailey, 1981).

Antlers as a Signal of Male Quality

Antlers may also provide a signal of genetic or phenotypic quality to females. Bubenik (1983a) reported that red deer, elk, caribou, and moose females were attracted to males with larger antlers during behavioral experiments, but do antlers act as a signal of male quality to female white-tailed deer? Genetic heterozygosity has been positively associated with antler size of white-tailed deer (Smith et al., 1983; Scribner et al., 1989; Scribner and Smith, 1990; Webb, 2009) and antler development is related to genes that may influence pathogen resistance (Ditchkoff et al., 2001a). Thus, antlers may serve as a visual index of a male deer's quality as a potential mate. In red deer, antler size was positively related to sperm quality and production (Malo et al., 2005), providing a phenotypic signal of male fertility.

Antlers may also serve as a visual indicator to other deer of a male's current condition and vigor. Antlers are a good example of a condition-dependent secondary sexual trait (Andersson, 1986) that reflects the age, health, and nutritional status of the individual (Kodric-Brown and Brown, 1984; Kruuk

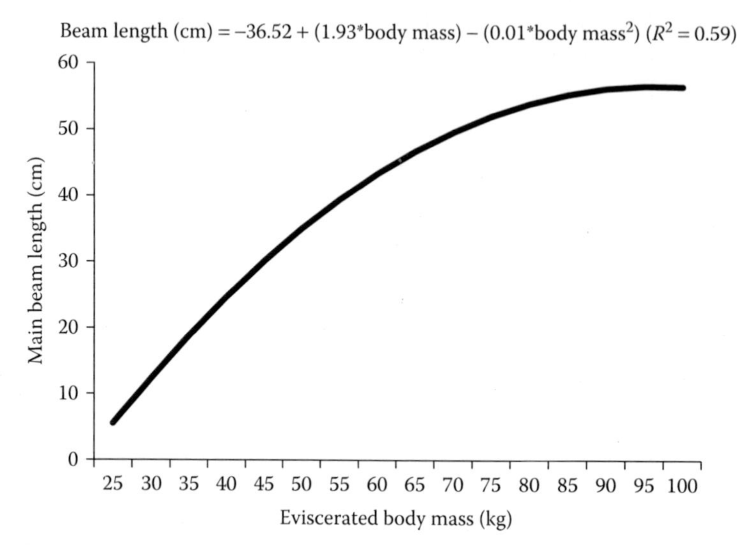

FIGURE 4.9 General relationship between antler size (main beam length) and body size (eviscerated body mass) of male white-tailed deer harvested in Mississippi from 1991 to 2007.

et al., 2002). Although genes control the limits of antler expression, an individual's phenotype (seen by females and other males) is governed by age and condition. Vanpé et al. (2007) demonstrated that antler size was related to phenotypic variation in roe deer and an honest indicator of male quality. Indeed, white-tailed deer antlers should be a reliable indicator of age because antler size generally covaries with body mass (Figures 4.9 and 4.10), but there are no studies demonstrating that deviations from this general antler–body allometric relationship influence dominance or reproductive success in white-tailed deer. That is, it is unknown if males with relatively larger antlers are afforded more breeding opportunities than those with relatively smaller antlers.

The role of antlers and their influence on lifetime reproductive success (i.e., fitness) may differ among cervids. In species like red deer and elk with a harem breeding system (Hirth, 2000; Wisdom and Cook, 2000), lifetime reproductive success is highly variable (Clutton-Brock, 1987) and intraspecific combat among males may yield far more reproductive dividends than it would for male white-tailed deer.

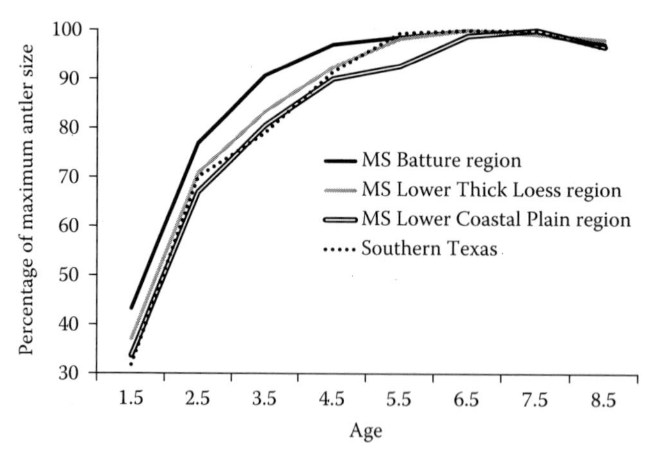

FIGURE 4.10 Age-specific growth rates for antler size (Boone and Crockett score) from four regions. Growth is expressed as a percentage of maximum size. Regions include Batture region of Mississippi (adjacent to Mississippi River), Lower Thick Loess region of Mississippi, Lower Coastal Plain of Mississippi, and southern Texas.

The tending-bond breeding behavior system that white-tailed deer exhibit (Hirth, 2000; Miller et al., 2003) may not result in significantly more breeding opportunities for victorious males. Undoubtedly, intraspecific combat occurs among male white-tailed deer (Marchinton and Hirth, 1984; Miller et al., 2003), but in these contests the victor may have gained courtship access to only a small number of females.

Antlers as a Weapon Used for Intraspecific Male–Male Combat

Antlers may have evolved in *Odocoileus* ancestors to counter head blows by an opposing male. Once ancestral deer reached sufficient size, using an antlerless head as a weapon by charging and ramming the opponent's flank could inflict severe damage to an opponent, as currently demonstrated by hummel red deer stags (Clutton-Brock, 1982). Horn-like head protuberances may have evolved to counter this behavior. Once antler-like structures appeared they facilitated the head-to-head battles witnessed today (Geist, 1966). Some ungulates developed structures for ramming (e.g., sheep), while most cervids developed structures to lock opponents together for pushing and wrestling (Geist, 1966). Prior to and during the breeding season, male white-tailed deer commonly spar or fight to establish dominance hierarchies (Bubenik, 1983a; Marchinton and Hirth, 1984; Barrette and Vandal, 1990; Miller et al., 2003). During combat for dominance, antlers are used as a structure to lock two males together allowing them to push and shove one another to determine the stronger of the two (Geist, 1966; Clutton-Brock, 1982; Lincoln, 1994; Figure 10.10). More often, one or more males will assume a subordinate role when in the company of another male, but if neither of the two males will subordinate to the other, the individuals will lock antlers and engage in an exhaustive contest of strength to determine dominance (Marchinton and Hirth, 1984; Miller et al., 2003). Through visual observation or sparring (Bubenik, 1983a; Barrette and Vandal, 1990), antlers may convey information regarding an individual's age and quality to other males without the risk of injury through combat.

Summary of "Why Antlers?"

On the basis of a review of many cervids, Clutton-Brock (1982) reported that evidence supports only the theory that antlers evolved primarily as weapons of aggression, but certain secondary functions may have co-evolved simultaneously. This view corresponds with the theory proposed by Berglund et al. (1996) regarding the dual utility of structures like antlers as "armaments and ornaments." That is, antlers may have evolved primarily as weapons used in intraspecific male–male combat, but secondarily antlers may also be a useful signal to females and thus sexually selected. Because antlers are physiologically costly to produce (Ullrey, 1983), once antlers reach sufficient size to use as an indicator of age and phenotypic quality, relatively larger antlers should be selected against. Alternatively, antlers may serve as an example of the handicap principle (Zahavi, 1975) for sexual selection where production of large antlers is an honest indicator of male quality because they are so costly to produce. Thus, males that can afford to produce relatively large antlers should be disproportionately selected by females.

Indeed, studies have demonstrated how antlers may be an indicator of male quality in white-tailed deer (Smith et al., 1983; Scribner et al., 1989; Scribner and Smith, 1990; Ditchkoff et al., 2001a; Webb, 2009), but currently there is no evidence that females select males with relatively large antlers. Due to the tending-bond mating strategy of white-tailed deer, female choice may be more passive by "choosing" the male that has availed himself through male–male competition and persistent courtship during estrus. Thus, we conclude antlers evolved primarily for intraspecific male–male combat in white-tailed deer, with a potential, yet undocumented secondary function as an honest signal of quality for female selection.

Why Are Antlers Replaced?

A physical structure like antlers appears to be beneficial and adaptive for white-tailed deer, but why have antlers and not horns? Several theories have been proposed to explain the annual antler regeneration and casting process, but is this process adaptive, or merely a physiological consequence of antlerogenesis?

Geist and Bromley (1978) proposed that antlers are cast following the breeding season so that predators cannot readily distinguish males that are likely in poor condition following the breeding season. However, Goss (1983) proposed this explanation is inconsistent with caribou, in which both sexes carry antlers and the shed-antlered male is easy to distinguish from females. Furthermore, if this signal to predators were negatively correlated to fitness of males we should see a similar process in bovids with horns.

Researchers have speculated the ability of antlers to be shed and renewed annually may be a response to antler breakage during combat (Goss, 1983; Prothero and Schock, 2002). In contrast, horns can be broken but will continue to grow and minimize the asymmetry caused by the breakage.

Another theory is based on allometric growth; that is, bony antler must be grown and shed annually to correspond with annual body growth (Goss, 1983). According to Bubenik (1983a), annual antlerogenesis is advantageous because the structure can convey the individual's annual status; thus, annual shedding of antlers is required for this process to track status.

Relationships between Antlers and Body Size

Among cervids, antler size generally increases with body size. Mature (five to seven years of age) males reared in the Mississippi State University research facility (n = 69) had 55 g of antler mass/kg body mass$^{0.75}$, or 3.8 g/kg body mass$^{1.35}$ (the exponent of 0.75 is used to represent metabolic mass and the exponent of 1.35 is used to make comparisons among smaller and larger bodied cervids; Geist and Bayer, 1988; Geist, 1998). On the antler size to body size ratio continuum among cervids, white-tailed deer have a ratio that is consistent with a species that has evolved in a forest environment, of moderate sexual dimorphism (Geist and Bayer, 1988), and a breeding group size of three to five individuals (Clutton-Brock et al., 1980).

Antler size is positively related to body mass (Figure 4.9), which increases with age (Scribner et al., 1989; Demarais et al., 2000; Miller et al., 2003). Maximum antler size is generally attained between five and seven years of age in free-ranging deer (Strickland and Demarais, 2000; Lewis, 2010). The rate of growth to attain maximum size may be influenced by diet quality, with males occupying poor habitats reaching maximum size one to two years later than males in areas of greater diet quality (Strickland and Demarais, 2000). Males from areas of greater available nutrition grow larger antlers on an absolute scale and also grow relatively larger antlers at younger ages (Figures 4.10 and 4.13).

A greater proportion of endogenous resources is allocated to antlers (and less to body mass) as a male ages. A measure of antler size (expressed as basal circumference multiplied by main beam length to approximate antler beam volume) can be compared to body size by dividing antler volume by eviscerated body mass, expressed on a metabolic scale (i.e., body mass$^{0.75}$; Figure 4.11). This measure of relative allocation of resources to antlers versus body shows how more resources are allocated to antlers as a male grows older. Similar relationships have been documented in moose (Stewart et al., 2000). At one year of age, only about 5 cm^3/kg body mass are allocated but this ratio increases to over 25 cm^3/kg body mass at six and seven years of age (Figure 4.11).

Within age classes the trend of greater allocation to antlers as body size increases holds; however, younger males do not allocate resources to antlers at the same rate relative to increasing body mass as older males (Figure 4.12). Younger males are likely allocating relatively more resources to general body development, like skeletal and muscle growth, before shifting proportionately more resources to antler growth.

Allocation of resources to antlers at the expense of body growth may be opportunistic when additional resources are available, and thus antlers may be a sensitive, plastic indicator of male physical condition. Variability associated with the ratio of antler size to body size is greatest in yearling males (Figure 4.11) and may by influenced by diet quality, birth date, or other maternal effects. The relative expression of antler size to body size also varies spatially. In Mississippi, relative antler development is greatest and less variable in regions of Mississippi with greater soil fertility and deer diet quality (Strickland and Demarais, 2000; Jones et al., 2008) and lesser and more variable in regions of lower diet quality

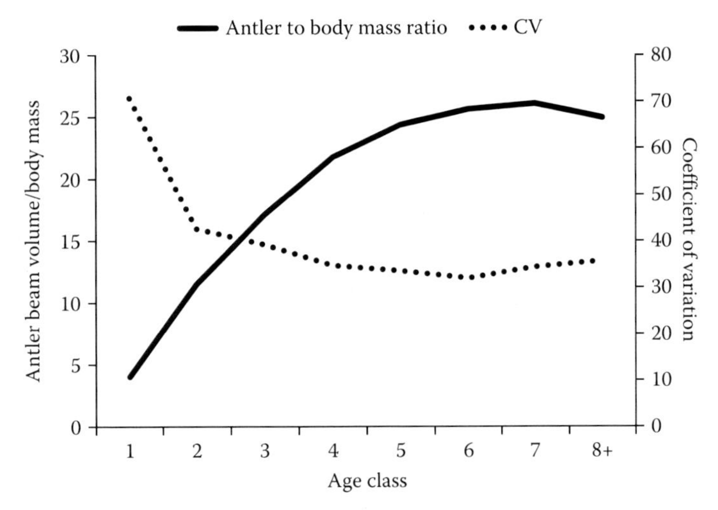

FIGURE 4.11 The ratio of antler volume relative to metabolic body mass (antler beam volume [cm^3]/metabolic body mass [$kg^{0.75}$]) increases with age. Antler size is expressed as an index of antler volume using main beam length accounting for basal circumference. Eviscerated body mass was converted into metabolic mass. Coefficient of variation (%) is the age-specific standard deviation divided by the mean.

(Figure 4.13). Thus, male white-tailed deer appear to capitalize on food resources and allocate a greater proportion to antlers when available, as demonstrated in red deer (Mysterud et al., 2005).

Sources of Variation in Antlers

Antler conformation and size in white-tailed deer are highly variable and depend generally on age of the animal, nutritional intake at all life stages, and genetic potential. Age and nutrition have direct, positive relationships with antler growth. Potential antler conformation and size are regulated by a deer's DNA, although debate continues over the relative heritability of specific antler features. Antler asymmetry and developmental abnormalities can confound the general effects of age, nutrition, and genetics on antler growth at any age and under any nutritional intake.

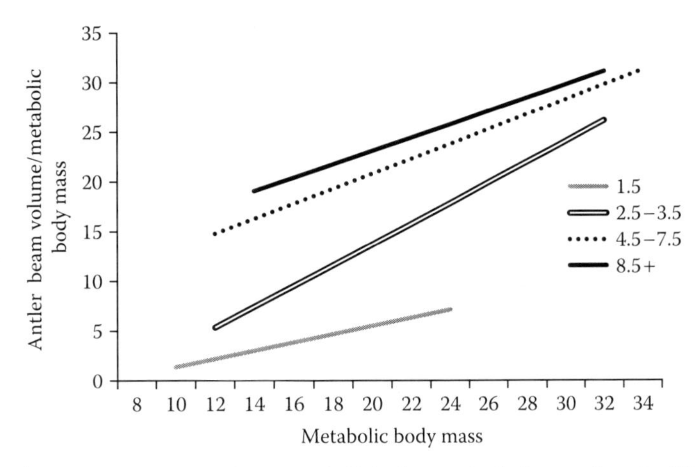

FIGURE 4.12 The ratio of antler volume relative to metabolic body mass (antler beam volume [cm^3]/metabolic body mass [$kg^{0.75}$]) increases with body size. Antler size is expressed as an index of antler volume using main beam length accounting for basal circumference. Eviscerated body mass was converted into metabolic mass. Older males allocate relatively more resources to antlers than do younger deer. Yearling males allocate the least amount of resources to antlers.

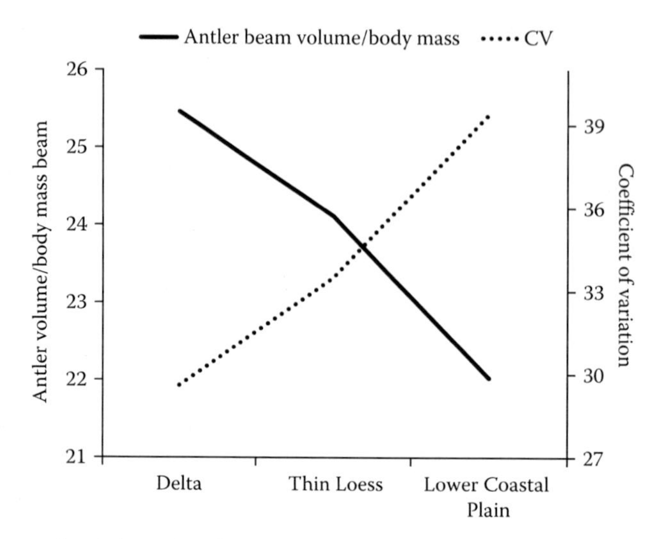

FIGURE 4.13 Differences in relative antler allocation among three soil regions of Mississippi with varying diet quality. Antler size is expressed as an index of antler volume using main beam length accounting for basal circumference, and eviscerated body mass was converted into metabolic mass (antler beam volume [cm^3]/metabolic body mass [$kg^{0.75}$]). Coefficient of variation (%) is the standard deviation divided by the mean. The Delta region has the greatest average diet quality, the Lower Thin Loess moderate quality, and Lower Coastal Plain the least diet quality. Allocation to antlers is more variable in the lower-quality soil regions.

Age

The first antlers are typically grown at one year of age, although 20–80% of buck fawns may develop hardened antler buttons at about eight months of age (Ozoga, 1988; Jacobson, 1995). This phenomenon is rarely seen by hunters because it usually happens after hunting season. In these cases, nutrition is adequate and birth dates are early enough to allow fawns to gain the critical body mass needed to reach sexual maturity and initiate antler growth.

Antler size increases annually until maximum antler development is reached at five to seven years of age (Figure 4.10). Based on averages of captive deer in Mississippi and wild deer in Texas, one-year-old bucks grew the equivalent of 28% of their ultimate maximum gross Boone and Crockett score (Jacobson, 1995; Lewis, 2010). The percentage increased to 62% at two years, 78% at three years, 92% at four years, and 99% at five years. Using antler mass as the measure of antler size with the captive deer revealed a similar story, but values were lower the first three years; only 10% of maximum at one year, 44% at two years, and 71% at three years. Bucks consuming less than optimal forage quality would be expected to take one or two years longer to reach their maximum antler size (Sauer, 1984; Strickland and Demarais, 2000).

Allowing bucks to grow older is a critical component of any deer management program designed to increase antler size. Just how old they need to get will depend on management goals. Of the three factors influencing antler characteristics, age is the simplest to manipulate and progress in this area is certainly most easily documented. Average antler size can be doubled merely by allowing bucks to age from one to two years. An additional 30% improvement can be obtained by letting deer age another year.

An individual's antler conformation or shape is normally fairly consistent throughout life (Figure 4.14). In some cases a peculiar yet consistent feature allows a unique wild deer to be identified each year with reasonable accuracy. However, variations such as an odd point or drop tine may alternate from one side to the other between years or be lost altogether. Prevalence of abnormal points typically increases with age. Bucks that live to an old age may produce smaller antlers after their prime and are more susceptible to annual environmental (nutritional) variation. The last antler set of bucks dying of old age may be quite abnormal (Goss, 1983) (Figure 4.15).

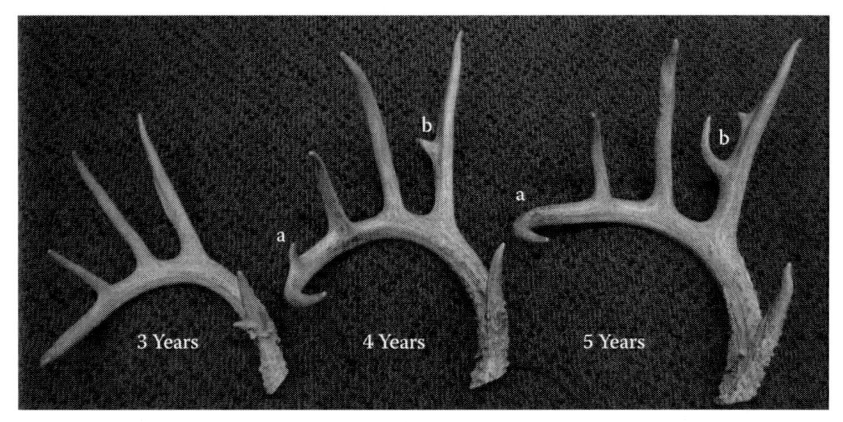

FIGURE 4.14 Antlers from the same deer at ages 3, 4, and 5 years showing consistency of general conformation, annual variation in presence of tines (a), and abnormal points increasing at older ages (b). (Photo by S. Demarais. With permission.)

In areas with extended breeding seasons, late birth dates can negatively affect antler development. The average birth date of forked-antlered yearlings was about one month earlier than spike-antlered yearlings in the wild in Alabama (Gray et al., 2002). In captivity in Mississippi, fawns born September/August were more often spikes and had fewer antler points at one year of age compared to fawns born in June (Jacobson, 1995). Although early- and late-born fawns had similar-sized antlers at maturity, late-born fawns were less able to express their genetic potential for antler development for one or more years. These types of effects would be expected only in more southern latitudes because there is less breeding date variation in northern latitudes.

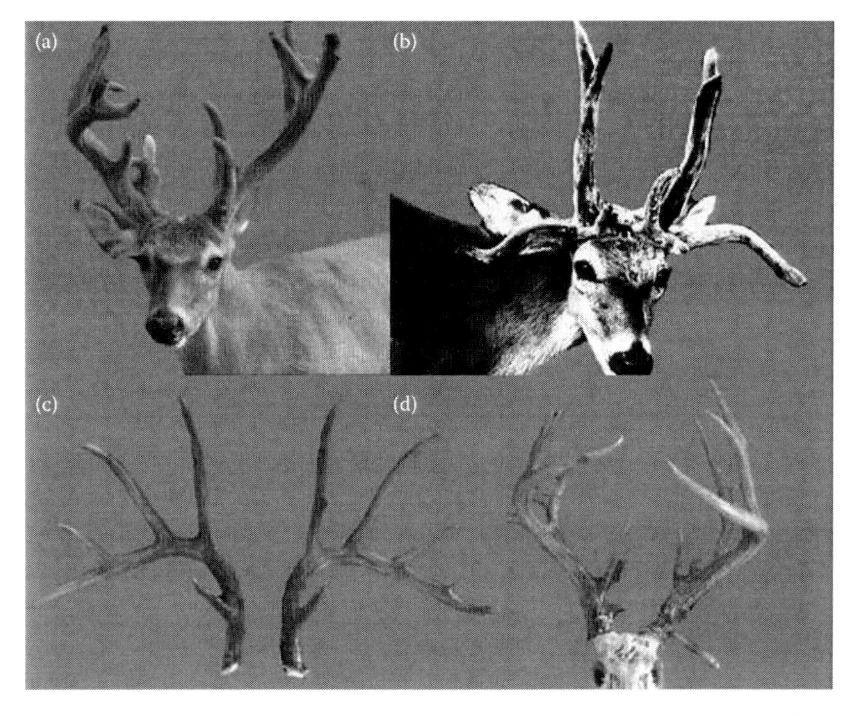

FIGURE 4.15 Some deer antlers exhibit annual variation in conformation, and a deer's last antler set may be quite abnormal. The same deer at 10 years (a) and at 14 years of age, just prior to his death (b). Another deer at maturity (c) and the year of his death (d). (Photo by S. Demarais. With permission.)

The accuracy of predicting antler characteristics of mature bucks based on yearling antler size has been debated. Number of points in yearling bucks has been reported to be a poor predictor and an excellent predictor of ultimate antler size (Jacobson, 1995; Ott et al., 1997). Increased variation from late-born fawns contributed to the former study's results. In a subsequent analysis of Jacobson's data, average Boone and Crockett score of yearling bucks was a reasonable predictor of antler size of the smallest and largest groups of yearling bucks, but was less accurate with "average" antler sizes. Accuracy increased greatly when using score at two years of age (Demarais, 1998, Figure 4.16). In a study of captive deer in Louisiana antler mass differed through three years of age when grouped as yearlings with spikes or with forks; a loss of significance at four years must be viewed cautiously as sample size decreased greatly at that point (Schultz and Johnson, 1992).

As with any sample from a population, there will be a mean value and the majority of values will be clustered within one or two standard deviations of the mean. There will also be animals that represent the full breadth of variation. It must be remembered that any generalization is based on the "measure of central tendency" or the mean values, and there always will be exceptions to the "rule" that represent variation in the population.

The first data on predictability of antler size using known-age animals from wild populations were reported recently (Koerth and Kroll, 2008). However, several design and analysis issues reduced the scope and validity of their conclusion that size of a buck's yearling antlers was a poor predictor of antler size at maturity (Demarais and Strickland, 2010). Their major problems were that their statistical test did not coincide with their specific conclusion and their methods were susceptible to measurement bias.

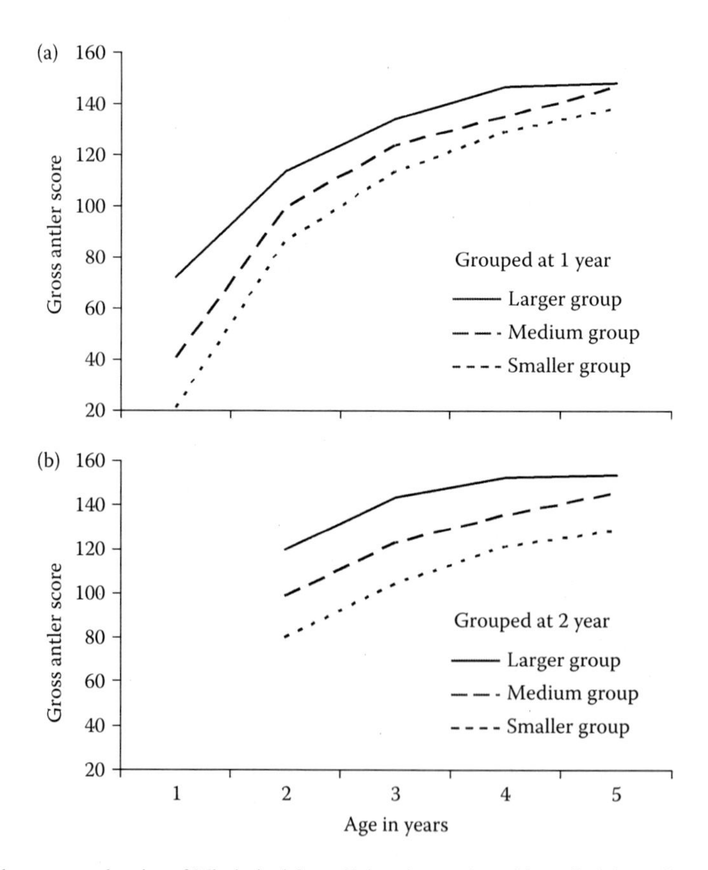

FIGURE 4.16 Subsequent antler size of Mississippi State University captive white-tailed deer when grouped at one year (a) or two years (b) of age into three groups based on relative antler size (gross Boone and Crockett score). On average, yearling antler size is a reasonable predictor of antler size at maturity, but selection would be more effectively applied to two-year-old males. (Adapted from Demarais, S. 1998. In *The Role of Genetics in White-Tailed Deer Management* (2nd edition), eds. K. A. Cearley and D. Rollins, 33–36. College Station, TX: Texas A&M University.)

Additionally, their results were applicable only to populations with similar culling and management programs.

Nutrition

Antler development is greatly affected by nutritional intake prior to and during growth. The annual casting and growth of new antlers allows the animal to respond to annual variation in resource availability. Although physiological mechanisms are rarely addressed in nutritional studies, one possible pathway involves the rate of healing after antler casting. The rate of production of the "wound epithelium" after antler casting might be slowed if nutrient deficiencies decrease healing rates. Another mechanism could be a direct impact on growth rates mediated via the supportive role of growth hormone and its relationship with IGF-I. These and other potential mechanisms should be evaluated in future research.

Adequate nutritional intake is required for males to fulfill their genetic potential for growth at all ages. The relationship between antler size and nutrition has long been known, but modern research to quantifying the impact of specific nutritional factors on antler development began in the 1950s. We now know that a number of nutritional components interact to generate the boney matrix of antlers, most importantly protein, energy, and minerals. A variety of experimental approaches have been employed to unravel the nutrition–antler mystery, but most experiments compare antler characteristics between a group fed "optimally" and one or more groups fed "sub-optimally."

Early studies on whitetails showed that fawns fed 16% protein from weaning until 1.5 years of age grew antlers much larger than fawns fed 4.5% or 9.5% protein (French et al., 1956). Antlers from the low-protein groups were barely measureable spikes, while the high-protein group grew antlers averaging 36 cm (14 in.) main beams and six points. This effect could have been due to retarded development of the pedicle and/or slower growth of the first antler, as both can be negatively impacted by the inadequate nutrition. Red deer calves fed 30% less food during November–May than an unrestricted group had delayed pedicle development and grew fewer antler points at one year of age (Suttie and Kay, 1983). In Michigan, whitetail fawns fed a diet simulating an early green-up with access to acorns grew twice as many antler points at one year of age as fawns fed a diet simulating late green-up (Ullrey, 1983). This relationship between diet quality and a buck's first set of antlers is important in management decisions as well as in understanding the nutrition and birth date interaction to be discussed in the section on "Genetics."

The level of protein in forage required for maximum antler development varies with age. Two-year-old whitetails fed a diet with 16% protein grew antlers almost twice as heavy as bucks fed 8% protein (Harmel et al., 1989). Adult whitetails fulfilled requirements for antler development with as little as 10% protein (Asleson et al., 1996). However, nutritional requirements for body growth generally take precedence over requirements for antler growth (French et al., 1956) and younger animals generally require greater levels of protein. For example, weaned fawns require up to 20% protein for optimum growth, which would include requirements for pedicle development (Ullrey et al., 1967). Therefore, nutritional management for optimum antler production at a variety of ages should strive for an average intake of at least 16% protein from spring through autumn.

Energy requirements are just as important as protein to antler development. Dietary energy restrictions decreased beam diameter and length and the number of points in one-year-old white-tailed deer (Ullrey, 1983). Habitat managers do not usually target specific energy levels in forages because energy requirements are typically met when protein content is adequate.

Little is known about the mineral requirements for antler growth in white-tailed deer, although we can assume by their presence in antlers that some quantity is required. Research has emphasized the macroelements, such as calcium and phosphorus.

Antler growth in white-tailed deer requires a net daily accumulation of 0.5–1.0 g calcium (Ullrey, 1983). Early researchers concluded that a diet of 0.7% calcium was required for antler growth in deer up to three years of age (French et al., 1956). The proportion of cortical bone in antlers increased when dietary calcium increased from 0.18% to 0.40%, and the conclusion was drawn that about 0.45% calcium would meet requirements for antler development (Ullrey, 1983).

Antler growth in white-tailed deer requires a net daily accumulation of 0.25–0.5 g phosphorus (Ullrey, 1983). Early work indicated that a diet of 0.6% phosphorus was required for antler growth up to three years of age (French et al., 1956). More quantitative work on adults (two years or older) reported a phosphorus requirement of 0.14% annually, varying seasonally from 0.16% during spring to 0.11% during summer (Grasman and Hellgren, 1993).

Much of the immediate demand for calcium and phosphorus during antler growth is met by mobilization from other skeletal sites (Bubenik, 1983b). This mobilization takes place even when deer have access to highly concentrated dietary minerals (Ullrey, 1983). A 23% reduction in rib bone mineral content during peak antler growth of mule deer was reduced to 3% shortly after growth was completed (Banks and Davis, 1966 cited Ullrey, 1983). Presumably, bone reserves used during antler growth are subsequently replaced by dietary intake (Ullrey, 1983).

Although our knowledge is minimal for macroelements, we know even less about the specific requirements for trace minerals, such as manganese, copper, and zinc. Selenium is present throughout the velvet antler, especially at the growing tip (Suttie and Fennessy, 1992).

Mineral requirements are met by the environment in most areas, but specific minerals may be limited. For example, phosphorus is generally deficient in the sandy soils of the southeastern coastal plain (Perry et al., 2008, p. 324). Deer can make up for low mineral content of their diet by eating soil, creating "deer licks." A relatively inexpensive homemade mineral supplement can be created from supplies available at most county cooperatives using two parts dicalcium phosphate and one part mineralized salt.

Our knowledge of vitamin requirements for antler growth in white-tailed deer is even less than for mineral requirements. Vitamin A deficiency during winter was reported in white-tailed deer (Youatt et al., 1976). There is likely some minimal requirement for vitamin A and its derivative, retinoic acid, because of their importance in cellular differentiation in the growing antler and their presence there in all stages of growth (Price and Allen, 2004). Treatment of the growing pedicle with retinoic acid increased the size of first antlers in fallow deer (Kierdorf and Bartoš, 1999). Vitamin D concentration was elevated during mineralization of white-tailed deer antlers, which is likely related to its involvement in calcium metabolism (Van Der Eems et al., 1988).

Genetics

Control over the potential for antler development is contained within an animal's genetic material. Antler size, conformation or shape, and number of points certainly vary within bounds controlled by an animal's DNA. Beyond these accepted generalizations lie significant areas for differences of interpretation and varying opinions over how best to apply knowledge of genetics to the management of white-tailed deer.

A fawn is the product of the combination of DNA from sire and dam. There is no reason to think that either sex makes a greater genetic contribution to antler size. However, the dam provides an important added contribution based on several maternal factors. Her inherent ability to support fetal growth and to produce adequate quality and quantity of milk contribute directly to offspring growth rates. Additionally, her ability to obtain adequate habitat indirectly affects her offspring's well-being. These nutrition-related factors acting during his first six to eight months of a fawn's life can confound expression of even the best genetic potential.

Nutritional factors influencing the initiation of pedicle growth during his first winter and antler growth the next spring and summer affect the expression of a young buck's genetic potential for antler development. Subsequent environmental variation and animal health affect expression of genetic potential at older ages. These varied causes of phenotypic variation make it difficult to evaluate genetic potential of individual animals. Evaluation of genetic potential is required to fully integrate genetic manipulation into management programs.

The degree of heritability of antler traits is another often debated antler topic because the only two attempts to study the topic generated different results. A Texas study reported relatively high heritabilities at one year of age for number of antler points (0.22–0.56), main beam length (0.47–0.70), basal circumference (0.80–0.89), and antler mass (0.71–0.86) and concluded that substantial genetic change could be expected from individual selection at the yearling age class if realistic selection differentials were used

(Williams et al., 1994). A Mississippi study reported heritabilities at the same age that were low (0.00–0.13) for number of antler points, main beam length, and antler mass and moderate for basal circumference (0.25) and noted a significant maternal influence. The authors concluded that genetic selection of yearling males could not alter genetic quality of a deer population (Lukefahr and Jacobson, 1998).

The problem with the heritability debate is that it mostly focuses on which study was "wrong." This approach is inappropriate for two reasons. First, there is no reason to believe that either of the studies is "wrong." Their respective results are most "correct" and therefore most applicable to deer populations similar to the studied populations, especially regarding reproductive patterns. Second, expression of genetic potential (i.e., phenotype) is affected by many external factors. One of these external factors is the difference in fawning date effects between the respective research pens. The Texas deer had an earlier fawning season and fawning date did not affect yearling antler size (Harmel et al., 1998). The Mississippi deer had later fawning dates that affected yearling antler development (Jacobson, 1995). This birth-date effect likely limited the clear expression of genetic potential of one-year-old deer and would have contributed to the significant "maternal influence" reported by Lukefahr and Jacobson (1998). Heritability was greater in males aged two years and older, indicating they had at least partially compensated for birth-date effects. The disparity between these two studies also may be related to the genetic composition or degree of genetic variability in the penned populations. For comparison, Kruuk et al. (2002) reported a heritability estimate of 0.33 for antler mass in free-ranging red deer, so heritability values for free-ranging white-tailed deer likely are somewhere between the greater estimates reported by Williams et al. (1994) and the lesser estimates reported by Lukefahr and Jacobson (1998). Finally, it is not realistic to expect the same results from two studies estimating a genetic component when the phenotypic expression is highly susceptible to external influences. Additional studies are needed to estimate heritability and to more completely quantify how genetic potential for antler development is affected by external factors.

Spatial Variation

Spatial scale is one of those ecological terms that southerners like to say is "clear as mud." It can have a host of potential contexts; from habitat selection within a home range to the extreme of latitudinal variation across the range of a species. This section will describe sources of antler variation at the landscape scale, at regional scales, and at the somewhat overlapping levels of subspecies and latitude. One must keep in mind that the basis for all these sources of variation lay within the previous two topics: variation related to nutrition and genetics.

The landscape context describes variation associated with a variety of factors, such as habitat types, stage of succession, and land use and management decisions; research on these topics abounds. Habitat quality at the landscape scale has been shown to influence several population-level phenotypic parameters in mule deer, roe deer, and moose (Sæther et al., 1996; Kie et al., 2002; Mysterud et al., 2002), but work is limited on antler development in white-tailed deer. A habitat suitability index was positively related to deer abundance and negatively related to antler size in Arkansas (Miranda and Porter, 2003), possibly reflecting the interaction of nutrition and density on expression of genetic potential. Animal density can influence relationships between habitat quality and phenotype, and so deer density relative to carrying capacity should be considered when accounting for variation in antler size. Antler size variation across the landscape was explained by the composition of land-use types in Mississippi; those that promoted growth of early succession communities with abundant herbaceous forage positively influenced deer habitat quality (i.e., nutritional intake) and antler size in forested landscapes (Strickland and Demarais, 2008).

Large-scale regional soil and land-use characteristics influence white-tailed deer population-level phenotype, with the cause likely related to variation in nutritional intake. Yearling beam diameters differed similarly among regions in West Virginia and New York, with smallest values being 76–77% of the largest values (Severinghaus et al., 1950; Gill, 1956). Antler size differed among soil resource regions of Mississippi, with antlers from the lowest-quality region only 79% as large as those from the highest-quality region (Strickland and Demarais, 2000). Growth rate varied among regions; deer in lower-quality soil resource regions may require one to two additional years for their antlers to reach maximum size.

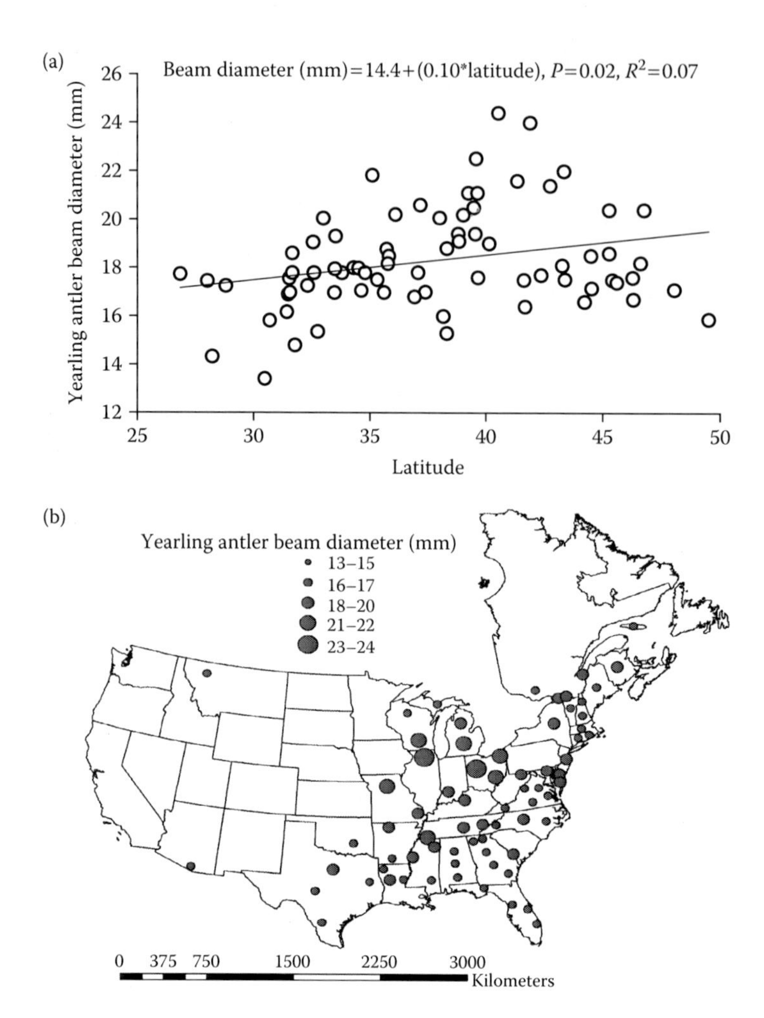

FIGURE 4.17 Regression comparing latitude and mean population antler beam diameter of yearling males ($n = 74$) harvested throughout North America (a), and locations of populations used for the analysis (b). Although a general trend of increasing antler size with increasing latitude exists, soil fertility and land use likely has a greater influence on population antler size.

Looking at the spatial distribution of Boone and Crockett record book bucks from North America might convince you that there are regions where deer grow larger antlers. There may be some ecological basis for this conclusion. There was a positive relationship with latitude that explained 22% of variation in basal diameter (ages not specified) at 21 locations from 26 to 46° N (Smith et al., 1983). In contrast, our comparison of average yearling antler beam diameter and latitude at 74 locations explained only 7% of the variation (Figure 4.17). There is no reason to expect antler size alone to be larger at more northern latitudes, but antler size is generally associated with body size (Figure 4.9). Relative body size of white-tailed deer increases with increasing latitude, with the exception that island deer are generally smaller than continental deer at the same latitude (Geist, 1998). Subspecies may have adaptations that are habitat based and independent of latitude. For example, subspecies occupying savannas of Venezuela and Texas have classic open-landscape adaptations, including larger, more complex antlers (Geist, 1998).

Antler Asymmetry

Smith et al. (1983) reported that antler asymmetry decreased with age in males from a South Carolina population. Similar relationships are found in comparisons of main beam lengths of bucks in Mississippi

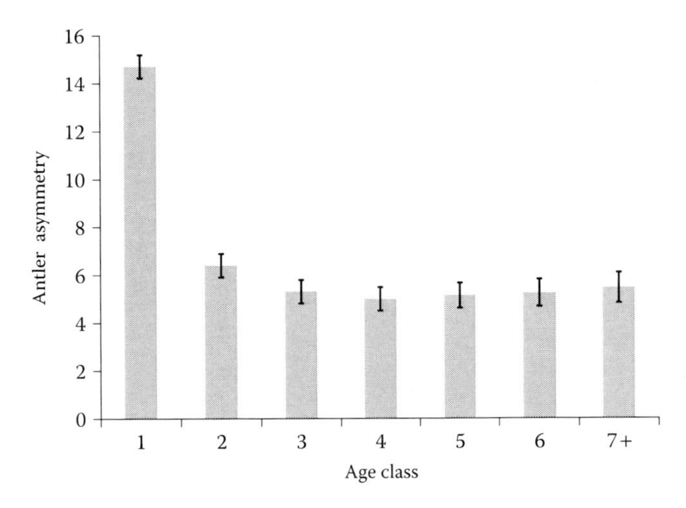

FIGURE 4.18 Antler asymmetry (% difference) of main beam length of hunter-harvested deer in Mississippi from 1991 to 2007. Error bars represent standard error.

(Figure 4.18), with the greatest asymmetry occurring in the yearling age class. Ditchkoff et al. (2001b) reported a general decline in antler asymmetry with increasing age in an Oklahoma population as well. They further documented that asymmetry declined with increasing antler size within all age classes except in males greater than five years old. However, upon further experimentation using a three-dimensional analysis, Ditchkoff and deFreese (2010) found no relationship between antler asymmetry and increasing age or antler size. The authors suggest that fluctuating asymmetry of white-tailed deer antlers may not be a reliable signal of male quality.

Developmental Abnormalities

A wide range of developmental oddities have been described for white-tailed deer antlers. A specific cause is hard to assign to an antler abnormality in wild deer, because there are so many potentially confounding and overlapping circumstances of which we know little. However, there are three general causes of antler abnormalities in "normal" males: injury, systemic conditions, and genetics (Goss, 1983). Various other antler abnormalities can be related to abnormal males and females.

The velvet antler tip is the single most important tissue influencing antler size and shape (Suttie and Fennessy, 1992). Although a buck is particularly protective of his growing antlers, the softness of the velvet tip makes it susceptible to accidental injury and thus damage-induced irregularities. A fractured beam or tine can heal and continue growing in an abnormal shape or direction (Figure 4.19).

Because the pedicle is the initial source of the regenerating antler, injury to this base can increase prevalence of injury-related abnormal antlers (Goss, 1983). Pedicle injuries can cause abnormal conformation, size, and asymmetry of an antler (Figure 4.20). Additionally, injuries to the pedicle and even the adjacent frontal bone have caused a range of variations, including accessory or extra antlers (Goss, 1983). Widespread management to increase age distribution has produced older-aged populations. Older bucks tend to fight more aggressively and have had more opportunity to damage their pedicles from intraspecific fights or accidents. Brain abscess in 20% of bucks three years and older examined in the Southeast (Davidson, 2006) and 35% of adult bucks in Maryland (Karns et al., 2009) indicates a significant rate of skull and pedicle damage. Therefore, prevalence of asymmetric antlers due to pedicle injury should increase when management promotes an older age structure.

Contralateral asymmetry is an injury-related antler abnormality where there is retarded development of the opposite antler following a significant injury to a hind leg. For example, a rear leg amputation resulted in a significantly reduced opposite antler for six subsequent years (Davis, 1983). There is no obvious explanation, but one possible mechanism is a modified blood supply due to an altered gait by the crippled deer (Goss, 1983). Most deer biologists have personal knowledge of excellent examples of contralateral

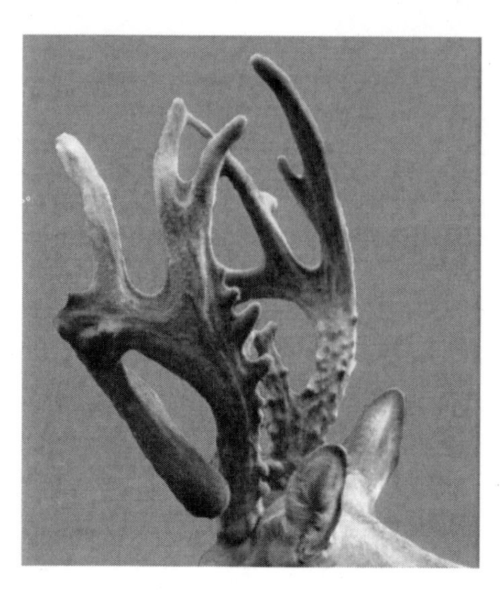

FIGURE 4.19 Example of a broken beam that maintained its blood supply, survived and hardened into an abnormal conformation. (Photo by B. Murphy and M. Hellickson. With permission.)

asymmetry. However, there are also cases where significant leg injuries have not caused developmental problems, and so the cause and effect are neither fully understood nor universally accepted.

Antlers still in velvet have been reported well beyond the seasonal norm, and may be maintained year-round. Antlers remain in velvet if testosterone levels do not reach the high levels needed to cause hardening and velvet shedding. Insufficient testosterone production usually is due to loss of or damage to testicles or to a birth defect where the testicles do not develop. Although they lack the normal pattern of hardening and casting, velvet growth is still limited to the normal period of March to August. Remaining in velvet throughout the fall and winter increases the likelihood of physical damage and such antlers are often thick, but short and blunted (Figure 4.21).

Antlered adult females have been reported throughout the whitetail's range; in most cases, antlers are short spikes still in velvet. Prevalence data are varied, including one in 4437 "bucks" that were actually antlered females harvested accidentally in Pennsylvania (Doutt and Donaldson, 1961 cited by Goss, 1983), one in 900 adult females in Michigan (Ryel, 1963), and one in 65 adult females in Alberta (Wishart, 1985). Hesselton and Hesselton (1982) generalized a frequency of one in every 1000–1100 adult females. Surprisingly, reports of antlered females have not increased now that either-sex harvest is ubiquitous.

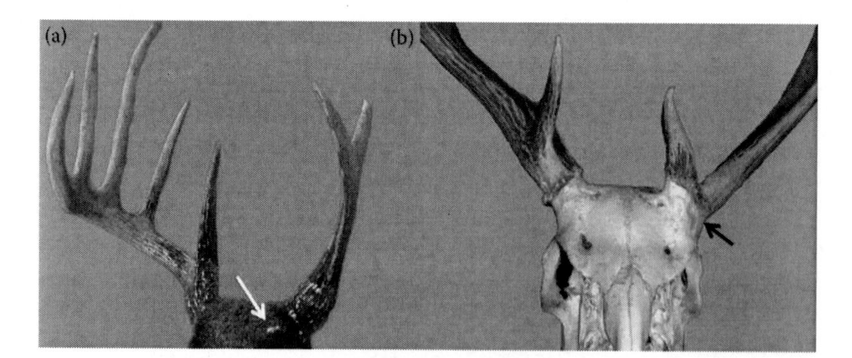

FIGURE 4.20 Injuries to the pedicle and frontal bone may cause abnormal conformation and asymmetry. A presumptive injury caused (a) asymmetry and growth of accessory antler (arrow) and (b) asymmetry similar to the one on the left, lack of a burr and malformed pedicle indicative of recovery from trauma (arrow; note the difference between the left and the right pedicle, the burr, and the shape of the frontal bone above the eye orbit). (Photo by S. Demarais. With permission.)

FIGURE 4.21 Antlers may remain in velvet permanently if testosterone production is insufficient to promote hardening and velvet shedding. (Photo by S. Demarais. With permission.)

Female cervids possess a latent capacity to grow antlers (Goss, 1983), and so some occurrence of antlered females is expected. Pedicles have developed in female deer after testosterone administration, and antlers later grew if the pedicle tip was injured (Goss, 1983). The case of freemartins (heifers partially masculinized due to sharing of testosterone in the uterus with their male twin) has been proposed as one possible explanation for antlered females, but it may not apply for cervids (Goss, 1983).

Adult "females" with hardened antlers have been reported, although less frequently than those with velvet antlers. These animals typically are males possessing female external genitalia. A male pseudohermaphrodite white-tailed deer harvested in November from South Carolina was described as having the external appearance of an antlered doe. The six-point antlers were in velvet; the vulva, clitoris, vagina, and cervix were the only female organs present and the testes were located in the body cavity (Scanlon et al., 1975). A similar animal occurred in Mississippi, with the exception that the antlers were hardened and the testicles were external to the abdominal cavity and subcutaneous to the nonfunctional teats (Figure 4.22).

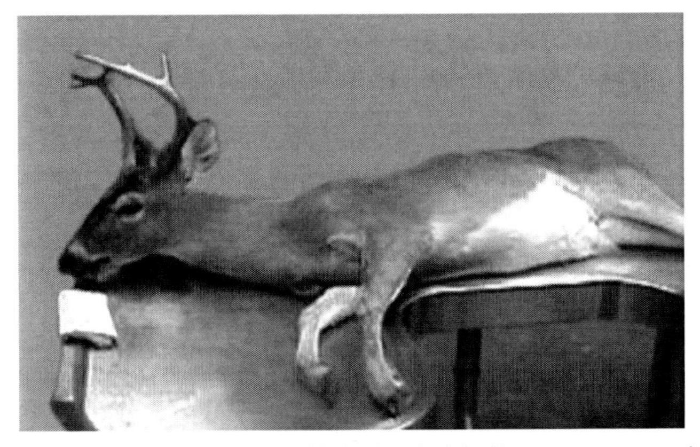

FIGURE 4.22 An "adult female white-tailed deer with hardened antlers" on necropsy proved to be a male pseudohermaphrodite. This animal possessed external female genitalia but no uterus or ovaries. The testes were located between the abdominal cavity and the teats. (Photo by S. Demarais. With permission.)

Antlers as a Tool of Management

The attractiveness of antlers as a target of management makes them an obvious choice as a parameter to monitor throughout the course of a management program. To be a valid part of a monitoring program samples must adequately represent the population; an adequate number of unbiased samples is critical. Antlers can be used as a "condition index" to measure herd health relative to habitat carrying capacity because they are responsive to nutritional intake. Condition indices are monitored over time as an indication of proximity to management goals. Monitoring of condition indices using an adequate number of unbiased samples is essential to establishment of a valid deer management program.

Random sampling is a basic requirement for extrapolating sampling results to a population. Complete randomness is nearly impossible outside the realm of lotteries, but efforts must be made to minimize potential biases to the composition of a sample. Where biases are inevitable, they must be documented and results extrapolated with appropriate qualifications. Harvest biases due to hunter selection and antler restrictions are common in deer management programs.

A sample is biased when hunters do not harvest the first available animal (in the strictest sense, sample bias is likely even when the first available animal is harvested due to behavioral factors). For example, harvest data could not be used to estimate the average antler size of two-year-old males within a population prior to and after instituting antler-based harvest criteria that differentially protect some two-year-old males. Although not ideal, the best alternative in this case would be to remove animals from the precriteria sample that did not meet the new harvest criteria and compare the resultant data sets. Another potential problem exists with evaluating antler size of animals within an age class that are harvested for different reasons. For example, management for antler improvement typically requires harvest of animals from both sides of the antler size frequency distribution (Figure 4.23). When analyzing the average antler size of harvested mature males, you would not group animals harvested as "trophies" with animals harvested as "cull" and "management" bucks because each of these categories represent different targets of management.

Adequate sample size is a second important requirement for a monitoring program. Natural variation within a population dictates adequate sample size for population-level inference. Sample size required for valid inference varies spatially and by variable and should be evaluated separately for each management program. In general, an increase in sample size will increase precision and accuracy, which leads to better management decisions. Simulated subsampling of antler beam length data exemplifies how the number of individuals sampled influences precision and accuracy of the average (Figure 4.24). With a sample size up to 30 individuals on this particular property, a 5-cm annual variation in average beam length could be due to the sample composition as easily as it could be due to management actions.

Antler growth responds annually to nutritional variation and so size within age class can be used to evaluate nutritional intake prior to and during the antler growth period. Yearling antlers are susceptible to nutritional impacts over a wider time period, because the added time for pedicle development also is susceptible to nutritional variation.

Monitoring population health relative to carrying capacity is important for sound deer management. In a monitoring program average values (i.e., phenotypic expression) are compared with some benchmark value. One obvious benchmark is the age-specific genetic potential. Not all management goals require fulfilling genetic potential; however, the information generated from such a comparison can be applied to any management context. Initial management goals should include steady improvement in antler size indices over time. If the goal is to maximize antler size within age classes, then the *de facto* requirement is to provide nutritional resources that allow each animal to fully express their genetic potential. If genetic potential is not known, then relative change can still be used as an index.

The effectiveness of antler-based selective harvest programs should be monitored using antler characteristics. The quantity and quality of various antler characteristics should respond to management actions. Proper sampling is the only way to determine the effectiveness of management actions. We present several examples that show how antler characteristics can be used as a management tool to predict and/or monitor response to management.

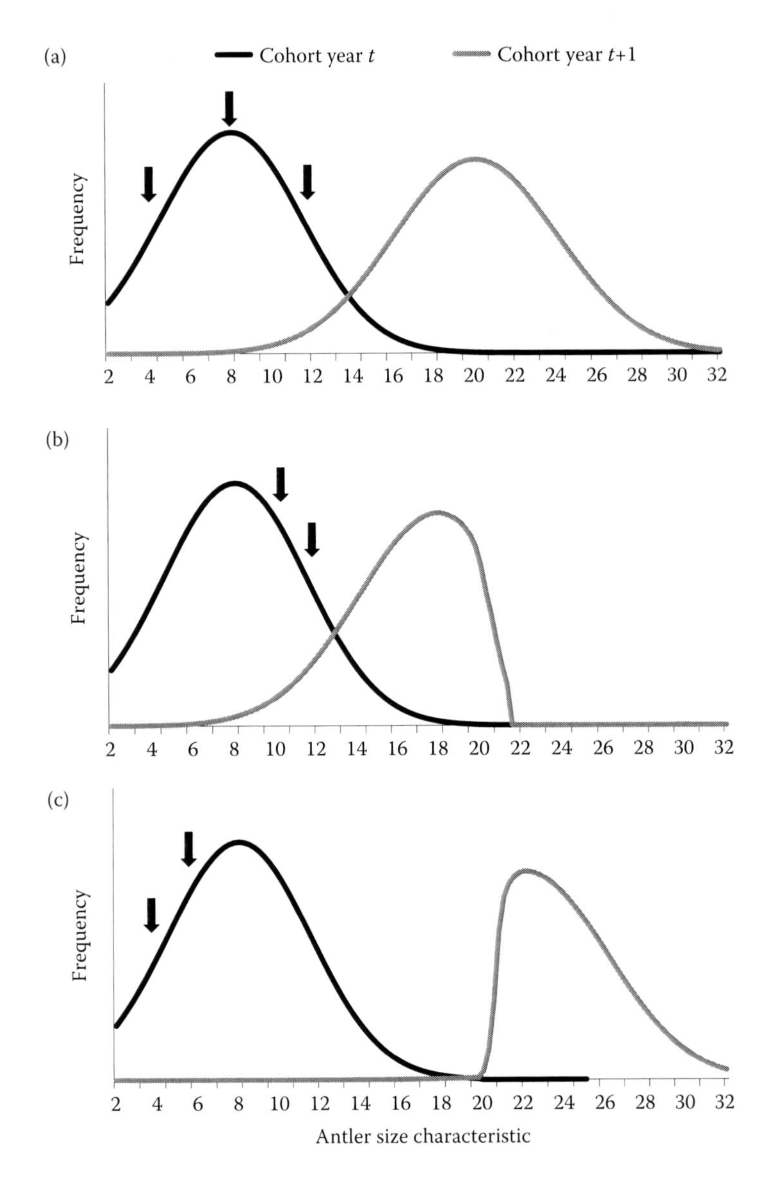

FIGURE 4.23 Theoretical effects of selective harvest on cohort antler characteristics assuming that antler characteristics of individuals are positively correlated between ages and the characteristic would increase annually due to increased age. (a) Uniform selection applied to an antler characteristic of the younger age class results in no change in shape of the distribution in the following year. (b) Selective harvest of males with the greatest antler development in the younger age class results in a negatively skewed distribution with a lesser mean. (c) Selective harvest of males with the least antler development in the younger age class results in a positively skewed distribution with a greater mean. Arrows indicate the portion of distribution that was targeted for selective removal in year *t*.

Strickland et al. (2001) used antler characteristics from 220 captive bucks to predict the effects of various antler-based selective harvest criteria on average antler size of the remaining cohort. They concluded that harvest restrictions that protected smaller-antlered, young males and permitted harvest of larger-antlered, young males reduced mean cohort antler size in subsequent years if the harvest rate of vulnerable males was high. Subsequently, they quantified Mississippi's four-total-point restriction effect on antler size of harvested deer on 22 public areas encompassing 240,000 ha. Antler size declined in at least one year class after the restriction across the range of soil resource regions; gross Boone and Crockett scores decreased 13–23 cm (5–9 in.) at two years of age and 25–43 cm (10–17 in.)

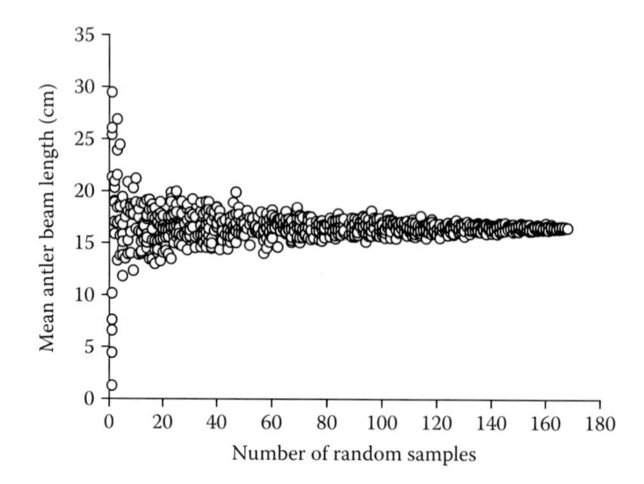

FIGURE 4.24 Natural variation requires adequate sample size to properly estimate how a variable responds to management activities. Sample size affects the precision and accuracy of antler size estimation as indicated from randomly subsampling from a "population" of 168 yearling bucks harvested on the Noxubee National Wildlife Refuge, Mississippi, in 1991. Even with a sample size of up to 30 individuals, the subsample means varied by about 5 cm.

at three years of age. The studied populations had high harvest rates (intense selection) and relatively stable interannual environment. The authors concluded that antler restrictions should be considered a short-term solution to age-structure problems because of the potential negative biological effects (Demarais et al., 2005).

Ashley et al. (1998) used antler measurements to monitor the effectiveness of a severe population reduction on average animal condition and expression of genetic potential. Deer from an island in Lake Erie, Ontario, Canada had some of the smallest antlers on record for the subspecies (*O. virginianus borealis*) and some believed that these animals were genetically programmed to be smaller than adjacent mainland deer. They reduced the estimated population by an amazing 85% using public hunts over a five-year period. Antler beam diameter increased by 93% in yearlings and 35% in two-year-olds in response to the significant population reduction. They concluded that deer in this population were not genetically smaller, but their growth had been restricted by environmental (i.e., nutritional) conditions.

Selective Harvest

Selective harvest involves the differential protection or harvest of a particular type of animal. In its most basic form, it may include selective preference toward females to improve an unbalanced sex ratio. More advanced management programs may incorporate selective harvest of males based on phenotypic expression of antler characteristics. The goal of such programs would be to increase prevalence of males with greater growth potential or decrease prevalence of males with lesser growth potential. Such efforts can involve selective harvest decisions under two scenarios: management of the "standing crop" and management of the "genetic composition" of a population. Standing crop involves managing the current deer population with the intent to improve the age structure or antler characteristics of the current population. Managing genetic composition is a much more complex process involving the management of current deer with the intent of altering genetic characteristics in a future deer population.

Selective Harvest to Improve Age Structure

Overall antler size and the morphological components therein (e.g., points, main beam length, etc.) are related to age (McCullough, 1982; Smith et al., 1983; Strickland and Demarais, 2000; Monteith et al., 2009). Thus, antlers can be used as a general indicator of age class, with a certain degree of error. In Mississippi, antler beam length and overall antler size (as indexed by Boone and Crockett score) has the greatest correlation with age while number of antler points has the weakest correlation (Table 4.1).

TABLE 4.1

Pearson's Correlation Coefficients (r) between Age and Antler Characteristics from Male White-tailed Deer Harvested in Mississippi from 1991 to 2007 ($n > 120{,}000$; $P < 0.01$)

	Circumferences[a]	Length[b]	Spread[c]	Points[d]	Score[e]
Age	0.66	0.71	0.66	0.60	0.72
Circumferences		0.84	0.79	0.75	0.88
Length			0.91	0.82	0.98
Spread				0.79	0.95
Points					0.86

[a] Basal circumference of antler beam taken approximately 2.54 cm above coronet.
[b] Length of main antler beam.
[c] Widest spread inside the main beams.
[d] Total number of antler points.
[e] Predicted Boone and Crockett score based on equations developed by Strickland and Demarais (2000).

Population age structure can be improved using a harvest restriction based on antler points. In Mississippi, a four-total-points restriction improved age structure of harvested bucks on public lands within six years (Strickland et al., 2001; Demarais et al., 2005).

Although antler characteristics are related to age, frequency distributions of these characteristics overlap among age classes; any particular measure may be possessed by a significant proportion of both younger and older age classes (Figure 4.25a–c). This overlap in the frequency distributions of antler characteristics by age class makes the use of antlers an inefficient tool for managing male age structure. Because younger and older males may possess similar antler measures, intense selection may disproportionately remove higher-quality young males and protect lower-quality older males, subsequently decreasing mean cohort antler size, as seen after five years of a statewide antler point restriction in Mississippi (Strickland et al., 2001; Demarais et al., 2005).

Selective Harvest to Improve Cohorts

Management of cohorts, or "standing crop," of a population can provide both positive and negative effects, depending on the approach to selective harvest (Strickland et al., 2001). Cohorts can be manipulated to improve antler development of surviving bucks if there is an excess of bucks within a population and you have the luxury and ability to selectively remove "inferior" animals. Removing these animals leaves more forage resources for bucks that have greater potential to grow larger antlers. Standing crop can and should be managed at any age when there is a need for control of overpopulation; failure to do so will perpetuate the existence of below-average mature males within the foraging and breeding environments.

Intensive cohort management may involve selective removal of younger animals. This approach is effective only if you can predict future antler size based on current antler development. The reliability of such prediction has been discussed previously. Although researchers disagree about the correlation between yearling and mature antlers and its management applications (Koerth and Kroll, 2008; Demarais and Strickland, 2010), most agree that the predictive relationship between a male's current set of antlers and his ultimate antler size at maturity increases from two to four years of age.

The distribution of antler characteristics (e.g., total antler points, beam length, inside spread, and an index of gross Boone and Crockett score) closely follows a bell-shaped curve, or normal distribution (Figure 4.25a–c). For a characteristic to be distributed normally, the most frequent observations occur in the middle of the distribution (average or median) and relatively fewer observations occur on the sides, or tails, of the distribution. This means there are very few individuals in a population that exhibit extremely large or very small values for a desired phenotypic characteristic. Extensive data collection over time will

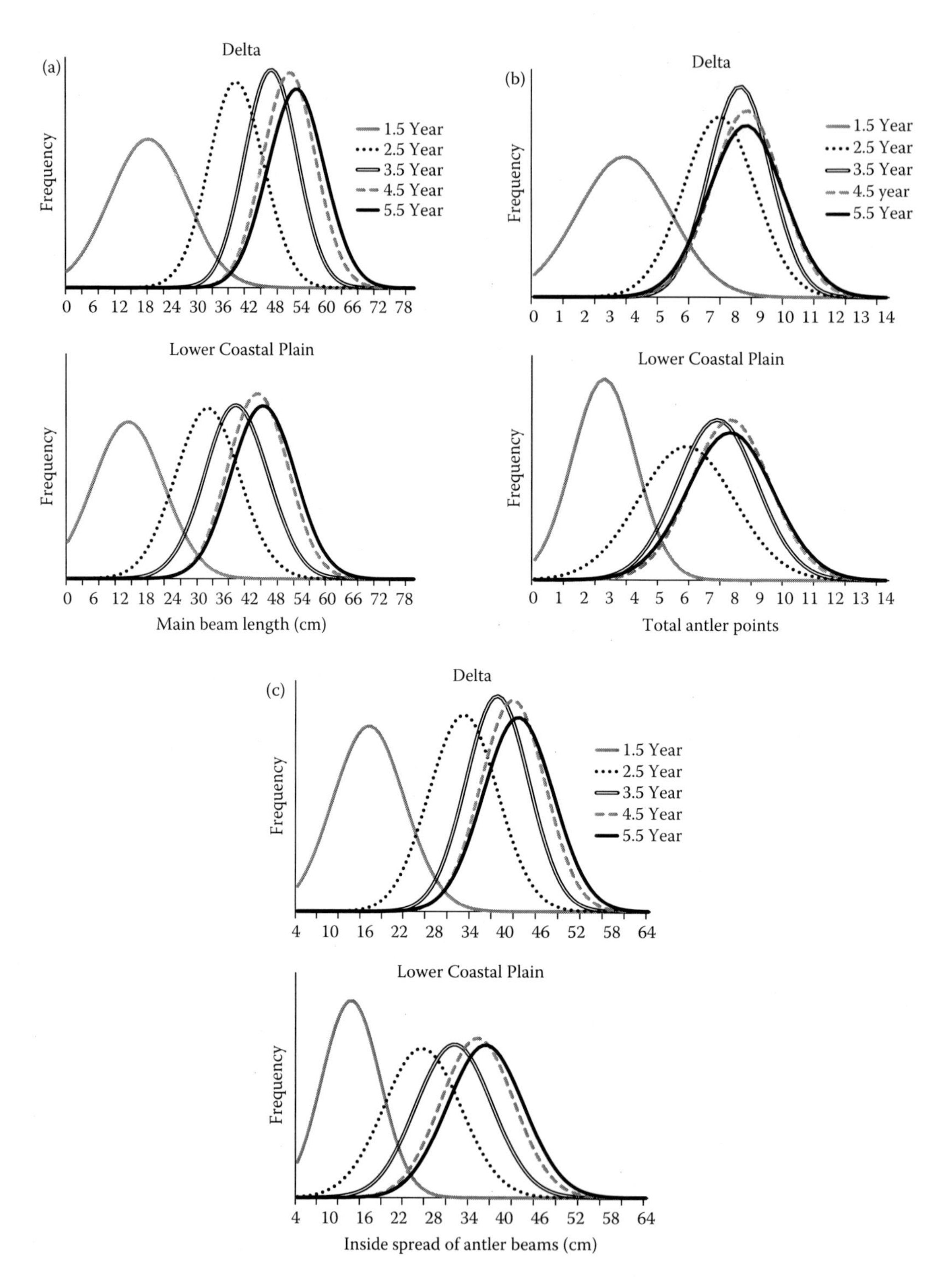

FIGURE 4.25 (a) Age-class-specific distributions of antler main beam length from male white-tailed deer harvested in Mississippi from 1991 to 2007. The Delta region is associated with greater deer diet quality than in the Lower Coastal Plain region. (b) Age-class-specific distributions of total antler points from male white-tailed deer harvested in Mississippi from 1991 to 2007. The Delta region is associated with greater deer diet quality than in the Lower Coastal Plain region. (c) Age-class-specific distributions of the inside spread of antler beams from male white-tailed deer harvested in Mississippi from 1991 to 2007. The Delta region is associated with greater deer diet quality than in the Lower Coastal Plain region.

yield site- and age-specific distributions of antler characteristics. These distributions can be used as the basis for formulating selective-harvest strategies to shape the distribution of antler characteristics of cohorts at maturity. How cohort phenotype is influenced by selective harvest is simply quantitative—the extent of directional selective pressure (Ridley, 1996) determines the extent of phenotypic change in the cohort.

Selective harvest to change cohort phenotype involves targeting harvest efforts at the portion of the distribution most related to management goals (Figure 4.23). Selective harvest focused on the left side of an age-specific distribution will improve the mean size of the antler characteristic in the cohort at older ages. Selective harvest of the right side of an age-specific distribution will decrease the average size of an antler characteristic in the cohort at older ages. Uniform selection across a distribution (i.e., random harvest of bucks) will result in a numerical change in the number of individuals in the cohort, but will not affect average size of the desired characteristic. The mean and associated variation of these age-specific distributions likely varies spatially, and so selective-harvest programs should be based on site-specific data.

Selection intensity (i.e., the proportion of the cohort removed) affects how cohort antler size is altered. In a simulation study, Strickland et al. (2001) determined that at least 50% of males that met a predefined antler size criterion in younger-aged cohorts (primarily at the yearling age class) had to be harvested to generate meaningful change in cohort phenotype at four years of age. Thus, negligible harvest rates should not impact subsequent cohort antler characteristics in any measurable way.

Persistence of a phenotypic change in a cohort is influenced by interannual environmental variation. Stochastic environmental events may affect cohort phenotype and mask the effects of selective harvest; conversely, a more stable environment makes selective harvest more effective and the response more measurable. Effective selective harvest also requires a long-term commitment. Any change in phenotypic distribution will start reverting to the original distribution the year that selective harvest ends.

For selective harvest strategies to be effective, hunters must be able to accurately judge age-specific antler development and selectively harvest inferior animals. The great degree of overlap in the distribution of antler characteristics (Figure 4.25a–c) makes it difficult to determine age using only antler characteristics and this ability to discern age-specific antler development is vital for selective-harvest strategies to improve cohort phenotype. The hunter must determine if an animal is a high-quality, younger-aged male, or a low-quality, older-aged male. Failure to do so will eliminate directional selection to improve phenotype.

The reverse approach, selective protection of inferior-antlered young bucks and removal of superior-antlered young bucks can negatively impact standing crop antler development (Strickland et al., 2001); this phenomenon is referred to as "high-grading." In another simulation using data from captive deer, we selectively removed all three-year-old bucks exceeding 129 gross Boone and Crockett score and recruited the remaining bucks to the next year. We again removed all four-year-old bucks exceeding 129 gross Boone and Crockett score and recruited the remainder another year. For comparison, we calculated the average score at four and five years without removing the larger-antlered bucks. The results were remarkable; average scores for the high-graded cohort's antlers were 9% smaller at four years and 18% smaller at five years.

Selective harvest criteria should incorporate a measure of antler size appropriate to the management goal. Many agency restrictions incorporate number of points, because points are easier for the hunter to judge. However, a single antler feature like points may not accurately represent the true size of an antler. Based on antler points, Jacobson (1995) concluded that yearling antler size was a poor predictor of size at older ages. In contrast, Demarais (1998) used the same database but a different measure of antler size, and concluded that yearling antler size was a reasonable predictor of subsequent antler size (Figure 4.16). This discrepancy in conclusions using the same database was due to their use of different descriptive variables. A single antler characteristic (e.g., number of antler points) does not provide as much information about total antler size, and thus predictive power, as a variable that incorporates several antler characteristics (e.g., Boone and Crockett score; Demarais and Strickland, 2010).

Selective Harvest to Improve Genetics

Managing the "genetics" of a population involves altering gene frequencies in a breeding population, such that there is an increase in genetic potential to grow larger antlers. Genetic composition can be manipulated only if you can judge the genetic potential of bucks and then increase the reproductive success of the superior animals. This type of selective-harvest approach would increase mean cohort antler size and population-level gene frequencies that code for production of relatively larger antlers.

Directional selection (Figure 4.23) for increasing gene frequencies that code for relatively larger antlers can only be successful if two criteria are fulfilled. First, the phenotypic expression of antler development at the time of selection accurately portrays the genetic potential for antler production. Second, the genetic potential for production of relatively large antlers must be passed to offspring, that is, antler characteristics must be heritable.

As described above, the interannual predictive relationship of antler characteristics in white-tailed deer is controversial. Several studies have demonstrated a positive correlation between a male's antlers when he is young and at maturity (Williams and Harmel, 1984; Schultz and Johnson, 1992; Ott et al., 1997; Demarais, 1998; Lewis, 2010) while others have not (Koerth and Kroll, 2008). Phenotypic expression of antler size at younger ages must be related to genetic potential at older ages for selective harvest to be effective. Factors that interfere with the expression of genetic potential also interfere with the application of genetic selection. These factors include late fawning dates (Jacobson, 1995) and any number of sources of environment variation, such as precipitation and temperature.

The effectiveness of selective-harvest programs to change the genetic composition of a population is influenced by other factors. Our inability to judge a female's genetic potential for antler development is a significant limitation because there is no directional selection for half of the population; thus, change in the genetic composition of females must come indirectly from the offspring of surviving sires. The tending mating system of white-tailed deer further complicates this scenario. Because breeding opportunity is spread throughout the male segment of the population (DeYoung et al., 2006, 2009), genetically "superior" males cannot sire a disproportionate number of offspring, slowing the potential change in antler-coding gene frequencies over time.

The last impediment to changing the genetic composition of free-ranging white-tailed deer populations is dispersal. Immigration of yearling males will perpetually introduce genetic diversity back into the managed population and dilute the gene pool, potentially negating any change through selection. This is especially true if the primary animals targeted for selection are yearling males. Emigration of yearling males from the managed population furthers the inefficiency.

Although studies have demonstrated a link between genetic properties, such as heterozygosity, and antler size in free-ranging populations (Smith et al., 1983; Scribner et al., 1984; Scribner and Smith, 1990), there are no documented cases of genetic manipulation in wild populations of white-tailed deer. Working in a captive deer facility, Lockwood et al. (2007) demonstrated that eight years of directional selection in a nutritionally stressed environment altered antler characteristics of yearling males in subsequent generations. They recorded a 12.9 cm (5.1 in.) increase in main beam length, an increase of 3.2 total antler points, and a 92.3 cm (36.4 in.) increase in gross Boone and Crockett score of yearling deer.

The multiple sources of variation affecting expression of genetic potential represent considerable challenges to impacting genetics under any management scenario with free-ranging deer or in large enclosed populations. Webb (2009) simulated the efficacy of selective harvest at altering genetic composition and increasing antler size in free-ranging populations. He accounted for variation in heritability estimates, lack of female selection, and dispersal. Using a heritability value of 0.35, antler points increased only 0.1–0.4 after 20 years of selective harvest, and that increase required an annual selective removal of 28–56% of males. Thus, given the severe limitations to success, manipulation of population-level genetics may not be a viable option under most white-tailed deer management scenarios.

Antler Restrictions as a Tool in Selective Harvest

Antler restrictions are commonly used to increase buck age structure based on the fact that antler size generally increases with age. By protecting younger bucks with antlers that do not meet certain criteria, an antler restriction can protect a particular age class of bucks. However, antler restrictions are site specific. That is, an antler restriction that works on one property may not work on another. So, using historic harvest data from an area is essential for developing effective antler restriction. No single approach to antler-based restrictions can be universally applied. Having no antler restriction is better than having a poorly designed one.

In special cases antler restrictions can be designed to protect higher-quality, younger-aged bucks and allow harvest of lower-quality bucks at any age and harvest of higher-quality, older-aged bucks. However, antler restrictions designed to differentially remove smaller-antlered young bucks must be developed with careful consideration of their potential impacts, especially over-harvest of an age class. The slot-limit approach to antler restrictions lets managers be more specific when protecting or targeting certain types of bucks for removal, similar to a fisheries slot-limit that protects medium-sized fish. A carefully constructed slot-limit antler restriction can allow removal of lower-quality bucks, both young and old, and protect higher-quality, younger bucks from harvest. But greater complexity requires more effort and education on the part of hunters; the most biologically valid antler restrictions succeed only if the hunter properly applies them.

Acknowledgments

We thank the following for contributing data for the latitudinal analysis of yearling antler size. Chris Cook, Alabama Department of Conservation and Natural Resources; Brad Miller, Arkansas Game and Fish Commission; Jim Heffelfinger, Arizona Game and Fish Department; Joe Rogerson, Delaware Department of Natural Resources and Environmental Control; Brigham Mason, Deseret Ranches of Florida; Matthew Chopp, Donald Francis, Cory Morea, Katherine Roscoe, Florida Fish and Wildlife Conservation Commission; Charles Killmaster, Georgia Department of Natural Resources; Tom Micetich, Illinois Department of Natural Resources; Kristina Brunjes and David Yancy, Kentucky Department of Fish and Wildlife Resources; Scott Durham, Louisiana Department of Wildlife and Fisheries; Brian Eyler, Maryland Department of Natural Resources; Brent Rudolph, Michigan Department of Natural Resources and Environment; Chad Dacus, William McKinley, Chris McDonald, Lann Wilf, Amy Blaylock, Mississippi Department of Wildlife, Fisheries and Parks; Lonnie Hansen, Missouri Department of Conservation; John Vore, Montana Department of Natural Resources and Conservation; Rod Cumberland, New Brunswick Department of Natural Resources; Kent Gustafson, New Hampshire Fish and Game Department; Evin Stanford, North Carolina Wildlife Resources Division; Michael Tonkovich, Ohio Department of Natural Resources; Harmon Weeks, Purdue University; Kenneth Gee, The Samuel Roberts Noble Foundation; Claude Daigle, Quebec Ministry of Natural Resources and Wildlife; Charles Ruth, South Carolina Department of Natural Resources; Mitch Lockwood, Texas Parks and Wildlife Department; Tennessee Wildlife Resources Agency; Don White, Jr., University of Arkansas at Monticello; Steeve Côté and Anouk Simard, Université Laval; Craig Harper, University of Tennessee; Shawn Haskell, Vermont Fish and Wildlife Department; Matt Knox, Virginia Department of Game and Inland Fisheries; Keith McCaffery and Robert Rolley, Wisconsin Department of Natural Resources. We would also like to thank Andrew Little for his assistance with manuscript preparation.

REFERENCES

Allen, S. P., M. Maden, and J. S. Price. 2002. A role for retinoic acid in regulating the regeneration of deer antlers. *Developmental Biology* 251:409–423.

Andersson, M. 1986. Evolution of condition-dependent sex ornaments and mating preferences: Sexual selection based on viability differences. *Evolution* 40:804–816.

Ashley, E. P., G. B. McCullough, and J. T. Robinson. 1998. Morphological responses of white-tailed deer to a severe population reduction. *Canadian Journal of Zoology* 76:1–5.

Asleson, M. A., E. C. Hellgren, and L. W. Varner. 1996. Nitrogen requirements for antler growth and maintenance in white-tailed deer. *Journal of Wildlife Management* 60:744–752.

Banks, W. J. and R. W. Davis. 1966. Observations on the relationship of antlerogenesis to bone morphology and composition in the Rocky Mountain mule deer (*Odocoileus hemionus hemionus*). *Anatomical Record* 154:312.

Barrell, G. K., R. Davies, and C. I. Bailey. 1999. Immunocytochemical localization of oestrogen receptors in perichondrium of antlers in red deer (*Cervus elaphus*). *Reproduction, Fertility and Development* 11:189–192.

Barrette, C. and D. Vandal. 1990. Sparring, relative antler size, and assessment in male caribou. *Behavioral Ecology and Sociobiology* 26:383–387.

Bartoš, L. and J. Hyánek. 1983. Social position in the red deer stag I: The effect on developing antlers. In *Antler Development in Cervidae*, ed. R. D. Brown, 451–461. Kingsville, TX: Caesar Kleberg Wildlife Research Institute.

Bartoš, L., D. Schams, and G. A. Bubenik. 2009. Testosterone, but not IGF1, prolactin, or cortisol, may serve as antler-stimulating hormone in red deer stags (*Cervus elaphus*). *Bone* 44:691–698.

Bartoš, L., D. Schams, U. Kierdorf et al. 2000. Cyproterone acetate reduced antler growth in surgically castrated fallow deer. *Journal of Endocrinology* 164:87–95.

Berglund, A., A. Bisazza, and A. Pilastro. 1996. Armaments and ornaments: An evolutionary explanation of traits of dual utility. *Biological Journal of the Linnean Society* 58:385–399.

Brown, R. D., C. C. Chao, and L. W. Faulkner. 1983. Hormone levels and antler development in white-tailed and sika fawns. *Comparative Biochemistry and Physiology A: Physiology* 75:385–390.

Bubenik, A. B. 1983a. The behavioral aspects of antlerogenesis. In *Antler Development in Cervidae*, ed. R. D. Brown, 389–449. Kingsville, TX: Caesar Kleberg Wildlife Research Institute.

Bubenik, G. A. 1983b. The endocrine regulation of the antler cycle. In *Antler Development in Cervidae*, ed. R. D. Brown, 73–107. Kingsville, TX: Caesar Kleberg Wildlife Research Institute.

Bubenik, G. A. 1990. Neuroendocrine regulation of the antler cycle. In *Horns, Pronghorns, and Antlers: Evolution, Morphology, Physiology, and Social Significance*, eds. G. A. Bubenik and A. B. Bubenik, 265–297. New York, NY: Springer-Verlag.

Bubenik, G. A. and A. B. Bubenik. 1978. The role of sex hormones in the growth of antler bone tissue: Influence of antiestrogen therapy. *Saugetierkundl Mitt* 40:284–291.

Bubenik, G. A., K. V. Miller, A. L. Lister, D. A. Osborn, L. Bartoš, and G. J. Van Der Kraak. 2005. Testosterone and estradiol concentrations in serum, velvet skin, and growing antler bone of male white-tailed deer. *Journal of Experimental Zoology* 303A:186–192.

Bubenik, G. A., E. Reyes, D. Schams, L. Bartoš, and F. Koerner. 2002. Effect of antiandrogen cyproterone acetate on the development of the antler cycle in southern pudu (*Pudu puda*). *Journal of Experimental Zoology* 292:393–401.

Clutton-Brock, T. H. 1982. The functions of antlers. *Behaviour* 79:108–125.

Clutton-Brock, T. H. 1987. Sexual selection in the cervidae. In *Biology and Management of the Cervidae*, ed. C. M. Wemmer, 110–122. Washington, DC: Smithsonian Institution Press.

Clutton-Brock, T. H., S. D. Albon, and P. Harvey. 1980. Antlers, body size, and breeding group size in the Cervidae. *Nature* 285:565–567.

Darwin, C. 1874. *The Descent of Man, and Selection in Relation to Sex* (2nd edition). New York, NY: A.L. Burt.

Davidson, W. R. 2006. *Field Manual of Wildlife Diseases in the Southeastern United States* (3rd edition). Athens, GA: Southeastern Cooperative Wildlife Disease Study.

Davis, T. A. 1983. Antler asymmetry caused by limb amputation and geo-physical forces. In *Antler Development in Cervidae*, ed. R. D. Brown, 223–230. Kingsville, TX: Caesar Kleberg Wildlife Research Institute.

Demarais, S. 1998. Managing for antler production: Understanding the age–nutrition–genetics interaction. In *The Role of Genetics in White-tailed Deer Management* (2nd edition), eds. K. A. Cearley and D. Rollins, 33–36. College Station, TX: Texas A&M University.

Demarais, S., K. V. Miller, and H. A. Jacobson. 2000. White-tailed deer. In *Ecology and Management of Large Mammals in North America*, eds. S. Demarais and P. R. Krausman, 601–628. Upper Saddle River, NJ: Prentice-Hall.

Demarais, S. and B. K. Strickland. 2010. White-tailed deer antler research: A critique of design and analysis methodology. *Journal of Wildlife Management* 74:193–197.

Demarais, S., B. K. Strickland, and L. E. Castle. 2005. Antler regulation effects on white-tailed deer on Mississippi public hunting areas. *Proceedings of the Annual Conference of the Southeastern Association of Fish and Wildlife Agencies* 59:1–9.

DeYoung, R. W., S. Demarais, K. L. Gee, R. L. Honeycutt, M. W. Hellickson, and R. A. Gonzales. 2009. Molecular evaluation of the white-tailed deer (*Odocoileus virginianus*) mating system. *Journal of Mammalogy* 90:946–953.

DeYoung, R. W., S. Demarais, R. L. Honeycutt, K. L. Gee, and R. A. Gonzales. 2006. Social dominance and male breeding success in captive white-tailed deer. *Wildlife Society Bulletin* 34:131–136.

Ditchkoff, S. S. and R. L. deFreese. 2010. Assessing fluctuating asymmetry of white-tailed deer antlers in a three-dimensional context. *Journal of Mammalogy* 91:27–37.

Ditchkoff, S. S., R. L. Lochmiller, R. E. Masters, S. R. Hoofer, and R. A. Van Den Bussche. 2001a. Major-histocompatibility-complex-associated variation in secondary sexual traits of white-tailed deer (*Odocoileus virginianus*): Evidence for good-genes advertisement. *Evolution* 55:616–625.

Ditchkoff, S. S., R. L. Lochmiller, R. E. Masters, W. R. Starry, and D. M. Leslie, Jr. 2001b. Does fluctuating asymmetry of antlers in white-tailed deer (*Odocoileus virginianus*) follow patterns predicted for sexually selected traits? *Proceedings of the Royal Society B* 268:891–898.

Doutt, J. K. and J. C. Donaldson. 1961. Antlered doe study. *Pennsylvania Game News* 30(11):23–25.

French, C. E., L. C. McEwen, N. D. Magruder, R. H. Ingram, and R. W. Swift. 1956. Nutrient requirements for growth and antler development in the white-tailed deer. *Journal of Wildlife Management* 20:221–232.

Geist, V. 1966. The evolution of horn-like organs. *Behaviour* 27:175–214.

Geist, V. 1994. Origin of the species. In *Deer*, eds. D. Gerlach, S. Atwater, and J. Schnell, 2–16. Mechanicsburg, PA: Stackpole Books.

Geist, V. 1998. *Deer of the World: Their Evolution, Behaviour, and Ecology*. Mechanicsburg, PA: Stackpole Books.

Geist, V. and M. Bayer. 1988. Sexual dimorphism in the Cervidae and its relation to habitat. *Journal of Zoology* 214:45–53.

Geist, V. and P. T. Bromley. 1978. Why deer shed antlers. *Zeitschrift für Säugetierkunde* 43:223–231.

Gill, J. 1956. Regional differences in size and productivity of deer in West Virginia. *Journal of Wildlife Management* 62:801–807.

Goss, R. J. 1968. Inhibition of growth and shedding of antlers by sex hormones. *Nature* 220:83–85.

Goss, R. J. 1983. *Deer Antlers: Regeneration, Function, and Evolution*. New York, NY: Academic Press.

Goss, R. J., A. Van Praagh, and P. Brewer. 1992. The mechanism of antler casting in the fallow deer. *Journal of Experimental Zoology* 264:429–436.

Grasman, B. T. and E. C. Hellgren. 1993. Phosphorus nutrition in white-tailed deer: Nutrient balance, physiological responses, and antler growth. *Ecology* 74:2279–2296.

Gray, W. N., II, S. S. Ditchkoff, M. K. Causey, and C. W. Cook. 2002. The yearling disadvantage in Alabama deer: Effect of birth date on development. *Proceedings Annual Conference Southeast Association of Fish and Wildlife Agencies* 56:255–264.

Guyton, A. C. and J. E. Hall. 2000. *Textbook of Medical Physiology* (10th edition). New York, NY: W. B. Saunders Co.

Harmel, D. E., W. E. Armstrong, E. R. Fuchs, E. L. Young, and K. D. McGinty. 1998. The Kerr Area penned deer research facility. In *The Role of Genetics in White-tailed Deer Management*, eds. K. Cearley and D. Rollins, 40–45. College Station, TX: Texas A&M University.

Harmel, D. E., J. D. Williams, and W. E. Armstrong. 1989. Effects of genetics and nutrition on antler development and body size of white-tailed deer. Federal Aid Report series No. 26. Projects W-56-D, W-76-R, W-109-R, and W-14-C. Texas Parks and Wildlife Department, Austin, Texas.

Hesselton, W. T. and R. M. Hesselton. 1982. White-tailed deer. In *Wild Mammals of North America*, eds. J. A. Chapman and G. A. Feldhamer, 878–901. Baltimore, MD: The John Hopkins University Press.

Hirth, D. H. 2000. Behavioral ecology. In *Ecology and Management of Large Mammals in North America*, eds. S. Demarais and P. R. Krausman, 175–191. Upper Saddle River, NJ: Prentice-Hall.

Jacobson, H. A. 1995. Age and quality relationships. In *Quality Whitetails: The How and Why of Quality Deer Management*, eds. K. V. Miller and R. L. Marchinton, 103–111. Mechanicsburg, PA: Stackpole Books.

Jacobson, H. A. and R. N. Griffin. 1983. Antler cycles of white-tailed deer in Mississippi. In *Antler Development in Cervidae*, ed. R. D. Brown, 15–22. Kingsville, TX: Caesar Kleberg Wildlife Research Institute.

Jacobson, H. A. and S. J. Waldhalm. 1992. Antler cycles of a white-tailed deer with congenital anophthalmia. In *The Biology of Deer*, ed. R. D. Brown, 520–524. New York, NY: Springer-Verlag.

Jones, P. D., S. Demarais, B. K. Strickland, and S. L. Edwards. 2008. Soil region effects on white-tailed deer forage protein content. *Southeastern Naturalist* 7:595–606.

Karns, G. R., R. A. Lancia, C. S. DePerno, M. C. Conner, and M. K. Stoskopf. 2009. Intracranial abscessation as a natural mortality factor for adult male white-tailed deer (*Odocoileus virginianus*) in Kent County, Maryland, USA. *Journal of Wildlife Disease* 45:196–200.

Kellert, S. R. and C. P. Smith. 2000. Human values toward large mammals. In *Ecology and Management of Large Mammals in North America*, eds. S. Demarais and P. R. Krausman 38–63. Upper Saddle River, NJ: Prentice-Hall.

Kie, J. G., T. Bowyer, M. C. Nicholson, B. B. Boroski, and E. R. Loft. 2002. Landscape heterogeneity at differing scales: Effects on spatial distribution of mule deer. *Ecology* 83:530–544.

Kierdorf, U. and L. Bartoš. 1999. Treatment of the growing pedicle with retinoic acid increased the size of first antlers in fallow deer (*Dama dama* L.). *Comparative Biochemistry and Physiology Part C: Pharmacology, Toxicology, and Endocrinology* 124:7–9.

Kierdorf, U. and H. Kierdorf. 2002. Pedicle and first antler formation in deer: Anatomical, histological, and developmental aspects. *Zeitschrift fur Jagdwissenschaft* 48:22–34.

Kierdorf, U., H. Kierdorf, and T. Szuwart. 2007. Deer antler regeneration: Cells, concepts, and controversies. *Journal of Morphology* 268:726–738.

Kierdorf, U., C. Li, and J. S. Price. 2009. Improbable appendages: Deer antler renewal as a unique case of mammalian regeneration. *Seminars in Cell and Developmental Biology* 20:535–542.

Kodric-Brown, A. and J. H. Brown. 1984. Truth in advertising: The kinds of traits favored by sexual selection. *American Naturalist* 124:309–323.

Koerth, B. H. and J. C. Kroll. 2008. Juvenile-to-adult antler development in white-tailed deer in South Texas. *Journal of Wildlife Management* 72:1109–1113.

Kruuk, L. E. B., J. Slate, J. M. Pemberton, S. Brotherstone, F. Guiness, and T. Clutton-Brock. 2002. Antler size in red deer: Heritability and selection but no evolution. *Evolution* 56:1683–1695.

Landete-Castillejos, T., J. A. Estevez, A. Martinez, F. Ceacero, A. Garcia, and L. Gallego. 2007a. Does chemical composition of antler bone reflect the physiological effort made to grow it? *Bone* 40:1095–1102.

Landete-Castillejos, T., A. Garcia, and L. Gallego. 2007b. Body weight, early growth, and antler size influence antler bone composition of Iberian red deer (*Cervus elaphus hispanicus*). *Bone* 40:230–235.

Lewis, J.S. 2010. Factors influencing antler size in free-ranging white-tailed deer and mark/recapture estimates of demographic traits. PhD dissertation, Texas A&M University–Kingsville.

Lewis, L. K. and G. K. Barrell. 1994. Regional distribution of estradiol receptors in growing antlers. *Steroids* 59:490–492.

Li, C., R. P. Littlejohn, I. D. Corson, and J. M. Suttie. 2003. Effects of testosterone on pedicle formation and its transformation to antler in castrated male, freemartin, and normal female red deer (*Cervus elaphus*). *General and Comparative Endocrinology* 131:21–31.

Li, C. and J. M. Suttie. 1994. Light microscopic studies of pedicle and early first antler development in red deer (*Cervus elaphus*). *Anatomical Record* 239:198–215.

Li, C., J. M. Suttie, and D. E. Clark. 2005. Histological examination of antler regeneration in red deer (*Cervus elaphus*). *The Anatomical Record Part A: Discoveries in Molecular, Cellular, and Evolutionary Biology* 282A:163–174.

Lincoln, G. A. 1972. The role of antlers in the behavior of red deer. *Journal of Experimental Zoology* 182:233–250.

Lincoln, G. A. 1973. Appearance of antler pedicles in early foetal life in red deer. *Journal of Embryology and Experimental Morphology* 29:431–437.

Lincoln, G. A. 1994. Teeth, horns and antlers: The weapons of sex. In *The Differences between the Sexes*, eds. R. V. Short and E. Balaban, 131–158. Cambridge: Cambridge University Press.

Lockwood, M. A., D. B. Frels, Jr., W. E. Armstrong, E. Fuchs, and D. E. Harmel. 2007. Genetic and environmental interaction in white-tailed deer. *Journal of Wildlife Management* 71:2732–2735.

Lukefahr, S. D. and H. A. Jacobson. 1998. Variance component analysis and heritability of antler traits in white-tailed deer. *Journal of Wildlife Management* 62:262–268.

Malo, A. F., E. R. S. Roldan, J. Garde, A. J. Soler, and M. Gomendio. 2005. Antlers honestly advertise sperm production and quality. *Proceedings of the Royal Society B* 272:149–157.

Marchinton, R. L. and D. H. Hirth. 1984. Behavior. In *White-tailed Deer: Ecology and Management*, ed. L. K. Halls, 129–168. Harrisburg, PA: Stackpole Books.

McCullough, D. R. 1982. Antler characteristics of George Reserve white-tailed deer. *Journal of Wildlife Management* 46:821–826.

McDonald, C. G., S. Demarais. T. A., Campbell, H. F. Janssen, V. G. Allen, and A. M. Kelly. 2005. Physical and chemical characteristics of antlers and antler breakage in white-tailed deer. *Southwestern Naturalist* 50:356–362.

Mech, L. D. 1984. Predators and predation. In *White-tailed Deer: Ecology and Management*, ed. L. K. Halls, 189–200. Harrisburg, PA: Stackpole Books.
Miller, K. V., R. L. Marchington, J. R. Beckwith, and P. B. Bush. 1985. Variations in density and chemical composition of white-tailed deer antlers. *Journal of Mammalogy* 66:693–701.
Miller, K. V., L. I. Muller, and S. Demarais. 2003. White-tailed deer. In *Wild Mammals of North America: Biology, Management and Conservation* (2nd edition), eds. G. A. Feldhamer, B. C. Thompson, and J. A. Chapman, 906–930. Baltimore, MD: John Hopkins University Press.
Miranda, B. R. and W. F. Porter. 2003. Statewide habitat assessment for white-tailed deer in Arkansas using satellite imagery. *Wildlife Society Bulletin* 31:715–726.
Monteith, K. L., L. E. Schmitz, J. A. Jenks, J. A. Delger, and R. T. Bowyer. 2009. Growth of male white-tailed deer: Consequences of maternal effects. *Journal of Mammalogy* 90:651–660.
Mysterud, A., R. Langvatn, N. G. Yoccoz, and N. C. Stenseth. 2002. Large-scale habitat variability, delayed density effects and red deer populations in Norway. *Journal of Animal Ecology* 71:569–580.
Mysterud, A., E. Meisingset, R. Langvatn, N. G. Yoccoz, and N. C. Stenseth. 2005. Climate-dependent allocation of resources to secondary sexual traits in red deer. *Oikos* 111:245–252.
Ott, J. R., S. A. Roberts, J. T. Baccus, D. E. Harmel, W. E. Armstrong, and E. Fuchs. 1997. Antler characteristics and body mass of spike- and fork-antlered yearling white-tailed deer at maturity. *Proceedings of the Annual Conference of the Southeastern Association of Fish and Wildlife Agencies* 51:400–413.
Ozoga, J. J. 1988. Incidence of "infant" antlers among supplementally fed white-tailed deer. *Journal of Mammalogy* 69:393–395.
Perry, D. A., R. Oren, and S. C. Hart. 2008. *Forest Ecosystems* (2nd edition). Baltimore, MD: Johns Hopkins University Press.
Price, J. and S. Allen. 2004. Exploring the mechanisms regulating regeneration of deer antlers. *Philosophical Transactions of the Royal Society of London B* 359:809–822.
Price, J. S., S. Allen, C. Faucheux, T. Althanian, and J. G. Mount. 2005. Deer antlers: A zoological curiosity or the key to understanding organ regeneration in mammals? *Journal of Anatomy* 207:603–618.
Price, J. S., B. Nichols, L. E. Lanyon, and S. P. Allen. 2002. Estrogen rapidly induces bone formation in regenerating deer antlers. *Bone* 30:10S
Price, J. S., B. O. Oyajobi, R. O. C. Oreffo, and R. G. G. Russell. 1994. Cells cultured from the growing antler tip of red deer antler express alkaline phosphatase and proliferate in response to insulin-like growth factor-I. *Journal of Endocrinology* 143:R9–R16.
Prothero, D. R. and R. M. Schoch. 2002. Where the deer and the antelope play. In *Horns, Tusks, and Flippers: The Evolution of Hoofed Mammals,* eds. D. R. Prothero and R. M. Schoch, 61–85. Baltimore, MD: John Hopkins University Press.
Putman, R. 1988. *The Natural History of Deer*. Ithaca, NY: Cornell University Press.
Ridley, M. 1996. *Evolution*. Cambridge: Blackwell Science.
Rolf, H. J. and A. Enderle. 1999. Hard fallow deer antler: A living bone till antler casting? *The Anatomical Record* 255:69–77.
Ryel, L. A. 1963. The occurrence of certain anomalies in Michigan white-tailed deer. *Journal of Mammalogy* 44:79–98.
Sadighi, M., S. R. Haines, A. Skottner, A. J. Harris, and J. M. Suttie. 1994. Effects of insulin-like growth factor-I (IGF-I) and IGF-II on the growth of antler cells *in vitro*. *Journal of Endocrinology* 143:461–469.
Sæther, B.-E., R. Andersen, O. Hjeljord, and M. Heim. 1996. Ecological correlates of regional variation in life history of the moose *Alces alces*. *Ecology* 77:1493–1500.
Sauer, P. R. 1984. Physical characteristics. In *White-tailed Deer Ecology and Management*, ed. L. K. Halls, 73–90. Harrisburg, PA: Stackpole Books.
Scanlon, P. F., D. F. Urbson, and J. A. Sullivan. 1975. A male pseudohermaphrodite white-tailed deer resembling an antlered doe. *Journal of Wildlife Diseases* 11:237–240.
Schultz, S. R. and M. K. Johnson. 1992. Antler development of captive Louisiana white-tailed bucks. *Proceedings of the Annual Conference of the Southeastern Association of Fish and Wildlife Agencies* 46:67–74.
Schultz, S. R., M. K. Johnson, S. E. Feagley, L. L. Southern, and T. L. Ward. 1994. Mineral content of Louisiana white-tailed deer. *Journal of Wildlife Diseases* 30:77–85.

Scribner, K. T. and M. H. Smith. 1990. Genetic variability and antler development. In *Horns, Pronghorns, and Antlers: Ecology, Morphology, Physiology, and Social Significance*, eds. G. A. Bubenik and A. B. Bubenik, 460–473. New York, NY: Springer-Verlag.

Scribner, K. T., M. H. Smith, and P. E. Johns. 1984. Age, condition and genetic effects on incidence of spike bucks. *Proceedings of the Annual Conference of the Southeastern Association of Fish and Wildlife Agencies* 38:23–32.

Scribner, K. T., M. H. Smith, and P. E. Johns. 1989. Environmental and genetic components of antler growth in white-tailed deer. *Journal of Mammalogy* 70:284–291.

Severinghaus, C. W., H. F. Maguire, R. A. Cookingham, and J. E. Tanck. 1950. Variations by age class in the antler beam diameters of white-tailed deer related to range conditions. *Transactions of the North American Wildlife Conference* 15:551–570.

Smith, M. H., R. K. Chesser, E. G. Cothran, and P. E. Johns. 1983. Genetic variability and antler growth in a natural population of white-tailed deer. In *Antler Development in Cervidae,* ed. R. D. Brown, 365–387. Kingsville, TX: Caesar Kleberg Wildlife Research Institute.

Snyder, D. L., R. L. Cowan, D. R. Hagen, and B. D. Schanbacher. 1983. Effects of pinealectomy on seasonal changes in antler growth and concentrations of testosterone and prolactin in white-tailed deer. *Biology of Reproduction* 29:63–71.

Stewart, K. M., R. T. Bowyer, J. G. Kie, and W. C. Gasaway. 2000. Antler size relative to body mass in moose: Tradeoffs associated with reproduction. *Alces* 36:77–83.

Strickland, B. K. and S. Demarais. 2000. Age and regional differences in antlers and mass of white-tailed deer. *Journal of Wildlife Management* 64:903–911.

Strickland, B. K. and S. Demarais. 2008. Influence of landscape composition and structure on antler size of white-tailed deer. *Journal of Wildlife Management* 72:1101–1108.

Strickland, B. K., S. Demarais, L. E. Castle et al. 2001. Effects of selective-harvest strategies on white-tailed deer antler size. *Wildlife Society Bulletin* 29:509–520.

Suttie, J. M. and P. F. Fennessy. 1992. Recent advances in the physiological control of velvet antler growth. In *Antler Development in Cervidae,* ed. R. D. Brown, 471–486. Kingsville, TX: Caesar Kleberg Wildlife Research Institute.

Suttie, J. M., P. D. Gluckman, J. H. Butler, P. F. Fennessy, I. D. Corson, and F. J. Laas. 1985. Insulin-like growth factor 1 (IGF-1): Antler stimulating hormone? *Endocrinology* 116:846–848.

Suttie, J. M. and R. N. B. Kay. 1983. The influence of nutrition and photoperiod on the growth of antlers of young red deer. In *Antler Development in Cervidae,* ed. R. D. Brown, 61–71. Kingsville, TX: Caesar Kleberg Wildlife Research Institute.

Suva L. J., G. A. Winslow, R. E. Wettenhall et al. 1987. A parathyroid hormone-related protein implicated in malignant hypercalcemia: Cloning and expression. *Science* 237:893–896.

Townsend, T. W. and E. D. Bailey. 1981. Effects of age, sex, and weight on social rank in penned white-tailed deer. *American Midland Naturalist* 106:92–101.

Ullrey, D. E. 1983. Nutrition and antler development in white-tailed deer. In *Antler Development in Cervidae,* ed. R. D. Brown, 49–59. Kingsville, TX: Caesar Kleberg Wildlife Research Institute.

Ullrey, D. E., W. G. Youatt, H. E. Johnson et al. 1973. Calcium requirements of weaned white-tailed deer fawns. *Journal of Wildlife Management* 37:187–194.

Ullrey, D. E., W. G. Youatt, H. E. Johnson, L. D. Fay, and B. L. Bradley. 1967. Protein requirements of white-tailed deer fawns. *Journal of Wildlife Management* 31:679–685.

Van Der Eems, K. L., R. D. Brown, and C. M. Gundberg. 1988. Circulating levels of 1,25 dihydroxyvitamin D, alkaline phosphatase, hydroxyproline, and osteocalcin associated with antler growth in white-tailed deer. *Acta Endocrinologica* 118:407–414.

Vanpé, C., J.-M. Gaillard, P. Kjellander et al. 2007. Antler size provides an honest signal of male phenotypic quality in roe deer. *American Naturalist* 169:481–493.

Webb, S. L. 2009. Movements, relatedness and modeled genetic manipulation of white-tailed deer. PhD dissertation, Mississippi State University.

Williams, J. D. and D. E. Harmel. 1984. Selection for antler points and body weight in white-tailed deer populations. *Proceedings of the Annual Conference of the Southeastern Association of Fish and Wildlife Agencies* 38:43–50.

Williams, J. D., W. F. Krueger, and D. H. Harmel. 1994. Heritabilities for antler characteristics and body weight in yearling white-tailed deer. *Heredity* 73:78–83.

Wisdom, M. J. and J. G. Cook. 2000. North American elk. In *Ecology and Management of Large Mammals in North America*, eds. S. Demarais and P. R. Krausman, 694–735. Upper Saddle River, NJ: Prentice-Hall.
Wishart, W. D. 1985. Frequency of antlered white-tailed does in Camp Wainwright, Alberta. *Journal of Wildlife Management* 49:386–388.
Wislocki, G. B. 1943. Studies on growth of deer antlers. In *Essays in Biology: In honor of Herbert M. Evans*, 631–653. Berkeley, CA: University of California Press.
Youatt, W. G., D. E. Ullrey, and W. T. Magee. 1976. Vitamin A concentration in livers of white-tailed deer. *Journal of Wildlife Management* 40:172–173.
Zahavi, A. 1975. Mate selection—A selection for a handicap. *Journal of Theoretical Biology* 53:205–214.

5

Population Dynamics

Charles A. DeYoung

CONTENTS

Perspective

White-tailed deer are an ancient, very adaptable species ranging from Canada in North America to Peru in South America and from the Atlantic to Pacific coasts (Baker, 1984) (Chapter 1). Across this broad range, whitetails are successful in a wide variety of habitats and climatic regimes. This adaptability requires that the population biology of the species is also flexible.

The conditions under which white-tailed deer evolved and persisted were vastly different from the past 12,000 years. The late Pleistocene extinctions resulted in the loss of 70% of the larger herbivores in North America (Martin, 1970), potential competitors of whitetails. These included ground sloths, mammoths, mastodons, extinct bison, native camels, and native horses. Heffelfinger (2006, p. 30) wrote: "Deer did not become the widespread, dominant ungulates we know today until the multitude of large grazing mammals disappeared, allowing them to expand their ranges and assume vacated niches." All species exceeding 1000 kg in adult body mass were lost in the Pleistocene extinctions in the Americas, along with three-fourth of the herbivores in the 100–1000 kg range (Owen-Smith, 1987).

Also, whitetails existed in the Pleistocene with larger predators than today, including large cats, canids, bears, and hyenas. Haynes (1983, p. 164) wrote: "During the late Pleistocene, North America was inhabited by more taxa of mammalian carnivores than in Recent times. These taxa included *Arctodus*, an enormous bear that has no modern counterpart …; *Panthera leo atrox*, a lion larger than modern African relatives; and *Canis dirus*, a wolf with massive, hyena-like dentition …"

Extinction of late Pleistocene megaherbivores was particularly noteworthy because of their outsized effect in shaping and modifying habitat (Owen-Smith, 1987). North American megafauna (1000 kg) in the late Pleistocene were Columbian mammoth, woolly mammoth, American mastodon, and Rusconi's ground sloth. There are examples of African elephants converting regions of savanna into grassland and both African and Asian elephants promoting the replacement of shade-tolerant trees with fast-growing trees, shrubs, and herbaceous vegetation (Owen-Smith, 1987). The former example would be disadvantageous to whitetails and the latter advantageous.

Whitetails migrate to winter deer yards as a means of surviving brutal winter weather in northern climes (Hoskinson and Mech, 1976). They respond to early successional stages in forested regions of eastern North America with rapid population growth (Russell et al., 2001). They occur in many semiarid and tropical habitats in relatively stable populations of modest or low density (DeYoung et al., 2008). Whitetails occupy a fragmented, although highly productive habitat in Illinois with the unusual adaptation whereby half the fawns and 20% of females disperse (Nixon et al., 1991). Clearly, the variety of white-tailed deer adaptation is remarkable.

Our understanding of the population dynamics of white-tailed deer has been strongly influenced by the landmark work of Dale McCullough (1979) on the George Reserve in Michigan. This study was conducted in an enclosure where the deer population was reduced in stages and then allowed to recover. Results showed density-dependent population behavior from low density to the point of population equilibrium (K carrying capacity). Since then, it has slowly become evident that George Reserve deer dynamics are not a template for ALL white-tailed deer populations, but are rather an important subset among many for this flexible species. DeYoung et al. (2008) suggested a way forward based largely on McCullough's extension of the role of density dependence in whitetail dynamics. They reinforced McCullough's statement that all whitetail populations sometimes show density dependence and sometimes do not. The relative mix of density-dependent and -independent population responses defines the population dynamics of an individual population.

Whereas many limiting factors simultaneously impinge on any white-tailed population, their interaction with the local food environment is fundamental to population dynamics. Populations in eastern North America commonly occupy forests where woody forage species have few defenses from grazing besides growing fast to get out of reach of deer. In these habitats, deer respond to forest disturbance with rapid population increase which can often lead to overbrowsing and thus damage or death of forage plants (Russell et al., 2001). Research on effects of white-tailed deer browsing and grazing on plants and plant communities in the more arid west and southwestern United States is lacking (Russell et al., 2001). However, plant–animal relationships within these habitats may differ greatly from those in more mesic habitats because forage plants have defenses such as thorns and unpalatable chemicals which limit damage by grazing deer. Browsing in these habitats may stimulate increased physical defense such as greater spine density (Cooper et al., 2003; Schindler et al., 2003) and increased chemical defenses (Bryant et al., 1992). Also, browsing in these habitats may stimulate shrub growth rather than impede it (Cooper et al., 2003), although compensatory growth may be muted by drought (Teaschner and Fulbright, 2007). Furthermore, in these habitats, favored forb species are often annual versus the more common perennials in forest habitats. Availability of forbs, like browse, is mediated by precipitation. During droughts, forb biomass may be negligible, whereas forbs are abundant in wet years (Grahmann, 2009). This, and slower plant growth overall, creates a very different population dynamic in white-tailed deer not present in the George Reserve study.

Much of the research on white-tailed deer has been conducted in areas of the eastern, Midwestern, and Lake States regions of the United States. Here, land-use changes and suburban living have created conditions for population growth resulting in deer being judged "overpopulated." Sport hunting is often used in a seldom successful attempt to "control" populations in such situations (Chapter 20). Because of the attention to research, including into population ecology, in northern and eastern portions of the species' range, these population dynamics are frequently assumed to be ubiquitous. Gavin et al. (1984, p. 32) wrote: "Generalizations about the population biology of white-tailed deer are difficult to derive. This may be due to the historical emphasis given to research and management of this species in the Midwest and Great Lakes States, a region with strong seasonality near the northern limits of *O. virginianus* range." Past emphasis on populations with high deer densities does a disservice to the flexibility of the species and leads to an unrealistic view. White-tailed deer should be defined less by the ability to "overpopulate" and more properly by their ability to thrive over a mind-boggling variety of situations. They are truly a cosmopolitan species and their adaptability and flexibility will be emphasized throughout this chapter which covers fecundity, mortality, dispersal, density dependence, population estimation, aging, and harvest.

Fecundity

White-tailed deer have a large reproductive potential, which is modified in each habitat by a complex milieu of physiological, environmental, and behavioral factors (Verme, 1969). McCullough (1979) estimated a potential rate of increase of 89% for a population at low density in a good habitat, whereas Downing and Guynn (1985) estimated 82%. In extremely favorable circumstances, doe fawns breed, some adult does have triplets (Ozoga, 1987), and the average number of fetuses/adult pregnant doe exceeds 2.00 (Haugen, 1975; Dapson et al., 1979) (Table 5.1).

The percent of female fawns that breed is a sensitive measure of a population's level of nutrition and a rare phenomenon in many habitats. In a sample of 12 studies (Table 5.1), the highest pregnancy rate for fawns was 29% (Haugen, 1975). However, about half of doe fawns in a southern Michigan study reached puberty and bred (Ozoga, 1987). Body mass attained by female fawns during the breeding season is a strong determinant of the ability to breed (Gaillard et al., 2000). Likewise, triplet litters are to be expected only when females receive excellent nutrition. In a supplementally fed Michigan enclosure, 14% of mature does had triplet litters at least once over a several-year period (Ozoga, 1987). One doe in this study averaged 2.67 fetuses/pregnancy over 12 years, whereas another, despite failing to conceive at 1.5 years, averaged 2.44 over 9 years.

The typical variability of white-tailed deer is evident in Table 5.1, with mature does in the fertile soils of the Midwestern United States averaging over 2.00 fetuses/pregnant doe (Nixon, 1971; Haugen, 1975). The diminutive Key Deer in southern Florida averaged relatively low 1.20 fetuses/mature doe (Folk and Klimstra, 1991), as did other Florida populations that varied from 1.31 to 1.38 (Richter and Labisky, 1985). Fetuses/mature doe were intermediate at 1.41–1.56 on northern ranges in Ontario, Canada (Mansell, 1974). Effects of highly variable precipitation in semiarid south Texas are evident in fetuses/mature doe that varied from 1.24 to 1.89 over several years in one study (Barron and Harwell, 1973), and from 1.11 to 1.94 in another (Kie and White, 1985). Obviously, the dynamics of populations reviewed in Table 5.1 vary considerably, as is typical for a species where populations persist in many habitats.

Mortality

Mortality in white-tailed deer can be divided into five age categories: Fetal (prenatal), neonatal (up to time of weaning), older fawns 0.5–1.0 years (postweaning), yearlings (1–2 years), and 2 years of age and older. Prenatal mortality is generally low, as white-tailed deer tend to carry embryos to term rather than abort or resorb. In Texas, losses of fertilized ova before implantation in yearlings and older does combined were 12% and less than 1% for older term embryos (Teer et al., 1965). Illinois research revealed that losses in the first six weeks of gestation were 4%, with losses for all ages of embryos at 12.9% (Roseberry and Klimstra, 1970). Another Texas study estimated preimplantation mortality of fertilized ova at 11% for yearlings and 9% for older does outside of an experimental area, with only three cases of intrauterine mortality among 948 embryos (Kie and White, 1985).

As is typical for all attributes of this species, neonatal mortality varies greatly among various habitats (Table 5.2). For fawns caught in the first few days after birth, mortality, at least in 1 year, has been as high

TABLE 5.1
Ovulation and Fetal Rates of Pregnant White-tailed Deer (Does Not Include Barren Does) in Various Studies

	Fawns[a]			Yearlings			≥2.5 Years			
Location	**OR**	**PP**	**FR/P**	**OR**	**PP**	**FR/P**	**OR**	**PP**	**FR/P**	**Study**
New York			1.00					78–92	1.20–1.70 (yearling and older lumped)	Morton and Cheatum (1946)
Michigan								86–89	1.11 (low nutrition)–1.96 (high nutrition)	Verme (1965)
Ohio	1.25		1.29	2.04		1.87	2.22		2.04	Nixon (1971)
South Texas							1.56–2.01	93	1.24–1.89	Barron and Harwell (1973)
Ontario		11	1.00		64	1.29		74–99	1.41–1.56	Mansell (1974)
Iowa–Nebraska	1.39	29	1.21	2.36	55	1.96	2.14–2.33	71–75	2.00–2.25	Haugen (1975)
South Carolina		6–8	1.00–1.04		88–96	1.36–1.67		94–97	1.56–2.27	Dapson et al. (1979)
Mississippi	1.50		1.00	1.62		1.40	1.78		1.66	Jacobson et al. (1979)
South Texas		4		1.00–2.00	88	0.67–1.80 ($n \geq 3$)	1.67–2.13	94	1.11–1.94	Kie and White (1985)
Florida		14	1.25		93	1.14		87–98	1.31–1.38	Richter and Labisky (1985)
Florida Keys		4	1.00		39	1.08		42	1.20	Folk and Klimstra (1991)
Minnesota		2			88	1.30		91–96	1.8	DelGiudice et al. (2007)

[a] OR = ovulation rate; PP = percent pregnant; FR/P = fetal rate per pregnant doe.

TABLE 5.2

Percent Deaths and Mortality Rates for Neonatal White-tailed Deer

Location	Age	Percent Deaths	Mortality Rate	Study
Texas	≤32 days	66.7		Cook et al. (1971)
Texas	≤2 months	10.0–90.0		Carroll and Brown (1977)
Oklahoma	≤3 months	90.0		Bartush and Lewis (1981)
Iowa	≤180 days	27.2		Huegel et al. (1985)
Illinois	"pre-hunt"	30.0		Nelson and Woolf (1987)
Oklahoma	≤45 days	36.2	0.45	Sams et al. (1996b)
New Brunswick	Jun–Sep		0.53	Ballard et al. (1999)
Oregon	≤7 months	58.7	0.86	Ricca et al. (2002)
Pennsylvania	≤34 weeks	48.6		Vreeland et al. (2004)
Michigan	≤127 days	9.3		Pusateri Burroughs et al. (2006)
Michigan	60, 180 days		0.19, 0.33	Hiller et al. (2008)

as 90% in the first 2–3 months in Texas and Oklahoma (Carroll and Brown, 1977; Bartush and Lewis, 1981). At the other extreme, researchers in Michigan (Pusateri Burroughs et al., 2006) reported losses in the first 127 days as only 9.3%. Nutrition and predation are common factors cited in neonatal fawn losses, along with disease and accidents.

Research on mortality of older fawns has not been as extensive versus neonatal (Table 5.3). The magnitude of mortality after about 6 months tends to be lower, but can still be substantial. Mortality rates for older female fawns tend to be less than those of males. Female fawns had a mortality rate of 5% in Illinois (Nixon et al., 1991) and 7% in Minnesota (Brinkman et al., 2004). An intermediate rate of 15% was found in Illinois (Etter et al., 2002) and West Virginia (Campbell et al., 2005). However, Van Deelen et al. (1997) found a mortality rate of 32% in Michigan. Mortality rates in older male fawns have varied from 12% in Illinois (Nixon et al., 1991) to 28% in Michigan (Van Deelen et al., 1997).

Mortality rates in yearling deer also differ by sex and are strongly influenced in many areas by greater hunting pressure on yearling bucks (Table 5.4). In a sample of studies, hunting mortality rates for yearling males has varied from 16% in Mississippi (Bowman et al., 2007) to 63% in West Virginia (Campbell et al., 2005). Nonhunting mortality in young bucks was negligible at 1.5% in a Mississippi study (Bowman et al., 2007). However, nonhunting mortality in yearling males is generally higher (Table 5.4), as much as 22% in an Illinois study (Etter et al., 2002). Total mortality rates in yearling bucks have been high in most studies, as high as 75% in West Virginia (Campbell et al., 2005). Hunting and nonhunting mortality rates tend to be lower in yearling females (Table 5.4). The highest total mortality rates were 37% in Illinois (Nixon et al., 1991) and Montana (Dusek et al., 1992).

TABLE 5.3

Mortality Rates for White-tailed Deer 0.5–1.0 Years of Age

	Males			Females			
Location	Hunting	Nonhunting	Total	Hunting	Nonhunting	Total	Study
Illinois	0.039	0.078	0.117		0.051[a]	0.051	Nixon et al. (1991)[b]
Michigan		0.28	0.28		0.32	0.32	Van Deelen et al. (1997)
Illinois					0.15	0.15	Etter et al. (2002)
Minnesota					0.07	0.07	Brinkman et al. (2004)
West Virginia		0.20	0.20		0.15	0.15	Campbell et al. (2005)

Poaching and wounding loss included in hunting mortality where applicable.

[a] Caused by automobile.

[b] 0.58–1.0 years.

TABLE 5.4

Mortality Rates for Yearling White-tailed Deer

	Males			Females			
Location	**Hunting**	**Nonhunting**	**Total**	**Hunting**	**Nonhunting**	**Total**	**Study**
Minnesota	0.38	0.21	0.59	0.05	0.15	0.20	Nelson and Mech (1986)
Illinois	0.385	0.220	0.605	0.0259	0.114	0.373	Nixon et al. (1991)
Montana				0.19[a]	0.18[b]	0.37	Dusek et al. (1992)
Michigan	0.47	0.16	0.63	0.12		0.12	Van Deelen et al. (1997)
Illinois					0.15	0.15	Etter et al. (2002)
West Virginia	0.63	0.12	0.75	0.09	0.05	0.14	Campbell et al. (2005)
Mississippi	0.164	0.015	0.179				Bowman et al. (2007)

Poaching and wounding loss included in hunting mortality where applicable.

[a] Mean of 2 years, River Bottom study area.

[b] Natural and Other combined, mean of 2 years, River Bottom study area.

Deer ≥2.0 years experience mortality rates that vary considerably depending on habitat, hunting pressure, and extent of older age structure (Table 5.5). Most studies have lumped older bucks into one common age class because of sample size or aging problems. However, when data are sufficient, a typical "U"-shaped mammalian mortality curve is apparent (DelGuidice et al., 2006). Figures 5.1 and 5.2 show male and female U-shaped mortality patterns from south Texas (Box 5.1).

BOX 5.1

Data for Figures 5.1 and 5.2 are based on 231 radio-collared deer from the Faith and Camaron ranches, south Texas, 1986–1990. Mortality rates for 0–3-month-old deer were approximated and not based on telemetry. Remaining mortality rates by age class were calculated with program MICROMORT. Transmitter days totaled 97,180 for 41 adult bucks, 43 adult does, and 26 fawns on the Faith Ranch and 45 adult bucks, 51 adult does, and 25 fawns on the Camaron Ranch.

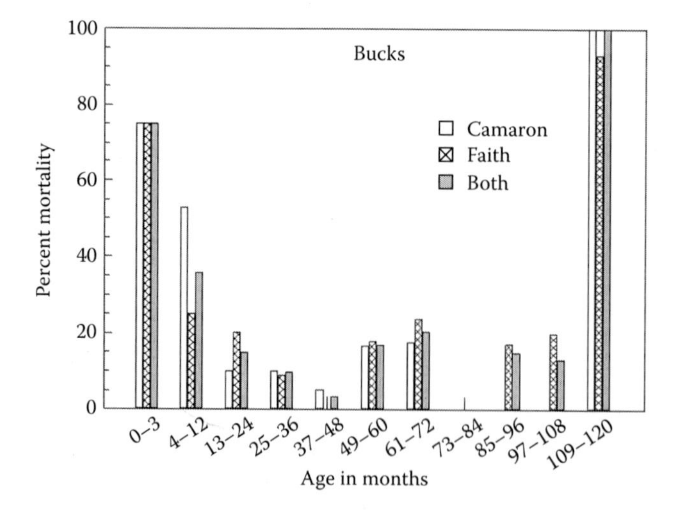

FIGURE 5.1 Mortality rates by age class for bucks on two study areas in south Texas. Box 5.1 describes methodology.

TABLE 5.5

Mortality Rates for White-tailed Deer at Least 2 Years of Age or Yearlings Grouped with Older Deer

		Males			Females			
Location	**Age (Years)**	**Hunting**	**Nonhunting**	**Total**	**Hunting**	**Nonhunting**	**Total**	**Study**
Minnesota	Adult	0.28	0.25	0.53	0.01	0.20	0.21	Nelson and Mech (1986)
Texas	≥1.5	0.05–0.12	0.08–0.23	0.16–0.35				DeYoung (1989b)
Minnesota	≥1.0	0.428	0.109	0.537	0.223	0.087	0.310	Fuller (1990)
Minnesota	≥10					0.56	0.56	Nelson and Mech (1990)
Illinois	≥2.0	0.553	0.059[a]	0.592	0.213	0.071[a]	0.284	Nixon et al. (1991)
Montana	≥2.0				0.224[b]	0.092[c]		Dusek et al. (1992)
Michigan	≥2.0	0.72		0.72	0.04	0.19	0.23	Van Deelen et al. (1997)
New Brunswick	≥2			0.43 –0.62[d]			0.08–0.52[e]	Whitlaw et al. (1998)
South Dakota and Wyoming	≥1.0						0.43	DePerno et al. (2000)
Oklahoma	1.5–2.5, 3.5–4.5, ≥5.5	0.133, 0.209, 0.133	0.027, 0.042, 0.133	0.16, 0.251, 0.263				Ditchkoff et al. (2001)
Illinois	Adult				0.05	0.12	0.17	Etter et al. (2002)
Nova Scotia	Adult	0.342	0.138	0.48	0.082	0.111	0.193	Patterson et al. (2002)
Oregon	Adult			0.26			0.27	Ricca et al. (2002)
Minnesota	>8 months				0.115	0.135	0.25	Brinkman et al. (2004)
West Virginia	≥2.0		0.73	0.73	0.04	0.08	0.12	Campbell et al. (2005)
Mississippi	≥2.5	0.297–0.442	0.064–0.184	0.367–0.556				Bowman et al. (2007)

Poaching and wounding loss included in hunting mortality where applicable.

[a] Automobile included in nonhunting mortality.

[b] Combined three study areas across five time periods.

[c] Combined three study areas across five time periods, Natural and Other combined.

[d] Two study areas, 1 year.

[e] Two study areas, 3 years.

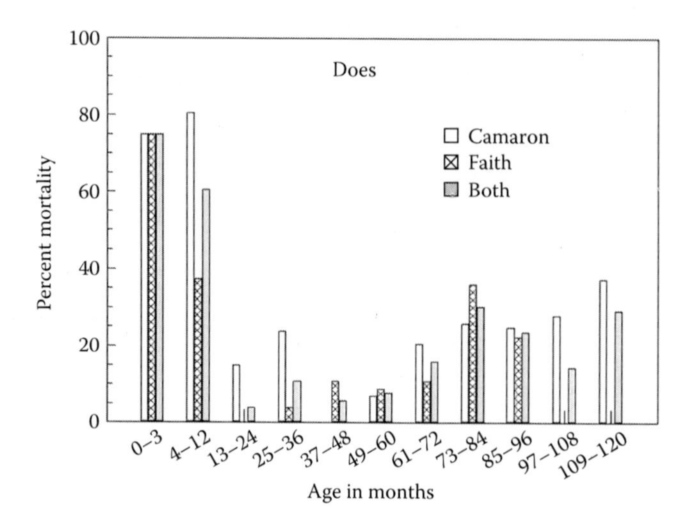

FIGURE 5.2 Mortality rates by age class for does on two study areas, south Texas. Methodology described in Box 5.1.

DeYoung (1989b) was unable to separate mortality rates among age classes for older bucks in an earlier data set from south Texas, probably because of insufficient sample size. Hunting mortality rates (females 0.21–0.22) in some studies (Fuller, 1990; Nixon et al., 1991) (Table 5.5) would be unsustainable in less productive habitats, as we will see later. Nonhunting mortality tends to be relatively low in most studies (Table 5.5), although rates can be substantial in old deer (Nelson and Mech, 1990) (Figures. 5.1, 5.2). In a review of ungulates, Gaillard et al. (2000) stated that prime-aged females are resistant to common sources of mortality that afflict other components of populations. Overall, survival of neonates is highly variable and often low, whereas adult survival is high and constant (Gaillard et al., 2000).

Dispersal

Dispersal is common in yearling male white-tailed deer but females disperse from some populations as well (Table 5.6). And, sometimes younger and older deer disperse. Dispersal is usually defined as a permanent movement from one home range to another (Kammermeyer and Marchinton, 1976) (Figure 5.3). The reason deer disperse is usually thought to be aggression from other deer. Dispersal was not related to population density in a Maryland study (Rosenberry et al., 1999). Orphaning was implicated in a Georgia study, where only 9.1% of orphans dispersed, versus 86.5% of nonorphans. Rosenberry et al. (2001) concluded that aggression related to sexual competition among yearling males was a factor motivating dispersal.

The percent of young bucks that disperse is commonly greater than 50% (Table 5.6). Dispersal distance is typically in the range of 4–10 km, but Nixon et al. (1991) reported distances of 45–50 km in Illinois. Percent of young females that disperse is typically lower versus males. No females dispersed in Minnesota (Nelson and Mech, 1984) and West Virginia (Campbell et al., 2004) populations. Contrast this with up to 50% of 10–12-month-old females that Nixon et al. (1991) found dispersing in Illinois. Table 5.6 provides a sense that female dispersal is not only lower, but also more variable from population to population versus male.

Amount of woody cover in a habitat seems to affect dispersal distance (Long et al., 2005) and dispersal rate (Nixon et al., 2007). In Pennsylvania, dispersal distance and rate were unaffected by population density (Long et al., 2005). Reduction in density of adult females and increase in density of breeding-age males did not affect the overall rate of dispersal, but seemed to influence the season in which it occurred (Long et al., 2008).

Mixing of deer due to dispersal has strong implications for population genetics. Also, dispersal rate and distance have strong implications for spread of chronic wasting disease and tick-borne human diseases

TABLE 5.6
Dispersal by White-tailed Deer in Various Studies

	Males			Females			
Location	Age	Percent Dispersing	Distance (km)	Age	Percent Dispersing	Distance (km)	Study
Georgia	1.5–2.5	50	Mean 4.4		5		Kammermeyer and Marchinton (1976)
Minnesota	By 17 months	70	Up to 9.6	Up to 3 years	0		Nelson and Mech (1984)
Eastern Illinois	10–12 months	50	45–50	10–12 months; Yearling	50; 20	45–50	Nixon et al. (1991)
Georgia	By 30 months	9.1% of orphans; 86.5% of nonorphans					Holzenbein and Marchinton (1992)
Northeastern Minnesota				Yearling	20	18–168	Nelson and Mech (1992)
Maryland	Yearling	70	Median 6				Rosenberry et al. (1999)
Northeastern Minnesota	1.0–1.5 years	64		1.0–1.5 years	20		Nelson (1993)
Maryland	Yearling	67					Rosenberry et al. (2001)
West Virginia				Fawns; >1 year	3.6; 0	15.4	Campbell et al. (2004)
New York				Yearlings and adults	<15		Porter et al. (2004)
Southwestern Minnesota				61 adults, 16 young	8	Mean 71.3	Brinkman et al. (2005)
Central Pennsylvania	>7 months	46	Mean 7				Long et al. (2005)
South Texas	Yearling	Area 1 = 68; Area 2 = 44	Area 1 = mean 4.4; Area 2 = mean 8.2				McCoy et al. (2005)
Ilinois	Fawns	25–58	28–44	Fawns	22–49	37–41	Nixon et al. (2007)

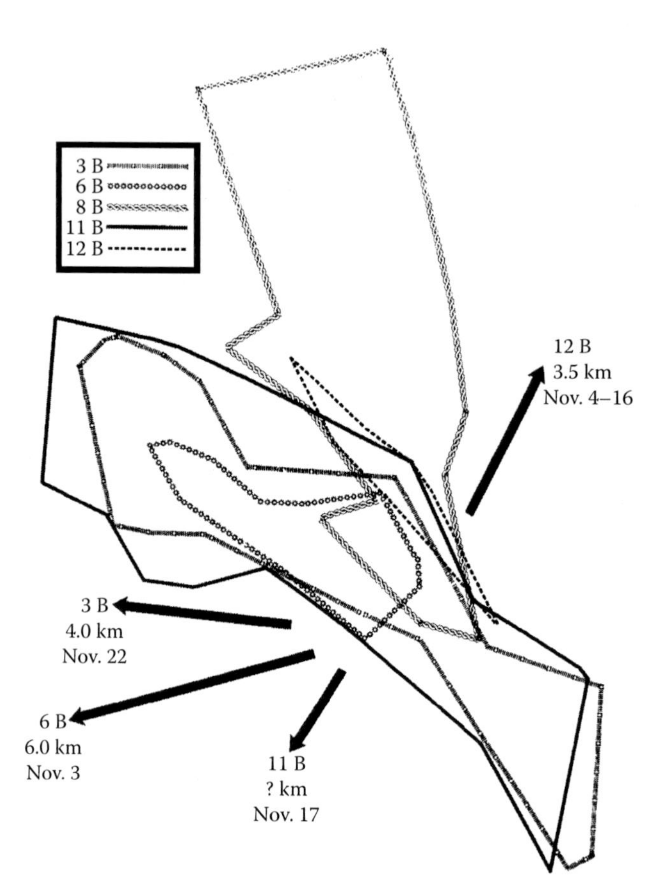

FIGURE 5.3 Overlapping home ranges of five Georgia bucks 1.5–2.5 years of age showing direction and distance of four that dispersed. (From Kammermeyer, K. E. and R. L. Marchinton. 1976. *Journal of Mammalogy* 57:776–778. Published by American Society of Mammalogists and Allen Press, Inc. With permission.)

(Long et al., 2005; Nixon et al., 2007). For population growth, the net sum of immigration and emigration is an important statistic and one that few managers know (Rosenberry et al., 1999). Accordingly, the net effect is frequently assumed to be zero, or dispersal is ignored altogether. The effect on population size could be significant if a pool of high-density deer is surrounded by lower densities. Rosenberry et al. (1999) speculated that the net loss from a Maryland population was about 5% of young males.

Density Dependence

There has long been strong interest in the net effect of the interplay of intrinsic and extrinsic factors on white-tailed deer populations (Kie et al., 2003). All populations are at times density dependent (DD) and at other times density independent (DI; McCullough, 1992; DeYoung et al., 2008). These are not competing models but interact in complex and variable pathways through time. In this chapter, emphasis will be on the net population response to limiting factors. Thus, a population determined to behave in a DD manner means that DD is the *net* effect of the various factors. Likewise, a DI population may have DD influences operating, but the net effect is DI. The *frequency* and *strength* of episodes of net DD versus net DI define local population behavior. Because white-tailed deer occur in such a variety of habitats, there is no generic model that is widely applicable. However, simple DD models, while useful in understanding population theory, are not generally suitable for application by managers because DD behavior is obscured by environmental factors (DeYoung et al., 2008). There are several types of density-dependent behavior that can be identified in animal populations (McCullough, 1992). This chapter will address density-dependent population responses caused by intraspecific competition due to food limitation.

Deer literature commonly considers primarily animal responses and treats climatic and habitat variables as nuisances to be overcome in modeling, an approach used because much of the deer literature is written by animal ecologists. McCullough (1990) considered population phenomena such as DD the "signal" and environmental factors as "noise." This emphasis on animal processes is not wrong, but is related to the specialty of the scientist. McCullough (1999, p. 1131) later stated: "... it is unsafe to assume that a given shape of the DD function is characteristic of a species because environmental conditions in specific situations can reshape the DD function. Shape of the DD function is influenced by scale of landscape over which the population occurs (and associated dispersal capability of the population), heterogeneity of the environment, quality and distribution of the food resources, and perhaps other variables not yet recognized." A climatologist or a plant/habitat specialist would emphasize the demographic effects of environment. Fryxell et al. (1991, p. 385) wrote: "... we may never be able to accurately predict long-term population trajectories of white-tailed deer until we understand the influence of foraging on plant physiological and demographic parameters and the influence of vegetation abundance and nutritional quality on deer demography."

Carrying Capacity and Habitat Considerations

There are many definitions of carrying capacity (Macnab, 1985). This discussion of density dependence will use nutritional carrying capacity (NCC; Bishop et al., 2009), defined as the equilibrium reached between deer and their food supply at the maximum sustainable population size (Macnab, 1985). NCC has also commonly been called K, ecological carrying capacity, in the deer literature (McCullough, 1979).

Density-dependent responses are generally expected from populations close to NCC (Macnab, 1985). Because of the wide variety of niches that white-tailed deer occupy across their range, getting close to NCC can be easy and swift for populations in some habitats (McCullough, 1979) or difficult and rare in others (DeYoung et al., 2008). Often overlooked by managers is the fact that NCC is a miserable place for a deer herd to be. Nutritional stress and mortality are high while fecundity is low and population growth is effectively zero. In habitats where forage plants have few defenses against grazing, whitetail populations are said to damage or destroy their habitat, giving rise to the descriptor "keystone species" (Waller and Alverson, 1997). Most managers consider populations at NCC as "overpopulated." Nevertheless, NCC is a useful benchmark on which to anchor concepts of population dynamics.

NCC is an abstract term usually defined by animal productivity. For example, if a population of white-tailed deer has no female fawns breeding and 0.9 fawns born/adult doe, it may be at or close to NCC (Downing and Guynn, 1985). However, there are habitat and environmental attributes unique to each situation, which vary the pathway through which each population arrives at NCC. Environmental influences (variable precipitation, severe and mild winters, etc.) on population trajectories require considering a forage-based, as opposed to a deer demography-based, NCC. The importance of this complexity cannot be overemphasized. These pathways must be considered to understand local population behavior. An abstract, demography-based NCC is useful for theory because it is difficult to conceptualize the range of actual complexity across whitetail distribution, but managers need some insight into how the local population gets there. Among the factors influencing population growth are the relative quantity of good-, medium-, and poor-quality forage; nature of plant defenses against grazing; and, degree of seasonal and annual variation in NCC. Predation may also influence population growth patterns (Kie et al., 2003, Chapter 8).

Little attention has been given by deer researchers to the relative quantity of nutritious forage versus medium- and poor-quality forage (Verme, 1969; Hobbs and Swift, 1985; Figure 5.4). This concept has an important role in density-dependent population behavior. The relative distribution of forage quality is unique to each habitat and is a continuously distributed variable across whitetail range. However, it is convenient to consider three discrete examples.

Type A populations are envisioned as similar to the classic George Reserve deer herd studied by McCullough (1979), in which frequent and regular density-dependent dynamics can be expected. Density-dependent responses are present (and may be evident) across a broad spectrum from low density to NCC. These populations occupy relatively stable climates giving rise to productive habitats (Figure 5.5).

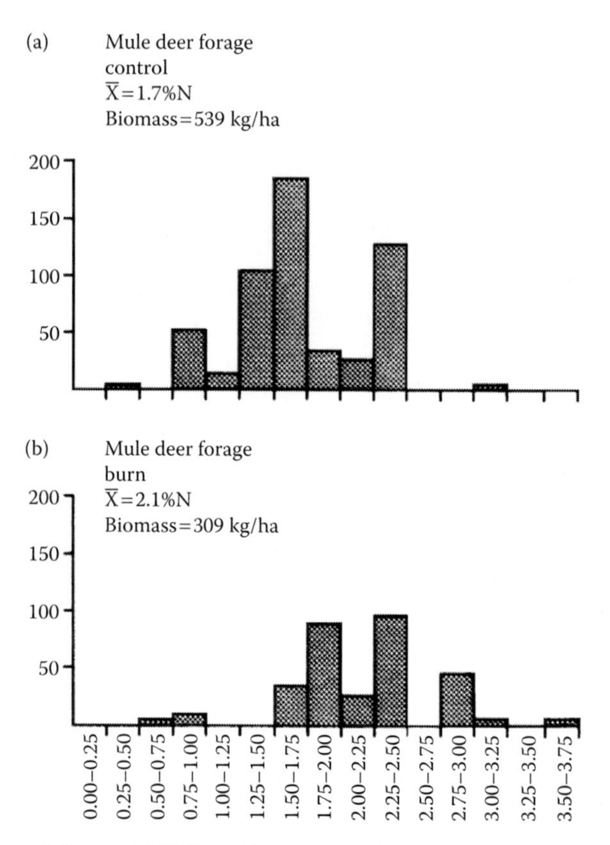

FIGURE 5.4 Distribution of nitrogen (gN/100 g DM) in forages of mule deer in (a) unburned control and (b) burned mountain shrub habitat, Colorado. (From Hobbs, N. T. and D. M. Swift. 1985. *Journal of Wildlife Management* 49:814–822. Copyright 1985, The Wildlife Society and John Wiley & Sons, Inc. This material reproduced with permission of John Wiley & Sons, Inc.)

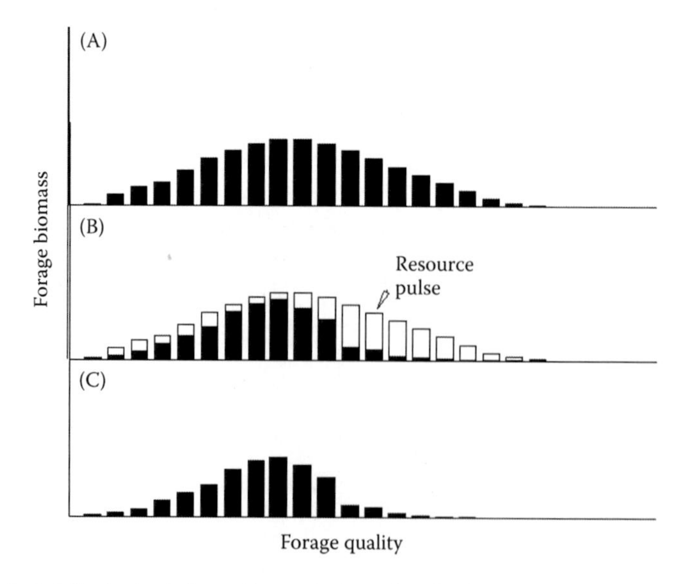

FIGURE 5.5 Hypothetical distribution in forage biomass by level of quality for three types of white-tailed deer populations.

The large reproductive potential of white-tailed deer frequently results in populations near, at, or above NCC. Harvesting sufficient females (male harvest may have little impact unless so severe as to impact breeding) from these populations results in increased juvenile survival and female fecundity that compensates for the lower number of foragers in the habitat. Type A roughly corresponds to Class I and II habitat described by Verme (1969) and Model A of McCullough (1999).

Type B populations occupy habitats where in most years there is a large supply of medium- and poor-quality food, sometimes with a thin veneer of high-quality food (Figure 5.5). These conditions provide for good deer productivity at very low density. However, above that, the forage sustains adults but does not provide for reproduction sufficient for population increase, except in occasional productive years such as above-average rainfall in semiarid habitats or mild winters in northern ranges (DeYoung et al., 2008) (Figure 5.5). The occasional productive years experienced by Type B populations have been described as resource pulses (Ostfeld and Keesing, 2000; Holt, 2008). In the southwestern United States, resource pulses are usually tied to El Nino–Southern Oscillation events (Quin et al., 1987). NCC may vary with the resource pulses in habitats of Type B populations but is more stable at high density for Type A and low density for Type C. Type B populations generally correspond to Class III habitat of Verme (1969) and Model B of McCullough (1999).

Type C populations occur in habitats with mostly medium- and poor-quality forage (Shea et al., 1992; Shea and Osborne, 1995; DeYoung et al., 2008; Jones et al., 2008) (Figure 5.5) that never get the occasional resource pulse like Type B. This situation was described by Shea and Osborne (1995) for some habitats in the southeastern United States. Poor nutrition is endemic in these situations which typically have poor-quality soils and relatively high annual precipitation. However, many states in the United States reported similar or equivalent areas of poor habitat (Shea and Osborne, 1995; DeYoung et al., 2008). Similar to habitats of Type B populations without a resource pulse, habitats of Type C populations can show good productivity at very low density, in response to a thin veneer of high-quality food. They exhibit a density-dependent response when they initially increase to somewhat higher densities, where population size then drifts up and down in response to a large amount of mediocre and poor-quality food with no density-dependent effects again unless they approach NCC. Type C is equivalent to Class IV of Verme (1969) and Model C of McCullough (1999).

Another important habitat attribute that acts in concert with the distribution of nutritional quality is the *defenses* against grazing of plants in the local habitat. This limiting factor varies among habitats. Frequently, the only defenses of most forage plants in habitats of Type A populations are to grow quickly, beyond the reach of deer. Type A populations are commonly associated with early forest seral stages. Following a forest disturbance, an abundance of nutritious forage is available within the reach of a foraging deer for a few years. However, as plant succession progresses, most of the good forage grows out of the reach of deer, or declines due to competition for sunlight with taller species. Alternatively, or concomitantly, the deer population may have grown so fast as to be at or above NCC. In these cases, Type A populations severely damage desirable forage plants by grazing, sometimes causing plant species to disappear from the community (Tremblay et al., 2005). The preponderance of research on Type A populations has led to a widespread impression that Type A populations are normal for white-tailed deer. However, as we will see, this is far from the case.

Habitats of Type B populations have quite different forage dynamics. Woody forage plants may have strong defenses against browsing, signified by thorns, plant architecture, and chemical compounds (e.g., tannins, alkaloids, and terpenes) that interfere with intake, digestion, metabolism, or palatability (Haukioja, 1980). Some plants exhibit compensatory growth in which grazing stimulates more growth than would have occurred in the absence of grazing (McNaughton, 1983; Teaschner and Fulbright, 2007). Annual precipitation variability is commonly high, with a CV greater than 25%, resulting in pulsed resource availability (Ostfeld and Keesing, 2000). In cold climates, occasional mild winters may produce a resource pulse. This happens when deer exit the winter in relatively good condition, which is further enhanced by early forage-plant green-up. The resource pulses cause NCC to vary through time. There is a large stock of medium- and poor-quality food present in sometimes complex mixtures of shrubs. High rainfall years not only promote nutritious stem tips and fruits on woody plants, but frequently encourage a strong suite of highly digestible annual forbs. Annual forbs are much more difficult to impact by grazing as long as there is some seed production. During dry years, the level of nutrition available to deer supports maintenance of adult deer, but does not provide for significant recruitment. Palatable forbs may be largely absent. However, during wet times, deer fecundity is high and mortality low. Whereas there still can be successional effects if habitats are disturbed, regrowth is not as rapid, nor is shading as big a concern as in Type A. Because woody forage plants have mechanical, chemical, and growth defenses against grazing, it is more difficult for Type B deer population to cause lasting damage to habitats.

Type B populations tend to vary in density through time, but it is difficult for them to approach NCC except in the rare event of several favorable years in a row. For example, this may have happened with Type B populations in south Texas during the 1970s, the wettest decade of the twentieth century in that region (DeYoung et al., 2008). Density-dependent population responses are less frequent, and/or density dependence is difficult to detect because influences are intertwined with extrinsic factors. NCC tends to vary through time with maximum levels depending on the strength and duration of resource pulses.

NCC may not be commonly reached in habitats of Type C populations. Once the small amount of quality food is eliminated or reduced, deer populations remain at low density and reduction in density does not usually result in a detectable compensatory response (Shea et al., 1992). There is typically a large amount of forage that provides maintenance level nutrition for adults, but little extra for producing recruits.

Lag effects involving DD population behavior are sometimes described in the literature (Fryxell et al., 1991). A variety of factors can cause time-lagged DD behavior (Fryxell et al., 1991), but delay in regrowth of grazed plants will be considered here. If a population near NCC is harvested, pressure on forage plants is reduced and some may then recover with new, nutritious growth. This enhanced nutrition then enhances productivity of the deer population, a DD response. However, forage plants have varying responses to grazing. In habitats of Type A populations, regrowth may be rapid if plant succession is not too far advanced. However, in arid, semiarid, or cold habitats of Type B populations, plant growth can be slow, and more nutritious forage may be delayed, thus causing a lag effect in deer productivity.

Another strong influence on DD behavior in deer populations is the temporal variability of NCC, both seasonally and annually. The mix of nutritious forage available varies seasonally among populations. In northern parts of whitetail range, populations migrate to deer yards to survive the winter (Hoskinson and Mech, 1976). Deer in these habitats depend on summer and fall seasons when enough nutritious forage is produced to allow deer to store enough fat to survive long, cold, snowy winters (Mautz, 1978). In semiarid, subtropical south Texas (Type B), late winter and early spring rains, when they occur periodically, promote a flush of cool season forbs. This burst of spring nutrition promotes a successful fawning season. If several wet springs occur in a row (rare), populations may approach NCC. During dry years, these populations are more like Type C.

Figure 5.6 shows NCC estimated roughly bimonthly on a south Texas study area using the tame deer technique (Box 5.2) (Strickland, 1998; Strickland et al., 2005).

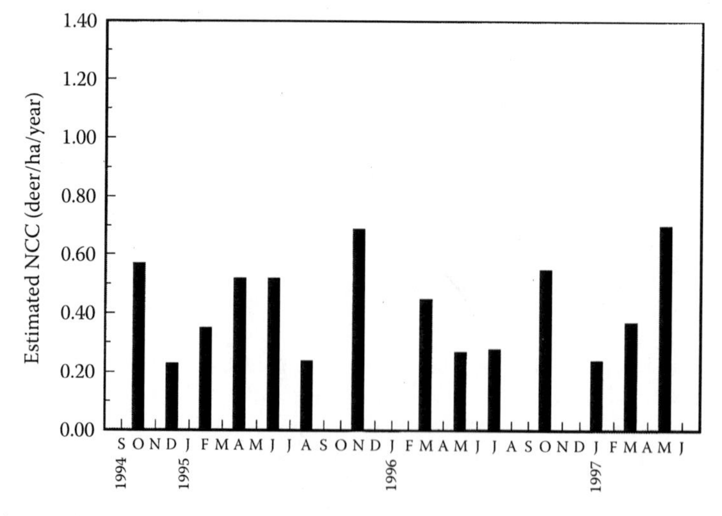

FIGURE 5.6 Estimates of nutritional carrying capacity (NCC) using the tame deer technique on the Faith Ranch, south Texas. See Box 5.2 for methodology.

BOX 5.2

Data for Figure 5.6 were collected using the tame deer technique on the Faith Ranch, Dimmit County, Texas, United States. This technique utilizes mass loss (or gain) of tame deer in small field enclosures to estimate digestible energy consumed by the deer. The equations developed by Strickland et al. (2005) were then used with the digestible energy estimates to estimate NCC. McCall et al. (1997) validated the tame deer technique by showing that it was correlated with forage-based techniques for estimating NCC. For details of the Faith Ranch study, see Draeger (1996) and Strickland (1998).

How does this seasonal rhythm result in an "overall NCC"? Should we measure NCC seasonally, annually, or averaged long term? Averaging across time is overly simplistic, because it does not capture the effect of segments of the seasonal rhythm on population behavior. However, for illustrative purposes, Table 5.7 shows averaged carrying capacity values (from Figure 5.6) compared with average corrected census values for the Faith Ranch, south Texas. This crude comparison suggests this population averaged about 38–54% of NCC during the sampling periods. DeYoung et al. (2008) used time-series analysis and failed to detect DD in this same population. However, they speculated there were occasional episodes of density dependence.

Zooming out to a broader view, annual precipitation variation or variation in winter severity, causing resource pulses, has a marked effect on population dynamics of white-tailed deer over about one-third of their range in the United States (DeYoung et al., 2008). Presumably, high variation in annual precipitation or winter severity leads to significant annual variation in NCC as well, causing Type B population dynamics.

This review of differing seasonal and annual dynamics shows how very different situations lead populations to NCC. Local dynamics are important to the manager, but an all-encompassing thought model is also instructive. DeYoung et al. (2008) showed how McCullough (1999) has produced the best unifying conceptual models to date (Figure 5.7). McCullough (1999) envisioned a plateau phase and a ramp phase in the relationship between population growth rate and population size. There is little evidence of DD in the plateau phase. Once a population enters the ramp phase, DD behavior becomes more evident. McCullough (1999) envisioned model C (Figure 5.7) only in rare populations on the periphery of deer range. However, DeYoung et al. (2008) hypothesized that populations that spend considerable time in the plateau phase were common across the range of white-tailed deer.

Before considering harvest of white-tailed deer populations, population estimation and deer aging techniques frequently used to support harvest decisions, will be covered.

TABLE 5.7

Comparison of Estimate of Nutritional Carrying Capacity (NCC, deer/ha) Obtained by Tame Deer Technique to Helicopter Surveys (deer/ha) on Faith Ranch, South Texas

NCC			Helicopter Survey			
n	**Mean (SE)**	**Time Period**	**n**	**Mean (SE)**	**Time Period**	**Percent**[a]
7	0.45 (0.07)	11 Oct 94–25 Nov 95	10	0.17 (0.02)	1983–1992	38
7	0.41 (0.06)	12 Mar 96–30 May 97	11	0.22 (0.03)	1987–1997	54

[a] Percent corrected helicopter survey was of NCC estimate.

Helicopter surveys were corrected by multiplying by a factor of 2.865, based on the percentage the raw counts were of a population estimate by mark-recapture (mean = 34.7%, $n = 72$, SE = 1.71).

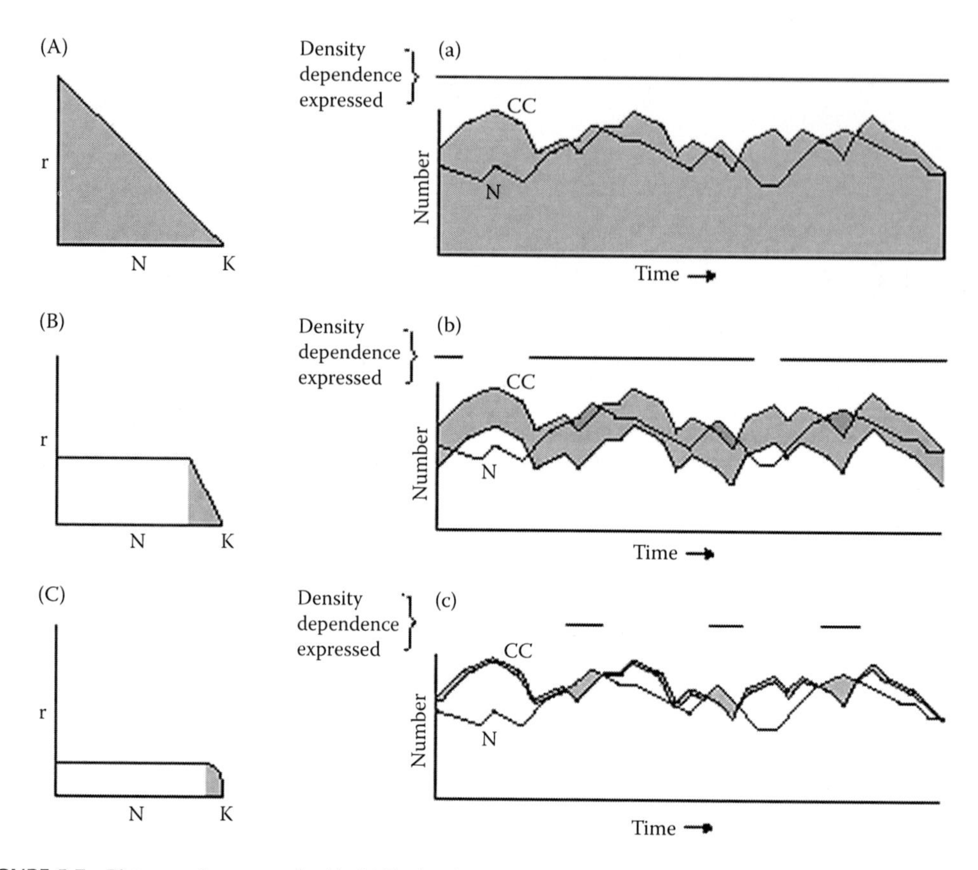

FIGURE 5.7 Plateau and ramp graphs (A, B, C) showing a range of deer population density-dependent responses and corresponding graphs (a, b, c) of carrying capacity variation in comparison with population level variations. (From McCullough, D. R. 1999. *Journal of Mammalogy* 80:1130–1146. Published by American Society of Mammalogists and Allen Press, Inc. With permission.)

Population Estimation

Reliable methods to estimate or index population size are integral to studying and managing white-tailed deer. Researchers and managers covet accurate, precise, and relatively inexpensive techniques for estimating population size. This goal has proven impossible to attain. Population estimates that can be demonstrated to be relatively accurate are usually imprecise. Most estimates are expensive to attain. Following is a review of selected commonly used population estimation techniques for white-tailed deer.

Pellet Group Counts

One of the oldest techniques for estimating population size for white-tailed deer is counting the number of fecal pellet groups in an area. Circular sampling plots are frequently employed with the results projected later to a larger area. A "daily defecation rate" is assumed, frequently in the 13–15 range (Eberhardt and Van Etten, 1956). Researchers in an early study in Michigan counted pellets accumulated since autumn leaf-fall (Eberhardt and Van Etten, 1956). However, other studies have employed physically clearing plots, then waiting a prescribed number of days for accumulation. Neff (1968) reviewed factors affecting pellet group surveys. In addition to good sampling protocol to attain adequate sample size, variations in defecation rate, observer bias, shape and size of sample plot, period of deposition, insect attack, and identification of species producing pellet groups affected surveys.

Factors that can affect defecation rate include range condition (moisture content varies), rate of feed intake, class of deer (fawns have higher rates versus adults), and abrupt changes in diet. In some habitats, dung beetle (superfamily *Scarabaeoidea*) activity is prevalent, resulting in the disappearance of pellets soon after deposition (Downing et al., 1965). Results of a Mississippi study revealed up to 100% loss of pellet groups on monthly surveys, with amount of precipitation a major factor (Wigley and Johnson, 1981). Penned whitetails consuming a commercial ration and alfalfa defecated about 20 pellet groups/day (Rollins et al., 1984). Tame, free-ranging deer produced an average of 34 pellet groups/day during October to April in Minnesota (Rogers, 1987) and 31 pellet groups/day during September to February in Georgia (Sawyer et al., 1990). Fuller (1991, p. 395) stated: "Given the number of factors that might affect pellet group deposition, counts of pellet groups probably do not unambiguously reflect deer abundance."

Pellet group counts are not commonly used in recent years. The technique is not likely to provide accurate population estimates. If factors enumerated by Neff (1968) are minimized, pellet group counts may detect relatively large changes in population size or trend.

Track Counts

Track counts to estimate deer population size are conducted by clearing dirt in a strip (frequently a road) and then counting the number of deer crossings during the subsequent 24 hours. Deer track counts have also been conducted in snow (D'Eon, 2001). In an early study in the southeastern United States, Tyson (1952) assumed that 1.6 deer tracks/km of road equaled 1 deer/km^2. A subsequent Georgia study (Downing et al., 1965) employed cleared plots instead of strips during summer. They estimated that 10–12 counts of 100–200 plots of 0.008 ha size were needed to detect a 20% population change. Harlow and Downing (1967) emphasized that large samples of tracks were needed because of high day-to-day variability in track crossings. Similarly, Minnesota researchers (Mooty and Karns, 1984) also reported high variability in track count samples. A Texas study (DeYoung et al., 1988) assessed track count population estimates by comparing to mark-recapture estimates of deer population size from repeated helicopter surveys. They reported that track counts produced very conservative estimates and failed to detect a population increase in one area. However, track counts did reflect a large difference in density between two study areas. A Florida study (Fritzen et al., 1995) also evaluated the Tyson (1959) method and found that deer population size was underestimated by about 20%. Mandujano and Gallina (1995) evaluated track counts against other census techniques in tropical dry forest in Jalisco, Mexico. They concluded that a conversion statistic was lacking for that habitat in order to estimate population size but track counts did index population trend over time.

Overall, track counts are not a reliable method of estimating white-tailed deer population size. If adequate sampling procedures are used, a large change in population size may be detected using track counts.

Infrared Thermal Imagery

Wildlife biologists have experimented with infrared thermal imagery to estimate white-tailed deer population size since the 1960s (Croon et al., 1968). The imaging equipment has usually been used from fixed-wing aircraft or helicopters. However, in recent years, truck-mounted or hand-held equipment has been used. Infrared scanning equipment senses electromagnetic radiation given off by objects and produces a television-like image of a deer if it emits more energy than background objects. However, screening objects like trees can partially or totally block a deer from being detected. A deer standing in the open against a snow background makes for ideal counting by infrared thermal imagery. Reliability of such a census declines when the background is near the energy given off by a deer or if the animal is obstructed by vegetation.

In Pennsylvania, detectability of deer from a fixed-wing aircraft was influenced by time-of-day, season, altitude, and wavelength sensitivity of infrared detectors (Graves et al., 1972). In South Dakota, also using a fixed-wing aircraft, thermal infrared equipment detected 88% of deer counted by ground personnel (Naugle et al., 1996). Spotlight counts in the same area underestimated deer density by 38% compared to infrared. Gill et al. (1997) evaluated ground surveys using a portable imager, in conjunction with distance sampling to estimate ungulate populations in the United Kingdom. Use of a hand-held sensor from a helicopter along transects in Florida resulted in counting 42% more deer versus visual observations from the helicopter (Havens and Sharp, 1998). In this study, thermal imaging was more accurate versus traditional helicopter surveys. Haroldson et al. (2003) evaluated aerial thermal imaging in an area of Missouri mostly covered by deciduous forest. They conducted thermal imaging from a fixed-wing aircraft and compared results to mark-resight estimates from conventional visual surveys by helicopter. They surveyed quadrats three times and infrared thermal imaging resulted in counts that varied from 31% to 89% of the population estimate by mark-resight. The CV for imaging surveys was 36%. They attributed much of the variation to observer bias and inconsistent thermal contrast between deer and the background. In spruce–fir forests, infrared imaging from a helicopter produced estimates that varied from 54% to 89% of reconstructed populations (Potvin and Breton, 2005).

There is a perennial confidence that improved imaging equipment will eventually be released into the commercial market by "the military." Until that time, white-tailed deer population estimates using infrared thermal imaging are best made in open or relatively open habitat with a cold background. A warm background and/or forest cover may result in deer population estimates with poor accuracy and precision.

Night Spotlighting

Utilizing light from a spotlight reflected off the tapetums of a deer's eyes, night spotlighting has been a common method of counting deer. Progulske and Duerre (1964) evaluated white-tailed deer population change using spotlight counts in South Dakota. Most deer were observed in the early hours of darkness, although bucks were not readily observable during summer. McCullough (1982) conducted bi-weekly spotlight surveys throughout the year on the George Reserve in Michigan and compared them with a base count estimated by population reconstruction. He estimated the area surveyed from a map and adjusted by habitat type. Spotlight counts were quite variable, ranging from less than 10% of the base count to above 50% in November and April. Bucks were typically underrepresented in counts, with highest buck counts in July. Fawns were greatly underrepresented until several months old, and did not approach the base count until 10 months old. Spotlight counts were hampered on the George Reserve by inability to spotlight one habitat type (tamarack swamp).

A south Texas study evaluated spotlight counts along a 16-km route counted 184 times during August to October (Fafarman and DeYoung, 1986). Researchers calculated a population estimate using an estimated visible area along the route determined by perpendicular measurements from the route. Such estimates are affected by the net difference between a positive bias in the visibility estimates and a negative bias due to missed deer (Evans, 1975). Contrary to the results of Progulske and Duerre (1964), fewest deer were observed in the first 13% of darkness. Spotlight counts later in the night resulted in population estimates that were 82–98% of corrected helicopter surveys. Fawn counts increased monthly from August to October, similar to results from the George Reserve (McCullough, 1982).

Attention-eliciting techniques, such as a police whistle or predator call have increased spotlight counts by 10% (Cypher 1991). Whipple et al. (1994) simulated deer with reflective thumb-tacks in western Texas. They used methods similar to Fafarman and DeYoung (1986) for population estimates. Observers tended to underestimate visibility along the count route in open habitat and overestimate in closed habitat.

Night spotlighting for population estimates may work reasonably well in open habitats but utility declines with thicker vegetation. Spotlight counts are also quite variable and counts are best repeated several times. Sex ratio estimates vary with time of the year and can also be quite variable. Fawn counts are nearly always conservative versus adults.

Surveys by Helicopter

Maneuverability, slow speed, lower-altitude capability, and good visibility have made the helicopter the usual choice for aerial counting of white-tailed deer. Rice and Harder (1977) marked deer with collars on an Ohio study area, then flew a survey by helicopter and subsequently estimated a population size using mark-recapture techniques. They stated (Rice and Harder, 1977, p. 205): "The method is best suited to enclosed or isolated areas with adequate snow cover and minimal canopy." Rice and Harder (1977) attained good precision with their techniques, but in Texas research (Beasom, 1979) CVs of repeated flights were 18%, 8%, and 13% for winter, spring, and autumn, respectively. In south Texas, it has been common to count 100 m on each side of the flight line. On average, 53% fewer deer are sighted at 50–100 m from the survey line, versus 0–50 m (Beasom et al., 1981). Another Texas study marked deer individually with collars and ear tags and then flew repeated surveys by helicopter (DeYoung, 1985). Subsequently, a mark-recapture population estimate was calculated and the raw survey count compared for accuracy. Autumn-count accuracy averaged 36% versus 65% for winter. Beasom et al. (1986) used similar methods also in south Texas and reported accuracy ranging from 28% to 40% for repeated flights in autumn. Accuracy was unaffected by sample intensity (percent of area surveyed) but CVs roughly doubled from about 30% to about 60% for 100% and 10% area coverage, respectively. A correction factor has been developed for helicopter surveys based on counting deer in strips out from the flight line (DeYoung et al., 1989). The correction procedure differed from mark-recapture population estimates by <6%. In a major improvement, mark-resight estimators have subsequently been developed for combining multiple surveys (White, 1996; McClintock and White, 2009; McClintock et al., 2009).

Counting deer from a helicopter against a snow background seems to enhance the percent counted (Stoll et al., 1991). In a Missouri study, 54% of marked deer counted over snow were sighted (Beringer et al., 1998). Potvin et al. (2004) used a double-counting technique with two independent observers and distance sampling methodology to estimate deer in plots from a helicopter in Quebec. Population estimates for large areas (583–19,043 km^2) had 90% confidence intervals that varied from 10% to 49% of the population estimate. They concluded that their techniques applied best to highly visible populations and would not work in closed-canopy habitats.

Surveys of deer populations by helicopter work best in habitats without dense woody cover. Surveys by helicopter are not suited for heavily forested habitats. Helicopter surveys tend to be moderately to highly variable from count-to-count and repeated counts of subsets of the habitat are most efficient in producing reasonable precision versus a single flight of a large portion. Accuracy depends on many factors and correction factors for visibility biases or distance sampling techniques work best.

Infrared Cameras

Use of infrared- or motion-triggered remote cameras to count deer is an increasingly popular technique, especially for smaller areas. Jacobson et al. (1997) evaluated the technique in Mississippi and developed a protocol for estimating population size. Their method was based on individual recognition of bucks based on antler characteristics and photographing all or nearly all bucks in an area over bait. Once the number of bucks is determined, does are estimated using the ratio of their photos versus bucks, and assuming equal sightability of the sexes. Fawns are estimated using the ratio of their photos versus does and again assuming equal sightability. Adding all classes gives a population estimate. At a camera density of 1/65 ha, they photographed 100% of individually marked deer in the area one year and 88% the next. A Texas study (Koerth et al., 1997) compared estimates from camera surveys (camera density 1/32.5 ha) to repeated helicopter surveys corrected as per DeYoung et al. (1989). The two techniques produced similar estimates for a number of bucks but the camera procedures estimated fewer does, and thus a lower population size. Another Texas study found that sex ratio varied considerably by month (Koerth and Kroll, 2000). Evaluation of camera census in Mississippi and Oklahoma at camera densities of 1/41 and 1/81 ha used marked deer (McKinley et al., 2006). Photographic recapture rates were 92% and 89% for adult males and females, respectively, on the Mississippi area. However, recapture rates were only 22% for adult males and 34% for adult females in Oklahoma. The lower rates in Oklahoma were attributed to abundant acorns, which made the corn bait less attractive. Importantly, little difference in recapture rates was found for males and females.

Roberts et al. (2006) and Watts et al. (2008) evaluated camera surveys of Key deer in southern Florida. Both studies used marked deer and did not bait camera sites, instead placing cameras on well-used trails and at watering sites. Roberts et al. (2006) recommended camera surveys over traditional road surveys for Key deer. Watts et al. (2008) recaptured on photos 84%, 70%, and 100% of marked individuals by day 10 on three areas. They also found no bias in sighting probabilities by sex.

Moore (2008) evaluated camera surveys in south Texas within 81-ha enclosures using marked deer. He used corn-baited camera sites at a density of 1/20.3 ha. Unlike some previous studies, Moore (2008) found a bias toward males in photographic recapture rate, but no change in bias across differing deer density. The bias toward more bucks photographed resulted in a tendency to underestimate population size using the methods of Jacobson et al. (1997).

Although no panacea, infrared remote cameras provide the best chance at a reasonable population estimate in forested habitats. Most camera counts of deer are conducted on relatively small properties. However, there is no reason that with good sampling design, they could not be used over extensive areas. Cost is always a concern when counting deer, and a limiting factor with remote cameras is the labor required to view thousands of photos. The differences among studies in whether there is a bias in photos by sex, and in photographic recapture rates of marked deer, may suggest that researchers will need to adapt methods on a regional basis.

Mark-Resight

Mark-resight models are a subset of traditional mark-recapture techniques and are primarily used by researchers (White, 1996; White and Shenk, 2001; McClintock et al., 2009; McClintock and White, 2009). Basically, some animals in a population are marked and then resighted in a subsequent survey. Thus, physical recapture is avoided in the surveys, reducing sampling effort and stress on animals. When different techniques are used for initial marking and follow-up resighting, bias due to "trap avoidance" is not present. Mark-resight models were initially based on radio-collared animals because of the necessity of knowing the exact number of marked individuals (White, 1996). However, techniques have been developed for situations where the number marked is unknown (Arnason et al., 1991). Another fundamental assumption of mark-resight models is that the population of interest is closed, although ways of getting around this assumption are available (McClintock et al., 2009).

Because of available software, particularly PROGRAM MARK, use of mark-resight techniques by researchers is likely to increase. This is particularly true for users of aerial survey and camera survey.

Deer Population Estimation in General

The above review covers the most used population estimation methods for white-tailed deer, but is by no means exhaustive. The goal of good accuracy and precision from an estimator is elusive, and seldom attained. With good research vetting and careful design, a reasonably accurate estimate with 90% confidence interval of 20% of the mean may be attainable at modest cost. As usual with white-tailed deer, the particular population estimation method that is best, and the sampling intensity required, varies widely over the species' range.

Estimating Deer Age

Like population estimation, aging white-tailed deer is a widely used technique for studying and managing populations. Aging white-tailed deer fetuses is frequently undertaken in management and research in order to approximate the breeding season of a population. Several early researchers established a chronology for backdating fetuses to the time of conception (Cheatum and Morton, 1946; Armstrong, 1950; Short, 1970). From these early studies, the crown-rump measurement emerged as the primary statistic for determining fetal age (Hamilton et al., 1985) (Figure 3.6).

Haugen and Speake (1958) developed criteria for aging young fawns, including hoof length. Hoof growth has subsequently been found to be the best predictor of age out of eight measurements taken on fawns up to 31-days old (Sams et al., 1996a).

Aging white-tailed deer by dentition has been the main technique for aging older fawns and adults since Severinghaus (1949) published his seminal paper. Tooth replacement and wear (TRW) continues to be widely used today, especially in management applications (Figure 5.8). There have been several evaluations of the technique (Ryel et al., 1961; Cook and Hart, 1979; Jacobson and Reiner, 1989; Hamlin et al., 2000; Gee et al., 2002; Meares et al., 2006). There is a tendency across studies to overage young animals (2 years of age) and underage older animals (at least 4 years of age)

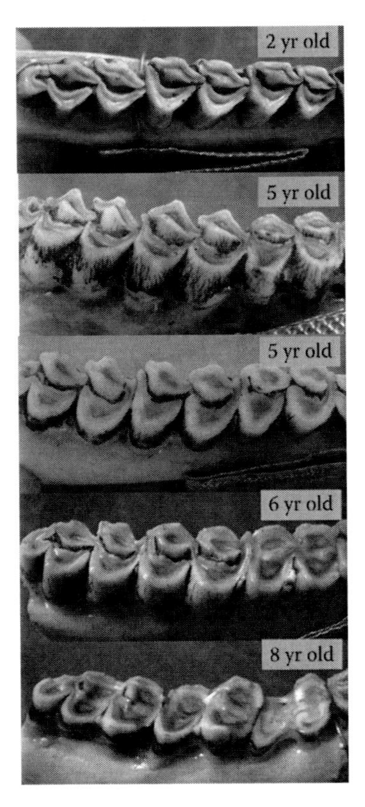

FIGURE 5.8 Known-age white-tailed deer jaws from south Texas showing variation in wear by age and within age. (Photos by J. Lewis. With permission.)

TABLE 5.8

Comparison of Age Class by Tooth Replacement and Wear versus Known or Cementum Annuli Aging for White-tailed Deer

	Wear Age (years)						
Wear Age Compared with	**2**	**3**	**4**	**5**	**6**	**7**	**Source**
Known age	+	=	–	–	–		Ryel et al. (1961)
Cementum age	+	–	–	–	–	–	Gilbert and Stolt (1970)
Cementum age	=	+	+	=	–		Brokx (1972)
Cementum age	–	+	=	–	–		Lockhard (1972)
Known age	+	+	+	=	=		Cook and Hart (1979)
Known age	=	–	–	– (≥5.5)			Jacobson and Reiner (1989)
Known age	+	+	–	–	–	–	Hamlin et al. (2000)
Known age	+	+	+	+	+	+	Gee et al. (2002)
Known age	+	=	–				Meares et al. (2006)
Score[a]	0.56	0.33	–0.22	–0.50	–0.57	–0.33	

Source: Adapted from DeYoung, C. A. 1989a. *Journal of Wildlife Management* 53:519–523.

[a] Score determined by assigning a 1 to a "+," 0 to =, and −1 to "–," summing and then dividing by the number of studies that evaluated each age.

(Table 5.8). If one assumes equivalent research quality, the bias appears to vary by study site. Texas research (Cook and Hart, 1979) found a tendency to overage through 4 years, then no bias at 5 and 6 years of age. In Mississippi, no tendency to overage was found, with no bias at 2 years, then tendency to underage thereafter (Jacobson and Reiner, 1989). Older fawns and yearlings are often assumed to be aged accurately by TRW, although low levels of miss-aging have been reported for yearlings (Cook and Hart, 1979).

Gee et al. (2002) recommended no more than three age classes be established when using TRW: fawn, yearling, and at least 2 years. Jacobson and Reiner (1989) and Hamlin et al. (2000) suggested grouping TRW ages in four classes: fawn, yearling, 2.5 years and at least 3.5 years. However, this seems to ignore useful information if there are a lot of older animals in the population, and whereas TRW may miss many, even most, ages at least 2.5 or 3.5 years, there is a correlation with age (Hamlin et al., 2000). Bias in TRW tends to "pile up" animals in middle-age classes (DeYoung, 1989a) (Figure 5.9). Correcting such a piled-up distribution can result in reasonable approximation of population age structure for management purposes (but see Gee et al. [2002], who felt variation in aging errors was too great to employ a correction). Recently, techniques have been developed for estimating age structure from inaccurate aging techniques if known-age comparisons are available (Conn and Diefenbach, 2007; Conn et al., 2008).

Ransom (1966) described the use of the cementum annuli technique to age molars of white-tailed deer, along with Gilbert (1966), who extended the technique to sectioning primary (I_1) incisors. Most biologists have subsequently used I_1 incisors, although I_2 incisors have been used to age live deer (DeYoung, 1989a). Some early evaluations of aging white-tailed deer by cementum annuli were not encouraging when compared against known-age specimens (Cook and Hart, 1979). However, subsequent studies have shown greater accuracy versus TRW (DeYoung, 1989a). Generally, accuracy of cementum annuli aging is greater in northern versus southern latitudes (DeYoung, 1989a; Matson, 2009). Importantly, when

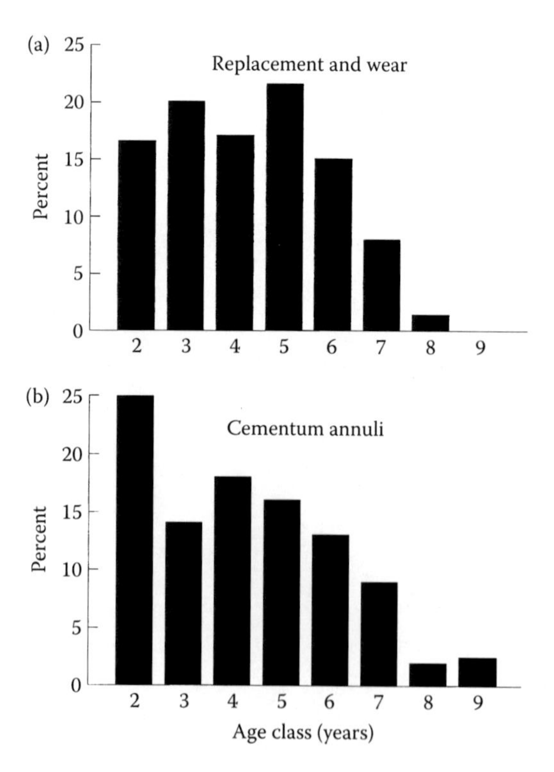

FIGURE 5.9 Age distributions for 199 male white-tailed deer from south Texas produced from two aging techniques. (From DeYoung, C. A. 1989a. *Journal of Wildlife Management* 53:521. Copyright 1989, The Wildlife Society and John Wiley & Sons, Inc. This material reproduced with permission of John Wiley & Sons, Inc.)

incorrect ages are assigned when using the cementum annuli technique, the misses are less biased than those with TRW (DeYoung, 1989a). Overall, cementum annuli analysis is more accurate and less biased than TRW, and is highly recommended for research work. Managers need to decide if the extra expense of cementum analysis is justified versus the information provided by TRW, as long as the limitations of the technique are understood. Tooth replacement and wear will continue to be widely used in management.

There has recently been much interest in aging live bucks using body characteristics, or aging on the hoof, because of the increase in state and private managers setting antler restrictions on buck harvest (Chapter 4). Antler restrictions typically have the goal of increasing buck age distribution in public management and increasing number of bucks with large antlers in private management. There are subjective criteria for aging bucks on the hoof (Demarais et al., 1999; Richards and Brothers, 2003), but relatively little research evaluation. DeYoung (1990) determined age by cementum annuli on two areas in south Texas and examined antler overlap among age classes. He concluded that when using antler size alone, it was not possible to distinguish among age classes 3.5 years and older. However, he stated that it may be possible to distinguish most bucks equal to or less than 2.5 years from those 3.5 years or older based on antler size. Hellickson et al. (2008) compared antler and various body measurements of bucks aged by TRW and concluded that the best option for aging on the hoof was a combination of antler and body characteristics. They concluded that chest or stomach girth, combined with antler size, were most useful for estimating buck age in the field. Specific criteria likely vary considerably among populations.

Harvest

Caughley (1985, p. 12) stated: "As of now we lack the knowledge to manage wildlife harvesting scientifically." He urged that designed, replicated experiments be conducted incorporating several harvest rates. Unfortunately, there has not been a lot of progress in the last 25 years. Controlled experiments at the population level on white-tailed deer are difficult, costly, and ideally long term. These are all excuses that Caughley (1985) rejected. The classic study by McCullough (1979), albeit unreplicated and unduplicated, has had a large influence on theory and management of white-tailed deer. This and subsequent thinking by McCullough (1984, 1990, 1992, 1999) have produced the best hypotheses for white-tailed deer population behavior, including harvest. However, these hypotheses have not been widely tested throughout whitetail range and have given rise to expectations of predictability of population behavior to harvest that are unrealistic for some populations. These concepts are not wrong until proven wrong, but the reality is that managers cannot safely employ simple predictive models in many habitats without risk (DeYoung et al., 2008). McCullough (1990, p. 534) stated: "...unambiguous demonstrations of DD as measured by population growth rate are few, and there are many contradictory results." This strongly suggests that most populations will not follow the predicted path of simple DD models, and that local models incorporating at least one environmental variable, along with DD behavior, need to be developed (DeYoung et al., 2008). Models should also consider the potential for time lags.

In a simple form, DD predictive models are something like Figure 5.10. As a population increases from minimum size to NCC, the harvest that it can yield at first increases rapidly with population growth. At some point, density-dependent feedback slows population growth, which in turn slows the harvest yield until it declines to zero at NCC. The peak of yield is termed maximum sustained yield (MSY).

As noted under the "Density Dependence" section of this chapter, the conceptual models of McCullough (1999) (Figure 5.7) are the best for representing white-tailed deer population behavior in a variety of habitats. McCullough (1999) uses a plateau and ramp in various configurations to conceptually model deer population behavior. Actually, McCullough (1999) suggested slower breeding ungulate species for some of his configurations. However, DeYoung et al. (2008) argued that whitetail populations may be represented by all McCullough's configurations in some parts of the species' range. Furthermore, the first configuration is all ramp, illustrative of the George Reserve population (McCullough, 1979). However, McCullough (1990) stated that this was an artifact of the small scale of the study area, and

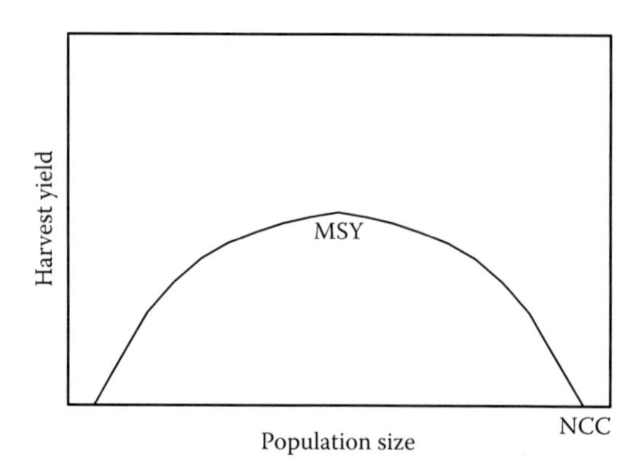

FIGURE 5.10 Hypothetical relationship of harvest yield across a range of population sizes for white-tailed deer.

that for a larger area with the same dynamics, a plateau would develop. When a population is on the plateau, no DD behavior is observed and any harvest is additive mortality. Only when a population is in the ramp phase do compensatory responses occur and allow a significant harvest through time. There is virtually nothing written about harvesting populations in the plateau phase. Some authors refer to harvesting a population on the left side of the parabola in Figure 5.10 as "unstable" meaning that the harvest is additive mortality and if significant, could cause a rapid and severe population decline (McCullough, 1979, 1984; Keyser et al., 2005). Determining population size relative to carrying capacity (Figure 5.10) is difficult because of expense and limitations of population monitoring techniques. Biologists commonly assume that whitetail populations are at "K" (NCC) without any real evidence this is the case (Ballard et al., 2001).

McCullough's (1999) models do not directly include environmental influences. Thus, they are only a part of the dynamics with which a deer manager must cope. McCullough (1984, p. 224) recognized the real-world problems with environmental variation and wrote: "If the environmental stochasticity is great relative to density-dependent responses of the population, an *ad hoc* strategy (for harvest) will be required. That is the biologist or manager will not be able to develop predictive models based on density-dependent response of the population and follow a more-or-less consistent management plan. Instead, management decisions will have to be made on a year-by-year basis in response to the environmental factors of the particular year."

A major problem hampering a scientific approach to deer harvesting (i.e., maximizing the efficiency of harvest at some density goal) is the lack of accurate and precise monitoring techniques. As reviewed in sections of this chapter on population estimation and estimating deer age, such monitoring techniques have limitations in most applications. Thus, assuming the development of a reasonably realistic model of population behavior, it is difficult to determine at a point in time where the population is within the model's limits of population growth.

So far in this discussion, the outlook appears gloomy for scientific deer harvest practices. This is not the case. Expectations have been unrealistically raised by concepts such as "maximum sustained yield" (MSY) (Figure 5.10). While a useful concept in discussing and understanding population theory, it is not particularly useful in the real world of deer managers. The issue is really *harvesting efficiency*. Modern deer harvest, in its many facets, is not efficient like MSY, but it works. Harvest produces no harm in most cases to local white-tailed deer populations and provides recreation through hunting and viewing for millions of people. Harvesting, at sufficient levels, reduces deer damage in other situations.

Compensatory population response is required for MSY. Another fallacy of MSY is that even if a manager could achieve it, it is not the goal in many management situations. MSY purports to be the point where the maximum number of individuals can be removed from a population. However, deer harvest in most applications is much more complex that just number harvested. Harvest goals for bucks versus does versus fawns, young bucks versus older bucks, ability to see a lot of deer, and reducing deer damage to a minimum are in the mix of common management goals that do not necessarily involve MSY. Harvest of prime-age does will have more impact on population productivity than harvest of bucks due to the polygynous mating system of white-tailed deer. This system of behavior results in some bucks fertilizing multiple does (DeYoung et al., 2009).

Population effects due to widening sex ratios caused by greater harvest rates on bucks have received little research attention in white-tailed deer. Possible effects include late breeding or no breeding by some does because bucks cannot find receptive does in a timely manner (White et al., 2001). For mule deer in Colorado, adjusting sex ratio from 1 buck:10 does to 1 buck:2.5 does was estimated to result in only 7.4 additional fawns:100 does (White et al., 2001).

Adaptive Management

Most managers follow a monitoring program and adjust harvest if a population is below or exceeds a goal. Managers in most local areas have experience with a harvest level that satisfies management goals and causes no harm to the population (unless population reduction is a goal). Managers also commonly use harvest data as well as population estimates or indices to guide harvest regulations. Harvest data such as deer sex, mass, age, antler measurements, and lactation rates are commonly collected and used to interpret population trend.

Harvest Efficiency

Controlled hunts in an enclosure in northern Michigan over seven years showed that fawns were most vulnerable to hunting and mature bucks least vulnerable (Van Etten et al., 1965). Researchers speculated that an annual harvest rate of 33% would have maintained the enclosed herd at a relatively constant level. Holsworth (1973) conducted a controlled hunt on Griffith Island, Ontario. He calculated an index of hunting efficiency as number of deer killed/hour = number of deer/259 ha times 0.006. Roseberry and Klimstra (1974) conducted a controlled hunt on Crab Orchard National Wildlife Refuge, Illinois. They calculated an index of hunter efficiency k using the equation $C(t) = kN(t)$, where $C(t)$ is the kill/hour at time t and $N(t)$ is the population density (deer/259 ha) at time t. The index declined at Crab Orchard from 0.0016 on the first day of hunting to 0.0006 on the last day. This is compared with an index value from Van Etten et al. (1965) of 0.004, and from Holsworth (1973) of 0.006. Thus, the Crab Orchard harvest was less efficient versus the other studies, which could at least partially be attributed to the method of hunting. Crab Orchard hunters were restricted to blinds whereas hunters in other studies were not.

Harvest Rate

Tables 5.3–5.5 show some of the great variation in harvest rate that has been reported among different ages and sexes of white-tailed deer. This variation is not surprising given the many different management goals and local population dynamics encompassed by these studies. However, the harvest rate that can be sustained by Type A populations versus Types B and C can be quite different.

The east-central Illinois deer herd studied by Nixon et al. (1991) appears to be a Type A population, although about 20% of the population migrated seasonally, suggesting limiting winter weather. The population occupied a highly fragmented but very productive habitat and had an unusually high rate of fawn and female dispersal. Dispersal served to replenish small islands of forest among extensive farm fields when deer populations in woodlots were decimated by hunting. Nixon et al. (1991) reported hunting-related annual mortality rates for adult bucks as 0.55 and does of 0.21 (Table 5.5). Nevertheless, the population was generally increasing before and during the study.

A declining deer population further north in Minnesota had only slightly lower mortality rates from hunting at 0.43 for bucks and 0.22 for does at least 1 year of age (Fuller, 1990) (Table 5.5). This population was subjected to more severe winters and wolf predation than that studied by Nixon et al. (1991). Fuller (1990) concluded that hunting was a significant factor in the decline of the Minnesota population and suggested reduction in female harvest.

BOX 5.3

Ranch replicates consisting of the Faith and Camaron ranches were separated by 125 km. Before the study, both ranches had received no harvest or incidental harvest for several years. Two sections of about 2000 ha were identified on each ranch. One section was harvested at a target rate of 15% and the other served as a control with no harvest. Population estimates were based on four or five repeated helicopter surveys in each of the four study years. Population estimates were made utilizing mark-recapture techniques the first two years and the method described by DeYoung et al. (1989) the final two years. Harvest treatment areas on each ranch were harvested by catching bucks, does, and fawns at random and translocating them off the areas.

The hunting mortality rates documented in Illinois (Nixon et al., 1991) and Minnesota (Fuller, 1990) would rapidly decimate Type B or C populations. Box 5.3 describes methods employed in a study of harvest on two Type B populations in south Texas. An area of about 2000 ha on the Faith Ranch was experimentally harvested at a rate of 10.6%, 16.5%, and 18.6% in 1987, 1988, and 1989, respectively. A similar area on the Camaron Ranch was harvested at a rate of 9.0%, 12.3%, and 27.0% over the same years. Each harvested area was compared with an unharvested area of similar size on the same ranch. Figure 5.11 shows trends in total population for the treatment and control areas across both ranches. Although the study was short, harvested portions declined versus unharvested portions. This study of Type B populations in south Texas shows that they were not able to withstand anywhere near the harvest that was sustained by the populations studied by Nixon et al. (1991). Again, the diversity of situations producing different population dynamics across the range of white-tailed deer is apparent.

Final Comments

White-tailed deer are a very flexible species that has expanded or increased in many empty Pliestocene niches. This expansion has also likely benefited from the late Pliestocene extinction of several large predators. The adaptability of the species is illustrated by its expansive range from Canada to Peru and from Atlantic to Pacific coasts. Because they occur in such a variety of situations, the population dynamics of white-tailed deer are also varied. Researchers and managers should evaluate local factors affecting

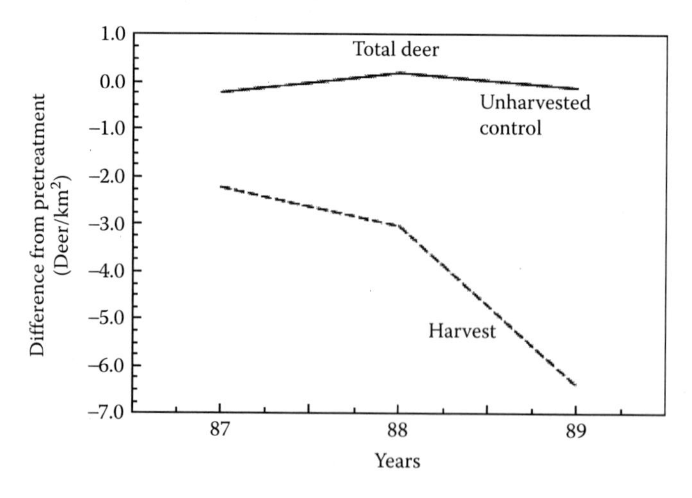

FIGURE 5.11 Comparison of population trend from repeated helicopter surveys for harvested and unharvested portions of the Faith and Camaron ranches, south Texas.

population dynamics and refrain from use of simple DD models in many habitats. However, all models should contain a DD term. All populations will at times be net DD and at other times net DI. The more DD, the greater the harvest a local population can likely stand. Frequently, DI populations should be harvested sparingly and carefully.

Acknowledgments

This chapter benefited from reviews by editor D. G. Hewitt, G. C. White, M. Festa-Bianchet, T. M. Fulbright, and K. M. Stewart.

REFERENCES

Armstrong, R. A. 1950. Fetal development of the northern white-tailed deer (*Odocoileus virginianus borealis* Miller). *American Midland Naturalist* 43:650–666.

Arnason, A. N., C. J. Schwarz, and J. M. Gerrard. 1991. Estimating closed population size and number of marked animals from sighting data. *Journal of Wildlife Management* 55:716–730.

Baker, R. H. 1984. Origin, classification and distribution. In *White-tailed Deer: Ecology and Management*, ed. L. K. Halls, 1–18. Harrisburg, PA: Stackpole Books.

Ballard, W. B., D. Lutz, T. W. Keegan, L. H. Carpenter, and J. C. deVos. 2001. Deer–predator relationships: A review of recent North American studies with emphasis on mule and black-tailed deer. *Wildlife Society Bulletin* 29:99–115.

Ballard, W. B., H. A. Whitlaw, S. J. Young, R. A. Jenkins, and G. J. Forbes. 1999. Predation and survival of white-tailed deer fawns in northcentral New Brunswick. *Journal of Wildlife Management* 63:574–579.

Barron, J. C. and W. F. Harwell. 1973. Fertilization rates of south Texas deer. *Journal of Wildlife Management* 37:179–182.

Bartush, W. S. and J. C. Lewis. 1981. Mortality of white-tailed deer fawns in the Wichita Mountains. *Proceedings of the Oklahoma Academy of Science* 61:23–27.

Beasom, S. L. 1979. Precision in helicopter censusing of white-tailed deer. *Journal of Wildlife Management* 43:777–780.

Beasom, S. L, J. C. Hood, and J. R. Cain. 1981. The effect of strip width on helicopter censusing of deer. *Journal of Range Management* 34:36–37.

Beasom, S. L., F. G. Leon, III, and D. R. Synatzske. 1986. Accuracy and precision of counting white-tailed deer with helicopters at different sampling intensities. *Wildlife Society Bulletin* 14:364–368.

Beringer, J., L. P. Hansen, and O. Sexton. 1998. Detection rates of white-tailed deer with a helicopter over snow. *Wildlife Society Bulletin* 26:24–28.

Bishop, C. J., G. C. White, D. J. Freddy, B. E. Watkins, and T. R. Stephenson. 2009. Effect of enhanced nutrition on mule deer population rate of change. *Wildlife Monographs* 172:1–28.

Bowman, J. L., H. A. Jacobson, D. S. Coggin, J. R. Heffelfinger, and B. D. Leopold. 2007. Survival and cause-specific mortality in adult male white-tailed deer managed under the Quality Deer Management paradigm. *Proceedings of the Annual Conference of the Southeastern Association of Fish and Wildlife Agencies* 61:76–81.

Brinkman, T. J., C. S. DePerno, J. A. Jenks, B. S. Haroldson, and R. G. Osborn. 2005. Movement of female white-tailed deer: Effects of climate and intensive row-crop agriculture. *Journal of Wildlife Management* 69:1099–1111.

Brinkman, T. J., J. A. Jenks, C. S. DePerno, B. S. Haroldson, and R. G. Osborn. 2004. Survival of white-tailed deer in an intensively farmed region of Minnesota. *Wildlife Society Bulletin* 32:726–731.

Brokx, P. A. 1972. Age determination of Venezuelan white-tailed deer. *Journal of Wildlife Management* 36:1060–1067.

Bryant, J. P., P. B. Reichardt, and T. P. Clausen. 1992. Chemically mediated interactions between woody plants and browsing mammals. *Journal of Range Management* 45:18–24.

Campbell, T. A., B. R. Laseter, W. M. Ford, and K. V. Miller. 2004. Feasibility of localized management to control white-tailed deer in forest regeneration areas. *Wildlife Society Bulletin* 32:1124–1131.

Campbell, T. A., B. R. Laseter, W. M. Ford, and K. V. Miller. 2005. Population characteristics of a central Appalachian white-tailed deer herd. *Wildlife Society Bulletin* 33:212–221.

Carroll, B. K. and D. L. Brown. 1977. Factors affecting neonatal fawn survival in southern-central Texas. *Journal of Wildlife Management* 41:63–69.

Caughley, G. 1985. Harvesting of wildlife: Past, present, and future. In *Game Harvest Management*, eds. S. L. Beasom and S. F. Robertson, 3–14. Kingsville, TX: Caesar Kleberg Wildlife Research Institute.

Cheatum, E. L. and G. H. Morton. 1946. Breeding season of the white-tailed deer in New York. *Journal of Wildlife Management* 10:249–263.

Conn, P. B., and D. R. Diefenbach. 2007. Adjusting age and stage distributions for misclassification errors. *Ecology* 88:1977–1983.

Conn, P. B. D. R. Diefenbach, J. L. Laake, M. A. Tement, and G. C. White. 2008. Bayesian analysis of wildlife age-at-harvest data. *Biometrics* 64:1170–1177.

Cook, R. L. and R. V. Hart. 1979. Ages assigned known-age Texas white-tailed deer: Tooth wear versus cementum analysis. *Proceedings of the Annual Conference of the Southeastern Association of Fish and Wildlife Agencies* 33:195–201.

Cook, R. S., M. White, D. O. Trainer, and W. C. Glazner. 1971. Mortality of young white-tailed deer fawns in south Texas. *Journal of Wildlife Management* 35:47–56.

Cooper, S. M., M. K. Owens, D. E. Spalinger, and T. F. Ginnett. 2003. The architecture of shrubs after defoliation and the subsequent feeding behavior of browsers. *Oikos* 100:387–393.

Croon, G. W., D. R. McCullough, C. E. Olson, Jr., and L. M. Queal. 1968. Infrared scanning techniques for big game censusing. *Journal of Wildlife Management* 32:751–759.

Cypher, B. L. 1991. A technique to improve spotlight observations of deer. *Wildlife Society Bulletin* 19:391–393.

Dapson, R. W., P. R. Ramsey, M. H. Smith, and D. F. Urbston. 1979. Demographic differences in contiguous populations of white-tailed deer. *Journal of Wildlife Management* 43:889–898.

DelGuidice, G. D., J. Fieberg, M. R. Riggs, M. C. Powell, and W. Pan. 2006. A long-term age-specific survival analysis of female white-tailed deer. *Journal of Wildlife Management* 70:1556–1568.

DelGiudice, G. D., M. S. Lenarz, and M. C. Powell. 2007. Age-specific fertility and fecundity in northern free-ranging white-tailed deer: Evidence for reproductive senescence? *Journal of Mammalogy* 88:427–435.

Demarais, S., D. Stewart, and R. N. Griffin. 1999. *A Hunter's Guide to Aging and Judging Live White-tailed Deer in the Southeast*. Mississippi State University Extension Service Publication 2206. Mississippi State, Mississippi.

D'Eon, R. G. 2001. Using snow-track surveys to determine deer winter distribution and habitat. *Wildlife Society Bulletin* 29:879–887.

DePerno, C. S., J. A. Jenks, S. L. Griffin, and L. A. Rice. 2000. Female survival rates in a declining white-tailed deer population. *Wildlife Society Bulletin* 28:1030–1037.

DeYoung, C. A. 1985. Accuracy of helicopter surveys of deer in south Texas. *Wildlife Society Bulletin* 13:146–149.

DeYoung, C. A. 1989a. Aging live white-tailed deer on southern ranges. *Journal of Wildlife Management* 53:519–523.

DeYoung, C. A. 1989b. Mortality of adult male white-tailed deer in south Texas. *Journal of Wildlife Management* 53:513–518.

DeYoung, C. A. 1990. Inefficiency in trophy white-tailed deer harvest. *Wildlife Society Bulletin* 18:7–12.

DeYoung, C. A., D. L. Drawe, T. E. Fulbright et al. 2008. Density dependence in deer populations: Relevance for management in variable environments. In *Wildlife Science: Linking Ecological Theory and Management Applications*, eds. T. E. Fulbright and D. G. Hewitt, 202–222. Boca Raton, FL: CRC Press.

DeYoung, C. A., F. S. Guthery, S. L. Beasom, S. P. Coughlin, and J. R. Heffelfinger. 1989. Improving estimates of white-tailed deer abundance from helicopter surveys. *Wildlife Society Bulletin* 17:275–279.

DeYoung, C. A., J. R. Heffelfinger, S. P. Coughlin, and S. L. Beasom. 1988. Accuracy of track counts to estimate white-tailed deer abundance. *Proceedings of the Annual Conference of the Southeastern Association of Fish and Wildlife Agencies* 42:464–469.

DeYoung, R. W., S. Demarias, K. L. Gee, R. L. Honeycutt, M. W. Hellickson, and R. A. Gonzales. 2009. Molecular evaluation of the white-tailed deer (*Odocoileus virginianus*) mating system. *Journal of Mammalogy* 90:946–953.

Ditchkoff, S. S., E. R. Welch, Jr., R. L. Lochmiller, R. E. Masters, and W. R. Starry. 2001. Age-specific causes of mortality among male white-tailed deer support mate-competition theory. *Journal of Wildlife Management* 65:552–559.

Downing, R. L. and D. C. Guynn, Jr. 1985. A generalized sustained yield table for white-tailed deer. In *Game Harvest Management*, eds. S. L. Beasom and S. F. Roberson, 95–103. Kingsville, TX: Caesar Kleberg Wildlife Research Institute.

Downing, R. L., W. H. Moore, and J. Kight. 1965. Comparison of deer census techniques applied to a known population in a Georgia enclosure. *Proceedings of the Southeastern Association of Game and Fish Commissions* 19:26–30.

Draeger, D. A. 1996. Predicting seasonal flux in white-tailed deer carrying capacity in south Texas: Root-plowed versus undisturbed sites. MS thesis, Texas A&M University—Kingsville.

Dusek, G. L., A. K. Wood, and S. T. Stewart. 1992. Spatial and temporal patterns of mortality among female white-tailed deer. *Journal of Wildlife Management* 56:645–650.

Eberhardt, L. and R. C. Van Etten. 1956. Evaluation of the pellet group count as a deer census method. *Journal of Wildlife Management* 20:70–74.

Etter, D. R., K. M. Hollis, T. R. Van Deelen et al. 2002. Survival and movements of white-tailed deer in suburban Chicago, Illinois. *Journal of Wildlife Management* 66:500–510.

Evans, W. 1975. Methods of estimating densities of white-tailed deer. PhD dissertation, Texas A&M University.

Fafarman, K. R. and C. A. DeYoung. 1986. Evaluation of spotlight counts of deer in south Texas. *Wildlife Society Bulletin* 14:180–185.

Folk, M. J. and W. D. Klimstra. 1991. Reproductive performance of female Key deer. *Journal of Wildlife Management* 55:386–390.

Fritzen, D. E., R. F. Labisky, D. E. Easton, and J. C. Kilgo. 1995. Nocturnal movements of white-tailed deer: Implications for refinement of track-count surveys. *Wildlife Society Bulletin* 23:187–193.

Fryxell, J. M., D. J. T. Hussell, A. B. Lambert, and P. C. Smith. 1991. Time lags and population fluctuations in white-tailed deer. *Journal of Wildlife Management* 55:377–385.

Fuller, T. K. 1990. Dynamics of a declining white-tailed deer population in north-central Minnesota. *Wildlife Monographs* 110:1–37.

Fuller, T. K. 1991. Do pellet counts index white-tailed deer numbers and population change? *Journal of Wildlife Management* 55:393–396.

Gaillard, J. M., M. Festa-Blanchet, N. G. Yoccoz, A. Loison, and C. Toigo. 2000. Temporal variation in fitness components and population dynamics of large herbivores. *Annual Review of Ecology and Systematics* 31:367–393.

Gavin, T. A., L. H. Suring, P. A. Vohs, Jr., and E. C. Meslow. 1984. Population characteristics, spatial organization, and natural mortality in the Columbian white-tailed deer. *Wildlife Monographs* 91:1–41.

Gee, K. L., J. H. Holman, M. K. Causey, A. N. Rossi, and J. B. Armstrong. 2002. Aging white-tailed deer by tooth replacement and wear: A critical evaluation of a time-honored technique. *Wildlife Society Bulletin* 30:387–393.

Gilbert, F. F. 1966. Aging white-tailed deer by annuli in the cementum of the first molar. *Journal of Wildlife Management* 30:200–202.

Gilbert, F. F. and S. L. Stolt. 1970. Variability in aging Maine white-tailed deer by tooth-wear characteristics. *Journal of Wildlife Management* 34:532–535.

Gill, R. M. A., M. L. Thomas, and D. Stocker. 1997. The use of portable thermal imaging for estimating deer population density in forested habitats. *Journal of Applied Ecology* 34:1273–1286.

Grahmann, E. D. 2009. Effects of three deer densities and supplemental feeding on vegetation in the western Rio Grande Plains of Texas. MS thesis, Texas A&M University—Kingsville.

Graves, H. B., E. D. Bellis, and W. M. Knuth. 1972. Censusing white-tailed deer by airborne thermal infrared imagery. *Journal of Wildlife Management* 36:875–884.

Hamilton, R. J., M. L. Tobin, and W. G. Moore. 1985. Aging fetal white-tailed deer. *Proceedings of the Annual Conference of the Southeastern Association of Fish and Wildlife Agencies* 39:389–395.

Hamlin, K. L., D. F. Pac, C. A. Sime, R. M. DeSimone, and G. L. Dusek. 2000. Evaluating the accuracy of ages obtained by two methods for Montana ungulates. *Journal of Wildlife Management* 64:441–449.

Harlow, R. F. and R. L. Downing. 1967. Evaluating the deer track census method used in the Southeast. *Proceedings of the Annual Conference of the Southeastern Association of Game and Fish Commissioners* 21:39–41.

Haroldson, B. S., E. P. Wiggers, J. Beringer, L. P. Hansen, and J. B. McAninch. 2003. Evaluation of aerial thermal imaging for detecting white-tailed deer in a deciduous forest environment. *Wildlife Society Bulletin* 31:1188–1197.

Haugen, A. O. 1975. Reproductive performance of white-tailed deer in Iowa. *Journal of Mammalogy* 56:151–159.

Haugen, A. O. and D. W. Speake. 1958. Determining age of young fawn white-tailed deer. *Journal of Wildlife Management* 22:319–321.

Haukioja, E. 1980. On the role of plant defences in the fluctuation of herbivore populations. *Oikos* 35:202–213.

Havens, K. J. and E. J. Sharp. 1998. Using thermal imagery in the aerial survey of animals. *Wildlife Society Bulletin* 26:17–23.

Haynes, G. 1983. A guide for differentiating mammalian carnivore taxa responsible for gnaw damage to herbivore limb bones. *Paleobiology* 9:164–172.

Heffelfinger, J. 2006. *Deer of the Southwest*. College Station, TX: Texas A&M University Press.

Hellickson, M. W., K. V. Miller, C. A. DeYoung, R. L. Marchinton, S. W. Stedman, and R. E. Hall. 2008. Physical characteristics for age estimation of male white-tailed deer in southern Texas. *Proceedings of the Annual Conference of the Southeastern Association of Fish and Wildlife Agencies* 62:40–45.

Hiller, T. L., H. Campa, III, and S. R. Winterstein. 2008. Survival and space use of fawn white-tailed deer in southern Michigan. *American Midland Naturalist* 159:403–412.

Hobbs, N. T. and D. M. Swift. 1985. Estimates of habitat carrying capacity incorporating explicit nutritional constraints. *Journal of Wildlife Management* 49:814–822.

Holsworth, W. N. 1973. Hunting efficiency and white-tailed deer density. *Journal of Wildlife Management* 37:336–342.

Holt, R. D. 2008. Theoretical perspectives on resource pulses. *Ecology* 89:671–681.

Holzenbein, S. and R. L. Marchinton. 1992. Emigration and mortality in orphaned male white-tailed deer. *Journal of Wildlife Management* 56:147–153.

Hoskinson, R. L. and L. D. Mech. 1976. White-tailed deer migration and its role in wolf predation. *Journal of Wildlife Management* 40:429–441.

Huegel, C. N., R. B. Dahlgren, and H. L. Gladfelter. 1985. Mortality of white-tailed deer fawns in south-central Iowa. *Journal of Wildlife Management* 49:377–380.

Jacobson, H. A., D. C. Guynn, R. N. Griffin, and D. Lewis. 1979. Fecundity of white-tailed deer in Mississippi and periodicity of corpora lutea and lactation. *Proceedings of the Annual Conference of the Southeastern Association of Fish and Wildlife Agencies* 33:30–35.

Jacobson, H. A., J. C. Kroll, R. W. Browning, B. H. Koerth, and M. H. Conway. 1997. Infrared-triggered cameras for censusing white-tailed deer. *Wildlife Society Bulletin* 25:547–556.

Jacobson, H. A. and R. J. Reiner. 1989. Estimating age of white-tailed deer: Tooth wear versus cementum annuli. *Proceedings of the Annual Conference of the Southeastern Association of Fish and Wildlife Agencies* 43:286–291.

Jones, P. D., S. Demarias, B. K. Strickland, and S. L. Edwards. 2008. Soil region effects on white-tailed deer forage protein content. *Southeastern Naturalist* 7:595–606.

Kammermeyer, K. E. and R. L. Marchinton. 1976. Notes on dispersal of male white-tailed deer. *Journal of Mammalogy* 57:776–778.

Keyser, P. D., D. C. Guynn, Jr., and H. S. Hill, Jr. 2005. Density-dependent recruitment patterns in white-tailed deer. *Wildlife Society Bulletin* 33:222–232.

Kie, J. G., R. T. Bowyer, and K. M. Stewart. 2003. Ungulates in western forests: Habitat relationships, population dynamics, and ecosystem processes. In *Mammal Community Dynamics in Western Coniferous Forests: Management and Conservation*, eds. C. Zabel and R. Anthony, 296–340. Baltimore, MD: The John Hopkins University Press.

Kie, J. G. and M. White. 1985. Population dynamics of white-tailed deer (*Odocoileus virginianus*) on the Welder Wildlife Refuge, Texas. *Southwestern Naturalist* 30:105–118.

Koerth, B. H. and J. C. Kroll. 2000. Bait type and timing for deer counts using cameras triggered by infrared monitors. *Wildlife Society Bulletin* 25:557–562.

Koerth, B. H., C. D. McKown, and J. C. Kroll. 1997. Infrared-triggered camera versus helicopter counts of white-tailed deer. *Wildlife Society Bulletin* 28:630–635.

Lockhard, G. R. 1972. Further studies of dental annuli for aging white-tailed deer. *Journal of Wildlife Management* 36:45–55.

Long, E. S., D. R. Diefenbach, C. S. Rosenberry, and B. D. Wallingford. 2008. Multiple proximate and ultimate causes of natal dispersal in white-tailed deer. *Behavioral Ecology* 19:1235–1242.

Long, E. S., D. R. Diefenbach, C. S. Rosenberry, B. D. Wallinford, and M. D. Grund. 2005. Forest cover influences dispersal distance of white-tailed deer. *Journal of Mammalogy* 86:623–629.

Macnab, J. 1985. Carrying capacity and related slippery shibboleths. *Wildlife Society Bulletin* 13:403–410.

Mandujano, S. and S. Gallina. 1995. Comparison of deer censusing methods in tropical dry forest. *Wildlife Society Bulletin* 23:180–186.

Mansell, W. D. 1974. Productivity of white-tailed deer on the Bruce Peninsula, Ontario. *Journal of Wildlife Management* 38:808–814.

Martin, P. S. 1970. Pleistocene niches for alien animals. *BioScience* 20:218–221.

Matson, G. 2009. Aging experience, accuracy, and precision. www.matsonslab.com. Accessed January 10, 2010.

Mautz, W. W. 1978. Sledding on a bushy hillside: The fat cycle in deer. *Wildlife Society Bulletin* 6:88–90.

McCall, T. C., R. D. Brown, and L. C. Bender. 1997. Comparison of techniques for determining the nutritional carrying capacity for white-tailed deer. *Journal of Range Management* 50:33–38.

McClintock, B. T. and G. C. White. 2009. A less field-intensive robust design for estimating demographic parameters with mark-resight data. *Ecology* 90:313–320.

McClintock, B. T., G. C. White, M. F. Antolin, and D. W. Tripp. 2009. Estimating abundance using mark-resight when sampling is with replacement or the number of marked individuals is unknown. *Biometrics* 65:237–246.

McCoy, J. E., D. G. Hewitt, and F. C. Bryant. 2005. Dispersal by yearling male white-tailed deer and implications for management. *Journal of Wildlife Management* 69:366–376.

McCullough, D. R. 1979. *The George Reserve Deer Herd.* Ann Arbor, MI: University of Michigan Press.

McCullough, D. R. 1982. Evaluation of night spotlighting as a deer study technique. *Journal of Wildlife Management* 46:963–973.

McCullough, D. R. 1984. Lessons from the George Reserve, Michigan. In *White-tailed Deer: Ecology and Management*, ed. L. K. Halls, 211–242. Harrisburg, PA: Stackpole Books.

McCullough, D. R. 1990. Detecting density dependence: Filtering the baby from the bathwater. *Transactions of the North American Wildlife and Natural Resources Conference* 55:534–543.

McCullough, D. R. 1992. Concepts of large herbivore population dynamics. In *Wildlife 2001: Populations*, eds. D. R. McCullough and R. H. Barrett, 967–984. London: Elsevier Applied Science.

McCullough, D. R. 1999. Density dependence and life-history strategies of ungulates. *Journal of Mammalogy* 80:1130–1146.

McKinley, W. T., S. Demarias, K. L. Gee, and H. A. Jacobson. 2006. Accuracy of the camera technique for estimating white-tailed deer population characteristics. *Proceedings of the Annual Conference of the Southeastern Association of Fish and Wildlife Agencies* 60:83–88.

McNaughton, S. J. 1983. Compensatory plant growth as a response to herbivory. *Oikos* 40:329–336.

Meares, J. M., B. P. Murphy, C. R. Ruth, D. A. Osborn, R. J. Warren, and K. V. Miller. 2006. A quantitative evaluation of the Severinghaus technique for estimating age of white-tailed deer. *Proceedings of the Annual Conference of the Southeastern Association of Fish and Wildlife Agencies* 60:89–93.

Moore, M. T. 2008. Evaluation of a camera census technique at three white-tailed deer densities. MS thesis, Texas A&M University–Kingsville.

Mooty, J. J. and P. D. Karns. 1984. The relationship between white-tailed deer track counts and pellet-group surveys. *Journal of Wildlife Management* 48:275–278.

Morton, H. and E. L. Cheatum. 1946. Regional differences in breeding potential of white-tailed deer in New York. *Journal of Wildlife Management* 10:242–248.

Naugle, D. E., J. A. Jenks, and B. J. Kernohan. 1996. Use of thermal infrared sensing to estimate density of white-tailed deer. *Wildlife Society Bulletin* 24:37–43.

Neff, D. J. 1968. The pellet-group count technique for big game trend, census, and distribution: A review. *Journal of Wildlife Management* 32:597–614.

Nelson, M. E. 1993. Natal dispersal and gene flow in white-tailed deer in northeastern Minnesota. *Journal of Mammalogy* 74:316–322.

Nelson, M. E. and L. D. Mech. 1984. Home-range and dispersal of deer in northeastern Minnesota. *Journal of Mammalogy* 65:567–575.

Nelson, M. E. and L. D. Mech. 1986. Mortality of white-tailed deer in northeastern Minnesota. *Journal of Wildlife Management* 50:691–698.

Nelson, M. E. and L. D. Mech. 1990. Weights, productivity, and mortality of old white-tailed deer. *Journal of Mammalogy* 71:689–691.

Nelson, M. E. and L. D. Mech. 1992. Dispersal in female white-tailed deer. *Journal of Mammalogy* 73:891–894.

Nelson, T. A. and A. Woolf. 1987. Mortality of white-tailed deer fawns in southern Illinois. *Journal of Wildlife Management* 51:326–329.

Nixon, C. M. 1971. Productivity of white-tailed deer in Ohio. *Ohio Journal of Science* 71:217–225.

Nixon, C. M., L. P. Hansen, P. A Brewer, and J. E. Chelsvig. 1991. Ecology of white-tailed deer in an intensively farmed region of Illinois. *Wildlife Monographs* 118:1–77.

Nixon, C. M., P. C. Mankin, D. R. Etter et al. 2007. White-tailed deer dispersal behavior in an agricultural environment. *American Midland Naturalist* 157:212–220.

Ostfeld, R. S. and F. Keesing. 2000. Pulsed resources and community dynamics of consumers in terrestrial ecosystems. *Trends in Ecology and Evolution* 15:232–237.

Owen-Smith, N. 1987. Pleistocene extinctions: The pivotal role of megaherbivores. *Paleobiology* 13:351–362.

Ozoga, J. J. 1987. Maximum fecundity in supplementally-fed northern Michigan white-tailed deer. *Journal of Mammalogy* 68:878–879.

Patternson, B. R., B. A. MacDonald, B. A. Lock, D. G. Anderson, and L. K. Benjamin. 2002. Proximate factors limiting population growth of white-tailed deer in Nova Scotia. *Journal of Wildlife Management* 66:511–521.

Porter, W. F., H. B. Underwood, and J. L. Woodard. 2004. Movement behavior, dispersal, and the potential for localized management of deer in a suburban environment. *Journal of Wildlife Management* 68:247–256.

Potvin, F. and L. Breton. 2005. From the field: Testing 2 aerial survey techniques on deer in fenced enclosures—visual double-counts and thermal infrared sensing. *Wildlife Society Bulletin* 33:317–325.

Potvin, F., L. Breton, and L. Rivest. 2004. Aerial surveys for white-tailed deer with the double-count technique in Quebec: Two 5-year plans completed. *Wildlife Society Bulletin* 32:1099–1107.

Progulske, D. R. and D. C. Duerre. 1964. Factors influencing spotlight counts of deer. *Journal of Wildlife Management* 28:27–34.

Pusateri Burroughs, J., H. Campa, III, S. R. Winterstein, B. A. Rudolph, and W. E. Moritz. 2006. Cause-specific mortality and survival of white-tailed deer fawns in southwestern lower Michigan. *Journal of Wildlife Management* 70:743–751.

Quinn, W., V. Neal, and S. De Mayolo. 1987. El Niño occurrences over the past four and a half centuries. *Journal of Geophysical Research* 92:14449–14461.

Ransom, A. B. 1966. Determining age of white-tailed deer from layers in cementum of molars. *Journal of Wildlife Management* 30:197–199.

Ricca, M. A., R. G. Anthony, D. H. Jackson, and S. A. Wolfe. 2002. Survival of Columbian white-tailed deer in western Oregon. *Journal of Wildlife Management* 66:1255–1266.

Rice, W. R. and J. D. Harder. 1977. Application of multiple aerial sampling to a mark-recapture census of white-tailed deer. *Journal of Wildlife Management* 41:197–206.

Richards, D. and A. Brothers. 2003. *Observing and Evaluating Whitetails*. Boerne, TX: Dave Richards Wilds of Texas Photography.

Richter, A. R. and R. F. Labisky. 1985. Reproductive dynamics among disjunct white-tailed deer herds in Florida. *Journal of Wildlife Management* 49:964–971.

Roberts, C. A., B. L. Pierce, A. W. Braden et al. 2006. Comparison of camera and road survey estimates for white-tailed deer. *Journal of Wildlife Management* 70:263–267.

Rogers, L. L. 1987. Seasonal changes in defecation rates of free-ranging white-tailed deer. *Journal of Wildlife Management* 51:330–333.

Rollins, D., F. C. Bryant, and R. Montandon. 1984. Fecal pH and defecation rates of eight ruminants fed known diets. *Journal of Wildlife Management* 48:807–813.

Roseberry, J. L. and W. D. Klimstra. 1970. Productivity of white-tailed deer on Crab Orchard National Wildlife Refuge. *Journal of Wildlife Management* 34:23–28.

Roseberry, J. L. and W. D. Klimstra. 1974. Differential vulnerability during a controlled deer harvest. *Journal of Wildlife Management* 38:499–507.

Rosenberry, C. S., M. C. Conner, and R. A. Lancia. 2001. Behavior and dispersal of white-tailed deer during the breeding season. *Canadian Journal of Zoology* 79:171–174.

Rosenberry, C. S., R. A. Lancia, and M. C. Conner. 1999. Population effects of white-tailed deer dispersal. *Wildlife Society Bulletin* 27:858–864.

Russell, F. L., D. B. Zippin, and N. L. Fowler. 2001. Effects of white-tailed deer (*Odocoileus virginianus*) on plants, plant populations and communities: A review. *American Midland Naturalist* 146:1–26.
Ryel, L. A., L. D. Fay, and R. C. Van Etten. 1961. Validity of age determination in Michigan deer. *Papers of the Michigan Academy of Science, Arts and Letters* 46:289–316.
Sams, M. G., R. L. Lochmiller, E. C. Hellgren, W. D. Warde, and L. W. Varner. 1996a. Morphometric predictors of neonatal age for white-tailed deer. *Wildlife Society Bulletin* 24:53–57.
Sams, M. G., R. L. Lochmiller, C. W. Qualls, Jr., D. M. Leslie, Jr., and M. E. Payton. 1996b. Physiological correlates of neonatal mortality in an overpopulated herd of white-tailed deer. *Journal of Mammalogy* 77:179–190.
Sawyer, T. G., R. L. Marchinton, and W. M. Lentz. 1990. Defecation rates of female white-tailed deer in Georgia. *Wildlife Society Bulletin* 18:16–18.
Schindler, J. R., T. E. Fulbright, and T. D. A. Forbes. 2003. Influence of thorns and tannins on white-tailed deer browsing after mowing. *Journal of Arid Environments* 55:361–377.
Severinghaus, C. W. 1949. Tooth development and wear as criteria of age in white-tailed deer. *Journal of Wildlife Management* 13:195–216.
Shea, S. M. T. A. Breault, and M. L. Richardson. 1992. Herd density and physical conditions of white-tailed deer in Florida flatwoods. *Journal of Wildlife Management* 56:262–267.
Shea, S. M., and J. S. Osborne. 1995. Poor quality habitats. In *Quality Whitetails: The Why and How of Quality Deer Management*, eds. K. V. Miller and R. L. Marchinton, 193–209. Mechanicsburg, PA: Stackpole Books.
Short, C. 1970. Morphological development and aging of mule and white-tailed deer fetuses. *Journal of Wildlife Management* 34:383–388.
Stoll, R. J., Jr., M. W. McClain, J. C. Clem, and T. Plageman. 1991. Accuracy of helicopter counts of white-tailed deer in western Ohio farmland. *Wildlife Society Bulletin* 19:309–314.
Strickland, B. K. 1998. Using tame white-tailed deer to index carrying capacity in south Texas. MS thesis, Texas A&M University—Kingsville.
Strickland, B. K., D. G. Hewitt, C. A. DeYoung, and R. L. Bingham. 2005. Digestible energy requirements for maintenance of body mass of white-tailed deer in southern Texas. *Journal of Mammalogy* 86:56–60.
Teaschner, T. B. and T. E. Fulbright. 2007. Shrub biomass production following simulated herbivory: A test of the compensatory growth hypothesis. In *Proceedings: Shrubland Dynamics—Fire and Water*, eds. R. E. Sosbee, R. E. Wester, D. B. Briton, and C. M. McArthur, 107–111. Fort Collins, CO: USDA Forest Service RMRS-P-47.
Teer, J. G., J. W. Thomas, and E. A. Walker. 1965. Ecology and management of white-tailed deer in the Llano Basin of Texas. *Wildlife Monographs* 15:1–62.
Tremblay, J., I. Thibault, C. Dussault, J. Huot, and S. D. Côté. 2005. Long-term decline in white-tailed deer browse supply: Can lichens and litterfall act as alternative food sources that preclude density-dependent feedbacks. *Canadian Journal of Zoology* 83:1087–1096.
Tyson, E. L. 1952. Estimating deer populations from tracks. *Proceedings of the Annual Conference of the Southeastern Association of Game and Fish Commissioners* 6:507–517.
Tyson, E. L. 1959. A deer drive vs. track census. *Transactions of the North American Wildlife and Natural Resources Conference* 24:457–464.
Van Deelen, T. R., H Campa III, J. B. Haufler, and P. D. Thompson. 1997. Mortality patterns of white-tailed deer in Michigan's Upper Peninsula. *Journal of Wildlife Management* 61:903–910.
Van Etten, R. C., D. F. Switzenberg, and L. Eberhardt. 1965. Controlled deer hunting in a square-mile enclosure. *Journal of Wildlife Management* 29:59–73.
Verme, L. J. 1965. Reproduction studies on penned white-tailed deer. *Journal of Wildlife Management* 29:74–78.
Verme, L. J. 1969. Reproductive patterns of white-tailed deer related to nutritional plane. *Journal of Wildlife Management* 33:881–887.
Vreeland, J. K., D. R. Diefenbach, and B. D. Wallingford. 2004. Survival rates, mortality causes, and habitats of Pennsylvania white-tailed deer fawns. *Wildlife Society Bulletin* 32:542–553.
Waller, D. M. and W. S. Alverson. 1997. The white-tailed deer: A keystone species. *Wildlife Society Bulletin* 25:217–226.
Watts, D. E., I. D. Parker, R. R. Lopez, N. J. Silvey, and D. S. Davis. 2008. Distribution and abundance of endangered Florida Key deer on outer islands. *Journal of Wildlife Management* 72:360–366.

Whipple, J. D., D. L. Rollins, and W. H. Schacht. 1994. A field simulation for assessing accuracy of spotlight deer surveys. *Wildlife Society Bulletin* 22:667–673.

White, G. C. 1996. NOREMARK: Population estimation from mark-resighting surveys. *Wildlife Society Bulletin* 24:50–52.

White, G. C., D. J. Freddy, R. B. Gill, and J. H. Ellenberger. 2001. Effect of adult sex ratio on mule deer and elk productivity in Colorado. *Journal of Wildlife Management* 65:543–551.

White, G. C. and T. M. Shenk. 2001. Population estimation with radio-marked animals. In *Radio Tracking and Animal Populations*, eds. J. Millspaugh and J. M. Marzluff, 329–350. San Diego, CA: Academic Press.

Whitlaw, H. A., W. B. Ballard, D. L. Sabine, S. J. Young, R. A. Jenkins, and G. J. Forbes. 1998. Survival and cause-specific mortality rates of adult white-tailed deer in New Brunswick. *Journal of Wildlife Management* 62:1335–1341.

Wigley, T. B. and M. K. Johnson. 1981. Disappearance rates for deer pellets in the Southeast. *Journal of Wildlife Management* 45:251–253.

6

Spatial Use of Landscapes

Kelley M. Stewart, R. Terry Bowyer, and Peter J. Weisberg

CONTENTS

Introduction

Knowledge of spatial needs of white-tailed deer is fundamental to effective management of their populations and habitats (Fulbright and Ortega-S, 2006). White-tailed deer, and other large mammals, require temporally and spatially diverse elements of habitat such as food, water, and cover, and these mammals can have significant effects on vegetation composition, community structure, and ecosystem processes, thereby acting as keystone species (Molvar et al., 1993; Wallis and de Vries, 1995; Hobbs, 1996; Bowyer et al., 1997; Simberloff, 1998; Kie et al., 2002, 2003). Landscape structure can affect habitat selection by deer, patterns of movement, and home-range size, but those effects also influence the developmental trajectory of landscapes (Kie et al., 2002). Spatial distributions, movement patterns, and the corresponding effects on landscapes, communities, or ecosystems may be intense but seasonal for migratory populations. Such effects may be of longer duration with fine-scale changes strongly influenced by density-dependent processes (*sensu* McCullough, 1979) in areas where white-tailed deer are resident. Large mammals tend to be vagile and often integrate landscape processes, such as nutrient cycling and rates of mineralization over large areas (Chapter 12) and herbivores often concentrate foraging and deposition of urine and feces in localized areas (McNaughton, 1985; Etchberger et al., 1988; Ruess and McNaughton, 1987, 1988; Day and Detling, 1990; Bowyer and Kie, 2006; Stewart et al., 2006). Because large mammals do not use habitats uniformly, changes in ecosystem processes resulting from movements of these mammals may lead to increases in patchiness within habitats, which ultimately influences successional changes and habitat selection in ecosystems inhabited by white-tailed deer and other large mammals. Moreover, resource selection by white-tailed deer and other large mammals can occur at both extremely large and fine scales, necessitating a hierarchical approach to understanding their behavior and ecology (Johnson et al., 2001, 2002; Bowyer and Kie, 2006). Moreover, the sexes of deer may behave quite differently to meet their nutritional and reproductive needs (Kie and Bowyer, 1999). Finally, understanding life-history characteristics of white-tailed deer and other large mammals requires that biologists consider entire landscapes rather than isolated patches of habitat for conservation and management (Kie et al., 2002, 2003).

The advent of Global Positioning System (GPS) technology for collecting fine-scale data on locations of animals and Geographic Information Systems (GIS) for analyzing spatial data allows for complex analysis of spatial distributions, resource selection, and movements of animals, including white-tailed deer. Those techniques allow for collection and interpretation of data at very fine scales, for example, examination of characteristics at birth-sites (Bowyer et al., 1998b, 1999), or very large scales, such as landscape level movements related to dispersal and migration (Long et al., 2005). Moreover, these data are often collected on very fine temporal as well as spatial scales, which also allow detailed analyses of life-history characteristics that are essential to understanding fitness consequences for animals.

Our purpose is to interpret information on white-tailed deer primarily related to movement patterns, spatial distributions, effects of landscape configuration, and spatial separation of sexes. We elucidate how landscape characteristics affect movement patterns, including home ranges, dispersal, and migration, among populations, and other aspects of the ecology of white-tailed deer, including sexual segregation. Landscape heterogeneity and spatial scale have profound affects on life-history characteristics of white-tailed deer and a more complete understanding of how those characteristics of the environment interact with life-history characteristics is important for understanding the ecology and effectively managing populations of white-tailed deer. This chapter is subdivided into three separate, but integrated topics: scale and movement patterns, landscape characteristics and fragmentation, and sexual segregation. We will conclude with an overview of how these topics relate to management at the landscape scale and set directions for future research.

Scale and Movement Patterns

Understanding how the spatial configuration and sizes of habitat patches affect the distribution of animals requires information about how individuals perceive and respond to variation in spatial extent within their environment (Turner, 1989; Wiens, 1989; Wiens and Milne, 1989; Kie et al., 2002). Life-history characteristics for large mammals, including white-tailed deer, are extremely scale sensitive, and sampling at multiple scales is an important key to understanding their life-history strategies (Bowyer and Kie, 2006). Indeed, failure to select the correct scale for spatial analyses can easily lead to misinterpretations of biological data and management errors (Powell, 1994; Bowyer et al., 1996; Kie et al., 2002). Moreover, the patchy nature of interactions of white-tailed deer and other ungulates with their environment, and their ability to move over long distances introduces scale sensitivity into many processes associated with these large herbivores (Gomper and Gittleman, 1991; Bowyer et al., 1996; Bowyer and Kie, 2006). Therefore, careful consideration of the scale at which spatial heterogeneity is defined and

FIGURE 6.1 (a) White-tailed deer in what appears to be open grassy habitat, viewed at a relatively small spatial scale, and (b) view of the same habitat at a larger spatial scale. *Note*: The larger scale indicates that those deer are using a habitat comprised of a matrix of shrubs and clearings. (Photos by T. E. Fulbright.)

measured is critical for examining resource selection or life-history characteristics of white-tailed deer (Kie et al., 2002; Boyce, 2006). For example, Figure 6.1a may lead to the conclusion that the animals are using grassland habitats, but a larger spatial scale, Figure 6.1b, indicates that the habitat is a mixed-shrub community with small clearings. Any assessment of habitat use or resource selection by white-tailed deer differs dramatically between those two spatial scales.

Large mammals do not use their environment uniformly (Hobbs, 1996; Kie et al., 2003; Stewart et al., 2006, 2009), and variation in use and selection of habitats among individuals or populations may be overemphasized or underemphasized if an inappropriate scale is used to examine those processes. We agree with Bowyer and Kie (2006) that study areas should be selected to be of sufficient size to accommodate sampling at a scale that is likely to be relevant to life-history characteristics of interest. An important difficulty, however, in choosing an appropriate scale at which to measure habitat selection based on a life-history characteristic is that some desirable characteristics may vary spatially or temporally among individuals in the same population (Bowyer and Kie, 2006). Different types of movement patterns by white-tailed deer occur at different spatial scales. Movement patterns at small spatial scales, such as habitat selection or movements within home ranges, may be influenced by availability of resources, heterogeneity of habitats, or distribution of conspecifics or predators (Phillips et al., 2004; Webb et al., 2009). Movements by white-tailed deer at large spatial scales, such as migration or dispersal, are likely influenced by geographic variation and structure of landscapes (Webb et al., 2009). Indeed, hypotheses relating to resource selection or movement patterns of white-tailed deer and other large mammals must be framed at the appropriate spatial scale to gain an understanding of how those large mammals use habitats and landscapes.

Home Range

A home range is simply the area used by an animal as it meets its needs for food, water, cover, social interactions, and caring for young (Burt, 1943). Home range serves as the most fundamental index to use of space by wildlife (Hemson et al., 2005), and perhaps no other ecological characteristic of vertebrates has been measured as often as home range (Kie et al., 2002). This observation holds for species of deer (Gruell and Papez, 1963; Eberhardt et al., 1984; Garrott et al., 1987; Pac et al., 1988; Nicholson et al., 1997; Kie et al., 2002, and many others). Sizes of home ranges of deer are highly variable (Nicholson et al., 1997). Indeed, variation in home-range size is related to a variety of factors including body size (McNab, 1963; Jenkins, 1981; Swihart et al., 1988), trophic level (Harestad and Bunnell, 1979), sex and age (Cederlund and Sand, 1994; Nicholson et al., 1997; Relyea et al., 2000), reproductive status (Bertrand et al., 1996), season (Nicholson et al., 1997), subspecies (Anderson and Wallmo, 1984), distribution of resources such as forage (Schoener, 1981; Ford, 1983; Tufto et al., 1996; Powell et al., 1997; Relyea et al., 2000) or water (Herbert and Krausman, 1986; Bowers et al., 1990), as well as intra- (Riley and Dood, 1984) and interspecific competition (Loft et al., 1993; Kie et al., 2002). Home ranges often are based on available resources; deer in areas with higher-quality resources (food, water, and cover) that are abundant and well distributed tend to have smaller home ranges than deer that occupy less productive areas (Marchinton and Hirth, 1984). Nonetheless, home ranges may shrink or grow seasonally as food or water change in availability, and home ranges can vary tremendously in size and shape with differences in the heterogeneity of habitats, and even among individuals within a single population. Indeed, in northern regions snow depth, population density, and low temperatures have the greatest influence on the daily activity of deer, and deer often minimize movements in response to severe weather conditions to conserve energy (Telfer, 1967; Verme, 1973; Tierson et al., 1985; Beier and McCullough, 1990; Parker et al., 1984; Brinkman et al., 2005).

Male white-tailed deer tend to have larger home ranges than females, presumably because they need more space to meet nutritional demands (Nicholson et al., 1997; Beier and McCullough, 1990; Barboza and Bowyer, 2000, 2001; Long et al., 2008). The mating season (rut) also affects home-range sizes; male white-tailed deer nearly doubled their home-range size with the onset of rut in autumn, presumably for increased access to females, although yearling males continuously maintained smaller home ranges than adults in all seasons (Nelson and Mech, 1981, 1984). In addition, home ranges of females tend to shrink near parturition, with corresponding reductions in mobility and social interactions (Ozoga et al., 1982; Bertrand et al., 1996; D'Angelo et al., 2004). Daily movements of white-tailed deer are greatest during the breeding season and least during late gestation, close to parturition (Marchinton and Hirth, 1984).

Within the home range there are often core areas where the animal spends a disproportionate amount of time, presumably because conditions are optimal in that area (Heffelfinger, 2006). In arid regions, home ranges tend to shrink during the dry season as animals remain closer to water and expand under more mesic conditions (Heffelfinger, 2006). Deer living in arid regions, however, generally have larger home ranges because of more widely distributed resources. Bed sites or water sources may be important components of core areas, particularly in arid regions where animals are dependent on the availability of water (Marchinton and Hirth, 1984).

Kie et al. (2002) demonstrated that the location of a home range among several populations of mule deer resulted from selection of habitat features occurring across a much greater area than the home range itself. Thus, selection for or against particular landscape features already had occurred in the process of deer deciding where to locate their home ranges; similar higher-level selection likely occurs in populations of white-tailed deer (Kie et al., 2002; Bowyer and Kie, 2006). Indeed, home ranges of white-tailed deer on the George Reserve in Michigan were small, which was thought to be a result of high interspersion (low contagion) of habitat types (Beier and McCullough, 1990; Kie et al., 2002).

TABLE 6.1

Home-Range Sizes (ha) for Resident White-tailed Deer

Location	Size (ha)	Habitat	Source
Illinois	99		Walter et al. (2009)
Michigan	134		Walter et al. (2009)
Nebraska	120		Walter et al. (2009)
Florida,	270	Pine–oak uplands	Bridges (1968)
Alabama	81	Bottomland hardwood	Byford (1970)
Texas	71	Mesquite chaparral	Hood (1971)
Wisconsin	178	Tamarack swamp/Ag	Larson et al. (1978)
Wisconsin	147		Walter et al. (2009)
Louisiana	232	Longleaf pine/scrub oak	Lewis (1968)
Georgia	342	Pine/hardwood	Marshall and Whittington (1969)
South Carolina	171	Pine/scrub hardwood	Sweeney (1970)

Source: Adapted from Marchinton, R. L. and D. H. Hirth. 1984. *White-tailed Deer Ecology and Management*, ed. L. K. Halls, 129–168. Harrisburg, PA: Stackpole Books.

Sizes of home ranges often vary seasonally. In southern regions, white-tailed deer are resident and occupy ranges year-round (Table 6.1). During winter, deer tend to minimize movement to conserve energy (Moen, 1976; Parker et al., 1984); thus, white-tailed deer generally have smaller home ranges during winter compared to summer, especially in highly seasonal environments (Table 6.2). In forested regions of New York (Tierson et al., 1985) and northern Minnesota (Nelson and Mech, 1981; Mooty et al., 1987) winter ranges were substantially smaller than summer ranges, in part because deer congregated in traditional wintering areas, called yards, during winter (Telfer, 1967; Rongstad and Tester, 1969). Nonetheless, in agricultural areas of southwest Minnesota, home-range size during winter was more than double the size of summer ranges (Nixon et al., 1991; Brinkman et al., 2005). In those agricultural lands, summer ranges were likely smaller because of abundant cover and nutritious forages throughout the landscape resulting from intensive farming (Brinkman et al., 2005). Thus, managers should keep in mind that deer densities fluctuate spatially and temporally especially in areas with intensive agriculture (Brinkman et al., 2005).

TABLE 6.2

Seasonal Home-Range Sizes (ha) for White-tailed Deer Populations

Location	Winter	Summer	Source
Quebec	129	2435	Lesage et al. (2000)
New Brunswick	370	277	Lesage et al. (2000)
Florida		22–26	Kilgo et al. (1998)
Florida		23	Labisky and Fritzen (1998)
Wisconsin	178		Larson et al. (1978)
Minnesota	43	69	Mooty et al. (1987)
Minnesota	44	83–319	Nelson and Mech (1981,1987)
Illinois		267	Nixon et al. (1991)
New York	135	225	Tierson et al. (1985)
Michigan	194–212	170–490	Van Deelen (1995)
Montana	634	326	Wood et al. (1989)

Source: Adapted from Lesage, L. et al. 2000. *Canadian Journal of Zoology* 78:1930–1940.

Habitat Selection

Although many ungulates, particularly white-tailed deer, are considered well adapted to habitat edges, life-history characteristics in deer, such as home-range size, are related to landscape patterns in more complex ways than simply a function of habitat edge (Kie et al., 2003). For example, the amount of habitat edge measured within 2000 m of the center of the home ranges for 80 mule deer (area larger than most home ranges) accounted for 27% of the variability in home-range size (Kie et al., 2002, 2003); the same likely applies to white-tailed deer. Moreover, spatial heterogeneity, measured at the same spatial scale accounted for 57% of variation in home-range size. If ungulates perceive potential habitats at scales greater than those they eventually choose as a home range, analysis of habitat selection within the home range may yield biased results because the home range already includes landscape attributes that the individual has selected (Kie et al., 2003; Bowyer and Kie, 2006).

Originally thought to be mostly a forest-dwelling animal, white-tailed deer have adapted to use nearly every ecosystem in North, Central, and much of South America, although disturbed vegetation is favored because of the high diversity of plants in those areas (Teer, 1996). Indeed, white-tailed deer often are associated with brushlands and forests throughout their range in North America; woody vegetation is often used for forage and cover (Teer, 1996). Disturbed communities that produce high amounts of forbs or browse often will support relative high densities of white-tailed deer, given similar sources and rates of mortality.

Important components of habitat for white-tailed deer vary across their distribution (Chapter 1). In northern and eastern ranges, deer are associated with forests and spend the winter in yards to avoid deep snow and mitigate cold temperatures (Telfer, 1967; Rongstad and Tester, 1969). Those populations tend to have separate winter and summer ranges; winter ranges are generally areas protected from deep snow (Teer, 1996). In western North America, white-tailed deer are associated with riparian zones, which are important components of habitat. In southern regions, where deer are resident, optimum habitat generally consists of clearings containing forage species such as forbs and grasses, interspersed in a woodland matrix.

In areas where they are sympatric with mule deer, white-tailed deer occupy more mesic areas such as riparian zones, and mule deer are associated with more xeric habitats (Teer, 1996). In the western Great Plains and northern Rocky Mountains mule deer generally occupy higher elevations and more xeric habitats than white-tailed deer in areas where they are sympatric (Krämer, 1973). In mountainous regions of the southwestern United States, such as Arizona, white-tailed deer are generally observed at higher elevations than mule deer (Anthony and Smith, 1977); lower elevations in Arizona are more xeric than the high-elevation habitats occupied by white-tailed deer. Although the two species appear to switch roles in Arizona, with white-tailed deer occurring at higher elevations than mule deer, the tendency of white-tailed deer to occupy more mesic habitats is consistent with observations in other studies and other geographic areas. Riparian zones also function as important habitats for foraging and bedding, provide screening and thermal cover, and serve as travel corridors for deer (Fulbright and Ortega-S, 2006).

Migration

Migration is defined as movement of animals periodically from one region or climatic zone to another for feeding or reproduction; or in general terms as movement of individuals between summer and winter ranges (Caughley and Sinclair, 1994). Migration behavior in white-tailed deer most likely evolved as a learned behavior as return movement (Sinclair, 1983; Nelson, 1998) to mediate variable climates or seasonal changes in forage quality and availability (Brower and Malcolm, 1991). White-tailed deer exhibit multiple types of migratory strategies; they may be nonmigratory (year-round residents), obligate migrators (migrating every year), or conditional or facultative migrators (migrating some years but not others; Nelson, 1995, 1998; Van Deelen et al., 1998; Sabine et al., 2002; Brinkman et al., 2005; Fieberg et al., 2008). All three strategies may be observed within the same population, although in general young of mothers that migrate are more likely to migrate than young of sedentary mothers (Nixon et al., 2008).

Migration between summer and winter ranges tends to be more pronounced where there are marked differences in seasonal weather patterns, such as northern or mountainous areas and particularly with increasing snow depth (Nelson and Mech, 1981; Marchinton and Hirth, 1984; Sabine et al., 2002; Fieberg et al., 2008; Nixon et al., 2008). Distances of migration among populations often vary by geographical

TABLE 6.3

Migration Distances of White-tailed Deer from Different Geographical Areas

Location	Distance (km)	Source
Illinois	7.3–15.9	Nixon et al. (2008)
North Michigan	Mean 38.6	Bartlet (1950)
Upper Peninsula, Michigan	Mean 13.8	Verme (1973)
North Central Minnesota	1.5–34.8, mean 9.4	Fieberg et al. (2008)
Southwestern Minnesota	Mean 10.1	Brinkman et al. (2005)
Northeast Minnesota	10.0–38.1, mean 20.7	Hoskin and Mech (1976)
South Dakota	Mean 23.2	Sparrowe and Springer (1970)

Source: Updated from Marchinton, R. L. and D. H. Hirth. 1984. *White-tailed Deer Ecology and Management*, ed. L. K. Halls, 129–168. Harrisburg, PA: Stackpole Books.

region or topographic variation (Table 6.3). McCullough (1985) reported that migration behavior may be genetically predisposed rather than only a learned behavior based on observations that distances of migration appear to be longer than necessary to avoid severe weather or locate nutritious forages, although the full range of weather conditions, including extremely severe winters may not have been observed in that study. In most northern areas, the timing of migration appears to be driven by climate, with fluctuations of temperature and snow depth exerting the strongest effects on seasonal movements of white-tailed deer (Verme, 1968, 1973; Ozoga and Gysel, 1972; Marchington and Hirth, 1984; Sandegren et al., 1985; Nelson, 1995; Brinkman et al., 2005). Nixon et al. (2008) reported that in agricultural areas, migration appears to be driven partly by changing availability of cover following harvest in autumn, causing deer to move to areas of permanent cover on winter range. Conversely during summer deer show high fidelity to summer ranges and used crop areas with high resource availability, particularly safe parturition sites (Nixon et al., 2008). For both mule deer and white-tailed deer, variability in climate appears to be the strongest determinant responsible for mixed-migration strategies (Nicholson et al., 1997; Brinkman et al., 2005).

In the north eastern and Midwestern United States, migratory deer concentrate in yards during winter and generally have relatively short migration distances (Dahlberg and Guettinger, 1956; Telfer, 1967; Rongstad and Tester, 1969; Nelson and Mech, 1981; Marchinton and Hirth, 1984). The primary use of yards by white-tailed deer is associated with cold temperatures rather than snow depth, although often the two are related (Marchinton and Hirth, 1984). Suitable forage also affects areas used for wintering by white-tailed deer (Hodgman and Bowyer, 1986). Deer may occupy yards from 12 to 20 weeks, depending on the severity of winter (Verme and Ozoga, 1971; Marchinton and Hirth, 1984). In addition, where wolf predation is an important source of mortality during winter, yards also appear to offer potential antipredator benefits (Nelson and Mech, 1992).

Dispersal

Natal dispersal is defined as permanent emigration of an individual from the birth site to an area of potential reproduction (Howard, 1960; Caughley, 2004). Dispersal is one of the least understood life-history traits, although dispersal is fundamental to our understanding of population dynamics, gene flow, and spread of diseases (Diefenbach et al., 2008). Natal dispersal may be caused by multiple processes. The three most common explanations for the ultimate causes of emigration typically relate increased fitness of dispersers to: (1) avoidance of inbreeding (Wolff et al., 1988; Pusey and Wolf, 1996), (2) reduced competition for mates (Dobson, 1979, 1982; Moore and Ali, 1984), or (3) reduced competition for resources (Murray, 1967; Dobson, 1979; Long et al., 2008). Field-based inquires addressing dispersal are rare for large mammals, although white-tailed deer are among the few species where dispersal has been addressed (Rosenberry et al., 1999; Long et al., 2005, 2008; Diefenbach et al., 2008).

In polygynous species, such as white-tailed deer (Hirth, 1977), dispersal tends to be sex biased, such that juvenile males are more likely to emigrate than females (Perrin and Mazalov, 2000; Hjeljord, 2001). Indeed for white-tailed deer, yearling males are the primary sex and age class that disperses, although female dispersal has been reported (Nelson and Mech, 1992; Nelson, 1993; Rosenberry et al., 1999; Purdue et al., 2000; Diefenbach et al., 2008). Natal dispersal in white-tailed deer is seasonal with emigration occurring most often in spring, when yearlings associate most closely with related adult females (Marchinton and Hirth, 1984; Holzenbein and Marchington, 1992), and during autumn when male yearlings associate most closely with other males (Hirth, 1977; Ozoga and Verme, 1985). Thus, spring dispersal appears to be driven by mother–young interactions, and autumn dispersal likely results from male–male competition during the mating season. At a local level, dispersal resulting from competition for limited resources, such as mates, may lead to increases in inclusive fitness by reducing competition with kin (Hamilton and May, 1977; Long et al., 2008). Yearling males with larger antlers were more likely to disperse than smaller yearling males, presumably because larger males were closer to being sexually mature than smaller ones (Shaw et al., 2003; McCoy et al., 2005).

Dispersal distances of white-tailed deer differ according to the ultimate cause of dispersal (Waser, 1987; Sutherland et al., 2000; Long et al., 2008). Short movements, immediately outside the home range, may be sufficient for escaping local mate competition, but when opposite-sex relatives are philopatric longer dispersal distances, outside the social group, may be necessary to escape inbreeding (Ronce et al., 2001; Long et al., 2008). Long et al. (2008) reported that dispersal distances of white-tailed deer in Pennsylvania, United States, were greater during spring than autumn, consistent with the idea that dispersal for avoidance of inbreeding would be longer than dispersal to reduce local mate competition.

Structure of some populations of white-tailed deer have been described as overlapping home ranges of related females that are arranged like rose petals (Porter et al., 1991). Home ranges of less related individuals radiate outward, such that genetic distance increases with linear distance from the center (Comer et al., 2005). Genetic structuring of male white-tailed deer is less well understood, but dispersal of males is common, about 50–80% disperse as yearlings (Long et al., 2005, 2008; McCoy et al., 2005). Thus, local populations of male white-tailed deer are less likely to be related than females in those populations (Purdue et al., 2000; Long et al., 2005).

Long et al. (2005) reported that dispersal in white-tailed deer was highly correlated with percentage of forested cover, and that forest cover explained 94% of the variation in average dispersal distance. Those authors posited that in less forested landscapes, deer may need to travel farther to find suitable patches of habitat. Habitat fragmentation has been observed to increase dispersal distances in birds (Matthysen et al., 1995) and small mammals (Wiggett and Boag, 1989), and the effects of habitat fragmentation on dispersal distances have not been fully addressed in white-tailed deer. Long et al. (2005) posited that the ability of deer and other animals to disperse farther in more suitable habitats may increase population sustainability and growth. Conversely, increasingly patchy and fragmented habitats, may increase risk of mortality during dispersal and lead to lower sustainability of populations.

Multiple causes of dispersal likely occur within deer populations, and are often caused by differing ultimate and proximate factors. Research addressing questions related to dispersal are rare, particularly for large mammals which are inherently difficult to study because of the necessity of incorporating landscape-level areas of habitat, and substantial cost of such research. Nonetheless, large mammals, especially white-tailed deer because of their wide distribution across North America, are appropriate models for examining interactions among dispersal, population dynamics, and ultimately gene flow among populations. Indeed, these large, vagile mammals likely integrate those processes across landscapes. To date, studies examining dispersal in white-tailed deer have not documented effects of density-dependent processes in dispersal behavior (Long et al., 2005). Interactions of density-dependent changes in physical condition and reproduction, with the corresponding effects on selection of sex ratios and ultimately their effects on dispersal may be difficult, but important to document.

Response to Landscape Heterogeneity and Land-Use Change

An important concern for many species is the effect of land-use change on habitat amount, quality, and connectivity. Because white-tailed deer are strong dispersers as well as habitat generalists, the effects of reducing connectivity (i.e., fragmentation) may be less important than for other species that are more dispersal or habitat limited. Fahrig (2007) differentiates among types of landscape structure with respect to patchiness of habitat and movement risk associated with the nonhabitat matrix, postulating that evolved adaptations of species to a particular landscape structural type should exhibit predictably different responses to human-caused landscape alterations. Species that evolved in landscapes where habitat is distributed in a patchy fashion, and where nonhabitat areas between patches create high risk of mortality during dispersal, are expected to be highly susceptible to habitat fragmentation effects that further isolate habitat patches. Conversely, species that evolved with a low-risk matrix are likely to have evolved a high movement probability overall (because potential benefits of movement outweigh costs), weak boundary response (low risk of mortality when leaving favorable habitat), and convoluted movement patterns within habitats but straight long-distance movements in the nonhabitat matrix. White-tailed deer fall into this latter category, and so are robust to habitat fragmentation as long as dispersal is possible.

Although white-tailed deer may not be particularly susceptible to negative effects of habitat fragmentation *per se*, being capable of long-range movements, they may be highly susceptible to increased mortality associated with movement through high-risk areas. The proportion of high-risk areas is likely to increase in many landscapes due to habitat loss and degradation of matrix quality. Collisions with vehicles on roads are a significant cause of mortality for white-tailed deer in the United States, as well as a public safety hazard in high-risk locations (Romin and Bissonette, 1996). Predictive models using GIS and spatial analysis show promise for proactive management to reduce deer–vehicle collisions using fencing, road realignment, or travel corridors such as highway underpasses (Finder et al., 1999; Clevenger et al., 2001; Malo et al., 2004).

An absence of strong boundary response results in white-tailed deer leaving safe habitat to travel through habitat with higher mortality risk. For example, ownership parcels in much of the eastern United States are smaller in size than the scale of landscape use by deer, resulting in high mortality rates during the hunting season (McCoy et al., 2005). Furthermore, because the probability of seasonal deer migrations is influenced by the arrangement of different habitat types that collectively satisfy annual resource requirements (Felix et al., 2007), land-use change may compound the problem of increased movement mortality where altered habitat mosaics or reduced habitat quality cause resident deer populations to become migratory or lead to reductions in population size.

Deer Use of Cultural Landscapes

There is evidence that white-tailed deer shift their landscape use during winter to energetically favorable areas, including those with human residents in regions with harsh winter weather and deep snow such as northern New York (Hurst and Porter, 2008). Deer often abandon historical winter yards for nearby residential areas, where smaller home ranges result from localized concentrations of resources. Low-density residential areas may allow for a more favorable energy balance than undeveloped winter ranges, leading to inevitable conflicts between habitat use of deer and human communities. Seasonal home-range size of white-tailed deer in the Midwestern United States was greatest for rural areas, intermediate for exurban areas, and least for suburban areas (Storm et al., 2007). Cultural landscapes in wooded regions are also characterized by an increase in forest edge density, which can be directly linked to increased vital attributes such as pregnancy rate in other edge-associated cervids such as sika (Miyashita et al., 2008).

Natural disturbances that increase forest heterogeneity also can improve habitat quality and thereby reduce home-range size, as reported for roe deer after severe windstorms in western Europe (Said and Servanty, 2005). Over time, repeated large-scale disturbances produce landscape mosaics of different successional stages. At the landscape scale, white-tailed deer require such mosaics to satisfy the different requirements of their annual life cycle, such as early successional stages for food resources during the growing season, early and middle stages for food resources during autumn and winter, middle stages for thermal cover at any time of the year, and late stages in areas where mast is important (Felix et al., 2004).

White-tailed Deer as Agents of Landscape Change

Much research has documented how large populations of ungulates can alter the structure and composition of vegetation communities through selective foraging, where certain plant species are favored over others (reviewed in Côté et al., 2004) (Chapter 12). For example, overabundance of white-tailed deer in a western

Pennsylvania forest increased relative dominance of invasive herbs (garlic mustard and Japanese stilt-grass) because of selective foraging on natives and creation of open patches (Knight et al., 2009). Selective effects of deer herbivory can also produce landscape- or regional-scale shifts in forest-species composition that can cascade to other wildlife species. Deer overpopulation is one of several ultimate causes implicated in oak decline in the eastern United States, where oak forests are gradually being replaced by maple species, which has implications for numerous species of game birds and small mammals that rely on acorn production (McShea et al., 2007). Such shifts in forest composition can lead to altered disturbance regimes and nutrient cycling (Chapter 12). The role of white-tailed deer for long-distance transport of seeds is highly significant and often underappreciated (Myers et al., 2004). Over a 1-year period of fecal pellet collection in New York, more than 70 plant species germinated from deer feces, with a mean of over 30 seeds germinated from each fecal pellet group. White-tailed deer also play a significant role in transport of nutrients between ecosystems or land cover types, particularly where they have high population levels. Where cropland and forest occur in a fine-grained mosaic in the eastern United States, diel movement of deer for foraging and concealment results in a net transport of nitrogen from nitrogen-rich cropland, where deer feed, to nitrogen-poor forest where deer bed down or take cover (Seagle, 2003).

Landscape Historical Perspective

Before Euro-American settlement, white-tailed deer were abundant over much of their range, occupying upland glades and riverine woodlands, with greatest abundance on islands and coastal wetlands (McCabe and McCabe, 1984). In mature, virgin forests white-tailed deer were primarily disturbance-associated species using forest edges created by windstorms and occasional wildfires (McCabe and McCabe, 1984). Canopy gap dynamics from tree falls created forest edge over small areas during periods separating infrequent large-scale disturbance. For example, in the temperate deciduous forest of the Great Smoky Mountains canopy gap disturbances affected about 1% of the landscape annually (Runkle, 1985). Native peoples of the northeastern United States likely used fire to manage vegetation, partly for the purpose of improving habitat for deer, an important food source, although such fire management may have been important only near native population centers (Russell, 1983; Clark and Royall, 1996). With Euro-American settlement came extensive agricultural development over much of the white-tailed deer's range, perhaps causing expansion of deer populations. Subsequently, large acreages of agricultural land were abandoned in the late nineteenth and early twentieth centuries, resulting in extensive reforestation (Orwig and Abrams, 1994; Motzkin et al., 1996). Over the past half-century there has been a pronounced decline in the extent of early successional forest across much of the eastern United States because of forest successional processes, suburban and exurban development, and fire suppression (Trani et al., 2001; Brown et al., 2005). For these regions, deer likely depend on residential areas and remaining agricultural fields to subsidize energy sources from forested landscapes (Hurst and Porter, 2008), and to sustain artificially high populations, often with a cost to adjacent natural areas (Horsley et al., 2003).

For white-tailed deer, the important issues associated with landscape heterogeneity and land-use change are not the negative effects of habitat fragmentation as commonly observed for other species that have patchy population structures or low movement rates. Such species are highly dependent on dispersal to recolonize unoccupied patches (Fahrig, 2003, 2007). Instead, problems associated with white-tailed deer more commonly result from locally concentrated landscape use associated with anthropogenic subsidies to their available energy sources. Consequences of overpopulated deer ranges include increased incidence of wildlife diseases such as mengial worm or bovine tuberculosis (Schmitt et al., 1997), increased transmission of zoonotic diseases to humans (e.g., Lyme disease; Barbour and Fish, 1993), increased conflict with human uses, negative effects on vegetation in core habitat areas, movement mortality associated with roads, and die-offs during severe drought or winter conditions.

Sexual Segregation

Sexual segregation, which involves the differential use of space or other resources by the sexes outside the mating season, is likely ubiquitous among dimorphic ruminants (Bowyer, 2004), including white-tailed deer (McCullough, 1979; Beier, 1987; McCullough et al., 1989; LaGory et al., 1991; Jenks et al., 1994; Kie and Bowyer, 1999; DePerno et al., 2003; Stewart et al., 2003; Monteith et al., 2007; Cooper et al., 2008). Clarifying reasons for this phenomenon is critical for wise management of deer, including the need to understand their distribution across the landscape. Indeed, spatial separation of sexes has far-reaching consequences for implementation of habitat management for deer (Chapters 14 through 21) as well as understanding their behavior (Chapter 10), population ecology (Chapters 5 and 8), and ultimately their effects on ecosystem structure and function (Chapter 12). There is much to be learned concerning the behavior and ecology of white-tailed deer from a more generalized understanding of sexual segregation in ruminants.

The concept of sexual segregation can be traced to Charles Darwin (Bowyer, 2004 for review). Interest among biologists in sexual segregation has not waned over time, and publications on this topic have increased exponentially in recent decades, with research on ruminants responsible for most advances in our understanding of this topic (Bowyer, 2004). Nonetheless, debate continues over the causes of sexual segregation (Miquelle et al., 1992; Main et al., 1996; Bleich et al., 1997; Bowyer, 2004; Main, 2008 for reviews). Clarifying causes and consequences of sexual segregation is a necessary first step in using this knowledge to implement management strategies to benefit white-tailed deer.

Sexual segregation has been documented for most dimorphic ruminants, but is absent in monomorphic species (Bowyer, 2004). This widespread occurrence indicates there are common causes among species (Bowyer, 2004), and necessitates a search for ultimate hypotheses with strong evolutionary underpinnings capable of explaining sexual segregation under the array of environmental conditions for which it has been described (Bowyer, 2004). Proposing a hodgepodge of proximal causes that operate under some circumstances but not others is neither a fruitful avenue for advancing our knowledge of sexual segregation nor a viable approach for helping develop management strategies for deer.

There is consensus that sexual dimorphism is an important key to understanding sexual segregation (Bowyer, 2004 for review). Male ruminants attaining body sizes larger than females relates directly to degree of polygyny, wherein natural selection has favored large males that are most effective in male–male competition for mates (Weckerly, 1998; Loison et al., 1999; Spaeth et al., 2001; Perez-Barberia et al., 2002). Models of sexual dimorphism based on competition between sexes proposed for other taxa are not sufficient to explain sexual dimorphism in mammals (Ralls, 1977; Bowyer, 2004). Moreover, sexual dimorphism and sex-ratio variation among ungulates span millions of years (Berger et al., 2001). More proximal hypotheses for ruminants that do not incorporate predictions stemming from sexual dimorphism, including related differences in life-history strategies of sexes, are not likely to be profitable lines of inquiry. Misunderstanding how and why sexual segregation occur holds potential for mismanaging deer.

Regrettably, no operational definition for sexual segregation has been widely accepted (Barboza and Bowyer, 2000). Failure to adequately conceptualize, describe, and define this phenomenon has hindered formalizing testable hypotheses for sexual segregation, as well as slowed the application of that knowledge to management of ruminants (Bowyer, 2004). The most recent review of sexual segregation in ungulates (Main, 2008) has done little to clarify these important issues. Our purpose is to rectify some problems related to definitions and hypotheses, and then focus on how our approach for understanding sexual segregation can improve and extend the existing knowledge of the ecology, and behavior, of white-tailed deer and their spatial use of landscapes. We also provide insights into how the various approaches for understanding sexual segregation relate to the management of white-tailed deer.

Describing, Detecting, and Defining Sexual Segregation

Describing, detecting, and defining sexual segregation is not a simple process. Bowyer (2004) listed some requirements for demonstrating the occurrence of sexual segregation based on types of social group or sex ratios: (1) seasonal variation in the degree of segregation must occur on an annual basis; and (2) the ratio of sexes must differ between areas or from the sex ratio of the whole population during the period of segregation. Methods used to categorize social groups or measure sex ratios, however, can yield varying results (Conradt, 1998a; Bonenfant et al., 2007; Monteith et al., 2007). Similarly, determining the degree of sexual segregation is strongly influenced by scale of measurement (Bowyer et al.,

1996, 2002; Bowyer and Kie, 2006). Both methods and scale of measurements will affect the interpretation of how the sexes of deer use the landscape. Indeed, failure to document sexual segregation may be methodological rather than biological (Bowyer, 2004). Moreover, variation among published studies in methods used for measuring sexual segregation, and the scale at which it was sampled, render a comparative approach for understanding this phenomenon ineffective and sometimes misleading (Bowyer, 2004). Those cautionary remarks, however, have yet to serve as a deterrent to making inappropriate comparisons among studies.

Group Composition and Sex Ratios

Changes in the types of social groups in which individuals occur (Hirth, 1977; Bowyer, 1984; Miquelle et al., 1992; Molvar and Bowyer, 1994; Bleich et al., 1997; Bowyer and Kie, 2004; Bowyer et al., 2007) have been used to assess periods of segregation and aggregation for ruminants. Indeed, the method developed by Hirth (1977 for definitions) for white tailed deer, wherein social groups are categorized as mixed sex, male only, female only, yearling, or young only, has been the basis for developing much of our understanding of social organization in ungulates. For species such as white-tailed deer, where many individuals may be solitary, including lone individuals as a group (despite traditional dictionary definitions) is critical to capture the complete range of sociality for that species (Bowyer, 2004). Failure to do so can omit numerous individuals from analyses (e.g., Conradt, 1998a), which can affect perceived patterns of segregation and aggregation on the landscape. Monteith et al. (2007) demonstrated that how individuals were categorized into social groups had considerable effect on interpreting the extent and timing of sexual segregation in white-tailed deer (Figure 6.2). In this instance, the solitary categorization of deer was superior in depicting the pattern of segregation around parturition, because most yearling females were pregnant and preparturient females became solitary near parturition (Monteith et al., 2007).

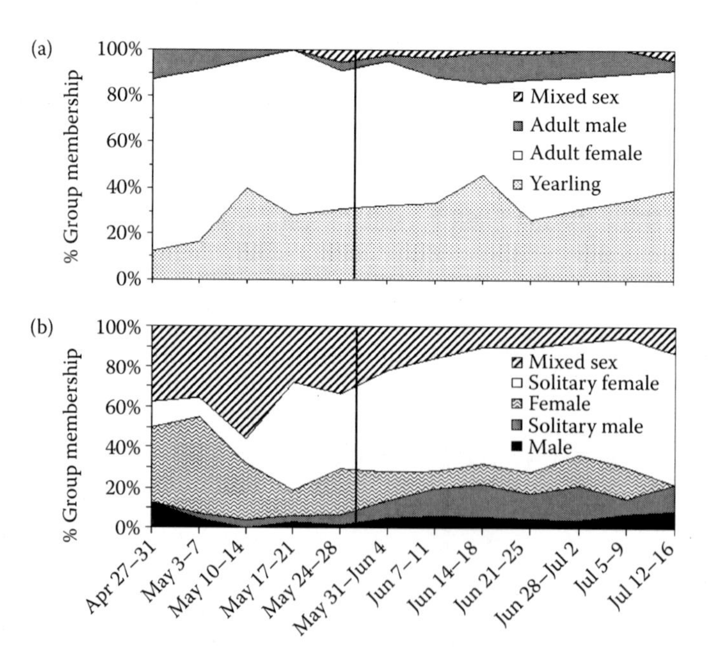

FIGURE 6.2 (a) Percentage of social groups of white-tailed deer categorized according to the method of Hirth (1977) and (b) a method of solitary categorization developed by Monteith et al. (2007). Solid vertical lines indicate peak parturition. The primary differences between methods are that the solitary categorization includes yearlings as adults, records solitary males and females as separate categories, and considers a group mixed-sex if it contains at least one male and one female whether they are yearling or adult, whereas the method of Hirth separates yearlings, includes solitary individuals in male or female groups, and only considers adults when classifying mixed-sex groups. (Modified from Monteith, K. L. et al. 2007. *Journal of Wildlife Management* 71:1712–1716.)

Moreover, the solitary method avoids the problem of categorizing large yearling females as adults. This outcome indicates that the method of Hirth (1977) is probably superior for gregarious species, as well as those where yearlings seldom reproduce, or for relatively unproductive populations of white-tailed deer, where reproduction by yearlings would be unlikely (Monteith et al., 2007). Determining the correct technique for assessing the timing of segregation for white-tailed deer will require information concerning the relationship of the population to carrying capacity (K), or the physical condition of adult and yearling females, which affects reproduction in those age classes (McCullough, 1979).

Changes in sex ratios also have been used to assess degree and timing of sexual segregation. Conradt (1998a) developed a statistical approach to examine changes in sex ratios over time. This technique has the advantage of providing a relatively nonarbitrary method for evaluating if and when sexual segregation occurs. Her metric for social segregation, however, does not include solitary individuals, which may limit its usefulness for some species, especially white-tailed deer. More recently, Bonenfant et al. (2007) developed an improved method based on the chi-square test to assess sex ratios during segregation and aggregation. This test, however, does not have a spatial component, and may require additional modifications to assess the patchy distribution of deer upon the landscape. Moreover, the method of Bonenfant et al. (2007) simply tests (with a yes or no answer) for segregation and aggregation relative to the expected sex ratio; however, this method does not provide a descriptive statistic indicating the degree of segregation. Another method to assess patterns of segregation and aggregation on the landscape was developed by Stewart et al. (2003) to study responses of male and female white-tailed deer to habitat manipulations. Those authors assessed how changes in proportions of males using various habitat manipulations changed during periods of aggregation and segregation, as well as how the proportion of males on treatments differed from the population as a whole. The best method for assessing sexual segregation in white-tailed deer likely will require considerable information and deliberation, including a detailed sampling design that is specifically focused on questions related to sexual segregation.

Scale

Species of ruminants may sexually segregate at markedly different scales (Bowyer, 2004). For example, male and female white-tailed deer partition space at relatively fine scales during segregation compared to other ruminants, with separation occurring at 1–4 ha on study areas where this has been assessed (McCullough et al., 1989; Kie and Bowyer, 1999; Stewart et al., 2003). Conversely, sexes of desert-dwelling bighorn sheep occupy mountain ranges separated by >15 km while spatially segregated (Bleich et al., 1997). Determining the best scale for detecting sexual segregation is important. Sampling at too large a scale would lead to all sampling units being categorized as sexually aggregated, whereas too small a scale would lead to the conclusion that most animals were segregated (Bowyer, 2004). Bowyer et al. (1996) offer methods to select the most appropriate scale for sampling sexual segregation, but the correct determination likely is unique to each species and study area.

Understanding the scale at which white-tailed deer are distributed on the landscape and determining the degree to which deer sexually segregate are not trivial and have implications for management activities, especially habitat manipulations (Stewart et al., 2003). Many aspects of the ecology of ruminants are scale sensitive (Kie et al., 2002; Boyce et al., 2003; Anderson et al., 2005; Maier et al., 2005). A hierarchical sampling approach using differing scales (Bowyer et al., 1996, 2001a; Bowyer and Kie, 2006; Wheatley and Johnson, 2009) would more likely detect sexual segregation than simply applying a scale that had worked for other species or the same species in a different environment.

Other issues related to sampling sexual segregation with quadrats or plots remain. First, basing sampling units on habitat patches is not advised. Those patches likely differ in size and shape, and effects of scale, and perhaps shape, would be introduced inadvertently into the analyses (Bowyer and Kie, 2006). Second, mixing habitat or vegetation types within sampling units could lead to biases in understanding sexual segregation, especially if one sex or the other strongly selected a particular habitat that varied in abundance across samples. Third, selecting criteria for determining whether a particular sample is segregated can be problematic. Polygynous ruminants, including white-tailed deer, often have sex ratios that are biased toward females (Hirth, 1977; McCullough, 1979; Kie and Bowyer, 1999; Stewart et al., 2003; Monteith et al., 2009). Thus, even a sex ratio that significantly favored females might not be evidence of sexual segregation, especially if that ratio did not differ from the overall ratio of males to females in the population. Bowyer et al. (1996, 2001a) adopted a value of ≥90% of one sex or the other to define a segregated quadrat, but other values are possible depending on the overall sex ratio of the population under study.

Niche-Based Approach for Understanding Sexual Segregation

Despite numerous publications on sexual segregation, no general definition for this phenomenon is widely accepted (Barboza and Bowyer, 2000). Traditionally, sexual segregation has been defined as the differential use of space, and sometime resources, by the sexes (Bowyer, 2004). We will return to other hypotheses, including those concerning differences in activity patterns by the sexes, in the following section. Herein, we consider how the ecology of the sexes might influence segregation. Mysterud (2000) subdivided ecological segregation into three broad categories: (1) spatial; (2) dietary; and (3) habitat. His approach provides a useful framework for understanding ecological segregation, but has the disadvantage of separating interconnected processes. Analyses based on these ecological subdivisions also run the risk of concluding that sexual segregation does not occur where that assessment is based on only a single category (e.g., space, diet, or habitat). Indeed, we believe that a more synthetic approach is needed. Such an approach is founded on a niche-based understanding of sexual segregation (Table 6.4), and proposes that the sexes of dimorphic ruminants behave as if they were separate but coexisting species (Kie and Bowyer, 1999; Bowyer, 2004; Bowyer and Kie, 2004; Shannon et al., 2006; Schroeder et al., 2010). As with other advances in knowledge related to large mammal ecology (*sensu* McCullough, 1979), white tailed deer have played the key role in development of the niche-based model for sexual segregation.

Briefly, research on the population dynamics of white-tailed deer was undertaken on the Welder Wildlife Refuge in South Texas, United States (Kie and White, 1985). This research design reduced, but not completely removed, coyotes from a 391-ha enclosure. Fewer coyotes resulted in a marked increase in deer within the enclosure (77 deer/km^2), compared to a more moderate density for the remainder of the refuge (39 deer/km^2). The sexes of deer at high density exhibited more spatial overlap than those at moderate density (Figure 6.3), but differed markedly in their dietary niches (Figure 6.4). Conversely, deer at moderate density showed more spatial separation of sexes than deer living at high density, but the sexes had dietary niches that were similar (Kie and Bowyer, 1999). These changes also were accompanied by differences in habitats used by the sexes at high and moderate densities (Kie and Bowyer, 1999). Results from this study conform to standard niche theory, wherein overlap on one niche axis is accompanied by

TABLE 6.4

A Niche-Based Model for Conceptualizing Patterns of Sexual Segregation in Polygynous Ruminants

Pattern of Segregation	Space	Diet	Habitat
I	+	–	+
II	–	+[a]	–[b]
III	–	–	+/–

Source: Adapted from Bowyer, R. T. and J. G. Kie. 2004. *Journal of Mammalogy* 85:498–504.

Note: + = Overlap of sexes; – = separation of sexes.

The model does not intend that overlap or separation be complete on any particular niche axis but indicates the general direction of outcomes. Other potential combinations of overlap and separation between sexes were judged to be inconsistent with niche theory (e.g., overlap in space, diet, and habitat), or otherwise infeasible (e.g., overlap in space but not habitat).

[a] Some differences in diet would be expected because of allometric and life-history differences between sexes.

[b] Sexes potentially could use the same habitat at different locations.

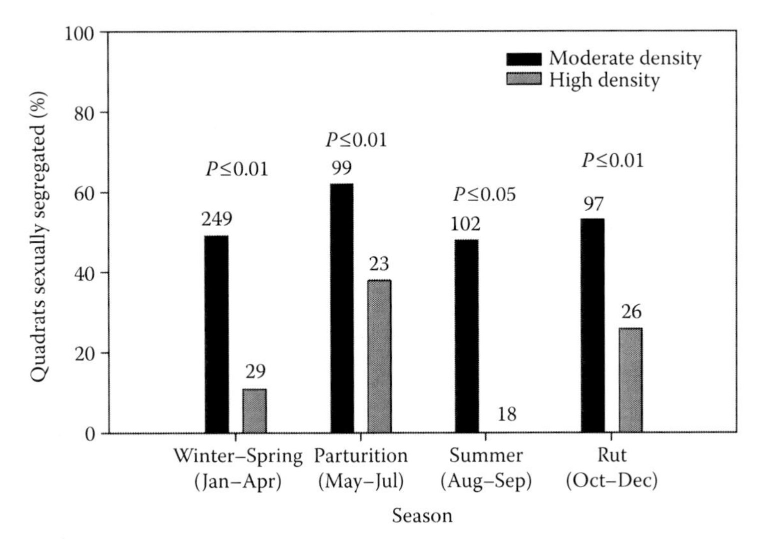

FIGURE 6.3 Percentage of quadrats that were sexually segregated on a monthly basis at moderate and high densities of white-tailed deer. (Redrawn from Kie, J. G. and R. T. Bowyer. 1999. *Journal of Mammalogy* 80:1004–1020.)

avoidance on another (Kie and Bowyer, 1999; Bowyer, 2004; Bowyer and Kie, 2004; Schroeder et al., 2010). These results also illustrate the importance of measuring multiple niche axes to fully understand the ecology of sexes and the best approaches for their management.

Hypotheses for Sexual Segregation

Although a plethora of hypotheses have been forwarded to elucidate why sexes of dimorphic ruminants segregate, many authors have proffered explanations that were too narrow or proximal and therefore, lacked broad applicability, or were subsequently rejected by empirical research. Proximal explanations for sexual segregation, such as the need of lactating females for water (Bowyer, 1984), or differential sensitivity of sexes to weather (Conradt et al., 2000) are unlikely to be the root cause of this phenomenon (Bowyer, 2004). Confusion over this point is sufficient that proximate and ultimate causes of sexual segregation, including failing to distinguish an hypothesis from a prediction, have been muddled in several recent publications (Neuhaus et al., 2005; Main, 2008; Ruckstuhl, 2007; Gregory et al., 2009). We will not discuss all hypotheses for sexual segregation or the reasons they were rejected herein—adequate treatments exist elsewhere (Miquelle et al., 1992; Main et al., 1996; Bleich et al., 1997; Bowyer, 2004). Several hypotheses remain, however, which still receive attention in the literature and require discussion relative to the spatial use of landscapes by deer. Moreover, an understanding of which hypotheses are not appropriate is as essential for wise management of white-tailed deer as delineating those hypotheses capable of explaining sexual segregation. Hypotheses should be capable of explaining the widespread occurrence of sexual segregation, not confuse cause and effect, and be independent and testable (Bowyer, 2004).

Reproductive Strategy Hypothesis

Main et al. (1996) presented predation risk (see the following section) as part of a broader and somewhat bewildering treatment under the reproductive-strategy hypothesis; unfortunately, many aspects of ungulate behavior can be viewed as a reproductive strategy (Bowyer, 2004). Indeed, Bowyer (2004) and Neuhaus et al. (2005) noted that the reproductive-strategy hypothesis was cast too broadly to provide specific predictions about sexual segregation, and consequently was difficult to test. Most hypotheses for sexual segregation, including the gastrocentric hypothesis and predation hypothesis, also involve life-history strategies (Bowyer, 2004). Indeed, the reproductive-strategy hypothesis is not independent

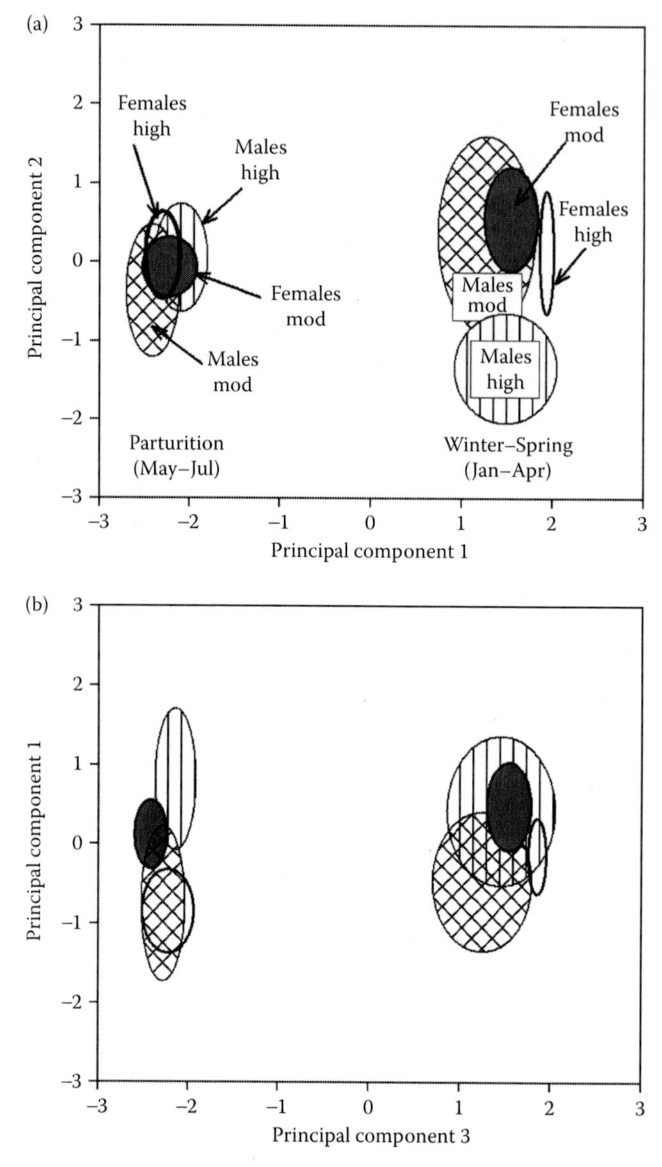

FIGURE 6.4 Principal components 1 and 2 (a) and principal components 1 and 3 (b) for diets of white-tailed deer at moderate and high densities. Ellipses are 95% CI, and represent the dietary niche for males and females. The first three principal components explained 66% of the variation in diet. (Redrawn from Kie, J. G. and R. T. Bowyer. 1999. *Journal of Mammalogy* 80:1004–1020.)

of other hypotheses for sexual segregation, which creates statistical difficulties for an unbiased test (this criticism is far more serious than hypotheses not being mutually exclusive). Ambiguity in the formulation of this hypothesis, and accompanying confusion concerning clear predictions, makes its value to either theory or management nil.

Activity-Budget Hypothesis

One widely investigated hypothesis is that differences in activity between sexes cause social segregation (Conradt, 1998b; Ruckstuhl, 1998). Studies of various ruminants, however, have revealed mixed outcomes as to whether activity patterns are related to formation and maintenance of social groups (Bowyer and Kie, 2004; Mooring and Rominger, 2004; Mooring et al., 2005; Michelena et al., 2006, and many others). One problem with this hypothesis is that it is not possible to disentangle whether differences in activity budgets cause social segregation or are simply correlated with segregation or another related variable such as sexual dimorphism (Bowyer, 2004). Moreover, a critical test of this hypothesis with a manipulative experiment is difficult if not impossible to obtain (Bowyer, 2004).

Understanding how differences in activities of sexes relate to the formation of social groups is of considerable theoretical interest, although specifically how this fits with other hypotheses related to the evolution of sociality is yet to be adequately explored (Bowyer, 2004). Indeed, hypotheses related to sociality include predation risk and clumped resources (Bowyer, 2001b for review); neither of these hypotheses has been evaluated with regard to activity patterns and their relationship to social segregation. Including social segregation with dietary, spatial, and habitat segregation was reasonable when it was believed that this process could affect the distribution of sexes on the landscape (Mysterud, 2000)—this assertion, however, has been rejected (Bowyer, 2004; Bowyer and Kie, 2004) and subsequently retracted (Neuhaus et al., 2005). To include this process under the general term sexual segregation may not be justified because of its inability to explain the spatial distribution of sexes. Nonetheless, the pattern of sexual aggregation and separation can influence how the extent and timing of sexual segregation are measured (see the previous section).

There is general agreement that social segregation cannot explain sexual segregation with respect to habitat or space (Bowyer, 2004; Neuhaus et al., 2005). Indeed, Kie and Bowyer (1999) demonstrated that there was substantial variation in the degree of spatial segregation in white-tailed deer with few concomitant changes in the proportion of single-sex or mixed-sex groups. As noted previously, the activity-budget hypothesis cannot explain the distribution of the sexes and their spatial use of landscapes, and consequently has only peripheral value in developing management strategies for white-tailed deer.

Social-Factors Hypothesis

Hypotheses for sexual segregation related to social factors are diverse. These have included the need for males to occur in same-sex groups to develop and assess fighting abilities and form dominance hierarchies. Reasons underpinning female associations might include learning places to forage, locations of water, and places to give birth. A social affinity of the sexes for one another has also been considered under this hypothesis (Bon et al., 2005). Bon et al. (2005) believed it was difficult to separate social from ecological reasons for sexual segregation. Nonetheless, the manipulative experiment of Kie and Bowyer (1999) involving white-tailed deer mentioned previously provided a critical test—there were large changes in spatial, dietary, and habitat separation between sexes without concomitant modifications in social groups. In addition, Neuhaus et al. (2005) characterized many of the social-factors hypotheses as not falsifiable, and consequently untestable.

One potential exception to the aforementioned problems with the role of social factors in causing sexual segregation relates to aggression between sexes or avoidance of one sex by the other (Weckerly, 2001; Weckerly et al., 2001). Aggression and cropping rates were related to types and sizes of social groups, which presumably led to spatial difference between sexes for North American elk (Weckerly et al., 2001). Mutual avoidance by the sexes, however, might result from differences in body size and may not be independent of ecological or physiological requirements of sexes (Weckerly et al., 2001; Bowyer, 2004). Indeed, no aspect of this hypothesis will result in long-term differences in use of space or diets between sexes (Bowyer, 2004). Temporal differences in the use of resources between sexes do not result in long-term displacement of one sex by the other. Understanding social-factor hypotheses does not lead to obvious outcomes for enhancing habitat or other aspects of the management of white-tailed deer because this hypothesis does not result in sexes consistently using different space.

Scramble-Competition Hypothesis

This hypothesis also has been cast as the sexual dimorphism-body size hypothesis (Main et al., 1996), although serious problems exist with the independence of this and other hypotheses for sexual segregation

(Bowyer, 2004). Clutton-Brock et al. (1987) initially proposed that females competitively excluded males, which ultimately resulted in sexual segregation (i.e., scramble competition—also termed indirect competition). That females exclude males is critical to this hypothesis, because an explanation is needed for why larger males do not displace smaller females from higher-quality areas, or use the same areas as females. Several investigators have rejected this hypothesis because patterns of sexual segregation did not follow logically from predictions of competitive exclusion, including that segregation was not marked during seasons when competition between sexes should have been most prominent (du Toit, 1995; Bleich et al., 1997), or that pronounced sexual segregation occurred in a population at extremely low density where intersexual competition was lax (Miquelle et al., 1992). Shannon et al. (2006) also rejected this hypothesis based on foraging behavior by the sexes of African elephants. Scramble competition also fails to provide a mechanism for explaining why the sexes should aggregate (Bowyer, 2004), a topic we will return to later. In addition, further studies by the research group who initially proposed the scramble-competition hypothesis have led to them recanting this idea, and to substantial doubts about its utility as an explanation for sexual segregation (Conradt et al., 1999, 2001).

Straightforward tests of scramble competition between sexes require a manipulative design (Kie and Bowyer, 1999). In an experiment involving foraging behavior in captive moose, Spaeth et al. (2004) demonstrated that previous foraging by females failed to substantially reduce forage intake by males. Those authors concluded that competition for forage could not lead to exclusion of males by females. One of the most important contributions to our understanding of sexual segregation came from another experimental manipulation of densities of white tailed deer (Kie and Bowyer, 1999). Changing deer density provided a clear-cut and unequivocal test of whether scramble competition could cause sexual segregation. A reduction in the degree of sexual segregation occurred with increasing population density (Figure 6.3). The scramble-competition hypothesis predicts that an increase in density would have intensified intersexual competition and resulted in increased sexual segregation. Outcomes from this manipulative experiment are antithetical to predictions from the scramble-competition hypothesis. Main (2008) recently tried to resurrect the scramble-competition hypothesis. That attempt ignored the evidence presented herein, and provided no new substantive evidence to support the contention that scramble competition caused sexual segregation. The weight of evidence overwhelmingly indicates that this hypothesis is not a viable explanation for sexual segregation. Developing management strategies for white-tailed deer around ideas involving scramble competition between sexes is not a fruitful endeavor.

Forage-Selection Hypothesis

The forage-selection hypothesis has largely replaced the sexual dimorphism-body size hypothesis in the literature on sexual segregation, because of problems with the latter hypothesis as described in the previous section. Regrettably, this is not an improvement. The forage-selection hypothesis is not a hypothesis, but at best is a prediction or observation. This "hypothesis" gives no indication of why the sexes should aggregate or segregate, it simply notes that they forage differently. This observation (the sexes forage differently) requires some explanation (i.e., hypothesis) for why they should do so. For dimorphic ruminants, we know of only two hypotheses that have the potential to explain differences in morphology and physiology between sexes that would lead to differences in foraging: the Bell–Jarman hypothesis and the gastrocentric hypothesis. Several authors have badly confused differences between hypotheses and predictions by subsuming these true hypotheses under the mantel of the forage-selection hypothesis (Ruckstuhl, 2007; Main, 2008; Gregory et al., 2009). Moreover, Main and du Toit (2005) maintained that these hypotheses related to sexual size dimorphism cannot explain why sexes partition space within a habitat. This contention is incorrect. Spatial separation of sexes within habitats has been documented for ruminants (Bowyer, 1984, 1986); both the Bell–Jarman and gastrocentric hypotheses can deal with this outcome because they make predictions concerning both forage quality and quantity, not just differences in habitat use. We suspect some confusion stems from not recognizing that segregation can occur with respect to diet, space, and habitat (Table 6.4). Moreover, the gastrocentric hypothesis was developed specifically to explain differences in both forage quality and quantity (Barboza and Bowyer, 2000, 2001), and does not require heterogeneous habitats to bring about sexual segregation (Bowyer, 2004).

Although differences in forage requirements of the sexes can have important management implications for white-tailed deer, understanding why those differences occur is essential for the wise management of these ungulates.

Bell–Jarman Hypothesis

The Bell–Jarman hypothesis, as articulated by Demment and Van Soest (1985), explains niche differentiation among species of ruminants (Bell, 1970; Jarman, 1974) based on scaling relationships of rumen and reticulum size (hereafter referred to as rumen size), and metabolic rate with body mass. Isometric scaling of the rumen (i.e., ~1.0) in relation to the allometric scaling of metabolic rate (i.e., ~0.75) with respect to body mass has been proposed to explain differences in digestive capabilities across species of ruminants. Larger species have proportionally larger rumens in relation to their respective metabolic rates than do smaller ruminants. This greater rumen capacity allows larger species to consume more food (they also require absolutely more food), and retain forage longer during fermentation than smaller species do (Demment and Van Soest, 1985; Clauss et al., 2007; Weckerly, 2010). These digestive differences have been proposed as the mechanism underpinning niche differentiation across a wide range of body sizes in ruminants (Demment and Van Soest, 1985).

Although the Bell–Jarman hypothesis was developed to explain niche partitioning among species of ruminants, this hypothesis also has been proposed as a possible explanation for sexual segregation in dimorphic ruminants (McCullough, 1979). Nonetheless, the full implications of understanding niche differences between sexes have only recently been elucidated (Kie and Bowyer, 1999; Bowyer, 2004; Bowyer and Kie, 2004, Table 6.4). Moreover, Barboza and Bowyer (2001) cautioned that morphological and physiological mechanisms that explained niche differences among species of ruminants might not be applicable to body size differences between sexes of the same species. The Bell–Jarman hypothesis cannot be tested critically by examining differences in foraging behavior or digestive efficiency (*sensu* Perez-Barberia et al., 2008) between sexes, because other competing hypotheses can lead to similar results under particular sets of circumstances (see the section on Gastrocentric Hypothesis). What is required is a test of the aforementioned scaling relationship between rumen size and body mass (data on metabolic rate in relation to body mass are well established and much less controversial). Tests of this hypothesis between sexes of ruminants has been equivocal, in part because of difficulties of controlling for seasonal and daily patterns of rumen fill, which is used as an index to rumen size (Weckerly, 2010, for review). A detailed study of rumen capacity and body mass for sexes of white tailed deer documented that relationship was allometric, and offered no support for the Bell–Jarman hypothesis being involved in sexual segregation (Weckerly, 2010). Another problem with extending the Bell–Jarman hypothesis to sexual segregation is that although it offers a potential explanation for why the sexes segregate, there is no explanation for why they aggregate (Bowyer, 2004). The need to acquire mates may be invoked, but sexual aggregation occurs outside the mating season in some species (Bowyer, 1984). Moreover, there is no explanation in the Bell–Jarman hypothesis for why sexual segregation tends to be concentrated around parturition (Bowyer, 2004). Clearly, the Bell–Jarman hypothesis is not a sufficient explanation for sexual segregation—other hypotheses are needed to understand this phenomenon.

Gastrocentric Hypothesis

The gastrocentric hypothesis (Barboza and Bowyer, 2000, 2001) offers an explanation for sexual segregation in ruminants based on an allometric model of metabolic requirements, minimal food quality, and digestive retention. Adult male deer are predicted to consume abundant forages high in fiber because ruminal capacity prolongs retention and thereby allows greater use of fiber for energy than in females (Figure 6.5). Smaller-bodied females are better at postruminal digestion of food than males, particularly when energy and protein requirements necessary for reproduction increase. These relationships are well documented for white-tailed deer (Jenks et al., 1994). Indeed, reproductive females also increase rumen size, and the length and width of rumen papillae over that of nonreproductive females (Zimmerman et al., 2006). High demand for absorption of nutrients during lactation additionally stimulates maternal

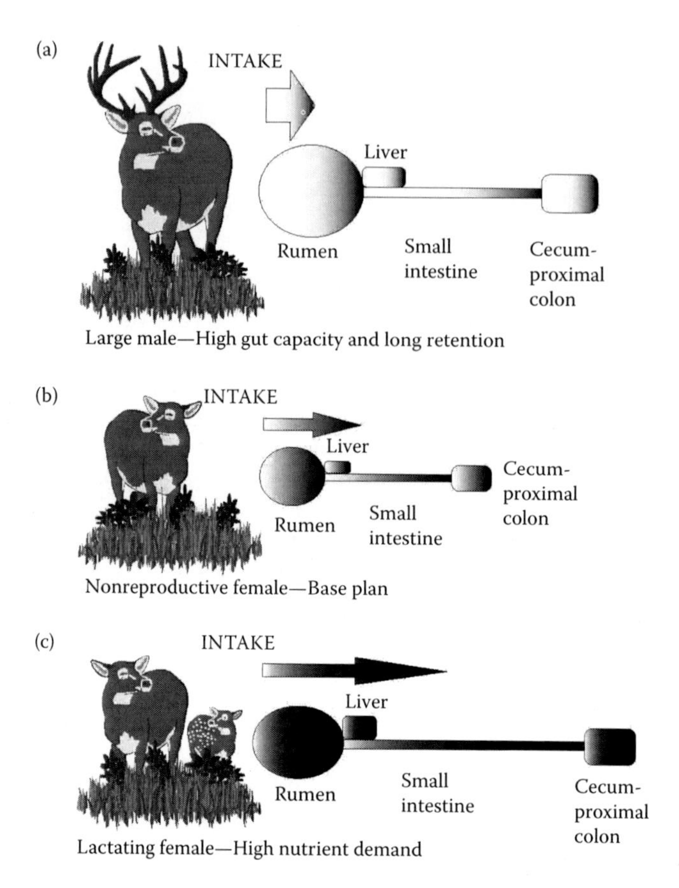

FIGURE 6.5 Model of intake and digestive function and nonreproducing females compared with large males and lactating females. Width of arrow reflects amount of food intake, length of arrows indicate rate of digesta passage, and shading indicates density of nutrients in food. Diagrams of the digestive tract are stippled to reflect potential changes in fibrosity of food for males and increases in postruminal size and function for lactating females. This figure is modified from the original to include new information (Zimmerman et al., 2006) documenting that the rumen of lactating females is larger and has greater papillae length and width compared to that of nonlactating females. (Modified from Barboza, P. S. and R. T. Bowyer. 2000. *Journal of Mammalogy* 81:473–489; figure redrawn by C. McVicars.)

investment in intestinal and hepatic tissue (Barboza and Bowyer, 2000). These increased nutrient demands in reproductive females, including the associated costs of remodeling their digestive systems to accommodate late gestation and lactation, reinforce differential use of habitats and forages and may lead to sexual segregation (Barboza and Bowyer, 2000, 2001).

The gastrocentric hypothesis has a number of advantages over other hypotheses forwarded to explain sexual segregation. There is no need to invoke competitive exclusion (e.g., scramble competition) of males by females to bring about spatial separation of sexes. The gastrocentric hypothesis offers a reason that males do not use areas of high quality occupied by females around the time of parturition. Males possess a larger ruminal volume than do females that can accommodate digestion of bulky fibers, whereas females on a similar diet may require more processing by chewing and rumination (Barboza and Bowyer, 2000, 2001). Slow rates of passage and a fibrous diet favor a cellulolytic microflora in the rumen of males. Females, however, have a complement of microbes more accustomed to faster passage of food and acquire a greater proportion of nutrients from cellular contents than from cell walls of plants (Barboza and Bowyer, 2000). Males would be prevented from switching between diets of differing quality quickly because such changes would disrupt ruminal fermentation, and risk excess production of gases and bloat, malabsorption and scouring (Gordon and Illius, 1996; Stevens and Hume, 1995; Van Soest, 1994; Barboza et al., 2009) (Figure 6.5). Males can obtain adequate nutrition on lower-quality and abundant foods than those required by reproductive females (Barboza and Bowyer, 2000). Likewise, sexual segregation can occur without invoking predation risk as a cause of this phenomenon (Barboza and Bowyer, 2000, 2001). Moreover, unlike the Bell–Jarman hypothesis, the gastrocentric model (Barboza and Bowyer, 2000, 2001) offers an explanation for why sexes aggregate and segregate on an annual basis (Bowyer, 2004). Care should be taken in extrapolating these ideas concerning forage selection to other unrelated taxa (Bowyer, 2004). Clearly, predictions from this hypothesis are uniquely related to ruminants. Finally, the gastrocentric hypothesis predicts differences in forage selection by the sexes, an outcome that unmistakably indicates that the forage-selection hypothesis is simply a prediction that emanates from the gastrocentric model.

The gastrocentric hypothesis has huge implications for management of habitat for white-tailed deer. This hypothesis places emphasis on the nutritional needs of sexes in relation to their reproductive strategies (Barboza and Bowyer, 2000, 2001). The gastrocentric model also dovetails nicely with the niche-based approach for understanding the divergent needs of sexes when they are spatially separated. This hypothesis provides an understanding of why the sexes should be treated as if they were different yet co-existing species (Bowyer, 2004), and specifically how management of habitat should be structured to benefit each sex, a topic we will return to later (see section "Habitat Manipulations and the Sexes" of this chapter).

Predation Hypothesis

The predation hypothesis predicts that the sexes will spatially segregate because of differential vulnerability of large males compared with females (especially those with young) to predation from large carnivores (Bowyer, 2004). The predation hypothesis, and changes in the distribution of sexes because of predation risk, was first clearly delineated by Bleich et al. (1997).

Risk of predation is sufficient but perhaps not necessary to explain sexual segregation (Bowyer, 2004). Bleich et al. (1997) clearly demonstrated that large male bighorn sheep inhabited areas with more predators that those occupied by females with young during segregation. Moreover, females occupied ranges of lower quality than did males, presumably trading-off forage for security (Bleich et al., 1997). Little doubt exists that the predation hypothesis can explain sexual segregation (Corti and Shackelton, 2002; Hay et al., 2008). Sexes of red deer, however, segregated under circumstances where no predators were present (Clutton-Brock et al., 1987), but care should be taken in assuming ungulates inhabiting predator-free areas do not exhibit adaptations for avoiding or eluding predators (Bowyer, 2004). Some responses of ungulates to large carnivores may wane quickly in the absence of predators, but others may persist for centuries or longer (Berger, 1999; Barten et al., 2001). For instance, white-tailed deer on Ossabaw Island, Georgia, United States, which have inhabited that island for centuries, still maintain classic anti-predator behaviors even though predators never occurred there (LaGory, 1987; LaGory et al., 1991).

Understanding how the sexes of white-tailed deer respond to risk of predation has far-reaching consequences for managing their habitats and population dynamics. Females, however, may not always use areas of lower-quality forage to avoid predators. Bowyer (1984) and Pierce et al. (2004) noted that areas with better food also could include concealment cover for females and young. Consequently, understanding the relationship between forage quality and risk of predation may not always require a trade-off (Bowyer et al., 1998a, 1999). Nonetheless, trade-offs between forage and risk of predation are common, and require consideration when crafting viable management plans for white-tailed deer.

Multiple Causations and Trade-Offs

One school of thought holds that numerous, multiple causes are necessary to explain sexual segregation (Neuhaus et al., 2005; Main, 2008). This conclusion, however, is not consistent with patterns of sexual segregation among taxa of dimorphic ruminants—sexual segregation is near ubiquitous (Bowyer, 2004); all potential causes would not be expected to operate for all species and all circumstances. If a laundry list of hypotheses was necessary to explain sexual segregation, there should be multiple examples where this phenomenon did not occur—such observations are nil for dimorphic ruminants. The best that can be hoped for by following this line of reasoning is to attribute a relative amount of explanatory power to each hypothesis (Bowyer, 2004). This procedure will not lead to clear-cut tests of why sexual segregation occurs, and presents statistical problems when hypotheses are not independent. Strategies based on multiple causations are likely to result in misdirected and unnecessary management activities for ruminants, including white tailed deer.

There may be no need to invoke more than two postulates to explain sexual segregation in dimorphic ruminants: the gastrocentric and predation hypotheses (Bowyer, 2004). Indeed, Long et al. (2009) used those two hypotheses to frame clear-cut tests of why the sexes selected habitats differently. This approach is also amenable to the niche-based model for understanding sexual segregation. Moreover, considering how these hypotheses interact (they are independent but not mutually exclusive) is necessary to clarify patterns of spatial separation of sexes on the landscape (Bowyer, 2004). Trade-offs between acquiring essential resources and avoiding predation are well documented for large herbivores (Festa-Bianchet, 1988; Berger, 1991; Molvar and Bowyer, 1994; Rachlow and Bowyer, 1998; Kie, 1999; Barten et al., 2001; Corti and Shackelton, 2002; Hay et al., 2008, and many others), and such a trade-off may result in females acquiring a lower-quality diet than that of males, as long as it satisfies requirements for reproduction (Bowyer, 2004). Greater vulnerability of smaller females and young to predation compared to larger males is well recognized as a cause of differences in habitat use and other behaviors between sex and age classes of ungulates (Bowyer, 1987; Bleich et al., 1997; Bleich, 1999). Failure to assess trade-offs can result in misinterpretation of outcomes from tests of sexual segregation. For instance, Main (2008) reviewed numerous studies to examine differences in forage quantity and quality (although these two measures are confounded in his review) between sexes. We previously cautioned against using a comparative approach to assess causes of sexual segregation. In this instance, the comparisons were inappropriate because potential effects of predation risk were not considered in tests for forage quality. Understanding trade-offs (or lack thereof) lies at the heart of management activities for the sexes of white-tailed deer.

Role of Sexual Segregation in Management

Despite an exponential increase in studies of sexual segregation in ruminants during the past two decades, and growing recognition that the niche requirements of sexes were sufficiently different that they should be managed as if they were separate species (Bowyer, 2004), this information has yet to find its way into management plans for species of dimorphic ruminants, including white-tailed deer. Indeed, we know of no management plans for ungulates that incorporate this knowledge. A few modern texts on wildlife management, as well as numerous publications in wildlife literature, have included information on sexual segregation (Bowyer, 2004), but this topic has yet to become firmly established in the discipline of range management. Clearly, sound management of ungulates requires better integration of knowledge from scientific publications into management practices. That knowledge involves both population dynamics and habitat manipulation.

Population Dynamics

McCullough (1979) documented that changing densities of white-tailed deer in relation to carrying capacity (K) influenced female nutrition via intraspecific competition for forage, resulting in changes in their physical condition, rates of pregnancy, and ultimately recruitment of young into the population. Moreover, when he examined effects of female density compared with that of males, number of females was the best predictor of recruitment—males markedly reduced the fit of the relationship predicting recruitment rate from population size (McCullough, 1979). Indeed, Kie and Bowyer (1999) reported that fat reserves of females were reduced more than those of males at high population density. What is less-well recognized is that this outcome results from partitioning of space or food by the sexes outside the mating season, so that competition among adult females and their young for resources is more intense than competition of those sex and age classes with spatially separated males (Bowyer, 2004). Males play a role in the population dynamics of ungulates (Mysterud et al., 2002;

Bowyer et al., 2007; Whiting et al., 2008 for reviews), but effects of female nutrition often overwhelm the influence of males on the productivity of populations. For populations near *K*, harvest of females does more to enhance successful reproduction than killing males (McCullough, 1979). These well-established relationships underpin the necessity of treating the sexes as if they were separate species for the management of populations (Kie and Bowyer, 1999; Bowyer et al., 2001b; Stewart et al., 2003; Bowyer, 2004).

Increasing populations of deer also influence the degree of sexual segregation, with sexes exhibiting more spatial overlap at higher than lower densities (Kie and Bowyer, 1999; Bowyer et al., 2002). This observation is important for management purposes, because one conclusion might be that sexual segregation was not occurring at high density and there was no need to manage habitats differently for the sexes. Again, this is the reason that understanding the niche-based model for sexual segregation (Table 6.4) is critical. Increasing overlap on the spatial (or habitat) niche axis could lead to differences on the dietary axis, which would be important for the management of forages for deer.

Sexual segregation also holds potential to bias estimates of population size and sex ratios of large herbivores. If sampling is unwittingly directed toward male or female ranges when the sexes are spatially separated, biased estimates of those population parameters would result (Bowyer, 2004). This potential bias may pose less of a problem for white-tailed deer than other species, because of the relatively fine scale at which sexes segregate (e.g., 1–4 ha). Even fine-scale differences in use of habitat, however, could lead to differences in visibility of deer, thereby affecting population and sex-ratio estimates. Moreover, there would be implications for the management of habitat and forages for the sexes of deer.

Habitat Manipulations and the Sexes

Bowyer et al. (2001a) reported that the mechanical crushing of willows to benefit moose resulted in a net reduction in winter range for females, even though suitable browse for moose was increased threefold. Mostly males occurred on what previously had been winter range for females. Ostensibly, females and young did not make full use of this manipulated area because it provided less concealment cover and made them more vulnerable to predators (Bowyer et al., 2001a). Whiting et al. (2010) noted that water sources were used differently by sexes of bighorn sheep when they were sexually segregated, and that the existing criterion for distance of bighorn reintroductions from water may be inadequate for successful establishment of populations. Stewart et al. (2006) recorded distributions of male and female North American elk separately to construct maps of population density necessary to understand the role of herbivory in influencing net above ground primary productivity (i.e., herbivore optimization). Recognizing that the sexes use space and habitat differently is becoming a critical element in the design of experiments to test hypotheses concerning ecological processes. The integration of this knowledge into management plans, however, is less common. Without considering the disparate needs of the sexes, danger exists of inadvertently managing habitat such that one sex benefits to the potential detriment of the other (Bowyer, 2004).

Sexes of white-tailed deer use habitats, space, and forages differently during periods of segregation (McCullough, 1979; Beier, 1987; McCullough et al., 1989; LaGory et al., 1991; Jenks et al., 1994; Kie and Bowyer, 1999; DePerno et al., 2003; Monteith et al., 2007; Cooper et al., 2008). How this knowledge relates to management practices, however, is less well studied. Stewart et al. (2003) conducted one of the few experiments relating habitat to sexual segregation in white-tailed deer. Those investigators applied five common habitat manipulations to rangelands in South Texas, USA, and monitored responses of the sexes to those treatments (Figure 6.6). Males occurred more often on treatments that reduced forbs and resprouts of shrubs, and had the greatest biomass of graminoids (Stewart et al., 2003). Those authors concluded that reducing forbs and shrubs likely created a plot that benefited males more than females. Males also occurred on more open areas than females, an outcome that has been reported previously (LaGory et al., 1991; Kie and Bowyer, 1999), and likely relates to males using areas with a greater risk of predation than those used by females and young. Males and females clearly respond differently to manipulation of habitats, but this information seldom is implemented in managing white-tailed deer.

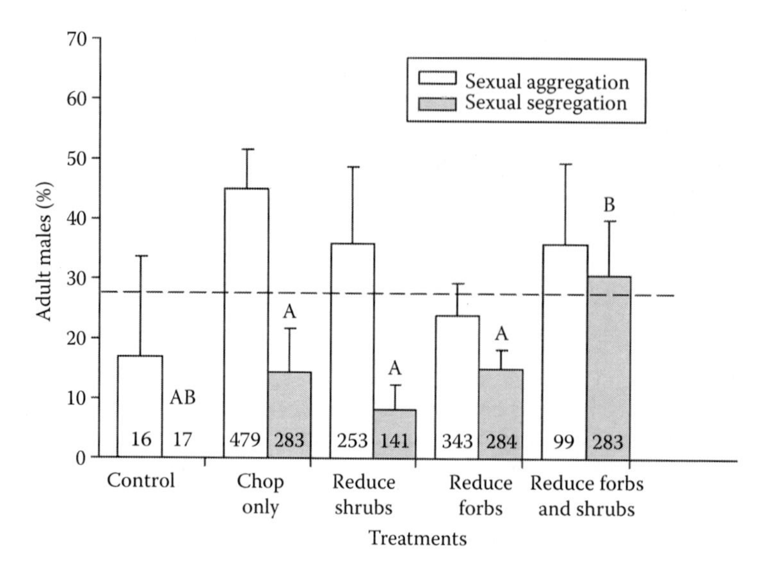

FIGURE 6.6 Percentage of male white-tailed deer (±SE) occurring on habitat treatments during sexual aggregation and segregation, South Texas, USA, 1995–1996. The dashed line represents the overall percentage of adult males observed in the study area. Letters above bars indicate differences in contrasts from repeated measures analysis of variance ($P < 0.03$), weighted by number of deer in each treatment (indicated on each bar). (From Stewart, K. M. et al. 2003. *Wildlife Society Bulletin* 31:1210–1217. With permission.)

Populations of white-tailed deer have sex ratios naturally biased toward females (McCullough, 1979; Kie and White, 1985; Kie and Bowyer, 1999), and density of females relative to K is critical in determining the productivity of populations (McCullough, 1979; Kie and White, 1985). Consequently, most management programs should emphasize manipulating habitat for females if allowable harvest is the primary goal. Nutrition of males (Asleson et al., 1997) also is important if producing large-antlered males (e.g., trophy management) is a management objective. Nevertheless, moose grow larger antlers in areas with lower densities of females than where females occur at higher density (Schmidt et al., 2007). Harvest of females should be a component of any management plan designed to improve habitat for females, because unchecked growth of females would soon reduce their nutritional condition and that of their offspring. Moreover, recent evidence for white-tailed deer indicates that nutrition of females is a primary determinant of male body and antler size, and that small males may never recover antler or body size even on a high nutritional plane (Monteith et al., 2009). Monteith et al. (2009) demonstrated that young born to smaller females from the Black Hills, South Dakota, USA, remained smaller and had smaller antler sizes throughout their lives compared to young born to larger females from eastern South Dakota, despite all deer being on a highly nutritious diet (Figures 6.7 and 6.8). Females required two generations to recover body size and give birth to males that ultimately had larger bodies and antlers (Figures 6.7 and 6.8). These results indicate that habitat manipulations for females are critical for producing large males, but will not have an immediate effect on size of male offspring. In addition, harvesting small males to improve antler size will not have the desired effect in the absence of management that results in females being in good physical condition.

Future Directions

One long-standing difficulty in testing hypotheses related to sexual segregation has been the dearth of manipulative experiments (Kie and Bowyer, 1999). Studies based mostly on correlations among variables related to sexual segregation run the risk of confusing cause and effect. Several studies have employed an experimental approach for this purpose (Kie and Bowyer, 1999; Stewart et al., 2003; Spaeth et al., 2004), but such studies are a minority among numerous research endeavors concerning sexual segregation (Bowyer, 2004). This manipulative approach is critical to gaining new knowledge about why the sexes

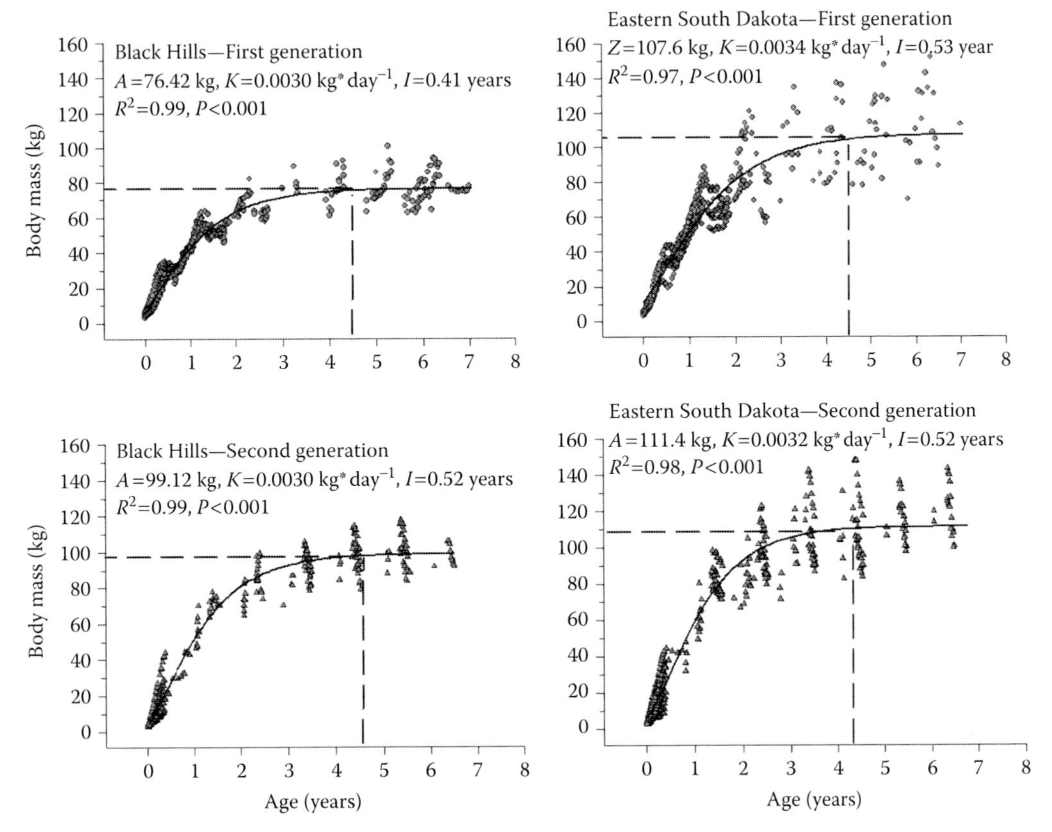

FIGURE 6.7 Body mass relative to age of male white-tailed deer raised in captivity (1997–2007). Study animals were composed of original cohorts (first generation) acquired as neonates from eastern South Dakota, USA. Second-generation young were born in captivity and sired from first-generation adults. Sample size indicates number of adults and the dashed line illustrates time to asymptotic body growth. (From Monteith, K. L. et al. 2009. *Journal of Mammalogy* 90:651–660. With permission.)

segregate, and for furthering our ability to manage white-tailed deer based on this important life-history characteristic.

There also is a need to state hypotheses for sexual segregation in a clear and concise manner. As noted previously, the forage-selection hypothesis is not really a hypothesis but at best a prediction. Moreover, none of the three general hypotheses proposed by Main et al. (1996) are independent (and hence not testable); all have outlived their usefulness for understanding sexual segregation (Bowyer, 2004). Formulating a study design to critically test these hypotheses as constructed is not possible, because they do not provide clear alternative explanations or testable predictions; these hypotheses need to be set aside if we are to advance our understanding of sexual segregation (Bowyer, 2004). More care is needed in not continually resurrecting old, rejected ideas or hypotheses if we are to further our knowledge of sexual segregation (Bleich et al., 1997; Bowyer, 2004). Indeed, many prominent hypotheses forwarded to explain sexual segregation in ruminants are not viable and have limited applied or theoretical value for white-tailed deer (Table 6.5). We maintain that the gastrocentric and predation hypotheses are sufficient to explain patterns of sexual segregation on the landscape, and provide clear and testable predictions—future research should focus on those hypotheses.

There is a clear need to understand how differential selection of space, habitat, and diets (Table 6.4) by the sexes relate to fitness consequences of such behaviors. We often assume that these behaviors are adaptive, but supportive data are rare. Farmer et al. (2006) related habitat selection to survivorship by the sexes of deer. Studies of other factors related to reproductive success resulting for differing strategies by the sexes await future research.

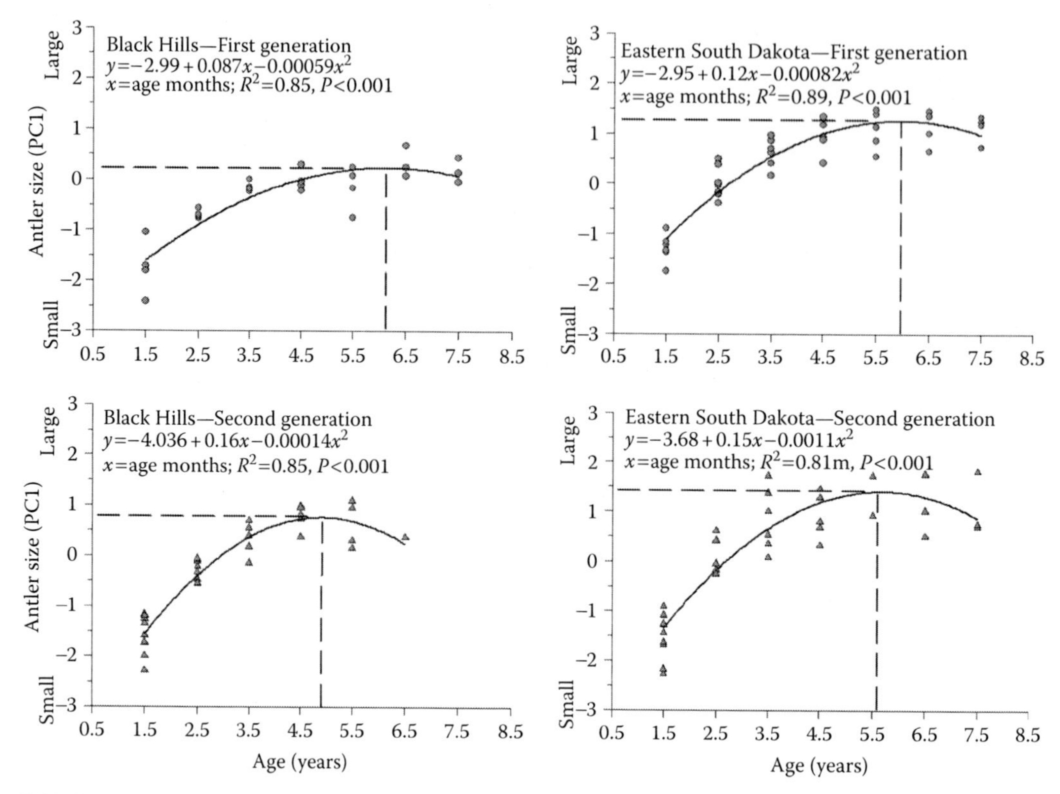

FIGURE 6.8 Size of antlers as indexed by principal component 1 of male white-tailed deer raised in captivity (1997–2007). Study animals were composed of original cohorts (first generation) acquired as neonates from eastern South Dakota (larger adult females) and the Black Hills (smaller adult females) in western South Dakota, USA. Second-generation young were born in captivity and sired from first-generation adults. Sample size indicates number of adults and the dashed line illustrates time to peak antler growth. (From Monteith, K. L. et al. 2009. *Journal of Mammalogy* 90:651–660. With permission.)

Sexual segregation may also hold import for understanding areas of population ecology that we have not discussed previously. For instance, how is interspecific competition among ruminants (*sensu* Stewart et al., 2002) influenced by smaller males of one species competing with larger females of another? We know of no model that considers such a possibility. Likewise, sexual segregation may influence predator–prey relationships (Bowyer, 2004). Sexes of ruminants exhibit differential vulnerability to predators—a male killed by a predator will have a disproportionally small effect on the dynamics of that prey population compared to the death of a female. In addition, a male, because of his larger size, would provide more food than a female or young to predators, and accordingly have a greater effect on the predator population (Bowyer et al., 2005). None of these considerations are dealt with in models of predator–prey dynamics.

Further research also is needed on how spatial patterns of sexual segregation and aggregation on the landscape influence patterns of gene flow (Bowyer, 2004; Rubin and Bleich, 2005). These data are likely critical when considering the management of rare or endangered populations, including documenting differential needs of the sexes with respect to translocation sites. Likewise, understanding how patterns of sexual segregation affect the spread of diseases and parasites requires additional research (Bowyer, 2004; Rubin and Bleich, 2005). Anthropogenic changes to the landscape, including housing developments, roads, power lines, energy developments, pipelines, aqueducts, and probably numerous other factors, may influence the manner in which the sexes of ruminants use space (Bleich et al., 1997; Bowyer, 2004; Rubin and Bleich, 2005). Understanding needs of the sexes should be a requisite component of mitigation plans. There remains considerable research to be conducted on this fascinating subject, which should aid the management of white-tailed deer.

TABLE 6.5

Overview of Extant Hypotheses Forwarded to Explain Sexual Segregation in Ruminants Including Their Relevance to the Landscape Ecology of White-tailed Deer (Applied Value)

Hypothesis	Viable?	Theoretical Value?	Applied Value?	Comments
Reproductive strategy	No	No	No	Not independent of other hypotheses; not testable
Activity budget	Probably not	Yes	No	Cannot explain spatial separation of sexes; conclusions based on correlations; mixed outcomes from empirical tests
Social factors	Maybe	Yes	No	Cannot explain spatial patterns of sexes over time; seldom demonstrated empirically
Scramble competition	No	No	No	Rejected logically and empirically on numerous occasions, including with manipulative experiments
Forage selection	No	No	No	Not an hypothesis, but an outcome or prediction from other hypotheses
Bell–Jarman	No	No	No	Rejected based on empirical evidence; cannot explain patterns of sexual aggregation and segregation
Gastrocentric	Yes	Yes	Yes	Explains patterns of sexual aggregation and segregation; explains why males do not displace females from better habitats; considerable empirical support
Predation	Yes	Yes	Yes	May be sufficient but not necessary to explain sexual segregation; overwhelming empirical support especially in relation to trade-offs

REFERENCES

Anderson, A. E. and O. C. Wallmo. 1984. *Odocoileus hemionus*: Mule deer. *Mammalian Species* 219:1–9.

Anderson, D. P., M. G. Turner, J. D. Forester et al. 2005. Scale-dependent summer resource selection by reintroduced elk in Wisconsin, USA. *Journal of Wildlife Management* 69:298–310.

Anthony, R. G. and N. Smith. 1977. Ecological relationships between mule deer and white-tailed deer in southeastern Arizona. *Ecological Monographs* 47:255–277.

Asleson, M. A., E. C. Hellgren, and L. W. Varner. 1997. Effects of seasonal protein restriction on antlerogensis and body mass of white-tailed deer. *Journal of Wildlife Management* 61:1098–1107.

Barbour, A. G. and D. Fish. 1993. The biological and social phenomenon of Lyme Disease. *Science* 260:1610–1616.

Barboza, P. S. and R. T. Bowyer. 2000. Sexual segregation in dimorphic deer: A new gastrocentric hypothesis. *Journal of Mammalogy* 81:473–489.

Barboza, P. S. and R. T. Bowyer. 2001. Seasonality of sexual segregation in dimorphic deer: Extending the gastrocentric model. *Alces* 37:275–292.

Barboza, P. S., K. L. Parker, and I. D. Hume. 2009. *Integrative Wildlife Nutrition*. Berlin: Springer-Verlag.

Bartlett, I. H. 1950. Michigan deer. Michigan Department of Conservation, Game Division, Lansing.

Barten, N. L., R. T. Bowyer, and K. J. Jenkins. 2001. Habitat use by female caribou: Tradeoffs associated with parturition. *Journal of Wildlife Management* 65:77–92.

Beier, P. 1987. Sex difference in quality of white-tailed deer diets. *Journal of Mammalogy* 68:323–329.

Beier, P. and D. R. McCullough. 1990. Factors influencing white-tailed deer activity patterns and habitat use. *Wildlife Monographs* 109:1–51.

Bell, R. H. V. 1970. The use of the herb layer by grazing ungulates in the Serengeti. In *Animal Populations in Relation to Their Food Resources*, ed. A. Watson, 111–123. Oxford: Blackwell.

Berger, J. 1991. Pregnancy incentives, predation constraints and habitat shifts: Experimental and field evidence for wild bighorn sheep. *Animal Behaviour* 41:61–77.

Berger, J. 1999. Anthropogenic extinction of top carnivores and interspecific animal behaviour: Implications of the rapid decoupling of a web involving wolves, bears, moose and ravens. *Proceedings of the Royal Society of London B* 266:2261–2267.

Berger, J., S. Dulamtseren, S. Cain, D. et al. 2001. Back-casting sociality in extinct species: New perspectives using mass death assemblages and sex ratios. *Proceedings of the Royal Society B* 268:131–139.

Bertrand, M. R., A. J. DeNicola, S. R. Beissinger, and R. K. Swihart. 1996. Effects of parturition on home ranges and social affiliations of female white-tailed deer. *Journal of Wildlife Management* 60:899–909.

Bleich, V. C. 1999. Mountain sheep and coyotes: Patterns of predator evasion in a mountain ungulate. *Journal of Mammalogy* 80:283–289.

Bleich, V. C., R. T. Bowyer, and J. D. Wehausen. 1997. Sexual segregation in mountain sheep: Resources or predation? *Wildlife Monographs* 134:1–50.

Bon, R., J.-L. Deneubourg, J.-F. Gerard, and P. Michelena. 2005. Sexual segregation in ungulates: From individual mechanisms to collective patterns. In *Sexual Segregation in Vertebrates: Ecology of Two Sexes*, eds. K. E. Ruckstuhl and P. Neuhaus, 180–199. Cambridge: Cambridge University Press.

Bonenfant, C., J.-M. Gaillard, S. Dray, A. Loison, M. Royer, and D. Chessel. 2007. Testing sexual segregation and aggregation: Old ways are best. *Ecology* 88:3202–3208.

Bowers, M. A., D. N. Welch, and T. G. Carr. 1990. Home range size adjustments in response to natural and manipulated water availability in the eastern chipmunk, *Tamias striatus*. *Canadian Journal of Zoology* 68:2016–2020.

Bowyer, R. T. 1984. Sexual segregation in southern mule deer. *Journal of Mammalogy* 65:410–417.

Bowyer, R. T. 1986. Habitat selection by southern mule deer. *California Fish and Game* 72:153–169.

Bowyer, R. T. 1987. Coyote group size relative to predation on mule deer. *Mammalia* 51:515–526.

Bowyer, R. T. 2004. Sexual segregation in ruminants: Definitions, hypotheses, and implications for conservation and management. *Journal of Mammalogy* 85:1039–1052.

Bowyer, R. T., V. C. Bleich, X. Manteca, J. C. Whiting, and K. M. Stewart. 2007. Sociality, mate choice, and timing of mating in American bison (*Bison bison*): Effects of large males. *Ethology* 113:1048–1060.

Bowyer, R. T. and J. G. Kie. 2004. Effects of foraging activity on sexual segregation in mule deer. *Journal of Mammalogy* 85:498–504.

Bowyer, R. T. and J. G. Kie. 2006. Effects of scale on interpreting life-history characteristics of ungulates and carnivores. *Diversity and Distributions* 12:244–257.

Bowyer, R. T., J. G. Kie, and V. Van Ballenberghe. 1996. Sexual segregation in black tailed deer: Effects of scale. *Journal of Wildlife Management* 60:10–17.

Bowyer, R. T., J. G. Kie, and V. Van Ballenberghe. 1998a. Habitat selection by neonatal black-tailed deer: Climate, forage, or risk of predation? *Journal of Mammalogy* 79:415–425.

Bowyer, R. T., D. R. McCullough, and G. E. Belovsky. 2001b. Causes and consequences of sociality in mule deer. *Alces* 37:371–402.

Bowyer, R. T., D. K. Person, and B. M. Pierce. 2005. Detecting top-down versus bottom-up regulation of ungulates by large carnivores: Implications for conservation of biodiversity. In *Large Carnivores and the Conservation of Biodiversity*, eds. J. C. Ray, K. H. Redford, R. S. Steneck, and J. Berger, 342–361. Covelo, CA: Island Press.

Bowyer, R. T., B. M. Pierce, L. K. Duffy, and D. A. Haggstrom. 2001a. Sexual segregation in moose: Effects of habitat manipulation. *Alces* 37:109–122.

Bowyer, R. T., K. M. Stewart, S. A. Wolfe et al. 2002. Assessing sexual segregation in deer. *Journal of Wildlife Management* 66:536–544.

Bowyer, R. T., V. Van Ballenberghe, and J. G. Kie. 1997. The role of moose in landscape processes: Effects of biogeography, population dynamics and predation. In *Wildlife and Landscape Ecology: Effects of Pattern and Scale*, ed. J. A. Bissonette, 265–287. New York, NY: Springer-Verlag.

Bowyer, R. T., V. Van Ballenberghe, and J. G. Kie. 1998b. Timing and synchrony of parturition in Alaskan moose: Long-term versus proximal effects of climate. *Journal of Mammalogy* 79:1332–1334.

Bowyer, R. T., V. Van Ballenberghe, J. G. Kie, and J. A. K. Maier. 1999. Birth-site selection by Alaskan moose: Maternal strategies for coping with a risky environment. *Journal of Mammalogy* 80:1070–1083.

Boyce, M. S. 2006. Scale for resource selection functions. *Diversity and Distributions* 12:269–276.

Boyce, M. S., J. S. Mao, E. H. Merrill et al. 2003. Scale and heterogeneity in habitat selection by elk in Yellowstone National Park. *Écoscience* 10:421–431.

Bridges, R. J. 1968. Individual white-tailed deer movement and related behavior during the winter and spring in northwestern Florida. MS thesis, University of Florida.

Brinkman, T. J., C. S. Deperno, J. A. Jenks, B. S. Haroldson, and R. G. Osborn. 2005. Movement of female white-tailed deer: Effects of climate and intensive row-crop agriculture. *Journal of Wildlife Management* 69:1099–1111.

Brower, L. P. and S. B. Malcolm. 1991. Animal migration: Endangered phenomena. *American Naturalist* 31:265–276.

Brown, D. G., K. M. Johnson, T. R. Loveland, and D. M. Theobald. 2005. Rural land-use trends in the conterminous United States, 1950–2000. *Ecological Applications* 15:1851–1863.

Burt, W. H. 1943. Territoriality and home range as applied to mammals. *Journal of Mammalogy* 24:346–352.

Byford, J. L. 1970. Movement responses of white-tailed deer to changing food supplies. *Proceedings of the Annual Conference of Southeast Association of Game and Fish Commissioners* 24:63–78.

Caughley, G. 2004. *Analysis of Vertebrate Populations*. Caldwell, ID: Blackburn Press.

Caughley, G. and A. R. E. Sinclair. 1994. *Wildlife Ecology and Management*. Boston: Blackwell Science Inc.

Cederlund, G. and H. Sand. 1994. Home-range size in relation to age and sex in moose. *Journal of Mammalogy* 75:1005–1012.

Clark, J. S. and P. D. Royall. 1996. Local and regional sediment charcoal evidence for fire regimes in presettlement north-eastern North America. *Journal of Ecology* 84:365–382.

Clauss, M., A. Schwarm, S. Ortmann, W. J. Streich, and J. Hummel. 2007. A case of non-scaling in mammalian physiology? Body size, digestive capacity, food intake, and ingesta passage in mammalian herbivores. *Comparative Biochemistry and Physiology A* 148:249–265.

Clevenger, A. P., B. Chruszcz, and K. E. Gunson. 2001. Highway mitigation fencing reduces wildlife–vehicle collisions. *Wildlife Society Bulletin* 29:646–653.

Clutton-Brock, T. H., G. R. Iason, and F. E. Guinness. 1987. Sexual segregation and density-related changes in habitat use in male and female red deer (*Cervus elaphus*). *Journal of Zoology London* 211:275–289.

Comer, C. E., J. C. Kilgo, G. J. D'Angelo, T. C. Glenn, and K. V. Miller. 2005. Fine-scale genetic structure and social organization in female white-tailed deer. *Journal of Wildlife Management* 69:332–344.

Conradt, L. 1998a. Measuring the degree of sexual segregation in group-living animals. *Journal of Animal Ecology* 67:217–226.

Conradt, L. 1998b. Could asynchrony in activity between the sexes cause intersexual social segregation in ruminants? *Proceedings of the Royal Society of London B* 265:1359–1363.

Conradt, L., T. H. Clutton-Brock, and F. E. Guinness. 2000. Sex differences in weather sensitivity can cause habitat segregation: Red deer as an example. *Animal Behaviour* 59:1049–1060.

Conradt, L., T. H. Clutton-Brock, and D. Thompson. 1999. Habitat segregation in ungulates: Are males forced into suboptimal foraging habitats through indirect competition by females? *Oecologia* 119:367–377.

Conradt, L., I. J. Gordon, T. H. Clutton-Brock, D. Thompson, and F. E. Guinness. 2001. Could the indirect competition hypothesis explain inter-sexual site segregation in red deer (*Cervus elaphus* L.)? *Journal of Zoology London* 254:185–193.

Cooper, S. M., H. L. Perotto-Baldivieso, M. K. Owens, M. G. Meek, and M. Figueroa-Pagen. 2008. Distribution and interaction of white-tailed deer and cattle in a semi-arid grazing system. *Agriculture Ecosystems and Environment* 127:85–92.

Corti, P. and D. M. Shackelton. 2002. Relationship between predation-risk factors and sexual segregation in Dall's sheep (*Ovis dalli dalli*). *Canadian Journal of Zoology* 80:2108–2117.

Côté, S. D., T. P. Rooney, J. P. Tremblay, C. Dussault, and D. M. Waller. 2004. Ecological impacts of deer overabundance. *Annual Review of Ecology and Systematics* 35:113–147.

D'Angelo, G. J., C. E. Comer, J. C. Kilgo, C. D. Drennan, D. A. Osborn, and K. V. Miller. 2004. Daily movements of female white-tailed deer relative to parturition and breeding. *Proceedings of the Southeastern Association of Fish and Wildlife Agencies* 58:292–301.

Dahlberg, B. L. and R. C. Guettinger. 1956. The white-tailed deer in Wisconsin. Wisconsin Conservation Department Technical Wildlife Bulletin 14. Madison, WI.

Day, T. A. and J. K. Detling. 1990. Grassland patch dynamics and herbivore grazing preference following urine deposition. *Ecology* 71:180–188.

Demment, M. W. and P. J. Van Soest. 1985. A nutritional explanation for body-size patterns of ruminant and nonruminant herbivores. *American Naturalist* 125:641–672.

DePerno, C. S., J. A. Jenks, and S. L. Griffin. 2003. Multidimensional cover characteristics: Is variation in habitat selection related to white-tailed deer sexual segregation. *Journal of Mammalogy* 84:1316–1329.

Diefenbach, D. R., E. S. Long, C. S. Rosenberry, B. D. Wallingford, and D. R. Smith. 2008. Modeling distribution of dispersal distances in male white-tailed deer. *Journal of Wildlife Management* 72:1296–1303.

Dobson, F. S. 1979. An experimental study of dispersal in the California ground squirrel. *Ecology* 60:1103–1109.

Dobson, F. S. 1982. Competition for mates and predominant juvenile male dispersal in mammals. *Animal Behaviour* 30:1183–1192.

du Toit, J. 1995. Sexual segregation in kudu: Sex difference in competitive ability, predation risk, or nutritional needs? *South African Journal of Wildlife Research* 25:127–132.

Eberhardt, L. E., E. E. Hanson, and L. L. Caldwell. 1984. Movement and activity patterns of mule deer in the sagebrush–steppe region. *Journal of Mammalogy* 65:404–409.

Etchberger, R. C., R. Mazaika, and R.T. Bowyer. 1988. White-tailed deer, *Odocoileus virginianus*, fecal groups relative to vegetation biomass and quality in Maine. *Canadian Field-Naturalist* 102:671–674.

Fahrig, L. 2003. Effects of habitat fragmentation on biodiversity. *Annual Review of Ecology, Evolution, and Systematics* 34:487–515.

Fahrig, L. 2007. Non-optimal animal movement in human-altered landscapes. *Functional Ecology* 21:1003–1015.

Farmer, C. F., D. K. Person, and R. T. Bowyer. 2006. Risk factors and survivorship of black-tailed deer in a managed forest landscape. *Journal of Wildlife Management* 70:1403–1415.

Felix, A. B., H. Campa, K. F. Millenbah, S. R. Winterstein, and W. E. Moritz. 2004. Development of landscape-scale habitat-potential models for forest wildlife planning and management. *Wildlife Society Bulletin* 32:795–806.

Felix, A. B., D. P. Walsh, B. D. Hughey, H. Campa, and S. R. Winterstein. 2007. Applying landscape-scale habitat-potential models to understand deer spatial structure and movement patterns. *Journal of Wildlife Management* 71:804–810.

Festa-Bianchet, M. 1988. Seasonal range selection in bighorn sheep: Conflicts between forage quality, forage quantity, and predator avoidance. *Oecologia* 75:580–586.

Fieberg, J., D. W. Kuehn and G. D. DelGiudice. 2008. Understanding variation in autumn migration of northern white-tailed deer by long-term study. *Journal of Mammalogy* 89:1529–1539.

Finder, R. A., J. L. Roseberry, and A. Woolf. 1999. Site and landscape conditions at white-tailed deer vehicle collision locations in Illinois. *Landscape and Urban Planning* 44:77–85.

Ford, R. G. 1983. Home range in a patchy environment: Optimal foraging predictions. *American Zoologist* 23:315–326.

Fulbright, T. E. and J. A. Ortega-S. 2006. *White-tailed Deer Habitat: Ecology and Management on Rangelands.* College Station, TX: Texas A&M University Press.

Garrott, R. A., G. C. White, R. M. Bartman, L. H. Carpenter, and A. W. Alldredge. 1987. Movements of female mule deer in northwest Colorado. *Journal of Wildlife Management* 51:634–643.

Gompper, M. E. and J. L. Gittleman. 1991. Home range scaling: Intraspecific and comparative trends. *Oecologia* 87:343–348.

Gordon, I. J. and A. W. Illius. 1996. The nutritional ecology of African ruminants: A reinterpretation. *Journal of Animal Ecology* 65:18–28.

Gregory, A. J., M. A. Lung, T. M. Gehring, and B. J. Swanson. 2009. The importance of sex and spatial scale when evaluating sexual segregation by elk in Yellowstone. *Journal of Mammalogy* 90:971–979.

Gruell, G. E. and N. J. Papez. 1963. Movements of mule deer in northeastern Nevada. *Journal of Wildlife Management* 27:414–422.

Hamilton, W. D. and R. M. May. 1977. Dispersal in stable habitats. *Nature* 269:578–581.

Harestad, A. S. and F. L. Bunnell. 1979. Home range and body weight—A re-evaluation. *Ecology* 60:389–402.

Hay, C. T., P. C. Cross, and P. J. Funston. 2008. Trade-offs of predation and foraging explain sexual segregation in African buffalo. *Journal of Animal Ecology* 77:850–858.

Heffelfinger, J. 2006. *Deer of the Southwest.* College Station, TX: Texas A&M University Press.

Hemson, G., P. Johnson, A. South, R. Kenward, R. Ripley, and D. MacDonald. 2005. Are kernels the mustard? Data from global positioning system collars suggests problems for kernel home-range analyses with least-squares cross-validation. *Journal of Animal Ecology* 74:455–463.

Herbert, J. J. and P. R. Krausman. 1986. Desert mule deer use of water developments in Arizona. *Journal of Wildlife Management* 50:670–676.
Hirth, D. H. 1977. Social behavior of white-tailed deer in relation to habitat. *Wildlife Monographs* 54:1–55.
Hjeljord, O. 2001. Dispersal and migration in northern forest deer—Are there unifying concepts? *Alces* 37:353–370.
Hobbs, N. T. 1996. Modification of ecosystems by ungulates. *Journal of Wildlife Management* 60:695–713.
Hodgman, T. P. and R. T. Bowyer. 1986. Fecal crude protein relative to browsing intensity by white-tailed deer on wintering areas in Maine. *Acta Theriologica* 31:347–353.
Holzenbein, S. and R. L. Marchinton. 1992. Spatial integration of maturing male white-tailed deer into the adult population. *Journal of Mammalogy* 73:326–334.
Hood, R. E. 1971. Seasonal variations in home range, diel movement, and activity patterns of white-tailed deer on the Rob and Bessie Welder Wildlife Refuge (San Patricio County, Texas). MS thesis. Texas A&M University, College Station, TX.
Horsley, S. B., S. L. Stout, and D. S. DeCalesta. 2003. White-tailed deer impact on the vegetation dynamics of a northern hardwood forest. *Ecological Applications* 13:98–118.
Hoskinson, R. L. and L. D. Mech. 1976. White-tailed deer migration and its role in wolf predation. *Journal of Wildlife Management* 40:429–441.
Howard, W. E. 1960. Innate and environmental dispersal of individual vertebrates. *American Midland Naturalist* 63:152–161.
Hurst, J. E. and W. F. Porter. 2008. Evaluation of shifts in white-tailed deer winter yards in the Adirondack region of New York. *Journal of Wildlife Management* 72:367–375.
Jarman, P. J. 1974. The social organization of antelope in relation to their ecology. *Behaviour* 48:215–266.
Jenks, J. A., D. M. Leslie, Jr., R. L. Lockmiller, and M. A. Melchiors. 1994. Variation in gastrointestinal characteristics of male and female white-tailed deer: Implications for resource partitioning. *Journal of Mammalogy* 75:1045–1053.
Jenkins, S. H. 1981. Common patterns in home range-body size relationships of birds and mammals. *American Naturalist* 118:126–128.
Johnson, C. J., K. L. Parker, and D. C. Heard. 2001. Foraging across a variable landscape: Behavioral decisions made by woodland caribou at multiple spatial scales. *Oecologia* 127:590–602.
Johnson, C. J., K. L. Parker, D. C. Heard, and M. P. Gillingham. 2002. A multiscale behavioral approach to understanding the movements of woodland caribou. *Ecological Applications* 12:1840–1860.
Kie, J. G. 1999. Optimal foraging and risk of predation: Effects on behavior and social structure in ungulates. *Journal of Mammalogy* 80:1114–1129.
Kie, J. G. and R. T. Bowyer. 1999. Sexual segregation in white-tailed deer: Density dependent changes in use of space, habitat selection, and dietary niche. *Journal of Mammalogy* 80:1004–1020.
Kie, J. G., R. T. Bowyer, M. C. Nicholson, B. B. Boroski, and E. R. Loft. 2002. Landscape heterogeneity at differing scales: Effects on spatial distribution of mule deer. *Ecology* 83:530–544.
Kie, J. G., R. T. Bowyer, and K. M. Stewart. 2003. Ungulates in western forests: Habitat requirements, population dynamics, and ecosystem processes. In *Mammal Community Dynamics: Management and Conservation in the Coniferous Forests of Western North America*, eds. C. J. Zabel and R. G. Anthony, 296–340. New York, NY: Cambridge University.
Kie, J. G. and M. White. 1985. Population dynamics of white-tailed deer (*Odocoileus virginianus*) on the Welder Wildlife Refuge, Texas. *Southwestern Naturalist* 30:105–118.
Kilgo, J. C., R. F. Labinski, and D. E. Fritzen. 1998. Influences of hunting on the behavior of white-tailed deer: Implications for conservation of the Florida panther. *Conservation Biology* 12:1359–1364.
Knight, T. M., J. L. Dunn, L. A. Smith, J. Davis, and S. Kalisz. 2009. Deer facilitate invasive plant success in a Pennsylvania forest understory. *Natural Areas Journal* 29:110–116.
Krämer, A. 1973. Interspecific behavior and dispersion of two sympatric deer species. *Journal of Wildlife Management* 37:572–573.
Labisky, R. F. and D. E. Fritzen. 1998. Spatial mobility of breeding female white-tailed deer in a low-density population. *Journal of Wildlife Management* 62:1329–1334.
LaGory, K. E. 1987. The influence of habitat and group characteristics on the alarm and flight response of white-tailed deer. *Animal Behaviour* 35:20–25.
LaGory, K. E., C. Bagshaw, and I. L. Brisbin. 1991. Niche differences between male and female white-tailed deer on Ossabaw Island, Georgia. *Applied Animal Behaviour Science* 29:205–214.

Larson, T. J., O. J. Rongstad, and F. W. Terbilcox. 1978. Movement and habitat use of white-tailed deer in southcentral Wisconsin. *Journal of Wildlife Management* 42:113–117.

Lesage, L., M. Crete, J. Huot, A. Dumont, and J.-P. Ouellet. 2000. Seasonal home range size and philopatry in two northern white-tailed deer populations. *Canadian Journal of Zoology* 78:1930–1940.

Lewis, D. M. 1968. Telemetry studies of white-tailed deer on Red Dirt Game Management Area, MS thesis, Louisiana State University and Texas A&M College.

Loft, E. R., J. G. Kie, and J. W. Menke. 1993. Grazing in the Sierra Nevada: Home range and space use patterns of mule deer as influenced by cattle. *California Fish and Game* 79:145–166.

Loison, A., J.-M. Gillard, C. Pélabon, and N. G. Yoccoz. 1999. What factors shape sexual size dimorphism in ungulates? *Evolutionary Ecology Research* 1:611–633.

Long, E. S., D. R. Diefenbach, C. S. Rosenberry, and B. D. Wallingford. 2008. Multiple proximate and ultimate causes of natal dispersal in white-tailed deer. *Behavioral Ecology* 19:1235–1242.

Long, E. S., D. R. Diefenbach, C. S. Rosenberry, B. D. Wallingford, and M. D. Grund. 2005. Forest cover influences dispersal distance of white-tailed deer. *Journal of Mammalogy* 86:623–629.

Long, R. A., J. L. Rachlow, and J. G. Kie. 2009. Sex-specific responses of North American elk to habitat manipulations. *Journal of Mammalogy* 90:423–432.

Maier, J. A. K., J. M. Ver Hoef, A. D. McGuire, R. T. Bowyer, L. Saperstein, and H. A. Maier. 2005. Distribution and density of moose in relation to landscape characteristics: Effects of scale. *Canadian Journal of Forest Research* 35:2233–2243.

Main, M. B. 2008. Reconciling competing ecological explanations for sexual segregation in ungulates. *Ecology* 89:693–704.

Main, M. B. and J. T. du Toit. 2005. Sex differences in reproductive strategies affect habitat choice in ungulates. In *Sexual Segregation in Vertebrates: Ecology of Two Sexes*, eds. K. E. Ruckstuhl and P. Neuhaus, 148–161. Cambridge: Cambridge University Press.

Main, M. B., F. W. Weckerly, and V. C. Bleich. 1996. Sexual segregation in ungulates: New directions for research. *Journal of Mammalogy* 77:449–461.

Malo, J. E., F. Suarez, and A. Diez. 2004. Can we mitigate animal–vehicle accidents using predictive models? *Journal of Applied Ecology* 41:701–710.

Marchinton, R. L. and D. H. Hirth. 1984. Behavior. In *White-tailed Deer Ecology and Management,* ed. L. K. Halls, 129–168. Harrisburg, PA: Stackpole Books.

Marshall, A. D. and R. W. Whittington. 1969. A telemetric study of deer home ranges and behavior of deer during managed hunts. *Proceedings of the Annual Conference of Southeast Association of Game and Fish Commissioners* 22:30–46.

Matthysen, E., F. Adriaensen, and A. A. Dhondt. 1995. Dispersal distances of nuthatches, *Sitta europaea*, in a highly fragmented forest habitat. *Oikos* 72:375–381.

McCabe, R. E. and T. R. McCabe. 1984. Of slings and arrows: An historical retrospective. In *White-tailed Deer Ecology and Management,* ed. L. K. Halls, 19–74. Harrisburg, PA: Stackpole Books.

McCoy, J. E., D. G. Hewitt, and F. C. Bryant. 2005. Dispersal by yearling male white-tailed deer and implications for management. *Journal of Wildlife Management* 69:366–376.

McCullough, D. R. 1979. *The George Reserve Deer Herd: Population Ecology of a K Selected Species*. Ann Arbor, MI: University of Michigan.

McCullough, D. R. 1985. Long range movements of large terrestrial mammals. In *Migration: Mechanisms and Adaptive Significance*, ed. M. A. Rankin, 444–465. Contributions in Marine Science 27. Austin, TX: University of Texas.

McCullough, D. R., D. H. Hirth, and S. J. Newhouse. 1989. Resource partitioning between sexes in white-tailed deer. *Journal of Wildlife Management* 53:277–283.

McShea, W. J., W. M. Healy, P. Devers et al. 2007. Forestry matters: Decline of oaks will impact wildlife in hardwood forests. *Journal of Wildlife Management* 71:1717–1728.

McNab, B. K. 1963. Bioenergetics and the determination of home range size. *American Naturalist* 97:133–140.

McNaughton, S. J. 1985. Ecology of a grazing ecosystem: The Serengeti. *Ecological Monographs* 55:259–294.

Michelena, P., S. Noel, J. G., J. Gautrais, J. F. Gerard, J. L. Deniubourg, and R. Bon. 2006. Sexual dimorphism, activity budget and synchrony in groups of sheep. *Oecologia* 148:170–180.

Miquelle, D. G., J. M. Peek, and V. Van Ballenberghe. 1992. Sexual segregation in Alaskan moose. *Wildlife Monographs* 122:1–57.

Miyashita, T., M. Suzuki, D. Ando, G. Fujita, K. Ochiai, and M. Asada. 2008. Forest edge creates small-scale variation in reproductive rate of sika deer. *Population Ecology* 50:111–120.
Moen, A. N. 1976. Energy conservation by white-tailed deer in the winter. *Ecology* 57:192–198.
Molvar, E. M. and R. T. Bowyer. 1994. Costs and benefits of group living in a recently social ungulate: The Alaskan moose. *Journal of Mammalogy* 75:621–630.
Molvar, E. M., R. T. Bowyer, and V. Van Ballenberghe. 1993. Moose herbivory, browse quality, and nutrient cycling in an Alaskan treeline community. *Oecologia* 94:472–479.
Monteith, K. L., L. E. Schmitz, J. A. Jenks, J. A. Delger, and R. T. Bowyer. 2009. Growth of male white-tailed deer: Consequences of maternal effects. *Journal of Mammalogy* 90:651–660.
Monteith, K. L., C. L. Sexton, J. A. Jenks, and R. T. Bowyer. 2007. Evaluation of techniques for categorizing group membership of white-tailed deer. *Journal of Wildlife Management* 71:1712–1716.
Moore, J. and R. Ali. 1984. Are dispersal and inbreeding avoidance related? *Animal Behavior* 32:94–112.
Mooring, M. S., D. D. Reisig, E. R. Osborn et al. 2005. Sexual segregation in bison: A test of multiple hypotheses. *Behaviour* 142:897–927.
Mooring, M. S. and E. M. Rominger. 2004. Is the activity budget hypothesis the holy grail of sexual segregation? *Behaviour* 141:521–530.
Mooty, J. J., P. D. Karns, and T. K. Fuller. 1987. Habitat use and seasonal range size of white-tailed deer in northcentral Minnesota. *Journal of Wildlife Management* 51:644–648.
Motzkin, G., D. Foster, A. Allen, J. Harrod, and R. Boone. 1996. Controlling site to evaluate history: Vegetation patterns of a New England sand plain. *Ecological Monographs* 66:345–365.
Murray, B. G. 1967. Dispersal in vertebrates. *Ecology* 48:975–978.
Myers, J. A., M. Vellend, S. Gardescu, and P. L. Marks. 2004. Seed dispersal by white-tailed deer: Implications for long-distance dispersal, invasion, and migration of plants in eastern North America. *Oecologia* 139:35–44.
Mysterud, A. 2000. The relationship between ecological segregation and sexual body size dimorphism in large herbivores. *Oecologia* 124:40–54.
Mysterud, A., T. Coulson, and N. C. Stenseth. 2002. The role of males in the dynamics of ungulate populations. *Journal of Animal Ecology* 71:907–915.
Nelson, M. E. 1993. Natal dispersal and gene flow in white-tailed deer in northeastern Minnesota. *Journal of Mammalogy* 74:316–322.
Nelson, M. E. 1995. Winter range arrival and departure of white-tailed deer in northeastern Minnesota. *Canadian of Journal of Zoology* 73:1069–1076.
Nelson, M. E. 1998. Development of migratory behavior in northern white-tailed deer. *Canadian Journal of Zoology* 76:426–432.
Nelson, M. E. and L. D. Mech. 1981. Deer social organization and wolf predation in northeastern Minnesota. *Wildlife Monographs* 11:1–53.
Nelson, M. E. and L. D. Mech. 1984. Home-range formation and dispersal of deer in northeastern Minnesota. *Journal of Mammalogy* 65:567–575.
Nelson, M. E. and L. D. Mech. 1987. Demes within a northeastern Minnesota deer population. In *Mammalian Dispersal Patterns: The Effects of Social Structure on Population Genetics*, eds. B. D. Chepko-Sade and Z. T. Halpin, 567–575. Chicago, IL: University of Chicago Press.
Nelson, M. E. and L. D. Mech. 1992. Dispersal in female white-tailed deer. *Journal of Mammalogy* 73:891–894.
Neuhaus, P., K. E. Ruckstuhl, and L. Conradt. 2005. Conclusions and future directions. In *Sexual Segregation in Vertebrates: Ecology of Two Sexes*, eds. K. E. Ruckstuhl and P. Neuhaus, 395–402. Cambridge: Cambridge University Press.
Nicholson, M. C., R. T. Bowyer, and J. G. Kie. 1997. Habitat selection and survival of mule deer: Tradeoffs associated with migration. *Journal of Mammalogy* 78:483–504.
Nixon, C. M., L. P. Hansen, P. A. Brewer, and J. E. Chelsvig. 1991. Ecology of white-tailed deer in an intensively farmed region of Illinois. *Wildlife Monographs* 118:1–77.
Nixon, C. M., P. C. Mankin, D. R. Etter et al. 2008. Migration behavior among female white-tailed deer in central and northern Illinois. *American Midland Naturalist* 160:178–190.
Orwig, D. A. and M. D. Abrams. 1994. Land-use history (1720–1992), composition, and dynamics of oak-pine forests within the Piedmont and coastal-plain of northern Virginia. *Canadian Journal of Forest Research* 24:1216–1225.

Ozoga, J. J. and L. W. Gysel. 1972. Response of white-tailed deer to winter weather. *Journal of Wildlife Management* 36:892–896.

Ozoga, J. J. and L. J. Verme. 1985. Comparative breeding behavior and performance of yearlings vs. prime-aged white-tailed bucks. *Journal of Wildlife Management* 49:364–372.

Ozoga, J. J., L. J. Verme, and C. S. Bienz. 1982. Parturition behavior and territoriality in white-tailed deer: Impacts on neonatal mortality. *Journal of Wildlife Management* 46:1–11.

Pac, H. I., W. F. Kasworm, L. R. Irby et al. 1988. Ecology of mule deer, *Odocoileus hemionus*, along the east front of the Rocky Mountains, Montana. *Canadian Field-Naturalist* 102:227–236.

Parker, K. L., C. T. Robbins, and T. A. Hanley. 1984. Energy expenditures for locomotion by mule deer and elk. *Journal of Wildlife Management* 33:366–379.

Perez-Barberia, F. J., I. J. Gordon, and M. Pagel. 2002. The origins of sexual dimorphism in body size of ungulates. *Evolution* 56:1276–1285.

Perez-Barberia, F. J., E. Perez-Fernandex, E. Robertson, and B. Alvarez-Enriquez. 2008. Does the Jarman–Bell principle at intra-specific level explain sexual segregation in polygynous ungulates? Sex differences in forage digestibility in Soay sheep. *Oecologia* 157:21–30.

Perrin, N. and V. Mazalov. 2000. Local competition, inbreeding, and the evolution of sex-biased dispersal. *American Naturalist* 155:116–127.

Phillips, M. L., W. R. Clark, S. M. Nusser, M. A. Sovada, and R. J. Greenwood. 2004. Analysis of predator movement in prairie landscapes with contrasting grassland composition. *Journal of Mammalogy* 85:187–195.

Pierce, B. M., R. T. Bowyer, and V. C. Bleich. 2004. Habitat selection by mule deer: Forage benefits or risk of predation? *Journal of Mammalogy* 68:533–541.

Porter, W. F., N. E. Mathews, H. B. Underwood, R. W. Sage, and D. F. Behrend. 1991. Social organization in deer—implications for localized management. *Environmental Management* 15:809–814.

Powell, R. A. 1994. Effects of scale on habitat selection and foraging behavior of fishers in winter. *Journal of Mammalogy* 75:349–356.

Powell, R. A., J. W. Zimmerman, and D. E. Seaman. 1997. *Ecology and Behavior of North American Black Bears: Home Ranges, Habitat and Social Organization*. London: Chapman & Hall.

Purdue, J. R., M. H. Smith, and J. C. Patton. 2000. Female philopatry and extreme spatial genetic heterogeneity in white-tailed deer. *Journal of Mammalogy* 81:179–185.

Pusey, A. E. and M. Wolf. 1996. Inbreeding avoidance in animals. *Trends in Ecology and Evolution* 11:201–206.

Rachlow, J. L. and R. T. Bowyer. 1998. Habitat selection by Dall's sheep (*Ovis dalli*): Maternal trade-offs. *Journal of Zoology London* 245:457–465.

Ralls, K. 1977. Sexual dimorphism in mammals: Avian models and unanswered questions. *American Naturalist* 122:917–938.

Relyea, R. A., R. K. Lawrence, and S. Demarais. 2000. Home range of desert mule deer: Testing the body-size and habitat-productivity hypotheses. *Journal of Wildlife Management* 64:146–153.

Riley, S. J. and A. R. Dood. 1984. Summer movements, home range, habitat use, and behavior of mule deer fawns. *Journal of Wildlife Management* 48:1302–1310.

Romin, L. A. and J. A. Bissonette. 1996. Deer–vehicle collisions: Status of state monitoring activities and mitigation efforts. *Wildlife Society Bulletin* 24:276–283.

Ronce, O., I. Olivieri, J. Clobert, and E. Danchin. 2001. Perspectives on the study of dispersal evolution. In *Dispersal*, eds. J. Clobert, E. Dranchin, A. A. Dhondt, and J. D. Nichols, 341–357. Oxford: Oxford University Press.

Rongstad, O. J. and J. R. Tester. 1969. Movements and habitat use of white-tailed deer in Minnesota. *Journal of Wildlife Management* 33:366–379.

Rosenberry, C. S., R. A. Lancia, and M. S. Conner. 1999. Population effects of white-tailed deer dispersal. *Wildlife Society Bulletin* 27:858–864.

Rubin, E. S. and V. C. Bleich. 2005. Sexual segregation: A necessary consideration in wildlife conservation. In *Sexual Segregation in Vertebrates: Ecology of Two Sexes*, eds. K. E. Ruckstuhl and P. Neuhaus, 379–391. Cambridge: Cambridge University Press.

Ruckstuhl, K. E. 1998. Foraging behaviour and sexual segregation in bighorn sheep. *Animal Behaviour* 137:361–377.

Ruckstuhl, K. E. 2007. Sexual segregation in vertebrates: Proximate and ultimate causes. *Integrative and Comparative Biology* 2:245–257.

Ruess, R. W. and S. J. McNaughton. 1987. Grazing and the dynamics of nutrient and energy regulated microbial processes in the Serengeti grasslands. *Oikos* 49:101–110.

Ruess, R. W. and S. J. McNaughton. 1988. Ammonia volatilization and the effects of large grazing mammals on nutrient loss from East African grasslands. *Oecologia* 77:550–556.

Runkle, J. R. 1985. Disturbance regimes in temperate forests. In *The Ecology of Natural Disturbance and Patch Dynamics*, eds. S. T. A. Pickett and P. White, 17–33. Orlando, FL: Academic Press.

Russell, E. W. B. 1983. Indian-set fires in the forests of the northeastern United States. *Ecology* 64:78–88.

Sabine, D. L., S. F. Morrison, H. A. Whitlaw, W. B. Ballard, G. J. Forbes, and J. Bowman. 2002. Migration behavior of white-tailed deer under varying winter climate regimes in New Brunswick. *Journal of Wildlife Management* 66:718–728.

Said, S. and S. Servanty. 2005. The influence of landscape structure on female roe deer home-range size. *Landscape Ecology* 20:1003–1012.

Sandegren, F., R. Bergstrom, and P. Y. Sweanor. 1985. Seasonal moose migration related to snow in Sweden. *Alces* 21:321–338.

Schmidt, J. I., J. M. Ver Hoef, and R. T. Bowyer. 2007. Antler size of Alaskan moose: Effects of population density, harvest intensity, and use of guides. *Wildlife Biology* 13:53–65.

Schmitt, S. M., S. D. Fitzgerald, T. M. Cooley et al. 1997. Bovine tuberculosis in free-ranging white-tailed deer from Michigan. *Journal of Wildlife Diseases* 33:749–758.

Schoener, T. W. 1981. Sizes of feeding territories among birds. *Ecology* 49:123–141.

Schroeder, C. A., R. T. Bowyer, V. C. Bleich, and T. R. Stephenson. 2010. Sexual segregation in Sierra Nevada bighorn sheep, *Ovis canadensis sierrae*: Ramifications for conservation. *Arctic, Antarctic, and Alpine Research* 42:476–489.

Seagle, S. W. 2003. Can ungulates foraging in a multiple-use landscape alter forest nitrogen budgets? *Oikos* 103:230–234.

Shannon, G., B. R. Page, K. J. Duffy, and R. Slotow. 2006. The role of foraging behaviour in the sexual segregation of the African elephant. *Oecologia* 150:344–354.

Shaw, J., R. Lancia, M. Conner, and C. Rosenberry. 2003. Why are the young bucks leaving? *Quality Whitetails* 10:30–34.

Simberloff, D. 1998. Flagships, umbrellas, and keystones: Is single species management passé in the landscape era? *Biological Conservation* 83:247–257.

Sinclair, A. R. E. 1983. The function of distance movements in vertebrates. In *The Ecology of Animal Movement*, eds. I. R. Swingland and P. J. Greenwood, 240–301. Oxford: Clarendon Press.

Spaeth, D. F., R. T. Bowyer, P. S. Barboza, and T. R. Stephenson. 2004. Sexual segregation in moose *Alces alces*: An experimental manipulation of foraging behaviour. *Wildlife Biology* 10:59–72.

Spaeth, D. F., K. J. Hundertmark, R. T. Bowyer, P. S. Barboza, and T. R. Stephenson, and R. O. Peterson. 2001. Incisor arcades of Alaskan moose: Is dimorphism related to sexual segregation? *Alces* 37:217–226.

Sparrowe, R. D. and P. F. Springer. 1970. Seasonal activity patterns of white-tailed deer in eastern South Dakota. *Journal of Wildlife Management* 34:420–431.

Stevens, C. E. and I. D. Hume. 1995. *Comparative Physiology of the Vertebrate Digestive System* (2nd edition). Cambridge: Cambridge University Press.

Stewart, K. M., R. T. Bowyer, J. G. Kie, N. J. Cimon, and B. K. Johnson. 2002. Temprospatial distributions of elk, mule deer, and cattle: Resource partitioning and competitive displacement. *Journal of Mammalogy* 83:229–244.

Stewart, K. M., R. T. Bowyer, J. G. Kie, B. L. Dick, and R. W. Ruess. 2009. Population density of North American elk: Effects on plant diversity. *Oecologia* 161:303–312.

Stewart, K. M., R. T. Bowyer, R. W. Ruess, B. L. Dick, and J. G. Kie. 2006. Herbivore optimization in North American elk: Consequences for theory and management. *Wildlife Monographs* 167:1–24.

Stewart, K. M., T. E. Fulbright, D. L. Drawe, and R. T. Bowyer. 2003. Sexual segregation in white-tailed deer: Responses to habitat manipulations. *Wildlife Society Bulletin* 31:1210–1217.

Storm, D. J., C. K. Nielsen, E. M. Schauber, and A. Woolf. 2007. Space use and survival of white-tailed deer in an exurban landscape. *Journal of Wildlife Management* 71:1170–1176.

Sutherland, G. D., A. S. Harestad, K. Price, and K. P. Lertzman. 2000. Scaling of natal dispersal distances in terrestrial birds and mammals. *Conservation Ecology* published online. http://www.consecol.org/vol4/iss1/art16, accessed February 7, 2011.

Sweeney, J. R. 1970. The effects of harassment by hunting dogs on the movement patterns of white-tailed deer on the Savannah River Plant, South Carolina. MS thesis, University of Georgia.

Swihart, R. K., N. A. Slade, and B. J. Bergstrom. 1988. Relating body size to the rate of home range use in mammals. *Ecology* 69:393–399.

Teer, J. G. 1996. The white-tailed deer: Natural history and management. In *Rangeland Wildlife*, ed. P. R. Krausman, 193–210. Denver, CO: Society for Range Management.

Telfer, E. S. 1967. Comparison of a deer yard and a moose yard in Nova Scotia. *Canadian Journal of Zoology* 45:485–490.

Tierson, W. C., G. F. Mattfeld, R. W. Sage, and D. F. Behrend. 1985. Seasonal movements and home ranges of white-tailed deer in the Adirondacks. *Journal of Wildlife Management* 49:760–769.

Trani, M. K., R. T. Brooks, T. L. Schmidt, V. A. Rudis, and C. M. Gabbard. 2001. Patterns and trends of early successional forests in the eastern United States. *Wildlife Society Bulletin* 29:413–424.

Tufto, J., R. Anderson, and J. Linnell. 1996. Habitat use and ecological correlates of home range size in a small cervid: The roe deer. *Journal of Animal Ecology* 65:715–724.

Turner, M. G. 1989. Landscape ecology: The effect of pattern on process. *Annual Review of Ecology and Systematics* 20:171–197.

Van Deelen, T. R. 1995. Seasonal migration and mortality of white-tailed deer in Michigan's Upper Peninsula. PhD dissertation, University of Michigan.

Van Deelen, T. R., H. Campa III, M. Hamady, and J. B. Haufler. 1998. Migration and seasonal range dynamics of deer using adjacent deer yards in northern Michigan. *Journal of Wildlife Management* 62:205–213.

Van Soest, P. J. 1994. *Nutritional Ecology of the Ruminant* (2nd edition). Ithaca, NY: Cornell University Press.

Verme, L. J. 1968. An index of winter weather severity for northern deer. *Journal of Wildlife Management* 32:566–574.

Verme, L. J. 1973. Movements of white-tailed deer in upper Michigan. *Journal of Wildlife Management* 37:545–552.

Verme, L. J. and J. J. Ozoga. 1971. Influence of winter weather on white-tailed deer in upper Michigan. In *Proceedings of Snow and Ice Symposium*, ed. A. O. Haugan, 16–28. Ames, IA: Iowa State University.

Wallis de Vries, M. F. 1995. Large herbivores and the design of large-scale nature reserves in western Europe. *Conservation Biology* 9:25–33.

Walter, W. D., K. C. VerCauteren, H. Campa III et al. 2009. Regional assessment on influence of landscape configuration and connectivity on range size of white-tailed deer. *Landscape Ecology* 24:1405–1420.

Waser, P. M. 1987. A model predicting dispersal distance distributions. In *Mammalian Dispersal Patterns: The Effects of Social Structure on Population Genetics*, eds. B. D. Chepko-Sade and Z. T. Halpin, 251–256. Chicago, IL: University of Chicago Press.

Webb, S. L., S. K. Riffell, K. L. Gee, and S. Demarais. 2009. Using fractal analyses to characterize movement paths of white-tailed deer and response to spatial scale. *Journal of Mammalogy* 90:1210–1217.

Weckerly, F. W. 1998. Sexual size dimorphism: Influence of mass and mating systems in the most dimorphic mammals. *Journal of Mammalogy* 79:33–52.

Weckerly, F. W. 2001. Are large male Roosevelt elk less social because of aggression? *Journal of Mammalogy* 82:414–422.

Weckerly, F. W. 2010. Allometric scaling of rumen-reticulum capacity in white-tailed deer. *Journal of Zoology London* 280:41–48.

Weckerly, F. W., M. A. Ricca, and K. P. Meyer. 2001. Sexual segregation in Roosevelt elk: Cropping rates and aggression in mixed-sex groups. *Journal of Mammalogy* 82:825–835.

Wheatley, M. and C. Johnson. 2009. Factors limiting our understanding of ecological scale. *Ecological Complexity* 6:150–159.

Whiting, J. C., R. T. Bowyer, and J. T. Flinders. 2008. Young bighorn (*Ovis canadensis*) males: Can they successfully woo females? *Ethology* 114:32–41.

Whiting, J. C., R. T. Bowyer, J. T. Flinders, V. C. Bleich, and J. G. Kie. 2010. Sexual segregation and use of water by bighorn sheep: Implications for conservation. *Animal Conservation* 13:541–548.

Wiens, J. A. 1989. Spatial scaling in ecology. *Functional Ecology* 3:385–397.

Wiens, J. A. and B. T. Milne. 1989. Scaling of 'landscapes' in ecology, or, landscape ecology from a beetle's perspective. *Landscape Ecology* 3:87–96.

Wiggett, D. R. and D. A. Boag. 1989. Intercolony natal dispersal in the Columbian ground squirrel. *Canadian Journal of Zoology* 67:42–50.

Wolff, J. O., K. I. Lundy, and R. Baccus. 1988. Dispersal, inbreeding avoidance, and reproductive success in white-footed mice. *Animal Behaviour* 36:456–465.

Wood, A. K., R. J. Mackie, and K. L. Hamlin. 1989. Ecology of sympatric populations of mule deer and white-tailed deer in a prairie environment. Montana Department of Fish, Wildlife, and Parks, Helena, Montana.

Zimmerman, T. J., J. A. Jenks, and D. M. Leslie, Jr. 2006. Gastrointestinal morphology of female white-tailed and mule deer: Effects of fire, reproduction, and feeding type. *Journal of Mammalogy* 87:598–605.

7

Diseases and Parasites

Tyler A. Campbell and Kurt C. VerCauteren

CONTENTS

Introduction

Wildlife biologists have long pursued understanding the ecology of diseases and parasites impacting white-tailed deer (e.g., see Whitlock, 1939), an important field of study because they can detrimentally affect deer populations, other wildlife, livestock, and humans (Davidson et al., 1981). Diseases and parasites of white-tailed deer, perhaps more than any other North American large mammal species, have received much attention in the literature and complete treatises have been devoted to the subject (e.g., see Davidson et al., 1981). In the last 20 years it has become necessary for wildlife biologists to incorporate disease concerns into the management of white-tailed deer (Figure 7.1). For example, at the federal level the United States Department of Agriculture (USDA), Animal and Plant Health Inspection Service is working to manage white-tailed deer diseases, some with implications for livestock health, including bluetongue (BT), bovine tuberculosis, cattle fever ticks, chronic wasting disease (CWD), and Johne's disease. Much of our knowledge stems from the exhaustive work with white-tailed deer diseases and parasites performed at the Southeastern Cooperative Wildlife Disease Study at the University of Georgia over nearly six decades (e.g., see Hayes et al., 1958). Recent advances in our understanding of the ecology of white-tailed deer diseases and parasites have also been made by state and federal agencies, and university scientists.

Numerous diseases and parasites cause morbidity and mortality in white-tailed deer. Altered deer behavior and reproductive success have also been noted (Matschke et al., 1984). White-tailed deer management programs should consider the significance of diseases and parasites early during the planning phases and throughout program implementation. Specifically, white-tailed deer biologists and managers would benefit by familiarizing themselves with the common infectious and parasitic diseases of deer, including viruses, bacteria, infectious prions, and parasites. Herein, the purpose is to provide a brief synopsis of these diseases and parasites and the chapter is organized into primary headings of: viral diseases, bacterial diseases, rickettsial diseases, CWD, and parasites. For a more detailed account of many of these infectious and parasitic agents readers should peruse Davidson et al. (1981), Samuel et al. (2001), and Williams and Barker (2001). For an easy-to-use and practical field guide for many white-tailed deer diseases and parasites, readers should see Davidson (2006). Furthermore, this chapter does not consider the morbidity and mortality factors of toxicosis, environmental contaminants, trauma, and weather-related phenomenon.

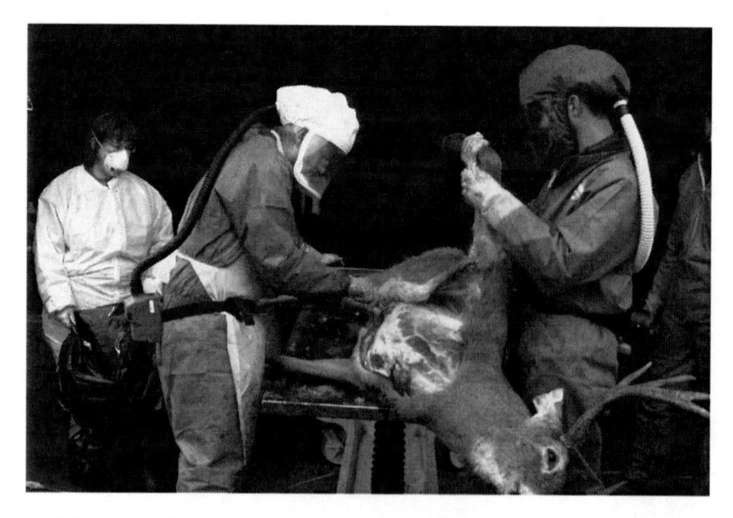

FIGURE 7.1 Personnel from the Michigan Department of Natural Resources and U.S. Department of Agriculture's Animal and Plant Health Inspection Service conducting necropsy on white-tailed deer under moderate biosecurity conditions. (Photo by Michigan Department of Natural Resources and Environment. With permission.)

Viral Diseases

Hemorrhagic Disease

Hemorrhagic disease (HD) in white-tailed deer populations is caused by orbiviruses, which occur in the family Reoviridae and are mostly double-stranded RNA viruses that are vector-borne (Roy, 1996). Of the >120 viral serotypes classified within 14 serogroups, only epizootic hemorrhagic disease (EHD) and BT viruses are associated with large-scale morbidity and mortality in white-tailed deer (Howerth et al., 2001). Two serotypes of EHD and six serotypes of BT occur in the United States, and all but BT serotype 2 (BT-2) have caused outbreaks of HD in white-tailed deer (Davidson, 2006; Murphy et al., 2006).

The clinical signs of HD are similar for both EHD and BT infections; however, there is much variability among deer populations and individuals may display chronic to peracute symptoms (Davidson, 2006). Although not completely understood, variation in clinical signs is likely related to differences in virus virulence (Howerth et al., 2001), innate host resistance (Gaydos et al., 2002b), cross-protection immunity (Gaydos et al., 2002a), and acquisition of maternal antibodies (Gaydos et al., 2002c). Furthermore, HD can be a slow and progressive (chronic form) disease resulting in changing clinical signs within infected individuals. In general, deer may exhibit depression, emaciation, facial swelling, fever, lameness, loss of appetite, reduced activity, and respiratory complications associated with HD (Howerth et al., 2001).

HD lesions are also variable and follow characterizations of being peracute, acute, and chronic. In the peracute form, animals often die rapidly and exhibit edema of the conjunctiva, head, lungs, neck, and tongue (Davidson, 2006). In the classic acute form of HD, peracute lesions often exist plus animals may display congestion or hemorrhages in the heart, intestines, and rumen, and necrosis on the dental pad, omasum, rumen, and tongue (Howerth et al., 2001). In addition to the above, deer displaying the chronic form of HD may have cracked or irregular hooves and loss of rumen papillae.

Wildlife hosts for HD include white-tailed deer and mule deer (Nettles and Stallknecht, 1992). However, HD has been detected in pronghorn, bighorn sheep, bison, elk, and mountain goats (Howerth et al., 2001). Livestock hosts include cattle and sheep. Cattle are vulnerable to both EHD and BT viruses, but only rarely exhibit clinical signs. Sheep are not susceptible to EHD virus, but BT viruses may cause severe morbidity and mortality (Davidson, 2006).

Numerous serologic tests for antibodies to HD viruses exist, including agar gel immunodiffusion, serum neutralization, and competitive enzyme-linked immunosorbent assays (Howerth et al., 2001). These tests may be used to confirm activity of viruses in deer populations or to determine probable cause of lesions. Molecular techniques may also be used in diagnostics, including reverse transcriptase polymerase chain reaction (PCR) (Shad et al., 1997). A suspected diagnosis can be determined with the presence of gross lesions, but confirmed diagnoses require virus isolation from morbid or dead deer (Davidson, 2006).

HD is transmitted by several species of *Culicoides* midges (Gibbs and Greiner, 1989). Although viruses associated with BT are primarily vectored by *C. sonorensis*, other species are important for transmission to white-tailed deer (Howerth et al., 2001; Davidson, 2006). Seasonal peaks in HD occur in late summer and early fall and correspond with life cycles of their vectors (Davidson, 2006). Lesions of chronic forms of HD may be detected into winter (Howerth et al., 2001).

Of viral diseases, HD is the most important agent impacting white-tailed deer herds (Nettles and Stallknecht, 1992). Nonetheless, population-level effects are not well documented (Fischer et al., 1995; Flacke et al., 2004, Gaydos et al., 2004). Risk factors associated with HD outbreaks in white-tailed deer populations are not clear and therefore management prescriptions for this disease have been elusive. Furthermore, funding for HD research has been limited in part because the disease is not zoonotic and human disease has not been reported with either EHD or BT viruses.

Cutaneous Fibromas

Fibromas and papillomas are hairless tumors or warts infecting white-tailed deer and other mammals that are caused by papillomaviruses. These double-stranded DNA viruses (Watson and Littlefield, 1960) are comprised of a nonenveloped icosahedral capsid structure 50–55 nm in diameter (Pfister, 1987).

FIGURE 7.2 Cutaneous fibromas on a white-tailed deer. (Photo by D. G. Hewitt. With permission.)

There are >125 papillomaviruses that infect >50 mammalian species, including six that have been found on Cervidae (Sundberg et al., 1997).

Fibromas infecting white-tailed deer are cutaneous and occur on the surface of the skin (Figure 7.2). Deer infected with fibromas seldom develop clinical disease. However, deer with heavy tumor loads may suffer from exhaustion and incapacitation due to tumors interfering with vision, respiration, food ingestion, and locomotion (Sundberg et al., 2001). Cutaneous fibromas are self-limiting and lesions normally regress, a process related to the development of antibodies and immunity (Ghim et al., 2000). Population-level impacts to white-tailed deer do not occur (Davidson, 2006).

Cutaneous fibromas may be found at any location on white-tailed deer and are usually smooth, but they may also be convoluted and resemble the head of a cauliflower (Davidson, 2006). Fibromas are commonly black, but may also be gray or white in color (Sundberg et al., 2001). The abundance and size of fibromas vary with severity of infection, but fibromas ranging from 1 to 10 cm in diameter are common.

Papillomaviruses infect species-specific mammalian hosts (Sundberg et al., 1997). Under natural conditions, the papillomaviruses that infect white-tailed deer are believed to only infect deer. However, experimental inoculation studies have demonstrated a white-tailed deer papillomavirus infection within hamsters (Koller and Olson, 1972). Cross-species transmission has not been demonstrated under natural conditions.

With a considerable amount of certainty, cutaneous fibromas can be diagnosed from gross observation of lesions (Sundberg et al., 2001). Histological determination is considered confirmatory and can be performed on tumor biopsies preserved in 10% buffered formalin (Davidson, 2006). Additional tests can be conducted using monoclonal and polyclonal antibodies to determine evidence of productive infections and to type papillomas through immunohistochemistry (Lim et al., 1990; Sundberg et al., 1996; Jenson et al., 1997).

Historically, sparring male white-tailed deer were hypothesized to spread cutaneous fibromas through direct contact with lesions (Friend, 1967). While infection via contact has not been verified, papillomavirus infections are believed to be transmitted from deer to deer by insect vectors and direct contact with abrasive fomites (Davidson, 2006).

More than 80 human papillomaviruses are known to exist (Sundberg et al., 1997). However, because papillomaviruses are specific to particular mammalian hosts, cutaneous fibromas from white-tailed deer are not infectious to humans. Furthermore, fibromas are only skin-deep and once skinned, harvested animals are suitable for consumption (Sundberg et al., 2001).

Arboviruses

Several arboviruses (arthropod-borne viruses), which replicate in blood-feeding arthropods such as mosquitoes, have been found in white-tailed deer (Yuill and Seymour, 2001). For example, West Nile virus was the cause of death in a white-tailed deer in Georgia (Miller et al., 2005) and eastern equine encephalitis infection within white-tailed deer has been found in Michigan (Schmitt et al., 2007) and Georgia (Tate et al., 2005). However, with the exception of vesicular stomatitis (VS), these arboviruses have not been determined to impact deer populations and are considered insignificant (Forrester, 1992). Lesions associated with VS include vesicles (or blisters) on the skin of the feet or mouth indistinguishable from lesions associated with foot-and-mouth disease (FMD). Because FMD is a highly contagious foreign animal disease, it is important to report any deer found with fluid-filled blisters on the mouth, tongue, muzzle, teats, and feet to animal health authorities and to not move the animal or carcass from the site (Davidson, 2006).

Other Viruses

White-tailed deer have been found with common cattle viruses including bovine viral diarrhea virus (Passler et al., 2007, 2009; Ridpath et al., 2007; Chase et al., 2008), malignant catarrhal fever (Forrester, 1992), and infectious bovine rhinotracheitis (Cantu et al., 2008). Under normal circumstances these viruses are not believed to significantly impact white-tailed deer populations. Other viruses that may infect white-tailed deer are rabies (Davidson, 2006), pseudorabies (Forrester, 1992), and parainfluenza virus (Forrester, 1992).

Bacterial Diseases

Anthrax

Anthrax is a highly lethal disease caused by the bacterium *Bacillus anthracis*, which belongs to the family Bacillaceae. *B. anthracis* is rod shaped and usually surrounded by a capsule. It forms spores and produces toxin. The vegetative form of this bacterium is susceptible to environmental degradation, while spores are very resistant and capable of surviving almost indefinitely under favorable conditions (200 ± 50 years) (de Vos, 2003). Anthrax spores are found on every continent except for Antarctica.

Depending on how *B. anthracis* enters the body of the host, the bacterium can cause cutaneous, respiratory, or gastrointestinal infections. The clinical disease is short in duration, sometimes resulting in sudden death of infected individuals. Clinical signs have not been reported in deer (Davidson and Nettles, 1997) though numerous in other animals, including dehydration, scouring, constipation, grinding teeth (Buxton and Barlow, 1994), fever, rapidly progressing debility, disorientation, respiratory distress (Cormack Gates et al., 2001), apoplectic seizures, depression, staggering gait (Van Ness, 1981), and swelling of the face, throat, and neck (de Vos, 2003). Infected individuals may also produce bloody discharges from orifices (Buxton and Barlow, 1994; Mackintosh et al., 2002) and subcutaneous edematous areas (Van Ness, 1981). Death can occur shortly after onset of clinical signs and several animals may die in close succession.

Animals with anthrax, a septicemic disease, may develop edema, hemorrhage, and necrotic lesions (de Vos, 2003). Sero-sanguinous fluid may be observed exuding from the nostrils and copious amounts of fluid may be present in the body cavity (Kellogg et al., 1970; Mackintosh et al., 2002). The blood is generally thick and dark, clots poorly, and flows freely from cut surfaces of infected carcasses (Cormack Gates et al., 2001).

Anthrax is most prevalent in wild and domestic mammalian herbivores, although birds and carnivores may also contract the disease (de Vos, 2003). Some mammalian carnivores and avian scavengers are resistant to the disease and act as carriers of *B. anthracis* spores (Cormack Gates et al., 2001). The bacteria can infect and be lethal to humans.

Anthrax is diagnosed by isolating *B. anthracis* from stained blood films (Buxton and Barlow, 1994; Mackintosh et al., 2002). Spores can easily be isolated from blood in the last few hours before death to one hour after (Cormack Gates et al., 2001). Because a carcass suspected to be infected with anthrax should not be opened, swabbings of mouth, nares, and anus can be used to diagnose the disease (Cormack Gates et al., 2001).

Anthrax epidemics in herbivores generally occur by ingesting food or water contaminated with *B. anthracis* spores (Choquette, 1970). Seasonally, these epidemics typically occur during dry summers following periods of heavy rain. These rain events promote runoff and areas of standing water accumulate *B. anthracis* spores. The dry weather that follows concentrates spores even more in these depressions, creating highly contaminated areas (Cormack Gates et al., 2001). Transmission can occur via cutaneous, respiratory, or gastrointestinal infections. Insect vectors, such as house flies and other arthropods, including biting flies, mosquitoes, and ticks may also transmit the disease (de Vos, 2003). Scavengers and predators may also disseminate the disease by feeding on and dismembering infected carcasses and spreading pathogens across the landscape (Choquette, 1970; de Vos, 2003).

Anthrax outbreaks in livestock can be controlled and even eradicated through quarantines and vaccinations, but applying these measures to free-ranging wildlife is currently impossible (Hugh-Jones and de Vos, 2002). A key to preventing or limiting new outbreaks is reducing soil contamination with *B. anthracis* spores; unfortunately, no current management practices to do so exist (Cormack Gates et al., 2001). Reducing animal densities in endemic areas will not prevent anthrax from occurring, but might limit the number of individuals exposed and thereby lessen mortality (Van Ness, 1981). In the event of an outbreak, easy to erect temporary fencing or other means to prevent infected animals from leaving the outbreak area and contacting other animals should be considered.

Anthrax is highly lethal in humans when left untreated. There are effective antibiotic therapies to treat the disease, but only when initiated immediately (Van Ness, 1981). Humans may become infected via inhalation, ingestion, or cutaneous exposure. Handling infected animals or infected animal products are common routes of human infection (Choquette, 1970; Van Ness, 1981). Early detection and treatment is critical in regulating incidence of the disease in humans (Choquette, 1970).

Dermatophilosis

Dermatophilosis is a skin disease caused by the bacterium *Dermatophilus congolensis*, of the family Dermatophilaceae. Dermatophilosis is distributed worldwide and is most common in ruminants (Salkin et al., 1983), but is most pronounced in tropical and subtropical climates (Roscoe et al., 1975). These bacteria are aerobic, Gram-positive, and nonacid-fast (Salkin and Gordon, 1981).

Clinical signs include areas of hair loss, thick scabs, emaciation (Leighton, 2001), a decline in overall physical activity, and potentially death (Salkin and Gordon, 1981). In white-tailed deer, fawns have been reported as more adversely affected than adults (Salkin and Gordon, 1981). Typical lesions consist of raised, matted hair tufts, held stiffly by an enveloping crust of shed epidermis and exudates (Leighton, 2001). These lesions or encrustations may become detached revealing pus and a red, inflamed dermis that bleeds (Leighton, 2001). Severe hair loss may also result and lesions may be found all over the affected animal, especially around the eyes, ears, and muzzle.

D. congolensis has a wide host range and is found in many parts of the world. Domestic animals are affected most frequently by the disease, but it also occurs in wild animals and humans (Salkin and Gordon, 1981). In North American wild animals in particular, the disease has been found in white-tailed deer, raccoons, woodchucks, striped skunks (Salkin et al., 1983), mule deer (Williams et al., 1984), rabbits, and rodents (Richard, 1981).

Presumptive diagnosis of *D. congolensis* depends on observations of gross lesions on infected animals and is verified by microscopic examinations of stained smears or histologic sections of scabs (Leighton, 2001). Under magnification, two parallel rows of Gram-positive bacteria form a hypha-like array (Leighton, 2001). The best place to collect bacteria needed for microscopic examination is the moist undersurface of freshly removed, uncontaminated scabs.

D. congolensis transmission may occur from direct contact or insect vectors (Salkin et al., 1983). Lesions or scabs on infected individuals contain *D. congolensis* bacteria and contact with them may be a route for direct disease transmission. Infected scabs may also fall off individuals and contaminate the environment, but the bacterium cannot persist for long periods outside a suitable host (Leighton, 2001). Insect vectors that have been implicated in transmitting the disease include mosquitoes, mange mites, ticks, and biting and nonbiting flies (Salkin and Gordon, 1981). Moisture facilitates release of the motile zoospore phase of the bacterium, thereby enhancing transmission (Williams et al., 1984).

Animals with *D. congolensis* can be treated successfully with antibiotics. Free-ranging wildlife may become a reservoir for the disease and potentially transmit dermatophilosis to domestic livestock and humans (Salkin and Gordon, 1981). In white-tailed deer it has been reported as self-limiting but may exhibit spontaneous remissions (Salkin and Gordon, 1981). Controlling biting insects is believed to reduce the disease in animal populations (Leighton, 2001).

Dermatophilosis is a contagious, zoonotic disease and human cases are usually associated with handling diseased animals. Lesions are typically found on hands and feet in humans (Hyslop, 1980). Infected humans usually heal with minimal treatment (Hyslop, 1980). When handling diseased animals, gloves and protective coveralls should be worn to prevent infection.

Brain Abscesses/Intracranial Abscesses

Brain abscesses are bacterial infections in the skull and brain. Although numerous genera of bacteria have been isolated, *Arcanobacterium pyogenes* is the primary bacteria found in brain abscesses of white-tailed deer (Davidson and Nettles, 1997; Baumann et al., 2001). *A. pyogenes* is a pyogenic, Gram-positive bacterium (Sneath et al., 1986).

Clinical signs include lack of coordination and fear, blindness, weakness, profound depression, emaciation, circling, single instances of torticollis, bilateral horizontal nystagmus, lameness, fever, and anorexia (Davidson et al., 1990). Lack of fear may result in wild animals walking toward humans and allowing handling or displaying the apparent desire to be handled (Debbie, 1965).

As bacterial infection develops in the brain a membrane forms around the area and a mass or lesion develops. The mass may vary in size and be filled with a creamy purulent material or pus (Debbie, 1965; Davidson and Nettles, 1997). Swelling and inflammation intensifies in response to infection. Signs of brain abscesses may also be evident by examination of skulls of infected individuals, often revealing erosion or pitting of cranial bones (Davidson et al., 1990). In white-tailed deer, lesions are often associated with antler pedicels of adult males (Davidson et al., 1990; Davidson and Nettles, 1997).

Brain abscesses occur in many species of mammals (Davidson et al., 1990), though rather infrequently in white-tailed deer (Davidson and Nettles, 1997; Baumann et al., 2001). When found in white-tailed deer, they have a strong bias toward adult males and incidence of the disease is seasonal, most likely due to velvet shedding, antler casting (Davidson and Nettles, 1997; Baumann et al., 2001), or a result of trauma associated with fighting with other males during the rut. Diagnosis includes opening brain cases to reveal the presence of abscesses or, if only skeletal remains are found, erosion or pitting of cranial bones (Davidson et al., 1990; Davidson and Nettles, 1997; Baumann et al., 2001; Karns et al., 2009).

It is not known whether brain abscesses can be transmitted between animals by direct contact (Davidson and Nettles, 1997). *A. pyogenes*, the primary bacteria found in brain abscesses of white-tailed deer, has been found in the nasopharyngeal mucosa of male deer suggesting that the direct contact of sparring males could transmit the disease among individuals (Karns et al., 2009).

Low prevalence rate in white-tailed deer suggests that overall this disease is not an important natural mortality source (Baumann et al., 2001). However, strong bias toward antlered males may affect management efforts to produce older-age males (Davidson and Nettles, 1997; Baumann et al., 2001).

Bovine Tuberculosis

Tuberculosis is a chronic, zoonotic bacterial disease found primarily in cattle, although it has a broad host range (Palmer et al., 2000). The causative agent is *Mycobacterium bovis*, a nonsporing, nonmotile, Gram-positive bacteria (Clifton-Hadley et al., 2001). Clinical signs of *M. bovis* infection may become apparent in weeks or take several years (Clifton-Hadley et al., 2001; de Lisle et al., 2002). Clinical signs include weight loss, swollen lymph nodes, discharging lymph node abscesses (Figure 7.3), coughing, and exercise intolerance (de Lisle et al., 2002; Mackintosh et al., 2002). Bovine tuberculosis is typically characterized by the formation of granulomas (Fitzgerald et al., 2000; de Lisle et al., 2002; Mackintosh et al., 2002). Tuberculosis granulomas have been observed in lungs, lymph nodes, and pericardium of white-tailed deer (Schmitt et al., 1997; O'Brien et al., 2001; de Lisle et al., 2002).

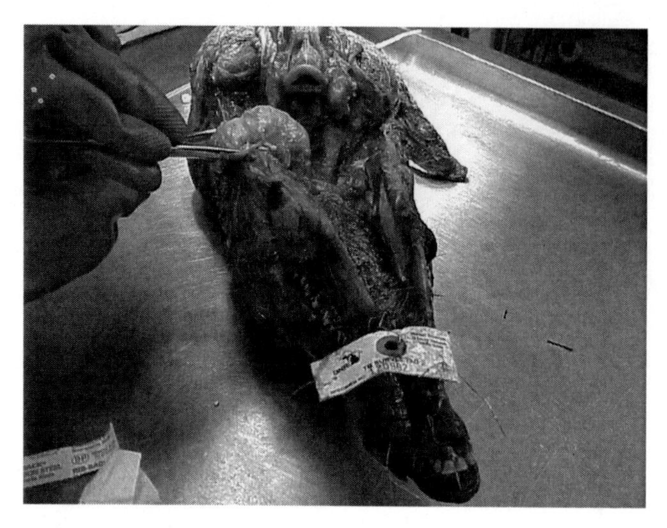

FIGURE 7.3 White-tailed deer head with swollen lymph node and abscesses. (Photo by Michigan Department of Natural Resources and Environment. With permission.)

White-tailed deer are the primary maintenance host of the disease in North America. Other wildlife species that can horizontally transmit *M. bovis* infection include brush-tailed possum, European badger, bison, and African buffalo (de Lisle et al., 2002). North American elk, red deer, fallow deer, Arabian oryx, dromedary camels, llamas, alpacas, and Asiatic water buffalo are other wildlife species that have been diagnosed with *M. bovis* (Hunter, 1996; Isaza, 2003). Additionally, coyotes can serve as sentinels for the disease in the environment (VerCauteren et al., 2008b).

Macroscopic examination of suspect tuberculosis lesions is frequently used to presumptively diagnose *M. bovis* infection postmortem (Figure 7.4) (Clifton-Hadley et al., 2001; de Lisle et al., 2002; Mackintosh et al., 2002). Histopathologic examination of suspect tuberculosis lesions and bacterial culture are the primary means used to diagnose *M. bovis* infection (de Lisle et al., 2002; Mackintosh et al., 2002). Another diagnostic test for deer includes a composite immune cell and antibody test (Mackintosh et al., 2002).

FIGURE 7.4 Tissue collection from white-tailed deer for presumptive diagnosis. (Photo by Minnesota Department of Natural Resources. With permission.)

Tuberculosis can be spread by either oral or respiratory routes (Mackintosh et al., 2002; Palmer et al., 2002). Routes of infection include contact with contaminated feed, mutual grooming, and inhaling infected droplets of fluid from the respiratory tract of infected individuals or the environment (Schmitt et al., 1997; Mackintosh et al., 2002; Palmer et al., 2004).

The presence of bovine tuberculosis in deer and subsequent transmission to cattle can severely impact local cattle industries. Restricting supplemental feeding and baiting of white-tailed deer is thought to reduce transmission of *M. bovis* bacteria (Schmitt et al., 1997; Palmer et al., 2001, 2004; O'Brien et al., 2002). Increasing harvest of deer to reduce densities and therefore the potential for disease transmission has also been practiced (O'Brien et al., 2006). The use of fencing, livestock protection of dogs, and specialized frightening devices can serve to reduce deer contact with cattle and feed meant for cattle (VerCauteren et al., 2006, 2008a; Seward et al., 2007).

M. bovis in humans is most commonly associated with consumption of raw, unpasteurized milk products (Isaza, 2003). Tuberculosis transmission from white-tailed deer to humans through consumption of infected meat, field dressing, or inhaling droplets from infected deer is unlikely, but can occur (Wilkins et al., 2003; de la Rua-Domenech, 2006).

Johne's Disease/Paratuberculosis

Johne's disease, also known as paratuberculosis, is a contagious, chronic, and often fatal infection that primarily affects small intestines of ruminants (Stabel, 1998; Manning, 2001). Johne's disease is caused by *Mycobacterium paratuberculosis* and is found worldwide. The bacterium is resistant to low pH, high temperature, and chemical agents, which leads to persistence in the environment (Manning, 2001).

Clinical symptoms include diarrhea, weight loss, decreased milk production, inappetence, and emaciation (Rosen, 1981; Stabel, 1998; Manning, 2001). Progression of the disease may be more rapid in deer than cattle (Williams, 2001). Typical gross lesions include thickening of intestinal walls in the posterior ileum and colon and enlargement of adjacent lymph nodes (Rosen, 1981; Manning, 2001; Isaza, 2003). Lesions in deer may not always be present and the intestinal wall can appear normal or only slightly edematous (Williams, 2001).

Johne's disease has been reported in domestic (e.g., cattle, goats, and sheep) and wild and captive ruminants (e.g., axis deer, fallow deer, red deer, reindeer, roe deer, sika deer, white-tailed deer, elk, moose, bison, aoudad, mouflon, and bighorn sheep) (Soltys et al., 1967; Libke and Walton, 1975; Davidson and Nettles, 1997; Manning, 2001).

There are many tests that will diagnose Johne's disease, each with their own advantages and disadvantages; but there is no single test that will diagnose all stages of the disease and subclinical carriers (Williams, 2001). The disease can be diagnosed from observing clinical signs, culture, immunological tests, or polymerase chain reaction (Williams, 2001).

M. paratuberculosis is predominately spread by ingesting food or water contaminated with infected feces (Mackintosh et al., 2002). Offspring may also become infected by ingesting contaminated milk (Williams, 2001; Mackintosh et al., 2002). Fecal shedding of *M. paratuberculosis* by apparently healthy animals may occur unknowingly for years and serve as a primary source of transmission (Williams and Barker, 2001).

With the exception of a few local populations, Johne's disease infrequently occurs in wild ruminants in North America (Williams, 2001) and white-tailed deer are not a major reservoir for the disease (Davidson and Nettles, 1997; Sleeman et al., 2009). Conversely, cattle producers incur large economic losses from Johne's disease and prevalence rates of 7–10% have been reported (Davidson and Nettles, 1997). Management of the disease may include minimizing contact between infected and uninfected individuals, moving or relocating individuals from only *M. paratuberculosis*-free herds or areas, and testing for *M. paratuberculosis* as part of a herd-health monitoring plan (Williams, 2001). Even though there are similarities between Johne's disease and Crohn's disease, a debilitating chronic inflammatory bowel disease found in people, no records exist of Johne's disease being transmitted to humans (Van Kruiningen, 2001).

Leptospirosis

Leptospirosis is caused by the spirochetal bacterium *Leptospira interrogans* (Bolin, 2003). These spirochete bacteria or leptospires are slender, Gram-negative aerobes, actively motile, and distributed worldwide (Roth, 1970; Bolin, 2003). There are 184 serovarieties or "serovars" of *Leptospira* belonging to 20 serogroups (Shotts, 1981). Naturally infected white-tailed deer rarely exhibit clinical leptospirosis signs (Davidson and Nettles, 1997). White-tailed deer experimentally inoculated subcutaneously with leptospirosis also showed no obvious clinical signs (Reid, 1994), but deer experimentally infected with serovariety *L. interrogans pomona* developed nephritis, hepatitis, and hemorrhages (Shotts, 1981). Hemorrhages, liver enlargement, edema, congestion in lungs, and hematuria were observed in cervids with leptospiral antibodies (Mackintosh et al., 2002).

Leptospirosis is a worldwide disease that can infect humans and many wild and domestic animals (Mackintosh et al., 2002). Different serovars are associated with particular host species, which are then important reservoirs of the disease (Bolin, 2003). Common wild and domestic animal maintenance hosts include raccoons, opossums, skunks, rats, mice, cattle, dogs, swine, and horses (Bolin, 2003). Diagnosis of leptospirosis is difficult and tests can be separated into those designed to detect antibodies and those designed to detect leptospires (Bolin, 2003). Examining body fluids and tissues with microscopy and fluorescent antibody techniques will provide confirmation of disease existence (Mackintosh et al., 2002).

Direct or indirect contact with infected urine, placental fluids, or milk will transmit leptospirosis (Bolin, 2003). Leptospiral organisms can survive outside the host for several months if protected from sunlight and conditions are damp and temperate (Mackintosh et al., 2002; Bolin, 2003). Vaccination can reduce the risk of exposure to leptospirosis (Mackintosh et al., 2002). If an animal is showing clinical signs attributable to *Leptospira*, appropriate treatment may be ampicillin and tetracyclines (Reid, 1994). Limited information suggests that white-tailed deer do not play a major role in the maintenance and spread of leptospirosis (Trainer et al., 1963; Shotts, 1981; Goyal et al., 1992). Leptospirosis is rarely found in white-tailed deer and risk of contracting the disease from deer is low (Shotts, 1981).

Salmonella

Bacteria of the family Enterobacteriacae and genus *Salmonella* are small nonspore-forming facultative Gram-negative rods that cause enteric or systemic infection. They are ubiquitous and highly adaptable, infecting a variety of vertebrates with little host preference. *Salmonella* bacteria can be responsible for both human and animal illness, most commonly manifesting as gastroenteritis. In certain *Salmonella* infections, for example, typhoid fever, systemic infections may also occur (Robinson, 1981). Salmonella bacteria have been generally classified into three species: *S. typhi*, *S. cholerasuis*, and *S. enteritidis*. The first two consist of single serotype, whereas the latter consists of more than 2400 serotypes (Ketz-Riley, 2003).

In deer, salmonellosis most commonly occurs in fawns. Clinical salmonellosis may manifest as primary enteritis and colitis, generalized infection (septicemia), or abortion (Clarke and Gyles, 1993). The disease occurs in all forms from peracute to chronic. Mild salmonellosis is characterized by mild gastroenteritis with vomiting and diarrhea. Diarrheic feces are often yellowish gray, spotted with blood, and have a foul odor (Davidson and Nettles, 1997). Anorexia, lethargy, fever, polydipsia, depression, recumbency, opisthotonos, and dehydration can occur (Mackintosh et al., 2002).

Gross lesions are not pronounced though edema may occur in mesenteric lymph nodes (Davidson and Nettles, 1997). In septicemic cases widespread hemorrhage on serous membranes, enlargement of the spleen and lymph nodes, and edema and congestion of organs such as the lung may be present. Microscopic lesions can include microvascular thrombosis in any tissue, necrosis in liver, spleen, and lymph nodes, and focal granulomas in various organs (Mörner, 2001).

Many *Salmonella* serotypes have been isolated from a wide variety of mammals and other vertebrates. Animals can be carriers of the bacteria and not have the disease salmonellosis. There is no evidence that deer serve as a reservoir for *Salmonella* (Robinson, 1981).

With living animals, culture of fresh fecal material is the most commonly used diagnostic method (Ketz-Riley, 2003). At necropsy, culture of fresh mesenteric lymph nodes has been the most consistent tissue for diagnostic confirmation (Robinson, 1981). Transmission of *Salmonella* occurs fecal orally, primarily via ingestion of contaminated food and water or through direct contact with infected and shedding animals (e.g., Ketz-Riley, 2003). The bacteria can be persistent in the environment and outbreaks are sometimes associated with riparian areas and periodic flooding. Deer are not considered to disseminate *Salmonella* to livestock.

Salmonella infections can cause significant mortality, morbidity, and economic losses among domestic animals (Robinson, 1981). Control of *Salmonella* infection in the wild is not feasible. To the extent that environmental contamination with sewage sludge, manure, or effluent from slaughterhouses contribute to the occurrence of *Salmonella* in wildlife (Murray, 1991), improved sanitation is probably the best way to reduce occurrence among wild mammals (Mörner, 2001).

In humans, *Salmonella* infections usually result in temporary gastrointestinal infection caused by contaminated food or drink. Cases of human salmonellosis have not been attributed to infected deer, but there is no reason to doubt that infections could be acquired from clinically affected deer (Robinson, 1981). Persons handling living or dead deer should practice common-sense sanitation.

Lyme Disease

Lyme disease is a zoonotic, tick-borne disease caused by the spirochetal bacterium, *Borrelia burgdorferi* (Brown and Burgess, 2001). *B. burgdorferi* bacteria are Gram-negative and 10–30 μm long (Shapiro and Gerber, 2000). White-tailed deer and other wildlife species typically do not exhibit clinical signs or lesions when infected with *B. burgdorferi* (Brown and Burgess, 2001). However, many species of mammals, birds, and reptiles play a role in maintaining the wildlife-tick cycle (Davidson and Nettle, 1997). White-tailed deer are primary hosts of adult black-legged deer ticks (*Ixodes scapularis*), but are not important reservoirs for the disease (Lane et al., 1991; Brown and Burgess, 2001). The role that white-tailed deer appear to play is maintaining and transporting infected ticks (Lane et al., 1991).

Lyme disease diagnosis in wildlife is problematic because the disease has little impact on reservoir species (Brown and Burgess, 2001). Clinical signs mimic other diseases and should not be used solely to diagnose infection with *B. burgdorferi* (Brown and Burgess, 2001). Various serologic tests (e.g., IFA, ELISA, and Western blots) have been used for detecting *B. burgdorferi* in wildlife, humans, and domestic animals (Brown and Burgess, 2001).

In eastern and midwestern United States, black-legged deer ticks are the vectors that transmit *B. burgdorferi* (Shapiro and Gerber, 2000). In western United States, western black-legged deer ticks (*I. pacificus*) transmit the disease (Shapiro and Gerber, 2000). The most likely tick stage during which transmission of Lyme disease occurs is the nymph stage (Shapiro and Gerber, 2000). It seems logical that if hosts of deer-ticks were removed from an area, then deer-tick numbers and potential for Lyme disease transmission would be reduced. Studies investigating this disease management approach on islands or other geographically isolated areas provide evidence that reducing deer densities does reduce numbers of host-seeking *I. scapularis* (Telford et al., 1988; Wilson et al., 1990; Jordan et al., 2007). Others have reported that reducing deer densities had no effect on nymph tick abundance and that deer densities would have to be reduced to very low numbers to impact disease transmission risk (Jordan et al., 2007). Scientists with the USDA, Agricultural Research Services have developed a passive topical treatment device to apply acaricide for controlling ticks feeding on white-tailed deer, called the "4-Poster" (Figure 7.5) (Pound et al., 2000). Throughout the northeastern United States, the "4-Poster" device has reduced the number of Ixodidae ticks infecting deer, and has been suggested as a tool to help control Lyme disease (Figure 7.6) (Carroll and Kramer, 2003; Carroll et al., 2009; Gatewood Hoen et al., 2009; Miller et al., 2009; Schulze et al., 2009).

For humans, Lyme disease is the most common vector-borne illness in the United States (Shapiro and Gerber, 2000). Typical early symptoms include fever, headache, fatigue, and a characteristic skin rash. Later phases of the disease may mimic other conditions, including rheumatoid arthritis, Bell's palsy, and neurologic impairment (Brown and Burgess, 2001).

FIGURE 7.5 White-tailed deer feeding from a passive topical treatment device that applies acaricide for controlling ticks called a "4-Poster." (Photo by U.S. Department of Agriculture's Agricultural Research Service.)

FIGURE 7.6 Biologists examining acaricide-impregnated rollers for signs of wear at a heavily used "4-Poster" in the northeastern United States. (Photo by U.S. Department of Agriculture's Agricultural Research Service.)

Rickettsial Diseases

Anaplasmosis

Anaplasmosis is a vector-borne, infectious, noncontagious disease caused by rickettsia belonging to the family Anaplasmataceae, genus *Anaplasma* (Kuttler, 1981; Davidson and Goff, 2001). Bacteria that cause the disease are Gram-negative and nonacid-fast (Davidson and Goff, 2001). Anaplasmosis causes destruction of red blood cells and is a worldwide disease of cattle, sheep, goats, and wild ruminants (Davidson and Goff, 2001). The *Anaplasma* genus was recently expanded to include three species that infect ruminants: *Anaplasma marginale* (cattle and deer), *Anaplasma ovis* (sheep, deer, and goats), and *Anaplasma centrale* (a less pathogenic organism) (Kocan et al., 2003).

Anaplasmosis is characterized by anemia as erythrocytes are parasitized and destroyed (Howe, 1981). As more red blood cells become infected, animals may develop rapid breathing, weakness (Kuttler, 1981), depression, inappetence, pale mucous membranes, rapid pulse, dehydration, thirst, and constipation (Howe, 1981). Most members of the deer family do not show clinical signs of anaplasmosis, even though they are susceptible to infection (Howe, 1981; Davidson and Nettles, 1997). Lesions are consistent with animals suffering from anemia and include thin watery blood, pale mucous membranes, enlargement of the spleen, liver, and gallbladder, generalized lymphadenopathy, and petechial hemorrhages in the endocardium (Howe, 1981; Kuttler, 1981).

In North American ruminants, where suitable vectors exist, anaplasmosis has been documented in wild white-tailed deer, mule deer, black-tailed deer, elk, bighorn sheep, and pronghorn antelope. Until recently, experimentally infecting nonruminant species with *Anaplasma* organisms was unsuccessful. The human anaplasmosis variant, *A. phagocytophilum* is now recognized to infect some rodent and shrew species, suggesting that these small mammals may also be important reservoirs of the disease (Woldehiwet, 2006).

Methods for diagnosis of anaplasmosis differ among animals with acute disease symptoms and symptoms from chronic or latent infections (Davidson and Goff, 2001). Animals with acute disease symptoms can be confirmed by microscopic examination of stained blood films, accompanied by serologic or molecular investigations (Davidson and Goff, 2001). Serologic demonstration of antibodies, with confirmation by animal subinoculation or molecular methods, is typically used for diagnosing anaplasmosis in wild ruminants with chronic or latent infections (Davidson and Goff, 2001). Molecular diagnostic procedures are used to detect anaplasmosis DNA in ticks and vertebrae hosts (Eriks et al., 1989; Kieser et al., 1990).

Ticks are the natural biologic vector capable of transmitting anaplasmosis; biting insects are capable of mechanically transmitting the disease, though less efficiently (Davidson and Goff, 2001). Numerous species of ticks (including members of the genera *Boophilus*, *Hyalomma*, *Amblyomma*, *Rhipicephalus*, *Dermacentor*, *Ixodes*, *Argus*, and *Haemophysalis*) have been shown to transmit anaplasmosis (Figure 7.7) (Kuttler, 1981). Transmission of the disease through biting insects must occur quickly in order for the infectious agent to remain viable on insect mouth parts (Davidson and Goff, 2001). Whether the disease becomes established in wild or domestic populations depends on the presence of suitable vectors in

FIGURE 7.7 Dorsal view of an engorged female lone star tick. (*Amblyomma americanum*; photo by U.S. Centers for Disease Control and Prevention.)

the area. Infected blood on surgical instruments and hypodermic needles can also transmit the disease (Kuttler, 1981).

Management of anaplasmosis can be controversial in areas where wildlife species known to carry the disease overlap with domestic livestock range. Cattle can be extremely susceptible to anaplasmosis, with mortality exceeding 50% (Kuttler, 1981). Wild ruminants, primarily black-tailed and mule deer, are considered asymptomatic carriers in western North America (Kuttler, 1981). Management of the disease in free-ranging wild ruminants is often logistically impossible. Furthermore, removing wild ruminants that are known carriers is seldom a consideration because they rarely display clinical signs of infection (Howe, 1981).

The three *Anaplasma* species that infect ruminants, *A. marginale*, *A. ovis*, and *A. centrale* do not affect humans (Davidson and Goff, 2001). Human anaplasmosis, caused by the bacteria *A. phagocytophilum*, can be transmitted to humans by the bite of deer ticks and western black-legged ticks.

Chronic Wasting Disease

Chronic wasting disease is a transmissible spongiform encephalopathy (TSE), a group of neurological diseases that include bovine spongiform encephalopathy, sheep scrapie, transmissible mink encephalopathy, and Creutzfeldt–Jakob disease in humans. The causative agents of TSEs are thought to be proteinaceous infectious agents called prions. Prions are abnormal, protease-resistant forms of cellular proteins coded for and normally synthesized in the central nervous system and lymphoid tissues (Williams and Miller, 2002). The accumulation of prion leads to neurodegeneration and ultimately death (Sigurdson, 2008).

Clinical signs of CWD are unspecific and subtle in early disease and become detectable a year to just days prior to death, with most infected animals surviving three to four months following the onset of clinical disease (Williams et al., 2001). The primary clinical signs of animals with CWD include weight loss that progresses to emaciation; excessive salivation; abnormal behavior, including loss of fear of human beings; and mild ataxia (Spraker, 2003). Animals often carry their head and ears low and may walk in repetitive patterns, appear depressed, and rouse easily. As the disease progresses, many affected animals display polydipsia and polyuria, increased salivation with resultant drooling, as well as lack of coordination, posterior ataxia, fine-head tremors, and a wide-based stance. Esophageal dilation, hyperexcitability, and syncope are occasionally observed. Death is inevitable (Williams and Miller, 2002).

Primary gross lesions in advanced cases of CWD include weight loss, emaciation, and loss of abdominal and subcutaneous adipose tissue. Adipose tissue behind the eye, around the spinal cord, within the bone marrow and joints, within the renal pelvis, and around the coronary vessels undergoes serous atrophy. Adrenal gland enlargement may be noted. Aspiration pneumonia is common (Spraker, 2003). Carcasses may be in poor nutritional state or emaciated, but may be in fair condition if the animal died of aspiration pneumonia or after only a short clinical course. Primary histological lesions are limited to the nervous system and are typical of spongiform encephalopathies. The lesions are characterized by a spongiform degeneration of the neuropil, vacuolar degeneration of neurons with neuronal loss, and mild astrocytosis. Lymphoid depletion of the tonsils, lymph nodes, and spleen may occur in terminal stages (Figure 7.8). Histopathological examination can be used to confirm secondary gross lesions, such as bronchiopneumonia, gastric ulceration and peritonitis, hypertrophy of the adrenal cortex, and serous atrophy of fat (Spraker, 2003).

Naturally occurring cases of CWD have been documented in white-tailed deer, mule deer, elk, and moose. Concern exists regarding potential transmission or adaptation to other species of wildlife and domestic livestock. Clinical diagnosis of CWD relies on the appearance of the aforementioned symptoms. Confirmatory diagnosis is by identification of abnormal prion protein in lymphoid and brain tissues by immunohistochemistry and by the appearance of spongiform changes on histopathological examination of brain tissues. Monoclonal antibodies specific for the prion protein are effective on fresh and formalin-fixed tissues and allow the detection of CWD. Biopsies of tonsil and rectal mucosa are being evaluated and used in some cases for antemortem-CWD surveillance (Spraker et al., 2006; Wolfe et al., 2007).

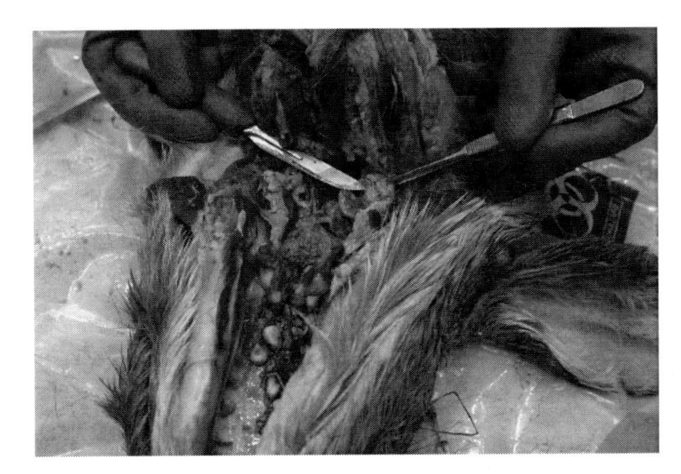

FIGURE 7.8 Inspection of white-tailed deer head for lymphoid depletion. (Photo by Michigan Department of Natural Resources and Environment. With permission.)

Although mechanisms are not completely understood, transmission of CWD occurs directly through contact with infected individuals and indirectly through contact with environments and fomites that have been contaminated by excretions of infected individuals prior to death and by their carcasses following death (e.g., Miller et al., 2004; Mathiason et al., 2006; Haley et al., 2009). Prions remain infectious in soil for over two years, suggesting that soil may serve as a reservoir for CWD prions (Seidel et al., 2007). Although interspecies transmission likely occurs within the Cervidae, CWD has not been transmitted by oral inoculation to species outside this family.

The presence of CWD in captive and free-ranging populations of white-tailed deer and other cervids is a serious management problem. There is no known treatment for animals affected with CWD, and it is considered 100% fatal once clinical signs develop. Eradication from free-ranging deer is unlikely due to CWD's long incubation period, subtle early clinical signs, extremely resistant infectious agent, environmental contamination, and multiple modes of transmission (e.g., Williams and Miller, 2002; Spraker, 2003; Sigurdson, 2008). Active surveillance aids in determining the distribution and prevalence of CWD and can be used to elucidate changes over time (Figure 7.9). Localized population reduction, regulating translocation of deer, and banning baiting and feeding have all been attempted to slow down the spread of CWD.

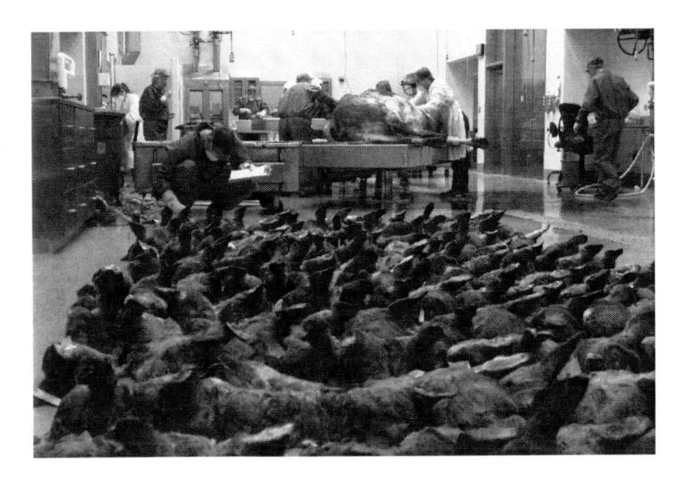

FIGURE 7.9 White-tailed deer heads waiting evaluation and testing during active surveillance activities for CWD. (Photo by Michigan Department of Natural Resources and Environment. With permission.)

No cases of human disease have been associated with CWD and there appears to be a robust species barrier. As CWD and other TSEs are incompletely understood, though, hunters harvesting deer in endemic areas should consider having carcasses tested, and, as with other diseases, it is not advisable to consume meat from infected deer. Additionally, common sense measures should be employed when handling and processing carcasses from endemic areas.

Parasites

Protozoa

Toxoplasmosis

Toxoplasma gondii, which has felids as definitive hosts and has many mammals as intermediate hosts, is the causative agent of toxoplasmosis (Dubey and Odening, 2001). This protozoa normally parasitizes hosts, including white-tailed deer (Lindsay et al., 1991), without causing clinical signs, although experimentally infected animals display diarrhea, weakness, and lethargy (Davidson, 2006). Lesions of toxoplasmosis in deer may include enlarged heart and lymph nodes, congestion of ocular vessels, abomasum, lungs, and spleen, and hemorrhages within the myocardium and internal abdominal wall (Davidson, 2006). Definitive diagnosis is conducted through histological, biological, and/or serological methods (Dubey and Odening, 2001). Transmission of *T. gondii* oocysts to deer occurs through the fecal (from felids)–oral route via environmental exposure (Davidson, 2006). Morbidity and mortality among deer due to toxoplasmosis are rare, however, seroprevalence rates are high (30–60%; Lindsay et al., 1991; Brillhart et al., 1994; Humphreys et al., 1995; Vanek et al., 1996). This highlights the importance of either freezing or thoroughly cooking venison because toxoplasmosis can negatively and severely impact human health if exposed.

Babesiosis

The causative agents of babesiosis are obligate intraerythrocytic protozoan parasites in the genus *Babesia* (Kocan and Waldrup, 2001). White-tailed deer are most commonly infected with *B. odocoilei*, which has been found in Texas, New Mexico, Oklahoma, Virginia, and Florida (Spindler et al., 1958; Emerson and Wright, 1968; Perry et al., 1985; Waldrup et al., 1989b), and is vectored by the deer tick (*I. scapularis*) (Waldrup et al., 1990). Numerous other ixodid ticks transmit *Babesia* spp., such as the southern cattle tick (*Rhipicephalus* [*Boophilus*] *microplus*) and the cattle tick (*Rhipicephalus* [*Boophilus*] *annualatus*), which are both one-host ticks (Kocan and Waldrup, 2001). Both of these tick species have been found on white-tailed deer (Cooksey et al., 1989; Cantu et al., 2007). Other species susceptible to *B. odocoilei* are elk, reindeer, caribou, desert bighorn sheep, and muskoxen (Schoelkopf et al., 2005). Clinical signs and mortality for free-ranging deer infected with *Babesia* spp. are rare; however, mortality is high for immunocompromised deer under experimental conditions (Kocan and Waldrup, 2001). Definitive diagnosis in species that are more susceptible to babesiosis than deer involves microscopically demonstrating the agent in whole blood (Kocan and Waldrup, 2001). *Babesia* spp. that infect white-tailed deer are not infectious to humans. However, given that bovine babesiosis (*B. bovis* and *B. bigemina*) has recently been demonstrated in white-tailed deer in northern Mexico (Cantu et al., 2007), there is great concern among livestock producers and government agencies about the role deer might play in the event these diseases were reintroduced into the United States (Figure 7.10).

Theileriosis

White-tailed deer may be infected with *Theileria cervi*, an intraerythrocytic protozoan parasite and the causative agent of theileriosis (Kreier et al., 1962). Clinical signs of *T. cervi* infecting free-ranging white-tailed deer are extremely rare and may include only mild-to-nonapparent anemia (Kocan and Waldrup, 2001). However, fever, anemia, emaciation, pale membranes, and overall debilitation has been observed in cases where *T. cervi* infection is concurrent with severe infestations of lone star ticks (*A. americanum*) in fawns (Barker et al., 1973; Kocan and Waldrup, 2001; Yabsley et al., 2005). In North America, *T. cervi*

FIGURE 7.10 Inspection of wild-caught white-tailed deer for cattle fever ticks (*Rhipicephalus* [*Boophilus*] spp.) that carry bovine babesiosis within the southern Texas border region. (Photo by U.S. Department of Agriculture's Animal and Plant Health Inspection Service.)

has been found in white-tailed deer, mule deer, sika deer, fallow deer, and elk (Davidson et al., 1985; Waldrup et al., 1989a) in regions with lone star ticks, its only known vector. Diagnosis is based on identification of intraerythrocytic piroplasms (Kocan and Waldrup, 2001). *T. cervi* does not infect humans and poses a minimal threat to white-tailed deer populations.

Trematodes

Liver Fluke

Fascioloides magna is the scientific name for liver flukes, which are also known as giant liver flukes, large American liver flukes, or deer flukes (Pybus, 2001). In white-tailed deer and other definitive hosts clinical signs of disease are uncommon. Liver fibrosis may occur in severe *F. magna* infestations (Davidson, 2006). In aberrant hosts, depression, lethargy, anorexia, and weight loss are common just prior to death (Foreyt, 1992, 1996). Lesions associated with liver fluke infestation are fibrous capsules in liver tissues. These capsules commonly contain ≥2 adult flukes. In aberrant hosts, lesions may involve severe liver damage including hemorrhage, necrosis, inflammation, and diffuse fibrosis (Foreyt and Leathers, 1980). Definitive hosts of *F. magna* are white-tailed deer, elk, caribou, black-tailed deer, mule deer, red deer, and fallow deer (Pybus, 2001). Dead-end hosts of *F. magna* are moose, sika deer, sambar, cattle, bison, yak, horse, swine, collared peccary, and llama (Pybus, 2001). Aberrant hosts of *F. magna* are domestic sheep, domestic goat, chamois, bighorn sheep, mouflon, and roe deer. Diagnosis of liver flukes is through identification of reddish-brown trematodes (8 × 3 cm) encapsulated in the liver (Pybus, 2001). The life cycle of *F. magna* is complex and involves snail intermediate hosts and deer definitive hosts that obtain metacercariae through consumption of vegetation (Pybus, 2001). Care should be taken when translocating elk, red deer, white-tailed deer, and fallow deer from enzootic to nonenzootic areas (Pybus, 2001). Anthelmintic treatment should occur in these situations as well as at game farms and where sheep and deer comingle (Davidson, 2006). Eating venison from animals with liver flukes poses no threat to human health (Davidson, 2006).

Nematodes

Large Lungworm

The large lungworm found in white-tailed deer is *Dictyocaulus viviparous*. Deer infected with *D. viviparous* rarely display clinical signs (Davidson, 2006). However, deer that are severely infected are often weak, underweight, and undergoing respiratory distress (Munro, 1988). Deer experiencing lungworm pneumonia are usually fawns, malnourished, and infected with high levels of other parasites (Bergstrom, 1975). As the name implies, lesions associated with large lungworm infection occur in the respiratory tract and may include excessive foamy mucous in air passages, bronchopneumonia, and lungs coated with fibrin (Anderson and Prestwood, 1981). *D. viviparous* have been found in numerous other wild and domestic ruminants, including mule deer, elk, moose, cattle, sheep, and goats (Anderson and Prestwood, 1981). Large lungworm infection can be diagnosed through identification of these up to 3.8 cm long white nematodes in the trachea and bronchi. Live animal tests are performed through microscopic examination of feces for larvae. The life cycle of *D. viviparous* is direct and deer obtain larvae while feeding on vegetation (Mason, 1985). Large lungworms present no human health risks. However, *D. viviparous* infections are a significant source of deer mortality on properties where deer densities exceed nutritional carrying capacity. As such, and particularly on properties where deer co-mingle with livestock, biologists should seek to maintain deer densities below carrying capacity (Davidson, 2006).

Large Stomach Worm

The scientific name for large stomach worm or barber pole worm is *Haemonchus contortus*. These gastrointestinal nematodes are characterized by a well-developed synlophe, prominent buccal tooth, and well-developed copulatory bursa in males (Hoberg et al., 2001). Most deer serve as hosts of the large stomach worm without demonstrating signs of disease (Davidson, 2006). In fact, prevalence rates approximate 100% in parts of the coastal plain of the southeastern United States (Prestwood and Pursglove, 1981). At high *H. contortus* intensities, haemonchosis may occur, which may result in death (Hoberg et al., 2001). Deer experiencing haemonchosis are usually fawns, weak, anemic, underweight, malnourished, and are often infected with many other parasites (Prestwood and Kellogg, 1971; Davidson et al., 1980; Forrester, 1992). Lesions associated with haemonchosis are accumulation of fluid in the submandibular region and pale mucous membranes and organs due to blood loss (Foreyt and Trainer, 1970; Prestwood and Purslove, 1981). *H. contortus* have been found in most wild and domestic ruminants occurring in North America (Davidson, 2006). Definitive diagnosis of haemonchosis using fecal analysis is challenging because eggs of *Haemonchus* are indistinguishable from other trichostrongylids (Sommer, 1996). However, the presence of large numbers of *H. contortus* in young animals that are weak and emaciated is presumptive evidence (Figure 7.11). Also, molecular markers have been used to partition trichostrongylids into species (Zarlenga et al., 1994; Lichtenfels et al., 1997). The life cycle of *H. contortus* is direct and deer obtain larvae while feeding on vegetation (Davidson, 2006).

Large stomach worms pose no human health risk. The occurrence of haemonchosis in white-tailed deer herds is indicative of populations that exceed nutritional carrying capacity (Davidson, 2006). Similar to other severe nematode infections, biologists should seek to maintain deer densities below carrying capacity to promote healthy herds of deer and other co-occurring ruminants (Davidson, 2006).

Meningeal Worm

The long-lived meningeal worm (*Parelaphostrongylus tenuis*) is common across the range of white-tailed deer, with the possible exception of the coastal plain region of the southeastern United States (Comer et al., 1991; Duffy et al., 2002). For example, recent reports have demonstrated *P. tenuis* within cervids in Saskatchewan, Manitoba, North Dakota (Wasel et al., 2003), Kentucky (Larkin et al., 2003), Michigan (Bender et al., 2005), and South Dakota (Jacques and Jenks, 2004). White-tailed deer infected with meningeal worms only rarely display clinical signs of disease (Lankester, 2001). In cases of massive infestations, white-tailed deer may show neurological signs, including partial paralysis, loss of motor function, or circling (Prestwood, 1970). *P. tenuis* infections within other native cervids cause devastating morbidity and mortality (Lankester, 2001). Lesions in white-tailed deer with clinical disease are inflammation of the cranial meninges, small (~1 mm) red spots scattered widely across the surface of lungs, and occasionally mild pneumonia (Lankester, 2001; Davidson, 2006). Definitive hosts for meningeal worms are white-tailed deer. Clinical disease is near certain when *P. tenuis* enters other native cervids, some exotic ungulates, and domestic sheep (Lankester, 2001; Davidson, 2006). To diagnose meningeal worm

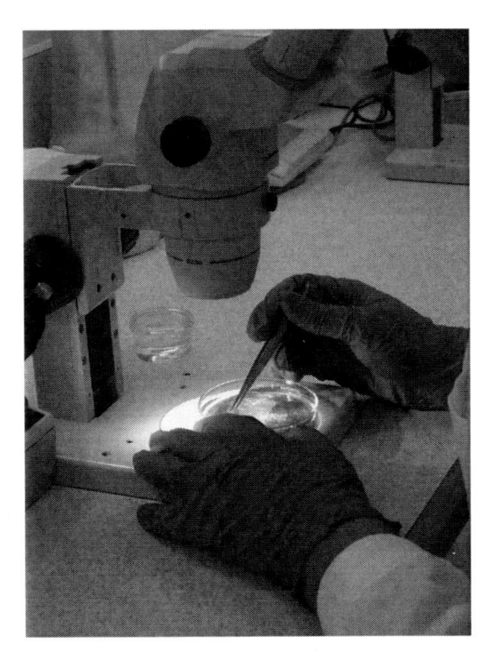

FIGURE 7.11 Personnel collecting large stomach worms (*Haemonchus contortus*) from a diluted sample of white-tailed deer fawn abomasal contents. (Photo by U.S. Department of Agriculture's Animal and Plant Health Inspection Service.)

infestation, adult worms must be identified from the central nervous system, which requires animals to be euthanized (Carreno and Lankester, 1993; Davidson, 2006). Efforts are being made to develop molecular diagnostic tools (Gajadhar et al., 2000). The life cycle of *P. tenuis* is complex and involves terrestrial mollusks as intermediate hosts and deer as definitive hosts that obtain infective third-stage larvae through inadvertent consumption of mollusks (Lankester, 2001). Meningeal worms present no human health risks. However, *P. tenuis* is a significant threat to all native cervids in North America other than white-tailed deer (Samuel et al., 1992; Lankester, 2001). Natural resource managers and biologists undertaking cervid translocation activities should consider and take preventative measures to prevent meningeal worm introductions into susceptible herds (Davidson, 2006).

Arterial Worm

As their common name implies, arterial worms (*Elaeophora schneideri*) reside in the arteries of cervids, including white-tailed deer (Pence, 1991). Arterial worms may cause elaeophorosis, within their hosts. The clinical signs of elaeophorosis in white-tailed deer are oral food compaction and associated facial swelling (Couvillion et al., 1986). White-tailed deer with only a few worms often present no gross lesions. However, with clinical disease, lesions may include coronary obstruction, plaque-like lesions on internal lining of carotids, and thickening of vessel walls (Anderson, 2001). Death may also occur (Titche et al., 1979). Natural hosts for *E. schneideri* are mule deer (Hibler et al., 1969), black-tailed deer (Weinmann et al., 1973), white-tailed deer (Prestwood and Ridgeway, 1972), elk (Hibler and Adcock, 1971), moose (Madden et al., 1991), domestic barbary sheep (Pence and Grey, 1981), and sika deer (Robinson et al., 1978), primarily in western North America (Anderson, 2001).

When white-tailed deer are seen with food compaction, elaeophorosis should be strongly suspected (Davidson, 2006). Confirmation of elaeophorosis is through identification of *E. schneideri* within arteries (Davidson, 2006) or through skin biopsy (Hibler and Adcock, 1971). Arterial worms pose no threat to human health. The life cycles of *E. schneideri* are indirect and require horseflies of the genera *Hybomitra* and *Tabanus* as intermediate hosts and for transmission to definitive hosts (Anderson, 2001). Population-level impacts of elaeophorosis in white-tailed deer have not been documented. However, because elaeophorosis is more often observed in older animals, management aimed at increasing the herd age-structure may also increase the occurrence of disease (Davidson, 2006).

Abdominal Worm

Setaria yehi is a large filarial worm, also known as the abdominal worm, often occurring in the abdominal and thoracic cavities of white-tailed deer (Forrester, 1992). The abdominal worm does not cause clinical disease in its hosts, which include most native cervids from North America and sambar deer (Davidson et al., 1987; Davidson, 2006). However, hunters may discover live or postmortem adult worms encysted on the surface of organs while eviscerating deer. Furthermore, infected deer often have mild fibrinous peritonitis (Prestwood and Pursglove, 1977). The life cycle of *S. yehi* is indirect and it uses mosquitoes as intermediate hosts and biological vectors for transmission back to deer (Forrester, 1992). Abdominal worms pose no human, wildlife, or livestock health risks. However, hunters who notice *S. yehi* may communicate concerns or questions to biologists and managers (Davidson, 2006).

Cestodes

Larval Tapeworm

White-tailed deer serve as intermediate hosts for several tapeworms, the most common being *Taenia hydatigena*. In white-tailed deer and other intermediate hosts, no clinical signs are reported (Jones and Pybus, 2001). Lesions in white-tailed deer are rare and include bladder-like larvae on or in visceral organs, such as liver tissue (Davidson, 2006). Given the global distribution of *T. hydatigena*, the dominant definitive and intermediate hosts vary by region (Jones and Pybus, 2001). Recent reports document larval *T. hydatigena* in white-tailed deer in Ontario (Addison et al., 1988), Alberta (Pybus, 1990), Ohio (Schurr et al., 1988), and Florida (Forrester and Rausch, 1990). In white-tailed deer, diagnosis is through identification of *T. hydatigena* larvae in tissues. The life cycle of *T. hydatigena* is indirect and involves herbivores as intermediate hosts and carnivores as definitive hosts (Jones and Pybus, 2001). Larval tapeworms present no human, wildlife, or livestock health risks. As with the abdominal worm, hunters who find larval tapeworms in tissues may communicate concerns or questions to biologists and managers (Davidson, 2006).

Arthropods

Ticks

Without question, ticks are the most significant ectoparasite of wild mammals, including white-tailed deer, because of their ability to harbor and transmit pathogens (Strickland et al., 1981; Forrester, 1992). The most common ticks that infect white-tailed deer are hard ticks of the genera *Amblyomma* (Figure 7.12), *Ixodes* (Figure 7.13), and *Dermacentor* (Davidson, 2006). Ticks of the genera *Boophilus* were once abundant across the range of white-tailed deer; however, with the exception of a small eradication zone in southern Texas, they are now eradicated from the United States (Allan, 2001b). Most deer with ticks show no signs of disease and often those that do are malnourished and have high internal parasite loads (Davidson, 2006). Massive tick infestations may result in local irritation at the site of feeding, blood loss, secondary infections, and mortality (Allan, 2001b). Fawns with heavy infestations may experience blindness and death (Davidson, 2006). Ticks commonly found on white-tailed deer also infect livestock and other wildlife. Diagnosis is through identification of ticks on the skin (Allan, 2001b). Ticks of the genera *Amblyomma*, *Ixodes*, and *Dermacentor* are all three-host ticks, requiring a blood meal from a different host to complete each life stage (i.e., larva, nymph, adult; Allan, 2001b). Several species of ticks that infect white-tailed deer will coinfect humans and may function as vectors of diseases (Davidson, 2006).

Ear Mites

White-tailed deer may be infected with ear mites (*Psoroptes cuniculi*). Clinical signs of ear mite infestation are circling, incoordination, scratching or shaking of the head, and secondary bacterial infections inside the ear (Scott et al., 2000). Lesions associated with severe ear mite infections include thick, firm debris in the external ear canal (Scott et al., 2000). In addition to white-tailed deer, ear mites have been

FIGURE 7.12 Dorsal view of an adult female lone star tick (*Amblyomma americanum*). (Photo by U.S. Centers for Disease Control and Prevention.)

found in mule deer, domestic rabbits, and goats (Davidson, 2006). Infestations by ear mites are diagnosed through acquisition and identification of *P. cuniculi* from ears (Davidson, 2006). Ear mites mature from eggs, to larvae, to adults on white-tailed deer, with transmission from deer-to-deer via direct contact (Scott et al., 2000). Ear mites pose no human health risks and are not considered overtly pathogenic to white-tailed deer unless heavily infested (Davidson, 2006).

FIGURE 7.13 Dorsal view of an adult female western black-legged tick (*Ixodes pacificus*). (Photo by U.S. Centers for Disease Control and Prevention.)

Demodectic Mange

White-tailed deer have been reported with demodectic mange or demodicidosis, which is caused by the mange mite *Demodex odocoilei*, in Oklahoma, Georgia, Virginia, South Dakota, and Texas (Kellogg et al., 1971; Carpenter et al., 1972; Jacques et al., 2001; Turner and Cano, 2008). Clinical signs and lesions in infested white-tailed deer usually do not develop (Davidson, 2006). With severe infestations, hair loss with thickening of the skin and secondary bacterial infections may develop (Desch and Nutting, 1974). *Demodex* is taxonomically complex because of similarities within the genera (Turner and Cano, 2008). However, *D. odocoilei* have only been found on white-tailed deer (Davidson, 2006). An infestation with demodectic mange is diagnosed through acquisition and identification of *D. odocoilei* from pustular lesions on skin. (Scott et al., 2000). *D. odocoilei* mature from eggs, to larvae, to adults on white-tailed deer, with transmission from deer-to-deer via direct contact or contact with areas used by infected animals (Scott et al., 2000; Davidson, 2006). Demodectic mange presents no human health risks and is not considered a problem for managers of white-tailed deer (Davidson, 2006).

Other Common Arthropods

Nasal bots (*Cephenemyia* spp.), louse flies (*Lipoptena* spp. and *Neolipoptena* spp.), and lice (sucking lice, *Solenopotes binipilosus*; chewing lice, *Tricholipeurus lipeuroides* and *T. parallelus*) are commonly found on white-tailed deer throughout their range (Colwell et al., 2008). Clinical signs and lesions for nasal bots, louse flies, and lice have not been reported (Allan, 2001a; Durden, 2001; Colwell, 2001). The above-mentioned ectoparasites only occur on white-tailed deer and other cervids (Davidson, 2006). Nasal bots, louse flies, and lice are diagnosed through recovery and identification of ectoparasites in the oral cavity (larvae), skin, and in the axillary and inguinal regions with little hair, respectively (Allan, 2001a; Colwell, 2001; Durden, 2001). Nasal bots, louse flies, and lice pose no threat to human health and safety (Davidson 2006).

Conclusions

Several recommendations can be made related to white-tailed deer diseases and parasites. First, it would behoove white-tailed deer biologists and managers to become knowledgeable in the above-mentioned diseases and parasites. This, in part, is needed in distinguishing potentially harmful diseases to humans, other wildlife, and livestock from those which are benign. For example, hunters that encounter white-tailed deer displaying unusual behavior or who observe parasites on or in their harvested animal frequently have questions related to the safety of handling carcasses or consuming venison. Second, it would be wise for biologists and managers who handle large numbers of animals and carcasses to forge partnerships with disease diagnostic laboratories, such as the Southeastern Cooperative Wildlife Disease Study at the University of Georgia or other university, state, or federal wildlife health laboratories. Such collaborations will facilitate appropriate sample collection, storage, and transport, ensure rapid and accurate diagnoses, and enable reliable information exchange. Lastly, biologists and managers considering translocation of cervids should follow the guidelines of Corn and Nettles (2001) to minimize the threat of introducing high-risk ectoparasites and infectious diseases, which are often spread through contact. These guidelines include evaluating the health status of source populations, quarantines, physical examination and diagnostic testing, restrictions on translocation of animals from certain geographic areas or populations, and prophylactic treatment (Corn and Nettles, 2001).

REFERENCES

Addison, E. M. J., J. Hoeve, D. G. Joachim, and D. J. McLachlin. 1988. *Fascioloides magna* (Trematoda) and (Cestoda) from white-tailed deer. *Canadian Journal of Zoology* 66:1359–1364.

Allan, S. A. 2001a. Biting flies (Class Insecta: Order Diptera). In *Parasitic Diseases of Wild Mammals* (2nd edition), eds. W. M. Samuel, M. J. Pybus, and A. A. Kocan, 18–45. Ames, IA: Iowa State Press.

Allan, S. A. 2001b. Ticks (Class Arachnida: Order Acarina). In *Parasitic Diseases of Wild Mammals* (2nd edition), eds. W. M. Samuel, M. J. Pybus, and A. A. Kocan, 72–106. Ames, IA: Iowa State Press.

Anderson, R. C. 2001. Filarioid nematodes. In *Parasitic Diseases of Wild Mammals* (2nd edition), ed. W. M. Samuel, M. J. Pybus, and A. A. Kocan, 342–356. Ames, IA: Iowa State Press.

Anderson, R. C. and A. K. Prestwood. 1981. Lungworms. In *Diseases and Parasites of White-tailed Deer*, eds. W. R. Davidson, F. A. Hayes, V. F. Nettles, and F. E. Kellogg, 266–317. Tallahassee, FL: Tall Timbers Research Station.

Barker, R. W., A. L. Hoch, R. G. Buckner, and J. A. Hair. 1973. Hematological changes in white-tailed deer fawns, *Odocoileus virginianus*, infested with *Theileria* infected lone star ticks. *Journal of Parasitology* 59:1091–1098.

Baumann, C. D., W. R. Davidson, D. E. Roscoe, and K. Beheler-Amass. 2001. Intracranial abscessation in white-tailed deer of North America. *Journal of Wildlife Diseases* 37:661–670.

Bender, L. C., S. M. Schmitt, E. Carlson, J. B. Haufler, and D. E. Beyer, Jr. 2005. Mortality of Rocky Mountain elk in Michigan due to meningeal worm. *Journal of Wildlife Diseases* 41:134–140.

Bergstrom, R. C. 1975. Prevalence of *Dictyocaulus viviparous* infection in Rocky Mountain elk in Teton County, Wyoming. *Journal of Wildlife Diseases* 11:40–44.

Bolin, C. A. 2003. Leptospirosis. In *Zoo and Wild Animal Medicine* (5th edition), eds. M. E. Fowler and R. E. Miller, 699–702. St. Louis, MO: Elsevier Science.

Brillhart, D. B., L. B. Fox, J. P. Dubey, and S. J. Upton. 1994. Seroprevalence of *Toxoplasma gondii* in wild mammals in Kansas. *Journal of the Helminthological Society of Washington* 61:117–121.

Brown, R. N. and E. C. Burgess. 2001. Lyme Borreliosis. In *Infectious Diseases of Wild Mammals* (3rd edition), eds. E. S. Williams and I. K. Barker, 435–454. Ames, IA: Iowa State University Press.

Buxton, D. and R. M. Barlow. 1994. Cardiovascular system: Anthrax. In *Management and Diseases of Deer* (2nd edition), eds. T. L. Alexander and D. Buxton, 151–152. London: The Veterinary Deer Society.

Cantu, A., J. A. Ortega-S., J. Mosqueda, Z. Garcia-Vazquez, S. E. Henke, and J. E. George. 2007. Immunologic and molecular identification of *Babesia bovis* and *Babesia bigemina* in free-ranging white-tailed deer in northern Mexico. *Journal of Wildlife Diseases* 43:504–507.

Cantu, A., J. A. Ortega-S., J. Mosqueda, Z. Garcia-Vazquez, S. E. Henke, and J. E. George. 2008. Prevalence of infectious agents in free-ranging white-tailed deer in northeastern Mexico. *Journal of Wildlife Diseases* 44:1002–1007.

Carpenter, J. W., J. C. Freeny, and C. S. Patton. 1972. Occurrence of *Demodex* Owen 1843 on a white-tailed deer from Oklahoma. *Journal of Wildlife Diseases* 8:112–114.

Carreno, R. A. and M. W. Lankester. 1993. Additional information on the morphology of the Elaphostrongylinae (Nematoda: Protostrongylidae) of North American Cervidae. *Canadian Journal of Zoology* 71:592–600.

Carroll, J. F. and M. Kramer. 2003. Winter activity of *Ixodes scapularis* (Acari: Ixodidae) and the operation of deer-targeted tick control devices in Maryland. *Journal of Medical Entomology* 40:238–244.

Carroll, J. F., J. M. Pound, J. A. Miller, and M. Kramer. 2009. Sustained control of Gibson Island, Maryland, populations of *Ixodes scapularis* and *Amblyomma americanum* (Acari: Ixodidae) by community-administered 4-Poster deer self-treatment bait stations. *Vector-Borne and Zoonotic Diseases* 9:417–421.

Chase, C. C. L., L. J. Braun, P. Leslie-Steen, T. Graham, D. Miskimins, and J. F. Ridpath. 2008. Bovine viral diarrhea virus multiorgan infection in two white-tailed deer in southeastern South Dakota. *Journal of Wildlife Diseases* 44:753–759.

Choquette, L. P. E. 1970. Anthrax. In *Infectious Diseases of Wild Mammals*, eds. J. W. Davis, L. H. Karstad, and D. O. Trainer, 256–266. Ames, IA: Iowa State University Press.

Clark, R. C. and C. L. Gyles. 1993. Salmonella. In *Pathogenesis of Bacterial Infections in Animals* (2nd edition), eds. C. L. Gyles and C. O. Thoen, 133–153. Iowa, IA: Iowa State University Press.

Clifton-Hadley, R. S., C. M. Sauter-Louis, I. W. Lugton, R. Jackson, P. A. Durr, and J. W. Wilesmith. 2001. *Mycobacterium bovis* infections. In *Infectious Diseases of Wild Mammals* (3rd edition), eds. E. S. Williams and I. K. Barker, 340–361. Ames, IA: Iowa State University Press.

Colwell, D. D. 2001. Bot flies and warble flies (Order Diptera: Family Oestridae). In *Parasitic Diseases of Wild Mammals* (2nd edition), eds. W. M. Samuel, M. J. Pybus, and A. A. Kocan, 46–71. Ames, IA: Iowa State Press.

Colwell, D. D., D. Gray, K. Morton, and M. Pybus. 2008. Nasal bots and lice from white-tailed deer in southern Alberta, Canada. *Journal of Wildlife Diseases* 44:687–692.

Comer, J. A., W. R. Davidson, A. K. Prestwood, and V. F. Nettles. 1991. An update on the distribution of *Parelaphostrongylus tenuis* in the southeastern United States. *Journal of Wildlife Diseases* 27:348–354.

Cooksey, L. M., R. B. Davey, E. H. Ahrens, and J. E. George. 1989. Suitability of white-tailed deer as hosts for cattle fever ticks (Acari: Ixodidae). *Journal of Medical Entomology* 26:155–158.

Cormack Gates, C., B. Elkin, and D. Dragon. 2001. Anthrax. In *Infectious Diseases of Wild Mammals* (3rd edition), eds. E. S. Williams, and I. K. Barker, 396–412. Ames, IA: Iowa State University Press.

Corn, J. L. and V. F. Nettles. 2001. Health protocol for translocation of free-ranging elk. *Journal of Wildlife Diseases* 37:413–426.

Couvillion, C. E., V. F. Nettles, C. A. Rawlings, and R. L. Joyner. 1986. Elaeophorosis in white-tailed deer: Pathology of the natural disease and its relation to oral food impactions. *Journal of Wildlife Diseases* 22:214–223.

Davidson, W. R. 2006. *Field Manual of Wildlife Diseases in the Southeastern United States* (3rd edition). Athens, GA: Southeastern Cooperative Wildlife Disease Study.

Davidson, W. R., J. L. Blue, L. B. Flynn, S. M. Shea, R. L. Marchinton, and J. A. Lewis. 1987. Parasites, diseases and health status of sympatric populations of sambar deer and white-tailed deer in Florida. *Journal of Wildlife Diseases* 23:267–272.

Davidson, W. R., J. M. Crum, J. L. Blue, D. W. Sharp, and J. H. Phillips. 1985. Parasites, diseases and health status of sympatric populations of fallow deer and white-tailed deer in Kentucky. *Journal of Wildlife Diseases* 21:153–159.

Davidson, W. R. and W. L. Goff. 2001. Anaplasmosis. In *Infectious Diseases of Wild Mammals* (3rd edition), eds. E. S. Williams and I. K. Barker, 455–462. Ames, IA: Iowa State University Press.

Davidson, W. R., F. A. Hayes, V. F. Nettles, and F. E. Kellogg. 1981. *Diseases and Parasites of White-tailed Deer*. Tallahassee, FL: Tall Timbers Research Station.

Davidson, W. R., M. B. McGhee, V. F. Nettles, and L. C. Chapell. 1980. Haemonchosis in white-tailed deer in the southeastern United States. *Journal of Wildlife Diseases* 16:499–508.

Davidson, W. R. and V. F. Nettles. 1997. *Field Manual of Wildlife Diseases in the Southeastern United States* (2nd edition). Athens, GA: Southeastern Cooperative Wildlife Disease Study.

Davidson, W. R., V. F. Nettles, L. E. Hayes, E. W. Howerth, and C. E. Couvillion. 1990. Epidemiologic features of an intracranial abscessation/suppurative meningoencephalitis complex in white-tailed deer. *Journal of Wildlife Diseases* 26:460–467.

Debbie, J. G. 1965. Brain abscess in a white-tailed deer (*Odocoileus virginianus*). *Bulletin of the Wildlife Disease Association* 1:3–4.

de la Rua-Domenech, R. 2006. Human *Mycobacterium bovis* infection in the United Kingdom: Incidence, risks, control measures and review of the zoonotic aspects of bovine tuberculosis. *Tuberculosis* 86:77–109.

de Lisle, G. W., R. G. Bengis, S. M. Schmitt, and D. J. O'Brien. 2002. Tuberculosis in free-ranging wildlife: Detection, diagnosis and management. *Revue Scientifique Et Technique* 21:317–334.

Desch, C. E. and W. B. Nutting. 1974. *Demodex odocoilei* from the white-tailed deer, *Odocoileus virginianus*. *Canadian Journal of Zoology* 52:785–789.

de Vos, V. 2003. Anthrax. In *Zoo and Wild Animal Medicine* (5th edition), eds. M. E. Fowler and R. E. Miller, 696–699. St. Louis, MO: Elsevier Science.

Dubey, J. P. and K. Odening. 2001. Toxoplasmosis and related infections. In *Parasitic Diseases of Wild Mammals* (2nd edition), eds. W. M. Samuel, M. J. Pybus, and A. A. Kocan, 478–519. Ames, IA: Iowa State Press.

Duffy, M. S., T. A. Greaves, N. J. Keppie, and M. D. B. Burt. 2002. Meningeal worm is a long-lived parasitic nematode in white-tailed deer. *Journal of Wildlife Diseases* 38:448–452.

Durden, L. A. 2001. Lice (Phthiraptera). In *Parasitic Diseases of Wild Mammals* (2nd edition), eds. W. M. Samuel, M. J. Pybus, and A. A. Kocan, 3–17. Ames, IA: Iowa State Press.

Emerson, H. R. and W. T. Wright. 1968. Isolation of a *Babesia* in white-tailed deer. *Bulletin of the Wildlife Disease Association* 4:142–143.

Eriks, I. S., G. H. Palmer, T. C. McGuire, D. R. Allred and A. F. Barbet. 1989. Detection and quantitation of *Anaplasma marginale* in carrier cattle by using a nucleic acid probe. *Journal of Clinical Microbiology* 27:279–284.

Fischer, J. R., L. P. Hanson, J. R. Turk, M. A. Miller, W. H. Fales, and H. S. Gosser. 1995. An epizootic of hemorrhagic disease in white-tailed deer (*Odocoileus virginianus*) in Missouri: Necropsy findings and population impact. *Journal of Wildlife Diseases* 31:30–36.

Fitzgerald S. D., J. B. Kaneene, K. L. Butler et al. 2000. Comparison of postmortem techniques for the detection of *Mycobacterium bovis* in white-tailed deer (*Odocoileus virginianus*). *Journal of Veterinary Diagnostic Investigation* 12:322–327.

Flacke, G. L., M. J. Yabsley, B. A. Hanson, and D. E. Stallknecht. 2004. Hemorrhagic disease in Kansas: Enzootic stability meets epizootic disease. *Journal of Wildlife Diseases* 40:288–293.

Foreyt, W. J. 1992. Experimental *Fascioloides magna* infections of mule deer (*Odocoileus hemionus hemionus*). *Journal of Wildlife Diseases* 28:183–187.

Foreyt, W. J. 1996. Susceptibility of bighorn sheep (*Ovis canadensis*) to experimentally induced *Fascioloides magna* infections. *Journal of Wildlife Diseases* 32:556–559.

Foreyt, W. J. and C. W. Leathers. 1980. Experimental infection of domestic goats with *Fascioloides magna*. *American Journal of Veterinary Research* 41:883–884.

Foreyt, W. J. and D. O. Trainer. 1970. Seasonal parasitism changes in two populations of white-tailed deer in Wisconsin. *Journal of Wildlife Management* 44:758–764.

Forrester, D. J. 1992. *Parasites and Diseases of Wild Mammals in Florida*. Gainesville, FL: University Press of Florida.

Forrester, D. J. and R. L. Rausch. 1990. Cysticerci (Cestoda: Taeniidae) from white-tailed deer, *Odocoileus virginianus*, in southern Florida. *Journal of Parasitology* 76:583–585.

Friend, M. 1967. Skin tumors in New York deer. *Bulletin of the Wildlife Disease Association* 3:102–104.

Gajadhar, A., T. Steeves-Gurnsey, J. Kendall, M. Lankester, and M. Steen. 2000. Differentiation of dorsal-spined elaphostrongyline larvae by polymerase chain reaction amplification of ITS-2 rDNA. *Journal of Wildlife Diseases* 36:713–723.

Gatewood Hoen, A., L. G. Rollend, M. A. Papero, J. F. Carroll, T. J. Daniels, T. N. Mather, T. L. Schulze, K. C. Stafford, III, and D. Fish. 2009. Effects of tick control by acaricide self-treatment of white-tailed deer on host-seeking tick infection prevalence and entomologic risk for *Ixodes scapularis*-borne pathogens. *Vector-Borne and Zoonotic Diseases* 9:431–438.

Gaydos, J. K., J. M. Crum, W. R. Davidson, S. S. Cross, S. F. Owen, and D. E. Stallknecht. 2004. Epizootiology of an epizootic hemorrhagic disease outbreak in West Virginia. *Journal of Wildlife Diseases* 40:383–393.

Gaydos, J. K., W. R. Davidson, F. Elvinger, E. W. Howerth, M. Murphy, and D. E. Stallknecht. 2002a. Cross-protection between epizootic hemorrhagic disease virus serotypes 1 and 2 in white-tailed deer. *Journal of Wildlife Diseases* 38:720–728.

Gaydos, J. K., W. R. Davidson, F. Elvinger, D. G. Mead, E. W. Howerth, and D. E. Stallknecht. 2002b. Innate resistance to epizootic hemorrhagic disease in white-tailed deer. *Journal of Wildlife Diseases* 38:713–719.

Gaydos, J. K., D. E. Stallknecht, D. Kavanaugh, R. J. Olson, and E. R. Fuchs. 2002c. Dynamics of maternal antibodies to hemorrhagic disease viruses (Reoviridae: Orbivirus) in white-tailed deer. *Journal of Wildlife Diseases* 38:253–257.

Ghim, S. J., J. Newsome, J. Bell, J. P. Sundberg, R. Schlegel, and A. B. Jenson. 2000. Spontaneous regressing oral papillomas induce systemic antibodies that neutralize canine oral papillomavirus. *Experimental and Molecular Pathology* 68:147–151.

Gibbs, E. P. J. and E. C. Greiner. 1989. Bluetongue and epizootic hemorrhagic disease. In *The Arboviruses: Epidemiology and Ecology* (Vol. 2), ed. T. P. Monath, 39–70. Boca Raton, FL: CRC Press.

Goyal, S. M., L. D. Mech, and M. E. Nelson. 1992. Prevalence of antibody titers to *Leptospira* spp. in Minnesota white-tailed deer. *Journal of Wildlife Diseases* 28:445–448.

Haley, N. J., D. M. Seelig, M. D. Zabel, and E. A. Hoover. 2009. Detection of CWD prions in urine and saliva of deer by transgenic mouse bioassay. *PLoS One* 4:e4848.

Hayes, F. A., W. E., Greer, and S. B. Shotts. 1958. A progress report from the Southeastern Cooperative Deer Disease Study. *Transactions of the North American Wildlife Conference* 23:133–136.

Hibler, C. P. and J. L. Adcock. 1971. Elaeophorosis. In *Parasitic Diseases of Wild Mammals*, eds. J. W. Davis and R. C. Anderson, 263–278. Ames, IA: Iowa State Press.

Hibler, C. P., J. L. Adcock, R. W. Davis, and Y. Z. Adbelbaki. 1969. Elaeophorosis in deer and elk in the Gila National Forest, New Mexico. *Bulletin of the Wildlife Disease Association* 5:27–30.

Hoberg, E. P., A. A. Kocan, and L. G. Rickard. 2001. Gastrointestinal strongyles in wild ruminants. In *Parasitic Diseases of Wild Mammals* (2nd edition), eds. W. M. Samuel, M. J. Pybus, and A. A. Kocan, 193–227. Ames, IA: Iowa State Press.

Howe, D. L. 1981. Anaplasmosis. In *Infectious Diseases of Wild Mammals* (2nd edition), eds. J. W. Davis, L. H. Karstad, and D. O. Trainer, 407–414. Ames, IA: Iowa State University Press.

Howerth, E. W., D. E. Stallknecht, and P. D. Kirkland. 2001. Bluetongue, epizootic hemorrhagic disease and other orbivirus-related diseases. In *Infectious Diseases of Wild Mammals* (3rd edition), eds. E. S. Williams and I. K. Barker, 77–97. Ames, IA: Iowa State University Press.

Hugh-Jones, M. E. and V. de Vos. 2002. Anthrax and wildlife. *Revue Scientifique Et Technique* 21:358–383.

Humphreys, J. G., R. L. Stewart, and J. P. Dubey. 1995. Prevalance of *Toxoplasma gondii* antibodies in sera of hunter-killed white-tailed deer in Pennsylvania. *American Journal of Veterinary Research* 56:172–173.

Hunter, D. L. 1996. Tuberculosis in free-ranging, semi free-ranging and captive cervids. *Revue Scientifique Et Technique* 15:171–181.

Hyslop, N. S. T. G. 1980. Dermatophilosis (Streptothricosis) in animals and man. *Comparative Immunology, Microbiology and Infectious Diseases* 2:389–404.

Isaza, R. 2003. Tuberculosis in all taxa. In *Zoo and Wild Animal Medicine* (5th edition), eds. M. E. Fowler and R. E. Miller, 689–696. St. Louis, MO: Elsevier Science.

Jacques, C. N. and J. A. Jenks. 2004. Distribution of meningeal worm (*Parelaphostrongylus tenuis*) in South Dakota. *Journal of Wildlife Diseases* 40:133–136.

Jacques, C. N., J. A. Jenks, M. B. Hildreth, R. J. Schauer, and D. D. Johnson. 2001. Demodicosis in a white-tailed deer (*Odocoileus virginianus*) in South Dakota. *Prairie Naturalist* 33:221–226.

Jenson, A. B., M. C. Jenson, L. Cowsert, S. J. Ghim, and J. P. Sundberg. 1997. Multiplicity of uses of monoclonal antibodies that define papillomavirus linear immunodominant epitopes. *Immunologic Research* 16:115–119.

Jones, A. and M. J. Pybus. 2001. Taeniasis and echinococcosis. In *Parasitic Diseases of Wild Mammals* (2nd edition), eds. W. M. Samuel, M. J. Pybus, and A. A. Kocan, 150–192. Ames, IA: Iowa State Press.

Jordan, R. A., T. L. Schulze, and M. B. Jahn. 2007. Effects of reduced deer density on the abundance of *Ixodes scapularis* (Acari: Ixodidae) and Lyme disease incidence in a northern New Jersey endemic area. *Journal of Medical Entomology* 44:752–757.

Karns, G. R., R. A. Lancia, C. S. de Perno, M. C. Conner, and M. K. Stoskopf. 2009. Intracranial abscessation as a natural mortality factor for adult male white-tailed deer (*Odocoileus virginianus*) in Kent County, Maryland, USA. *Journal of Wildlife Diseases* 45:196–200.

Kellogg, F. E., T. P. Kistner, R. K. Strickland, and R. R. Gerrish. 1971. Arthropod parasites collected from white-tailed deer. *Journal of Medical Entomology* 8:495–498.

Kellogg, F. E., A. K. Prestwood, and R. E. Noble. 1970. Anthrax epizootic in white-tailed deer. *Journal of Wildlife Diseases* 6:226–228.

Ketz-Riley, C. J. 2003. Salmonellosis and shigellosis. In *Zoo and Wild Animal Medicine* (5th edition), eds. M. E. Fowler and R. E. Miller, 686–689. St. Louis, MO: Elsevier Science.

Kieser, S. T., I. S. Eriks, and G. H. Palmer. 1990. Cyclic rickettsia during persistent *Anaplasma marginale* infection in cattle. *Infection and Immunity* 58:1117–1119.

Kocan, K. M, J. de la Fuents, A. A. Guglielmone, and R. D. Melendez. 2003. Antigens and alternatives for control of *Anaplasma marginale* infection in cattle. *Clinical Microbiology Reviews* 16:698–712.

Kocan, A. A. and K. A. Waldrup. 2001. Piroplasms (*Theileria* spp., *Cytauxzoon* spp., and *Babesia* spp.). In *Parasitic Diseases of Wild Mammals* (2nd edition), eds. W. M. Samuel, M. J. Pybus, and A. A. Kocan, 524–536. Ames, IA: Iowa State Press.

Koller, L. D. and C. Olson. 1972. Attempted transmission of warts from man, cattle, and horses and of deer fibroma to selected hosts. *Journal of Investigative Dermatology* 58:366–368.

Kreier, J. P., M. Ristic, and A. M. Watrach. 1962. *Theileria* sp. in a deer in the United States. *American Journal of Veterinary Research* 23:657–662.

Kuttler, K. L. 1981. Anaplasmosis. In *Diseases and Parasites of White-tailed Deer*, eds. W. R. Davidson, F. A. Hayes, V. F. Nettles, and F. E. Kellogg, 126–135. Tallahassee, FL: Tall Timbers Research Station.

Lane, R. S., J. Piesman, and W. Burgdorfer. 1991. Lyme Borreliosis: Relation of its causative agent to its vectors and hosts in North America and Europe. *Annual Review of Entomology* 36:587–609.

Lankester, M. W. 2001. Extrapulmonary lungworms of cervids. In *Parasitic Diseases of Wild Mammals* (2nd edition), eds. W. M. Samuel, M. J. Pybus, and A. A. Kocan, 228–278. Ames, IA: Iowa State Press.

Larkin, J. L., K. J. Alexy, D. C. Bolin et al. 2003. Meningeal worm in a reintroduced elk population in Kentucky. *Journal of Wildlife Diseases* 39:588–592.

Leighton, F. A. 2001. Dermatophilosis. In *Infectious Diseases of Wild Mammals* (3rd edition), eds. E. S. Williams and I. K. Barker, 489–491. Ames, IA: Iowa State University Press.

Libke, K. G. and A. M. Walton. 1975. Presumptive paratuberculosis in a Virginia white-tailed deer. *Journal of Wildlife Diseases* 11:552–553.
Lichtenfels, J. R., E. P. Hoberg, and D. S. Zarlenga. 1997. Systematics of gastrointestinal nematodes of domestic ruminants: Advances 1992–1995, and proposals for future research. *Veterinary Parasitology* 72:225–245.
Lim, P. S., A. B. Jenson, L. Cowsert et al. 1990. Distribution and specific identification of papillomavirus major capsid protein epitopes by immunocytochemistry and epitope scanning of synthetic peptides. *Journal of Infectious Diseases* 162:1263–1269.
Lindsay, D. S., B. L. Blagburn, J. P. Dubey, and W. H. Mason. 1991. Prevalence and isolation of *Toxoplasma gondii* from white-tailed deer in Alabama. *Journal of Parasitology* 77:62–64.
Mackintosh, C., J. C. Haigh, and F. Griffin. 2002. Bacterial diseases of farmed deer and bison. *Revue Scientifique Et Technique* 21:249–263.
Madden, D. J., T. R. Spraker, and W. J. Adrian. 1991. *Elaeophora schneideri* in moose (*Alces alces*) from Colorado. *Journal of Wildlife Diseases* 27:340–341.
Manning, E. J. B. 2001. *Myobacterium avium* subspecies *Paratuberculosis:* A review of current knowledge. *Journal of Zoo and Wildlife Medicine* 32:293–304.
Mason, P. C. 1985. Biology and control of the lungworm *Dictyocaulus viviparous* in farmed red deer in New Zealand. *Royal Society of New Zealand Bulletin* 22:119–121.
Mathiason C. K., J. G. Powers, S. J. Dahmes et al. 2006. Infectious prions in the saliva and blood of deer with chronic wasting disease. *Science* 314:133–136.
Matschke, G. H., K. A. Fagerstone, R. F. Harlow et al. 1984. Population influences. In *White-tailed Deer Ecology and Management*, ed. L. K. Halls, 169–188. Harrisburg, PA: Stackpole Books.
Miller, D. L., Z. A. Radi, C. Baldwin, and D. Ingram. 2005. Fatal West Nile virus infection in a white-tailed deer (*Odocoileus virginianus*). *Journal of Wildlife Diseases* 41:246–249.
Miller, M. W., E. S. Williams, N. T. Hobbs, and L. L. Wolfe. 2004. Environmental sources of prion transmission in mule deer. *Emerging Infectious Diseases* 10:1003–1006.
Miller, N. J., W. A. Thomas, and T. N. Mather. 2009. Evaluating a deer-targeted Acaricide applicator for area-wide suppression of blacklegged ticks, *Ixodes scapularis* (Acari: Ixodidae), in Rhode Island. *Vector-Borne and Zoonotic Diseases* 9:401–406.
Mörner, T. 2001. Salmonellosis. In *Infectious Diseases of Wild Mammals* (3rd edition), eds. E. S. Williams and I. K. Barker, 505–507. Ames, IA: Iowa State University Press.
Munro, R. 1988. Pulmonary parasites: Pathology and control. In *Management and Health of Farmed Deer*, ed. H. W. Reid, 27–42. Norwell, MA: Kluwer Academic Publishers.
Murphy, M. D., B. A. Hanson, E. W. Howerth, and D. E. Stallknecht. 2006. Molecular characterization of epizootic hemorrhagic disease virus serotype 1 associated with a 1999 epizootic in white-tailed deer in the eastern United States. *Journal of Wildlife Diseases* 42:616–624.
Murray, C. J. 1991. Salmonellae in the environment. *Revue Scientifique Et Technique* 10:765–785.
Nettles, V. F. and D. E. Stallknecht. 1992. History and progress in the study of hemorrhagic disease of deer. *Transactions of the North American Wildlife and Natural Resources Conference* 57:499–516.
O'Brien, D. J., S. D. Fitzgerald, T. J. Lyon et al. 2001. Tuberculous lesions in free-ranging white-tailed deer in Michigan. *Journal of Wildlife Diseases* 37:608–613.
O'Brien, D. J., S. M. Schmitt, J. S. Fierke et al. 2002. Epidemiology of *Mycobacterum bovis* in free-ranging white-tailed deer, Michigan, USA, 1995–2000. *Preventative Veterinary Medicine* 54:47–63.
O'Brien D. J., S. M. Schmitt, S. D. Fitzgerald, D. E. Berry, G. J. Hickling. 2006. Managing the wildlife reservoir of *Mycobacterium bovis*: The Michigan, USA, experience. *Veterinary Microbiology* 112:313–323.
Palmer, M. V., W. R. Waters, and D. L. Whipple. 2002. Aerosol exposure of white-tailed deer (*Odocoileus virginianus)* to *Mycobacterium bovis. Journal of Wildlife Diseases* 39:817–823.
Palmer, M. V., W. R. Waters, and D. L. Whipple. 2004. Shared feed as a means of deer-to-deer transmission of *Mycobacterium bovis. Journal of Wildlife Diseases* 40:87–91.
Palmer, M. V., D. L. Whipple, J. B. Payeur et al. 2000. Naturally occurring tuberculosis in white-tailed deer. *Journal of the American Veterinary Medical Association* 216:1921–1924.
Palmer, M. V., D. L. Whipple, and W. R. Waters. 2001. Experimental deer-to-deer transmission of *Mycobacterium bovis. American Journal of Veterinary Research* 62:692–696.
Passler, T., P. H. Walz, S. S. Ditchkoff et al. 2009. Cohabitation of pregnant white-tailed deer and cattle persistently infected with bovine viral diarrhea virus results in persistently infected fawns. *Veterinary Microbiology* 134:262–267.

Passler, T., P. H. Walz, S. S. Ditchkoff, M. D. Givens, H. S. Maxwell, and K. V. Brock. 2007. Experimental persistent infection with bovine viral diarrhea virus in white-tailed deer. *Veterinary Microbiology* 122:350–356.

Pence, D. B. 1991. Elaeophorosis in wild ruminants. *Bulletin of the Society for Vector Ecology* 16:149–160.

Pence, D. B. and G. G. Gray. 1981. Elaeophorosis in Barbary sheep and mule deer from the Texas panhandle. *Journal of Wildlife Diseases* 17:49–56.

Perry, B. D., D. K. Nickols, and E. S. Cullon. 1985. *Babesia odocoilei* Emerson and Wright, 1970, in white-tailed deer, *Odocoileus virginianus*, in Virginia. *Journal of Wildlife Diseases* 21:149–152.

Pfister, H. 1987. Papillomaviruses: General description, taxonomy, and classification. In *The Papovaviridae* (volume 2), eds. N. P. Salzman and P. M. Howley, 1–38. New York, NY: Plenum Press.

Pound, J. M., J. A. Miller, J. E. George, and C. A. Lemeilleur. 2000. The '4-Poster' passive topical treatment device to apply acaricide for controlling ticks (Acari: Ixodidae) feeding on white-tailed deer. *Journal of Medical Entomology* 37:588–594.

Prestwood, A. K. 1970. Neurologic disease in a white-tailed deer massively infected with meningeal worm (*Pneumostrongylus tenuis*). *Journal of Wildlife Diseases* 6:84–86.

Prestwood, A. K. and F. E. Kellogg. 1971. Naturally occurring haemonchosis in a white-tailed deer. *Journal of Wildlife Diseases* 7:133–134.

Prestwood, A. K. and S. R. Pursglove. 1977. Prevalence and distribution of *Setaria yehi* in southeastern white-tailed deer. *Journal of the American Veterinary Medical Association* 171:933–935.

Prestwood, A. K. and S. R. Pursglove. 1981. Gastrointestinal nematodes. In *Diseases and Parasites of White-tailed Deer*, eds. W. R. Davidson, F. A. Hayes, V. F. Nettles, and F. E. Kellogg, 318–349. Tallahassee, FL: Tall Timbers Research Station.

Prestwood, A. K. and T. R. Ridgeway. 1972. Elaeophorosis in white-tailed deer of the southeastern USA: Case report and distribution. *Journal of Wildlife Diseases* 8:233–236.

Pybus, M. J. 1990. Survey of hepatic and pulmonary helminths of wild cervids in Alberta, Canada. *Journal of Wildlife Diseases* 26:453–459.

Pybus, M. J. 2001. Liver flukes. In *Parasitic Diseases of Wild Mammals* (2nd edition), eds. W. M. Samuel, M. J. Pybus, and A. A. Kocan, 121–149. Ames, IA: Iowa State Press.

Reid, H. W. 1994. Leptospirosis. In *Management and Diseases of Deer* (2nd edition), eds. T. L. Alexander and D. Buxton, 143–144. London: The Veterinary Deer Society.

Richard, J. L. 1981. Dermatophilosis. In *Infectious Diseases of Wild Mammals*, eds. J. W. Davis, L. H. Karstad, and D. O. Trainer, 339–346. Ames, IA: Iowa State University Press.

Ridpath, J. F., C. S. Mark, C. C. L. Chase, A. C. Ridpath, and J. D. Neill. 2007. Febrile response and decrease in circulating lymphocytes following acute infection of white-tailed deer fawns with either a BVDV1 or a BVDV2 strain. *Journal of Wildlife Diseases* 43:653–659.

Robinson, R. M. 1981. Salmonellosis. In *Diseases and Parasites of White-tailed Deer*, eds. W. R. Davidson, F. A. Hayes, V. F. Nettles, and F. E. Kellogg, 155–160. Tallahassee, FL: Tall Timbers Research Station.

Robinson, R. M., L. P. Jones, T. J. Galvin, and G. M. Harwell. 1978. Elaeophorosis in sika deer in Texas. *Journal of Wildlife Diseases* 14:137–141.

Roscoe, D. E., R. C. Lund, M. A. Gordon, and I. F. Salkin. 1975. Spontaneous dermatophilosis in twin white-tailed deer fawns. *Journal of Wildlife Diseases* 11:398–401.

Rosen, R. N. 1981. Miscellaneous bacterial and mycotic diseases. In *Diseases and Parasites of White-tailed Deer*, eds. W. R. Davidson, F. A. Hayes, V. F. Nettles, and F. E. Kellogg, 175–192. Tallahassee, FL: Tall Timbers Research Station.

Roth, E. E. 1970. Leptospirosis. In *Infectious Diseases of Wild Mammals*, eds. J. W. Davis, L. H. Karstad, and D. O. Trainer, 293–303. Ames, IA: Iowa State University Press.

Roy, P. 1996. Orbiviruses and their replication. In *Fields Virology* (3rd edition), eds. B. N. Fields, D. M. Knipe, and P. M. Howley, 1709–1734. Philadelphia, PA: Lippencott-Raven.

Salkin, I. F. and M. A. Gordon. 1981. Dermatophilosis. In *Diseases and Parasites of White-tailed Deer*, eds. W. R. Davidson, F. A. Hayes, V. F. Nettles, and F. E. Kellogg, 168–173. Tallahassee, FL: Tall Timbers Research Station.

Salkin, I. F., M. A. Gordon, and W. B. Stone. 1983. Cutaneous granules associated with dermatophilosis in a white-tailed deer. *Journal of Wildlife Diseases* 19:361–363.

Samuel, W. M., M. J. Pybus, and A. A. Kocan. 2001. *Parasitic Diseases of Wild Mammals* (2nd edition). Ames, IA: Iowa State Press.

Samuel, W. M., M. J. Pybus, D. A. Welch, and C. J. Wilke. 1992. Elk as a potential host for meningeal worm: Implications for transmission. *Journal of Wildlife Management* 56:629–639.
Schmitt, S. M., T. M. Cooley, S. D. Fitzgerald et al. 2007. An outbreak of eastern equine encephalitis virus in free-ranging white-tailed deer in Michigan. *Journal of Wildlife Diseases* 43:635–644.
Schmitt, S. M., S. D. Fitzgerald, T. M. Cooley et al. 1997. Bovine tuberculosis in free-ranging white-tailed deer from Michigan. *Journal of Wildlife Diseases* 33:749–758.
Schoelkopf, L., C. E. Hutchinson, K. G. Bendele et al. 2005. New ruminant hosts and wider geographic range identified for *Babesia odocoilei* (Emerson and Wright 1970). *Journal of Wildlife Diseases* 41:683–690.
Schulze, T. L., R. A. Jordan, R. W. Hung, and C. J. Schulze. 2009. Effectiveness of the 4-Poster passive topical treatment device in the control of *Ixodes scapularis* and *Amblyomma americanum* (Acari: Ixodidae) in New Jersey. *Vector-Borne and Zoonotic Diseases* 9:389–400.
Schurr, K., F. Rabalais, and W. Terwilliger. 1988. *Cysticercus tenuicollis*: A new state record for Ohio. *Ohio Journal of Science* 88:104–105.
Scott, D. W., W. H. Miller, and C. E. Griffin. 2000. *Muller and Kirk's Small Animal Dermatology* (6th edition). Philadelphia, PA: Elsevier.
Seidel, B., A. Thomzig, A. Buschmann et al. 2007. Scrapie agent (Strain 263K) can transmit disease via the oral route after persistence in soil over years. *PLoS One* 5:1–8.
Seward, N. W., G. E. Phillips, J. F. Duquette, and K. C. VerCauteren. 2007. A frightening device for deterring deer from cattle feed. *Journal of Wildlife Management* 71:271–276.
Shad, G., W. C. Wilson, J. O. Meacham, and J. F. Evermann. 1997. Bluetongue virus isolation detection: A safer reverse-transcriptase polymerase chain reaction for prediction of viremia in sheep. *Journal of Veterinary Diagnostic Investigation* 9:118–124.
Shapiro, E.D. and M. A. Gerber. 2000. Lyme disease. *Clinical Infectious Diseases* 31:533–542.
Shotts, E. B. 1981. Leptospirosis. In *Diseases and Parasites of White-tailed Deer*, eds. W. R. Davidson, F. A. Hayes, V. F. Nettles, and F. E. Kellogg, 138–147. Tallahassee, FL: Tall Timbers Research Station.
Sigurdson, C. J. 2008. A prion disease of cervids: Chronic wasting disease. *Veterinary Research* 39:41. http://www.vetres.org
Sleeman, J. M., E. J. B. Manning, J. H. Rohm et al. 2009. Johne's disease in a free-ranging white-tailed deer from Virginia and subsequent surveillance for *Myobacterium avium* subspecies *paratuberculosis*. *Journal of Wildlife Diseases* 45:201–206.
Sneath, P. H., N. S. Mair, and M. E. Sharpe. 1986. *Bergey's Manual of Systematic Bacteriology* (Vol. 2). Baltimore, MD: Williams & Wilkins.
Soltys, M. A., C. E. Andress and A. L. Fletch. 1967. Johne's disease in a moose (*Alces alces*). *Bulletin of the Wildlife Disease Association* 3:183–184.
Sommer, C. 1996. Digital image analysis and identification of eggs from bovine parasitic nematodes. *Journal of Helminthology* 70:143–151.
Spindler, L. A., R. W. Allen, L. S. Diamond, and J. C. Lotz. 1958. *Babesia* in white-tailed deer. *Journal of Protozoology* 5:8.
Spraker, T. R. 2003. Spongiform encephalopathy. In *Zoo and Wild Animal Medicine* (5th edition), eds. M. E. Fowler and R. E. Miller, 741–745. St. Louis, MO: Elsevier Science.
Spraker, T. R., T. L. Gidlewski, A. Balachandran, K. C. VerCauteren, L. Creekmore, and R. D. Munger. 2006. Detection of PrPCWD in postmortem rectal lymphoid tissues in Rocky Mountain elk (*Cervus elaphus nelsoni)* infected with chronic wasting disease. *Journal of Veterinary Diagnostic Investigation* 18:553–557.
Stabel, J. R. 1998. Johne's disease: A hidden threat. *Journal of Dairy Science* 81:283–288.
Strickland, R. K., R. R. Gerrish, and J. S. Smith. 1981. Arthropods. In *Diseases and Parasites of White-tailed Deer*, eds. W. R. Davidson, F. A. Hayes, V. F. Nettles, and F. E. Kellogg, 363–389. Tallahassee, FL: Tall Timbers Research Station.
Sundberg, J. P., S. J. Ghim, M. Van Ranst, and A. B. Jenson. 1997. Nonhuman papillomaviruses: Host range, pathology, epitope conservation, and new vaccine approaches. In *Spontaneous Animal Tumors: A Survey*, eds. L. Rossi, R. Richardson, and J. Harshbarger, 33–40. Milon: Press Point di Abbiategrasso.
Sundberg, J. P., M. Van Ranst, R. D. Burk, and A. B. Janson. 1996. The nonhuman (animal) papillomaviruses: Host range, epitope conservation, and molecular diversity. In *Human Papillomavirus Infections in Dermatology and Venereology*, eds. G. Gross and G. von Krogh, 47–68. Boca Raton, FL: CRC Press.

Sundberg, J. P., M. Van Ranst, and A. B. Jenson. 2001. Papillomavirus infections. In *Infectious Diseases of Wild Mammals* (3rd edition), eds. E. S. Williams and I. K. Barker, 223–231. Ames, IA: Iowa State University Press.

Tate, C. M., E. W. Howerth, D. E. Stallknecht, A. B. Allison, J. R. Fischer, and D. G. Mead. 2005. Eastern equine encephalitis in free-ranging white-tailed deer (*Odocoileus virginianus*). *Journal of Wildlife Diseases* 41:241–245.

Telford, S. R. III, T. N. Mather, S. I. Moore, M. L. Wilson, and A. Spielman. 1988. Incompetence of deer as reservoirs of the Lyme disease spirochete. *American Journal of Tropical Medicine and Hygiene* 39:105–109.

Titche, A. R., A. K. Prestwood, and C. P. Hibler. 1979. Experimental infections of white-tailed deer with *Elaeophora schneideri*. *Journal of Wildlife Diseases* 15:273–280.

Trainer, D. O., R. P. Hanson, E. P. Pope, and E. A. Carbrey. 1963. The role of deer in the epizootiology of leptospirosis in Wisconsin. *American Journal of Veterinary Research* 24:159–167.

Turner, J. C. and J. Cano. 2008. Demodectic mange in a white-tailed deer from Walker County, Texas. *Journal of Medical Entomology* 45:572–575.

Vanek, J. A., J. P. Dubey, P. Thulliez, M. R. Riggs, and B. E. Stromberg. 1996. Prevalence of *Toxoplasma gondii* antibodies in hunter-killed white-tailed deer (*Odocoileus virginianus*) in four regions of Minnesota. *Journal of Parasitology* 82:41–44.

Van Kruiningen, H. 2001. Lack of support for a common etiology in Johne's disease of animals and Crohn's disease in humans. *Inflammatory Bowel Diseases* 5:183–191.

Van Ness, G. B. 1981. Anthrax. In *Diseases and Parasites of White-tailed Deer*, eds. W. R. Davidson, F. A. Hayes, V. F. Nettles, and F. E. Kellogg, 161–167. Tallahassee, FL: Tall Timbers Research Station.

VerCauteren K. C., T. C. Atwood, T. J. DeLiberto et al. 2008b. Sentinel-based surveillance of coyotes to detect bovine tuberculosis, Michigan. *Emerging Infectious Diseases* 14:1862–1869.

VerCauteren, K. C., M. J. Lavelle, and S. E. Hygnstrom. 2006. Fences and deer-damage management: A review of designs and efficacy. *Wildlife Society Bulletin* 34:191–200.

VerCauteren, K. C., M. J. Lavelle, and G. E. Phillips. 2008a. Livestock protection dogs for deterring deer from cattle and feed. *Journal of Wildlife Management* 72:1443–1448.

Waldrup, K., A. E. Collisson, S. E. Bentsen, C. K. Winkler, and G. G. Wagner. 1989a. Prevalence of erythrocytic protozoa and serologic reactivity to selected pathogens in deer in Texas. *Preventive Veterinary Medicine* 7:49–58.

Waldrup, K., A. A. Kocan, R. W. Barker, and G. G. Wagner. 1990. Transmission of *Babesia odocoilei* in white-tailed deer (*Odocoileus virginianus*) by *Ixodes scapularis* (Acari: Ixodidae). *Journal of Wildlife Diseases* 26:390–391.

Waldrup, K., A. A. Kocan, T. Quereshi, D. S. Davis, D. Baggett, and G. G. Wagner. 1989b. Serologic prevalence and isolation of *Babesia odocoilei* among white-tailed deer (*Odocoileus virginianus*) in Texas and Oklahoma. *Journal of Wildlife Diseases* 25:194–201.

Wasel, S. M., W. M. Samuel, and V. Crichton. 2003. Distribution and ecology of meningeal worm, *Parelaphostrongylus tenuis* (Nematoda), in northcentral North America. *Journal of Wildlife Diseases* 39:338–346.

Watson, J. D. and J. W. Littlefield. 1960. Some properties of DNA from Shope papillomavirus. *Journal of Molecular Biology* 2:161–165.

Weinmann, C. J., J. R. Anderson, W. M. Longhurst, and G. Connolly. 1973. Filarial worms of Columbian black-tailed deer in California. 1. Observations in the vertebrate host. *Journal of Wildlife Diseases* 9:213–220.

Whitlock, S. C. 1939. The prevalence of disease and parasites of white-tailed deer. *Transactions of the North American Wildlife Conference* 4:244–249.

Wilkins, M. J., P. C. Bartlett, B. Frawley, D. J. O'Brien, C. E. Miller, and M. L. Boulton. 2003. *Mycobacterium bovis* (bovine TB) exposure as a recreational risk for hunters: Results of a Michigan hunter survey, 2001. *The International Journal of Tuberculosis and Lung Disease* 7:1001–1009.

Williams, E. S. 2001. Paratuberculosis and other mycobacterial diseases. In *Infectious Diseases of Wild Mammals* (3rd edition), eds. E. S. Williams and I. K. Barker, 361–371. Ames, IA: Iowa State University Press.

Williams, E. S. and I. K. Barker. 2001. *Infectious Diseases of Wild Mammals* (3rd edition). Ames, IA: Iowa State Press.

Williams, E. S., J. K. Kirkwood, and M. W. Miller. 2001. Transmissible spongiform encephalopathies. In *Infectious Diseases of Wild Mammals* (3rd edition), eds. E. S. Williams and I. K. Barker, 392–301. Ames, IA: Iowa State University Press.

Williams, E. S. and M. W. Miller. 2002. Chronic wasting disease in deer and elk in North America. *Revue Scientifique Et Technique* 21:305–316.

Williams, E. S., A. C. Pier. and R. W. Wilson. 1984. Dermatophilosis in a mule deer, *Odocoileus hemionus* (Rafinesque), from Wyoming. *Journal of Wildlife Diseases* 20:236–238.

Wilson, M. L., A. M. Ducey, T. S. Litwin, T. A. Gavin, and A. Spielman. 1990. Microgeographic distribution of immature *Ixodes dammini* ticks correlated with that of deer. *Medical and Veterinary Entomology* 4:151–159.

Woldehiwet, Z. 2006. *Anaplasma phagocytophilum* in ruminants in Europe. *Annals of the New York Academy of Sciences* 1078:446–460.

Wolfe, L. L, T. R. Spraker, L. Gonzalez et al. 2007. PrPCWD in rectal lymphoid tissue of deer (*Odocoileus* spp.). *Journal of General Virology* 88:2078–2082.

Yabsley, M. J., T. C. Quick, and S. E. Little. 2005. Theileriosis in a white-tailed deer (*Odocoileus virginianus*) fawn. *Journal of Wildlife Diseases* 41:806–809.

Yuill, T. M. and C. Seymour. 2001. Arbovirus infections. In *Infectious Diseases of Wild Mammals* (3rd edition), eds. E. S. Williams and I. K. Barker, 98–118. Ames, IA: Iowa State University Press.

Zarlenga, D. S., F. Stringfellow, M. Nobary, and J. R. Lichtenfels. 1994. Cloning and characterizations of ribosomal RNA genes from 3 species of *Haemonchus* (Nematoda: Trichostronglyoidea) and identification of PCR primers for rapid differentiation. *Experimental Parasitology* 78:28–36.

8

Predator–Prey Relationships

Warren Ballard

CONTENTS

The relationships of prey species such as deer and their predators have been and continue to be highly controversial among scientists and lay persons alike. Several recent reviews of predation and its effects on ungulate populations have helped to shed some light on this topic and bring new findings to the forefront. Reviews have been conducted by Mech (1984) for white-tailed deer, Ballard and Van Ballenberghe (1998a) for moose, Bergerud (1983, 1988) and Bergerud et al. (1984) for caribou, and Ballard et al. (2001) for mule and black-tailed deer. Here I provide an update of scientific studies on white-tailed deer since Mech's (1984) summary.

The principal predators of white-tailed deer include coyotes, black bears, bobcats, foxes, grizzly bears, mountain lions, and wolves. Some of these predators focus primarily on fawns, others on adults, and many predators consume both age classes of white-tailed deer. Predators can be generally classified into two types; obligate carnivores are those that must consume flesh for sustenance and include wolves and cats, while facultative predators consume flesh and a wide variety of other foods such as vegetative material. Bears, coyotes, and foxes fall into the latter category.

Historical Perspective on White-tailed Deer Abundance

McCabe and McCabe (1984) (also see Chapter 11 for a more detailed account) provided an account of anecdotal historical records of white-tailed deer numbers and distribution prior to and after European settlement. Many biologists consider today's abundance of white-tailed deer to be one of the best examples of wildlife management successes, while others argue that it is simply a reflection of the species' ability to respond to anthropogenic-induced changes in the environment. In many respects, both perspectives may be correct.

McCabe and McCabe (1984) provided numerous accounts of deer abundance at the time of European settlement. However, they also show that deer distribution was not even and there appeared to be wide discrepancies in numbers depending on time period and area. Deer in northern climates of North America apparently experienced dramatic fluctuations in population size due to severe winters and predation. McCabe and McCabe (1984) found some indications that deer were more numerous before European colonization than today. Deer populations declined to low numbers in the late 1800s due to excessive human exploitation. The advent of hunting seasons and bag limits and an agricultural shift to the Great Plains helped reverse this trend. After the establishment of hunting regulations, the expansion of human settlement in the eastern half of the continent with moderate amounts of agriculture and forestry undoubtedly helped create favorable conditions for whitetails by setting back plant succession, reducing frequency of fires, and reducing or extirpating predators. Currently, in many areas of eastern North America, and in urban situations in particular, deer are perceived by many individuals as a nuisance species and a threat to ecosystem diversity. Deer survival in urban areas is typically high due to a lack of hunting and natural predators, and collisions with automobiles are often the largest source of mortality (Etter et al., 2002).

A different historical situation may have existed in the western half of North America. McCabe and McCabe (1984) indicated that there was general consensus among early explorers that white-tailed deer were not abundant in the western flatlands and deserts. Kay (1995) and Kay et al. (2000) speculated that hunting by Native Americans and predation kept ungulates at low numbers in the Rocky Mountains.

Fires were also a frequent occurrence in these areas. When wolves were abundant it is likely that deer existed between the boundaries of wolf territories. A similar prey distribution appeared to exist in relation to the distribution of Native American Indian tribes. Changes in vegetation and predator populations allowed white-tailed deer populations to expand. Currently, deer are now expanding their distributions into urban areas all across North America.

Definitions

Our understanding of the effects of predation on wildlife has been confounded in large part by the complexities involved with predator–prey relationships, and also by inconsistent use of terminology, such as control, regulation, and limiting factors (Messier, 1991; Sinclair, 1991; Skogland, 1991; Boutin, 1992; Dale et al., 1994; Van Ballenberghe and Ballard, 1994; Ballard et al., 2001). In this chapter, I use the definitions provided by Messier (1991). I consider the terms control and regulation as being synonymous. Limiting factors are any factors that impede the population rate of growth and can include both density-dependent and density-independent factors. Regulating factors are a subset of limiting factors and are density dependent. Populations that start to reach equilibrium with their habitat are depressed by regulating factors. Thus, regulating factors are density-dependent processes.

Bartmann et al. (1992) defined the terms compensatory and additive mortality. Compensatory mortality factors generally replace other mortality factors so that total mortality remains similar. Generally this occurs to the right of the inflection point on the logistic population curve (Figure 8.1). Additive mortality, on the other hand, is an addition to total mortality and there is no compensation. Such mortality is more prominent to the left of the inflection point of the logistic curve. However, this view only represents a theoretical approach. In fact, these types of mortality probably occur on both sides of the population curve, but vary in magnitude depending on where the population occurs on the growth curve, particularly in highly variable environments (Mackie et al., 1998).

For this chapter, I define habitat carrying capacity as forage carrying capacity (K) (MacNab, 1985). At this population level births generally equal deaths, and the population does not increase, but theoretically is in equilibrium with its habitat. Deer populations near K are particularly susceptible to environmental influences such as severe winter weather or drought. Another misuse of terms has involved the use predation rate and kill rate (Messier, 1995a,b). Predation rate is derived from the functional response (i.e., kill rate per predator in relation to prey density) and the predator's numerical response (i.e., changes in predator population in response to prey density) (Messier, 1995a,b).

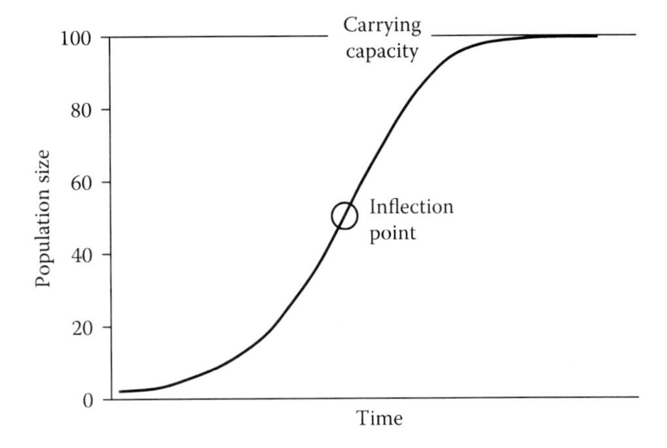

FIGURE 8.1 Growth of deer populations through time can be described by a logistic growth model in which the net increase in population size is more each successive year until the population reaches 1/2 of the carrying capacity at which time density dependent factors, such as food limitation, cause population growth rate to decline. The population size with the greatest annual growth rate is the inflection point and the population size at which the population ceases to grow is the carrying capacity.

Deer Population Dynamics

Ballard et al. (2001) reviewed deer population dynamics and that discussion remains pertinent to our understanding of the impacts of predation on white-tailed deer. Population levels are a function of births, deaths, immigration, and emigration and their relationship to habitat carrying capacity (MacNab, 1985). Unfortunately, in most cases biologists cannot determine habitat carrying capacity with any degree of certainty and, consequently, must rely on indicators of when a population is below or above carrying capacity. In theory, deer at carrying capacity should have lower weights and lower productivity, and mortality should be high. A population at this point is stable with stable environmental conditions. Where a population occurs on the population growth curve in theory determines that population's performance and how mortality factors are classified.

Theoretical population biology suggests that at the inflection point or to the right of the logistic population growth curve, mortality factors change from primarily additive to primarily compensatory. However, ungulate populations never conform entirely to this population model because the density-dependent function is not linear but curvilinear (Dale McCullough, University of California, personal communication). As previously mentioned, additive mortality is an additional source of mortality, whereas compensatory mortality consists of mortality causes that simply replace each other. This concept has implications for attempting to implement predator control to reduce deer mortality. For example, if a population is at habitat carrying capacity and managers attempt to reduce predator-caused mortality with control programs, the deer population is likely not to respond, because other forms of mortality will replace predation. Consequently, predator control will have little to no impact on the deer population in such situations. On the other hand, if a population is well below habitat carrying capacity and predation is a significant mortality factor, then predator control could result in a deer population increase.

The studies I reviewed suggest that in most cases white-tailed deer pregnancy rates and twinning rates (as indicated by fetal counts and ultrasound) (Figure 8.2) are relatively high, except when populations are at or above habitat carrying capacity or when severe winters or droughts have occurred. In the latter cases, productivity may be unchanged, but recruitment rates may decline because fawns may have low birth weights and be in poor condition, or the neonates simply succumb to a stressful environment. Consistently high productivity indicates that most deer populations produce adequate numbers of deer to allow a population increase, but failure to increase is almost entirely due to mortality factors. Other factors such as deer age and condition and characteristics of vegetation cover also impact these relationships.

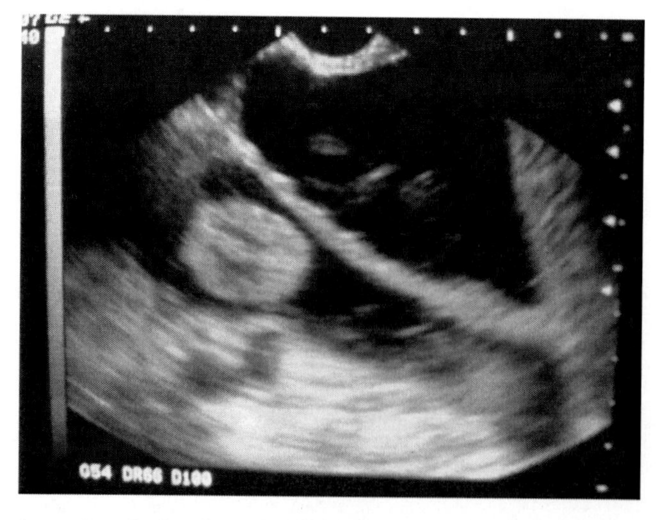

FIGURE 8.2 Recent advances in technology have allowed biologists to use portable ultrasound machines to noninvasively estimate pregnancy and twinning rates from live deer. This image shows a cotyledon of pregnancy which indicates the deer is pregnant with at least one fawn. Often twins and triplets can be identified. (Photo by S. Haskell. With permission.)

It may be important to consider the age or sex of deer prey if predation is selective in this regard. Deer population growth rate is most sensitive to changes in adult female survival (Gaillard et al., 2000; Haskell and Ballard, 2007). However, annual adult female survival is a relatively consistent vital rate for deer. Juvenile recruitment tends to be the more variable fitness component in deer populations that influences growth rate, except when harvest of females is high.

Factors Affecting Prey Vulnerability

A variety of factors influence a predator's success at making a kill. In general, predators take young, old, weak, and otherwise infirmed individuals, which could be a form of compensatory mortality if death was imminent (Mech, 1970; Mech and Boitani, 2003; and many others). However, there are many exceptions to this general rule, and the nature of both predator and prey animals needs to be considered. Severe winters (as determined by snow depth, density, and hardness; winter duration; and extreme cold) and drought can impact a predator's selection of prey. These factors impact physical condition of deer and make individuals more vulnerable than normal to predation. But factors such as deep snow can also make prey in good condition vulnerable by impeding their movements, but not the movements of predators (e.g., crusted snow may carry the weight of predators). Mech (2007) indicated that during 1984–2002, marrow fat of fawns killed by wolves in Minnesota was lower than that of deer killed by accidental causes, suggesting that poor condition predisposed fawns to predation. However, there is considerable evidence that many predators kill fawns randomly in terms of the health of individual fawns. Drought can have a similar impact by reducing vegetative cover, which can adversely affect nutrition or hiding cover, thus making deer more susceptible to predation (Kucera, 1988; Taylor, 1996) (Figure 8.3). In west-central Texas, bobcats killed both healthy and unhealthy fawns seemingly at random (Haskell, 2007).

A number of other factors also determine what sex and age class of animal might be susceptible to predation (Table 8.1). These include the species of prey (some species are preferred over others), sex (males more susceptible during rut; females more susceptible during parturition), age class (younger and older individuals more susceptible than prime age individuals), nutritional condition (animals in poorer condition more susceptible), weight (lighter individuals more often taken), disease and parasites (render animals more susceptible due to poor condition or behavioral differences), injuries or other abnormalities (these factors affect condition and behavior), defensiveness (some individuals unlikely to mount strong defense), parental (offspring of older parents less likely to be killed) or grandparental care (offspring of malnourished parents

FIGURE 8.3 Drought can have many impacts on fawn survival. After wet years (such as in this photo) fawns have better hiding cover which improves their chances of avoiding predators. (Fawn is in middle of the photo; photo by S. Haskell. With permission.)

TABLE 8.1

Antipredator Characteristics and Behavior of Wolf Prey Species

Trait/Behavior	Species	References
Physical traits		
Size	Moose	Mech (1966b)
	Bison	Carbyn et al. (1993)
	Muskoxen	Gray (1987)
Weapons		
Antlers/horns	Male ungulates	Nelson and Mech (1981)
	Some females	Mech and Peterson (2003)
Hooves	All ungulates	Mech and Peterson (2003)
Cryptic coloration	Most ungulate young	Lent (1974)
Speed/agility	Pronghorn	Kitchen (1974)
	Hares	Mech (unpublished data)
	Blackbuck	Jhala (1993)
Lack of scent	Deer neonates	Severinghaus and Cheatum (1956)
Behavior		
Birth synchrony	Most ungulates	Estes (1966), Rutberg (1987), Ims (1990), Adams and Dale (1998)
Hiding	Deer neonates	Walther (1961), Lent (1974), Carl and Robbins (1988)
	Pronghorn neonates	Lent (1974), Carl and Robbins (1988)
Following	Caribou neonates	Walther (1961), Lent (1974)
	Goat neonates	Lent (1974)
	Sheep neonates	Lent (1974)
	Moose neonates	Lent (1974)
Aggressiveness	All ungulates	Mech and Peterson (2003)
Grouping	Caribou	Bergerud et al. (1984)
	Elk	Darling (1937), Hebblewhite and Pletscher (2002)
	Muskoxen	Gray (1987), Heard (1992)
	Bison	Carbyn et al. (1993)
	Deer (winter)	Nelson and Mech (1981)
	Pronghorn	Kitchen (1974), Berger (1978)
	Sheep	Berger (1978)
	Goats	Holroyd (1967)
	Hares	Mech (unpublished data)
Vigilance	All species	Dehn (1990), Laundré et al. (2001)
	Deer	Mech (1966a)
Vocalizations	Deer	Schaller (1967), Hirth and McCullough (1977), LaGory (1987)
	Sheep	Berger (1978)
Visual signals	Deer	Smythe (1970, 1977), Bildstein (1983), LaGory (1986), Caro et al. (1995)
	Elk	Guthrie (1971)
	Sheep	Berger (1978)
	Muskoxen	Gray (1987)
	Arctic hares	Mech (unpublished data)
Landscape use		
Migration	Caribou	Banfield (1954)
	Deer	Nelson and Mech (1981)
	Elk	Schaefer (2000)
	General	Fryxell et al. (1988)

TABLE 8.1 (continued)

Antipredator Characteristics and Behavior of Wolf Prey Species

Trait/Behavior	Species	References
Nomadism	Caribou	Bergerud et al. (1984)
	Muskoxen	Gray (1987)
	Bison	Roe (1951)
	Saiga	Bannikov et al. (1967)
Spacing		
Away	Caribou	Bergerud et al. (1984), Ferguson et al. (1988), Adams et al. (1995)
	Deer	Hoskinson and Mech (1976); Mech (1977a,b); Nelson and Mech (1981)
	Moose	Edwards (1983), Stephens and Peterson (1984)
Out	Deer	Nelson and Mech (1981)
	Moose	Mech et al. (1998)
	Caribou	Bergerud et al. (1984)
Escape features		
Water	Deer	Nelson and Mech (1981)
	Moose	Peterson (1955), Mech (1966b)
	Caribou	Crisler (1956)
	Elk	Cowan (1947), Carbyn (1974)
	Beavers	Mech (1970)
Steepness	Sheep	Murie (1944), Sumanik (1987)
	Goats	Rideout (1978), Fox and Streveler (1986)
Shorelines	Caribou	Bergerud (1985), Stephens and Peterson (1984)
Burrows	Wild boar	Grundlach (1968)

Source: From Mech, L. D. and R. O. Peterson. 2003. In *Wolves: Behavior, Ecology, and Conservation*, eds. L. D. Mech and L. Boitani, 131–160. Chicago, IL: University of Chicago Press. With permission.

and grandparents more likely to be killed), and season (Mech and Peterson, 2003). All of the above factors singly or, more commonly, in combination can influence predation's effect on a particular prey population.

Perceived Risk of Predation and Habitat Avoidance

Prey populations have evolved with predator populations and have made adjustments to counter predation risk. Mech and Peterson (2003) summarized a large number of studies which identified ungulate adaptations to predation risk from wolves, but these same adaptations likely apply to all types of predation on ungulates. They divided these adaptations into physical traits, behavior, and landscape-use patterns (Table 8.2). With regard to white-tailed deer they listed traits such as antlers and hooves for defense, and cryptic coloration and hiding behavior of neonates to avoid predation. Behavioral adaptations to predation for white-tailed deer included the following: synchrony of births, hiding, aggressiveness, grouping (herds or yarding behavior), vigilance, vocalizations, and visual signals (Table 8.2). Deer have also altered their use of landscapes and habitat through several mechanisms. Migration is one response perhaps in relation to predation, but also in relation to severe winters (i.e., deep snows increase vulnerability); generally northern white-tailed deer or deer in areas with extreme weather migrate to traditionally used winter yards, whereas southern deer often do not. Deer can separate themselves from predation risk by occupying areas not occupied by predators (termed spacing away). Deer frequently use water bodies as an escape mechanism. Of the above factors, yarding behavior has received considerable attention.

Reasons why white-tailed deer use deer yards have been the focus of several studies. Deer yards frequently do not contain abundant food resources, so other factors must be responsible for their use. Several authors have indicated that in areas where deer use yards those deer have higher survival rates

TABLE 8.2

Prey Characteristics That May Determine Vulnerability to Wolves

Characteristic	Remarks	References
Species	Some indication that in multiprey systems, certain species may be "preferred" to others, but no definite	Cowan (1947), Mech (1966a), Carbyn (1974, 1983), Potvin et al. (1988), Huggard (1993), Weaver (1994), Kunkel et al. (1999)
Sex	Males killed most often around the rut	Nelson and Mech (1986b), Mech et al. (1995)
Age	Calves and fawns and old animals most often taken	Summarized by Mech (1970, 1995)
Nutritional condition	Individuals in poor condition most often taken	Summarized by Mech (1970), Mech et al. (1998, 2001), Seal et al. (1978), Kunkel and Mech (1994)
Weight	Lighter individuals most often taken	Peterson (1977), Kunkel and Mech (1994), Adams et al. (1995)
Disease	Diseased animals most often taken	Summarized by Mech (1970) and Mech et al. (1998)
Parasites	Hydatid cysts and winter ticks may predispose prey	Summarized by Mech (1970) and Mech et al. (1998)
Injuries, abnormalities	Injured or abnormal individuals most often taken	Summarized by Mech (1970) and Mech et al. (1998), Mech and Frenzel (1971), Landis (1998)
Parental or grandparental condition	Offspring of malnourished mothers or grandmothers most often taken	Peterson (1977), Mech and Karns (1977), Mech et al. (1991)
Defensiveness	Aggressive individuals taken less often	Mech (1966b, 1988), Haber (1977), Peterson (1977), Nelson and Mech (1993), Mech et al. (1998)
Parental age	Offspring of older parents taken less often	Mech and McRoberts (1990)

Source: From Mech, L. D. and R. O. Peterson. 2003. In *Wolves: Behavior, Ecology, and Conservation*, eds. L. D. Mech and L. Boitani, 131–160. Chicago, IL: University of Chicago Press. With permission.

than nonyarding deer (Messier and Barrette, 1985; Nelson and Mech, 1991). Messier and Barrette (1985) indicated that because deer yards contained many "runways," the trail network enhanced deer's ability to escape coyotes. However, Whitlaw et al. (1998) found no differences in survival rates or predator-caused mortality rates between yarding and nonyarding deer populations. Other factors such as ease of movement, lower snow depths, and thermal cover might be responsible for their use. Clearly, further research is needed to clarify the reasons for yarding behavior.

Recently, considerable research has focused on perceived risk of predation and how it may alter habitat use by ungulates, and subsequently alter ecosystems (Lima, 1988; Lima and Dill, 1990; Altendorf et al., 2001). Berger et al. (2001) demonstrated a cascade effect from the elimination of grizzly bears and wolves from the Greater Yellowstone Ecosystem. They identified three primary results: (1) eruption of moose which depend on riparian habitats, (2) alteration of riparian vegetation structure and density due to ungulate use, and (3) reductions of avian neotropical migrants. They suggested that restoration options included accepting the current situation with less biological diversity, replacing natural predation with human harvest, or allowing and encouraging top predators to recolonize areas where they have been extirpated. Ripple and Beschta (2003, 2004) came to similar conclusions and documented increases in cottonwood and willow communities as a result of elk avoiding riparian areas in response to presence of wolves. Creel et al. (2007) suggested that elk have altered their behavior in relation to wolf re-colonization by altering patterns of aggregation, habitat selection, vigilance, foraging, and sensitivity to environmental conditions. They suggested that without consideration of the indirect impacts of predation, declines in reproduction might be falsely blamed on bottom-up limitation due to resources.

O'Brien et al. (2010) documented an increased use by mule deer of areas with the greatest vegetative cover in a 246-ha enclosure following a wildfire after coyotes were introduced. How deer perceive predation risk can be important when attempting to understand and mitigate undesirable habitat avoidance caused by human disturbance (Frid and Dill, 2002; Sawyer et al., 2006; Haskell and Ballard, 2008).

Predator–Deer Case Histories

Connolly (1978) reviewed 45 studies concerning predator–ungulate relationships and concluded that a selective review of the literature could reinforce any view concerning the importance of predation. Ballard et al. (2003) reviewed 13 case histories for mule deer, nine case histories for white-tailed deer, and two case histories for black-tailed deer to increase understanding of mule deer–predator relationships. Here I repeat and update white-tailed deer studies and report on new studies that have occurred since the mid-1970s with specific focus on predation. There are numerous other studies that are not reviewed, where predation was not a significant mortality factor or the study lacked documentation of whether predation was an important factor. I focus on deer–predator studies that are based upon radio-telemetry data or long-term studies with adequate scientific documentation (Figure 8.4). A number of studies that I report were only two to three years duration and, consequently, their usefulness in understanding predator–prey relationships is limited in relation to long-term studies.

South-Central British Columbia

Robinson et al. (2002) examined survival and causes of mortality of adult sympatric mule and white-tailed deer during 1997–2000. White-tailed deer were three times more abundant than mule deer. Fetal rates between deer species were similar (1.83 and 1.78 for white-tailed deer and mule deer, respectively). Recruitment rates were similar between species (56 and 38 fawns:100 does for white-tailed and mule deer, respectively), but white-tailed deer survival rates were higher than mule deer survival rates (i.e., 0.81 for white-tailed deer vs. 0.72 for mule deer). Predation by cougars was the largest source of mortality for both deer species; the predation rate was higher on mule deer (0.17) than on white-tailed deer (0.09). Annual finite rate of growth was higher for white-tailed deer (λ=1.02) than for mule deer (λ=0.88). The authors suggested that the differences in survival and predation rates were caused by apparent competition for habitat between the two deer species. They offered managers' suggestions for increasing mule deer numbers, such as increasing cougar harvests following severe winters and a gradual reduction in white-tailed deer numbers to prevent prey switching.

Colorado

Whittaker and Lindzey (1999) studied the causes and timing of early fawn mortality (i.e., less than 30 days of age) in a sympatric population of mule deer and white-tailed deer during 1991–1992. They found

FIGURE 8.4 Use of vaginal implant transmitters and development of expandable fawn collars have allowed biologists to study early fawn survival. (Photo by S. Haskell. With permission.)

that survival was greater for mule deer (0.66) than for white-tailed deer (0.34) fawns and that 79% of mortalities for each species were due to coyote predation. They attributed the lower survival rate for white-tailed deer to differences in timing of births between the two deer species. White-tailed deer birthed an average of 8–10 days earlier than mule deer. Whittaker and Lindzey (1999) speculated that because of these differences and because mule deer outnumbered white-tailed deer (4:1), mule deer were afforded protection via predator swamping strategies. All deer appeared to be in good physical condition.

Southern Illinois

Rohm et al. (2007) examined causes of fawn mortality during 2000–2004 in southern Illinois in or near areas studied earlier by Nelson and Woolf (1987). Overall fawn survival was 0.59 and predation was the leading cause of mortality (i.e., 64%), with coyotes accounting for 56% of predation mortalities. The second largest cause of mortality was abandonment (8%), although for 24% of mortalities the cause of death could not be determined. Bobcats and domestic dogs each accounted for about 7% of the mortalities. The authors suggested that predator-caused mortalities were probably underestimated. Their survival estimates were lower than those of Nelson and Woolf (1987), and they speculated that this difference was due to increase in coyote and bobcat populations during the 20-year interval between the two studies. Fawn survival was best explained by fawn age and landscape and forest characteristics. They indicated that areas inhabited by surviving fawns had forest patches next to nonforest patches and contained more edge habitats. They speculated that these habitats were areas where coyotes were less successful at locating and killing fawns. Their study was the first to relate macro habitat parameters to fawn survival.

Maine

Long et al. (1998) investigated causes of white-tailed deer fawn mortality in an island population where coyotes had recently become reestablished and hunting was prohibited. Annual survival rate of fawns was 0.26 and the predator-caused mortality rate was 0.52. Coyote predation accounted for 47% of all mortalities. Mortality related to drowning and human related (nonhunting) were also important sources of mortality (mortality rate = 0.24 and 0.14, respectively). They reported that the survival rate was higher for fawns in residential areas on the edges of coyote territories than elsewhere and concluded that this deer population was limited by coyote predation and mortalities associated with humans.

Southwestern Lower Michigan

Burroughs et al. (2006) examined fawn survival in a study area that was about 50% agriculture during 2001–2002. Survival to pre-hunt was 0.91 and 0.92 in 2000 and 2001, respectively. About 7% of the fawns each year died from legal hunting and vehicle collisions. Coyotes appeared to be abundant. Their study had one of the higher, if not the highest, fawn survival rates in North America, which they attributed to habitat quality and land cover and use.

North-Central Minnesota

DelGiudice et al. (2002) examined white-tailed deer survival, cause-specific mortality, and winter severity during 1991–1996 in north-central Minnesota in an area where wolf populations recently had become reestablished. Based on 153 radiocollared female deer (at least 0.6 years old), 84% of mortality occurred during November–May each year. Lifetime mortality rates were estimated at 43.3%, 33.1%, and 23.6% for mortalities due to hunting, wolf predation, and miscellaneous causes, respectively. They included estimates of age-specific hazard functions and survival, but then in a follow-up that included 76 radiocollared neonates and 302 females at least 0.6 years old, they expanded their analyses to 1991–2003 (DelGiudice et al., 2006). Risk of mortality was high at birth, declined precipitously through two years of age, was low and stable until six years of age, and then progressively increased through 17 years of age. Additional evidence indicated that the phase of stable, low mortality risk for female deer was prolonged to 10 years of age by mild winters, associated low wolf predation, and reduced hunting pressure (DelGiudice et al.,

2006). Mean percentages of winter (December 1 to May 31) deaths caused annually by wolves was 58%. Within 5–11 years after wolf reestablishment, wolf predation had become the leading cause of natural mortality for female deer, but its importance in relation to hunting was dependent on winter severity. DelGiudice et al. (2006) suggested that managers should consider environmental conditions (notably severe winters) and their interactions with predation and age distributions when setting hunting seasons.

Carstensen et al. (2009) studied causes of fawn mortality during 2001–2002. Fawn survival to 12 weeks of age was 0.47. Predation was the largest cause of death (86%) with black bears and bobcats having the greatest impact. Black bears accounted for 36% and 29% of mortalities during 2001 and 2002, respectively, while bobcats accounted for 9% and 47% of mortalities during 2001 and 2002, respectively. Wolf predation accounted for only 5% of the deaths, although it is likely that wolves were responsible for at least a portion of the 29% neonate mortality assigned a cause of "unknown" or "unknown predator." The authors reported that even though winter 2000–2001 was about 3.5 times more severe than winter 2001–2002, they did not detect a relationship between winter severity and birth and blood characteristics, but lower body mass, smaller body measurements, and elevated serum urea nitrogen and tumor necrosis factor-α were associated with fawns that died within one week of birth. Black bears were the most numerous predators in their study area, and they attributed increased bobcat predation to increases in the bobcat population.

Northeastern Minnesota

Kunkel and Mech (1994) studied the timing and causes of white-tailed deer fawn mortality within the Superior National Forest from May to October during 1989 and 1990. Nine of 21 neonate fawns died, and seven deaths occurred before June 28. All deaths were due to predation, five from wolves and four from black bears. They found that fawns from does greater than four years of age were heavier and had better survival rates than fawns from younger mothers.

Nelson and Mech (1986a) reported survival and cause-specific mortality rates for 209 white-tailed deer from 1973 through winter 1983–1984, also in the Superior National Forest. Annual survival was 0.31 for fawns, 0.80 for yearling females, 0.41 for yearling males, 0.79 for adult females, and 0.47 for adult males. Of 85 deaths, 44 were attributed to wolf predation, 22 to hunting, 12 to probable wolf kills, and the remaining seven to miscellaneous causes. Only one deer starved, and Mech and Karns (1977) indicated that this deer population was well below the levels imposed by habitat and winter weather alone. Most mortality in all age cohorts occurred during January through April and was due to wolf predation. Although few fawns died of malnutrition, Mech and Karns (1977) suspected that fawns that died were predisposed to death, because they were in poorer condition at the time of capture than were surviving fawns. However, these authors indicated that fate of fawns in the absence of wolf predation was debatable. Highest wolf-kill rates occurred during winters of deep snow with wolf predation limiting the recruitment of yearling deer (Nelson and Mech, 1986b).

Southwestern Minnesota

Brinkman et al. (2004) examined fawn and adult survival rates in an intensively farmed area during 2001–2002. Annual survival rate of adults and fawns older than seven months of age was 0.75. Hunting was the largest source of mortality (43% of mortalities). Neonate fawn survival in this population was relatively high, averaging 0.84. Four of six neonate mortalities were caused by predators (two coyotes and two unknown predators). They attributed the high survival to low predator densities, good fawn hiding cover, and high nutritional condition of adult females.

Montana

Dusek et al. (1992) determined survival and cause-specific mortality rates for 154 adult female radiocollared deer in three habitats during 1980–1989. Hunting was the largest cause of mortality in all areas. Of 99 deaths, only eight (8%) were attributed to natural causes other than predation. They concluded that harvest-related mortality had little effect on natural mortality rates and, therefore, hunting was an additive form of mortality in their study areas. If compensatory mortality existed in these populations, they suggested that it would probably occur in juvenile age classes.

New Brunswick

Ballard et al. (1999) and Whitlaw et al. (1998) reported survival and cause-specific mortality rates of radiocollared fawn and adult white-tailed deer, respectively, during 1994–1997. Fawn survival was lowest during summer (0.47), increased during autumn (0.86) and early winter (0.95), and then declined during late winter (0.76) and spring (0.81). Predation by coyotes was the largest source of annual cause-specific mortality (0.37), followed by black bear predation (0.14), predation by domestic dogs (0.08), unknown causes (0.07), predation by bobcats (0.06), and other causes (0.04).

Predation by coyotes was also the largest natural mortality factor among adult deer, with most mortality occurring during winter (Whitlaw et al., 1998). These authors indicated that eastern coyotes appeared to have replaced wolves as major predators of deer in northeast North America where wolves were extirpated. They suggested that predator management, along with continued provision of adequate winter habitats and mild winters, would be necessary to maintain and increase that deer population.

Nova Scotia

Patterson and Messier (2000) examined factors influencing predation on white-tailed deer by coyotes during winter. Deer abundance, body condition, distribution, vulnerability, snowshoe hare abundance, and coyote group size all affected coyote killing rates of white-tailed deer. Increased vulnerability of deer during late winter appeared to be a factor in coyotes switching from hares to deer.

Oklahoma

Stout (1982) indicated that effects of coyote predation were site specific. He reduced coyote densities and studied effects of the reduction on white-tailed deer fawn mortality during 1977–1980 relative to findings of studies by Garner (1976), Garner et al. (1976), and Bartush (1978), which indicated that 88–90% of neonate fawns died and 88–97% of fawn mortalities were caused by coyotes.

Initial deer conception rates appeared high, ranging from 1.6 to 2.1 corpora lutea/adult doe, but fawn losses from birth to November were also high. To test the hypothesis that reductions in coyote numbers could increase fawn survival, coyotes were killed by trapping and aerial-shooting from helicopters in three areas during midwinter to midspring prior to parturition by deer. Fawn:doe ratios were then compared between pre- and postcoyote removal.

Coyote densities apparently were significantly reduced, but pre- and postremoval coyote densities were not reported. Fawn recruitment from birth to mid-September during the first year of coyote removal increased by 262%, 92%, and 167% on the three areas, respectively. Prior to coyote removal in 1976, fawn:doe ratios averaged 0.37, but after coyote removal they averaged 0.94. Untreated study areas exhibited no increases in fawn recruitment.

Fawn mortality was greatest during the first four to eight weeks of life and coyote removal reduced this mortality. Stout (1982) indicated that coyote removal during late winter and early spring was effective for reducing fawn mortality. However, he indicated that disease in young does may partially offset increases in fawn survival. He concluded that coyote removal significantly increased fawn survival, but effects of the increase on herd dynamics were not understood due to interactions that were difficult to identify and study.

Western Oregon

Ricca et al. (2002) examined survival and causes of mortality for adult and fawn Columbian white-tailed deer during 1996–1998. Annual adult survival rates were relatively low over the three-year period (0.74). Eight adults died from emaciation and disease, and 92% of the carcasses they examined (12 of 30 deaths) were in poor condition. Fawn survival was low (0.14) and most mortality occurred during the first

1.5 months of life. Predation was the leading cause of fawn mortality (57%) and the following predators were responsible: bobcats (six fawns), fox (one fawn), domestic dog (one fawn), and unknown predator (11 fawns). The second leading cause of death was from abandonment ($n = 6$, 16%) followed by disease ($n = 4$, 11%), fence entanglement ($n = 1$, 3%), and unknown causes ($n = 5$, 14%). The authors suggested that the population had responded to density-dependent factors and that recruitment was so low that the population could not grow and expand into new habitats.

Central Pennsylvania

Vreeland et al. (2004) examined the causes of mortality of deer fawns during 2000–2001 in both forested and agricultural landscapes. Fawn survival at nine weeks of age was 72% in agricultural areas and 57% in forested areas; the same trend was observed at 34 weeks of age (i.e., 53% and 38% in agricultural areas and forested areas, respectively). Predation was responsible for 46% of mortalities. Coyotes accounted for 37% and black bears for 33% of predation events. Unidentified predators and bobcats accounted for the remainder of predation mortalities. Other natural causes accounted for 27% of mortalities. They indicated that in spite of the fawn survival rates they measured, the deer population continued to grow.

King Ranch, Texas

Beasom (1974a) conducted one of the first predator removal experiments on white-tailed deer by contrasting fawn:doe ratios and deer densities between a 2186-ha treated and a 2186-ha untreated area. Pretreatment deer densities within the areas were estimated at 7.4–14.7/km^2, and apparently the deer population was below habitat carrying capacity, because deer were in good condition (Beasom, 1974a). Autumn coyote densities were estimated at 1.5–2.3/km^2 (Knowlton, 1972) and bobcats were thought to be common (Beasom, 1974a).

Predator numbers were reduced in the treated area from February through June in 1971 and 1972 with the use of steel traps, M-44s, strychnine-treated meat and egg baits, and shooting (Beasom, 1974a). A total of 129 coyotes and 66 bobcats were removed during 1971 and 59 coyotes and 54 bobcats were removed during 1972. Predator track counts indicated that predator densities were similar between treated and untreated areas prior to predator reductions, but after predator removal, track counts suggested near elimination of predators in the treatment area. When the experiment ended, predator numbers appeared to rebound (based upon track counts) to pre-removal levels within six months.

August fawn:doe ratios in the treatment area were 0.47 and 0.82 in 1971 and 1972, respectively, while in the untreated area, fawn:doe ratios averaged 0.12 and 0.32 in 1971 and 1972, respectively. Fawn mortality was 74% and 61% higher in the untreated area than in the treated area during 1971 and 1972, respectively. Deer density within the treatment area increased from 15.6 to 19.6 deer/km^2, whereas density in the untreated area remained essentially unchanged from 8.0 to 7.8 deer/km^2 between 1971 and 1972, respectively. Beasom (1974a) indicated that fawn:doe ratio data suggested there may have been an influence of predator removal out to 23 km, but data were highly variable. There were clearly higher ratios out to 3.2 km from the treatment area. It should also be noted that poisons were used. Beasom (1974a) concluded that there was potential for producing many more deer on a sustained yield basis through the use of predator removal, but there would have to be adequate hunter harvests to prevent overabundance. Another experimental removal of predators was conducted in south Texas with similar results (Guthery and Beasom, 1977). Beasom (1974b) and Guthery and Beasom (1977) also monitored the effects of predator removal on small mammal and lagomorph populations and found no effect.

Southern Texas

Heffelfinger et al. (1990) tested the hypothesis that adult male (greater than four years of age) mortality would be lower following coyote control than without coyote control. The hypothesis was based on the observation that adult male white-tailed deer suffered high rates of natural mortality (20–23%; DeYoung, 1989) during the post-rut period. During January 1987 through June 1989, 97 radiocollared male deer were monitored. During this time period, only eight males died; all were either killed or scavenged by coyotes. Natural mortality rates between coyote removal areas (9.3%) and nonremoval areas (11.9%) were not different. Neither coyote nor deer densities before or after coyote control were reported by Heffelfinger et al. (1990), but scent-station surveys indicated that coyote populations were reduced by 81% (Heffelfinger, 1989). Heffelfinger (1989) indicated that the coyote treatment did not increase the number of bucks available for harvest. He further commented that more intensive coyote-control efforts may have produced a different result, but such treatment would have been impractical given methods they had available at that time (e.g., traps, M-44s, hunting, incidental shooting, and aerial-gunning). A total of $28,000 was spent controlling coyotes, and yet no positive results were documented in terms of increased bucks available for harvest, total number of deer, or number of mature bucks (Heffelfinger, 1989, p. 85).

South-Central Texas

Carroll and Brown (1977) studied the causes and timing of neonate fawn mortality in two counties of south-central Texas for two-month periods during 1971–1973. Deer densities were 41 and 10 deer/km^2 in the Lavaca and Gonzales County study areas, respectively. Natality rates were estimated to be 1.50 fawns/doe. Sixty fawns were radiocollared in each study area over the three-year period. In the Lavaca study area, 37% of fawns died, and 45% of mortalities were attributed to coyote predation. In the Gonzales study area, 45% of fawns died, with coyotes and bobcats being responsible for 55% and 11% of mortalities, respectively. They indicated that there were substantial differences between study areas. In the Gonzales area, 50% of adult deer died in 1970 due to starvation from drought conditions. Consequently, they attributed the 90% loss of fawns in 1971 to lack of ground cover and poor nutrition (presumably deer were above K at that point). They indicated that the wetter conditions in 1972 and 1973 resulted in improved fawn survival (65% and 90%, respectively). Fawn-at-heel counts indicated that there were a few fawn mortalities after August. They concluded that these high-density deer populations seemed to sustain moderate fawn mortality.

Welder Wildlife Refuge, Texas

Teer et al. (1991) summarized studies of white-tailed deer on Welder Wildlife Refuge (WWR) from 1954 through 1990. First, fawn survival to weaning appeared related to precipitation during gestation. Kie and White (1985) found no relationship between conception rates and deer densities of 20–60 deer/km^2, but they did find lower conception rates when deer densities peaked at 84 deer/km^2. Teer et al. (1991) indicated that deer densities appeared to stabilize at about 34 ± 10 deer/km^2 from 1962 through 1990.

Coyote densities in and near WWR have varied over the years from 0.2 to a high of 2.3/km^2 in favorable habitat. Coyote diets varied by year, with a pronounced shift to deer fawns during fawning season. As fruits became abundant, coyote diets shifted to fruit. Deer then became an important coyote food item during late winter and early spring.

Cook et al. (1971) were the first biologists to study causes of fawn mortality with the aid of radiotelemetry. Seventy-two percent (58 of 81) of radiocollared fawns died within 60 days of birth; 93% of those died during the first month of life. Predation accounted for 83% of losses. Seventeen percent died as a result of starvation, disease, or accidents.

West-Central Texas

Haskell (2007) studied survival and causes of mortality of adult female white-tailed deer in a sympatric white-tailed deer and mule deer population during 2000–2007. This deer population was at or above habitat carrying capacity. Coyotes and other large predators were largely absent from this study site. Adult white-tailed deer had higher survival rates during relatively wet years (0.92) than during dry years (0.82). From about 300 deer-years there were 31 adult mortalities, of which 39% were due to disease or starvation, 39% from unknown causes, 10% from poaching, 6% from vehicle collisions, 3% from hunting, and 3% from bobcat predation. Top-down limitation was negligible.

Fawn survival was studied during 2004–2006 with 170 fawns (106 mule deer and 64 white-tailed deer) captured and many from known birth sites by using vaginal implant transmitters in adult females. Nine percent of fawns died within 25 m of birth sites, and all deaths but one were due to sickness-starvation and determined not to be caused by researchers. Annual survival rates of white-tailed deer fawns ranged from 0.12 to 0.58 and were greater in relatively wet years when adult females weighed more than in a dry year. For both species combined, 47% of mortality was due to sickness-starvation, and about 47% was due to predation. All of the predation, except for four fawns (two fox predation and two other), was caused by bobcats that took about 31% of the annual fawn crop (Figure 8.5).

Mortality rate of white-tailed deer fawns due to sickness-starvation increased during the dry year, but mortality rate due to predation remained consistently high (about 40%) regardless of weather. Bobcats appeared abundant, perhaps due to the absence of coyotes (Henke and Bryant, 1999). It was found that even though large predators were absent, white-tailed deer females still separated their young fawns and stayed distant from them, which was a relatively unsuccessful strategy compared to more aggressive mule deer females that kept fawns nearby. This behavioral phenomenon has direct implications for predator-control efforts if reduction of large predators releases populations of smaller predators.

The question of whether or not directly additive mortality can be indirectly compensatory is worth considering (Linnell et al., 1995). For example, bobcats killed healthy fawns with relatively heavy thymus gland and body weights, and so such an event would seem to be directly additive (Haskell, 2007). However, bobcat predation on fawns was the only significant top-down limiting factor for these populations of deer living a tenuous existence near a habitat carrying capacity that fluctuated with rain. Thus, had bobcat predation been reduced, would other mortality factors related to sickness-starvation have increased? If so, then bobcat predation of apparently healthy fawns could be viewed as indirectly compensatory in a long-term population-level context.

Midwestern United States

Several studies of neonatal white-tailed deer fawns have been conducted in mixed agricultural and forested habitats in the East and Midwest. Huegel et al. (1985) indicated that fawn mortality in these areas was relatively low in comparison with semiarid areas of the Southwest. For example, McGinnes and Downing (1969) observed 8% fawn mortality in a Virginia enclosure, while Ozoga et al. (1982) reported mortality of 2–23% in a Michigan enclosure. Schultz (1982) reported 15% of fawns died in Minnesota, and Huegel et al. (1985) reported 27% died in Iowa, suggesting that growing conditions in the Midwest provided better fawning habitat than elsewhere in North America, which made fawns less susceptible

FIGURE 8.5 When other predator species have been eliminated, bobcats can be an important source of mortality. This photo shows a typical bobcat fawn mortality site in west-central Texas. (Photo by S. Haskell. With permission.)

to predation. Nelson and Woolf (1987) examined the causes of fawn mortality in southern Illinois and found that pre-hunt mortality rates for fawns were 29%, 38%, and 21% in 1980, 1981, and 1982, respectively. Sixty-nine percent of mortalities were due to predation by coyotes and domestic dogs followed by disease (13%), starvation (6%), and unknown factors (13%). They suggested that generally predation rates on fawns were lower in the Midwest.

Washington

Wielgus et al. (2007) recently examined cougar, white-tailed and mule deer, and human relationships in four areas of Washington during 1997–2007. They suggested that increases in the number of white-tailed deer resulted in increased cougar numbers and increased predation on mule deer. There was an interaction between hunting of cougars by humans, conflicts with humans, and deer populations. They suggested that such interactions would result in declines in the cougar population and a potential population explosion in white-tailed deer.

Summary

Wolves, coyotes, black bears, bobcats, and cougars all have been identified as important predators of deer fawns. Wolves and cougars are important predators of adult deer, and in some cases, coyotes are also important predators of adult deer. Where wolves are relatively abundant, coyote populations are usually low and wolves constitute a significant source of deer mortality of both fawns and adults. Black bears have been identified as a significant source of deer fawn mortality in a number of jurisdictions. All reports of significant black bear predation for white-tailed deer occur in the eastern and midwest portions of North America. Coyotes appear to be significant predators of deer fawns and adults when wolves are not present. Some late-winter scavenging by northeastern coyotes can be confused with predation, but adult deer are usually most vulnerable at that time. Bobcats also can be important predators of deer fawns, but are not important predators of adult deer. Mountain lions can be significant predators of fawn and adult deer in western states and provinces. There was no evidence of grizzly bears being a significant predator of white-tailed deer, although this may be due to lack of study and little overlap in the distribution of grizzly bears and white-tailed deer. Most mortality due to predation occurs on neonates usually within the first two months of life; however, in northern latitudes significant mortality may also occur during winter when fawns and older deer are also particularly vulnerable. Predator-control efforts may have little impact when a population is near habitat carrying capacity, because predation may be viewed as compensatory. In many jurisdictions predation is an added benefit to wildlife managers trying to prevent white-tailed deer overabundance amid human-social factors that limit efficacy of deer population control by regulated hunting.

Virtually all of the studies I reviewed were conducted in Canada and the United States. Although white-tailed deer are widely distributed in Mexico, I found only two studies that addressed predation. Tanner (1966) indicated that deer populations were originally controlled by predators (e.g., mostly mountain lions). Ezcurra and Gallina (1981) indicated that predator removal did not reverse a declining trend in deer numbers in the La Michlia area. Clearly, more study and dissemination of results are needed from Mexico to understand predator–prey relationships in this region.

Enclosed or Isolated Deer Population Studies

Carrera-Trevino (2008) indicated that the best evidence for the effects of weather come from predator-free islands or enclosures where resources ultimately limit ungulate populations and population parameters clearly respond to weather patterns. However, in multiple predator–prey systems partitioning cause and effect relationships among species make it difficult to identify contributions from all variables without planned experiments. Several studies of deer in enclosures and on predator-free islands have increased our understanding of the effects of weather and predators.

Arizona

During 1971, a 259-ha predator-proof enclosure (Walnut Canyon Predator Proof Enclosure) was constructed on the Three Bar Wildlife Area in central Arizona in an effort to determine the causes of mule deer population decline in the 1960s (Smith and LeCount, 1979). In Arizona, October–April rainfall (prior to fawning) and winter forb yields were strongly correlated with winter fawn:doe ratios (Smith and LeCount, 1979). However, within the predator-free enclosure, such correlations were not apparent. The deer population within the enclosure increased rapidly with annual finite rates of increase ranging from 1.0 to 1.6 from 1970–1971 through 1975–1976. After 1975–1976, the herd declined when deer densities reached 18 deer/km^2. The deer population outside the enclosure was thought to be stable with densities ranging from 3 to 5 deer/km^2 during the study. Fawn:doe ratios during 1970–1971 through 1975–1976 (prior to decline in enclosure) averaged 80 fawns:100 does in the enclosure and 50 fawns:100 does outside the enclosure. Researchers indicated that predators were abundant outside the enclosure. Although rainfall had profound effects on fawn:doe ratios, predation also had a large impact. In spite of these differences, the deer population outside the enclosure increased when favorable weather conditions returned.

Carrera-Trevino (2008) resumed the enclosure studies and expanded them beginning in 1997 with a population of 17 deer (6.6 deer/km^2) in the enclosure, which increased to 89 deer (34.4 deer/km^2) by 2006. During the same time period deer populations outside the enclosure remained stable at about 2 deer/km^2. Fawn:doe ratios in the enclosed population increased from 16 fawns:100 does to 108 fawns:100 does. The increase occurred during one of the worst drought periods in Arizona history.

From 2004 to 2006, Carrera-Trevino (2008) measured body condition, productivity, and quality of diets inside and outside the enclosure. The best model to predict fawn:doe ratios when predators were present was a model with drought index as a single variable, but this relationship was not evident inside the enclosure where predators were absent. He found no differences in body weights and number of fetuses/doe inside versus outside the enclosure. There was evidence that deer inside the enclosure were in better condition, and deer inside the enclosure were consuming a diet higher in energy. Protein levels in available vegetation inside and outside the enclosure were not different. He also found no differences in plant species diversity and composition inside versus outside the enclosure. This suggested that even at the high deer density of 34.4 deer/km^2, the population was not at carrying capacity although there may have been some density-dependent increase in fawn mortality due to overcrowding and social interactions among deer. Carrera-Trevino (2008) suggested the enclosure experiment indicated that low fawn survival due to predation was the most important factor affecting mule deer populations in central Arizona and that deer outside the enclosure were well below habitat carrying capacity.

Angel Island, California

This 2.2-km^2 predator-free island occurs in San Francisco Bay. Black-tailed deer apparently were extirpated by Europeans, but were reintroduced in 1915 by the military and were hunted until 1955 when the island was made into a State Park (McCullough, 1997). Between 1964 and 1984 at least four peaks in deer population size were observed; population estimates ranged from about 75 animals (34.1 deer/km^2) in 1977 to 275 animals (125 deer/km^2) in 1980, and then from 44 individuals (20.0 deer/km^2) in 1981 to 260 (118 deer/km^2) in 1984.

George Reserve Deer Herd, Michigan

The George Reserve predator-free, white-tailed deer herd is the most thoroughly studied deer herd in the world (McCullough, 1997) and provides useful insights of how a deer population responds to lack of predation. The herd exists in a 4.4-km^2 enclosure. This population grew from six individuals (1.4 deer/km^2) in 1928 to about 222 (50.5 deer/km^2) in 1935 (McCullough, 1997). The deer population was reduced in 1935, and there were no other causes of mortality. Culling continued through 1975, and the population was reduced to 10 animals (2.3 deer/km^2). Between 1975 and 1981, the population increased from 10 to

212 deer (48.2 deer/km^2). McCullough (1997) indicated that the high deer densities impacted vegetation, although evidence suggested that growth rates of the deer herd had not been negatively affected and deer habitat carrying capacity had not been lowered.

North Manitou Island, Michigan

This 59.7-km^2 apparently predator-free island located in Lake Michigan had white-tailed deer introduced to it in 1926 (Case, 1982; Case and McCullough, 1987: from McCullough, 1997). Deer were artificially fed, numbers became high despite hunting, and substantial damage to woody vegetation occurred. Administration of the Island was turned over to the National Park Service in 1977, and hunting was terminated. A massive deer-die off occurred in 1977–1978, with an estimated population prior to the die-off at 1879 deer (31.5 deer/km^2). The population again began to grow immediately and reached 2080 individuals (34.8 deer/km^2) in 1981, and then crashed (76% reduction) again in 1981–1982. The population again began to immediately increase and was stabilized by hunting.

Welder Wildlife Refuge, Texas

Following the studies of Cook et al. (1971), a predator-proof enclosure encompassing 391 ha was constructed. Coyote populations were reduced and fawn survival immediately increased by 30% over areas outside the enclosure (Teer et al., 1991). Deer increased until food became the limiting factor, and then deer densities declined. Increased deer densities were accompanied by increased parasite loads.

Teer et al. (1991) indicated that the largest changes in deer numbers on WWR occurred during drought years. They speculated that increased ground cover during wet years was responsible for reduced fawn mortality. They indicated their studies fell short of demonstrating that coyotes had "controlled" deer numbers since 1954, but that coyote control could be used to manage deer numbers. Teer et al. (1991, p. 559) concluded: "Control of coyotes need not be a management strategy when numbers are not cropped by hunting or natural means. Conversely, control of coyotes can be a management strategy when there is adequate habitat and deer numbers need to be increased for greater productivity."

Summary

Enclosure studies indicate that predators are important in controlling deer densities and may reduce population fluctuations. This suggests that human harvests could be substantially higher in such areas in contrast to areas where predators are abundant outside enclosures. Enclosure studies may force managers to rethink the concept of habitat carrying capacity and how deer populations respond following population crashes. McCullough (1997) indicated that most deer biologists would consider a population density of 39 deer/km^2 to be extremely high. The enclosure studies presented here suggest that deer can attain extremely high densities, which damage vegetation, but do not appear to diminish habitat carrying capacity. The traditional view has been that once a population exceeds its long-term habitat carrying capacity, the deer cause such damage that subsequent habitat carrying capacity is reduced. Results from enclosure and isolated populations suggest this accepted dogma may not be the case in all situations, particularly where vegetation can respond quickly to reduced herbivory. No doubt vegetation damage and changes occur, but population performance can be similar before and after such cycles. McCullough (1997) also pointed out that in populations that periodically increase and crash, rates of population increase were similar before and after crashes, and that these peaks appeared to be cyclic at about four-year intervals.

Although a number of the above studies suggest damage and changes in vegetative structure following deer population highs, the study conducted by Carrera-Trevino (2008) suggested that vegetation species composition and biomass were similar at high and low densities of deer. Obviously, there is a threshold of deer densities where serious changes to vegetation structure and composition occur. Additional research is needed to clarify at what densities these changes begin to occur. However, in spite of the vegetation damage and changes, ultimately deer populations continue to perform in similar fashion.

Predator Species Accounts

Wolves

Wolves are widely distributed in the northern hemisphere and are probably the best studied of the carnivore species (Figure 8.6). Both gray wolves and red wolves prey on white-tailed deer in addition to other ungulate species and other prey species. Little is known about impacts of red wolves on white-tailed deer. Based on a sample of 1890 scats, Phillips et al. (2003) estimated that about 25% of prey biomass was composed of white-tailed deer, which occurred in about 43% of red wolf scats. However, the proportion of white-tailed deer remains in scats resulting from scavenging as opposed to predation was unknown. Until recently, studies of gray wolf predation of white-tailed deer had largely been conducted in Canada, Minnesota, and Montana, because wolves had largely been exterminated from most of the continuous United States by the early twentieth century and prior to the advent of modern radiotelemetry.

Wolves recolonized Glacier National Park in Montana within the last few decades, and Kunkel et al. (1999) examined diets of wolves and mountain lions in this area. White-tailed deer comprised 83% and 87% of wolf and cougar kills, respectively. Elk and moose comprised more of the wolf's diet than of the cougar's diet. They suggested that as prey populations decline, carrying capacity for cougars may also decline, especially if wolf appropriation and consumption of cougar kills were significant. There was potential for exploitative and interference competition in this ecosystem. Apparently both deer and elk populations had declined (Kunkel, 1997). Kunkel et al. (2004) indicated that where white-tailed deer and elk were sympatric, wolves selectively killed elk. Numbers of wolves and elk harvested by humans were positively correlated with proportions of deer killed by wolves annually. Excluding severe winters, white-tailed deer, elk, and moose appeared equally vulnerable to wolf predation. Arjo et al. (2002) examined dietary overlap between wolves and coyotes in northwestern Montana and determined that winter wolf diets contained more deer, whereas coyote diets contained more small mammals. Wolves selected a larger proportion of adult deer than coyotes during summer, whereas coyotes selected more juveniles. The authors speculated that the differential use of prey resources facilitates coexistence of the two predator species.

Wolves were translocated to the Greater Yellowstone Area in 1995. Since that time wolf populations have substantially increased. One objective of ongoing studies was to determine the impact of wolf predation on ungulate prey. Smith et al. (2004) determined that the ability to locate wolf kills using ground and aerial surveys was about 73%, suggesting that estimates of kill rates may be low. Kill rates of ungulates in late winter (2.2 kills/wolf/month) were higher than those in early winter (1.6 kills/wolf/month), suggesting increased vulnerability during late winter. The primary prey of wolves during winter

FIGURE 8.6 Wherever wolves overlap the distribution of white-tailed deer they are frequently a significant cause of white-tailed deer mortality. (Photo by W. Ballard. With permission.)

was elk. Their study and recent studies (Smith et al., 2009) indicated that white-tailed deer were not important prey for wolves in this area, probably because elk are more abundant than white-tailed deer. During 2008, biologists detected 576 probable or possible kills, which were composed of 80% elk, 4% bison, 3% deer, 2% coyotes, 2% wolves, and other species (Smith et al., 2009). It is doubtful that wolves influence white-tailed deer population dynamics in the Greater Yellowstone Area due to abundant alternative prey resources.

Mexican gray wolves were translocated to southeastern Arizona and southwestern New Mexico during 1998. Wolf numbers as of 2003 were estimated at 50–60 individuals (Mexican Wolf Blue Range Adaptive Management Oversight Committee and Interagency Field Team 2005). Human harvest and government removals have limited wolf population growth thus far. Based on historical records it was thought that deer would be the principal prey species (Brown, 1983; Parsons, 1998). However, elk are the principal prey of reintroduced Mexican wolves. Historically, elk did not exist in the project area and are now common and locally abundant. Of 72 confirmed or probable ungulate kills, elk composed 90%, whereas only one deer kill was documented. The authors estimated prey availability at 32% elk and 68% deer, suggesting that wolves selected elk. Similarly, according to scat analyses, during winter elk comprised about 73% frequency of occurrence (Reed et al., 2006), and during summer elk were about 80% frequency of occurrence in the diet (Merkle et al., 2010). Carrera et al. (2004) compared diets between Mexican wolves and coyotes and found that both predators were feeding on elk. Although not definitive, competition was possible, because both species had similar diets. It was unknown what proportion of the diets was due to scavenging. The current low population of Mexican wolves probably has minimal impact on white-tailed deer.

Mech and Peterson (2003) provided a detailed account of gray wolf–prey relationships. Although several studies have indicated selection of some ungulate prey species over other species, this apparent selection could also be the result of differential vulnerability due to a wide variety of factors as described in the previous section.

Wolf-kill rates of white-tailed deer during winter have ranged from 0.5 to 6.8 kg/wolf/day (Mech and Peterson, 2003). Because most studies were conducted in late winter in northern environments, they concluded that published estimates of kills were probably maximum rates, because prey species would be in their poorest condition of the year and would be most vulnerable to predation. When kill rates are translated to actual numbers of white-tailed deer killed per wolf, estimates range from 15 to 19 adult-sized deer per year, assuming that about 20% of the diet was composed of other prey items. However, these rates can only be considered rough estimates, as it is difficult to estimate the weight of some prey classes due to growth of young deer and declining seasonal condition, which is usually lowest during late winter when most studies have been conducted. Also, few data exist on summer kill rates, because kills are difficult to document without snow.

Mech and Peterson (2003, p. 145) stressed the importance of seasonal prey vulnerability to wolf predation and provided an example for white-tailed deer: "Throughout the year, fawns are vulnerable as a class, although not every individual is (Kunkel and Mech, 1994); during summer, adults are rarely taken, so fawns form most of the wolf's diet. In fall, adult bucks—occupied with fighting and the rut instead of eating—become vulnerable, and finally during late winter and spring, when pregnant does reach the nadir of their condition (DelGiudice et al., 1991), they become more vulnerable (Nelson and Mech, 1986b)." However, it must be pointed out that there is considerable variation among seasons and species of predator and their impacts (Carstensen et al., 2009). Although the above example is for white-tailed deer, a similar pattern may exist for other ungulates that are wolf prey species.

Wolves have been known to kill white-tailed deer in excess of their immediate needs with little to no immediate consumption of the prey. Kruuk (1972) termed this type of predation as "surplus-killing." Little information exists on whether surplus-killing occurs during summer, but during winter, deep snow and poor nutritional condition of deer appeared to be linked to the occurrence of surplus-killing (DelGiudice, 1998). Increased incidence of surplus-killing of white-tailed deer in Minnesota could be predicted when snow depths exceeded 70 cm for four to eight weeks. Although surplus-killing is an interesting phenomenon, it is relatively rare in most ungulate populations. Its role in limiting ungulate populations is probably insignificant, but localized impacts could occur.

Clearly, wolves can be a significant limiting factor for white-tailed deer populations. Arguments continue on whether such predation is simply a limiting factor or whether it is regulatory (Eberhardt, 1997;

Messier and Joly, 2000). However, only in a few situations are wolves the sole predator. The combined effects of multiple predator species may be more instructive than examining single predators.

Coyotes

Coyote predation on white-tailed deer has received considerable attention and is a significant source of white-tailed deer mortality in many studies (Figure 8.7). Coyote predation was the leading cause of fawn mortality in eight of the 18 fawn mortality studies that I reviewed. The eastern coyote has replaced the gray wolf as the principal predator of white-tailed deer in northeastern portions of the continent when wolves are absent (Matthews and Porter, 1992; Ballard et al., 1999). Generalizations about the impacts of coyote predation have been difficult, because of the facultative nature of this predator. Coyotes are opportunistic scavengers and predators, and can survive on a variety of foods, with diets ranging from entirely plant to entirely animal material. Several studies have attempted to evaluate the impacts of coyote predation on white-tailed deer.

Enclosure and coyote-control program studies indicate that under some circumstances coyote removal can benefit deer populations. However, benefit:cost ratios and long-term benefits have not been reported. Also, there could be some negative consequences associated with predator control (see subsequent sections).

Similar to other predators, coyotes do not always prey on deer and in several areas they are not significant predators of white-tailed deer. Much remains to be learned about coyote predation. Similar to other predator species, why do coyotes constitute a significant source of mortality in some places and not in others? Availability of alternate prey appears to be a major factor (Patterson and Messier, 2000; and many others). Additional research on coyote behavior, long-term impacts of control measures, and benefit:cost ratios are needed.

Bears

Ballard (1992) reviewed the effects of bear predation on moose and, more recently, Zager and Beecham (2006) reviewed the role of bears as predators of ungulates in North America. Black bear predation has been documented as a significant source of white-tailed deer fawn mortality in several areas of North America, particularly in the eastern portion of white-tailed deer distribution (Figure 8.8). However,

FIGURE 8.7 Where wolves have been extirpated, coyote can be a significant source of fawn and adult white-tailed deer mortality. (Photo by G. Andrejko. With permission.)

FIGURE 8.8 In many areas of North America, particularly in eastern North America, black bears have been identified as an important cause of white-tailed deer fawn mortality. (Photo by D. G. Hewitt. With permission.)

Vreeland et al. (2004) indicated that there were anecdotal accounts from other areas such as Michigan, Vancouver, and Alberta that black bears preyed on fawns. Bear predation is almost entirely focused on neonates during the first two months of life. There are numerous other studies in the midwest and western portions of the continent where black bear predation was probably an insignificant source of fawn mortality. Black bears undoubtedly kill some adult deer, but I found no reports to indicate that they are a significant source of mortality for adult deer.

Much is unknown about the nature of bear predation. Zager and Beecham (2006) indicated that we do not know if bear predation is the proximate or ultimate cause of mortality. Similar to arguments presented here, they concluded that success of attempts to reduce bear predation would likely depend on the prey population's relationship to habitat carrying capacity. Other factors such as predator and prey density, weather conditions, and human influences on the environment are other important considerations.

Two experiments attempted to reduce black bear densities to increase ungulate survival. Stewart et al. (1985) attempted to remove all black bears from two study areas in Saskatchewan. Moose calf:cow ratios increased and were twice as high as those in areas where bears were not removed. In subsequent years when bears were not removed, ratios declined to pre-removal levels. However, proportions of yearlings in the herd increased from 5% to 24% following bear reduction. They concluded that bear removal had not provided lasting relief from bear predation on moose, but the short-term nature of this study precluded firm conclusions.

Keech (2005) (cited in Boertje et al., 2010) translocated black bears from a 1368-km^2 study area in interior Alaska to determine if low moose calf survival could be improved. Eighty-one black bears and nine grizzly bears were transplanted during 2003 and 34 black bears and one grizzly bear were transplanted from the area in 2004. Keech (2005) concluded that removing a majority of the bears resulted in decreased numbers of calves being killed by bears and an increase in moose calf survival. Long-term impacts of this experiment and differences in adult survival were not reported, although the moose population did increase.

Black bears can be a significant cause of ungulate neonate mortality, but this is not always the case. Why do black bears prey on neonates in some areas and not in others? Are all bears equally responsible for predation? Limited evidence suggests that adult male bears prey more on ungulates than other sex and age classes of bears (Ballard, 1992), but this has not been investigated for white-tailed deer. Would bear reduction programs result in more ungulates for harvest? Much more research is needed to elucidate the nature and impacts of black bear predation on ungulate populations in general and white-tailed deer in particular.

There is no evidence that grizzly bears constitute significant mortality to white-tailed deer. This may be due to the limited overlap in distribution of grizzly bears and white-tailed deer. Also, grizzly bears occur in multipredator and multiprey systems, and white-tailed deer are often not the most numerous prey species.

Mountain Lions

Predation by mountain lions on deer has not received as much research effort as other predator species (Figure 8.9). Logan et al. (1996) conducted a long-term study on lion predation on mule deer and the effects of lion removal. They concluded that drought was a major factor affecting the deer population and that lion predation was a major mortality factor, but that habitat quality and quantity were the main driving forces. Deer appear to be the primary prey of mountain lions, but predation on other species can also be significant and depress secondary species (Knopff and Boyce, 2007). Individual lions may specialize on certain species or types of prey so that mere overlap of lion home ranges with prey species is not a good indicator of a predator population's impact on prey species (Knopff and Boyce, 2007). Thus, lion control may not have the desired impact on some prey populations.

Interactions among wolves and other predators with mountain lions have gained attention recently. Wolves appear to usurp lion kills and could result in higher kill rates by lions. More research will help describe this apparent pattern.

A number of western states allow or sanction reductions in lion populations to protect endangered ungulates such as bighorn sheep and other species (Rominger, 2007). Rominger (2007) indicated that many lion populations were subsidized by cattle and other exotic ungulates, resulting in increased predation pressure on native ungulates. Much more research is necessary to evaluate the impacts of lion predation on all ungulates including white-tailed deer.

Bobcats and Foxes

Little is known about the nature of bobcat or fox predation on populations of white-tailed deer or other ungulates. Bobcats can constitute a significant source of mortality on neonates as evidenced by this review, but factors determining when bobcats will become a significant source of mortality are unknown. Undoubtedly, the availability of alternate prey influences this relationship. In other cases, both bobcats and foxes may prey on neonates that are predisposed to die from other factors, such as disease and starvation. This may have been the case in west-central Texas, but apparently healthy fawns also were killed by other predators (Haskell, 2007). Neither bobcat nor fox appear to be significant predators of adult deer.

FIGURE 8.9 Recent advances in GPS technology have allowed biologists to better understand prey species selection and kill rates by mountain lions. Here, Dr. Kerry Nicholson attaches a GPS collar to a mountain lion in Arizona. (Photo by B. Buckley. With permission.)

More research is needed to determine conditions under which such predation occurs and whether or not it should be managed to increase white-tailed deer populations. Lynx surely kill some white-tailed deer, but I found no evidence that they are a significant predator of deer.

Regulation versus Limitation of Deer Populations

The presence of multiple predator species greatly influences the ecosystems they inhabit and their prey species (Bergerud, 1988; Berger and Wehausen, 1991; Hatter and Janz, 1994; Van Ballenberghe and Ballard, 1994; McNay and Voller, 1995; Boertje et al., 1996; Wehausen, 1996; Ballard and Van Ballenberghe, 1998a,b; Ballard et al., 2001). The number of predator species also has large implications for whether predation is solely a limiting factor or a regulating factor. Undoubtedly, the impacts of a single predator species are different from those of multiple predator species.

Debate continues on whether predation is a regulating or limiting factor of ungulate populations. A number of studies have provided data, arguments, and models of how predation could limit or regulate ungulate numbers (Boutin, 1992; Van Ballenberghe and Ballard, 1994; McRoberts et al., 1995; Eberhardt, 1997, 2000; Ballard and Van Ballenberghe, 1998a,b; Messier and Joly, 2000; Eberhardt et al., 2003; Mech and Peterson, 2003). The models being debated have focused on four basic themes: low-density equilibria, multiple stable states, stable-limit cycles, and recurrent fluctuations (Van Ballenberghe and Ballard, 1994; Ballard and Van Ballenberghe, 1998a,b; Ballard et al., 2001). In all of these cases, predation is a significant limiting factor and in some cases could be regulatory. Ballard and Van Ballenberghe (1998a,b) and Ballard et al. (2001) provide visual conceptual models of ungulate numbers under each scenario. Evidence from northern latitudes involving wolves, bears, and moose suggests the low-density equilibria model, where ungulate densities would be held at low levels for extended periods of time, with little surplus available for hunters (Boertje et al., 2010). Virtually all of this debate focuses on northern latitudes where wolves and bears are the predominant predators and moose and caribou are the principal prey. Only some localized debate has occurred about effects of predation on white-tailed deer. Perhaps the largest reason is that, within the range of white-tailed deer, anthropogenic factors have greatly changed the relationships between predators and prey and their habitats (Ballard et al., 2001). Indeed, predator species have been extirpated from many areas of their former range, facultative carnivores such as bears and coyotes have numerically increased and expanded their distributions, and habitat and grazing schemes have been greatly modified due to anthropogenic causes. In such situations, any of the proposed models are possible and warrant further study. In any case, whether predation is a limiting or regulating factor, it is clear that predation can greatly modify management strategies. Understanding why prey populations fluctuate is important in deciding how to manage predator–prey relationships. However, if predation is a limiting factor and reductions in limiting factors achieve higher ungulate populations and higher public harvests, then managers may have all of the information needed to manage predator–prey relationships. In other cases, human harvest may not be able to limit deer numbers and other solutions will have to be sought.

Predator Control, Ecosystem Effects, and Public Perceptions

A number of predator-reduction experiments have been conducted to improve ungulate survival and recruitment, and ultimately result in higher human harvests. Techniques used have included ground-shooting, trapping, aerial-gunning from both helicopter and fixed-wing aircraft, and, in some cases, poison. However, most of these experiments, other than enclosure studies, have been conducted on northern ungulates, such as moose and caribou. These studies clearly indicate that calf and yearling recruitment can be enhanced through predator-reduction programs involving wolves and bears when these species have been identified as significant limiting factors and ungulate populations were well below habitat carrying capacity. These programs have resulted in larger harvests of moose, which was the objective of the programs (Boertje et al., 2010). However, a number of differences exist between the northern programs and those that have been tested on white-tailed deer.

Examinations of urbanized and suburban areas clearly indicate that reductions in predator numbers along with habitat improvements result in higher deer densities. These areas are characterized by extirpation of many predators, changes in predator species (obligate to facultative), reductions in (or no) hunting, and changes in habitat to earlier successional stages that favor deer. If management for harvest were a priority, then managers would be harvesting annual increments or reducing deer populations with harvest. However, in many situations, due to concerns about public safety and public sentiment, harvest opportunities are restricted or eliminated, and deer become overpopulated and are viewed by many as nuisances. In these cases, managers are forced to develop alternative deer population control techniques such as translocation and sterilization.

There are no government-sanctioned predator-reduction programs that are aimed at controlling predator numbers to increase white-tailed deer. However, on many private properties, particularly in western states, predator-control programs are conducted routinely to benefit white-tailed deer. These programs utilize ground-shooting, aerial-gunning from both helicopters and fixed-wing aircraft, trapping, and poison. Many programs are conducted by private individuals. Such programs can and do result in deer population increases that may or may not result in the desired result of increased harvests of deer. On private properties deer harvest levels are often restricted and conservative and can lead to undesired results, such as overpopulation. Perhaps, a good example of this is on the Edwards Plateau, Texas, where extensive predator control has been associated with high populations of deer and other herbivores, such as lagomorphs and livestock. In many cases, these populations are at or above carrying capacity, and instead of predation or hunting being the leading cause of death, many deer succumb to disease and starvation (Haskell, 2007). The problem is not predation, but an overpopulation of deer, which can also result in undesired consequences.

Ecosystem Effects

For example, a number of studies have documented changes in rodents, lagomorphs, and mesocarnivores as a result of predator-control programs. Henke and Bryant (1999) evaluated the effects of coyote control on sympatric populations of rodents, lagomorphs, raptors, and mammalian mesopredators on four sites (two treatment and two control) in west Texas during a two-year period. Coyote densities were reduced from 0.12 to 0.06 coyotes/km^2. They found that rodent species richness and diversity decreased on the treatment sites; however, rodent density and biomass, black-tailed jackrabbit density, and badger, bobcat, and gray fox abundance increased on the treatment sites. The richness, diversity, and density of cottontail rabbits and raptor species were not different.

Other investigators have reported similar changes in lower fauna due to coyote-control programs (Clark, 1972; Wagner and Stoddard, 1972; Keith et al., 1977; Harris and Silva-Lopez, 1992; and many others). Other than these unintended ecological consequences, reductions in coyote numbers could result in large increases in black-tailed jackrabbit populations resulting in overuse of forage and greater competition with livestock (Wagner, 1988). Undoubtedly, other impacts probably occur from control of coyotes and other facultative predators. Ecosystem impacts from control of obligate carnivores have not received the same attention as that of facultative carnivores.

Public Perceptions and Benefit:Cost Ratios

Managers have a poor understanding of how various stakeholders view predator-reduction programs. There appears to be a wide discrepancy in public support for predator-reduction programs to benefit huntable species among regions, species of predator controlled, and a large number of socio-demographic factors such as education, age, and sex (Kellert, 1984, 1985). Although all these factors are important in determining public attitudes, there do appear to be large differences in public opinion based on the species being managed and the region where such actions are being conducted. For example, predator reductions of charismatic carnivores, such as wolves and grizzly bears, appear to evoke strong opposition from some stakeholders (Stephenson et al., 1995; Boertje et al., 2010), but for other species, the public is unconcerned, displays little opposition, or is uninformed about such actions. Coyote reduction, for instance, occurs across the western United States, but elicits little in the way of public response, except in some states that have large urban populations which have banned some forms of control (Ballard et al., 2001). Kamler et al. (2003) killed 227 coyotes during 2000 by aerial-shooting from fixed-wing aircraft to test whether such reductions would enhance swift fox populations, and virtually no protests were reported or received. This same situation appears to be occurring in private land states where landowners routinely control coyotes and other predators with a variety of techniques with little or no public comment.

Whether managers should attempt to control or manipulate predator populations to favor white-tailed deer as public policy has not been an issue and should be dependent on the biological circumstances involved, methods used, objectives of the program, benefit:cost ratios, and public acceptance of such management policies. Managers have little information to determine public acceptance of predator control to enhance wildlife even when biologically justified. Messmer et al. (1999) examined public attitudes concerning the control of mesopredators to enhance avian populations. Their results suggested that the interested public of the United States may be willing to support predator control under specific situations and when used as a "surgical procedure." Respondents were not convinced that human needs were more important than a predator's right to exist. However, there was little support for eliminating predator control as a management tool.

Decker et al. (2006) examined some of the factors that affected public acceptance or rejection of Alaska's predator-control programs to reduce wolf and bear predation to increase moose populations for increased harvests. Interestingly, they found that support for such programs was influenced by whether human hunting was for subsistence or recreation. Predator-control programs were less likely to be supported if predation was "perceived to be less severe with respect to human needs" (Decker et al., 2006, p. 426). A majority of participants indicated that when subsistence issues were involved they preferred lethal to nonlethal predator-control methods. The authors indicated that people make anthropocentric, contest-mediated distinctions about what is acceptable management even when they are not personally affected. Even people far removed from management actions based their decisions on the perceived impacts on other people. They concluded that public acceptability of control methods was often determined by how the situation was presented by news media, interest groups, and agencies. Perhaps the take-home message from their study is that local dependence on resources and responsibility for economic impacts may have a lot to do with public acceptance of management actions. Clearly, managers lack information on the reasons for public acceptance or rejection of management programs to increase big game populations. In some situations public acceptance may be attainable, and in others acceptance may be impossible.

A glaring deficiency of all predator management programs has been lack of monitoring (i.e., documentation of whether program objectives were met and whether the actions had a favorable benefit:cost ratio both ecologically and economically). The only study that has reported a benefit:cost ratio involved coyote reduction and pronghorn population performance. As a result of coyote reductions from helicopter-gunning, Smith et al. (1986) documented a 400% increase in pronghorn populations at Anderson Mesa, Arizona. They provided a benefit:cost ratio of 1.92 when coyote populations were reduced at two-year intervals. Such analyses should be required for all predator management programs designed to enhance white-tailed deer and other ungulate species.

Should Managers Control Predators to Increase White-tailed Deer?

Whether managers choose to initiate predator management programs to increase white-tailed deer populations depends on many important factors. Ballard et al. (2001) in discussing this same issue with regard to mule deer listed the factors as follows: public acceptance, management goals, scale, methodology, biological relevance, and perhaps, most importantly, relationship of the deer population to habitat carrying capacity. Predation has been identified as a significant limiting factor over large geographic areas, and it would be impossible to manage predators on such large scales. In other areas predation is not a significant factor, and in such cases, predation management is unwarranted. Success of predator management will depend on the deer population's relationship to habitat carrying capacity, with more success when the deer population is well below habitat carrying capacity and no success when at carrying capacity. Intensive predator management on relatively small areas (i.e., less than 1000 km^2, but greater than 200 km^2) appears to be more effective and feasible than wide-scale efforts, and may have more public support if viewed as "surgical management" rather than over large areas (Ballard et al, 2001).

Methods of predator reduction are also limited as ground-shooting and trapping are largely ineffective. Aerial-shooting from either helicopter or fixed-wing aircraft appears to be an effective management tool, but again benefit:cost ratios have not been evaluated. Use of poison is an effective tool, but also harms non-target species. Its use has been banned by federal legislation, but it is still used on some private ranches.

Ballard et al. (2001) identified characteristics of situations when predator control was effective and ineffective. These criteria are relevant to white-tailed deer management and are repeated here.

Successful programs:

- Predator control was implemented when the deer populations were below habitat carrying capacity.
- Predation was identified as a limiting factor.
- Control efforts reduced predator populations enough to yield results (e.g., expected to be approximately 70% of a local coyote population).
- Control efforts were timed to be most effective (just prior to predator or prey reproduction).
- Control took place at a focused scale (generally <1000 km^2).

Unsuccessful programs:

- When deer populations were at or near habitat carrying capacity
- When predation was not a key limiting factor
- Where control failed to reduce predator populations sufficiently to be effective
- Where control efforts were on large-scale areas

Before initiating predator management the following parameters should be determined for each program:

- A management plan needs to be in place that identifies the following:
    - Current status of deer populations and population objective desired from the predator-control project
    - Desired removal goals for the predator species
    - Timing and method of removal efforts
    - Scale of removal effort
    - What other limiting factors may be playing a role in depressing the deer population
- An adaptive management plan that sets monitoring criteria for evaluation of both predator and prey populations and identifies thresholds when predator control will cease or be modified.

Research Needs

Ballard et al. (2001) also identified aspects of predator control requiring additional research to clarify or determine the value of such programs. Much of the research conducted so far has been short term and has suffered from poor experimental design. Human-social issues have rarely been examined. Needs include the following:

- Initiate experimental, long-term research, particularly on coyote, mountain lion, and black bear predation to clarify the role of predation on deer. An experimental approach is necessary whereby deer population performance in relation to predator density is monitored in both manipulated and un-manipulated areas. The study period should be long enough so that both severe and favorable weather conditions occur. Confounding factors such as other predators, human harvests, densities of buffer species, and habitat conditions must also be measured to allow appropriate interpretation of study results.

- Well-designed research to measure social attitudes toward various aspects of predator-control programs is important. As a minimum, research should develop a better understanding of variation in public attitudes toward methods of predator control, timing of predator control, prey species thresholds that would warrant implementation of a predator-control program, and scale of a predator-control program.
- Research should include an analysis of benefit:cost ratios of any control efforts.

Conclusions

Predation by wolves, coyotes, mountain lions, bobcats, and, in some cases, foxes can be significant source of mortality on white-tailed deer neonates and can be an important limiting factor. Wolves, coyotes, and mountain lions can also have significant effects on adult white-tailed deer. Impacts of predation on white-tailed deer populations are determined by type of carnivore (facultative or obligate), alternate prey resources, vulnerability of deer, condition of habitat and deer, and the deer population's relationship to habitat carrying capacity. Predation on white-tailed deer can be a significant limiting factor, but it is unknown whether predation regulates white-tailed deer populations. Anthropogenic changes in the landscape make a number of limitation and regulation scenarios possible. In many states and provinces, humans are the predators that limit deer populations by taking the appropriate number of adult female deer annually, but this can be difficult to achieve in places where deer are highly productive or social factors prevent effective hunting strategies (e.g., large blocks of private restricted lands, urban and suburban firearms ordinances, and social stigma for shooting female deer). Results from predator-free enclosures and island populations without predators demonstrate the potential effects of predation that can significantly impede population growth. Predator management has been used to increase some prey populations, but the benefit and costs of such operations and resultant increases in harvest have not been thoroughly examined. Manipulative studies need to be better designed to answer important questions. Many questions remain about the role of predation, particularly for facultative carnivores such as coyote and bears. Management and research on predator–prey relationships remains controversial and this will continue into the future.

Acknowledgments

I thank M. Butler, A. Escobar, and R. Dillon for assistance with literature searches and preparation of tables. R. Carrera, S. Haskell, D. Hewitt, and H. Whitlaw provided valuable input and criticism. I am especially grateful for the critical and constructive reviews provided by G. DelGiudice, P. Krausman, and D. McCullough. These pioneers in deer biology improved this manuscript.

REFERENCES

Adams, L. G. and B. W. Dale. 1998. Timing and synchrony of parturition in Alaskan caribou. *Journal of Mammalogy* 79:287–294.

Adams, L. G., B. W. Dale, and L. D. Mech. 1995. Wolf predation on caribou calves in Denali National Park, Alaska. In *Ecology and Conservation of Wolves in a Changing World*, eds. L. N. Carbyn, S. H. Fritts, and D. R. Seip, 245–260. Edmonton, Canada: Canadian Circumpolar Institute.

Altendorf, K. B., J. W. Laundré, C. A. López González, and J. S. Brown. 2001. Assessing effects of predation risk on foraging behavior of mule deer. *Journal of Mammalogy* 82:430–439.

Arjo, W. M., D. H. Pletscher, and R. R. Ream. 2002. Dietary overlap between wolves and coyotes in northwestern Montana. *Journal of Mammalogy* 83:754–766.

Ballard, W. B. 1992. Bear predation on moose: A review of recent North American studies and their management implications. *Alces Supplement* 1:162–176.

Ballard, W. B., D. Lutz, T. W. Keegan, L. H. Carpenter, and J. C. deVos, Jr. 2001. Deer–predator relationships: A review of recent North American studies with emphasis on mule and black-tailed deer. *Wildlife Society Bulletin* 29:99–115.

Ballard, W. B., D. Lutz, T. W. Keegan, L. H. Carpenter, and J. C. deVos, Jr. 2003. Deer–predator relationships: A review of recent North American studies with emphasis on mule and black-tailed deer. In *Mule Deer Conservation: Issues and Management Strategies,* eds. J. C. deVos, Jr., M. R. Conover, and N. E. Headrick, 177–218. Logan, UT: Jack H. Berryman Institute Press.

Ballard, W. B. and V. Van Ballenberghe. 1998a. Predator–prey relationships. In *Ecology and Management of North American Moose*, eds. A. W. Franzmann and C. C. Schwartz, 247–273. Washington, DC: Smithsonian Institution Press.

Ballard, W. B. and V. Van Ballenberghe. 1998b. Moose–predator relationships: Research and management needs. *Alces* 34:91–105.

Ballard, W. B., H. A. Whitlaw, S. J. Young, R. A. Jenkins, and G. J. Forbes. 1999. Predation and survival of white-tailed deer fawns in northcentral New Brunswick. *Journal of Wildlife Management* 63:574–579.

Banfield, A. W. F. 1954. Preliminary investigation of the barren-ground caribou. Part I. Former and present distribution, migrations, and status. *Wildlife Management Bulletin Series 1*, no. 10A. Canadian Wildlife Service.

Bannikov, A. G., L. V. Zhirnov, L. S. Lebedeva, and A. A. Fandeev. 1967. *Biology of the Saiga.* [Translated from the Russian by M. Fleischmann.] Washington, DC: U.S. Department of the Interior and National Science Foundation.

Bartmann, R. M, G. C. White, and L. H. Carpenter. 1992. Compensatory mortality in a Colorado mule deer population. *Wildlife Monographs* 121:1–39.

Bartush, W. S. 1978. Mortality of white-tailed deer fawns in the Wichita Mountains, Comanche County, Oklahoma, Part II. MS thesis, Oklahoma State University.

Beasom, S. L. 1974a. Relationships between predator removal and white-tailed deer net productivity. *Journal of Wildlife Management* 38:854–859.

Beasom, S. L. 1974b. Intensive short-term predator removal as a game management tool. *Transactions of the North American Wildlife and Natural Resources Conference* 39:230–240.

Berger, J. 1978. Group size, foraging, and antipredator ploys: An analysis of bighorn sheep decisions. *Behavior, Ecology, and Sociobiology* 4:91–99.

Berger, J. and J. D. Wehausen. 1991. Consequences of mammalian predator–prey disequilibrium in the Great Basin Desert. *Conservation Biology* 5:244–248.

Berger, J., P. B. Stacey, L. Bellis, and M. P. Johnson. 2001. A mammalian predator–prey imbalance: Grizzly bear and wolf extinction affect avian neotropical migrants. *Ecological Applications* 11:947–960.

Bergerud, A. T. 1983. The natural population control of caribou. In *Symposium on the Natural Regulation of Wildlife Populations*, 14–61. Forest, Wildlife and Range Experiment Station, University of Idaho, Moscow.

Bergerud, A. T. 1985. Antipredator strategies of caribou: Dispersion along shorelines. *Canadian Journal of Zoology* 63:1324–1329.

Bergerud, A. T. 1988. Caribou, wolves and man. *Trends in Ecology and Evolution* 3:68–71.

Bergerud, A. T., H. E. Butler, and D. R. Miller. 1984. Antipredator tactics of calving caribou: Dispersion in mountains. *Canadian Journal of Zoology* 62:1566–1575.

Bergerud, A. T., R. D. Jakimchuk, and D. R. Carruthers. 1984. The buffalo of the north: Caribou (*Rangifer tarandus*) and human developments. *Arctic* 37:7–22.

Bildstein, K. L. 1983. Why white-tailed deer flag their tails. *American Naturalist* 121:709–715.

Boertje, R. D., P. Valkenburg, and M. E. McNay. 1996. Increase in moose, caribou, and wolves following wolf control in Alaska. *Journal of Wildlife Management* 60:474–489.

Boertje, R. D., M. A. Keech, and T. F. Paragi. 2010. Science and values influencing predator control for Alaska moose management. *Journal of Wildlife Management* 74:917–928.

Boutin, S. 1992. Predation and moose population dynamics: A critique. *Journal of Wildlife Management* 56:116–127.

Brinkman, T. J., J. A. Jenks, S. DePerno, B. S. Haroldson, and R. G. Osborn. 2004. Survival of white-tailed deer in an intensively farmed region of Minnesota. *Wildlife Society Bulletin* 32:726–731.

Brown, D. E., ed. 1983. *The Wolf in the Southwest*. Tucson, AZ: The University of Arizona Press.

Burroughs, J. P., H. Campa, III, S. R. Winterstein, B. A. Rudolph, and W. E. Moritz. 2006. Cause-specific mortality and survival of white-tailed deer fawns in southwestern lower Michigan. *Journal of Wildlife Management* 70:743–751.

Carbyn, L. N. 1974. *Wolf Predation and Behavioral Interactions with Elk and Other Ungulates in an Area of High Prey Diversity*. Edmonton, Alberta, Canada: Canadian Wildlife Service.

Carbyn, L. N. 1983. Wolf predation on elk in Riding Mountain National Park, Manitoba. *Journal of Wildlife Management* 47:963–976.

Carbyn, L. N., S. M. Oosenbrug, and D. W. Anions. 1993. Wolves, bison and the dynamics related to the Peace Athabaska Delta in Canada's Wood Buffalo National Park. *Circumpolar Research Series*, No. 4. Canadian Circumpolar Institute, Edmonton, Canada: University of Alberta.

Carl, G. and C. T. Robbins. 1988. The energetic cost of predator avoidance in neonatal ungulates: Hiding versus following. *Canadian Journal of Zoology* 66:239–246.

Caro, T. M., L. Lombardo, A. W. Goldizen, and M. Kelly. 1995. Tail-flagging and other antipredator signals in white-tailed deer: New data and synthesis. *Behavioral Ecology* 6:442–450.

Carrera, R., W. Ballard, P. Gipson et al. 2004. Comparison of Mexican wolf and coyote diets in Arizona and New Mexico. *Journal of Wildlife Management* 72:376–381.

Carrera-Trevino, R. 2008. Population dynamics of desert mule deer (*Odocoileus hemionus eremicus*) in central Arizona. PhD dissertation, Texas Tech University.

Carroll, B. K. and D. L. Brown. 1977. Factors affecting neonatal fawn survival in southern-central Texas. *Journal of Wildlife Management* 41:63–69.

Carstensen, M., G. D. DelGiudice, B. A. Sampson, and D. W. Kuehn. 2009. Survival, birth characteristics, and cause-specific mortality of white-tailed deer neonates. *Journal of Wildlife Management* 73:175–183.

Case, D. J. 1982. The white-tailed deer of North Manitou Island, Michigan. MS thesis, University of Michigan.

Case, D. J. and D. R. McCullough. 1987. The white-tailed deer of North Manitou Island, Michigan. *Hilgardia* 55:1–57.

Clark, F. W. 1972. Influence of jackrabbit density on population change. *Journal of Wildlife Management* 36:343–356.

Connolly, G. E. 1978. Predators and predator control. In *Big Game of North America*, eds. J. L. Schmidt and D. L. Gilbert, 369–394. Harrisburg, PA: Statckpole Books.

Cook, R. S., M. White, D. O. Trainer, and W. C. Glazener. 1971. Mortality of young white-tailed deer fawns in south Texas. *Journal of Wildlife Management* 35:47–56.

Cowan, I. M. 1947. The timber wolf in the Rocky Mountain national parks of Canada. *Canadian Journal of Research* 25:139–174.

Creel, S., D. Christianson, S. Liley, and J. A. Winnie, Jr. 2007. Predation risk affects reproductive physiology and demography of elk. *Science* 315:960.

Crisler, L. 1956. Observations of wolves hunting caribou. *Journal of Mammalogy* 37:337–346.

Dale, B. W., L. G. Adams, and R. T. Bowyer. 1994. Functional response of wolves preying on barren-ground caribou in a multiple-prey ecosystem. *Journal of Animal Ecology* 63:644–652.

Darling, F. F. 1937. *A Herd of Red Deer*. Oxford: Oxford University Press.

Decker, D. J., C. A. Jacobson, and T. L. Brown. 2006. Situation-specific "impact dependency" as a determinant of management acceptability: Insights from wolf and grizzly bear management in Alaska. *Wildlife Society Bulletin* 34:426–432.

Dehn, M. M. 1990. Vigilance for predators: Detection and dilution effects. *Behavioral Ecology and Sociobiology* 26:337–342.

DelGiudice, G. D. 1998. Surplus killing of white-tailed deer by wolves in northcentral Minnesota. *Journal of Mammalogy* 79:227–235.

DelGiudice, G. D., J. Fieberg, M. R. Riggs, M. C. Powell, and W. Pan. 2006. A long-term age-specific survival analysis of female white-tailed deer. *Journal of Wildlife Management* 70:1556–1568.

DelGiudice, G. D., L. D. Mech, and U. S. Seal. 1991. Gray wolf density and its association with weights and hematology of pups from 1970 to 1988. *Journal of Wildlife Diseases* 27:630–636.

DelGiudice, G. D., M. C. Riggs, P. Joly, and W. Pan. 2002. Winter severity, survival, and cause-specific mortality of female white-tailed deer in north-central Minnesota. *Journal of Wildlife Management* 66:698–717.
DeYoung, C. A. 1989. Mortality of adult male white-tailed deer in south Texas. *Journal of Wildlife Management* 53:513–523.
Dusek, G. L., A. K. Wood, and S. T. Stewart. 1992. Spatial and temporal patterns of mortality among female white-tailed deer. *Journal of Wildlife Management* 56:645–650.
Eberhardt, L. L. 1997. Is wolf predation ratio-dependent. *Canadian Journal of Zoology* 75:1940–1944.
Eberhardt, L. L. 2000. Reply: Predator–prey ratio dependence and regulation of moose populations. *Canadian Journal of Zoology* 78:511–513.
Eberhardt, L. L., R. A. Garrott, D. W. Smith, P. J. White, and R. O. Peterson. 2003. Assessing the impact of wolves on ungulate prey. *Ecological Applications* 13:776–783.
Edwards, J. 1983. Diet shifts in moose due to predator avoidance. *Oecologia* 60:185–189.
Estes, R. D. 1966. Behavior and life history of the wildebeest (*Connochaetes taurinus* Burchell). *Nature* 212:999–1000.
Etter, D. R., K. M. Hollis, T. R. Van Deelen et al. 2002. Survival and movement of white-tailed deer in suburban Chicago, Illinois. *Journal of Wildlife Management* 66:500–510.
Ezcurra, E. and S. Gallina. 1981. Biology and population dynamics of white-tailed deer in northwestern Mexico. In *Deer Biology, Habitat Requirements, and Management in Western North America,* eds. P. F. Ffolliott and S. Gallina, 79–108. Mexico: Instituto DeEcologia, A. C.
Ferguson, S. H., A. T. Bergerud, and R. Ferguson. 1988. Predation risk and habitat selection in the persistence of a remnant caribou population. *Oecologia* 76:236–245.
Fox, J. L. and G. P. Streveler. 1986. Wolf predation on mountain goats in southeastern Alaska. *Journal of Mammalogy* 67:192–195.
Frid, A. and L. Dill. 2002. Human-caused disturbance stimuli as a form of predation risk. *Conservation Ecology* 6:11 (online http://www.consecol.org/vol6/iss1/art11).
Fryxell, J. M., J. Greever, and A. R. E. Sinclair. 1988. Why are migratory ungulates so abundant? *Ecological Society American Bulletin* 69:140.
Gaillard, J. M., M. Festa-Bianchet, N. G. Yoccoz, A. Loison, and C. Toigo. 2000. Temporal variation in fitness components and population dynamics of large herbivores. *Annual Review of Ecology and Systematics* 31:367–393.
Garner, G. W. 1976. Mortality of white-tailed deer fawns in the Wichita Mountains, Oklahoma. PhD dissertation, Oklahoma State University.
Garner, G. W., J. A. Morrison, and J. C. Lewis. 1976. Mortality of white-tailed deer fawns in the Wichita Mountains, Oklahoma. *Proceedings Annual Conference Southeastern Association Fish and Wildlife Agencies* 13:493–506.
Gray, D. R. 1987. *The Muskoxen of Polar Bear Pass*. Markham, Canada: Fitzhenry and Whiteside.
Grundlach, H. 1968. Brutvorsorge, Brutpflege, Verhaltensontogenese and Tagesperiodik beim europaischen Wildschwein (*Sus scrofa* L.). *Zoological Tierpsychology* 25:955–995.
Guthery, F. S. and S. L. Beasom. 1977. Responses of game and nongame wildlife to predator control in south Texas. *Journal of Range Management* 30:404–409.
Guthrie, R. D. 1971. A new theory of mammalian rump patch evolution. *Behaviour* 38:132–145.
Haber, G. C. 1977. Socio-ecological dynamics of wolves and prey in a subarctic ecosystem. PhD dissertation, University of British Columbia.
Harris, L. D. and G. Silva-Lopez. 1992. Forest fragmentation and the conservation of biological diversity. In *Conservation Biology: The Theory and Practice of Nature Conservation Preservation and Management,* eds. P. L. Fedler and S. K. Jain, 197–238. New York, NY: Chapman & Hall.
Haskell, S. P. 2007. Ecology of sympatric deer species in west-central Texas: Methodology, reproductive biology, and mortality and antipredator strategies of adult females and fawns. PhD dissertation, Texas Tech University.
Haskell, S. P. and W. B. Ballard. 2007. Modeling the Western Arctic caribou during a positive growth phase: Potential effects of wolves and radiocollars. *Journal of Wildlife Management* 71:619–627.
Haskell, S. P. and W. B. Ballard. 2008. Annual re-habituation of calving caribou to oilfields in northern Alaska: Implications for expanding development. *Canadian Journal of Zoology* 86:627–637.
Hatter, J. W. and D. W. Janz. 1994. The apparent demographic changes in black-tailed deer associated with wolf control in northern Vancouver Island. *Canadian Journal of Zoology* 72:878–884.

Heard, D. C. 1992. The effect of wolf predation and snow cover on musk–ox group size. *American Naturalist* 139:190–204.

Hebblewhite, M. and D. H. Pletscher. 2002. Effects of elk group size on predation by wolves. *Canadian Journal of Zoology* 80:800–809.

Heffelfinger, J. R. 1989. The effects of intensive coyote control on post-rut mortality of buck white-tailed deer. MS thesis, Texas A&I University.

Heffelfinger, J. R., S. L. Beasom, and C. A. DeYoung. 1990. The effects of intensive coyote control on post-rut mortality of male white-tailed deer. In *Managing Wildlife in the Southwest*, eds. P. R. Krausman and N. S. Smith, 35–45. Phoenix, AZ: Arizona Chapter of the Wildlife Society.

Henke, S. E. and F. C. Bryant. 1999. Effects of coyote removal on the faunal community in western Texas. *Journal of Wildlife Management* 63:1066–1081.

Hirth, D. H. and D. R. McCullough. 1977. Evolution of alarm signals in ungulates with special reference to white-tailed deer. *American Naturalist* 111:31–42.

Holroyd, J. D. 1967. Observations of Rocky Mountain goats on Mount Wardle, Kootenay National Park, British Columbia. *Canadian Field Naturalist* 81:1–22.

Hoskinson, R. L. and L. D. Mech. 1976. White-tailed deer migration and its role in wolf predation. *Journal of Wildlife Management* 40:429–441.

Huegel, C. N., R. B. Dahlgren, and H. L. Gladfelter. 1985. Mortality of white-tailed deer fawns in south-central Iowa. *Journal of Wildlife Management* 49:377–380.

Huggard, D. J. 1993. Prey selectivity of wolves in Banff National Park. I. Prey species. *Canadian Journal of Zoology* 71:130–139.

Ims, R. A. 1990. The ecology and evolution of reproductive synchrony. *Trends in Ecology and Evolution* 5:135–140.

Jhala, Y. V. 1993. Predation on blackbuck by wolves in Velavadar National Park, Gujarat, India. *Conservation Biology* 7:874–881.

Kamler, J. F., W. Ballard, R. L. Gilliland, P. R. Lemons II, and K. Mote. 2003. Impacts of coyotes on swift foxes in northwestern Texas. *Journal of Wildlife Management* 67:317–323.

Kay, C. E. 1995. Aboriginal overkill and native burning: Implications for modern ecosystem management. *Western Journal of Applied Forestry* 10:120–126.

Kay, C. E., B. Patton, and C. A. White. 2000. Historical wildlife observations in Canadian Rockies: Implications for ecological integrity. *Canadian Field-Naturalist* 114:561–583.

Keech, M. A. 2005. Factors limiting moose at low density in Unit 19D East, and response of moose to wolf control. *Alaska Department of Fish and Game*. Federal Aid in Wildlife Restoration Final Research Performance Report. Grants W-27-5 and W-33-1 through W-33-3. Project 1.58. Juneau, Alaska.

Keith, L. B., A. W. Todd, C. J. Brand, R. S. Adamick, and D. H. Rusch. 1977. An analysis of predation during a cyclic fluctuation of snowshoe hares. *International Congress of Game Biologists* 13:151–175.

Kellert, S. R. 1984. Affective, cognitive, and evaluative perceptions of animals. In *Behavior and the Natural Environment*, eds. I. Altman and J. Wohlwill, 241–267. New York, NY: Plenum Press.

Kellert, S. R. 1985. Public perceptions of predators, particularly the wolf and coyote. *Biological Conservation* 31:167–189.

Kie, J. G. and M. White. 1985. Population dynamics of white-tailed deer (*Odocoileus virginianus*) on the Welder Wildlife Refuge, Texas. *Southwestern Naturalist* 30:105–118.

Kitchen, D. W. 1974. Social behavior and ecology of the pronghorn. *Wildlife Monographs* 38:1–96.

Knopff, K. H. and M. S. Boyce. 2007. Prey specialization by individual cougars in multiprey systems. *Proceedings of the North American Wildlife and Natural Resources Conference* 72:194–210.

Knowlton, F. F. 1972. Preliminary interpretations of coyote population mechanisms with some management implications. *Journal of Wildlife Management* 36:369–383.

Kruuk, H. 1972. Surplus killing by carnivores. *Journal of Zoology (London)* 166:233–244.

Kucera, T. E. 1988. Ecology and population dynamics of mule deer in the eastern Sierra Nevada, California. PhD dissertation, University of California, Berkeley.

Kunkel, K. E. 1997. Predation by wolves and other large carnivores in northwestern Montana and southeastern British Columbia. PhD dissertation, University of Montana.

Kunkel, K. E. and L. D. Mech. 1994. Wolf and bear predation on white-tailed deer fawns in northeastern Minnesota. *Canadian Journal of Zoology* 72:1557–1565.

Kunkel, K. E. and D. H. Pletscher. 1999. Species-specific population dynamics of cervids in a multipredator ecosystem. *Journal of Wildlife Management* 63:1082–1093.

Kunkel, K. E., D. H. Pletscher, D. K. Boyd, R. Ream, and M. W. Fairchild. 2004. Factors correlated with foraging behavior of wolves in and near Glacier National Park, Montana. *Journal of Wildlife Management* 68:167–178.

Kunkel, K. E., T. K. Ruth, D. H. Pletscher, and M. G. Hornocker. 1999. Winter prey selection by wolves and cougars in and near Glacier National Park, Montana. *Journal of Wildlife Management* 63:901–910.

LaGory, K. E. 1986. Habitat, group size, and the behaviour of white-tailed deer. *Behaviour* 98:168–179.

LaGory, K. E. 1987. The influence of habitat and group characteristic on the alarm and flight response of white-tailed deer. *Animal Behavior* 35:20–25.

Landis, R. 1998. *Yellowstone Wolves—Predation.* Trailwood-Landis Films, Gardiner, Montana.

Laundré, J. W., L. Hernández, and K. B Altendorf. 2001. Wolves, elk, and bison: Reestablishing the "landscape of fear" in Yellowstone National Park, U.S.A. *Canadian Journal of Zoology* 79:1401–1409.

Lent, P. C. 1974. Mother–infant relationships in ungulates. In *The Behaviour of Ungulates and its Relation to Management*, eds. V. Geist and F. Walther, New Serial 24, Vol. 1, 14–55. Gland, Switzerland: International Union Conservation Nature Publication.

Lima, S. L. 1988. Vigilance during the initiation of daily feeding in dark-eyed juncos. *Oikos* 53:12–16.

Lima, S. L. and L. M. Dill. 1990. Behavioral decisions made under the risk of predation: A review and prospectus. *Canadian Journal of Zoology* 68:619–640.

Linnell, J. D. C., R. Aanes, and R. Andersen. 1995. Who killed Bambi? The role of predation in the neonatal mortality of temperate ungulates. *Wildlife Biology* 1:209–223.

Logan, K. A., L. L. Sweanor, T. K. Ruth, and M. G. Hornocker. 1996. Cougars of the San Andres Mountains, New Mexico. Final Report to the New Mexico Department of Fish and Game. Hornocker Wildlife Institute, Moscow.

Long, R. A., A. F. O'Connell, Jr., and D. J. Harrison. 1998. Mortality and survival of white-tailed deer *Odocoileus virginianus* fawns on a north Atlantic coastal island. *Wildlife Biology* 4:237–247.

Mackie, R. J., D. F. Pac, K. L. Hamlin, and G. L. Dusek. 1998. Ecology and management of mule deer and white-tailed deer in Montana. Montana Fish, Wildlife and Parks, Wildlife Division, Federal Aid to Wildlife Restoration Report, Project W-120-R. Helena, Montana.

MacNab, J. 1985. Carrying capacity and related slippery shibboleths. *Wildlife Society Bulletin* 13:403–410.

Matthews, N. E. and W. F. Porter. 1992. Maternal defense behavior in white-tailed deer. In *Ecology and Management of the Eastern Coyote*, ed. A. H. Boer, 123–140. Fredericton: University of New Brunswick, Wildlife Research Unit.

McCabe, R. E. and T. R. McCabe. 1984. Of slings and arrows: A historical retrospection. In *White-tailed Deer Ecology and Management*, ed. L. H. Halls, 19–72. Harrisburg, PA: Stackpole Books.

McCullough, D. R. 1997. Irruptive behavior in ungulates. In *The Science of Overabundance: Deer Ecology and Population Management*, eds. W. J. McShea, H. B. Underwood, and J. H. Rappole, 69–98. Washington, DC: Smithsonian Institution Press.

McGinnes, B. S. and R. L. Downing. 1969. Fawn mortality in a confined Virginia deer herd. *Proceedings of Southeast Association of Game and Fish Commissioners* 23:188–191.

McNay, R. S. and J. M. Voller. 1995. Mortality causes and survival estimates for adult female Columbian black-tailed deer. *Journal of Wildlife Management* 59:138–146.

McRoberts, R. E., L. D. Mech, and R. O. Peterson. 1995. The cumulative effect of consecutive winter's snow depth on moose and deer populations: A defense. *Journal of Animal Ecology* 64:131–135.

Mech, L. D. 1966a. Hunting behavior of timber wolves in Minnesota. *Journal of Mammalogy* 47:347–348.

Mech, L. D. 1966b. The wolves of Isle Royale. *U.S. National Park Service Fauna Series 7.*

Mech, L. D. 1970. *The Wolf: The Ecology and Behavior of an Endangered Species.* Garden City, NY: Natural History Press.

Mech, L. D. 1977a. Population trend and winter deer consumption in a Minnesota wolf pack. In *Proceedings of the 1975 Predator Symposium*, eds. R. L. Phillips and C. Jonkel, 55–83. Montana Forest and Conservation Experiment Station, University of Montana, Missoula.

Mech, L. D. 1977b. Wolf-pack buffer zones as prey reservoirs. *Science* 198:320–321.

Mech, L. D. 1984. Predators and predation. In *White-tailed Deer Ecology and Management*, ed. L. H. Halls, 189–200. Harrisburg, PA: Stackpole Books.

Mech, L. D. 1988. *The Arctic Wolf: Living with the Pack*. Stillwater, MN: Voyageur Press.

Mech, L. D. 2007. Femur-marrow fat of white-tailed deer fawns killed by wolves. *Journal of Wildlife Management* 71:920–923.

Mech, L. D. and L. Boitani, eds. 2003. *Wolves: Behavior, Ecology, and Conservation*. Chicago, IL: University of Chicago Press.

Mech, L. D. and L. D. Frenzel, Jr., eds. 1971. Ecological studies of the timber wolf in Northeastern Minnesota. USDA Forest Service Research Paper NC-52. North Central Forest Experiment Station, St. Paul, Minnesota.

Mech, L. D. and P. D. Karns. 1977. Role of the wolf in a deer decline in the Superior National Forest. U. S. Forest Service Research Paper NC-148. St. Paul, Minnesota.

Mech, L. D. and R. E. McRoberts. 1990. Survival of white-tailed deer fawns in relation to maternal age. *Journal of Mammalogy* 71:465–467.

Mech, L. D. and R. O. Peterson. 2003. Wolf–prey relations. In *Wolves: Behavior, Ecology, and Conservation*, eds. L. D. Mech and L. Boitani, 131–160. Chicago, IL: University of Chicago Press.

Mech, L. D., L. G. Adams, T. J. Meier, J. W. Burch, and B. W. Dale. 1998. *The Wolves of Denali*. Minneapolis, MN: University of Minnesota Press.

Mech, L. D., T. J. Meier, J. W. Burch, and L. G. Adams. 1995. Patterns of prey selection by wolves in Denali National Park, Alaska. In *Ecology and Conservation of Wolves in a Changing World*, eds. L. N. Carbyn, S. H. Fritts, and D. R. Seip, 231–243. Edmonton, Canada: Canadian Circumpolar Institute.

Mech, L. D., M. E. Nelson, and R. E. McRoberts. 1991. Effects of maternal and grandmaternal nutrition on deer mass and vulnerability to wolf predation. *Journal of Mammalogy* 72:146–151.

Mech, L. D., D. W. Smith, K. M. Murphy, and D. R. MacNulty. 2001. Winter severity and wolf predation on a formerly wolf-free elk herd. *Journal of Wildlife Management* 64:998–1003.

Merkle, J. A., P. R. Krausman, D. W. Stark, J. K. Oakleaf, and W. B. Ballard. 2010. Summer diet of the Mexican gray wolf. *Southwestern Naturalist* 54: 480–524.

Messier, F. 1991. The significance of limiting and regulating factors on the demography of moose and white-tailed deer. *Journal of Animal Ecology* 60:377–393.

Messier, F. 1995a. Trophic interactions in two northern wolf-ungulate systems. *Wildlife Research* 22:131–146.

Messier, F. 1995b. On the functional and numerical responses of wolves to changing prey density. In *Ecology and Conservation of Wolves in a Changing World*, eds. L. N. Carbyn, S. H. Fritts, and D. R. Seip, 187–198. Edmonton, Canada: Canadian Circumpolar Institute, University of Alberta.

Messier, F. and C. Barrette. 1985. The efficiency of yarding behaviour by white-tailed deer as an antipredator strategy. *Canadian Journal of Zoology* 63:785–789.

Messier, F. and D. O. Joly. 2000. Comment: Regulation of moose populations by wolf predation. *Canadian Journal of Zoology* 78:506–510.

Messmer, T. A., M. W. Brunson, D. Reiter, and D. G. Hewitt. 1999. United States public attitudes regarding predators and their management to enhance avian recruitment. *Wildlife Society Bulletin* 27:75–85.

Mexican Wolf Blue Range Adaptive Management Oversight Committee and Interagency Field Team. 2005. Mexican wolf Blue Range reintroduction project 5-year review. Unpublished report to the U.S. Fish and Wildlife Service Region 2. Albuquerque, New Mexico.

Murie, A. 1944. The wolves of Mount McKinley. *U.S. National Park Service Fauna Series*, no. 5.

Nelson, M. E. and L. D. Mech. 1981. Deer social organization and wolf predation in northeastern Minnesota. *Wildlife Monographs* 77:1–53.

Nelson, M. E. and L. D. Mech. 1986a. Mortality of white-tailed deer in northeastern Minnesota. *Journal of Wildlife Management* 50:691–698.

Nelson, M. E. and L. D. Mech. 1986b. Relationship between snow depth and gray wolf predation on white-tailed deer. *Journal of Wildlife Management* 50:471–474.

Nelson, M. E. and L. D. Mech. 1991. Wolf predation risk associated with white-tailed deer movements. *Canadian Journal of Zoology* 69:2696–2699.

Nelson, M. E. and L. D. Mech. 1993. Prey escaping wolves, *Canis lupus*, despite close proximity. *Canadian Field-Naturalist* 107:245–246.

Nelson, T. A. and A. Woolf. 1987. Mortality of white-tailed deer fawns in southern Illinois. *Journal of Wildlife Management* 51:326–329.

O'Brien, C. S., P. R. Krausman, H. M. Boyd, W. B. Ballard, S. C. Cunningham, and J. C. deVos, Jr. 2010. Influence of coyotes on habitat use by mule deer following a wildfire. *California Game and Fish* 96:7–22.

Ozoga, J. J., L. J. Verme, and C. S. Bienz. 1982. Parturition behavior and territoriality in white-tailed deer: Impact on neonatal mortality. *Journal of Wildlife Management* 46:1–11.
Parsons, D. R. 1998. "Green fire" returns to the southwest: Reintroduction of the Mexican wolf. *Wildlife Society Bulletin* 26:799–807.
Patterson, B. R. and F. Messier. 2000. Factors influencing killing rates of white-tailed deer by coyotes in eastern Canada. *Journal of Wildlife Management* 64:721–732.
Peterson, R. L. 1955. *North American Moose*. Toronto: University of Toronto Press.
Peterson, R. O. 1977. Wolf ecology and prey relationships on Isle Royale. *U.S. National Park Service Scientific Monograph Series 11*, Washington, D.C.
Phillips, M. K., V. G. Henry, and B. T. Kelly. 2003. Restoration of the red wolf. In *Wolves: Behavior, Ecology, and Conservation*, eds. L. D. Mech and L. Boitani, 272–288. Chicago, IL: University of Chicago Press.
Potvin, F., H. Jolicoeur, and J. Huot. 1988. Wolf diet and prey selectivity during two periods for deer in Quebec: Decline versus expansion. *Canadian Journal of Zoology* 66:1274–1279.
Reed, J. E., W. B. Ballard, P. S. Gipson et al. 2006. Diets of free-ranging Mexican gray wolves in Arizona and New Mexico. *Wildlife Society Bulletin* 34:1127–1133.
Ricca, M. A., R. G. Anthony, D. H. Jackson, and S. A. Wolfe. 2002. Survival of Columbian white-tailed deer in western Oregon. *Journal of Wildlife Management* 66:1255–1266.
Rideout, C. B. 1978. Mountain goat. In *Big Game of North America: Ecology and Management*, eds. J. L. Schmidt and D. L. Gilbert, 149–159. Harrisburg, PA: Stackpole Books.
Ripple, W. J. and R. L. Beschta. 2003. Wolf reintroduction, predation risk, and cottonwood recovery in Yellowstone National Park. *Forest Ecology and Management* 184:299–313.
Ripple, W. J. and R. L. Beschta. 2004. Wolves and the ecology of fear: Can predation risk structure ecosystems. *BioScience* 54:755–766.
Robinson, H. S., R. B. Wielgus, and J. C. Gwilliam. 2002. Cougar predation and population growth of sympatric mule deer and white-tailed deer. *Canadian Journal of Zoology* 80:556–568.
Roe, F. G. 1951. *The North American Buffalo: A Critical Study of the Species in its Wild State*. Toronto: University of Toronto Press.
Rohm, J. H., C. K. Nielsen, and A. Woolf. 2007. Survival of white-tailed deer fawns in southern Illinois. *Journal of Wildlife Management* 71:851–860.
Rominger, E. M. 2007. Culling mountain lions to protect ungulate populations–some lives are more sacred than others. *Transactions of the North American Wildlife and Natural Resources Conference* 72:186–193.
Rutberg, A. T. 1987. Adaptive hypotheses of birth synchrony in ruminants: An interspecific test. *American Naturalist* 130:692–710.
Sawyer, H., R. M. Nielson, F. Lindzey, and L. L. McDonald. 2006. Winter habitat selection of mule deer before and during development of a natural gas field. *Journal of Wildlife Management* 70:396–403.
Schaefer, C. L. 2000. Spatial and temporal variation in wintering elk abundance and composition, and wolf response on Yellowstone's Northern Range. MS thesis, Michigan Technological University.
Schaller, G. B. 1967. *The Deer and the Tiger: A Study of Wildlife in India*. Chicago, IL: University of Chicago Press.
Schultz, J. H. 1982. Mortality and movements of white-tailed deer (*Odocoileus virginianus* Zimmerman) fawns in southeastern Minnesota. MS thesis, Mankato State University.
Seal, U. S., M. E. Nelson, L. D. Mech, and R. L. Hoskinson. 1978. Metabolic indicators of habitat differences in four Minnesota deer populations. *Journal of Wildlife Management* 42:746–754.
Severinghaus, C. W. and E. L. Cheatum. 1956. Life and times of the white-tailed deer. In *The Deer of North America*, ed. W. P. Taylor, 57–186. Harrisburg, PA: Stackpole Books.
Sinclair, A. R. E. 1991. Science and the practice of wildlife management. *Journal of Wildlife Management* 55:767–773.
Skogland, T. 1991. What are the effects of predators on large ungulate populations? *Oikos* 61:401–411.
Smith, D. W., T. D. Drummedr, K. M. Murphy, D. S. Guernsey, and S. B. Evans. 2004. Winter prey selection and estimation of wolf kill rates in Yellowstone National Park, 1995–2000. *Journal of Wildlife Management* 68:153–166.
Smith, D. W., D. R. Stahler, E. Albers et al. 2009. Yellowstone project: Annual report, 2008. National Park Service, Yellowstone Center for Resources, YCR-2008–01. Yellowstone National Park, Wyoming.
Smith, R. H. and A. LeCount. 1979. Some factors affecting survival of desert mule deer fawns. *Journal of Wildlife Management* 43:657–665.

Smith, R. H., D. J. Neff, and N. G. Woolsey. 1986. Pronghorn response to coyote control—a benefit:cost analysis. *Wildlife Society Bulletin* 14:226–231.

Smythe, N. 1970. On the existence of "pursuit invitation" signals in mammals. *American Naturalist* 104:491–494.

Smythe, N. 1977. The function of mammalian alarm advertising: Social signals or pursuit invitation? *American Naturalist* 111:191–194.

Stephens, P. W. and R. O. Peterson. 1984. Wolf avoidance strategies of moose. *Holarctic Ecology* 7:239–244.

Stephenson, R. O., W. B. Ballard, C. A. Smith, and K. Richardson. 1995. Wolf biology and management in Alaska, 1981–1992. In *Ecology and Conservation of Wolves in a Changing World*, eds. L. N. Carbyn, S. H. Fritts, and D. R. Seip, 43–54. Edmonton, Canada: Canadian Circumpolar Institute, University of Alberta.

Stewart, R. R., E. H. Kowal, R. Beaulieu, and T. W. Rock. 1985. The impact of black bear removal on moose calf survival in east-central Saskatchewan. *Alces* 21:403–418.

Stout, G. G. 1982. Effects of coyote reduction on white-tailed deer productivity on Fort Sill, Oklahoma. *Wildlife Society Bulletin* 10:329–332.

Sumanik, R. S. 1987. Wolf ecology in the Kluane Region, Yukon Territory. MS thesis, Michigan Technological University.

Tanner, J. T. 1966. Effects of population density on growth rates of animal populations. *Ecology* 47:733–745.

Taylor, T. J. 1996. Condition and reproductive performance of female mule deer in the central Sierra Nevada. California Department of Fish and Game, Bishop, CA.

Teer, J. G., D. L. Drawe, T. L. Blankenship et al. 1991. Deer and coyotes: The Welder experiments. *Transactions of the North American Wildlife and Natural Resources Conference* 56:550–560.

Van Ballenberghe, V. and W. B. Ballard. 1994. Limitation and regulation of moose populations: The role of predation. *Canadian Journal of Zoology* 72:2071–2077.

Vreeland, J. K., D. R. Diefenbach, and B. D. Wallingford. 2004. Survival rates, mortality causes, and habitats of Pennsylvania white-tailed deer fawns. *Wildlife Society Bulletin* 32:542–553.

Wagner, F. H. 1988. *Predator Control and the Sheep Industry*. Claremont, CA: Regina Books.

Wagner, F. H. and L. C. Stoddart. 1972. Influence of coyote predation on black-tailed jackrabbit populations in Utah. *Journal of Wildlife Management* 36:329–342.

Walther, F. 1961. Einige Verhaltensbeobachtungen am Bergwild des Georg von Opel Freigeheges. *Jahrb. G. v. Opel Freigehege. Tierforsch.* 1960–1961:53–89.

Weaver, J. L. 1994. Ecology of wolf predation amidst high ungulate diversity in Jasper National Park, Alberta. PhD dissertation, University of Montana.

Wehausen, J. D. 1996. Effects of mountain lion predation on bighorn sheep in the Sierra Nevada and Granite Mountains of California. *Wildlife Society Bulletin* 24:471–479.

Whitlaw, H. A., W. B. Ballard, D. L. Sabine, S. J. Young, R. A. Jenkins, and G. J. Forbes. 1998. Survival and cause-specific mortality rates of adult white-tailed deer in New Brunswick. *Journal of Wildlife Management* 62:1335–1341.

Whittaker, D. G. and F. G. Lindzey. 1999. Effect of coyote predation on early fawn survival in sympatric deer species. *Wildlife Society Bulletin* 27:256–262.

Wielgus, R. B., H. S. Robinson, and H. S. Cooley. 2007. Effects of white-tailed deer expansion and cougar hunting on cougar, deer and human interactions. *Transactions of North American Wildlife and Natural Resources Conference* 72:211–216.

Zager, P. and J. Beecham. 2006. The role of American black bears and brown bears as predators of ungulates in North America. *Ursus* 17:95–108.

9

Interactions with Other Large Herbivores

Jonathan A. Jenks and David M. Leslie, Jr.

CONTENTS

Introduction

White-tailed deer are widely distributed in North and South America (Branan et al., 1985; Smith, 1991; Miller et al., 2003; Weber and Gonzalez, 2003; Chapter 1 of this book) and occur sympatrically with numerous potential competitors, both wild and domestic. In northern and western regions of North America across a wide variety of habitats, white-tailed deer can occur sympatrically with moose, elk, and mule deer. Increasingly, white-tailed deer are exposed to potential competition with exotic ungulates (e.g., feral hogs, nilgai, and European fallow deer), domestic cattle, sheep, and goats.

Competition occurs when two species use or consume a common resource that is in short supply. Gause (1932) stated that two species sharing the same resources cannot coexist (competitive exclusion principle). The principle implies that if two noninterbreeding species (Species A and B) are adapted to the same ecological niche, and if they occur in the same geographical area, and if Species A reproduces at a faster rate than Species B, then Species B will ultimately be displaced and become extinct (Hardin, 1960). Hence, when competition is persistent and resources remain limited in supply, one species increases (wins) and another decreases (loses). Nevertheless, Cole (1960) questioned the justification of using such a conservative definition of competition in ecological systems where variability in factors affecting competition result in maintenance of multispecies communities. Shaw (1985, p. 121) further clarified the principle of competitive exclusion by adding that "the probability that two or more competing species could coexist in the same area was inversely proportional to their niche overlap." Such clarifications of the principle allow for the existence of "active" competition where two species can occur sympatrically but affect one another by consuming available vegetation (exploitation), aggressively defending food patches (interference), or through the indirect effects of predators (apparent effects). These examples of active competition are in contrast to niche partitioning in which species adapt to competition by developing divergent traits, thereby allowing the species to co-occupy habitats without the negative interactions associated with competition. Consequently, niche partitioning assumes that species respond more to adaptation-linked environmental factors than to potential competitors (Jenkins and Wright, 1988).

For competition to be active, a shared resource must be in short supply. Resource availability generally becomes limited during winter in northern climates when plants become dormant (Mautz, 1978) and during summer in southern climates when drought conditions limit plant growth (McMahan, 1964). During these seasons, competition for resources, space, or habitat can cause population decline. Competition for resources can be direct, such as when one species displaces another (Loft et al., 1991; Cooper et al., 2008) or when removal of resources makes them unavailable to a competitor (Jenks et al., 1996), or indirect, when competition is mediated through other species in the community (Lawlor, 1979). Competition also can be intense when populations increase or decrease over relatively short periods of time, or cumulative when changes in plant availability due to forage use result in long-term shifts from high- to low-quality resources for competitors, with concomitant population decline. For example, long-term foraging by elk at high density could modify vegetative succession by shifting plant availability from a mixture of high-quality foods to those of variable quality (Leslie et al., 1984; Wisdom and Cook, 2000). Such a shift would be beneficial to an intermediate feeder (*sensu* Hofmann, 1988) (Figure 9.1) such as elk but detrimental to a concentrate feeder such as white-tailed deer.

An understanding of the adaptations responsible for food selection is necessary to document competition (Hanley, 1982). For competition to exist, two or more wild ungulates must be sympatric and theoretically have comparable physiologies resulting in selection for similar foods and thus habitats. White-tailed deer differ physiologically from most other wild ungulates (Hofmann, 1988, 1989). Except for mule deer and pronghorn, white-tailed deer have a smaller body size, smaller rumen size, and more selective foraging patterns, and they require foods that contain proportionally greater amounts of cell solubles (e.g., lipids, sugars, pectin, starch; Van Soest, 1982) than other wild ruminants in North America. Competition should be minimal to nonexistent for species pairs that diverge the most in these

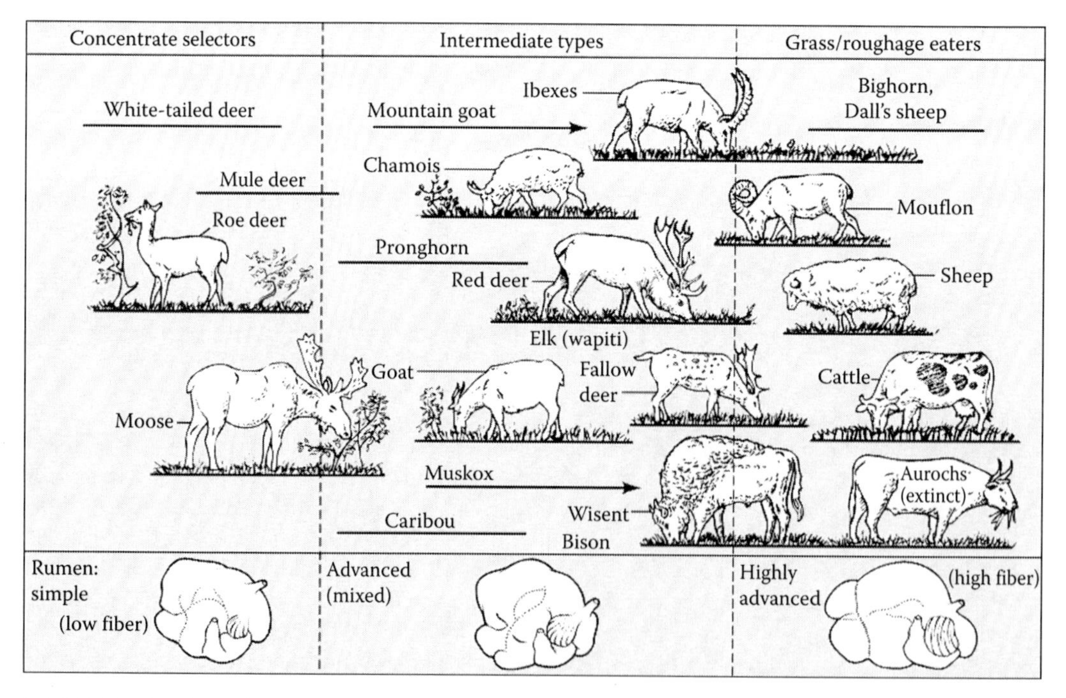

FIGURE 9.1 Positions of European and North American ruminant species within the system of feeding types. Ability to ferment fiber and complexity of the rumen increase from left to right. (Adapted from Hofmann, R. R. 1988. In *The Ruminant Animal: Digestive Physiology and Nutrition*, ed. D. C. Church, 14–43. Englewood Cliffs, NJ: Prentice-Hall and by permission from R. R. Hofmann; Willis, D. W., C. G. Scalet, and L. D. Flake. 2009. *Introduction to Wildlife and Fisheries: An Integrated Approach*. New York, NY: W. H. Freeman and Company.)

characteristics (e.g., white-tailed deer and American bison), but it would be intense between similar species pairs (e.g., white-tailed and mule deer).

Many studies have attempted to document competition between two or more ungulates. Methods used to assess competitive interactions have included assessment of dietary overlap (Leslie et al., 1984, 1987; Gogan and Barrett, 1995; Jenks et al., 1996) or a change in population size in response to the presence of a potential competitor (Harmel, 1980). Diet information can suggest potential for competition but provides little insight on divergence along other niche axes (Leslie et al., 1987; Putman, 1996). Furthermore, high or low dietary overlap might or might not indicate competitive interactions, depending on resource availability, season, or presence and abundance of potential competitors. Thus, to be a useful index of competition, diets must show a shift in components when potentially competing species are sympatric compared to when they occur separately (Putman, 1996).

Not all interactions among species result in detrimental competition; some can be beneficial. Facilitation generally occurs when one species enhances habitat for a second species (Gordon, 1988; McNaughton, 1976). For example, it has been hypothesized that zebra remove senescent grasses, which enhances new vegetative growth consumed by wildebeest in Africa (Bell, 1971; McNaughton, 1976; Jarman and Sinclair, 1979). Although facilitative interactions among ungulates are controversial (Sinclair and Norton-Griffiths, 1982; Stelfox et al., 1986; de Boer and Prins, 1990; Arsenault and Owen-Smith, 2002), long-term modifications by selective removal of particular plants can alter landscapes and improve availability of forages, allowing cohabitation by potential competitors (McNaughton, 1979).

Relative to white-tailed deer, Jenks et al. (1996) suggested that cattle consumption of broomsedge bluestem enhanced forb growth in clearcuts in southern pine forests. White-tailed deer that occurred sympatrically with cattle consumed more forbs than deer that occurred in an area without cattle, suggesting facilitation can occur when cattle and deer co-occupy southern clearcuts. Other potential outcomes for sympatric ungulates include neutralism (no impact on either species), direct interference (both species affected negatively), amensalism (one species affected negatively with no effect on second species), commensalism (one species benefits with no effect on the second species), and protocooperation (both species benefit) (Odum, 1971; Kingery et al., 1996).

Competitive interactions not only involve different species (interspecific competition) but also occur within species (intraspecific competition). Intraspecific competition generally occurs when populations approach theoretical carrying capacity and resources become limited. When this occurs, density-dependent mechanisms (e.g., reduced food availability, space limitations, lower reproduction, lower survival, lower resistance to disease and parasites; Sams et al., 1996, 1998) result in population stabilization or decline. Mysterud (2000) argued that density of individuals relative to resource abundance determines the strength of competitive interactions. Intraspecific and interspecific competitive interactions can occur simultaneously and result in unpredictable outcomes; therefore, a species assemblage in an area can increase system complexity and difficulty of interpretation. Furthermore, aggressive management of a species would result in populations maintained below ecological carrying capacity, reducing intraspecific and interspecific competitive interactions.

It is important to note that our understanding of competitive interactions among white-tailed deer and other ungulates comes from studies conducted largely in the absence of major predators. Recent work in the Intermountain West of the United States attests to the pronounced impacts such a predator can have on behavior and population levels of ungulates such as elk (Ripple and Beschta, 2004; Fortin et al., 2005; Wright et al., 2006). For example, apparent competition occurs when a predator preferentially kills one of the potentially competing prey species in an area (Holt, 1977). Predation could influence competitive interactions because an increase in a prey species could result in an increase in predators that could subsequently reduce a competing prey species. For example, differential preference by mountain lion for mule deer over white-tailed deer could result in an increase in white-tailed deer and a competitive disadvantage to mule deer (Robinson et al., 2002). Robinson (2007) noted that in the absence of predation by mountain lions, both white-tailed and mule deer increased, but with predation white-tailed deer were three times more abundant than mule deer despite similar reproductive rates for the two prey species. Apparent competition also was hypothesized for sympatric populations of mule deer and white-tailed deer affected by coyote predation (Whittaker and Lindzey, 1999). In this study, white-tailed deer fawns were born 8–10 days earlier than those of mule deer, which provided an advantage to mule deer through predator swamping. It must be noted, however, that even in systems where predation affects competitors equally, a variety of outcomes are possible (Holt, 1985). Therefore, historical information on the relationships between all species (predators and prey) must be known to adequately predict the outcome of competitive interactions (Noy-Meir, 1981).

Internal parasites and diseases can act in a similar fashion to large carnivores relative to their effects on competition in wild ruminant populations (Price et al., 1988). Such parasite-mediated competition has been hypothesized to explain declines in moose populations sympatric with white-tailed deer in areas that contain the parasite, meningeal worm (*Parelaphostrongylus tenuis*) (Anderson, 1972; Saunders, 1973; Schmitz and Nudds, 1994). In addition, cattle occupying ranglands can carry Bovine Viral Diarrhea Virus (BVDV), which can negatively affect white-tailed deer populations (Hibler, 1981). For example, Zimmerman (2004) documented that 17% of white-tailed deer sampled in the southern Black Hills in areas with cattle were exposed to the disease and Wolf et al. (2008) determined that prevalence (based on antibody titers) ranged from 35% to 41% in free-ranging white-tailed deer in southern Minnesota. The disease can cause blindness in adults and fawns (Hibler, 1981), stillborn fawns, and congenital hypotrichosis in fawns (Zimmerman et al., 2004); deer populations in close proximity to cattle or their feces can be affected. Thus, additive effects of predators, parasites, and diseases further complicate the understanding of competitive interactions and highlight the need for comprehensive field data to assess such interactions.

Large Herbivore Characteristics that Affect Competitive Interactions

Characteristics that influence the potential for competitive interactions among species are multidimensional and relate to the various axes that describe the niche of a species. These characteristics include but are not limited to feeding adaptations including physiology (Hofmann, 1988), body size (Demment and Van Soest, 1985), and time of the year (Thill et al., 1987).

Hofmann (1988) developed three classes of ruminant based on gastrointestinal anatomy; concentrate selector, intermediate feeder, and grass-roughage feeder (Figure 9.1). He classified white-tailed deer as a concentrate selector because it has a relatively small rumen and feeds every few hours to maintain rumen fill. In contrast, Hofmann (1988, 1989) classified cattle as grass-roughage feeders because of their relatively large rumen and habit of consuming sufficient forage to fill the rumen (meal eating) followed by a subsequent period of loafing and ruminating. Intermediate feeders, such as elk, have anatomical characteristics and feeding patterns intermediate between concentrate selectors and grass-roughage feeders.

As mentioned, the more similar two ungulate species are with regard to body size and digestive morphology—whether wild, domestic, or exotic—the greater the potential for competition because of their adaptations to foods with similar structural and chemical characteristics. For example, white-tailed deer are adapted to diets typified by browse (active growth of shrubs and trees) whereas domestic cattle are adapted to diets dominated by grasses (high in cellulose). These adaptations are related to external (body size) and internal (gastrointestinal) anatomical characteristics of the species. Four morphological characteristics of ungulates influence forage preferences and can aid in the assessment of competitive interactions: body size, type of digestive system, rumino-reticular volume to body weight ratio, and mouth size (Hanley, 1982; Hoffman, 1988) (Figure 9.2). Considering these four factors, Hanley and Hanley (1982) stressed that investigation of selective foraging patterns of ungulates provides insight on dietary overlap as an indicator of competition and food resource partitioning in sympatric ungulate species. Under most circumstances (cf. Leslie et al., 1984, 1987), elk and deer are generally thought to have different forage preferences and thus little dietary overlap (Nelson et al., 1982) because of their different body sizes. Deer, with their small body sizes and rumen volumes, can exploit browse diets with high cell solubles and lignin from woody habitats (Hanley and Hanley, 1982). In contrast, elk have larger body sizes and rumen volumes, which enable elk to use diets high in cellulose from grassland habitats (Hanley and Hanley, 1982).

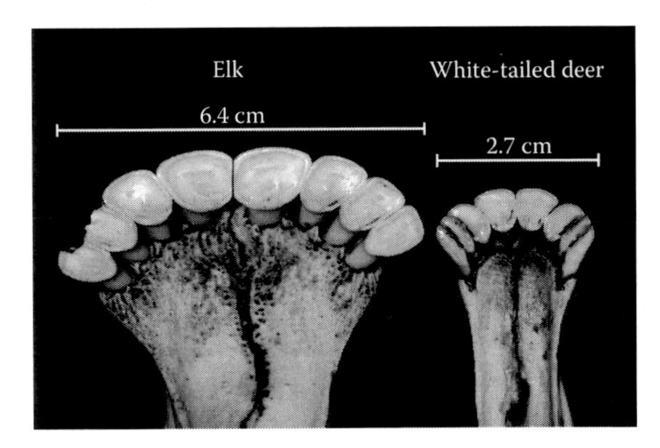

FIGURE 9.2 Comparison of incisor breadth (width of incisors) of white-tailed deer and elk. Reduced incisor breadth of white-tailed deer increases diet selectivity but reduces diet diversity compared to species like elk or cattle.

Demment and Van Soest (1985) noted that as body size increases, relative gut capacity remains constant despite a decrease in weight-relative metabolic rate. Thus, characteristics such as body size are directly related to the physiology of the species; as body size increases, rate of metabolism per kilogram of body weight declines (Kleiber's law, Kleiber, 1947). The long retention time made possible by the relationship between gut size and body size (same gut volume per unit body mass as body size increases) coupled with the low metabolic rate per unit body mass allow large herbivores to consume poor-quality food. Furthermore, large herbivores must consume poor-quality food because in most environments they would not be able to harvest sufficient amounts of high-quality food to stay alive. In some cases, morphological characteristics enable selection of high-quality forage, despite the constraints of large body size. For example, hartebeest are able to coexist with roan antelope in parts of Africa because they have adapted facial characteristics (long and thin muzzle) that allow them to consume sufficient amounts of freshly grown grasses that have a limited availability (Schuette et al., 1998).

Small ruminants have higher rates of metabolism and smaller rumens that result in consumption of food items that ferment quickly to replace nutrients in a timely manner (Henke et al., 1988). Despite lower metabolic rates of large ruminants, these species need to consume greater absolute amounts of food to fuel their greater size. As body sizes of ungulates diverge, competitive interactions might be minimized because one ungulate might respond differently than another to environmental gradients in forage availability.

There are conditions that can promote competition between sympatric ungulates even when species exhibit vast differences in body sizes. For example, a cow's rumen may contain 16 kg of wet material dominated by grasses, whereas white-tailed deer rumen fill averages about 1 kg and is represented largely by browse and forbs (Parra, 1978; Jenks et al., 1994; Zimmerman et al., 2006). These adaptations have resulted in estimates of dietary overlap between these species that range from over 60% (Ortega et al., 1997) to less than 25% (Martinez et al., 1997), suggesting that competitive interactions may vary in intensity. Furthermore, when cattle consume even small amounts of forages preferred by deer during seasons when such forages are in limited supply, the potential for competition is increased. As a simplified example, if cattle consumed 10% browse, that would represent 1.6 kg of potential forage lost to a deer, or more than the total amount of forage in one deer rumen. Consequently, factors such as density, physiological and nutritional status, animal age, plant composition, grazing intensity, and season (Kingery et al., 1996) can interact to create a unique set of conditions either promoting or demoting competitive interactions.

Morphological adaptations of concentrate selectors, such as white-tailed deer, allow quick fermentation of starches and other cell solubles (Van Soest, 1982). As a result, the symbiotic microbial complex in the gastrointestinal tract of white-tailed deer is relatively simple and dominated by amylolytic bacteria, such as *Bacteroides amylophilus*, and pectinolytic bacteria, such as *Butyrivibrio fibrisolvens* (Yokoyama and Johnson, 1988). Protozoa tend to be limited to entodiniomorphs (*Entodinium*, *Diplodinium*) because their high rate of reproduction allows them to maintain colonies in the rumen. Cellulose fermentation, in contrast, requires additional time and a more diverse microbial complex, and it is limited in deer because of the small size of the rumen and lower gastrointestinal tract (i.e., cecum-colon; Hofmann, 1989) (Figure 2.7). In contrast, rumens of grass-roughage feeders are dominated by cellolotytic bacteria (*Bacteroides succinogenes* and *Butyrivibrio fibrisolvens*) and contain holotric protozoa (*Isotricha* and *Dasytricha*)—protozoa species that require extended time (24–48 h) to reproduce and use sugars fermented from cellulose. Among ungulate species, these patterns are not categorical but represent a continuum from concentrate selector to intermediate feeder to grass-roughage feeder. Individual species can adjust their strategy—some more than others—to seasonal changes in forage characteristics and availability. Such adaptations can reduce competitive interactions.

In contrast to grass-roughage feeders (Goetsch and Galyean, 1983; Huston et al., 1986), concentrate selectors such as white-tailed deer have relatively fast transit times of digesta (Mautz and Petrides, 1971; Milchunas et al., 1978). Consequently, under what might be considered "natural conditions" where grass-roughage feeders (like cattle) move through an area (either by herding or nomadic movements) consuming forage, competitive interactions with resident concentrate selectors (like deer) likely would be minimal. However, when grass-roughage feeders are confined to areas at relatively high densities and supplementally fed with hay or concentrates, competitive interactions would be enhanced (Jenks et al., 1996).

The nomenclature of Hofmann (1988) and Demment and Van Soest (1985) help to explain how variation in body size and associated rumen characteristics affect dietary choice and potential competition. Yet, other characteristics of gastrointestinal anatomy of ruminants also affect their nutritional ecology (Hofmann, 1988, 1989). In addition to rumen size, omasum size and structure, cecum length, and length of the lower gastrointestinal tract vary by species relative to body size (Hofmann, 1989). These gastrointestinal characteristics of ruminants have evolved in response to nutrient availability, and their species-specific structures are best adapted to absorb nutrients needed to promote growth and reproduction. Functional adaptations of these characteristics (e.g., relatively small cecum in grass-roughage feeders; Hofmann, 1988) also are adapted to consumption of various forage categories (i.e., grass, browse, forb).

Despite differences among ruminants directly related to body size, relative differences in gastrointestinal characteristics also occur. For example, the ratio of length of small intestine to large intestine ranges from 3.7 to 4.2:1 in grass-roughage feeders but only 2.3–2.6:1 in concentrate selectors (Hofmann, 1988). Even within concentrate selectors differences in gastrointestinal characteristics occur (Jenks et al., 1994); papillae density, rumen digesta weight, and intestinal length were greater for mule deer than sympatric white-tailed deer collected in the southern Black Hills of South Dakota (Zimmerman et al., 2006). Such differences in morphology represent specific adaptations that interplay to minimize potential competitive interactions.

Season-related conditions that promote competitive interactions are more related to energy availability than to protein availability due to the nitrogen-recycling adaptations of ruminants (Schaefer et al., 1978; Lapierre and Lobley, 2001). In species such as white-tailed deer, up to 90% of protein needs during winter can be met by recycling nitrogen; however, energy requirements must come from stores (subcutaneous and organ lipids) or forage. When energy stores are low and forage availability is limited, deer might consume forages of low palatability (e.g., bark, pine needles) (Jenks et al., 1996) or alter behavior (limit movement) (Mautz, 1978) to reduce energy expenditure and maintain core body temperature. Competitive interactions that reduce forage availability or limit the use of habitats needed to promote energy conservation will negatively impact species. Such effects ultimately could reduce resources available for survival and reproduction (fat stores) with a concomitant decline in population size and distribution.

Maturity of plant material can affect the nutrient value to wild ruminants. Under some circumstances, a potential competitor can either enhance or diminish nutritional components for competitors. For example, compared to an ungrazed control, early-summer sheep grazing improved both autumn and winter browse quality in redstem ceanothus, ninebark, rose, and snowberry (Alpe et al., 1999). In contrast, late-summer sheep grazing reduced both autumn and winter browse quality of redstem ceanothus and ninebark, species known to be consumed by white-tailed deer (Thilenius and Hungerford, 1967). Late-summer grazing reduced autumn browse quality in snowberry and rose but improved browse quality of rose during winter. Thus, timing of foraging by potential competitors could affect resources available to a competitor even though competitors are not occupying the same habitat concurrently.

Limited food resources generally occur during winter when plants are dormant or during summer when drought conditions prevail. Limited forage availability can result in potentially competing species consuming forages for which they are not adapted. Hence, grass-roughage feeders consume more browse than optimal and concentrate selectors consume more grasses. White-tailed deer occurring sympatrically with heavily and moderately stocked cattle had more than 40% and more than 20% short leaf and loblolly pine in their diets, respectively, but deer in an area without cattle consumed less than 10% pine (Jenks et al., 1996). Cattle under heavy stocking consumed more than 20% browse and conifer during winter, but those occupying a moderately stocked area consumed less than 10% of those forage items (Jenks et al., 1996). In addition, deer co-occupying areas with cattle contained more digesta in their rumens, which suggested consumption of lower-quality diets (Jenks et al., 1994). These seemingly maladaptive behaviors, caused by a competitive reduction in food resources, can result in bulk limitation in concentrate selectors (deer) and digestive upset in grass-roughage feeders (cattle). Both for deer and cattle, secondary compounds in conifers also can kill rumen microbes (Oh et al., 1967) and potentially overwhelm detoxification systems in the liver (Freeland and Janzen, 1974; Dearing et al., 2005). Bulk limitation occurs when digestible energy of forage is less than 2.17 kcal/g, restricting digesta passage and nutrient assimilation (Ammann et al., 1973; Jenks and Leslie, 1989; Figure 3.3).

Other Factors Associated with Competitive Interactions

Other factors associated with competitive interactions among white-tailed deer and other ungulates are indirectly related to forage acquisition and include behavior, activity patterns, climate, and use of habitats (Anthony and Smith, 1977; Leslie et al., 1984, 1987; Lingle, 2003). White-tailed deer and elk have similar forage use in Glacier National Park, United States, and deer appear to be at a disadvantage because of their limited reach and inability to traverse deep snow (Singer, 1979; Jenkins and Wright, 1988). Structural features in a habitat can affect forage availability, acquisition possibilities, and competitive outcomes. In old-growth temperate forests in Olympic National Park, United States, Columbia black-tailed deer, congeneric with white-tailed deer, and Roosevelt elk appear to avoid competitive interactions for food because downed trees create enclaves of habitat that only deer can access (Leslie et al., 1984).

Differential habitat use by sympatric ruminants can minimize competitive interactions (Kerney and Gilbert, 1976; Anthony and Smith, 1977; Wiggers and Beasom, 1986; Smith, 1987; Hester et al., 1999). In Texas, high densities of desert mule deer were associated with 43% woody cover, whereas white-tailed deer were associated with woody cover of 63% (Wiggers and Beasom, 1986). A similar relationship occurs between Columbian white-tailed deer and Columbian black-tailed deer in western Oregon; highest densities of white-tailed deer occurred in areas with ≥50% woodland cover (Smith, 1981, 1987). Where white-tailed deer and moose co-occur in Ontario, Canada, deer distribution is influenced more by cover characteristics of the habitat, but moose distribution is influenced most by forage availability (Kerney and Gilbert, 1976). These dynamics indicate that differential habitat use, relative to its structural and forage characteristics, and patch size (Hester et al., 1999) influence intensity of competitive interactions among species.

Behavior also affects competitive interactions among sympatric species. For example white-tailed deer were significantly more likely than mule deer to occur in relatively small groups of females, whereas mule deer groups were more likely composed of both sexes during winter (Lingle, 2003). Mule deer groups also were more stable than groups of white-tailed deer. Compton et al. (1988) observed that the presence of cattle was the second most important factor influencing distribution of white-tailed deer in riparian areas along the Yellowstone River in Montana; deer were observed more often in areas where cattle did not occur. Presence of potential competitors can result in modified feeding patterns with, for example, deer spending less time in areas inhabited by other species (Kie et al., 1991).

Activity patterns are ultimately linked to digestive physiology but can be modified by the presence of other species, particularly predators (Nelson and Mech, 1981; Beier and McCullough, 1990) and potential competitors (Mackie, 1970; Kie et al., 1991; Loft et al., 1991). Cederlund (1989) estimated that roe deer in Sweden differed from moose in having active bouts more evenly distributed over the day. Generally, both species were most active during sunrise and sunset. Reduced use of habitats can most often be attributed to limited availability of forage or cover, but dominance of a competitor can cause shifts in activity away from otherwise preferred habitats. Differences in activity patterns during periods of limited forage availability could diminish potential competition. Such a pattern could help explain low trophic and ecological overlap between white-tailed deer and moose (Jenkins and Wright, 1988) and white-tailed deer and mule deer (Anthony and Smith, 1977), and also habitat shifts of deer linked to the presence of cattle (Cohen et al., 1989; Cooper et al., 2008).

Within populations of white-tailed deer, high dominance rank is critical for successful competition for limited resources (Smith, 1991; Taillon and Côté, 2006). Even among deer fawns, dominance hierarchies can affect forage consumption (Taillon and Côté, 2007). Generally, males are dominant over females, and older animals are dominant over younger animals. These relationships likely contribute to density-dependent mortality (e.g., younger animals being more susceptible to winter mortality) in large mammalian populations. Shi and Dunbar (2006) found that time of the year (but not sex or habitat type) had a significant effect on feeding interaction rates of feral goats, and the proportion of interactions involving intense conflict was highest in November when forage availability was beginning to decline. Their results indicated that the initiator won most feeding encounters, with adult males being dominant over females. The ability to win conflicts increased with age both for males and females (Shi and Dunbar, 2006).

Effects of Density

Interspecific and intraspecific competition for food increases as population density of white-tailed deer increases (Leberg and Smith, 1993; Jenks et al., 1996; Patterson and Power, 2002). The threshold at which population growth is in equilibrium with its environment is affected by a variety of factors including the presence of other species that consume the same or similar forages and can be regulatory (density-dependent processes that ultimately maintain populations at equilibriums) (Murray, 1982; Fowler, 1987) in nature (Messier, 1991). At the extreme, competition would lead to competitive exclusion (Gause, 1932; Hardin, 1960). Growth of white-tailed deer is slower at high population density due to energetic costs of competition for mates and the use of marginal habitats by males (Leberg and Smith, 1993). Forage competition alone results in lower body mass of deer (Lesage et al., 2001). These relationships have also been documented experimentally by placing white-tailed deer in enclosures with potential competitors (McMahan, 1964; Harmel, 1980, 1992). Furthermore, reduction in forage quality due to increased density represents evidence for exploitative competition for food (McMahan, 1964).

Density-dependent mechanisms that affect populations of large ungulates generally affect young of the year and/or reproduction in subsequent years (Sams et al., 1996, 1998; Dumont et al., 2000). Jenks and Leslie (2003) documented reduced reproduction in white-tailed deer populations exposed to cattle stocking from an adjusted 2.2 fawns/adult female to 1.1 fawns/adult female. Accumulated effects from reduced reproduction would include an increase in population age structure and decline in theoretical carrying capacity of habitat (population density). Furthermore, adult animals would show signs of malnutrition (Huerta-Patricio et al., 2005) and potentially shift forage selection patterns from what would be expected at moderate density (Kie and Bowyer, 1999). Parasite loads increase with deer density (Eve and Kellogg, 1977; Demarais et al., 1983; Sams et al., 1996); abomasal parasites can affect deer nutritional status via the use of consumed nutrients and reduction of forage intake and have been linked to white-tailed deer density. Because intraspecific density-dependence results in similar consequences for populations as competitive interactions do, it becomes necessary to assess all potential causes of population decline prior to invoking competition.

Evidence for and against Competition

Putman (1996:9) provided a revised overview (based on criteria of Wiens, 1989) of evidence necessary to document competition.

Evidence is considered weak:

a. Population trends and particularly southern states like Texas.
b. There is overlap in resource use.
c. There is some indication of intraspecific competition.

Evidence is considered indicative:

a. There is some indication that one species has reduced the availability of a limiting resource for a second species.

Evidence is considered strong:

a. Other possible explanations are inconsistent with the patterns of resource use and trends in populations.

Because of the high number of niche parameters that can allow coexistence (e.g., differences in activity, body size, or gastrointestinal characteristics), documentation of active competition is complex. Generally, three conditions are necessary for interspecific exploitation competition: overlap in habitat use and diet and limited availability of resources (de Boer and Prins, 1990; Mysterud, 2000). Most studies have examined the potential for competition using circumstantial evidence (one niche axis) and either implied that competition was occurring or that it was negligible (Table 9.1). Variables used to assess competition, such as diet overlap, can vary in their meaning and thus, complicate conclusions derived from field observations. For example, high dietary overlap might imply there is adequate forage to support coexistence of the two species, but it also could imply that other forages are unavailable and the sympatric species were competing for limited forage (Leslie et al., 1984; Jenkins and Wright, 1988; Mysterud, 2000). In contrast, low diet overlap could indicate that competition is occurring (or had occurred), and diets diverged to allow coexistence.

Some of the first experiments designed to document potential for diet competition between white-tailed deer and other species were conducted in Texas during the 1960s. Numbers of bites per forage species were recorded for a tame white-tailed deer, goats, sheep, and cattle (McMahan, 1964), and it was concluded that competition for grass between deer and domestic species would only occur when grass was succulent (e.g., spring and early summer) and other forages were unavailable. Another study was conducted in pastures stocked with deer and livestock in the Edwards Plateau Region; domestic species (sheep, goats, and cattle) were more efficient at competing for native forage than white-tailed deer (McMahan and Ramsey, 1965). Pastures stocked with livestock had higher adult deer mortality and lower fawn survival compared to pastures stocked with only deer.

In studies from the 1970s and 1980s, white-tailed deer were paired in enclosures with fallow deer, chital, and sika deer for six–eight years (Harmel, 1980, 1992). In all experiments, numbers of one of the two species (either white-tailed deer or the exotic) declined to, or approached, zero (extinction). When white-tailed deer were enclosed with chital or sika deer, numbers of white-tailed deer declined, but fallow deer declined to extinction when they were paired with white-tailed deer. Conclusions from these studies indicate that direct competition for forage within the enclosures resulted in the documented change in numbers of animals. Hofmann (1985) classified both sika and fallow deer as intermediate to grass-roughage feeders. Pappillae surface enlargement factor is considerably less for sika deer (4.76) (Fraser, 1996) than for white-tailed deer (16.4) (Zimmerman et al., 2006), which would indicate that sika deer are better at fermenting fibrous forages (Fraser, 1996). In contrast, fallow deer had lower cellulolytic activity than red deer (Prins and Geelen, 1971), which would suggest adaptations more similar to concentrate selectors, and might enhance competition between fallow and white-tailed deer. Consequently, differences in absorptive capacity help to explain the outcomes of previous studies designed to document competition between white-tailed deer and other species.

Research was conducted in Louisiana using tame white-tailed deer and cattle (Thill, 1984; Thill and Martin, 1986; Thill et al., 1987). Study sites included forested and clearcut stands of loblolly shortleaf pine. Moderate diet overlap was observed between deer and cattle in winter (45.6%) and early spring (31.9%), but diet overlap during other seasons was low, and diets were seen as complimentary (i.e., independent) (Thill, 1984). Burning affected diets more than grazing except during winter when diet overlap averaged 31% on recently burned habitat and 35% on older burned habitat (Thill and Martin, 1986). Deer diets on ungrazed pastures were higher in digestibility during spring, autumn, and winter, and selectivity by deer was higher but foraging efficiency lower on ungrazed pastures (Thill et al., 1987). Winter rosettes of some

TABLE 9.1

Species, Methods Used, and Outcomes of Studies that Have Evaluated Competitive Interactions between White-tailed Deer and other Species

Species	Deer Source	Methods Used	Season	Region	Findings	Publication
Domestic						
Cattle	Captive	Observations	Annual	Texas	Competition	McMahan (1964)
		Vegetation	Annual	Texas	Competition	McMahan and Ramsey (1965)
		Diet digestibility	Annual	Louisiana	No effect	Thill et al. (1987)
		Dietary overlap	Annual	Louisiana	Potential	Thill (1984)
					Potential	Thill and Martin (1986)
					Potential	Thill and Martin (1989)
	Wild	Distribution	Annual	Montana	Potential	Compton et al. (1988)
		Body condition	Annual	Oklahoma/Arkansas	Competition	Jenks and Leslie (2003)
		Observations	February	Texas	Competition	Michael (1967)
		Habitat use	Annual	Mexico	Competition	Galindo-Leal et al. (1993)
		Dietary overlap	Annual	Mexico	Potential	Gallina (1993)
			Summer/Fall	Idaho	No effect	Kingery et al. (1996)
			August/January	Mexico	Minimal	Martinez et al. (1997)
			October/July	Texas	Potential	Ortega et al. (1997)
Sheep	Captive	Observations	Annual	Texas	Competition	McMahan (1964)
		Vegetation	Annual	Texas	Competition	McMahan and Ramsey (1965)
Goats	Captive	Observations	Annual	Texas	Competition	McMahan (1964)
		Vegetation	Annual	Texas	Competition	McMahan and Ramsey (1965)
Feral Hogs	Wild	Dietary overlap	Fall	Texas	Potential	Taylor and Hellgren (1997)
Wild						
Mule deer	Wild	Observations	Annual	Alberta/Saskatchewan	No effect	Kramer (1973)
				Oregon	Competition	Smith (1987)
		Dietary overlap	Annual	Arizona	Competition	Anthony and Smith (1977)
		Home range use	Annual	Texas	Potential	Brunzes (2004)
		Habitat	Annual	Montana	Competition	Martinka (1968)
Sika deer	Wild	Nutrition	July	Virginia	Competition	Davidson and Crow (1983)
Red deer	Wild	Grazing time	May	Czech Republic	Positive effect	Bartos et al. (2002)
Elk	Wild	Habitat selection	Winter/Spring	Montana	Potential	Jenkins and Wright (1988)
		Dietary overlap	Summer/Fall	Idaho	Potential	Kingery et al. (1996)
Moose	Wild	Habitat selection	Winter/Spring	Montana	No effect	Jenkins and Wright (1988)
		Habitat use	Annual	Ontario	No effect	Kearney and Gilbert (1976)
		Dietary overlap	Winter	Maine	Limited	Ludewig and Bowyer (1985)
		Observations	Annual	Michigan	Competition	Messier (1991)

forbs and grasses were more abundant where grazing had removed taller bunch grasses (which might suggest facilitation) (Bryant et al., 1979; Jenks et al., 1996), but browse was less abundant on grazed pastures. Winter diets of white-tailed deer from grazed pastures also contained less browse and more forbs and grasses than on ungrazed pastures. Thill and Martin (1989) concluded that even moderate cattle stocking could significantly reduce the availability of deer forages, suggesting competition during winter.

Singer (1979) reported 54% dietary overlap between elk and white-tailed deer during winter in Glacier National Park. The browse component was critical to the high overlap, comprising 88–100% of each species' diet; however, the mechanism for niche separation was provided by snow depth (Singer, 1979). Stewart et al. (2002) observed strong resource partitioning in the use of slope, elevation, and vegetation among Rocky Mountain elk, mule deer, and cattle in northwestern United States. They also reported competitive displacement of elk after the addition of cattle. Others have provided evidence indicating that habitat selection by cattle could cause competition with white-tailed deer for forage in the Black Hills, South Dakota (Pase, 1958; Kranz and Linder, 1973). In Montana, Dusek (1975) hypothesized that competition between mule deer and cattle could occur in late summer and early autumn in overgrazed areas. Grass and grass-like plants that were abundant in the diet of cattle were insignificant in diets of mule deer, suggesting that competition was limited throughout most of the year. The presence of cattle reduced the nutritional condition of white-tailed deer (e.g., reproductive rate, morphometric and biochemical indices) in pine–broomsedge forests in southeastern Oklahoma (Jenks and Leslie, 2003).

Anthony and Smith's (1977) evaluation of interactions between white-tailed deer and mule deer in two localities in southeastern Arizona still represents one of the most comprehensive assessments of niche dynamics and competitive interactions among North American ungulates. They calculated seasonal coefficients of competition with multidimensional analyses of distribution, habitat selection, and diet and demonstrated that the potential for competition was greatest during the most resource-limited time of year and lowest when resources were most abundant. In one area, the two deer species were spatially allopatric on a relatively small landscape scale and occupied different elevational gradients with a buffer zone between; white-tailed deer used the highest elevational areas. Despite such disparities in areas of use, diet overlap was relatively high, ranging from 55% in May–July to 61% in November–January. In the other area, the two species were sympatric, and diet overlap was even higher: 74% from April to mid-July and 84% in November–March. In all behavioral interactions observed during the study, mule deer were dominant over white-tailed deer. Most (48%) of the recorded behavioral encounters between the two species across populations occurred during the forage-limited periods, and in 50% of the interactions, mule deer aggressively asserted dominance over white-tailed deer for a choice food item or a bedding site. Anthony and Smith (1977, p. 255) concluded that competitive exclusion resulted in distributional disparities in one population and that direct competition occurred in the other population—the latter resulting from "vegetational changes [due to climate change], livestock overgrazing and/or range fire suppression."

Exotic Ungulates and White-tailed Deer

In various parts of the United States, particularly southern states like Texas and Florida, white-tailed deer are sympatric with exotic wild ungulates that have been introduced mainly for sport and recreation (Presnall, 1958; Ramsey 1969; Ables and Ramsey, 1972; Feldhamer and Armstrong, 1993). Under such circumstances, competition for food and space (Sanders, 1963; Feldhamer and Armstrong, 1993) and sharing of parasites and diseases may exist (Davidson and Crow, 1983; Davidson et al., 1987; Richardson and Demarais, 1992). Unfortunately, few of these situations have been studied in depth, but some research has been conducted where introductions of African and Asian ungulates on large private ranches were particularly common in the mid-1900s (Presnall, 1958; Ables and Ramsey, 1972). Initially, such exotic species were confined by fencing, but some have escaped confinement and established feral populations that may compete with native white-tailed deer, other wildlife, and livestock. Two notable studies on the interactions of white-tailed deer and exotic ungulates, both of Asian origin, come from south Texas and a barrier island off the Florida panhandle.

The nilgai is a large antelope native largely to India (Schaller, 1967; Berwick 1974; Leslie, 2008). It was introduced in south Texas in the 1920s (Leslie, 2008) and became free-ranging by the 1930s (Sheffield et al., 1983). Differences in size and feeding preferences appear to minimize competition among nilgai, other exotics, and domestic ungulates, but when forage availability is low in south Texas, nilgai may compete with white-tailed deer, particularly for herbaceous vegetation (Sheffield, 1983; Sheffield et al., 1983). White-tailed deer appear to minimize competition for food with nilgai by selecting small plant parts, particularly forbs, and supplemental items such as mushrooms, lichens, moss, and fruits (Sheffield et al., 1983). Nevertheless, although forbs were a higher percentage of the diets of white-tailed deer (60%) than nilgai (25%) or cattle (3%), the greater body mass of nilgai indicated that they consumed a greater absolute amount of forbs (1.0 kg dry weight/day) than white-tailed deer (0.8 kg dry weight/day) (Sheffield et al., 1983).

Based on absolute amounts of consumption, the forbs, browse, and grass eaten by one nilgai in a given period of time is equivalent to that eaten by 1.3, 3.0, and 8.0 white-tailed deer, respectively (Sheffield et al., 1983). Percent *in vitro* dry, organic matter, and cell-wall digestibilities in nilgai (70.9, 69.4, 34.6) are higher than white-tailed deer (57.9, 57.4, 20.2) but comparable to domestic cattle (69.5, 68.3, 33.6) and goats (68.3, 67.4, 33.8) (Priebe et al., 1987). In India, nilgai harbor many ecto- and endoparasites (Leslie, 2008) and the potential for shared parasites with white-tailed deer exists (Sheffield et al., 1983), albeit no shared pathogenic effects or disease transmission has been reported between the two species. In another study in Texas, browse made up 80.2% of the annual diet of exotic greater kudu, suggesting competitive interactions with browse-preferring white-tailed deer that would be exacerbated under poor range conditions caused by seasonal droughts or heavy livestock use (Gray et al., 2007).

The sambar is an Asian cervid with a wide geographical distribution and mass comparable to North American elk (Leslie, 2011). The sambar occurs from India, Nepal, and Sri Lanka through southern and southeastern Asia to the Pacific Coast and Taiwan (Schaller, 1967; Geist, 1998) and has been introduced in the United States (Richardson, 1972), New Zealand, Australia, South Africa, and Brazil (Leslie, 2011). The sambar was introduced on St. Vincent Island, off the Florida panhandle, in 1908 and shares what is now St. Vincent National Wildlife Refuge with white-tailed deer (Lewis et al., 1990). The sambar and white-tailed deer partition habitat and food resources; in particular, the sambar uses freshwater habitats and eats associated aquatic vegetation (11.1–56.2% aquatic vegetation in seasonal diets) to a much greater extent than white-tailed deer (1.5–4.4%) (Shea et al., 1990). Sambars consume a greater variety of forages than white-tailed deer, which eat more browse throughout the year and select small specialized forage items (Shea et al., 1990). Dietary overlap between the two species is relatively low at 41% in summer, 40% in autumn, 41% in winter, and 30% in spring, and most of the overlap occurred among the most abundant browse species (Shea et al., 1990). Sambar and white-tailed share most nematode and arthropod parasites identified in cervids on St. Vincent Island, without obvious detriment (Davidson et al., 1987; Flynn et al., 1990). Relatively little change to population numbers of each species since sambar became established suggests that whatever competition might occur between them does not affect coexistence (Flynn et al., 1990).

White-tailed deer have been introduced in New Zealand where they interact with other non-native ungulates (e.g., Wardle et al., 2001; Forsyth et al., 2002; Bellingham and Allan, 2003), and in Europe where translocations in Bulgaria and the British Isles have been less successful than those in Finland and Yugoslavia (Bartos et al., 2002). For example, the success of white-tailed deer introductions initiated in the 1930s in Finland is reflected in more than 8,000 traffic collisions between 1989 and 1997 (Haikonen and Summala, 2001). In the Czech Republic, white-tailed deer coexist with native red deer, fallow deer, and roe deer, but population productivity and dispersal of white-tailed deer suggest negative consequences of competitive interactions for food resources (Baccus et al., 1985; Feldhamer and Armstrong, 1993; Bartos et al., 2002; Homolka et al., 2008). Bartos et al. (2002) did not detect any adverse behavioral interactions among white-tailed deer and native ungulates in the Czech Republic that would influence competitive interactions; in fact, interspecific cooperative antipredator behavior, particularly among white-tailed deer and fallow deer, was noted. The authors did not rule out competition for forage in limiting white-tailed deer productivity or dispersal.

Examples of Methods to Assess Competition

Forage Availability

Standing crop of forage has been used to assess competitive interactions. When compared with food habits of sympatric species, change in forage availability linked to forage removal by a competitor is assumed to negatively affect other species in sympatry. Methods available to assess forage availability include the use of transects (Ekblad et al., 1993), plots (Thill, 1984), and key forage plants (Cole, 1956). Methods to assess food habits include the use of esophageal fistulas, direct observations, and microhistological analysis of rumen contents or fecal samples (Bryant et al., 1979; Warren and Krysl, 1983; Jenks et al., 1996; Shi and Dunbar, 2006) (Chapter 3). Bryant et al. (1979) used several techniques to assess competitive interactions among white-tailed deer, sheep, Angora goats, and Spanish goats. Forage availability was determined by sampling 0.25-m^2 plots using a 10-cm grid along random transects (modified from Durham and Kothmann, 1977 and Bryant, 1977). Diets of the four species were determined by collecting samples via esophageal fistulas for the sheep and goats and by observing deer and determining feeding minutes. Results of the study indicated that livestock avoided consumption of forbs, which were utilized by deer, and that there were periods of competition, particularly during autumn and winter when competitors mainly consumed browse.

Diet Overlap

Numerous techniques to quantify resource overlap have been developed; many have been used to assess potential competition despite the fact that used alone they may not provide appropriate biological insights (Hurlbert, 1978; Abrams, 1980). Likely the simplest and most used index is Czekanowski's (Schoener, 1970), which is illustrated below.

$$PS = \sum_{i=1} \min(p_i q_i) * 100,$$

where PS=percent similarity, p_i=proportion of ith species in diet of ungulate A, and q_i=proportion of ith species in diet of ungulate B. A number of other indices, including those attributed to Morisita and Pianka, are available for evaluating dietary or niche overlap (see reviews of Horn, 1966 and Hurlbert, 1978).

Zimmerman (2004) used Czekanowski's index to compare diet overlap among white-tailed deer, mule deer, and elk occurring sympatrically in the southern Black Hills of South Dakota. At the time of the study, white-tailed deer and mule deer were believed to have declined in the presence of an expanding elk population. Overlap in winter diets for the three species ranged from 42% for mule deer–elk to 56% for elk–white-tailed deer; overlap between white-tailed deer and mule deer was intermediate at 52%. Selection of shrubs caused the highest diet overlap among the three species and ranged from 21% for mule deer–elk to 30% for white-tailed deer–mule deer.

Diet Quality

Indices used to assess the dietary quality of large mammals sympatric with white-tailed deer include forage digestibility, nutrient concentrations, and fecal characteristics (e.g., fecal nitrogen and phosphorus) that are correlated with diet quality. Direct indices include crude protein and *in vitro* digestibility of diets. Ekblad et al. (1993) evaluated potential competition between Angora goats and white-tailed deer in Texas using control (no stocking of goats), a traditional stocking rate (2 goats/ha), twice the traditional stocking rate (4 goats/ha), and three times the traditional stocking rate (6 goats/ha). They used bite tactics (type of bite: nip, frontal prehension with lips and teeth; jaw, lateral prehension with molars), diet composition, selection ratios (% forage species in diet—% forage species available), and diet quality (crude protein [%] and *in vitro* digestible organic matter [%]) to assess potential competition. Because comparisons across species are affected by physiology, Ekblad et al. (1993) assessed potential competitive effects of stocking by assessing changes in indices by species (deer and goats). They documented changes in bite tactics (goats increased nip bites with increased stocking) and relative consumption of plant species (blackbrush declined in deer diets with the stocking of goats in June and August but not January). Other methods, including a browse volume–weight method (Lubbering et al., 1991) and comparative yield/dry weight rank method (Friedel et al., 1988) that assess change in vegetation, and change in selection ratio (Durham and Kothmann, 1977) also were used.

Fecal nitrogen includes undigested dietary nitrogen (fiber bound), and water soluble, bacterial, and endogenous nitrogen (Arman et al., 1975; Leslie and Starkey, 1985; Leslie et al., 2008). Despite issues related to index variability due to effects of tannins (Hobbs, 1987; Leslie and Starkey, 1987; Robbins et al., 1987), the use of fecal nitrogen to compare populations has remained steady, but it should be emphasized that interspecific comparisons are perilous (Leslie et al., 2008). Jenks et al. (1996) used the index to assess competitive interactions between white-tailed deer and cattle under continuous grazing. As with the study by Ekblad et al. (1993), potential competition was assessed by evaluating change in diet quality via fecal nitrogen within species relative to change in stocking rate of cattle: stocking rates included no stocking, moderate stocking (1 AMU/18 ha), and heavy stocking (1 AMU/3 ha). Deer using areas with cattle had lower fecal nitrogen than did deer on the control area in winter (February), but fecal nitrogen was lower for deer with no stocking than for deer on areas with cattle during summer (August) and autumn (October). The index also was used to compare diet quality of cattle under moderate and heavy stocking. Fecal nitrogen was again lowest in winter; fecal nitrogen of cattle under heavy stocking was lower than for cattle under moderate stocking.

Due to potential factors affecting fecal nitrogen (e.g., secondary compounds), fecal diaminopimelic acid (DAPA) has been suggested as an alternative technique for indexing diet quality (Leslie et al., 1989). Fecal DAPA is a component of bacterial cell walls and, because it is correlated with fecal nitrogen (Leslie et al., 1989) and dietary digestible energy (Nelson and Davitt, 1984), could provide insight into competition between white-tailed deer and sympatric ungulates. Tissue and fecal stable isotopes of nitrogen and carbon also are likely to provide novel ways to assess effects of competitive interactions among white-tailed deer and other large mammalian herbivores (Ben-David et al., 2001; Walter and Leslie, 2008; Walter et al., 2009). Liver isotopes of white-tailed deer and mule deer in the Black Hills of South Dakota elucidated differences in forage acquisition (Walter et al., 2009) between the two species relative to a large wildfire that altered habitat composition and paralleled concomitant changes in digestive morphology (Zimmerman, 2004; Zimmerman et al., 2006).

Nutritional Condition

If competition is prevalent, an expectation would be that reduced availability and/or quality of forage would result in reduced nutritional condition with a concomitant reduction in long-term fitness. The long-term outcome of this relationship would be reduced population size and potential extinction (Hardin, 1960). Therefore, change in nutritional condition could index vegetation change. Numerous indices have been developed to assess nutritional condition, including body mass (Kie et al., 1983), kidney fat index (Riney, 1955), femur fat (Mech and Delgiudice, 1985), and blood urea nitrogen (Kirkpatrick et al., 1975).

Warren and Krysl (1983) evaluated potential competition between white-tailed deer and exotic ungulates by evaluating nutritional condition of deer. Deer were collected on two areas: one area where the exotic ungulates (axis deer, sika deer, fallow deer, blackbuck antelope, Barbado sheep, and aoudad) were considered common and another area where they were considered rare. Body mass and kidney fat were used to assess nutritional condition of white-tailed deer on the two areas. Body mass of deer collected from the two areas did not differ, but deer obtained from the area where exotics were considered common were older. Warren and Krysl (1983) interpreted the similar body masses of deer as indication of competition because the older age classes were expected to have heavier body mass. Jenks and Leslie (2003) also used body mass and other indices of nutritional condition to evaluate potential competition between white-tailed deer and cattle. They documented greater body mass, kidney fat, and reproductive rate for deer under no cattle stocking than for deer collected from areas with moderate to heavy stocking of cattle.

Home Range and Movements

Interactions between species can be documented using locations of animals from traditional telemetry systems or those that involve collection of geographic positioning systems (GPS) data. Home ranges and movements could be compared prior to and after an introduction of a competitor. Thus, effect of a competitor would be documented via changes in home range size or dimensions, or movements away from areas selected prior to introduction of the potential competitor. Cohen et al. (1989) used traditional telemetry methods to evaluate white-tailed deer responses to cattle presence and absence under continuous and short-duration grazing systems. White-tailed deer traveled 35% more between radio fixes in areas with short duration than continuous grazing, and they avoided cattle concentrations under short-duration grazing (Cohen et al., 1989).

Use of GPS technology allows for microsite evaluation of interactions between species because of the large number of locations that can be obtained over a short period of time. Cooper et al. (2008) used GPS to evaluate animal-to-animal interactions among six adult white-tailed deer and nine adult cross-bred Angus cows in Texas; collars were programmed to take locations every 5 min. They assessed deer-cattle interactions, habitat use, and distance to anthropogenic features. Deer generally passed cattle at an average distance of 53 m, and there was no evidence that deer changed course to avoid cattle.

Conclusions

When white-tailed deer occur sympatrically with other large herbivores, interactions can vary from direct competition to a complementary (i.e., facilitation) relationship. White-tailed deer populations can be affected by vegetative response to herbivory and successional changes that result in reduced availability and quality of preferred forage. In situations involving successional changes in vegetation, responses of white-tailed deer populations may be more related to temporal effects than to competitive interactions between species. In addition, climate, such as drought conditions, reduces forage availability and diversity, which can promote competitive interactions between white-tailed deer and other species. Consequences of punctuated changes that favor or disfavor white-tailed deer, such as intense hunter harvest, could also shift wild ruminant communities and result in a focal point for competition between or among species.

Interactions between white-tailed deer and domestic species are more likely to be competitive due to anthropogenic manipulations that are designed to maximize livestock production (Smith, 1991; Figure 9.3). Even when domestic species differ considerably from deer relative to body size, physiology, or forage and habitat preferences, competition can occur because as preferred forages are removed, individuals will consume unusual forages to maintain gut fill and maximize nutrient acquisition. These situations might not be beneficial to domestic species because reduced forage availability may cause consumption of plants that are poisonous (Laycock, 1978) resulting in decreased reproduction (e.g., pine needle abortion; Gardner et al., 1999). It therefore behooves ranchers, farmers, and wildlife managers to critically evaluate relationships between white-tailed deer and domestic species to document potential competitive interactions and to adjust deer densities via harvest or manipulate domestic species density to provide optimal conditions to sustain deer populations.

Most studies of competition involving white-tailed deer suffer from lack of replication or pseudoreplication (Hurlbert, 1984). Researchers studying relationships between white-tailed deer and other species must contend with primary niche axes directly involved with interactions to provide firm conclusions. An optimal design should be implemented such as a before-after/control-impact design (Morrison et al. 2008). Using this design, animal numbers could be manipulated to provide pre- and postinteraction data to assess relationships between or among species. Questions should be asked such as what resource is believed to be limiting? Are species competing for forage, space, or some combination of both? How will interactions be assessed? What are the indices that will be used to evaluate the objectives? For example, if forage is believed to be limiting, will diets be determined or will interactions be assessed based on radiocollared animals or condition indices? Finally, what analyses will be used to test objectives and what are the implications of the study? Studies based on a solid experimental design will result in clear

FIGURE 9.3 White-tailed deer and cattle foraging together in improved pasture in southern Texas. (Photo by A. Ortega. With permission.)

conclusions and will improve our understanding of competitive interactions between white-tailed deer and other species.

REFERENCES

Ables, E. D. and C. W. Ramsey. 1972. Indian mammals on Texas rangelands. *Journal of the Bombay Natural History Society* 71:18–25.

Abrams, P. 1980. Some comments on measuring niche overlap. *Ecology* 61:44–49.

Alpe, J. J., J. L. Kingery, and J. C. Mosely. 1999. Effects of summer sheep grazing on browse nutritive quality in autumn and winter. *Journal of Wildlife Management* 63:346–354.

Ammann, A. P., R. L. Cowan, C. L. Mothershead, and E. R. Baumgardt. 1973. Dry matter and energy intake in relation to digestibility in white-tailed deer. *Journal of Wildlife Management* 37:195–201.

Anderson, R. C. 1972. The ecological relationships of meningeal worm and native cervids in North America. *Journal of Wildlife Diseases* 8:304–310.

Anthony, R. G. and N. S. Smith. 1977. Ecological relationships between mule deer and white-tailed deer in southeastern Arizona. *Ecological Monographs* 47:255–277.

Arman, P., D. Hopcraft, and I. McDonald. 1975. Nutritional studies on East African herbivores. 2. Losses of nitrogen in the feces. *British Journal of Nutrition* 33:265–276.

Arsenault, R. and N. Owen-Smith. 2002. Facilitation versus competition in grazing herbivore assemblages. *Oikos* 97:313–318.

Baccus, J. T., D. E. Harmel, and W. E. Armstrong. 1985. Management of exotic deer in conjunction with white-tailed deer. In *Game Harvest Management*, eds. S. L. Beasom and S. F. Robertson, 213–226. Kingsville, TX: Caesar Kleberg Wildlife Research Institute.

Bartos, L., D. Vankova, K. V. Miller, and J. Siler. 2002. Interspecific competition between white-tailed, fallow, red, and roe deer. *Journal of Wildlife Management* 66:522–527.

Beier, P. and D. R. McCullough. 1990. Factors influencing white-tailed deer activity patterns and habitat use. *Wildlife Monographs* 109:1–51.

Bell, R. H. V. 1971. A grazing ecosystem in the Serengeti. *Scientific American* 225:86–93.

Bellingham, P. J. and C. N. Allan. 2003. Forest regeneration and the influences of white-tailed deer (*Odocoileus virginianus*) in cool temperate New Zealand rain forests. *Forest Ecology and Management* 175:71–86.

Ben-David, M., E. Shochat, and L. G. Adams. 2001. Utility of stable isotope analysis in studying forage ecology of herbivores: Examples from moose and caribou. *Alces* 37:1–14.

Berwick, S. H. 1974. The community of wild ruminants in the Gir Forest ecosystem, India. PhD dissertation, Yale University.

Branan, W. V., M. C. M. Werkhoven, and R. L. Marchington. 1985. Food habits of brocket and white-tailed deer in Suriname. *Journal of Wildlife Management* 49:972–976.

Brunzes, K. J. 2004. Ecology of sympatric mule deer and white-tailed deer in west-central Texas. Ph.D. dissertation, Texas Tech University.

Bryant, F. C. 1977. Botanical and nutritive content in diets of sheep, Angora goats, Spanish goats and white-tailed deer grazing a common pasture. PhD dissertation, Texas A&M University.

Bryant, F. C., M. M. Kothmann, and L. B. Merrill. 1979. Diets of sheep, Angora goats, Spanish goats, and white-tailed deer under excellent range conditions. *Journal of Range Management* 32:412–417.

Cederlund, G. 1989. Activity patterns in moose and roe deer in a north boreal forest. *Ecography* 12:39–45.

Cohen, W. E., D. L. Drawe, F. C. Bryant, and L. C. Bradley. 1989. Observations on white-tailed deer and habitat response to livestock grazing in south Texas. *Journal of Range Management* 42:361–365.

Cole, G. F. 1956. *The Pronghorn Antelope—Its Range, Use and Food Habits in Central Montana with special reference to alfalfa.* Montana Game and Fish Department and Montana State College Agricultural Experiment Station Bulletin Number 516.

Cole, L. C. 1960. Competitive exclusion. *Science* 132:348–349.

Compton, B. B., R. J. Mackie, and G. L. Dusek. 1988. Factors influencing distribution of white-tailed deer in riparian habitats. *Journal of Wildlife Management* 52:544–548.

Cooper, S. M., H. L. Peroto-Baldivieso, M. K. Owens, M. G. Meek, and M. Figueroa-Pagan. 2008. Distribution and interaction of white-tailed deer and cattle in a semi-arid grazing system. *Agriculture, Ecosystems, and Environment* 127:85–92.

Davidson W. R., J. L. Blue, L. B. Flynn, S. M. Shea, R. L. Marchinton, and J. A. Lewis. 1987. Parasites, diseases and health status of sympatric populations of sambar deer and white-tailed deer in Florida. *Journal of Wildlife Diseases* 23:267–272.

Davidson, W. R. and C. B. Crow. 1983. Parasites, diseases, and health status of sympatric populations of sika deer and white-tailed deer in Maryland and Virginia. *Journal of Wildlife Diseases* 19:345–348.

Dearing, M. D., W. J. Foley, and S. McLean. 2005. The influence of plant secondary metabolites on the nutritional ecology of herbivorous terrestrial vertebrates. *Annual Review of Ecology and Evolutionary Systematics* 36:169–189.

de Boer, W. F. and H. H. T. Pins. 1990. Large herbivores that strive mightily but eat and drink as friends. *Oecologia* 82:264–274.

Demarais, S., H. A. Jacobson, and D. C. Guynn. 1983. Abomasal parasites as a health index for white-tailed deer in Mississippi. *Journal of Wildlife Management* 47:247–252.

Demment, M. W. and P. J. Van Soest. 1985. A nutritional explanation of body-size patterns of ruminant and non-ruminant herbivores. *American Naturalist* 125:641–672.

Dumont, A., M. Crete, J. P. Ouellet, J. Hout, and J. Lamoureux. 2000. Population dynamics of northern white-tailed deer during mild winters: Evidence of regulation by food competition. *Canadian Journal of Zoology* 78:764–776.

Durham, A. J., Jr. and M. M. Kothmann. 1977. Forage availability and cattle diets on the Texas coastal prairie. *Journal of Range Management* 27:15–20.

Dusek, G. L. 1975. Range relations of mule deer and cattle in prairie habitat. *Journal of Wildlife Management* 39:605–616.

Ekblad, R. L., J. W. Stuth, and M. K. Owens. 1993. Grazing pressure impacts on potential foraging competition between Angora goats and white-tailed deer. *Small Ruminant Research* 11:195–208.

Eve, J. H. and F. E. Kellogg. 1977. Management implications of abomasal parasites in southeastern white-tailed deer. *Journal of Wildlife Management* 41:169–177.

Feldhamer, G. A. and W. E. Armstrong. 1993. Interspecific competition between four exotic species and native artiodactyls in the United States. *Transactions of the North American Wildlife and Natural Resources Conference* 58:468–478.

Flynn, L. B., S. M. Shea, J. C. Lewis, and R. L. Marchinton. 1990. Biology of sambar deer on St. Vincent National Wildlife Refuge, Florida: Part III. Population statistics, health and habitat use. *Bulletin of Tall Timbers Research Station* 25:63–107.

Forsyth, D. M., D. A. Coomes, G. Nugent, and G. M. J. Hall. 2002. Diet and diet preferences of introduced ungulates (Order: Artiodactyla) in New Zealand. *New Zealand Journal of Zoology* 29:323–343.

Fraser, K. W. 1996. Comparative rumen morphology of sympatric sika deer (*Cervus nippon*) and red deer (*C. elaphus scoticus*) in the Ahimanawa and Kaweka Ranges, central North Island, New Zealand. *Oecologia* 105:160–166.

Freeland, W. J. and D. H. Janzen. 1974. Strategies in herbivory by mammals: The role of plant secondary compounds. *American Naturalist* 108:269–289.

Friedel, M. H., G. N. Bastin, and G. F. Griffin. 1988. Range assessment and monitoring in arid lands: The derivation of functional group to simplify vegetation data. *Journal of Environmental Management* 27:85–97.

Fortin, D., H. L. Beyer, M. S. Boyce, D. W. Smith, T. Duchesne, and J. S. Mao. 2005. Wolves influence elk movements: Behavior shapes a trophic cascade in Yellowstone National Park. *Ecology* 86:1320–1330.

Fowler, C. W. 1987. A review of density dependence in populations of large mammals. *Current Mammalogy* 1:401–441.

Galindo-Leal, C., A. Morales G., and M. Weber R. 1993. Distribution and abundance of Coues deer and cattle in Michilia biosphere Reserve, Mexico. *Southwestern Naturalist* 38:127–135.

Gallina, S. 1993. White-tailed deer and cattle diets at la Michilia, Durango, Mexico. *Journal of Range Management* 46:487–492.

Gardner, D. R., K. E. Panter, and L. F. James. 1999. Pine needle abortion in cattle: Metabolism of iscupressic acid. *Journal of Agriculture and Food Chemistry* 47:2891–2897.

Gause, G. F. 1932. Experimental studies on the struggle for existence: 1. Mixed population of two species of yeast. *Journal of Experimental Biology* 9:389–402.

Geist, V. 1998. *Deer of the World: Their Evolution, Behaviour, and Ecology*. Mechanicsburg, PA: Stackpole Books.

Goetsch, A. L. and M .L. Galyean. 1983. Ruthenium phenanthroline, dysprosium and ytterbium as particulate markers in beef steers fed an all-alfalfa hay diet. *Nutrition Reports International* 27:171–178.

Gogan, P. J. P. and R. H. Barrett. 1995. Elk and deer diets in a coastal prairie–scrub mosaic, California. *Journal of Range Management* 48:327–335.

Gordon, I. J. 1988. Facilitation of red deer grazing by cattle and its impact on red deer performance. *Journal of Applied Ecology* 25:1–10.

Gray, S. S., T. R. Simpson, J. T. Baccus, R. W. Manning, and T. W. Schwertner. 2007. Seasonal diet and foraging preference of greater kudu *Tragelaphus strepsiceros* in the Llano Uplift of Texas. *Wildlife Biology* 13:75–83.

Haikonen, H. and H. Summala. 2001. Deer–vehicle crashes: Extensive peak at 1 hour after sunset. *American Journal of Preventive Medicine* 21:209–213.

Hanley, T. A. 1982. The nutritional basis for food selection by ungulates. *Journal of Range Management* 35:146–151.

Hanley, T. A. and K. A. Hanley. 1982. Food resource partitioning by sympatric ungulates on Great Basin rangeland. *Journal of Range Management* 35:152–158.

Hardin, G. 1960. The competitive exclusion principle. *Science* 131:1292–1297.

Harmel, D. E. 1980. *The Influence of Exotic Artiodactyls on White-tailed Deer Performance and Survival.* Texas Parks and Wildlife Department, Performance Report: Job No. 20, Federal Aid to Wildlife Restoration W-109-R-3. Austin, TX.

Harmel, D. E. 1992. *The Influence of Fallow Deer and Aoudad Sheep on White-tailed Deer Production and Performance.* Texas Parks and Wildlife Department, Performance Report: Job No. 20, Federal Aid to Wildlife Restoration W-109-R-3. Austin, TX.

Henke, S. E., S. Demarais, and J. A. Pfister. 1988. Digestive capacity and diets of white-tailed deer and exotic ruminants. *Journal of Wildlife Management* 52:595–598.

Hester, A. J., I. J. Gordon, G. J. Baillie, and E. Tappin. 1999. Foraging behaviour of sheep and red deer within natural heather/grass mosaics. *Journal of Applied Ecology* 36:133–146.

Hibler, C. P. 1981. Diseases. In *Mule and Black-Tailed Deer of North America*, ed. O. C. Wallmo, 129–155. Lincoln, NE: University of Nebraska Press.

Hobbs, N. T. 1987. Fecal indices to dietary quality: A critique. *Journal of Wildlife Management* 51:317–320.

Hofmann, R. R. 1985. Digestive physiology of deer: Their morphophysiological specialization and adaptation. *Bulletin of the Royal Society of New Zealand* 22:393–407.

Hofmann, R. R. 1988. Anatomy of the gastro-intestinal tract. In *The Ruminant Animal: Digestive Physiology and Nutrition*, ed. D. C. Church, 14–43. Englewood Cliffs, NJ: Prentice-Hall.

Hofmann, R. R. 1989. Evolutionary steps of ecophysiological adaption and diversification of ruminants: A comparative view of their digestive system. *Oecologia* 78:443–457.
Holt, R. D. 1977. Predation, apparent competition, and the structure of prey communities. *Theoretical Population Biology* 12:197–229.
Holt, R. D. 1985. Density-independent mortality, non-linear competitive interactions, and species coexistence. *Journal of Theoretical Biology* 116:479–493.
Homolka, M., M. Heroldova, and L. Bartos. 2008. White-tailed deer winter feeding strategy in area shared with other deer species. *Folia Zoologica* 57:283–293.
Horn, H. S. 1966. Measurement of "overlap" in comparative ecological studies. *American Naturalist* 100:419–424.
Huerta-Patricio, E., K. D. Cameron, G. N. Cameron, and R. A. Medellin. 2005. Conservation implications of exotic game ranching in the Texas Hill Country. In *Contribuciones Mastozoologicas en Homenaje a Bernardo Villa*, eds. V. Sanchez-Cordero and R. A. Medellin, 237–252. Conabio, Mexico: Instituto de Ecologiam, UNAM.
Hurlbert, S. H. 1978. The measurement of niche overlap and some relatives. *Ecology* 59:67–77.
Hurlbert, S. H. 1984. Pseudoreplication and the design of ecological field experiments. *Ecological Monographs* 54:187–211.
Huston, J. E., B. S. Rector, W. C. Ellis, and M. L. Allen. 1986. Dynamics of digestion in cattle, sheep, goats, and deer. *Journal of Animal Science* 62:208–215.
Jarman, P. J. and A. R. E. Sinclair. 1979. Ungulate feeding strategy and resource partitioning. In *Serengeti*, eds. A. R. E. Sinclair and M. Norton-Griffiths, 131–163. Chicago, IL: University of Chicago Press.
Jenkins, K. J. and R. G. Wright. 1988. Resource partitioning and competition among cervids in the northern Rocky Mountains. *Journal of Applied Ecology* 25:11–24.
Jenks, J. A. and D. M. Leslie, Jr. 1989. Digesta retention of winter diets in white tailed deer (*Odocoileus virginianus*) fawns in Maine, USA. *Canadian Journal of Zoology* 67:1500–1504.
Jenks, J. A. and D. M. Leslie, Jr. 2003. Effect of domestic cattle on the condition of female white-tailed deer (*Odocoileus virginianus*) in southern pine-bluestem forests, USA. *Acta Theriologica* 48:131–144.
Jenks, J. A., D. M. Leslie, Jr., R. L. Lochmiller, M. A. Melchiors, and F. T. McCollum. 1994. Variation in gastrointestinal characteristics of male and female white-tailed deer: Implications for resource partitioning. *Journal of Mammalogy* 75:1045–1053.
Jenks, J. A., D. M. Leslie, Jr., R. L. Lochmiller, M. A. Melchiors, and F. T. McCollum. 1996. Competition in sympatric white-tailed deer and cattle populations in southern pine forests of Oklahoma and Arkansas, USA. *Acta Theriologica* 41:287–306.
Kearney, S. R. and F. F. Gilbert. 1976. Habitat use by white-tailed deer and moose on sympatric range. *Journal of Wildlife Management* 40:645–657.
Kie, J. G. and R. T. Bowyer. 1999. Sexual segregation in white-tailed deer: Density-dependent changes in use of space, habitat selection, and dietary niche. *Journal of Mammalogy* 80:1004–1020.
Kie, J. G., C. J. Evans, E. R. Loft, and J. W. Menke. 1991. Foraging behavior by mule deer: The influence of cattle grazing. *Journal of Wildlife Management* 55:665–674.
Kie, J. G., M. White, and D. L. Drawe. 1983. Condition parameters of white-tailed deer in Texas. *Journal of Wildlife Management* 47:583–594.
Kingery, J. L., J. C. Mosley, and K. C. Bordwell. 1996. Dietary overlap among cattle and cervids in northern Idaho forests. *Journal of Range Management* 49:8–15.
Kirkpatrick, R. L., D. E. Buckland, W. A. Abler, P. F. Scanlon, J. B. Whelan, and H. E. Burkhart. 1975. Energy and protein influences on blood urea nitrogen of white-tailed deer fawns. *Journal of Wildlife Management* 39:692–698.
Kleiber, M. 1947. Body size and metabolic rate. *Physiological Reviews* 27:511–541.
Kramer, A. 1973. Interspecific behavior and dispersion of two sympatric deer species. *Journal of Wildlife Management* 37:288–300.
Kranz, J. J. and R. L. Linder. 1973. Value of Black Hills forest communities to deer and cattle. *Journal of Range Management* 26:263–265.
Lapierre, H. and G. E. Lobley. 2001. Nitrogen recycling in the ruminant: A review. *Journal of Dairy Science* 84(E. Supplement):E223–E236.
Lawlor, L. R. 1979. Direct and indirect effects of n-species competition. *Oecologia* 43:355–364.
Laycock, W. A. 1978. Coevolution of poisonous plants and large herbivores on rangelands. *Journal of Range Management* 31:335–342.

Leberg, P. L. and M. H. Smith. 1993. Influence of density on growth of white-tailed deer. *Journal of Mammalogy* 74:723–731.

Lesage, L. L., M. Crete, J. Huot, and J. P. Ouellet. 2001. Evidence for a trade-off between growth and body reserves in northern white-tailed deer. *Oecologia* 126:30–41.

Leslie, D. M., Jr. 2011. *Rusa unicolor* (Artiodactyla: Cervidae). *Mammalian Species* 43:1–30.

Leslie, D. M., Jr. 2008. *Boselaphus tragocamelus* (Artiodactyla: Bovidae). *Mammalian Species* 813:1–16.

Leslie, D. M., Jr., R. T. Bowyer, and J. A. Jenks. 2008. Facts from feces: Nitrogen has measured up as a nutritional index for mammalian herbivores. *Journal of Wildlife Management* 72:1420–1438.

Leslie, D. M., Jr., J. A. Jenks, M. Chilelli, and G. A. Lavigne. 1989. Nitrogen and diaminopimelic acid in deer and moose feces. *Journal of Wildlife Management* 53:216–218.

Leslie, D. M., Jr. and E. E. Starkey. 1985. Fecal indices to dietary quality of cervids in old-growth forests. *Journal of Wildlife Management* 49:142–146.

Leslie, D. M., Jr. and E. E. Starkey. 1987. Fecal indices to dietary quality: A reply. *Journal of Wildlife Management* 51:321–325.

Leslie, D. M., Jr., E. E. Starkey, and B. G. Smith. 1987. Forage acquisition by sympatric cervids along an old-growth sere. *Journal of Mammalogy* 68:430–434.

Leslie, D. M., Jr., E. E. Starkey, and M. Vavra. 1984. Elk and deer diets in old-growth forests in western Washington. *Journal of Wildlife Management* 48:762–775.

Lewis, J. C., L. B. Flynn, R. L. Marchinton, S. M. Shea, and E. M. Marchinton. 1990. Biology of sambar deer on St. Vincent National Wildlife Refuge, Florida: Part I. Introduction, study area, and literature review. *Bulletin of Tall Timbers Research Station* 25:1–12.

Lingle, S. 2003. Group composition and cohesion in sympatric white-tailed deer and mule deer. *Canadian Journal of Zoology* 81:1119–1130.

Loft, E. R., J. W. Menke, and J. G. Kie. 1991. Habitat shifts by mule deer: The influence of cattle grazing. *Journal of Wildlife Management* 55:16–26.

Lubbering, J., J. Stuth, E. Mungall, and W. Sheffield. 1991. An approach for strategic planning of stocking rates for exotic and native ungulates. *Applied Animal Behaviour Science* 29:483–488.

Ludewig, H. A. and R. T. Bowyer. 1985. Overlap in winter diets of sympatric moose and white-tailed deer in Maine. *Journal of Mammalogy* 66:390–392.

Mackie, R. J. 1970. Range ecology and relations of mule deer, elk, and cattle in the Missouri River Breaks, Montana. *Wildlife Monographs* 20:1–79.

Martinez, A. M., V. Molina, F. Gonzalez S., J. S. Marroquin, and J. Navar CH. 1997. Observations of white-tailed deer and cattle diets in Mexico. *Journal of Range Management* 50:253–257.

Martinka, C. J. 1968. Habitat relationships of white-tailed and mule deer in northern Montana. *Journal of Wildlife Management* 32:558–565.

Mautz, W. W. 1978. Sledding on a bushy hillside: The fat cycle in deer. *Wildlife Society Bulletin* 6:88–90.

Mautz, W. W. and G. A. Petrides. 1971. Food passage rate in the white-tailed deer. *Journal of Wildlife Management* 35:723–731.

McMahan, C. A. 1964. Comparative food habits of deer and three classes of livestock. *Journal of Wildlife Management* 28:798–808.

McMahan, C. A. and C. W. Ramsey. 1965. Response of deer and livestock to controlled grazing in central Texas. *Journal of Range Management* 18:1–7.

McNaughton, S. J. 1976. Serengeti migratory wildebeest: Facilitation of energy flow by grazing. *Science* 191:92–94.

McNaughton, S. J. 1979. Grazing as an optimization process: Grass–ungulate relationships in the Serengeti. *American Naturalist* 113:691–703.

Mech, L. D. and G. D. Delgiudice. 1985. Limitations of the marrow-fat technique as an indicator of body condition. *Wildlife Society Bulletin* 13:204–206.

Messier, F. 1991. The significance of limiting and regulating factors on the demography of moose and white-tailed deer. *Journal of Animal Ecology* 60:377–393.

Michael, E. D. 1967. Behavioral interactions of deer and some other mammals. *Southwestern Naturalist* 12:156–162.

Milchunas, D. G., M. L. Dyer, O. C. Wallmo, and D. E. Johnson. 1978. *In-Vivo/In-Vitro Relationships of Colorado Mule Deer Forages*. Colorado Division of Wildlife, Special Report No. 43. Denver, Colorado.

Miller, K. V., L. I. Muller, and S. Demarais. 2003. White-tailed deer. In *Wild Mammals of North America*, eds. G. A. Feldhamer, B. C. Thompson, and J. A. Chapman, 906–930. Baltimore, MD: The John Hopkins University Press.
Morrison, M. S., W. M. Block, M. D. Strickland, B. A. Collier, and M. J. Peterson. 2008. *Wildlife Study Design*. New York, NY: Springer Science+Business Media, LLC.
Murray, B. G., Jr. 1982. On the meaning of density dependence. *Oecologia* 53:370–373.
Mysterud, A. 2000. Diet overlap among ruminants in Fennoscandia. *Oecologia* 124:130–137.
Nelson, J. R. and B. B. Davitt. 1984. A regional cooperative DAPA research and development program. In *Proceedings of the Western States and Provinces Elk Workshop*, ed. R. W. Nelson, 148–163. Edmonton, Canada: Alberta Fish and Game Department.
Nelson, J. R., J. W. Thomas, and D. E. Toweill. 1982. Relationships of elk and other large herbivores. In *Elk of North America*, eds. J. W. Thomas and D. E. Toweill, 415–441. Harrisburg, PA: Stackpole Books.
Nelson, M. E. and L. D. Mech. 1981. Deer social organization and wolf predation in northeastern Minnesota. *Wildlife Monographs* 77:1–53.
Noy-Meir, I. 1981. Theoretical dynamics of competitors under predation. *Oecologia* 50:1432–1439.
Odum, E. P. 1971. *Fundamentals of Ecology*. Philadelphia: W. B. Saunders Company.
Oh, H. K., T. Sakai, M. B. Jones, and W. M. Longhurst. 1967. Effects of various essential oils isolated from Douglas fir needles upon sheep and deer rumen microbial activity. *Applied Microbiology* 15:777–784.
Ortega, I. M., S. Soltero-Gardea, F. C. Bryant, and D. L. Drawe. 1997. Evaluating grazing strategies for cattle: Deer and cattle food partitioning. *Journal of Range Management* 50:622–630.
Parra, R. 1978. Comparison of foregut and hindgut fermentation in herbivores. In *The Ecology of Arboreal Folivores*, ed. G. G. Montgomery, 205–230. Washington, DC: Smithsonian Institution Press.
Pase, C. P. 1958. Herbage production and composition under immature ponderosa pine stands in the Black Hills. *Journal of Range Management* 11:238–243.
Patterson, B. R. and V. A. Power. 2002. Contributions of forage competition, harvest, and climate fluctuation to changes in population growth of northern white-tailed deer. *Oecologia* 130:62–71.
Presnall, C. C. 1958. The present status of exotic mammals in the United States. *Journal of Wildlife Management* 22:45–50.
Price, P. W., M. Westoby, and B. Rice. 1988. Parasite-mediated completion: Some predictions and tests. *American Naturalist* 131:544–555.
Priebe, J. C., R. D. Brown, and D. Swakon. 1987. Comparative *in vitro* digestive efficiencies of cattle, goats, nilgai antelope, and white-tailed deer. *Texas Journal of Science* 39:341–348.
Prins, R. A. and M. J. H. Geelen. 1971. Rumen characteristics of red deer, fallow deer, and roe deer. *Journal of Wildlife Management* 35:673–680.
Putman, R. J. 1996. *Competition and Resource Partitioning in Temperate Ungulate Assemblies*. London: Chapman & Hall.
Ramsey, C. W. 1969. Texotics. *Texas Parks and Wildlife Bulletin* 49:1–46.
Richardson, M. L. and S. Demarais. 1992. Parasites and condition of coexisting populations of white-tailed and exotic deer in south-central Texas. *Journal of Wildlife Diseases* 28:485–489.
Richardson, W. A., II 1972. A natural history survey of the sambar deer (*Cervus unicolor*) on the Powderhorn Ranch, Calhoun County, Texas. MS thesis, Texas A&M University.
Riney, T. 1955. Evaluating condition of free-ranging red deer (*Cervus elaphus*), with special reference to New Zealand. *New Zealand Journal of Science and Technology* 36B:429–463.
Ripple, W. J. and R. L. Beschta. 2004. Wolves, elk, willows, and trophic cascades in the upper Gallatin range of southwestern Montana, USA. *Forest Ecology and Management* 200:161–181.
Robbins, C. T., S. Mole, A. E. Hagerman, and T. A. Hanley. 1987. Role of tannins in defending plants against ruminants: Reduction in dry matter digestion? *Ecology* 68:1606–1615.
Robinson, H. S. 2007. Cougar demographics and resource use in response to mule deer and white-tailed deer densities: A test of the apparent competition hypothesis. PhD dissertation, Washington State University.
Robinson, H. S., R. B. Wielgus, and J. C. Gwilliam. 2002. Cougar predation and population growth of sympatric mule deer and white-tailed deer. *Canadian Journal of Zoology* 80:556–568.
Sams, M. G., R. L. Lochmiller, C. W. Qualls, Jr., and D. M. Leslie, Jr. 1998. Sensitivity of condition indices to changing density in a white-tailed deer population. *Journal of Wildlife Diseases* 34:110–125.

Sams, M. G., R. L. Lochmiller, C. W. Qualls, Jr., D. M. Leslie, Jr., and M. E. Payton. 1996. Physiological correlates of neonatal mortality in an overpopulated herd of white-tailed deer. *Journal of Mammalogy* 77:181–190.

Sanders, C. L., Jr. 1963. Habitat preferences of the white-tailed deer and several exotic ungulates in south Texas. *Ecology* 44:803–806.

Saunders, B. P. 1973. Meningeal worm in white-tailed deer in northwestern Ontario and moose population densities. *Journal of Wildlife Management* 37:327–330.

Schaefer, A. L., B. A. Young, and A. M. Chimwano. 1978. Ration digestion and retention times of digesta in domestic cattle (*Bos taurus*), American bison (*Bison bison*), and Tibetan yak (*Bos grunniens*). *Canadian Journal of Zoology* 56:2355–2358.

Schaller, G. B. 1967. *The Deer and the Tiger: A Study of Wildlife in India*. Chicago, IL: University of Chicago Press.

Schmitz, O. J. and T. D. Nudds. 1994. Parasite-mediated competition in deer and moose: How strong is the effect of meningeal worm on moose? *Ecological Applications* 4:91–103.

Schoener, T. W. 1970. Non-synchronous spatial overlap of lizards in patchy habitats. *Ecology* 51:408–418.

Schuette, J. R., D. M. Leslie, Jr., R. L. Lochmiller, and J. A. Jenks. 1998. Dietary overlap between hartebeest and roan antelope in Burkina Faso, West Africa: Support of the "long-faced" hypothesis. *Journal of Mammalogy* 79:426–436.

Shaw, J. H. 1985. *Introduction to Wildlife Management*. New York, NY: McGraw-Hill Book Company.

Shea, S. M., L. B. Flynn, R. L. Marchinton, and J. C. Lewis. 1990. Biology of sambar deer on St. Vincent National Wildlife Refuge, Florida: Part II. Social behavior, movement ecology, and food habits. *Bulletin of Tall Timbers Research Station* 25:13–62.

Sheffield, W. J. 1983. Food habits of nilgai antelope in Texas. *Journal of Range Management* 36:316–322.

Sheffield, W. J., B. A. Fall, and B. A. Brown. 1983. *The Nilgai Antelope in Texas*. Kleberg Studies in Natural Resources, Texas A&M University, College Station, TX.

Shi, J. and R. I. M. Dunbar. 2006. Feeding competition within a feral goat population on the Isle of Rum, NW Scotland. *Journal of Ethology* 24:117–124.

Sinclair, A. R. E. and M. Norton-Griffiths. 1982. Does competition or facilitation regulate migrant ungulate populations in the Serengeti? A test of hypotheses. *Oecologia* 53:364–369.

Singer, F. J. 1979. Habitat partitioning and wildfire relationships of cervids in Glacier National Park, Montana. *Journal of Wildlife Management* 43: 437–444.

Smith, W. P. 1981. Status and habitat use of Columbia white-tailed deer in Douglas County, Oregon. Ph.D. dissertation, Oregon State University, Corvallis.

Smith, W. P. 1987. Dispersion and habitat use by sympatric Columbian white-tailed deer and Columbian black-tailed deer. *Journal of Mammalogy* 68:337–347.

Smith, W. P. 1991. *Odocoileus virginianus*. *Mammalian Species* 388:1–13.

Stelfox, J. G., D. G. Peden, H. Epp, R. J. Hudson, S. W. Mbugua, J. L. Agatsiva, and C. L. Amuyunzu. 1986. Herbivore dynamics in southern Narok, Kenya. *Journal of Wildlife Management* 50:339–347.

Stewart, K. M., R. T. Bowyer, J. G. Kie, N. J. Cimon, and B. K. Johnson. 2002. Temporospatial distributions of elk, mule deer, and cattle: Resource partitioning and competitive displacement. *Journal of Mammalogy* 83:229–244.

Taillon, J. and S. D. Côté. 2006. The role of previous social encounters and body mass in determining social rank: An experiment with white-tailed deer. *Animal Behaviour* 72:1103–1110.

Taillon, J. and S. D. Côté. 2007. Social rank and winter forage quality affect aggressiveness in white-tailed deer fawns. *Animal Behaviour* 74:265–275.

Taylor, R. B. and E. C. Hellgren. 1997. Diet of feral hog in the western south Texas plains. *Southwestern Naturalist* 42:33–39.

Thilenius, J. F. and K. E. Hungerford. 1967. Browse use by cattle and deer in northern Idaho. *Journal of Wildlife Management* 31:141–145.

Thill, R. E. 1984. Deer and cattle diets on Louisiana pine-hardwood sites. *Journal of Wildlife Management* 48:788–798.

Thill, R. E. and A. Martin, Jr. 1986. Deer and cattle diet overlap on Louisiana Pine-bluestem range. *Journal of Wildlife Management* 50:707–713.

Thill, R. E. and A. Martin, Jr. 1989. Deer and cattle diets on heavily grazed pine-bluestem range. *Journal of Wildlife Management* 53:540–548.

Thill, R. E., A. Martin, Jr., H. F. Morris, Jr., and E. D. McCune. 1987. Grazing and burning impacts on deer diets on Louisiana pine-bluestem range. *Journal of Wildlife Management* 51:873–880.
Van Soest, P. J. 1982. *Nutritional Ecology of the Ruminant*. Ithaca, NY: Comstock Publishing Associates, Cornell University Press.
Walter, W. D. and D. M. Leslie, Jr. 2008. Stable isotope ratio analysis to differentiate temporal diets of a free-ranging herbivore. *Rapid Communications in Mass Spectrometry* 23:2190–2194.
Walter, W. D., T. J. Zimmerman, D. M. Leslie, Jr., and J. A. Jenks. 2009. Dietary response of sympatric deer to fire using stable isotope analysis of liver tissues. *Wildlife Biology in Practice* 5:128–135.
Wardle, D. A., G. M. Barker, G. W. Yeates, K. I. Bonner, and A. Ghani. 2001. Introduced browsing mammals in New Zealand natural forests: Aboveground and belowground consequences. *Ecological Monographs* 71:587–614.
Warren, R. J. and L. J. Krysl. 1983. White-tailed deer food habits and nutritional status as affected by grazing and deer-harvest management. *Journal of Range Management* 36:104–109.
Weber, M. and S. Gonzalez. 2003. Latin American deer diversity and conservation: A review of status and distribution. *Écoscience* 10:443–454.
Whittaker, D. G. and F. G. Lindzey. 1999. Effect of coyote predation on early fawn survival in sympatric deer species. *Wildlife Society Bulletin* 27:256–262.
Wiens, J. A. 1989. *The Ecology of Bird Communities. Vol. 2: Processes and Variations*. London: Cambridge University Press.
Wiggers, E. P. and S. L. Beasom. 1986. Characterization of sympatric or adjacent habitat of 2 deer species in west Texas. *Journal of Wildlife Management* 50:129–134.
Willis, D. W., C. G. Scalet, and L. D. Flake. 2009. *Introduction to Wildlife and Fisheries: An Integrated Approach*. New York, NY: W. H. Freeman and Company.
Wisdom, M. J. and J. G. Cook. 2000. Elk. In *Ecology and Management of Large Mammals in North America*, eds. S. Demarais and P. R. Krausman, 27–37. Upper Saddle River, NJ: Prentice-Hall.
Wolf, K. N., C. S. DePerno, J. A. Jenks et al. 2008. Selenium status and antibodies to selected pathogens in white-tailed deer (*Odocoileus virginianus*) in southern Minnesota. *Journal of Wildlife Diseases* 44:181–187.
Wright, G. J., R. O. Peterson, D. W. Smith, and T. O. Lemke. 2006. Selection of northern Yellowstone elk by gray wolves and hunters. *Journal of Wildlife Management* 70:1070–1078.
Yokoyama, M. T. and K. A. Johnson. 1988. Microbiology of the rumen and intestine. In *The Ruminant Animal: Digestive Physiology and Nutrition*, ed. D. C. Church, 125–144. Englewood Cliffs, NJ: Prentice-Hall.
Zimmerman, T. J. 2004. Effects of fire on the nutritional ecology of selected ungulates in the southern Black Hills, South Dakota. MS thesis, South Dakota State University.
Zimmerman, T. J., J. A. Jenks, L. D. Holler, C. N. Jacques, and W. W. Morlock. 2004. Congenital hypotrichosis in a white-tailed deer fawn from South Dakota. *Journal of Wildlife Diseases* 40:145–149.
Zimmerman, T. J., J. A. Jenks, and D. M. Leslie, Jr. 2006. Gastrointestinal morphology of female white-tailed and mule deer: Effects of fire, reproductive stage, and feeding type. *Journal of Mammalogy* 87:598–605.

10

White-tailed Deer Behavior

Randy W. DeYoung and Karl V. Miller

CONTENTS

Introduction

Viewing animals and their activities in a natural setting is rewarding for both biologists and the public. People relate to animal behaviors because behaviors have human analogs, which can both help and hinder interpretation of the behaviors. As a science, behavioral ecology applies ecological and evolutionary principles to understand animal behavior (Krebs and Davies, 1997). Thus, behavioral ecologists seek scientific answers to the "why" questions that occur to amateur and professional observers alike: why does an animal engage in a specific behavior and what are the short- and long-term consequences of the behavior?

The behavioral ecology of the white-tailed deer can be divided into annual and seasonal components. Foraging, daily movements, social interactions, and avoidance of predators are activities that occur throughout the year. Seasonal behaviors include response to changing weather, habitat conditions, and life history components such as breeding and parturition. Many seasonal behaviors are cued by changes in photoperiod.

Early behavioral studies of whitetails focused on captive or exploited populations and relied primarily on visual observations. Since the review by Marchinton and Hirth (1984), advances in technology have greatly expanded our understanding of whitetail behavior. Improvements in radiotelemetry, global positioning system (GPS) technology, remote photography, computing power, and genetic markers have collectively aided in the refinement of existing knowledge, and in some cases prompted re-evaluation of long-held assumptions. Despite these advances, many outstanding questions about deer behavior remain.

White-tailed deer foraging and social behaviors are exceedingly plastic, enabling the species to exist in a variety of habitats ranging from tropical forests of South America to sub-boreal forests of Canada (Demarais et al., 2000; Chapter 1). This adaptability also allows whitetails to thrive in human-altered landscapes including areas with intensive agriculture or urban/suburban development (Chapter 20). However, behavioral plasticity limits generalizations about whitetail behavior and complicates behavior studies because of varied behavioral responses in differing environmental and social contexts. Nevertheless, generalizations are unavoidable and can convey usual or expected patterns of behavior.

Some social behaviors of whitetails are primitive in comparison to other species of large mammals. Behaviors are influenced by selection over evolutionary timescales, as evidenced by the ability to recover phylogenetic relationships from an analysis of behaviors alone (Cap et al., 2002). Historically, whitetails were widespread, but likely not locally numerous to the point of overpopulation, unlike in many modern landscapes. During the Pleistocene, whitetails lived in a landscape populated by large specialist herbivores and a diverse community of predators. Whitetails are habitat opportunists and historically exploited ephemeral habitats rather than climax communities (Geist, 1998). The white-tailed deer were able to become locally numerous only after the terminal Pleistocene extinctions brought about the demise of the large specialists and their predators (Wolverton et al., 2009). Whitetails tend to be poor competitors and perform poorly in the presence of exotic deer species (e.g., sika, axis, and fallow deer) that can exist and prosper on low-quality diets that will not sustain productive populations of whitetails (Chapter 9). One can easily observe the results of behaviors honed by selection over millennia, where localized populations of whitetails exploited high-quality forages in small groups in or near heavy vegetative cover to avoid predators. In contrast, many contemporary populations of white-tailed deer exist in high-density or heavily exploited conditions.

Social Organization

White-tailed deer are social animals and display behavioral similarities to other species of ungulates. The social organization of white-tailed deer is centered on formation of matriarchal social groups

FIGURE 10.1 The matriarchal group, consisting of an adult doe and several generations of her female offspring, is the basic social unit of the white-tailed deer. (Photo by T. Kirkland. With permission.)

(Figure 10.1) composed of an adult doe and several generations of her female offspring (Hawkins and Klimstra, 1970; Hirth, 1977; Mathews and Porter, 1993). Members of female social groups typically associate throughout the year (Nelson and Mech, 1984; Aycrigg and Porter, 1997), with the exception of fawning season. In some cases, all individuals in social groups may not be close relatives (Comer et al., 2005; Miller et al., 2010b), but the occurrence and rate of nonrelatives in social groups of wild deer is difficult to verify without modern genetic techniques. In contrast, bucks are solitary during the breeding season, but otherwise aggregate into temporary bachelor groups of nonrelated individuals after dispersing from their natal group at about 1.5 years of age (Hirth, 1977). Evolution of sexual segregation has been widely debated, but recent consensus centers on differences between the sexes in dietary needs, with predation a secondary factor (Barboza and Bowyer, 2000; Chapter 6).

Mathews (1989) proposed a useful conceptual model of spatial and social organization of female white-tailed deer. Social groups are composed of female relatives, where the home ranges of one or more juveniles overlap that of an older female relative analogous to the petals of a rose. This theory of social group structure implies that the spatial area occupied by a social group will expand outward with each new member, eventually forming new groups via fissioning of existing groups (Porter et al., 1991). However, the rose petal analogy of doe social groups was developed in a low-density, migratory deer population, where mature does and several generations of their female relatives were present. The composition and spatial overlap of doe social groups in high-density or exploited (harvested) deer populations are less certain. As density increases, doe groups must either overlap in space or home ranges must contract markedly (Comer et al., 2005; Miller et al., 2010b). Outside of the fawning season, does are not known to be territorial, so it is likely that the range of doe groups must overlap in space, if not in time.

White-tailed deer do not form large herds. In comparison to other species of ungulates, white-tailed deer are termed a "small-group" species, and are typically less gregarious than other North American cervids, with the exception of moose (De Vos et al., 1967). Social group sizes across the whitetail's range are poorly documented, but long-term studies of marked deer have estimated doe social group sizes of 2–12 individuals (Nelson and Mech, 1981; Mathews, 1989; Miller et al., 2010b). Temporary aggregations of several smaller groups may occur at feeding or watering sites, giving the false impression that whitetail groups can be large. Furthermore, large aggregations of deer occur in northern environments as deer seek thermal cover from the elements. However, visual observations are often a misleading indicator of wild deer behavior because deer spend much of their time in dense cover, where continuous observation is seldom possible.

Group size and composition in ungulate species are shaped by forage distribution and the species' predator avoidance strategy. Whitetails typically are found in small groups due to dietary specialization and their predator avoidance strategy of hide and flee. Low-quality forages are typically abundant and comprise the vast majority of forage available to herbivores. Herbivores adapted to exploit abundant and predictable, yet low-quality forages (e.g., grasses) can exist in larger groups. In contrast, white-tailed deer are relatively small ruminants classified as browsers, or concentrate selectors, adapted to eating forage high in protein and energy (Hofmann, 1989; Chapters 2 and 3) (Figure 9.1). High-quality forages are typically patchy and unpredictable, and may be highly seasonal in availability. For example, mast and forbs provide high-quality forage, but appear at peak nutritional quality for only a few weeks or months of the year. As nutrient requirements rise, diet quality must increase, and individuals in large groups compete for high-quality forages. Furthermore, a large group could not move cohesively in thick vegetative cover.

The Home Range

Home range is the geographic area used by an individual during normal daily activities of food gathering, mating, and rearing young (Burt, 1943). Therefore, the home range must contain sufficient food, water, and vegetative cover required by the animal on a daily and annual basis. The annual home range size of whitetails varies dramatically among regions and habitat types (Table 10.1), ranging from less than 100 ha (Leach and Edge, 1994; Cornicelli et al., 1996; Storm et al., 2007) to more than 1000 ha

TABLE 10.1

Home Range Size (ha) for White-tailed Deer in Various Parts of Its Range by Sex and Age Class

Location	Animal (Sex/Age Class)	N Deer	Home Range (ha)	Estimator	References
Florida Everglades	Males	10	700 ± 140 SE	95% MCP	Sargent (1992)
		23	290 ± 40 SE	95% MCP	
Florida	Females (yearling)	7	2458	95% harmonic mean	Labisky et al. (1991)
	Females (adult)	6–16	275–420	95% harmonic mean	
	Males	5	701	95% harmonic mean	
Illinois	Female	7	50.8 ± 23 SE	MCP	Cornicelli et al. (1996)
Illinois	Female (fawning)	26	53.0 ± 5.2 SE	95% fixed kernel	Storm et al. (2007b)
	Females (winter)		90.6 ± 9.7 SE	95% fixed kernel	
Maryland	Male (autumn)	15	299.4 ± 31.3	95% adaptive kernel	Tomberlin (2007)
	Male (autumn)	9	386 ± 129 SE	95% adaptive kernel	Karns (2008)
Mexico	Females	6	206 ± 13	95% MCP	Gallina et al. (1997)
	Males	7	205 ± 14	95% MCP	
Michigan	Female	10–15	45	MCP	Beier and McCullough (1990)
	Male	17–19	142	MCP	
Michigan	Females (winter)	42–51	730–1859	95% adaptive kernel	Van Deelen et al. (1998)
	Females (summer)	37–49	1255–3037	95% adaptive kernel	
Minnesota	Females (winter)	30	85.3 ± 5.8 SD	90% adaptive kernel	Grund et al. (2002)
	Females (spring)	20	144.0 ± 7.5 SD	90% adaptive kernel	
	Females (summer)	19	50.4 ± 6.8 SD	90% adaptive kernel	
	Females (fall)	44	92.9 ± 11.4 SD	90% adaptive kernel	
Minnesota	Female	130	310 ± 30 SE	95% fixed kernel	Brinkman et al (2005)
Montana	Female (adult)	11	71 ± 18.4 SE	MCP	Leach and Edge (1994)
	Female (yearling)	5	91 ± 30.4 SE	MCP	
Nebraska	Females	14	170 (CI = 38)	95% harmonic mean	VerCauteren and Hygnstrom (1998)
New York	Female (summer)	64	221 ± 19.0 SE	100% CP	Tierson et al. (1985)
	Female (winter)	45	132 ± 18.3 SE	100% CP	
	Male (summer)	34	233 ± 23.4 SE	100% CP	
	Male (winter)	12	150 ± 31.6 SE	100% CP	
Quebec	Male (summer)	5–14	1144–1247	95% MCP	
	Female (summer)	13–24	910–2812	95% MCP	
	Male (winter)	4–7	193–272	95% MCP	
	Female (winter)	7–13	102–112	95% MCP	
South Dakota	M/F (summer)	49	920 ± 100 SE	95% fixed kernel	Grovenburg et al. (2009)
	M/F (winter)	27	1020 ± 120 SE	95% fixed kernel	
	All	24	437 ± 38.4 SE	95% adaptive kernel	Kernohan et al. (2002)
Texas	Female	27	84 ± 8 SE	95% HR ellipse	Inglis et al. (1979)
	Male	14	139 ± 37 SE	95% HR ellipse	
Texas	Male (<2 years old)	6	922 ± 328 SE	95% fixed kernel	Hellickson et al.(2008)
	Male (3–4 years old)	12	499 ± 130 SE	95% fixed kernel	
	Male (5–6 years old)	23	553 ± 118 SE	95% fixed kernel	
Washington	Female (adult)	18	103.6 ± 16.5 SE		Gavin et al. (1984)
		7	208.6 ± 24.6 SE		
West Virginia	Female (winter)	110	133 ± 12 SE	95% fixed kernel	Campbell et al. (2004a)
	Female (summer)	90	100 ± 8 SE	95% fixed kernel	
Wisconsin	Female	15	178 ± 102 SD	100% MCP	Larson et al. (1978)

(Lesage et al., 2000). Typically, uniformly distributed habitats containing a good mix of food, cover, and water are associated with smaller home ranges (Marchinton and Hirth, 1984).

Deer do not make equal use of the area encompassing their annual range. Half of a deer's use of its home range (50% core area) (Favreau, 2005) may be focused on only 12–17% of the entire area (Campbell et al., 2004a; Tomberlin, 2007).

Animal requirements change seasonally, as does the distribution of forage and other critical habitat features. Thus, the size and location of home ranges may differ throughout the year. In general, home range size varies inversely with population density and habitat productivity (Sanderson, 1966). Male deer typically have annual home ranges that are two to three times larger than that of females, but may exhibit sex-specific differences in response to habitat productivity (Relyea et al., 2000). Studies that monitor deer home range and movements typically detect some degree of individuality in home range and movements (Nelson and Mech, 1999; Campbell et al., 2004a). Furthermore, many deer make occasional excursions outside of seasonal home ranges (Inglis et al., 1979).

In nonmigratory deer herds, adults of both sexes display site fidelity to seasonal home ranges among years (Beier and McCullough, 1990; Hellickson et al., 2008). Does often use the same or adjacent areas as their female relatives; home ranges may be occupied by several generations of females and their female offspring. Most bucks disperse and establish seasonal home ranges distinct from the natal range, but bucks are typically faithful to seasonal home ranges thereafter. Young bucks tend to have larger home ranges than prime-aged and senescent bucks, although home range size may vary among individuals (Webb et al., 2007; Hellickson et al., 2008). Migratory deer also use the same seasonal ranges in successive years, albeit with greater fidelity to summer range than winter range (Tierson et al., 1985; Aycrigg and Porter, 1997; Nelson and Mech, 1999). In some regions, deer are facultative migrators and may not visit a distinct winter range unless prompted by inclement weather.

Home range size varies throughout the year in response to physiologic status and external influences, including disturbance and availability of seasonal forages. During rut, adults of both sexes undertake movements outside of their seasonal home range (Karns, 2008; Kolodzinski et al., 2010). Bucks may increase the size of their home range prior to or during rut (Nelson and Mech, 1981; Beier and McCullough, 1990; Tomberlin, 2007; but see Hellickson et al., 2008). However, does may also increase home range size during rut in low-density populations (Labisky and Fritzen, 1998). Doe home ranges are typically smallest during and after parturition. Does will segregate from other deer prior to giving birth, and may completely isolate their core home range from conspecifics (Schwede et al., 1993). Selection of suitable fawning sites is an important factor in fawn mortality due to predation (Ozoga and Verme, 1986b) and does may display fidelity to their postpartum range (Schwede et al., 1993). Fawns gradually expand their range after four to eight weeks of age as they begin to follow the dam (Jackson et al., 1972; Ozoga et al., 1982).

Deer home ranges vary in response to the availability of seasonal forages. Emergence of hard mast, such as acorns, can influence deer home range size and movements in portions of the eastern United States (McShea and Schwede, 1993). In agricultural areas, the emergence, maturation, and harvest of crops alters availability of food and cover, affecting deer use and home range size. Home range centers in Nebraska shifted an average of 174 m closer to cornfields when corn was in the tasseling–silking stage and 157 m away from fields after harvest (Vercauteren and Hyngstrom, 1998). Home range size increased 32% after corn harvest. Another Nebraska study reported that home range size decreased from 400 to 200 ha following a 44% reduction in cropland acreage on the study area (Walter et al., 2009).

Deer use natural or artificial mineral licks within their home range, but may travel several kilometers outside of their seasonal home range to visit a mineral source (Wiles and Weeks, 1986; Campbell et al., 2004b). Deer also change activity and population structure in response to artificial bait sites (Blanchong et al., 2006; Campbell et al., 2006).

Daily and Seasonal Patterns of Movement

Deer must allocate their time and use of available habitat for predation avoidance, foraging, and social interactions. Although white-tailed deer are crepuscular, with peaks of daily activity occurring at dawn and dusk (Kammermeyer and Marchinton, 1977; Ockenfels and Bissonette, 1982; Rouleau et al., 2002; Webb et al., 2010), activity budgets and habitat use patterns may vary with age, sex, time of day, season, and weather conditions. Sunset appears to be the period of greatest activity (Hosey, 1980; Beier and McCullough, 1990); lowest activity generally occurs during the predawn early morning, late morning, and late evening after sunset (Kammermeyer and Marchinton, 1977; Hosey, 1980; Ivey and Causey, 1981). Diurnal patterns of activity also have been reported (Hosey, 1980; Holzenbein and Schwede, 1989; Naugle et al., 1997; Hellickson, 2002). Other studies have reported a distinct crepuscular pattern during some seasons and a lack of any pattern (Demarais et al., 1989), or a shift to a single peak (Beier and McCullough, 1990) during other seasons. Naugle et al. (1997) reported that deer changed from a diurnal activity pattern to a crepuscular pattern between years. Minor activity peaks have also been reported at midday and midnight (Michael, 1970). Plasticity in behavioral responses to environmental conditions certainly would be expected as deer adjust their activities to avoid dangerous situations, maintain physical comfort, adjust to changing foraging opportunities, optimize energy conservation, and respond to social and physiological demands for reproduction. Diel patterns of activity are similar between males and females, except during early autumn when males are more active than females at night and females more active than males during daytime (Beier and McCullough, 1990).

Seasonal differences in activity rates between males and females reflect differences in social organization, reproductive behavior, physiological demands for gestation and lactation, and region. Females are, on average, active for 1.5 h/day longer than males during January through July (Beier and McCullough, 1990) and spend more time in heavy cover, presumably bedded (Zagata and Haugen, 1974). Increased metabolic demands due to gestation and lactation require a higher-quality diet and presumably increased foraging time. In northern ranges, reduced movement during winter is common for both sexes (Hoskinson and Mech, 1976; Beier and McCullough, 1990), and deer activity becomes more diurnal in response to lower temperatures (Ozoga and Gysel, 1972).

Activity rates also vary among regions. In southern Texas, greatest activity occurs during January and again during September–October (Michael, 1970; Hellickson, 2002). In contrast, greatest activity in southern Michigan occurs during May and October, with lowest activity during midwinter (Beier and McCullough, 1990). Researchers have even reported varying activity rates among individual deer (Hosey, 1980; Hellickson, 2002).

Among environmental variables, temperature appears to have the greatest impact on deer activity across all seasons. Deer observations during nightly spotlight counts were directly related to temperature and negatively correlated with cloud cover, precipitation, and dew (Progulske and Duerre, 1964). In a Michigan study, deer activity rates during spring, summer, and autumn were greatest between 10°C and 16°C (Beier and McCullough, 1990). Activity rates declined in direct response to temperatures above or below this range. Responses to changes in temperature may be different at more southerly latitudes. In Oklahoma, males during winter moved more when temperatures were lowest, whereas females during summer moved greater distances when temperatures were normal to high and less when temperatures were coolest (Webb et al., 2010).

Winds apparently have little effect on activity (Michael, 1970; Newhouse, 1973), although deer tend to select habitats that reduce wind speeds on windy days (Zagata and Haugen, 1974), particularly during winter (Ozoga and Gysel, 1972; Beier and McCullough, 1990). Movement in response to precipitation appears unpredictable, although deer often remain inactive during heavy rainfall or during snowstorms accompanied by high winds (Severinghaus and Cheatum, 1956; Michael, 1970).

Questions remain as to how deer activity responds to changes in lunar phase or the amount of moonlight. Kammermeyer (1975) reported increased diurnal and nocturnal movements around the time of the full moon and Newhouse (1973) observed increased use of open habitats on moonlit nights. However, most research has reported little or no effects of moon phase or the amount of moonlight on deer (Michael, 1970; Zagata and Haugen, 1974; Beier and McCullough, 1990; Webb et al., 2010).

Using GPS technology, Webb et al. (2010) documented fine-scale movements of deer in Oklahoma. They reported that male daily movements were greater during the rut (7363 ± 364 m) than postrut (6156 m ± 260). Female daily movements were greater following parturition (3357 m ± 91) than during (2902 m ± 107) or before parturition (2682 ± 121 m). Activity rates, as measured by daily distance moved, also vary according to annual rainfall patterns in Mexico and may average more than 7 km/day in years of abundant rainfall but less than 4 km/day during years of normal rainfall (Bello et al., 2004).

Unusual, short-term excursions to areas outside of the normal range have been reported for both male and female deer, primarily during autumn (Figure 10.2). Tomberlin (2007) reported that 58% of a sample of adult males made excursions during the breeding season and Karns (2008) reported similar behaviors in 53% of collared males, with a smaller number of extraneous movements occurring before and after the breeding season. Hawkins and Klimstra (1970) suggested that males travel outside their normal home range to search for receptive females. Alternately, males may be tending an estrous doe as she moves back to her core area (Holzenbein and Schwede, 1989).

Dispersal

Dispersal movements are an important but little-appreciated behavior of white-tailed deer. Dispersal is defined as a permanent movement from the natal range to a new home range some distance away. Dispersal of one sex is a common behavior in vertebrates and is believed to be a behavioral mechanism to avoid inbreeding. In white-tailed deer, juvenile or physically immature bucks disperse from their natal area, whereas does are typically philopatric. Early studies of white-tailed deer used radiotelemetry to establish that most dispersal was done by yearling bucks. Rates of doe dispersal are typically low (2–20%; Hawkins and Klimstra, 1970; Nelson, 1993; but see Nixon et al., 1991, 2007), while dispersal rates of bucks may exceed 70% (Rosenberry et al., 2001). Habitat features may influence dispersal distances, where bucks disperse farther in more open or fragmented habitats than in forested or dense habitats (Nixon et al., 1991; Long et al., 2005; McCoy et al., 2005; Diefenbach et al., 2008). Dispersal distances of males may exceed 30–40 km in agricultural habitats of the midwestern United States (see Table 5.6), where vegetative cover is fragmented and patchily distributed (Nixon et al., 1991, 2007).

Although dispersal appears to be an innate behavior for many mammals, the actual behavioral stimulus or cue for dispersal is less certain. Male dispersal behavior is probably genetically based. Differences in seasonal timing of buck dispersal prompted two hypotheses for the behavioral stimulus, commonly referred to as mate competition and maternal aggression. The hypotheses are not mutually exclusive, but provide a view into the underlying motivations for dispersal in whitetails. Some bucks disperse prior to the rut (Rosenberry et al., 2001; Shaw et al., 2006), suggesting social pressure from other bucks, or perhaps competition for mates, as behavioral cues for dispersal. Other studies found that most young bucks dispersed during spring, implicating social pressure from the dam or female relatives as the cue for

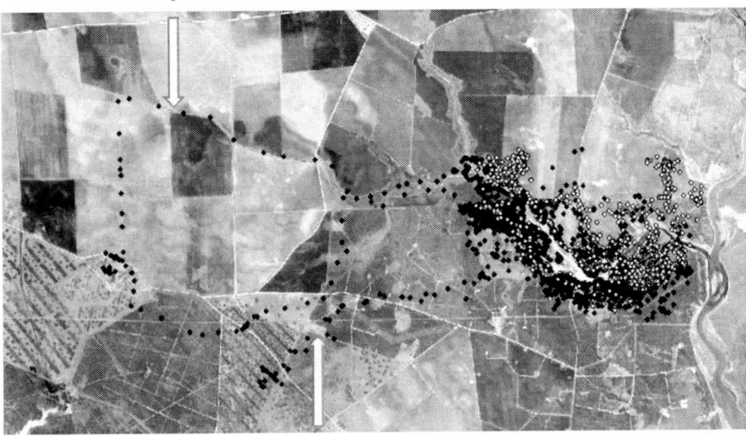

FIGURE 10.2 Excursions outside a buck's home range during the rut, as revealed by GPS telemetry. The buck made at least two relatively brief excursions of more than 10 miles. (Figure credit: Aaron Foley. With permission.)

dispersal. One of the few experimental studies of dispersal supported the maternal aggression hypothesis, as buck fawns orphaned during their first year of life were less likely to disperse than nonorphaned fawns (Holzenbein and Marchinton, 1992a,b). A recent large-scale study provided a balanced view of causal factors. Both mate competition and maternal aggression may cue dispersal in the same population, depending on population demographic conditions, such as buck age structure and sex ratio (Long et al., 2008). The stimuli for dispersal result in differing behavioral responses, such that bucks dispersing during spring (inbreeding avoidance dispersal) move greater distances on average than fall dispersers (mate competition dispersal) (Long et al., 2008).

Doe dispersal is relatively rare throughout the whitetail's range. However, doe dispersal occurs in some regions and has important implications for population dynamics in areas such as the agricultural matrix of the midwestern United States. More than 45% of does dispersed in an Illinois study, with dispersal distances exceeding 30 km (Nixon et al., 2007). Overall, high rates of doe dispersal appear to be a response to lack of habitats with sufficient vegetative cover during spring. However, doe fawns whose dam displayed migratory or long-range movements were more likely to disperse (Nixon et al., 2007). Therefore, learning or other behavioral interactions may act as contributing stimuli for doe dispersal.

Seasonal Migrations

In northern portions of the whitetail's range, deer often undertake pronounced seasonal migrations in response to snow depth and cold weather (Verme and Ozoga, 1971; Hoskinson and Mech, 1976). Migrations to wintering habitat generally are less than 16 km, but migrations in excess of 90 km have been reported from the Upper Peninsula of Michigan (Doepker and Ozoga, 1991). Migratory movements are relatively short and straight, spanning about two to seven days, depending on distance, with little wandering (Nelson and Mech, 1981).

Migratory patterns and seasonal home ranges of migratory deer are traditional. Fawns learn migration routes from their mothers but may also rely on female relatives because related does aggregate during autumn (Nelson and Mech, 1981; Tierson et al., 1985). Most yearling does establish home ranges near their natal areas and continue the migration patterns of their mothers. Deer in individual deer yards tend to represent distinct subpopulations or demes (Nelson and Mech, 1987) with adult deer from each yard occupying largely exclusive summer ranges. As a result, the behavior-based population separation may persist for years or decades (Nelson and Mech, 1999). Even females who have dispersed and established new summer ranges generally will continue migration to their natal winter ranges (Nelson, 1998).

Deer yards are often located in areas of closed-canopy coniferous forest which provide thermal cover, reduced wind velocity, and decreased snow depths compared to adjacent areas. In addition, yarding likely serves as an antipredator strategy, as the greater number of runways in high-deer-density areas and reduced snow depth enhance escape opportunities (Messier and Barrette, 1985). Winter yards may support from a few deer to several hundred. The 930 km^2 Mead Deer Yard in Upper Michigan supported an estimated 43,000 deer during the winter of 1987 (Ozoga, 1995).

The stimulus prompting deer to seek protective cover during winter is cold temperature (Ozoga and Gysel, 1972) or snow depth and seasonal cues (Sabine et al., 2002). In areas where winter weather varies annually, migration may be conditional and may depend on the onset of limiting conditions. For example, in southwestern Minnesota some deer remained sedentary during winter (15%), others were facultative migrators (35%), whereas the remainder were obligate migrators (Brinkman et al., 2005). About 20% of marked yearling and adult does migrated seasonally in Illinois (Nixon et al., 1991).

Once deer are in the yarding area, snow depth governs movements and habitat use. In areas of heavy snowfall, travel within yards often is confined to well-used trails that minimize energy expenditure. Mature northern white cedar is the preferred cover species in deer yards because it provides a narrow thermal range, firm snow pack, and significantly reduced wind speeds. In addition, it is a preferred and nutritious winter browse (Ozoga, 1995). Unfortunately, the thick coniferous cover and heavy deer use combine to limit available browse in deer yards. Deer in yards largely exist on fat reserves, and severe or prolonged winters result in high mortality from starvation and exposure. As a result, deer may minimize activity and movements to conserve energy during severe winter conditions.

Deer within yarding areas associate with one another frequently throughout the winter. Social hierarchies that are established via threats and other aggressive interactions help minimize social stress, although social strife may increase during late winter (Ozoga, 1972).

Deer leave yarding areas for summer ranges as soon as snowmelt advances sufficiently to allow ease of travel and daytime temperatures are consistently above freezing (Nelson and Mech, 1986). Spring migration is more direct and faster than autumn migration (Rongstad and Tester, 1969). As with autumn migration, fawns must be led back to their ancestral grounds by their dam or other adult female. Fawns orphaned overwinter may wander aimlessly during spring and summer and never return to their ancestral summer range (Nelson and Mech, 1986).

Although migrations are most pronounced on northern ranges, seasonal shifts in activity centers have been reported in other regions. In Illinois, 19.6% of adult does on a study area during winter migrated an average of 13 km away during spring and returned in autumn. In Nebraska, 13% of a marked population of does migrated seasonally (VerCauteren, 1998). Kammermeyer and Marchinton (1976) reported seasonal shifts onto a refuge in northwestern Georgia in response to habitat conditions and hunting pressure. Movements in response to seasonal flooding are common in southern river swamps (Byford, 1970; Joanen et al., 1985).

Predator Avoidance

Animal Predators

White-tailed deer use cryptic coloration, vigilance, and keen senses to minimize predation risk. Whitetails hide in thick cover, lying down or standing quietly, often allowing a predator to approach closely (Geist, 1998). However, once detected, they can use explosive speed to outdistance predators. Whitetails are fast runners for short distances, but are not adapted to outrunning predators in long high-speed chases. Because they inhabit dense vegetation, they do not require the extreme stamina and speed of open habitat specialists, such as pronghorn. Alternately, whitetails may opt to sneak away from potential predators.

Whitetails appear able to assess predation risks in their environment and weigh the costs and benefits of various behavioral options (Ozoga, 1995). As an example, habitats may differ in forage availability and predation vulnerability. Therefore, the whitetail must evaluate trade-offs in energy gain versus predation risk in determining where and when to forage. Behavioral modifications may include varying group size in relation to habitat or season, aggressive defense, escape behavior, and altering foraging patterns.

White-tailed deer group size varies according to season and habitat type. Group size tends to be larger in open habitats, allowing individual deer to be less vigilant. In a coastal Georgia study, deer in an open pasture spent more time feeding, less time moving, and less time being alert than did deer in an adjacent forested area (Lagory, 1986). Individual vigilance is negatively correlated with group size, suggesting that deer exploit loose aggregations to maximize foraging time by decreasing individual vigilance (Lark and Slade, 2008). In northern climates, whitetails congregate during winter in sheltered coniferous forests. Although these deer yards help conserve energy, their ultimate function may be a predator defense mechanism. Shallower snow depths, a well-traveled trail system, and social groupings allow greater mobility and improve chances of escaping predators (Nelson and Mech, 1981).

Whitetails are excellent swimmers and readily enter water to escape pursuit by predators (Figure 1.18). In a series of experimental chases by hunting dogs, three radiocollared deer were observed to run through water in swamps or streams on 40 occasions, resulting in the dogs losing the trail in all but two chases (Sweeney et al., 1971). Similarly, Barkalow and Keller (1950) observed a young deer successfully elude trained hunting dogs by entering a ditch and remaining submerged with the exception of her nostrils, forehead, and eyes.

Unlike mule deer, whitetails do not tend to change group size or formation in response to risks of coyote predation (Lingle, 2001). Whereas mule deer form large, cohesive groups and may aggressively defend conspecifics from a coyote attack (Lingle and Pellis, 2002), whitetails prefer to flee. Whitetails also maintain smaller group sizes than sympatric mule deer in open habitats, suggesting that differences in predator avoidance behaviors between the two species may influence group sizes (Lingle, 2003).

FIGURE 10.3 Newborn fawns adopt a hiding strategy during the first few weeks of life. Fawns remain isolated from their siblings and the dam, except for brief nursing bouts. The characteristic spotting pattern helps camouflage the fawn. (Photo by L. Thomas. With permission.)

Fawns are vulnerable to most medium and large predators, including coyotes, bobcats, and black bears (Chapter 8). Newborn fawns are "hiders" (Lent, 1974), remaining inactive and virtually undetectable until the dam returns to groom and nurse it once or twice daily (Figure 10.3). Most predator-induced mortality on fawns occurs during the first few weeks of life before fawns are able to outrun or evade predators (Vreeland et al., 2004). Separation of twin fawns soon after parturition (Schwede et al., 1994a,b), mother–infant separation, and considerable movement between consecutive fawn bedding sites help minimize predation. Experienced does appear more likely than younger does to lead their fawns to dense cover in response to disturbance, distress calls by fawns other than their own, or predator odors. In addition, does are hostile toward other deer at parturition (McCullough, 1979; Ozoga et al., 1982) and care-soliciting vocalizations by fawns may evoke defense of the fawn against potential predators (Garner and Morrison, 1980) or humans (Smith, 1987; Grovenburg et al., 2009). Because a doe's fawn-rearing skills improve with experience, older mothers are more successful than younger ones, especially in areas where predation seriously impacts neonatal survival (Ozoga and Verme, 1986a; Mech and McRoberts, 1990; Nixon and Etter, 1995). Interestingly, in urban or suburban habitats, where deer have become habituated to human presence, simple close association between humans and fawns may promote attacks toward humans (Hubbard and Nielsen, 2009).

Response to Hunting

Wild whitetails perceive humans as a potential threat or predator. However, whitetail responses to hunting can be variable depending on hunting pressure, deer population densities, and habitat type. Behavioral differences among deer also occur related to the age, sex, and prior experiences of the individual deer. Some studies indicate that deer remain within established home ranges in response to hunting (Marshall and Whittington, 1968; Sargent, 1992), while hunting can cause an increase in daily movement and changes in home range use elsewhere (Sparrow and Springer, 1970; Root et al., 1988; Naugle et al., 1997; VerCauteren and Hyngstrom, 1998; Kilpatrick and Lima, 1999; Williams et al., 2008). In Maryland, disturbance by hunters produced a behavioral response in only half of the encounters documented. In those situations, flight responses were temporary, and no changes in daily habits or shifts in use of home ranges or core areas occurred (Karns, 2008).

The average distance of deer to the nearest road and the amount of nocturnal activity were greater during the hunting season than during the nonhunting season in a Florida study (Kilgo et al., 1998). In addition, deer avoided clear-cuts, regenerating pine stands, and other open habitats and preferred the dense cover of swamps and mature forests. Similarly, areas of dense vegetative cover functioned as refugia for deer during controlled hunts in Connecticut (Kilpatrick et al., 2002). In the absence of dense cover, deer flushed by hunters may move long distances and may often leave their annual home ranges (Pilcher and Wampler, 1981; VerCauteren and Hygnstrom, 1998).

Deer may perceive and use nonhunted refuge areas during the hunting season if such areas are located within or adjacent to their home ranges (Zagata and Haugen, 1973; Kammermeyer and Marchinton, 1976), but do not appear to make long-distance movements to refuge areas. Proximity of refugia may limit the effectiveness of hunting in reducing deer populations, particularly in surburban and exurban areas (Brown et al., 2000; Storm et al., 2007a,b) (Chapter 20). Sex-specific differences in the response to hunting pressure may also occur. Does may move greater distances and cover larger spatial areas during hunting than nonhunting seasons, particularly if daily hunter activity exceeds 0.45 h/ha (Root et al., 1988). However, the critical level of disturbance likely varies among habitats. In contrast, males may not alter their movements in response to hunting (Root et al., 1988) or may variably increase (Van Etten et al., 1965) or decrease movements (Autry, 1967).

When pursued by hunting dogs, deer display high fidelity to seasonal and annual home ranges. Although they may leave their annual range, virtually all return to their normal range and resume normal activity patterns within 24 h (Sweeney et al., 1971; D'Angelo et al., 2003). When harassed by hunting dogs, deer exhibit characteristic behaviors to escape pursuit including holding, circuitous movements, distance running (Sweeney et al., 1971), periods of inactivity, and backtracking (D'Angelo et al., 2003).

Communication

White-tailed deer communicate through a series of chemical, audible, and visual cues. The design and spatial range of these modes of communication reflect the whitetail's evolutionary history and adaptations for living in heavy vegetative cover. Visual cues are mediated by posture and employed at close range. Likewise, whitetails have fewer and quieter vocalizations than other species of ungulates. In dense habitats, whitetails rely more on scent than on visual displays for communication, especially for locating conspecifics during the breeding season.

Olfactory Communication

Olfaction likely is the most important route of social communication in white-tailed deer. Chemical communication is accomplished through scent deposition on environmental objects (signposts) and through self-anointing via urine or glandular secretions. Whitetails possess a variety of specialized skin glands including the nasal, forehead, preorbital, tarsal, metatarsal, interdigital, and preputial glands (Figure 10.4). The forehead, tarsal, and interdigital glands are known to produce socially significant odors, but little is known of the communicative significance of the other glands.

FIGURE 10.4 In addition to vocal and visual displays, white-tailed deer communication relies on chemical cues produced by a variety of scent glands. The significance of forehead, tarsal, and interdigital glands secretions in social communication has been described, but the function or social significance of the other glands remains speculative. Other olfactory cues used for social communication include urine of both sexes and chemical signals produced in the urogenital region of the female. (Photo by T. Kirkland. With permission.)

The interdigital glands are pockets between the primary digits of all four feet (Quay, 1959). Because deer have been observed tracking other deer and even backtracking themselves, these glands likely are responsible for leaving an olfactory trail and conveying home range familiarity. Additionally, chemical analyses of secretions from the interdigital glands indicate differences in concentrations of some volatile constituents related to age and dominance rank, suggesting that social status likewise may be conveyed (Gassett et al., 1996).

The forehead gland is most active during the breeding season in both sexes; greatest activity is observed in dominant males (Atkeson and Marchinton, 1982). Forehead gland secretions are used in marking rubs and potentially overhanging branches at scrape sites (Moore and Marchinton, 1974). Social grooming among male deer is concentrated on the forehead region during the breeding season. Thus, chemical signals from the forehead region likely aid in individual recognition (Gassett et al., 1997).

All age classes and both sexes of whitetails rub their tarsal glands together while urinating over them in a behavior called rub-urination (Figure 10.5). Urinary products, glandular secretions, and bacteria on the tarsal tuft interact to produce a unique semiochemical profile (Alexy et al., 2003) that may facilitate conspecific recognition, communicate social and reproductive status (Marchinton and Hirth, 1984), aid in mate selection, and possibly serve as a priming pheromone for the female reproductive system (Miller et al., 1991). Black-tailed deer fawns as young as two days old have been observed urinating over their tarsal glands (Müller-Schwarze, 1971), and this behavior likely is similar among whitetail fawns. Does rub-urinate at least once per day, typically when they rise from a nocturnal bed (Sawyer et al., 1993). Does, fawns, and bucks during the nonreproductive period lick the excess urine off of the tarsal gland. Frequency of rub-urination increases in mature males during the rut (Miller et al., 1987; Marchinton et al., 1990). Because bucks no longer lick the excess urine the gland becomes darkly stained and produces a strong musky odor, although changes in urinary constituents likely contribute to the strong odor (Miller et al., 1998). Deer investigate the tarsal gland of conspecifics and often will rub-urinate in socially stressful situations.

Visual Communication

Whitetails use subtle visual cues to communicate dominance status, aggressive threats, and alarm response to perceived threats. Whereas threats signal aggressive intention, dominance displays proclaim status, serve to intimidate potential rivals, and potentially attract mates. Low-level threats are signaled by an ear-drop display in which the ears are laid back along the neck, coupled with a direct stare by the

FIGURE 10.5 Deer of all age classes and both sexes commonly urinate over the tarsal glands. The interactions of urine, glandular secretions, and bacterial fauna likely produce a unique scent profile for each individual. The increased frequency of rub-urination and changes in urinary constituents produce the characteristic dark stains and strong scent of rutting bucks. (Photo by T. Kirkland. With permission.)

FIGURE 10.6 Whitetails communicate through a variety of visual cues when individuals are in close proximity. Low-level threats are signaled by an ear-drop coupled with a direct stare by the aggressor. (Photo by T. Kirkland. With permission.)

aggressor (Figure 10.6). The head-high and head-low threats are more aggressive and signal intent to strike or chase (Hirth, 1973). The head-high threat is used predominantly by does and by bucks during the velvet antler phase and signals intent to rear and flail the adversary with front hooves. The head-low threat signals intent to chase and strike with a foreleg. Males also use the head-low thread during the breeding season to signal a willingness to engage antlers with an opponent.

Dominance displays by a rutting buck include the ear-drop, piloerection (erection of body hairs), a broadside stance (sidling), and flaring of the preorbital gland and nostrils (Hirth, 1977) (Figure 10.7). Piloerection increases perceived body size, while the ear-drop and flaring nostrils signal aggressive intent. Beads of moisture appear on the muzzle, and often there is noticeable drooling.

FIGURE 10.7 Dominance interactions among bucks often involve visual displays. The ear-drop, direct stares at a rival, flaring of the nostrils and preorbital gland, and erection of body hair signal aggressive intent. (Photo by T. Kirkland. With permission.)

When fleeing, whitetails often expose the white underside of their tails. Tail flagging is performed by all ages and both sexes. The frequency of flagging does not appear to differ between doe groups composed of female relatives and buck groups composed of nonrelatives. Therefore, tail flagging likely serves as a signal that helps maintain group cohesion for antipredator benefits (Hirth and McCullough, 1977). In addition, tail flagging may inform an approaching predator that it has been detected (Bildstein, 1983).

Vocal Communication

Several vocalizations have been identified and categorized based on tonality and social context: alarm and distress calls, agonistic calls, maternal–neonatal calls, and mating calls (Atkeson et al., 1988). Alarm calls include the characteristic snort produced by an alerted or fleeing deer, particularly adults. It is an intense, unvoiced sound of variable tonality produced by forcibly expiring air through the nasal passages. Snorts may be accompanied by a foot stomp that produces a thumping sound and serves to alert members of the social group to possible danger (Caro et al., 1995). Snorts often are elongated and given in a series when danger has been perceived, but the associated risk has not been identified. Snorts are of shorter duration and sharper during flight. Deer most commonly snort in circumstances of little risk to the giver and more frequently within groups of related individuals (Hirth and McCullough, 1977).

The bawl, or distress call, is produced by physically distressed or traumatized deer of both sexes and all ages. It is a voiced call of high tonality and the pitch declines with age (Atkeson et al., 1988). Richardson et al. (1983) suggested that bawls by fawns elicit maternal defense. They reported that the bawls of fawns were sufficiently distinct to allow individual recognition by the doe. However, Lingle et al. (2007) reported that whitetail does approached recordings of distress calls from nonfamilial fawns, but only when their own fawns were hidden.

Agonistic calls include the grunt, the grunt–snort, and the grunt–snort–wheeze. The grunt may be used by either sex throughout the year and often is accompanied by intention postures. The grunt–snort is emitted during progressively intense agonistic interactions between males, primarily during the breeding season. The most intense agonistic vocalization is the grunt–snort–wheeze given primarily by dominant bucks and consists of a low guttural grunt followed by one or more rapid nonvoiced snorts caused by rapid expulsion of air through open nostrils, followed by a drawn-out wheezing expulsion of air through pinched nostrils (Atkeson et al., 1988). Both the grunt–snort and the grunt–snort–wheeze are typically accompanied by aggressive postures including pinned ears, sidling, and piloerection.

Grunt calls are employed in a variety of contexts, including grunts given by bucks tending an estrous doe, cohesive calls given by does when separated from a group, and maternal grunts directed at fawns (Richardson et al., 1983). Fawns may produce a high-pitched mew sound in response to the maternal grunt, probably to solicit attention from the dam. The fawn bleat is a more demanding, care-soliciting vocalization and can be heard by humans up to 100 m away. Fawns periodically emit a nursing whine while suckling, which may help the dam identify the fawn, solidify the maternal bond, and elicit additional maternal attention, such as grooming.

Mating Behaviors

The suite of behaviors displayed during the breeding season includes some of the most widely recognized and interesting behaviors for hunters and nature enthusiasts. Sexual maturity occurs at a relatively young age in white-tailed deer compared to other species of large mammals. Bucks are sexually mature by 1.5 years of age (Sauer, 1984), although captive bucks have become physiologically capable of breeding prior to one year of age. Sexual maturity in does is dependent on body mass. Doe fawns can breed at six to eight months of age under conditions of excellent nutrition, but breed at 1.5 years of age in most populations. Thus, deer of all age classes could be sexually mature in their first year and may participate in rutting behavior.

In temperate regions, rutting activity is mediated by seasonal changes in photoperiod (Goss, 1983) (Chapter 2). However, in tropical regions of Central and South America, deer reproduce year-round although there may be seasonal peaks associated with regional weather patterns (Brannan and Marchinton,

1987). In northern regions, shortening day length triggers a chain of hormonal events initiated by the pineal gland which ensures that conception occurs during autumn, so fawns are born in spring or early summer. Photoperiod changes more abruptly in northern than in southern latitudes. A short, precisely timed breeding season is necessary in northern climates, so fawns have optimal nutrition and ample time for growth prior to harsh winter weather. However, whitetails display considerable geographic variation in the timing of breeding, particularly in southern temperate regions of the United States (McDowell, 1970; Richter and Labisky, 1985) (Figure 16.1). In the southern United States, seasonal climatic fluctuations are less severe and reproductive timing is less critical. Therefore, the changing photoperiod apparently opens a reproductive window during which breeding can occur.

Exactly when breeding occurs varies according to a variety of factors including herd demographics (Miller and Ozoga, 1997), genetic origin (Jacobson, 1994a), maternal age and condition, and possibly physiological cues passed from mother to daughter. In the southeastern United States, peak breeding varies widely among populations and may occur in September in portions of South Carolina, in November in Georgia, and from December through January in portions of Alabama and Mississippi. In areas of Florida and Texas, timing of the breeding season can vary dramatically among locations less than 80 km apart (Richter and Labisky, 1985; Traweek et al., 1996) (see Figure 16.1). In some cases, environmental factors shape rut timing. For example, historical flooding regimes have apparently selected for an early rut during August in some Florida populations to avoid fawning during floods (Richter and Labisky, 1985). However, high variability in rut timing between geographically proximate populations has defied easy explanation in other areas of the southeastern United States.

Many areas of the southeastern United States were restocked with deer from various regions of the whitetail's range and stocking sources comprised multiple subspecies. Thus, variable timing of reproduction in this region may be related to the genetic source of these populations. Stock sources spanned latitudes from Mexico to Wisconsin and eastward to North Carolina (McDonald and Miller, 2004). Restoration programs had a lasting impact on the genetic variation of deer populations in some regions, whereas native stocks recovered in other regions (Leberg and Ellsworth, 1999). However, if rut timing was only due to genetic determination based on stocking source, differences in rut timing likely would gradually erode in continuous populations. Therefore, maintenance of local breeding dates may be influenced by a physiological clock or other priming mechanism mediated as a maternal effect. Low rates of doe dispersal in the southeastern United States combined with a maternal influence on breeding dates likely preserve breeding dates similar to the source stock. A maternal effect is plausible in that maternal effects are an important source of phenotypic variation in quantitative traits in ungulates (Wilson et al., 2005). The influence of a maternal effect on timing of breeding is supported by a study of captive deer in which researchers experimentally altered the natural timing of breeding in two captive does. The does subsequently maintained a breeding schedule asynchronous with the normal period of estrus (Osborn et al., 2000).

The degree of breeding synchrony within populations also varies across the whitetail's range and is influenced by environmental and demographic factors. Births are timed to coincide with peak nutritional conditions, and synchrony of births serves to reduce predation on fawns. In general, does are receptive for mating during the estrous period which typically lasts about 24 h. Estrus may be prolonged more than 48 h if the doe remains unbred (Knox et al., 1988). Age and body condition influence timing of estrus for individual does. Adult does in good condition breed early in the rut, whereas first-time mothers (fawns or yearlings) mate later (Ozoga and Verme, 1986b). In northern portions of the whitetail's range, such as in New York (Cheatum and Morton, 1946) and Virginia (McGinnes and Downing, 1977), most does are bred over a relatively brief three-week rutting period due to more severe environmental and photoperiodic constraints. In tropical regions, whitetail breeding may last six months or more (Brannan and Marchinton, 1987). In the midwestern United States, breeding may be prolonged due to late breeding by doe fawns (Nixon et al., 1991), while herd demographics likely influence the duration of the breeding season in the southeastern United States (Jacobson et al., 1979; Miller et al., 1991). In exploited populations with male-biased harvest and skewed sex ratios, peak breeding may occur up to one month later and the breeding season may be prolonged compared to demographically balanced populations (Jacobson, 1992; Miller et al., 1995). The prolonged rut and later dates of conception in exploited populations may be due to an insufficient number of bucks to ensure that all does are bred during their first estrus. However, there is evidence that the presence of physically mature bucks may promote earlier breeding, perhaps due to scent or other chemical stimuli (Verme et al., 1987; Miller et al., 1991).

Phases of the Rut

Increasing testosterone concentrations in late summer trigger velvet shedding and behavioral changes among bucks in preparation for the rut (Figure 10.8). Almost as soon as the velvet is shed, bucks within bachelor groups begin sparring with each other (Hirth, 1977). All bucks engage in sparring, although yearling bucks initiate more sparring bouts (Ozoga and Verme, 1985), likely due to behavioral immaturity and uncertainty about their status within the dominance hierarchy. Older bucks appear to tolerate the sparring advances of these younger males. Sparring bouts begin with one buck approaching another, lowering his head, and presenting his antlers. Antler presentations during sparring are rarely accompanied by signals of aggression such as the ear-drop, piloerection, and aggressive vocalizations. Sparring matches may last from as little as a few seconds to several minutes, and often conclude amicably as the bucks resume feeding.

Shortly after velvet shedding, sparring matches are relatively mild encounters. These bouts of "skill sparring" (Figure 10.9) apparently do not have a clear winner or loser (Ozoga, 1994). However, the intensity of sparring encounters increases as testosterone concentrations rise preceding the rut. "Demonstrative sparring" bouts are more aggressive encounters and usually there is a clear winner and loser. These demonstrative sparring bouts serve as tests of strength and help to assess or establish dominance rank among bucks.

Either preceding or following a sparring match, the smaller and presumably subordinate buck often will groom the forehead region of his sparring partner. This behavior may allow individuals to associate the outcome of these encounters with the odor of their opponent's forehead gland secretions on antler rubs or at scrape sites (Forand and Marchinton, 1989) and maintain familiarity among bucks after bachelor groups dissolve during the rut.

Violent dominance fights are much less common than sparring matches and occur primarily during the breeding season between two unfamiliar bucks or over breeding rights to an estrous doe. Dominance fights often are preceded by dominance displays. These aggressive signals begin with a directed stare and pinned-back ears. If neither buck retreats, the display intensifies. Piloerection increases perceived body size, and the aggressor will angle toward his opponent in a stiff-legged broadside gait referred to as sidling. These aggressive displays are often accompanied by flared preorbitals, pawing, and the grunt–snort–wheeze vocalization.

FIGURE 10.8 An increase in testosterone concentrations during late summer triggers the maturation of the antlers and shedding of the antler velvet. Bucks often use small trees or shrubs to aid in removing the velvet. (Photo by T. Kirkland. With permission.)

FIGURE 10.9 Sparring matches often involve young bucks or a young and mature partner. Sparring allows bucks to test their strength and aid in establishing dominance before the rut. Shortly after velvet shedding, sparring matches are relatively mild encounters. However, the intensity of sparring increases as testosterone concentrations continue to rise preceding the rut. (Photo by T. Kirkland. With permission.)

At this point, one of the adversaries commonly breaks eye contact, steps away, and may resume feeding. More rarely, neither buck acknowledges defeat and they may slowly circle each other before coming together in a violent antler crash. During fights bucks drop low to the ground and attempt to forcibly push the opponent backward or throw him off balance to gain an advantage (Figure 10.10). Many fights are relatively brief, lasting only a minute or two before the loser breaks contact and retreats. However, fights between equally matched opponents can be much longer. Retreating bucks are pursued for only a short

FIGURE 10.10 Fighting between equally matched mature bucks is relatively uncommon. However, encounters between unfamiliar bucks during the peak of the rut or in contests over an estrous doe may be very intense. (Photo by J. Feild. With permission.)

FIGURE 10.11 Broken antlers and wounds to the head are a common result of dominance fights. (Photo by T. Kirkland. With permission.)

distance, or not at all, and usually the victor does not attempt to gore the body of the retreating buck. Fatal, or even severe, injuries are relatively uncommon, although broken antlers, torn ears, and gouges around the eyes and the neck are quite common (Figure 10.11).

Signpost Communication

Bucks communicate through the use of visual cues that accompany and draw attention to olfactory cues. The use and importance of these signposts vary throughout the annual cycle. Communal licking branches, where bucks deposit scent presumably from the forehead or preorbital gland are an important medium for communication during spring and summer (Marchinton et al., 1990). Following velvet shedding and throughout the breeding season, bucks make antler rubs by debarking small trees and shrubs with their antlers and anointing them with secretions from the forehead gland. When making a rub, a buck often is very deliberate, occasionally stopping to smell or lick the rubbed area. Occasionally, a rubbing bout will escalate into a vigorous thrashing of the tree that likely serves as a mock fight. Bucks select highly aromatic species of trees for rubbing such as pines, cherries, and Eastern red cedar presumably because they are more apparent to other deer (Benner and Bowyer, 1988; Oehler et al., 1995; Miller and Marchinton, 1999).

Rubs are made continually throughout the breeding season (Ozoga and Verme, 1985) (Figure 10.12) and rub density can exceed 1400 rubs/km^2 (Miller and Marchinton, 1999). On average, bucks may make 300–400 rubs each fall (Miller et al., 1987), although behaviorally immature yearling bucks make relatively few rubs early during the breeding season and overall only produce about 50% as many rubs as mature males (Ozoga and Verme, 1985). Other bucks and does respond to antler rubs (less than three days old) by smelling, licking, or rubbing them with their forehead (Sawyer et al., 1989).

Certain types of rubs may serve special purposes. "Traditional rubs" are larger-diameter trees that are rubbed in successive years (Woods, 1994). Several individual bucks may interact with these rubs, either by smelling the rubbed area or by rubbing on the exposed area themselves. Traditional rubs tend to be uncommon in heavily exploited deer herds with a young male age structure. What role these traditional rubs play is still speculative, but clearly they are important in communication of presence or status among bucks.

In producing a scrape, a buck approaches a branch hanging just above his head. Often, licking branches that have been marked throughout the summer and early fall become scrape sites during the breeding

FIGURE 10.12 Following velvet shedding and throughout the breeding season, bucks make antler rubs by debarking small trees and shrubs with their antlers and anointing them with secretions from the forehead gland. When making a rub, a buck may stop to smell or lick the rubbed area. Occasionally, a rubbing bout will escalate into a vigorous thrashing of the tree that likely serves as a mock fight. (Photo by T. Kirkland. With permission.)

season (Marchinton et al., 1990). However, bucks will utilize any branch of sufficient height (Ozoga, 1989a). The buck often mouths the branch and rakes it with his antlers. Judging from the buck's behavior, marking the overhanging branch appears pleasurable, and he sometimes seems almost oblivious to his surroundings. Clearly, the buck is leaving some type of scent on this overhanging branch, although the exact source remains ambiguous. Likely the forehead gland is involved in marking the limb, but other potential scent sources include the preorbital gland, the nasal gland, and even saliva. After the overhead limb is marked, the buck paws away the leaves directly below the limb, likely leaving scent from the interdigital gland in the pawed area. The area cleared of leaves varies, but a 1-m-diameter circle is common. The buck then steps forward and urinates over the tarsal glands (Figure 10.13) while rubbing them together, allowing urine to flow into the pawed area (Moore and Marchinton, 1974). This urine leaves a persistent strong odor and may stain the soil dark even after it has dried.

Although some scrapes may appear shortly after velvet shedding, peak scraping activity invariably occurs about two to three weeks in advance of peak breeding (Moore and Marchinton, 1974; Ozoga 1989b; Miller and Marchinton, 1999) (Figure 10.14). Scrape sites may be defended in the presence of a dominant male (Marchinton and Atkeson, 1985); however, scrapes may be visited and re-marked by several males during the breeding season (Dasher et al., 1998). There is some evidence that even mature bucks vary greatly in the amount of scraping that they do. Age is important, but a buck's testosterone levels, social position, experience, and behavioral maturity also interact to promote scraping behavior (Marchinton et al., 1990).

Scraping activity is highly influenced by population demographics, as most scrapes are made by mature males. Yearling bucks may make only 15% as many scrapes as produced by older males (Ozoga and Verme, 1985). However, individual mature bucks vary in the number and frequency of scraping activity. Scrapes apparently serve to advertise the presence and social status of bucks, but may be a communicative mechanism between the sexes as well. Scrapes, and perhaps antler rubs, may assist in mate selection as well as act as a source of priming pheromones that assist in termination of seasonal anestrus (Miller et al., 1987). In populations with a high percentage of does and a young age structure among bucks, the communicative role of rubs and scrapes may be less significant (Marchinton and Hirth, 1984).

FIGURE 10.13 In producing a scrape, a buck mouths an overhanging branch (a) and rakes it with his antlers. Sources of scent left on the branch may include the forehead gland, the preorbital gland, or perhaps saliva. The buck then paws away the leaves directly below the limb (b), likely leaving scent from the interdigital gland, then steps forward and rub-urinates urine to flow into the pawed area (c). This urine leaves a persistent strong odor and may stain the soil dark even after it has dried. (Photos (a) and (b) by T. Kirkland. With permission. Photo (c) by R. DeYoung.)

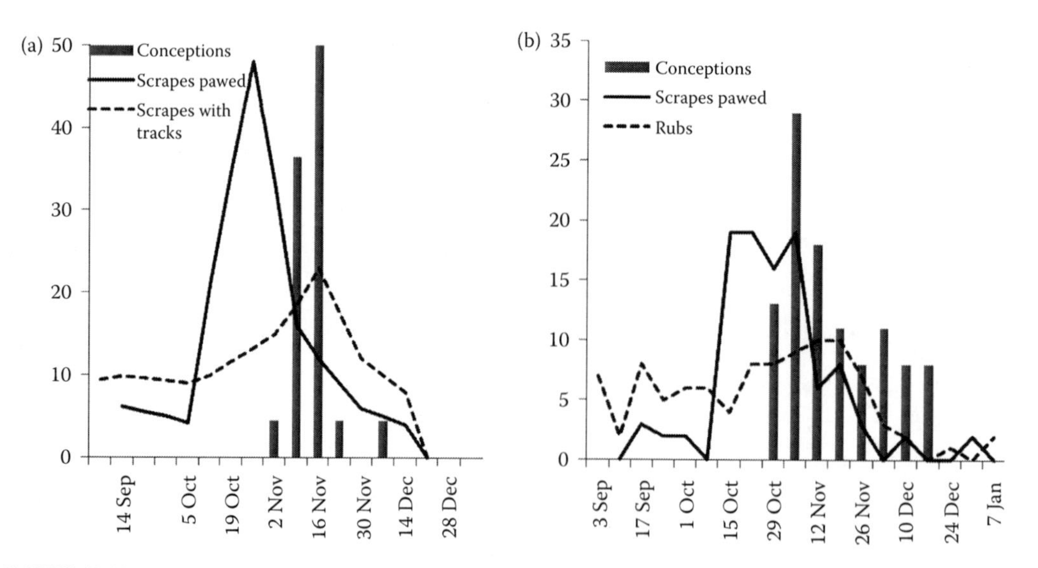

FIGURE 10.14 Temporal distribution of white-tailed deer signposts and conception dates in Michigan (a) and Georgia (b). (Data adapted from Ozoga, J. J. 1989b. *Journal of Mammalogy* 70:633–636 and Miller, K. V. and R. L. Marchinton. 1999. *Canadian Field-Naturalist* 113:519–521.)

Tending and Copulation

Behavioral ecologists characterize white-tailed deer as a species with a "tending-bond" mating system (Hirth, 1977). Mating is a one-on-one interaction between an individual buck and doe, rather than a group or harem. Bucks display no evidence of territoriality; the doe is defended, not a fixed location or physical resource.

Bucks are physiologically and behaviorally ready to breed long before females are receptive. Adult males increase movement and activity from summer to the breeding season with a subsequent decrease following the breeding season (Tomberlin, 2007). These movements typically are characterized by males covering large portions of their home range with continuous movement and returning to the point of origin within 8–30 h (Tomberlin, 2007), consistent with Brown's (1974) dominant floater description.

During the pre-rut, or courtship phase, bucks begin to investigate females for signs of estrus, but the does normally avoid bucks. In most regions, the courtship phase begins four to eight weeks after velvet is shed, although this can be quite variable in the southern United States. The behavioral stimulus for courtship initiation is speculative, but may result from elevated testosterone concentrations in the male or perhaps pheromonal signals from the female (Marchinton and Hirth, 1984). During this phase, bachelor groups dissolve as mature males increase their movements and home range sizes in efforts to locate estrous does (Beier and McCullough, 1990). During the courtship and subsequent breeding phases, some bucks may shift their seasonal home ranges to an area disjunct from their summer ranges, whereas others may simply increase their home range size (Tomberlin, 2007).

Bucks detect estrus via behavioral and olfactory cues. Nonreceptive does flee at the approach of a buck and may run in tight circles around or in front of the buck before fleeing. Does being chased often stop to urinate. The courting buck approaches the spot where the doe urinated, takes a sample of the urine into his mouth, and performs the flehmen behavior (Figure 10.15). Through this process, urine is drawn into the vomeronasal organ located in the buck's upper palate where urinary compounds of low volatility are analyzed by the accessory olfactory system (Müller-Schwarze, 1994). It was long believed that this behavior provides the buck with information about the doe's reproductive status. However, anosmic males can detect estrous females by behavioral cues whereas males with occluded vomeronasal organs are attracted to estrous females even in the absence of behavioral cues (Gassett et al., 1998). Vomerofaction may mediate physiological changes that ensure males and females are simultaneously in peak reproductive condition, but this concept is speculative (Miller and Marchinton, 1994). Volatile

FIGURE 10.15 The courting buck may sample urine from a doe by performing the flehmen behavior which introduces urine into the vomeronasal organ. Although it was once believed that this behavior allowed the buck to determine the doe's reproductive status, vomerofaction may mediate physiological changes that ensure males and females are simultaneously in peak reproductive condition. (Photo by D. Hewitt. With permission.)

compounds produced in the female's reproductive tract apparently are a source of sexually attractive odors (Whitney et al., 1991), although behavioral cues are equally important.

Activity of does increases just prior to conception (Ozoga and Verme, 1975), suggesting that females may play an active role in breeding activities. Even in high-density herds with balanced sex ratios, females might still need to search for prospective mates because the relative abundance of mature males to reproductively mature females may be low. If most males are preoccupied with receptive females, then females entering estrus may be forced to engage in mate-searching behaviors. Subsequently, does may increase activity or expand their home range during the rut (Labisky and Fritzen, 1998), including large excursions outside of their home range at the estimated time of conception (Holzenbein and Schwede, 1989; D'Angelo et al., 2004). In an Alabama study, does increased activity during the rut, but total distance moved decreased (Ivey and Causey, 1988). Many early radiotelemetry studies of female movements during the rut only infrequently reported breeding excursions by does at the time of conception. However, this infrequent reporting likely was due to the short duration of these movements and the relatively infrequent sampling in very high frequency (VHF) telemetry studies. Recently, a study using GPS technology reported that nine of 10 females displayed excursive behavior (Kolodzinski et al., 2010) and suggested that these excursions are common as females may be engaging in discrete selection for the most reproductively fit breeding partner.

As a doe approaches estrus, she and her suitor separate from other deer and form a tending bond. Because a doe is attractive to bucks for a day or more before she is willing to stand for copulation, a series of bucks may be displaced from the doe by larger or more dominant bucks. Behaviorally inexperienced yearling bucks may vigorously pursue does that are approaching estrus, but mature bucks court females in a highly ritualized fashion.

A mature buck tests a doe's receptivity with a courtship approach consisting of short, choppy steps while holding his head low to the ground and his neck outstretched (Brown and Hirth, 1979). If receptive, the doe allows the buck to approach closely. Prior to breeding, the buck will smell and lick the doe's tarsal gland as well as her urogenital region (Figure 10.16). Copulation begins with the buck resting his chin on her haunches. If receptive, she exhibits a mating stance characterized by extreme immobility and rigidity with ears laid backward. The male rises to her back and clasps the female with his forelegs. Intromission occurs with a single minor pelvic oscillation and tail twitching and is terminated with a hard penile thrust that often knocks the doe out from under him (Warren et al., 1978). Following copulation the female assumes a postcopulatory posture often accompanied by abdominal contractions and opening and closing of the vulva.

Unless displaced, tending bucks remain with does for 24–72 h preceding and during their estrous period. Often the pair will isolate themselves, typically in open habitats such as overgrown fields, clearcuts, or open woods. A doe will only be receptive for mating during the estrous period which typically

FIGURE 10.16 Volatile compounds produced in the female's reproductive tract apparently are a source of sexually attractive odors, although behavioral cues are equally important. (Photo by T. Kirkland. With permission.)

lasts about 24 h, although estrus may be prolonged for at least 48 h if the doe remains unbred (Knox et al., 1988). Generally, older and more experienced does breed first; young subordinate does breeding for the first time (as fawns or yearlings) mate later (Ozoga and Verme, 1986a).

Parturition and Fawn Development

Social isolation is essential for proper mother–infant bond formation and is accomplished in deer through a form of territorial behavior that lasts four to six weeks around the time of parturition (Ozoga et al., 1982; Bertrand et al., 1996). Dominant does often maintain these exclusive fawning territories for several years, although primiparous yearling or two-year-old does may establish fawning sites adjacent to their mother's territory. Three-year-old does established new fawning areas away from the family group. Because matriarchal does acquire the best fawning territory, adult does (and their associated two-year-old daughters) have the greatest fawn-rearing success (Mech and McRoberts, 1990; Nixon and Etter, 1995).

The doe's aggressive behavior associated with defense of her fawning territory differs from the dominance displays used at other times of the year. Instead of posturing and intent behaviors, territorial defense involves overt aggression as the doe attempts to evict the intruder from her fawn-rearing territory (Ozoga, 1996). Often she approaches with a head-high threat, but when within 6–9 m she rushes her adversary kicking with her front feet. These chases may extend 100 m in length.

One to two days prior to parturition, doe movements increase as they begin a restless pacing, often with their tail elevated. In a study of penned deer, the allantochorion ruptured about 4 h before birth of the first fawn and labor contractions lasted an average of 124 min before birth (Townsend and Bailey, 1975). In contrast, observations of a wild doe reported a time period of 28 min between a pregnant doe lying down and the fawn first appearing (Michael, 1964). The birthing process proceeds quickly after the forelegs appear, averaging 17 min. The interval between birth of the first fawn and the second can vary from 8 to 57 min (Townsend and Bailey, 1975).

Immediately after delivery does lick the fawn vigorously to remove all of the membranes and to dry the fawn. This forceful licking is important to clean the fawn quickly and perhaps to stimulate standing and nursing. This initial grooming is also important in imprinting the dam with her fawn as well as to reduce odors that may attract predators. Fawns will attempt to rise shortly after birth and be capable of standing within 20–30 min, although second fawns usually take longer to stand than first fawns (Townsend and Bailey, 1975). Does invariably consume the entire afterbirth after it is expelled, presumably to minimize attractiveness of the birth site to predators as well as to gain a nutritional benefit (Marchinton and Hirth, 1984).

Nursing begins almost immediately after standing, once the fawn locates the teat, although initial nursing may also occur while the doe is lying down. Does remain at the birth site with their fawns for approximately 3 h before moving them to separate bedding sites, although does may leave fawns at the birth site for 24 h after parturition (Bartush and Lewis, 1978). Siblings remain isolated during the first three to four weeks after which the doe guides them to a communal bed site (Schwede et al., 1994b). During the first few weeks of life, mothers associate with their fawns only briefly, primarily at sunrise and sunset. At this time, fawns are active for only about 8% of the day (Jackson et al., 1972). Young fawns react to potential danger by remaining hidden and freezing in a prone position, either curled or outstretched with their chin and neck on the ground. Accompanying this freezing posture, the fawn's heart rate may decrease by 11–68% depending on the perceived stimuli and age of the fawn (Jacobsen, 1979). By the second week of life, fawns are more agile and are equally likely to run as to freeze when alarmed (White et al., 1972), and by the third week fawns typically will run from danger.

As fawns approach one month of age, daily activity increases and twin fawns start to appear together. At this time male fawns are more active than female fawns (18% vs. 12% of the time) (Jackson et al., 1972). Male fawns also associate less and more loosely with their mothers than do female fawns (Schwede et al., 1994a).

Fawns nurse only two or three times daily during the first couple of weeks, but the frequency of nursing bouts increases to five or six as the fawns age. During the first couple of weeks, the mother grooms the fawn while it is nursing, particularly around the perineal area to stimulate urination and

FIGURE 10.17 During their first couple of weeks, fawns nurse only two or three times daily, but the frequency of nursing bouts increases as the fawns age. Mothers often groom young fawns while nursing, particularly around the perineal area, to stimulate urination and defecation. (Photo by T. Kirkland. With permission.)

defecation (Figure 10.17). Nursing bouts at first are terminated by the fawn, but by two or three weeks of age the doe may begin terminating nursing bouts (Gauthier and Barrette, 1985) by simply walking away. Does may lead the fawn to a new bedding area, although the fawn selects a bedding site without apparent assistance from the mother. After nursing, the fawn simply wanders 12–25 m from its mother to a bedding site. A study in Iowa indicated that fawn bed sites contained denser vegetation than surrounding areas, although ambient temperature apparently influenced the fawns' selection of cover density and aspect (Huegel et al., 1986). On cool days fawn bed sites were found more often on slopes facing the sun in relatively open habitats, whereas on warm days fawns bedded in dense habitat on slopes facing away from the sun. Does locate their bedded fawns either by odor or by returning to the area where the fawn is bedded and emitting a soft maternal grunt to which the fawn responds by running to the dam.

During the first couple of weeks of life, the dam remains within about 90 m of her resting fawn. If threatened, the fawn may bleat or bawl, eliciting a defensive response from the dam. However, does apparently do not begin defending fawns until they are approximately four days old (Grovenburg et al., 2009), and fawn defense may continue until the fawn is two weeks old and capable of fleeing from dangerous situations. In addition, male fawns are more likely to be defended than female fawns. Types and intensity of defensive behaviors vary according to the condition of the female and their fawns and among age classes (Smith, 1987). Does may defend their fawns from other deer and predators (Ozoga and Verme, 1986b; Mech and McRoberts, 1990) as well as humans (Hubbard and Nielsen, 2009).

By three weeks of age fawns begin consuming vegetation and by six weeks the does end most nursing bouts. Home range size of fawns and their mothers increases over the first three months of life and by week nine the fawn's home range approaches that of its mother. By 10–12 weeks of age fawns are as active as their mothers and they are functionally weaned. Does will reject most nursing attempts at this time.

Play is an important part of fawn development that helps develop locomotor skills, adult behavior, and survival instincts. By the second week of life, nursing bouts are often followed by locomotor play (Hirth, 1973). A fawn may run a short distance from its mother before stopping abruptly and dashing back while bucking and dodging. These episodes may be repeated several times, and occasionally the doe will join the frivolity. Locomotor play is primarily solitary, even after siblings begin traveling together with their dam. However, play bouts can stimulate play in other fawns, and at times evolve into games of chase. Social play becomes more common later in the summer as fawns interact socially with siblings and other fawns to develop skills necessary for establishing their place in the social order (Jacobson, 1994b). Fawn play mimics adult behavior and they may engage in mock fights, mimic scent marking, and practice aggressive postures. By five months of age they have established their social position with other fawns.

Reproductive Success and Mating Tactics

Mating strategies and number of fawns sired by individual bucks are difficult to study in white-tailed deer because not all individuals or copulations can be observed. The most intensive observational study of mating behaviors in wild deer tallied four copulations in three years of observations (Hirth, 1977). Recently, genetic parentage and GPS technology have revolutionized the study of mating behaviors and it is now possible to accurately assess patterns of mating success and mating strategies. A series of studies in captive (DeYoung et al., 2002, 2006) and wild populations of whitetails (Sorin, 2004; DeYoung et al., 2009) revealed that individual bucks sire relatively few fawns per year in wild populations, with successful bucks producing nine or fewer offspring (DeYoung, 2004; Sorin, 2004; DeYoung et al., 2009). Overall, physically mature bucks (aged 3.5 years and older) are more likely to be successful sires (Figure 10.18), and collectively sire about 70% of fawns in populations with a natural age structure (Sorin, 2004; DeYoung et al., 2009). Young, physically immature bucks (1.5 and 2.5 years of age) sire the remaining 30% of fawns, even in populations where mature bucks are present. In exploited populations, high rates of buck harvest result in a younger age structure and female-biased sex ratio, resulting in more equitable breeding opportunities for bucks of all age classes (DeYoung et al., 2009). Furthermore, does may mate with more than one buck during a single estrous period as studies of captive and wild deer found up to 25% of twins had more than one sire (DeYoung et al., 2002, 2006; Sorin, 2004).

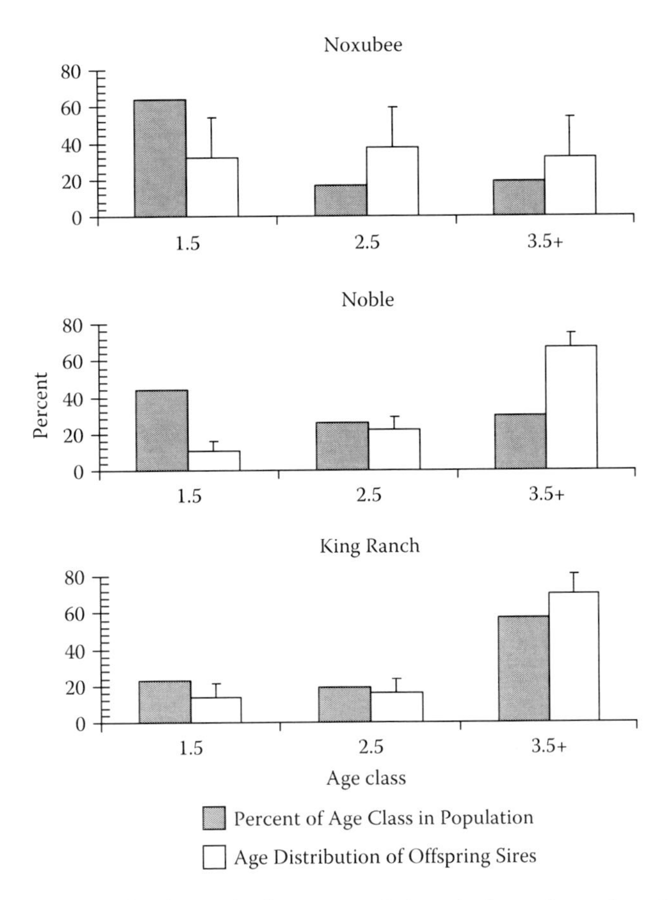

FIGURE 10.18 The distribution of buck reproductive success relative to buck age class and population age structure in populations managed under differing harvest plans. Reproductive success was estimated based on paternity assignment using DNA microsatellite loci. Bars indicate 95% bound on error of estimation, where more samples reflect greater confidence in the stated proportions. (Data from DeYoung, R. W. et al. 2009. *Journal of Mammalogy* 90:946–953.)

Collectively, the breeding behaviors of whitetails, including a wide distribution of mating success among bucks, ability of young bucks to breed, and promiscuous mating by does, are different from expected based on previous ecological and behavioral studies of white-tailed deer (Hirth, 1977; McCullough, 1979; Marchinton and Hirth, 1984). Previous studies based on visual observations concluded that most fawns were sired by a small number of dominant bucks. However, visual observations can be highly misleading for paternity assignment in ungulates, even in relatively open habitats (Hogg and Forbes, 1997; Pemberton et al., 1999) because all individuals cannot be continuously monitored in the best conditions, and rare or cryptic tactics are underrepresented.

Overall, an age-related pattern of dominance is consistent with increased reproductive success for bucks aged 3.5 years and older. White-tailed deer dominance hierarchies are influenced by age, body size, and aggressiveness (Townsend and Bailey, 1981; Miller et al., 1987) in which the top ranks in male dominance hierarchies are usually held by mature, large-bodied bucks. Bucks attain physical maturity and maximum body size at approximately four years of age (Sauer, 1984; Strickland and Demarais, 2000), and are at the peak of physical condition and dominance during prime age (about four to seven years old). However, the wide distribution of reproductive success among individuals and age classes of bucks indicates that opportunities to exercise dominance may be limited. Studies of captive deer have revealed that dominance is associated with breeding success, but dominant bucks could not monopolize breeding, as evidenced by repeated occurrences of multiple paternity and siring of offspring by subordinates (DeYoung et al., 2002, 2006). This pattern occurred even though in captive settings the dominant buck could keep all does in view and quickly interfere with any breeding attempts by subordinate bucks. Furthermore, dominance ranks may be fluid or unstable during the rut (Walock, 1997; DeYoung et al., 2006). Thus, the ability of dominant bucks to suppress breeding of subordinates in free-ranging populations may be limited to interactions over individual does.

The wide distribution of male reproductive success involves more than dominance interactions among individual bucks. Some bucks may successfully employ alternative mating tactics that do not rely on dominance, as observed in other species of large mammals (Hogg and Forbes, 1997). Group size and distribution of does across the landscape are influenced by habitat pattern and in turn influence the distribution of male breeding success in white-tailed deer. The dense vegetative cover and patchy distribution of resources in many white-tailed deer habitats inhibits the formation and maintenance of large groups (Demarais et al., 2000). Therefore, mating strategies where bucks attempt to monopolize groups of does or territories (lekking, territoriality, and harem defense) are not viable for white-tailed deer.

Does may remain in estrus for 48 h or more if not bred immediately, a behavior thought to ensure fertilization in low-density populations (White et al., 1995). Mating may curtail estrus duration to about 24 h (White et al., 1995), although successful fertilizations have occurred 49–95 h after onset of estrus (Verme and Ozoga, 1981). The spatial dispersion of does within populations combined with temporal synchrony of estrus limits the number of estrous does an individual buck can locate and breed, allowing more males access to mating opportunities and restricting the success of individual bucks (e.g., Say et al., 2001). Thus, whitetail bucks may face limited mating opportunities relative to opportunities available to a dominant buck in a harem or lek mating system.

Mating tactics are difficult to quantify because most male–male and male–female interactions are not observed. Young males do not directly challenge mature bucks, but rely on persistence or alternative tactics for success. Young bucks participate in mating chases at similar frequencies to adult bucks, and may loiter in the vicinity of mature bucks engaged in tending does (Hirth, 1977). The tending behavior is thought to be a means to ensure paternity through repeated mating and defense of an estrous doe. Young bucks do not tend does in age-structured populations, perhaps reflecting low dominance rank or lack of experience (Hirth, 1977). Mature bucks tending does can be displaced by a more dominant buck, although it is possible that the original tending buck had already mated with the doe. Multiple paternity in white-tailed deer may occur when a tending buck is displaced by another (DeYoung et al., 2002; Sorin, 2004), or through the use of alternative mating tactics by subdominant bucks (e.g., Hogg and Forbes, 1997; Preston et al., 2003).

The physical and behavioral attributes that equate to mating success in whitetails have yet to be established. Fighting is critical to male mating success in polygynous cervids such as red deer (Clutton-Brock et al., 1988), where only mature, large-antlered males in good physical condition are able to defend harems and gain access to females (Kruuk et al., 2000). Breeding success in white-tailed deer is not skewed to a few bucks and no clear association between breeding success and antler size is apparent. Therefore, fighting may be less important to the reproductive success of individual whitetail bucks. During times when many does are in concurrent estrous, competition among bucks may be limited to a subset of males that has located the same estrous female; subordinates may not be forced to challenge dominants for breeding rights. If this reasoning is correct, young bucks in age-structured populations might be successful more often during peak rut.

Experience and body condition may also influence a buck's success both in knowing where to locate estrous does and in having sufficient energy reserves to engage in courtship activities. Bucks eat little during the rut and rely on energy stored during the pre-rut to fuel the majority of their rut movements. Bucks may lose 30% of their body mass during the rut and experience their highest natural mortality during the post-rut period (DeYoung, 1989; Ditchkoff et al., 2001). Therefore, bucks entering the rut in good condition may be able to expend more time engaging in breeding activities. Conversely, young bucks may have fewer energy reserves to expend.

Most research on white-tailed deer mating tactics has centered on factors affecting reproductive success of bucks. The potential role of female choice remains obscure. Some have suggested an age or productivity bias in mating effort, where mature bucks avoid mating with young does (Sorin, 2004) or prefer to mate with does that did not recruit offspring in the current year (Margulis, 1993). Young does have less experience and fewer physical resources, and may have suboptimal fawning territories, possibly affecting offspring fitness. Although compelling, convincing evidence for age bias in mating is lacking and was not observed in wild deer (DeYoung et al., 2009). Further study of male mating preferences in relation to female age or reproductive status is warranted.

Discrete female choice has not been convincingly demonstrated in cervids, although the potential role of female choice in mating cannot be discounted. Does increase their movements and activity prior to estrus (Ivey and Causey, 1981; Beier and McCullough, 1990), even in captive herds (Ozoga and Verme, 1975). Telemetry studies have documented unusual doe movements outside of the doe's seasonal home range during the rut (Kolodzinski et al., 2010; Figure 10.19), and observational studies have documented does seemingly inviting pursuit (Hirth, 1977). Although behavioral evidence is lacking, does likely are not passive participants in the mating process. A doe faces a trade-off in that the long-term fitness advantage to her offspring gained by mating with the fittest male may be outweighed by the disadvantages of not conceiving during her first estrous. Thus, does that solicit mates could ensure both competition among bucks for breeding rights and early conception. Mating with multiple bucks could result in potential genetic benefits for the offspring if male quality varies. However, does also might mate with multiple bucks to avoid harassment, or as a simple by-product of male–male competition and displacement (Jennions and Petrie, 2000; Chapman, 2006). Finally, much of whitetail communication is based on scent, and it is possible that scent carries some information about genetic quality or compatibility that does use as a cue for preferential mating.

Consideration of Behavior in Management

Management is the art of applying knowledge gained through scientific research to change the status of a population and satisfy human goals. Thus, reliable knowledge of animal behavior can increase the effectiveness of management strategies (Porter et al., 1991) and offer more realistic predictions of population responses to management actions. Accordingly, many have advocated an increased consideration of behaviors in management of wildlife (Côté, 2003; Festa-Bianchet and Apollonio, 2003).

Successful management relies on assumptions about the potential response of populations or individuals to management actions. However, current understanding of white-tailed deer behavior, although rapidly expanding, appears to be only marginally adequate for applying behavior-based management strategies. The existing knowledge base of whitetail behaviors is built primarily on studies done in exploited or captive populations and therefore may not be representative of deer behavior in other contexts (Miller and Ozoga, 1997). Studies of signpost behavior emphasize the limitations of captive research.

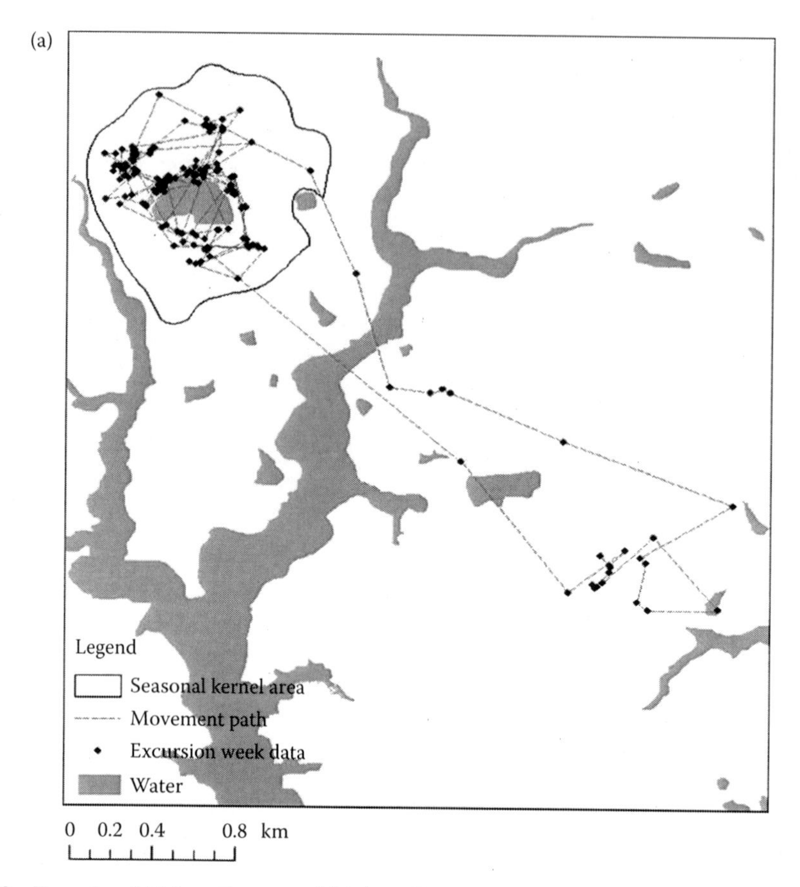

FIGURE 10.19 Examples of (a) long-distance and (b) short-distance excursions outside of the seasonal home range of two adult female white-tailed deer during the breeding season. Individual points represent observations during the week in which the excursion occurred. (Data from Kolodzinski, J. J. et al. 2010. *American Midland Naturalist* 163:366–373.)

For example, Alexy et al. (2001) used remote cameras to demonstrate that bucks of all age classes used the same scrapes in free-ranging conditions, contrary to observations of captive deer. In addition, much knowledge of deer behaviors is based on anecdotal evidence or a limited number of visual observations. Management based on mistaken assumptions or knowledge of animal behavior out of context will be ineffective at best. Nevertheless, behavior-based management has been proposed and evaluated in a variety of contexts.

Management Based on the Social Behavior of Does

The most prominent behavior-based management strategy is termed localized management (Porter et al., 1991). Localized management relies on conceptual models of doe social group behavior and is intended for use in fine-scale management of whitetail populations at the level of doe social groups. Porter et al. (1991) proposed that removal of an entire doe social group may create a spatial void that may take several generations to fill. The localized management approach is based on an extensive study of doe group fidelity and movement patterns of free-ranging deer, and compliments similar research done in other regions (Nelson and Mech, 1987). However, deer exhibit substantial behavioral plasticity across their range and research conducted on these migratory herds may not be representative of behaviors in all populations. Furthermore, the localized management concept was developed in areas with low-density deer populations and low rates of doe dispersal. Thus, the

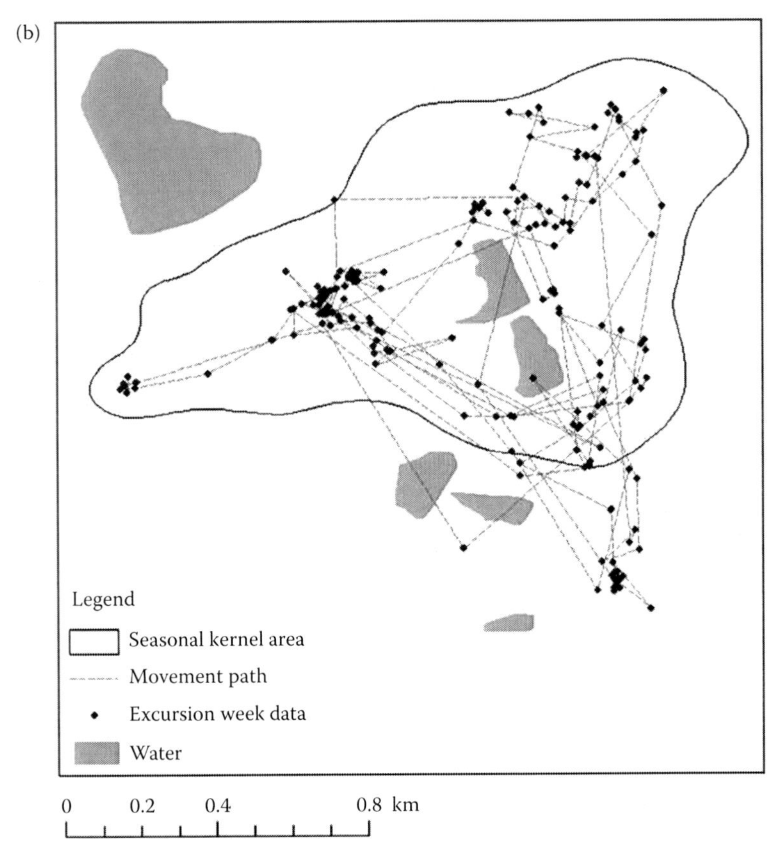

FIGURE 10.19 *continued.*

efficacy of localized management in areas with high doe dispersal rates (Nixon et al., 1991, 2007), in high-density populations (Miller et al., 2010b), or in exploited populations (Comer et al., 2005) appears limited.

Management strategies designed to mitigate or contain disease outbreaks, such as chronic wasting disease, also rely on assumptions about doe social behavior. For example, managers have advocated increasing doe harvest rates in local areas to limit contact among individuals, and disruption of social groups may effectively reduce deer-to-deer contact rates (Figure 10.20). A doe population with a well-developed age structure is a requisite for the expected sociospatial behavior in white-tailed deer (Aycrigg and Porter, 1997). Studies of exploited populations of white-tailed deer, where rates of doe harvest are high, have revealed a low degree of spatial and genetic structuring (Scribner et al., 1997; Comer et al., 2005). Disruption of family groups due to harvest may limit formation of persistent, cohesive social groups (Comer et al., 2005) and promote association of unrelated does (Williams et al., 2008). Thus, increased doe harvest may promote dispersal or association between nonrelatives. Furthermore, the conceptual model of doe social structure implies that spatial overlap among social groups is minimal. At high population densities, however, home range sizes decrease markedly and social groups overlap in space and time (Laseter, 2004). Furthermore, local reductions in density might promote immigration into the managed area in high-density populations (Miller et al., 2010a). Social group structure likely is reduced in agricultural areas of the midwestern United States, where habitat fragmentation apparently results in high rates of dispersal among juvenile and adult does (Nixon et al., 1991). Overall, the effects of intense harvest on doe dispersal and social group dynamics are fertile ground for future research.

FIGURE 10.20 Appropriate antlerless harvest is critical to maintaining deer densities within the carrying capacity of the habitat. In addition, harvest strategies based on an understanding of deer social ecology may have utility in reducing deer damage and containing outbreaks of diseases such as chronic wasting disease. (Photo by K. V. Miller. With permission.)

Mating Behavior and Selective Harvest

One common attempt to incorporate behavior into management involves the effects of harvest on the breeding distribution of bucks. Managers have voiced concern over potential negative effects of male-biased harvest strategies on population fitness (Ryman et al., 1981; Rhodes and Smith, 1992; Harris et al., 2002; Festa-Bianchet, 2003) and have proposed harvest strategies to manipulate the distribution of mating to improve population attributes (e.g., antler size; Rollins, 1998; Chapter 15).

Harvest management may unintentionally affect breeding by overharvest of mature, large-antlered bucks. Changes in buck age structure due to intense harvest could reduce population fitness over time because dominance-based breeding becomes a "scramble" competition, allowing less-fit bucks to breed (Ozoga and Verme, 1985). The recent increase in intensive deer management has led some to recommend protecting mature, large-antlered bucks to increase population antler size, assuming that large-antlered individuals will sire many offspring (Rollins, 1998). However, breeding is not exclusively determined by social dominance or antler size and dominance ranks may not be predictable or stable. Successful bucks sire few offspring per year or during a lifetime. Furthermore, the factors that determine buck mating success in wild deer are not clearly understood. Finally, yearling dispersal and rut-related movements of adult males will impact harvest strategies aimed at altering genetic characteristics. Clearly, the subject warrants additional research attention.

Dispersal

Dispersal is easily overlooked despite its importance to management and dynamics of deer populations. Dispersal behavior has important implications for disease and overpopulation in areas of the midwestern United States due to high rates of dispersal of both sexes (Nixon et al., 2007). In terms of harvest management, managers often assume that immigration and emigration are balanced. However, the deer herd often uses the landscape at spatial scales larger than contiguous landholdings. For instance, landholdings spanning more than 10,000 ha are needed to retain 50% of yearling bucks in semiarid habitats (Webb et al., 2007).

The scale of deer dispersal can complicate management if harvest practices differ among adjacent landholdings. Uneven rates of harvest may promote a dispersal imbalance of bucks, where emigration of young bucks is not balanced by immigration (Rosenberry et al., 2001). The large spatial scale of deer dispersal movements relative to property size provides a serious obstacle that can only be overcome through social management (Chapter 15). Management cooperatives have been successful in open or continuous populations dominated by small landholdings. Cooperatives are voluntary associations in which adjacent landowners agree to manage the pooled area under the same practices and harvest guidelines. Many states in the southeastern United States have legislated county- or state-wide cooperatives through implementation of restrictive buck harvest based on antler size (Chapter 4; Figure 11.2). In these areas a legal buck must meet minimum antler or age criteria.

Use of Behaviors in Hunting

Hunters have made informal use of deer behaviors since the time when indigenous Americans first encountered whitetails. Careful observation combined with a trial and error approach resulted in hunting strategies that consider the behavioral characteristics of deer. Hunters recognize feeding or bedding areas and travel corridors as areas where deer congregate. They also learn to avoid detection by watching the wind, minimizing movements, and using cover for concealment and to break up their outline.

The most recognizable and effective behaviors used by hunters are the imitation of deer-audible communication or vocalizations. Dominance fights among bucks may attract others to the vicinity, although the reason bucks congregate during fights is not known. Hunters take advantage of this behavior by mimicking the sound of a fight using cast antlers or antlers removed from harvested deer (Figure 10.21). As would be expected, loud antler rattling attracts the greatest number of bucks, but the duration of rattling is not important (Hellickson et al., 2009). Mature bucks respond more often post-rut, whereas young and mid-aged bucks respond more often during the pre-rut and rut. Antler rattling appears to be effective in most regions of the whitetail range, given a reasonable sex ratio and buck age structure.

FIGURE 10.21 Behavior-based hunting strategies include mimicking deer scent communication and vocalizations. Because dominance fights among bucks may attract other bucks, antler rattling has become an increasingly popular hunting tactic, particularly in areas with a more natural male age structure. (Photo by Mossy Oak. With permission.)

Hunters often imitate deer grunt vocalizations as a method of attracting deer. Calls that mimic the tending grunt of a courting buck have been marketed for several decades and are effective in some situations. Similarly, scents are widely marketed that purportedly mimic the glandular secretions or urine of estrous does or of bucks, although evidence of the attractiveness of these odors to conspecifics remains tenuous (Whitney et al., 1991).

REFERENCES

Alexy, K. J., J. W. Gassett, D. A. Osborn, and K. V. Miller. 2001. Remote monitoring of scraping behaviors of a wild population of white-tailed deer. *Wildlife Society Bulletin* 29:873–878.

Alexy, K. J., J. W. Gassett, D. A. Osborn, S. M. Russell, and K. V. Miller. 2003. Bacterial fauna of the tarsal tufts of white-tailed deer (*Odocoileus virginianus*). *American Midland Naturalist* 149:237–240.

Atkeson, T. D. and R. L. Marchinton. 1982. Forehead glands in white-tailed deer. *Journal of Mammalogy* 63:613–617.

Atkeson, T. D., R. L. Marchinton, and K. V. Miller. 1988. Vocalizations of white-tailed deer. *American Midland Naturalist* 120:194–200.

Autry, D. C. 1967. Movements of white-tailed deer in response to hunting on Crab Orchard National Wildlife Refuge. MS thesis, Southern Illinois University.

Aycrigg, J. L. and W. F. Porter. 1997. Sociospatial dynamics of white-tailed deer in the central Adirondack Mountains, New York. *Journal of Mammalogy* 78:468–482.

Barboza, P. S. and R. T. Bowyer. 2000. Sexual segregation in dimorphic deer: A new gastrocentric hypothesis. *Journal of Mammalogy* 81:473–489.

Barkalow, F. S. and W. E. Keller. 1950. Escape behavior of the white-tailed deer. *Journal of Wildlife Management* 14:246–247.

Bartush, W. S. and J. C. Lewis. 1978. Behavior of whitetail does and fawns during the parturition period. *Proceedings of the Southeastern Association of Fish and Wildlife Agencies* 32:246–255.

Beier, P. and D. R. McCullough. 1990. Factors influencing white-tailed deer activity patterns and habitat use. *Wildlife Monographs* 109:1–51.

Bello, J., S. Gallina, and M. Equihua. 2004. Movements of the white-tailed deer and their relationship with precipitation in northeastern Mexico. *Interciencia* 29:357–361.

Benner, J. M. and R. T. Bowyer. 1988. Selection of trees for rubs by white-tailed deer in Maine. *Journal of Mammalogy* 69:624–627.

Bertrand, M. R., A. J. DeNicola, S. R. Beissinger, and R. K. Swihart. 1996. Effects of parturition on home ranges and social affiliations of female white-tailed deer. *Journal of Wildlife Management* 60: 899–909.

Bildstein, K. L. 1983. Why white-tailed deer flag their tails. *American Naturalist* 121:709–715.

Blanchong, J. A., K. T. Scribner, B. K. Epperson, and S. R. Winterstein. 2006. Changes in artificial feeding regulations impact white-tailed deer fine-scale spatial genetic structure. *Journal of Wildlife Management* 70:1037–1043.

Brannan, W. V. and R. L. Marchinton. 1987. Reproductive ecology of white-tailed and red brocket deer in Suriname. In *Biology and Management of the Cervidae*, ed. C. M. Wemmer, 344–351. Washington, DC: Smithsonian Institution Press.

Brinkman, T. J., C. S. Deperno, J. A. Jenks, B. S. Haroldson, and R. G. Osborn. 2005. Movement of female white-tailed deer: Effects of climate and intensive row-crop agriculture. *Journal of Wildlife Management* 69:1099–1111.

Brown, B. A., Jr. 1974. Social organization in male groups of white-tailed deer. In *The Behaviour of Ungulates and its Relation to Management*, eds. V. Geist and F. Walther, Publication 24, 436–446. Morges, Switzerland: International Union for the Conservation of Nature.

Brown, B. A. and D. H. Hirth. 1979. Breeding behavior in white-tailed deer. *Proceedings of the First Welder Wildlife Foundation Symposium* 1:83–95.

Brown, T. L., D. J. Decker, S. J. Riley et al. 2000. The future of hunting as a mechanism to control white-tailed deer populations. *Wildlife Society Bulletin* 28:797–807.

Burt, W. H. 1943. Territoriality and home range concepts as applied to mammals. *Journal of Mammalogy* 24:346–352.

Byford, J. L. 1970. Movements and ecology of white-tailed deer in a logged floodplain habitat. PhD dissertation, Auburn University.

Campbell, T. A., C. A. Langdon, B. R. Laseter, W. M. Ford, J. W. Edwards, and K. V. Miller. 2006. Movements of female white-tailed deer to bait sites in West Virginia, USA. *Wildlife Research* 33:1–4.

Campbell, T. A., B. R. Laseter, W. M. Ford, and K. V. Miller. 2004a. Feasibility of localized management to control white-tailed deer in forest regeneration areas. *Wildlife Society Bulletin* 32:1124–1131.

Campbell, T. A., B. R. Laseter, W. M. Ford, and K. V. Miller. 2004b. Unusual white-tailed deer movements to a gas well in the central Appalachians. *Wildlife Society Bulletin* 32:983–986.

Cap, H., S. Aulagnier, and P. Deleporte. 2002. The phylogeny and behavior of Cervidae (Ruminantia Pecora). *Ethology, Ecology & Evolution* 14:199–216.

Caro, T. M., L. Lombardo, A. W. Goldizen, and M. Kelly. 1995. Tail-flagging and other antipredator signals in white-tailed deer: New data and synthesis. *Behavioral Ecology* 6:442–450.

Chapman, T. 2006. Evolutionary conflicts of interest between males and females. *Current Biology* 16:R744–R754.

Cheatum, E. L. and G. H. Morton. 1946. Breeding season of white-tailed deer in New York. *Journal of Wildlife Management* 10:249–263.

Clutton-Brock, T. H., S. D. Albon, and F. E. Guinness. 1988. Reproductive success in male and female red deer. In *Reproductive Success: Studies of Individual Variation in Contrasting Breeding Systems*, ed. T. H. Clutton-Brock, 325–343. Chicago, IL: University of Chicago Press.

Comer, C. E., J. C. Kilgo, G. J. D'Angelo, T. C. Glenn, and K. V. Miller. 2005. Fine-scale genetic structure and social organization in female white-tailed deer. *Journal of Wildlife Management* 69:332–344.

Cornicelli, L., A. Woolf, and J. L. Roseberry. 1996. White-tailed deer use of a suburban environment in southern Illinois. *Transactions of the Illinois State Academy of Science* 89:93–103.

Côté, I. M. 2003. Knowledge of reproductive behavior contributes to conservation programs. In *Animal Behavior and Wildlife Conservation*, eds. M. Festa-Bianchet and M. Apollonio, 77–92. Washington, DC: Island Press.

D'Angelo, G. J., C. E. Comer, J. C. Kilgo, C. D. Drennan, D. A. Osborn, and K. V. Miller. 2004. Daily movements of female white-tailed deer relative to parturition and breeding. *Proceedings of the Annual Conference of the Southeastern Association of Fish and Wildlife Agencies* 58:292–301.

D'Angelo, G. J., J. C. Kilgo, C. E. Comer, C. D. Drennan, D. A. Osborn, and K. V. Miller. 2003. Effects of controlled dog hunting on movements of female white-tailed deer. *Proceedings of the Annual Conference of the Southeastern Association of Fish and Wildlife Agencies* 57:317–325.

Dasher, K. A., J. W. Gassett, D. A. Osborn, and K. V. Miller. 1998. Remote monitoring of scraping behaviour in a wild population of white-tailed deer. *Biology of Deer, Kaposvar, Hungary* 4:34 (abstract).

Demarais, S., K. V. Miller, and H. A. Jacobson. 2000. White-tailed deer. In *Ecology and Management of Large Mammals in North America*, eds. S. Demarais and P. R. Krausman, 601–628. Upper Saddle River, NJ: Prentice-Hall.

Demarais, S., R. E. Saiglin, M. T. Pollock, and D. G. Whittaker. 1989. Environmental effects on activity patterns of mature male white-tailed deer. *Proceedings of the Southeast Deer Study Group* 12:20–21.

De Vos, A., P. Brokx, and V. Geist. 1967. A review of social behavior of the North American cervids during the reproductive period. *American Midland Naturalist* 77:390–417.

DeYoung, C. A. 1989. Mortality of adult male white-tailed deer in South Texas. *Journal of Wildlife Management* 53:513–518.

DeYoung, R. W. 2004. Effects of social and population characteristics on the reproductive success of male white-tailed deer. PhD dissertation, Mississippi State University.

DeYoung, R. W., S. Demarais, K. L. Gee, R. L. Honeycutt, M. W. Hellickson, and R. A. Gonzales. 2009. Molecular evaluation of the white-tailed deer (*Odocoileus virginianus*) mating system. *Journal of Mammalogy* 90:946–953.

DeYoung, R. W., S. Demarais, R. A. Gonzales, R. L. Honeycutt, and K. L. Gee. 2002. Multiple paternity in white-tailed deer (*Odocoileus virginianus*) revealed by DNA microsatellites. *Journal of Mammalogy* 83:884–892.

DeYoung, R. W., S. Demarais, R. L. Honeycutt, K. L. Gee, and R. A. Gonzales. 2006. Social dominance and male breeding success in captive white-tailed deer. *Wildlife Society Bulletin* 34:131–136.

Diefenbach, D. R., E. S. Long, C. S. Rosenberry, B. D. Wallingford, and D. R. Smith. 2008. Modeling distribution of dispersal distances in male white-tailed deer. *Journal of Wildlife Management* 72:1296–1303.

Ditchkoff, S. S., E. R. Welch, Jr., R. L. Lochmiller, R. E. Masters, and W. R. Starry. 2001. Age-specific mortality of adult, male white-tailed deer supports mate competition theory. *Journal of Wildlife Management* 65:552–559.

Doepker, R. V. and J. J. Ozoga. 1991. Wildlife value of northern white-cedar. In *Northern White-Cedar in Michigan*, ed. D. O. Lantagne, 15–34. Michigan State University Report 512.

Favreau, J. 2005. Standard movement terms and definitions. PhD dissertation, North Carolina State University.

Festa-Bianchet, M. 2003. Exploitative wildlife management as a selective pressure for the life history evolution of large mammals. In *Animal Behavior and Wildlife Conservation*, eds. M. Festa-Bianchet and M. Apollonio, 191–207. Washington, DC: Island Press.

Festa-Bianchet, M. and M. Apollonio. 2003. *Animal Behavior and Wildlife Conservation*. Washington, DC: Island Press.

Forand, K. J. and R. L. Marchinton. 1989. Patterns of social grooming in adult white-tailed deer. *American Midland Naturalist* 122:357–364.

Gallina, S., S. Mandujano, J. Bello, and C. Delfin. 1997. Home-range size of white-tailed deer in northeastern Mexico. In *Proceedings of the Deer/Elk Workshop*, ed. J. C. deVos, Jr., 47–50. Phoenix, AZ: Arizona Game and Fish Department.

Garner, G. W. and J. A. Morrison. 1980. Observations of interspecific behavior between predators and white-tailed deer in southwestern Oklahoma. *Journal of Mammalogy* 61:126–130.

Gassett, J. W., K. A. Dasher, D. A. Osborn, and K. V. Miller. 1998. What the nose knows: Detection of oestrus by male white-tailed deer. *Biology of Deer*, Kaposvar, Hungary 4:35 (abstract).

Gassett, J. W., D. P. Wiesler, A. G. Baker et al. 1996. Volatile compounds from interdigital gland of male white-tailed deer (*Odocoileus virginianus*). *Journal of Chemical Ecology* 22:1689–1696.

Gassett, J. W., D. P. Wiesler, A. G. Baker et al. 1997. Volatile compounds from the forehead region of male white-tailed deer (*Odocoileus virginianus*). *Journal of Chemical Ecology* 23:569–578.

Gauthier, D. and C. Barrette. 1985. Suckling and weaning in captive white-tailed and fallow deer. *Behaviour* 94:128–149.

Gavin, T. A., L. H. Suring, P. A. Vohs, Jr., and E. C. Meslow. 1984. Population characteristics, spatial organization, and natural mortality in the Columbian white-tailed deer. *Wildlife Monographs* 91:1–41.

Geist, V. 1998. *Deer of the World*. Mechanicsburg, PA: Stackpole Books.

Goss, R. J. 1983. *Deer Antlers: Regeneration, Function, and Evolution*. New York, NY: Academic Press.

Grovenburg, T. W., J. A. Jenks, C. N. Jacques, R. W. Klaver, and C. C. Swanson. 2009. Aggressive defense behavior by free-ranging white-tailed deer. *Journal of Mammalogy* 90:1218–1223.

Grund, M. D., J. B. McAninch, and E. P. Wiggers. 2002. Seasonal movements and habitat use of female white-tailed deer associated with an urban park. *Journal of Wildlife Management* 66:123–130.

Harris, R. B., W. A. Wall, and F. W. Allendorf. 2002. Genetic consequences of hunting: What do we know and what should we do? *Wildlife Society Bulletin* 30:634–643.

Hawkins, R. E. and W. D. Klimstra. 1970. A preliminary study of the social organization of the white-tailed deer. *Journal of Wildlife Management* 34:460–464.

Hellickson, M. W. 2002. Age-specific physical characteristics, activity, and behavior patterns of male white-tailed deer in southern Texas. PhD dissertation, University of Georgia.

Hellickson, M. W., T. A. Campbell, K. V. Miller, R. L. Marchinton, and C. A. DeYoung. 2008. Seasonal ranges and site fidelity of adult male white-tailed deer (*Odocoileus virginianus*) in southern Texas. *Southwestern Naturalist* 53:1–8.

Hellickson, M. W., K. V. Miller, R. L. Marchinton, C. A. DeYoung, and R. E. Hall. 2009. Behavioral responses of male white-tailed deer to antler rattling. *Proceedings of the Annual Conference of the Southeastern Association of Fish and Wildlife Agencies* 63:7–10.

Hirth, D. H. 1973. Social behavior of white-tailed deer in relation to habitat. PhD dissertation, University of Michigan.

Hirth, D. H. 1977. Social behavior of white-tailed deer in relation to habitat. *Wildlife Monographs* 53:1–55.

Hirth, D. H. and D. R. McCullough. 1977. Evolution of alarm signals in ungulates with special reference to white-tailed deer. *American Midland Naturalist* 111:31–42.

Hofmann, R. R. 1989. Evolutionary steps of ecophysiological adaptation and diversification of ruminants: A comparative view of their digestive system. *Oecologia* 78:443–457.

Hogg, J. T. and S. H. Forbes. 1997. Mating in bighorn sheep: Frequent male reproduction via a high-risk "unconventional" tactic. *Behavioral Ecology and Sociobiology* 41:33–48.

Holzenbein, S. and R. L. Marchinton. 1992a. Emigration and mortality of orphaned male white-tailed deer. *Journal of Wildlife Management* 56:147–153.

Holzenbein, S. and R. L. Marchinton. 1992b. Spatial integration of maturing male white-tailed deer into the adult population. *Journal of Mammalogy* 73:326–334.

Holzenbein, S. and G. Schwede. 1989. Activity and movements of female white-tailed deer during the rut. *Journal of Wildlife Management* 53:219–223.

Hosey, A. G., Jr. 1980. Activity patterns and notes on behavior of male white-tailed deer during rut. MS thesis, Auburn University.

Hoskinson, R. L. and L. D. Mech. 1976. White-tailed deer migration and its role in wolf predation. *Journal of Wildlife Management* 40:429–441.

Hubbard, R. D. and C. K. Nielsen. 2009. White-tailed deer attacking humans during the fawning season: A unique human–wildlife conflict on a university campus. *Human–Wildlife Conflicts* 3:129–135.

Huegel, C. N., R. B. Dahlgren, and H. L. Gladfelter. 1986. Bedsite selection by white-tailed deer fawns in Iowa. *Journal of Wildlife Management* 50:474–480.

Inglis, J. M., R. E. Hood, B. A. Brown, and C. A. DeYoung. 1979. Home range of white-tailed deer in Texas coastal prairie brushland. *Journal of Mammalogy* 60:377–389.

Ivey, T. L. and M. K. Causey. 1981. Movements and activity patterns of female white-tailed deer during rut. *Proceedings of the Southeastern Association of Fish and Wildlife Agencies* 35:149–166.

Ivey, T. L. and M. K. Causey. 1988. Social organization among white-tailed deer during the rut. *Proceedings of the Annual Conference of the Southeastern Association of Fish and Wildlife Agencies* 42:266–271.

Jackson, R. M., M. White, and F. F. Knowlton. 1972. Activity patterns of young white-tailed deer fawns in south Texas. *Ecology* 53:262–270.

Jacobsen, N. K. 1979. Alarm bradycardia in white-tailed deer fawns (*Odocoileus virginianus*). *Journal of Mammalogy* 60:343–349.

Jacobson, H. A. 1992. Deer condition response to changing harvest strategy, David Island, Mississippi. In *The Biology of Deer*, ed. R. D. Brown, 48–55. New York, NY: Springer-Verlag.

Jacobson, H. A. 1994a. Reproduction. In *Deer*, eds. D. Gerlach, S. Atwater, and J. Schnell, 98–108. Mechanicsburg, PA: Stackpole Books.

Jacobson, H. A. 1994b. Fawn behavior. In *Deer*, eds. D. Gerlach, S. Atwater, and J. Schnell, 135–141. Mechanicsburg, PA: Stackpole Books.

Jacobson, H. A., D. C. Guynn, Jr., R. N. Griffin, and D. Lewis. 1979. Fecundity of white-tailed deer in Mississippi and periodicity of corpora lutea and lactation. *Proceedings of the Southeastern Association of Fish and Wildlife Agencies* 33:30–35.

Jennions, M. D. and M. Petrie. 2000. Why do females mate multiply? A review of the genetic benefits. *Biological Reviews of the Cambridge Philosophical Society* 75:21–64.

Joanen, T., L. McNease, and D. Richard. 1985. The effects of winter flooding on white-tailed deer in southwestern Louisiana. *Proceedings of the Louisiana Academy of Sciences* 48:109–115.

Kammermeyer, K. E. 1975. Movement-ecology of white-tailed deer in relation to a refuge and hunted area. MS thesis, University of Georgia.

Kammermeyer, K. E. and R. L. Marchinton. 1976. The dynamic aspects of deer populations utilizing a refuge. *Proceedings of the Annual Conference of the Southeastern Association of Game and Fish Commissioners* 29:466–475.

Kammermeyer, K. E. and R. L. Marchinton. 1977. Seasonal changes in circadian activity of white-tailed deer. *Journal of Wildlife Management* 41:315–317.

Karns, G. R. 2008. Impact of hunting on adult male white-tailed deer behavior. MS thesis, North Carolina State University.

Kernohan, B. J., J. A. Jenks, and D. E. Naugle. 2002. Localized movements and site fidelity of white-tailed deer in the Northern Great Plains. *Prairie Naturalist* 34:1–12.

Kilgo, J. C., R. F. Labisky, and D. E. Fritzen. 1998. Influences of hunting on the behavior of white-tailed deer: Implications for conservation of the Florida panther. *Conservation Biology* 12:1359–1364.

Kilpatrick, H. J. and K. K. Lima. 1999. Effects of archery hunting on movement and activity of female white-tailed deer in an urban landscape. *Wildlife Society Bulletin* 27:433–440.

Kilpatrick, H. J., A. M. LaBonte, and J. T. Seymour. 2002. A shotgun-archery deer hunt in a residential community: Evaluation of hunt strategies and effectiveness. *Wildlife Society Bulletin* 30:478–486.

Knox, W. M., K. V. Miller, and R. L. Marchinton. 1988. Recurrent estrous cycles in white-tailed deer. *Journal of Mammalogy* 69:384–386.

Kolodzinski, J. J., L. V. Tannenbaum, L. I. Muller et al. 2010. Excursive behaviors of female white-tailed deer during estrus at two mid-Atlantic sites. *American Midland Naturalist* 163:366–373.

Krebs, J. R. and N. B. Davies. 1997. The evolution of behavioural ecology. In *Behavioral Ecology: An Evolutionary Approach* (4th edition), eds. J. R. Krebs and N. B. Davies, 3–12. Malden, MA: Blackwell Science.

Kruuk, L. E. B., T. H. Clutton-Brock, J. Slate, J. M. Pemberton, S. Brotherstone, and F. E. Guinness. 2000. Heritability of fitness in a wild mammal population. *Proceedings of the National Academy of Sciences of the USA* 97:698–703.

Labisky, R. F. and D. E. Fritzen. 1998. Spatial mobility of breeding female white-tailed deer in a low-density population. *Journal of Wildlife Management* 62:1329–1334.

Labisky, R. F., D. E. Fritzen, and J. C. Kilgo. 1991. Population ecology and management of white-tailed deer on the Osceola National Forest, Florida. Final Report to the Florida Game and Fresh Water Fish Commission, University of Florida.

Lagory, K. E. 1986. Habitat, group size, and the behavior of white-tailed deer. *Behaviour* 98:168–179.

Lark, A. M. and N. A. Slade. 2008. Variation in vigilance in white-tailed deer (*Odocoileus virginianus*) in northeastern Kansas. *American Midland Naturalist* 159:67–74.

Larson, T. J., O. J. Rongstad, and F. W. Terbilcox. 1978. Movement and habitat use of white-tailed deer in southcentral Wisconsin. *Journal of Wildlife Management* 42:113–117.

Laseter, B. R. 2004. Sociospatial characteristics and genetic structure of white-tailed deer in the central Appalachians of West Virginia. PhD dissertation, University of Georgia.

Leach, R. H. and W. D. Edge. 1994. Summer home range and habitat selection by white-tailed deer in the Swan Valley, Montana. *Northwest Science* 68:31–36.

Leberg, P. L. and D. L. Ellsworth. 1999. Further evaluation of the genetic consequences of translocations of southeastern white-tailed deer populations. *Journal of Wildlife Management* 63:327–334.

Lent, P. C. 1974. Mother–infant relationships in ungulates. In *The Behaviour of Ungulates and Its Relation to Management*, eds. V. Geist and F. Walther, Publication 24, 14–55. Morges, Switzerland: International Union for the Conservation of Nature Publication 24.

Lesage, L., M. Crête, J. Huot, A. Dumont, and J. Ouellet. 2000. Seasonal home range size and philopatry in two northern white-tailed deer populations. *Canadian Journal of Zoology* 78:1930–1940.

Lingle, S. 2001. Anti-predator strategies and grouping patterns in white-tailed deer and mule deer. *Ethology* 107:295–314.

Lingle, S. 2003. Group composition and cohesion in sympatric white-tailed deer and mule deer. *Canadian Journal of Zoology* 81:1119–1130.

Lingle, S. and S. M. Pellis. 2002. Fight or flight? Antipredator behavior and the escalation of coyote encounters with deer. *Oecologia* 131:154–164.

Lingle, S., D. Rendall, W. F. Wilson, R. W. DeYoung, and S. M. Pellis. 2007. Altruism and recognition in the antipredator defence of deer: 2. Why mule deer help nonoffspring fawns. *Animal Behaviour* 73:907–916.

Long, E. S., D. R. Diefenbach, C. S. Rosenberry, and B. D. Wallingford. 2008. Multiple proximate and ultimate causes of natal dispersal in white-tailed deer. *Behavioral Ecology* 19:1235–1242.

Long, E. S., D. R. Diefenbach, C. S. Rosenberry, B. D. Wallingford, and M. D. Grund. 2005. Forest cover influences dispersal distance of white-tailed deer. *Journal of Mammalogy* 86:623–629.

Marchinton, R. L. and T. D. Atkeson. 1985. Plasticity of socio-spatial behaviour of white-tailed deer and the concept of facultative territoriality. In *The Biology of Deer Production*, eds. P. F. Fennessy and K. R. Drew, 375–377. Bulletin 22. Wellington, New Zealand: The Royal Society of New Zealand.

Marchinton, R. L. and D. H. Hirth. 1984. Behavior. In *White-tailed Deer: Ecology and Management*, ed. L. K. Halls, 129–168. Harrisburg, PA: Stackpole Books.

Marchinton, R. L., K. L. Johansen, and K. V. Miller. 1990. Behavioural components of white-tailed deer scent marking: Social and seasonal effects. In *Chemical Signals in Vertebrates V*, eds. D. W. Macdonald, D. Müller-Schwarze, and S. E. Natynczuk, 295–310. Oxford: Oxford University Press.

Margulis, S. W. 1993. Mate choice in Rocky Mountain mule deer bucks (*Odocoileus hemionus hemionus*): Is there a preference for does without fawns? *Ethology, Ecology and Evolution* 5:115–119.

Marshall, A. D. and R. W. Whittington. 1968. A telemetric study of deer home ranges and behavior of deer during managed hunts. *Proceedings of the Annual Conference of the Southeastern Association of Game and Fish Commissioners* 22:30–46.

Mathews, N. E. 1989. Social structure, genetic structure and anti-predator behavior of white-tailed deer in the central Adirondacks. PhD dissertation, State University of New York, Syracuse.

Mathews, N. E. and W. F. Porter. 1993. Effect of social structure on genetic structure of free-ranging white-tailed deer in the Adirondack Mountains. *Journal of Mammalogy* 74:33–43.

McCoy, J. E., D. G. Hewitt, and F. C. Bryant. 2005. Dispersal by yearling male white-tailed deer and implications for management. *Journal of Wildlife Management* 69:366–376.

McCullough, D. R. 1979. *The George Reserve Deer Herd: Population Ecology of a K-Selected Species*. Ann Arbor, MI: University of Michigan Press.

McDonald, J. S. and K. V. Miller. 2004. *A History of White-tailed Deer Restocking in the United States, 1878–2004*. Bogart, GA: Quality Deer management Association.

McDowell, R. D. 1970. Photoperiodism among breeding white-tailed deer (*Odocoileus virginianus*). *Transactions of the Northeast Section of the Wildlife Society* 27:19–38.

McGinnes, B. S. and R. L. Downing. 1977. Factors affecting the peak of white-tailed deer fawning in Virginia. *Journal of Wildlife Management* 41:715–719.

McShea, W. J. and G. Schwede. 1993. Variable acorn crops: Responses of white-tailed deer and other mast consumers. *Journal of Mammalogy* 74:999–1006.

Mech, L. D. and R. E. McRoberts. 1990. Survival of white-tailed deer fawns in relation to maternal age. *Journal of Mammalogy* 71:465–567.

Messier, F. and C. Barrette. 1985. The efficiency of yarding behavior by white-tailed deer as an antipredator strategy. *Canadian Journal of Zoology* 63:785–789.

Michael, E. D. 1964. Birth of white-tailed deer fawns. *Journal of Wildlife Management* 28:171–173.

Michael, E. D. 1970. Activity patterns of white-tailed deer in South Texas. *Texas Journal of Science* 21:417–428.

Miller, B. F., T. A. Campbell, B. R. Laseter, W. M. Ford, D. A. Osborn, and K. V. Miller. 2010a. A test of localized management for reducing white-tailed deer browsing in central Appalachian forest regeneration sites. *Journal of Wildlife Management* 74:370–378.

Miller, B. F., R. W. DeYoung, T. A. Campbell, B. R. Laseter, W. M. Ford, and K. V. Miller. 2010b. Fine-scale genetic and social structuring in a central Appalachian white-tailed deer herd. *Journal of Mammalogy* 93:681–689.

Miller, K. V. and R. L. Marchinton. 1994. Deer talk: Sounds, smells, and postures. In *Deer*, eds. D. Gerlach, S. Atwater, and J. Schnell, 158–168. Mechanicsburg, PA: Stackpole Books.

Miller, K. V. and R. L. Marchinton. 1999. Temporal distribution of rubbing and scraping by a high-density white-tailed deer, *Odocoileus virginianus*, population in Georgia. *Canadian Field-Naturalist* 113:519–521.

Miller, K. V., R. L. Marchinton, K. J. Forand, and K. L. Johansen. 1987. Dominance, testosterone levels, and scraping activity in a captive herd of white-tailed deer. *Journal of Mammalogy* 68:812–817.

Miller, K. V., R. L. Marchinton, and W. M. Knox. 1991. White-tailed deer signposts and their role as a source of priming pheromones: A hypothesis. In *Global Trends in Wildlife Management*, eds. B. Bobek, K. Perzanowski, and W. L. Regelin, 455–458. Krakow-Warszawa, Poland: Swait Press.

Miller, K. V., R. L. Marchinton, and J. J. Ozoga. 1995. Deer sociobiology. In *Quality Whitetails: The Why and How of Quality Deer Management*, eds. K. V. Miller and R. L. Marchinton, 118–128. Mechanicsburg, PA: Stackpole Books.

Miller, K. V., B. Jemiolo, J. W. Gassett, I. Jelinek, D. Wiesler, and M. Novotny. 1998. Putative chemical signals from white-tailed deer (*Odocoileus virginianus*): Social and seasonal effects on urinary volatile excretion in males. *Journal of Chemical Ecology* 24:673–683.

Miller, K. V. and J. J. Ozoga. 1997. Density effects on deer sociobiology. In *The Science of Overabundance: Deer Ecology and Population Management*, eds. W. J. McShea, H. B. Underwood, and J. H. Rappole, 136–150. Washington, DC: Smithsonian Institution Press.

Moore, W. G. and R. L. Marchinton. 1974. Marking behavior and its social function in white-tailed deer. In *The Behaviour of Ungulates and Its Relation to Management*, eds. V. Geist and F. Walther, Publication 24, 447–456. Morges, Switzerland: International Union for the Conservation of Nature.

Müller-Schwarze, D. 1971. Pheromones in black-tailed deer (*Odocoileus hemionus columbianus*). *Animal Behaviour* 19:141–152.

Müller-Schwarze, D. 1994. The senses of deer. In *Deer*, eds. D. Gerlach, S. Atwater, and J. Schnell, 58–65. Mechanicsburg, PA: Stackpole Books.

Naugle, D. E., J. A. Jenks, B. J. Kernohan, and R. R. Johnson. 1997. Effects of hunting and loss of escape cover on movements and activity of female white-tailed deer, *Odocoileus virginianus*. *Canadian Field-Naturalist* 111:595–600.

Nelson, M. E. 1993. Natal dispersal and gene flow in white-tailed deer in northeastern Minnesota. *Journal of Mammalogy* 74:316–322.

Nelson, M. E. 1998. Development of migratory behavior in northern white-tailed deer. *Canadian Journal of Zoology* 76:426–432.

Nelson, M. E. and L. D. Mech. 1981. Deer social organization and wolf predation in northeastern Minnesota. *Wildlife Monographs* 77:1–53.

Nelson, M. E. and L. D. Mech. 1984. Home range formation and dispersal of deer in northeastern Minnesota. *Journal of Mammalogy* 65:567–575.

Nelson, M. E. and L. D. Mech. 1986. Relationship between snow depth and gray wolf predation on white-tailed deer. *Journal of Wildlife Management* 50:471–474.

Nelson, M. E. and L. D. Mech. 1987. Demes within a northeastern Minnesota deer population. In *Mammalian Dispersal Patterns*, eds. B. D. Chepko-Sade and Z. T. Halpin, 27–40. Chicago, IL: University of Chicago Press.

Nelson, M. E. and L. D. Mech. 1999. Twenty-year home-range dynamics of a white-tailed deer matriline. *Canadian Journal of Zoology* 77:1128–1135.

Newhouse, S. J. 1973. Effects of weather on behavior of white-tailed deer of the George Reserve, Michigan. MS thesis, University of Michigan.

Nixon, C. M. and D. Etter. 1995. Maternal age and fawn rearing success for white-tailed deer in Illinois. *American Midland Naturalist* 133:290–297.

Nixon, C. M., L. P. Hansen, P. A. Brewer, and J. E. Chelsvig. 1991. Ecology of white-tailed deer in an intensively farmed region of Illinois. *Wildlife Monographs* 118:1–77.

Nixon, C. M., P. C. Mankin, D. R. Etter et al. 2007. White-tailed deer dispersal behavior in an agricultural environment. *American Midland Naturalist* 157:212–220.

Ockenfels, R. A. and J. A. Bissonette. 1982. Estimates of white-tailed deer activity levels in Oklahoma. *Proceedings of the Annual Conference of the Southeastern Association of Fish and Wildlife Agencies* 36:892–896.

Oehler, M. W., Sr., J. A. Jenks, and R. T. Bowyer. 1995. Antler rubs by white-tailed deer: The importance of trees in a prairie environment. *Canadian Journal of Zoology* 73:1383–1386.

Osborn, D. A., J. W. Gassett, K. V. Miller, and W. R. Lance. 2000. Out-of-season breeding of captive white-tailed deer. *Theriogenology* 54:611–619.

Ozoga, J. J. 1972. Aggressive behavior of white-tailed deer at winter cuttings. *Journal of Wildlife Management* 36:861–868.

Ozoga, J. J. 1989a. Induced scraping activity in white-tailed deer. *Journal of Wildlife Management* 3:877–880.

Ozoga, J. J. 1989b. Temporal pattern of scraping behavior in white-tailed deer. *Journal of Mammalogy* 70:633–636.

Ozoga, J. J. 1994. *Whitetail Autumn*. Minocqua, WI: Willow Creek Press.

Ozoga, J. J. 1995. *Whitetail Winter*. Minocqua, WI: Willow Creek Press.

Ozoga, J. J. 1996. *Whitetail Spring*. Minocqua, WI: Willow Creek Press.

Ozoga, J. J. and L. W. Gysel. 1972. Response of white-tailed deer to winter weather. *Journal of Wildlife Management* 36:892–896.

Ozoga, J. J. and L. J. Verme. 1975. Activity patterns of white-tailed deer during estrus. *Journal of Wildlife Management* 39:679–683.

Ozoga, J. J. and L. J. Verme. 1985. Comparative breeding behavior and performance of yearling vs. prime-age white-tailed bucks. *Journal of Wildlife Management* 49:364–372.

Ozoga, J. J. and L. J. Verme. 1986a. Initial and subsequent maternal success of white-tailed deer. *Journal of Wildlife Management* 50:122–124.

Ozoga J. J. and L. J. Verme. 1986b. Relation of maternal age to fawn-rearing success in white-tailed deer. *Journal of Wildlife Management* 50:480–486.

Ozoga, J. J., L. J. Verme, and C. S. Bienz. 1982. Parturition behavior and territoriality in white-tailed deer: Impact on neonatal mortality. *Journal of Wildlife Management* 46:1–11.

Pemberton, J. M., D. W. Coltman, J. A. Smith, and J. G. Pilkington. 1999. Molecular analysis of a promiscuous, fluctuating mating system. *Biological Journal of the Linnean Society* 68:289–301.

Pilcher, B. K. and G. E. Wampler. 1981. Hunting season movements of white-tailed deer on Fort Sill Military Reservation. *Proceedings of the Annual Conference of the Southeastern Association of Fish and Wildlife Agencies* 35:142–148.

Porter, W. F., N. E. Mathews, H. B. Underwood, R. W. Sage, Jr., and D. F. Behrend. 1991. Social organization in deer: Implications for localized management. *Environmental Management* 15:809–814.

Preston, B. T., I. R. Stevenson, and K. Wilson. 2003. Soay rams target reproductive activity towards promiscuous females' optimal insemination period. *Proceedings of the Royal Society of London, Series B* 270:2073–2078.

Progulske, D. R. and D. C. Duerre. 1964. Factors influencing spotlighting counts of deer. *Journal of Wildlife Management* 22:184–192.

Quay, W. B. 1959. Microscopic structure and variation in the cutaneous glands of the deer, *Odocoileus virginianus*. *Journal of Mammalogy* 40:144–128.

Relyea, R. A., R. K. Lawrence, and S. Demarais. 2000. Home range of desert mule deer: Testing the body-size and habitat-productivity hypotheses. *Journal of Wildlife Management* 64:146–153.

Rhodes O. E., Jr. and M. H. Smith. 1992. Genetic perspectives in wildlife management: The case of large herbivores. In *Wildlife 2001: Populations*, eds. D. R. McCullough and R. H. Barrett, 985–996. New York, NY: Elsevier Science.

Richardson, L. W., H. A. Jacobson, R. J. Muncy, and C. J. Perkins. 1983. Acoustics of white-tailed deer (*Odocoileus virginianus*). *Journal of Mammalogy* 64:245–252.

Richter, A. R. and R. F. Labisky. 1985. Reproductive dynamics among disjunct white-tailed deer herds in Florida. *Journal of Wildlife Management* 49:964–971.

Rollins, D., ed. 1998. *Proceedings of the Symposium on the Role of Genetics in White-tailed Deer Management.* Texas Agricultural Extension Service, College Station.

Rongstad, O. J. and J. R. Tester. 1969. Movements and habitat use of white-tailed deer. *Journal of Wildlife Management* 33:366–379.

Root, B. G., E. K. Fritzell, and N. F. Giessman. 1988. Effects of intensive hunting on white-tailed deer movement. *Wildlife Society Bulletin* 16:145–151.

Rosenberry, C. S., M. C. Conner, and R. A. Lancia. 2001. Behavior and dispersal of white-tailed deer during the breeding season. *Canadian Journal of Zoology* 79:171–174.

Rouleau, I., M. Crête, and J. Ouellet. 2002. Contrasting the summer ecology of white-tailed deer inhabiting a forested and an agricultural landscape. *Écoscience* 9:459–469.

Ryman, N., R. Baccus, C. Reuterwall, and M. H. Smith. 1981. Effective population size, generation interval, and potential loss of genetic variability in game species under different hunting regimes. *Oikos* 36:257–266.

Sabine, D. L., S. F. Morrison, H. A. Whitlaw, W. B. Ballard, G. J. Forbes, and J. Bowman. 2002. Migration behavior of white-tailed deer under varying winter climate regimes in New Brunswick. *Journal of Wildlife Management* 66:718–728.

Sanderson, G. C. 1966. The study of mammal movements: A review. *Journal of Wildlife Management* 30:215–235.

Sargent, R. A., Jr. 1992. Movement ecology of adult male white-tailed deer in hunted and non-hunted populations in the wet prairie of the everglades. MS thesis, University of Florida.

Sauer, P. R. 1984. Physical characteristics. In *White-tailed Deer: Ecology and Management*, ed. L. K. Halls, 73–90. Harrisburg, PA: Stackpole Books.

Sawyer, T. G., R. L. Marchinton, and K. V. Miller. 1989. Response of female white-tailed deer to scrapes and antler rubs. *Journal of Mammalogy* 70:431–433.

Sawyer, T. G., K. V. Miller, and R. L. Marchinton. 1993. Patterns of urination and rub-urination in female white-tailed deer. *Journal of Mammalogy* 74:477–479.

Say, L., D. Pontier, and E. Natoli. 2001. Influence of oestrus synchronization on male reproductive success in the domestic cat (*Felis catus* L.). *Proceedings of the Royal Society of London, B. Biological Sciences* 268:1049–1053.

Schwede, G., H. Hendrichs, and W. McShea. 1993. Social and spatial organization of female white-tailed deer, *Odocoileus virginianus*, during the fawning season. *Animal Behaviour* 45:1007–1017.

Schwede, G., H. Hendrichs, and C. Wemmer. 1994a. Early mother–young relations in white-tailed deer. *Journal of Mammalogy* 75:438–445.

Schwede, G., H. Hendrichs, and C. Wemmer. 1994b. Sibling relations in young white-tailed deer fawns *Odocoileus virginianus*. *Mammalia* 58:175–181.

Scribner, K. T., M. H. Smith, and R. K. Chesser. 1997. Spatial and temporal variability of microgeographic genetic structure in white-tailed deer. *Journal of Mammalogy* 78:744–755.

Severinghaus, C. W. and E. L. Cheatum. 1956. Life and times of the white-tailed deer. In *The Deer of North America*, ed. W. P. Taylor, 57–186. Harrisburg, PA: Stackpole Books.

Shaw, J. C., R. A. Lancia, M. C. Conner, C. S. Rosenberry. 2006. Effect of population demographics and social pressures on white-tailed deer dispersal ecology. *Journal of Wildlife Management* 70:1293–1301.

Smith, W. P. 1987. Maternal defense behavior in Columbian white-tailed deer: When is it worth it? *American Naturalist* 130:310–316.

Sorin, A. B. 2004. Paternity assignment for white-tailed deer (*Odocoileus virginianus*): Mating across age classes and multiple paternity. *Journal of Mammalogy* 85:356–362.

Sparrow, R. D. and P. F. Springer. 1970. Seasonal activity patterns of white-tailed deer in eastern South Dakota. *Journal of Wildlife Management* 34:420–431.

Storm, D. L., C. K. Nielsen, E. M. Schauber, and A. Woolf. 2007a. Deer–human conflict and hunter access in an exurban landscape. *Human–Wildlife Conflicts* 1:53–59.

Storm, D. J., C. K. Nielsen, E. M. Schauber, and A. Woolf. 2007b. Space use and survival of white-tailed deer in an exurban landscape. *Journal of Wildlife Management* 71:1170–1176.

Strickland, B. K. and S. Demarais. 2000. Age and regional differences in antlers and body mass of white-tailed deer. *Journal of Wildlife Management* 64:903–911.

Sweeney, J. R., R. L. Marchinton, and J. M. Sweeney. 1971. Responses of radio-monitored white-tailed deer chased by hunting dogs. *Journal of Wildlife Management* 35:707–716.

Tierson, W.C., G. F. Mattfeld, R.W. Sage, Jr., and D. F. Behrend. 1985. Seasonal movements and home ranges of white-tailed deer in the Adirondacks. *Journal of Wildlife Management* 49:760–769.

Tomberlin, J. W. 2007. Movement, activity, and habitat use of adult male white-tailed deer at Chesapeake Farms, Maryland. MS thesis, North Carolina State University.

Townsend, T. W. and E. D. Bailey. 1975. Parturitional, early maternal, and neonatal behavior in penned white-tailed deer. *Journal of Mammalogy* 56:347–362.

Townsend, T. W. and E. D. Bailey. 1981. Effects of age, sex, and weight on social rank in penned white-tailed deer. *American Midland Naturalist* 106:92–101.

Traweek, M., S. Waldroup, J. Williams, and E. L. Young. 1996. The rut in white-tailed deer. Texas Parks and Wildlife http://www.tpwd.state.tx.us/huntwild/hunt/planning/rut_whitetailed_deer/#map.

Van Deelen, T. R., H. Campa, III, M. Hamady, and J. B. Haufler. 1998. Migration and seasonal range dynamics of deer using adjacent deeryards in northern Michigan. *Journal of Wildlife Management* 62:205–213.

Van Etten, R. C., D. F. Switzenberg, and L. Eberhardt. 1965. Controlled deer hunting in a square-mile enclosure. *Journal of Wildlife Management* 29:59–73.

VerCauteren, K. C. 1998. Dispersal, home range fidelity, and vulnerability of white-tailed deer in the Missouri River Valley. PhD dissertation, University of Nebraska.

VerCauteren, K. C. and S. E. Hygnstrom. 1998. Effects of agricultural activities and hunting on home ranges of female white-tailed deer. *Journal of Wildlife Management* 62:280–285.

Verme, L. J. and J. J. Ozoga. 1971. Influence of winter weather on white-tailed deer in Upper Michigan. In *Snow and Ice in Relation to Wildlife and Recreation Symposium*, ed. A. O. Haugen, 16–28. Ames, IA: Iowa State University.

Verme, L. J. and J. J. Ozoga. 1981. Sex ratio of white-tailed deer and the estrus cycle. *Journal of Wildlife Management* 45:710–715.

Verme, L. J., J. J. Ozoga, and J. T. Nellist. 1987. Induced early estrus in penned white-tailed deer. *Journal of Wildlife Management* 51:54–56.

Vreeland, J. K., D. R. Diefenbach, and B. D. Wallingford. 2004. Survival rates, mortality causes, and habitats of Pennsylvania white-tailed deer fawns. *Wildlife Society Bulletin* 32:542–553.

Walock, S. C. 1997. Behavioral changes in a captive population of white-tailed deer after sterilization of does by tubal ligation. MS thesis, Mississippi State University.

Walter, W. D., K. C. VerCauteren, J. M. Gilsdorf, and S. E. Hygnstrom. 2009. Crop, native vegetation, and biofuels: Response of white-tailed deer to changing management priorities. *Journal of Wildlife Management* 73:339–344.

Warren, R. J., R. W. Vogelsang, R. L. Kirkpatrick, and P. F. Scanlon. 1978. Reproductive behavior of captive white-tailed deer. *Animal Behaviour* 26:179–183.

Webb, S. L., K. L. Gee, B. K. Strickland, S. Demarais, and R. W. DeYoung. 2010. Measuring fine-scale white-tailed deer movements and environmental influences using GPS collars. *International Journal of Ecology*, published online: DOI /10.1155/2010/459610.

Webb, S. L., D. G. Hewitt, and M. W. Hellickson. 2007. Scale of management for mature male white-tailed deer as influenced by home range and movements. *Journal of Wildlife Management* 71:1507–1512.

White, L. M., D. A. Hosack, R. J. Warren, and R. A. Fayrer-Hosken. 1995. Influence of mating on duration of estrus in captive white-tailed deer. *Journal of Mammalogy* 76:1159–1163.

White, M., F. F. Knowlton, and W. C. Glazener. 1972. Effects of dam-newborn fawn behavior on capture and mortality. *Journal of Wildlife Management* 36:897–906.

Whitney, M. D., D. L. Forster, K. V. Miller, and R. L. Marchinton. 1991. Sexual attraction in white-tailed deer. In *The Biology of Deer*, ed. R. D. Brown, 327–333. New York, NY: Springer-Verlag.

Wiles, G. J. and H. P. Weeks, Jr. 1986. Movements and use patterns of white-tailed deer visiting natural licks. *Journal of Wildlife Management* 50:487–496.

Williams, S. C., A. J. DeNicola, and I. M. Ortega. 2008. Behavioral responses of white-tailed deer subjected to lethal management. *Canadian Journal of Zoology* 86:1358–1366.

Wilson, A. J., D. W. Coltman, J. M. Pemberton, A. D. J. Overall, K. A. Byrne, and L. E. B. Kruuk. 2005. Maternal genetic effects set the potential for evolution in a free-living vertebrate population. *Journal of Evolutionary Biology* 18:405–414.

Wolverton, S., R. L. Lyman, J. H. Kennedy, and T. W. La Point. 2009. The terminal Pleistocene extinctions in North America, hypermorphic evolution, and the dynamic equilibrium model. *Journal of Ethnobiology* 29:28–63.

Woods, G. R. 1994. Studies on traditional rubs and the human dimensions associated with quality deer management. PhD dissertation, Clemson University.

Zagata, M. D. and A. O. Haugen. 1973. Pilot Knob State Park: A winter deer haven. *Iowa State Journal of Research* 47:199–217.

Zagata, M. D. and A. O. Haugen. 1974. Influence of light and weather on observability of Iowa deer. *Journal of Wildlife Management* 38:220–228.

Part III

Management

11

Management History

Kip P. Adams and R. Joseph Hamilton

CONTENTS

Game management is the art of making land produce sustained annual crops of wild game for recreational use (Leopold, 1933). Aldo Leopold made that statement in his book, *Game Management*, and it is a fitting beginning to a chapter on the management history of white-tailed deer. McCabe and McCabe (1997) later stated that wildlife management is the practice of applying biological principles to wildlife systems while keeping in balance with societal conditions and expectations. These statements have fundamental differences in their application to white-tailed deer management. While both philosophies were applied to deer populations of their respective time periods, they achieved varying degrees of success. However, one constant during human habitation of North America with respect to deer management is that humans, either directly through exploitation or indirectly through habitat alteration, have been the all-pervasive influence on wildlife resources (Peek, 1986).

According to Halls (1978), no wildlife species exceeds the white-tailed deer in long-time influence on the entire field of conservation. Hunting laws have been enacted, conservation policies promulgated, wildlife restoration begun, and public attitudes toward environmental issues swayed by ecological principles developed in the long history of whitetail management. Deer of the genus *Odocoileus* are the most important wild ungulates in North America in both numbers and economic value (McCullough, 1987), with whitetails more prominent than mule deer or black-tailed deer. Whitetail hunters comprise 81% of all hunters 16 years of age and older (U.S. Fish and Wildlife Service, 2006) which highlights the importance of deer hunting and emphasizes that whitetails heavily influence the hunting industry.

The history of white-tailed deer in the United States is one of near tragedy and triumph (Trefethen, 1975). White-tailed deer management in North America can be considered the greatest wildlife success story and the greatest failure. Whitetail populations have been in balance with the habitat, grossly overabundant, and extirpated. These varying densities have occurred throughout the whitetail's range and across major time periods. In general, there are five distinct phases in the history of white-tailed deer management in North America.

Phase I (0–1800): Native American Exploitation

White-tailed deer were assumed to be quite abundant, especially toward the end of this time period, and many Native American tribes depended on them for subsistence (McDonald and Miller, 2004). Whitetails were a prime source of food and clothing (Halls, 1978), and antlers were used as ornaments, chipping tools, headdresses, and awls; sinews were used for bowstrings, fishing lines, and thread; and brains were used for tanning hides (Calhoun and Loomis, 1974).

Indians hunted deer year-round and although this phase is referred to as Native American Exploitation, whitetails were still numerous by 1800. An analysis by McCabe (1982) suggested that up to 949 adult deer were necessary to support a band of 100 North American Indians annually. Another analysis using assumptions for the size of whitetail range, Indian density across that area, the amount of venison consumed per Indian, and other variables, estimated Indians in Canada and the United States may have harvested 4.6–6.4 million whitetails annually (McCabe and McCabe, 1984). In 2008, hunters in Canada and the United States harvested less than 7 million whitetails. The similarity in deer harvests emphasizes that humans have had a major impact on deer populations for nearly the past 2000 years, and possibly for much longer. Deer remains identified in an archeological site in Texas were dated at 18,000 years ago during the Last Glacial Maximum (Lewis, 2009). Evidence for harvest of white-tailed deer since humans first arrived in North America argues against modern antihunting statements that deer herds were principally managed by nonhuman predators and other natural agents prior to European man's involvement. The whitetail population in 2008 could have supported a larger harvest, but the similarity of the 2008 harvest to McCabe and McCabe's (1984) estimate suggests deer populations during this phase were substantial. If we assume an extremely aggressive removal rate of 50% of the standing population, McCabe and McCabe's estimated harvest suggests a minimum standing population of 9–13 million whitetails. However, removing a third of the herd is a more realistic sustainable harvest rate, and suggests a deer population of 9–19 million deer (Figure 11.1).

Whitetails also provided footwear and other apparel, currency, and commerce (McCabe and McCabe, 1997). The relative lack of material goods has led some historians to describe Native Americans as living "lightly on the land." This assertion may be true, but by 1000, Native American trade relationships covered the continent; mother-of-pearl from the Gulf of Mexico has been found in Manitoba, and Lake Superior copper has been found in present-day Louisiana (Mann, 2005).

Mann (2005) asserts the current view of North America prior to Columbus' arrival in 1492 was that of a thriving and diverse place, with a tumult of languages, trade, and culture. Humans were spread throughout the continent, and agriculture occurred in as much as two-thirds of the present-day continental United States with large swathes of the Southwest terraced and irrigated. He continues that much of this world vanished after Columbus as a result of disease, and estimates the human population declined by 20% from Columbus' arrival to the mid-1600s. Some species, including whitetails, may have benefited from reduced Native American populations. Native Americans routinely relied on the whitetail resource (Halls, 1978; McCabe and McCabe, 1997, 1984; McDonald and Miller, 2004), and reduced harvest pressure could have allowed deer populations to experience an "ecological release" and increase in size. Thus, deer populations may have increased by around 1500 (Figure 11.1).

The corresponding decline through 1800 may have resulted from a combination of heavy harvest pressure by surviving Native Americans and less habitat management (e.g., prescribed burning) that resulted in reduced habitat quality and the ability to support fewer deer. Although there were fewer Native Americans, harvest pressure was high as commerce between the Americas and Europe developed. Native Americans used white-tailed deer as a resource to trade for European goods, and for nearly 100 years beginning in the late 1600s Native Americans helped supply the 100,000 deer hides exported annually from the port of Charles Town, South Carolina, to European leather factories (Kane and Keaton, 1993). The latter part of this phase was characterized by massive harvests by Indians

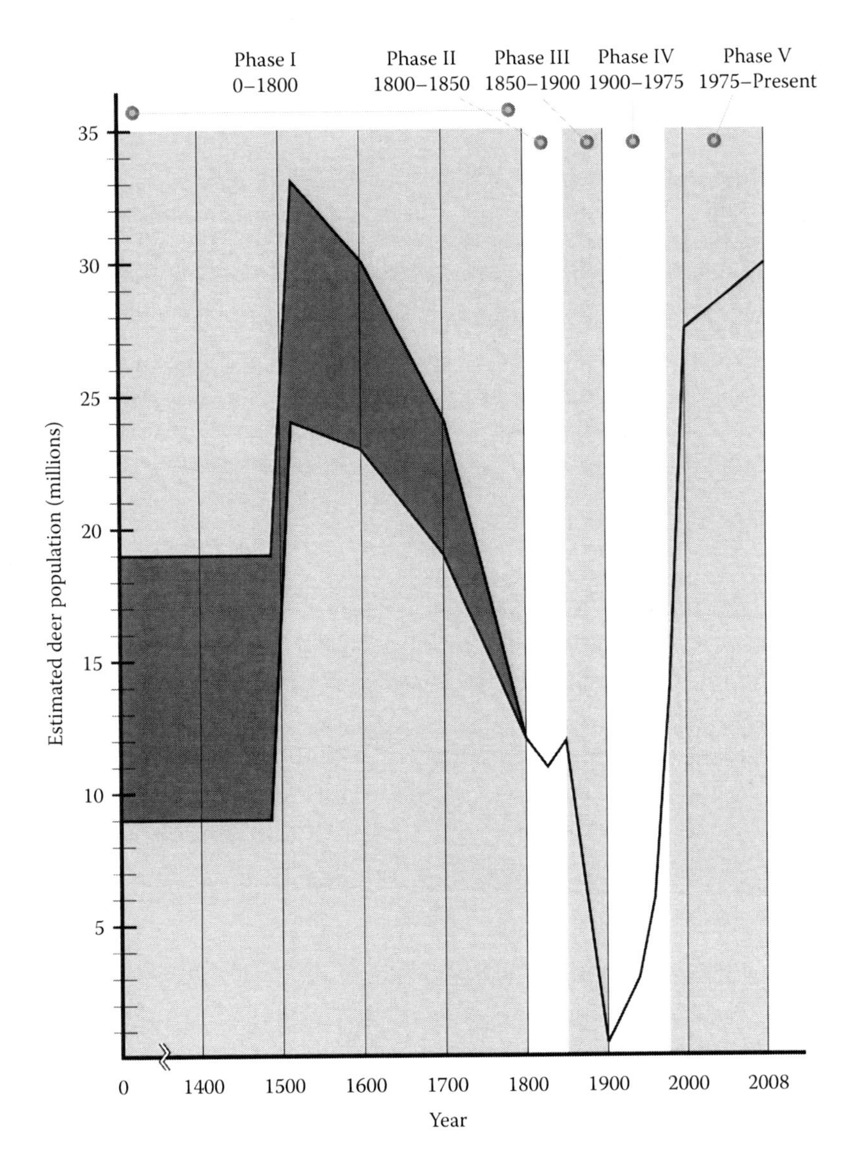

FIGURE 11.1 Estimated white-tailed deer population in the United States and Canada, 0–2008. The population is designated by screens from 0 to 1800 to show estimated numerical ranges. Estimated population from 0 to 1500 based, in part, on analysis from McCabe and McCabe (1984), population from 1500 to 1980 adapted from McCabe and McCabe (1984), and population from 1980 to 2008 from the Quality Deer Management Association (2009).

seeking to obtain trader geegaws, metalwares, guns, alcohol, textiles, and promises (McCabe and McCabe, 1984).

The deer population at the beginning of this phase is unknown, although McCabe and McCabe's (1984) estimate of 23–34 million in 1500 is probably liberal. The phase ended with approximately 12 million deer. Depletion of white-tailed deer was evident to settlers as early as the mid-seventeenth century, and so some states enacted legislation to protect deer and other wildlife. Rhode Island passed legislation in 1646 prohibiting hunting, New Jersey outlawed the export of deer skins from deer killed by Native Americans in 1679, and other colonies enacted similar laws (McDonald and Miller, 2004). Similarly, Virginia banned harvest of female deer in 1738, and New York forbade the use of hounds in deer hunting in 1788.

Interestingly, the assumed "depletion" of deer in the 1600s and 1700s may actually have been a return to normalcy for whitetail populations. Native Americans purposefully reduced numbers of deer and other animals that competed with them for hard and soft mast and maize (Mann, 2005). Mann (2005) also provides evidence that bison, elk, passenger pigeon (and likely whitetail) populations were much lower until disease (from Columbus' arrival) greatly reduced populations of Native Americans. His evidence includes bones (or lack of) from these species at Native American middens.

While Native Americans may or may not have intentionally manipulated deer populations, their actions certainly met the criteria for deer management. They harvested deer annually across all age classes, and modified habitat in ways preferred by deer. Native Americans routinely used fire to modify habitat for their benefit, such as maintaining it in an early successional stage to promote soft mast production (Stewart, 1951). Native Americans intentionally burned thousands of square kilometers of whitetail habitat annually which helped sustain deer populations in many areas (Trefethen, 1970). These habitat improvements at least partially helped offset the high harvest rate Native Americans applied to deer populations. However, there was apparently great variation in time, place, degree of skill, and foresight regarding the use of fire both as process and tool (McCabe and McCabe, 1984). Whitetails, as well as Native Americans, clearly benefited from this practice.

Indians also used fire as a hunting technique. While they were adept at stalking deer and possessed more hunting ability than the average European (Jenness, 1932), communal hunting was more important and provided greater amounts of meat (Carr, 1895). Communal hunts during autumn and early winter had the added advantage of providing high-quality hides that Native Americans used for leggings, shanks, dresses, breechclouts, moccasins, sashes, shirts, robes, skirts, headware, and mittens (McCabe and McCabe, 1984). They burned prairies and woods to drive deer to hunters waiting with weapons; a strategy similar to modern deer or black bear drives. Autumn fire drives were most productive as vegetation was conducive to burning and deer were in prime condition. This hunting technique modified habitat as well as ensured that deer from a range of age classes were harvested. Ironically, by harvesting deer of both sexes and all age classes and manipulating habitat, it can be argued that Native Americans managed whitetails better than "educated" Europeans centuries later.

Phase II (1800–1850): Moderate Recovery

This short phase essentially marked the end of the 300-year decline in whitetail populations and included a modest increase in deer numbers. European settlements invaded the continental interior and nearly all Native American influence on the landscape within whitetail range was terminated (McCabe and McCabe, 1984). Massive harvests at the end of *Phase I* dropped precipitously and whitetail numbers began rebounding (Figure 11.1). Recovery of whitetail populations was further promoted because, by 1800, skins had lost prominence in the hide trade as deer numbers were low in and around European settlements (Phillips, 1961). While Indian populations and impacts declined, the number of Europeans in North America were increasing at an astonishing rate (McCabe and McCabe, 1984), and this set the stage for European exploitation in *Phase III.*

Phase III (1850–1900): European Exploitation

Extensive exploitation of deer and their habitat by European settlers supplying markets for deer meat and skins in new settlements and Europe resulted in a great increase in hunting pressure and the greatest period of whitetail exploitation and population decline (McDonald and Miller, 2004). Coupled with extensive timber harvest, repeated burning of cutover forests, market hunting, agricultural development, and a lack of game laws, the whitetail experienced difficult times (Halls, 1978).

Theodore Roosevelt was an avid hunter and outdoorsman and saw firsthand the decimation of our wildlife resources. In 1887, he founded the Boone and Crockett Club, and the Club's Fair Chase statement was the first document outlining a code of conduct and ethics for sportsmen. This statement later became the cornerstone of today's game laws. In an era of destruction a statement of hunter ethics was a glimmer of hope for the future.

With respect to whitetails, the intensity of harvest was magnified because there were no more wilderness sanctuaries where whitetails could thrive beyond the reach of hunters, and positive land-use impacts on habitat were prevented or negated by the continual influx of people (McCabe and McCabe, 1984). Gone were the massive deer populations of the past 300 years. Gone were the forests. Gone was the tie between deer and their consumers. Whitetails were extirpated from many areas during this phase and truly faced extinction. By the end of the nineteenth century, whitetail populations were at their lowest number in recorded history, ranging from 300,000 in 1890 to 500,000 in 1900 (Figure 11.1) (Downing 1987).

Phase IV (1900–1975): Protection and Recovery

As mentioned in *Phase I*, some states enacted legislation to protect deer and other wildlife as early as the mid-seventeenth century. However, it was nearly 200 years later (1878) before the first game wardens were hired in California and New Hampshire; so many protective measures were unenforceable. As a result, year-round deer hunting continued in most areas (McDonald and Miller, 2004).

The Lacey Act of 1900 provided the first real protection for wildlife. Signed into law by President William McKinley on May 25, 1900, this Act prohibited interstate transport of illegally killed wildlife. It was the first federal law protecting wildlife and it helped to effectively bring an end to market hunting.

Also beginning in the late 1890s and continuing through the early 1900s, most state wildlife agencies were formed. Initially they were designed primarily to establish and enforce game laws, and thus game wardens were the first employees. In 1933, the University of Wisconsin published Aldo Leopold's *Game Management*. Leopold became the first professor of game management in the United States and a new profession was born. Wildlife biologists then arrived on the scene as early as the 1940s, and began gaining popularity in the 1950s.

Another key protection mechanism was the beginning of the National Wildlife Refuge System in 1903 when President Theodore Roosevelt designated Florida's Pelican Island as the first wildlife refuge. The National Wildlife Refuge System, managed by the U.S. Fish and Wildlife Service, includes a system of public lands and waters set aside to conserve America's fish, wildlife, and plants. As of 2009, the system had grown to more than 60 million hectares, 550 National Wildlife Refuges and other units of the Refuge System, and 37 wetland conservation districts (U.S. Fish and Wildlife Service, 2009a).

Formation of the U.S. Forest Service followed in 1905 and the National Park Service in 1916. While these agencies did not necessarily manage specifically for deer, the habitats they incorporated and the legislation they fostered (e.g., Multiple-Use Sustained Yield Act, Wilderness Act, National Forest Management Act, National Environmental Policy Act) were crucial for whitetails. The Lacey Act provided federal protection to white-tailed deer and the National Refuge System, U.S. Forest Service, and National Park Service provided critical habitat. Combined, these measures set the stage for the recovery of whitetails.

The next windfall for white-tailed deer, and all wildlife, was the Federal Aid in Wildlife Restoration Act. Popularly known as the Pittman–Robertson Act, this is the world's premiere wildlife conservation effort. In 1937, at the urging of organized sportsmen, state wildlife agencies, and the firearms and ammunition industries, Congress extended the life of an existing 10% tax on ammunition and firearms used for sport hunting, and earmarked the proceeds to be distributed to the states for wildlife restoration (U.S. Fish and Wildlife Service, 2009b). The Pittman–Robertson Act provided funding for restoration, rehabilitation, and improvement of wildlife habitat, wildlife management research, and distribution of information produced by each project. The Act was amended in 1970 to provide funding for hunter education and the development, operation, and maintenance of public shooting ranges. By 2009, the Act provided an 11% Federal excise tax on sporting arms, ammunition, and archery equipment, and a 10% excise tax on handguns. These funds are apportioned to each state on the basis of total area and number of licensed hunters.

Not all whitetails were doing well, however, despite these important legislative efforts. Key deer, a distinct subspecies of white-tailed deer, became landlocked on the western end of the Florida Keys

10,000–15,000 years ago (Chapter 1). Loss of habitat and unregulated hunting in the late 1800s and early 1900s reduced Key deer populations to 25 individuals by the late 1940s. Fortunately, J. N. "Ding" Darling, Chief of the U.S. Bureau of Biological Survey (precursor to the U.S. Fish and Wildlife Service) recognized the Key deer's precarious situation and began campaigning for their protection. The Boone and Crockett Club helped by offering to pay a warden to patrol the Keys. This funding and law enforcement commissions from the Bureau of Biological Survey and State of Florida helped Key deer populations. A real victory came in 1957 when the National Key Deer Refuge was established. Deer–vehicle collisions are now the greatest mortality source for Key deer (Lopez, 2001). Projects to reduce these accidents include the management of roadside vegetation making it less attractive to deer and multimillion dollar construction projects to create highway underpasses (Williams, 2008a). The Key deer population was stable at approximately 800 animals in 2008 (Williams, 2008a), and given continued development, the population will likely never be large enough to sustain hunter harvest.

The final act of protection during this phase benefited Columbia River white-tailed deer the westernmost subspecies of whitetails found in southwestern Washington and northwestern Oregon (Chapter 1). By 1900, these deer were extirpated from most of their range due to habitat loss and hunting, and by 1930 they were believed to persist in a single place near Roseburg, Oregon (Williams, 2008b). In the late 1930s, the U.S. Bureau of Biological Survey located a new herd of Columbia River whitetails along the lower Columbia River of southwestern Washington. This subspecies has a strong preference for riparian habitat and thus suffered from periodic floods, land-clearing for agriculture, and illegal harvest (Williams, 2008b). This site-specific habitat affinity is why the establishment of the Julia Butler Hansen Refuge, a 2500-hectare tract in southwestern Washington and northwestern Oregon in 1972 to protect the Columbia River whitetail, was so crucial. The Roseburg, Oregon population has a brighter history than the Columbia River population. The Roseburg population has more intact habitat and has thrived to the point that it was reopened to hunting in 2005 after a closure of 27 years (Williams, 2008b). As of 2009, the Columbia River Whitetail was an endangered but recovering subspecies.

Concurrent with improved protection for deer through legislation and state law enforcement personnel, many habitats began recovering and providing high-quality forage and cover. This set the stage for the second half of *Phase IV—Recovery*. Initial deer-restoration efforts in some states began as early as the late 1800s (Iowa, New York, North Carolina, and Vermont), but these efforts were small and usually consisted of a single stocking or translocation of deer (McDonald and Miller, 2004). During the 1930s to 1950s, many state agencies were seriously involved in restocking programs for white-tailed deer, and they moved over 46,000 animals; Texas alone restocked over 16,000 deer (McDonald and Miller, 2004). Most restoration efforts were completed by the late 1960s and early 1970s, but some remained active into the 1980s (Idaho, Louisiana, Mississippi, North Carolina, South Carolina, and Tennessee) and 1990s (Alabama, Arkansas, Georgia, Kentucky, Texas, Virginia, and West Virginia) (McDonald and Miller, 2004).

As deer numbers increased, state wildlife agencies were careful to protect their breeding stock. Hence, the management approach now viewed as "Traditional Deer Management" emerged. This approach allowed antlered bucks of any size or age class to be harvested while female deer were protected. This harvest strategy allowed for maximum recreational opportunity for sportsmen while concurrently allowing for deer population growth.

Many restocking efforts coincided with forest regeneration that followed extensive clearcutting in the eastern United States. Many restocked deer populations living in ideal habitat grew exponentially. Population growth was good during the *Recovery* phase, but rapid growth combined with an ingrained tradition of not harvesting female deer set the stage for overabundant deer herds, extensive forested habitat degradation, and decades of conflict among state wildlife agencies, sportsmen, and legislators.

During the latter half of the twentieth century, many people moved from cities to suburban or rural areas. This sign of affluence affected whitetail populations through habitat loss, limited hunter access, and increased deer–vehicle collisions resulting from commuters traveling to and from work during peak activity times for whitetails—early morning and late afternoon. Many of these new suburban and rural residents retained their urban views toward wildlife and hunting. This change in landowners' attitude toward wildlife, and hunting in particular, changed the management options open to wildlife agencies seeking to manage deer populations as an increased number of landowners disapproved of hunting and often denied access to hunters.

By the 1960s, state public hunting lands (e.g., wildlife management areas) were established throughout the East, but primarily in the Southeast, and the burgeoning whitetail populations provided additional hunting opportunities. Deer hunting on Federal lands (National Forest lands, National Wildlife Refuges) also grew in popularity during this period. Timber companies began acquiring vast acreages of forested land to provide raw materials for new paper companies, primarily in the South. Leasing of these lands by deer hunters began a new era in North American hunting. In this new era, loose-knit groups of neighborhood hunters with irregular hunting schedules on a variety of properties were being replaced by organized groups and clubs hunting specific tracts of land and paying for the privilege to do so.

Hound hunting for white-tailed deer declined in the late 1960s and early 1970s. This traditional method of hunting deer was being replaced by still hunting. Terminology is important in that still hunting in the South referred to stationary hunting from stands either in trees, free-standing, or on the ground. Still hunting in other regions, particularly in the Northeast and Midwest, was synonymous with stalking.

Increasing whitetail populations (Figure 11.1) eventually gave rise to the need for antlerless deer harvests to control high deer densities. However, deer hunters continued their reluctance to harvest antlerless deer, a throwback to the early years of buck-only harvests. Deer populations continued to spread and increase in density as a result of traditional harvest regimes, including heavy pressure on bucks and light or no harvest of antlerless deer.

In the early years of deer management, state wildlife agencies were obliged to manage for quantity instead of quality. Hunters responded to this approach of maximum opportunity by overharvesting the male segment while protecting females. It was not unusual for state deer harvests to be comprised of at least 75% antlered bucks, and an extremely high percentage (in some instances 80% or more) of yearling bucks in the annual harvest. In most regions, deer populations increased exponentially with this protection of the female segment. Traditional deer management remained steadfast through the 1970s in most of North America.

Phase V (1975–2009): A Changing Paradigm

By the late 1960s and early 1970s, deer herds across much of the whitetail's range were showing ill signs of traditional deer management. Many herds were more abundant than the habitat could support, and adult sex ratios and buck age structures were skewed. Traditional management worked well to recover deer populations, but many deer herds had recovered, and now needed management to control their growth. Despite the problems associated with over abundant deer herds, the tremendous recovery effort made possible by sportsmen is worth noting. Recovery of white-tailed deer populations is among the greatest success stories in the history of wildlife management, and it would not have been possible if not for the efforts of sportsmen.

By this point in time whitetails were likely more numerous in the United States and Canada than they had been prior to European man's arrival (Figure 11.1). The problem was that hunters had been instructed for decades that it was wrong to harvest female deer, and so it was not easy to convince them that doe harvests were now necessary.

New England, interior Midwestern, northern plains, and western states were exceptions to the general pattern of overabundant deer herds with skewed sex ratios and age structures. New England's severe winters kept deer herds in check. Interior Midwestern, northern plains, and western states recovered deer populations later than other regions, and thus were not yet seeing the ill effects of overly successful restoration efforts.

One major change with respect to management programs was a philosophical shift in hunters from mere consumers to becoming deer managers. Quality deer management (QDM) was an emerging paradigm (Green and Stowe, 2000), and by the end of this phase, the average hunter was far more knowledgeable about deer biology, management, and habitat improvement than his/her predecessors. Hunters had more access to deer literature and educational programs, were more likely to own land, and were more engaged in wildlife management and land stewardship. All these qualities bode well for the future of successful deer management programs.

Deer Management Strategies

Deer managers generally recognize three distinct management approaches: traditional, quality, and trophy deer management. The aforementioned traditional deer management strategy, often referred to as "maximum buck harvest management," was used by every state in the United States and some continue to use it today. This strategy may work when the deer herd is below what the habitat can support or a management program's objective, but fails when the herd equals or exceeds these densities.

In 1975, Texas wildlife biologists Al Brothers and Murphy E. Ray, Jr. published their landmark book, *Producing Quality Whitetails* (Brothers and Ray, 1975), which gave rise to QDM. In contrast to traditional deer management, QDM was designed to protect young bucks from harvest while shifting the harvest pressure to female deer, thus producing healthy deer herds in balance with existing habitat conditions. QDM is first and foremost about having the biologically appropriate number of deer for the habitat. QDM also improves age structure by protecting most yearling bucks, allowing them to reach older age classes. This approach includes the four cornerstones of herd management, habitat management, hunter management, and herd monitoring. Texas was the obvious home for a new paradigm in deer management as favorable conditions included large property sizes, family ownership, deer management as a priority, the presence of professionally trained managers and biologists, and economic feasibility. QDM was originally perceived as a practice primarily for Texans, but this novel approach of "total deer herd management" was destined to become a new paradigm for whitetail managers and hunters throughout North America.

In 1982, South Carolina wildlife biologists Gerald Moore and Joe Hamilton invited Al Brothers to be the keynote speaker at the Fifth Annual Meeting of the Southeast Deer Study Group in Charleston, South Carolina. Brothers accepted the invitation on the condition that the audience would consist of half local deer hunters and half professionals. His request was fulfilled, and the audience of over 300 people included more than 150 local deer hunters. In his opening comments, Brothers first chastised the biologists for spending too much time talking to each other and too little talking to hunters. Brothers then chastised the hunters for exploiting bucks and protecting does for far too many years. He asked the group to begin practicing "total deer herd management" by affording the buck segment of the population the same degree of protection currently given the female segment while reducing female numbers in an effort to balance deer populations with their habitat capacity. Brothers' message was taken to heart and South Carolina deer hunters were primed for change. Their 4.5-month deer-hunting season and NO limit on the number of bucks harvested throughout most of the state's coastal region had created an unbalanced sex ratio and a very young buck age structure. Hamilton et al. (1991) tracked changes in hunter attitudes and their effect on harvest trends in this region. The region's known minimum deer harvest increased by 357% from 14,280 in 1972 to 69,305 in 1989. During this period the antlerless harvests increased by 2324% from 1282 to 31,078. As a result, the male-to-female harvest ratio shifted from 10.1:1 in 1972 to 1.3:1 in 1989. Furthermore, hunter selectivity increased steadily, reflected by a decline in occurrence of male fawns in the antlerless harvest from 16.2% in 1982 to 9.7% in 1989. In Game District 5 (a seven-county area in the lower coastal plain) only 52% of approximately 14,000 antlered males harvested in 1989 were yearlings, compared to the traditional occurrence of 75–90% yearlings (Hamilton et al. 1991). The immediate impacts of these changes in harvest were improved deer herd health (e.g., body weights and antler development) and additional recreational opportunities (Hamilton et al., 1991). According to Hamilton and his colleagues, related benefits included enhanced hunter satisfaction and the development of integrated resource management involving wildlife, forestry, and agriculture. These harvest trends were attributed to changes in hunter attitudes resulting from personal contact with wildlife biologists, peer pressure, a progressive regulatory program (a unique antlerless deer quota program), and growing participation in QDM.

Trophy deer management (TDM) is the approach in which only fully mature bucks, generally greater than 5.5 years of age, with high-scoring antlers are harvested and does are harvested to maintain low deer density and optimum nutrition for the remaining animals. In some applications of TDM, bucks that are not fully mature are harvested if their antlers are smaller than some property-specific criteria. While biologically sound, TDM is not practical in much of the white-tailed deer's range and the strategy is

viewed negatively by much of the hunting and nonhunting public (Duda et al., 1998). See Chapter 15 for a complete discussion on TDM.

Urban Deer Management

One final management challenge that became increasingly important during *Phase V* in the history of deer management was the concept of urban/suburban deer management. Urban/suburban deer management was among the most pressing challenges for state wildlife agencies and local municipalities in the early 2000s. Many states struggled to reduce or maintain deer populations at appropriate levels in rural areas even with the willing aid of hunters. Some deer populations were difficult to control even with property access, proficient hunters, and favorable hunting seasons and bag limits. In urban/suburban areas, restricted property access, landowner unfamiliarity with hunting, safety zones, and weapons discharge ordinances made deer management exceedingly difficult.

One approach often touted for use in urban and similar environments, such as National Parks, is immunocontraception. This management strategy uses birth control rather than hunters to control the deer population. The approach has received much publicity because it is nonlethal and has the potential to regulate deer populations in urban/suburban areas that are closed to hunting. Immunocontraception is a birth control method that uses the deer's immune system to prevent pregnancy (Chapter 2). This is the commonest method of inducing infertility in deer, and much research has been conducted over the past four decades to develop an effective contraceptive that can be used on free-ranging deer herds (Fagerstone et al., 2002).

Unfortunately much confusion surrounds the status of fertility control agents. The public has a general misunderstanding regarding the availability and practicality of immunocontraceptive vaccines. Despite misperceptions, overabundant deer herds cannot be controlled solely with immunocontraceptive vaccines. Successful fertility control may limit population growth, but it does little to reduce the existing population. There are also misconceptions about vaccine availability and effectiveness. Prior to October 2009, all research on vaccines was conducted under investigative permits, as there were no vaccines authorized for use on free-ranging deer herds. However, effective October 2009, GonaCon™ was the first contraceptive vaccine registered for use in free-ranging white-tailed deer herds by the Environmental Protection Agency. GonaCon is an immunocontraceptive vaccine for bucks and does developed by scientists at the U.S. Department of Agriculture's Wildlife Services' National Wildlife Research Center. It is a single-shot multiyear vaccine that is effective for two to four years in deer. GonaCon works in the hypothalamus portion of the brain to disrupt the body's reproductive processes. This means female deer do not come into estrous and thus will not be bred. Bucks treated with GonaCon have suppressed testosterone production which reduces rutting behavior, neck swelling, and muscular growth. Treated bucks resemble does with antlers (Killian et al., 2005).

Fertility control in deer is a rapidly advancing technology. However, even with current advancements, immunocontraception is expensive, with estimated costs ranging from $500 to $1000/deer (due mainly to deer capture and handling costs). Because annual mortality rates for urban/suburban deer populations are low, an estimated 70% or more of the does in a population need to be treated to limit or stop herd growth (Warren, 2000). This approach's effectiveness and practicality are limited to enclosed or localized herds rather than free-ranging populations. Urban/suburban deer management is a difficult challenge for state wildlife agencies and local municipalities, and given our country's rate of development and increasing human population, these deer management arenas will necessitate additional time and resources in the future.

Antler Restriction Programs

Antler restrictions are a strategy to protect a specific age class (generally 1.5-year-old) or classes of bucks. Antler restrictions first appeared in Texas in the 1920s to protect spike-antlered deer, but it was not until the late 1980s that they were used in the southeast, and the early 2000s in the northeastern United States. The first experiment with antler regulations on a wildlife management area (WMA) in the southeast took place in Mississippi in the late 1980s (Thomas, 2004). In 1993, Dooly County, Georgia became the first county in the United States to establish countywide antler restrictions requiring bucks to have an outside spread of at least 15 in. (38 cm) to be legal, and in 1995 Mississippi was the first state to implement statewide antler regulations, requiring bucks to have a total of four or more antler points. These programs paved the way for other states and by 2004 five states had statewide antler regulations, four states had antler regulations at a county or deer management unit level, and four states were discussing antler regulations.

Arkansas implemented a statewide restriction in 1998 that required bucks to have at least three antler points on one side. The percentage of yearling bucks in the harvest dropped from 49% prior to the restriction to 14% in the 2003–2004 season (Thomas, 2004). Georgia had already experimented with antler regulations and in 2002 the state required at least one of the two bucks in the bag limit to have four points on one side. The restriction reduced the buck harvest more than 30% during the first year. Michigan adopted rules in 1998 that required at least one buck of the two-buck bag limit to have four points on one side. The restriction protected some bucks, but even a decade later Michigan's buck harvest exceeded 60% yearlings, a yearling buck harvest rate among the highest in the nation.

In 2002, Pennsylvania made national news by implementing an antler point restriction. The majority of the state required three antler points on one side while the western third required four points on one side. Junior (16 years and younger), senior (65 years and older), and disabled hunters as well as military personnel were exempt from the restrictions. The regulations were designed to protect 50–75% of the yearling bucks. The additional antler point required in the western third of the state was necessary because habitat quality was higher in that region. This was a monumental undertaking in a state with 1 million deer hunters where about 80% of the antlered bucks harvested had been yearlings. In some areas of the state, over 90% of the annual buck harvest was comprised of yearlings. A change was needed because less than 1% of bucks in Pennsylvania reached maturity (Gary Alt, Pennsylvania Game Commission, personal communication). Antler point restrictions reduced yearling harvest rates to about 50%. The number of bucks 2.5 years and older in the population and in the harvest increased as did hunter satisfaction with the restrictions. According to 2007 hunter survey data, approximately 75% of hunters with an opinion on antler restrictions supported the new regulations (Pennsylvania Game Commission, 2009a).

By 2004, antler regulations were also being implemented in certain counties or deer management units in Georgia, Louisiana, Michigan, and Texas. In addition to Dooly County, eight other Georgia counties had regulations. One required a 15-in. (38 cm) outside spread (like Dooly County), while the other seven required four antler points on at least one side (Thomas, 2004). In 2002, Louisiana required bucks to have six or more total antler points in three southern Louisiana parishes. Michigan experimented with antler regulations in nine deer management units. Three required bucks to have at least two points on one side and six required at least three points on one side.

Also in 2002, Texas Parks and Wildlife Department (TPWD) implemented regulations in six counties in the oak prairie region of southeastern Texas (Carroll et al., 2010). The regulations required bucks to have a minimum inside spread of 13 in. (33 cm), or six or more antler points on an antler, or have at least one unbranched antler (e.g., a spike). This regulation differed from all others mentioned in that it attempted to protect only the larger-antlered young bucks. By 2006, the Texas "slot limit" approach spread to include 61 counties and had gained high levels of hunter and landowner acceptance. The proportion of bucks less than 3.5 years old in the harvest dropped from 79% (prior to the experimental regulation) to less than 30% during the 2004–2007 hunting seasons. Thus, harvest of bucks at least 3.5 years old increased from 20% prior to 2002 to more than 70% during the 2004–2007 hunting seasons (Carroll et al., 2010). For the 2009–2010 hunting season, TPWD extended the slot limit for bucks to 52 additional counties so that the regulations would apply to about 45% of the state. This technique was applied to areas, or counties, with a history of overharvesting young bucks. After the initial three-year experiment, the antler-restriction regulation was modified to delete the category of bucks with at least six antler points on one side. The bag limit was two bucks. Although only one buck could have a 13-in. (33 cm) or greater inside spread, both bucks could have at least one unbranched antler (i.e., spikes). The TPWD also increased antlerless bag limits and provided a longer season in certain areas to address some resource concerns and provide additional hunting opportunity.

It is clear that in the early 2000s antler restrictions were a hot topic among deer hunters. Whether you loved or hated them, you could be sure your state wildlife agency had discussed them. In fact, 22 states had some form of antler restrictions implemented in 2008 (Figure 11.2). Eight states (Alabama, Arkansas, Delaware, Georgia, Michigan, Mississippi, Pennsylvania, and Vermont) had statewide restrictions, at least for one buck, while 14 states used them in some wildlife management areas, units, regions, or military bases. The most commonly used restriction was the number of antler points. Sixteen states employed this technique, and depending on the state, the number varied from two to four points on a side. Three states used an antler spread restriction; West Virginia required harvested bucks to have an outside spread of 14 in. (35 cm) while Delaware and Kentucky required 15 in. (38 cm).

South Carolina and Texas used a combination of antler points and spread, and Mississippi used a combination of antler points, spread, and main beam length. These combination approaches allowed hunters to harvest bucks that met one of the two or three antler criteria. Combination approaches are generally more biologically sound, flexible, and preferred to single restriction strategies.

In 2009, the Mississippi Wildlife, Fisheries and Parks Department removed the antler point criterion (the least biologically sound) from its combination restriction. Their innovative approach now required a buck to meet either an antler spread or main beam length requirement. Research in Mississippi (Strickland and Demarais, 2008) demonstrated soil region differences with regard to antler development. Wildlife biologists divided the state on the basis of soil quality and assigned antler spread and beam length minimums accordingly (Figure 11.3). In the deer management zone with the best soil, a legal buck was required to have a minimum inside antler spread of 12 in. (30 cm) *or* a beam length of 15 in. (38 cm).

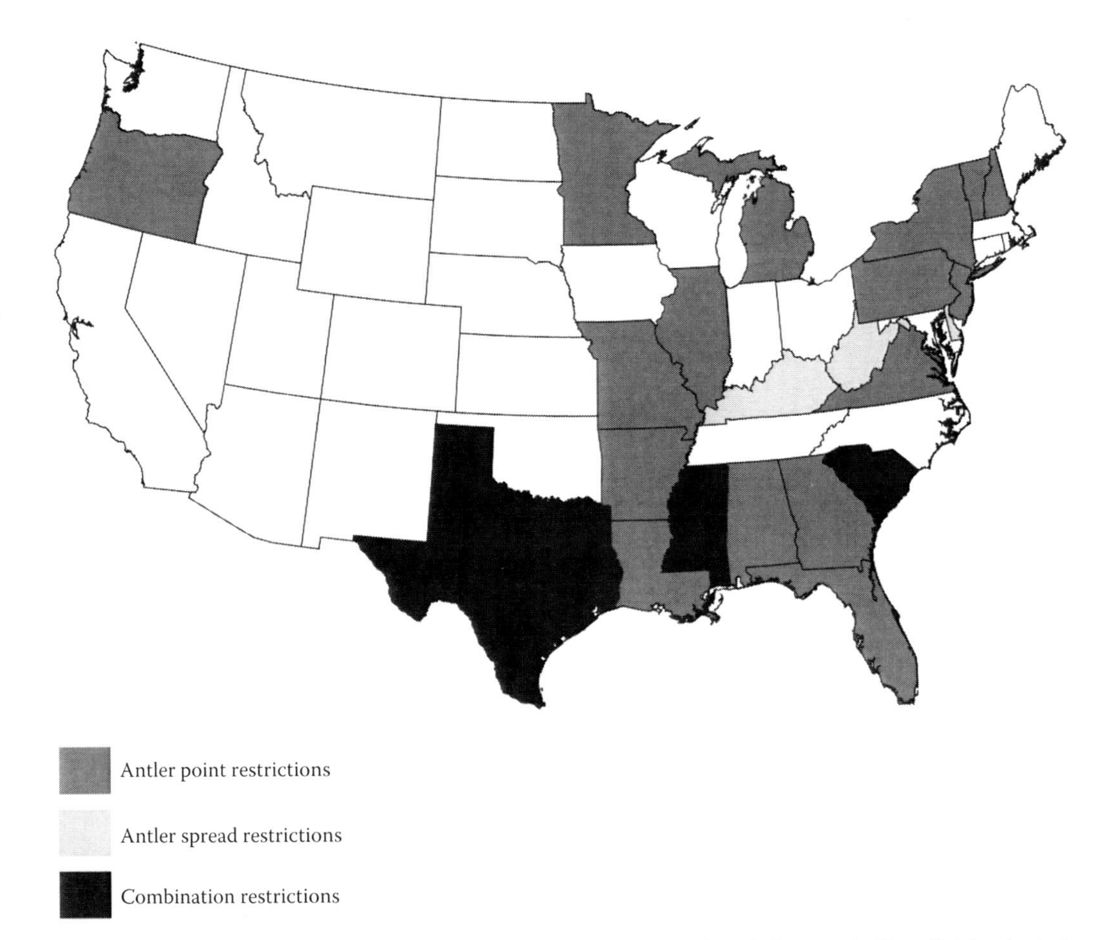

FIGURE 11.2 Twenty-two states in the United States that had antler restrictions for harvest of white-tailed deer bucks in 2008. Sixteen states employed antler point restrictions, while three states used antler-spread restrictions, and three used a combination of antler point and spread and/or main beam length.

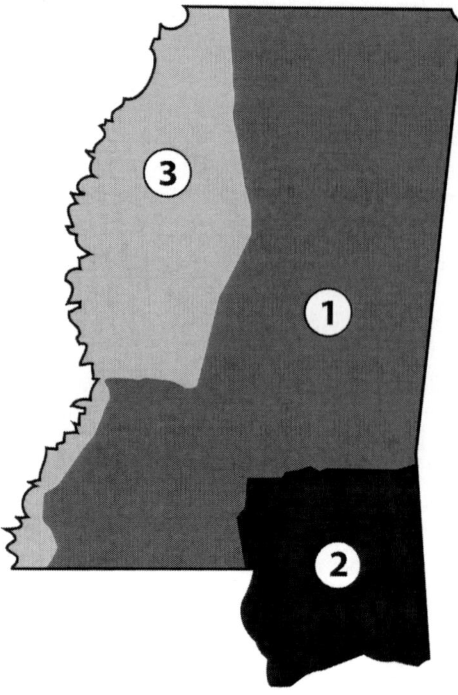

Legal bucks			
Zone	Inside spread		Main beam
1	10 in. (25.5 cm)	or	13 in. (33 cm)
2	10 in. (25.5 cm)	or	13 in. (33 cm)
3	12 in. (30.5 cm)	or	15 in. (38 cm)

FIGURE 11.3 Three zones of soil productivity (zone 1—medium productivity; zone 2—low productivity; zone 3—high productivity) in Mississippi and the corresponding antler criteria required for bucks to be legal for harvest in each zone during the 2009 deer hunting season.

In the other two deer management zones, a majority of the state, the required antler criteria were a 10-in. (25 cm) inside spread *or* a 13-in. (33 cm) beam length. These antler restrictions were designed to protect nearly 100% of yearling bucks.

Deer management in the early 2000s differed vastly from that of a decade or more removed. Twenty-first century hunters were more knowledgeable on deer biology and management than ever and demanded progressive management programs from their state agencies. This was in complete contrast to their counterparts in the *Protection and Recovery* phase that harvested young bucks at every opportunity and demanded low antlerless harvest rates to ensure high deer densities.

Culling

No discussion on deer management, especially regarding bucks, would be complete without a discussion on culling. No other aspect of white-tailed deer management has created more controversy, both among laymen and management professionals, as the practice of culling to improve genetics (Kroll, 1994). Culling is the selective removal of specific individuals in an attempt to improve the quality of the remaining population. Culling has been most often applied to whitetails in the form of removing bucks with inferior antler traits as a strategy to improve antler quality.

Use of this strategy was widely advocated in Texas since at least the 1980s based on research from the Kerr Wildlife Management Area suggesting antler quality could be improved by removing spike-antlered yearling bucks (Harmel, 1983; Newsom, 1984; Lockwood et al., 2007). Later, others cautioned against selectively removing yearlings with spike or few antler points, arguing a yearling bucks' antlers were more a reflection of birth date and nutrition than genetics (Lukefahr and Jacobson, 1998).

This issue has been a hot and contentious topic at many Southeast and West Texas Deer Study Group meetings. Culling has been discussed at annual Northeast Deer Technical Committee meetings, the Midwest Deer Study Group meetings, and at countless camps and hunt clubs. To this point, most culling discussions focused on the impacts of removing the theoretically "poorest" individuals. However, designed protection of yearling bucks in Mississippi in the late 1990s inadvertently caused many of the largest individuals in the yearling age class to be harvested. This "culling" resulted in high-grading yearling bucks on state hunting lands and produced smaller antlers in the corresponding older age classes (Strickland et al., 2001). This unintended result fueled the theory that antler genetics could be impacted by culling. However, it is important to note that antler genetics were not shown to be impacted in Mississippi. Rather, antler size was smaller as a result of the harvest restriction (bucks were required to have at least four total antler points) mainly protecting deer born later in the year. Many of the earlier-born bucks grew antlers consisting of four or more points, thus they were eligible for harvest during the hunting season. Average antler size of harvested bucks two years old and older was smaller after implementation of antler restrictions compared to antler size of the same-aged bucks before antler restrictions. Smaller antlers in older cohorts can be explained by harvest of bucks with the largest antlers when they were yearlings. Harvested yearling bucks may have had large antlers for many reasons other than genetics, such as having been born early in the year, being a single fawn raised by an experience mother, being disease free, or having been raised in exceptionally good habitat.

Culling has been and continues to be a confusing topic. Culling proponents favor research suggesting yearling spikes would always be smaller than their branch-antlered counterparts. Culling opponents favor research suggesting a buck's first set of antlers were not a good indicator of his antler growth potential. The argument raged even as research in south Texas on the King Ranch suggested that aggressively culling the smallest antlered bucks in all age classes for six consecutive years on a free-ranging deer herd at the 10,000-acre scale had little impact on antler quality (M. W. Hellickson, Orion Wildlife Management Services, unpublished data).

The vast majority of the 22 states selectively harvesting bucks in 2008 had programs designed to protect yearling bucks rather than selectively remove "inferior" individuals. However, the culling argument is far from over, as evidence from discussions by researchers and managers at professional deer meetings, as well as by hunters at camps and on Web forums (Thomas, 2007). In this chapter some deer management programs, especially those in Texas, strongly advocate the practice of culling while others discourage its use.

Deer Harvest Trends

The Quality Deer Management Association (QDMA, 2001) produced a deer density map for the contiguous United States using data provided by state wildlife agencies from 1994 to 1999 on the relative deer density per county. The QDMA (2008) updated the map with information from 2001 to 2005 and noticeable changes had occurred since the initial map was produced (Figure 11.4). The maps also included tables with harvest data by sex and age class (Tables 11.1 and 11.2), and QDMA updated that data in their 2010 Whitetail Report (Adams et al., 2010). A few states did not provide data and some did not provide all of the requested information, but the submitted information provided for meaningful comparisons among states and between surveys. These data show a significant shift in many deer management programs and draw attention to the new direction many state agencies were headed.

Yearling Buck Harvest Trends

A significant indication of the spread of the QDM philosophy was the percentage of yearling bucks in the harvest. From 1989 to 2008, this percentage declined from an average of 62% to 41%. During this same time period the percentage of 2.5-year-olds increased from 23% to 32% and 3.5-year-olds or older increased from 15% to 29%. Some states made tremendous advances, such as Pennsylvania which dropped from 81% to 52% yearling bucks in the harvest. Wisconsin dropped from 74% to 53%,

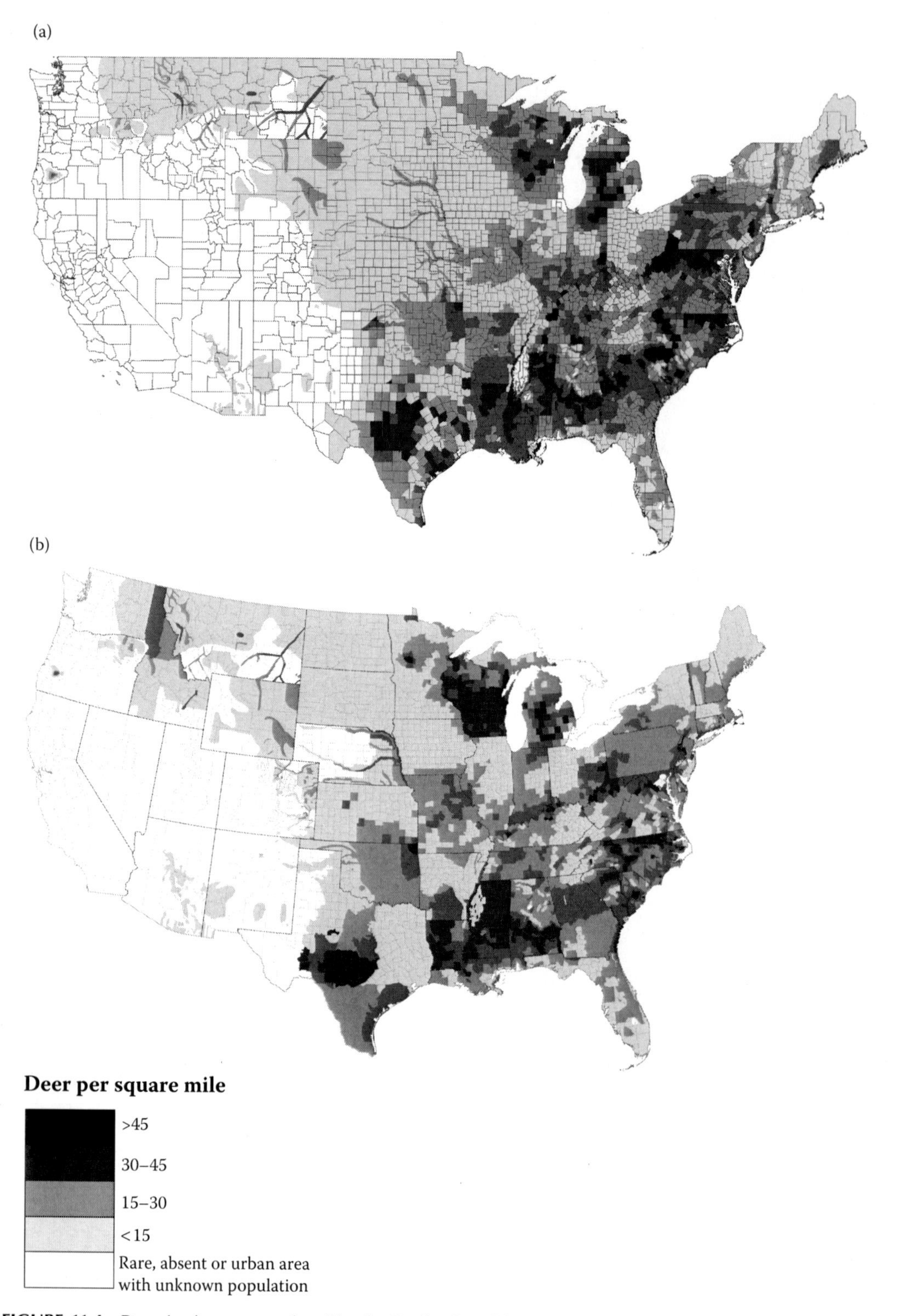

FIGURE 11.4 Deer density maps produced by the Quality Deer Management Association showing white-tailed deer density by county for the United States in 1999 (a) and 2005 (b). Data were provided by each state's wildlife management agency as relative deer density per county.

TABLE 11.1

Estimated Antlered and Antlerless Deer Harvest by State in 1989, 1999, and 2008

	Estimated Harvest					
	Antlered Deer			Antlerless Deer		
State	1989	1999	2008	1989	1999	2008
Alabama	159,000	191,156	[a]	83,000	224,536	[a]
Arizona	5650	3550	5080	58	36	0
Arkansas	68,262	66,024	93,375	44,817	117,745	74,963
Colorado	640	655	[a]	530	555	[a]
Connecticut	3855	5293	5892	4908	5722	6790
Delaware	1777	[a]	3771	1229	[a]	10,105
Florida	[a]	88,961	[a]	[a]	32,756	[a]
Georgia	165,842	150,788	159,567	156,821	253,272	239,350
Idaho	[a]	11,760	13,610	[a]	5360	6149
Illinois	27,560	36,649	71,813	28,360	48,066	117,088
Indiana	40,500	46,400	50,845	38,800	53,200	78,903
Iowa	46,602	55,251	51,710	51,326	66,384	90,484
Kansas	[a]	37,526	41,462	[a]	58,692	39,028
Kentucky	37,303	49,436	54,936	32,846	45,791	65,674
Louisiana	123,500	267,500	87,010	66,500	139,100	71,190
Maine	17,009	19,289	13,564	13,251	12,184	7497
Maryland	22,520	37,821	34,725	21,245	39,839	65,712
Massachusetts	3650	4840	5582	3100	4700	5620
Michigan	253,220	265,878	248,350	199,270	271,695	241,573
Minnesota	77,132	93,448	96,000	52,419	72,168	126,000
Mississippi	186,337	[a]	132,167	72,856	[a]	148,687
Missouri	78,875	100,265	99,957	90,297	117,595	182,162
Montana	26,161	25,600	[a]	17,688	18,627	[a]
Nebraska	17,800	25,700	36,235	10,100	15,600	32,397
New Hampshire	5122	6819	6390	2116	3884	4526
New Jersey	23,100	22,000	18,399	25,500	53,400	34,859
New Mexico	[a]	[a]	137	[a]	[a]	0
New York	99,589	125,392	105,747	82,290	130,567	117,232
North Carolina	94,700	119,600	85,051	53,500	82,400	91,246
North Dakota	23,370	27,336	33,963	26,589	28,773	57,577
Ohio	37,000	58,000	89,962	53,000	68,000	162,055
Oklahoma	28,078	52,577	59,449	10,263	30,147	45,820
Oregon	[a]	[a]	815	[a]	[a]	63
Pennsylvania	170,000	194,000	122,410	219,000	184,000	213,440
Rhode Island	635	1173	1055	450	631	1210
South Carolina	57,866	137,505	119,346	46,422	169,949	129,432
South Dakota	22,123	20,036	33,413	23,891	19,285	30,459
Tennessee	67,512	78,811	93,873	41,250	64,686	70,540
Texas	256,045	239,859	340,159	221,446	180,328	279,491
Utah	[a]	[a]	[a]	[a]	[a]	[a]
Vermont	8390	11,900	9539	840	7880	7452
Virginia	[a]	96,544	112,207	[a]	91,160	144,175
Washington	[a]	[a]	[a]	[a]	[a]	[a]
West Virginia	82,000	109,000	86,914	32,000	122,000	76,689
Wisconsin	165,916	206,207	138,507	194,414	291,463	313,378
Wyoming	5889	3665	8304	2874	1699	6488
Total	**2,510,530**	**3,094,214**	**2,771,291**	**2,025,266**	**3,133,875**	**3,395,504**

Source: Adapted from Quality Deer Management Association. 2001. *White-tailed Deer Density Map—1999*. Bogart, GA: Quality Deer Management Association and Adams, K. P., R. J. Hamilton, and M. D. Ross. 2010. *QDMA's Whitetail Report*. Bogart, GA: Quality Deer Management Association.

[a] Data not available.

TABLE 11.2

Percent of Antlered Deer Harvest by Age Class and State in 1989, 1999, and 2008

	Percent Harvest								
	1.5 Years			2.5 Years			3.5+ Years		
State	**1989**	**1999**	**2008**	**1989**	**1999**	**2008**	**1989**	**1999**	**2008**
Alabama	[a]	[a]	25	[a]	[a]	35	[a]	[a]	40
Arizona	[a]	[a]	[a]	[a]	[a]	[a]	[a]	[a]	[a]
Arkansas	61	[a]	13	24	[a]	38	15	[a]	49
Colorado	[a]	[a]	[a]	[a]	[a]	[a]	[a]	[a]	[a]
Connecticut	[a]	[a]	40	[a]	[a]	[a]	[a]	[a]	[a]
Delaware	63	[a]	53	25	[a]	28	12	[a]	19
Florida	48	35	[a]	31	39	[a]	22	26	[a]
Georgia	61	47	45	25	32	32	15	21	23
Idaho	[a]	17	[a]	[a]	[a]	[a]	[a]	[a]	[a]
Illinois	43	38	41	22	26	[a]	12	11	[a]
Indiana	69	63	40	24	25	40	7	11	20
Iowa	[a]	[a]	[a]	[a]	[a]	[a]	[a]	[a]	[a]
Kansas	51	49	17	[a]	[a]	34	[a]	[a]	49
Kentucky	68	[a]	41	25	[a]	38	7	[a]	21
Louisiana	70	55	24	20	30	22	10	15	54
Maine	43	48	37	25	25	23	32	27	40
Maryland	55	54	62	21	23	[a]	7	7	[a]
Massachusetts	48	48	48	29	26	29	23	26	23
Michigan	73	66	61	18	22	25	9	12	14
Minnesota	[a]	[a]	67	[a]	[a]	23	[a]	[a]	10
Mississippi	70	50	18[b]	20	30	22[b]	10	20	60[b]
Missouri	63	61	22(58)[c]	[a]	[a]	54(31)[c]	[a]	[a]	24(11)[c]
Montana	25	17	[a]	23	17	[a]	52	63	[a]
Nebraska	71	58	34	20	29	[a]	9	13	[a]
New Hampshire	47	46	45	28	27	26	24	27	29
New Jersey	73	71	64	21	23	[a]	6	6	[a]
New Mexico	[a]	[a]	[a]	[a]	[a]	[a]	[a]	[a]	[a]
New York	69	69	62	20	22	26	11	9	12
North Carolina	65	50	39[b]	24	33	39[b]	11	17	22[b]
North Dakota	[a]	[a]	[a]	[a]	[a]	[a]	[a]	[a]	[a]
Ohio	65	65	50	25	25	32	10	10	18
Oklahoma	66	50	27	22	30	32	12	20	41
Oregon	[a]	[a]	[a]	[a]	[a]	[a]	[a]	[a]	[a]
Pennsylvania	81	80	52	19	20	35	[a]	[a]	13
Rhode Island	[a]	[a]	38	[a]	[a]	27	[a]	[a]	35
South Carolina	54	48	59	22	21	23	13	16	18
South Dakota	53	55	[a]	29	30	[a]	18	15	[a]
Tennessee	78	56	44	18	33	40	4	11	16
Texas	[a]	30	27	[a]	28	19	[a]	42	54
Utah	[a]	[a]	[a]	[a]	[a]	[a]	[a]	[a]	[a]
Vermont	50	55	15	25	22	59	25	23	26
Virginia	[a]	51	37	[a]	33	37	[a]	17	26
Washington	[a]	[a]	[a]	[a]	[a]	[a]	[a]	[a]	[a]
West Virginia	[a]	62	[a]	[a]	30	[a]	[a]	8	[a]
Wisconsin	74	68	53	19	22	[a]	7	10	[a]
Wyoming	[a]	[a]	[a]	[a]	[a]	[a]	[a]	[a]	[a]
Average	**62**	**57**	**41**	**23**	**28**	**32**	**15**	**19**	**29**

Source: Adapted from Quality Deer Management Association. 2001. *White-tailed Deer Density Map—1999*. Bogart, GA: Quality Deer Management Association and Adams, K. P., R. J. Hamilton, and M. D. Ross. 2010. *QDMA's Whitetail Report*. Bogart, GA: Quality Deer Management Association.

[a] Data not available.

[b] Data from check stations and DMAP.

[c] Data from antler-point-restriction counties (non-antler-point-restriction counties).

and Arkansas dropped from 61% to a nationwide low of 13% yearlings in the antlered buck harvest. Pennsylvania had historically led the nation in this category, but was now around the national average with at least seven other states harvesting a higher percentage of yearlings. This included neighboring Maryland, New Jersey, and New York; three of only five states reporting that yearling bucks constituted over 60% of the antlered buck harvest.

Only two states regressed in this statistic as South Carolina and Maryland both harvested a higher percentage of yearlings in 2008 than in 1989. South Carolina increased from 54% to 59% yearlings in the antlered buck harvest. South Carolina was below the national average in 1989 and was still doing fairly well in 2008, especially considering that more than half of the state had a 4.5 month season with no buck limit. It is noteworthy that approximately 1.2 million ha are involved in voluntary QDM programs and that nearly 32,000 ha of public lands employ antler point restrictions, but it is the stronghold of the traditional deer management philosophy by South Carolina's dog hunters and others that retards progress on a statewide basis.

Maryland was below the national average in 1989 and 1999 but far above it in 2008. The percent of yearling bucks in the antlered buck harvest increased from 55% to 62% from 1989 to 2008 in Maryland, while the national average dropped from 62% to 41%.

In New Hampshire the percentage of yearling bucks in the antlered harvest increased from 47% to 51% from 1989 to 2005. Then, in 2005 the New Hampshire Fish and Game Department acted to reverse this trend. Objective 2-1 of New Hampshire's 2006–2015 Big Game Management Plan stated, "Manage regional deer populations to ensure that yearling males do not exceed 50% of the adult male population." From 2004 to 2006, the percentage of yearling bucks had exceeded 50% of the buck harvest in one of the state's wildlife management units (WMU). Therefore, the Department organized an ad-hoc deer advisory committee to determine the preferred strategies for reducing the yearling harvest rate in that WMU, and they implemented an antler point restriction in 2007. This component of their deer management plan was arguably one of the most progressive QDM procedures implemented by any state agency, and the percentage of yearlings in the buck harvest dropped to 24% in that WMU the following year and to a statewide average of 45% by 2008.

In 2008, Texas led the nation in antlered buck harvest by harvesting over 340,000 bucks and only 27% of them were yearlings. Michigan was second with over 248,000 bucks, but 61% of them were yearlings. This is an example of two states with distinct philosophical differences with respect to buck harvest. Both states have high hunter numbers, but Texas has historically placed more focus on moving bucks into the older age classes, while Michigan has placed less emphasis on protecting yearling bucks.

Older Bucks in the Harvest

Vermont increased the percentage 2.5-year-old bucks in the harvest from 25% in 1989 to 59% in 2008, and Tennessee increased from 18% to 40% during this time period with voluntary QDM practices by hunters. Pennsylvania also made tremendous gains as the Keystone State improved from 19% to 35% 2.5-year-old bucks in the antlered harvest, which in actual numbers represents an increase from about 32,000 to nearly 43,000 2.5-year-old antlered bucks harvested.

With respect to bucks 3.5 years and older, Mississippi led the nation with 60% of their buck harvest reaching this age category in 2008, nearly a sixfold increase since 1989. Fifty-four percent of the antlered buck harvest in Texas reached this age class, and 50% did in Louisiana. The new paradigm in deer management was creating dramatic results. On average, far more yearling bucks were making it through the hunting season and surviving to reach 2.5, 3.5, or even 4.5 years of age. In addition to creating a balanced and healthier deer herd, these bucks were providing tremendous opportunities and excitement for hunters, photographers, and other deer enthusiasts.

Antlerless Harvest Trends

A milestone in deer management was obtained in 1999 because that year marked the first time in modern history hunters in the United States harvested more antlerless deer than antlered bucks. In 1999, hunters harvested approximately 6.2 million whitetails, with antlered bucks accounting for slightly less than half. In 2008 hunters harvested approximately the same number of deer (within 1%), but antlered bucks accounted for only 45% of the total. From 1999 to 2005, buck harvest as a percentage of the total harvest declined while the antlerless harvest increased nearly 10%.

Antlerless deer harvests are more difficult to compare across states and years as some states may be aggressively reducing populations while others may be seeking to stabilize or even grow deer herds. From 1989 to 2008, one of the most notable statistics was the increase in antlerless harvest in the Midwest. Illinois, Indiana, Iowa, Kentucky, Ohio, and Missouri increased their antlerless harvests by an average of 136% from less than 300,000 to nearly 700,000. Illinois and Ohio led the list by increasing their antlerless harvests from 28,360 to 117,088 and 53,000 to 162,055, respectively.

Overall, Wisconsin led the nation in this category in 2008 by harvesting over 313,000 antlerless deer (Table 11.1). Wisconsin confirmed the presence of chronic wasting disease (CWD) in 2002 and had been aggressively attempting to reduce the state's deer population. For comparison, Pennsylvania's antlerless harvest was 27% higher in 2005 than in 1999, and in 2005 the state was stabilizing (vs. reducing) its deer herd. Pennsylvania's 2005 antlerless harvest was 38% lower than the 2003 harvest when the state was aggressively reducing the deer herd.

In 2005, 64% of states shot at least as many antlerless deer as antlered bucks, while 36% shot more bucks (Figure 11.5). This is vastly different than during the *Protection and Recovery* phase when antlerless harvests comprised a small percentage of the total deer harvest. For example, in 1907, does were given complete protection in Pennsylvania. In 1923, the Pennsylvania Game Commission was given the authority to establish antlerless seasons. The Commission allotted 100 antlerless licenses that year and nine deer were taken (Pennsylvania Game Commission, 2009b). Hunters took 25,095 antlerless deer in 1928, 83,963 in 1949 (0 in 1948), 79,376 in 1968, and 65,808 antlerless deer in 1975 (Pennsylvania Game Commission, 2002). In contrast, during the current Phase, hunters took an estimated 216,492 antlerless deer

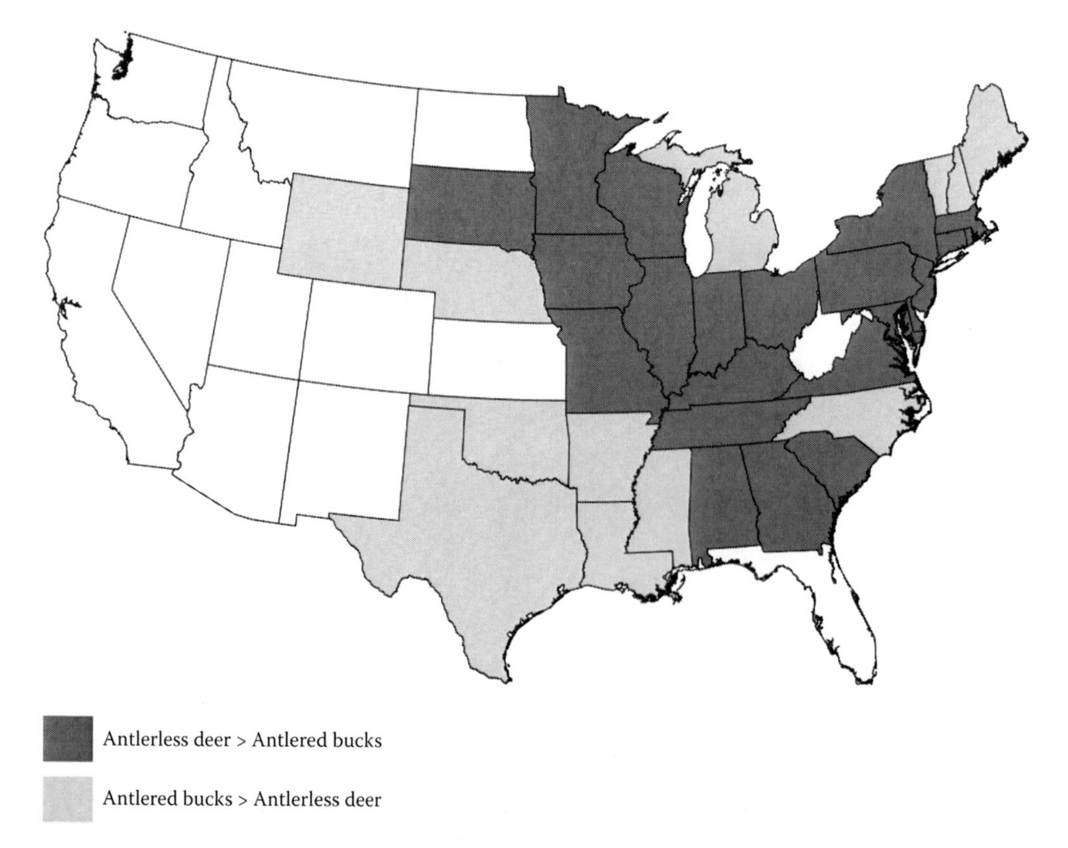

FIGURE 11.5 States in the United States that harvested more antlerless deer than antlered bucks and the states that harvested more antlered bucks than antlerless deer in 2005.

in 1988, 233,890 in 2005, and 213,440 antlerless deer in 2008 (Pennsylvania Game Commission, 2009b). From nine in their first season to over 200,000 antlerless deer a century later is remarkable. Similarly, Wisconsin's antlerless harvest climbed from about 30,000 in 1964 to 400,000 in 2000 (Wisconsin Department of Natural Resources, 2009a), before declining to 310,000 in 2008 (Wisconsin Department of Natural Resources, 2009b).

In 2008, antlerless deer comprised 73% of the total deer harvest in Delaware. Antlerless harvest ranged from 65% to 69% in Maryland, New Jersey, and Wisconsin. States such as Maine, New Hampshire, and Vermont shot fewer than 50% antlerless deer. This is not surprising in northern New England where lower deer densities combined with severe winter weather allow for successful deer management programs with reduced antlerless harvests. However, states with productive deer herds like Arkansas, Michigan, Nebraska, and West Virginia should likely have harvests comprised of more than 50% antlerless deer rather than the 45–49% they averaged in 2008.

Habitat Influence on Population Size

Habitat quality is a vital component of a deer herd's productivity and directly relates to the harvest intensity a population can withstand. During the *Protection and Recovery* phase many deer herds were blessed with regenerating forests and farmsteads. This early successional habitat provided abundant forage and cover, and deer herds flourished. Near the end of that phase and entering *Phase V* many forests were maturing, and where silvicultural treatments were not applied, habitat quality diminished rapidly for whitetails.

While many states were actively trying to reduce deer populations, expanded seasons and increased antlerless licenses were not the sole culprits for fewer deer in some areas. Habitat quality was also a factor, and in forested environments habitat quality is related to the tree species present and their age classes. A range of age classes is important because mature closed-canopy forests only produce a fraction of the forage that early-successional forests provide (Stransky, 1980). Availability of selected forages for deer will increase 5–10 times following a timber harvest and/or forest management practice (Lashley, 2009). Given that the average deer eats about 1000 kg of forage/year on a fresh weight basis, it is clear that early successional habitats provide a lot more forage and can sustain many more deer than mature forests.

Mature forests are also important components of deer habitat. Mature forests provide critical cover for deer and, depending on the species present, may provide hard and soft mast in addition to browse. Mast is important to deer but its availability varies seasonally and annually. The point is that deer herds are more productive in areas with a variety of habitat types and diversity of age classes.

According to USDA Forest Service (2008) data from the 1980s to the 2000s the United States gained over 55.6 million ha of forestland. This was a 28% increase and most of it was in the Midwest and West. Ironically, even with the increase in forestland, the percentage of early successional forestland declined from 22% to 18%. During the last 10–20 years, New York, Pennsylvania, and Michigan lost 200,000 to 360,000 ha of early successional forest and some southern states, notably Mississippi and Alabama, lost 680,000 to 1.8 million ha.

Of particular interest in the above list were "key" deer states like Michigan, New York, and Pennsylvania. Hunters in those states had enjoyed high deer densities in the past, and such densities were directly influenced by the amount of early successional habitat. As these states reduced area in early successional forest, they also reduced the number of deer and other wildlife those habitats could support. Expanded seasons and increased antlerless licenses were partly responsible for reduced deer sightings by hunters, but in some instances the loss of early successional habitat may have been an equally important factor for areas having fewer whitetails.

Other Factors Affecting White-tailed Deer Management

Many state and national hunting and conservation organizations lent support to white-tailed deer during the 1900s and 2000s. Some assisted state wildlife agencies with restocking efforts, some lobbied for protective measures such as closed seasons and restricted antlered or antlerless harvests, and others supported research and supplemental feeding programs. These were valiant efforts and they are important to the whitetail's management history. One organization in particular, QDMA, played a prominent role during *Phase V*.

The QDMA was formed in 1988 as a national nonprofit wildlife conservation organization with a mission to ensure the future of white-tailed deer, wildlife habitat, and our hunting heritage. The QDMA was an educational organization focusing on providing unbiased, scientific information on deer biology and management and habitat improvement to hunters, managers, and researchers. The QDMA worked with hunters, landowners, and natural resource professionals to balance deer herds with the habitat and balance the adult sex ratios and age structures. In short, QDMA promoted the QDM philosophy by educating hunters and managers on its merits through hundreds of annual seminars and workshops, and via *Quality Whitetails*, QDMA's flagship publication. The QDMA's ongoing commitment to education and stewardship was formally recognized in 2001 when it received the prestigious "Group Achievement Award" from The Wildlife Society—an international professional organization of nearly 10,000 wildlife professionals. The QDMA is the only white-tailed deer organization ever to receive this coveted award. As of 2009, the QDMA's membership included more than 3000 of the nation's leading wildlife and forestry professionals. As such, QDMA was widely regarded as the most respected and influential white-tailed deer organization in the United States.

Throughout most of the whitetail's range for the past 200 years, ever-changing patterns of man's use of the land, predation, disease, climate cycles, and disturbances such as fire, wind, and flood created conditions that either favored or precluded healthy deer populations. Vast changes occurred in hardwood stands throughout the Appalachians due to demand by foundries and industrial logging. By the end of the 1860s, the iron industry had rendered tens of thousands of hectares treeless. Expansion of railways and industrial logging during the late 1800s and early 1900s enabled additional clearing of hardwoods across vast areas (Davis, 2003).

The longleaf pine ecosystem, one of the most extensive and biologically diverse ecosystems in North America, once covered 37 million hectares from southeastern Virginia southward through Florida and westward to east Texas. This forest type was especially important to white-tailed deer, and it once produced copious amounts of natural forage, due to frequent naturally occurring and man-induced burns. However, it now covers less than 1.6 million hectares; a 96% decline (Earley, 2004).

The boll weevil (*Anthonomus grandis*), migrated into the United States in 1892 from Mexico and had infested all cotton-growing areas by the 1920s. Until the arrival of the boll weevil, cotton fields had consumed vast areas of prime deer habitat throughout the South. Farmers became bankrupt and workers migrated to industrial centers for jobs. Then, the Great Depression and the infamous Dust Bowl era occurred in the 1930s. On the heels of this disastrous period came World War II, which turned thousands of men to protecting their country and away from recreational activities, especially deer hunting.

During this time, white-tailed deer also encountered a new adversary in some areas, the screwworm (*Cochliomyia hominivorax*). Migrating from Mexico into the southern United States from California across to the Carolinas in the early twentieth century, this parasite controlled deer population growth in many regions for a quarter century until it was eradicated in the Southeast in 1959 and over a decade later in the Gulf Coast states (Novy, 2009). Another exotic, invasive parasite from Mexico, the cattle-fever tick (*Boophilus microplus*), caused problems from Texas to Florida beginning in the 1930s (Chapter 7). White-tailed deer in Florida numbered approximately 20,000 in the late 1930s and deer were nearly extirpated in south Florida in an effort to eradicate tick-borne diseases (Schaefer and Main, 1997).

With the fall of "King" cotton, corn and soybeans became the cash crops for Southern farmers. Then, in 1978, the boll weevil eradication program in the United States allowed full-scale cotton cultivation to continue. Again, cotton replaced corn and soybeans throughout much of the South, and once again dramatically reduced food supplies for the whitetail.

Whitetail populations throughout the Southeast were compromised during the 1980s and 1990s when marginally productive agricultural land was taken out of production and replaced with pine plantations. This practice initially created vast areas of early successional natural foods for whitetails, but ultimately resulted in a marked decrease in available foods due to crown closure within the loblolly pine plantations.

Dam building beginning in the early 1900s inundated large areas of productive bottomland habitat. Riparian corridors had long been the conduit for expanding whitetail populations, and the construction of hydroelectric reservoirs consumed thousands of square kilometers of prime habitat throughout the whitetail's range.

The resilient, adaptable white-tailed deer has weathered these "storms" and many others for centuries. Essentially unfettered from its journey through time, this species has reached a milestone. Management of the whitetail is on the cusp of a new paradigm, an approach divergent of traditional means and one that benefits the species, its habitats, and those who revere the animal. Managed with an emphasis on age structure, sex ratio, and density, the whitetail has gained new prominence as a game animal. There will be future storms; some probably driven by the emotions of nonsportsmen and some related to predation by coyotes or competition from feral swine, but the white-tailed deer has an educated, dedicated, and enthusiastic cadre of supporters to ensure its well-being.

REFERENCES

Adams, K. P., R. J. Hamilton, and M. D. Ross. 2010. *QDMA's Whitetail Report.* Bogart, GA: Quality Deer Management Association.

Brothers, A. and M. E. Ray, Jr. 1975. *Producing Quality Whitetails.* Laredo, TX: Fiesta Publishing Company.

Calhoun, J. and F. Loomis. 1974. *Prairie Whitetails.* Springfield, IL: Illinois Department of Conservation.

Carr, L. 1895. *The Food of Certain American Indians and Their Methods of Preparing it.* Worcester, MA: Charles Hamilton Printer.

Carroll, B. K., R. Jurries, G. Pleasant et al. 2010. *Antler Restrictions in Texas.* PWD RP W7000–1686. Texas Parks and Wildlife. Austin, TX.

Davis, D. E. 2003. *Where There Are Mountains: An Environmental History of the Southern Appalachians.* Athens, GA: University of Georgia Press.

Downing, R. L. 1987. Success story: White-tailed deer. In *Restoring America's Wildlife*, eds. H. Kallman, C. P. Agee, W. R. Gogorth, and J. P. Linduska, 45–57. Washington, DC: U.S. Fish and Wildlife Service.

Duda, M. D., S. J. Bissell, and K. C. Young. 1998. *Wildlife and the American Mind: Public Opinions On and Attitudes Toward Fish and Wildlife Management.* Harrisonburg, VA: Responsive Management.

Earley, L. S. 2004. *Looking for Longleaf.* Chapel Hill, NC: University of North Carolina Press.

Fagerstone, K. A., M. A. Coffey, P. D. Curtis. 2002. Immunocontraceptive vaccines. In *Wildlife Fertility Control*, ed. R. Kirkpatrick, 17–19. Bethesda: The Wildlife Society Technical Review 02-2.

Green, D. and J. P. Stowe, Jr. 2000. Quality deer management: Ethical and social issues. *Human Dimensions of Wildlife* 5:62–71.

Halls, L. K. 1978. White-tailed deer. In *Big Game of North America: Ecology and Management*, eds. J. L. Schmidt and D. L. Gilbert, 43–65. Harrisburg, PA: Stackpole Books.

Hamilton, R. J., W. M. Knox, and L. O. Rogers. 1991. An evaluation of antlerless deer harvest trends in the South Carolina Coastal Plain, 1972–1989. *Paper Presented at the 14th Annual Meeting of the Southeast Deer Study Group*, Baton Rouge.

Harmel, D. E. 1983. Effects of genetics on antler quality and body size in white-tailed deer. In *Antler Development in Cervidae*, ed. R. D. Brown, 339–348. Kingsville, TX: Caeser Kleberg Wildlife Research Institute.

Jenness, D. 1932. *The Indians of Canada.* Bulletin 65. Ottawa, Canada: National Museum of Canada.

Kane, S. and R. Keaton. 1993. *Beneath These Waters: Archaelogical and Historical Studies of 11,500 Years along the Savannah River.* Atlanta, GA: National Park Service.

Killian, G. J., D. Wagner, and L. A. Miller. 2005. Observations on the use of the GNRH vaccine Gonacon™ in male white-tailed deer (*Odocoileus virginianus*). *Proceedings of the 11th Wildlife Damage Management Conference* 11:256–263.

Kroll, J. C. 1994. Genetics and culling. In *A Practical Guide to Producing and Harvesting White-tailed Deer*, ed. J. C. Kroll, 235–249. Nacogdoches, TX: Stephen F. Austin State University.

Lashley, M. A. 2009. Deer forage available following silvicultural treatments in upland hardwood forests and warm-season plantings. MS thesis, University of Tennessee.

Leopold, A. 1933. *Game Management.* Madison, WI: University of Wisconsin Press.

Lewis, C. R. 2009. An 18,000 year old occupation along Petronila Creek in Texas. *Bulletin of the Texas Archeological Society* 80:15–50.

Lockwood, M. A., D. B. Frels, Jr., W. E. Armstrong, E. Fuchs, and D. E. Harmel. 2007. Genetic and environmental interaction in white-tailed deer. *Journal of Wildlife Management* 71:2732–2735.

Lopez, R. R. 2001. Population ecology of Florida Key deer. PhD dissertation, Texas A&M University.

Lukefahr, S. D. and H. A. Jacobson. 1998. Variance component analysis and heritability of antler traits in white-tailed deer. *Journal of Wildlife Management* 62:262–268.

Mann, C. C. 2005. *1491: New Revelations of the Americas before Columbus*. New York, NY: Vintage Books.

McCabe, R. E. 1982. Elk and Indians: Historical values and perspectives. In *Elk of North America: Ecology and Management*, eds. J. W. Thomas and D. E. Toweill, 60–123. Harrisburg, PA: Stackpole Books.

McCabe, R. E. and T. R. McCabe. 1984. Of slings and arrows: An historical retrospection. In *White-tailed Deer Ecology and Management*, ed. L. K. Halls, 19–72. Harrisburg, PA: Stackpole Books.

McCabe, T. R. and R. E. McCabe. 1997. Recounting whitetails past. In *The Science of Overabundance: Deer Ecology and Population Management*, eds. W. J. McShea, H. B. Underwood, and J. H. Rappole, 11–26. Washington, DC: Smithsonian Institution Press.

McCullough, D. R. 1987. The theory and management of *Odocoileus* populations. In *Biology and Management of the Cervidae*, ed. C. M. Wemmer, 535–548. Washington, DC: Smithsonian Institution Press.

McDonald, J. S. and K. V. Miller. 2004. *A History of White-tailed Deer Restocking in the United States 1878 to 2004*. Bogart, GA: Quality Deer Management Association.

Newsom, J. D. 1984. Coastal plain. In *White-tailed Deer Ecology and Management*, ed. L. K. Halls, 367–380. Harrisburg, PA: Stackpole Books.

Novy, J. E. 2009. Screwworm control and eradication in the southern United States of America. FAO Corporate Document Repository. http://www.fao.org/DOCREP/U4220T/U4220T0A.HTM. Accessed November 12, 2009.

Peek, J. M. 1986. *A Review of Wildlife Management*. Englewood Cliffs, NJ: Prentice-Hall.

Pennsylvania Game Commission. 2002. *Pennsylvania Deer Harvests and Road Kills 1915–2001*. Harrisburg, PA: Pennsylvania Game Commission.

Pennsylvania Game Commission. 2009a. http://www.portal.state.pa.us/portal/server.pt/community/deer/11949. Accessed January 31, 2011.

Pennsylvania Game Commission. 2009b. http://www.portal.state.pa.us/portal/server.pt?open=514&objID=625882&mode=2. Accessed January 31, 2011.

Phillips, P. C. 1961. *The Fur Trade*. Norman, OK: University of Oklahoma Press.

Quality Deer Management Association. 2001. Estimated antlered and antlerless deer harvest and percent antlered deer harvest by age class and state in 1989 and 1999. *White-tailed Deer Density Map—1999*. Bogart, GA: Quality Deer Management Association.

Quality Deer Management Association. 2008. Estimated antlered and antlerless deer harvest and percent antlered deer harvest by age class and state in 2005. *White-tailed Deer Density Map—2005*. Bogart, GA: Quality Deer Management Association.

Quality Deer Management Association. 2009. White-tailed deer population size. *Unpublished Data on Estimated Number of White-tailed Deer in the United States*. Bogart, GA: Quality Deer Management Association.

Schaefer, J. and M. B. Main. 1997. *Florida's White-tailed Deer*. SS-WEC-11. Gainesville, FL: University of Florida.

Stewart, O. C. 1951. Burning and natural vegetation in the United States. *Georgia Review* 41:317–320.

Stransky, J. J. 1980. Forage produced by clearcutting and site preparation in East Texas. *Paper Presented at the Third Annual Meeting of the Southeast Deer Study Group*, Nacogdoches, TX.

Strickland, B. K. and S. Demarais. 2008. Influence of landscape composition and structure on antler size of white-tailed deer. *Journal of Wildlife Management* 72:1101–1108.

Strickland, B. K., S. Demarais, L. E. Castle et al. 2001. Effects of selective-harvest strategies on white-tailed deer antler size. *Wildlife Society Bulletin* 29:509–520.

Texas Parks and Wildlife Department. 2009. http://www.tpwd.state.tx.us/huntwild/wild/game_management/deer/antler_restrictions. Accessed November 2, 2009.

Thomas, L. 2004. Rules to hunt by. *Quality Whitetails* 11(2):14–22.

Thomas, L. 2007. Mis-management bucks. *Quality Whitetails* 14(2):58–64.

Trefethen, J. B. 1970. The return of the white-tailed deer. *American Heritage* 21:97–103.

Trefethen, J. B. 1975. *The Wild Sheep in Modern North America*. New York, NY: The Winchester Press.

U.S.D.A. Forest Service. 2008. Inventory and analysis mapmaker 3.0 program. http://fia.fs.fed.us/tools-data/other/default.asp. Accessed December 8, 2008.

U.S. Fish and Wildlife Service. 2006. *2006 National Survey of Fishing, Hunting, and Wildlife-Associated Recreation*. Washington, DC: U.S. Fish and Wildlife Service.

U.S. Fish and Wildlife Service. 2009a. http://www.fws.gov/refuges. Accessed June 2, 2009.

U.S. Fish and Wildlife Service. 2009b. http://www.fws.gov/southeast/federalaid/pittmanrobertson.html. Accessed June 2, 2009.

Warren, R. J. 2000. *Fertility Control in Urban Deer: Questions and Answers*. Gainesville, FL: American Archery Council.

Williams, L. 2008a. Lessons of whitetail conservation: The Key deer. *Quality Whitetails* 15(1):80–86.

Williams, L. 2008b. Lessons of whitetail conservation: The Columbian whitetail. *Quality Whitetails* 15(2):44–51.

Wisconsin Department of Natural Resources. 2009a. http://dnr.wi.gov/org/land/wildlife/HUNT/deer/buckfawn.pdf. Accessed November 12, 2009.

Wisconsin Department of Natural Resources. 2009b. http://dnr.wi.gov/org/land/wildlife/harvest/reports/deerharv08.pdf. Accessed January 31, 2011.

12

Impacts on Ecosystems

Steeve D. Côté

CONTENTS

The abundance and distribution range of white-tailed deer in North America (Tilghman, 1989; Conover, 1997; Waller and Alverson, 1997; Rooney, 2001; Côté et al., 2004), together with other cervid species worldwide (Akashi and Nakashizuka, 1999; Cooke and Farrell, 2001; Martin and Baltzinger, 2002; Darimont et al., 2005), have been increasing rapidly over the last few decades. The situation is particularly alarming in areas where there are no natural predators or where predators have been extirpated (Horsley et al., 2003; Potvin et al., 2003). Overabundant white-tailed deer populations have led to major impacts on preferred plant species (Rooney, 2001; Côté et al., 2004), affecting competitive interactions between plants and favoring some species over others (Huntly, 1991; Côté et al., 2004). Overbrowsing by deer has modified forest succession in many ecosystems, from temperate deciduous forests to boreal forests (Alverson and Waller, 1997; Healy et al., 1997; Horsley et al., 2003; Potvin et al., 2003). Because of the nature of their impact on ecosystems, white-tailed deer have been referred to as ecosystem engineers (Côté et al., 2004; Baiser et al., 2008). By modifying vegetation structure in various habitats, white-tailed deer at high population densities can have numerous direct and indirect effects on other organisms (Huntly, 1991; Rooney and Waller, 2003), such as small mammals (McShea, 2000), birds (deCalesta, 1994; McShea and Rappole, 2000), and invertebrates (Barrett and Stiling, 2007, see also Allombert et al., 2005b). Given that white-tailed deer population numbers are still increasing in North America, mainly due to improved habitat conditions, decreasing hunting pressure, and milder climate, impacts on ecosystems are likely to become increasingly frequent. This chapter addresses and discusses the impacts of overabundant white-tailed deer populations on ecosystems, from individual responses of organisms to community level processes.

Overview of Causes of High Deer Densities

From population lows early in the nineteenth century, many white-tailed deer populations have now increased, often to levels impacting their habitat (Côté et al., 2004; Chapter 11). Local population densities greater than 15 deer/km^2 are now common. There are several reasons possibly explaining the current high abundance of white-tailed deer populations, the most influential being the reduced or extirpated populations of large predators concomitant with decreased hunting pressure, improved habitat conditions with agricultural lands or fallow lands in proximity to forests, and milder climate particularly at northern latitudes (Chapter 1).

Many populations of white-tailed deer, especially in the eastern United States, were released from predator regulation over a century ago, notably with the near extirpation of timber wolves and the extirpation of cougars. In absence of predation, populations increased to high densities and then became controlled by the vegetation. Populations on predator-free islands such as introduced white-tailed deer on Anticosti Island in eastern Canada have also increased dramatically in the 1900s (Côté et al., 2008). Introductions and re-introductions contributed greatly to the recovery of white-tailed deer populations and now to the issues of overabundance. In parallel with the reduction of predator numbers, hunting regulations became stricter at the turn of the century to encourage recovery of deer populations (Diefenbach et al., 1997; McShea et al., 1997a; Brown et al., 2000). In more recent years, however, the problem has reversed and hunting pressure has decreased, mainly through a decrease of interest by people or for social beliefs (Brown et al., 2000; Enck et al., 2000; Riley et al., 2003). Although more than 10 million people in the United States alone hunt white-tailed deer, interest has decreased in recent years, particularly through lower recruitment of young hunters. Despite some effort to increase harvest of female deer in recent years, hunting pressure is still gradually decreasing.

White-tailed deer have a high behavioral phenotypic plasticity (Simard et al., 2008) and can thus live in close proximity to human infrastructure. This characteristic, joined to the increasing abundance of fallow lands in proximity to forest woodlots, created good habitat conditions for white-tailed deer at large spatial scales. Abundant vegetation in abandoned agricultural lands and crops in cultivated lands, as well as silvicultural activities located near forest cover, created suitable habitat with interspersed forage and cover for white-tailed deer (Alverson et al., 1988; Porter and Underwood, 1999; Massé and Côté, 2009). Young planted trees and early successional forested landscapes provide abundant forage. This situation has generated several conflict situations with farmers growing crops or orchards, for example (Chapter 13).

Finally, climate has been warming worldwide, particularly at northern latitudes (Parmesan and Yohe 2003). One immediate consequence of global warming is the more frequent occurrence of mild winters. Winter harshness had been a limiting factor for temperate and northern white-tailed deer populations, but the limiting effect of winter is now decreasing. Because deer populations can be very productive, a series of mild winters can lead to large increases in population sizes (Loison et al., 1999). It is currently unknown whether white-tailed deer populations are now more or less abundant than before European colonization, but the evidence available today suggests that current white-tailed deer numbers are unprecedented (McCabe and McCabe, 1997, Chapter 11).

History of Deer Impacts on Ecosystems

The first evidences of deer impacts on ecosystems appeared in the 1930s (Leopold, 1933). People first noticed impacts on vegetation, particularly damage to trees. One of the first and more convincing experiments performed was to build exclosures to demonstrate the effects of deer browsing (Figure 12.1) (Webb et al., 1956; Stoeckeler et al., 1957). Other studies followed in the 1970s that demonstrated persistent

FIGURE 12.1 A large deer exclosure 3 years after timber harvesting on Anticosti Island, Quebec, Canada. The ground is completely covered by fireweeds. (Photo by S. D. Côté. With permission.)

negative impacts of deer on vegetation and on forest composition and succession (Harlow and Downing, 1970; Anderson and Loucks, 1979; McCullough, 1979). Soon after, impacts on other components of ecosystems such as nutrient cycling and negative effects on other animal species were reported (Casey and Hein, 1983; DeGraaf et al., 1991). The frequency of impacts has gradually increased, so that many examples have now been documented (reviews in McShea et al., 1997a; Russell et al., 2001, Côté et al., 2004).

Impacts on Vegetation

Nutrient Cycling and Energy Transfer

By altering competitive interactions between plant species and modifying plant community composition and forest succession, deer may impact ecosystem processes such as energy transfer, nutrient cycling, and development of soil (Hobbs, 1996; Paine, 2000). Deer at high densities consume a large proportion of the available plant biomass, especially in unproductive systems such as forest understories where vegetation growth rates are generally low (Bråthen and Oksanen, 2001). Deer browsing at high density should thus reduce productivity and nutrient cycling (Côté et al., 2004).

High densities of deer may interfere with nutrient cycling through their impacts on vegetation litterfall. Forage consumption by abundant white-tailed deer decreases the shrub layer in forested ecosystems and thus the contribution of shrub leaves to litterfall (Ritchie et al., 1998). Return of nutrients to the decomposition pathway is thus compromised depending on the intensity of deer browsing. In addition, the effects of deer trampling, rubbing antlers on trees, defecation, and urination can impact ecosystem processes (Côté et al., 2004; Vavra et al., 2007).

Defecation and urination by deer normally increase nitrogen cycling by re-distributing nitrogen on the landscape and thus increasing its availability for plants (Bardgett and Wardle, 2003; Singer and Schoenecker, 2003). The positive effects of feces, however, may be small compared to the negative effects of browsing (Pastor and Naiman, 1992). Furthermore, in areas where nitrogen is a limiting factor, conversion of more productive nitrogen-rich deciduous forests into coniferous forests with plants that have low nitrogen content should decrease nutrient cycling because the quality and availability of litter for decomposers would decline (Pastor and Naiman, 1992; Ritchie et al., 1998; Bardgett and Wardle, 2003). Similarly, replacement of deciduous plants by lower-quality coniferous plants may reduce the rate of energy transfer between vegetation and deer. A reduction in diet quality may increase intake rate in white-tailed deer (Taillon et al., 2006), and thus magnify impacts on vegetation.

Forest Understory

At high density, deer browsing has a direct impact on plants in the forest understory by the removal of a high proportion of plant biomass. Deer at high-density impact both preferred and nonpreferred plant species mainly by affecting competitive relationships between species. By reducing the size and abundance of preferred species, deer often advantage nonpreferred species.

Preferred Plants

Deer browsing at high density impacts the growth, reproduction, and eventually survival of preferred plant species such as certain grasses and sedges, and many forb and shrub species of the understory (Augustine and Frelich, 1998; Ruhren and Handel, 2003; Côté et al., 2004; Mudrak et al., 2009; Thiemann et al., 2009). The initial impact of deer browsing on plants is generally on growth (Kraft et al., 2004). By consuming part of the plant biomass, deer browsing may reduce plant growth. Tolerance to herbivory, however, is highly variable and depends on a variety of factors including species and plant individual characteristics (Canham et al., 1994; Hochwender et al., 2000), the degree of browsing (Saunders and Puettmann, 1999), current and past environmental conditions (Canham et al., 1994; Cronin and Hay, 1996), and composition of the plant community, that is, of competitive interactions between species. Some plant species, however, are tolerant to a certain extent of browsing and browsing may even increase productivity in some cases (Paige and Whitham, 1987; Hobbs, 1996). Certain species, such as grasses, are highly tolerant to browsing because they hide their meristem at or below ground level (Augustine and McNaughton, 1998). In general, slow-growing understory and shade-tolerant herbs are the least tolerant to repeated and heavy deer browsing and are thus more vulnerable to deer browsing (Côté et al., 2004). Forbs flowering in spring and early summer in regions where deer are abundant and few other plants are available may be particularly vulnerable to browsing (Rooney and Waller, 2001; Augustine and deCalesta, 2003; Webster et al., 2005). Finally, the effects of deer browsing on plant abundance may not be direct or linear (e.g., Eschtruth and Battles, 2009; Stewart et al., 2009). Some studies have reported little to moderate impact when a species was common but rapid extirpation when the species was rare (Augustine et al., 1998). The impact of deer browsing appears therefore not to be proportional to deer density and the risk of disappearance of a plant population may thus increase when the population is gradually reduced (Côté et al., 2004).

Many plant species have developed protective responses to browsing. Some species have evolved mechanic responses such as thick leaves, spines, or thorns. For example, a study on resource allocation in the shrub *Damnacanthus indicus* (Rubiaceae) showed that both spine thickness and density were higher in areas with deer browsing than in areas with no deer (Takada et al., 2001). Other plant species show a chemically induced response to deer browsing and produce increasing concentrations of secondary metabolites such as total phenols or terpenes when browsed (Vourc'h et al., 2001). Young leaves of shrubs such as birch or blueberries are often preferred by deer, but are avoided when concentrations of phenolic compounds increase (Manseau et al., 1996; Côté, 1998). If deer density remains high for an extended period of time, however, white-tailed deer can feed on virtually all shrub species. On Anticosti Island, for example, all shrub species are browsed by deer, even species reported to be unpalatable to deer (Pellerin et al., 2006).

In some instances, impacts on plant growth may be negligible or subtle, but reproduction may be affected (Augustine and Frelich, 1998). The number of flowers and eventually fruits on plants has been reported to decrease at high deer density. Deer often select larger individual plants to feed on and plants repeatedly browsed may suffer reduction in growth and reproduction (Anderson, 1994; Rooney and Waller, 2001; Knight, 2003, 2004; Kraft et al., 2004). The proportion of small nonreproductive plants in the populations then increases (see Knight, 2004 for an example with white trillium). For example, on Anticosti Island, the number of flowers and fruits of cloudberry was only 20% of that on Mingan Islands, a group of nearby islands devoid of deer (Pellerin et al., 2006). In a long-term deer exclusion experiment in sugar maple-dominated forests of southern Québec, red trillium plants flowered and fruited more frequently in the absence of deer browsing than in control plots (Collard, 2009). In some cases, individual plants may not produce flowers for several growth seasons after being browsed by deer (Whigham, 1990).

Browsing by deer may extirpate nontolerant species (Rooney and Dress, 1997). The presence and abundance of certain preferred species can thus be used as an indication of deer density. For example, *Clintonia* and *Trillium* have been used as indicator plants of deer density (Balgooyen and Waller, 1995; Augustine and Frelich, 1998). Certain species develop different growth forms when browsed extensively and their terminal lead or apical bud is consumed. Altered growth forms are particularly obvious for shrubs heavily browsed during winter (e.g., yellow birch, mountain maple) where axillary buds produce many branches. The presence and abundance of these different growth forms can also be used to estimate deer density.

Nonpreferred Plants

Deer can also affect nonpreferred plant species in two opposite ways. First, they may directly consume nonpreferred species when palatable preferred species have been extirpated and thus have a negative impact (Côté et al., 2004). Second, and more interestingly, deer browsing can have a positive impact on nonpreferred species. The consumption of preferred species by deer reduces the abundance or size of preferred species and thus their competitive abilities. Reduced competition by species consumed by deer may have a positive impact on growth and reproduction of other less palatable species. For example, unpreferred species such as grasses, thistles, and ferns generally increase at high deer density on Anticosti Island (Casabon and Pothier, 2008; Beguin et al., 2009). In several areas in the United States, forb cover declined at higher deer densities, while grass cover increased (Brooks and Healy, 1988; Rooney, 2009; Mudrak et al., 2009). Studies in Pennsylvania in the United States have also shown that the hay-scented fern was advantaged at high deer density because deer do not consume it but reduce the abundance of its competitor, the thorny shrub Allegheny blackberry (Horsley and Marquis, 1983; George and Bazzaz, 1999). Increased penetration of light in the understory of overbrowsed forests may also, in certain situations, increase plant diversity (Risenhoover and Maass, 1987; Côté et al., 2004).

Invasive Species

Many studies have shown that heavy deer browsing was correlated with declines in native species and increases in exotic plant abundance (Rooney and Dress, 1997; Rooney et al., 2002; Wiegmann and Waller, 2006; Eschtruth and Battles, 2009; Knight et al., 2009). Similar to the situation with nonpreferred species, sustained heavy deer browsing on native preferred plants may favor establishment and propagation of invasive and exotic plant species (Vavra et al., 2007). Invasives are often highly competitive species that may proliferate because of the reduced competition induced by deer browsing on preferred plants (Eschtruth and Battles, 2009). In some situations, heavy deer browsing may thus favor propagation of exotic weeds if those species are avoided by deer (Vavra et al., 2007). Exotic plants are often avoided by herbivores because they are unpalatable or are not detected by herbivores because they are absent from their traditional diet (Lankau et al., 2004). In secondary forests, exotic plants may form extensive monocultures and replace a diverse understory layer of herbs and tree saplings (e.g., Gorchov and Trisel, 2003; Stinson et al., 2006). Deer are also known to consume the inflorescences and seeds of some noxious exotic plants (Rachich and Reader, 1999), and thus exotic plants can comprise over 80% of the seedlings germinating from deer fecal pellets (Myers et al., 2004). A mean of greater than 30 seeds germinated per fecal pellet group in a study conducted in central New York State (Myers et al., 2004). In addition, it was estimated that a deer herd in southern Connecticut containing approximately 23 deer/km^2 had the potential to disperse 390–1046 viable exotic seeds/day/km^2 (Williams and Ward, 2006). Thus, deer could be a major dispersal agent for introduced plants through both endozoochory and epizoochory (Vellend, 2002; Williams and Ward, 2006; Vavra et al., 2007; Williams et al., 2008).

Alternatively, in some situations, deer browsing on invasive species may limit deer impacts on native forest herbs and at the same time reduce progression of invasive species (Parker et al., 2006; Rossell et al., 2007). The role of deer browsing on propagation of invasive plant species, however, remains little understood and deserves more attention.

Tree Regeneration

Trees are more vulnerable to deer browsing during their young life-stages than when adults. For example, deer consume seeds, seedlings, and saplings. Several studies have reported strong deer impacts on growth and density of seedlings from both deciduous (Healy et al., 1997; Engelmark et al., 1998; Horsley et al., 2003) and coniferous species (Frelich and Lorimer, 1985; Anderson and Katz, 1993; Alverson and Waller, 1997; Rooney et al., 2000, 2002; Potvin et al., 2003). In many forests, the shrub layer is open and the density of saplings is low. Deer normally prefer seedlings and saplings of deciduous trees and increased tree harvesting intensity has been proposed to reduce the overall browsing pressure in deciduous forests and ensure regeneration (Kalen, 2005). Mean browsing rates decline when the amount of early successional habitat increases (Miller et al., 2009). In West Virginia, a recent study indicated that when 14% of the area was composed of stands younger than 10 years old, browsing rates declined below 5% in both regeneration sites and adjacent mature forests (Miller et al., 2009). The presence of highly preferred species may act as a buffer to reduce the browsing pressure on deciduous tree species. For example, the high availability of blackberry in West Virginia protected tree saplings from heavy browsing by white-tailed deer by providing preferred forage that reduced dependence on deciduous tree species (Miller et al., 2009). While providing abundant browse by forest management appears a viable option to decrease herbivory, production of increased forage may augment the carrying capacity of stands and stimulate deer population growth, thus possibly exacerbating future problems associated with deer browsing (Kramer et al., 2006; Miller et al., 2009).

The impact of deer browsing may last for long periods and it may take as much as 70 years at low deer density for slow-growing species such as hemlock to recover (Anderson and Katz, 1993). Other species, however, such as balsam fir may recover more rapidly when deer density is reduced (Tremblay et al., 2007). In general, coniferous species may be more vulnerable to deer browsing than deciduous ones because they allocate a greater proportion of their resources to leaves and do not retranslocate nutrients to stems and roots to the same extent as deciduous species (Ammer, 1996; Côté et al., 2004).

Woody species also produce constitutive or induced chemical and physical defenses against deer browsing. Generally, defenses increase with the extent of deer browsing, but often decrease when plants grow beyond the reach of deer (Bryant and Raffa, 1995). In certain situations, seedlings and saplings growing on islands without deer have been reported to evolve lower concentrations of secondary plant metabolites than those growing at high deer density (Vourc'h et al., 2001). However, tree species are likely to lose the evolutionary arms race with deer in situations where large reproducing trees are not subjected to deer browsing and thus produce seedlings that may not be adapted to heavy browsing. This situation is likely common in North America because deer density has mainly increased in the last 50 years while most reproducing trees present today became established more than 50 years ago, and are thus beyond the reach of deer and their selection effect.

Forest Succession

In classical models of forest succession, the effects of deer browsing on plant density are proportional to deer density. In slightly more complex models, a threshold exists beyond which sustained deer browsing induces a rapid decrease in plant abundance (Scheffer et al., 2001). The common point of these models is that the effects of deer browsing are reversible. A third model exists in which the effects of deer browsing are not readily reversible when the browsing pressure is relaxed and the system is thus maintained in an alternate stable state even at low deer density (Scheffer et al., 2001; Côté et al., 2004). At this point, a long period at low deer density and interventions (e.g., silvicultural treatments) favoring the return of preferred species are needed for the forest to recover (Scheffer et al., 2001).

Similarly to plants from the understory, deer browsing on preferred tree seedlings may reduce abundance of dominant species locally and advantage competitive tree species. For example, in an old-growth beech–maple forest in Pennsylvania, browse tolerant and nonpreferred species (e.g., beech, black cherry, and sugar maple) increased in abundance at the expense of preferred browse species such as oak, ash, and red maple in a deer-exclusion experiment (Long et al., 2007). When deer browsing is sustained, unpreferred species receive a competitive advantage because deer reduce the abundance of preferred species, and the disturbance regime of the forest can be greatly modified (Huntly, 1991; Hobbs, 1996).

The normally subordinate species may then become dominant and shift forest succession to other alternate states (Côté et al., 2004). Deer can accelerate forest succession if they impact species dominating early successional stages or favor establishment of plants from late seral stages (Hobbs, 1996; Seagle and Liang, 2001). Alternatively, deer browsing can slow succession if sustained deer browsing negatively affects colonization and survival of species from late successional stages (Hobbs, 1996; Ritchie et al., 1998).

Several examples exist of compositional shifts in dominant tree species induced by deer browsing in temperate and boreal forests. A classic example is the replacement of eastern hemlock by sugar maple in many temperate forests of the Midwest induced by sustained heavy deer browsing (Anderson and Loucks, 1979; Frelich and Lorimer, 1985; Alverson and Waller, 1997; Rooney et al., 2000). A similar situation is occurring in Pennsylvania forests where mixed hardwoods are gradually replaced by black cherry (Tilghman, 1989; Horsley et al., 2003). These dynamics, however, are rarely linear as shown in several ecosystems, including temperate and boreal forests (Pastor et al., 1993; Augustine et al., 1998).

As discussed earlier, heavy deer browsing favors establishment and growth of the hay-scented fern because deer avoid it. Once the fern is established, however, a reduction of deer browsing by controlled management or other factors may not be sufficient to return to the original state in which Allegheny blackberry or other species colonize openings. In this situation, deer browsing would have shifted the forest understory to an alternate stable state that cannot be colonized by the original dominant plant species of the understory when browsing is relaxed (Stromayer and Warren, 1997). A similar situation may also be occurring on Anticosti Island where balsam fir forests are quickly disappearing due to heavy browsing by white-tailed deer (Potvin et al., 2003; Côté et al., 2008). Deer browse the seedlings so that most do not survive more than four years after a perturbation increases light in the understory (Potvin et al., 2003; Casabon and Pothier, 2007). Fir browse has almost disappeared on the island (Tremblay et al., 2005) and fir stands are being gradually replaced by white spruce stands. After a perturbation such as a cutblock, the ground layer is quickly covered by grasses and thistles and the only tree species regenerating is white spruce (Casabon and Pothier, 2008). Thus, heavy deer browsing leads to open parkland forests dominated by white spruce that may represent an alternate stable state where balsam fir regeneration may not be possible even when deer density is reduced (Figure 12.2). Other examples of new savanna-type ecosystems replacing oak forests have also been reported (Healy et al., 1997).

Hard Mast Seeds

Mast seeds are a critical resource for many animals, especially rodents and even black bears (Otsfeld et al., 1996). The impacts of deer browsing in the shrub layer may greatly reduce the abundance of mast seeds. Species producing hard mast seeds such as oak and beech are often targeted by deer, particularly

FIGURE 12.2 Open parkland forest dominated by white spruce and grasses on Anticosti Island, Quebec, Canada. (Photo by S. D. Côté. With permission.)

during winter. At high density, deer browsing reduces abundance of deciduous shrubs of mast-producing species and eventually recruitment of saplings into adult reproducing trees. As explained in earlier sections, deer browsing could induce compositional shifts in the dominant species of the forest and initiate replacement of mast seed species by species that do not produce hard mast seeds. For example, Healy et al. (1997) have shown that certain oak forests were replaced by savanna-type systems at high deer density in northeastern United States.

Impacts on Other Animals

By foraging selectively, deer affect the growth and survival of many herb, shrub, and tree species, thereby modifying patterns of relative plant abundance and vegetation dynamics. Deer also exert cascading effects on other animals both by competing directly for resources with other herbivores and omnivores and by indirectly modifying the composition and physical structure of habitats of both invertebrates and vertebrates (Van Wieren, 1998; Côté et al., 2004; Suominen and Danell, 2006). By modifying species abundance and diversity, deer populations at high density can modify trophic interactions among species (Ostfeld et al., 1996; McShea and Rappole, 2000; Smit et al., 2001). Effects on interactions within the food web may be particularly important in ecosystems where several species of herbivores and omnivores co-habit such as in western and eastern North America.

At low density, deer browsing has a low impact on vegetation and other organisms. On the other hand, at high density, deer impact can be considerable. At intermediate density, however, deer browsing may increase vegetation productivity and habitat structure so that a greater abundance and diversity of both plants and animals can occur. This phenomenon has been termed the intermediate disturbance hypothesis (Grime, 1973). According to this hypothesis, we should expect positive effects of deer on other organisms in the ecosystems at intermediate density. Maximum diversity within a forest stand often appears to occur at intermediate deer density (deCalesta and Stout, 1997; van Wieren, 1998; Fuller, 2001; Rooney and Waller, 2003; Suominen et al., 2003).

Invertebrates

Deer browsing modifies the composition and structure of plant communities. By affecting the growth form and the relative proportions of the different plant species available, deer structure the habitat for other species, particularly invertebrates (Stewart, 2001). Many invertebrates depend on vegetation cover and several species of invertebrates (e.g., pollinators) are associated with particular plant species, so that the impact of deer density on their ecology may be considerable (Miyashita et al., 2004). For example, high deer density may disrupt associations between pollinators and plants by modifying the relative abundance of flowers from different species (Vázquez and Simberloff, 2003). In affecting both the rate and nature of return of dead material to the ground, deer may also affect the decomposition cycle in the forest and the whole suite of invertebrates involved in decomposition of plant material.

There has been a plethora of studies on the relationships between ungulate density and invertebrates in the last decade but surprisingly few in North America with white-tailed deer (see Côté et al., 2004 for a review; Barrett and Stiling, 2007). For example, several studies on moose and reindeer have been conducted in Europe (Danell and Huss-Danell, 1985; Suominen et al., 1999a,b, 2003, 2008). The results are mixed; many studies found deer browsing had a negative impact on the abundance and diversity of invertebrates (Baines et al., 1994; Bailey and Whitham, 2002; den Herder et al., 2004; Miyashita et al., 2004; Suominen et al., 2008). Other studies, however, did not detect any clear impact of deer density on forest invertebrates (Brooks, 1999; Suominen et al., 1999b; Barrett and Stiling, 2007). Some studies supported the intermediate disturbance hypothesis with higher density and diversity of insects at intermediate deer density (Danell and Huss-Danell, 1985; Suominen et al., 1999a, 2003). Finally, some authors have reported positive effects of deer herbivory on the abundance of invertebrates (Suominen et al., 1999b, 2003; Greenwald et al., 2008). In addition, deer herbivory had a positive indirect effect on the abundance of other species such as red-backed salamanders, garter snakes, and gastropods in a National Park in Ohio (Greenwald et al., 2008).

The effect of herbivore population density on forest invertebrates may also depend on site productivity. For example, the response of spider richness to moose density was positive at productive sites and negative at unproductive sites in a controlled browsing experiment conducted in northern Sweden (Suominen et al., 2008). There is still much work to be conducted to assess the effects of white-tailed deer on forest invertebrates in North America; studies on Anticosti Island, for instance, are now being conducted. In another island ecosystem, in Haida Gwaii on the west coast of British Columbia (Canada), Allombert et al. (2005b) found that the abundance of forest insects decreased to less than 15% of that originally present following introduction of black-tailed deer.

Avian Communities

The ecology and life-history of forest birds are strongly related to the vertical structure of the forest. By affecting the structure and cover of different vegetation strata in the forest, particularly that of the forest floor, deer at high density may affect the abundance and diversity of forest bird communities (McShea and Rappole, 1997). For example, the abundance of small birds dependent on the shrub layer was 93% lower on islands where black-tailed deer have been introduced than on islands free of deer on Haida Gwaii (Allombert et al., 2005a). A few studies on the impact of white-tailed deer browsing on species richness and abundance of songbirds have been conducted, and most have found negative effects of deer (Côté et al., 2004). In a controlled grazing experiment in the northern hardwoods of Pennsylvania that included four simulated deer densities, deCalesta (1994) reported nonlinear declines of 27% and 37% in species richness and abundance, respectively, for intermediate canopy nesters between the lowest and highest deer densities. He found no effect of deer density on ground and canopy nesters, however. Surprisingly, DeGraaf et al. (1991) found no relation between deer density and the abundance and diversity of ground-feeding species, but lower abundance and diversity of canopy feeders at high deer density. It thus seems that deer browsing can also impact canopy species and not just species of the shrub layer. Using an exclosure experiment in oak-dominated mixed hardwoods of Virginia, McShea and Rappole (2000) reported an increase in the abundance of ground nesting species and intermediate canopy nesters after deer exclusion, but no effect on bird diversity because of species replacement.

Small Mammals

The ecology of many small mammals is tightly tied to plant cover in the understory. By interfering with the shrub layer and decreasing plant cover and also plant biomass, deer browsing may affect small mammals by reducing both their food supply and protective cover (Smit et al., 2001). There have been comparatively fewer studies of deer impacts conducted on small mammals than on invertebrates and birds. In the first study relating white-tailed deer density to the abundance of small mammals, Brooks and Healy (1988) reported that red-backed voles in Massachusetts were much more abundant in stands of low deer density than at high deer density, but white-footed mice were almost twice as abundant in stands of high deer density as in stands of low deer density. Abundance of short-tailed shrews also declined with increasing deer density (Brooks and Healy, 1988). McShea (2000) studied small mammal abundance in relationship with deer density in oak-dominated mixed hardwoods of Virginia in the United States. He found higher eastern chipmunk and white-footed mouse abundance in deer exclosures than under natural deer density, but there was an interaction with the previous acorn crop (see below). In an experimental study in Oregon, Moser and Witmer (2000) found higher abundance, species richness, and diversity of small mammals in ungrazed sites than in sites grazed by elk, but no effects for birds. In another experimental study controlling white-tailed deer density on Anticosti Island (Québec, Canada), preliminary results indicate deer mouse abundance is not affected by deer density ranging from 0 to >55 deer/km^2 (Côté et al., unpublished data).

Other Mast Consumers

By reducing the deciduous shrub layer in the forest, deer browsing at high density inevitably reduces recruitment of species producing mast seeds. Mast seeds and fruits are rich in lipids or soluble carbohydrates and represent a high-energy resource for animals. Seeds are the staple food of many rodent

FIGURE 12.3 The last black bear observed on Anticosti Island (Quebec, Canada) in June 1996. Note the low body condition of the animal. (Photo by J.-F. Boudreau. With permission.)

species in various ecosystems (Otsfeld et al., 1996). If this food source is reduced or disappears, the impacts on rodent populations may be major. Some specialist species may even become extirpated. There is, however, little research that has addressed these questions although there is a growing need to better understand the importance and complexity of habitat modifications induced by deer browsing on other organisms in the forest. In a study on eastern chipmunk and white-footed mouse, McShea (2000) found a higher abundance of both species when deer were excluded in low-mast years, but no difference occurred in good-mast years.

Bears in deciduous forests also rely on mast seeds as a food source during the autumn hyperphagy period to build up body reserves for the winter (Rogers, 1976). In Maine, black bear females entered their den six weeks earlier on average during low mast years because of food shortages (Schooley et al., 1994). In another study, the risk of mortality during the hunting season for females was twice higher during years of low oak and hazelnut production than during years of high mast productivity (Noyce and Garshelis, 1997). The production of mast seeds, therefore, has many management and conservation implications for black bears.

In the boreal forest of Anticosti Island, browsing by white-tailed deer is thought to have led to the extirpation of the black bear population from the island within 50–70 years of deer introduction (Figure 12.3) (Côté, 2005). Black bears were abundant and widely distributed on the large island of nearly 8000 km^2 and relied mainly on berries during autumn to accumulate body reserves for the winter because there was virtually no hard mast species present on the island (Côté et al., 2008). Introduced deer likely increased to over 50,000 individuals within 30 years of release and deer browsing almost eliminated the shrub layer (Tremblay et al., 2005). It was estimated that 0.28 berry/m^2 on average were available on the island in the early 2000s, which was about 235 times lower than the threshold necessary for black bears to maintain body mass (Pelchat and Ruff, 1986; Côté, 2005). Alternative explanations for the decline of black bears were examined, but in the absence of other convincing evidence, it was concluded that white-tailed deer likely extirpated black bears by exploitative competition.

Impacts of Disease Transmission

White-tailed deer are opportunist and behaviorally plastic animals that can occur in several types of habitats, from hardwood and mixed wood forest interiors to agro-forested landscapes. In many regions, deer have become accustomed to humans and urban and semiurban white-tailed deer populations are now common throughout the United States and Canada. Through their proximity with human settlements, white-tailed deer at high density could be the transmission vector of zoonoses and other diseases (Côté et al., 2004). For example, deer at high density can transmit bovine tuberculosis

(*Mycobacterium bovis*), although its incidence has so far remained relatively rare in North America, except perhaps in Michigan where concerns have been higher (Schmitt et al., 1997; Miller et al., 2003). Deer are also known to increase the transmission of tick-borne zoonoses by favoring the maintenance of abundant tick (*Ixodes* spp.) populations (Ostfeld et al., 1996; Wilson and Childs, 1997). The most important and well-known tick-borne disease associated with high densities of white-tailed deer in North America is Lyme disease (Telford III, 2002). The incidence of Lyme disease has been reported to be related to deer density in eastern United States (Telford III, 2002; Wilson et al., 1990). White-tailed deer can serve as an alternative host to cattle fever ticks (*Rhipicephalus* [*Boophilus*] spp.) which can transmit *Babesia* spp. parasites and cause babesiosis in cattle (Chapter 7). A massive effort by the U.S. Department of Agriculture to eradicate cattle fever ticks from the United States has been complicated in Florida during the 1930s and in southern Texas during the past two decades because of white-tailed deer maintaining populations of cattle fever ticks (Pound et al., 2010). White-tailed deer could also transmit chronic wasting disease, a spongiform encephalopathy, probably originating from farm-raised animals (Williams et al., 2002). Chronic wasting disease can be transmitted within and among cervid species (Gross and Miller, 2001), but transmission to humans appears unlikely (Raymond et al., 2000).

Impacts of Management to Increase or Maintain Deer Densities

The control of deer densities has interested biologists and wildlife managers for over a century (Leopold, 1933). Several studies have focused on factors limiting and regulating deer populations (McCullough, 1979; Sæther, 1997; Dumont et al., 2000; Riley et al., 2003) and, in parallel to these studies, a whole suite of management tools have been developed for increasing or maintaining deer densities (Porter and Underwood, 1999; Côté et al., 2004). As of today, however, much less effort has been put in developing management strategies to reduce overabundant deer populations (Côté et al., 2004).

Several management practices developed for increasing or maintaining deer densities can accentuate deer impacts on vegetation and other organisms. First, if successful, management practices such as predator control or more restrictive hunting regulations may allow deer densities to increase dramatically and thus accentuate the direct impact of deer on ecosystems (Côté et al., 2004). Predator control has been used extensively in North America to increase populations of ungulates (Boertje et al., 1996; Valkenburg et al., 2004). Although the role of predation in regulating deer populations remains uncertain (e.g., Peterson, 1999), deer will likely benefit from lowered densities of predators, even in situations where the practice is targeting another species of ungulate. The loss of predators in an ecosystem may allow a dramatic increase in large herbivores and flip the system to an irreversible alternate state (Côté et al., 2004). Reduced hunting pressure either through more restrictive hunting regulations or a lower number of hunters (Enck et al., 2000) may also allow deer populations to irrupt to very high densities and thus exacerbate their impact on ecosystems.

Second, practices such as supplemental feeding or deer proof fences may alter the foraging behavior and space use of deer and increase their impact locally (Murden and Risenhoover, 1993; Bartoskewitz et al., 2003; Timmons et al., 2010). Supplemental feeding, for instance, concentrates deer in certain areas where resources are heavily utilized and their impact thus increases (Figure 12.4) (Cooper et al., 2006). When supplemented, deer can avoid low-quality forages and selectively consume remaining palatable plant species (Murden and Risenhoover, 1993). Deer populations are thus maintained at densities artificially higher than the carrying capacity of ecosystems and negative impacts on the habitat may occur. Similarly, deer proof fences concentrate animals in certain areas and may also help maintain high deer densities by decreasing road-kill accidents. As long as these techniques will be employed, we can expect increasing impacts of high deer density on ecosystems, at least locally.

Some solutions to reduce deer density such as more liberal hunting regulations, localized management, fertility-control programs, and even sharp-shooting in semiurban areas have been proposed and used effectively in some instances (McNulty et al., 1997; McShea et al., 1997b; Campbell et al., 2004; Côté et al., 2004). Most techniques, however, are costly and only effective at small spatial and temporal scales. The most efficient technique applicable at large scale remains sport hunting, but the decreasing

FIGURE 12.4 Emergency supplementary feeding of white-tailed deer in southern Quebec, Canada. (Photo by R. Charest. With permission.)

population of hunters makes it a real challenge to use this management strategy throughout the range of overabundant white-tailed deer (Brown et al., 2000; Giles and Findlay, 2004).

Conclusion

Deer browsing at high density can impact growth, reproduction, and survival of many herbs, shrubs, and tree species. Deer have the capacity to modify competitive relationships between plant species and thus their relative abundance and dynamics. By affecting forest composition, deer browsing also alters nutrient cycling and eventually forest dynamics and succession, sometimes leading to alternative stable states that appear difficult to reverse. These states often also include cascading effects on other invertebrates and vertebrates in the ecosystems.

A major gap in our knowledge of deer impacts on ecosystems is a better understanding of the effect of density. Deer impacts may be linear but more likely increase at a certain threshold density (Côté et al., 2004). The threshold density allowing the maintenance of a particular ecosystem and its properties may vary depending on the productivity and biodiversity of the system. Scientists urgently need to identify the different deer threshold densities in various ecosystems as well as the duration of the recovery periods. One of the most powerful approaches to achieve this goal is the use of controlled browsing experiments in different ecosystems where density is experimentally manipulated in large enclosures in natural habitat and the responses of plants and other organisms are followed through long time periods (Figure 12.5) (Hester et al., 2000; Côté et al., 2008). This approach has already started to produce very robust results amenable to management (Horsley et al., 2003; Tremblay et al., 2007).

Once the deer density (and its uncertainty) compatible with forest regeneration and maintenance of all other ecosystem processes is known, the next step is to move from traditional wildlife management to ecosystem management (Christensen et al., 1996). If the goal is to reduce the impact of deer on ecosystems, we must consider white-tailed deer in relation to all other ecosystem components and include in management decisions the responses of these ecosystem constituents (including predators) to different deer densities (Tester et al., 1997). Because deer are "ecosystem engineers," their effects are far-reaching and likely include impacts that we have not yet been able to identify and monitor. Once a target deer density is agreed on, one of the biggest challenges for managers and conservationists would be to convince hunters of the need to reduce deer density (Diefenbach et al., 1997). Many hunting communities have been reluctant to new programs aimed at reducing deer density such as antlerless deer hunting (Riley et al., 2003). A deep modification in hunting philosophy emphasizing the ecological role of hunters may

FIGURE 12.5 Vegetation responses after a 7-year controlled browsing experiment in the boreal forest of Anticosti Island, Quebec, Canada. (Photo by É. Cardinal. With permission.)

be required to answer the problem of deer impacts on ecosystems in the context of ecosystem management because hunting still remains the best tool to manage wildlife populations.

Acknowledgments

My research on white-tailed deer is funded by the Natural Sciences and Engineering Research Council of Canada (NSERC)-Produits forestiers Anticosti Industrial Research Chair, Université Laval, and the Ministère des Ressources naturelles et de la Faune du Québec. I thank D. Hewitt for comments on a previous draft of the chapter, C. Hins for help with the literature review, and S. de Bellefeuille for formatting the references.

REFERENCES

Akashi, N. and T. Nakashizuka. 1999. Effects of bark-stripping by sika deer (*Cervus nippon*) on population dynamics of a mixed forest in Japan. *Forest Ecology and Management* 113:75–82.

Allombert, S., A. J. Gaston, and J.-L. Martin. 2005a. A natural experiment on the impact of overabundant deer on songbird populations. *Biological Conservation* 126:1–13.

Allombert, S., S. Stockton, and J.-L. Martin. 2005b. A natural experiment on the impact of overabundant deer on forest invertebrates. *Conservation Biology* 19:1917–1929.

Alverson, W. S. and D. M. Waller. 1997. Deer populations and the widespread failure of hemlock regeneration in northern forests. In *The Science of Overabundance: Deer Ecology and Population Management*, eds. W. J. McShea, H. B. Underwood, and J. H. Rappole, 280–297. Washington, D.C.: Smithsonian Institution Press.

Alverson, W. S., D. M. Waller, and S. L. Solheim. 1988. Forests too deer: Edge effects in northern Wisconsin. *Conservation Biology* 2:348–358.

Ammer, C. 1996. Impact of ungulates on structure and dynamics of natural regeneration of mixed mountain forests in the Bavarian Alps. *Forest Ecology and Management* 88:43–53.

Anderson, R. C. 1994. Height of white-flowered trillium (*Trillium grandiflorum*) as an index of deer browsing intensity. *Ecological Applications* 4:104–109.

Anderson, R. C. and A. J. Katz. 1993. Recovery of browse-sensitive tree species following release from white-tailed deer *Odocoileus virginianus* Zimmerman browsing pressure. *Biological Conservation* 63:203–208.

Anderson, R. C. and O. L. Loucks. 1979. White-tail deer (*Odocoileus virginianus*) influence on structure and composition of *Tsuga canadensis* forests. *Journal of Applied Ecology* 16:855–861.

Augustine, D. J. and D. deCalesta. 2003. Defining deer overabundance and threats to forest communities: From individual plants to landscape structure. *Écoscience* 10:472–486.

Augustine, D. J. and L. E. Frelich. 1998. Effects of white-tailed deer on populations of an understory forb in fragmented deciduous forests. *Conservation Biology* 12:995–1004.

Augustine, D. J., L. E. Frelich, and P. A. Jordan. 1998. Evidence for two alternate stable states in an ungulate grazing system. *Ecological Applications* 8:1260–1269.

Augustine, D. J. and S. J. McNaughton. 1998. Ungulate effects on the functional species composition of plant communities: Herbivore selectivity and plant tolerance. *Journal of Wildlife Management* 62:1165–1183.

Bailey, J. K. and T. G. Whitham. 2002. Interactions among fire, aspen, and elk affect insect diversity: Reversal of a community response. *Ecology* 83:1701–1712.

Baines, D., R. B. Sage, and M. M. Baines. 1994. The implications of red deer grazing to ground vegetation and invertebrate communities of Scottish native pinewoods. *Journal of Applied Ecology* 31:776–783.

Baiser, B., J. L. Lockwood, D. La Puma, and M. F. J. Aronson. 2008. A perfect storm: Two ecosystem engineers interact to degrade deciduous forests of New Jersey. *Biological Invasions* 10:785–795.

Balgooyen, C. P. and D. M. Waller. 1995. The use of *Clintonia borealis* and other indicators to gauge impacts of white-tailed deer on plant communities in northern Wisconsin, USA. *Natural Areas Journal* 15:308–318.

Bardgett, R. D. and D. A. Wardle. 2003. Herbivore-mediated linkages between aboveground and belowground communities. *Ecology* 84:2258–2268.

Barrett, M. A. and P. Stiling. 2007. Relationships among Key deer, insect herbivores, and plant quality. *Ecological Research* 22:268–273.

Bartoskewitz, M. L., D. G. Hewitt, J. S. Pitts, and F. C. Bryant. 2003. Supplemental feed use by free-ranging white-tailed deer in southern Texas. *Wildlife Society Bulletin* 31:1218–1228.

Beguin, J., D. Pothier, and M. Prévost. 2009. Can the impact of deer browsing on tree regeneration be mitigated by shelterwood cutting and strip clearcutting. *Forest Ecology and Management* 257:38–45.

Boertje, R. D., P. Valkenburg, and M. E. McNay. 1996. Increases in moose, caribou, and wolves following wolf control in Alaska. *Journal of Wildlife Management* 60:474–489.

Bråthen, K. A. and J. Oksanen. 2001. Reindeer reduce biomass of preferred plant species. *Journal of Vegetation Science* 12:473–480.

Brooks, R. T. 1999. Residual effects of thinning and high white-tailed deer densities on northern redback salamanders in southern New England oak forests. *Journal of Wildlife Management* 63:1172–1180.

Brooks, R. T. and W. M. Healy. 1988. Response of small mammal communities to silvicultural treatments in eastern hardwood forests of West Virginia and Massachusetts. In *Management of Amphibians, Reptiles, and Small Mammals in North America*, eds. R. C. Szaro, K. E. Severson, and D. R. Patton, 313–318. Fort Collins, CO: USDA Forest Service.

Brown, T. L., D. J. Decker, S. J. Riley et al. 2000. The future of hunting as a mechanism to control white-tailed deer populations. *Wildlife Society Bulletin* 28:797–807.

Bryant, J. P. and K. F. Raffa. 1995. Chemical antiherbivore defense. In *Plant Stems: Physiology and Functional Morphology*, ed. B. L. Gartner, 365–381. San Diego, CA: Academic Press.

Campbell, D., G. M. Swanson, and J. Sales. 2004. Comparing the precision and cost-effectiveness of faecal pellet group count methods. *Journal of Applied Ecology* 41:1185–1196.

Canham, C. D., J. B. McAninch, and D. M. Wood. 1994. Effects of the frequency, timing, and intensity of simulated browsing on growth and mortality of tree seedlings. *Canadian Journal of Forest Research* 24:817–825.

Casabon, C. and D. Pothier. 2007. Browsing of tree regeneration by white-tailed deer in large clearcuts on Anticosti Island, Quebec. *Forest Ecology and Management* 253:112–119.
Casabon, C. and D. Pothier. 2008. Impact of deer browsing on plant communities in cutover sites on Anticosti Island. *Écoscience* 15:389–397.
Casey, D. and D. Hein. 1983. Effects of heavy browsing on a bird community in deciduous forest. *Journal of Wildlife Management* 47:829–836.
Christensen, N. L., A. M. Bartuska, J. H. Brown et al. 1996. The report of the Ecological Society of America Committee on the scientific basis for ecosystem management. *Ecological Applications* 6:665–691.
Collard, A. 2009. Réponses des plantes de sous-bois au retrait expérimental du cerf de Virginie dans les forêts du sud du Québec. MSc thesis, Université Laval.
Conover, M. R. 1997. Monetary and intangible valuation of deer in the United States. *Wildlife Society Bulletin* 25:298–305.
Cooke, A. S. and L. Farrell. 2001. Impact of muntjac deer (*Muntiacus reevesi*) at Monks Wood National Nature Reserve, Cambridgeshire, eastern England. *Forestry* 74:241–250.
Cooper, S. M., M. K. Owens, R. M. Cooper, and T. F. Ginnett. 2006. Effect of supplemental feeding on spatial distribution and browse utilization by white-tailed deer in semi-arid rangeland. *Journal of Arid Environments* 66:716–726.
Côté, S. D. 1998. *In vitro* digestibilities of summer forages utilized by the Rivière George caribou herd. *Arctic* 51:48–54.
Côté, S. D. 2005. Extirpation of a large black bear population by introduced white-tailed deer. *Conservation Biology* 19:1668–1671.
Côté, S. D., C. Dussault, J. Huot, F. Potvin, J.-P. Tremblay, and V. Viera. 2008. High herbivore density and boreal forest ecology: White-tailed deer on Anticosti Island. In *Lessons from the Islands: Introduced Species and what they Tell us about How Ecosystems Work. Proceedings from the Research Group on Introduced Species 2002 Symposium,* Queen Charlotte City, Queen Charlotte Islands, British Columbia, eds. A. J. Gaston, T. E. Golumbia, J. L. Martin and, S. T. Sharpe, 154–161. Ottawa, Canada: Canadian Wildlife Service, Environment Canada.
Côté, S. D., T. P. Rooney, J. P. Tremblay, C. Dussault, and D. M. Waller. 2004. Ecological impacts of deer overabundance. *Annual Review of Ecology, Evolution, and Systematics* 35:113–147.
Cronin, G. and M. E. Hay. 1996. Susceptibility to herbivores depends on recent history of both the plant and animal. *Ecology* 77:1531–1543.
Danell, K. and K. Huss-Danell. 1985. Feeding by insects and hares on birches earlier affected by moose browsing. *Oikos* 44: 75–81.
Darimont, C. T., P. C. Paquet, T. E. Reimchen, and V. Crichton. 2005. Range expansion by moose into coastal temperate rainforests of British Columbia, Canada. *Diversity and Distributions* 11:235–239.
deCalesta, D. S. 1994. Effect of white-tailed deer on songbirds within managed forests in Pennsylvania. *Journal of Wildlife Management* 58:711–718.
deCalesta, D. S. and S. L. Stout. 1997. Relative deer density and sustainability: A conceptual framework for integrating deer management with ecosystem management. *Wildlife Society Bulletin* 25:252–258.
DeGraaf, R. M., W. M. Healy, and R. T. Brooks. 1991. Effects of thinning and deer browsing on breeding birds in New England oak woodlands. *Forest Ecology and Management* 41:179–191.
den Herder, M., R. Virtanen, and H. Roininen. 2004. Effects of reindeer browsing on tundra willow and its associated insect herbivores. *Journal of Applied Ecology* 41:870–879.
Diefenbach, D. R., W. L. Palmer, and W. K. Shope. 1997. Attitudes of Pennsylvania sportsmen towards managing white-tailed deer to protect the ecological integrity of forests. *Wildlife Society Bulletin* 25:244–251.
Dumont, A., M. Crête, J.-P. Ouellet, J. Huot, and J. Lamoureux. 2000. Population dynamics of northern white-tailed deer during mild winters: Evidence of regulation by food competition. *Canadian Journal of Zoology* 78:764–776.
Enck, J. W., D. J. Decker, and T. L. Brown. 2000. Status of hunter recruitment and retention in the United States. *Wildlife Society Bulletin* 28: 817–824.
Engelmark, O., A. Hofgaard, and T. Arnborg. 1998. Successional trends 219 years after fire in an old *Pinus sylvestris* stand in northern Sweden. *Journal of Vegetation Science* 9:583–592.
Eschtruth, A. K. and J. J. Battles. 2009. Acceleration of exotic plant invasion in a forested ecosystem by a generalist herbivore. *Conservation Biology* 23:388–399.

Frelich, L. E. and C. G. Lorimer. 1985. Current and predicted long-term effects of deer browsing in hemlock forests in Michigan, USA. *Biological Conservation* 34:99–120.

Fuller, R. J. 2001. Responses of woodland birds to increasing numbers of deer: A review of evidence and mechanisms. *Forestry* 74:289–298.

George, L. O. and F. A. Bazzaz. 1999. The fern understory as an ecological filter: Emergence and establishment of canopy-tree seedlings. *Ecology* 80:833–845.

Giles, B. G. and C. S. Findlay. 2004. Effectiveness of a selective harvest system in regulating deer populations in Ontario. *Journal of Wildlife Management* 68:266–277.

Gorchov, D. L. and D. E. Trisel. 2003. Competitive effects of the invasive shrub, *Lonicera maackii* (Rupr.) Herder (Caprifoliaceae) on the growth and survival of native tree seedlings. *Plant Ecology* 166:13–24.

Greenwald, K. R., L. J. Petit, and T. A. Waite. 2008. Indirect effects of a keystone herbivore elevate local animal diversity. *Journal of Wildlife Management* 72:1318–1321.

Grime, J. P. 1973. Competitive exclusion in herbaceous vegetation. *Nature* 242:344–347.

Gross, J. E. and M. W. Miller. 2001. Chronic wasting disease in mule deer: Disease dynamics and control. *Journal of Wildlife Management* 65:205–215.

Harlow, R. F. and R. L. Downing. 1970. Deer browsing and hardwood regeneration in the southern Appalachians. *Journal of Forestry* 68:298–300.

Healy, W. M., D. S. deCalesta, and S. L. Stout. 1997. A research perspective on white-tailed deer overabundance in the northeastern United States. *Wildlife Society Bulletin* 25:259–263.

Hester, A. J., L. Edenius, R. M. Buttenschøn, and A. T. Kuiters. 2000. Interactions between forests and herbivores: The role of controlled grazing experiments. *Forestry* 73:381–391.

Hobbs, N. T. 1996. Modification of ecosystems by ungulates. *Journal of Wildlife Management* 60:695–713.

Hochwender, C. G., R. J. Marquis, and K. A. Stowe. 2000. The potential for and constraints on the evolution of compensatory ability in *Asclepias syriaca*. *Oecologia* 122:361–370.

Horsley, S. B. and D. A. Marquis. 1983. Interference by weeds and deer with Allegheny hardwood reproduction. *Canadian Journal of Forest Research* 13:61–69.

Horsley, S. B., S. L. Stout, and D. S. deCalesta. 2003. White-tailed deer impact on the vegetation dynamics of a northern hardwood forest. *Ecological Applications* 13:98–118.

Huntly, N. 1991. Herbivores and the dynamics of communities and ecosystems. *Annual Review of Ecology and Systematics* 22:477–503.

Kalen, C. 2005. Deer browsing and impact on forest development. *Journal of Sustainable Forestry* 21:51–62.

Knight, T. M. 2003. Floral density, pollen limitation and reproductive success in *Trillium grandiflorum*. *Oecologia* 137:557–563.

Knight, T. M. 2004. The effect of herbivory and pollen limitation on a declining population of *Trillium grandiflorum*. *Ecological Applications* 14:915–928.

Knight, T. M., J. L. Dunn, L. A. Smith, J. Davis, and S. Kalisz. 2009. Deer facilitate invasive plant success in a Pennsylvania forest understory. *Natural Areas Journal* 29:110–116.

Kraft, L. S., T. R. Crow, D. S. Buckley, E. A. Nauertz, and J. C. Zasada. 2004. Effects of harvesting and deer browsing on attributes of understory plants in northern hardwood forests, Upper Michigan, USA. *Forest Ecology and Management* 199:219–230.

Kramer, K., G. W. T. A. Groot Bruinderink, and H. H. T. Prins. 2006. Spatial interactions between ungulate herbivory and forest management. *Forest Ecology and Management* 226:238–247.

Lankau, R. A., W. E. Rogers, and E. Siemann. 2004. Constraints on the utilisation of the invasive Chinese tallow tree *Sapium sebiferum* by generalist native herbivores in coastal prairies. *Ecological Entomology* 29:66–75.

Leopold, A. 1933. *Game Management*. New York, NY: Charles Scribner's Sons.

Loison, A., R. Langvatn, and E. J. Solberg. 1999. Body mass and winter mortality in red deer calves: Disentangling sex and climate effects. *Ecography* 22:20–30.

Long, Z. T., T. H. Pendergast IV, and W. P. Carson. 2007. The impact of deer on relationships between tree growth and mortality in an old-growth beech–maple forest. *Forest Ecology and Management* 252:230–238.

Manseau, M., J. Huot, and M. Crête. 1996. Effects of summer grazing by caribou on composition and productivity of vegetation: Community and landscape level. *Journal of Ecology* 84:503–513.

Martin, J. L. and C. Baltzinger. 2002. Interaction among deer browsing, hunting, and tree regeneration. *Canadian Journal of Forest Research* 32:1254–1264.

Massé, A. and S. D. Côté. 2009. Habitat selection of a large herbivore at high density and without predation: Trade-off between forage and cover? *Journal of Mammalogy* 90:961–970.

McCabe, T. R. and R. E. McCabe. 1997. Recounting whitetales past. In *The Science of Overabundance: Deer Ecology and Population Management*, eds. W. J. McShea, H. B. Underwood, and J. H. Rappole, 11–26. Washington, D.C.: Smithsonian Institution Press.

McCullough, D. R. 1979. *The George Reserve Deer Herd*. Ann Harbor, MI: Michigan University Press.

McNulty, S. A., W. F. Porter, N. E. Mathews, and J. A. Hill. 1997. Localized management for reducing white-tailed deer populations. *Wildlife Society Bulletin* 25:265–271.

McShea, W. J. 2000. The influence of acorn crops on annual variation in rodent and bird populations. *Ecology* 81:228–238.

McShea, W. J., S. L. Monfort, S. Hakim et al. 1997b. The effect of immunocontraception on the behavior and reproduction of white-tailed deer. *Journal of Wildlife Management* 61:560–569.

McShea, W. J. and J. H. Rappole. 1997. Herbivores and the ecology of forest understory birds. In *The Science of Overabundance: Deer Ecology and Population Management*, eds. W. J. McShea, H. B. Underwood, and J. H. Rappole, 298–309. Washington, D.C.: Smithsonian Institution Press.

McShea, W. J. and J. H. Rappole. 2000. Managing the abundance and diversity of breeding bird populations through manipulation of deer populations. *Conservation Biology* 14:1161–1170.

McShea, W. J., H. B. Underwood, and J. H. Rappole, editors. 1997a. *The Science of Overabundance: Deer Ecology and Population Management*. Washington, D.C.: Smithsonian Institution Press.

Miller, B. F., T. A. Campbell, B. R. Laseter, W. M. Ford, and K. V. Miller. 2009. White-tailed deer herbivory and timber harvesting rates: Implications for regeneration success. *Forest Ecology and Management* 258:1067–1072.

Miller, R., J. B. Kaneene, S. D. Fitzgerald, and S. M. Schmitt. 2003. Evaluation of the influence of supplemental feeding of white-tailed deer (*Odocoileus virginianus*) on the prevalence of bovine tuberculosis in the Michigan wild deer population. *Journal of Wildlife Diseases* 39:84–95.

Miyashita, T., M. Takada, and A. Shimazaki. 2004. Indirect effects of herbivory by deer reduce abundance and species richness of web spiders. *Écoscience* 11:74–79.

Moser, B. W. and G. W. Witmer. 2000. The effects of elk and cattle foraging on the vegetation, birds, and small mammals of the Bridge Creek Wildlife Area, Oregon. *International Biodeterioration and Biodegradation* 45:151–157.

Mudrak, E. L., S. E. Johnson, and D. A. Waller. 2009. Forty-seven year changes in vegetation at the Apostle Islands: Effects of deer on the forest understory. *Natural Areas Journal* 29:167–176.

Murden, S. B. and K. L. Risenhoover. 1993. Effects of habitat enrichment on patterns of diet selection. *Ecological Applications* 3:497–505.

Myers, J. A., M. Vellend, S. Gardescu, and P. L. Marks. 2004. Seed dispersal by white-tailed deer: Implications for long-distance dispersal, invasion, and migration of plants in eastern North America. *Oecologia* 139:35–44.

Noyce, K. V. and D. L. Garshelis. 1997. Influence of natural food abundance on black bear harvests in Minnesota. *Journal of Wildlife Management* 61:1067–1074.

Otsfeld, R. S., C. G. Jones, and J. O. Wolff. 1996. Of mice and mast: Ecological connections in eastern deciduous forests. *Bioscience* 46:323–330.

Paige, K. N. and T. G. Whitham. 1987. Overcompensation in response to mammalian herbivory: The advantage of being eaten. *American Naturalist* 129:407–416.

Paine, R. T. 2000. Phycology for the mammalogist: Marine rocky shores and mammal-dominated communities - How different are the structuring processes? *Journal of Mammalogy* 81:637–648.

Parker, J. D., D. E. Burkepile, and M. E. Hay. 2006. Opposing effects of native and exotic herbivores on plant invasions. *Science* 311:1459–1461.

Parmesan, C. and G. Yohe. 2003. A globally coherent fingerprint of climate change impacts across natural systems. *Nature* 421: 37–32.

Pastor, J., B. Dewey, R. J. Naiman, P. F. McInnes, and Y. Cohen. 1993. Moose browsing and soil fertility in the boreal forests of Isle Royale National Park. *Ecology* 74:467–480.

Pastor, J. and R. J. Naiman. 1992. Selective foraging and ecosystem processes in boreal forests. *American Naturalist* 139:690–705.

Pelchat, B. O. and R. L. Ruff. 1986. Habitat and spatial relationships of black bears in the boreal mixedwood forest of Alberta. *International Conference on Bear Research and Management* 6:81–92.

Pellerin, S., J. Huot, and S. D. Côté. 2006. Long term effects of deer browsing and trampling on the vegetation of peatlands. *Biological Conservation* 128:316–326.

Peterson, R. O. 1999. Wolf-moose interaction on Isle Royale: The end of natural regulation? *Ecological Applications* 9:10–16.

Porter, W. F. and H. B. Underwood. 1999. Of elephants and blind men: Deer management in the U.S. National Parks. *Ecological Applications* 9:3–9.

Potvin, F., P. Beaupré, and G. Laprise. 2003. The eradication of balsam fir stands by white-tailed deer on Anticosti island, Québec, Canada: A 150-year process. *Écoscience* 10:487–495.

Pound, J. M., J. E. George, D. M. Kammlah, K. H. Lohmeyer, and R. B. Davey. 2010. Evidence for role of white-tailed deer (Artiodactyla: Cervidae) in epizootiology of cattle ticks and southern cattle ticks (Acari:Ixodidae) in reinfestations along the Texas/Mexico border in South Texas: A review and update. *Journal of Economic Entomology* 103:211–218.

Rachich, J. and R. J. Reader. 1999. An experimental study of wetland invasibility by purple loosestrife (*Lythrum salicaria*). *Canadian Journal of Botany* 77:1499–1503.

Raymond, G. J., A. Bossers, L. D. Raymond et al. 2000. Evidence of a molecular barrier limiting susceptibility of humans, cattle and sheep to chronic wasting disease. *EMBO Journal* 17:4425–4430.

Riley, S. J., D. J. Decker, J. W. Enck, P. D. Curtis, T. B. Lauber, and T. L. Brown. 2003. Deer populations up, hunter populations down: Implications of interdependence of deer and hunter population dynamics on management. *Écoscience* 10:455–461.

Risenhoover, K. L. and S. A. Maass. 1987. The influence of moose on the composition and structure of Isle Royale forests. *Canadian Journal of Forest Research* 17:357–364.

Ritchie, M. E., D. Tilman, and J. M. H. Knops. 1998. Herbivore effects on plant and nitrogen dynamics in oak savanna. *Ecology* 79:165–177.

Rogers, L. 1976. Effects of mast and berry crop failures on survival, growth, and reproductive success of black bears. *Transactions of the North American Wildlife and Natural Resources Conference* 41:431–438.

Rooney, T. P. 2001. Deer impacts on forest ecosystems: A North American perspective. *Forestry* 74:201–208.

Rooney, T. P. 2009. High white-tailed deer densities benefit graminoids and contribute to biotic homogenization of forest ground-layer vegetation. *Plant Ecology* 202:103–111.

Rooney, T. P. and W. J. Dress. 1997. Species loss over sixty-six years in the ground-layer vegetation of Heart's Content, an old-growth forest in Pennsylvania USA. *Natural Areas Journal* 17:297–305.

Rooney, T. P., R. J. McCormick, S. L. Solheim, and D. M. Waller. 2000. Regional variation in recruitment of hemlock seedlings and saplings in the upper Great Lakes, USA. *Ecological Applications* 10:1119–1132.

Rooney, T. P., S. L. Solheim, and D. M. Waller. 2002. Factors affecting the regeneration of northern white cedar in lowland forests of the Upper Great Lakes region, USA. *Forest Ecology and Management* 163:119–130.

Rooney, T. P. and D. M. Waller. 2001. How experimental defoliation and leaf height affect growth and reproduction in *Trillium grandiflorum*. *Journal of the Torrey Botanical Society* 128:393–399.

Rooney, T. P. and D. M. Waller. 2003. Direct and indirect effects of deer in forest ecosystems. *Forest Ecology and Management* 181:165–176.

Rossell, C. R. Jr, S. Patch, and S. Salmons. 2007. Effects of deer browsing on native and non-native vegetation in a mixed oak-beech forest on the Atlantic coastal plain. *Northeastern Naturalist* 14:61–72.

Ruhren, S. and S. N. Handel. 2003. Herbivory constrains survival, reproduction, and mutualisms when restoring nine temperate forest herbs. *Journal of the Torrey Botanical Society* 130:34–42.

Russell, F. L., D. B. Zippin, and N. L. Fowler. 2001. Effects of white-tailed deer (*Odocoileus virginianus*) on plants, plant populations and communities: A review. *American Midland Naturalist* 146:1–26.

Sæther, B.-E. 1997. Environmental stochasticity and population dynamics of large herbivores: A search for mechanisms. *Trends in Ecology and Evolution* 12:143–149.

Saunders, M. R. and K. J. Puettmann. 1999. Effects of overstory and understory competition and simulated herbivory on growth and survival of white pine seedlings. *Canadian Journal of Forest Research* 29:536–546.

Scheffer, M., S. Carpenter, J. A. Foley, C. Folke, and B. Walker. 2001. Catastrophic shifts in ecosystems. *Nature* 413:591–596.

Schmitt, S. M., S. D. Fitzgerald, T. M. Cooley et al. 1997. Bovine tuberculosis in free-ranging white-tailed deer from Michigan. *Journal of Wildlife Diseases* 33:749–758.

Schooley, R. L., C. R. McLaughling, G. J. Matula, Jr., and W. B. Krohn. 1994. Denning chronology of female black bears: Effects of food, weather, and reproduction. *Journal of Mammalogy* 75:466–477.

Seagle, S. and S. Liang. 2001. Application of a forest gap model for prediction of browsing effects on riparian forest succession. *Ecological Modelling* 144:213–229.

Simard, M. A., S. D. Côté, R. B. Weladji, and J. Huot. 2008. Feedback effects of chronic browsing on life-history traits of a large herbivore. *Journal of Animal Ecology* 77:678–686.

Singer, F. J. and K. A. Schoenecker. 2003. Do ungulates accelerate or decelerate nitrogen cycling? *Forest Ecology and Management* 181:189–204.

Smit, R., J. Bokdam, J. den Ouden, H. Olff, H. Schot-Opschoor, and M. Schrijvers. 2001. Effects of introduction and exclusion of large herbivores on small rodent communities. *Plant Ecology* 155:119–127.

Stewart, A. J. A. 2001. The impact of deer on lowland woodland invertebrates: A review of the evidence and priorities for future research. *Forestry* 74:259–270.

Stewart, K. M., R. T. Bowyer, J. G. Kie, B. L. Dick, and R. W. Ruess. 2009. Population density of North American elk: Effects on plant diversity. *Oecologia* 161:303–312.

Stinson, K. A., S. A. Campbell, J. R. Powell et al. 2006. Invasive plant suppresses the growth of native tree seedlings by disrupting belowground mutualisms. *PLoS Biology* 4:727–731.

Stoeckeler, J. H., R. O. Strothmann, and L. W. Krefting. 1957. Effect of deer browsing on reproduction in the northern hardwood-hemlock type in northeastern Wisconsin. *Journal of Wildlife Management* 21:75–80.

Stromayer, K. A. K. and R. J. Warren. 1997. Are overabundant deer herds in the eastern United States creating alternate stable states in forest plant communities? *Wildlife Society Bulletin* 25:227–234.

Suominen, O. and K. Danell. 2006. Effects of large herbivores on other fauna. In *Large Herbivore Ecology, Ecosystem Dynamics and Conservation*, eds. K. Danell, R. Bergström, P. Duncan, and J. Pastor, 383–412. Cambridge: Cambridge University Press.

Suominen, O., K. Danell, and R. Bergström. 1999a. Moose, trees, and ground-living invertebrates: Indirect interactions in Swedish pine forests. *Oikos* 84:215–226.

Suominen, O., K. Danell, and J. P. Bryant. 1999b. Indirect effects of mammalian browsers on vegetation and ground-dwelling insects in an Alaskan floodplain. *Écoscience* 6:505–510.

Suominen, O., J. Niemelä, P. Martikainen, P. Niemelä, and I. Kojola. 2003. Impact of reindeer grazing on ground-dwelling Carabidae and Curculionidae assemblages in Lapland. *Ecography* 26:503–513.

Suominen, O., I.-L. Persson, K. Danell, R. Bergström, and J. Pastor. 2008. Impact of simulated moose densities on abundance and richness of vegetation, herbivorous and predatory arthropods along a productivity gradient. *Ecography* 31:636–645.

Taillon, J., D. G. Sauvé, and S. D. Côté. 2006. The effects of decreasing winter diet quality on foraging behavior and life-history traits of white-tailed deer fawns. *Journal of Wildlife Management* 70:1445–1454.

Takada, M., M. Asada, and T. Miyashita. 2001. Regional differences in the morphology of a shrub *Damnacanthus indicus*: An induced resistance to deer herbivory? *Ecological Research* 16:809–813.

Telford III, S. R. 2002. Deer tick-transmitted zoonoses in the eastern United States. In *Conservation Medicine: Ecological Health in Practice*, eds. A. Aguirre, R. S. Ostfeld, G. M. Tabor, C. House, M. C. Pearl, 310–324. New York, NY: Oxford University Press.

Tester, J. R., A. M. Starfield, and L. E. Frelich. 1997. Modeling for ecosystem management in Minnesota pine forests. *Biological Conservation* 80:313–324.

Thiemann, J. A., C. R. Webster, M. A. Jenkins, P. M. Hurley, J. H. Rock, and P. S. White. 2009. Herbaceous-layer impoverishment in a post-agricultural southern Appalachian landscape. *American Midland Naturalist* 162:148–168.

Tilghman, N. G. 1989. Impacts of white-tailed deer on forest regeneration in northwestern Pennsylvania. *Journal of Wildlife Management* 53:524–532.

Timmons, G. R., D. G. Hewitt, C. A. DeYoung, T. E. Fulbright, and D. A. Draeger. 2010. Does supplemental feed increase selective foraging in a browsing ungulate? *Journal of Wildlife Management* 74:995–1002.

Tremblay, J.-P., J. Huot, and F. Potvin. 2007. Density-related effects of deer browsing on the regeneration dynamics of boreal forests. *Journal of Applied Ecology* 44:552–562.

Tremblay, J.-P., I. Thibault, C. Dussault, J. Huot, and S. D. Côté. 2005. Long-term decline in white-tailed deer browse supply: Can lichens and litterfall act as alternate food sources that preclude density-dependent feedbacks? *Canadian Journal of Zoology* 83:1087–1096.

Valkenburg, P., M. E. McNay, and B. W. Dale. 2004. Calf mortality and population growth in the Delta caribou herd after wolf control. *Wildlife Society Bulletin* 32:746–756.

Van Wieren, S. E. 1998. Effects of large herbivores upon the animal community. In *Grazing and Conservation Management*, eds. M. F. WallisDeVries, J. P. Bakker, S. E. Van Wieren, 185–214. Boston: Kluwer Academic Publishers.

Vavra, M., C. G. Parks, and M. J. Wisdom. 2007. Biodiversity, exotic plant species, and herbivory: The good, the bad, and the ungulate. *Forest Ecology and Management* 246:66–72.

Vázquez, D. P. and D. Simberloff. 2003. Changes in interaction biodiversity induced by an introduced ungulate. *Ecology Letters* 6:1077–1083.

Vellend, M. 2002. A pest and an invader: White-tailed deer (*Odocoileus virginianus* Zimm.) as a seed dispersal agent for honeysuckle shrubs (*Lonicera* L.). *Natural Areas Journal* 22:230–234.

Vourc'h, G., J. L. Martin, P. Duncan, J. Escarré, and T. P. Clausen. 2001. Defensive adaptations of *Thuja plicata* to ungulate browsing: A comparative study between mainland and island populations. *Oecologia* 126:84–93.

Waller, D. M. and W. S. Alverson. 1997. The white-tailed deer: A keystone herbivore. *Wildlife Society Bulletin* 25:217–226.

Webb, W. L., R. T. King, and E. F. Patric. 1956. Effect of white-tailed deer on a mature northern hardwood forest. *Journal of Forestry* 54:391–398.

Webster, C. R., M. A. Jenkins, and J. H. Rock. 2005. Long-term response of spring flora to chronic herbivory and deer exclusion in Great Smoky Mountains National Park, USA. *Biological Conservation* 125: 297–307.

Whigham, D. F. 1990. The effect of experimental defoliation on the growth and reproduction of a woodland orchid, *Tipularia discolor*. *Canadian Journal of Botany* 68:1812–1816.

Wiegmann, S. M. and D. M. Waller. 2006. Fifty years of change in northern upland forest understories: Identity and traits of "winner" and "loser" plant species. *Biological Conservation* 129:109–123.

Williams, E. S., M. W. Miller., T. J. Kreeger, R. H. Kahn, and E. T. Thorne. 2002. Chronic wasting disease of deer and elk: A review with recommendations for management. *Journal of Wildlife Management* 66:551–563.

Williams, S. C. and J. S. Ward. 2006. Exotic seed dispersal by white-tailed deer in southern Connecticut. *Natural Areas Journal* 26:383–390.

Williams, S. C., J. S. Ward, and U. Ramakrishnan. 2008. Endozoochory by white-tailed deer (*Odocoileus virginianus*) across a suburban/woodland interface. *Forest Ecology and Management* 255:940–947.

Wilson, M. L. and J. E. Childs. 1997. Vertebrate abundance and the epidemiology of zoonotic diseases. In *The Science of Overabundance: Deer Ecology and Population Management*, eds. W. J. McShea, H. B. Underwood, and J. H. Rappole, 224–248. Washington, D.C.: Smithsonian Institution Press.

Wilson, M. L., A. M. Ducey, T. S. Litwin, T. A. Gavin, and A. Spielman. 1990. Microgeographic distribution of immature *Ixodes dammini* ticks correlated with that of deer. *Medical and Veterinary Entomology* 4:151–159.

13

Impacts of Deer on Society

Michael R. Conover

CONTENTS

Any entity, including deer, has both positive and negative values. Its positive values can be defined as its attributes that provide benefits and enjoyment to someone while its negative values are those attributes that cause economic losses, heartache, or problems for someone (Giles, 1978, 1990). Paradoxically, white-tailed deer provide both more benefits and cause more problems for humans than any other wildlife species in North America. Positive values of deer include the pleasure that people experience from watching or hunting them or the satisfaction of just knowing that deer populations are doing well. Negative values of deer include threats to human health or safety, economic losses, and nuisance problems. The goal of deer management is to increase the net value of deer, which can be defined as the positive values minus the negative values.

In the United States and Canada, wildlife belongs to society and is managed by government for society's benefit. It is the responsibility of elected officials to determine how best to achieve this. If they fail to do so or make poor decisions, then voters may elect new officials during the next election. Of course, government has grown too complex for elected officers to vote on all matters, so they have delegated specific tasks to government employees to perform on their behalf. In the case of white-tailed deer, management authority resides with each state's wildlife department, which makes decisions about how wildlife should be managed to meet the goals of society.

Decisions about how to manage deer are often controversial because opinions about deer are diverse. For a fruit grower, deer may have a negative net value because they consume his or her produce, reducing farm income. For a deer hunter, deer may have a positive value because of enjoyment derived from hunting them. Because deer impact growers and hunters in different ways, a hunter may want more deer while fruit growers may seek fewer deer. Because of such divided opinions, decisions about deer management can best be made after considering the many positive and negative ways deer affect members of society.

In this chapter, I use the scientific literature to estimate both the positive and negative values that deer provide to society. When possible, I define these values in terms of United States dollars. For this study, I corrected for inflation by converting the original dollar estimates provided in prior studies into the value of a United States dollar during 2008, using the Consumer Price Index of the United States Bureau of Labor Statistics. Most economic studies do not differentiate between mule deer and white-tailed deer, so this study likewise considers both species as a single category.

Placing a dollar value on the positive values of deer is difficult, but one way to do this is to determine how much money people spend in deer-related activities. The enjoyment (positive value) that people receive from the activity has to be higher than the cost required to engage in the activity, otherwise people would not do it. The actual value, or enjoyment, that people receive from engaging in the activity may be several times higher than the amount of money they expend. Summing the amount of money people spend to engage in deer-related activities underestimates the true positive value of deer, but it does provide a defensible conservative estimate of value (Conover, 2001).

Positive Values of Deer

Existence, Empathic, and Historic Values

People gain a sense of satisfaction or feel joy from the knowledge that white-tailed deer are thriving even if they never see one. This sense of satisfaction is referred to as the deer's existence value (Conover, 2002).

Empathy can be defined as the emotional and intellectual ability to project one's own consciousness into that of another person or animal. It is a uniquely human ability. Deer have a positive empathy value when people derive satisfaction, pleasure, or a sense of fulfillment from their empathetic experience with deer (Conover and Conover, 2003).

The historical value of deer is the satisfaction that people experience knowing that deer populations have not changed over time (Conover and Conover, 2003). This positive value arises from the important role deer have played in both Native American history and American colonial and frontier history. People are nostalgic and most think fondly about the past. Because deer populations thrived in North America in the past, people want deer populations to continue to thrive.

Recreational Value: Hunting

Many people are avid about wildlife and report that watching wildlife, hunting, or fishing provides them with more enjoyment than any other recreational activity. During 2006, people in the United States spent $122 billion, equal to 1% of the nation's gross domestic product, to engage in wildlife recreation (U.S. Fish and Wildlife Service, 2006). Of this amount, $64 billion was spent on equipment, $37 billion was trip-related expenses (food, lodging, and transportation), and $21 billion was spent to buy licenses and lease land. Annual expenditures by recreational activity were $42 billion for fishing, $23 billion for hunting, $12 billion for gear used for both fishing and hunting, and $46 billion for watching wildlife (U.S. Fish and Wildlife Service, 2006).

During 2006, 12.5 million people hunted wildlife in the United States. Most hunters (10.7 million) pursued big game (deer and elk) and spent $12 billion doing so (U.S. Fish and Wildlife Service, 2006). An indication of how much the average big-game hunter enjoys the sport is that he or she spent 15 days and $1121 doing it during 2006.

Recreational Value: Wildlife Viewing

During 2006, 71 million United States residents spent time watching wildlife (U.S. Fish and Wildlife Service, 2006). Wildlife watchers include people who spend time feeding, photographing, or observing

FIGURE 13.1 Many people value seeing and photographing white-tailed deer. (Photo by D. G. Hewitt.)

wildlife (Figure 13.1). Many wildlife watchers visit public parks to enjoy wildlife. They also maintain plantings and natural areas around their home for the benefit of wildlife. During 2006, people spent $46 billion watching wildlife. This amount includes $13 billion spent on wildlife-viewing trips and $23 billion spent on equipment for wildlife watching (e.g., binoculars, cameras, bird food, and camping equipment). They also spent $7 billion to lease and buy land where they can view wildlife and $2 billion on plantings to attract wildlife.

During 2006, 23 million people took trips to observe, feed, or photograph wildlife. Of these, 13 million travelers, or 70%, observed or photographed large land mammals, including deer (U.S. Fish and Wildlife Service, 2006). Deer are among the most awe-inspiring mammals of North America, owing to their size and grace. If deer account for 10% of the pleasure people receive from viewing wildlife, then the recreational value of viewing deer is $4.6 billion.

Negative Values of Deer

Ecological Value

The ecological value of deer is the role that deer play in shaping their ecosystem. The ecological value of deer is positive when the changes they cause are beneficial from a human perspective. For instance, cougars prey heavily upon deer, and if deer were unavailable, many cougars would starve. Hence, part of the value that people attach to having a healthy cougar population (whether positive or negative) can be attributed to deer.

Deer also shape their ecosystem by their herbivory. Again, this value can be positive if deer herbivory improves the habitat from the perspective of people and negative if their herbivory degrades the habitat. Over-browsing by deer can damage understory vegetation of forests by reducing the density of tree seedlings and forbs and increasing the density of grass and ferns (Marquis and Brenneman, 1981; Warren, 1991). Deer herbivory on young trees can cause palatable species, such as hemlock, to be eliminated from the forest canopy. High deer densities also impact ecosystems indirectly by altering primary production, litter quantity and quality, nutrient cycling, and fire regimes (Hobbs, 1996). Not surprisingly, these changes adversely affect some species of birds and small mammals while favoring others (Warren, 1991; deCalesta, 1994; McShea and Rappole, 2000).

Deer–Vehicle Collisions

Deer pose a significant hazard for motorists in the United States and Canada. In Utah, 4% of all accidents reported to police were deer–vehicle collisions (DVCs) (Bissonette et al., 2008); 16% of accidents in Michigan were DVCs (Marcoux and Riley, 2010). The National Highway Traffic Safety Administration (2000) estimated that 4% of the 6.1 million motor vehicle accidents reported to police during 2000 in the United States resulted from a DVC, that is, 247,000 accidents. Romin and Bissonette (1996) determined that >500,000 DVCs were reported to state authorities during 1991. However, many DVCs go unreported. During 1988, 41% of DVCs in Tompkins County, New York, were reported to insurance companies; 52% were reported to police. Romin and Bissonette (1996) and Marcoux and Riley (2010) estimated that half of all DVCs are reported to state authorities, indicating that there must be greater than 1.0 million DVCs annually in the United States.

Another way to assess the frequency of DVCs is to determine the proportion of drivers who are involved in DVCs annually. During 1988 and 1990, 3% and 4% of the population, respectively, of Tompkins County, New York, were involved in a DVC (Decker et al., 1990; Stout et al., 1993). In Michigan, 18% of drivers reported having a DVC during a 5-year period, or 3.6% each year (Marcoux and Riley, 2010). There were 191 million drivers in the United States during 2008 (U.S. Department of Transportation, 2009). If 3% of them are involved annually in DVCs, then there are 5.7 million DVC each year. Hence, it seems safe to assume that there are more than 1 million DVCs annually in the United States.

Bissonette et al. (2008) estimated that vehicles involved in a DVC in Utah during 2001 sustained an average of $1434 in damages ($1748 in 2008 after correcting for inflation). In Michigan, vehicles involved in DVCs sustained $2300 in damages (Marcoux and Riley, 2010). If the mean of these two estimates ($2024) is multiplied by the 500,000 reported DVCs, the total vehicle repair bill in the United States is $1.0 billion. A DVC involving heavy damage is more likely to be reported than one with little damage. In Tomkins County, New York, damage sustained by vehicles in unreported DVCs was about half the level of damage sustained by vehicles in reported DVCs (Decker et al., 1990). Assuming that a vehicle in an unreported DVC sustains 50% of the damage of a vehicle in a reported DVC, the total amount of damage to vehicles involved in DVCs equals $1.5 billion ($1.0 billion for reported DVCs and $0.5 billion for unreported DVCs).

In Michigan, the 60,000 reported DVCs resulted in more than 1880 human injuries (3.1% of DVCs) and seven human fatalities (0.01% of DVCs) (Marcoux and Riley, 2010). In Alabama, 8% of 27,780 DVCs from 1994 to 2003 resulted in a human injury and 0.1% in a human fatality (Hussain et al., 2007). In Utah, investigating highway patrolmen can classify people involved in an automobile accident as having died, sustained an obvious wound, possible injury, and not injured. For people involved in DVCs, 0.01% were killed, 3% had an obvious wounds, and 2.2% had a possible injury (Bissonette et al., 2008). Four hundred forty-eight of the victims (2.1%) were taken to the hospital and then released while 44 (0.2%) were hospitalized overnight as a result of DVCs. If the 3% injury rate and 0.01% fatality rate are extrapolated to the >500,000 reported DVCs in the United States each year, there are 15,000 human injuries and 50 human fatalities from DVCs annually. U.S. Centers for Disease Control (2004) reported that 26,647 people were injured annually during 2001 and 2002 in vehicle collisions with an animal (most of which were with a deer).

In Utah, medical costs for people involved in DVCs and taken to the hospital averaged $2037 per individual. If medical costs in Utah are similar to those in the rest of the United States, then the 26,647 people injured annually (U.S. Centers for Disease Control, 2004) would result in hospital bills for $54 million. The U.S. Department of Transportation (2002) and Bissonette et al. (2008) recommend using $3.0 million for the value of a human life. Hence, 50 human fatalities from DVCs are valued at $150 million. Thus, the total cost of DVCs (for both mule deer and white-tailed deer) equals $1.7 billion annually ($1.5 billion for vehicle damage, $54 million for medical care, and $150 million for human fatalities).

Much attention has been focused on bird–aircraft collisions after US Airways Flight 1549 was ditched into the Hudson River during January 2009 (Dolbeer, 2009). Although aircraft usually collide with birds, deer on airfields have become more common in recent years (De Vault et al., 2008; VerCauteren et al., 2009). Any aircraft–deer collision poses a hazard to human health and can be extremely expensive. This problem is especially serious at small general aviation airports and private runways (De Vault et al., 2008).

Physical Attacks on People

White-tailed deer occasionally attack people. As one example, 13 people on the campus of Southern Illinois University–Carbondale were attacked by deer during 2005 and 2006 (Hubbard and Nielsen, 2009). The attacks occurred at multiple locations on campus; injuries to the victims included bruises, broken bones, and skin lacerations requiring stitches. All attacks occurred during the fawning season (May and June) and probably involved does attacking people that the does thought posed a threat to their fawns.

Elsewhere, attacks on humans have occurred during the rutting season by bucks that were raised since they were fawns by humans, then released once they became too big to be a pet. Presumably, these bucks had imprinted on humans during a critical period and viewed humans as rivals.

Zoonotic Diseases

Zoonotic diseases are human diseases for which the virus or bacteria that cause the disease use another animal either as a reservoir or vector. I have focused on those zoonotic diseases that are vectored by ticks of the genus *Ixodes* because the adult ticks require large mammals for a blood meal. In the midwestern and eastern United States, white-tailed deer serve as the primary host for adult *Ixodes* and are key to ticks' survival (Kilpatrick and LaBonte, 2007; Stafford, 2007).

Lyme Disease

Lyme disease is caused by a spirochetal bacterium, *Borrelia burgdorferi*, identified in the United States during the 1980s. The disease is transmitted to humans in the midwestern and eastern United States through the bite of the black-legged tick (*Ixodes scapularis*), which is also called the deer tick (Lane et al., 1991; Stafford, 2007). The number of Lyme disease cases reported to health officials increased from 8257 during 1993 to 27,444 during 2007 (U.S. Centers for Disease Control 2009a). Among people who sought medical attention for Lyme disease, 68% had a rash, 33% arthritis, 8% facial palsy, 4% radiculopathy, and 1% meningitis or encephalitis (U.S. Centers for Disease Control, 2009a). Although Lyme disease is not fatal, it can cause a stillbirth if the mother is infected during pregnancy.

Granulocytic Anaplasmosis

Human granulocytic anaplasmosis (HGA) was first detected during 1990 when a person in Wisconsin died two weeks after being bitten by a tick. This disease is caused by a rickettsial bacteria, *Anaplasma phagocytophilum*, that invades leucocytes: a type of white blood cell (Dumler et al., 2005). At least seven deaths have been attributed to HGA since 1990 (Dumler et al., 2005). The number of reported cases of HGA in humans has increased from 100 to 200 during the 1990s to over 700 during 2005 (Stafford, 2007), but many infections go undetected; in some parts of the United States, as many as 36% of humans have been infected (Dumler et al., 2005). The disease organism is transmitted to humans in the eastern United States by the black-legged tick.

Babesiosis

This disease is caused by the microscopic parasite, *Babesia microti*. The parasite is vectored to humans through the bite of the black-legged tick, the same species that vectors Lyme disease and HGA to humans (Stafford, 2007; U.S. Centers for Disease Control, 2009b). The parasite is endemic in the upper midwestern and northeastern United States.

After being transmitted to a person, *B. microti* invade red blood cells where they reproduce asexually and emerge to infect other blood cells. The red blood cells are destroyed and, not surprisingly, hemolytic anemia may occur. Many infected people feel fine and are asymptomatic; others develop malaria-like symptoms, such as fever, body aches, chills, nausea, and fatigue. When high levels of red blood cells are infected, complications can occur such as acute respiratory failure, congestive heart failure, and renal failure. Babesiosis can be life-threatening, especially in victims who are elderly, have a weak immune system, or lack a spleen. The disease is diagnosed through microscopic examination of red blood cells for the parasite and by polymerase chain reaction assay for DNA from *B. microti*. The number of people who contract babesiosis annually is unknown because many cases go undetected. What is known is that the number of reported cases has increased in recent years (Stafford, 2007; U.S. Centers for Disease Control, 2009b).

Economic Losses to Agricultural Production

During 1994, most United States agricultural producers (53%) experienced problems with deer; this was far more than number of problems caused by any other wildlife species (Conover, 1998). Agricultural producers reported spending a mean of 43.6 h and $1002 that year trying to prevent wildlife damage. When extrapolated to the nation's 2,088,000 agricultural producers, the amounts exceed 90 million hours and $2 billion (Conover, 1998). Despite efforts to prevent wildlife damage, annual damage of wildlife to agricultural producers was $2 billion in 1994 (Figure 13.2). Due to inflation, $4 billion during 1994 (the amount spent to prevent wildlife damage plus the amount of wildlife damage), was $5.75 billion during 2008. Deer are likely responsible for more than 10% or $575 million of the wildlife damage to agriculture.

Economic Losses to Timber Industry

White-tailed deer injure small trees when they browse them or rub off the tree's bark with their antlers. Doing so reduces the trees' growth rate and results in a loss of timber productivity because the injured trees will take more years to grow to merchantable size. In the 6,500,000-ha Allegheny forest of Pennsylvania, annual timber losses from deer amounted to $102/ha (in 2008 dollars, adjusted for inflation) or more than $500 million (Marquis, 1981; Marquis and Brenneman, 1981; Tilghman, 1989). Yet, this region contains only a small proportion (14%) of North America's oak–hickory forests (Gedney and Van Sickle, 1979). Conover (2002) estimated that white-tailed deer losses to commercial timber throughout the Midwestern and eastern United States were $1.6 billion annually.

Deer Damage to United States Households

Deer can be a nuisance when their herbivory damages landscape plants around homes and destroys gardens. During 1993, 61% of metropolitan residents in the United States reported that they had experienced

FIGURE 13.2 White-tailed deer cause millions of dollars of damage to agriculture every year. (Photo by L. Thomas, Jr., Quality Deer Management Association. With permission.)

a problem caused by wildlife during the previous year; damages per household averaged $64 despite the average household spending $32 in an attempt to alleviate the problem (Conover, 1997). When extrapolated to the 60 million metropolitan households in the United States, losses totaled $5.7 billion ($3.8 billion in damages and $1.9 billion in prevention). More than 10% of metropolitan residents reported that they experienced problems during the previous year caused by mice, squirrels, raccoons, moles, pigeons, starlings, and skunks. Only 4% of metropolitan residents complained of having a deer problem (Conover, 1997). If deer were responsible for 4% of the wildlife damage to metropolitan households, then deer caused $228 million in losses during 1993, an amount equal to $328 million in 2008 after adjusting for inflation.

There are an additional 34 million U.S households outside of metropolitan areas (Conover, 2002). These rural and small city residents should face even greater losses to wildlife than those of metropolitan residents because wildlife densities are lower in metropolitan areas. However, if we assume that rural and small city households suffer at least as much damage as metropolitan households, then wildlife damage to non-metropolitan households equaled $3.2 billion during 2008 (Conover, 2002). The amount of this attributable to deer equals $186 million, if deer cause 4% of the wildlife problems in non-metropolitan areas.

Effect of Deer Densities on Positive and Negative Values

Even when a deer population is low, it still has high existence, historic, and recreational values. Concomitantly, negative values of deer are minimal because deer are not numerous enough to cause many problems. As the deer population increases, its positive values also increase but at an ever-slowing rate. The reason for this is because most people receive more enjoyment from seeing their first deer than they receive from seeing their thousandth deer. Likewise, deer hunters get more satisfaction from shooting their first deer than their hundredth.

The negative values of deer also increase as deer numbers increase. For some deer problems, the relationship between deer densities and deer damage is linear and proportional. That is, a reduction in deer numbers causes an equal decrease in the severity of deer damage. For example, DeNicola and Williams (2008) found that a 54% decrease in deer densities resulted in a 49% decrease in the frequency of DVCs, while a 72% decrease in deer densities caused a 75% decrease in DVCs. Other problems caused by deer, such as damage to unpalatable crops, are minor until deer densities approach carrying capacity, and then they increase exponentially in both severity and frequency (Figure 13.3).

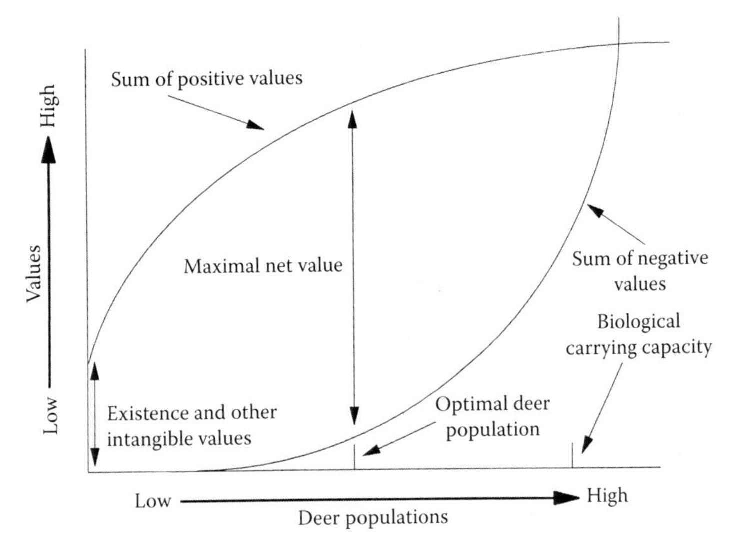

FIGURE 13.3 Positive and negative values of white-tailed deer as a function of deer density.

TABLE 13.1

Positive, Negative, and Net Values of Deer (Both White-tailed and Mule Deer) on an Annual Basis (Values are Reported on the Inflation-Adjusted Value of a United States Dollar during 2008)

Positive values	
Ecological benefits	Unknown
Existence, empathic, and historical values	Unknown
Recreational value—Deer hunting	$12.0 billion
Recreational value—Deer viewing	$4.6 billion
Total positive values	>$16.6 billion
Negative values	
Ecological damage	Unknown
Deer–vehicle collisions	$1.7 billion
Physical attacks of people	Minor (<$1 million)
Zoonotic diseases	Unknown
Losses to agricultural production	$0.6 billion
Losses to timber industry	$1.6 billion
Damage to households	$0.5 billion
Total negative values	>$4.4 billion
Net value of deer (annual profit for society)	$12.2 billion

Net Value of Deer

To determine the net value of deer, the negative value of deer (>$4.4 billion) must be subtracted from their positive value (>$16.6 billion). Doing so indicates that deer have a net positive value of $12.2 billion (Table 13.1). This amount can be viewed as the value that deer bring to society on an annual basis. In this regard, it is similar to the annual profit of a company or the annual yield on a savings bond. Further, evidence that the positive values of deer greatly outweigh the negative values is the finding that few people, including those that have been in a DVC, want lower deer populations (Stout et al., 1993, Storm et al., 2007).

The optimal deer population is one where the positive value exceeds negative value by the greatest amount (i.e., net value is greatest) (Figure 13.3). This optimal deer population is usually below an area's carrying capacity. If a deer population is going to be maintained below the carrying capacity, then society must intervene by either increasing the deer population's mortality rate or decreasing its birth rate. Although reducing the fertility of deer populations may be an option in the future, society currently lacks the means to do so (Curtis et al., 2008; Miller et al., 2008, 2009; Rutberg and Naugle, 2008). Hence, society's only option at present is to try to increase the mortality rate of deer. Efforts to do so have been successful either by using recreational hunters (Conover, 2001) or sharp shooters (DeNicola and Williams, 2008). However, these practices need to be continued annually, or the deer population will rebound to its prior level.

REFERENCES

Bissonette, J. A., C. A. Kassar, and L. J. Cook. 2008. Assessment of costs associated with deer–vehicle collisions: Human death and injury, vehicle damage, and deer loss. *Human–Wildlife Conflicts* 2:17–27.

Conover, M. R. 1997. Wildlife management by metropolitan residents in the United States: Practices, perceptions, costs, and values. *Wildlife Society Bulletin* 25:306–311.

Conover, M. R. 1998. Perceptions of American agricultural producers about wildlife on their farms and ranches. *Wildlife Society Bulletin* 26:597–604.

Conover, M. R. 2001. Effect of hunting and trapping on wildlife damage. *Wildlife Society Bulletin* 29:521–532.

Conover, M. R. 2002. *Resolving Human–Wildlife Conflicts: The Science of Wildlife Damage Management*. New York, NY: Lewis Publishers.

Conover, M. R. and D. O. Conover. 2003. Unrecognized values of wildlife and the consequences of ignoring them. *Wildlife Society Bulletin* 31:843–848.

Curtis, P. D., M. E. Richmond, L. A. Miller, and F. W. Quimby. 2008. Physiological effects of gonadotropin-releasing hormone immunocontraception on white-tailed deer. *Human–Wildlife Conflicts* 2:68–69.

deCalesta, D. S. 1994. Effect of white-tailed deer on songbirds within managed forests in Pennsylvania. *Journal of Wildlife Management* 58:711–718.

Decker, D. J., K. M. Loconti Lee, and N. A. Connelly. 1990. Incidence and costs of deer-related vehicular accidents in Tompklins County, New York, NY: Insights on an increasingly important aspect of deer management. Human Dimensions Research Unit Series 89-7. Cornell University, Ithaca, New York, USA.

DeNicola, A. J. and S. C. Williams. 2008. Sharpshooting suburban white-tailed deer reduces deer–vehicle collisions. *Human–Wildlife Conflicts* 2:28–33.

De Vault, T. L., J. E. Kubel, D. J. Glista, and O. E. Rhodes, Jr. 2008. Mammalian hazards at small airports in Indiana: Impact of perimeter fencing. *Human–Wildlife Conflicts* 2:225–239.

Dolbeer, R. A. 2009. Birds and aircraft: Fighting for airspace in ever more crowded skies. *Human–Wildlife Conflicts* 3:165–166.

Dumler, J. S., K-S Choi, J. C. Garcia-Garcia, N. S. Barat, D. S. Scorpio, J. W. Garyu, D. J. Grab, and J. S. Bakken. 2005. Human granulocytic anaplasmosis and *Anaplasma phagocytophilum*. *Emerging Infectious Diseases* 11:1828–1834.

Gedney, D. R. and C. Van Sickle. 1979. Geographic context of forestry. In *Forest Resource Management: Decision-Making Principles and Cases*, eds. W. A. Duerr, D. E. Teeguarden, N. B. Christiansen, and S. Guttenberg, 301–318. Philadelphia, PA: W. B. Saunders.

Giles, R. H., Jr. 1978. *Wildlife Management*. San Francisco, CA: W. H. Freeman.

Giles, R. H., Jr. 1990. A new focus for wildlife resource managers. *Journal of Forestry* 88:21–26.

Hobbs, N. T. 1996. Modification of ecosystems by ungulates. *Journal of Wildlife Management* 60:695–713.

Hubbard, R. D. and C. K. Nielsen. 2009. White-tailed deer attacking humans during the fawning season: A unique human–wildlife conflict on a university campus. *Human–Wildlife Conflicts* 3:129–135.

Hussain, A., J. B. Armstrong, D. B. Brown, and J. Hogland. 2007. Land-use pattern, urbanization, and deer–vehicle collisions in Alabama. *Human–Wildlife Conflicts* 1:89–96.

Kilpatrick, H. J. and A. M. LaBonte. 2007. *Managing Urban Deer in Connecticut*. Hartford, CT: Connecticut Department of Environmental Protection.

Lane, R. S., J. Piesman, and W. Burgdorfer. 1991. Lyme borreliosis: Relation of its causative agent to its vectors and hosts in North America and Europe. *Annual Review of Entomology* 36:587–609.

Marcoux, A. and S. J. Riley. 2010. Driver knowledge, beliefs, and attitudes about deer–vehicle collisions in southern Michigan. *Human–Wildlife Conflicts* 4:47–55.

Marquis, D. A. 1981. *Effect of Deer Browsing on Timber Production in Allegheny Hardwood Forests of Northwestern Pennsylvania*. Northeastern Forest Experiment Station, Broomall, Pennsylvania.

Marquis, D. A. and R. Brenneman. 1981. *The Impact of Deer on Forest Vegetation in Pennsylvania*. Northeastern Forest Experiment Station, Broomall, Pennsylvania.

McShea, W. J. and J. H. Rappole. 2000. Managing the abundance and diversity of breeding bird populations through manipulation of deer populations. *Conservation Biology* 14:1161–1170.

Miller, L. A., K. A. Fagerstone, D. C. Wagner, and G. J. Killian. 2009. Factors contributing to the success of a single-shot, multiyear PZP immunocontraceptive vaccine for white-tailed deer. *Human–Wildlife Conflicts* 3:93–102.

Miller, L. A., J. P. Gionfriddo, J. C. Rhyan, K. A. Fagerstone, D. C. Wagner, and G. J. Killian. 2008. GnRH immunocontraception of male and female white-tailed deer. *Human–Wildlife Conflicts* 2:93–101.

National Highway Traffic Safety Administration. 2000. Analysis of light vehicle crashes and pre-crash scenarios based on the 2000 General Estimates System. U.S. Department of Transportation Publication DOT-VNTSC-NHTSA-02–04. Washington, DC, USA.

Romin, L. A. and J. A. Bissonette. 1996. Deer–vehicle collisions: Status of state monitoring activities and mitigation efforts. *Wildlife Society Bulletin* 24:276–283.

Rutberg, A. T. and R. E. Naugle. 2008. Deer–vehicle collision trends at a suburban immunocontraception site. *Human–Wildlife Conflicts* 2:60–67.

Stafford, K. C., III. 2007. *Tick Management Handbook*. Connecticut Agricultural Experiment Station, Bulletin No. 1010, New Haven, CT, USA.

Storm, D. J., C. K. Nielsen, E. M. Schauber, and A. Woolf. 2007. Deer–human conflict and hunter access in an exurban landscape. *Human–Wildlife Conflicts* 1:53–59.

Stout, R. J., R. C. Stedman, D. J. Decker, and B. A. Knuth. 1993. Perceptions of risk from deer-related vehicle accidents: Implications for public preferences for deer herd size. *Wildlife Society Bulletin* 21:237–249.

Tilghman, N. G. 1989. Impacts of white-tailed deer on forest regeneration in northwest Pennsylvania. *Journal of Wildlife Management* 53:524–532.

U.S. Centers for Disease Control. 2004. Nonfatal motor–vehicle animal crash-related injuries, United States, 2001–2002. *Morbidity and Mortality Weekly Report* 53:675–678.

U.S. Centers for Disease Control. 2009a. Lyme disease transmission. http://www.cdc.gov/ncidod/dvbin/lyme/ld_transmission.htm. Accessed June 5, 2009.

U.S. Centers for Disease Control. 2009b. Babesiosis fact sheet. http://www.cdc/babesiosis/factsheet.html. Accessed June 5, 2009.

U.S. Department of Transportation. 2002. *Revision of Departmental Guidance: Treatment of Value of Life and Injuries*. U.S. Department of Transportation, Washington, DC, USA.

U.S. Department of Transportation. 2009. Licensed drivers—our nation's highways. http://www.fhwa.dot.gov/ohim/onh00/onh2p4.htm. Accessed August 26, 2009.

U.S. Fish and Wildlife Service. 2006. National survey of fishing, hunting and wildlife-associated recreation. U.S. Department of the Interior and U.S. Department of Commerce, Washington, DC, USA.

VerCauteren, K. C., N. W. Seward, M. J. Lavelle, J. W. Fischer, and G. E. Phillips. 2009. Deer guards and bump gates for excluding white-tailed deer from fenced resources. *Human–Wildlife Conflicts* 3:145–145.

Warren, R. J. 1991. Ecological justification for controlling deer populations in eastern national parks. *Transactions of the North American Wildlife and Natural Resources Conference* 56:56–66.

14

Extensive Management

Lonnie Hansen

CONTENTS

Introduction

The recovery of North American white-tailed deer populations is one of wildlife management's greatest success stories. In every state and province within the range of the white-tailed deer, numbers have gone from abundance prior to European settlement, to historical lows between 1850 and the early 1900s, to probable record numbers in recent years (McDonald and Miller, 2004) (Chapter 11). The decline and then recovery occurred at a time when humans were impacting the landscape at scales never before experienced. Throughout much of the white-tailed deer's range, forests were cut, most accessible land was grazed or farmed, and humans were scattered on small parcels of the rural landscape (Figure 14.1). The decrease in deer resulted partly from the human-induced environmental changes. However, overexploitation as a result of local consumption and market hunting probably played a major role.

How did recovery occur in spite of such an immense change in the landscape and what management elements produced successful restoration programs in every state? The white-tailed deer is highly adaptable to human activities and in spite of what humans did to the landscape, the white-tailed deer tolerated or even took advantage of the changes and flourished. Even as adaptable as the white-tailed deer is, however, population recovery would not have occurred without a change in public attitudes toward wildlife.

FIGURE 14.1 Humans drastically altered the landscape throughout much of the range of the white-tailed deer in the late nineteenth and early twentieth centuries. (Photo courtesy of the Missouri Department of Conservation. With permission.)

Early efforts to stem the decline in deer populations through regulation generally failed because of the lack of public support. In the early twentieth century, attitudes toward wildlife shifted from a utilitarian to a more conservation-oriented emphasis; citizens recognized their impacts on wildlife and began supporting efforts to protect and restore many wildlife species (McCabe and McCabe, 1984). The stage was set for recovery of many species when state agencies were formed and modern day conservation began.

What role did conservation agencies play in the recovery of white-tailed deer populations? Early agencies varied greatly in how they managed deer-hunting seasons and harvests and the amount of resources dedicated to research, habitat management, and enforcement of regulations. Although timing of the recovery may have varied, deer population restoration occurred in every state within the range of the white-tailed deer in spite of these large management differences. It is clear that there was a common element in agency deer management that produced the successful recovery. This element was protection from overexploitation. Deer season regulations became restrictive in every state, especially related to doe harvest, and enforcement efforts were increased.

One might argue that early management was relatively simple for agencies because the objective was to grow deer populations and this could be accomplished by protecting does. As deer populations grew and problems associated with this growth developed, such as crop damage and deer–vehicle accidents, management has become more complex. More deer demographic information is required when managing for stable or reduced deer populations rather than growing populations. Human interests beyond hunters also must be considered as deer and human populations grow and deer increasingly impact human activities. Agencies must address a whole new arena of human dimensions when making management decisions.

Most agency programs have evolved since the early years (Figure 14.2). We know much more about white-tailed deer biology and population dynamics, hunting opportunities have expanded, human dimensions play a large role in most agency management decisions, and many challenges to traditional management have presented themselves, such as urban deer problems and declining hunter numbers.

Although agency deer population management goal statements may read differently, most include emphases on maintaining deer numbers at levels that are consistent with the land to support deer,

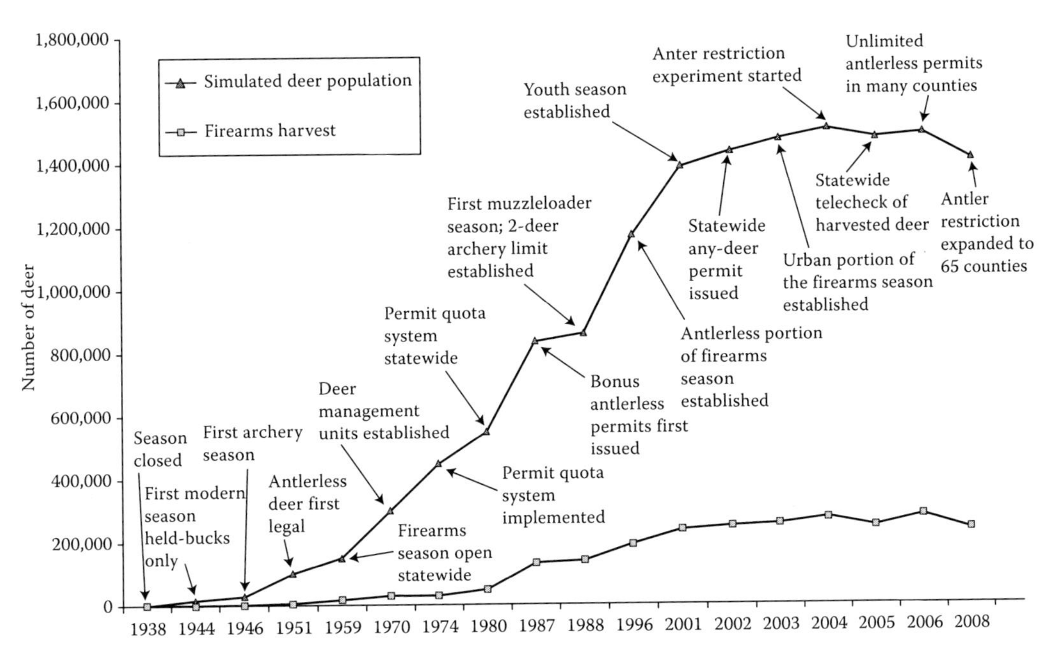

FIGURE 14.2 Deer population estimates, firearms deer harvest, and highlights of deer management in Missouri, 1938–2008. Although timing and regulation type vary considerably among agencies, most follow similar trends in deer season liberalizations and deer population increases.

providing deer hunters and viewers with acceptable levels of opportunity, and minimizing conflicts between humans and deer. As a result, deer management on the scale faced by agencies involves a combination of economic, political, philosophical, and biological considerations (Roseberry and Woolf, 1991). Agencies must maintain a clear set of: (1) deer management goals and objectives; (2) reliable data on deer population demographics; and (3) a conceptual framework for decision making. Every state in some way addresses these issues; most in very different ways. In this chapter I will describe the management processes and constraints encountered by agencies charged with managing deer in highly variable landscapes in which white-tailed deer exist in North America. Many factors must be considered with the foundation of agency deer management rooted in the North American Model of Wildlife Conservation (NAMWC).

North American Model of Wildlife Conservation

The NAMWC is based on early European law which viewed wild animals as belonging to no one when alive, much like the wind and oceans, but becoming property of an individual once captured or killed (Bean and Rowland, 1997). Key components of the NAMWC presented and interpreted by Geist et al. (2001) are provided in the following with a discussion of the challenges to agencies in incorporating and maintaining these components in deer management programs given today's changing recreational, land-management, social, and economic climate.

1. Wildlife as Public Trust Resources—Agency powers to oversee wildlife for the benefit of all were established in the early twentieth century (Geist et al., 2001) with general agreement that wildlife does not belong to the individual but to all citizens. This concept is easiest to apply on publicly owned land because government agencies control access to these properties and therefore can allocate use of their resources to all citizens. However, on private land, which constitutes a high percentage of the landscape in most states, public ownership of wildlife is largely theoretical (Allen, 1974) because landowners control access and thus use of the resources.

Historically, much private land was open to deer hunters who gained access free-of-charge, often through family and friends (Mozumder et al., 2007). Although this continues to be the norm in most states (Cordell et al., 1999), leasing property for hunting has increased over time (Adams et al., 2000). Increasing privatization (individual control of a resource) acts contrary to the NAMWC principle that deer belong to all citizens, especially on those properties purchased for personal hunting of deer or managed for landowner income through leasing the property for hunting deer. However, potential ecological benefits may result from land dedicated to hunting. Benson (1992) argued that privatization can be an incentive to manage habitat for deer and other wildlife which otherwise would not happen; more economically but not ecologically sound land management practices otherwise could prevail.

2. Elimination of markets for wildlife—Markets for wild animal products has a tainted history because it led to the near demise of many North American species, including white-tailed deer (Figure 14.3). This unregulated killing is a history lesson ingrained in most conservationists' memories and rightly so. Also, deer ranching for economic gain and the implications for disease spread and genetic consequences are a concern. The market issue today related to deer and other wildlife is varied and not clear cut. For example, products from wild furbearers are accepted commodities but sale of other wild game products such as venison from wild white-tailed deer is prohibited by most agencies. Yet in most states, antlers, hides, and other body parts can be bought and sold. Allowing the sale of venison from wild deer is unacceptable to most deer biologists because of the negative historical consequences of market hunting. However, predicted social and landscape changes will challenge the ability of agencies to manage deer (Brown et al., 2000). Marketing of deer, for example, through the sale of venison, may be a future consideration to facilitate population control (Nugent and Choquenot, 2004).
3. Allocation of wildlife by law—Conservation agencies have historically had the authority to establish deer-hunting regulations. These regulations are applied on a broad scale such as a county or deer management unit (DMU) and are intended to maintain deer densities at levels that are in the best interest of all citizens. As deer populations have increased, the number of stakeholders impacted by deer has increased. Changing social climates have forced agencies to measure public sentiment and incorporate public interests into deer management programs. Successful agencies do this; those that do not risk loss of credibility and authority to set and enforce biologically sound deer-hunting regulations (Jacobson and Decker, 2006). Already, ballot initiatives and referenda have threatened agency authority; this will likely continue (Williamson, 1998). Privatization of wildlife also threatens agency authority to regulate,

FIGURE 14.3 Market scene in the late nineteenth century when wildlife was an important commodity. Marketing wildlife resulted in the extinction or near extinction of many North American wildlife species. (Photo courtesy of the Missouri Department of Conservation. With permission.)

usually indirectly by compromising the intended objective of a regulation, potentially deterring the ability of an agency to manage deer populations (Collier and Krementz, 2006).

4. Wildlife can only be killed for a legitimate purpose—Geist et al. (2001) stated that killing wildlife for frivolous reasons is prohibited by law; legitimate purposes for killing wildlife include food, fur, and self-defense or property protection. Benson (1992) suggested that strict interpretation of this concept is contrary to policy and hunting practice in North America because there is a significant recreational component to hunting. Many interpret Geist et al. (2001) to include recreational hunting as long as there is no waste of the wildlife taken. Wanton waste rules have been established by many agencies to address this.
5. Science is the proper tool for discharge of wildlife policy—This concept states that science-based decisions should drive wildlife management. Most agency deer biologists are trained to collect, interpret, and apply deer ecological and demographic information in making management decisions. However, because of the impacts white-tailed deer can have on human activities, human-associated issues have become increasingly important in decisions agencies make concerning deer management, sometimes overriding science-based decision making.
6. Democracy of hunting—The basic concept is that all citizens have an equal opportunity to hunt. In practice this is most easily accomplished on public land. Because landowners control hunter access, hunting for all on private land is more problematic. This is especially a problem for white-tailed deer because in most states private land is unavailable to all citizens for deer hunting. Lessees, owners purchasing property specifically for hunting deer, or family and friends of the landowner often have exclusive deer-hunting rights. States have developed various programs to promote deer-hunting opportunities on private land with mixed success.

These components of the NAMWC form the philosophical basis of most agency wildlife management programs but there is considerable variability in interpretation and adherence to the principles. This is especially true when applied to species such as the white-tailed deer, which generate considerable interest from all segments of the public. As a result, there is disagreement among white-tailed deer biologists and managers about the management conflicts and opportunities associated with the progressive privatization of the deer resource. This will probably increase in the future as social, land ownership, and economic patterns related to deer evolve.

Deer Population Assessment

Hunting accounts for a significant proportion of mortality of white-tailed deer, especially males, throughout much of their range (Ditchkoff et al., 2001b; Nixon et al., 2001; Hansen and Beringer, 2003; DelGiudice et al., 2006). As a result, annual harvests influence deer population size and thus provide basic information useful in assessing population status (Roseberry and Woolf, 1991; Skalski et al., 2007). It follows that monitoring deer harvest and deer population size are integral parts of most agency deer management programs. The spatial scale of deer population monitoring varies among agencies but usually ranges from smaller than a county up to multiple counties. I will refer to the level at which an agency regulates as a DMU. Monitoring deer populations at the scale of a DMU presents significant challenges and has been the object of considerable research on methods of collecting harvest information and determining population size.

Deer Harvest Determination

Annual deer harvests provide information on population status and serve as the basis for establishing harvest regulations. Basic information collected on harvested deer varies among agencies and may include gender, age, location (usually DMU) and date of kill, number of days hunted, harvest method (e.g., rifle, bow, muzzleloader), type of permit used, and a hunter identifier. Methods of collecting harvest information vary among states (Rupp et al., 2000). Direct methods, most commonly mandatory in-person

FIGURE 14.4 For many agencies, deer checking stations serve as a valuable means of collecting data from harvested deer and enforcing deer-hunting regulations. (Photo courtesy of the Missouri Department of Conservation. With permission.)

checking, require the deer to be taken to a station where it is observed by a checker and information on the deer is recorded (Figure 14.4). Mandatory checking not only serves as an enumeration of legal harvest, but also is a regulation enforcement tool and a convenient vehicle for collecting additional biological data, such as antler measurements, weights, parasites, and tissue samples. Disadvantages of this method include agency expense (Rupp et al., 2000; Hansen et al., 2006), inconvenience and expense for the successful hunter, and potentially poor compliance (Rosenberry et al., 2004; Kilpatrick et al., 2005; Hansen et al., 2006). Indirect methods of measuring harvest include mandatory reporting by mail, internet, or telephone which provides an enumeration of harvest. Estimates of harvest are obtained through mail and telephone surveys of randomly selected hunters (Skalski et al., 2006). Indirect methods are less expensive for the agency and more convenient for the hunter. However, these methods may be biased if the hunter does not respond or is not truthful in what they report (Taylor et al., 2000) and require other methods for collecting biological information. Mandatory telechecking (check deer by phone or internet) is becoming an increasingly popular method for reporting harvested deer in many states.

Deer Population Indices, Estimates, and Simulations

Knowing the status of deer populations is essential for tracking the effectiveness of management programs and setting harvest regulations. Deer population assessments fall into various categories most commonly including indices of relative abundance, estimates of actual abundance, and model simulations.

Many techniques for assessing deer population status have been developed to monitor local population trends, generally on publically owned properties. Indices include pellet group counts (Neff, 1968; Patterson and Power, 2002), track/trail counts (Daniel and Frels, 1971; McCaffery, 1976; Fritzen et al., 1995; Forsyth et al., 2007), browse surveys (Morellet et al., 2001), physical condition models (Shea et al., 1992; Keyser et al., 2006), hunter observations/perceptions (Winchcombe and Ostfeld, 2001a), habitat-based models (Rutledge et al., 2008), and spotlight counts (Progulske and Duerre, 1964; Wood et al., 1985; Storm et al., 1992; Winchcombe and Ostfeld, 2001b). Methods of deriving population estimates from some of these techniques also have been developed (Fafarman and DeYoung, 1986; Mandujano, 2005; Forsyth et al., 2007) although their validity has been challenged (Collier et al., 2007). Methods of estimating deer population size and characteristics on a local scale have included use of aerial (DeYoung, 1985; Beringer et al., 1998; Potvin and Breton, 2005), infrared (Naugle et al., 1996;

Haroldson et al., 2003; Drake et al., 2005), and trail camera censuses (Jacobson et al., 1997; Koerth and Kroll, 2000; Roberts et al., 2006; Slade, 2009).

The scale at which agencies apply deer regulations is too large for practical application of most of the indices described above, although a few states use them on a broader scale. For example, Iowa uses extensive aerial surveys as indices of DMU-level deer population trends (Iowa Deer Study Advisory Committee, 2009) and county-level deer population estimates are made from multiple roadside infrared counts in Tennessee (D. Ratajczak, Tennessee Wildlife Resources Agency, personal communication). Following are techniques most commonly used by agencies to assess population status on a broad scale.

Population Indices

Although most agencies use deer population indices extensively to track deer population trends, the validity of their use has been questioned (Anderson, 2001, 2003). Concern centers mostly on often unrealistic assumptions about the probability of detecting the deer or sign left by the deer. Depending on the survey method, many factors can affect the probability of detection. If a project uses more than one observer, as most multiyear deer projects do, individual variability in factors such as training, experience, conscientiousness, and ability to concentrate can affect the probability of detection. Environmental variables such as habitat type, season of year, precipitation, and temperature also can affect probability of detection. Convenience sampling in which data are collected along existing transects (e.g., road, power line) may not provide a representative sample of the deer population. Differential detectability of gender and age classes of deer can affect the calculated index. Finally, because deer studies often occur over long periods of time during which indices are determined annually, changes in vegetation and landscape due to succession or human disturbance can affect probability of detection. Although Anderson (2001) pointed out these problems relative to indices, Engeman (2003) countered that these issues also may be a problem with any wildlife study, including estimates of population size.

Deer–Vehicle Accidents

Deer–vehicle accidents are a major source of concern for all agencies because of the costs in damage, human injuries, and loss of deer and human life throughout the range of the white-tailed deer (Conover et al., 1995; Bissonette et al., 2008) but they also may serve as an index of deer population size. The number of deer–vehicle accidents may be dependent on numerous factors including deer population size, traffic volume and speed, and landscape characteristics (Hubbard et al., 2000; Farrell and Tappe, 2007; Grovenburg et al., 2008). McCaffery (1973) found that buck harvest and the number of deer–vehicle accidents adjusted for traffic volume were highly correlated in Wisconsin. However, this trend was most useful in assessing statewide not DMU-level trends. Problems associated with using deer–vehicle accidents as an index include consistency of data collection over time, annual variation in factors that may affect deer–vehicle accidents such as food distribution and abundance (e.g., acorn production) (Ryan et al., 2004), and timing of the opening of the firearms deer-hunting season (Sudharsan et al., 2006). The number of deer–vehicle accidents standardized for traffic volume may best represent statewide deer population trends; use for DMU-level trends may be limited in most states. This statewide deer population index based on deer–vehicle accidents may be most useful to agencies in justifying harvest liberalizations or restrictions and highlighting deer management successes and not for prescribing specific DMU-level harvests.

Bowhunter Index

Archery hunting is popular throughout the range of the white-tailed deer (Figure 14.5). Participants are dedicated, given long seasons and liberal bag limits, and spend considerable time hunting. Bowhunter observations of wildlife over a period of years may provide useful population indices (Winchcombe and Ostfeld, 2001a). As part of a number of agency furbearer programs, cooperating archery deer hunters are asked to maintain a log of their observations of furbearers while deer hunting. Information generally recorded includes hunting location, date, number of hours hunted, and number of furbearers seen while hunting (Frawley, 2004). White-tailed deer are included as a recordable species in most of these surveys;

FIGURE 14.5 Deer bowhunters are provided with long seasons and liberal harvest opportunities. Bowhunter observations of deer while hunting can provide useful indices of deer population size and demographics. (Photo by the Missouri Department of Conservation. With permission.)

some include deer gender and age categories. At the end of the designated recording period, diary forms are submitted to an agency biologist and deer observation rates are calculated.

Factors other than deer population size can affect observation rates. Annual variation in food availability on a broad scale, especially acorns (Feldhamer, 2002; Ryan et al., 2004), can affect harvest which is likely correlated with deer observation rates. Time trends (Anderson, 2001) that involve increasing hunter experience also could affect observation rates. Many participants are long-term cooperators whose skills and familiarity with the places they hunt probably improve over time, independent of deer abundance.

As with deer–vehicle accidents, archer observations may serve as a general population index (Winchcombe and Ostfeld, 2001a). Downing et al. (1965) compared five census techniques and concluded that hunter observations could provide accurate index values. Lancia et al. (1996) used data from hunter diaries to estimate the number of bucks, using a catch per unit effort model. Depending on the number and distribution of participants, population indices on a DMU or statewide basis may be tallied.

Hunter and Landowner Perceptions of Deer Population Status

Hunter and landowner-based indices of deer population size can be generated from random surveys of landowners and deer hunters (Brown and Decker, 1979; Stoll and Mountz, 1983; Miller and Shelton, 2000; West and Parkhurst, 2002). Generally these mail or telephone surveys are not conducted on a scale appropriate for DMU-level inferences; they are limited to broader groups of DMUs or statewide trends. Annual variation or trends in hunter effort per deer killed, hunter success at harvesting one or more deer, or hunter and landowner perceptions of deer population status may reflect deer abundance although these indices suffer from the same problems as bowhunter observation indices. In addition, landowner attitudes and perceptions of deer can be affected by current crop-growing conditions and market prices for their products.

Antlered Buck Harvest Trends

Annual harvest, especially antlered buck harvest (Figure 14.6), has been used commonly as an index of population size and often serves as the "known" population index to which other indices are compared (McCaffery, 1976; Brown and Decker, 1979; West and Parkhurst, 2002). Assumptions that must be met

FIGURE 14.6 Harvest of antlered deer has served as an index of deer population for many agencies. (Photo by L. Hansen.)

for the antlered harvest to be representative of the population include stable hunter effort and hunter selectivity, and constant deer reproductive and mortality rates. Environmental factors also can affect harvest, such as distribution and abundance of natural food supplies (Ryan et al., 2004), weather (Hansen et al., 1986), and stage of crop harvest, especially corn (Hansen et al., 1986). Unfortunately, all of these factors can vary annually, potentially affecting overall deer harvest and its composition rendering use of the antlered harvest as an index of population size questionable (Millspaugh et al., 2009).

Herd Composition

Agencies often attempt to determine the sex and age composition of the deer herd (Roseberry and Woolf, 1991; Kubisiak et al., 2001; Bender, 2006). Although criticized for use in establishing trends in population size (McCullough, 1994), sex and age ratios may provide insights into mortality rates and productivity, and may help agencies track the success of specific deer management programs (Bender, 2006). Effort required to collect adequate data to make DMU inferences is often not possible for most agencies; data are often pooled across DMUs creating biases if the pooled DMUs are not homogenous (Millspaugh et al., 2007). There also is often high variance and a lack of a random sampling design associated with collection of herd composition data in many states (Rabe et al., 2002). Rabe et al. (2002) described an ideal big-game survey as one that uses a single method for all parameter estimates, directly estimates population size, uses random sampling, and estimates survival of females of breeding age. These may be most easily achieved for western ungulates because of the characteristics of the landscape and may not easily be applied to white-tailed deer.

Habitat-Based Indices

Roseberry and Woolf (1998) found that habitat suitability indices explained 81% of the variation in county deer population densities in Illinois (Figure 14.7). However, they suggested that deer herds in most states are regulated primarily by harvest and not habitat. Roseberry and Woolf (1991) concluded that habitat-based population estimates may not be useful for predicting population size or the relationship between deer abundance and ecological carrying capacity where populations are maintained at low densities through harvest. A stem count index used to appraise white-tailed deer habitat suitability in Texas was considered a reliable measure of deer density (Rutledge et al., 2008). However, its application for measuring broad-scale deer populations may be limited.

FIGURE 14.7 Habitat suitability can explain variation in deer densities but often harvest maintains deer numbers below levels that would be supported by the habitat. (Photo by L. Hansen.)

Deer Population Estimates Based on Sex and Age Distribution in Harvest

For most agencies estimated or enumerated deer harvest is the primary source of information on the status of DMU-level deer populations. As a result, use of data collected from harvested deer to estimate deer populations has been the subject of considerable research (Roseberry and Woolf, 1991; Williams et al., 2001; Skalski et al., 2005). Analysis methodologies and output vary considerably. Roseberry and Woolf (1991) categorized output in terms of the information provided (abundance, population parameters, or both), type of estimate (absolute or index), and time frame (retrospective or predictive). They suggested that agencies concentrate on annual collections of data that include total harvest, effort expended, sex ratio of the harvest, and sex-specific age structure. They recommended that managers use a variety of techniques to estimate deer abundance and concluded that the best techniques for estimating a known population on a public area in Illinois were the DeLury catch per unit effort (DeLury, 1947; Eberhardt, 1960) and population reconstruction methods (Fry, 1949; Woolf and Harder, 1979).

Most agencies that develop broad-scale population estimates from harvest data use some form of reconstruction (Davis et al., 2007; Mattson and Moritz, 2008; Millspaugh et al., 2009); often a sex–age–kill (SAK) model variation (Millspaugh et al., 2007). Millspaugh et al. (2009) evaluated the robustness of the SAK to stochastic effects such as variation in survival and recruitment, and changes in hunter harvest strategies or agency regulations influencing sex and age distribution of the harvest. Violations of the requirement of a stable age distribution and stationary population also were assessed. Millspaugh et al. (2009) demonstrated that differential buck harvest rates could dramatically impact SAK estimates. Unfortunately, increasing interest among hunters in taking bucks in older age classes and some agency regulations (e.g., antler restrictions) have produced differential harvest rates, with 1.5-year-old bucks taken at a lower rate than older bucks (Chapter 11). As a result, Millspaugh et al. (2009) suggested that use of SAK is questionable unless independent estimates of male recovery rates are available. Pennsylvania has addressed differential harvest rates of their antlered bucks by determining survival rates of bucks by large-scale monitoring of radiocollared deer (D. R. Dieffenbach, Cooperative Fish and Wildlife Research Unit, Pennsylvania State University, personal communication). Most states, however, do not have the resources to conduct such extensive studies.

Millspaugh et al. (2009) recommended using likelihood-based statistical population reconstruction methods as described by Skalski et al. (2007). Statistical population reconstruction is an approach to analyzing age-at-harvest data that can be updated as more refined or additional demographic data are collected. It provides a framework upon which a variety of demographic studies can be linked for the purpose of estimating wildlife populations at broad scales. Data requirements include annual harvests by age class and at least one other demographic parameter. Auxiliary demographic data might include the population indices described above.

Population Simulation Models

Accounting-type population simulation models have been used by many agencies to track population size and composition and guide deer management (Walters and Gross, 1972; Medin and Anderson, 1979; McCullough, 1984; Roseberry and Woolf, 1991; Roseberry, 1995; Xie et al., 1999; Grund and Woolf, 2004). These models can help determine factors that most affect deer population change so that research can be targeted (Medin and Anderson, 1979; Grund and Woolf, 2004) and to evaluate the impacts of various harvest strategies or population perturbations on population change (McCullough, 1984; Fischer et al., 1995; Starfield, 1997; Riley et al., 2002b; Keyser et al., 2006). Most deer population accounting models are data demanding and it is generally not practical for agencies to collect survival and reproductive data annually for input; relatively constant values for these must be assumed. In parts of the white-tailed deer range, most mortality is hunting-caused and reproductive rates do not vary much (Hansen et al., 1996; Nixon et al., 2001; Hansen and Beringer, 2003; Brinkman et al., 2004; Burroughs et al., 2006) but extreme weather, predators (Nelson and Mech, 1986; Van Deelen et al., 1997; Ginnett and Young, 2000; DelGiudice et al., 2006), and disease outbreaks (Nettles and Stallknecht, 1992; Fischer et al., 1995; Beringer et al., 2000) can cause annual variation that confounds model output. Grund and Woolf (2004) simulated the deer population on a large public area in Minnesota and found that model estimates were sensitive to small changes in adult female summer survival. They concluded that because it is not practical to measure annual survival rates, an accounting model should not be used exclusively to monitor long-term deer populations; other methods of population assessment should also be incorporated. Perhaps recommendations by Starfield (1997) concerning model development and use are most appropriate for agencies. He recommended a pragmatic approach to model building that emphasizes simplicity and problem-oriented development. Starfield (1997) proposed that researchers and managers cooperate in the development and implementation of models and that by so doing, improved communication and cooperation may result.

Deer Population Goals

All agencies, formally or informally, set goals for their deer populations that may be based on a multitude of biological and social factors (Carpenter, 2000). These include deer condition, deer population demographics, habitat characteristics, and stakeholder interests. Information used to establish these goals varies greatly among agencies and often is driven by traditions, the regulatory system, and landscape characteristics.

Carrying capacity can be defined in many ways but all imply a long-term stability of deer numbers, vegetation biomass, and community species composition (Sinclair, 1997). Some carrying capacities are biologically based, others socially or economically based. For white-tailed deer, most are related to the number of deer that a particular stakeholder feels can be accommodated. Many terms have been used to describe carrying capacity, and there has been considerable confusion about various aspects of carrying capacity (MacNab, 1985). For purposes here I will define ecological carrying capacity [ECC; also coined K carrying capacity by McCullough (1979)] as the maximum number of animals that can be sustained by a landscape over an extended time. Maximum sustainable yield [MSY or I carrying capacity of McCullough (1979)] is the population level relative to ECC where the rate of population change is greatest and thus the potential yield is greatest (McCullough, 1979; Skalski et al., 2007). ECC and MSY are most relevant in areas where deer populations have a history or the potential of reaching or exceeding ECC such as in the southeastern United States and some northern and eastern states and provinces where deer are dependent on limited natural food bases and/or where extreme conditions during parts of the year, especially winter and early spring, can be significant stressors (Mautz, 1978; Potvin and Huot, 1983; Fryxell et al., 1991; Shea and Osborne, 1995; Ginnett and Young, 2000; Patterson and Power, 2002; Williamson, 2003). White-tailed deer populations associated with agricultural areas, especially those areas without severe winters (e.g., much of the lower Midwest), are generally held well below ECC and MSY because factors other than deer demography, especially human social and economic issues, are more likely to dictate deer population goals. The role of these human dimension issues in deer management has expanded geographically and increased in importance as human and deer populations have increased (Woolf and Roseberry, 1998).

Biological Impacts on Carrying Capacity

Determining deer ECC and MSY at a DMU level can be problematic because both can vary considerably geographically and temporally (Carpenter, 2000) and most agencies do not have biological databases relating deer population size to ECC or MSY on a broad scale. Most efforts to determine the relationship between deer densities and carrying capacity have either evaluated food abundance relative to deer nutritional requirements, measured status of the flora, especially browsing intensity, or determined deer physical or demographic characteristics.

Attempts to determine carrying capacity using forage availability and quality, and nutritional requirements (Potvin and Huot, 1983; McCall et al., 1997; Miller and Wentworth, 2000) are labor intensive and not practical in most areas for establishing DMU-level carrying capacity, especially where most of the landscape is in private ownership and hunter distribution and harvest vary. More commonly, deer physical characteristics (Severinghaus et al., 1950; Eve and Kellogg, 1977; Williamson, 2003) are utilized for DMU-level determination of the relationship between the current deer population size and ECC. Deer physical characteristics such as antler development in 1.5-year-old males, body weight, and parasite loads are assumed to be correlated with herd condition which is an indicator of the herd status relative to ECC (Eve and Kellogg, 1977; Williamson, 2003).

Regressing some correlate of deer population condition (e.g., fawn recruitment) against population size has been proposed as a means of establishing ECC (McCullough, 1984). Wisconsin Division of Natural Resources used a ratio of the pre-hunt deer population to the post-hunt population the previous year as a measure of deer population growth (lambda) and thus an index of recruitment (Kubisiak et al., 2001). They assumed that lambda was an indicator of the deer population status relative to ECC and regressed lambda over time (dependent variable) against the post-hunt deer population (independent variable). The resulting equation could then be solved for lambda = 1 to determine the deer population size at ECC. ECC varied greatly among DMUs in Wisconsin, from 6 to >34 deer/km^2 (Rolley, 1995). Differences were attributed to variability in habitat type and food abundance associated with agricultural activities. In Wisconsin, the concept of ECC was most relevant in the northern forest region; in the southern agriculture zone social and not biological issues played a key role in management decisions (Rolley, 1995). Validity of this innovative procedure is contingent on accurate deer population estimates which are questionable for the SAK procedure used in Wisconsin, especially at the DMU level (Millspaugh et al., 2009). Temporal variation in ECC also can confound estimates of ECC. These issues exemplify the difficulty of managing deer populations in relation to ECC and illustrate why most states have no formal procedure for measuring ECC or MSY and managing at these levels.

White-tailed deer can have significant impacts on herbaceous plant communities (Russell et al., 2001; Côté et al., 2004; Killmaster et al., 2007) although such impacts have received less attention than other management considerations by agencies making DMU-level deer-management decisions. Protection of sensitive plant species generally requires deer populations to be maintained substantially below ECC. For example, densities of 15 deer/km^2 in Missouri slowed but did not stop population growth of American ginseng (Farrington et al., 2009). In West Virginia, McGraw and Furedi (2005) projected that populations ranging from 15 to 49 deer/km^2 would cause ginseng populations to decline. Farrington et al. (2009) concluded that ginseng can tolerate browsing by deer at moderate population sizes but may decline at browsing levels produced by high deer populations.

White-flowered trillium appears to be especially sensitive to deer browsing (Anderson, 1994; Augustine and Frelich, 1998; Knight, 2004). The number of trillium plants in flower decreases with increased browsing and plant height decreases over time (Anderson, 1994). Anderson (1994) recommended that densities <6 deer/km^2 should be maintained in northeastern Illinois forests to minimize impacts on trillium (Figure 14.8). Knight (2004) suggested that trillium is a good indicator species of deer browsing because it is a widespread, long-lived, preferred food of white-tailed deer, and browsing effects are easily quantified. Augustine and Jordan (1998) demonstrated significant deer impacts on forest forbs, especially those that are palatable and at low densities. They found that availability of alfalfa and row crops nearby could reduce browsing on forest plants.

Deer also can significantly impact regeneration of woody plants (Russell et al., 2001; Côté et al., 2004). Most studies of deer browsing effects on forest regeneration have been done in the Great Lakes and Allegheny Plateau regions although impacts have been evident in other parts of North America (Stromayer and Warren, 1997). Deer exclusion studies have been a common procedure for demonstrating the effects of deer browsing (Côté et al., 2004). Tilghman (1989) manipulated deer densities in large exclosures (densities ranged from 0 to 31 deer/km^2) and found that when deer densities were high, browsing reduced the diversity and impacted size and composition of tree seedlings after a clear cut. Tilghman (1989) concluded that deer populations should be maintained at densities <7 deer/km^2 to allow forest regeneration. Varying deer densities in large exclosures in Pennsylvania demonstrated that deer effects on forest composition are linearly correlated with deer density and are cumulative (Horsley et al., 2003). Horsley et al. (2003) concluded that densities >8 deer/km^2 would impact abundance of many tree species in areas of continuous forest, although they suggested that deer impacts might be reduced in forests associated with agricultural crops because of the alternative food source. Stromayer and Warren (1997) proposed that deer can create alternate stable states in forest ecosystems by suppressing advanced regeneration required to restock sites following removal of the overstory, such as occurs in timber harvest.

Establishing deer population densities that minimize forest regeneration impacts can be problematic because deer impacts may vary with forest type and food availability on adjacent landscapes and these differences may occur at scales smaller than most agency DMU-level management. In addition, threshold levels of deer impacts are not available in most situations so it is difficult to set deer population goals that minimize browsing impacts. Rooney and Waller (2003) indicated that although there are many examples demonstrating deer impacts on forest species, few have attempted to develop indicators of

FIGURE 14.8 *Trillium* sp. may serve as a good indicator species of deer browsing. An abundance of trillium in bloom indicates that deer densities are probably less than 5/km^2. (Photo courtesy of G. P. Fleming, the Virginia Natural Heritage Program. With permission.)

browsing that managers can use to assess deer population status relative to impacts on forest regeneration. Browse lines in high deer density areas are prominent but are a "post mortem" or "lagging" indicator (Rooney and Waller, 2003).

Frelich and Lorimer (1985) used percentage of sugar maple seedlings browsed as an index of browsing severity in Michigan and found a significant negative relationship between the browsing index and the abundance of eastern hemlock seedlings. Mladenoff and Stearns (1993) cautioned against making regional conclusions (e.g., DMU level) about deer impacts on forest regeneration that are based on a stand-scale. They concluded that deer browsing was not the critical element affecting hemlock forest regeneration on a regional scale in the upper Great Lakes region. Rooney and Waller (2003), however, used a broad-scale, multifactor approach to evaluate the effect of deer on regeneration of eastern hemlock and northern white-cedar and found that seedling regeneration declined linearly with deer browsing intensity.

So what are the consequences of deer browsing on ecological integrity and how should agencies respond? Waller and Alverson (1997) warned of the immediate and long-term ecological costs of deer overabundance. They challenged deer managers to develop research programs in which deer and their ecological impacts are monitored and to develop management protocols that enhance the ability to manage deer populations to achieve desired levels. Agency response to these issues is difficult because of the considerable variability in deer impacts on ecosystems and the lack of indicators of deer impacts that can be used to proactively manage. Deciding the appropriate density at which to manage deer is further confounded by large differences between deer densities that minimize ecological impacts, especially of sensitive plant species, and densities desired by many constituencies, especially hunters (Diefenbach et al., 1997). An additional, probably unresolvable issue is knowing the relationship between current deer population densities and densities under which plant communities evolved (Sinclair, 1997). Obviously, white-tailed deer have played a significant evolutionary role in ecosystem development but we do not know how far outside that role deer are in modern times when there are so many other factors affecting our forests. Currently, most agencies do not manage for ecological integrity on a DMU scale, but will dedicate areas, usually publicly owned, in which deer populations are maintained at levels that ensure ecological integrity (Killmaster et al., 2007).

Human Social Impacts on Carrying Capacity

Diverse stakeholder interests and input have had an increasingly large impact on many agency deer management programs, even in states where biological issues can significantly influence deer population status (Williamson, 2003). Ellingwood and Spignesi (1986, p. 42) defined the term cultural carrying capacity (CCC) as "the maximum number of deer that can compatibly coexist with a local human population." Other names have been given to a similar concept, most recently "wildlife acceptance capacity" (Decker and Purdy, 1988; Lischka et al., 2008). However, Minnis and Peyton (1995) described wildlife acceptance capacity as a narrower concept than CCC that deals with individual stakeholders while CCC includes a broader scope involving multiple value systems.

Moving CCC from a general concept to a tool agencies can apply quantitatively in deer management decisions is challenging. Minnis and Peyton (1995, p. 30) provided a framework to move CCC "beyond a notion to the status of a comprehensive, structured, and evolving model." They defined a complex array of biological and social factors that affect human perceptions of deer and their abundance and recommended that agencies target these perceptions to resolve conflict. Lischka et al. (2008) compared the acceptance capacity for white-tailed deer of three stakeholder groups (hunter, farmer, and rural resident who neither hunted nor farmed) in Michigan. They found that whether a person hunted or farmed influenced their perception of the impacts from deer, but that the classification as a farmer or hunter was of limited value in predicting acceptance capacity of deer. They concluded that defining stakeholder groups on their perceptions of deer impacts would more accurately reflect how readily they accepted deer. Thus they suggested that agencies should focus on constituents' perceptions of impacts for resolving deer and human conflicts. In addition to deer population management, Lischka et al. (2008) suggested that agencies could improve public perceptions of deer by managing the frequency of interactions and how they are perceived.

On a DMU scale, many factors can affect acceptance capacity for deer because of the diversity of stakeholder groups and their attitudes toward deer (Decker and Purdy, 1988). Even within stakeholder groups there is variability in age, gender, education, occupation, and experience with deer that can affect attitudes toward deer. Attempts have been made to quantitatively relate deer densities with attitudes toward deer (Brown and Decker, 1979; Decker and Brown, 1982; Stoll and Mountz, 1983; West and Parkhurst, 2002). In New York, Brown and Decker (1979) determined the optimum antlered buck harvest (a measure of deer population size) to be that at which the proportion of farmers who wanted the deer population to remain the same was high, the proportion who wanted a decrease was as low as possible, and the proportion who wanted an increase to be at least as high as those wanting a decrease. They recommended that where these deer population goals have been achieved, farmers who continued to complain because they were intolerant of any damage, adjacent to high deer density areas, or raising crops especially vulnerable to deer, be given special permits to kill deer. West and Parkhurst (2002) found that agriculture producers' attitudes toward deer were correlated with deer density (as reflected in antlered buck harvest) although there was too much variability to be of predictive value. They suggested that relating deer density to CCC for individual stakeholder groups (e.g., row crop farmers, orchardists, cattle farmers) would produce better quantitative relationships (Figure 14.9). Stoll and Mountz (1983) asked in a random survey of landowners if they wanted the deer population to increase, remain the same, or decrease. They then regressed annual buck harvest on the proportion of landowners who responded "increase" and on the proportion responding "decrease." Stoll and Mountz (1983) suggested that where the regression lines intersected represented the appropriate deer population level (as reflected in buck harvest). Christoffel and Craven (2000) surveyed woodland owners about their perceptions of white-tailed deer and deer impacts on their woodlands. Most were deer hunters and were concerned about deer health but less than 10% noticed a decline in any plant group and most were not concerned about the ecological or economic impacts of deer on their property.

Assessing public attitudes toward deer should be a part of all agency deer management programs to qualitatively or quantitatively establish DMU-level satisfaction with deer population status. Randomly distributed attitude surveys of constituents provide the best measure of public opinions about management issues but public meetings, although often biased by a vocal few (Johnson et al., 1993), provide an opportunity for agencies to communicate directly with citizens. Some agencies set permit quotas to achieve specific harvests that produce desired populations (Wisconsin Department of Natural Resources, 2001). However, most use available deer population information, along with annual input from constituents and political influences to establish DMU status relative to overall public interests (deer density too high,

FIGURE 14.9 Deer in a soybean field can impact production and lead to landowner dissatisfaction. Note difference in soybean growth in small area protected by a fence. (Photo courtesy of D. Nicks. With permission.)

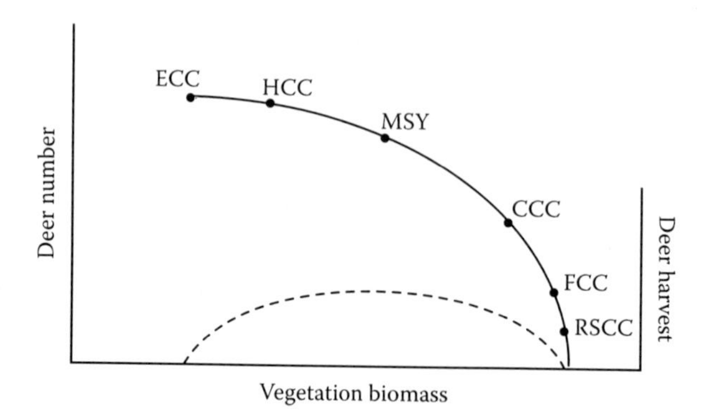

FIGURE 14.10 Graph showing points on a deer population continuum (solid line) representing ecological carrying capacity (ECC), hunter dictated carrying capacity (HCC), maximum sustainable yield (MSY), cultural carrying capacity (CCC), timber management carrying capacity (FCC), and rare species carrying capacity (RSCC). Deer harvest required to maintain constant deer numbers and vegetation biomass is shown by dashed line. (Adapted from Sinclair, A. R. E. 1997. In *The Science of Overabundance: Deer Ecology and Population Management*, eds. W. J. McShea, H. B. Underwood, and J. H. Rappole, 380–394. Washington, DC: Smithsonian Institution Press. With permission.)

about right, or too low; for example, Virginia Department of Game and Inland Fisheries, 2007) and set harvest quotas accordingly.

Deer population goals and how they are established vary among agencies and may depend on many factors, including land-use characteristics (e.g., forest, agriculture, urban, rural), human population density, stakeholder perceptions and attitudes, land ownership patterns (private, public, size of ownership), and future considerations such as trends in deer hunter numbers, ownership patterns, urban sprawl, and evolving hunter interests. Sinclair (1997) defined the relative deer population density at which various carrying capacities might occur (Figure 14.10). At the ECC, no deer harvest could occur without affecting deer numbers and thus moving deer populations away from ECC. Maximum harvest could occur at the MSY, but in some landscapes hunters are not satisfied with deer observation rates at these levels. Thus, hunter-desired deer abundance generally occurs somewhere between the MSY and ECC although this may vary considerably depending on habitat quality. Unfortunately, these deer densities often exceed the desires of other stakeholders such as farmers, timber managers, and ecologists interested in protecting sensitive plant and animal species. Appropriate population goals are confounded by the interests of these diverse stakeholders. The reality is that there is no level of deer abundance that will satisfy all constituents. Conceptually, the most satisfactory level may be where deer populations are tolerable by all constituencies but not optimal for any (Figure 14.11).

The influence of constituencies in the process of setting a deer population goal can vary according to the legal system in which regulations are established, economics of human activities that occur on the landscape and how deer impact these activities, traditions especially relating to hunting, and current status of deer populations and management. As a result agencies approach goal setting in many different ways. In areas with extreme conditions (such as severe winters) or where deer populations have the potential to exceed the ability of the landscape to support them, population goals may be biologically rather than socially based. However, as deer populations have increased, agencies have become more sensitive to stakeholder interests and increasingly set goals based on social rather than biological issues.

Harvest Management

There is considerable variation in agency deer harvest management programs depending on land-use characteristics, ownership, deer abundance, human social factors, and traditions. Providing maximum hunting opportunity while at the same time managing deer populations at levels that are in the best interest of citizens is a common agency goal.

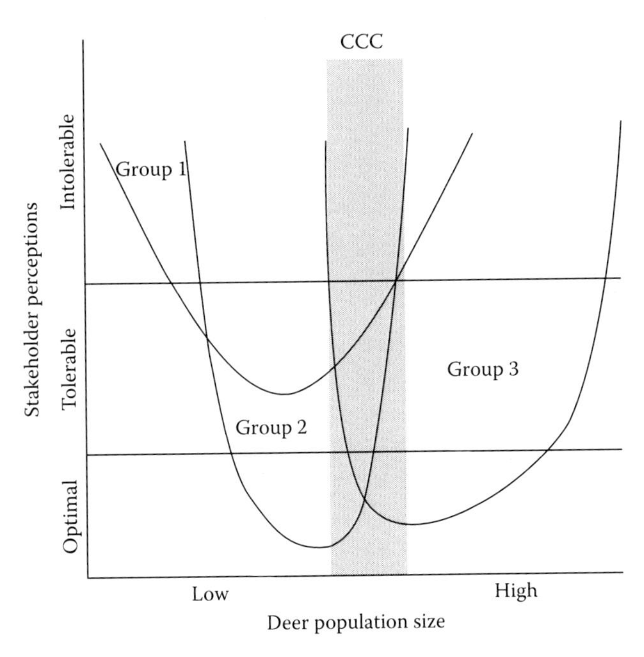

FIGURE 14.11 Perceptions of three hypothetical stakeholder groups concerning deer abundance. The cultural carrying capacity (CCC) represents the range of deer population size that all groups find tolerable or optimal. (Adapted from Conover, M. R. 2001. *Resolving Human–Wildlife Conflicts: The Science of Wildlife Damage Management.* New York, NY: Lewis Publishers. With permission.)

Regulations

All agency deer regulations include broad-based DMU-level regulations and often more restrictive regulations directed at public areas. Some agencies, especially in the east and southeast where leasing for deer hunting is common, also incorporate regulations that address management at the level of individual properties or groups of properties.

Public Land Management

Private and public land present different management challenges for agencies, primarily because public land often receives considerably more hunting pressure than private land (Root et al., 1988). Historically in Missouri and some other states, DMU-level harvest regulations have been applied to private and most public land equally. Considerable liberalizations in regulations to address increasing deer populations on private land (e.g., unlimited availability of antlerless permits) were initially applied to private and public land. The result was declining deer populations on most public areas. Missouri firearms deer hunter respondents to random mail questionnaires were less successful, felt there were fewer deer, and were less satisfied if they hunted mostly or entirely on public land compared to those hunting private land (Table 14.1). As a follow-up, hunters on selected public lands with a variety of deer-hunting regulations were contacted to determine their success and opinion about the areas for deer hunting. Results indicated poor success and dissatisfaction with the experience if hunting areas were open to statewide regulations; aerial surveys of deer confirmed lower deer densities where statewide regulations were applied (Hansen and Beringer, 2007). As a result, more restrictive deer-hunting regulations that included method (modern firearms, muzzleloader, archery) or bag limit restrictions were implemented on many Missouri public areas although access was not limited on most areas. Some agencies do not impose special regulations on public areas. This may be especially true in areas where public hunting land makes up a high percentage of the landscape and deer bag limits are low such as in the western fringe of the white-tailed deer range. Other agencies manage harvest and hunter densities by requiring special permits to hunt public areas. Often, a lottery-style permit drawing is held.

TABLE 14.1

Results of Random Survey of Missouri Firearms Deer Hunters in 2008

Hunting Activity and Perceptions	Percentage of Respondents	
	Public Land Hunters	Private Land Hunters
Harvested at least one deer	35.5	49.9
Harvested an antlered buck	16.5	24.8
Feeling that there were too few deer	58.7	40.1
Hunting strategy—harvest first legal deer seen	55.7	37.7
Rated season as good or excellent	32.7	48.9
Perception that there were more deer now than five years ago in place they hunt	10.7	14.7
In favor of a 4-point antler restriction	51.4	62.3
Could have taken more antlerless deer but decided not to	29.1	45.5

A total of 9000 surveys were mailed yielding 4394 usable responses. Public land hunters were those who responded they hunted mostly or entirely on publically owned land. Private land hunters were those who responded that they hunted mostly or entirely on privately owned land.

Some form of a "managed" or "controlled" hunt is another method of regulating hunter numbers and/or harvest on sensitive public areas. These hunts usually involve a process in which a person has to apply for a limited quota hunt. Bag limits are generally set separately from the statewide bag (Hansen and Beringer, 1997; McDonald et al., 2007). These hunts are usually associated with deer population control, often in situations where limited access and special hunter requirements are desirable, such as urban settings (Kilpatrick et al., 1997, 2004). Success at controlling small- to large-scale deer numbers have been achieved under the managed hunt format (Dohm and Brown, 1998; McDonald et al., 2002, 2007; Suchy et al., 2002).

Private Land Management

For most agencies, the greatest deer management challenges occur on private land. Most agency programs have historically relied on DMU-level regulations to achieve deer management goals, but increasingly, agencies have incorporated programs to address property-specific deer management (Guynn et al., 1983), especially where land ownerships are large and leasing is prevalent.

In the early deer population growth phase of the mid-twentieth century, hunters either were restricted to taking antlered bucks or where permits were valid for either sex, not everyone got to hunt each year. If any antlerless harvest was allowed, quotas often were used to allocate either-sex or antlerless permits. As deer numbers increased, so did opportunities to harvest antlerless deer, and often the proportion of the harvest that is antlerless has increased (Figure 14.12; Chapter 11). Hunting and harvest opportunities have increased such that all get to participate in some form of deer hunting if desired, with the exception of a few situations where politically motivated regulations limit nonresident participation. In recent years, with increasing interest in older, larger bucks and significant problems in places with deer overabundance, management efforts by many agencies have reversed with buck harvest being more restricted than antlerless harvest, stimulating development of innovative deer management programs (Miller and Marchinton, 1995) (Chapter 15).

Antlered Buck Bag Limits

Buck-only regulations were routine in many states and provinces during the period of white-tailed deer population restoration; high buck harvest rates often resulted. As deer populations increased, harvesting a deer has become somewhat routine in many areas. As a result, hunters have become more selective, especially in the bucks they harvest. Many agencies have tried to accommodate the accelerated interest in older-aged bucks through a diversity of regulations designed to protect younger bucks. In the southeast, where antlered buck limits generally have exceeded those in other parts of the white-tailed deer's

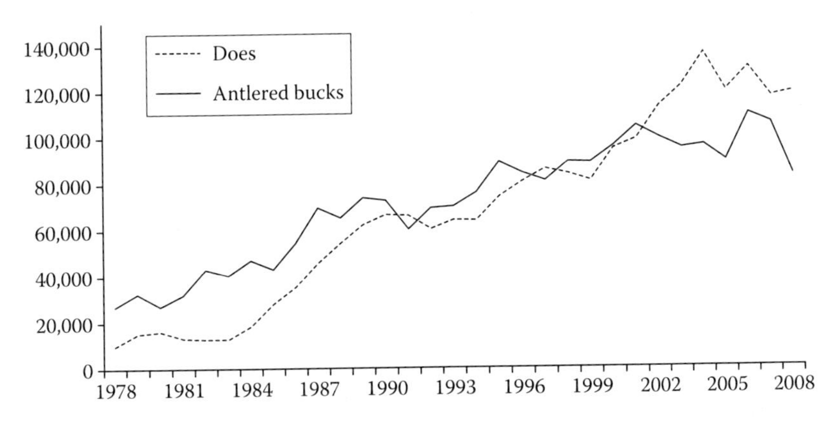

FIGURE 14.12 Statewide doe and antlered buck harvest in Missouri from 1978 to 2008.

range, efforts to reduce limits have often met with considerable opposition. In Texas, a buck permit quota reduced buck harvest by 59% on land ownerships less than 259 ha and did not change buck:doe ratios because doe harvest also declined (Gore et al., 1985). The number and percentage of 4.5-year-old bucks in the harvest increased but antler quality declined because of increasing deer densities. The number of hunters and number of days hunted dropped as a result of the regulation (Gore et al., 1985). Concern also has been expressed about the impacts of a buck quota on lease hunting value in Texas (Huggins et al., 2005).

Selective harvest based on characteristics of the buck have been the public's preferred method of reducing harvest of young bucks; some form of antler restriction being the most popular (Durkin, 2004). Antler restrictions can take on many forms including minimum number of antler points (one exception in some states: deer with at least one side with a single point are legal to minimize potential high-grading effect), antler beam width, or combinations of the two. Some agencies have a restriction only on part of the antlered buck bag limit. Key issues with any type of selective harvest regulation include enforceability, effectiveness at achieving desired objectives, negative long-term biological consequences, and public acceptance.

Antler point restrictions (APR) have been the most popular antler restriction because they are enforceable and relatively easy for hunters to follow (Figure 14.13). APR were first imposed on elk and mule deer in western states (Biederbeck et al., 2001; Bender et al., 2002). In general, the western experience was disappointing because primary objectives, such as increased recruitment of young and production of older-aged males, were not achieved unless very restrictive regulations were implemented (Weigand and Mackie, 1987). Although male deer or elk lived one year longer, they were subsequently cropped excessively (Biederbeck et al., 2001). In some locations considerable illegal harvest of sublegal bucks also occurred (Winchell and Buck, 1993). One aspect of the western experience which is different throughout much of the range of the white-tailed deer is that harvests of female deer and elk were tightly restricted in most areas with the antler restrictions; hunters generally had a permit that allowed them to take only an antlered animal. As a result, instead of the harvest pressure shifting from a male to a female deer or elk, as it might have if the hunter had a permit to take an antlerless animal, all harvest pressure was shifted to legal males. The result was excessive pressure on legal males with few surviving beyond the age at which they became legal. This may be less likely to occur in much of the white-tailed deer range because hunters generally have the option to take an antlerless deer.

Results of studies of the impact of APR on white-tailed deer buck harvest age structure have been mixed. Strickland et al. (2001) compared antler size of deer taken from Mississippi wildlife management areas before and after implementation of a regulation that required a buck to have a minimum of four total points before being harvested. They found antler size of 2.5- and 3.5-year-old males declined in more fertile regions where yearling growth was optimal but not in other less fertile regions. They suggested that there may have been a "highgrading" effect in the more fertile sites. Although Strickland et al. (2001) saw increasing proportions of 2.5 year and older bucks in the harvest, they were not able to determine if these older bucks increased in the population.

How to count points

An antlered deer must have at least four points on one side to be taken. Each of the following counts as a point:

- An antler point, if it is at least 1-inch long
- The brow tine, if it is at least 1-inch long
- The end of the main beam
- Any broken tine that is at least 1-inch long

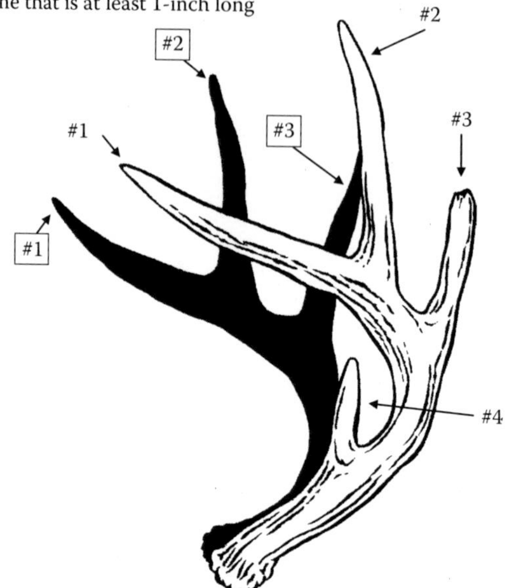

FIGURE 14.13 Example of an antler point restriction. In Missouri, a minimum of four points on at least one side is required in some counties. (Photo courtesy of the Missouri Department of Conservation. With permission.)

Harvest prior to and after implementation of an APR in New Jersey indicated that the number of 3.5 year and older bucks increased but the number of 2.5-year-old bucks unexpectedly decreased (J. Penkala, New Jersey Division of Fish and Wildlife, personal communication). Harvest of 1.5-year-old bucks decreased as expected. After four years of the APR the ratio of the reduction in harvest of 1.5-year-old bucks to the increase in older buck harvest was 6.2 to 1, a large reduction in total buck harvest opportunity for a small increase in adult buck harvest.

In Missouri, APR (minimum of four points on at least one side) were implemented to shift harvest pressure from antlered to antlerless deer (Hansen, 2007). After a four-year study comparing harvest sex and age structure in control and APR counties, there was an overall decrease in the number of antlered bucks taken by hunters but an increase in 2.5, 3.5, and 4.5+ year-old males (Figure 14.14). Doe harvest increased in areas with moderate deer densities which compensated for the reduction in buck harvest. In APR counties with high deer densities, doe harvest did not increase and overall harvest decreased (Hansen, 2007).

The long-term impacts of selective harvest on antler development and other characteristics of white-tailed deer have been hotly debated; research has not resolved the issue. Concern is that some antler restrictions, especially APR, allow hunters to take the best 1.5-year-old bucks and protect the poorer ones. If genetic potential for antler development is expressed at 1.5 years of age, then there could be long-term effects on antler development as well as other essential biological factors (Ditchkoff et al., 2001a). Potential impacts of selective harvest have been demonstrated in other hunted species (Ginsberg and Milner-Gulland, 1994; Singer and Zeigenfuss, 2002; Coltman et al., 2003; Milner et al., 2007). A key element with white-tailed deer is whether protecting small-antlered bucks produces a high-grading effect. There is not yet a clear resolution on this with some research indicating that small-antlered 1.5-year-old bucks have a genetic basis (Williams et al., 1994; Lockwood et al., 2007); other results suggest other factors are most important in 1.5 -year-old antler development (Lukefahr and Jacobsen, 1998;

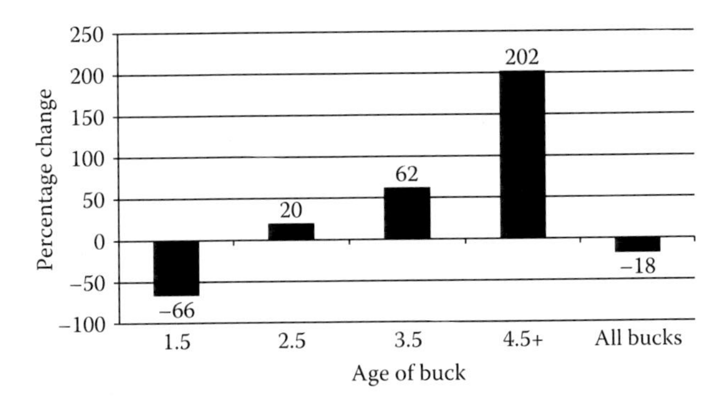

FIGURE 14.14 Percent change in the harvest of bucks, by age and for all bucks, that can be attributed to an antler restriction in Missouri that required a buck to have at least 4 points on one side to be legal.

Koerth and Kroll, 2008). Complexities produced by differential harvest pressure across the landscape (Thomas et al., 1976; Foster et al., 1997; Harden et al., 2005; Blanchong et al., 2008), varied distribution and movement of deer (Gould and Jenkins, 1993; Nixon et al., 1994, 2007; Diefenbach et al., 2008; Long et al., 2008), the contribution of the doe to antler development (Lukefahr and Jacobsen, 1998; Wilson et al., 2005), a mating system in which a diversity of buck age classes breed (Sorin, 2004; DeYoung et al., 2009), and the lack of a relationship between breeding success and antler or body size in males 2.5 years and older (Shaw, 2005) further confound the issue. Nevertheless, regulations such as minimum antler spread regulations which protect a higher proportion of 1.5-year-old bucks or APR that allow spike bucks to be taken have been implemented to reduce the risk of high-grading.

Selective harvest regulations have been popular with the hunting public but "trophy" hunting has generally been deemed unacceptable by the nonhunting public (Kellert, 1996). This nonhunting public sentiment has generated agency concern about implementing regulations intended to produce larger bucks. Perceptions that management for older bucks may lead to privatization of the deer resource (Green and Stowe, 2000) and difficulty in achieving adequate doe harvests also have produced some agency reluctance to implement selective harvest regulations.

Antlerless Deer Harvest

Most agency attempts to increase antlerless harvest have involved increased availability of antlerless permits or lengthened seasons. Giles and Findlay (2004) found that antlerless harvest was related to the number of antlerless permits issued only when less than 40% of hunters possessed them. They concluded that in Ontario, density-dependent mortality was more of a factor in population control than hunting. Van Deelen et al. (2006) found that extra days of firearms antlerless hunting in October produced increased antlerless harvest by both firearms and archery deer hunters in Wisconsin. They cautioned that limits on the number of deer that hunters are willing to take could ultimately reduce the effectiveness of further liberalizations of antlerless harvests.

Efforts to promote use of antlerless permits by requiring that hunters take an antlerless deer prior to taking an antlered deer (earn-a-buck) have produced significant increases in antlerless harvest and may be the best management tool available to manage deer numbers. In New Jersey, antlerless harvests were increased the first year of an earn-a-buck regulation by 138% over the previous year (Ferrigno et al., 2002). Holsman and Perchenik (2006) found that nearly half of surveyed hunters listed the earn-a-buck regulation in Wisconsin as the primary reason they harvested their first deer. They concluded that earn-a-buck regulations motivated hunters to take an antlerless deer but would not be a motivation to use additional antlerless permits even if more antlered bucks could be taken. Earn-a-buck regulations in Wisconsin have produced an antlerless to antlered harvest ratio of 3:1 (K. Warnke, Wisconsin Department of Natural Resources, personal communication). Unfortunately, earn-a-buck regulations are unpopular with hunters and create enforcement issues, especially where mandatory in-person checking is not required. In DMUs with deer hot (high deer density and damage problems) and cold (low deer density) spots on private land created by variable hunter access, an earn-a-buck system can further polarize deer abundance by unnecessary reduction of deer numbers in the cold spots and not solving hot spots where hunting is not allowed.

Long firearms seasons are the norm in the southeastern states; short firearms seasons are more typical in midwestern states. Up to a point, especially when deer are scarce or poor conditions for hunting reduce early-season harvests, a longer firearms deer season or addition of antlerless portions may increase antlerless harvest. Other times, when deer are abundant and antlerless harvest opportunities are already liberal, extension of existing seasons or changing opening dates has minimal impact other than perhaps redistributing the harvest (Murphy, 1965). In Missouri, addition of an antlerless portion of the firearms season that occurred after the main firearms season, resulted in a shift in timing but not a significant increase in antlerless harvest (Hansen, 2001). It was concluded that in some years, especially those when poor hunting conditions occur during the main firearms season, an antlerless extension could improve antlerless harvest. However, because some hunters delay their antlerless harvest until the antlerless portion, poor conditions during that portion (which were more likely to occur because it was shorter and later) could actually reduce overall antlerless harvest.

Promoting antlerless harvest early in the firearms season may produce the best harvests (Figure 14.15). Deer are most vulnerable then (Roseberry and Klimstra, 1974) and hunters are "fresh" and ready to take a deer; the odds of harvesting a deer are greatest early and decline as the season progresses. Also, the earlier hunters take their first deer, the more likely they will hunt again and harvest another deer. Allowing antlerless permits to be used over an extended time may encourage hunters to delay harvest because there is a lot of time to hunt; ultimately hunters may never take an antlerless deer because it becomes too difficult or they never get around to it. Conversely, restricting opportunity and requiring a hunt-specific permit (e.g., restrict time when an antlerless permit can be used) early in or before the main firearms season may encourage early harvest of antlerless deer, ensuring that harvest occurs (Haroldson et al., 2007; M. Tonkovich, Ohio Department of Natural Resources, personal communication). For agencies that already have traditionally long firearms seasons, many of the timing options for improving antlerless harvests are not useful. Hunter education through groups such as the Quality Deer Management Association which encourage antlerless harvest as part of responsible deer management is promoted by many agencies to encourage antlerless harvest.

FIGURE 14.15 One of the biggest challenges for agencies throughout the range of the white-tailed deer is to attain adequate harvest of antlerless deer. Numerous incentives to take antlerless deer have been implemented by agencies with mixed success. (Photo by L. Hansen.)

Baiting and Feeding Deer

Feeding, which may include food plots or food imported from outside sources, and baiting are used extensively in many areas throughout the range of the white-tailed deer (Figure 12.4). Attitudes toward feeding and baiting deer are mixed; some passionately support, others oppose. A primary reason people feed deer is to supplement limited natural food, whether perceived or real, to increase survival, or improve antler development. Baiting takes on additional objectives such as improved potential for a safe, clean shot and increased deer sightings and hunting success.

Baiting and feeding can impact deer use of the landscape and therefore harvests. Van Deelen et al. (2006) found that use of bait to hunt deer in Wisconsin tended to increase archery harvest but decreased firearms harvest. Van Deelen et al. (2006) concluded that negligible impacts of baiting on hunter efficiency and increased disease risks and environmental and social costs associated with baiting made legalization of baiting questionable. Bartelt et al. (2003) found that hunters using bait were slightly more efficient than nonbaiters. They suggested, however, that harvest efficiency of nonbaiters may have been reduced because properties that baited attracted deer away from nonbaited properties, making the overall change in harvest as a result of baiting negligible. In South Carolina baiting was allowed in some counties but not in others; all had similar deer densities (Ruth and Shipes, 2004). Harvest was 33% greater and effort per deer 6% lower where baiting was prohibited. Ruth and Shipes (2004) suggested that hunters became too dependent on the bait and that deer behavior changed at baited sites as a result of disturbance, making it more difficult to harvest deer. Legal baiting has been shown to change deer movement patterns and behavior (Cooper et al., 2002), generate deer hunter conflict (Michigan Department of Natural Resources, 1999), and produce ethical concerns within the hunting community and general public (Brown and Cooper, 2006).

Generally, discussion of problems associated with feeding deer involves imported supplemental food although many of these problems pertain to food plots. Feeding deer may impact deer demographics and exacerbate problems with deer overabundance (Brown and Cooper, 2006; Smith et al., 2007) such as overbrowsing of native plants (Schmitz, 1990; Doenier et al., 1997; Williamson, 2000). Increasing concerns about disease in deer, especially chronic wasting disease and bovine tuberculosis, have led many agencies to consider or to implement feeding bans. These bans generally involve imported supplemental food and not food plots.

Baiting and feeding issues are complex and confounding for many agencies to address because of the difficulty of defining what each entails, enforcement problems, hunter conflicts generated, and general public perceptions, which all interact with traditions, economics, and emotions associated with baiting and feeding. Most agencies allow or even promote food plots for deer but discourage supplemental feeding. There is considerable diversity among agencies relative to baiting. Some allow baiting with few conflicts (L. Fox, Kansas Department of Wildlife and Parks, personal communication); for others, the baiting issue has generated considerable conflict among hunters and often in hunter–agency relationships (McCaffery, 2000). In most situations where baiting is a tradition, compromises in which both social and biological considerations drive decisions concerning baiting may be appropriate and necessary (Rudolph et al., 2006).

Use of Dogs to Hunt Deer

Using dogs to hunt deer is a long-held tradition in parts of the white-tailed deer's range, especially the southeastern United States. General styles of hunting deer with dogs include the traditional method in which hunters are posted and dogs are released in cover where deer are jumped and chased past the hunters. A more modern method involves use of vehicles and radios to coordinate posting of hunters to intercept deer being run by hounds (Virginia Hound-Hunting Technical Committee, 2008). From a deer-management standpoint, using hounds to hunt deer is generally more efficient than stand hunting (Marchinton et al., 1970; Novak et al., 1991; Virginia Hound-Hunting Technical Committee, 2008). Where deer population control is desired, allowing use of hounds may increase hunter success. However, use of hounds where deer densities are below desired levels could inhibit management goals (Campo et al., 1987).

Although hunting deer with dogs can be more efficient and hunting-related wounding losses higher than still hunting, negative biological factors (reduced reproduction/recruitment and indirect hunting-related mortality) associated with hound hunting are negligible (Campo et al., 1987). Social issues are the primary concerns agencies must address in hunting deer with dogs. Hunting deer with dogs requires large properties and changing land use and ownership have reduced the number of such properties; trespass has become an important conflict for many landowners. Societal attitudes often view use of hounds to hunt deer as unfair and there is increasing concern about objectionable hunter behavior, especially the use of trucks and radios to track deer movement and using roads as shooting lanes. Conflicts associated with hunting deer with dogs has stimulated some states to implement more restrictive regulations pertaining to the use of dogs to hunt deer (Steffen et al., 1983) or prohibit the practice (Campo and Spencer, 1991).

Weapon Type to Take Deer

Hunter harvest efficiency can be strongly influenced by weapon type (Figure 14.16) (Kilpatrick et al., 2002; Weckerly et al., 2005). Agency allocation of hunting opportunity often is based on these efficiency differences. For example, in Missouri, archers account for only about 15% of the total deer harvest but are allocated 112 days to hunt compared to 11 days for modern firearms hunters (although there are additional days for firearm youth and antlerless hunts). Bag limits for archers also are often more liberal than for firearms hunters. Agency consideration of weapon type for deer hunting generally includes allocation of deer-hunting opportunities, deer population objectives, and public sentiments toward the method.

Deer Damage Permits

Most agencies have some system for handling deer problems that have not been solved through hunting during the statewide hunting seasons. These may be associated with urban or rural settings; urban issues will be discussed in Section "Urban, Suburban, and Exurban Deer Management." Most deer-related problems in rural areas are associated with crop damage. Circumstances that generate these problems often involve specialty crops that are highly vulnerable to depredation or crops impacted by deer that originate from neighboring properties. Landowners who refuse to use hunting as a tool to manage deer populations may or may not be considered for deer damage permits depending on the circumstances

FIGURE 14.16 Hunting method and weapon type affect harvest and must be taken into consideration when agencies set deer-hunting regulations. (Photo courtesy of the Missouri Department of Conservation. With permission.)

and agency. Provisions in policy or regulations that address deer problems usually involve shooting deer in-season (Erickson and Giessman, 1989) or out-of-season (Horton and Craven, 1997). To be considered for the permits generally requires a field visit by agency staff to assess the damage and determine if issuance of damage permits is warranted. Restrictions on use of the permits vary widely. Some crop damage permittees must follow statewide deer-hunting methods restrictions; other states have special provisions that may be outside of the season framework such as hunting over bait or with spotlights at night. Some agencies allow only antlerless deer to be shot; others any deer. Some agencies allow the shooter to keep the deer; others do not but most require that the deer be utilized. In many cases issuance of these permits is an appeasement to the landowner; usually, the deer overpopulation and crop damage problems are not solved (Horton and Craven, 1997). Issuance of crop damage permits can generate complaints from neighbors and local hunters who often exaggerate the degree to which the permits are used and the impact on their hunting opportunities. The process for issuing the permits also can require considerable staff time (Erickson and Giessman, 1989).

Landscape Factors Affecting Harvest

Many landscape factors affect deer harvest on private land. In Illinois, the most important factor influencing deer harvest vulnerability, defined as the proportion of white-tailed deer harvested per unit of hunting effort, was the proportion of the landscape in forest cover; vulnerability was highest in counties with small, highly fragmented forests (Foster et al., 1997). Harden et al. (2005) used similar estimates of harvest vulnerability to determine if increased development of rural landscapes reduced hunting pressure and thus harvest vulnerability in Illinois counties. In Illinois there is a 274-m buffer around human dwellings in which hunting is not allowed, encompassing 31% of the total land area in Illinois. Harden et al. (2005) found that as rural development increased, thus protecting more deer habitat, harvest vulnerability decreased (Figure 14.17). They reported that although many hunters were unaware of the restricted buffer zone, hunter activity around human dwellings was reduced. Harden et al. (2005) concluded that as human populations expand, the proportion of the landscape open to hunting will decline

FIGURE 14.17 Development in rural landscapes can create more refuge from hunting. Deer vulnerability to harvest may decrease, thus confounding management of deer numbers. (Photo courtesy of the Missouri Department of Conservation. With permission.)

and that in some regions and perhaps entire states, traditional hunting will no longer be effective at controlling deer populations.

Refuges may play a positive deer management role in some settings. In parts of Illinois under intensive row-crop agriculture (<5% forested), deer vulnerability to harvest was high in the small habitat patches existing in this landscape (Nixon and Hansen, 1992; Hansen et al., 1997). Deer in small habitat patches open to hunting were often depopulated during hunting seasons but were "restocked" by deer originating in refuges.

In agricultural areas the stage of crop harvest, especially corn, may influence deer harvests (Hansen et al., 1986). Deer use standing corn for cover (Nixon et al., 1991; VerCauteren and Hygnstrom, 1998) and when crop harvest is late, deer may be less vulnerable to harvest. Farmers also may not have the time to hunt deer because they are too busy harvesting crops.

Deer abundance and activity in oak-dominated forests have been associated with acorn production (Feldhamer et al., 1989; Wentworth et al., 1992; Carlock et al., 1993; McShea and Schwede, 1993; Ryan et al., 2004). Ryan et al. (2004) found that hard mast abundance, especially acorns, had a negative relationship with total, archery, and antlerless white-tailed deer harvest but not antlered buck harvest. They suggested that deer were scattered and less vulnerable when acorn crops were large and attributed the lack of a relationship between buck harvest and acorn crop to the large number of hunters afield, ensuring bucks were encountered and harvested. The firearms hunting season coinciding with the rut also may make bucks vulnerable in spite of abundant mast. Wentworth et al. (1992) found acorn availability and age structure of the harvest to be related in the southern Appalachians and suggested that acorn availability affected reproduction and/or fawn survival.

Weather during the deer-hunting season can affect hunter activity and thus deer harvest (Fobes, 1945; Van Etten et al., 1965; Mechler, 1970; Curtis et al., 1972; Hansen et al., 1986). In Virginia, the amount of precipitation on the first day of the hunt correlated negatively with total harvest (Mechler, 1970), although mean daily temperature and total precipitation were negatively and positively related to deer observability, respectively (Curtis et al., 1972). Hansen et al. (1986) found that daily harvest of deer during the Illinois firearms season was affected by the amount and duration of rainfall but that this did not affect overall harvest. As deer populations have increased throughout the range of the white-tailed deer, hunting seasons have expanded and although weather might affect harvest during some portion of the season, impacts on overall harvest in recent times may be less likely because hunter effort and harvest can be shifted to other parts of the season.

Hunter Numbers and Willingness to Harvest Deer

As agency management efforts continue to shift from deer population growth to control, hunter numbers, willingness to harvest antlerless deer, and access to private land for hunting will affect the success of deer management programs. Considerable discussion has revolved around the current and future feasibility of managing deer populations through hunting (Witmer and deCalesta, 1991; Woolf and Roseberry, 1998; Brown et al., 2000; Enck et al., 2000; Riley et al., 2003; Van Deelen and Etter, 2003; Holsman and Petchenik, 2006). Brown et al. (2000) provided a discouraging assessment of the future effectiveness of hunting in regulating deer populations. They described key issues faced by agency deer managers including increasing deer populations, decreasing hunter numbers, more restrictive hunter access to private property, deer adaptability to developed areas allowing them to thrive in settings where hunting may not be feasible, and possible erosion of the social acceptability of hunting. Brown et al. (2000, p. 800) accurately described most state season frameworks as "applied across broad scales that include a mixture of public and private lands, unrestricted and restricted access to deer, and local governments with various hunting and firearms ordinances." The complexity of the landscape across which deer regulations are applied usually precludes precise application of harvest and targeting areas with unique deer management goals. Brown et al. (2000) contended that where deer are abundant and hunter numbers are low, an increasing scenario across the range of the white-tailed deer, the ability to control deer populations through recreational hunting may become limited.

Riley et al. (2003) compared the 1997 deer harvests in New York with those required to maintain the deer population at densities that would not inhibit forest regeneration (deCalesta, 1994). They found that the level of harvest was far below that needed to achieve target populations and that without an appreciable gain in hunter numbers, harvest effectiveness, or willingness to kill more antlerless deer, methods other than hunting would need to be incorporated into the deer management programs.

Holsman and Petchenik (2006) evaluated deer hunter harvest behavior in an area of Wisconsin where deer depopulation was attempted to control chronic wasting disease. They defined characteristics of hunters that affected what and how many deer they harvested including type of hunter (buck or antlerless preference), harvest threshold (the maximum number of deer a hunter was willing to take for his or her own use), efficiency, attitude toward the Wisconsin DNR's eradication program, hunter access to private land, and hunting effort. Holsman and Petchenik (2006) found that large deer harvests were dependent on having an adequate number of hunters who were willing to take antlerless deer and who had a good place to hunt. In the Wisconsin scenario, hunter effort was not a significant determinant of deer harvest because some hunters spent considerable time hunting but were selective (e.g., buck hunters) or hunted in poor locations. Holsman and Petchenik (2006) concluded that hunter harvest decisions and therefore the number of deer taken depended most heavily on intrinsic attitudes that could not be affected by agency attempts to encourage hunters to take more deer. The authors suggested that agencies would be best served by managing the segment of hunters most willing to harvest deer rather than taking a broad approach (e.g., longer season) as a means of increasing harvest.

Ward (2005) also suggested that deer overabundance problems might be more efficiently solved if agencies distinguished and targeted high-productivity hunters to receive the majority of antlerless permits. He defined serious and casual Pennsylvania hunters by their willingness to pay for antlerless permits (high, medium, and low demand for antlerless permits). Only about 10% of hunters fit into the high demand category. The more serious hunters were willing to pay more for antlerless permits and this was correlated with their antlerless harvest. Casual hunters were not willing to pay much for an antlerless permit and did not harvest many if any antlerless deer. Ward (2005) suggested that permit allocation inefficiency in Pennsylvania (also true of many other states with limited quotas on antlerless permits) occurred because casual hunters would obtain antlerless permits but not use them. As a result, serious hunters would receive fewer antlerless permits. Ward (2005) recommended an antlerless permit pricing system that would attract serious hunters but not casual hunters, and thus allocate more antlerless permits into the hands of hunters who would use them. The process would increase the price for the first or second antlerless permit to target serious hunters. Ward (2005) predicted that this permit system would have positive economic and harvest impacts. He did acknowledge, however, that if a higher price for antlerless permits was charged, then a significant number of hunters might stop deer hunting or become concerned about other hunters killing too many deer. Another option which most agencies have adopted is to issue an unlimited number of antlerless permits.

Sometimes hunter mistrust of the agency's deer population estimates and harvest goals may affect what hunters will harvest. Wisconsin hunters have historically questioned agency estimates of deer population size, stimulating a technical review by an outside group of experts in an effort to add credibility to their methods (Millspaugh et al., 2007). A survey of deer hunters was conducted after the review (Holsman, 2007) to assess hunter perceptions and understanding of deer populations and Wisconsin's method of estimating deer numbers. Holsman (2007) concluded that the technical review was not likely to change hunter skepticism over the agency's deer estimates because hunters use their own field observations to establish their perceptions of deer population status. He suggested that hunters and biologists define credibility differently. For hunters the quality of science is least important; trust in the agency, fairness, and shared values affect their perceptions. Challenging the accuracy of the agency's deer estimates was considered a surrogate for the real issue: hunter preferences for deer densities that differed from agency goals.

Hunter Recruitment and Retention

Peyton (2000) described major social changes that will affect hunter numbers and impact deer management programs. Hunting competes with many other demands on youth and adult time. Deer hunter numbers are at best stable and most are decreasing throughout the range of the white-tailed deer (U.S. Fish and Wildlife Service, 2006). These decreasing trends portend serious implications for many agency programs (game and nongame) supported by revenues generated by deer license sales and potentially a

FIGURE 14.18 Hunter recruitment and retention are key elements of most agency deer management programs. (Photo by L. Hansen.)

decreasing need for programs intended to provide hunting recreation (Enck et al., 2000; Frawley, 2006). Declining deer hunter numbers also will have significant impacts on the ability to manage deer. Not only will there be fewer hunters to attain the necessary harvests but the average age of deer hunters is increasing and older deer hunters tend to take fewer deer (L. P. Hansen, unpublished data).

Agencies are addressing hunter recruitment and retention in a variety of ways (Figure 14.18). Most involve an effort to remove barriers to hunting or to provide special opportunities. Some barriers such as hunter education programs and age restrictions are being addressed by reducing the age at which a person can start hunting and making required hunter education programs more accessible (e.g., online opportunities). Many states have apprenticeship programs in which adults can hunt in the presence of hunter education certified mentors for one or more seasons without taking hunter education. The goal is to expose these individuals to hunting without initially going through the rigor of hunter education training. Many states provide special hunting opportunities for youth, either on public hunting areas or statewide. These usually involve a separate hunting season in which only youth can participate. Many agencies also are trying to attract persons into existing seasons by liberalizing hunting opportunities. For example, increasingly, crossbow hunting opportunities have been expanded; in some cases being allowed during archery seasons. States that have done this see increasing participation in archery seasons, including younger and older hunters (Tonkovich and Cartwright, 2002), although the impact on total hunter numbers is questionable.

Hunter Access on Private Land

A key element and limitation of agency management programs is access to private land for hunting deer. If hunting is not allowed or is severely limited, it will be difficult if not impossible to achieve deer population management objectives (Figure 14.19). Deer overabundance on public hunting lands is generally not a problem because they receive considerable deer-hunting pressure although there may be exceptions to this on large public tracts (Diefenbach et al., 2005).

FIGURE 14.19 Access to deer hunting can affect hunter recruitment and retention and create challenges to agencies trying to manage deer numbers on private land. (Photo by L. Hansen.)

Although most hunting is done on private land free-of-charge (Cordell et al., 1999), there is an increasing trend toward purchase of land or leasing to hunt deer (Mozumder et al., 2007). The average amount spent per hunter for hunting access varies considerably among states ranging from $4 per year in West Virginia to $872 per year in Alaska (Mozumder et al., 2007). Although leasing land for deer hunting provides economic benefits to the landowner, may make some private land more accessible to hunters, and may result in improved habitat management, it can negatively impact hunters not able or willing to invest in a lease. Some of these hunters will quit hunting deer. Where much land is leased for hunting, many hunters can lose access to traditional hunting areas. In Illinois, most hunters depend on private land for deer hunting but, because of increased leasing, 48% of surveyed hunters had lost access in the previous 5 years (Miller and Vaske, 2003). Although increased leasing may result in the loss of some hunters, it may recruit and retain others because of quality of the experience. Many are willing to pay for a better quality experience. For example, Zhang et al. (2004) estimated that in Alabama, persons were willing to pay more than double the average lease payment to hunt deer. The net effect on hunter retention and recruitment is not known.

Agency efforts to improve access to hunting on private lands have varied considerably. In most cases agencies promote private land hunting to ensure hunter opportunity and responsible deer management. Some agencies, especially in the southeast and west, have promoted or at least accommodated leasing for deer hunting by providing landowners or lessees special privileges such as supplemental antlerless permits that promote management or facilitate the leasing process (Leal and Grewell, 1999). Generally, attempts to link hunters with landowners to provide free hunting opportunities have been unsuccessful. In Missouri, a landowner–hunter link program designed to put hunters into contact with landowners who had experienced deer damage produced many interested hunters but practically no landowner participants. Landowners were reluctant to allow free access to persons they did not know even though an effort was made to familiarize the landowner with the hunter. A similar but more aggressive program in Illinois (Access Illinois) was successful but evolved into a leasing program (Miller and Vaske, 2003). Walk-in private land access programs in which the agency pays the landowner a fee to allow free hunter access have been popular and have provided additional opportunity, especially in western states where land ownerships are often large and landowners often are absentee. Most agencies, however, cannot compete with lease fees for deer hunting generally charged by private landowners and this system has limited application in most states or provinces.

Texas has the most extensive lease hunting program in North America (Adams et al., 2000). Hunting in Texas has evolved from traditional hunting participation by family and friends to an extensive leasing system that is driven by management for trophy deer. A concern in the Texas system is that economics is driving deer management and as hunter numbers decline, agency funding sources will decrease and there will be less public acceptance of hunting. Efforts to promote hunting participation have resulted (Adams et al., 2004).

Urban, Suburban, and Exurban Deer Management

White-tailed deer are highly adaptable to human disturbance and often flourish close to human population centers (Figure 14.20). One of the most significant challenges to agency deer managers is the increasing impact of urbanization on the ability to manage deer numbers (Chapter 20). Deer management is not only a challenge in urban and suburban areas but increasingly in developed areas outside of city limits that are mixed with working farms and undeveloped land (exurban). Recently, human population increases in exurban areas have exceeded those of all other developments (Storm et al., 2007). Many of the methodologies for assessing deer population status and limitations associated with deer management in rural environments can be applied to urbanized environments. Further complicating deer management in urbanized areas, however, are variable attitudes toward deer and deer management, generating public scrutiny of traditional deer population control methodologies (Curtis and Richmond, 1992; Decker et al., 1996; Stout et al., 1996). Many agencies have responded to increasing urban deer problems by developing urban deer management plans that may include urban deer zones with special regulations to promote control (Messmer et al., 1997).

FIGURE 14.20 The white-tailed deer is adaptable to human activities and can thrive in suburban settings. Management of deer numbers can be difficult because of the lack of access to hunting. (Photo courtesy of the Missouri Department of Conservation. With permission.)

The popular and scientific literature on white-tailed deer and their management in urbanized areas is extensive and nearly every state in the range of the white-tailed deer is faced with urban deer issues. Traditional methods of control have been effective in selected cases (Hansen and Beringer, 1997; Dohm and Brown, 1998; Kilpatrick et al., 2002; Suchy et al., 2002) but increasingly, agencies have attempted nontraditional means of controlling deer numbers (DeNicola et al., 1997; Doerr et al., 2001; DeNicola and Williams, 2008).

Controlled shooting of deer, often coined sharpshooting, has been used effectively in urban settings (Deblinger et al., 1995; Jones and Witham, 1995; Stradtmann et al., 1995; Doerr et al., 2001; DeNicola and Williams, 2008). DeNicola et al. (1997) reported costs ranging from $91 to $260/deer removed by sharpshooters. Doerr et al. (2001) compared controlled hunts, opportunistic sharpshooting, and sharpshooting over bait. Costs per deer killed were similar (although using police officers to shoot deer over bait increased costs). They concluded that sharpshooting over bait was the most efficient and adaptable method for urban settings. Sharpshooting can be effective, safe, and discrete even on small areas and to some urbanites may be more acceptable than hunting because of the perceived safety issues and lack of a recreational aspect (Doerr et al., 2001).

Trap and translocation has been used as a nonlethal means of controlling locally abundant deer (Jones and Witham, 1990; Ishmael and Rongstad, 1984; Beringer et al., 2002). Costs ranged greatly from $72/deer in a small-scale Wisconsin program (Ishmael et al., 1995) to $387/deer in Missouri (Beringer et al., 2002) and $1251/deer in Illinois (Drummond, 1995). Capture myopathy and increased deer vulnerability to hunting and other mortality factors significantly reduced annual survival rates of translocated deer (Jones and Witham, 1990; Beringer et al., 2003). Trap and translocation is not now considered a feasible method of control in most states because movement of deer to other sites may contribute to existing deer problems at release sites and because of increasing concerns about disease transmission, especially chronic wasting disease and bovine tuberculosis.

Fertility control as a means of managing wildlife populations is often a highly favored method to urbanites because the animal does not have to be killed (Curtis et al., 1993; Warren, 2000). Many research efforts to maintain deer populations at desired levels using fertility control have failed (Warren, 2000; Fagerstone et al., 2002). Fagerstone et al. (2002) cited major hurdles to the effective use of fertility control to include delivery of the contraceptive, public and conservation agency acceptance of fertility control as a valid method, and lack of commercialization of vaccines and baits. Warren (1995) encouraged wildlife professionals to be active in research and management efforts relating to contraception because of increasing situations where traditional deer management is not practical for controlling deer populations.

Resolving urban deer problems without affecting deer numbers has been attempted with mixed success. Deer deterrents such as chemical repellents (Swihart and Conover, 1990; Wagner and Nolte, 2001), fencing (Rosenberry et al., 2001; VerCauteren et al., 2006), scare devices (Beringer et al., 2003; Gilsdorf et al., 2004), compensation programs (Wagner et al., 1997), dogs (Beringer et al., 1994), and efforts to raise public tolerance thresholds of deer (Rutberg, 1997) may have positive effects. However, unless deer populations are naturally stable, the benefits are generally short-term.

Agency White-tailed Deer Management Outlook

Until recently, deer management has been based on the precept that science-based and biologically sound decisions should drive deer management programs. This requires as a basis some knowledge of deer population status and distribution, population limiting factors, and how the population will respond to perturbations, especially hunting. Although prediction is ideal, much management is adaptive and agencies increasingly need to develop monitoring programs in which impacts of hunting or other management efforts can be measured so that knowledge is gained from these efforts. Unfortunately, most agencies do not have adequate financial or personnel resources to measure and monitor the demographic parameters necessary to accurately and precisely measure population change at the DMU level. Agencies must focus on collection of information that is practical and will guide management most efficiently (Skalski et al., 2005).

Most white-tailed deer management occurs on private land. Therefore, hunting effort and harvest are not evenly distributed across the landscape. Purchasing and leasing of land for deer hunting is increasing and will challenge agencies' abilities to ensure responsible deer management occurs on a local scale. Unfortunately, management by agencies generally involves a smoothing of deer density peaks and valleys within DMU (Porter and Underwood, 1999). Areas of high and low deer abundance do and always will occur on private land and will generate criticism of agency deer management abilities. Agency consultations with landowners on a scale smaller than the DMU, as occurs in parts of the south and east, increasingly will be a part of agency deer management focus throughout the range of the white-tailed deer.

Deer managers trained in biology sometimes struggle with the human dimensions aspects of deer management (Green et al., 1997) but human societal issues will continue to be an equal partner with biological issues in modern-day deer management. Distancing of urbanites from their rural roots, public intolerance of government, private property rights, increasing interest in nonconsumptive orientations, and diverse attitudes toward deer will all challenge agency ability to manage deer. Diefenbach and Palmer (1997) proposed that the science of deer management should drive the social aspects. However, the wide array of stakeholders and their increasing expectation that they be involved in deer management decisions necessitates integration of social and biological considerations. Success or failure of future deer management programs may depend on how agencies approach societal change. Deer population management success and continued agency control over the deer management decision-making process may depend on whether an agency resists change and continues as is or embraces change and adapts management programs that incorporate stakeholder input into deer management decisions (Riley et al., 2002a; Jacobson and Decker, 2006). Unfortunately, resistance to change is common in agencies and the hunting public because management traditions often are engrained, creating a growing gap between agencies and the norms of society (Jacobson and Decker, 2006). The key to agency deer management successes will be to balance public interests with sound biological information in making management decisions.

REFERENCES

Adams, C. E., R. D. Brown, and B. J. Higginbotham. 2004. Developing a strategic plan for future hunting participation in Texas. *Wildlife Society Bulletin* 32:1156–1165.

Adams, C. E., N. Wilkins, and J. L. Cooke. 2000. A place to hunt: Organizational changes in recreational hunting, using Texas as a case study. *Wildlife Society Bulletin* 28:788–796.

Allen, D. L. 1974. *Our Wildlife Legacy.* New York, NY: Funk & Wagnalls.

Anderson, D. R. 2001. The need to get the basics right in wildlife field studies. *Wildlife Society Bulletin* 29:1294–1297.

Anderson, D. R. 2003. Response to Engeman: Index values rarely constitute reliable information. *Wildlife Society Bulletin* 31:288–291.

Anderson, R. C. 1994. Height of white-flowered trillium (*Trillium grandiflorum*) as an index of deer browsing intensity. *Ecological Applications* 4:104–109.

Augustine, D. J. and L. E. Frelich. 1998. Effects of white-tailed deer on populations of an understory forb in fragmented deciduous forests. *Conservation Biology* 12:995–1004.

Augustine, D. J. and P. A. Jordan. 1998. Predictors of white-tailed deer grazing intensity in fragmented deciduous forests. *Journal of Wildlife Management* 62:1076–1085.

Bartelt, G., J. Pardee, and K. Thiede. 2003. Environmental impact statement of rules to eradicate chronic wasting disease in Wisconsin's free-ranging white-tailed deer herd. Wisconsin Department of Natural Resources PUB-SS-980. Madison, WI.

Bean, M. J. and M. J. Rowland. 1997. *The Evolution of National Wildlife Law.* Westport, CT: Praeger.

Bender, L. C. 2006. Uses of herd composition and age ratios in ungulate management. *Wildlife Society Bulletin* 34:1225–1230.

Bender, L. C., P. E. Fowler, J. A. Bernatowicz, J. L. Musser, and L. E. Stream. 2002. Effects of open-entry spike-bull, limited-entry branched-bull harvesting on elk composition in Washington. *Wildlife Society Bulletin* 30:1078–1084.

Benson, D. E. 1992. Commercialization of wildlife: A value-added incentive for conservation. In *The Biology of Deer*, ed. R. D. Brown, 539–553. New York, NY: Springer-Verlag.

Beringer, J., L. P. Hansen, J. Demand, J. A. Sartwell, M. Wallendorf, and R. Mange. 2002. Efficacy of translocation to control urban deer in Missouri: Costs, efficiency, and outcome. *Wildlife Society Bulletin* 30:767–774.

Beringer, J., L. P. Hansen, R. A. Heinen, and N. F. Giessman. 1994. Use of dogs to reduce damage by deer to a white pine plantation. *Wildlife Society Bulletin* 22:627–632.

Beringer, J., L. P. Hansen, and O. Sexton. 1998. Detection rates of white-tailed deer with a helicopter over snow. *Wildlife Society Bulletin* 26:24–26.

Beringer, J., L. P. Hansen, and D. E. Stallknecht. 2000. An epizootic of hemorrhagic disease in white-tailed deer in Missouri. *Journal of Wildlife Diseases* 36:588–591.

Beringer, J., K. C. VerCauteren, and J. J. Millspaugh. 2003. Evaluation of an animal-activated scarecrow and a monofilament fence for reducing deer use of soybean fields. *Wildlife Society Bulletin* 31:492–498.

Biederbeck, H. H., M. C. Boulay, and D. H. Jackson. 2001. Effects of hunting regulations on bull elk survival and age structure. *Wildlife Society Bulletin* 29:1271–1277.

Bissonette, J. A., C. A. Kassar, and L. J. Cook. 2008. Assessment of costs associated with deer–vehicle collisions: Human death and injury, vehicle damage, and deer loss. *Human–Wildlife Conflicts* 2:17–27.

Blanchong, J. A., M. D. Samuel, K. T. Scribner, B. V. Weckworth, J. A. Langenberg, and K. B. Filcek. 2008. Landscape genetics and the spatial distribution of chronic wasting disease. *Biology Letter* 4:130–133.

Brinkman, T. J., J. A. Jenks, C. S. DePerno, B. S. Haroldson, and R. G. Osborn. 2004. Survival of white-tailed deer in an intensively farmed region of Minnesota. *Wildlife Society Bulletin* 32:726–731.

Brown, R. D. and S. M. Cooper. 2006. The nutritional, ecological, and ethical arguments against baiting and feeding white-tailed deer. *Wildlife Society Bulletin* 34:519–524.

Brown, T. L. and D. J. Decker. 1979. Incorporating farmers' attitudes into management of white-tailed deer in New York. *Journal of Wildlife Management* 43:236–239.

Brown, T. L., D. J. Decker, S. J. Riley et al. 2000. The future of hunting as a mechanism to control white-tailed deer populations. *Wildlife Society Bulletin* 28:797–807.

Burroughs, J. P., H. Campa III, S. R. Winterstein, B. A. Rudolph, and W. E. Moritz. 2006. Cause-specific mortality and survival of white-tailed deer fawns in southwestern Lower Michigan. *Journal of Wildlife Management* 70:743–751.

Campo, J. J. and G. E. Spencer. 1991. Regulatory response to deer hunting with dogs in eastern Texas. *Proceedings of the Annual Conference of the Southeastern Association of Fish and Wildlife Agencies* 45:235–240.

Campo, J. J., G. E. Spencer, and B. Ortego. 1987. White-tailed deer hunting with dogs in east Texas. *Proceedings of the Annual Conference of the Southeastern Association of Fish and Wildlife Agencies* 41:404–409.

Carlock, D. M., K. E. Kammermeyer, L. E. McSwain, and E. J. Wentworth. 1993. Deer movements in relation to food supplies in Southern Appalachians. *Proceedings of the Annual Conference of the Southeastern Association of Fish and Wildlife Agencies* 47:16–23.

Carpenter, L. H. 2000. Harvest management goals. In *Ecology and Management of Large Mammals in North America*, eds. S. Demarais and P. R. Krausman, 192–213. Upper Saddle River, NJ: Prentice-Hall.

Christoffel, R. A. and S. R. Craven. 2000. Attitudes of woodland owners toward white-tailed deer and herbivory in Wisconsin. *Wildlife Society Bulletin* 28:227–234.

Collier, B. A., S. S. Ditchkoff, J. B. Raglin, and J. M. Smith. 2007. Detection probability and sources of variation in white-tailed deer spotlight surveys. *Journal of Wildlife Management* 71:277–281.

Collier, B. A. and D. G. Krementz. 2006. White-tailed deer management practices on private lands in Arkansas. *Wildlife Society Bulletin* 34:307–313.

Coltman, D. W., P. O'Donoghue, J. T. Jorgenson, J. T. Hogg, C. Strobeck, and M. Festa-Blanchet. 2003. Undesirable evolutionary consequences of trophy hunting. *Nature* 426:655–658.

Conover, M. R. 2001. *Resolving Human–Wildlife Conflicts: The Science of Wildlife Damage Management*. New York, NY: Lewis Publishers.

Conover, M. R., W. C. Pitt, K. K. Kessler, T. J. DuBow, and W. A. Sanborn. 1995. Review of human injuries, illnesses, and economic losses caused by wildlife in the United States. *Wildlife Society Bulletin* 23:407–414.

Cooper, S. M., R. M. Cooper, M. K. Owens, and T. F. Ginnett. 2002. Effect of supplemental feeding on use of space and browse utilization by white-tailed deer. In *Land Use for Water and Wildlife*, eds. D. Forbes and G. Piccinni, 31–32. Texas Agricultural Research and Extension Center, UREC-02–031. Uvalde, TX.

Cordell, H. K., C. Betz, J. Bowker et al. 1999. *Outdoor Recreation in American Life: A National Assessment of Demand and Supply Trends*. Champaign, IL: Sagamore Publishing.

Côté, S. D., T. P. Rooney, J. P. Tremblay, C. Dussault, and D. M. Waller. 2004. Ecological impacts of deer overabundance. *Annual Review of Ecology, Evolution and Systematics* 35:113–147.

Curtis, P. D., D. J. Decker, R. J. Stout, M. E. Richmond, C. A. Loker. 1993. Human dimensions of contraception in wildlife management. *USDA National Wildlife Research Center Symposium: Contraception in Wildlife Management*. http://digitalcommons.unl.edu/nwrccontraception/7. Accessed September 17, 2009.

Curtis, P. D. and M. E. Richmond. 1992. Future challenges of suburban white-tailed deer management. *Transactions of the North American Wildlife and Natural Resources Conference* 58:102–116.

Curtis, R. L., H. S. Mosby, and C. T. Cushwa. 1972. The influence of weather on hunter–deer contacts in Western Virginia. *Transactions of the North American Wildlife and Natural Resources Conference* 37:282–285.

Daniel, W. S. and D. B. Frels. 1971. *A Track-Count Method for Censusing White-tailed Deer*. Technical Series Number 7, Texas Parks and Wildlife Department. Austin, TX.

Davis, M. L., J. Berkson, D. Steffen, and M. K. Tilton. 2007. Evaluation of accuracy and precision of Downing population reconstruction. *Journal of Wildlife Management* 71:2297–2303.

Deblinger, R. D., D. W. Rimmer, J. J. Vaske, and G. Vecellio. 1995. Efficiency of controlled, limited hunting at the Crane Reservation, Ipswich, Massachusetts. In *Urban Deer: A Manageable Resource? Proceedings of a Symposium at the 55th Midwest Fish and Wildlife Conference*, ed. J. B. McAninch, 82–86. St. Louis, MO: North Central Section of the Wildlife Society.

deCalesta, D. S. 1994. Impact of white-tailed deer on songbirds within managed forests in Pennsylvania. *Journal of Wildlife Management* 58:711–718.

Decker, D. J. and T. L. Brown. 1982. Fruit growers' vs. other farmers' attitudes toward deer in New York. *Wildlife Society Bulletin* 10:150–155.

Decker, D. J., C. C. Krueger, R. A. Baer, Jr., B. A. Knuth, and M. E. Richmond. 1996. From clients to stakeholders: A philosophical shift for fish and wildlife management. *Human Dimensions in Wildlife* 1:70–82.

Decker, D. J. and K. G. Purdy. 1988. Toward a concept of wildlife acceptance capacity in wildlife management. *Wildlife Society Bulletin* 16:53–57.

DelGiudice, G. D., J. Fieberg, M. R. Riggs, M. Carstensen Powell, and W. Pan. 2006. A long-term age-specific survival analysis of female white-tailed deer. *Journal of Wildlife Management* 70:1556–1568.

DeLury, D. B. 1947. On the estimation of biological populations. *Biometrics* 3:145–167.

DeNicola, A. J., S. J. Weber, C. A. Bridges, and J. L. Stokes. 1997. Nontraditional techniques for management of overabundant deer populations. *Wildlife Society Bulletin* 25:496–499.

DeNicola, A. J. and S. C. Williams. 2008. Sharpshooting suburban white-tailed deer reduces deer–vehicle collisions. *Human–Wildlife Conflicts* 2:28–33.

DeYoung, C. A. 1985. Accuracy of helicopter surveys of deer in South Texas. *Wildlife Society Bulletin* 13:146–149.

DeYoung, R. W., S. Demarais, K. L. Gee, R. L. Honeycutt, M. W. Hellickson, and R. A. Gonzales. 2009. Molecular evaluation of the white-tailed deer (*Odocoileus virginianus*) mating system. *Journal of Mammalogy* 90:946–953.

Diefenbach, D. R., J. C. Finley, A. E. Luloff et al. 2005. Bear and deer hunter density and distribution on public land in Pennsylvania. *Human Dimensions of Wildlife* 10:201–212.

Diefenbach, D. R., E. S. Long, C. S. Rosenberry, B. D. Wallingford, and D. R. Smith. 2008. Modeling distribution of dispersal distances in male white-tailed deer. *Journal of Wildlife Management* 72:1296–1303.

Diefenbach, D. R. and W. L. Palmer. 1997. Deer management: Marketing the science. *Wildlife Society Bulletin* 25:378–381.

Diefenbach, D. R., W. L. Palmer, and W. K. Shope. 1997. Attitudes of Pennsylvania sportsmen towards managing white-tailed deer to protect the ecological integrity of forests. *Wildlife Society Bulletin* 25:244–251.

Ditchkoff, S. S., R. L. Lochmiller, R. E. Masters, S. R. Hoofer, and R. A. Van Den Bussche. 2001a. Major-histocompatibility-complex-associated variation in secondary sexual traits of white-tailed deer (*Odocoileus virginianus*): Evidence for good-genes advertisement. *Evolution* 55:616–625.

Ditchkoff, S. S., E. R. Welch Jr., R. L. Lochmiller, R. E. Masters, and W. R. Starry. 2001b. Age-specific causes of mortality among male white-tailed deer support mate-competition theory. *Journal of Wildlife Management* 65:552–559.

Doenier, P. N., G. D. DelGiudice, and M. R. Riggs. 1997. Effects of winter supplemental feeding on browse consumption by white-tailed deer. *Wildlife Society Bulletin* 25:235–243.

Doerr, M. L., J. B. McAninch, and E. P. Wiggers. 2001. Comparison of 4 methods to reduce white-tailed deer abundance in an urban community. *Wildlife Society Bulletin* 29:1105–1113.

Dohm, L. and E. Brown. 1998. Public partnerships for urban deer management in Missouri. *Transactions of the North American Wildlife and Natural Resources Conference* 63:239–243.

Downing, R. L., W. H. Moore, and J. Kight. 1965. Comparison of deer census techniques applied to a known population in a Georgia enclosure. *Proceedings of the Annual Conference of the Southeastern Association of Fish and Wildlife Agencies* 19:26–30.

Drake, D., C. Aquila, and G. Huntington. 2005. Counting a suburban deer population using forward-looking infrared radar and road counts. *Wildlife Society Bulletin* 33:656–661.

Drummond, F. 1995. Lethal and non-lethal deer management at Ryerson Conservation Area Northeastern Illinois. In *Urban Deer: A Manageable Resource? Proceedings of a Symposium at the 55th Midwest Fish and Wildlife Conference*, ed. J. B. McAninch, 105–109. St. Louis, MO: North Central Section of the Wildlife Society.

Durkin, P. 2004. Antler restrictions spell conservation. Buck or doe? Antler restrictions spell controversy. *American Hunter* 32:28–32.

Eberhardt, L. L. 1960. *Estimation of Vital Characteristics of Michigan Deer Herds.* Michigan Department of Conservation, Game Division Report 2282. East Lansing, MI.

Ellingwood, M. R. and J. V. Spignesi. 1986. Management of an urban deer herd and the concept of cultural carrying capacity. *Transactions of the Northeast Deer Technical Committee* 22:42–45.

Enck, J. W., D. J. Decker, and T. L. Brown. 2000. Status of hunter recruitment and retention in the United States. *Wildlife Society Bulletin* 28:817–824.

Engeman, R. M. 2003. More on the need to get the basics right: Population indices. *Wildlife Society Bulletin* 31:286–287.

Erickson, D. W. and N. F. Giessman. 1989. Review of a program to alleviate localized deer damage. *Wildlife Society Bulletin* 17:544–548.

Eve, J. H. and F. E. Kellogg. 1977. Management implications of abomasal parasites in southeastern white-tailed deer. *Journal of Wildlife Management* 41:169–177.

Fafarman, K. R. and C. A. DeYoung. 1986. Evaluation of spotlight counts of deer in South Texas. *Wildlife Society Bulletin* 14:180–185.

Fagerstone, K. A., M. A. Coffey, P. D. Curtis et al. 2002. *Wildlife Fertility Control.* The Wildlife Society Technical Review 02–2. Bethesda, MD.

Farrell, M. C. and P. A. Tappe. 2007. County-level factors contributing to deer–vehicle collisions in Arkansas. *Journal of Wildlife Management* 71:2727–2731.

Farrington, S. J., R. M. Muzika, D. Drees, and T. M. Knight. 2009. Interactive effects of harvest and deer herbivory on the population dynamics of American ginseng. *Conservation Biology* 23:719–728.

Feldhamer, G. A. 2002. Acorns and white-tailed deer: Interrelationships in forest ecosystems. In *Oak Forest Ecosystems: Ecology and Management for Wildlife*, eds. W. J. McShea and W. M. Healy, 215–223. Baltimore, MD: Johns Hopkins University Press.

Feldhamer, G. A., T. P. Kilbane, and D. W. Sharp. 1989. Cumulative effect of winter on acorn yield and deer body weight. *Journal of Wildlife Management* 53:292–295.

Ferrigno, D. M., S. E. Martka, J. Powers, and D. Roberts. 2002. Results of an experimental "earn-a-buck" regulation in New Jersey's agricultural and suburban deer management zones. In *Proceedings of the First National Bowhunting Conference*, ed. R. J. Warren, 62–63. Comfrey, MN: Archery Manufacturers and erchants Organization.

Fischer, J. R., L. P. Hansen, J. R. Turk, M. A. Miller, W. H. Fales, and H. S. Gosser. 1995. An epizootic of hemorrhagic disease in white-tailed deer (*Odocoileus virginianus*) in Missouri: Necropsy findings and population impact. *Journal of Wildlife Diseases* 31:30–36.

Fobes, C. B. 1945. Weather and the kill of white-tailed deer in Maine. *Journal of Wildlife Management* 9:76–78.

Forsyth, D. M., R. J. Barker, G. Morriss, and M. P. Scroggie. 2007. Modeling the relationship between fecal pellet indices and deer density. *Journal of Wildlife Management* 71:964–970.

Foster, J. R., J. L. Roseberry, and A. Woolf. 1997. Factors influencing efficiency of white-tailed deer harvest in Illinois. *Journal of Wildlife Management* 61:1091–1097.

Frawley, F. J. 2004. *2003 Archer's Furbearer Population Index.* Michigan Department of Natural Resources Wildlife Division Report No. 3415. Lansing, MI.

Frawley, F. J. 2006. *Demographics, Recruitment, and Retention of Michigan Hunters: 2005 Update.* Michigan Department of Natural Resources Wildlife Division Report No. 3462. Lansing, MI.

Frelich, L. E. and C. G. Lorimer. 1985. Current and predicted long-term effects of deer browsing in hemlock forests in Michigan, USA. *Biological Conservation* 34:99–120.

Fritzen, D. E., R. F. Labisky, D. E. Easton, and J. C. Kilgo. 1995. Nocturnal movements of white-tailed deer: Implications for refinement of track-count surveys. *Wildlife Society Bulletin* 23:187–193.

Fry, F. E. J. 1949. Statistics of a lake trout fishery. *Biometrics* 5:26–67.

Fryxell, J. M., D. J. T. Hussel, A. B. Lambert, and P. C. Smith. 1991. Time lags and population fluctuations in white-tailed deer. *Journal of Wildlife Management* 55:377–385.

Geist, V., S. P. Mahoney, and J. F. Organ. 2001. Why hunting has defined the North American model of wildlife conservation. *Transactions of the North American Wildlife and Natural Resources Conference* 66:175–185.

Giles, B. G. and C. S. Findlay. 2004. Effectiveness of a selective harvest system in regulating deer populations in Ontario. *Journal of Wildlife Management* 68:266–277.

Gilsdorf, J. M., S. E. Hygnstrom, K. C. VerCauteren, G. M. Clements, E. E. Blankenship, and R. M. Engeman. 2004. Evaluation of a deer-activated bio-acoustic frightening device for reducing deer damage in cornfields. *Wildlife Society Bulletin* 32:515–523.

Ginnett, T. F. and E. L. B. Young. 2000. Stochastic recruitment in white-tailed deer along an environmental gradient. *Journal of Wildlife Management* 64:713–720.

Ginsberg, J. R. and E. J. Milner-Gulland. 1994. Sex-biased harvesting and population dynamics in ungulates: Implications for conservation and sustainable use. *Conservation Biology* 8:157–166.

Gore, H., W. F. Harwell, M. D. Hobson, and W. J. Williams. 1985. Buck permits as a management tool in South Texas. In *Game Harvest Management*, eds. S. L. Beasom and S. F. Roberson, 149–163. Kingsville, TX: Caesar Kleberg Wildlife Research Institute.

Gould, J. H. and K. J. Jenkins. 1993. Seasonal use of conservation reserve program lands by white-tailed deer in east-central South Dakota. *Wildlife Society Bulletin* 21:250–255.

Green, D., G. R. Askins, and P. D. West. 1997. Public opinion: Obstacle or aid to sound deer management? *Wildlife Society Bulletin* 25:367–370.

Green, D. and J. P. Stowe Jr. 2000. Quality deer management: Ethical and social issues. *Human Dimensions of Wildlife* 5:62–71.

Grovenburg, T. W., J. A. Jenks, R. W. Klaver et al. 2008. Factors affecting road mortality of white-tailed deer in eastern South Dakota. *Human–Wildlife Conflicts* 2:48–59.

Grund, M. D. and A. Woolf. 2004. Development and evaluation of an accounting model for estimating deer population sizes. *Ecological Modeling* 180:345–357.

Guynn, D. C. Jr., S. P. Mott, W. D. Cotton, and H. A. Jacobson. 1983. Cooperative management of white-tailed deer on private lands in Mississippi. *Wildlife Society Bulletin* 11:211–214.

Hansen, L. P. 2001. *Effect of the Addition of an Antlerless Portion of the Firearms Season on Antlerless Harvest.* Jefferson City, MO: Missouri Department of Conservation.

Hansen, L. P. 2007. *Results of the Pilot Antler Point Restriction*. Jefferson City, MO: Missouri Department of Conservation.

Hansen, L. P. and J. Beringer. 1997. Managed hunts to control white-tailed deer populations on urban public areas in Missouri. *Wildlife Society Bulletin* 25:484–487.

Hansen, L. P. and J. Beringer. 2003. Survival of rural and urban white-tailed deer in Missouri. *Proceedings of the Annual Conference of the Southeastern Association of Fish and Wildlife Agencies* 57:326–336.

Hansen, L. P. and J. Beringer. 2007. *Deer Management Evaluation on Conservation Areas–2005–2006.* Jefferson City, MO: Missouri Department of Conservation.

Hansen, L. P., J. Beringer, and J. H. Schulz. 1996. Reproductive characteristics of female white-tailed deer in Missouri. *Proceedings of the Annual Conference of the Southeastern Association of Fish and Wildlife Agencies* 50:357–366.

Hansen, L. P., C. M. Nixon, and J. Beringer. 1997. Role of refuges in the dynamics of outlying deer populations. In *The Science of Overabundance: Deer Ecology and Population Management*, eds. W. J. McShea, H. B. Underwood, and J. H. Rappole, 327–345. Washington, DC: Smithsonian Institution Press.

Hansen, L. P., C. M. Nixon, and F. Loomis. 1986. Factors affecting daily and annual harvest of white-tailed deer in Illinois. *Wildlife Society Bulletin* 14:368–376.

Hansen, L. P., M. Wallendorf, and J. Beringer. 2006. A comparison of deer and turkey harvest data collection methods in Missouri. *Wildlife Society Bulletin* 34:1356–1361.

Harden, C. D., A. Woolf, and J. Roseberry. 2005. Influence of exurban development on hunting opportunity, hunter distribution, and harvest efficiency of white-tailed deer. *Wildlife Society Bulletin* 33:233–242.

Haroldson, B. S., M. Grund, and L. Cornicelli. 2007. *Minnesota Deer Status Report.* Report presented at the 31st Midwest Deer and Wild Turkey Study Group, Jackson, OH.

Haroldson, B. S., E. P. Wiggers, J. Beringer, L. P. Hansen, and J. B. McAninch. 2003. Evaluation of aerial thermal imaging for detecting white-tailed deer in a deciduous forest environment. *Wildlife Society Bulletin* 31:1188–1197.

Holsman, R. H. 2007. *What Wisconsin Deer Hunters Think about SAK, Deer Populations and DNR Credibility: Results of a 2006 Statewide Survey.* Stevens Point, WI: University of Wisconsin.

Holsman, R. H. and J. Petchenik. 2006. Predicting deer hunter harvest behavior in Wisconsin's chronic wasting disease eradication zone. *Human Dimensions of Wildlife* 11:177–189.

Horsley, S. B., S. L. Stout, and D. S. deCalesta. 2003. White-tailed deer impact on the vegetation dynamics of a northern hardwood forest. *Ecological Applications* 13:98–118.

Horton, R. R. and S. R. Craven. 1997. Perceptions of shooting-permit use for deer damage abatement in Wisconsin. *Wildlife Society Bulletin* 25:330–336.

Hubbard, M. W., B. J. DaNeilsen, and R. A. Schmitz. 2000. Factors influencing the location of deer–vehicle accidents in Iowa. *Journal of Wildlife Management* 64:707–713.

Huggins, J. G., R. Stevens, B. Unruh, and B. DeVille. 2005. Impact of buck harvest limits on lease hunting value. *Wildlife Society Bulletin* 33:749–750.

Iowa Deer Study Advisory Committee. 2009. *A Review of Iowa's Deer Management Program.* Iowa Department of Natural Resources, Des Moines, IA.

Ishmael, W. E. and O. J. Rongtad. 1984. Economics of an urban deer-removal program. *Wildlife Society Bulletin* 12:394–398.

Ishmael, W. E., D. E. Katsma, and T. A Isaac. 1995. Chronology of the deer management debate in River Hills, Wisconsin. In *Urban Deer: A Manageable Resource? Proceedings of a Symposium at the 55th Midwest Fish and Wildlife Conference*, ed. J. B. McAninch, 66–74. St. Louis, MO: North Central Section of the Wildlife Society.

Jacobson, C. A. and D. J. Decker. 2006. Ensuring the future of state wildlife management: Understanding challenges for institutional change. *Wildlife Society Bulletin* 34:531–536.

Jacobson, H. A., J. C. Kroll, R. W. Browning, B. H. Koerth, and M. H. Conway. 1997. Infrared-triggered cameras for censusing white-tailed deer. *Wildlife Society Bulletin* 25:547–556.

Johnson, K. N., R. L. Johnson, D. K. Edwards, and C. A. Wheaton. 1993. Public participation in wildlife management: Opinions from public meetings and random surveys. *Wildlife Society Bulletin* 21:218–225.

Jones, J. M. and J. H. Witham. 1990. Post-translocation survival and movement of metropolitan white-tailed deer. *Wildlife Society Bulletin* 18:434–441.

Jones, J. M. and J. H. Witham. 1995. Urban deer problem-solving in northeast Illinois: An overview. In *Urban Deer: A Manageable Resource? Proceedings of a Symposium at the 55th Midwest Fish and Wildlife Conference*, ed. J. B. McAninch, 58–65. St. Louis, MO: North Central Section of the Wildlife Society.

Kellert, S. R. 1996. *The Value of Life: Biological Diversity and Human Society.* Washington, DC: Island Press.

Keyser, P. D., D. C. Guynn, H. S. Hill, and W. M. Knox. 2006. Relative density-physical condition models: A potential application for managing white-tailed deer populations. *Wildlife Society Bulletin* 34:1113–1120.

Killmaster, C. H., D. A. Osborn, R. J. Warren, and K. V. Miller. 2007. Deer and understory plant responses to a large-scale herd reduction on a Georgia state park. *Natural Areas Journal* 27:161–168.

Kilpatrick, H. J., A. M. LaBonte, and J. S. Barclay. 2005. Factors affecting harvest-reporting rates for white-tailed deer. *Wildlife Society Bulletin* 33:974–980.

Kilpatrick, H. J., A. M. LaBonte, J. S. Barclay, and G. Warner. 2004. Assessing strategies to improve bowhunting as an urban deer management tool. *Wildlife Society Bulletin* 32:1177–1184.

Kilpatrick, H. J., A. M. LaBonte, and J. T. Seymour. 2002. A shotgun-archery deer hunt in a residential community: Evaluation of hunt strategies and effectiveness. *Wildlife Society Bulletin* 30:478–486.

Kilpatrick, H. J., S. M. Spohr, and G. G. Chasko. 1997. A controlled deer hunt on a state-owned coastal reserve in Connecticut: Controversies, strategies, and results. *Wildlife Society Bulletin* 25:451–456.
Knight, T. M. 2004. The effects of herbivory and pollen limitation on a declining population of *Trillium grandiflorum*. *Ecological Applications* 14:915–928.
Koerth, B. H. and J. C. Kroll. 2000. Bait type and timing for deer counts using cameras triggered by infrared monitors. *Wildlife Society Bulletin* 28:630–635.
Koerth, B. H. and J. C. Kroll. 2008. Juvenile-to-adult antler development in white-tailed deer in South Texas. *Journal of Wildlife Management* 72:1109–1113.
Kubisiak, J. F., K. R. McCaffery, W. A. Creed, T. A. Heberlein, R. C. Bishop, and R. E. Rolley. 2001. *Sandhill Whitetails: Providing New Perspective for Deer Management.* Madison, WI: Bureau of Integrated Science Services.
Lancia, R. A., J. W. Bishir, M. C. Conner, and C. S. Rosenberry. 1996. Use of catch-effort to estimate population size. *Wildlife Society Bulletin* 24:731–737.
Leal, D. R. and J. B. Grewell. 1999. *Hunting for Habitat: A Practical Guide to State-Landowner Partnerships.* Bozeman, MT: Political Economy Research Center.
Lischka, S. A., S. J. Riley, and B. A. Rudolph. 2008. Effects of impact perception on acceptance capacity for white-tailed deer. *Journal of Wildlife Management* 72:502–509.
Lockwood, M. A., D. B. Frels Jr., W. E. Armstrong, E. Fuchs, and D. E. Harmel. 2007. Genetic and environmental interaction in white-tailed deer. *Journal of Wildlife Management* 71:2732–2735.
Long, E. S., D. R. Diefenbach, C. S. Rosenberry, and B. D. Wallingford. 2008. Multiple proximate and ultimate causes of natal dispersal in white-tailed deer. *Behavioral Ecology* 19:1235–1242.
Lukefahr, S. D. and H. A. Jacobson. 1998. Variance component analysis and heritability of antler traits in white-tailed deer. *Journal of Wildlife Management* 62:262–268.
MacNab, J. 1985. Carrying capacity and related slippery shibboleths. *Wildlife Society Bulletin* 13:403–410.
Mandujano, S. 2005. Track count calibration to estimate density of white-tailed deer (*Odocoileus virginianus*) in Mexican dry tropical forest. *Southwestern Naturalist* 50:223–229.
Marchinton, R. L., A. S. Johnson, J. R. Sweeney, and J. M. Sweeney. 1970. Legal hunting of white-tailed deer with dogs: Biology, sociology and management. *Proceedings of the Annual Conference of the Southeastern Association of Game and Fish Commissioners* 24:74–89.
Mattson, K. M. and W. E. Moritz. 2008. Evaluating differences in harvest data used in the sex–age–kill deer population model. *Journal of Wildlife Management* 72:1019–1025.
Mautz, W. W. 1978. Nutrition and carrying capacity. In *Big Game of North America*, eds. J. L. Schmidt and D. L. Gilbert, 321–348. Harrisburg, PA: Stackpole Books.
McCabe, R. E. and T. R. McCabe. 1984. Of slings and arrows: An historical retrospection. In *White-tailed Deer Ecology and Management*, ed. L. K. Halls, 19–72. Harrisburg, PA: Stackpole Books.
McCaffery, K. R. 1973. Road-kills show trends in Wisconsin deer populations. *Journal of Wildlife Management* 37:212–216.
McCaffery, K. R. 1976. Deer trail counts as an index to populations and habitat use. *Journal of Wildlife Management* 40:308–316.
McCaffery, K. R. 2000. *Deer Baiting and Feeding Issue.* Adaptation from presentation at a joint Midwest and Northeast Deer Study Groups meeting, Hillman, MI.
McCall, T. C., R. D. Brown, and L. C. Bender. 1997. Comparison of techniques for determining the nutritional carrying capacity for white-tailed deer. *Journal of Range Management* 50:33–38.
McCullough, D. R. 1979. *The George Reserve Deer Herd: Population Ecology of a K-Selected Species.* Ann Arbor, MI: University of Michigan Press.
McCullough, D. R. 1984. Lessons from the George Reserve, Michigan. In *White-tailed Deer: Ecology and Management*, ed. L. K. Halls, 211–242. Harrisburg, PA: Stackpole Books.
McCullough, D. R. 1994. What do herd composition counts tell us? *Wildlife Society Bulletin* 22:295–300.
McDonald, J. E. Jr., D. E. Clark, and W. A. Woytek. 2007. Reduction and maintenance of a white-tailed deer herd in central Massachusetts. *Journal of Wildlife Management* 71:1585–1593.
McDonald, J. E. Jr., W. A. Woytek, and R. D. Deblinger. 2002. Can we overcome the largest obstacle to deer management? A Massachusetts case study. In *Proceedings of the First National Bowhunting Conference*, ed. R. J. Warren, 108–116. Comfrey, MN: Archery Manufacturers and Merchants Organization.
McDonald, J. S. and K. V. Miller. 2004. *A History of White-tailed Deer Restocking in the United States 1878–2004.* Bogart, GA: The Quality Deer Management Association.

McGraw, J. B. and M. A. Furedi. 2005. Deer browsing and population viability of a forest understory plant. *Science* 307:920–922.
McShea, W. J. and G. Schwede. 1993. Variable acorn crops: Responses of white-tailed deer and other mast consumers. *Journal of Mammalogy* 74:999–1006.
Mechler, J. L. 1970. Factors influencing the white-tailed deer harvest in Virginia, 1947–1967. MS thesis, Virginia Polytechnical Institute and State University.
Medin, D. E. and A. E. Anderson. 1979. Modeling the dynamics of a Colorado mule deer population. *Wildlife Monographs* 68:1–77.
Messmer, T. A., L. Cornicelli, D. J. Decker, and D. G. Hewitt. 1997. Stakeholder acceptance of urban deer management techniques. *Wildlife Society Bulletin* 25:360–366.
Michigan Department of Natural Resources. 1999. *Deer Baiting Issues in Michigan.* Michigan Department of Natural Resources Wildlife Division issue review paper 5. Lansing, MI.
Miller, C. A. and P. Shelton. 2000. Perceptions of white-tailed deer abundance and management among hunters and landowners in Illinois. In *The Ninth Wildlife Damage Management Conference Proceedings*, eds. M. C. Brittingham, J. Kays, and R. McPeake, 264–268. State College, PA.
Miller, C. A. and J. J. Vaske. 2003. Individual and situational influences on declining hunter effort in Illinois. *Human Dimensions of Wildlife* 8:263–276.
Miller, K. V. and R. L. Marchinton. 1995. *Quality Whitetails: The Why and How of Quality Deer Management.* Mechanicsburg, PA: Stackpole Books.
Miller, K. V. and J. M. Wentworth. 2000. Carrying capacity. In *Ecology and Management of Large Mammals in North America*, eds. S. Demarais and P. R. Krausman, 140–155. Upper Saddle River, NJ: Prentice-Hall.
Millspaugh, J. J., M. S. Boyce, D. R. Diefenbach, L. P. Hansen, K. Kammermeyer, and J. R. Skalski. 2007. *An Evaluation of the SAK Model as Applied in Wisconsin.* Madison, WI: Wisconsin Department of Natural Resources.
Millspaugh, J. J., J. R. Skalski, R. L. Townsend et al. 2009. An evaluation of sex–age–kill (SAK) model performance. *Journal of Wildlife Management* 73:442–451.
Milner, J. M., E. B. Nilsen, and H. P. Andreassen. 2007. Demographic side effects of selective hunting in ungulates and carnivores. *Conservation Biology* 21:36–47.
Minnis, D. L. and R. B. Peyton. 1995. Cultural carrying capacity: Modeling a notion. In *Urban Deer: A Manageable Resource? Proceedings of a Symposium at the 55th Midwest Fish and Wildlife Conference*, ed. J. B. McAninch, 19–34. St. Louis, MO: North Central Section of the Wildlife Society.
Mladenoff, D. J. and F. Stearns. 1993. Eastern hemlock regeneration and deer browsing in the northern Great Lakes region: A re-examination and model simulation. *Conservation Biology* 7:889–900.
Morellet, N., S. Champely, J. M. Gaillard, P. Ballon, and Y. Boscardin. 2001. The browsing index: New tool uses browsing pressure to monitor deer populations. *Wildlife Society Bulletin* 29:1243–1252.
Mozumder, P., C. M. Starbuck, R. P. Berrens, and S. Alexander. 2007. Lease and fee hunting on private lands in the U.S.: A review of the economic and legal issues. *Human Dimensions of Wildlife* 12:1–14.
Murphy, D. A. 1965. Effects of various opening days on deer harvest and hunting pressure. *Proceedings of the Annual Conference Southeastern Association of Fish and Wildlife Agencies* 19:141–146.
Naugle, D. E., J. A. Jenks, and B. J. Kernohan. 1996. Use of thermal infrared sensing to estimate density of white-tailed deer. *Wildlife Society Bulletin* 24:37–43.
Neff, D. J. 1968. The pellet-group count technique for big game trend, census, and distribution: A review. *Journal of Wildlife Management* 32:597–614.
Nelson, M. E. and L. D. Mech. 1986. Mortality of white-tailed deer in northeastern Minnesota. *Journal of Wildlife Management* 50:691–698.
Nettles, V. F. and D. E. Stallknecht. 1992. History and progress in the study of hemorrhagic disease of deer. *Proceedings of the North American Wildlife and Natural Resources Conference* 57:499–516.
Nixon, C. M. and L. P. Hansen. 1992. Habitat relationships and population dynamics of deer in the intensively farmed Midwestern United States. In *The Biology of Deer*, ed. R. D. Brown, 22–29. New York, NY: Springer-Verlag.
Nixon, C. M., L. P. Hansen, P. A. Brewer et al. 1994. Behavior, dispersal, and survival of male white-tailed deer in Illinois. *Illinois Natural History Survey Biological Notes* 139:1–29.
Nixon, C. M., L. P. Hansen, P. A. Brewer et al. 2001. Survival of white-tailed deer in intensively farmed areas of Illinois. *Canadian Journal of Zoology* 79:581–588.

Nixon, C. M., L. P. Hansen, P. A. Brewer, and J. E. Chelsvig. 1991. Ecology of white-tailed deer in an intensively farmed region of Illinois. *Wildlife Monographs* 118:1–77.

Nixon, C. M., P. C. Mankin, D. R. Etter et al. 2007. White-tailed deer dispersal behavior in an agricultural environment. *American Midland Naturalist* 157:212–220.

Novak, J. M., K. T. Scribner, W. D. DuPont, and M. H. Smith. 1991. Catch-effort estimation of white-tailed deer population size. *Journal of Wildlife Management* 55:31–38.

Nugent, G. and D. Choquenot. 2004. Comparing cost-effectiveness of commercial harvesting, state-funded culling, and recreational deer hunting in New Zealand. *Wildlife Society Bulletin* 32:481–492.

Patterson, B. R. and V. A. Power. 2002. Contributions of forage competition, harvest, and climate fluctuation to changes in population growth of northern white-tailed deer. *Oecologia* 130:62–71.

Peyton, R. B. 2000. Wildlife management: Cropping to manage or managing to crop? *Wildlife Society Bulletin* 28:774–779.

Porter, W. F. and H. B. Underwood. 1999. Of elephants and blind men: Deer management in the U.S. National Parks. *Ecological Applications* 9:3–9.

Potvin, F. and L. Breton. 2005. Testing 2 aerial survey techniques on deer in fenced enclosures-visual double-counts and thermal infrared sensing. *Wildlife Society Bulletin* 33:317–323.

Potvin, F. and J. Huot. 1983. Estimating carrying capacity of a white-tailed deer wintering area in Quebec. *Journal of Wildlife Management* 47:463–475.

Progulske, D. R. and D. C. Duerre. 1964. Factors influencing spotlighting counts of deer. *Journal of Wildlife Management* 28:27–34.

Rabe, M. J., S. S. Rosenstock, and J. C. deVos. 2002. Review of big-game survey methods used by wildlife agencies of the western United States. *Wildlife Society Bulletin* 30:46–52.

Riley, S. J., D. J. Decker, L. H. Carpenter et al. 2002a. The essence of wildlife management. *Wildlife Society Bulletin* 30:585–593.

Riley, S. J., D. J. Decker, J. W. Enck, P. D. Curtis, T. B. Lauber, and T. L. Brown. 2002b. Deer populations up, hunter populations down: Implications for interdependence of deer and hunter population dynamics on management. *Écoscience* 10:455–461.

Riley, S. J., W. F. Siemer, D. J. Decker, L. H. Carpenter, J. F. Organ, and L. T. Berchielli. 2003. Adaptive impact management: An integrative approach to wildlife management. *Human Dimensions of Wildlife* 8:81–95.

Roberts, C. W., B. L. Pierce, A. W. Braden et al. 2006. Comparison of camera and road survey estimates for white-tailed deer. *Journal of Wildlife Management* 70:263–267.

Rolley, R. 1995. Deer carrying capacity. In *Deer Population Goals and Harvest Management Environmental Assessment*, eds. W. J. Vander Zouwen and D. K. Warnke, 52–53. Madison, WI: Wisconsin Department of Natural Resources.

Rooney, T. P. and D. M. Waller. 2003. Direct and indirect effects of white-tailed deer in forest ecosystems. *Forest Ecology and Management* 181:165–176.

Root, B. G., E. K. Fritzell, and N. F. Giessman. 1988. Effects of intensive hunting on white-tailed deer movement. *Wildlife Society Bulletin* 16:145–151.

Roseberry, J. L. 1995. *Illinois Deer Harvest and Modeling Program (IDHAMP). Users Guide and Reference Manual.* Cooperative Wildlife Research Laboratory, Southern Illinois University Carbondale.

Roseberry, J. L. and W. D. Klimstra. 1974. Differential vulnerability during a controlled deer harvest. *Journal of Wildlife Management* 38:499–507.

Roseberry, J. L. and A. Woolf. 1991. A comparative evaluation of techniques for analyzing white-tailed deer harvest data. *Wildlife Monographs* 117:1–59.

Roseberry, J. L. and A. Woolf. 1998. Habitat-population density relationships for white-tailed deer in Illinois. *Wildlife Society Bulletin* 26:252–258.

Rosenberry, C. S., D. R. Diefenbach, and B. D. Wallingford. 2004. Reporting-rate variability and precision of white-tailed deer harvest estimates in Pennsylvania. *Journal of Wildlife Management* 68:860–869.

Rosenberry, C. S., L. I. Muller, and M. C. Conner. 2001. Movable, deer-proof fencing. *Wildlife Society Bulletin* 29:754–757.

Rudolph, B. A., S. J. Riley, G. J. Hickling, B. J. Frawley, M. S. Garner, and S. R. Winterstein. 2006. Regulating hunter baiting for white-tailed deer in Michigan: Biological and social considerations. *Wildlife Society Bulletin* 34:314–321.

Rupp, S. P., W. B. Ballard, and M. C. Wallace. 2000. A nationwide evaluation of deer hunter harvest survey techniques. *Wildlife Society Bulletin* 28:570–578.

Russell, F. L., D. B. Zippen, and N. L. Fowler. 2001. Effects of white-tailed deer (*Odocoileus virginianus*) on plants, plant populations and communities: A review. *American Midland Naturalist* 146:1–26.
Rutberg, A. T. 1997. Lessons from the urban deer battlefront: A plea for tolerance. *Wildlife Society Bulletin* 25:520–523.
Ruth, C. R. Jr. and D. A. Shipes. 2004. Potential negative effects of baiting on regional white-tailed deer harvest rates in South Carolina: A state with conflicting baiting laws. South Carolina Department of Natural Resources. Columbia, SC.
Rutledge, J., T. Bartoskewitz, and A. Cain. 2008. *Stem Count Index: A Habitat Appraisal Method for South Texas*. Austin, TX: Texas Parks and Wildlife.
Ryan, C. W., J. C. Pack, W. K. Igo, J. C. Rieffenberger, and A. B. Billings. 2004. Relationship of mast production to big-game harvests in West Virginia. *Wildlife Society Bulletin* 32:786–794.
Schmitz, O. J. 1990. Management implications of foraging theory: Evaluating deer supplemental feeding. *Journal of Wildlife Management* 54:522–532.
Severinghaus, C. W., H. F. Maguire, R. A. Cookingham, and J. E. Tanck. 1950. Variations by age class in the antler beam diameters of white-tailed deer related to range conditions. *Transactions of the North American Wildlife Conference* 15:551–568.
Shaw, J. C. 2005. Implications of quality deer management on population demographics, social pressures, dispersal ecology, and the genetic mating system of white-tailed deer at Chesapeake Farms, Maryland. PhD dissertation, North Carolina State University.
Shea, S. M., T. A. Breault, and M. L. Richardson. 1992. Herd density and physical condition of white-tailed deer in Florida flatwoods. *Journal of Wildlife Management* 56:262–267.
Shea, S. M. and J. S. Osborne. 1995. Poor quality habitats. In *Quality Whitetails: The Why and How of Quality Deer Management*, eds. K. V. Miller and R. L. Marchinton, 193–209. Harrisburg, PA: Stackpole Books.
Sinclair, A. R. E. 1997. Epilogue: Carrying capacity and the overabundance of deer. In *The Science of Overabundance: Deer Ecology and Population Management*, eds. W. J. McShea, H. B. Underwood, and J. H. Rappole, 380–394. Washington, DC: Smithsonian Institution Press.
Singer, F. J. and L. C. Zeigenfuss. 2002. Influence of trophy hunting and horn size on mating behavior and survivorship of mountain sheep. *Journal of Mammalogy* 83:682–698.
Skalski, J. R., J. J. Millspaugh, and K. E. Ryding. 2006. The impact of hunter postseason questionnaire design on big game harvest estimation. *Wildlife Society Bulletin* 34:329–337.
Skalski, J. R, K. E. Ryding, and J. J. Millspaugh. 2005. *Wildlife Demography: Analysis of Sex, Age, and Count Data*. San Diego, CA: Academic Press.
Skalski, J. R., R. L. Townsend, and B. A. Gilbert. 2007. Calibrating statistical population reconstruction models using catch-effort and index data. *Journal of Wildlife Management* 71:1309–1316.
Slade, J. 2009. Utility of trail camera users to assess deer population sex and age structure. MS thesis, University of Central Missouri.
Smith, J. R., R. A. Sweitzer, and W. F. Jensen. 2007. Diets, movements, and consequences of providing wildlife food plots for white-tailed deer in central North Dakota. *Journal of Wildlife Management* 71:2719–2726.
Sorin, A. B. 2004. Paternity assignment for white-tailed deer (*Odocoileus virginianus*): Mating across age classes and multiple paternity. *Journal of Mammalogy* 85:356–362.
Starfield, A. M. 1997. A pragmatic approach to modeling for wildlife management. *Journal of Wildlife Management* 61:261–270.
Steffen, D. E., D. M. Lewis, and P. J. Strong. 1983. The incidence and implications of road hunting during the dog and no-dog deer seasons in Mississippi. *Proceedings of the Annual Conference of the Southeastern Association of Fish and Wildlife Agencies* 37:513–518.
Stoll, R. J. Jr. and G. L. Mountz. 1983. Rural landowner attitudes toward deer and deer populations in Ohio. *Ohio Fish and Wildlife Report. 10*, Columbus, OH.
Storm, D. J., C. K. Nielsen, E. M. Schauber, and A. Woolf. 2007. Deer–human conflict and hunter access in an exurban landscape. *Human–Wildlife Conflicts* 1:53–59.
Storm, G. L., D. F. Cottam, R. H. Yahner, and J. D. Nichols. 1992. A comparison of 2 techniques for estimating deer density. *Wildlife Society Bulletin* 20:197–203.
Stout, R. J., D. J. Decker, B. A. Knuth, J. C. Proud, and D. H. Nelson. 1996. Comparison of three public-involvement approaches for stakeholder input into deer management decisions: A case study. *Wildlife Society Bulletin* 24:312–317.

Stradtmann, M. L., J. B. McAninch, E. P. Wiggers, and J. M. Parker. 1995. Police sharpshooting as a method to reduce urban deer populations. In *Urban Deer: A Manageable Resource? Proceedings of a Symposium at the 55th Midwest Fish and Wildlife Conference*, ed. J. B. McAninch, 117–122. St. Louis, MO: North Central Section of the Wildlife Society.

Strickland, B. K., S. Demarais, L. E. Castle et al. 2001. Effects of selective-harvest strategies on white-tailed deer antler size. *Wildlife Society Bulletin* 29:509–520.

Stromayer, K. A. K. and R. J. Warren. 1997. Are overabundant deer herds in the eastern United States creating alternate stable states in forest plant communities? *Wildlife Society Bulletin* 25:227–234.

Suchy, W. J., D. L. Garner, and W. R. Clark. 2002. Using bowhunting to successfully reduce deer numbers in an urban area in Iowa. In *Proceedings of the First National Bowhunting Conference*, ed. R. J. Warren, 40–44. Comfrey, MN: Archery Manufacturers and Merchants Organization.

Sudharsan, K., S. J. Riley, and S. R. Winterstein. 2006. Relationship of autumn hunting season to the frequency of deer–vehicle collisions in Michigan. *Journal of Wildlife Management* 70:1161–1164.

Swihart, R. K. and M. R. Conover. 1990. Reducing deer damage to yews and apple trees: Testing Big Game Repellent®, Ropel®, and soap as repellents. *Wildlife Society Bulletin* 18:156–162.

Taylor, C. E., D. L. Otis, H. S. Hill Jr., and C. R. Ruth. 2000. Design and evaluation of mail surveys to estimate deer harvest parameters. *Wildlife Society Bulletin* 28:717–723.

Thomas, J. W., J. D. Gill, J. C. Pack, W. M. Healy, and H. R. Sanderson. 1976. Influence of forestland characteristics on spatial distribution of hunters. *Journal of Wildlife Management* 40:500–506.

Tilghman, N. G. 1989. Impacts of white-tailed deer on forest regeneration in northwestern Pennsylvania. *Journal of Wildlife Management* 53:524–532.

Tonkovich, M. J. and M. E. Cartwright. 2002. Evaluation of the use of crossbows for deer hunting in Ohio and Arkansas. In *Proceedings of the First National Bowhunting Conference*, ed. R. J. Warren, 31–39. Comfrey, MN: Archery Manufacturers and Merchants Organization.

U.S. Fish and Wildlife Service. 2006. *National Survey of Fishing, Hunting and Wildlife-Associated Recreation.* U.S. Department of the Interior and U.S. Department of Commerce, Washington, DC.

Van Deelen, T. R., H. Campa III, J. B. Haufler, and P. D. Thompson. 1997. Mortality patterns of white-tailed deer in Michigan's Upper Peninsula. *Journal of Wildlife Management* 61:903–910.

Van Deelen, T. R., B. Dhuey, K. R. McCaffery, and R. E. Rolley. 2006. Relative effects of baiting and supplemental antlerless seasons on Wisconsin's 2003 deer harvest. *Wildlife Society Bulletin* 34:322–328.

Van Deelen, T. R. and D. R. Etter. 2003. Effort and the functional response of deer hunters. *Human Dimensions of Wildlife* 8:97–108.

Van Etten, R. C., D. F. Switzenberg, and L. Eberhardt. 1965. Controlled deer hunting in a square-mile enclosure. *Journal of Wildlife Management* 29:59–73.

VerCauteren, K. C. and S. E. Hygnstrom. 1998. Effects of agricultural activities and hunting on home ranges of female white-tailed deer. *Journal of Wildlife Management* 62:280–285.

VerCauteren, K. C., M. J. Lavelle, and S. Hygnstrom. 2006. Fences and deer-damage management: A review of designs and efficacy. *Wildlife Society Bulletin* 34:191–200.

Virginia Department of Game and Inland Fisheries. 2007. *Virginia Deer Management Plan 2006–2015.* Virginia Department of Game and Inland Fisheries Wildlife Information Publication Number 07–1. Richmond, VA.

Virginia Hound-Hunting Technical Committee. 2008. *Hunting with Hounds in Virginia: A Way Forward.* Richmond, VA: Virginia Department of Game and Inland Fisheries.

Wagner, K. K. and D. L. Nolte. 2001. Comparison of active ingredients and delivery systems in deer repellents. *Wildlife Society Bulletin* 29:322–330.

Wagner, K. K., R. H. Schmidt, and M. R. Conover. 1997. Compensation programs for wildlife damage in North America. *Wildlife Society Bulletin* 25:312–319.

Waller, D. M. and W. S. Alverson. 1997. The white-tailed deer: A keystone herbivore. *Wildlife Society Bulletin* 25:217–226.

Walters, C. J. and J. E. Gross. 1972. Development of big game management plans through simulation modeling. *Journal of Wildlife Management* 36:119–128.

Ward, K. J. 2005. Reducing deer overabundance by distinguishing high-productivity hunters: Revealed-preference, incentive-compatible licensing mechanisms. Paper presented at the American Agricultural Economics Association annual meeting, Series number 135580. Providence, RI.

Warren, R. J. 1995. Should wildlife biologists be involved in wildlife contraception research and management? *Wildlife Society Bulletin* 23:441–444.

Warren, R. J. 2000. Overview of fertility control in urban deer management. *Proceedings of the Annual Conference of the Society for Theriogenology and the American College of Theriogenologists.* November 28–December 2, 2000. San Antonio, TX.

Weckerly, F. W., M. L. Kennedy, and S. W. Stephenson. 2005. Hunter–effort–harvest–size relationships among hunt types of white-tailed deer. *Wildlife Society Bulletin* 33:1303–1311.

Weigand, J. P. and R. J. Mackie. 1987. What's working and what's not: An overview to approaches to management for quality hunting. *Proceedings of the Western Association of Fish and Wildlife Agencies* 67:69–76.

Wentworth, J. M., A. S. Johnson, P. E. Hale, and K. E. Kammermeyer. 1992. Relationships of acorn abundance and deer herd characteristics in the Southern Appalachians. *Southern Journal of Applied Forestry* 16:5–8.

West, B. C. and J. A. Parkhurst. 2002. Interactions between deer damage, deer density, and stakeholder attitudes in Virginia. *Wildlife Society Bulletin* 30:139–147.

Williams, B. K., J. D. Nichols, and M. J. Conroy. 2001. *Analysis and Management of Animal Populations.* San Diego, CA: Academic Press.

Williams, J. D., W. F. Krueger, and D. H. Harmel. 1994. Heritabilities for antler characteristics and body weight in yearling white-tailed deer. *Heredity* 73:78–83.

Williamson, S. J. 1998. Origins, history, and current use of ballot initiatives in wildlife management. *Human Dimensions of Wildlife* 3:51–59.

Williamson, S. J. 2000. *Feeding Wildlife—Just Say No!* Washington, DC: Wildlife Management Institute.

Williamson, S. J. 2003. *White-tailed Deer Harvest Management and Goal Setting in the Northeast.* Washington, DC: Wildlife Management Institute.

Wilson, A. J., D. W. Coltman, J. M. Pemberton, A. D. J. Overall, K. A. Byrne, and L. E. B. Kruuk. 2005. Maternal genetic effects set the potential for evolution in a free-living vertebrate population. *Journal of Evolutionary Biology* 18:405–414.

Winchcombe, R. J. and R. S. Ostfeld. 2001a. Bowhunter observations versus spotlighting as an index to deer abundance. *Northeast Wildlife* 56:39–48.

Winchcombe, R. J. and R. S. Ostfeld. 2001b. Indexing deer numbers with spotlighting: A long-term study of a managed deer population. *Northeast Wildlife* 56:31–38.

Winchell, C. S. and S. G. Buck. 1993. Illegal harvest of spike bucks during a regulated mule deer hunt. *California Fish and Game* 78:153–159.

Wisconsin Department of Natural Resources. 2001. *Management Workbook for White-tailed Deer.* Madison, WI: Bureaus of Wildlife Management and Integrated Science Services.

Witmer, G. W. and D. S. deCalesta. 1991. The need and difficulty of bringing the Pennsylvania deer herd under control. *Proceedings of the Eastern Wildlife Damage Control Conference* 5:130–137.

Wood, G. W., J. R. Davis, and G. R. Askew. 1985. An evaluation of 7 years of spotlight-count data on a coastal South Carolina plantation. In *Game Harvest Management*, eds. S. L. Beasom and S. F. Roberson. Kingsville, TX: Caesar Kleberg Wildlife Research Institute.

Woolf, A. and J. D. Harder. 1979. Population dynamics of a captive white-tailed deer herd with emphasis on reproduction and mortality. *Wildlife Monographs* 67:1–53.

Woolf, A. and J. L. Roseberry. 1998. Deer management: Our profession's symbol of success or failure? *Wildlife Society Bulletin* 26:515–521.

Xie, J. H., R. Hill, S. R. Winterstein et al. 1999. White-tailed deer management options model (DeerMOM): Design, quantification, and application. *Ecological Modeling* 124:121–130.

Zhang, D., A. Hussain, and J. B. Armstrong. 2004. Willingness to pay for hunting leases in Alabama. *Southern Journal of Applied Forestry* 28:21–27.

15

Management on Private Property

Harry A. Jacobson, Charles A. DeYoung, Randy W. DeYoung, Timothy E. Fulbright, and David G. Hewitt

CONTENTS

Introduction

Most big game species in North America are found in western portions of the continent that were sparsely settled by European immigrants, less desirable for agriculture, and therefore largely in public ownership. The model for managing big game species in the western United States is for habitat management to occur at low intensity on government property, most landowners to tolerate presence of the game species on private property, and state agencies to manage populations using public hunting and data from game surveys. Although this paradigm applies to white-tailed deer management in some areas (Chapter 14), a different approach has developed in many portions of the white-tailed deer's range because much of the area is in private ownership, whitetails are relatively sedentary, and property sizes are sufficiently small

FIGURE 15.1 Many deer management programs on private land in North America are designed to produce trophy bucks. To be consistently successful, managers must apply population and habitat management techniques in a manner consistent with their local conditions and personal situation. (Photo by J. Feild. With permission.)

that landowners can afford the scale of management necessary to influence deer on their property. These circumstances have given rise to a large number of people from all walks of life that manage white-tailed deer and the species' habitat. Aldo Leopold recognized wildlife and habitat management as the culmination in the development of a conservation ethic because effective husbandry requires a sense of responsibility, an understanding of the ecological system, and a desire to positively influence that system (Leopold, 1966). This chapter surveys practices used by landowners and deer enthusiasts on private property to meet their white-tailed deer management goals. These practices, as appropriate and legal, are applied on a continuum from captive deer in small pens to properties of thousands of hectares. Such site-specific management practices contrast with management at the level of the state agency which, by necessity must be extensive in nature (Chapter 14).

The first step in any wildlife management program is to define the management objective(s). Objectives of deer management on private property vary from reducing human–deer conflicts, through a desire to have deer present for nonconsumptive reasons, to various goals related to harvest of bucks. Management practices discussed in this chapter are those used to meet hunting goals, with a focus on management for harvest of large-antlered deer (Figure 15.1). This focus was chosen because most private-land deer management is motivated by hunting programs and many techniques used to meet trophy harvest objectives can be applied to meet other deer management goals. The chapter is not exhaustive and will compliment other excellent texts on management of white-tailed deer on private property (e.g., Miller and Marchinton, 1995; Fulbright and Ortega-S, 2006). Furthermore, this chapter will explore rapidly developing approaches to manage the genetic potential for a deer herd to produce large-antlered bucks.

Population Dynamics

An important component of deer management is understanding the dynamics of a deer population. In simplest form, population dynamics can be expressed as

$$\text{population change} = \text{births} - \text{deaths} + \text{immigration} - \text{emigration}.$$

The manager's task is to appreciate these four components of population change and alter them if necessary to meet management goals. To be successful, managers must be able to control the population, establish management goals, evaluate the population relative to goals, and then apply appropriate management techniques to achieve the goals.

Control of Population

Intensive management of white-tailed deer populations typically occurs on private lands and concentrates on increasing buck age and antler size. A fundamental prerequisite for intensive population management is control over the population on the property of interest (Berger et al., 2004). This means that the property has to be large enough so that at least a portion of the population is resident. Alternatively, smaller properties can be managed if neighbors either do not hunt or have similar management goals. If no deer are resident on a property and neighbors heavily harvest bucks, there is little a manager can do as far as population management. Habitat management, considered later in this chapter, may provide an attractive and secure refuge, regardless of property size.

The minimum size property that can be managed varies greatly over the range of white-tailed deer. Home range size, movements, and dispersal are major factors affecting the outcome of any deer management program, but are magnified under intensive management. Because male deer generally have home ranges twice as large as females (Demarais et al., 2000) (Chapter 6), male home range size influences the scale of most management operations. Mean buck home range size varied from 139 ha in south Texas (Inglis et al., 1979) and 150 ha in New York (Tierson et al., 1985) to 1182 ha in Quebec (Lesage et al., 2000). Deer inhabiting areas prone to flooding, seasonal changes in cropland cover, or drought have the largest home ranges (Sparrowe and Springer, 1970; Herriman, 1983; Labisky et al., 1991; Morgan, 1995; Demarais et al., 2000). Bucks also commonly take exploratory trips outside their home range (Inglis et al., 1979), which may take them outside of management control.

Seasonal migration and dispersal are additional factors complicating management efforts. Seasonal migrations are most prevalent in northern deer herds with deep snow depths where deer seek out traditional yarding areas during winter (Hoskinson and Mech, 1976; Nelson and Mech, 1986). Seasonal movements to and from yarding areas average 9–39 km (Chapter 6). To the extent migration is necessary for deer survival and reproduction, and because migratory movements are large relative to most property sizes, managers in areas where deer migrate will have less control over the system and be limited in the management goals they can realistically achieve.

Both male and female deer may disperse from their natal range, with fawns and yearlings showing the highest dispersal rates (Chapters 5, 6, and 10). Dispersal of yearling bucks may necessitate management units of 10,000–25,000 ha to ensure a large proportion of yearling bucks are subjected to consistent harvest criteria (Morgan, 1995; McCoy et al., 2005). Because such management units are large relative to most parcels of private property and deer do not respect property boundaries, multiple landowners must agree on management objectives either informally or formally through a management cooperative. Antler point restrictions enforced on a county or state-wide basis seek consistent harvest over large areas (Chapters 4 and 11), albeit with less precision than possible on a single property.

Properties less than home range size can be managed if bordered by larger unhunted areas or protected by barriers such as large rivers or busy highways. Still another way for a small property to gain management control over a deer population is to create or join a wildlife management association or co-op. Wildlife co-ops allow small landowners to join together and decide upon common management goals. Texas, like many states, has an active program for creating and maintaining wildlife co-ops, with over 150 active associations (Berger et al., 2004).

In the absence of large management units or cooperative management, and where legal, landowners with small properties may gain control over the spatial extent of management by constructing a game-proof fence (Brothers and Ray, 1975). Game-proof fences, also termed "high fences," consist of woven-wire fencing about 2.4 m in height. Fencing as a management tool is not available or socially acceptable in all areas, usually due to concerns about de facto ownership of a public resource or violation of fair chase. Furthermore, fencing is expensive to construct and maintain, especially in rocky terrain or flood-prone areas. Nevertheless, a game-proof fence properly constructed and maintained is highly effective at preventing most ingress and egress (VerCauteren et al., 2006; Webb et al., 2009). The full scope of the social and biological implications of game fencing is beyond this chapter; Demarais et al. (2002) provide a comprehensive review.

Establishing Objectives

After determining that management control is sufficient for the property to be intensively managed, the next step is to establish goals. The balanced scorecard (Kaplan and Norton, 1996; Norreklit, 2000) is one way to establish and monitor goals for intensive deer management. Developed for the business world, the balanced scorecard allows consideration of financial and nonfinancial measures and targets, and provides for monitoring progress. The balanced scorecard has been adapted to natural resource management on private lands by Dunn et al. (2006). "Balance" in the name comes from considering competing parts of a management program, including things such as landowner satisfaction and life-long learning (Table 15.1). Using the balanced scorecard involves (1) establishing a vision, (2) identifying basic components of the management effort, (3) identifying measures for tracking progress, (4) creating a plan to meet goals, and (5) tracking progress toward goals and vision (Dunn et al., 2006) (Table 15.1).

Evaluating Status of the Population

With control over the deer population and goals established, the next step in intensive management is to evaluate the status of the deer population in relation to goals. This means determining if population size, sex ratio, and buck age structure are consistent with management goals. Clues to population status come from many sources, including a survey, data collected from harvested deer, and experiences of other managers.

There are many ways to estimate deer populations (Chapter 5), but a common method used on smaller private properties is to use trail cameras. Population size can be estimated using the method described by Jacobson et al. (1997). Trail cameras are deployed for 14 days using corn as bait (some states may restrict baiting). Number of bucks is determined from photos using unique antler characteristics.

TABLE 15.1

Example of Completed Balanced Scorecard for Intensive Deer Management on a Private Property

Strategic Objectives	Goal	Actual (Current Year)
Vision and strategy: Intensive deer management on the property will be a gratifying experience for the owner and family members. Trophy bucks will be harvested annually, providing an enjoyable recreational experience for family and friends.		
Control of deer population		
1. Buy additional land	100 acres	No
2. Form co-op with neighbors	2000 acres	No, some progress
Financial resources		
1. Pay off land loan	10 years	On track
2. Save from outside business to support management practices	$20,000/year	Exceeded by $2500
Status of deer population		
1. Hold population steady with doe harvest	Conduct camera census; Harvest necessary does	Yes
2. Increase level of nutrition with food plots	Plant 3% of land to summer food plots; 2% to winter plots	Yes
3. Harvest only mature trophy bucks and culls	Harvest 3 bucks/year >150 B&C; harvest 5 bucks/year <125 B&C	No
Owner's quality of life		
1. Obtain enjoyment from deer management	Spend 10 weekends and 1 week of vacation on property	No
2. Practice lifelong learning through attending deer conferences and seminars	Attend Southeast Deer Study Group	Yes

Source: Data from Dunn, B. H. et al. 2006. *Using the Balanced Scorecard for Ranch Planning and Management: Setting Strategy and Measuring Performance.* South Dakota State University, Brookings.

The number of doe photos divided by number of buck photos and multiplied by the number of bucks provides an estimate of number of does. Similarly, number of fawn photos divided by number of doe photos multiplied by the estimated number of does provides an estimate of number of fawns. Summing estimated bucks, does, and fawns provides an estimate of population size. Moore (2008) determined that the Jacobson et al. (1997) method underestimated population size by about 30%. Regardless of whether used as a population estimate or index, this camera method is a useful tool for determining population status (McKinley et al., 2006).

Trail cameras are also useful for estimating buck age structure. Antler size alone is not a good method for separating most buck age classes (DeYoung, 1990). A combination of antler size and physical characteristics such as stomach girth improves estimates of age on the hoof (Hellickson et al., 2008).

Data collected from harvested deer provide additional clues to deer population status, especially whether the population is at a level of nutrition consistent with meeting goals and objectives. Age, weight, and antler measurements are commonly recorded by intensive managers. Insight into the reproductive status of a population comes from observation of corpora lutea, lactation, and fetal counts (Jacobson et al., 1979; Harder and Kirkpatrick, 1996) (Figure 3.6).

Finally, anecdotal experiences of other intensive deer managers in a region are very helpful in determining the status of a population. Most regions have a current or historical example of a quality deer population consistent with goals set using the balanced scorecard. Qualitative comparison of metrics from a local population to such a "model" population provides important clues to status. Managers must recognize that each of the measures described in this section are potentially biased, but if used in combination with one another, the composite data provide the necessary information to guide management decisions and gauge outcomes.

Adjusting a Population to Goal

Reducing the Population

Intensive deer managers in productive habitats may have to reduce the size of a population and continue to apply harvest pressure to maintain an optimum size consistent with their goal. The quickest and most effective way to reduce a population is to harvest does. Doe harvest not only reduces forage intake of the population represented by the animals removed, but also removes the recruits they would have produced. Harvesting bucks on the other hand has little effect, within limits, on a population's impact on the habitat.

Growing the Population

Less productive habitats commonly give intensive managers a different challenge. Fawn survival and recruitment can be low in these habitats for a variety of reasons. Managers may wish to increase such populations because producing trophy bucks is partly a numbers game. If only 10% of the mature buck population reaches a desired quality, the larger the population, the larger the number of bucks that will be represented by 10%. Increasing such populations commonly involves managing for better nutrition through habitat manipulation or supplemental feed such as food plots or pelleted feed (where legal). Habitat manipulation, however, may not increase deer foods in areas where plant growth is limited by soil infertility, salinity, or low rainfall; in fact, soil disturbance by mechanical means may reduce production of deer foods in such habitats. Food plots may also fail to produce expected results in less productive habitats.

Sex Ratio

The desired buck:doe ratio of an intensively managed population may depend largely on fawn survival (Figure 15.2). Low average fawn survival makes a wider sex ratio such as 1:3 more desirable because more females are needed to produce fawns to maintain a constant population size. However, if the habitat

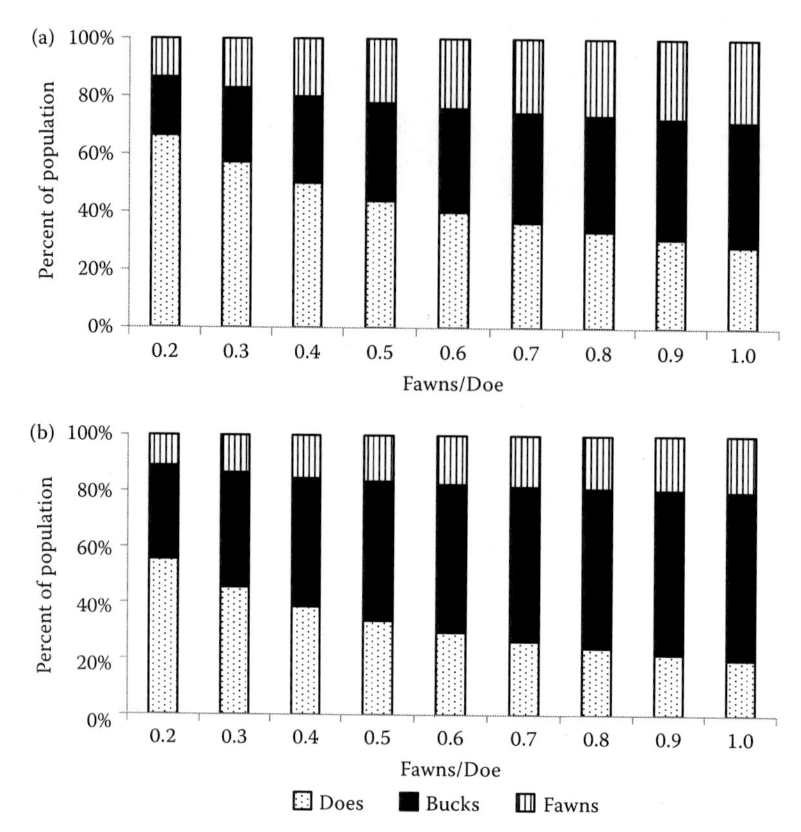

FIGURE 15.2 Change in composition of a hypothetical white-tailed deer population held constant in size, with no natural mortality, and harvest of bucks equal to or greater than (a) 3 years of age or (b) 6 years of age, as annual fawn recruitment varies.

is productive and fawn survival is high, some managers aim for more bucks than does because fewer does in the population can produce adequate fawns to maintain a population size goal.

Age Structure

Because intensive management commonly aims for older-aged bucks with larger antlers, buck age structure is important. An age structure with mature bucks requires minimal harvest of young bucks to allow them to grow older. Some managers have a goal of harvesting 3-year-old or older bucks, while intensive operations may not harvest good-quality bucks until 6 years of age or even older.

The term "stacking" was developed by Stedman (2009) to describe the management practice of carrying trophy bucks in a population to an older age. By not harvesting the best bucks until they are estimated to be seven to eight years old, a large number of mature bucks are stockpiled in the population. Then, harvest of several trophy bucks per year can begin and be sustained though time. In addition to harvest, stacking provides more opportunities to view and photograph trophy bucks. One disadvantage of stacking is that some bucks may be lost to natural mortality before they reach 7 to 8 years old.

Culling

If trophy bucks are desired, only a small percentage will have large antlers even when mature. Thus many intensive managers cull less-desirable bucks. Culling is done for two reasons. First, managers may remove bucks with small antlers to provide more resources for the remaining animals, a management action discussed in more detail in Chapter 4. The second reason to cull is that removing less-desirable individuals will let the better-quality animals do most of the breeding, potentially improving genetic traits for large antlers in the population. Culling to improve genetic characteristics of a deer population is discussed further in the section "Managing Genetic Potential for Antler Growth" of this chapter and in Chapter 4.

Fawn Production

Fawns, particularly neonates, are susceptible to predation and there are situations in which predators may influence fawn production to an extent that managers might consider actions to reduce predator impacts (Chapter 8). Reducing the density of predators is an option, but successful predator control must be implemented annually, often at considerable expense. Furthermore, effects of intensive predator control programs on production of white-tailed deer fawns are varied (Ballard et al., 2001). Another option would be habitat management to reduce fawn susceptibility to predation by increasing hiding cover. Studies demonstrating the effect of habitat management on susceptibility of fawns to predation are lacking.

Production of fawns is nutritionally demanding (Chapter 3) and therefore manipulating nutritional resources is a viable approach to increase fawn production (Zaiglin and DeYoung, 1989; Bishop et al., 2008). A higher nutritional plane can increase pregnancy rates and litter sizes, particularly in young deer (Verme, 1969; Ozoga and Verme, 1982), and improve fawn survival after birth (Murphy and Coates, 1966; Langenau and Lerg, 1976). Thus, deer managers wishing to improve fawn production should carefully evaluate the nutritional plane of the herd and take action to improve nutrition should it be found lacking.

Adult Survival

For managers, survival data on the adult segment of the deer population is more difficult to obtain than are estimates of juvenile survival and recruitment. Most studies on radio-collared deer generally have shown lower mortality for females than males (Dusek et al., 1989; Nixon et al., 1991, 2001), primarily due to greater harvest intensity of males than females. Survival rates of female deer are important to the extent that they influence population growth and also to the extent that nonharvest mortality influences opportunity for recreational or subsistence harvest of females. However, management of trophy deer requires minimizing mortality of young bucks so they reach greater than 4 years of age (DeYoung, 1990). Maximum antler development for captive deer is not reached until 5 to 8 years old (Jacobson, 1995, 1998) and observations on known age deer in wild populations support this (Koerth and Kroll, 2008). However, in areas where harvest of bucks less than 5 years of age is permitted, only 12–20% of yearlings may survive to 5 years of age (Ditchkoff et al., 2001; Bowman et al., 2007). In contrast, where males are protected from harvest, 52% of yearling bucks may survive to 6 years of age (Webb et al., 2007). Thus, a means of identifying and protecting younger males until they reach mature age classes is necessary to produce bucks with large antlers (DeYoung, 1990). Infrared-triggered camera surveys provide one means of estimating buck age (Kroll, 1996; Richards and Brothers, 2003; Demarais et al., 2004). Researchers have recently demonstrated that age can be correctly assigned to live white-tailed deer using morphometric ratios. Flinn et al. (2010) correctly assigned age to 84% of wild deer in 1-, 2-, 3-, and 4-year-and-older age classes from game camera photographs. Antler measurements can also be used to protect younger age class bucks from harvest. By using voluntary antler beam length restrictions of 38 cm, bucks harvested in 2.5 and older age classes increased from 13–23% to 73–75% in four years without any appreciable change in total buck harvest on an 8500 ha island in the Mississippi River (Jacobson, 1992).

Another means of changing age structure of adult bucks is to establish harvest quotas which are population based to restrict total buck harvest to a given percentage of the total buck population. If, for example, there is a known population of 100 bucks on a management unit, then assigning a harvest quota of 20 adult bucks should theoretically allow 80 bucks to move into the next age class. Whether or not this is successful will depend on other factors such as natural mortality rates and immigration or emigration in the management unit.

Nutrition and Habitat Management

Nutrition is an important regulator of reproduction, recruitment, survival, body growth, and antler size and therefore of great interest to deer managers (Chapter 3). Nutritional condition is often related to soil fertility, with the most fertile soils producing heavier-bodied and larger-antlered deer than low fertility soils (Short, 1969; Jacobson et al., 1977; Strickland and Demarais, 2000). Body weights and antler measurements were strongly correlated with soil P, Ca, Mg, K, pH, and organic matter for deer collected from 23 management areas in Mississippi and soil P was a useful predictor of physiological condition (Jacobson, 1985). This relationship had more to do with soil fertility and plant nutrient content than mineral requirements of deer because protein levels and production of important browse species was highest in those areas with the highest soil nutrients (Jones et al., 2008, in press).

When harvest data and habitat evaluation suggest nutrition may be limiting a population, managers may improve nutrition by reducing herbivore density, improving habitat, planting food plots, or providing supplemental feed. Each approach has positive and negative aspects, making it reasonable to consider the alternatives in the order listed. Furthermore, managers should be aware of the time necessary for results of a management action to fully materialize. Some improvement from better nutrition can occur in one to two years, but more dramatic improvements may occur over generations. Geist (1986) reviewed pre-World War II research on red deer and roe deer by Franz Vogt that showed body size and antler quality can be greatly enhanced through generational carryover effects of nutrition. Nutrition seems to operate through some maternal or perhaps epigenetic effect and can improve antler size progressively up to four generations. Similar effects have recently been demonstrated for white-tailed deer (Monteith et al., 2009; Figures 6.7 and 6.8). These findings show that projects to enhance nutrition should be long term and that full benefit may not be apparent for a decade or more.

Reducing Herbivore Density

White-tailed deer diet quality, body condition, reproduction, and antler qualities may be related to density of other herbivores, both livestock (Bryant et al., 1981; Ortega et al., 1997; Chapter 9) and the deer themselves (Kie et al., 1980; Keyser et al., 2005a,b). For example, long-term data from nine management areas across the southeastern United States show significant relationships between density and physical parameters of deer, with yearling male dressed body weight and antler characteristics showing the most consistent relationship (Keyser et al., 2005b). Jacobson (1992) demonstrated body weight, antler size, fawn survival, and conception dates all improved with increasing doe harvest in Mississippi. Both Jacobson (1992) and Keyser et al. (2005b) showed a two-year lag time from population reduction to increased condition, thus managers should not expect immediate effects of reduced deer density. Furthermore, there may be little relationship between deer density and nutritional condition in nutrient-poor environments with large amounts of poor-quality food (Shea and Osborn, 1995; DeYoung et al., 2008).

Habitat Management

Intensive white-tailed deer population management is often accompanied by intensive habitat management. Intensive habitat management involves manipulating forage nutritional quality or abundance, water availability, or vegetation cover to benefit white-tailed deer. The objective of manipulation may be to improve habitat quality so that white-tailed deer reproduction, survival, or body growth is increased (Van Horne, 1983). Habitat management may also attempt to increase the amount of usable space for deer by identifying portions of the landscape avoided by deer and converting these areas to suitable habitat (Hiller et al., 2009). Agricultural planting (see supplemental feeding below) may be practiced to improve abundance and nutritional quality of forage and cover.

Habitat management techniques commonly include forest and range management practices, such as forest thinning, clear cutting, controlled burning, fertilization, roller-chopping, mowing, aerating, chaining, cabling, railing, and herbicide treatments (Yoakum et al., 1980; Kammermeyer and Thackston, 1995; Fulbright and Ortega-S., 2006). The intensive deer habitat manager must have a thorough knowledge of native vegetation and cover conditions for the vegetation communities on their property. Descriptions of major habitat types are provided by Harlow and Guynn (1994) and in Halls (1984). Habitat management is truly an art that requires ability to alter vegetation succession to favor increased deer production. This requires ability to create interspersion and juxtaposition of plant communities which are most favorable to the white-tailed deer. Proper interspersion requires presence of all habitat elements within a deer's home range, including forage, hiding and thermal cover, and water. Balancing habitat manipulation is desirable so that one habitat element is not increased at the expense of causing another element to become limiting. Clearing forest or shrubland to increase herbaceous plants to improve deer nutritional status, for example, may be detrimental if deer respond by avoiding the area because inadequate woody cover remains following treatment.

Habitat manipulation is often directed at increasing herbaceous plants, particularly forbs, but woody plants are also a valuable habitat component. Woody plants provide browse, mast, and hiding cover, and provide cover to reduce heat and cold stress (Mysterud and Østbye, 1999). Adequate thermal cover may be particularly important during winter in northern climates and during summer in southern climates. It is extremely important that the deer manager has an in-depth knowledge and understanding of the ecology and management of native deer forage plants for their area. Many publications list native species of forage used by deer in different geographic regions (e.g., Warren and Hurst, 1981; Jacobson, 1994; Fulbright and Ortega-S., 2006).

Vegetation dynamics and successional processes may vary depending on site productivity. A habitat management practice that increases forbs for white-tailed deer in mesic environments in the southeastern United States may not be as effective in semiarid regions because of limited rainfall (Fulbright, 1999; Fulbright et al., 2008). Soil fertility also affects plant responses to habitat manipulation. For example, successional processes may vary with site productivity. On some sites, succession may consist of a temporal sequence leading from lower to higher successional communities following disturbance. In less productive sites, disturbance may result in a new, relatively stable plant community that does not exhibit successional trends (Fulbright and Ortega-S., 2006).

Habitat manipulation may not improve habitat quality if the existing landscape already meets nutritional, cover, and water needs of deer. Improving habitat quality, however, is not always the objective of habitat manipulation in intensive deer management. For example, woody plant clearing is sometimes done primarily to improve visibility of deer to hunters. Because habitat management is an art, there is potential for management decisions to be influenced by human values and perceptions. Park or savannah-like landscapes consisting of woody plants scattered in grassland receive the highest scores when people are tested for landscape preferences (Ulrich, 1986). Humans prefer landscapes that look "cared for" (Nassauer, 1995). A thinned forest or shrub-dominated landscape cleared in a pattern leaving patches or strips of brush surrounded by grassland is aesthetically appealing to many people. Just because a treatment looks good or facilitates hunting, however, does not mean it benefits white-tailed deer.

Supplemental Feeding

Supplemental agriculture production is a common practice to improve nutrition of white-tailed deer. Agricultural plantings may increase production of forage up to 100-fold over the same area of native vegetation (Kammermeyer and Thackston, 1995; Hehman and Fulbright, 1997), which not only increases carrying capacity, but can enhance diet quality of deer in nutrient-limited habitats. Agricultural food plot techniques have been described in detail for dry land environments by Fulbright and Ortega-S. (2006) and for more mesic regions by Kammermeyer et al. (2006) and Harper (2008).

In environments where conditions are unsuitable for agriculture and where it is legal to do so, many landowners provide supplemental feed in the form of whole grain, by-products of other agricultural production such as whole cottonseed and rice brand, or pelleted feed (Perkins, 1991; Kammermeyer and Thackston, 1995; Bartoskewitz et al., 2003). Supplemental feeding has given rise to a substantial economic industry based on the manufacture and sale of feed and the equipment for its distribution.

There is little question that supplemental feeding can improve nutritional condition in some habitats. Supplemental feeding may increase body weight, antler production, and fawn production (Ozoga and Verme, 1982; Zaiglin and DeYoung, 1989; Bartoskewitz et al., 2003). However, supplemental feed sites may also increase incidence of disease (Miller et al., 2003) and habitat degradation (Murden and Risenhoover, 1993; Brown and Cooper, 2006; Cooper et al., 2006; but see Timmons et al., 2010 for description of more nuanced effects). Awareness of supplemental feed effects on deer ecology, life history, and natural selection is increasing (Mysterud, 2010). The pros and cons of supplemental feeding have been the topic of numerous publications (Williamson, 2000; Brown and Cooper, 2006), symposia (Ramsey, 1996), and position statements by The Wildlife Society and state wildlife agencies (e.g., The Wildlife Society, 2007; Virginia Department of Game and Inland Fisheries, 2008; Maine Department of Inland Fisheries and Wildlife, 2010). All facets of a providing supplemental feed should be evaluated if such a program is being considered.

Feeding whole grains such as corn or other high-energy rations is not without hazard, particularly for deer in pens with few foraging choices. Under some conditions overeating high-energy rations can lead to ruminal acidosis or polioencephalomalacia (National Research Council, 2007). Diets with high calcium levels can lead to milk fever and mortality of does that are recently postpartum. Polioencephalomalacia of several deer and milk fever of two postpartum does were both diagnosed in the late 1970s in the captive deer research facility of Mississippi State University when an enriched diet with high calcium, greater than 20% crude protein, and low fiber levels was fed for a short period of time to captive deer (H. A. Jacobson, unpublished data). Following those incidents and veterinary recommendations, the diet of Mississippi State captive deer was changed to 16% protein and greater than 17% crude fiber.

Ullrey et al. (1971) documented good deer health, growth, and reproduction when fed a pelleted diet with 17.6% crude protein, 2% crude fat, 40% neutral detergent fiber, and a vitamin–mineral premix. During over 30 years of experience feeding supplemental rations to thousands of both captive and free-ranging wild deer, Jacobson (unpublished observations) has seen very good body condition and productivity and no nutritional problems reported from rations with 61–84% total digestible nutrients, 16–18% crude protein, 2–3.5% crude fat, 15–22% crude fiber, 0.8–1.2% Ca, 0.4–0.8% P, 0.4–0.68% salt, and a standard vitamin pre-mix.

Disease Management

An often overlooked aspect of deer management is disease management. Disease may severely impact deer populations, with mortality rates as high as 90% in isolated cases (Hayes, 1981). In some instances, there is nothing a manager can do but document the disease-induced mortality. However, some common sense management concepts can reduce the risk of disease. First, reducing deer density may reduce the incidence of disease. Eve (1981) showed positive relationships of deer density with abomasal parasites burdens and some infectious diseases such as anthrax. Other diseases and parasites may be largely independent of host densities, but may be related to habitat type. Numbers and species of external parasites have been linked directly to habitat with mature hardwoods having high prevalence of mosquitoes but low numbers of ticks and tabanids, whereas immature pine habitats have high numbers of tabanids and ticks but low numbers of mosquitoes (Jacobson, 1986). Other plant communities such as mature pines, mature pine hardwoods, and immature hardwoods all had different levels of external parasite species and numbers. By interspersing plant communities, the manager provides opportunity for evasion of external parasites in different seasons of the year.

Ticks can be particularly detrimental to white-tailed deer. Mortality of white-tailed deer fawns, in some cases up to 30%, has been attributed to tick infections (Emerson, 1969; Bolte et al., 1970). Ticks and other parasites can be controlled by habitat manipulation. Jacobson and Hurst (1979) demonstrated controlled burning effectively limited deer tick infestation of wild turkey poults. Periodic liming and disking of food plots to promote sunlight exposure and desiccation is another disease control measure. White-tailed deer are secondary hosts for cattle fever ticks and the incidence of infection is greater in areas with shrubs and exotic grasses such as buffelgrass (Cantu-C et al., 2009).

Another potential disease problem faced by intensive deer managers is disease transmission at sites of population concentration such as supplemental feeding stations or bait sites. Concern has been expressed about potential parasitic disease and infectious disease exposure at such sites (Miller et al., 2003; Palmer et al., 2004; Palmer and Whipple, 2006). Managers who choose to provide supplemental feed or bait should insure feeding stations are periodically moved and related equipment regularly sanitized to prevent disease transmission.

Managing Genetic Potential for Antler Growth

Intensive management is prompted in part by economic opportunity offered to landowners by hunting. Accordingly, managers have invested much time and resources in managing population density, habitat and forage quality, age structure, and sex ratio. Recently, the management community has placed increasing emphasis on genetic potential, to the extent that genetic explanations may receive the credit or blame for performance in terms of antler size in managed populations (see Rollins, 1998). Genetic attributes of deer populations are a persistent feature of popular media, including hunting magazines and television programs. One phrase constantly repeated is that large-antlered bucks are the product of "age, nutrition, and genetics." Nutrition and age are clearly required for bucks to attain their maximum antler and body size, as mature bucks in good condition have the greatest phenotypic development. Genetic potential is important in a conceptual sense, but harder to gauge in the field, unlike age and nutritional condition. Nevertheless, management strategies targeted at genetic characteristics of deer populations are widespread and seem to be accepted by many hunters and managers without question. Managers employ several strategies to influence the genetic potential for antler growth. Among these approaches are selective harvest, transplanting deer from another herd (Texas only), temporarily holding deer in a pen to control breeding (Texas only), and releasing captive-bred deer.

Selective Harvest

Culling or selective harvest is a viable management tool for use in choosing which individuals to invest resources and management. For instance, managers in productive habitats may elect to harvest bucks with poor antler development to aid in managing deer density or simply to invest resources in bucks with better potential for antler development. Culling to influence a cohort of bucks differs in scope and intent from culling to affect genetic characteristics of a population. The premise of selective harvest is to protect bucks with desirable traits and remove bucks with undesirable traits, resulting in increased population antler size in successive generations. Antler traits are heritable and should respond to selection. Studies of captive herds demonstrate that selection in a controlled environment can increase cohort antler size over six to eight generations, even under limiting nutrition (Lockwood et al., 2007). Evidence for improved physical attributes as a result of selective harvest in free-ranging deer is largely anecdotal; controlled studies are lacking. Therefore, the main questions are whether selection can be achieved in wild deer, under what conditions selection is feasible, what gain can be expected, and what is the cost to achieve the gain in antler size?

The effects of management on genetic characteristics of deer populations depend on trait heritability and the direction, intensity, and duration of selection. Estimation of antler trait heritability is difficult for wild deer because accurate estimates require hundreds to thousands of pedigreed individuals. Published estimates of heritability of antler characteristics in white-tailed deer are derived from two relatively small populations of captive deer and differ markedly in the estimate of trait heritability due to different approaches to the estimation of heritability (Williams et al., 1994; Lukefahr and Jacobson, 1998). Heritability estimates reported by Lukefahr and Jacobson (1998) are similar to antler trait heritability in wild and farmed red deer (Kruuk et al., 2002). If antler traits in white-tailed deer exhibit moderate heritability, selection could affect mean trait values in a population (Lukefahr and Jacobson, 1998), albeit with less efficiency than if heritability values are uniformly high (cf. Williams et al., 1994).

A key assumption of using selective harvest to increase antler size in a deer population is that bucks with desirable antler characters will sire most or all fawns produced. However, studies of genetic parentage in deer clearly indicate that breeding is distributed among many different bucks. Although mature bucks (3.5 years and older) sire about 70% of offspring in populations with reasonable sex ratio and age structure, young bucks (1.5- and 2.5-years-old) collectively sire the remaining 30% of fawns (Sorin, 2004; DeYoung et al., 2009). Breeding is widespread and not monopolized by large-antlered bucks. Without the ability to predict which wild bucks will breed or how many offspring they will sire, managers must exercise a great deal of control over a population to ensure that only desirable bucks breed. Therefore, one must reduce competition for mates by removing most or all bucks with undesirable antler traits, preferably before the rut, to increase the efficiency of selection.

The scale of management greatly affects harvest intensity and thus the efficiency of selective harvest. Stedman (1998) termed the scale-intensity relationship for selection the "corral continuum." On one extreme, imagine a deer pen or captive facility where a manager has complete control of which bucks and does breed and which of their offspring are allowed to breed. In a pen, a manager can derive some information about genetic potential for antler development through pedigree records, enabling selection on does as well as bucks. At the other extreme is a large, free-ranging population in which pedigrees are unavailable and the manager has little ability to select dams for progeny antler potential or to select bucks and does for breeding. In a practical sense, intense selection on bucks may be possible in small, fenced properties. However, selective harvest becomes increasingly difficult and labor-intensive as the spatial scale of management increases (Figure 15.3). Unfortunately, many misconceptions exist as to what culling or selective harvest can achieve. In the short term, culling simply removes individuals with undesirable phenotypic development; genetic characteristics of the population are unchanged. Removal of individuals erases their breeding contribution, with the potential for selection to affect population genetic characteristics over time.

If selection is feasible, the next critical question is how much time is required to achieve genetic change. Researchers have not addressed this question directly, but a critical evaluation of the biological

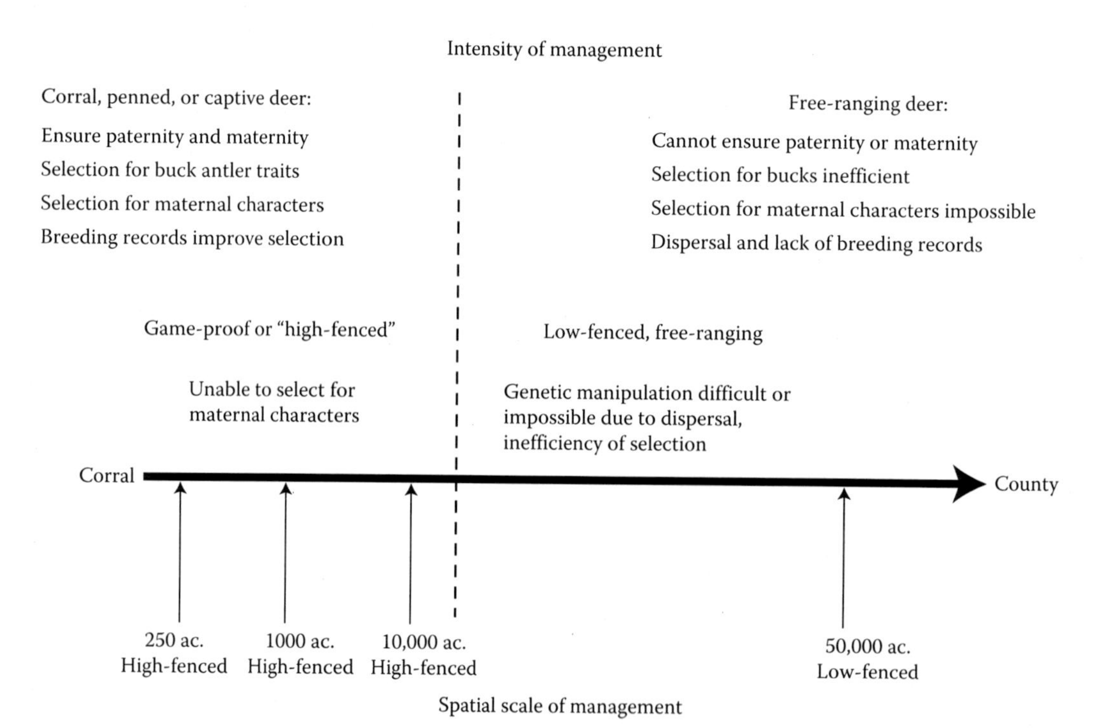

FIGURE 15.3 The corral to county concept, illustrating the difficulty of achieving intensive management and genetic manipulation along a continuum of spatial scales from the corral to county-scale management. As spatial scale of management increases, the achievable management intensity and control decreases. Control of breeding, breeding records, and control of dispersal are key components of genetic selection. Without these tools, selection to manipulate genetic characteristics of a deer population becomes difficult or impossible.

TABLE 15.2

Hypothetical Timeline for a Culling or Selective Harvest Program Aimed at Altering the Genetic Attributes of a Population

Timeline	Management	Result
Year 1	Begin culling	
Year 2	Fawns sired as a result of first year of culling are born	
Year 6	Fawns sired as a result of first year of culling are physically mature	Results of first year of culling may be seen
Year 11	Fawns sired as a result of fifth year of culling are physically mature	Results of fifth year of culling may be seen

Assume that all undesirable bucks are taken before the rut each year. Culling or selective harvest aimed at changing genetic characteristics is a long-term, intensive undertaking.

factors suggests that selective harvest or culling requires a multi-decade effort to achieve genetic change (Table 15.2). A long time is necessary because the effects of the first year of selection will not be apparent until the progeny reach physical maturity, about six years after the initiation of selection. During this time, one must maintain selection and refrain from harvesting desirable bucks until their full breeding potential is exhausted or diminished.

Overall, changing the genetic characteristics of a wild deer population through culling involves a long-term commitment to management. First, a manager must ensure control of the deer population, perhaps by building and maintaining a deer-proof fence. Second, bucks and does must have good nutrition to express their genetic potential. Improving diet quality may require decreasing livestock densities or other forms of production, habitat manipulation, establishment of food plots, a feeding program, or some combination. Third, fawn recruitment each year must be high and consistent or managers risk overharvesting the population, which may require many years to recover unless fawn production improves. Fourth, most or all undesirable bucks should be removed each year before the rut, which assumes that hunters can identify culls "on the hoof" and not make errors by culling bucks with good potential but poor current-year growth. Finally, desirable bucks should not be harvested until they have had many opportunities to breed, at least seven years of age. If a manager cannot maintain all of the above practices consistently for a 10–20-year period (many years, but only a few deer generations), there will be little or no chance to affect genetic characteristics of the population. Put in perspective, culling for genetic change is a long-term investment.

Managers should not attempt to affect genetic characteristics of a population without careful consideration of biological aspects of an intensive management plan. Often, seemingly simple management decisions involve a series of logical premises. Consider the question of whether to initiate selective harvest by culling bucks with spike-antlers. Before a decision, managers should answer several key biological questions (Table 15.3); if the answer is "no" to any of the questions, then management will have little effect and may waste time and resources that could be better applied elsewhere.

Just as selective harvest has been proposed as a means to increase antler size, there is concern that selective harvest of the largest, fastest-growing individuals may cause a decline in antler size or other traits of interest (Harris et al., 2002). Researchers focusing on phenotypic characters found that several decades of selective harvest may have decreased mean horn and body size in a population of bighorn sheep (Coltman et al., 2003). Bighorn sheep are a highly polygynous species where male mating success is associated with dominance, age, and horn size (Hogg and Forbes, 1997; Coltman et al., 2002). Rams with heritable phenotypic characters that normally confer increased mating success (e.g., horn size, early maturation) were taken by hunters prior to peak breeding age (about eight years), leaving less-fit individuals to breed (Coltman et al., 2003). A similar situation seems to have occurred in Asian elephants, where ivory poaching has increased the frequency of tuskless males in some areas (Kurt et al., 1995). Finally, several studies have documented a change in allozyme frequencies in hunted populations of white-tailed deer and red deer (Scribner et al., 1985; Hartl et al., 1991, 1995, 2003), though the ramifications for fitness or physical traits are unknown.

TABLE 15.3

Guidelines for Culling or Selective Harvest Plan of Yearling Bucks with Below-Average Antler Development, such as Spike Antlers

Biological Limiting Factor	Answer	Management Outcome
Do environmental conditions allow bucks to express potential for antler development? Is deer density appropriate, adequate nutrition available, and are fawning dates early and synchronous?	⇨ No	Culling yearlings is not a viable management practice
⇩ Yes to all		
Is recruitment (survival of fawns) high?	⇨ No	Culling yearlings is not a viable management practice
⇩ Yes		
Do less than 40% of yearling bucks have poor-quality or spike antlers?	⇨ No	Culling yearlings is not a viable management practice
⇩ Yes		
Can management affect sufficient spatial area to ensure consistent application of harvest plans?	⇨ No	Culling yearlings is not a viable management practice
⇩ Yes		
Can hunters judge age of bucks "on the hoof" and consistently apply harvest criteria?	⇨ No	Culling yearlings is not a viable management practice
⇩ Yes		
Are hunters dissatisfied with antler development of mature bucks?	⇨ No	Culling yearlings is not a viable management practice
⇩ Yes		
Culling or selective harvest of yearling bucks is a viable management practice		

Source: Adapted from Strickland, B. K. and S. Demarais. 2007. *Using Antler Restrictions to Manage for Older-Aged Bucks.* Mississippi State University Extension Service Publication 2427.

Note: Culling can be a useful management tool, but is ineffective if the many biological limiting factors affecting antler development are not addressed.

The myriad of factors that limit the effectiveness of selective harvest in improving antler size of white-tailed deer also likely protect the species from negative genetic effects of selective harvest of trophy deer. A review of sport harvest effects on wild ungulates suggests that demographic, behavioral, and life history factors govern population response to selective harvest. In polygynous species, variance in male reproductive success is high and the physical attributes which confer increased mating success, such as antler size and body mass, are selected by hunters. Thus, the degree of polygyny within populations may influence the sensitivity of genetic response to male-biased harvest. For instance, sport harvest significantly affected population structure of European roe deer, which have a highly polygynous territorial mating system (Kurt et al., 1993). Harvested populations of Spanish red deer had less microsatellite diversity than an unharvested population (Martinez et al., 2002). In contrast, the large proportion of white-tailed deer males that participate in breeding help protect the species from negative genetic effects of trophy harvest. Furthermore, selection properly applied might be used to alter or improve population averages for heritable physical traits such as antlers, while avoiding unintentional negative trait evolution (Mysterud and Bischof, 2010).

Translocation of Wild Deer

Translocation of white-tailed deer was a widely used technique during the twentieth century to reestablish and bolster deer populations extirpated or nearly extirpated from habitat changes and over-harvest (McDonald and Miller, 1993). Currently, translocation of wild deer in the United States occurs on a large scale only in Texas. Translocation is also legal in Mexico. In Texas, a permit is required along with a site inspection by state biologists to insure deer are not being released into an area that does not have suitable habitat conditions or is overpopulated with deer (Texas Parks and Wildlife Department, 2010a). The landowner receiving deer must pay the costs of capture, but no transaction can involve sale of the animals themselves. The objectives of deer translocation in Texas are landowners wanting to reduce deer populations on properties from which deer are captured and, for the release sites, landowners wanting to increase deer density on their property, or more commonly to obtain deer from new genetic sources which they feel will improve the quality of their deer herds. In Mexico, deer may be captured on ranches with high deer populations and translocated to areas with fewer deer. Landowners who obtain capture permits are also allowed to sell the animals in these transactions. Additionally, because of low deer numbers in some areas it has been illegal to harvest antlerless deer in much of Mexico and landowners with high deer populations have to resort to capture of deer to prevent overpopulation (Martinez et al., 1997).

Historical records of deer restocking have value in evaluating the potential impacts of translocations. For instance, researchers have shown with genetic studies that not all restockings were successful (Ellsworth et al., 1994; Leberg and Ellsworth, 1999). In some areas, populations clearly benefited from releases, while in other areas multiple releases were necessary to establish viable populations. In still other cases, remnant groups of native deer appear to have recovered. Circumstantial evidence indicates that deer from northern populations (e.g., Wisconsin), which are valued today by some managers for their large body and antler size, performed poorly in southern environments. Researchers conducting a genetic study of deer restoration in Mississippi found no similarity among populations that received northern deer, while southern and recovered native stocks clearly had a major impact on multiple deer populations (DeYoung et al., 2003). Based on evidence from captive and wild deer, it appears northern deer are highly susceptible to diseases that have less effect on southern stocks, such as epizootic hemorrhagic disease, or bluetongue, which may explain the apparent lack of success of northern deer in the southeastern United States.

Pens to Control Breeding

In Texas, landowners may obtain a permit to capture up to 20 does from their property before the breeding season and place them with a single buck in a pen which must be at least 2.4 ha in size (Texas Parks and Wildlife Department, 2010b). Does placed in these pens under the authority of this Deer Management Permit (DMP) must be released back to their capture area, either as bred does, or along with any fawns produced, prior to the next breeding season. Landowners are also permitted to capture a single buck or use a buck from a permitted captive facility in these pens. This allows landowners interested in enabling individual bucks with desirable antler qualities to sire many fawns and thereby increase the probability of influencing the genetic composition of the population.

Release of Captive Stock

Captive breeding of white-tailed deer is controversial. Geist (1992) outlined potential negative consequences of the deer farming and hunting preserve industries. Chief among these are threats to North American Conservation model, the risk of disease introduction, and genetic pollution of native deer. On the same platform as Geist, Benson (1992) argued that, contrary to being a major threat to wildlife conservation in North America, commercial value for game-enabled land currently being used for wildlife recreation to compete with other land uses, thus maintaining the land's role in wildlife conservation. Despite the diverse attitudes toward captive breeding of white-tailed deer, the practice has grown dramatically during the past two decades.

In 2002 the number of known white-tailed deer farms in North America exceeded 4000 and housed an estimated 164,000 deer (Jacobson, 2002). At that time 15 states, six Canadian providences, and Mexico allowed permitted facilities for captive production of deer. Today, there are likely greater numbers of captive deer-breeding facilities. For example, during 2002 Texas had 430 deer-breeding facilities with 17,289 animals (Jacobson, 2002); in 2010, there were 1271 deer-breeder permits reported in Texas, with 81,351 white-tailed deer in captivity (Mitch Lockwood, Texas Parks and Wildlife, personal communication). Researchers conducting a survey in 2007 reported 7828 cervid farms in the United States, many of which housed white-tailed deer (Frosch et al., 2008). Although many captive breeding operations probably have a hobby component because the owners obtain personal satisfaction from raising and observing deer, the captive deer industry generates economic activity particularly important in rural areas. In Texas, the captive cervid industry supported 7335 jobs and generated $652 million in economic activity (Frosch et al., 2008).

Although the highest financial return for today's deer farmer is from selling to other breeders, the driving force behind the captive white-tailed deer industry is to produce trophy deer that eventually will be harvested by sport hunting. While some deer raised in captivity are released for harvest on hunting preserves, there is a substantial contingent of private landowners who wish to release captive deer to improve the genetic potential for antler production of deer on their property. During 2009–2010, 15,384 captive deer were released at 1034 sites in Texas (Mitch Lockwood, Texas Parks and Wildlife, personal communication). No information is known on releases in other states.

For release of captive deer to influence the genetic composition of a deer population, the released deer must survive, breed, and recruit offspring. Furthermore, their potential for antler production must be better than the native deer in that particular environment. Unfortunately, there is little information on the ability of captive deer releases to change genetic structure of existing deer populations, nor on other consequences of such releases. There are data, however, that can be interpreted to indicate care should be exercised when releasing deer not native to the area.

Deer from different geographic sources have different body, reproductive, and disease-resistance characteristics. Researchers during a 1988–1995 study investigated differences between Michigan and Mississippi deer from the same sources being raised in research facilities in both respective states. The study was designed for researchers to evaluate both phenotypic and physiologic characteristics of the parents and crossbred offspring in both locations (Jacobson and Lukefahr, 1998). Traits such as body weight of crossbred offspring were much lower in the northern environment because of the impact of environmental extremes impacting fawns born in that environment when compared to the same animal crosses born in the southern environment. Reproductive timing, molting, antler casting, and velvet shedding were both genetically and environmentally linked (Jacobson and Lukefahr, 1998). All of these occurred much earlier for Michigan deer than Mississippi deer in both northern and southern latitudes; and crossbred offspring were intermediate to that of the parents. An outbreak of hemorrhagic disease in 1995 in the Mississippi State University facilities effectively terminated the study. At that time, 65% of the purebred northern deer, 43% of the first-generation (50% northern and 50% southern) crosses, and 21% of the 25% northern and 75% southern crosses died of the disease. In contrast, only 12% of the pure southern deer died during the epizootic.

Little research has been conducted on the effects of captive deer releases. Some data are available from observations made by Jacobson (unpublished) of marked captive deer released to the wild on a number of large (>800 ha) ranches enclosed with high fences in Texas (Figures 15.4 and 15.5). Captive deer released

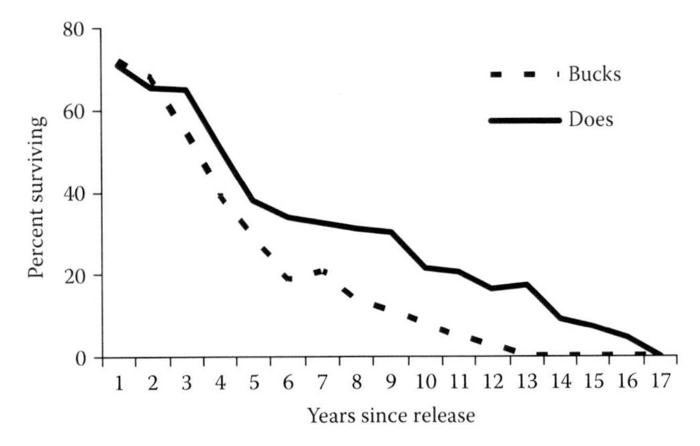

FIGURE 15.4 Percent survival of captive-bred white-tailed deer, primarily fawns, released onto private property in Texas. (Data from H. A. Jacobson, unpublished data.)

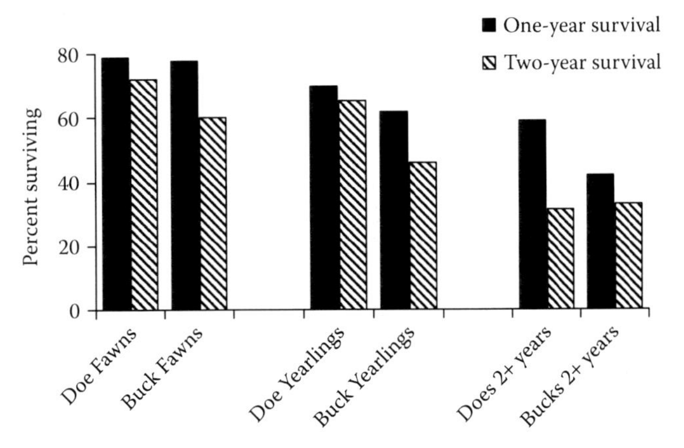

FIGURE 15.5 Percent survival for one or two years of captive-raised white-tailed deer released as fawns, yearlings, or at least two years old on to private property in Texas. (Data from H. A. Jacobson, unpublished data.)

on these ranches were mostly fawns at the time of release, almost all from southern genetic sources. These animals had not been bottle-reared and had little human contact other than the periodic placement of feed and water within their rearing pen. All animals were marked with ear tags or freeze brands that allowed individual identification following their release. The deer were released on intensively managed ranches being managed by certified wildlife biologists. Release properties had supplemental agriculture food plots for deer, deer densities appropriate to the habitat, and protected the released animals from harvest. Survival was monitored with the aid of infrared-triggered cameras used during annual census at each of the ranches.

Under the conditions of release, first year survival of both bucks and does was slightly over 70%. However, despite not being subjected to harvest mortality, by the sixth year after release, only 19% of the bucks and 34% of the does were still alive. Does had higher survival than bucks and at least 30% of the does released survived in the wild greater than nine years, with at least one surviving 16 years. These data demonstrate captive-reared deer can adapt and survive under conditions of intensive deer management on large Texas ranches. In particular, released does seem more likely than bucks to contribute genetic material to native deer populations. Under different conditions and with deer released from less locally adopted genetic sources, survival could be expected to be lower than those reported on here. There are no published data on the effectiveness of such introductions in changing phenotypic and genetic traits of animals in release sites over time. Research on the effects of the practice is necessary to provide guidance to managers considering introductions as a management tool.

Introduction of deer from populations not adapted to the release environment should be carried out with great caution based on the above information. Although body size and antlers may be larger in the animals being introduced, these traits may not convey genetic fitness in the environment to which they are being introduced. Although it is apparent to most people that a small-bodied deer will be at an extreme disadvantage in a cold northern environment, few people may consider that the opposite is also true and a large body is very inefficient in a hot, humid southern environment. Additionally, genetic traits such as disease resistance or reproductive timing may carry through to offspring, with detrimental consequences. We know little about other physiological traits, such as adaptations that may allow an animal adapted to a specific environment to consume local browse species that might otherwise be toxic. The end result of introducing animals that have not undergone selective pressures in the environment to which they are introduced will probably be self-limiting, as selection pressures favor endemic animals.

Final Thoughts on Genetic Management

Intensive management practices that may circumvent issues of culling intensity have recently become available in some states. For instance, the DMP program, described above, allows managers more control over which bucks are breeding, which might accelerate the selection process. The primary disadvantage of DMP is the cost of building and stocking a series of high-fenced enclosures, the risk of deer dying as a result of capture or in the pens, and unknown effectiveness.

Releasing captive deer with the genetic potential to produce large antlers is another technique to influence a population's genetic potential for antler development. An interpretation of the sparse data available is that captive deer do not perform well in the wild, especially if released at two years of age or older (Figures 15.4 and 15.5). Furthermore, many different bucks sire fawns in wild populations; thus, the breeding potential of a single or small number of captive bucks in a wild population is questionable. Captive does have little experience with predators and may recruit fewer fawns than wild deer. The ancestry of captives is also important; deer with northern ancestry do not perform well in southern climates (Jacobson and Lukefahr, 1998). The best potential for success with captive deer may be in using a captive buck to breed wild does in a DMP program. Nevertheless, many unknowns about the effectiveness of releasing captive deer into the wild remain and the practice is legal in only a few areas. Thus, buying or raising captive deer for release into the wild is expensive, inefficient, and largely unproven.

It is difficult to evaluate the effect of selective harvest or introduction in a conventional sense because many other management actions are often implemented simultaneously (e.g., habitat improvements, balancing sex ratio and age structure, planting food plots or providing supplemental feed). The concurrent use of many management actions makes it difficult to determine the relative contribution of each. The social or cognitive aspects of management should not be discounted; many managers have strong beliefs in the efficacy of specific techniques and look for confirmation rather than critically evaluate the strategy. For instance, the amount of feed consumed, area planted in food plots or subjected to brush control, or sex ratio and age structure can be easily measured. Yet, the one factor that cannot be measured, "genetics," may receive most of the credit for any improvements in antler size or most of the blame for lack of response. Managers often claim genetic change if large-antlered bucks appear within a few years of implementing selective harvest. Such claims do not consider the logical progression of selective harvest management; any mature buck harvested within four years of implementing a selective harvest program must have been present before selection began. Similarly, south Texas has long been renowned for producing large deer, with the so-called "trophy genetics." In reality, south Texas contains large ranches, where harvest can be strictly controlled. The primary reason large-antlered bucks are not harvested in many other regions of the United States is that hunters harvest too many young bucks; deer management is less complex than many wish to believe. Once harvest is controlled, many areas of the United States and elsewhere have the potential to produce large-antlered bucks under proper management.

Culling and intensive management techniques aimed at genetic change are relatively new and little information is available beyond anecdotal reports. Culling, DMP, trapping and transplanting wild deer, or releasing captive deer might affect genetic potential for antler development in some situations. Selective breeding has produced tremendous change in controlled, captive environments. However, the required intensity and duration of management to achieve genetic change in wild deer are unknown. Furthermore, the potential gain in antler size and cost–benefit ratio for a specified gain in antler growth are unknown. The biggest disadvantage to most of the intensive management techniques discussed here is cost and unknown effectiveness.

Overall, "bad genetics" is something most managers do not need to be concerned about. The recent focus on genetics and heritability has unfortunately obscured the fact that a large proportion of phenotypic variation is due to nongenetic causes, such as the environment (Figure 15.6), which may be affected by management. Several recent studies further illustrate how genetic and environmental influences interact to produce phenotypic traits in cervids, especially cohort and maternal effects. Individuals from a cohort born in a good year may experience faster growth, reach adult size quicker, and amass greater nutrient reserves (Pettorelli et al., 2002). The end result is greater survival, better physical development, and higher reproduction (Rose et al., 1998). Maternal effects are specific to the mother, usually her level of nutrition during pregnancy and lactation. Maternal effects are often considered a form of environmental variation, but are highly individualized, and as a consequence less is known about maternal effects in wild populations (Räsänen and Kruuk, 2007). The influence of maternal effects on physical development in white-tailed deer was illustrated in a recent study where deer from differing stock sources were reared in a common environment (Monteith et al., 2009). If interactions between environment and heritable genetic traits are common, then deer condition and physical development are products of current conditions as well as past conditions for one or more generations.

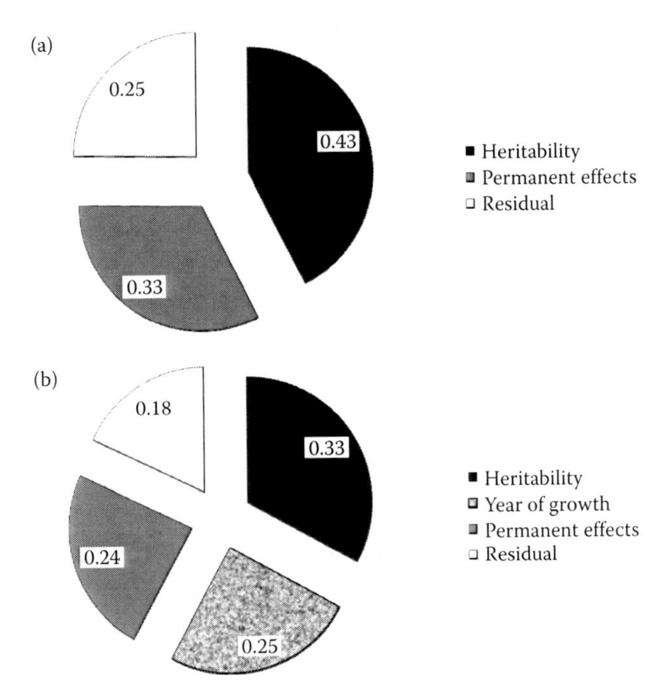

FIGURE 15.6 Variance component and heritability estimates for antler mass in prime-aged cervid males: (a) captive white-tailed deer (Based on data from Lukefahr, S. D. and H. A. Jacobson. 1998. *Journal of Wildlife Management* 62:262–268.) and (b) free-ranging red deer (Based on data from Kruuk, L. E. B. et al. Antler size in red deer: Heritability and selection but no evolution. *Evolution*. 2002. 56:1683–1695. Copyright Wiley-VCH Verlag GmbH & Co. KGaA. With permission.). Heritability estimates were derived using an animal model approach, where heritability was estimated as additive genetic variance divided by total phenotypic variance. Permanent effects include nonadditive genetic and environmental effects. Environmental conditions in year of antler growth accounted for 25% of variation in free-ranging red deer. Residual variation is the remaining variation among individuals not attributed to other sources.

Putting the Pieces Together: A Case Study

Intensive deer management requires integration of many different wildlife management tools, is a multi-year process, and involves many interrelated activities to be successful. Many management practices are available, but not all are appropriate, necessary, or effective for a given property. There are examples of intensive deer management reaching successful conclusion while providing substantial outdoor recreation and conservation benefit to both game and nongame wildlife resources.

One such example of intensive deer management is a 1518 ha privately owned sportsman's club, Concordia Rod and Gun Club, located in the Mississippi Delta in northern Mississippi. This hunting club has been intensively managed for white-tailed deer beginning in 1995. Concordia has a unique blend of plant communities, with most of the club's land consisting of impressive forests with a high percentage of pecans and other important mast-bearing species of timber, including water oak, willow oak, swamp chestnut oak, honey locust, and persimmon. It is bordered by agriculture to the north and forest land to the south. About 3% of the timberland has been placed in wildlife food plots which are interspersed throughout the property. This combination, along with the highly fertile alluvial soils, makes Concordia a haven for wildlife and provides excellent hunting opportunities for waterfowl, turkey, mourning dove, gray and fox squirrels, swamp rabbits, and of course deer. It also has a variety of other game and nongame wildlife including black bear, otter, coyotes, bobcats, and gray fox. It is a bird-watchers paradise and its oxbow lakes are teeming with fish. The club also has a long and storied history of recreational hunting for deer, bear, waterfowl, and squirrels. There are 16 club memberships, but each member can take several guests on their membership and hunting is provided for 50–75 individual hunters who visit Concordia with the members.

Harvest of deer on Concordia since 1995 has averaged 73 antlerless deer and 20.5 antlered bucks annually. Doe harvest has ranged from a low of 14 does harvested in 2003 to a high of 118 taken in 2000. Antlered buck harvest has ranged from a low of nine in 2000 to a high of 33 in 2008. Fawn bucks mistakenly harvested as does constituted 8% of the antlerless harvest in the years 1995–2001, but after hunter education efforts on aging deer in the field, fawn bucks became less than 3% of the antlerless harvest during 2002–2009. Despite harvesting does at a rate of 2.6 does to every buck over a 15 year period, hunter observations during 2009 showed a buck:doe ratio of 1:1.5. These data illustrate the difficulty of reaching optimum herd structure for sustained harvest of mature bucks without having control of dispersal and immigration, or being able to account for unknown buck mortality.

All deer harvest on Concordia was regulated annually based on camera census and observed recruitment success indicated by lactation rates the previous year and hunter observation of fawn:doe ratios. Because of low fawn recruitment in 2002, 2003, and 2008, doe harvest was reduced substantially in the 2003, 2006, and 2008 seasons. Low fawn recruitment years were related to high water and flood conditions from the Mississippi River and also to documented hemorrhagic disease outbreaks. Harvest goals for each year since 1995 were set on the basis of recommendations from a certified wildlife biologist.

In 1998 the club decided on a management goal of increasing the size of harvested bucks beyond the 130 gross B&C score, which at the time was the antler size of the largest bucks harvested. Before 1998, the club enforced harvest restrictions requiring bucks have a minimum of eight points and a 40-cm (16-in.) spread, and that before a member could harvest a second buck, they and their guests had to harvest five does. Beginning in 1998, the club began to collect more intensive data on all deer harvested. They were also given several recommendations to change their annual planting of food plots from that of winter small grain and summer soybeans, to a more diverse planting of legumes, clovers, and small grain. Additionally, they were requested to initiate a timber management program that included periodic selective timber harvest which would provide improved timber growth, income to the club, and better conditions for deer and other wildlife. Following these recommendations, the club initiated a forest management plan and began a food planting program that included a high diversity of legumes which made highly palatable high protein and energy forage available throughout the year. They also initiated a camera census in 1998. In 1998, harvest data showed the club was taking antlered bucks which were primarily three and four years old and contained a disproportionate number of bucks with 10 or more antler points. At that time club members were asked to refrain from harvesting bucks with 10 or more main frame points and members followed the restriction from 1998 through 2000. Beginning in 2001, club members agreed that they could harvest one ten point main frame deer per membership, but once a membership harvested such a buck, they would be required to wait two years before harvesting another one. During the 2001 hunting season, success was realized with harvest of one 14-point buck that had a gross score of 183 Boone and Crocket points and a 12-point buck with a gross score of 178 Boone and Crockett points. Net typical scores for these bucks were below 170. During the next three years, the club harvested bucks scoring in the high 140s, but nothing of the caliber of the 2001 harvest.

Beginning in 2002 the club initiated a culling program by agreeing to shoot bucks which members could identify as being older than two and having less than eight points. Club members were also asked to be more aggressive on harvesting mature bucks with less than 10 points. In 2005, the club again harvested bucks scoring above 160 gross Boone and Crockett and has continued to harvest such bucks in each of the following hunting seasons. In 2006, the club began a more intensive culling program with bonus bucks that were older than four years of age and had gross Boone and Crockett scores less than 135, or older than five years of age that scored below 145, given to each membership. In that same year, the 1620 ha club adjoining Concordia with over 100 memberships retained the consulting biologist working with Concordia and began a trophy deer management program on that property.

In 2007, Concordia reached its ultimate management goal with harvest of a 12-point buck by Dr. Paul Warrington scoring 180 gross Boone and Crockett and a net score of 176 6/8 entitling it to entry as a record book buck, which was the first official entry of a buck with a net score above 170 in this region of the Mississippi Delta since 1952 (Figure 15.7). Another buck was harvested in the same year that scored 169, but which had a broken main beam tip which brought its score below 170. In 2008, the club again recorded a net record book 10-point buck taken by Jeremy McMahan having a gross Boone and Crockett score of 180 6/8 and a net score of 177 2/8. In 2009 the club failed to harvest a buck above 170, but two were taken above 160.

The above example demonstrates the role of intensive management preserving a valuable wildlife resource. Without the economic incentive to manage intensively, much of the forested wildlife habitat in the Mississippi Delta would be lost to other land uses, primarily large-scale agriculture or high intensity, short term rotation cutting of hardwood forests. Intensive deer management in the case of Concordia and neighboring hunting clubs throughout the Mississippi Delta support high populations of both game and nongame species and even though they are private lands, provide hunting recreation to large numbers of hunters that otherwise would not have the opportunity for recreation. As a side note, at the same time as Concordia began its intensive deer management program black bear sightings became more common on the property. Concordia and neighboring hunting clubs are now seeing a return and recovery of black bear populations that were extirpated from the northern Mississippi Delta in the early 1900s.

FIGURE 15.7 Hunters in the Mississippi Delta show the results of several years of good habitat and harvest management which not only benefited white-tailed deer but many other wildlife species.

Acknowledgments

The authors appreciate reviews of an earlier version of this chapter by B. Strickland and M. Tonkovich. S. Stedman provided valuable insight on techniques and philosophies to manage white-tailed deer on private land.

REFERENCES

Ballard, W. B., D. Lutz, T. W. Keegan, L. H. Carpenter, and J. C. Devos, Jr. 2001. Deer–predator relationships: A review of recent North American studies with emphasis on mule and black-tailed deer. *Wildlife Society Bulletin* 29:99–115.

Bartoskewitz, M. L., D. G. Hewitt, J. S. Pitts, and F. C. Bryant. 2003. Supplemental feed use by free-ranging white-tailed deer in southern Texas. *Wildlife Society Bulletin* 31:1218–1227.

Benson, D. E. 1992. Commercialization of wildlife: A value-added incentive for conservation. In *The Biology of Deer*, ed. R.D. Brown, 539–553. New York, NY: Springer-Verlag.

Berger, M., L. McMurry, C. Richardson et al. 2004. A guide for wildlife management associations and co-ops. Texas Parks and Wildlife Department. Austin, TX.

Bishop, C. J., G. C. White, D. J. Freddy, B. E. Watkins, and T. R. Stephenson. 2008. Effect of enhanced nutrition on mule deer population rate of change. *Wildlife Monographs* 172:1–28.

Bolte, J. R., J. A. Hair, and J. Fletcher. 1970. White-tailed deer mortality following tissue destruction induced by one star ticks. *Journal of Wildlife Management* 34:546–552.

Bowman, J. L., H. A. Jacobson, D. S. Coggin, J. R. Heffelfinger, and B. D. Leopold. 2007. Survival and cause-specific mortality of adult male white-tailed deer managed under the Quality Deer Management paradigm. *Proceedings of the Annual Conference of Southeastern Association of Fish and Wildlife Agencies* 61:76–81.

Brothers, A. and M. E. Ray, Jr. 1975. *Producing Quality Whitetails.* Laredo, TX: Wildlife Services.

Brown, R. D. and S. M. Cooper. 2006. The nutritional, ecological, and ethical arguments against baiting and feeding white-tailed deer. *Wildlife Society Bulletin* 34:519–524.

Bryant, F. C., C. A. Taylor, and L. B. Merrill. 1981. White-tailed deer diets from pastures in excellent and poor range condition. *Journal of Range Management* 34:193–200.

Cantu-C., A., J. A. Ortega-S., Z. García-Vázquez, J. Mosqueda, S. E. Henke, and J. E. George. 2009. Epizootiology of *Babesia bigemia* in free-ranging white-tailed deer in northeastern Mexico. *Journal of Parasitology* 95:536–542.

Coltman, D. W., M. Festa-Bianchet, J. T. Jorgenson, and C. Strobek. 2002. Age-dependent sexual selection in bighorn rams. *Proceedings of the Royal Society of London B* 269:165–172.

Coltman, D. W., P. O'Donoghue, J. T. Jorgenson, J. T. Hogg, C. Strobek, and M. Festa-Bianchet. 2003. Undesirable evolutionary consequences of trophy hunting. *Nature* 426:655–658.

Cooper, S. M., M. K. Owens, R. M. Cooper, and T. F. Ginnett. 2006. Effect of supplemental feeding on spatial distribution and browse utilization by white-tailed deer in semi-arid rangeland. *Journal of Arid Environments* 66:716–726.

Demarais, S., R. W. DeYoung, L. J. Lyon, E. S. Williams, S. J. Williamson, and G. J. Wolfe. 2002. *Biological and Social Issues Related to Confinement of Wild Ungulates*. Wildlife Society Technical Review 02–3. Bethesda, MD.

Demarais, S., K. V. Miller, and H. A. Jacobson. 2000. White-tailed deer. In *Ecology and Management of Large Mammals in North America*, eds. S. Demarais and P. R. Krausman, 601–628. Upper Saddle River, NJ: Prentice-Hall.

Demarais, S., D. Stewart, and R. F. Griffin. 2004. *A Hunter's Guide to Aging and Judging Live White-tailed Deer in the Southeast.* FWRC Publication WF 113 and MSU-ES Publication 2206. Mississippi State University, Mississippi State, MS.

DeYoung, C. A. 1990. Inefficiency in trophy white-tailed deer harvest. *Wildlife Society Bulletin* 18:7–12.

DeYoung, C. A., D. L. Drawe, T. E. Fulbright et al. 2008. Density dependence in deer populations: Relevance for management in variable environments. In *Wildlife Science: Linking Ecological Theory and Management Applications*, eds. T. E. Fulbright and D. G. Hewitt, 203–222. Boca Raton, FL: CRC Press-Taylor & Francis Group.

DeYoung, R. W., S. Demarais, K. L. Gee, R. L. Honeycutt, M. W. Hellickson and R. A. Gonzales. 2009. Molecular evaluation of the white-tailed deer (*Odocoileus virginianus*) mating system. *Journal of Mammalogy* 90:946–953.

DeYoung, R. W., S. Demarais, R. L. Honeycutt, A. P. Rooney, R. A. Gonzales, and K. L. Gee. 2003. Genetic consequences of white-tailed deer (*Odocoileus virginianus*) restoration in Mississippi. *Molecular Ecology* 12:3237–3252.

Ditchkoff, S. S., E. R. Welch Jr., R. L. Lochmiller, R. E. Masters, and W. R. Starry. 2001. Age-specific causes of mortality among male white-tailed deer support mate-competition theory. *Journal Wildlife Management* 65:552–559.

Dunn, B. H., R. N. Gates, J. Davis, and A. Arzeno. 2006. *Using the Balanced Scorecard for Ranch Planning and Management: Setting Strategy and Measuring Performance*. South Dakota State University, Brookings.

Dusek, G. L., R. J. Mackie, J. D. Herriges, and B. B. Compton. 1989. Population ecology of white-tailed deer along the lower Yellowstone River. *Wildlife Monographs* 104:1–68.

Ellsworth, D. L., R. L. Honeycutt, N. J. Silvy, M. H Smith, J. W. Bickham, and W. D. Klimstra. 1994. White-tailed deer restoration to the southeastern United States: Evaluating genetic variation. *Journal of Wildlife Management* 58:686–697.

Emerson, H. R. 1969. A comparison of parasitic infections of white-tailed deer (*Odocoileus virginianus*) from central and east Texas. *Bulletin of the Wildlife Disease Association* 5:137–139.

Eve, J. H. 1981. Management implications of disease. In *Diseases and Parasites of White-tailed Deer*, eds. W. R. Davidson, F. A. Hayes, V. F. Nettles, and F. E. Kellogg, 413–423. Tallahassee, FL: Tall Timbers Research Station miscellaneous publication number 7.

Flinn, J. J., S. Demarias, H. A. Jacobson, and P. D. Jones. 2010. Aging deer using morphometric measurements. *Proceedings of the Southeast Deer Study Group Meeting* 33:39 (abstract).

Frosch, B. J., D. P. Anderson, and J. L. Outlaw. 2008. Economic impact of deer breeding operations in Texas. In *Southern Agricultural Economics Association Annual Meeting*. Dallas, TX.

Fulbright, T. E. 1999. Response of white-tailed deer foods to discing in a semiarid habitat. *Journal of Range Management* 52:346–350.

Fulbright, T. E. and J. A. Ortega-S. 2006. *White-tailed Deer Habitat: Ecology and Management on Rangelands* College Station, TX: Texas A&M University Press.

Fulbright, T. E., J. A. Ortega-S., A. Rasmussen, and E. J. Redeker. 2008. Applying ecological theory to habitat management: The altering effects of climate. In *Wildlife Science: Linking Ecological Theory and Management Application*, eds. Fulbright, T. E. and D. G. Hewitt, 241–258. Boca Raton, FL: CRC Press.

Geist, V. 1986. Super antlers and pre-World War II European research. *Wildlife Society Bulletin* 14:91–94.

Geist, V. 1992. Deer ranching for products and paid hunting: Threat to conservation and biodiversity by luxury markets. In *The Biology of Deer*, ed. R. D. Brown, 554–561. New York, NY: Springer-Verlag.

Halls, L. K., editor. 1984. *White-tailed Deer Ecology and Management*. Harrisburg, PA: Stackpole Books.

Harder, J. D. and R. L. Kirkpatrick, R. L. 1996. Physiological methods in wildlife research. In *Research and Management Techniques for Wildlife and Habitats*, ed. T. A. Bookhout, 275–306. Bethesda, MD: The Wildlife Society.

Harlow, R. F. and D. C. Guynn. 1994. Whitetail habitats and range. In *Deer*, eds. D. Gerlach, S. Atwater, and J. Schnell, 169–173. Harrisburg, PA: Stackpole Books.

Harper, C. A. 2008. *A Guide to Successful Wildlife Food Plots: Blending Science with Common Sense*. Knoxville, TN: University of Tennessee Institute of Agriculture.

Harris, R. B., W. A. Wall, and F. W. Allendorf. 2002. Genetic consequences of hunting: What do we know and what should we do? *Wildlife Society Bulletin* 30:634–643.

Hartl, G. B., F. Klein, R. Willing, M. Apollonio, and G. Lang. 1995. Allozymes and the genetics of antler development in red deer (*Cervus elaphus*). *Journal of Zoology (London)* 237:83–100.

Hartl, G. B., G. Lang, F. Klein, and R. Willing. 1991. Relationship between allozymes, heterozygosity and morphological characters in red deer (*Cervus elaphus*), and the influence of selective hunting on allele frequency distribution. *Heredity* 66:343–350.

Hartl, G. B., F. Zachos, and K. Nadlinger. 2003. Genetic diversity in European red deer (*Cervus elaphus* L.): Anthropogenic influences on natural populations. *Comptus Rendus Biologie* 326 (Suppl.):37–42.

Hayes, F. A. 1981. Preface. In *Diseases and Parasites of White-tailed Deer*, eds. W. R. Davidson, F. A. Hayes, V. F. Nettles, and F. E. Kellogg, vii–xii. Tallahassee, FL: Tall Timbers Research Station miscellaneous publication number 7.

Hehman, M. W. and T. E. Fulbright. 1997. Utilization of warm-season food plots by white-tailed deer. *Journal of Wildlife Management* 61:1108–1115.

Hellickson, M. K., K. V. Miller, C. A. DeYoung, R. L. Marchinton, S. W. Stedman, and R. E. Hall. 2008. Physical characteristics for age estimation of male white-tailed deer in southern Texas. *Proceedings of the Annual Conference of the Southeastern Association of Fish and Wildlife Agencies* 62:40–45.

Herriman, K. R. 1983. Hunting season movements of male white-tailed deer on Davis Island. MS thesis. Mississippi State University.

Hiller, T. L., H. Campa, III, and S. R. Winterstein. 2009. Estimation and implications of space use for white-tailed deer management in southern Michigan. *Journal of Wildlife Management* 73:201–209.

Hogg, J. T. and S. H. Forbes. 1997. Mating in bighorn sheep: Frequent male reproduction via a high-risk "unconventional" tactic. *Behavioral Ecology and Sociobiology* 41:33–48.

Hoskinson, R. L. and L. D. Mech. 1976. White-tailed deer migration and its role in wolf predation. *Journal of Wildlife Management* 40:429–441.

Inglis, J. M., R. E. Hood, B. A. Brown, and C. A. DeYoung. 1979. Home range of white-tailed deer in Texas coastal prairie brushland. *Journal of Mammalogy* 60:377–389.

Jacobson, H. A. 1985. *Investigation of Phosphorus in the Nutritional Ecology of White-tailed Deer*. Federal Aid in Fish and Wildlife Restoration Completion Report W-48–31, Study XXIII. Mississippi Department of Wildlife Conservation.

Jacobson, H. A. 1986. Wilderness and animal disease relationships. In *Wilderness and Natural Areas in the Eastern United States: A Management Challenge*, eds. D. L. Kulhavy and R. N. Conner, 94–97. Nacogdoches, TX: Southern Forest Experiment Station.

Jacobson, H. A. 1992. Deer condition response to changing harvest strategy, Davis Island, Mississippi. In *Biology of Deer*, ed. R. D. Brown, 48–55. New York, NY: Springer-Verlag.

Jacobson H. A. 1994. Feeding behavior. In *Deer*, eds. D. Gerlach, S. Atwater, and J. Schnell, 192–202. Harrisburg, PA: Stackpole Books.

Jacobson, H. A. 1995. Age and quality relationships. In *Quality Whitetails: The Why and How of Quality Deer Management*, eds. K. V. Miller and R. L. Marchinton, 103–111. Mechanicsburg, PA: Stackpole Books.

Jacobson, H. A. 2002. The history of white-tailed deer and its commerce. In *Proceedings of 2002 North American Deer Farmers annual conference and the World Deer Farming Congress III*, 43–54.

Jacobson, H. A. 1998. Culling as a management practice, "the dark side." In *The Role of Genetics in White-tailed Deer Management. Proceedings of a Symposium*, ed. D. Rollins, 72–77. College Station, TX: Texas A&M University.

Jacobson, H. A., D. Coggin, S. Bothum et al. 1997. Tag returns from deer captured in Mississippi. *Proceedings of the Southeast Deer Study Group Meeting* 20:29 (abstract).

Jacobson, H. A., D. C. Guynn, L. F. Castle, and E. J. Hackett. 1977. Relationships between soil characteristics and body weights, antler measurements, and reproduction of white-tailed deer in Mississippi. *Proceedings Northeast-Southeast Section Deer Study Group Meeting* 1:46–55.

Jacobson, H. A., D. C. Guynn, R. N. Griffin, and D. L. Lewis. 1979. Fecundity of white-tailed deer in Mississippi and periodicity of corpora lutea and lactation. *Proceedings Annual Conference Southeastern Association Fish and Wildlife Agencies* 33:30–35.

Jacobson, H. A. and G. A. Hurst. 1979. Prevalence of parasitism by *Amblyomma americanum* on wild turkey poults as influenced by prescribed burning. *Journal Wildlife Diseases* 15:43–47.

Jacobson, H. A., J. C. Kroll, R. W. Browning, B. H. Koerth, and M. H. Conway. 1997. Infrared-triggered cameras for censusing white-tailed deer. *Wildlife Society Bulletin* 25:547–556.

Jacobson, H. A. and S. D. Lukefahr, 1998. Genetics research on captive white-tailed deer at Mississippi State University. In *The Role of Genetics in White-tailed Deer Management. Proceedings of a Symposium*, ed. D. Rollins, 46–59. College Station, TX: Texas A&M University.

Jones, P. D., S. Demarais, B. K. Strickland, and S. L. Edwards. 2008. Soil region effects on white-tailed deer forage protein content. *Southeastern Naturalist* 7:595–606.

Jones, P. D., B. K. Strickland, S. Demarais, B. J. Rude, S. L. Edwards, and J. P. Muir. 2010. Soils and forage quality as predictors of white-tailed deer *Odocoileus virginianus* morphometrics. *Wildlife Biology* 16:430–439.

Kammermeyer, K., K. V. Miller, and L. Thomas Jr, editors. 2006. *Quality Food Plots: Your Guide to Better Deer and Better Hunting*. Bogart, GA: Quality Deer Management Association.

Kammermeyer, K. E. and R. Thackston. 1995. Habitat management and supplemental feeding. In *Quality Whitetails: The Why and How of Quality Deer Management*, eds. K. V. Miller and R. L. Marchinton, 129–154. Mechanicsburg, PA: Stackpole Books.

Kaplan, R. S. and D. P. Norton. 1996. *Using the Balanced Scorecard as a Strategic Management System*. Cambridge, MA: Harvard Business Review Product 4126.

Keyser, P. D., D. C. Guynn, Jr., and H. S. Hill, Jr. 2005a. Density-dependent recruitment patterns in white-tailed deer. *Wildlife Society Bulletin* 33:222–232.

Keyser, P. D., D. C. Guynn, Jr., and H. S. Hill, Jr. 2005b. Population density: Physical condition relationships in white-tailed deer. *Journal of Wildlife Management* 69:356–365.

Kie, J. G., D. L. Drawe, and G. Scott. 1980. Changes in diet and nutrition with increased herd size in Texas white-tailed deer. *Journal of Range Management* 33:28–34.

Koerth, B. H. and J. C. Kroll. 2008. Juvenile to adult antler development in South Texas. *Journal of Wildlife Management* 72:1109–1113.

Kroll, J. C. 1996. *Aging and Judging Trophy White-tailed Deer*. Nacogdoches, TX: Center for Applied Studies in Forestry, Stephen F. Austin University.

Kruuk, L. E. B., J. Slate, J. M. Pemberton, S. Brotherstone, F. Guinness, and T. H. Clutton-Brock. 2002. Antler size in red deer: Heritability and selection but no evolution. *Evolution* 56:1683–1695.

Kurt, F., G. B. Hartl, and R. Tiedemann 1995. Tuskless bulls in Asian elephant (*Elephas maximus* L.): History and population genetics of a man-made phenomenon. *Acta Theriologica* 40 (Suppl. 3):125–143.

Kurt, F., G. B. Hartl, and F. Volk. 1993. Breeding strategies and genetic variation in European roe deer *Capreolus capreolus* populations. *Acta Theriologica* 38 (Suppl. 2):187–194.

Labisky, R. F., M. C. Boulay, R. A. Sargent, K. E. Miller, and J. M. Zultowsky. 1991. *Population Dynamics of White-tailed Deer in Big Cypress National Preserve*. Annual Progress Report Cooperative Agreement CA-5000–78007. Department of Wildlife and Range Sciences, University of Florida, Gainesville, FL.

Langenau, E. E. and J. M. Lerg. 1976. The effects of winter stress on maternal and neonatal behavior in penned white-tailed deer. *Applied Animal Ethology* 2:207–223.

Leberg, P. L. and D. L. Ellsworth. 1999. Further evaluation of the genetic consequences of translocations on southeastern white-tailed deer populations. *Journal of Wildlife Management* 63 327–334.

Leopold, A. 1966. *A Sand County Almanac With Essays on Conservation from Round River*. New York, NY: Ballantine Books.

Lesage, L., M. Crete, J. Huot, A. Dumont, and J. Ouellet. 2000. Seasonal home range size and philopatry in two northern white-tailed deer populations. *Canadian Journal of Zoology* 78:1930–1940.

Lockwood, M. A., D. B. Frels, Jr., W. E. Armstrong, E. Fuchs, and D. E. Harmel. 2007. Genetic and environmental interaction in white-tailed deer. *Journal of Wildlife Management* 71:2732–2735.

Lukefahr, S. D. and H. A. Jacobson. 1998. Variance component analysis and heritability of antler traits in white-tailed deer. *Journal of Wildlife Management* 62:262–268.

Maine Department of Inland Fisheries and Wildlife. 2010. *Supplemental Feeding of White-tailed Deer During Winter*. http://www.state.me.us/ifw/wildlife/species/deer/feeding_deer.htm. Accessed July 23, 2010.

Martinez, A., D. G. Hewitt, and M. C. Correa. 1997. Managing overabundant white-tailed deer in northern Mexico. *Wildlife Society Bulletin* 25:430–432.

Martinez, J. G., J. Carranza, J. L. Fernandez-Garcia, and C. B. Sanchez-Prieto. 2002. Genetic variation of red deer populations under hunting exploitation in southwestern Spain. *Journal of Wildlife Management* 66:1273–1282.

McCoy, J. E., D. G. Hewitt, and F. C. Bryant. 2005. Dispersal by yearling male white-tailed deer and implications for management. *Journal of Wildlife Management* 69:366–376.

McDonald, J. S. and K. V. Miller. 1993. *A History of White-tailed Deer Restocking in the United States 1878 to 1992*. Research Publication 93–1. Bogart, GA: Quality Deer Management Association.

McKinley, W. T., S. Demarais, K. L. Gee, and H. A. Jacobson. 2006. Accuracy of the camera technique for estimating white-tailed deer population characteristics. *Proceedings of the Annual Conference of the Southeastern Association of Fish and Wildlife Agencies* 60:83–88.

Miller, K. V. and R. L. Marchinton, eds. 1995. *Quality Whitetails: The Why and How of Quality Deer Management*. Mechanicsburg, PA: Stackpole Books.

Miller, R., J. B. Kaneene, S. D. Fitzgerald, and S. M. Schmitt. 2003. Evaluation of the influence of supplemental feeding of white-tailed deer (*Odocoileus virginianus*) on the prevalence of bovine tuberculosis in the Michigan wild deer population. *Journal of Wildlife Diseases* 39:84–95.

Monteith, K. L., L .E. Schmitz, J. A. Jenks, J. A. Delger, and R. T. Bowyer. 2009. Growth of male white-tailed deer: Consequences of maternal effects. *Journal of Mammalogy* 90:651–660.
Moore, M. T. 2008. Evaluation of a camera census technique at three white-tailed deer densities. MS thesis, Texas A&M University-Kingsville, TX.
Morgan, Z. R. 1995. Male white-tailed deer daytime home range size and shape in relation to age and physiographic region in Mississippi. MS thesis, Mississippi State University, MS.
Murden, S. B. and K. L. Risenhoover. 1993. Effects of habitat enrichment on patterns of diet selection. *Ecological Applications* 3:497–505.
Murphy, D. A. and J. A. Coates. 1966. Effects of dietary protein on deer. *Transactions of the North American Wildlife and Natural Resources Conference* 31:129–139.
Mysterud, A. 2010. Still walking on the wild side? Management actions as steps towards "semi-domestication" of hunted ungulates. *Journal of Applied Ecology* 47:920–925.
Mysterud, A. and R. Bischof. 2010. Can compensatory culling offset undesirable evolutionary consequences of trophy hunting? *Journal of Animal Ecology* 79:148–160.
Mysterud, A. and E. Ostbye. 1999. Cover as a habitat element for temperate ungulates: Effects on habitat selection and demography. *Wildlife Society Bulletin* 27:385–394.
Nassauer, J. I. 1995. Culture and changing landscape structure. *Landscape Ecology* 10:229–237.
National Research Council. 2007. *Nutrient Requirements of Small Ruminants: Sheep, Goats, Cervids, and New World Camelids*. Washington, DC: National Academies Press.
Nelson, M. E. and L. D. Mech. 1986. Relationship between snow depth and gray wolf predation on white-tailed deer. *Journal of Wildlife Management* 50:471–474.
Nixon, C. M., L. P. Hansen, P. A. Brewer et al. 2001. Survival of white-tailed deer in intensively farmed areas of Illinois. *Canadian Journal Zoology* 79:581–588.
Nixon, C. M, L. P. Hansen, P. A. Brewer, J. E. Chelsvig. 1991. Ecology of white-tailed deer in an intensively farmed region of Illinois. *Wildlife Monographs* 118:1–77.
Norreklit, H. 2000. The balance on the balanced scorecard—A critical analysis of some of its assumptions. *Management Accounting Research* 11:65–88.
Ortega, I. M., S. Soltero-Gardea, D. L. Drawe, and F. C. Bryant. 1997. Evaluating grazing strategies for cattle: Nutrition of cattle and deer. *Journal of Range Management* 50:631–637.
Ozoga, J. J. and L. J. Verme. 1982. Physical and reproductive characteristics of a supplementally-fed white-tailed deer herd. *Journal of Wildlife Management* 46:281–301.
Palmer, M. V., W. R. Waters, and D. L. Whipple. 2004. Shared feed as a means of deer-to-deer transmission of *Mycobacterium bovis*. *Journal of Wildlife Diseases* 40:87–91.
Palmer, M. V. and D. L. Whipple. 2006. Survival of *Mycobacterium bovis* on feedstuffs commonly used as supplemental feed for white-tailed deer (*Odocoileus virginianus*). *Journal of Wildlife Diseases* 42:853–858.
Perkins, J. R. 1991. Supplemental feeding. Texas Parks and Wildlife Department, Austin, TX.
Pettorelli, N., J.-M. Gaillard, G. Van Laere et al. 2002. Variations in adult body mass in roe deer: The effects of population density at birth and of habitat quality. *Proceedings of the Royal Society of London B* 269:747–753.
Ramsey, C.W., ed. 1996. *Supplemental Feeding for Deer: Beyond Dogma*. College Station, TX: Texas A&M University.
Räsänen, K. and L. E. B. Kruuk. 2007. Maternal effects and evolution at ecological time-scales. *Functional Ecology* 21:408–421.
Richards, D. and A. Brothers. 2003. *Observing and Evaluating Whitetails*. Boerne, TX: Dave Richards Wilds of Texas Photography.
Rollins, D., ed. 1998. *The Role of Genetics in White-tailed Deer Management. Proceedings of a Symposium*. Texas Agricultural Extension Service, Texas A&M University.
Rose, K. E., T. H. Clutton-Brock, and F. E. Guinness. 1998. Cohort variation in male survival and lifetime breeding success in red deer. *Journal of Animal Ecology* 67:979–986.
Scribner, K. T., M. C. Wooten, M. H. Smith, and P. E. Johns. 1985. Demographic and genetic characteristics of white-tailed deer populations subjected to still or dog hunting. In *Game Harvest Management*, eds. S. L. Beasom and S. F. Roberson, 197–212. Kingsville, TX: Caesar Kleberg Wildlife Research Institute.
Shea, S. M. and J. S. Osborne 1995. Poor quality habitats. In *Quality Whitetails: The Why and How of Quality Deer Management*, eds. K. V. Miller and R. L. Marchinton, 193–209. Mechanicsburg, PA: Stackpole Books.

Short, H. L. 1969. Physiology and nutrition of deer in southern upland forests. In *Proceedings of the White-tailed Deer in Southern Forest Habitat Symposium*, 14–18. USDA Southern Forest Experiment Station, Nacogdoches, TX.

Sorin, A. B. 2004. Paternity assignment for white-tailed deer (*Odocoileus virginianus*): Mating across age classes and multiple paternity. *Journal of Mammalogy* 85:356–362.

Sparrowe, R. D. and P. F. Springer 1970. Seasonal activity patterns of white-tailed deer in eastern South Dakota. *Journal of Wildlife Management* 34:420–431.

Stedman, S. 1998. One manager's perspective on culling. In *The Role of Genetics in White-tailed Deer Management. Proceedings of a Symposium*, ed. D. Rollins, 94–96. Texas Agricultural Extension Service, Texas A&M University, TX.

Stedman, S. W. 2009. The magic of stacking. *Deer Associates eNews*, October 2009. http://ckwri.tamuk.edu/fileadmin/user_upload/docs/Deer_Research/eNews_Archives/The_Magic_of_Stacking_October_2009.pdf. Accessed September 7, 2010.

Strickland, B. K. and S. Demarais. 2000. Age and regional differences in antlers and mass of white-tailed deer. *Journal of Wildlife Management* 64:903–911.

Strickland, B. K. and S. Demarais. 2007. *Using Antler Restrictions to Manage for Older-Aged Bucks*. Mississippi State University Extension Service Publication 2427. Mississippi State, Mississippi.

Texas Parks and Wildlife Department. 2010a. *Trap, Transport and Transplant Permit Program.* http://www.tpwd.state.tx.us/business/permits/land/wildlife_management/trap_transport/. Accessed August 21, 2010.

Texas Parks and Wildlife Department. 2010b. *Deer management permits*. http://www.tpwd.state.tx.us/business/permits/land/wildlife_management/deer_management/index.phtml. Accessed August 21, 2010.

The Wildlife Society. 2007. *Final TWS Position Statement: Baiting and Supplemental Feeding of Game Wildlife Species*. http://joomla.wildlife.org/documents/positionstatements/42-Baiting%20and%20Feeding.pdf. Accessed July 26, 2010.

Tierson, W. C., G. F. Mattfeld, R. W. Sage, and D. F. Behrend. 1985. Seasonal movements and home ranges of white-tailed deer in the Adirondacks. *Journal of Wildlife Management* 49:760–769.

Timmons, G. R., D. G. Hewitt, C. A. DeYoung, T. E. Fulbright, and D. A. Draeger. 2010. Does supplemental feed increase selective foraging in a browsing ungulate? *Journal of Wildlife Management* 74:995–1002.

Ullrey, D. E., H. E. Johnson, W. G. Youatt et al. 1971. A basal diet for deer nutrition research. *Journal of Wildlife Management* 35:57–62.

Ulrich, R. S. 1986. Human responses to vegetation and landscapes. *Landscape and Urban Planning* 13:29–44.

Van Horne, B. 1983. Density as a misleading indicator of habitat quality. *Journal of Wildlife Management* 47:893–901.

VerCauteren, K. C., M. J. Lavelle, and S. Hygnstrom. 2006. Fences and deer-damage management: A review of designs and efficacy. *Wildlife Society Bulletin* 34:191–200.

Verme, L. J. 1969. Reproductive patterns of white-tailed deer related to nutritional plane. *Journal of Wildlife Management* 33:881–887.

Virginia Department of Game and Inland Fisheries. 2008. *Effective September 1, Feeding Deer Will Be Illegal in Virginia*. http://www.dgif.virginia.gov/news/release.asp?id=186. Accessed July 23, 2010.

Warren, R. C. and G. A. Hurst. 1981. *Ratings of Plants in Pine Plantations As White-tailed Deer Food.* Information Bulletin 18. Mississippi Agricultural and Forestry Experiment Station. Mississippi State, MS.

Webb, S. L., K. L. Gee, S. Demarais, B. K. Strickland, and R. W. DeYoung. 2009. Efficacy of a 15-strand high-tensile electric fence to control white-tailed deer movements. *Wildlife Biology in Practice* 5:45–57.

Webb, S. L., D. G. Hewitt, and M. W. Hellickson. 2007. Survival and cause specific mortality of mature white-tailed deer. *Journal of Wildlife Management* 71:555–558.

Williams, J. D., W. F. Krueger, and D. H. Harmel. 1994. Heritabilities for antler characteristics and body weight in yearling white-tailed deer. *Heredity* 73:78–83.

Williamson, S. 2000. *Feeding Wildlife ... Just Say No*. Washington, DC: Wildlife Management Institute.

Yoakum, J., W. P. Dasmann, H. R. Saanderson, C. M. Nixon, and H. S. Crawford. 1980. Habitat improvement techniques. In *Wildlife Management Techniques Manual* (4th edition), ed. S. D. Schemnitz, 329–404. Washington, DC: The Wildlife Society.

Zaiglin, R. E. and C. A. DeYoung. 1989. Supplemental feeding of free-ranging deer in south Texas. *Texas Journal of Agriculture and Natural Resources* 3:39–41.

16

Managing White-tailed Deer: Eastern North America

Duane R. Diefenbach and Stephen M. Shea

CONTENTS

The ability of the white-tailed deer to adapt to and thrive in a wide variety of habitats requires different approaches to managing this species. Variation in both the population dynamics of the species and in the social and political factors that influence how humans value white-tailed deer also necessitate different approaches. Consequently, providing an overview of the management of white-tailed deer in eastern North America runs the gamut from a species limited by snow depths in the remote boreal forests of Canada to a species inhabiting metropolitan areas. In this chapter, we address issues that influence deer management among ecoregions that have similar deer population characteristics and related management concerns.

Management of Subspecies

As many as 11 subspecies of white-tailed deer have been identified in eastern North America (Halls, 1984), but restoration efforts in the early and mid-twentieth century resulted in as many as seven different subspecies being introduced to some areas (Miller et al., 2003). Consequently, the subspecific status of white-tailed deer populations in North America is uncertain. Also, given that most subspecific classifications are based on morphometric characteristics, which can be influenced by environmental conditions, detailed molecular genetic analyses likely will be needed to better describe the genetic structure of white-tailed deer in eastern North America (Miller et al., 2003).

Provincial or state agencies responsible for managing deer give little consideration of their subspecific status (but see section of this chapter on breeding chronology), except for the endangered Key deer (*O. virginianus clavium*) (Chapter 1). The range of Key deer is composed of 17 islands in the lower Florida keys that total 9836 ha (Harveson et al., 2004). About 75% of the population exists on Big Pine and No Name keys (Lopez, 2001) and population size on these two keys was estimated to be 453–517 deer in 2001 (Lopez et al., 2004). The subspecies has not been hunted since 1939, but the population continued to decline through the 1950s. In 1957 the National Key Deer Refuge was established in part to protect this subspecies, and in 1967 Key deer were listed as endangered by the U.S. Fish and Wildlife Service.

Recovery of the Key deer is addressed in the Habitat Conservation Plan established in 1986 by the U.S. Fish and Wildlife Service for Big Pine and No Name keys. The primary source of mortality (~50%) is vehicle collisions, and population recovery has focused on protecting critical habitat from human development and reducing deer–vehicle collisions. Deer density on Big Pine and No Name keys is nearing carrying capacity, such that translocations to other keys may be required. Establishing Key deer populations on other islands would lessen the risk of extinction due to hurricanes or other catastrophic events.

Climate Change

During the twenty-first century it is predicted that throughout the range of the white-tailed deer in eastern North America average surface temperatures will increase 2.5–4.0°C (Intergovernmental Panel on Climate Change, 2007). Furthermore, in more northern latitudes precipitation is predicted to increase by 10–20%, including greater winter snowfall, and precipitation events are expected to occur less frequently but with greater intensity (Post and Stentheth, 1999). Consequently, agricultural landscapes at northern latitudes are expected to be influenced by longer growing seasons, more frequent short-term droughts, and shorter duration of snow cover, which are expected to adversely affect perennial forage crops used by deer (Bélanger et al., 2002). At more southerly latitudes it is expected that water shortages and droughts will become more common. Less snowfall and warmer temperatures could potentially reduce winter mortality of deer and possibly allow the species' distribution to expand northward (Chapter 1).

Forest vegetation changes are unlikely to have major effects on deer populations because the species currently inhabits all forest vegetation types. However, it is possible that climate change could influence key population parameters such that harvest management systems may need to be responsive to climatically induced changes. In northern latitudes more frequent fires and insect outbreaks will likely shift forest age structure to younger age classes that provide more forage for deer. However, the distribution of many plant species is expected to shift northward and some changes could have adverse effects on deer. For example, eastern hemlock, a species that provides deer with thermal cover and protection from deep snows, is expected to disappear from most of the United States. Also, other ungulates (e.g., moose) may be less adaptable to climate change (Bowyer et al., 1998) and could reduce interspecific competition with white-tailed deer (Jenkins and Wright, 1987). Climate-induced environmental conditions could cause population dynamics to shift between density-dependent and density-independent controlling factors.

Probably the most direct influence of climate change related to deer will be with respect to disease. In eastern Canada, ticks responsible for Lyme disease are predicted to spread through the region within 10–20 years (Bourque and Simonet, 2008), and white-tailed deer are an important overwinter host for the tick vector (Rand et al., 2004). Epizootic hemorrhagic disease (EHD), endemic to the southeastern United States, was first confirmed in 2008 in Pennsylvania (W. Cottrell, Pennsylvania Game Commission, personal communication). Warmer temperatures and short-term droughts interrupted by intense rainfall events will create ideal conditions for reproduction and survival of the midge (*Culicoides* spp.) that transmits EHD.

Effects of Deer on Forest Ecosystems

At certain densities, herbivory by white-tailed deer can have measurable effects on plant species composition and abundance and can reduce or prevent tree regeneration (Côté et al., 2004, Chapter 12). Despite disagreement concerning the relative importance of deer and other factors (e.g., acid deposition, suppressed fire regimes, etc.) on forest ecosystems (e.g., Abrams, 1992; Sharpe and Drohan, 1999), deer clearly affect tree species abundance and composition (Horsley et al., 2003).

Latham et al. (2005) reviewed evidence in support of the hypothesis that deer are the primary factor influencing forested ecosystems in Pennsylvania and proposed an adaptive management approach for managing deer from an ecosystem perspective. To date, no such approach has been attempted, although Latham et al. (2005) outlined how such an endeavor should be approached on state forest lands in Pennsylvania. In fact, the only state agency responsible for managing deer that uses measures of forest tree regeneration to assess habitat quality when making deer harvest recommendations is the Pennsylvania Game Commission (Shissler and Grund, 2009).

White-tailed deer were introduced to Anticosti Island, Quebec, in 1896–1897. The island is almost 8000 km^2 in size and deer densities exceed 20 deer/km^2. Consequently, dramatic changes have occurred in plant species composition and diversity and regeneration of balsam fir is difficult (Sauve and Côté, 2007). Research to test if different silvicultural treatments can offset the effects of deer browsing at high deer densities has failed to identify suitable prescriptions (Casabon and Pothier, 2008; Beguin et al., 2009a,b).

On Cumberland Island, Georgia, deer populations increased following the reduction of free-ranging, nonnative herbivores (cattle, horses, and hogs) and by 1979 an estimated density of 40 deer/km^2 occurred on the island (Nelms, 1999). In turn, regeneration of live oak, the island's dominant tree species, was severely reduced (Lieske et al., 1990). In 1988 and 1989, 30 bobcats were reintroduced to the island (Diefenbach et al., 1993) and white-tailed deer were a primary prey species (Baker et al., 2001). By 1998–1999, observations of bobcat use of deer as a primary prey species, a decline in deer abundance and an average 11 kg increase in body mass, and an increase in oak regeneration indicated that bobcat predation on deer caused a trophic cascade effect on the island (Nelms, 1999; Diefenbach et al., 2009).

The primary cause of mortality in white-tailed deer in eastern North America is human related: hunting mortality followed by vehicle collision mortality. Except in Canada where gray wolf populations are established, there are few effective predators of adult deer in eastern North America. Consequently, if hunting is ineffective in areas of high human density, or is banned (e.g., national parks), or hunters are opposed to increased harvest rates on antlerless deer, deer densities can increase to levels that affect vegetation diversity, composition, and abundance. The challenge for deer managers is to balance stakeholder desires and needs when managing white-tailed deer across hundreds of square kilometers where most land is privately owned and landowners make the ultimate decision regarding deer management (i.e., whether to permit hunting and how many deer to harvest).

Land-Use Changes

Large-scale land-use changes have, and will continue to have, profound impacts on deer density, distribution, and management across eastern North America. Many recent land-use changes have been caused by economic impacts on development, agriculture, forestry, and land ownership.

Development

The coastal plain of eastern North America with its sea ports, cities, and industrial centers has traditionally incurred higher rates of development and human population growth than other regions. The U.S. Census Bureau (2000) estimates that 82% of the U.S. population lives in urban and suburban areas and that the population will grow by 46% over the next 40 years. Recently, population growth in the Southeast has far exceeded that of the Northeast. Rapid commercial, industrial, and residential growth in both regions has caused loss and fragmentation of deer habitat. However, many areas of the eastern United States sustain some of the highest densities of white-tailed deer in the nation (Adams et al., 2009).

Economic growth has stimulated land values in this region, especially for property in close proximity to urban centers; therefore, income from the real-estate market has become competitive with other historical economic activities on many properties. Many property owners whose primary income was from farming, livestock, or forest products have sold or developed their land. Therefore, the demand for real estate has indirectly increased the rate of development and urban sprawl in this region. Housing density restrictions of many municipal comprehensive plans, especially in rural and suburban areas, have created an interspersion of low-density or clustered housing developments where deer and people share the same communities (Figure 14.20).

These land-use changes in the urban–wildland interface have heightened many social, economic, and management issues related to urban and suburban deer populations. Increased urban sprawl, road density, and fragmentation of deer habitat has led to more human–deer conflicts (vehicle collisions, damage to landscaping, etc.) and efforts to control deer population growth. Deer population management philosophies and objectives have evolved with land-use changes and led to a transition from deer management based solely on biological carrying capacity to management favoring cultural carrying capacity in highly developed areas.

Habitat loss and fragmentation caused by development also has led to displacement of hunters from traditional hunting grounds on private lands to public hunting areas or other adjacent private lands open to hunting. Hunter displacement has increased competition and conflict among hunters in some areas of Florida (C. Morea, Florida Fish and Wildlife Conservation Commission, personal communication).

Agriculture

Over the past 30 years, developments in agricultural technology and genetics have led to improved methodology and efficiency in farming practices. These changes have caused significant changes to the quality, quantity, and juxtaposition of deer habitat and to deer density in some areas. Adverse economic conditions in agriculture markets also have had profound impacts on farming and ranching in eastern North America.

Improvements in agricultural machinery, genetic seed stock, irrigation, herbicides, pesticides, and fertilizers have improved harvest efficiency and yields especially for row crops. Increased crop yields, both foreign and domestic, also have affected prices and profit margins. Small family-owned farms and ranches that could not compete with larger farming and cattle operations were bought by corporations. Large-scale farming and cattle operations optimized efficiency and increased acreage of clean farming and improved pasture. These changes reduced the amount of deer habitat and cover in many areas that traditionally had greater interspersion of habitats. Although clean farming and harvest practices have reduced waste grain available to deer, farming continues to enhance the seasonal nutritional plane of deer in many areas. However, deer have become more concentrated in the remaining areas that contain adequate cover. These areas often suffer from seasonal overpopulation, habitat degradation, and increased competition from hunters.

Market volatility also has led to variability in crop production. In many areas of the Southeast, corn and soybean crops have been replaced by cotton. Other farmlands have been converted to pine plantations and some farmers have removed lands from crop production by taking advantage of government programs such as the Conservation Reserve Program and Wildlife Habitat Improvement Program (Alig et al., 1990). These changes in the agricultural landscape have altered the carrying capacity, productivity, and density of white-tailed deer in some locations.

Forestry

Forest ownership in the eastern United States, especially in the Southeast, continues to change dramatically. Establishment of the Employee Retirement Income Security Act in 1974 and other similar state legislation thereafter encouraged institutional investors to diversify portfolios. Recognizing that timberlands were selling at bargain prices due to declines in the wood products market, timberland investment management organizations (TIMOs) emerged to organize and manage partnerships of institutional investors. This change in timberland ownership is evident in the fact that TIMOs now account for the majority of timber transactions in the United States (Block and Sample, 2001).

This ownership change also changed forest management philosophy from one with a primary focus of adhering to mill fiber agreements to one focused on shareholder needs. Timberland investors have placed more emphasis on financial returns of timberlands, which has led to the increase in forest fragmentation as holdings are sold for commercial, residential, and other developments (Alig et al., 1990).

Although these ownership changes have caused increased fragmentation and loss of habitat, they also have benefited deer habitat quality. During the 1990s, industrial and TIMO owners of forest lands began to adapt silviculture and harvest management practices to diversify their wood products. Reduction in canopy closure through selective harvest strategies has created higher-quality habitats for white-tailed deer, especially in pine plantations of the Southeast. However, timber harvest management on some public lands has been restricted by state and federal regulations including the Endangered Species Act and National Environmental Policy Act. In some cases, these regulations have directly or indirectly affected and reduced the amount of deer habitat management on public lands.

Land Ownership

As discussed previously, our nation's economic policies and human population growth have caused further fragmentation and diversity of ownership of lands especially those in close proximity to urban centers. As municipalities incorporate more rural lands along the urban–wild land interface, deer management issues become more complicated. Often layers of community, city, and county hunting restrictions are imposed in addition to state regulations. In some cases, especially in northeastern states, restrictions preclude access and methods necessary to sufficiently control deer populations in urban areas (Conover, 1995; Bishop et al., 1999; DeNicola et al., 2000; Kilpatrick and LaBonte, 2003). A 2009 survey of state and provincial wildlife agencies in the Northeast identified deer population control and urban/suburban deer management as the two most important deer management issues in that region. As habitat fragmentation, urbanization, management attitudes, and perceptions of hunting change with land ownership, so will wildlife managers' ability to properly manage deer populations in many areas.

As land ownership has diversified, so have landowner hunting and wildlife management philosophies. Economic and liability issues have caused an increase in the number and acreage of private properties that lease hunting rights to hunt clubs (Mozumder et al., 2007). Leasing hunting rights on private lands has been common in the Southeast for over 50 years. However, privatization issues including the increase in acreage of private lands placed under high-fence and hunting lease rates are major concerns of state wildlife agencies in the Southeast (Table 16.1). Due to self-limitations on hunting pressure, hunting leases typically do not accommodate as many hunters as open private or public lands. Therefore, privatization can cause increased hunting pressure on public lands and competition for and inflation of hunting lease rates on private lands. Land ownership patterns and differing views on hunting leases have limited privatization issues and concerns in the Northeast.

Disease

Only three diseases at this time have any significant influence on management of white-tailed deer in eastern North America: hemorrhagic disease, Lyme disease, and chronic wasting disease (CWD). Hemorrhagic disease can be caused either by EHD viruses or by bluetongue viruses of the genus *Orbivirus*. In southeastern United States the disease occurs annually and highly visible outbreaks are common. Severe outbreaks are more likely to occur in more northern regions (e.g., Virginia and Pennsylvania) because a smaller proportion of the population has developed antibodies to the virus.

Little can be done to control outbreaks of hemorrhagic disease, and management responses may involve reducing deer harvest rates in subsequent years. The disease is transmitted by a biting midge (*Culicoides* spp.) and environmental conditions in late summer (warm temperatures and exposed muddy areas from low river and lake water levels) allow midge populations to increase. Below-freezing temperatures during

TABLE 16.1

Legality of Fencing to Restrict Movements of White-tailed Deer in Eastern United States, Number of Fenced Properties, and Total Area, 2009

State or Province	Are High Fences Legal?	No. of High Fences	Area (ha)
Alabama	Yes	104	32,031
Arkansas	Yes		
Connecticut	Yes		
Delaware	No		
Florida	Yes	90	
Georgia	Yes		
Kentucky	Yes	80	
Louisiana	Yes		
Maine	Yes	11	891
Maryland	No		
Massachusetts	No		
Missouri	Yes	36	10,931
Mississippi	Yes	111	32,422
New Brunswick	No		
New Jersey			
New York	Yes		
New Hampshire	Yes	1	10,121
North Carolina	Yes	0	0
Oklahoma	Yes		
Pennsylvania[a]	Yes	1139	
Quebec	Yes		
South Carolina	Yes	28	5848
Texas	Yes		
Virginia	Yes	3	404
Vermont	Yes[b]	2	607
West Virginia	Yes	4	850

[a] Does not include captive deer operations that do not sell deer.
[b] No native species of cervids can be fenced.

autumn kill midges and stop transmission of the disease. Climate change is expected to allow this disease to spread northward (see section on Climate Change).

Ticks commonly infest white-tailed deer but rarely affect deer populations or individual animals. However, deer can serve as hosts of tick species known to be vectors of human diseases, such as Lyme disease and ehrlichioses. With respect to Lyme disease, deer are the overwinter host of the deer tick (*Ixodes* spp.) and are important to sustaining tick populations. Reducing deer populations can reduce abundance of ticks and incidence of Lyme disease in humans (Stafford et al., 2003). Consequently, some states specifically manage some deer populations to reduce the risk of Lyme disease for humans by maintaining populations at lower densities than otherwise could be sustained (S. A. Christensen, Massachusetts Division of Wildlife, personal communication).

In eastern North America, CWD, a spongiform encephalopathy caused by a prion, has been detected in New York and West Virginia. The disease was thought to have been introduced to a New York captive deer herd from a hunter-killed deer carcass from the western United States. However, eradication of the privately owned deer herd and subsequent monitoring of harvested and culled free-ranging deer suggest that the disease did not become established. In West Virginia, however, CWD was first detected in 2005 in Hampshire County and as of May 2009 a total of 39 deer have tested positive for the disease (West Virginia Department of Natural Resources, 2010). The disease continues to spread, including an infected deer detected in Virginia in 2010, and containment is unlikely.

Factors that risk introduction of CWD include concentrations of farmed or captive cervids, translocating cervids from CWD-affected regions, and permitting transport of hunter-killed cervid carcasses from CWD-infected areas (Samuel et al., 2002). In addition, factors that risk amplification of the disease once introduced include high cervid population densities, low abundance of large predators, and areas where free-ranging cervids are artificially concentrated (baiting, feeding, water development, and other human-related habitat modifications).

Most states and provinces allow captive cervid herds (Table 16.1) and transportation of live and harvested deer across state lines. Furthermore, many residents of eastern North America hunt cervids in states and provinces where CWD occurs and bring back infected carcasses. Some state wildlife agencies, such as New York, Virginia, and Vermont, have strict regulations or prohibitions regarding importation of live deer or carcasses. However, in other states, such as Pennsylvania, regulation of captive deer herds is under the authority of the Department of Agriculture and there is no regulation on purchase of white-tailed deer if there is no intent to sell animals. Furthermore, many states have no regulations regarding feeding of deer, which is considered a risk factor because it concentrates deer and increases the probability of disease transmission among individuals. Because most states and provinces in eastern North America exhibit one or more risk factors, prevention of the introduction of CWD to free-ranging deer herds in eastern North America will be a challenge.

Breeding Chronology

Across most white-tailed deer range in North America, deer are considered short-day breeders whose reproductive cycle is triggered by change in photoperiod (Chapter 2). As the daily duration of darkness increases during autumn, the amount of melatonin secreted by the pineal gland increases, triggering increased secretion of reproductive hormones, which stimulates breeding behavior, initiating the rut. Breeding chronology and synchrony of deer in northern latitudes is likely a product of natural selection optimizing survival and reproduction with respect to average weather conditions and forage availability. Breeding is more synchronous in temperate regions of North America than the southeastern United States (Harlow and Jones, 1965; Williams et al., 1995).

We created a breeding chronology map using information from reproductive studies, supplemented with information from state and provincial deer biologists (Figure 16.1). The map illustrates 30-day intervals during which most breeding occurred. Although the scale of the map does not capture all variation in breeding chronology within regions, it illustrates differences among regions. Compared with the majority of whitetail range in North America, a much later and wider variation in breeding chronology is evident in the Southeast, especially in Texas, Louisiana, Mississippi, Alabama, Florida, Georgia, and South Carolina. In this region, peak breeding ranges from summer through late winter compared to a one-month breeding period during autumn in temperate regions. Florida has the greatest disparity in breeding dates, which ranges over nine months from July in southern portions to March in northwest portions of the state (C. Morea, Florida Fish and Wildlife Conservation Commission, personal communication).

The reason for variation of breeding chronology in the Southeast is unclear. Some biologists suggest a genetic link caused by variation in origin of deer translocations during the early 1900s. However, this theory is not supported well by data or trends in breeding chronology. Although herd age structure and sex ratio are known to affect the timing of breeding of some herds, demographic differences do not explain the variation evident across the Southeast. In northern portions of the whitetail's range the timing of breeding is critical to fawn survival. However, in more southern latitudes with mild winters the timing of breeding probably does not greatly affect survival of fawns. Therefore, natural selection would not favor early breeders. Instead, subtle regional or local differences in other environmental factors that influence breeding condition or fawn survival (e.g., rainfall, forage availability, and range quality) may have a greater influence on the timing of breeding. However, no major differences in environmental factors are evident across 300 km of northern Florida where the breeding period varies by five months at the same latitude.

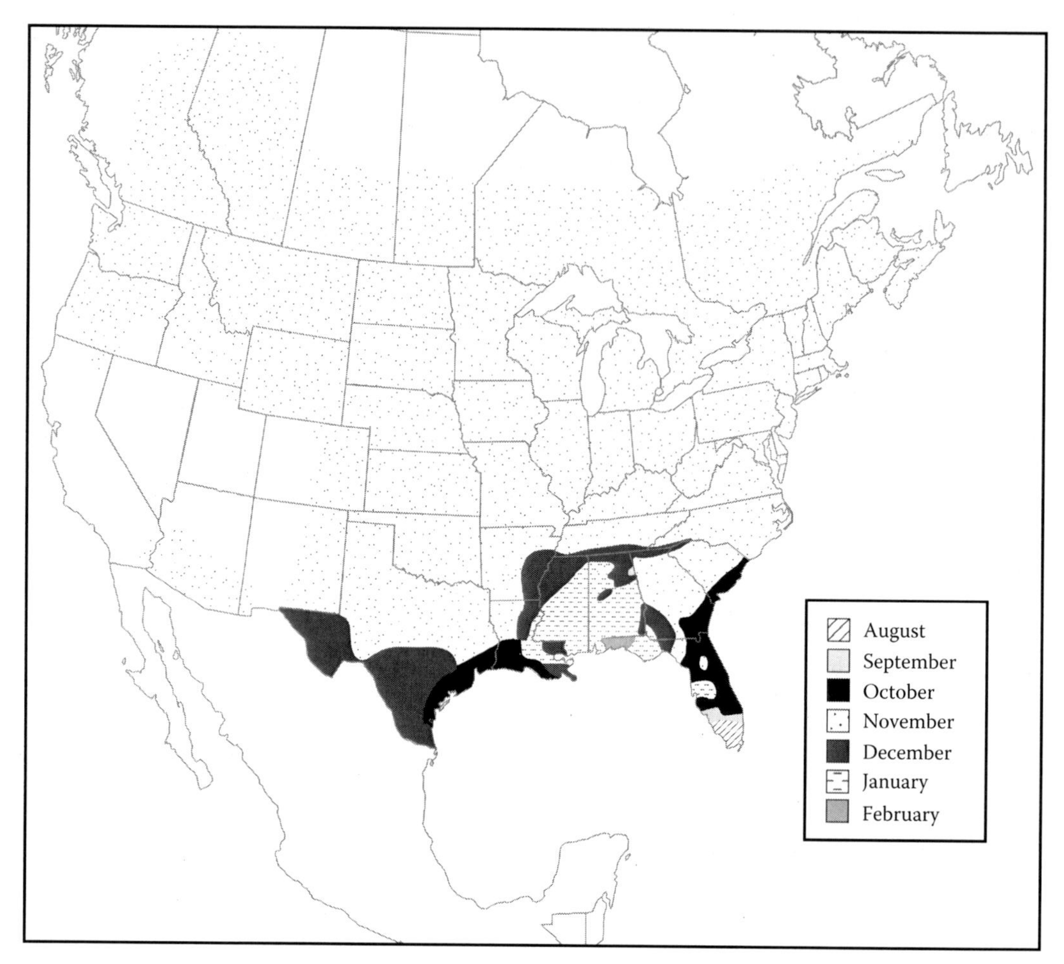

FIGURE 16.1 Breeding chronology of white-tailed deer in North America.

Although photoperiod is the primary factor influencing timing of the breeding season, some herds in the Southeast may exhibit differences in timing based on other environmental factors. However, the timing of breeding could be passed from doe to fawn (Miller, 2002). Therefore, without limiting factors that influence the timing of breeding, timing may become ingrained in the herd and passed on through successive generations.

Regardless of the reason for the diversity in breeding chronology in the Southeast, it has an important influence on deer biology and management in the region. Late breeding and fawning periods in some southern states affect antler mass, body size, and reproductive potential of yearling deer. Antler and body mass of yearling bucks from herds in Alabama and Florida where late breeding occurs are smaller than deer born earlier in the year (Shea et al., 1992a; Gray et al., 2002). Most yearling bucks in these states that are late born have spike antlers (Shea et al., 1992a; Cook and Gray, 2003). Therefore, antler size of most yearling bucks in these areas is not indicative of the buck's genetic potential for antler mass. Consequently, culling yearling bucks as a genetic harvest management strategy in regions with late breeding is not recommended. Physical condition parameters of yearling deer are less reliable in evaluating herd condition and population density when a large proportion of fawns are born late. Furthermore, late birth also delays reproductive maturity and productivity of fawn and yearling does. Late born fawns in this region rarely breed and yearling does have extremely low *in utero* productivity values (Harlow and Jones, 1965; Cook and Gray, 2003). Poor fawn and yearling doe productivity in these areas diminishes herd recruitment and therefore reduces the need for aggressive antlerless harvest. The effects of late birth are further exacerbated by nutritional deficiencies in some lower Gulf coastal plain habitats (Shea et al., 1992b; Shea and Osborne, 1995; Cook and Gray, 2003).

Asynchronous breeding chronology of herds in some southeastern states also complicates agency management when setting hunting seasons. It is difficult for states with variable conception and fawning periods to synchronize hunting seasons so that they occur prior to antler casting or when fawns are weaned. Subsequently, states with the most asynchronous rut, Florida, Georgia, Louisiana, and South Carolina, have the most diverse hunting season dates and zones to accommodate irregular breeding seasons. These states have deer hunting seasons that extend from four to seven months of the year.

Population Management

Management of deer populations is the responsibility of each state or province and deer abundance goals are established based on a combination of environmental conditions that determine how many deer can be sustained long term (environmental carrying capacity) as well as the number of deer deemed acceptable to society (social carrying capacity). From a biological perspective, social carrying capacity should never exceed environmental carrying capacity. In the northeastern United States, management agencies identified social carrying capacity as the most frequent basis for establishing deer density goals for management units (Williamson, 2003). However, in northern regions density-independent factors (e.g., climate and predation) are often the most important factors influencing deer density goals and corresponding management strategies.

Ideally, deer populations are managed according to land-based areas that have similar environmental conditions and are easily identifiable by the general public based on landscape features and political boundaries. Furthermore, management units need to be large enough for resource managers to be able to collect sufficient biological data to monitor changes in deer population characteristics (e.g., abundance, reproductive rates, age structure, etc.). Some agencies still delineate their management units solely according to political boundaries, which usually correspond to the county. Political boundaries often correspond well with physiographic conditions but can be too small to collect sufficient biological data to monitor deer population changes. Consequently, biological data from groups of counties may be used to monitor deer populations, even though hunting regulations (season length, bag limits, antlerless permit allocations, etc.) are promulgated by county. For example, prior to establishing 22 wildlife management units (WMUs) Pennsylvania managed deer according to management units defined by the Commonwealth's 65 counties. However, those 65 counties were grouped into about the same number of WMUs currently used when evaluating biological data for making recommendations for seasons and bag limits.

Population Monitoring

Population monitoring by state and provincial agencies primarily relies upon making inferences about abundance or trends using hunter harvest data (Table 16.2). Models often are used to estimate abundance such as the sex–age–kill (SAK) model (Eberhardt, 1960), Lang and Wood (1976) model, and Downing (1980) method. Such methods require strong assumptions about population structure or sources of mortality, lack measures of precision, and have been criticized for sensitivity to violation of assumptions (Millspaugh et al., 2009). Consequently, most agency biologists use a variety of information to assess population status (Table 16.2). Typically, biologists collect data from harvested deer to estimate the number of deer harvested, hunter success rates, age–sex structure of the harvest, and physical condition of deer (body mass, antler beam diameter, and number of antler points). In addition, some agencies may count embryos in road-killed deer to monitor reproductive rates or conduct sighting surveys to estimate fawn:doe ratios.

Harvest Strategies

Harvest bag limits and hunting methods generally become increasingly liberal from north to south because of a trend in increasing deer recruitment in warmer climates. However, comparing harvest regulations among states and provinces is difficult because there are many ways to regulate the number of

TABLE 16.2

Method of Estimating White-tailed Deer Abundance, Population Estimates, and Deer Population and Hunter Trends, Eastern United States, 2009

State or Province	Method of Estimating Deer Abundance	Deer Population Estimate	Deer Population Trend[a]		Trend in No. of Hunters[a]
			Urban/ Suburban	Rural	
Alabama		1,750,000	+	0	–
Arkansas		800,000	+	0	0
Connecticut	Aerial winter deer surveys	64,000	+	0	0
Delaware	Aerial infrared survey	42,000	+	0	0
Florida			0	0	0
Georgia	Age reconstruction	1,161,153	0	0	–
Kentucky	SAK	1,000,000	+	0	–
Louisiana			0	0	0
Maine	SAK	199,600	0	–	–
Maryland	Downing method; Lang and Wood method	229,000	+	0	–
Massachusetts	Age reconstruction	93,000	0	0	–
Missouri	Accounting model	1,400,000	+	0	0
Mississippi	Age reconstruction	1,750,000	0	0	0
New Brunswick	Aerial surveys and a deer model utilizing rate of increase	80,000	+	0	0
New Jersey					
New York	SAK statewide and buck kill/sq mile for trends in mgt units	940,000	+	+	0
New Hampshire	SAK	90,000	+	+	–
North Carolina	Downing reconstruction	1,250,000	+	+	–
Oklahoma			+	+	–
Pennsylvania	Modified SAK method for trend analysis		0	0	–
Quebec	Aerial survey	367,000	+	0	+
South Carolina	Downing method; Lang and Wood method; population reconstruction	780,000	+	–	0
Texas	Distance sampling within each management unit	4,000,000	+	0	0
Virginia	Combination of reconstruction and SAK	1,000,000	+	0	–
Vermont	MARK removal model, age data for sex ratios, and other population modeling	140,000	+	+	+
West Virginia	Buck kill per square mile during traditional firearms season and population reconstruction	700,000	+	0	–

[a] + = increasing; – = decreasing; 0 = stable.

antlerless deer harvested. Factors that influence the number of deer harvested include the type of sporting arm used, number of hunters, length of season, land ownership, and accessibility to land (whether hunting is allowed or physical accessibility). Consequently, completely different harvest regulations could result in similar percentages of deer harvested. In northern states and provinces the modern firearm deer season is relatively short (two weeks). The number of antlerless deer harvest permits is limited and allocated to hunters based on a lottery or first-come, first-served basis. For example, in Quebec and

New Brunswick, generally a hunter is allowed to harvest one deer/year, regardless whether they possess an antlerless permit, and some management units are completely closed to deer hunting. Such restrictive harvest regulations are required because recruitment into the population is low because of high winter mortality and low reproductive rates. In Maine and New Hampshire hunters are allowed to harvest one deer/year, although in some management units an antlered deer and an antlerless deer (with permit) can be harvested. Because New York encompasses a greater range of habitat types and large differences in winter weather conditions, some management units are antlered deer only (or by permit) whereas other parts of the state allow harvest of antlered deer with the purchase of a hunting license and additional antlerless deer can be harvested with an antlerless permit. In Pennsylvania, where there are more than 750,000 deer hunters, antlerless permits are issued by management unit and distributed on a first-come, first-served basis. New Jersey has fewer hunters and so the number of antlerless permits are not limited but can only be used in certain seasons and with specific sporting arms.

South of Pennsylvania, hunting seasons become extremely liberal in comparison to northern states. In parts of Maryland, the season bag limit is two antlered and 10 antlerless deer and the various hunting seasons begin in September and extend through January. In parts of South Carolina there is no limit on the number of antlered deer that can be harvested and on days that antlerless deer can be harvested (about 12 days/season); the bag limit is two deer/day. Florida has a daily bag limit of two bucks/day for a season that lasts more than four months. In Alabama three antlered deer can be harvested per year and during the antlerless season (over two months long) two antlerless deer/day can be harvested.

Liberal bag limits and season lengths of southern states are partly due to state wildlife agencies allowing private landowners more control in managing deer populations on their lands. Hunting rights on most private lands in the Southeast are leased to private hunt clubs. Many states provide technical assistance programs for private lands and issue antlerless deer permits and tags specific to those lands that can be used in addition to the regular antlerless deer permits or licenses issued for the management unit. Extended season length is necessary in many ecological communities in the Southeast where thick non-deciduous cover makes it difficult to attain harvest objectives. Hunting leases also have vested and financial interest in sound deer management and carefully prescribe harvest limits and restrictions, often more limiting than those required by state law. Liberal season and bag limit regulations of the Southeast better enable landowners to practice quality deer management (QDM) by providing sufficient season length and structure to meet antlerless harvest goals and selectivity in buck harvest. Subsequently, although some southeastern states have more liberal bag limits and season lengths, the average age of bucks harvested exceeds that of many northeastern states with more traditional season structures (Table 11.2).

Some poor- or suboptimal-quality habitats of the mountainous and coastal plain regions of eastern North America require more conservative harvest management prescriptions (Shea and Osborne, 1995). Soil fertility in these ecosystems often is low and adversely affects forage quality and abundance. Where forage quality and abundance is low, so is deer carrying capacity. These areas require more conservative antlerless harvests because deer populations are more susceptible to, and take longer to recover from, overharvest. Some lower coastal plain habitats in Florida have high quantities of low-quality deer forage and can support high deer densities. However, low productivity contributes to slow population growth and herds may not exhibit density-dependent responses to herd reduction (Shea et al., 1992b). In some regions of the Southeast, habitat quality varies widely within small geographic areas, which requires deer managers to be site specific in habitat and herd evaluations and management prescriptions.

Hunting Methods

Methods of hunting and harvesting deer vary in efficiency, traditions, and social acceptance. Perhaps the best example is the use of dogs to hunt deer. In northern states this method of hunting generally is not accepted by hunters or society because there never was a dog hunting tradition and, consequently, the method is considered contrary to the standards of fair chase. Southern states (Virginia, North Carolina, South Carolina, Georgia, Florida, Alabama, Mississippi, Louisiana, and Arkansas), in contrast, have a long tradition of using dogs to hunt deer, especially in the Atlantic and Gulf Coast plains where large tracts of dense vegetation and swamps reduce accessibility to deer. However, hunter–landowner conflicts are increasing because dog hunting requires large tracts of land, which are increasingly scarce (see section of this chapter on land use changes). Almost all states that allow hunting with dogs have added restrictions in recent years (e.g., closing areas or requiring permits and minimum acreages) because of conflicts with landowners and less acceptance of the hunting practice by society. Texas banned hunting with dogs in 1991 and most states where hunting with dogs is legal are involved in stakeholder mediation efforts or legislation.

From a deer management perspective, however, the method of harvesting deer is less important than how it affects hunter success rates. In northern regions where the harvest of antlerless deer is strictly regulated to prevent overharvest, hunting methods are limited. For example, the number of hunters who may participate in a deer drive may be limited, such as in Pennsylvania where no more than 25 people can participate. Muzzleloader and archery hunting have lower hunter participation and success rates than modern rifle hunting. Consequently, archery and muzzleloader seasons are generally longer, may allow harvest of an antlerless deer without a special permit, or may be the only season in which antlerless deer can be harvested. The use of bait is generally illegal in northern regions, except perhaps in urban and suburban areas where firearms are considered unsafe and greater effectiveness of archery hunting is needed to reduce deer abundance. In contrast, southern regions provide longer seasons, larger bag limits, and fewer restrictions on hunting methods.

Changes in Harvest Strategies

Interest in QDM began in the southeastern United States and has greatly influenced how deer hunters, landowners, and natural resource agencies manage deer. Most states in the Southeast have adopted some form of QDM in their statewide, regional, or WMU regulations. Throughout the eastern United States, an increasing number of states have implemented regulations that restrict harvest of antlered deer with the intent to increase the number of older-aged bucks in the population (Figure 11.2). Pennsylvania was the first northeastern state to implement regulations that required antlered deer legal for harvest have three or four points on at least one antler (depending on WMU). Since then other northeastern states have implemented similar regulations either statewide or on a management unit basis. More importantly, some states have seen hunters voluntary reduce harvest rates on younger antlered deer and increase harvest of antlerless deer. These changing harvest strategies have important implications for how biologists monitor deer population abundance (see section of this chapter on population monitoring) because several methods assume that harvest rates are similar across age classes of bucks. Consequently, an increasing number of older bucks in the harvest may not reflect overall reduced harvest rates but simply less harvest pressure on younger age classes.

An important factor causing management agencies to consider longer season lengths, larger bag limits, and different hunting methods (e.g., use of bait) is number of hunters that pursue white-tailed deer. During 2006, 5% of the U.S. population participated in deer hunting, but participation rates in most regions of the eastern United States were generally at or below average (3–5%, except 8% in Mississippi, Alabama, Kentucky, and Tennessee; U.S. Fish and Wildlife Service, 2006). Lower participation rates in the eastern United States may partly be explained by the fact that many people live in metropolitan areas (with greater than one million people) and a smaller percentage of these residents hunt (3%). However, little change in participation in deer hunting has occurred since 1991. Nationally, about the same number of hunters pursued deer in 1991 (10,277,000 hunters) as in 2006 (10,062,000 hunters) and since 1996 the number of days hunters spent afield has remained at just over 132 million. However, deer hunters have a limit to the number of deer they are willing to harvest (Brown et al., 2000) and may not be interested in hunting where deer need to be harvested. Deer populations are stable or increasing in both rural and urban areas, but more likely to be increasing in urban areas (Table 16.2). Consequently, management agencies have had to consider new ways to increase hunter participation and success rates in urban areas (e.g., Kilpatrick and Walter, 1999).

Urban Population Management

Between 1951 and 2001, metropolitan areas in North America have more than doubled in size and more than 90% of urban growth in the United States and Canada has created suburbs. Managing deer in urban, suburban, and exurban areas involves a wide array of stakeholders with diverse opinions on the problem and solution (Messmer et al., 1997) and requires management agencies to take a different approach than used for traditional deer management based on recreational hunting. Problems between deer and humans in urban areas usually involve deer–vehicle collisions (resulting in vehicle damage and human injuries), browsing damage to home landscaping, and tick-borne diseases. Deer–human conflicts in urban areas have increased dramatically in the eastern United States because deer adapt well to urban habitats, the pace of land development has increased dramatically in the past several decades, and deer in urban areas have relatively high reproductive rates and low mortality rates (Moriarty, 2006). Furthermore, either traditional recreational hunting is considered unsafe or firearm discharge ordinances preclude use of modern firearms. State agencies most successful in dealing with deer–human conflicts generally have required communities to identify acceptable methods to resolving these problems, within some broadly defined boundaries, rather than dictating the solution (Kilpatrick and Walter, 1997).

Potential methods of reducing deer abundance in urban areas include fertility control, recreational hunting (primarily archery hunting), culling by sharpshooters usually on a fee basis, live trapping and relocation, and trapping and euthanasia. Fertility control has been applied on an experimental basis in localized areas and may only be applicable to populations of fewer than 100 females (Rudolph et al., 2000; Warren, 2000). In Ithaca, New York, a study is underway to evaluate a combination of culling and fertility control to reduce deer–human conflicts (P. Curtis, Cornell University, personal communication). Culling by sharpshooters is one of the more cost-effective methods of reducing deer density and has been applied by the National Park Service (Frost et al., 1997) and local communities (DeNicola and Williams, 2008). Archery hunting can also be an effective method of population reduction, but requires substantial planning and coordination to be successful (Kilpatrick and Walter, 1999; Kilpatrick et al., 2004). Trapping deer can be expensive and relocation of deer is generally not an alternative given the risk of spreading disease and reduced survival of translocated deer (Cromwell et al., 1999); consequently, this approach is generally discouraged by management agencies.

Successful deer management in urban areas will require community-based solutions in which citizens take the initiative in solving deer–human conflicts (Curtis and Hauber, 1997; Lund, 1997). Natural resource agencies responsible for managing deer will have to be perceived as a source for objective information for resolving problems and helping communities negotiate regulatory issues. Deer problems in urban areas will increase as human development continues to transform landscapes from rural to suburban habitats. However, providing information about solution alternatives can promote consensus among stakeholders with respect to the preferred alternative for reducing deer–human conflicts (Lauber and Knuth, 2004). Furthermore, research in New York has indicated that, in general, lethal methods of deer population control are socially acceptable (Loker et al., 1999).

High Fences

Many consider the use of high fences the epitome of privatization of wildlife in North America. Privatization issues related to hunting have become more apparent, especially in eastern North America, over the past 30 years. Of the 26 states and provinces in eastern North America that responded to a 2009 deer management survey, fencing of white-tailed deer was legal in all but five states, all of which were in the Northeast (Table 16.1). Pennsylvania was the only northeastern state to report numerous high-fence operations. The use of high fences to facilitate trophy management of whitetails has increased, especially in the southeastern states, where Texas, Mississippi, Alabama, Florida, and Kentucky have extensive high-fence operations. The milder weather of most southern states is similar to home climates of many species of exotic big game that are introduced to many high-fenced areas, which increases season length and revenue opportunity for commercial hunting operations.

Outside of the occasional escapes, the use of high fencing creates a closed population which advocates claim improves the effectiveness in controlling deer population levels, sex ratio, age structure, and genetics. The smaller the enclosed area, the more precise managers can manipulate these variables. High fences also control immigration and emigration of deer, and control access and illegal harvest. However, the Boone and Crockett Club and Pope and Young Club have determined that hunting behind high fences, irrespective of the acreage, violates fair chase and do not allow deer killed within enclosures to be eligible for their record books. The increase in the number and size of high-fence enclosures may also have detrimental impacts on other species of wildlife, especially some mega-fauna species. The number and juxtaposition of high-fence operations could potentially impede the natural movements and range of black bears and cougars in Florida.

State fish and wildlife agencies also must consider the effect of fencing operations on regulatory aspects of wildlife populations. High-fence operations have created challenges regarding the regulatory authority among state fish and wildlife and agriculture agencies. This issue also has heightened the debate on whether fish and wildlife resources are the property of the state or private landowner and further challenges the North American Conservation Model (Organ and Mahoney, 2007). The high-fence issue will continue to be the catalyst for continued debate of wildlife privatization issues during the next several decades.

Habitat Management

White-tailed deer are generalists that can use a variety of habitat types to secure adequate food, water, and cover. Although forest is almost always an important habitat component, contiguous stands of mature forest rarely provide all the necessary habitat types required by white-tailed deer (Miller, 2001). Therefore, manipulation of habitat via timber management, prescribed burning, and establishment of food plots are options for improving habitat quality and diversity.

Wintering Areas

In the northern range of white-tailed deer, the most important habitat management tool is protection and maintenance of deer wintering areas, commonly termed "deer yards." Wintering areas have reduced snow cover and warmer microclimates (warmer temperatures and reduced wind) than other habitat types. Consequently, these areas are used by the same deer each winter and annual movements to traditional wintering areas can be continued by offspring (Verme, 1973; Drolet, 1976). Characteristics of wintering areas that provide the greatest benefit to deer are conifer stands with >70% crown closure located in areas with southern exposures and with browse nearby (Gill, 1957; Weber et al., 1983).

The importance of protecting wintering areas was recognized in the 1950s and 1960s when unregulated logging destroyed many wintering areas (Boer, 1978). Since that time, legislation and agency regulations (Boer, 1978) protect wintering areas and identify specific guidelines for protection and maintenance of these areas (e.g., Vermont Department of Fish and Wildlife, 1999). In general, wintering areas should have 50–60% of the area in conifer. Timber harvest should occur in autumn and winter because logging operations make deer forage accessible (fallen tree tops and tree lichens) and skid trails improve deer mobility. Although logging during summer is recommended if soil scarification is needed to regenerate hemlock, spruce, and fir (Frank and Bjorkbom, 1973), in general, spruce and fir should be selectively cut and hemlock and cedar should be protected.

Although wintering areas can be long established, they are not permanent and change over time. Boer (1992) identified 99 wintering areas in New Brunswick in 1975 and found that 42 of these were vacant 13 years later. In addition, wintering areas <50 ha were more likely to be vacant. Hurst and Porter (2008) found that deer wintering in the Adirondacks of New York changed their movement behavior to include residential areas. Residential areas may provide advantages of reduced energetic costs and access to high-quality habitat interspersed with potential food sources. As such, wildlife managers may need to consider residential areas and adapt management strategies accordingly as well as consider increased deer–human conflicts.

Forest Management

Deer are generalists and can use a wide variety of habitats to obtain the necessary resources to survive and reproduce. Consequently, habitat management for deer in eastern North America is generally synonymous with forest management because much of the nonforested habitat is primarily land intensively managed by humans (e.g., for growing food crops). Forest management activities that benefit deer should provide a mix of forest ages (early successional, mid-successional, and mature forest) in both a spatial and a temporal context. Early successional habitat, which is composed of regenerating forest stands less than six to eight years old, provides food for deer in the form of woody browse, forbs, and soft mast. Mid-successional age and mature forests provide less browse and forbs, but produce more hard mast. Forest and timber harvest management programs can greatly influence the diversity, quantity, quality, and juxtaposition of deer habitat forage resources.

Timber harvest rotation cycles and harvest methods vary by forest type, but should be designed to ensure that no single age class of forest will dominate at any given point in the future. For example, research in Pennsylvania has indicated that in oak–hickory forests early successional habitat can support 23 deer/km^2 during winter, mid successional can support 2 deer/km^2, and mature forest can support 8 deer km^2 (Diefenbach et al., 1997). Mixed oak forests in Pennsylvania can be intensively managed on a 100-year rotation (Storm et al., 2003) and faster growing tree species, such as aspen and some conifers, can be harvested on a 30- or 40-year rotation.

Although natural forest regeneration following clear or shelterwood cuts is typical in northeastern North America and portions of the hardwood forests and natural pinelands in the Southeast, intensive silvicultural management activities and tree planting of even aged plantations are the norm in at least 15% of southern forests (Sheffield and Dickson, 1998). Over 31 million acres of pine–oak uplands, sand hills, and flatwoods of the Southeast are intensively managed for loblolly, slash, sand, and longleaf pine. Site preparation activities (chopping, bedding, herbicide application, and burning) and plantation age may cause temporal limitations or shifts in deer forage quantity, quality, and diversity, but limitations of plantation size and distribution ensures adequate juxtaposition of forage resources. However, these activities have greatly reduced the hard mast component of many southern forests (Miller, 2001). Thinning operations should be conducted as soon as economically feasible in order to increase sunlight penetration to forest floor which is critical to maintaining deer forage production. Conversion of much of the South's longleaf pine ecosystem to loblolly and sand pine plantations has reduced the carrying capacity, nutritional plane, and deer density in some areas. Many forestry agencies of the southeastern states have developed silviculture best management practices that provide methodologies to protect wildlife habitat quality and other natural resources (Florida Department of Agriculture and Consumer Service, 2008).

Most pine ecosystems of the south require periodic fire to maintain species diversity, habitat quality, and population levels of many plant and wildlife species (Stanturf et al., 2002). Prescribed burning of these ecosystems is the most cost-effective way to maintain habitat quality and nutrition for white-tailed deer. Burning frequency and timing are prescribed based on site-specific conditions and goals but typically a burning interval of three to five years is most effective at increasing browse production, palatability, and nutrition (Miller, 2001). Most states require permits to conduct prescribed burns, which must be executed by trained professionals.

Habitat Quality

There are many areas of poor- or suboptimal-quality deer habitat across the United States and Canada (Shea and Osborne, 1995). Many of these areas occur in eastern North America and include lower coastal plain, sandhills, and mountain regions of the Southeast and some mountain and coastal regions of the Northeast. Climatic conditions and habitat quality most likely contributed to the genetic variation in white-tailed deer populations, especially in the Southeastern United States where suboptimal range is more pronounced. The smaller body and antler size and lower reproductive performance of deer in this region compared to more temperate regions may be attributed to these factors. Poor soil fertility of shallow shale soils of mountain regions and deep sandy and acidic loam soils of coastal regions cause nutritive values of deer forage to be relatively low. The low nutritional plane of these regions has been reported to be a limiting factor in deer reproductive and growth potential (Short, 1975; Wood and Tanner, 1985; Petrick et al., 1994; Shea and Osborne, 1995).

It is difficult to identify when deer populations approach or exceed carrying capacity in poor-quality habitats (Shea and Osborne, 1995). Some herds in extremely low-quality habitat exhibit poor relationships between physical condition and reproductive performance and population density

(Osborne et al., 1992; Shea et al., 1992a; Petrick et al., 1994). These herds often demonstrate limited response to herd reduction; therefore, habitat management and food plots can benefit deer by compensating for nutritional deficiencies.

Food Plots

Food plots are areas seeded with agronomic plant species or native forages and treated with prescribed fire, fertilizer, and selective herbicides with the intent to provide quality forage to improve the physical condition of deer. Food plots have become a popular habitat management tool with public and private landowners, in part, because they increase the opportunity to view and harvest deer (Harper, 2006). However, whether food plots can actually increase deer physical condition, and whether they are economically feasible is questionable. Johnson et al. (1987) demonstrated that quality of forage on upland pine and pine–hardwood sites in the southern United States limited deer production and Edwards et al. (2004) found that over a 10-year economic planning horizon it was cost effective to produce high-quality natural deer forages on treated food plots in loblolly pine stands in north-central Mississippi.

However, Hehman and Fulbright (1997) found that food plots in Texas did not replace browsing on native species such that any artificial increase in carrying capacity due to food plots could increase browsing intensity on native plants. Johnson et al. (1987) questioned whether food plots could improve growth of free-ranging white-tailed deer. Furthermore, Johnson and Dancak (1993) failed to find improved body condition of deer on an area in Louisiana with an extensive system of food plots.

Consequently, food plots should be used only as a supplement to management of habitat and natural forages and not a substitute for them (Miller, 2001). In northeastern North America, the greatest period of nutritional stress is late winter when deer metabolism increases but vegetation is still dormant. In these regions, food plots can probably best benefit deer by providing cool-season forages that initiate spring growth sooner than other plant species (e.g., clover). In southeastern North America, both winter and late summer can be periods of nutritional stress. Food plots planted with warm-season species can provide quality forage during late summer because these plants grow better in hot, dry conditions. A variety of cool-season plant species as well as root and tuber crops (e.g., beets) can provide quality forage during winter.

Challenges for Deer Management in Eastern North America

We identified several interrelated issues of managing white-tailed deer in eastern North America that we believe will be continuing challenges for natural resource agencies responsible for this species. First, use of recreational hunting as the primary tool to manage deer populations may become more difficult in some locations either because of changes in numbers and attitudes of deer hunters or because of reduced effectiveness of recreational hunting in some landscapes. Hunter numbers continue to decline as a combination of factors such as reduced recruitment of youth, the aging hunter population, and a greater proportion of people living in urban and suburban areas. In addition, even in areas where the number of hunters is stable, deer populations may have increased such that hunters are unwilling or not interested in harvesting the number of deer needed to control the population through recreational hunting (Brown et al., 2000; Bhandari et al., 2006).

Second, privatization of deer and access to land for hunting will create more challenges. Changes in land ownership are resulting in reduced land for hunting, such as through loss of leased hunting property, but mostly through land development. Privatization of wildlife challenges the North American model of wildlife management and will likely limit the authority and options available to state and provincial governments for managing this resource for its citizens. Third, deer management in urban and suburban areas will require greater involvement of a wider array of stakeholders to reach consensus on decisions for deer population management. This approach is essential for success but will require more resources in both personnel and funds for natural resource agencies that have traditionally been funded by hunting license fees. Consequently, deer management in urban areas will not necessarily generate revenue but will be required to meet the agency's mission.

Fourth, we envision disease issues will be of greater concern in eastern North America. CWD is already endemic in West Virginia and spreading. Many states have huge numbers of cervids in hundreds of captive facilities that are relatively unregulated (Table 16.1), and transport of free-ranging deer killed by hunters is difficult to monitor and control. The consequence of this situation is that transport of deer and deer carcasses and escape of captive cervids increases the risk of disease spread.

Fortunately, the white-tailed deer is a resilient and adaptable species that is likely to persist despite any changes that humans make to the landscape. The challenge for natural resource agencies will be to manage this species to meet the goals and desires of society.

REFERENCES

Abrams, M. D. 1992. Fire and the development of oak forests. *BioScience* 42:346–353.

Adams, K. P., R. J. Hamilton, and M. D. Ross. 2009. *QDMA's Whitetail Report*. Bogart, GA: Quality Deer Management Association.

Alig, R. J., W. G. Hohenstein, B. C. Murray, and R. G. Haight. 1990. Area changes in timberland by ownership and forest type in the United States by region and state, 1952–1987, with projections to 2040. General Technical Report SE-64, Southeastern Forest Experiment Station, U.S. Forest Service. Asheville, NC.

Baker, L. A., R. J. Warren, D. R. Diefenbach, W. E. James. 2001. Prey selection by reintroduced bobcats (*Lynx rufus*) on Cumberland Island, Georgia. *American Midland Naturalist* 145:80–93.

Beguin, J., D. Pothier, and M. Prévost. 2009a. Can the impact of deer browsing on tree regeneration be mitigated by shelterwood cutting and strip clearcutting? *Forest Ecology and Management* 257:38–45.

Beguin, J., M. Prévost, D. Pothier, and S. D. Côté. 2009b. Establishment of natural regeneration under severe browsing pressure from white-tailed deer after group seed-tree cutting with scarification on Anticosti Island. *Canadian Journal of Forest Research* 39:596–605.

Bélanger, G., P. Rochette, Y. Castonguay, A. Bootsmab, D. Mongrain, and D. A. J. Ryan. 2002. Climate change and winter survival of perennial forage crops in eastern Canada. *Agronomy Journal* 94:1120–1130.

Bhandari, P., R. C. Stedman, A. E. Luloff, J. C. Finley, and D. R. Diefenbach. 2006. Effort versus motivation: Factors affecting antlered and antlerless deer harvest success in Pennsylvania. *Human Dimensions of Wildlife* 11:423–436.

Bishop, P., J. Glidden, M. Lowery, and D. Riehlman. 1999. *A Citizen's Guide to the Management of White-tailed Deer in Urban and Suburban New York*. Albany, NY: New York Department of Environmental Conservation.

Block, N. E. and V. A. Sample. 2001. *Industrial Timberland Divestitures and Investments: Opportunities and Challenges in Forestland Conservation*. Washington, DC: Pinchot Institute for Conservation.

Boer, A. 1978. Management of deer wintering areas in New Brunswick. *Wildlife Society Bulletin* 6:200–205.

Boer, A. H. 1992. Transience of deer wintering areas. *Canadian Journal of Forest Research* 22:1421–1423.

Bourque, A. and G. Simonet. 2008. Quebec. In *Impacts to Adaptation: Canada in a Changing Climate*, eds. D. S. Lemmen, F. J. Warren, J. Lacroix, and E. Bush, 171–226. Ottawa, Canada: Government of Canada.

Bowyer, R. T., V. Van Ballenberghe, and J. G. Kie. 1998. Timing and synchrony of parturition in Alaskan moose: Long-term versus proximal effects of climate. *Journal of Mammalogy* 79:1332–1344.

Brown, T. L., D. J. Decker, S. J. Riley et al. 2000. The future of hunting as a mechanism to control white-tailed deer populations. *Wildlife Society Bulletin* 28:797–807.

Casabon, C. and D. Pothier. 2008. Impact of deer browsing on plant communities in cutover sites on Anticosti Island. *Écoscience* 15:389–397.

Conover, M. R. 1995. What is the urban deer problem and where did it come from? In *Urban Deer: A Manageable Resource? Proceedings of the Symposium of the North Central Section*, ed. J. B. McAnich, 11–18. The Wildlife Society.

Cook, C. and B. Gray. 2003. *Biology and Management of White-tailed Deer in Alabama*. Montgomery, AL: Alabama Department of Conservation and Natural Resources.

Côté, S. D., T. P. Rooney, J.-P. Tremblay, C. Dussault, and D. M. Waller. 2004. Ecological impacts of deer overabundance. *Annual Review of Ecology, Evolution, and Systematics* 35:113–147.

Cromwell, J. A., R. J. Warren, and D. W. Henderson. 1999. Live-capture and small-scale relocation of urban deer on Hilton Head Island, South Carolina. *Wildlife Society Bulletin* 27:1025–1031.

Curtis, P. D. and J. R. Hauber. 1997. Public involvement in deer management decisions: Consensus versus content. *Wildlife Society Bulletin* 25:399–403.

DeNicola, A. J. and S. C. Williams. 2008. Sharpshooting suburban white-tailed deer reduces deer–vehicle collisions. *Human–Wildlife Conflicts* 2:28–33.

DeNicola A. J., K. C. VerCauteren, P. D. Curtis, and S. E. Hyngstrom. 2000. *Managing White-tailed Deer in Suburban Environments*. Ithaca, NY: Cornell Cooperative Extension.

Diefenbach, D. R., L. A. Baker, W. E. James, R. J. Warren, and M. J. Conroy. 1993. Reintroducing bobcats to Cumberland Island, Georgia. *Restoration Ecology* 1:241–247.

Diefenbach, D. R., L. A. Hansen, R. J. Warren, M. J. Conroy, and M. G. Nelms. 2009. Restoration of bobcats to Cumberland Island, Georgia, USA: Lessons learned and evidence for the role of bobcats as keystone predators. In *Iberian Lynx ex situ Conservation: An Interdisciplinary Approach*, eds. A. Vargas, C. Breitenmoser, and U. Breitenmoser, 423–435. Madrid, Spain: Fundación Biodiversidad.

Diefenbach, D. R., W. R. Palmer, and W. K. Shope. 1997. Attitudes of Pennsylvania sportsmen towards managing white-tailed deer to protect the ecological integrity of forests. *Wildlife Society Bulletin* 25:244–251.

Downing, R. L. 1980. Vital statistics of animal populations. In *Wildlife Management Techniques Manual* (4th edition), ed. S. D. Schemnitz, 247–267. Washington, DC: The Wildlife Society.

Drolet, C. A. 1976. Distribution of movements of white-tailed deer in southern New Brunswick in relation to environmental factors. *Canadian Field Naturalist* 90:123–126.

Eberhardt, L. L. 1960. *Estimation of Vital Characteristics of Michigan Deer Herds*. Michigan Department of Conservation Game Division Report 2282. Lansing, MI.

Edwards, S. L., S. Demarais, B. Watkins, and B. K. Strickland. 2004. White-tailed deer forage production in managed and unmanaged pine stands and summer food plots in Mississippi. *Wildlife Society Bulletin* 32:739–745.

Florida Department of Agriculture and Consumer Service. 2008. *Silviculture Best Management Practices*. Tallahassee, FL: Florida Department of Agriculture and Consumer Services.

Frank, R. M. and J. C. Bjorkbomn. 1973. *A Silvicultural Guide for Spruce–Fir in the Northeast*. General Technical Report NE-6, Northeastern Forest Experiment Station, U.S. Forest Service. Upper Darby, PA.

Frost, H. C., G. L. Storm, M. J. Batcheller, and M. J. Lovallo. 1997. White-tailed deer management at Gettysburg National Military Park and Eisenhower National Historic Site. *Wildlife Society Bulletin* 25:462–469.

Gill, J. D. 1957. Review of deeryard management 1956. Game Division Bulletin 5, Maine Department of Inland Fish and Game,. Orono, ME.

Gray, W. N., S. S. Ditchkoff, M. K. Causey, and C. W. Cook. 2002. The yearling disadvantage in Alabama deer: Effect of birth date on development. *Proceedings Annual Conference of the Southeastearn Association Fish and Wildlife Agencies* 56:255–264.

Halls, L. K., ed. 1984. *White-tailed Deer: Ecology and Management*. Harrisburg, PA: Stackpole Books.

Harlow, R. F. and F. K. Jones Jr., eds. 1965. The white-tailed deer in Florida. Technical Bulletin 9, Florida Game and Fresh Water Fish Commission. Tallahassee, FL.

Harper, C. A. 2006. Initial considerations in developing a food plot program. In *Quality Food Plots*, eds. K. Kammermeyer, K. V. Miller, and L. Thomas, 30–45. Bogart, GA: Quality Deer Management Association.

Harveson P. M., R. R. Lopez, N. J. Silvy, and P. A. Frank. 2004. Source–sink dynamics of Florida key deer on Big Pine Key, Florida. *Journal of Wildlife Management* 68:909–915.

Hehman, M. W. and T. E. Fulbright. 1997. Use of warm-season food plots by white-tailed deer. *Journal of Wildlife Management* 61:1108–1115.

Horsley, S. B., S. L. Stout, and D. S. deCalesta. 2003. White-tailed deer impact on the vegetation dynamics of a northern hardwood forest. *Ecological Applications* 13:98–118.

Hurst, J. E. and W. F. Porter. 2008. Evaluation of shifts in white-tailed deer winter yards in the Adirondack region of New York. *Journal of Wildlife Management* 72:367–375.

Intergovernmental Panel on Climate Change. 2007. *Climate Change 2007: Synthesis Report*. http://www.ipcc.ch/pdf/assessment-report/ar4/syr/ar4_syr.pdf. Accessed July 24, 2009.

Jenkins, K. J. and R. G. Wright. 1987. Dietary niche relationships among cervids relative to winter snowpack in northwestern Montana. *Canadian Journal of Zoology* 65:1397–1401.

Johnson, M. K. and K. D. Dancak. 1993. Effects of food plots on white-tailed deer in Kisatchie National Forest. *Journal of Range Management* 46:110–114.

Johnson, M. K., B. W. Delany, S. P. Lynch et al. 1987. Effects of cool-season agronomic forages on white-tailed deer. *Wildlife Society Bulletin* 15:330–339.

Kilpatrick, H. J. and A. M. LaBonte. 2003. Deer hunting in a residential community: The community's perspective. *Wildlife Society Bulletin* 31:340–348.

Kilpatrick, H. J. and W. D. Walter. 1997. Urban deer management: A community vote. *Wildlife Society Bulletin* 26:388–391.

Kilpatrick, H. J. and W. D. Walter. 1999. A controlled archery deer hunt in a residential community: Cost, effectiveness, and deer recovery rates. *Wildlife Society Bulletin* 27:115–123.

Kilpatrick, H. J., A. M. LaBonte, J. S. Barclay, and G. Warner. 2004. Assessing strategies to improve bowhunting as an urban deer management tool. *Wildlife Society Bulletin* 32:1177–1184.

Lang, L. M. and G. W. Wood. 1976. Manipulation of the Pennsylvania deer herd. *Wildlife Society Bulletin* 4:159–165.

Latham, R. E., J. Beyea, M. Benner et al. 2005. *Managing Deer from an Ecosystem Perspective: Pennsylvania Case Study*. Harrisburg, PA: Deer Management Forum for Audubon Pennsylvania and Pennsylvania Habitat Alliance.

Lauber, T. B. and B. A. Knuth. 2004. Effects of information on attitudes toward suburban deer management. *Wildlife Society Bulletin* 32:322–331.

Lieske, S., J. E. Hoeldtke, and S. P. Bratton. 1990. *Live Oak Regrowth Monitoring Data Base*. Technical Report 57, National Park Service, Cooperative Park Studies Unit, University of Georgia. Athens, GA.

Loker, C. A., D. J. Decker, and S. J. Schwager. 1999. Social acceptability of wildlife management actions in suburban areas: 3 cases from New York. *Wildlife Society Bulletin* 27:152–159.

Lopez, R. R. 2001. Population ecology of Florida Key deer. PhD dissertation, Texas A&M University.

Lopez, R. R., N. J. Silvy, B. L. Pierce, P. A. Frank, M. T. Wilson, and K. M. Burke. 2004. Population density of the endangered Florida key deer. *Journal of Wildlife Management* 68:570–575.

Lund, R. C. 1997. A cooperative, community-based approach for the management of suburban deer populations. *Wildlife Society Bulletin* 25:488–490.

Messmer, T. A., L. Cornicelli, D. J. Decker, and D. G. Hewitt. 1997. Stakeholder acceptance of urban deer management techniques. *Wildlife Society Bulletin* 25:360–366.

Miller, K. V. 2001. White-tailed deer. In *Wildlife of Southern Forests*, ed. J. G. Dickson, 95–107. Blaine, WA: Hancock House Publishers.

Miller, K. V. 2002. A year in the life of the white-tailed deer: What times the rut? *Georgia Outdoor News* December:120–121.

Miller, K. V., L. I. Muller, and S. Demarais. 2003. White-tailed deer (*Odocoileus virginianus*). In *Wild Mammals of North America: Biology, Management, and Conservation* (2nd edition), eds. G. A. Feldhamer, B. C. Thompson, and J. A. Chapman, 906–930. Baltimore, MD: Johns Hopkins University Press.

Millspaugh, J. J., J. R. Skalski, R. L. Townsend et al. 2009. An evaluation of sex–age–kill (SAK) model performance. *Journal of Wildlife Management* 73:442–451.

Moriarty, J. 2006. Distribution, abundance, and management considerations of resident Canada geese and urban white-tailed deer. In *Urban Wildlife Management*, eds. C. E. Adams, K. J. Lindsey, and S. J. Ash, 287–303. Boca Raton, FL: CRC Press.

Mozumder, P., C. M. Starbuck, R. P. Berrens, and S. Alexander. 2007. Lease and fee hunting on private lands in the U.S.: A review of the economic and legal issues. *Human Dimensions of Wildlife* 12:1–14.

Nelms, M. G. 1999. Deer herd trends, bobcat food habits, and vegetation change on Cumberland Island, Georgia following bobcat reintroduction. MS thesis, University of Georgia.

Organ J. and S. Mahoney. 2007. The future of public trust: The legal status of the Public Trust Doctrine. *The Wildlife Professional* 1:18–22.

Osborne, J. S., A. S. Johnson, P. E. Hale, R. L. Marchinton, C. V. Vansant, and J. M. Wentworth. 1992. Population ecology of the Blackbeard Island white-tailed deer. Tall Timbers Research Station Bulletin 26. Tallahassee, FL.

Petrick, C. J., R. E. Vanderhoof, and S. M. Shea. 1994. Relationship of *in utero* productivity to population indices of white-tailed deer in the Florida sandhills. *Annual Meeting of the Southeast Deer Study Group* 17:30.

Post, E. and N. C. Stenseth. 1999. Climatic variability, plant phenology, and northern ungulates. *Ecology* 80:1322–1339.

Rand, P. W., C. Lubelczyk, M. S. Holman, E. H. Lacombe, and R. P. Smith, Jr. 2004. Abundance of *Ixodes scapularis* (Acari: Ixodidae) after the complete removal of deer from an isolated offshore island, endemic for Lyme disease. *Journal of Medical Entomology* 44:779–784.

Rudolph, B. A., W. F. Porter, and H. B. Underwood. 2000. Evaluating immunocontraception for managing suburban white-tailed deer in Irondequoit, New York. *Journal of Wildlife Management* 64:463–473.

Samuel, M. D., D. O. Joly, M. A. Wild et al. 2002. Surveillance strategies for detecting chronic wasting disease in free-ranging deer and elk. U.S. Geological Survey, National Wildlife Health Center. Madison, WI.

Sauvé, D. G. and S. D. Côté. 2007. Winter forage selection in white-tailed deer at high density: Balsam fir is the best of a bad choice. *Journal of Wildlife Management* 71: 911–914.

Sharpe, W. J. and J. R. Drohan, editors. 1999. *The Effects of Acidic Deposition on Pennsylvania's Forests.* University Park, PA: Environmental Resources Research Institute, The Pennsylvania State University.

Shea, S. M. and J. S. Osborne. 1995. Poor quality habitats. In *Quality Whitetails*, eds. K. V. Miller and R. L. Marchinton, 193–209. Mechanicsburg, PA: Stackpole Books.

Shea, S. M., T. A. Breaut, and M. L. Richardson. 1992a. Relationship of birth date and physical development of yearling white-tailed deer in Florida. *Proceedings Annual Conference of the Southeastearn Association Fish and Wildlife Agencies* 46:159–166.

Shea, S. M., T. A. Breault, and M. L. Richardson. 1992b. Herd density and physical condition of white-tailed deer in Florida Flatwoods. *Journal of Wildlife Management* 56:262–267.

Sheffield, R. M. and J. G. Dickson. 1998. The South's forest land on the hot seat to provide more. *North American Wildlife and Natural Resources Conference* 63:316–331.

Shissler, B. and M. Grund. 2009. *Managing Deer in the Commonwealth: A Study of Pennsylvania and Other States.* Washington, DC: Pinchot Institute for Conservation.

Short, H. L. 1975. Nutrition of southern deer in different seasons. *Journal of Wildlife Management* 39:321–329.

Stafford, K. C., III, A. J. DeNicola, and H. J. Kilpatrick. 2003. Reduced abundance of *Ixodes scapularis* (Acari: Ixodidae) and the tick parasitoid *Ixodiphagus hookeri* (Hymenoptera: Encyrtidae) with reduction of white-tailed deer. *Journal of Medical Entomology* 40:642–652.

Stanturf, J. A., D. D. Wade, T. A. Waldrop, D. K. Kennard, and G. L. Achtemeier. 2002. Background paper: Fire in southern forest landscapes. In *Southern Forest Resource Assessment*, eds. D. N. Wear and J. G. Greis, 607–622. Asheville: U.S. Forest Service, Southern Research Station.

Storm, G. L., W. L. Palmer, and D. R. Diefenbach. 2003. Ruffed grouse response to management of mixed oak and aspen communities in central Pennsylvania. Pennsylvania Game Commission Grouse Research Bulletin Number 1. Harrisburg, PA.

U.S. Census Bureau. 2000. Population change and distribution 1990–2000. http://www.census.gov/main/www/cen2000.html. Accessed May 25, 2010.

U.S. Fish and Wildlife Service. 2006. *2006 National Survey of Fishing, Hunting, and Wildlife-Associated Recreation.* Washington, DC: U.S. Fish and Wildlife Service.

Verme, L. J. 1973. Movements of white-tailed deer in upper Michigan. *Journal of Wildlife Management* 37:545–552.

Vermont Department of Fish and Wildlife. 1999. *Guidelines for the Review and Mitigation of Impacts to White-tailed Deer Winter Habitat in Vermont.* Waterbury, VT: Agency of Natural Resources.

Warren, R. J. 2000. Overview of fertility control in urban deer management. In *Proceedings of the 2000 Annual Conference of the Society for Theriogenology*, 237–246. Nashville, TN: Society for Theriogenology.

Weber, S. J., W. W. Mautz, J. W. Lanier, and J. E. Wiley, III. 1983. Predictive equations for deeryards in northern New Hampshire. *Wildlife Society Bulletin* 11:331–338.

West Virginia Department of Natural Resources. 2010. Chronic wasting disease. http://www.wvdnr.gov/hunting/ChronicWaste.shtm. Accessed May 25, 2010.

Williams, W. J., S. E. Wardroup, and M. S. Traweek. 1995. *White-tailed Deer Breeding Chronology and Reproduction.* Texas Parks and Wildlife Department project number 95: Federal Aid Project number W-127-R-3. Austin, TX.

Williamson, S. J. 2003. White-tailed deer harvest management and goal setting in the northeast. A Report to the Northeast Deer Technical Committee and the Northeast Wildlife Administrators Association, Wildlife Management Institute. Washington, DC.

Wood, J. M. and G. W. Tanner. 1985. Browse quality response to forest fertilization and soils in Florida. *Journal of Range Management* 38:432–435.

17

Managing White-tailed Deer: Midwest North America

Kurt C. VerCauteren and Scott E. Hygnstrom

CONTENTS

It easily can be said that the white-tailed deer is the most important species of wildlife in the midwestern United States. Approximately 10 million white-tailed deer inhabit the Midwest, which is more than any other region in North America. Further, many of the most impressive whitetails have come from the Midwest (Figure 17.1). Dramatic changes have occurred on the landscape since European settlement that influenced populations of white-tailed deer in this region. Timber harvest in the north and east and agricultural tillage and residential development throughout the region have had both positive and negative impacts on deer numbers. White-tailed deer were over-harvested through unregulated subsistence and market hunting in the 1800s and were nearly extirpated from many areas in the Midwest (Gladfelter, 1984; Menzel, 1984). Populations rebounded over time, however, to current record highs, due to an interested

FIGURE 17.1 (a) Milo Hanson World's Record typical white-tailed deer, 213–6/8 points, Biggar, Saskatchewan. (b) James Jordan. Number 2 typical white-tailed deer, 206–1/8 points from Burnett County, Wisconsin. (c) World's Record nontypical, 333 7/8 points, St. Louis County, Missouri. (d) Tony Lovstuen. Number 3 nontypical, 307 5/8 points from Monroe County, Iowa. (Images courtesy of Boone and Crockett Club [www.booneandcrockettclub.com]. With permission.)

public and the diligent efforts of management agency personnel. White-tailed deer have also expanded their range westward and now occupy areas of the Great Plains that they never have before.

Public interests in whitetails are high, as over 4 million deer hunters and untold numbers of deer watchers take to the field every year in the Midwest. An entire culture has grown out of sport hunting of white-tailed deer, especially in the upper Midwest where it is not uncommon for schools to be closed on opening day of the firearm deer season. Traditional meat poles that were once the center of attention every autumn in towns across the upper Midwest still exist and are used in some communities. Deer camps, where generations of hunters congregate every autumn, are still common across the North Woods (Willging, 2008), carrying on the hunting tradition and keeping the interest in wildlife, the outdoors, and our natural world alive. Public concerns also are high, thus crop damage and deer–vehicle collisions (DVCs) are major drivers in the management of local deer numbers. Managers are also concerned about the impacts of high populations of white-tailed deer on plant communities, ecological succession, and forest regeneration. The times for those interested in white-tailed deer in the Midwest are better than ever and the outlook is good, but even greater diligence and effort will be required to properly manage populations of white-tailed deer in the future.

Climatic and Geographic Description of the Region

Climate is the most significant over-riding environmental factor that affects the distribution and population levels of white-tailed deer in North America. North–south and east–west gradients exist in the Midwest for solar radiation, temperature, and moisture. The Midwestern range of white-tailed deer in North America occurs across a diverse array of subregions that includes all or parts of 14 states and three provinces (Figure 17.2). Low temperatures and deep snow in the north and western alpine regions limit the distribution of white-tailed deer. Highly adverse weather conditions such as drought, tornadoes, and rain-associated floods may have direct impacts on local populations of white-tailed deer. Climate in the Midwest also affects the distribution, density, and physio-chemical features of plants and plant communities which in turn, directly affect deer. Predictions of impending climate change could impact plant communities and agricultural production in the future, which will also directly affect deer in the Midwest.

Northern Plains Subregion

The Northern Plains Subregion is the northwestern extent of white-tailed deer range in the Midwest (Figure 17.2). Elevations range from 200 m on the rolling plains to about 1000 m in southwestern South Dakota (Petersen, 1984). Wooded draws, lowlands, and floodplains are preferred habitats of white-tailed

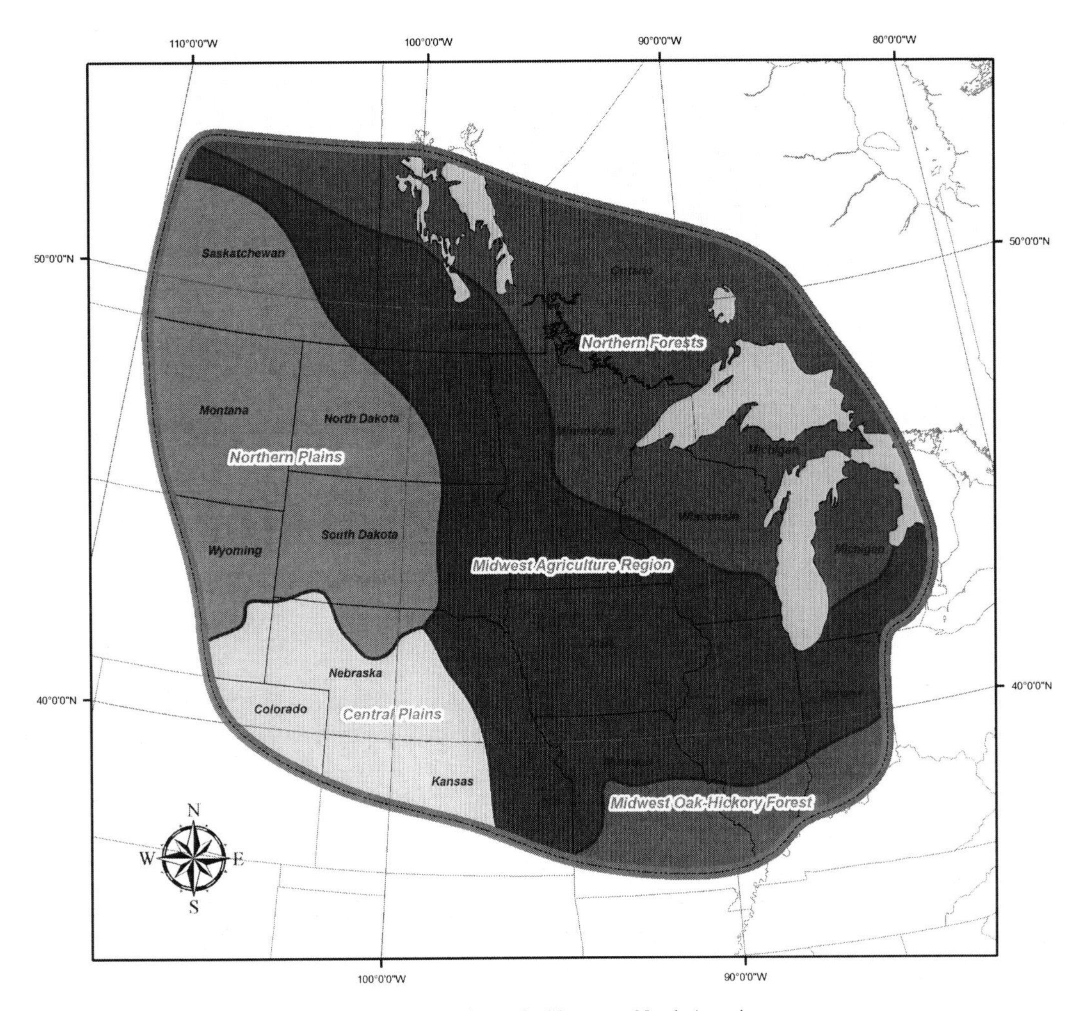

FIGURE 17.2 Geographic boundaries and ecoregions of midwestern North America.

deer (Petersen, 1984). Although less prevalent, mountainous terrain is found in parts of eastern Wyoming and southwestern South Dakota. White-tailed deer range overlaps with mule deer and elk in many locations across the Northern Plains. Common trees in the lowlands include green ash, American elm, and boxelder with Rocky Mountain juniper and ponderosa pine in draws and uplands (Petersen, 1984). Shrubland vegetated with western snowberry, silver buffaloberry, and common chokecherry provides valuable year-round cover and food (Petersen, 1984). Cattle ranching is the most prevalent land use in the region though dry-land and irrigated agriculture has increased in recent decades. Primary crops include alfalfa, winter wheat, corn, and sunflowers. Coal and uranium mining are relatively common and energy development including oil, gas, hydro, and wind are becoming more prominent across the landscape.

Average winter temperatures in the far northwest range from –20–9°C in January to 9–23°C in July (Saskatoon, Saskatchewan; Pearce and Smith, 1984). Average winter temperatures in the far southeast range from –13–0°C in January to 19–59°C in July (Rapid City, South Dakota) with a growing season of 90–240 days/year (Harlow and Guynn, 1994). The subregion receives 25–38 cm of rainfall annually in the northwest to 46–58 cm in the southeast (Harlow and Guynn, 1994).

Midwest Agricultural Subregion

A large central section of midwestern white-tailed deer range consists of the Midwest Agricultural Subregion (Figure 17.2), which covers much of what once made up the mixed- and tall-grass prairie ecosystems of the United States. The prairie ecosystem contributed to deep and fertile soils that enable the agricultural productivity seen today. Agricultural crops are nearly unlimited and make up the majority of the diets of white-tailed deer in the Midwest Agricultural Subregion (Gladfelter, 1984; Nixon et al., 1991). Primary crops produced include corn and soybeans although winter wheat, sorghum, alfalfa, and oats are also grown. Most land is privately owned with over 75% currently in crop production. Permanent cover is extensively fragmented and deer must adapt to the dramatic seasonal changes in available cover and food associated with harvest of crops (Nixon et al., 1991). Riparian woodlands and wooded uplands provide year-round cover that is especially vital during autumn and winter after crop harvest (Figure 17.3). Common trees species include red and white oak, sugar maple, American

FIGURE 17.3 Highly fragmented landscapes in the Midwest provide excellent habitat for white-tailed deer, especially along riparian areas where blocks of woodland and adjacent cropfields are sources of cover and food. (Photo by G. Clements. With permission.)

beech, paper birch, American elm, green ash, and cottonwood. The topography is low, flat, and gently rolling.

Annual rainfall within the Midwest Agricultural Subregion averages 64 cm in the west to 152 cm in the east. Snow cover is rarely substantial enough to present detrimental effects on winter survival as can occur in northern parts of the Midwest. Average northern temperatures in Winnipeg, Manitoba range from −25–14°C in January to 13–26°C in July (Pearce and Smith, 1984). Farther south in Kansas City, Missouri, temperatures range from −6–3°C in January to 21–32°C in July (Pearce and Smith, 1984) and an average of 160 growing days/year occur (Harlow and Guynn, 1994).

Northern Forests/Great Lakes Subregion

The northeastern reaches of white-tailed deer range in the Midwest comprise the Northern Forests/Great Lakes Subregion (Figure 17.2). Topography is generally flat to rolling and hilly with elevations from near sea level to 350 m. The majority of the subregion consists of managed deciduous and coniferous forests. Farmland makes up less than 25% of the land area. Predominant tree species include quaking aspen, paper birch, red maple, jack pine, black and white spruce, northern white cedar, balsam fir, American beech, and eastern hemlock. Important deer foods include northern white cedar, red maple, hemlock, American mountain ash, alternate leaf dogwood, yellow birch, mountain maple, and serviceberry (Harlow and Guynn, 1994).

About a third of the land is publicly owned and managed for timber production and recreation. Privately owned lands such as hunt clubs, weekend homes, and vacation resorts are also common. In the mid-1930s hunt clubs began purchasing inexpensive marginal lands. One particular hunt club formed in Michigan in 1938 encompassed >11,000 contiguous hectares and approximately 250 other such clubs originated thereafter (O'Brien et al., 2006). Privately owned captive cervid facilities are also common in this subregion with white-tailed deer being the most common species raised. Mining of iron, copper, and nickel is relatively common, yet involves a small percentage of the landmass (Blouch, 1994).

Precipitation ranges from 64 to 81 cm of rain and 127 to 508 cm of snow (Harlow and Guynn, 1994). Average temperatures range from −14–0°C in January to 17–28°C in July with a growing season of 80–135 days/year. Extreme winter conditions define the northern extent of white-tailed deer range. Seasonal movements to yarding areas where thermal cover exists are common in response primarily to deep snow and secondarily to cold temperatures (Dahlberg and Guettinger, 1956). The extent and quality of winter range is the main limiting factor for deer populations in this subregion (Dahlberg and Guettinger, 1956). Gray wolves have become more common here and contribute substantially to deer mortality (Nelson and Mech, 1986a), particularly during periods when snow inhibits mobility of deer (Nelson and Mech, 1986b).

Midwestern Oak and Hickory Forest Subregion

The southeastern reaches of white-tailed deer range in the Midwest are made up of the Midwestern Oak and Hickory Forest Subregion (Figure 17.2). Topography is largely unglaciated, rolling, and hilly with elevations from near sea level along the Mississippi River to over 600 m in moderately rugged mountains (Torgerson and Porath, 1984). Deciduous forests once covered much of this subregion; however, millions of hectares have been cleared for a variety of purposes (Torgerson and Porath, 1984). Cropland is relatively common with corn, soybeans, and wheat serving as high-quality sources of food for deer. Common tree species include hickory, oak, gum, cypress, and pine. In addition to agricultural crops, other important foods for deer include dogwood, redbud, serviceberry, sumac, blueberry, strawberry bush, elderberry, spice bush, farkleberry, black haw, deciduous holly, yaupon, and oak mast (Harlow and Guynn, 1994).

Weather itself is not a limiting factor for deer in this subregion although it can impact the abundance of foods such as acorns, which deer may rely upon for building energy reserves to sustain themselves through winter (Torgerson and Porath, 1984). Precipitation ranges from 89 to 127 cm annually (Harlow and Guynn, 1994). Average temperatures range from −4–3°C during January to 24–27°C in July with a growing season of 170–220 days/year.

Central and Southern Plains Subregion

The southwestern extent of white-tailed deer range in the Midwest is made up of the Central and Southern Plains Subregion (Figure 17.2). Topography is generally flat to rolling and hilly with elevations from near sea level to over 1800 m. Much of this subregion was once under short- to mixed-grass prairie with occasional riparian corridors providing habitat for white-tailed deer. Rugged terrain of limestone escarpments with dramatic geological formations are forested with ponderosa pine and eastern red cedar providing diversity throughout much of this subregion where mule deer, elk, and white-tailed deer ranges overlap. Although much of this landscape is grazed by livestock, a considerable amount has been converted to the production of crops such as corn and alfalfa through the use of center-pivot irrigation (Menzel, 1984). Common tree species in wooded riparian areas include cottonwood, green ash, burr oak, and eastern red cedar. Important foods for deer include snowberry, rose, grape, western soapberry, cottonwood, juniper, scrub oak, and agricultural crops (Harlow and Guynn, 1994).

Precipitation ranges from 28 to 89 cm of rain (Harlow and Guynn, 1994). Average temperatures range from −7–7°C in January to 18–29°C in July with a growing season of 120–220 days/year. Adverse weather conditions rarely occur long enough to significantly impact deer survival.

White-tailed deer have steadily expanded their range westward in recent history. Human manipulation and control of rivers in the Great Plains has enabled permanent wooded cover to become established along riparian corridors and center-pivot irrigation has allowed for crop production, especially corn. These two human-induced factors are the primary reasons whitetails extended their range westward and now thrive in many areas of the Great Plains.

Deer Impacts in the Midwest

The popularity of hunting and viewing white-tailed deer has created year-round entertainment for participants and an entire industry in the Midwest. In the past 25 years, the number of deer hunters has increased in the region and they clearly enjoy spending their leisure time and money on activities related to deer. Hunter participation within the Midwest (6–12%) exceeds the national average (5%) (U.S. Fish and Wildlife Service, 2006). Besides pursuing deer during hunting seasons, many people spotlight deer in summer and collect shed antlers in late winter. In spring, some put out mineral blocks or plant foodplots for deer and they tend these plots throughout the growing season. With the advent of trail cameras, scouting for deer has become a year-round endeavor for many whitetail enthusiasts.

Willingness of hunters to invest money to harvest deer has also been a boon to the agricultural community and local economies. Farmers and ranchers earn extra income from their land by leasing hunting rights. They are also selling agricultural products like corn and root vegetables to be used for baiting and feeding deer (Dorn and Mertig, 2005; O'Brien et al., 2006), though laws associated with these activities are tightening. In 2006 19.2 million Americans fed wildlife (other than birds) and 9.6 million maintained plantings for wildlife (U.S. Fish and Wildlife Service, 2006). In Michigan, prior to the bovine tuberculosis (TB)-related ban on feeding deer, over 50% of survey respondents fed deer (Dorn and Mertig, 2005) and it was a $50 million market in Michigan in the early 1990s (Winterstein et al., 1995). Some people provide feed to increase the opportunity to simply view wildlife while others provide supplemental food to decrease the potential for starvation during harsh winters. Others provide feed (bait) strategically to increase their hunting success or in an attempt to keep deer from leaving a property to seek food sources elsewhere (Garner, 2001; O'Brien et al., 2006). Prior to the bovine TB-related deer baiting ban in Michigan, 72% of nonresidents and 87% of residents hunted deer over bait (Dorn and Mertig, 2005). Feeding sites in northern Michigan have been observed to contain tons of food items such as corn, root vegetables, and hay (O'Brien et al., 2002). The occurrence of chronic wasting disease and bovine TB in several states that once allowed baiting and feeding (i.e., Minnesota, Wisconsin, Michigan) has prompted bans and restrictions on placement of food for deer to minimize potential for disease transmission (O'Brien et al., 2006). In the Midwest, hunting directly over bait is unregulated in just Kansas and Ontario and some form of regulated baiting is allowed in five states and provinces (Michigan, North Dakota, Wisconsin, Wyoming, and Saskatchewan) while it is completely banned in the remaining 10 states and provinces.

Several other economic benefits are associated with deer and deer hunting. Four of the largest companies that deal in retail and catalogue sales of outdoor equipment (Bass Pro Shops Incorporated, Cabella's Incorporated, Gander Mountain Company, and Scheels All Sports) are based in the Midwest. Combined, they account for over $8 billion/year in retail sales and 40,000 employees (Answers.com, 2010). Whitetails Unlimited is a Wisconsin-based organization established in 1982 with over 90,000 members across North America that has spent over $50 million on educational programs, habitat conservation, and preservation of the hunting tradition to benefit white-tailed deer and other wildlife (www.whitetailsunlimited.com). *Deer and Deer Hunting* magazine was established in Wisconsin 1977 to provide practical and comprehensive information to hunters of white-tailed deer, especially in the upper Midwest. Current circulation is about 120,000 and the organization has branched out to include a television show, website, videos, books, and other information (www.deeranddeerhunting.com). Another growing and thriving venture made possible by the importance of deer to hunters is outfitting. Outfitters and guides service hunters by gaining access to land, setting up stands, providing food and lodging, and taking care of the needs of their hunting clientele, who are willing to pay for the service and the opportunity to hunt in prime areas without the time and effort associated with finding and setting up their own hunting grounds. In some areas, the presence of deer increases the value of land considerably. Once categorized as "agricultural land," such sites are now marketed to hunters at much higher prices as "recreational property."

The popularity of white-tailed deer across the Midwest is not limited to those deer that are free ranging. The captive deer industry also has grown substantially in the last 25 years in the Midwest. Over 7800 facilities occur across the nation, of which roughly 40% are in midwestern states (Frosch et al., 2008). The market is driven primarily by the demand for opportunities to harvest large bucks, albeit behind high fences. Captive facilities range from small breeding operations of just a few hectares to large "hunt" properties of several thousand hectares. In addition to income derived from harvest of deer, the captive industry also profits from selling breeding stock, antlers, antler velvet, urine, photographic opportunities, venison, and other deer-derived products and deer-related activities. It is not legal to sell venison from deer harvested in the wild in the United States, but marketing of commercially produced venison is legal. Sale and consumption of commercially produced venison in the United States, though, is so low that it is not tracked by the U.S. Department of Agriculture.

While many positive attributes are associated with white-tailed deer, many negative impacts occur as well (Chapter 13). Deer cause more agricultural damage than any other species of wildlife in the United States (Conover, 2002). In the Midwest, they are responsible for about a third of the damage caused to corn by wildlife, which amounted to over $30 million in 1993 (Wywialowski, 1996) (Figure 17.4). Some states and provinces in the region provide support to agricultural producers that experience damage from deer. Aid may be compensation for crops damaged (Wagner et al., 1997), provision of damage abatement tools (i.e., frightening devices, fence), or advice on how to alleviate damage (Craven

FIGURE 17.4 Extensive deer damage in cornfields in late-June leads to significant economic impacts for some midwestern farmers. Note that the plants should be over head-high. (Photo by S. Hygnstrom. With permission.)

and Hygnstrom, 1994; Hygnstrom et al., 2008b). After crop harvest, damage by deer continues to be a threat with consumption of stored feed, though one study showed amounts consumed to be minimal (VerCauteren et al., 2003).

Another real and significant concern is the potential for disease transmission from deer to livestock via direct contact as well as contamination of crops and feed (VerCauteren et al., 2003). Deer also play a role in transmission of zoonotic diseases (those that affect humans and wildlife). Lyme disease was first observed as arthritic conditions of children in Lyme, Connecticut in the early-1970s, but it was not actually identified until 1981 (Centers for Disease Control and Prevention [CDC], 2010). Since then, it has expanded across the country (Hoen et al., 2009). White-tailed deer are not effective reservoirs of Lyme disease but are hosts and vectors of black-legged ticks (*Ixodes scapularis*) that carry it (Brown and Burgess, 2001). Over 20,000 human cases of Lyme disease have occurred annually across the United States since 2002, with highest prevalence occurring in the Northeast and Midwest (CDC, 2010). In some midwestern states, such as Minnesota and Wisconsin, the range of black-legged ticks is expanding and the number of reported cases of Lyme disease in humans is increasing (Bacon et al., 2008). In addition, white-tailed deer carry various agents that can contaminate fruits and vegetables destined for human consumption (Rice, 2009), such as *Escherichia coli* 0157:H7 (Renter et al., 2001) and *Salmonella* (Renter et al., 2006).

Collisions between deer and vehicles cause considerable property damage and are a serious threat to human health and safety. Roughly 1.5 million DVCs occur annually across the United States, resulting in over $1 billion in damage, at least 200 human fatalities, at least 29,000 injuries to humans, and more than 650,000 deer fatalities (Conover, 1997) (Chapter 13). Illinois, Iowa, Michigan, Minnesota, and Wisconsin experienced a 22.7% increase in DVCs from 1993 to 2002, peaking at nearly 123,000 (Knapp, 2005). An average of 21 fatalities and 4715 injuries to humans occurred annually in these states during the study (Knapp, 2005). Annual vehicle-related mortalities of deer in the Midwest ranged from a low of 987 in Wyoming to a high of 76,626 in Wisconsin between 1982 and 1991 (Romin and Bissonette, 1996). In addition, 712 deer–aircraft (civilian) collisions occurred between 1990 and 2007 (Dolbeer and Wright, 2009).

Interestingly, deer attacks on humans have been documented in recent years in the Midwest. On 13 occasions people were attacked by deer on an Illinois campus (Hubbard and Nielson, 2009). The incidences were the result of many contributing factors including fawning period, relatively high deer densities, habituation to humans, and restriction of hunting (Hubbard and Nielson, 2009). Aggression toward humans by female white-tailed deer may occur when there is a perceived threat toward fawns (Grovenburg et al., 2009), as occurs when researchers capture fawns for tagging. Human-raised deer also attack people, likely because of habituation to humans.

White-tailed deer in the Midwest are considered a "keystone species," based on their affects on other species and the overall community structure in which they exist (Waller and Alverson, 1997; Rooney and Waller, 2003). Overabundant populations of deer can have deleterious impacts on entire biotic communities, impacting flora and fauna (Waller and Alverson, 1997; Wisdom et al., 2006). In 1947, Aldo Leopold, the "Father of Wildlife Management," predicted the impact that overabundant populations of deer could have on their habitat (Leopold et al., 1947). Deer populations can detrimentally impact both natural and induced forest regeneration (Waller and Alverson, 1997). They can damage trees at any developmental stage, and even at maturity trees continue to be shaped and scraped by deer browsing and rubbing. Ornamental trees and shrubs are particularly susceptible to deer damage, with nursery stock sustaining particularly high levels of economic loss (Hygnstrom et al., 2009). Production of Christmas trees has proven economically impractical in some areas without protection from browsing deer (Beringer et al., 1994). Deer not only inflict economic damage to plantings, but also may deplete particular tree species from entire areas (Côté et al., 2004). Overabundant deer are detrimentally affecting Canada yew and associated plant communities on the Apostle Islands in northern Wisconsin (Beals et al., 1960; Allison, 1990). Overabundant deer can actually lower the carrying capacity for their own species by removing preferred plant species and leaving only less-preferred species to prevail. Some unpalatable ferns, for example, have the potential to interfere with forest regeneration by outcompeting shrub and tree seedlings (Horsley et al., 2003; Côté et al., 2004).

High densities and expanding ranges of white-tailed deer increase interactions with other cervids including moose, elk, and mule deer in some areas of the Midwest. The interspecies interaction can lead to a greater risk of disease transmission, competition for resources, and, in the case of mule deer, hybridization. White-tailed deer and mule deer live sympatrically in the western half of the Midwest region. Competition for resources between the two species has been hypothesized to be a contributing factor in the decline of mule deer populations throughout their range. Studies suggest that while there may be some common exploitation of food resources, little direct competition occurs between the two species (Martinka, 1968; Krämer, 1973; Swenson et al., 1983). Exploitation of common food resources was most common during winter when food availability was lowest (Martinka, 1968). Differences in preferred habitats minimized most of the interactions between the two species during the remainder of the year (Martinka, 1968). Interestingly, we observed that indirect contact between mule deer and white-tailed deer in western Nebraska was higher during summer (mean overlap = 79%) than winter (mean overlap = 53%), a response to smaller home ranges and greater niche segregation during winter (W. D. Walter, USDA-National Wildlife Research Center, unpublished data).

Mule deer and white-tailed deer are able to interbreed and produce reproductively viable offspring (Krämer, 1973; Wishart, 1980). The potential for hybridization coupled with declining mule deer populations in parts of their range has led to some concerns about the negative effects hybridization may have on mule deer populations. Male white-tailed deer are more likely to breed with female mule deer than are male mule deer likely to breed female white-tailed deer, due to the persistence of whitetail bucks and more approachable manner of mule deer does (Wishart, 1980; Stubblefield et al., 1986). In addition, white-tailed deer generally produce more fawns/doe than do mule deer. The western portion of the Midwest was historically dominated by mule deer, but habitat is transitioning into wooded cover in some areas, which favors white-tailed deer. As whitetail populations increase in these areas, hybridization may increase. Recommendations to minimize the risk of continued hybridization include reversing habitat succession and increasing harvest of white-tailed deer (Stubblefield et al., 1986; Wiggers and Beasom, 1986). Some state wildlife agencies have attempted to protect less robust populations of mule deer by suppressing white-tailed deer through increased harvest.

Landscape Influences

Landscape

One might ask why white-tailed deer are distributed across the Midwest landscape as they are. In general, white-tailed deer are found throughout the Midwest, but their abundance is greatly influenced by the amount of forested cover and agricultural food available (VerCauteren and Hygnstrom, 2004; Walter et al., 2010). In addition, their home ranges, seasonal movements and use areas, and daily movements are influenced by the environment that surrounds them (VerCauteren and Hygnstrom, 1994; Hygnstrom et al., 2008a). White-tailed deer classically are considered an "edge species" (Swift, 1946; Williamson and Hirth, 1985; Alverson et al., 1988). They do best in mixed landscapes where cover and food are juxtaposed and readily available. Much of the Midwest provides excellent habitat because it consists of a mix of forest and agricultural lands (Zwank et al., 1979; Compton et al., 1988; VerCauteren and Hygnstrom, 1998; Figure 17.3). Lowest densities of deer in the Midwest are found in the heavily forested area of the northern Great Lakes region, the western Great Plains that is dominated by grasslands, the "Corn Belt" where wooded cover is sparse, and the inner city where there isn't a tree behind which to hide and 8-cylinder predators from Detroit abound. Several environmental and anthropomorphic factors have shaped the Midwest that influence the distribution and abundance of white-tailed deer across the landscape.

Climate and Land-Use

The Midwest landscape follows an east–west gradient in precipitation that influences plant communities. High rainfall in the east supports lush deciduous forests with dense and diverse understories and ground covers. Low rainfall in the west limits distributions of plants to include drought-tolerant species generally associated with arid shortgrass prairies. Expansion of agricultural crops in the Midwest followed timber harvest in the east, conversion of prairie in the center, and irrigation development in the west. High densities of white-tailed deer are found in the agro-forest complexes found throughout the Midwest. Large areas of predominantly agricultural land, however, do not sustain large populations of white-tailed, due to the lack of permanent cover. Large grassland areas of the western Midwest have relatively low densities of white-tailed deer because forest and agricultural covertypes are limited.

Ecological Succession

Agro-forest ecosystems can support high densities of white-tailed deer, but foraging activities by deer can have serious and far-reaching effects on plant and animal communities (deCalesta, 1994a,b). White-tailed deer can pose a direct threat to forest structure and composition (Rooney, 2001; Russell et al., 2001). In the early 1990s, Diamond (1992) noted that overabundant deer were degrading habitat at Fontenelle Forest, Nebraska and causing "reverse succession." Persistent browsing by deer was preventing regeneration of climax canopy species, which were being replaced by mid-level and invasive species. Diversity of woody plants in the forest understory declined, but grasses increased (Gubanyi et al., 2008). Alverson et al. (1988) suggested that repeated browsing by white-tailed deer has prevented regeneration of Canada yew, eastern hemlock, and white cedar in northern Wisconsin. Others have reported, however, that although extremely high densities of deer and associated browsing negatively affected many forest species, including eastern hemlock, a multitude of factors influence hemlock regeneration in the upper Great Lake region, including climate, disturbance, hemlock life history, ecosystem processes, and historical land use (Mladenoff and Stearns, 1993).

Timber Harvest

Forests and forestry practices in the Midwest have had a huge impact on populations of deer historically, especially in the northern, eastern, and Great Lakes regions where climax forest types dominated the landscape prior to European settlement. Biological carrying capacity of these old-growth forests for deer was relatively low and densities of deer were much lower than they are today. Extensive timber harvest and associated wildfires in the late-1800s and early-1900s led to the widespread regression of seral stages, expansion of agriculture, and the resultant creation of edge. Populations of white-tailed deer responded and peaked in the 1930s and 1940s at 14 deer/km^2 in Wisconsin and other Great Lakes states (Swift, 1946; Alverson et al., 1988). Through the twentieth century, eastern deciduous and southern boreal forests regrew and now provide high-quality cover for white-tailed deer.

Habitat management is needed in northern forests where deer numbers are relatively low due to seral stages nearing climax, low biological carrying capacity, and winter severity. Lesage et al. (2000) maintain that the limiting factor for white-tailed deer in their northern range is adequate wintering areas, and that these areas are highly vulnerable to insect epidemics, forest fires, and logging. In specific situations when food resources are limiting, the potential for considerable winter mortality exists. In northern reaches of the Midwest, over-browsed winter range can be detrimental to local populations when weather conditions prompt deer to congregate in areas that are deficient in quality food sources (Dahlberg and Guettinger, 1956; Van Deelen et al., 1997). Losses due directly to starvation are observable while delayed impacts, such as reduced neonatal survival, can be equally, if not more devastating (Blouch, 1994). Critical wintering areas (i.e., deeryards) should be protected from timber harvest to maintain current deer populations and provide sources for future populations at northern latitudes (Van Deelen et al., 1998; Lesage et al., 2000; Morrison et al., 2003). Millions of hectares of public and commercial forests in the North are managed for the production of timber and other forest products, which benefits white-tailed deer. In large blocks of forests, agencies created forest openings to increase edge and forage production (McCaffery, 1984). In addition, yarding areas were protected from timber harvest to sustain these critical winter habitats.

Agricultural Expansion

Removal of much of the native forests in the eastern and northern portions of the Midwest resulted in a landscape that was available for conversion to agriculture (Ramankutty and Foley, 1999; Turner and Rabalais, 2003) and the untilled prairie was a sea of opportunity to early farmers and settlers.

Early settlement precipitated small farms and fields with a relatively high degree of edge. The crosscut and horse were replaced by the chainsaw and tractor as increased mechanization led to more land converted from forests and grasslands to agriculture throughout the Midwest. Diverse small fields gave way to large fields, monocultures, and simple rotations of corn and soybeans. Much of the native tallgrass prairie in Illinois, Iowa, Kansas, and Nebraska fell to the plow. The Midwest became known as the "Cornbelt" and "America's breadbasket" in the mid- to late-1900s.

Agricultural crops in the Midwest compliment the annual cycle of nutrient requirements of white-tailed deer (Mautz, 1978). Demand for digestible energy, protein, and other nutrients is high during spring when deer are recovering from winter food shortages and pregnant does are nearing parturition. Winter and spring wheat, as well as alfalfa, are used extensively by deer during spring green-up (Vecellio et al., 1994; Frost et al., 1997), when the winter snows subside, because these crops are available, palatable, and provide a highly needed source of nutrition at a time when other foods are lacking. Nutrient requirements also are high during summer and early autumn as does nurse fawns and bucks produce antlers. Both sexes also must accumulate fat reserves for winter and be healthy going into the fall breeding season. Agricultural crops such as corn and soybeans are abundant and high in carbohydrates, protein, and other nutrients during the growing season. Deer eat corn at all stages of growth, especially during the silking–tasseling stage in early- to mid-summer and when ears are ripe in autumn (VerCauteren and Hygnstrom, 1998). Deer eat soybeans, especially during the early stages of growth up to flowering and pod set (Nixon et al., 1970; deCalesta and Schwendeman, 1978). During winter, deer experience a negative energy balance because of low temperatures, snow depths, and limited forage resources (Mautz, 1978). Waste grain in fields after harvest provides a forage base for deer in agricultural areas throughout the winter (Figure 17.5). A plethora of specialty crops also are highly preferred and eaten year-round by white-tailed deer, including nursery crops, fruit trees, vineyards, vegetables, ornamentals, and organic farm produce.

Today, many areas of the Midwest are so intensively farmed that populations of white-tailed deer are relatively low. Where agriculture exceeds about 75% of the landscape, populations of white-tailed deer decline, but deer thrive in areas where at least 25% of the landscape exists in forest. West of the 100th meridian the Midwest is too arid for extensive forest growth and sustained dryland farming. Instead, shortgrass and mixed-grass prairie dominate the landscape. Since the mid-1900s, irrigation has led the

FIGURE 17.5 White-tailed deer use agricultural crops in the Midwest to satisfy their nutritional requirements throughout the year which has contributed to overabundance of white-tailed deer, a problem in many areas of the Midwest. Concerns about crop damage and deer–vehicle collisions offset the demand for high densities of deer for hunting and viewing, a concept biologists call "sociological carrying capacity." (Photo by G. Clements. With permission.)

way to expansion of agricultural crops westward, especially corn and soybeans. Human interests in western land and water led to the damming, channelization, diversion, and stabilization of rivers and streams in the west for flood control, irrigation, and power development. Stabilization led to expansion of western riparian forests and increased continuity of riparian forested corridors. Recent colonization by white-tailed deer has occurred in the western Great Plains, primarily along forested riparian areas (Kufeld and Bowden, 1995; Colorado Division of Wildlife, 2008) such as the Missouri, Platte, and Republican Rivers and their tributaries. In addition, suppression of fire in the grasslands, because of agriculture and human development, often has resulted in the increase of woody vegetation.

Rangeland is an important component of American agriculture. Over 60% of the western landscape consists of grasslands, most of which are grazed by livestock. These rangelands are not very suitable for white-tailed deer and typically forested cover is lacking, except along riparian areas. White-tailed deer are browsers first and do not do as well where grazing and grasslands predominate. Deer tend to avoid cattle and so are often displaced from the most desirable areas in rangeland–forested cover and watering areas.

Conversion of cropland to grassland through government agricultural programs, such as the Soil Bank Program in the 1950s and 1960s led to dramatic changes in the midwestern landscape. More recently, the Conservation Reserve Program, established by the Federal Food Security Act of 1985, resulted in conversion of up to 16 million hectares of cropland to grassland and other conservation plantings. Grasslands are not ideal habitat for deer, but they can provide food, cover, and vast areas of undisturbed space.

Development

Concrete and blacktop are not good deer habitat. However, in recent times, deer have acclimated to urban sprawl and suburban development (Cornicelli et al., 1996; Hygnstrom and VerCauteren, 1999; DeNicola et al., 2000; Etter et al., 2002; Grund et al., 2002) and more recently exurban development (Storm et al., 2007). Urbanites enjoy the aesthetic benefits of trees, grass, flowers, ornamentals, gardens, and ponds. Many "green spaces" (i.e., parks, arboretums, wetlands) create a mosaic with edge, providing sufficient food, water, and shelter for deer. Deer have moved into developed areas and development has moved into traditional deer habitat. The human population of Michigan is expected to increase 12% between 1990 and 2020, which equates to a 63–87% increase in the amount of land devoted to urban use (Smyth, 1995). Hunting is often restricted in developed areas due to concerns about public safety (DeNicola et al., 2000). Populations of deer can increase to extremely high levels in developed areas where deer are not actively managed through hunting or some other intervention.

Human Demographics

Archaeological evidence indicates that sizable communities of Native Americans and extensive agriculture existed in the Midwest long before Columbus brushed the shores of the North American continent (Mann, 2005). It appears that rampant diseases associated with early European explorers and settlers led to dramatic declines in these prehistoric societies. According to recent history (within the last 400 years), the Midwest went from periods of relatively low densities of Native Americans, settled and nomadic, to the high densities of our urbanized and diverse culture of today. Exploration of the Midwest by the French started in the mid-1600s. The most notable explorers, Meriwether Lewis and William Clark, traversed the Midwest in 1801–1804 by way of the Missouri River and brought back stories of the untamed and bountiful landscape to the then President Thomas Jefferson (Moulton, 2003). The Mormon Trail, Oregon Trail, and other routes followed, bisecting the western Midwest and provided settlers access to many areas that had little evidence of human development. The Homestead Act of 1862 encouraged hundreds of thousands of European settlers to fill the landscape of the Midwest and Great Plains (Cross, 1995). By law, they had to build homes and break the land to raise crops. Railroads eventually crossed the Midwest and provided a steady stream of meat from deer, elk, and bison eastward to feed a growing nation. The market hunting era (i.e., unrestricted hunting for commercial use) led to decimation of many populations of wildlife in the Midwest in the late 1800s. By 1900, only 50 deer were left in all of Nebraska (Menzel, 1984). In 1930, about 1.6 million people occupied the state of Nebraska and surprisingly, the human population is about the same today. The difference, however, is the distribution of those people. In the 1930s, people were scattered throughout the state, living on homesteads and in the small communities of rural Nebraska. After World War II, a shift of people occurred from the farms and rural areas to the bright lights of the big cities. Cities such as Chicago, Des Moines, Detroit, Kansas City, Milwaukee, Minneapolis, Omaha, and St. Louis grew at remarkable rates. The shift of people from rural to urban areas left large rural landscapes relatively vacant, both spatially and culturally (Berry, 1977).

Refugia

In the urban areas, city planners protected greenspaces from development. City parks, arboretums, wooded corridors, and wetlands were relatively undeveloped and deer colonized these areas where hunting was restricted or prohibited (Figures 14.19 and 14.20). The 1980s saw a movement of people out of the cities and into the nearby country. Development of 1–20 ha sized lots and ranchettes became popular and these periurban areas often provide ample food and cover for deer and other wildlife. In addition, absentee landowners purchased blocks of prime deer habitat, often for hunting and other recreational purposes. Hunting typically is limited to family or friends, if at all, in these areas, so mortality rates of local deer herds are much lower than surrounding areas. These refugia led to expanding populations of deer that have often exceeded carrying capacity and caused significant problems for adjacent landowners. On the other hand, wildlife refuges, such as those managed by the U.S. Fish and Wildlife Service, play an important role in sustaining populations of white-tailed deer in the Midwest, once adjacent croplands are cleared during harvest (Hansen et al., 1991; VerCauteren, 1998; Walter et al., 2009). VerCauteren (1998) found densities of deer on DeSoto National Wildlife Refuge (DNWR) in the Missouri River Valley of eastern Nebraska to be nearly four times greater than on surrounding agricultural lands (19 vs. 5 deer/km^2). Organized public hunts are conducted on DNWR, yet densities of deer exceed those of surrounding areas. Areas where hunting is not permitted and suitable deer habitat exists, such as suburban developments and larger city parks, allow deer densities to soar (DeNicola et al., 2000).

Predators

In most areas of the Midwest, nearly all mortality of white-tailed deer can be attributed to human-related causes, primarily hunting and DVCs. Disease, predation, starvation, and occasionally old age encompass the remaining mortality factors. The wide distribution and increase in abundance of white-tailed deer throughout the Midwest have aided in recolonization and increased densities of predators, primarily coyotes and gray wolves.

Coyotes are the most widely distributed predator–scavenger in the Midwest (Figure 8.7). Although relatively small, they have the ability to take even healthy white-tailed deer and probably consume more deer than the other predators. Coyotes commonly prey upon young fawns and old, sick, or injured individuals. Coyote predation was reported as the leading cause of fawn mortality in Illinois (69%) (Nelson and Woolf, 1987), and Iowa (54%) (Huegel et al., 1985). However, Bryan (1980) and Schulz (1982) reported little or no mortality caused by predation in Missouri and Minnesota, respectively. White-tailed deer are especially important to coyotes during fawning, during and after hunting seasons, and in late winter (Huebschman et al., 1997). White-tailed deer are vulnerable to predators during late winter due to the prolonged effects of winter weather and reduced forage.

Populations of gray wolves in Minnesota, Wisconsin, and Michigan have increased in recent years and are expanding into areas that have been unoccupied by wolves since the early 1980s. Recolonization in Michigan and Wisconsin likely occurred from a large and stable population of wolves in northeastern Minnesota (Fuller et al., 1992; Wydeven et al., 1995). As their range has continued to expand in recent years, states such as Illinois, Indiana, Nebraska, and South Dakota have documented the presence of wolves: these isolated incidents likely reflect transient or dispersing individuals that have moved south from Minnesota or Wisconsin. Predation of white-tailed deer by wolves contributes significantly to mortality in some locations such as northern Minnesota (Nelson and Mech, 1986a,b; Kunkel and Mech, 1994), yet less so in Wisconsin and Michigan where wolf populations are still in recovery

(Blouch, 1994; Van Deelen et al., 1997). The frequency of kills by a wolf pack varies depending on pack size, deer density, snow conditions, and abundance of alternative prey. In Minnesota, where wolves eat white-tailed deer almost exclusively, estimated kill rates range from 15 to 20 deer/wolf/year. Deer are the primary prey of wolves, so it is reasonable to assume that wolves compete with hunters for deer in the area. The extent to which wolf predation and human hunting actually compete, however, may depend on the intensity and compensatory nature of each. Nelson and Mech (2000) found inconsistent relationships between wolf numbers and harvest of male white-tailed deer. They suggest that competition between hunters and wolves is probably greatest in areas of poor quality habitat where deer densities are low.

Black bears also prey on white-tailed deer, primarily young fawns that cannot yet escape search and pursuit (Kunkel and Mech, 1994; Carstensen et al., 2006). Compared to wolf predation though, which is a continual threat to fawns, bear predation is generally a minor contributor to fawn mortality (Nelson and Mech, 1986a). In most studies, mortality of fawns due to black bears was low (Ozoga et al., 1982; Ballard et al., 1999); however, Vreeland et al. (2004) reported that black bears were responsible for 31% of all mortality by predators.

Mountain lions were extirpated from most of the Midwest in the early 1900s but they have been observed throughout the region with increasing frequency in the last decade (The Cougar Network, 2010). Mountain lions can prey on white-tailed deer of any sex–age class, but lion numbers currently are so low that they have little impact on populations of white-tailed deer. Bobcats also have been documented to kill fawns (Ballard et al., 1999; Vreeland et al., 2004; Carstensen et al., 2006). They are not, however, considered an important predator of white-tailed deer.

Disease

In the Midwest, a few diseases of deer have state and federal biologists seeking solutions (Figure 17.6). Bovine TB became established in a five-county area in the northern portion of the Lower Peninsula of Michigan in the 1990s. The disease was likely introduced by cattle and is now maintained in free-ranging deer. High densities of deer and supplemental feeding likely contributed to the establishment and spread of bovine TB in Michigan (O'Brien et al., 2002, 2006). Deer densities in the core of the bovine TB area of northern Michigan were 19–23 deer/km^2 (Schmitt et al., 1997). The Michigan Department of Natural Resources adjusted hunting regulations to increase harvest and reduce deer density. Although much higher densities of deer have been documented elsewhere in the United States, concentrating deer over limited food supplies provides the means needed for a disease outbreak to occur (O'Brien et al., 2002). More reasonable goals of 8–15 deer/km^2 in areas of the Midwest, such as Missouri, represent populations that satisfy needs of the general public while still maintaining a healthy population (Hansen and Beringer, 1997).

FIGURE 17.6 Diseases such as epizootic hemorrhagic disease-bluetongue (EHD-BT) have caused significant mortality in white-tailed deer in localized areas in the Midwest. (Photo by G. Clements. With permission.)

Chronic wasting disease (CWD) is another disease of white-tailed deer, occurring in free-ranging and captive populations, that is currently found in some parts of the Midwest. It also affects mule deer, elk, and moose. The disease typically leads to illness and death in infected individuals within 18 months (Chapter 7). Although prevalence is usually low, up to 35% prevalence has been observed in some free-ranging populations of white-tailed and mule deer (Williams, 2005; Farnsworth et al., 2006; Joly et al., 2006). Mobility of deer may play a role in transmission of CWD across the landscape (Frost et al., 2009; Clements et al., 2011). Levels of prevalence up to 79% have been documented in captive white-tailed deer facilities (Keane et al., 2008). Eradication of CWD in free-ranging populations may not be possible without eradicating the host (Wasserberg et al., 2008) and the disease agent can be remarkably persistent in the environment (Miller et al., 2004). Computer-based models that facilitate prediction of disease spread and impact on deer populations are contributing information for developing management plans (e.g., Frost et al., 2009). One model shows the importance of managed harvest in CWD-infected herds in Wisconsin and predicts that CWD may eventually compete with recreational hunting (Wasserberg et al., 2008). Unfortunately, CWD appears self-sustaining in deer populations, a treatment for the disease has not been identified, and the disease is spreading across the landscape.

Hemorrhagic disease (HD) is the most important viral disease affecting populations of white-tailed deer in the United States: it is seasonal, cyclic, and herd immunity is highly variable (Davidson, 2006). Hemorrhagic disease, caused by EHD and BT virus is common in the Midwest (Beringer et al., 2000) and occasionally has substantial impacts on local populations of deer (Gaydos et al., 2004). Although other species are susceptible to HD, white-tailed deer experience the highest mortality (Matschke et al., 1984). It is an arthropod-borne disease and outbreaks typically occur during late summer and early fall in association with Culicoid midges. It is not uncommon for hundreds of deer to fall victim to the disease when an outbreak occurs (Matschke et al., 1984). Global warming has been suggested as playing a role in its distribution and spread, especially into northern latitudes (Gibbs, 1992; Sleeman et al., 2009). Management agencies can do little to prevent HD and local populations of deer typically rebound quickly, so harvest regulations often are not modified following an outbreak.

White-tailed deer and moose are both hosts to a parasitic meningeal worm (*Parelaphostrongylus tenuis*). Eggs of this parasite are expelled in the feces of deer and are ingested by gastropods that come into contact with the fecal matter. Deer and moose ingest the metacercarial stage when they inadvertently consume infected gastropods while feeding (Anderson and Lankester, 1974). Worms mature and migrate into the central nervous system, often coming to rest in the meninges in deer, where they have a limited effect on overall health (Alibasogulu et al., 1961; Anderson, 1963; Anderson et al., 1966). In moose, however, the worms more often migrate throughout the brain, leading to radical changes in behavior and death. Until the 1980s, researchers hypothesized that moose populations were in decline because of increasing deer densities and resultant higher infection rates in moose from meningeal worm (Lenarz, 2010). Recent research, however, indicates that meningeal worms are not a major cause of mortality in moose populations (Whitlaw and Lankester, 1994; Murray et al., 2006). Further, increased deer numbers have not led to an increase in the prevalence of meningeal worm in the environment (Thomas and Dodds, 1988; Bogaczyk et al., 1993).

Recent History of Deer in the Midwest

In the early 1800s, populations of white-tailed deer across the Midwest were in recovery following extensive harvests for trade by indigenous people. However, by the mid-1800s, market hunting became a lucrative endeavor, thus populations of white-tailed deer again were on the decline. Reports from across the upper Midwest include details such as "in the fall of 1877 Ivory Livermore and three companions hunted on the North Fork of the Eau Claire River and killed for the market 118 deer and 26 bears from September 1–December 1" (Young, 1956). A single shipment of deer meat from Litchfield, Minnesota in December 1877 carried 5443 kg of venison and was destined for Boston, Massachusetts (Young, 1956). The advent of repeating rifles and the westward spread of railroads provided means for more efficient harvest and shipping, intensifying the decline of white-tailed deer. Prior to 1900 and the passage of the Lacey Act, restrictive game laws were only considered a nuisance (Young, 1956). The Lacey Act provided means for prosecuting individuals involved in transport of illegally taken wild game across state borders and combined with scarcity of deer resulted in the beginning of the end of market hunting for white-tailed deer (McCabe and McCabe, 1984).

Several factors contributed to the steady rise in deer populations throughout the twentieth century, including widespread predator suppression, a focus on hunting bucks, decreasing hunter access, moderate climates (Côté et al., 2004), and high-intensity agriculture, abandonment of small farms, forest protection, and suburban development. It was estimated that white-tailed deer numbered between 24 and 33 million around 1500 (McCabe and McCabe, 1984; VerCauteren, 2003). Surveys of deer biologists from state wildlife agencies suggest the current population of white-tailed deer in the United States is about 28.5 million (Figure 11.1), with over a third (about 10 million) in the Midwest (Table 17.1).

When populations began to rebound, they did so quickly and a review of over-populated deer ranges conducted in the 1940s revealed several locations across the Midwest where high densities were leading to habitat damage or starvation of deer (Leopold et al., 1947). Decades later, habitat loss across the Midwest associated with suburban sprawl and increased agricultural development created concerns over the potential for population declines (Hansen and Beringer, 1997). Concerns were unjustified; populations continued to climb as deer readily adapted and even thrived in the newly fragmented landscape. High-quality agricultural crops were now in greater abundance, reducing reliance on natural forage and enabling deer populations to increase (Figure 17.5). Deer are increasingly managed at social carrying capacity, with target densities based on landowner tolerances relative to damage inflicted by deer (Hansen and Beringer, 1997). Increased disease, DVCs, and damage to agricultural crops, landscapes, gardens, and habitat all serve to lower sociological carrying capacity in many areas across the Midwest (DeNicola et al., 2000). Ecological carrying capacity, K, the number of deer the habitat can support, is still relevant although often not a key component in developing management strategies for deer. For example, food often is the limiting factor in northern reaches of the Midwest, which in the past led to supplemental feeding. Supplemental feeding is a tradition in parts of this region and not only allows populations to exceed what their habitat can sustain, but also facilitates disease transmission (O'Brien et al., 2006).

TABLE 17.1

Number of White-tailed Deer and Deer Hunters and Harvest Data, by State, in the Midwestern United States

	Number of White-tailed Deer	Number of Deer Hunters	More Hunters Now Than 25 Years Ago?	Number of White-tailed Deer Harvested/Year	Maximum Number of White-tailed Deer a Hunter Can Harvest/Year
Colorado	15,000	90,000[a]	No	1826	2
Illinois	775,000	275,000	Yes	195,000	Unlimited antlerless
Indiana	550,000	229,000	Yes	130,000	~400
Iowa	340,000	180,650	Yes	140,000	Unlimited antlerless
Kansas	568,000	115,000	Yes	77,500	7
Manitoba	125,000	34,000	No	18,329	4
Michigan	1,700,000	735,000	Yes	480,000	Unlimited in some areas
Minnesota	1,000,000	500,000	Yes	340,000	Unlimited in some areas
Missouri	1,400,000	500,000	Yes	300,000	Unlimited antlerless
Montana	250,000	150,000[a]	Yes	65,000	11
Nebraska	300,000	60,000	Yes	68,600	Unlimited antlerless
North Dakota	500,000	110,000	Yes	100,000	~20
Ontario	450,000	185,000	Yes	85,000	6
Saskatchewan	355,000	53,325	Similar	32,793	4
South Dakota	200,000	81,500	Yes	74,000	45
Wisconsin	1,500,000	670,000	Similar	450,000	Unlimited antlerless
Wyoming	57,027	26,143	Yes	14,792	2

[a] Includes both white-tailed and mule deer.

Current Management

Deer are the economic engine of most state wildlife agencies in the Midwest, as revenue generated from the sale of hunting permits typically supports most state agency personnel and management programs, including those for nongame species. State and provincial wildlife agencies primarily manage, or manipulate the size of deer populations by regulating hunter harvest. Agencies typically have limited jurisdiction to manage deer habitat on a broad scale, but have advisory input to state, provincial, and federal land management agencies. As such, wildlife agencies directly influence one component of deer mortality, hunter harvest, but may only indirectly influence habitat factors that contribute to both survival and recruitment. Hunter harvest is a powerful management tool, however, that can impact both survival and recruitment through density-dependent processes. Hunter harvest also can be used to alter the sex–age structure of deer populations to achieve management goals.

By necessity, management of white-tailed deer is evolving and dynamic in most midwestern jurisdictions, transitioning from an emphasis on recovering deer herds from historic overexploitation to that of sustaining deer populations in balance with ecological and sociological carrying capacities. Current regulations across the Midwest suggest a need to reduce white-tailed deer on at least localized scales. All midwestern states currently allow hunters to take antlerless deer and many have liberal seasons of extended length that allow harvest of several antlerless deer (Table 17.1). Several states have "urban" hunts to reduce numbers of deer in urban/suburban areas using hunting under specific regulations (Hansen and Beringer, 1997; Doerr et al., 2001). Other sensitive areas such as military bases, industrial parks, wildlife refuges, and national parks (Fulton et al., 2004) have begun using managed hunts to control white-tailed deer populations.

Wisconsin, Illinois, South Dakota, Colorado, and Saskatchewan will be highlighted here as examples of how and why management philosophies and techniques need to vary among jurisdictions to achieve goals. Deer management and hunting in Wisconsin has a long and storied history (Swift, 1946; Wisconsin Department of Natural Resources [WDNR], 2009d). Illinois is particularly interesting because a task force appointed by the General Assembly recently recommended that deer management quotas be determined based on the number of DVCs by county (Joint Task Force, 2008) and because it is one of the last jurisdictions that continues to manage CWD by aggressively reducing the density of local deer herds. South Dakota removed antler-point requirements in the Black Hills in favor of reduced numbers of deer permits to increase numbers of bucks, unlike Missouri, where antler-point restrictions imposed in some units require hunters to harvest bucks with a minimum number of antler points, to increase numbers of mature bucks and balance sex ratios. Colorado allows liberal harvest of white-tailed deer in its eastern plains to reduce the number of white-tailed deer relative to sympatric and declining mule deer. Saskatchewan was the only Midwestern jurisdiction that depended primarily on aerial surveys to estimate the size of deer populations (Saskatchewan Ministry of Environment [SME], 2009). These jurisdictions represent the broad diversity in philosophy, goals, and objectives that exist today in midwestern deer management.

Wisconsin

Millspaugh et al. (2009) described Wisconsin's deer management process as the most comprehensive and transparent among comparable states, and credited it for the variety and quantity of biological data acquired as well as for the thoroughness with which these data are used. Wisconsin has 133 deer management units (DMU) (WDNR, 2009a). Population goals for DMUs are set every three years and antlerless harvest quotas and permit numbers are set annually (WDNR, 2001). Population goals are set with consideration of both ecological and sociological carrying capacities. Extensive public input has been included in the goal-setting process for decades. In forested DMUs, goals are approximately 60–65% of ecological carrying capacity (WDNR, 2009b) to ensure conservative harvest where environmental uncertainty causes highly variable winter mortality, particularly in the northern region of the state (WDNR, 2001). In largely agricultural and urbanized DMUs, population goals are less than 50% of ecological carrying capacity, reflecting reduced sociological carrying capacity due to crop damage and the risk of DVCs (WDNR, 2001). Despite intensive management of deer in Wisconsin, the overall trend in the statewide postharvest population has progressively exceeded annual goals since 1986 (Rolley, 2008; WDNR, 2009c) and deer numbers were 38% over goal in 2008.

The size of prehunt and posthunt deer populations are estimated statewide and by DMU using the sex–age–kill (SAK) model, which relies heavily on data from mandatory hunter registration and inspection of harvested deer (Eberhardt, 1960; Creed et al., 1984; Skalski et al., 2005; Millspaugh et al., 2009). In DMUs where the SAK model is inappropriate because of temporally inconsistent harvest rates of bucks, population estimates are based on accounting-type population models and/or aerial surveys (Rolley, 2008). Post-hunt populations are projected forward to estimate population size prior to the next hunting season, which is used along with various subjective considerations to set annual antlerless harvest quotas. The number of antlerless permits to be issued is determined using DMU-specific nonlinear models of estimated antlerless harvest as a function of number of permits (WDNR, 2001). To reduce crop damage and manage CWD, Wisconsin has also used DMU-specific opportunities and incentives (e.g., early antlerless-only seasons, unlimited antlerless permits, and inexpensive or free antlerless licenses) and special requirements (e.g., earn-a-buck, which requires hunters to harvest an antlerless deer before harvesting a buck) to increase harvests beyond those attainable under regular buck and antlerless permitting (Van Deelen, 2010). The quota-setting process is intensive and highly data driven, yet open to informed subjective modification, reflecting both the science and art of deer management.

Finding CWD in three deer harvested in south-central Wisconsin in 2001 (WDNR, 2009a) complicated deer management in the state. Beginning in 2002, CWD has been managed in portions of the state using combinations of special reduced population goals, increased quotas and numbers of permits, special collection periods, exceptional methods of shooting deer, and additional or extended harvest seasons (WDNR, 2002; Wisconsin Natural Resource Board [WNRB], 2002; WDNR, 2009a,b). In 2002, a core "Eradication Zone" (enclosing all cases of CWD-positive harvested deer) was established with a goal of 0 deer/km^2 and a surrounding buffer area or "Management Zone" with a radius of 64 km from the center of the Eradication Zone was established with a goal of 26 deer/km^2 (WDNR, 2002; WNRB, 2002; Wisconsin Legislative Audit Bureau, 2006). Objectives were to eradicate CWD from the area where it was known to exist and to reduce risk of further spread by reducing densities of deer in the surrounding area. In mid-2003, CWD was detected in a wild deer outside the original Eradication Zone near the border with Illinois, which led to the designation of a second Eradication Zone (both renamed as "Disease Eradication Zones," or DEZs) and an expanded "Herd Management Zone" (HRZ, formerly Management Zone) for the 2004 hunting season (WDNR, 2004). Population goals were relaxed to a target of <13 deer/km^2 for DEZs in mid-2004 (WNRB, 2004). A CWD Stakeholder Advisory Group (CWDSAG) was established in 2007 to seek public input and provide recommendations for future management directions (CWDSAG 2008). Subsequently, DEZs and the HRZ were combined into a single CWD Management Zone, 2008 goals were revised to DMU-specific densities of 26–62 deer/km^2, and season structure in the CWD Management Zone was shortened to better align with those of other over-goal DMUs in the state (WNRB, 2008a,b,c). These changes reflected the evolving recognition that CWD cannot be eradicated from Wisconsin using existing technology under prevailing sociological constraints (WDNR, 2009a; Garner et al., 2009). Saskatchewan also discontinued their CWD eradication program recently (SME, 2009a).

Estimates of deer densities in CWD units indicated a stable or slightly declining trend since 2002 (WDNR, 2009a). However, the number of deer in CWD units were never reduced to near goal in 2002–2007 and remained more than two-fold greater than the revised higher goal for 2008 (Garner et al., 2009). Wisconsin hunters have the opportunity to kill more deer but chose not to, partly because they remain unconvinced of risks posed by CWD (Holsman and Cooney, 2007; WDNR, 2009a). Unless hunters can be convinced that disease management is a higher priority than traditional factors that motivate them, it will be difficult to achieve harvest goals established for management of CWD.

Illinois

The Illinois Department of Natural Resources (IDNR) estimates fecundity, mortality, and age structure of white-tailed deer from annual or periodic surveys (Millspaugh et al., 2006). These data are used to estimate the size of county-specific populations of deer when analyzed within an accounting-type population model that can be simultaneously calibrated against various indices. The IDNR uses rates of DVCs, deer depredation complaints, hunter harvest data, bowhunter observations, and spotlight surveys to generate indices for the deer population. Beginning in 2009, Illinois began using rates of DVCs (number of collisions and deer-related accidents/1.6 billion vehicular km traveled) as the objective for measuring status of deer populations at county and statewide levels (Joint Task Force, 2008). Initial objectives were determined as the midpoint between minimum and maximum annual DVCs during 1994–2007 (Joint Task Force, 2008). Biological data and modeling continue to be considered, along with tolerable levels of DVCs when setting harvest quotas and season structures, as well as for assessing whether county DVC goals are reasonable. In general, counties well above DVC goals are managed using liberal antlerless harvest during all firearm deer seasons (Joint Task Force, 2008). In addition, late-winter antlerless seasons are used to increase harvest in counties above or near DVC goals. The approach in Illinois to managing deer is novel because efficacy of deer management is assessed directly against the number of DVCs.

Illinois shares a CWD outbreak area with Wisconsin. As of January 2010, IDNR continues aggressive management, using both public hunting and agency sharpshooting, to reduce deer herds in CWD counties irrespective of DVC levels. Liberal quotas (exceeding demand) during archery and firearm seasons have been in place since 2003 in CWD counties (Shelton and Hulin, 2004, 2005, 2006; Shelton and McDonald, 2008a,b, 2009). Beginning in January 2006, a special late CWD season concurrent with late-winter seasons in some non-CWD counties was established in four CWD counties. During CWD seasons, hunters can use unfilled permits from previous firearm seasons, and/or purchase inexpensive additional antlerless-only permits (Shelton and Hulin, 2006). Following public hunting seasons, sharpshooters have annually removed deer from CWD counties since January 2004 to further reduce deer populations and to increase CWD surveillance, particularly in areas inaccessible to public hunters. Sharpshooting has focused on areas of winter deer concentrations on or near CWD-infected properties. Helicopter surveys have been used to count deer in CWD management areas during winter and identify wintering areas for sharpshooting. Preliminary results suggested that sharpshooting sustained over multiple years had reduced local densities of deer when culling intensity was above a minimum threshold (not quantified) (Shelton and McDonald, 2009). Shelton and McDonald (2009) also observed that overall patterns of CWD distribution and intensity changed little during 2005–2008, and that declines in prevalence have occurred in fawns, yearlings, and adult female deer in sharpshooting areas. Based on recommendations of the Joint Task Force (2008), late winter deer season and the special CWD season were expanded from three to seven days, beginning in 2009 (Shelton and McDonald, 2009). It appears that IDNR will continue to aggressively manage CWD through liberal hunting regulations and focused use of sharpshooting to reduce densities of deer.

South Dakota

White-tailed deer and mule deer occur throughout South Dakota (South Dakota Game, Fish and Parks [SDGFP], 2009i). Management objectives and practices are specific to deer hunting units, that correspond to counties east of the Missouri River (East River Units, SDGFP, 2009e) and subdivisions of counties or combined portions of multiple counties west of the Missouri River (West River Units, SDGFP, 2009g; Black Hills units, SDGFP, 2009b). Statewide, white-tailed deer are more abundant than mule deer and are most common east of the Missouri River (SDGFP, 2009f), while mule deer are most common west of the river (SDGFP, 2009i).

The SDGFP conducts several types of annual surveys that provide information for modeling deer populations (SDGFP, 2009i). Species-sex composition and spatio-temporal distribution of the harvest, hunter success rates, days of hunting recreation, and hunter satisfaction are estimated from hunter report cards that are randomly allocated to deer license holders by management unit. Hunter success rates are used to estimate numbers of deer harvested. Each September (sometimes into October) a minimum of 30 independent classification surveys per management unit are used to estimate species-specific age–sex structure (fawn:doe and buck:doe ratios) of prehunt populations. Additional population surveys may be conducted for select areas to acquire information on reproduction and other factors. Data are used in models to estimate survival and to simulate population sizes and trends (SDGFP, 2009i).

Primary goals of big game management in South Dakota are to manipulate populations toward the highest possible level consistent with ecological and sociological carrying capacity, utilize surplus animals, and provide recreational and hunting opportunities for the public (SDGFP, 2009i). Population objectives in East and West River Units primarily are determined by tolerance of deer by landowners, while objectives in Black Hills Units primarily are determined by habitat conditions with increasing consideration of landowner tolerance in specific areas. In a 2006–2007 survey of landowner attitudes regarding the number of white-tailed deer on their land, 47% reported that deer numbers were acceptable, 17% thought too few deer existed, and 36% thought there were too many deer (Gigliotti, 2007). Hunter harvest is managed to address landowner tolerance. Liberal numbers of permits for hunting white-tailed deer have been available since 2001, reflecting increasing populations. Archery, youth antlerless, mentored youth, general muzzleloader, and landowner permits had unlimited quotas in 2008, while muzzleloader permits in certain areas and firearm season permits had limited quotas (SDGFP, 2009f). In 2009 a variety of permits reflecting variable bag limits were liberally offered to both residents and nonresidents (SDGFP, 2009a,b,d,e,g,h). For example, any deer (either species or sex), any antlerless deer (either species), any white-tailed deer, and antlerless white-tailed deer permits were available individually or in combination (including one–three antlerless permits), depending on season and management unit. Archery and muzzleloader hunters were allowed to harvest up to five deer. Rifle hunters were allowed to acquire up to five permits during three different drawings (nonresidents limited to third drawing) and unlimited numbers of leftover permits were available after the third drawing. In addition, East and West River permits not filled during the regular seasons converted to antlerless permits during a late season. Individuals can hunt and harvest deer in all seasons (archery, muzzleloader, and rifle) with appropriate permits.

Most white-tailed deer in South Dakota are harvested during rifle seasons in East and West River Units, although archery seasons account for substantial harvest as well (SDGFP, 2009f). Since 2001 (2008 latest year data available), white-tailed and mule deer numbers in East River and West River Units, combined, have remained relatively stable (SDGFP, 2009f). Managers are concerned, however, that hunters focus too much on harvesting bucks and not enough on harvesting antlerless deer (SDGFP, 2009c). From 2001 to 2008, thousands of unsold firearm permits were available in the East and West River Units (SDGFP, 2001–2009). Deer hunter success rates remained relatively stable during this period, ranging between 50–59% and 50–56% for rifle seasons in the East and West River Units, respectively. Liberalized availability of permits has brought populations of white-tailed deer closer to goals in some units recently and permits there were more limited in 2009. Currently, deer managers are trying to increase the number of white-tailed deer in the Black Hills, stabilize numbers in several East River Units, and reduce numbers in most of the western half of the state (T. Benzon, South Dakota Department of Game, Fish and Parks, personal communication).

Colorado

White-tailed deer in Colorado primarily occur on the eastern plains and foothills. None were year-round residents prior to 1948; although small groups of white-tailed deer had been reported earlier in eastern Colorado (Hunter, 1948, cited in Kufeld and Bowden, 1995). In 1964, two groups of white-tailed deer were introduced in northeastern Colorado: 20 near Weldona along the South Platte River and 22 in the foothills northwest of Fort Collins (Anonymous, 1965; cited in Kufeld and Bowden, 1995). Growth of the white-tailed deer population was facilitated by changing habitats along river bottoms in eastern Colorado, including increased woody vegetation and irrigated agriculture (Kufeld and Bowden, 1995). White-tailed deer in eastern Colorado appear to be more mobile than those in other parts of the United States and are common in some areas that were previously dominated by mule deer (Kufeld and Bowden, 1995). Wildlife managers of the Colorado Division of Wildlife (CDOW) and interest groups are concerned about interspecific competition and hybridization between these species (Stratman, 2006, 2007). Colorado deer hunters traditionally have preferred mule deer and mule deer are more vulnerable to rifle hunters due to preferences for open habitats. Therefore, harvests under either-species licensing tend to be skewed toward mule deer (Stratman, 2006, 2007).

Big game are managed by the CDOW to achieve population and sex-ratio objectives within Data Analysis Units (DAUs) Stratman, 2006, 2007). Each DAU consists of one or more Game Management

Units (GMUs), and approximately encompasses all of the seasonal ranges of big game herds with minimal interchange among adjacent herds (Stratman, 2006, 2007). Objectives are set for 10-year intervals (Stratman, 2006, 2007) and season structures are set for five-year intervals (CDOW, 2009c). Concerns and ideas of land management agencies and the public are integrated into planning processes for DAUs (Stratman, 2006, 2007). Deer population sizes for DAUs are projected using computer models parameterized with harvest data from hunter surveys, sex and age composition data from field surveys, and mortality factors such as wounding loss and winter severity approximated from field observations (Vieira, 2007). Ground surveys are typically used to estimate sex–age structures (Stratman, 2006, 2007). Model-projected population levels are compared to objectives to determine the harvest structure required to achieve objectives.

All deer hunting permits in Colorado have been issued by drawing since 1999 (CDOW, 2000). Deer permits are not species specific in GMUs, but where white-tailed deer are perceived to be encroaching on mule deer, white-tailed deer only (WTO) permits are issued in sufficient quantities to increase harvest of white-tailed deer without risk of overharvesting mule deer (Stratman, 2006, 2007). Special WTO seasons were held during 2003–2005 in conjunction with established deer seasons to evaluate efficacy of selective harvest for managing species ratios. Desired outcomes were achieved and separate WTO seasons were discontinued in favor of integrating WTO permits into existing season structures. Harvest management of white-tailed deer has become more aggressive over time. In 2004, 130 antlerless permits were allotted across three GMUs during the Late Plains WTO season. Either-sex permits were added in 2005, with a total quota of 135 in the same GMUs (CDOW, 2005). By 2009, 2230 WTO (antlerless or either-sex) permits were allocated across various seasons, in 29 GMUs (CDOW, 2009b). The deer hunting seasons in 2010–2014 will continue with those management options and also may include experimental season-choice private landowner permits and over-the-counter either-sex permits in select GMUs (CDOW, 2009c).

Although CWD was originally discovered in Colorado and exists in cervid populations across much of the state (CDOW, 2009a), it has not been a specific management issue for white-tailed deer. Other than monitoring for CWD and informing the public of disease status, management of CWD by the CDOW has focused primarily on foothills mule deer (Conner et al., 2007; Vieira, 2007).

Saskatchewan

White-tailed deer are at the northern edge of their range in Saskatchewan and as such are limited by winter severity (SME, 2008b). They are the most abundant big game species, however, with a total population of over 350,000 (SME, 2008b). The Saskatchewan Ministry of Environment manages hunting seasons for white-tailed deer in the southern two-thirds of the province (SME, 2008b). The priority for permit allocation is: species conservation, treaty and aboriginal rights, Saskatchewan resident hunters, Canadian resident hunters, and nonresident hunters. In farmland areas of southern Saskatchewan, landowner tolerance of crop damage and depredation to stored hay by deer are considered when establishing population objectives. The SME has an informal set of objectives for managing white-tailed deer that includes maintaining a province-wide winter population of 325,000 + 10% with an autumn herd structure of greater than 40 bucks:100 does:90 fawns (Arsenault, 2005).

Saskatchewan is divided into 42 white-tailed deer management units (WDMUs) and each is comprised of at least one of 76 wildlife management zones (WMZs) (Arsenault, 2005; SME, 2008b). Population modeling and status assessments are conducted at the WDMU level and harvest management is implemented at the WMZ level.

Population estimates are obtained during winter for select WDMUs, WMZs, or subdivisions of WMZs using aerial surveys with a precision target of no more than 20% (Arsenault, 2005; SME, 2008b, 2009). The aerial surveys are spatially and temporally sparse (SME, 2009) and trends in abundance of surveyed units typically are applied to adjacent units that are not surveyed (Arsenault, 2005). The sex–age structure of the population in each WMZ is estimated annually through the Cooperative Deer Management Survey, where citizens and SME staff quantify numbers of deer observed by sex, age, antler configuration, and habitat during September–November (Arsenault, 2005; SME, 2008b). Surveys of hunter harvest provide data on the sex–age structure of the population of harvested deer and hunter effort (SME, 2008b). Spotlight surveys are conducted in 13 WMZs, using consistent methods and time of year, to develop indices of population abundance and structure (SME, 2009). Data from these surveys are included in population models to predict deer numbers within WDMUs (Arsenault, 2005). Population status is evaluated annually, based on deer abundance and population structure; weather severity; habitat conditions; results from biological sampling and necropsies; and field reports from the general public, landowners, and SME staff (Arsenault, 2005). Harvest goals are based on perceived effects of management strategies versus other environmental factors on deer numbers relative to long-term average levels (Arsenault, 2005).

In 2009, hunters were allowed to harvest one adult buck, adult doe, or fawn (either-sex permit) in any WMZ, except three large WMZs in the far north where no deer seasons were established. In four farmland WMZs and in all forested WMZs, where recent winters had reduced deer herds, no additional permits were available. However, in the remaining southern WMZs, residents were allowed an additional permit to harvest one antlerless deer. In a further reduced subset of southern WMZs, hunters were allowed a second additional permit to harvest two antlerless deer (up to four deer). All permits were offered over-the-counter, could be purchased in any order, and were valid for archery, muzzleloader, and rifle seasons, where archery seasons began on September 1 and rifle seasons ended December 7. Nonresidents of Saskatchewan also were allowed one over-the-counter either-sex permit, for a limited number of WMZs with somewhat reduced seasons compared to those available for residents. In-season depredation permits and Big Game Management Licenses also are available to provide greater flexibility for directing hunting pressure and reducing deer numbers where deer damage is a concern (Arsenault, 2005; SME, 2007a).

Chronic wasting disease was first discovered in Saskatchewan's wild mule deer in 2000 and in wild white-tailed deer in 2002 (SME, 2010b). Herd reduction areas (HRAs) were established where CWD was found and free CWD control permits were issued to reduce herd sizes and provide samples for CWD testing (SME, 2002). In 2004, a formal statement by the SME on the CWD control program included herd reduction and sampling of deer in known CWD affected areas, planning for future sampling to monitor spread, and continuous intensive herd reduction in infected areas (SME, 2004). The number and size of HRAs increased as CWD was detected in additional areas of the province (SME, 2004, 2005, 2006, 2007b). By 2008, the focus of management shifted from CWD eradication to province-wide sampling to determine distribution of CWD, along with monitoring and managing prevalence of CWD in areas where it was known to exist at that time (SME, 2008a). Despite several years of aggressive harvest management using cooperative efforts of hunters and landowners, it was concluded that CWD could not be eradicated from Saskatchewan's herds of wild deer and elk (SME, 2010a). Herd reduction efforts previously initiated to eradicate CWD were discontinued in 2009 (SME, 2010a). However, population goals will be reduced on WMZs in which CWD is endemic and collection of samples from hunter harvest will continue to monitor prevalence and distribution of the disease.

Research Needs

The thriving population of white-tailed deer in the Midwest presents many challenges and questions that wildlife biologists must address. Overabundant deer populations threaten human health and safety and have economic and ecological ramifications as related to, for example, increases in human injuries and fatalities from DVCs, damage to crop fields and natural ecosystems, and transmission of zoonotic diseases (Chapters 12 and 13). Conversely, deer are also recognized as a valuable natural resource and positively impact society in the forms of wildlife viewing, hunting, venison, and ecosystem diversity (Conover, 1997). For these reasons and others, timely research is a high priority so that wildlife managers are provided the knowledge and tools they need to optimally manage deer (Figure 17.7).

Research is needed to develop alternative methods to estimate white-tailed deer population densities, sex ratios, and age structures. Current methods include drive and spot-light counts, aerial surveys, infrared cameras, pellet counts, mark-recapture techniques, and harvest-based estimates (Roberts et al., 2005; Skalski et al., 2005; Yamamura et al., 2008). As deer populations are often at or above ecological and sociological carrying capacities (Waller and Alverson, 1997), improved means of monitoring populations are needed by wildlife managers responsible for setting harvest quotas, maximizing recreational opportunities, minimizing DVCs and damage caused by deer, and maintaining sustainable populations of white-tailed deer at tolerable levels.

FIGURE 17.7 More research is needed to answer questions about population monitoring, harvest management, damage management, diseases, and human dimensions. Field work involves capturing deer in cagetraps (a) and by chemical immobilization (b), ratio-telemetry equipment (c), and radio-marked animals (d). (Photos by G. Clements. With permission.)

Harvest management strategies have been the cornerstone of deer management across North America, but hunting has failed to control populations of white-tailed deer in many areas of the Midwest. Impediments to effective herd management include overly conservative population goals, limited hunter access, public and private refugia, adequate behavior and motivation of hunters, declining numbers of hunters, lack of effective alternative control methods, and our lack of understanding of natural and human-modified systems. We need to apply the research principles associated with adaptive management to evaluate the impacts of current management goals, objectives, and techniques to enable managers to improve management programs in the future (Walters, 1986; Lancia et al., 1996).

Populations of deer in urban/suburban landscapes have increased across North America. Further research is needed to develop new techniques that are socially acceptable, yet effective at alleviating conflicts in these environments. Highly controlled hunting has been effective in urban situations (Butfiloski et al., 1997) but concerns about public safety and acceptance need to be addressed (Kilpatrick et al., 2002, 2007). Sharpshooting is effective in reducing numbers of deer in a highly selective manner, but public acceptance is a concern (DeNicola et al., 1997, 2008). More research is needed on methods of population reduction in these environments, as it appears that urban/suburban areas are the wave front of all our current concerns regarding publicly acceptable herd control.

Prevention and control of deer damage will always be a research need as long as deer adversely affect society. Habitat modification, exclusion, repellents, frightening devices, and population reduction (i.e., legal harvest, depredation permits, fertility control, capture, and removal) are current tools available to wildlife managers to reduce damage caused by deer, but not all of these tools are effective or fully developed and new tools and techniques are needed. Recent advances in fertility control have led to the development of GonaCon, a fertility control agent for white-tailed deer that is registered by the U.S. Environmental Protection Agency (2010; Miller et al., 2008; National Wildlife Research Center, 2010). Application, delivery, side effects, and cost-effectiveness remain a concern for this and other methods of fertility control.

Research is needed to develop an improved understanding of how disease and parasites affect midwestern deer populations. Both can cause morbidity and mortality in deer, sometimes impacting local populations significantly. Deer can also serve as reservoirs and vectors, transmitting disease to other wildlife, livestock, and humans. As diseases such as CWD and bovine TB are becoming more prominent in the Midwest, additional research is needed to devise strategies for managing and controlling them over the long term. As the global community continues to expand, we anticipate that foreign animal diseases (FADs) such as foot and mouth disease will be an even greater concern and problem in the future. Current research and development of surveillance procedures, epidemiological models, and management strategies will aid in control of future outbreaks of FADs.

Outlook

Throughout this chapter, we have identified how the beliefs, attitudes, and opinions of various publics impact deer management in the Midwest. In essence, many aspects of white-tailed deer management are actually more aptly directed at management of human activities than deer. The cornerstone of deer management is the use of hunter harvest to attain population levels that are low enough to minimize public concern associated with deer damage (e.g., agricultural crops, landscapes, plant communities, vehicles) and human health (e.g., human death and injuries from DVCs, Lyme disease) while also providing ample opportunities for hunting and viewing deer.

Over 10 years ago, McCabe and McCabe (1997) pointed out that populations of white-tailed deer in North America were at all-time historic highs of 16–17 million. More recent estimates put the nationwide population at about 30 million (VerCauteren, 2003). By necessity, management of white-tailed deer has evolved from establishing and allowing herd expansion to controlling and reducing population growth. New tools are needed to help reduce densities of deer and manage them at population goals. To date, agencies have primarily used public hunting to manage and control deer populations, but the ability of hunting to control deer populations has come into question (Brown et al., 2000). Brown et al. (2000) identified several factors that may limit hunting as a control mechanism for white-tailed deer populations in the future, including: (1) continued increase in deer distribution and abundance, (2) stable or declining number of hunters, (3) reduced access to private hunting lands that act as refuges during hunts, (4) overabundant urban and suburban deer populations, and (5) diminished social acceptability of hunting. Human dimensions also play an important role when dealing with issues such as controlling disease in deer. For example, recent outbreaks of CWD in the Midwest were followed by liberalized hunting to facilitate rapid population reductions within and around the locality of the outbreak. Even though it had been theorized that this was the most appropriate method for containing the disease, minimizing spread, and protecting overall herd health, resentment existed among certain interest groups. Supplemental feeding and baiting have become commonplace in many areas across the Midwest, and especially in areas where deer densities are relatively high. Biologists have recognized that these activities facilitate disease transmission and habitat degradation and therefore should be curtailed. Enacting restrictions on these activities, however, have been met with considerable resistance (O'Brien et al., 2006).

Recreational hunting is and will remain the primary mechanism to control populations of white-tailed deer, though means to increase hunter harvest need to be identified and explored. In many areas even the most liberal of hunting seasons are not effectively managing populations and the need to decrease deer densities is urgent for several reasons (i.e., DVCs, habitat/landscaping destruction, intense crop damage, disease control). Currently, most hunters only harvest one–two deer/year (which is the number their families will consume), not taking nearly as many deer as they could legally harvest (Table 17.1). To increase harvest, agencies and hunters are developing venison dispersal programs (e.g., the Nebraska Deer Exchange program [http://www.ngpc.state.ne.us/hunting/programs/deerexchange/] and Hunters for the Hungry [e.g., http://www.nrahq.org/hunting/hunterhungry.asp]) where hunters are provided means to donate deer that they harvest to others interested in consuming venison (Figure 17.8). Development of a new culture that facilitates and accepts the ample harvest and distribution of deer from hunters to others in a community may serve to increase local deer harvest and reduce their negative impacts. Also, outreach and extension efforts to develop, recruit, and retain new hunters will become more important.

FIGURE 17.8 Co-author Scott Hygnstrom donates two harvested antlerless white-tailed deer to an interested recipient through Nebraska's Deer Exchange Program, one of the many programs in the Midwest designed to increase deer harvest, reduce deer densities, provide food for the needy, and improve hunter–public relations. (Photo by S. Vantassel. With permission.)

Programs to encourage harvesting deer as a means to obtain "green" and locally grown lean protein should be considered.

A more radical alternative to decrease local deer densities would be a highly regulated form of licensed commercial harvest of deer for sale of venison and other deer products (Thogmartin, 2006). A hypothetical version may look like this: after completing state agency sponsored training, a deer harvester purchases a license from the agency to harvest a number of does (harvesters would be quite competent so the number of male fawns harvested could be minimal). Harvest would occur where dictated by the agency and the harvester would be given a quota of the maximum number of deer he or she is allowed to take. The harvester, then, can sell the deer he harvests for profit. An analogy would be timber harvests on state-owned lands. It is common practice in Midwestern states for state agencies to manage their forests and achieve their goals by contracting with private timber harvesters who pay the agency for the privilege to harvest (and therefore manage) the timber and then sell and profit directly from its sale. Our paradigm would allow for entrepreneurship and small-business opportunity while benefiting agricultural producers, home owners, motorists, and others. Note that commercial hunting would be dramatically different from "market hunting" because it would be highly regulated. The program would need to be controlled and dynamic. As densities of deer approach management goals, harvesters would stop taking deer and move to the next designated area. Commercial harvesting would have to achieve additive mortality in local areas to reduce densities to pre-determined levels set by the state agency. No loss of opportunity for hunters would occur. Recreational hunting would continue to be the primary means of managing deer populations.

Another realization is that more value can be derived from white-tailed deer than generally is obtained. Deer are a renewable natural resource and hunters and consumers of deer can use more of deer carcasses than just the meat for culinary enjoyment and antlers for fond memories. The hides, bones, and other wastes from processing could be put to better use than landfill material in the form of leather products, fertilizer, repellents, and other value-added products.

In the near future conversion of agricultural lands from traditional row crops into biomass crops could have significant impacts on deer and other wildlife species (Bies, 2006; Tilman et al., 2006; Walter et al., 2009). As global demand and prices for energy increase, alternative sources of energy (i.e., biofuels) will become increasingly important (Energy Information Administration, 2009). From 2006 to 2030, world consumption of marketed energy is projected to increase 44% (Energy Information Administration, 2009). To meet this demand, biofuels produced from biomass (i.e., forestry residues, agricultural crops and residues,

FIGURE 17.9 The outlook for white-tailed deer in the Midwest is very promising—We have healthy and abundant populations throughout much of the region and human interest may be at an all-time high. (Photo by G. Clements. With permission.)

wood, livestock, and municipal wastes) are being used to create ethanol and biodiesel (Bies, 2006). Current ethanol production is primarily from corn grain. New technology in cellulosic digestion may also enable use of corn husks and stalks and switchgrass residues as energy sources. Walter et al. (2009) reported that decreasing the amount of crops by 44% in Desoto National Wildlife Refuge, Iowa and converting these fields to native grasses reduced size of resident deer home ranges, but percentages of crops in home ranges increased. Increased biofuel production in the Midwest could impact deer negatively by limiting forage in corn fields after harvest and removing hiding and thermal cover by harvesting switchgrass.

Urban and suburban areas will continue to grow in size as the human population in the United States continues to increase. Approximately 64% of the United States population lived in urban areas in 1950. Today, more than 80% lives in urban areas (Hobbs and Stoops, 2002) and there is a trend for urbanites to disperse from urban centers to suburbia. A principal result of suburban growth (i.e., urban sprawl) is more vehicular traffic. Complex road networks are needed to connect large tracts of low-density neighborhoods and suburban residents to cities. People living in suburbs often must drive to work, putting them at risk to DVCs. Urbanization may also create and enhance deer habitat compared to previously rural areas. Notably, as residential acreages and subdivisions increase in extent, so could refugia for suburban deer populations and potential conflicts.

To sum, as Gladfelter (1984) pointed out in the first compilation of white-tailed deer research and management, white-tailed deer are flourishing in the Midwest (Figure 17.9). He could not have been more correct and it would have been difficult to predict just how successful deer would be early into the twenty-first century. Our predecessors did an excellent job of restoring and fostering reestablishment and expansion of white-tailed deer. Now it is our duty to manage our wonderful and dear deer resource as best possible. We have the expertise and continue to learn more, but the scientific aspects of managing deer are only a component. We must work toward increasing the understanding of our publics and politicians about the complexities associated with deer management, for their support and wherewithal will be required into the future as our management efforts will inevitably need to be intensified.

Acknowledgments

We are indebted to M. Lavelle, J. Fischer, G. Phillips, T. Ruby, A. Hildreth, and G. Clements for assisting with a variety of aspects of this chapter. We also thank T. Benzon, C. Bishop, L. Cornicelli, L. Fox, K. Hams, L. Hansen, P. Shelton, A. Schmidt, and other deer biologists from the Midwest for their assistance and insights.

REFERENCES

Alibasogulu, M. D., C. Kradel, and H. W. Dunne. 1961. Cerebral nematodasis in Pennsylvania deer (*Odocoileus virginianus*). *Cornell Veterinarian* 54: 431–441.

Allison, T. D. 1990. The influence of deer browsing on the reproductive biology of Canada yew (*Taxus canadensis* Marsh.). *Oecologia* 83:523–529.

Alverson, W. S., D. M. Waller, and S. L. Solheim. 1988. Forest too deer: Edge effects in northern Wisconsin. *Conservation Biology* 2:348–358.

Anderson, R. C. 1963. The incidence, development, and experimental transmission of *Pneumostronylys tenuis* Dougherty (Metastrongyloidea: Protostrongylidea) of the meninges of white-tailed deer (*Odocoileus virginianus* Borealis) in Ontario. *Canadian Journal of Zoology* 41:775–792.

Anderson, R. C. and M. W. Lankester. 1974. Infectious and parasitic diseases and arthropod pests of moose in North America. *Naturaliste Canadien* 101:23–50.

Anderson, R. C., M. W. Lankester, and V. R. Strelive. 1966. Further experimental studies of *Pneumostrongylus tenuis* in cervids. *Canadian Journal of Zoology* 44:851–861.

Anonymous. 1965. More whitetails for Colorado. *Colorado Outdoors* 14:5.

Answers.com. 2010. *ReferenceAnswers*. http://www.answers.com. Accessed May 18, 2010.

Arsenault, A. A. 2005. *Status and Management of Wildlife in Saskatchewan, 2002 and 2003*. Saskatchewan Environment Resource Stewardship Branch Technical Report 2005–2. Regina, Saskatchewan.

Bacon, R. M., K. J. Kugeler, and P. S. Mead. 2008. Surveillance for Lyme disease—United States, 1992–2006. *Centers for Disease Control Surveillance Summaries* 57:1–9.

Ballard, W. B., H. A. Whitlaw, S. J. Young, R. A. Jenkins, and G. J. Forbes. 1999. Predation and survival of white-tailed deer fawns in north central New Brunswick. *Journal of Wildlife Management* 63:574–579.

Beals, E. W., G. Cottam, and R. J. Vogl. 1960. Influence of deer on vegetation of the Apostle Islands. *Journal of Wildlife Management* 24:68–80.

Beringer, J., L. P. Hansen, R. A. Heined, and N. F. Geissman. 1994. Use of dogs to reduce damage by deer to a white pine plantation. *Wildlife Society Bulletin* 22:627–632.

Beringer, J., L. P. Hansen, and D. E. Stallknecht. 2000. An epizootic of hemorrhagic disease in white-tailed deer in Missouri. *Journal of Wildlife Diseases* 36:588–591.

Berry, W. 1977. *The Unsettling of America: Culture and Agriculture*. San Francisco, CA: Sierra Club Books.

Bies, L. 2006. The biofuels explosion: Is green energy good for wildlife? *Wildlife Society Bulletin* 34:1203–1205.

Blouch, R. I. 1994. Northern Great Lakes states and Ontario forests. In *White-tailed Deer: Ecology and Management*, ed. L. K. Halls, 391–410. Harrisburg, PA: Stackpole Books.

Bogaczyk, B. A., W. B. Krohn, and H. C. Gibbs. 1993. Factors affecting *Parelaphostrongylus tenuis* in white-tailed deer (*Odocoileus virginianus*) from Maine. *Journal of Wildlife Diseases* 29: 266–272.

Brown, R. N. and E. C. Burgess. 2001. Lyme Borreliosis. In *Infectious Diseases of Wild Mammals*, eds. E. S. Williams and I. K. Barker, 435–454. Ames, IA: Iowa State Press.

Brown, T. L., D. J. Decker, S. J. Riley et al. 2000. The future of hunting as a mechanism to control white-tailed deer populations. *Wildlife Society Bulletin* 28:797–807.

Bryan, D. A. 1980. White-tailed deer fawn mortality, home range, and habitat utilization in east central Missouri. MS thesis, University of Missouri.

Butfiloski, J. W., D. I. Hall, D. M. Hoffman et al. 1997. White-tailed deer management in a coastal Georgia residential community. *Wildlife Society Bulletin* 25:491–495.

Carstensen, M., G. D. Delgiudice, B. A. Simpson, and D. W. Kuehn. 2006. Survival, birth characteristics, and cause-specific mortality of white-tailed deer neonates. *Journal of Wildlife Management* 73:175–183.

Centers for Disease Control and Prevention. 2010. *Lyme Disease*. http://www.cdc.gov/ncidod/dvbid/lyme/ld_UpClimbLymeDis.htm. Accessed May 18, 2010.

Clements, G. M., S. E. Hygnstrom, J. M. Gilsdorf, D. M. Baasch, M. J. Clements, and K. C. VerCauteren. 2011. Movements and home range of male white-tailed deer in riparian habitat: Implications for management of infectious diseases. *Journal of Wildlife Management*. In press.

Colorado Division of Wildlife. 2000. *Annual Report*. http://wildlife.state.co.us/About/Reports/2000WildlifeRelatedRecreation.htm. Accessed February 10, 2010.

Colorado Division of Wildlife. 2005. *Final Regulations—Chapter 2—Big Game. Final as Approved by Wildlife Commission—5 May 2005*. http://wildlife.state.co.us/NR/rdonlyres/20A5EC3F-7A31–49CE-AE49–49ED3D533EA6/0/Ch2BigGame.pdf. Accessed February 10, 2010.

Colorado Division of Wildlife. 2008. *History of White-tailed Deer only (WTO) Licenses in Colorado.* http://wildlife.state.co.us/NR/rdonlyres/9B0CCFCB-0CF6–4FE4-AD36–8D852D42A3C9/0/WhitetailedDeerHuntingBackgroundInfoSep2008.pdf. Accessed April 29, 2010.

Colorado Division of Wildlife. 2009a. *Chronic Wasting Disease in Colorado, 2006–2008.* http://wildlife.state.co.us/NR/rdonlyres/763F5731-F895–4D52–9F27–2B8D5BE91175/0/CO_CWDreport_06082.pdf. Accessed February 9, 2010.

Colorado Division of Wildlife. 2009b. *Final Regulations—Chapter 2—Big Game. Final as Approved by Wildlife Commission—7 May 2009.* http://wildlife.state.co.us/NR/rdonlyres/F8FA1A12-C832–4D32–9E99-D5FFFFD65E8D/0/Ch2LicenseNumbersAsApproved.pdf. Accessed February 10, 2010.

Colorado Division of Wildlife. 2009c. *Five Year Big Game Season Structure 2010–2014. Final as Approved by Wildlife Commission–10 Sep 2009.* http://wildlife.state.co.us/NR/rdonlyres/DB92C226–7245–47B2-ACCB-265AC0335DBC/0/Final20102014BGSS.pdf. Accessed February 10, 2010.

Compton, B. R., R. J. Mackie, and G. L. Dusek. 1988. Factors influencing distribution of white-tailed deer in riparian habitats. *Journal of Wildlife Management* 52:544–548.

Conner, M. M., M. W. Miller, M. R. Ebinger, and K. P. Burnham. 2007. A meta-BACI approach for evaluating management intervention on chronic wasting disease in mule deer. *Ecological Applications* 17:140–153.

Conover, M. R. 1997. Monetary and intangible valuation of deer in the United States. *Wildlife Society Bulletin* 25:298–305.

Conover, M. R. 2002. *Resolving Human–Wildlife Conflicts: The Science of Wildlife Damage Management.* New York, NY: Lewis Publishers.

Cornicelli, L. A., A. Woolf, and J. L. Roseberry. 1996. White-tailed deer use of a suburban environment in southern Illinois. *Transactions of the Illinois State Academy of Science* 89:93–103.

Côté, S. D., T. P. Rooney, J. Tremblay, C. Dussault, and D. M. Waller. 2004. Ecological impacts of deer overabundance. *Annual Reviews of Ecology, Evolution, and Systematics* 35:113–147.

Craven, S. R. and S. E. Hygnstrom. 1994. Deer. In *Prevention and Control of Wildlife Damage*, eds. S. E. Hygnstrom, R. M. Timm, and G. E. Larson, D25–D40. Lincoln, NE: University of Nebraska Cooperative Extension.

Creed, W. A., F. Haberland, B. E. Kohn, and K. R. McCafferty. 1984. Harvest management: The Wisconsin experience. In *White-tailed Deer: Ecology and Management*, ed. L. K. Halls, 243–260. Harrisburg, PA: Stackpole Books.

Cross, C. F. 1995. *Go West Young Man!: Horace Greeley's Vision for America.* Albuquerque, NM: University of New Mexico Press.

CWD Stakeholder Advisory Group. 2008. Summary and recommendations to Secretary Matt Frank, WI DNR and the WI Natural Resources Board. http://dnr.wi.gov/org/land/wildlife/whealth/issues/cwd/doc/SAGreport.pdf. Accessed February 19, 2010.

Dahlberg, B. L. and R. C. Guettinger. 1956. *The White-tailed Deer in Wisconsin.* Madison, WI: Wisconsin Conservation Department.

Davidson, W. R. 2006. *Field Manual of Wildlife Diseases in the Southeastern United States* (3rd edition). Athens, GA: Southeastern Cooperative Wildlife Disease Study.

deCalesta, D. S. 1994a. Deer and diversity in Allegheny hardwood forests: Managing an unlikely challenge. *Landscape and Urban Planning* 28:47–53.

deCalesta, D. S. 1994b. Effect of white-tailed deer on songbirds within managed forests in Pennsylvania. *Journal of Wildlife Management* 58:711–718.

deCalesta, D. S. and D. B. Schwendeman. 1978. Characterization of deer damage to soybean plants. *Wildlife Society Bulletin* 6:250–253.

DeNicola, A. J., D. R. Etter, and T. Almendinger. 2008. Demographics of non-hunted white-tailed deer populations in suburban areas. *Human–Wildlife Conflicts* 2:102–109.

DeNicola, A. J., K. C. VerCauteren, P. D. Curtis, and S. E. Hyngstrom. 2000. *Managing White-tailed Deer in Suburban Environments.* Ithaca, NY: Cornell Cooperative Extension.

DeNicola, A. J., S. J. Weber, C. A. Bridges, and J. L. Stokes. 1997. Nontraditional techniques for management of overabundant deer populations. *Wildlife Society Bulletin* 25:496–499.

Diamond, J. 1992. Must we shoot deer to save nature? *Natural History* 8:2–8.

Doerr, M. L., J. B. McAninch, and E. P. Wiggers. 2001. Comparison of 4 methods to reduce white-tailed deer abundance in an urban community. *Wildlife Society Bulletin* 29:1105–1113.

Dolbeer, R. A. and S. E. Wright. 2009. Safety and management systems: How useful will the FAA National Wildlife Strike Database be? *Human–Wildlife Conflicts* 3:167–178.

Dorn, M. L. and A. G. Mertig. 2005. Bovine tuberculosis in Michigan: Stakeholder attitudes and implications for eradication efforts. *Wildlife Society Bulletin* 33:539–552.

Eberhardt, L. L. 1960. *Estimation of Vital Characteristics of Michigan Deer Herds*. Michigan Department of Conservation Game Division Report 2282. East Lansing, MI.

Energy Information Administration. 2009. *International Energy Outlook 2009*. Report # DOE/EIA-0484. U.S. Department of Energy, Washington, DC.

Etter, D. R., K. M. Hollis, T. R. Van Deelen. 2002. Survival and movements of white-tailed deer in suburban Chicago, Illinois. *Journal of Wildlife Management* 66:500–510.

Farnsworth, M. L., J. A. Hoeting, N. T. Hobbs, and M. W. Miller. 2006. Linking chronic wasting disease to mule deer movement scales: A hierarchical Bayesian approach. *Ecological Applications* 16:1026–1036.

Frosch, B. J., D. P. Anderson, and J. L. Outlaw. 2008. Economic impact of deer breeding operations in Texas. *Proceedings of the Southern Agricultural Economics Association Annual Meeting*. Dallas, TX.

Frost, C. J., S. E. Hygnstrom, A. J. Tyre et al., 2009. Probabilistic movement model with emigration simulates movements of deer in Nebraska, 1990–2006. *Ecological Modelling* 220:2481–2490.

Frost, H. C., G. L. Storm, M. J. Batcheller, and M. J. Lovallo. 1997. White-tailed deer management at Gettysburg National Historic Park and Eisenhower National Historic Site. *Wildlife Society Bulletin* 25:462–469.

Fuller, T. K., W. E. Berg, G. L. Radde, M. S. Lenarz, and G. B. Joselyn. 1992. A history and current estimate of wolf distribution and numbers in Minnesota. *Wildlife Society Bulletin* 20:42–55.

Fulton, D. C, J. T. Bruskotter, and L. Cornicelli. 2004. Assessing level of support for alternative deer management strategies in northwest Minnesota. Minnesota Cooperative Fish and Wildlife Research Unit Unpublished Project Report. St. Paul, Minnesota.

Garner, D. L., S. Dunwoody, D. Joly, D. O'Brien, M. J. Peterson, and M. J. Pybus. 2009. *External Panel Review of "A Plan for Managing Chronic Wasting Disease in Wisconsin: The Next Five Years." 8 Dec 2009*. http://dnr.wi.gov/org/land/wildlife/WHEALTH/issues/cwd/doc/External_Review.pdf. Accessed February 19, 2010.

Garner, M. S. 2001. Movement patterns and behavior at winter feeding and fall baiting stations in a population of white-tailed deer infected with bovine tuberculosis in the northeastern lower peninsula of Michigan. PhD dissertation, Michigan State University.

Gaydos, J. K., J. M. Crum, W. R. Davidson, S. S. Cross, S. F. Owen, and D. E. Stallknecht. 2004. Epizootiology of an epizootic hemorrhagic disease outbreak in West Virginia. *Journal of Wildlife Diseases* 40:383–393.

Gibbs, E. P. J. 1992. Epidemiology of orbiviruses–bluetongue: Towards 2000 and the search for patterns. In *Bluetongue, African Horse Sickness, and Related Orbiviruses*, eds. T. E. Walton and B. I. Osburn, 65–75. Boca Raton, FL: CRC Press.

Gigliotti, L. M. 2007. *Wildlife on Private Lands Status Report 2007*. HD-11–07.AMS. South Dakota Game, Fish and Parks. Pierre, SD.

Gladfelter. H. L. 1984. Midwest agricultural region. In *White-tailed Deer: Ecology and Management*, ed. L. K. Halls, 427–440. Harrisburg, PA: Stackpole Books.

Grovenburg, T. W., J. A. Jenks, C. N. Jacques, R. W. Klaver, and C. C. Swanson. 2009. Aggressive defensive behavior by free-ranging white-tailed deer. *Journal of Mammalogy* 90:1218–1223.

Grund, M. D., J. B. McAnnich, and E. P. Wiggers. 2002. Seasonal movements and habitat use of female white-tailed deer associated with an urban park. *Journal of Wildlife Management* 66:123–130.

Gubanyi, J. A., J. A. Savidge, S. E. Hygnstrom, K. C. VerCauteren, G. W. Garabrandt, and S. P. Korte. 2008. Deer impact on vegetation in natural areas in southeastern Nebraska. *Natural Areas Journal* 28:121–129.

Hansen, L. and J. Beringer. 1997. Managed hunts to control white-tailed deer populations on urban public areas. *Wildlife Society Bulletin* 25:484–487.

Hansen, L. P., C. M. Nixon, and J. Beringer. 1991. Role of refuges in the dynamics of outlying deer populations. In *The Science of Overabundance: Deer Ecology and Population Management*, eds. W. J. McShea, H. B. Underwood, and J. H. Rappole, 327–345. Washington, DC: Smithsonian Institution Press.

Harlow, R. F. and D. C. Guynn Jr. 1994. Whitetail habitats and ranges. In *Deer*, eds. D. Gerlach, S. Atwater, and J. Schnell, 169–173. Mechanicsburg, PA: Stackpole Books.

Hobbs, F. and N. Stoops. 2002. *Demographic Trends in the 20th Century: Census 2000 Special Reports*. U.S. Census Bureau, Series CENSR-4. Washington, DC.

Hoen, A. G., G. Margos, S. J. Bent et al. 2009. Phylogeography of *Borrelia burgdorferi* in the eastern United States reflects multiple independent Lyme disease emergence events. *Proceedings of the National Academy of Sciences, USA* 106:15013–15018.

Holsman, R. H. and E. E. Cooney. 2007. *Five Years Into the CWD Eradication Effort: Which Parts Do Hunters Support?* College of Natural Resources, University of Wisconsin-Stevens Point, WI.

Horsley, S. B., S. L. Stout, and D. S. deCalesta. 2003. White-tailed deer impact on the vegetation dynamics of a northern hardwood forest. *Ecological Applications* 13:98–118.

Hubbard, R. D. and C. K. Nielsen. 2009. White-tailed deer attacking humans during the fawning season: A unique human–wildlife conflict on a university campus. *Human–Wildlife Conflicts* 3:129–135.

Huebschman, J., S. E. Hygnstrom, and J. A. Gubany. 1997. Coyote food habits at DeSoto National Wildlife Refuge, Nebraska. *Prairie Naturalist* 29:100–109.

Huegel, C. N., R. B. Dahlgren, and H. L. Gladfelter. 1985. Mortality of white-tailed deer fawns in south-central Iowa. *Journal of Wildlife Management* 49:377–380.

Hunter, G. N. 1948. *History of White-tailed Deer in Colorado*. Denver, CO: Colorado Game and Fish Department.

Hygnstrom, S. E., S. R. Groepper, K. C. VerCauteren et al. 2008a. Literature review of mule deer and white-tailed deer movements in western and Midwestern landscapes. *Great Plains Research* 18:219–231.

Hygnstrom, S. E., P. D. Skelton, S. J. Josiah et al. 2009. White-tailed deer browsing and rubbing preferences for trees and shrubs that produce non-timber forest products. *Hortechnology* 19:204–211.

Hygnstrom, S. E., S. M. Vantassel, B. D. Trindle, and K. C. VerCauteren. 2008b. *Managing Deer Damage in Nebraska*. Lincoln, NE: University of Nebraska-Lincoln Extension NebGuide.

Hygnstrom, S. E. and K. C. VerCauteren. 1999. *Final Report: Ecology of White-tailed Deer in the Gifford Point-Fontenelle Forest Area, Nebraska*. Lincoln, NE: University of Nebraska-Lincoln.

Joint Task Force. 2008. *Report of the Joint Task Force on Deer Population Control as Required by House Joint Resolution 65*. Illinois Department of Natural Resources. Springfield, IL.

Joly, D. O., M. D. Samuel, J. A. Langenberg et al. 2006. Spatial epidemiology of chronic wasting disease in Wisconsin white-tailed deer. *Journal of Wildlife Diseases* 42:578–588.

Keane, D. P., D. J. Barr, P. N. Bochsler et al. 2008. Chronic wasting disease in a Wisconsin white-tailed deer farm. *Journal of Veterinary Diagnostic Investigations* 20:698–703.

Kilpatrick, H. J., A. M. Labonte, and J. S. Barclay. 2007. Acceptance of deer management strategies by suburban homeowners and bowhunters. *Journal of Wildlife Management* 71:2095–2101.

Kilpatrick, H. J., A. M. LaBonte, and J. T. Seymour. 2002. A shotgun-archery deer hunt in a residential community: Evaluation of hunt strategies and effectiveness. *Wildlife Society Bulletin* 30:478–486.

Knapp, K. K. 2005. *Statewide and Upper Midwest Summary of Deer–Vehicle Crash and Related Data from 1993 to 2003*. Deer–Vehicle Crash Information Clearinghouse Report DVCIC - 03. http://www.deercrash.com/10year/regional.htm. Accessed February 19, 2010.

Krämer, A. 1973. Interspecific behavior and dispersion of two sympatric deer species. *Journal of Wildlife Management* 37: 288–300.

Kufeld, R. C. and D. C. Bowden. 1995. *Mule Deer and White-tailed Deer Inhabiting Eastern Colorado Plains River Bottoms*. Colorado Division of Wildlife Technical Publication No. 41. Fort Collins, CO.

Kunkel, K. E. and L. D. Mech. 1994. Wolf and bear predation on white-tailed deer fawns in northeastern Minnesota. *Canadian Journal of Zoology* 72:1557–1565.

Lancia, R. A., C. E. Braun, M. W. Collopy et al. 1996. ARM! For the future: Adaptive resource management in the wildlife profession. *Wildlife Society Bulletin* 24:436–442.

Lenarz, M. S. 2010. A review of the ecology of *Parelaphostrongylus tenuis* in relation to deer and moose in North America. In *Summaries of Wildlife Research Findings*, eds. M. W. DonCarlos, R. O. Kimmel, J. S. Lawrence, and M. S. Lenarz, 70–75. St. Paul, MN. Minnesota Department of Natural Resources.

Leopold, A., L. K. Sowls, and D. L. Spencer. 1947. A survey of over-populated deer ranges in the United States. *Journal of Wildlife Management* 11:162–177.

Lesage, L., M. Crete, J. Huot, A. Dumont, and J. Ouellet. 2000. Seasonal home range size and philopatry in two northern white-tailed deer populations. *Canadian Journal of Zoology* 78:1930–1940.

Mann, C. C. 2005. *1491: New Revelations of the Americas Before a Person Named Columbus*. New York, NY: Borzoi Books, Alfred C. Knopf.

Martinka, C. J. 1968. Habitat relationships of white-tailed and mule deer in northern Montana. *Journal of Wildlife Management* 32: 558–565.

Matschke, G. H., K. A. Fagerstone, R. F. Harlow et al. 1984. Population influences. In *White-tailed Deer: Ecology and Management*, ed. L. K. Halls, 169–188. Harrisburg, PA: Stackpole Books.

Mautz, W. W. 1978. Sledding on a brushy hillside: The fat cycle in deer. *Wildlife Society Bulletin* 6:88–90.

McCabe, R. E. and T. R. McCabe. 1984. Of slings and arrows: An historical retrospection. In *White-tailed Deer: Ecology and Management*, ed. L. K. Halls, 19–72. Harrisburg, PA: Stackpole Books.

McCabe, T. R. and R. E. McCabe. 1997. Recounting whitetails past. In *The Science of Overabundance: Deer Ecology and Population Management,* eds. W. J. McShea, H. B. Underwood, and J. H. Rappole, 11–26. Washington, DC: Smithsonian Press.

McCaffery, K. 1984. Fat deer laugh at winter. *Wisconsin Natural Resources* 8:17–19.

Menzel, K. E. 1984. Central and southern plains. In *White-tailed Deer: Ecology and Management*, ed. L. K. Halls, 449–456. Harrisburg, PA: Stackpole Books.

Miller, L. A., J. P. Gionfriddo, K. A. Fagerstone, J. C. Rhyan, and G. J. Killian. 2008. The single-shot GnRH immunocontraceptive vaccine: Comparison of several GnRH preparations. *American Journal of Reproductive Immunology* 60:214–223.

Miller, M. W., E. S. Williams, N. T. Hobbs, and L. L. Wolfe. 2004. Environmental sources of prion transmission in mule deer. *Emerging Infectious Diseases* 10:1003–1006.

Millspaugh, J. J., M. S. Boyce, D. R. Diefenbach, L. P. Hansen, K. Kammermeyer, and J. R. Skalski. 2006. *An Evaluation of the SAK Model as Applied in Wisconsin. Draft.* http://dnr.wi.gov/org/land/wildlife/hunt/deer/SAKreport.pdf. Accessed February 18, 2010.

Millspaugh, J. J., J. R. Skalski, R. L. Townsend et al. 2009. An evaluation of sex–age–kill (SAK) model performance. *Journal of Wildlife Management* 73:442–451.

Mladenoff, D. J. and F. Stearns. 1993. Eastern hemlock regeneration and deer browsing in the northern Great Lakes region: A re-examination and model simulation. *Conservation Biology* 7:889–900.

Morrison, S. F., G. J. Forbes, S. J. Young, S. Lusk. 2003. Within-yard habitat use by White-tailed deer at varying winter severity. *Forest Ecology and Management* 172:173–182.

Moulton, G. E. 2003. *The Lewis and Clark Journals* (abridged edition). Lincoln, NE: University of Nebraska Press.

Murray, D. L., E. W. Cox, W. B. Ballard et al. 2006. Pathogens, nutritional deficiency, and climate influences on a declining moose population. *Wildlife Monographs* 166:1–30.

National Wildlife Research Center. 2010. *GonaCon™ New GNRH Single Shot*. U.S. Department of Agriculture-Animal Plant Health Inspection Service-Wildlife Services-National Wildlife Research Center http://www.aphis.usda.gov/wildlife_damage/nwrc/research/reproductive_control/gonacon.shtml. Accessed May 20, 2010.

Nelson, M. E. and L. D. Mech. 1986a. Mortality of White-tailed deer in northeastern Minnesota. *Journal of Wildlife Management* 50:691–698.

Nelson, M. E. and L. D. Mech. 1986b. Relationship between snow depth and gray wolf predation on White-tailed deer. *Journal of Wildlife Management* 50:471–474.

Nelson, M. E. and L. D. Mech. 2000. Do wolves affect White-tailed buck harvest in northeastern Minnesota? *Journal of Wildlife Management* 64:129–136.

Nelson, T. A. and A. Woolf. 1987. Mortality of White-tailed deer fawns in southern Illinois. *Journal of Wildlife Management* 51:326–329.

Nixon, C. M., L. P. Hansen, P. A. Brewer, and J. E. Chelsvig. 1991. Ecology of White-tailed deer in an intensively farmed region of Illinois. *Wildlife Monographs* 118:1–77.

Nixon, C. M., M. W. McClain, and K. R. Russell. 1970. Deer food habits and range characteristics in Ohio. *Journal of Wildlife Management* 34:870–886.

O'Brien, D. J., S. M. Schmitt, J. S. Fierke et al. 2002. Epidemiology of *Mycobacterium bovis* in free-ranging White-tailed deer, MI, USA, 1995–2000. *Preventive Veterinary Medicine* 54:47–63.

O'Brien, D. J., S. M. Schmitt, S. D. Fitzgerald, D. E. Berry, and G. J. Hickling. 2006. Managing the wildlife reservoir of *Mycobacterium bovis*: The Michigan, USA, experience. *Veterinary Microbiology* 112:313–323.

Ozoga, J. J., L. J. Verme, and C. S. Bienz. 1982. Parturition behavior and territoriality in White-tailed deer: Impact on neonatal mortality. *Journal of Wildlife Management* 46:1–11.

Pearce, E. A. and C. G. Smith. 1984. *World Weather Guide*. London: Hutchinson.

Petersen, L. E. 1984. Northern plains. In *White-tailed Deer: Ecology and Management*, ed. L. K. Halls, 441–448. Harrisburg, PA: Stackpole Books.

Ramankutty, N. and J. A. Foley. 1999. Estimating historical changes in global land cover: Croplands from 1700 to 1992. *Global Biochemical Cycles* 13:997–1027.

Renter, D. G., D. P. Gnad, J. M. Sargeant, and S. E. Hygnstrom. 2006. Prevalence and serovars of *Salmonella* in feces of free-ranging White-tailed deer in Nebraska. *Journal of Wildlife Diseases* 42:699–703.

Renter, D. G., J. M. Sargeant, S. E. Hygnstrom, J. Hoffman, and J. Gillespie. 2001. *Escherichia coli* O157:H7 in free-ranging deer in Nebraska. *Journal of Wildlife Diseases* 37:755–760.

Rice, D. H. 2009. Produce contamination by other wildlife. In *The Produce Contamination Problem: Causes and Solutions*, eds. G. M. Sapers, E. Solomon, and K. R. Matthews, 143–156. Burlington, MA: Academic Press.

Roberts, C. W., B. L. Pierce, A. W. Braden et al. 2005. Comparison of camera and road survey estimates for White-tailed deer. *Journal of Wildlife Management* 70:263–267.

Rolley, R. E. 2008. *White-tailed Deer Population Status*. Wisconsin Department of Natural Resources. http://dnr.wi.gov/org/land/wildlife/harvest/reports/wtaildeerpop08.pdf. Accessed February 18, 2010.

Romin, L. A. and J. A. Bisonette. 1996. Deer–vehicle collisions: Status of state monitoring activities and mitigation efforts. *Wildlife Society Bulletin* 24:276–283.

Rooney, T. P. 2001. Deer impacts on forest ecosystems: A North American perspective. *Forestry* 74:201–208.

Rooney, T. P. and D. M. Waller. 2003. Direct and indirect effects of White-tailed deer in forest ecosystems. *Forest Ecology and Management* 181:165–176.

Russell, F. L., D. B. Zippin, and N. L. Fowler. 2001. Effects of White-tailed deer (*Odocoileus virginianus*) on plants, plant populations and communities: A review. *American Midland Naturalist* 149:1–26.

Saskatchewan Ministry of Environment. 2002. Another Case of Chronic Wasting Disease in Wild. News release–16 Dec. http://www.gov.sk.ca/news?newsId=970f2f18-4900-4601-a38f-d6870c10feb2. Accessed February 10, 2010.

Saskatchewan Ministry of Environment. 2004. Saskatchewan Revises the Chronic Wasting Disease Control Program. News Release–13 Sep. http://www.gov.sk.ca/news?newsId=37008a57-f0e9-4d62-8c6b-e6300b2ebddf. Accessed February 10, 2010.

Saskatchewan Ministry of Environment. 2005. Saskatchewan Announces Fall 2005 Chronic Wasting Disease Control Program. News Release–29 Aug. http://www.gov.sk.ca/news?newsId=53f7b432-813f-47ae-b1d3-098c2b3a598f. Accessed February 10, 2010.

Saskatchewan Ministry of Environment. 2006. Chronic wasting disease 2006 management program. http://www.swf.sk.ca/pdfs/2006cwd_program.pdf. Accessed February 10, 2010.

Saskatchewan Ministry of Environment. 2007a. More Chances to Hunt Deer in Saskatchewan. Newsline, 4 Oct 2007. http://www.environment.gov.sk.ca/Default.aspx?DN=33ce5aa3-343e-423b-be67-40e0f1e40b6e. Accessed February 10, 2010.

Saskatchewan Ministry of Environment. 2007b. Saskatchewan's 2007 Chronic Wasting Disease Control Program. Newsline, Nov. 8, 2007. http://www.environment.gov.sk.ca/Default.aspx?DN=dae600a5-799d-4ff5-a3af-e99e9ee13e73. Accessed February 10, 2010.

Saskatchewan Ministry of Environment. 2008a. Chronic Wasting Disease Management Program. http://www.environment.gov.sk.ca/adx/aspx/adxGetMedia.aspx?DocID=1756,300,254,94,88,Documents&MediaID=981&Filename=2008+CWD+Program+Information.pdf&l=English. Accessed February 10, 2010.

Saskatchewan Ministry of Environment. 2008b. Saskatchewan Game Report 2007–08. http://www.environment.gov.sk.ca/adx/aspx/adxGetMedia.aspx?DocID=1968,254,94,88,Documents&MediaID=1061&Filename=Saskatchewan+Game+Report.pdf&l=English. Accessed February 10, 2010.

Saskatchewan Ministry of Environment. 2009. Wildlife surveys. http://www.environment.gov.sk.ca/adx/aspx/adxGetMedia.aspx?DocID=f8d120f6-1811-49f2-be0e-f09140bed4bd&MediaID=1500&Filename=Wildlife+Surveys+2009.pdf&l=English. Accessed February 10, 2010.

Saskatchewan Ministry of Environment. 2010a. Hunters' and Trappers' Guide. http://www.environment.gov.sk.ca/adx/aspx/adxGetMedia.aspx?DocID=655681 1e-477c-4a45-2a8-6d10cacbbf44&MediaID=3430&Filename=2010+Hunters'+and+Trappers'+Guide.pdf&l=English. Accessed February 10, 2010.

Saskatchewan Ministry of Environment. 2010b. Summary of the provincial CWD sampling effort 1997–2009. http://www.environment.gov.sk.ca/adx/aspx/adxGetMedia.aspx?DocID=1766,300,254,94,88,Documents&MediaID=2106&Filename=Summary+of+the+Provincial+CWD+Sampling+Effort+1997-2009.pdf&l=English. Accessed February 10, 2010.

Schmitt, S. M., S. D. Fitzgerald, T. M. Cooley et al. 1997. Bovine tuberculosis in free-ranging White-tailed deer from Michigan. *Journal of Wildlife Diseases* 33:749–758.

Schulz, J. H. 1982. Mortality and movements of White-tailed deer (*Odocoileus virginianus Zimmerman*) fawns in southeastern Minnesota. MA thesis, Mankato State University, Minnesota.

Shelton, P. and A. Hulin. 2004. *Illinois Chronic Wasting Disease Summary: Results for the 2003–2004 Surveillance Season*. Springfield, IL: Illinois Department of Natural Resources, Forest Wildlife Program.

Shelton, P. and A. Hulin. 2005. *Illinois Chronic Wasting Disease: 2004–2005 Surveillance/Management Summary*. Springfield, IL: Illinois Department of Natural Resources, Forest Wildlife Program.

Shelton, P. and A. Hulin. 2006. *Illinois Chronic Wasting Disease: 2005–2006 Surveillance/Management Summary*. Springfield, IL: Illinois Department of Natural Resources, Forest Wildlife Program.

Shelton, P. and P. McDonald. 2008a. *Illinois Chronic Wasting Disease: 2006–2007 Surveillance/Management Summary*. Springfield, IL: Illinois Department of Natural Resources, Forest Wildlife Program.

Shelton, P. and P. McDonald. 2008b. *Illinois Chronic Wasting Disease: 2007–2008 Surveillance/Management Summary*. Springfield, IL: Illinois Department of Natural Resources, Forest Wildlife Program.

Shelton, P. and P. McDonald. 2009. *Illinois Chronic Wasting Disease: 2008–2009 Surveillance/Management Summary*. Springfield, IL: Illinois Department of Natural Resources, Forest Wildlife Program.

Skalski, J. R., K. E. Ryding, and J. J. Millspaugh. 2005. *Wildlife Demography: Analysis of Sex, Age, and Count Data*. Burlington, MA: Elsevier Academic Press.

Sleeman, J. M., J. E. Howell, W. M. Knox, and P. J. Stenger. 2009. Incidence of hemorrhagic disease in White-tailed deer is associated with winter and summer climatic conditions. *EcoHealth* Published online: DOI 10.1007/s10393-009-0220-6.

Smyth, P. 1995. *Patterns on the Land: Our Choices—Our Future*. Lansing, MI: Lansing Printing.

South Dakota Game, Fish and Parks (SDGFP). 2009a. *Black Hills Deer, Fall Turkey*. Pierre, SD: South Dakota Game, Fish, and Parks application pamphlet.

South Dakota Game, Fish and Parks (SDGFP). 2009b. Black Hills Deer Seasons. https://www.sdgfp.info/Publications/BlackHillsDeer_unitmap.pdf. Accessed February 16, 2010.

South Dakota Game, Fish and Parks (SDGFP). 2009c. Central Regional Advisory Panel—January 2009 Summary. http://www.sdgfp.info/wildlife/rap/archive/Central_jan09.htm. Accessed February 16, 2010.

South Dakota Game, Fish and Parks (SDGFP). 2009d. East River Deer, East River Landowner Deer, Refuge Deer, Muzzleloader Deer. Pierre, SD: South Dakota Game, Fish, and Parks application pamphlet.

South Dakota Game, Fish and Parks (SDGFP). 2009e. East River Deer Units. http://www.sdgfp.info/Publications/EastRiverDeer_unitmap.pdf. Accessed February 16, 2010.

South Dakota Game, Fish and Parks (SDGFP). 2009f. South Dakota Game Report: 2008 Annual Report. Pierre, SD: South Dakota Game, Fish, and Parks report number 2009-03.

South Dakota Game, Fish and Parks (SDGFP). 2009g. West River Deer Units. http://www.sdgfp.info/Publications/WestRiverDeer_unitmap.pdf. Accessed February 16, 2010.

South Dakota Game, Fish and Parks (SDGFP). 2009h. West River Deer, West River Landowner Deer. Pierre, SD: South Dakota Game, Fish, and Parks application pamphlet.

South Dakota Game, Fish and Parks (SDGFP). 2009i. Wildlife Survey Manual: 2009–2015. Pierre, South Dakota: South Dakota Game, Fish, and Parks Project Number W-95-R.

South Dakota Game, Fish and Parks (SDGFP). 2001–2009. South Dakota Game Reports: Big Game Harvest Projections. http://www.sdgfp.info/Wildlife/hunting/Harvest/Projections.htm. Accessed February 16, 2010.

Storm, D. J., C. K. Nielsen, E. M. Schauber, and A. Woolf. 2007. Space use and survival of White-tailed deer in an exurban landscape. *Journal of Wildlife Management* 71:1170–1176.

Stratman, M. 2006. Arikaree Deer Herd Management Plan, Data Analysis Unit D-55. Fort Collins, CO: Colorado Division of Wildlife.

Stratman, M. 2007. South Tablelands Deer Herd Management Plan, Data Analysis Unit D-54. Fort Collins, CO: Colorado Division of Wildlife.

Stubblefield, S. S., R. J. Warren, and B. R. Murphy. 1986. Hybridization of free-ranging White-tailed and mule deer in Texas. *Journal of Wildlife Management* 50: 688–690.

Swenson, J. E., S. J. Knapp, and H. J. Wentland. 1983. Winter distribution and habitat use by mule deer and White-tailed deer in Southeastern Montana. *Prairie Naturalist* 15: 97–112.

Swift, E. 1946. *A History of Wisconsin Deer*. Madison, WI: Wisconsin Conservation Department.

The Cougar Network. 2010. *Big Picture Map*. http://thecougarnetwork.org/bigpicture.html. Accessed May 20, 2010.

Thogmartin, W. 2006. Why not consider the commercialization of deer harvests? *Bioscience* 56:957.

Thomas, J. E. and G. D. Dodds. 1988. Brainworm, *Parelaphostrongylus tenuis*, in moose, *Alces alces*, and White-tailed deer, *Odocoileus virginianus*, of Nova Scotia. *Canadian Field-Naturalist* 102: 639–642.

Tilman, D, J. Hill, and C. Lehman. 2006. Carbon-negative biofuels from low-input high diversity grassland biomass. *Science* 314:1598–1600.

Torgerson, O. and W. R. Porath. 1984. Midwest oak/hickory forest. In *White-tailed Deer: Ecology and Management*, ed. L. K. Halls, 411–426. Harrisburg, PA: Stackpole Books.

Turner, R. E. and N. N. Rabalais. 2003. Linking landscape and water quality in the Mississippi river basin for 200 years. *Bioscience* 53:563–572.

U.S. Environmental Protection Agency. 2010. *Pesticide Fact Sheet*. http://www.epa.gov/opprd001/factsheets/gonacon.pdf. Accessed May 20, 2010.

U.S. Fish and Wildlife Service. 2006. *National Survey of Fishing, Hunting and Wildlife-Associated Recreation*. Washington, DC: U.S. Department of the Interior and U.S. Department of Commerce.

Van Deelen, T. R., H. Campa, M. Hamady, and J. B. Haufler. 1998. Migration and seasonal range dynamics of deer using adjacent deeryards in northern Michigan. *Journal of Wildlife Management* 62:205–213.

Van Deelen, T. R., H. Campa, J. B. Haufler, and P. D. Thompson. 1997. Mortality patterns of White-tailed deer in Michigan's Upper Peninsula. *Journal of Wildlife Management* 61:903–910.

Van Deelen, T. R., B. J. Dhuey, C. N. Jacques, K. R. McCaffery, R. R E. Rolley, and K. Warnke. 2010. Effects of earn-a-buck on special antlerless-only seasons on Wisconsin's deer harvests. *Journal of Wildlife Management* 74:1693–1700.

Vecellio, G. M., R. H. Yahner, and G. L. Storm. 1994. Crop damage by deer at Gettysburg Park. *Wildlife Society Bulletin* 22:89–93.

VerCauteren, K. C. 1998. Dispersal, home-range fidelity, and vulnerability of White-tailed deer in the Missouri River Valley. PhD dissertation, University of Nebraska, Lincoln.

VerCauteren, K. C. 2003. The deer boom: Discussions on population growth and range expansion of the White-tailed deer. In *Bowhunting Records of North American White-tailed Deer* (2nd edition), eds. G. Hisey and K. Hisey, 15–20. Chatfield, MN: The Pope and Young Club.

VerCauteren, K. C. and S. E. Hygnsytrom. 1994. Movements of White-tailed deer in the eastern Great Plains relative to environmental conditions—A review. *Great Plains Research* 4:117–132.

VerCauteren, K. C. and S. E. Hygnsytrom. 1998. Effects of agricultural activities and hunting on home ranges of female White-tailed deer. *Journal of Wildlife Management* 62:280–285.

VerCauteren, K. C. and S. E. Hygnstrom. 2004. White-tailed deer. In *Encyclopedia of the Great Plains*, ed. D. Wishart, 642–643. Lincoln, NE: University of Nebraska Press.

VerCauteren, K., M. Pipas, P. Peterson, and S. Beckerman. 2003. Stored-crop loss due to deer consumption. *Wildlife Society Bulletin* 31:578–582.

Vieira, M. 2007. *Red Feather-Poudre Canyon Deer Herd Management Plan, Data Analysis Unit D-4*. Fort Collins, CO: Colorado Division of Wildlife.

Vreeland, J. K., D. R. Diefenbach, and B. D. Wallingford. 2004. Survival rates, mortality causes, and habitats of Pennsylvania White-tailed deer fawns. *Wildlife Society Bulletin* 32:542–553.

Wagner, K. K., R. H. Schmidt, and M. R. Conover. 1997. Compensation programs for wildlife damage in North America. *Wildlife Society Bulletin* 25:312–319.

Waller, D. M. and W. S. Alverson. 1997. The White-tailed deer: A keystone herbivore. *Wildlife Society Bulletin* 25:217–226.

Walter, W. D., K. C. VerCauteren, H. Campa, III et al. 2010. Regional assessment on influence of landscape configuration and connectivity on range size of White-tailed deer. *Landscape Ecology* 24:1405–1420.

Walter, W. D., K. C. VerCauteren, J. M. Gilsdorf, and S. E. Hygnstrom. 2009. Crop, native vegetation, and biofuels: Response of White-tailed deer to changing management priorities. *Journal of Wildlife Management* 73:339–344.

Walters, C. 1986. *Adaptive Management of Renewable Resources*. New York, NY: Macmillan Press.

Wasserberg, G., E. E. Osnas, R. E. Rolley, and M. D. Samuel. 2008. Host culling as an adaptive management tool for chronic wasting disease in White-tailed deer: A modelling study. *Journal of Applied Ecology* 46: 457–466.

Whitlaw, H. A. and M. W. Lankester. 1994. The co-occurrence of moose, White-tailed deer, and *Parelaphostrongylus tenuis* in Ontario. *Canadian Journal of Zoology* 72: 819–825.

Wiggers, E. P. and S. L. Beasom. 1986. Characterization of sympatric or adjacent habitats of 2 deer species in west Texas. *Journal of Wildlife Management* 50: 129–134.

Williams, E. S. 2005. Chronic wasting disease. *Veterinary Pathology* 42:530–549.

Williamson, S. J. and D. H. Hirth. 1985. An evaluation of edge use by White-tailed deer. *Wildlife Society Bulletin* 13:252–257.

Willging, R. C. 2008. *On the Hunt: The History of Deer Hunting in Wisconsin*. Madison, WI: Wisconsin Historical Society Press.

Winterstein, S. R., H. Campa, III, and K. F. Millenbah. 1995. *Status and Potential of Michigan Natural Resources: Wildlife*. Michigan Agricultural Experiment Station Special Report No. 75. East Lansing, MI.

Wisconsin Department of Natural Resources. 2001. *Management Workbook for White-tailed Deer* (2nd edition). Madison, WI: Wisconsin Department of Natural Resources PUB-WM-355–2001.

Wisconsin Department of Natural Resources. 2002. Chronic wasting disease management emergency rule. Correspondence/memorandum to Natural Resources Board members from Darrell Bazzell, 3 Jun 2002. File Ref: 2300. Madison, WI.

Wisconsin Department of Natural Resources. 2004. Deer hunting regulations. http://dnr.wi.gov/org/land/wildlife/regs/Deer04Regs1–35.pdf, http://dnr.wi.gov/org/land/wildlife/regs/Deer04CWDZones.pdf. Accessed February 19, 2010.

Wisconsin Department of Natural Resources. 2009a. *A Plan for Managing Chronic Wasting Disease in Wisconsin: the Next Five Years. Draft*. Madison, WI: Wisconsin Department of Natural Resources WM-482–2008 (rev.7–09).

Wisconsin Department of Natural Resources. 2009b. *Chronology of Wisconsin Gun Deer Hunting: From Closed Seasons to Antlerless Permits*. News Release 9 Nov 2009. http://dnr.wi.gov/news/BreakingNews_Print.asp?id=1492. Accessed January 6, 2010.

Wisconsin Department of Natural Resources. 2009c. *Deer Population Goals*. http://dnr.wi.gov/org/land/wildlife/hunt/deer/popgoal.htm. Accessed December 31, 2009.

Wisconsin Legislative Audit Bureau. 2006. *An Evaluation, Chronic Wasting Disease, Department of Natural Resources*. Madison, WI: Wisconsin Legislative Audit Bureau report 06–13.

Wisconsin Natural Resources Board. 2002. *Natural Resources Board Minutes, 25 June 2002*. http://dnr.wi.gov/org/nrboard/minutes/M02/06.pdf. Accessed February 19, 2010.

Wisconsin Natural Resources Board. 2004. *Administrative Rule Order WM-08-04, Clearinghouse Rule 04-020*. http://www.legis.state.wi.us/cr_final/04–020.pdf. Accessed February 19, 2010.

Wisconsin Natural Resources Board. 2008a. *Administrative Rule Order WM-05-08, Clearinghouse Rule 08-013*. http://www.legis.state.wi.us/cr_final/08–013.pdf. Accessed February 19, 2010.

Wisconsin Natural Resources Board. 2008b. *Natural Resources Board Minutes, 22–23 Apr 2008*. http://www.dnr.wi.gov/org/nrboard/2008/April/04–08-NRB-Minutes.pdf. Accessed February 19, 2010.

Wisconsin Natural Resources Board. 2008c. *Request Adoption of WM-05-08 Relating to Management of Chronic Wasting Disease. Natural Resources Board Agenda Item 3.B.1*. http://www.legis.state.wi.us/cr_final/08–013.pdf. Accessed February 19, 2010.

Wisdom, M. J., M. Vavra, J. M. Boyd, M. A. Hemstrom, A. A. Ager, and B. K. Johnson. 2006. Understanding ungulate herbivory-episodic disturbance effects on vegetation dynamics: Knowledge gaps and management needs. *Wildlife Society Bulletin* 34:283–292.

Wishart, W. D. 1980. Hybrids of white-tailed and mule deer in Alberta. *Journal of Mammalogy* 61:716–720.

Wydeven, A. P., T. K. Fuller, W. Weber, and K. MacDonald. 1995. The potential for wolf recovery in the northeastern United States via dispersal from southeastern Canada. *Wildlife Society Bulletin* 26:776–784.

Wywialowski, A. P. 1996. Wildlife damage to field corn in 1993. *Wildlife Society Bulletin* 24:264–271.

Yamamura, K., H. Matsuda, H. Yokomizo et al. 2008. Harvest-based Bayesian estimation of sika deer populations using state-space models. *Population Ecology* 50:131–144.

Young, S. P. 1956. The deer, the Indians and the American pioneers. In *The Deer of North America*, ed. W. P. Taylor, 1–27. Harrisburg, PA: Stackpole Books.

Zwank, P. J., R. D. Sparrowe, W. R. Porath, and O. Torgerson. 1979. Utilization of threatened bottomland habitats by white-tailed deer. *Wildlife Society Bulletin* 7:226–232.

18

Managing White-tailed Deer: Western North America

Timothy E. Fulbright

CONTENTS

White-tailed deer are most abundant east of the Rocky Mountains in North America; however, their populations are increasing and their range is expanding in parts of western North America. Unique subspecies of white-tailed deer inhabit the forests and river bottoms of the Pacific Northwest, the Sky Island mountain ranges of the southwest, and shrublands of the south-central United States and northern Mexico. Although they occupy distinctly different environments, these subspecies share many commonalities in ecology and management.

Geographic Description of the Region

White-tailed deer in three geographic regions of North America are the subject of this chapter. These regions are (1) the northern Rocky Mountains and Pacific Northwest in southwestern Alberta, western Montana, and Wyoming west to the Pacific Ocean; (2) the scattered mountain ranges of the Madrean archipelago in Arizona and New Mexico popularly referred to as the Sky Islands; and (3) the rangelands of southeastern Colorado and western Kansas south to southern and western Texas.

The diverse landscape of the Pacific Northwest ranges from mountains dominated by coniferous forest to semiarid steppes. Major mountain ranges include the northern Rocky Mountains, the Coast Mountains, the Cascade Range, the Olympic Mountains, Columbia Mountains, and the Blue Mountains of northeast Oregon and southeast Washington. The Columbia is the longest river, stretching from British Columbia through Washington and forming part of the border between Washington and Oregon before flowing into the Pacific Ocean. The Columbia Plateau stretches from Oregon north to Washington, lying east of the Coast and Cascade Mountain ranges. The Interior Plateau in British Columbia is a northward extension of the Columbia Plateau situated between the Coast range, Hazelton Mountains, and Cascade Range on the west and the Caribou and Monashee Mountains on the east.

The Madrean archipelago lies between the Rocky Mountains on the east and Sierra Nevada Mountains on the west (Warshall, 1995). The landscape of the region consists of a roughly parallel series of valleys and mountains. Increasing altitude in the mountains results in an ascending series of biotic communities, ranging from desert communities at low elevations to oak woodland and coniferous forest at higher elevations. White-tailed deer occupy the oak woodland and coniferous forests in Arizona, New Mexico, and western Texas.

White-tailed deer habitat in the southern Great Plains of western Kansas, eastern Colorado and New Mexico, and Texas Panhandle consists largely of bottomlands along streams and rivers. The Texas Rolling Plains and the Edwards Plateau lying to the south of the Great Plains support higher deer densities than the southern Great Plains, with the Edwards Plateau region supporting the highest deer densities of the geographic areas considered in this chapter (Fulbright and Ortega-S., 2006; Figure 18.1). The thornscrub

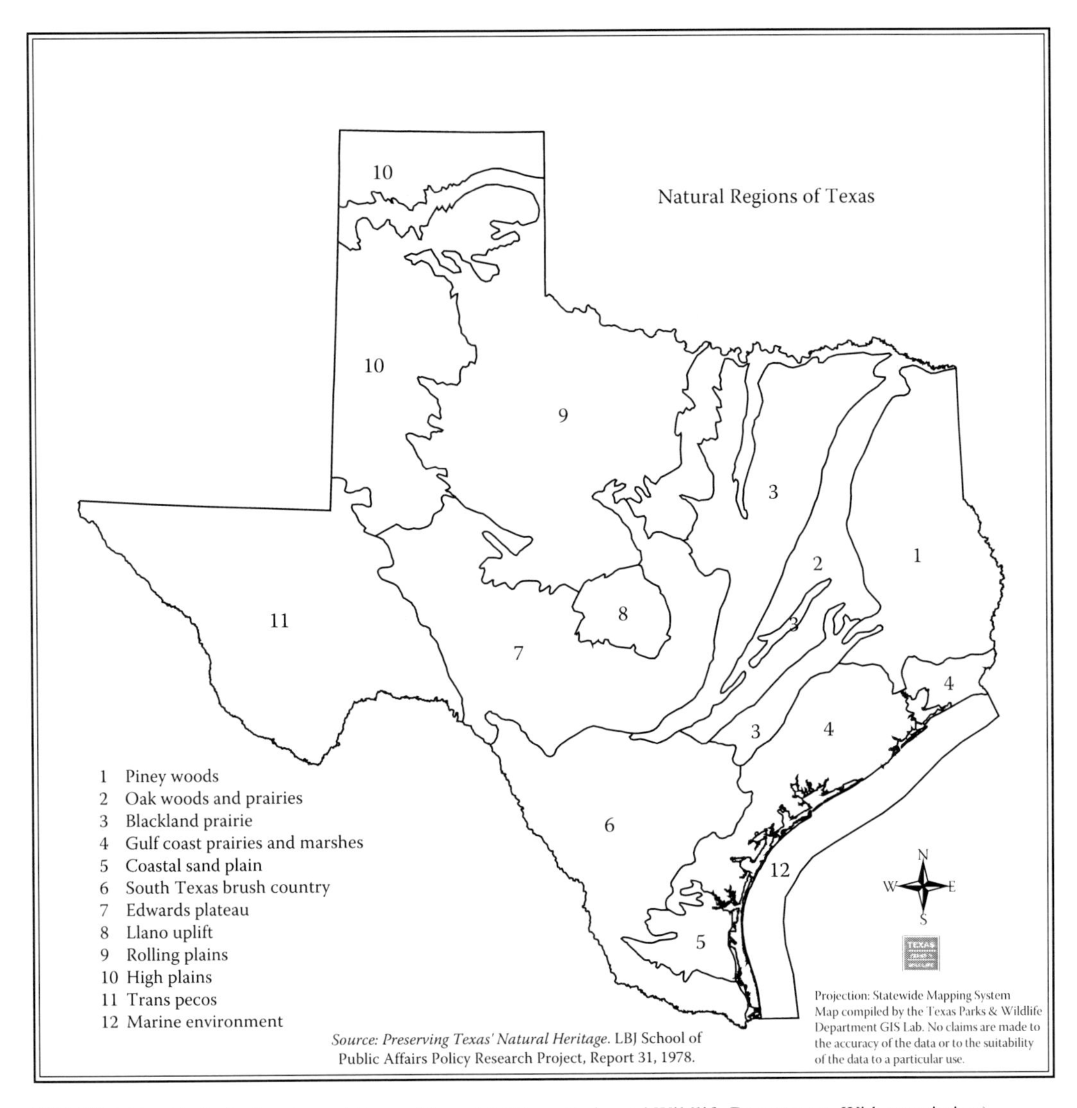

FIGURE 18.1 Ecological regions of Texas. (Copyright Texas Parks and Wildlife Department. With permission.)

region of the South Texas Plains also supports dense white-tailed deer populations. The western part of the range of Texas white-tailed deer (*Odocoileus virginianus texanus*) in Texas consists of a transition zone where the shortgrass prairies of the southern Great Plains, live oak and juniper woodland of the Edwards Plateau, and thornscrub of the South Texas Plains merge with Chihuahuan Desert shrubland.

Subspecies of Deer across the Region

Five subspecies of white-tailed deer occur in the western geographic region discussed in this chapter (Figure 1.7a). Columbian (*O. v. leucurus*) and northwest white-tailed deer (*O. v. ochrourus*) are found in the Pacific Northwest. Two geographically isolated subpopulations of Columbian white-tailed deer remain from a subspecies that was once distributed from the south end of Puget Sound in Washington to the Umpqua River drainage in southern Oregon (Suring and Vohs, 1979; Gavin and May, 1988). These subpopulations are in Douglas County, Oregon, and in and around the U.S. Fish and Wildlife Service Julia Butler Hansen Refuge for the Columbian White-tailed Deer in southwestern Washington.

These subpopulations are about 300 km from the nearest white-tailed deer populations in Oregon and Washington (Gavin and May, 1988). Genetic distance between Columbian white-tailed deer in Douglas County, Oregon, and northwest white-tailed deer may not be adequate for the Douglas County population to be classed as a distinct subspecies (Gavin and May, 1988). Genetic distance is greater between lower Columbia River population of Columbian white-tailed deer and northwest white-tailed deer, but it is unclear whether or not these differences are sufficient to consider them separate subspecies. Research using mitochondrial and microsatellite DNA to clarify genetic relationships between Columbian and northwest white-tailed deer is ongoing (A. J. Piaggio, USDA National Wildlife Research Center, personal communication). Hybridization occurs between Columbian white-tailed deer from on or near the Julia Butler Hansen National Wildlife Refuge and Columbian black-tailed deer (Gavin and May, 1988).

Northwest white-tailed deer occur in parts of British Columbia and Alberta south to Montana and west to northeastern Oregon and eastern Washington; occasional sightings of white-tailed deer have been reported in Utah (McClure et al., 1997; Rue, 2004; Lyman, 2006). White-tailed deer are more abundant in northern than in southern Idaho. In Montana, hybridization between white-tailed deer and mule deer is minimal (Cronin et al., 1988).

Carmen Mountains (*O. v. carminis*) and Coues whitetails (*O. v. couesi*) occur in the Sky Island mountains in southwestern United States and adjacent northern Mexico. Coues white-tailed deer range from southwestern New Mexico west to central Arizona and south into Mexico. Hybrids of Coues white-tailed deer and desert mule deer have been documented, but only about 2% of free-ranging Coues white-tailed deer are hybrids (Day, 1980; Lopez, 2006). Carmen Mountains white-tailed deer are found in the mountains of Texas west of the Pecos River (Trans-Pecos) and south into northern Mexico (Heffelfinger, 2006). Craniums of Carmen Mountains white-tailed deer in the Davis Mountains north of Big Bend National Park are closer in size to Texas white-tailed deer than to Carmen Mountains specimens from Big Bend National Park (Krausman et al., 1978). Larger ears and smaller body size distinguish Carmen Mountains white-tailed deer from Texas white-tailed deer.

Texas white-tailed deer inhabit the south-central United States and northern Mexico, ranging from western Kansas and eastern Colorado south through western Oklahoma and eastern New Mexico to western and southern Texas. Texas white-tailed deer and desert mule deer hybridize in areas where their ranges overlap (Derr, 1991; Ballinger et al., 1992). In western Texas, hybrids averaged 5.6% of the deer examined (Stubblefield et al., 1986). Hybridization is primarily between male mule deer and female white-tailed deer (Carr et al., 1986).

Prominent Ecological Influences

Columbian White-tailed Deer

Regional Climate and Habitat Preferences

Columbian white-tailed deer originally occupied river valleys and surrounding foothills dominated by shrubs (Suring and Vohs, 1979; Smith, 1985, 1987). Extant remnant populations occupy bottomlands and islands in the lower Columbia River and oak woodland in the Umpqua River Basin in Oregon (Smith, 1987). The range of Columbian white-tailed deer is characterized by relatively mild temperatures. Average monthly temperatures at the Julia Butler Hansen Refuge, for example, range from 1.7°C in winter to 18.5°C in summer (Suring and Vohs, 1979). Annual precipitation averages 84 cm in the Umpqua Basin and 254 cm at the Julia Butler Hansen Refuge (Suring and Vohs, 1979; Smith, 1987).

Plant communities receiving the highest use by Columbian white-tailed deer in the Julia Butler Hansen Refuge in Washington are discontinuous sitka spruce forest and opened-canopied forest dominated by western red cedar, red alder, and sitka spruce (Suring and Vohs, 1979). In the Umpqua River Basin, deer prefer grass–shrub, oak savanna, open oak, closed oak, riparian, and conifer plant communities (Smith, 1987). Highest deer densities occur in areas with greater than 50% woody plant canopy cover. Fawns in the Douglas County, Oregon, subpopulation prefer oak–madrone woodland and riparian areas (Ricca et al., 2003). Conifer-dominated areas are avoided by fawns.

Interactions with Other Herbivores

Conversion of natural bottomlands to cultivation in Oregon extirpated Columbian white-tailed deer from most of their historic range (Smith, 1987). Loss of woody plant communities allowed Columbian black-tailed deer to occupy what was formerly white-tailed deer habitat. Columbian white-tailed deer and Columbian black-tailed deer prefer different habitats, reducing spatial overlap between the two species (Smith, 1987). Where the two species overlap, some degree of interspecific avoidance appears to exist.

Columbian white-tailed deer in the lower Columbia River avoid pastures containing cattle (Gavin et al., 1984). Two to three months after cattle are removed, deer prefer the previously grazed pastures over pastures cut for hay possibly because grazed pastures have greater plant species diversity. Deer select areas with little or no use by livestock for fawning in preference to areas with livestock (Smith, 1981).

Predators

Predation is a minor cause of mortality in adult Columbian white-tailed deer; however, it is the major cause of mortality in fawns (Gavin et al., 1984; Ricca et al., 2002). Coyotes are the major natural predator in the lower Columbia River, whereas bobcats are the major natural predator in southwestern Oregon.

Food Habits

Diets of Columbian white-tailed deer in southwestern Washington during 1972–1977 consisted of 59% grasses, 17% forbs, and 24% browse (Gavin et al., 1984). Foxtail appeared to be the major grass genus in deer diets.

Northwest White-tailed Deer

Regional Climate and Habitat Preferences

White-tailed deer habitat in the northern Rocky Mountains occurs at elevations ranging from 300 to 2000 m (Gerlach et al., 1994). The climate of the region is cool temperate, and average annual precipitation within the range of northwest white-tailed deer varies from less than 50 to more than 90 cm. In British Columbia, white-tailed deer are most common in valley bottoms and along rivers (Hatter et al., 2000). Riparian areas and cultivated fields are preferred summer habitats in British Columbia, with grasslands being important immediately after spring thaw. In early winter, white-tailed deer may occupy aspen, cottonwood, and willow communities. As snow depth increases, coniferous forest is used more heavily. White-tailed deer prefer areas with snow less than 40 cm deep during winter (Poole and Mowat, 2005).

White-tailed deer occur from stream and river bottoms at lower elevations to coniferous forests at higher elevations in Montana and Idaho (Jenkins and Wright, 1988; Pauley et al., 1993). In the northern Rocky Mountains, white-tailed deer prefer mature forest communities and avoid lodgepole pine, deciduous forest, and nontimbered plant communities during January to May (Jenkins and Wright, 1988). Preferred plant communities include Douglas fir, spruce–aspen, and lowland spruce. White-tailed deer preference for forest communities increases with forest maturity.

White-tailed deer in northern Idaho select sites with 5–9% slopes at low elevations during early and late winter (Pauley et al., 1993). Deer primarily use Douglas fir and lodgepole pine communities with trees 7.5–23 cm diameter at breast height and mature timber stands on valley floors during early and late winter when snow is less than 30 cm deep. They may also use mature Douglas fir stands on adjacent slopes during winter. Old-growth western red cedar and western hemlock stands along rivers are heavily used during mid-winter when snow accumulation exceeds 40 cm in depth. Forest openings are avoided during winter because snow accumulation is too deep. In the Blue Mountains of northeastern Oregon, northwest white-tailed deer are primarily associated with riparian and agricultural areas, but also use adjacent slopes (D. G. Whittaker, Oregon Department of Game and Fish, personal communication).

Interactions with Other Herbivores

White-tailed deer diets were more species diverse than cattle or elk diets in a grand fir/queencup beadlily community in northern Idaho (Kingery et al., 1996). Diet overlap between cattle and white-tailed deer is minimal in the coniferous forests of northern Idaho (Thilenius and Hungerford, 1967; Kingery et al., 1996). Substantial diet and spatial overlap may exist between white-tailed deer and elk depending on season of the year and plant community (Jenkins and Wright, 1988; Kingery et al., 1996).

Mule deer and white-tailed deer are primarily associated with different plant communities during much of the year in north-central Montana (Martinka, 1968). The two species may compete for western serviceberry during mild winters. During severe winters, spatial overlap may exist between the two deer species in bunchgrass, pine–bunchgrass, and shrub communities. In British Columbia, mule deer and white-tailed deer maintain spatial separation during winter (Hudson et al., 1976).

Predators

Predators of northwest white-tailed deer include gray wolves, bears, bobcats, cougars, and coyotes (Hatter et al., 2000; Squires and Ruggiero, 2007). Cougar predation is the primary cause of mortality in British Columbia (Robinson et al., 2002). White-tailed deer are more abundant than mule deer in south-central British Columbia and in northeast Washington; however, the rate of predation by cougars is higher on mule deer than on white-tailed deer, which may be contributing to a decline in mule deer populations (Robinson et al., 2002; Cooley et al., 2008). Cougars kill white-tailed deer at sites that are steeper, have more mature trees, and have greater woody plant canopy cover than sites where wolves kill deer (Kunkel et al., 2003).

Wolves in northwestern Montana and southeastern British Columbia selectively kill elk and moose over deer (Kunkel et al., 2004). Wolves select deer when elk abundance is low and increasing snow depth makes deer more vulnerable to wolf predation.

White-tailed deer are less vulnerable to predation by coyotes than are mule deer in habitats with more level terrain (Lingle, 2002). White-tailed deer move downhill and away from slopes in response to presence of coyotes, whereas mule deer move uphill. Female mule deer defend fawns more aggressively than female white-tailed deer (Lingle et al., 2005). Coyotes avoid attacking fawns when mule deer females are present, but are not deterred by the presence of a female white-tailed deer.

Food Habits

In British Columbia, white-tailed deer diets are composed mostly of browse during summer and winter, with major species eaten including redstem ceanothus, choke cherry, western serviceberry, Oregon grape, snowbrush, and cottonwood seedlings (Hatter et al., 2000). Irrigated crops such as alfalfa are also eaten. Grasses and forbs are the major diet items during spring.

White-tailed deer diets in the northern Rocky Mountains during winter are primarily composed of conifers and deciduous shrubs (Thilenius and Hungerford, 1967; Jenkins and Wright, 1988). Douglas fir and Oregon grape were two of the most consistently important winter forage items during a 60-year study occurring from the 1940s through the 1990s (Dusek et al., 2005). In addition to Oregon grape, other low-growing evergreen shrub species are also eaten (Jenkins and Wright, 1988). Grasses and grasslikes compose less than 5% of diets. Lichens are also important winter forage (Ward and Marcum, 2005). Consumption of lichens increases during severe winters with deep snow.

Shrubs in white-tailed deer diets ranged from 43% in early summer to 39% in early autumn in a grand fir/queencup beadlily community in northern Idaho (Kingery et al., 1996). Forbs composed from 25% of diets during early summer to 51% of diets in early autumn, while grasses and grasslikes composed from 17% of diets during early summer to 3% in early autumn.

Spotted knapweed is a composite from Eurasia that has invaded almost 2 million ha in Idaho and Montana (Wright and Kelsey, 1997). Infestation of the plant appears to have little effect on carrying capacity of white-tailed deer habitat.

Carmen Mountains White-tailed Deer

Regional Climate and Habitat Preferences

Carmen Mountains white-tailed deer habitat is characterized by relatively low rainfall ranging from 33 to 51 cm/year; with less precipitation in the foothills, where the subspecies occasionally occurs, than the mountains, where the deer are more abundant (Krausman and Ables, 1981). Snow is rare within the range of distribution of the subspecies, but freezing temperatures occur during winter. Carmen Mountains white-tailed deer are most abundant above 1373 m elevation, but they occasionally occur down to 900 m (Krausman et al., 1978).

Habitat of Carmen Mountains white-tailed deer is characterized by dense woodland dominated by oaks, junipers, or other woody plants and presence of surface water (Krausman and Ables, 1981). Grasslands with sotol are also used by the deer, but deer are less abundant in more open habitats than in woody plant communities. Carmen Mountains white-tailed deer prefer to bed in dense vegetation. Ephemeral drainages dominated by dense vegetation including oak, pinyon pine, juniper, mountain mahogany, sumac, ash, Texas persimmon, and desert willow are favored by the deer in areas where vegetation is relatively sparse.

The ecosystem occupied by Carmen Mountains white-tailed deer in Big Bend National Park is relatively undisturbed by humans (Krausman and Ables, 1981). Although white-tailed deer are often considered a "mid-successional" species, this deer subspecies occupies habitats that are in advanced stages of ecological succession.

Interactions with Other Herbivores

In the Chisos Mountains, spatial overlap between Carmen Mountains white-tailed deer and desert mule deer occurs from about 1220 to 1464 m in elevation (Krausman and Ables, 1981). Plant species composition of the diets of the two deer species also overlaps considerably (Krausman, 1978).

Predators

Desert mule deer and Carmen Mountains white-tailed deer are the primary prey of cougars in Big Bend National Park (Krausman and Ables, 1981). Bobcats and coyotes in the National Park consume primarily rodents and lagomorphs; however, deer may be more heavily predated when lagomorph populations are low (Leopold and Krausman, 1986). Vehicle collisions are a minor cause of mortality (Krausman and Ables, 1981).

Food Habits

Browse and succulents, primarily lechuguilla and pricklypear, are the major foods of Carmen Mountains white-tailed deer (Krausman, 1978; Krausman and Ables, 1981). Forbs follow browse and succulents in importance, with grasses consumed primarily when they are producing new growth.

Coues White-tailed Deer

Regional Climate and Habitat Preferences

Coues white-tailed deer are most abundant at elevations between 1220 and 2134 m, although they can be found at elevations down to 800 m and up to 3000 m (Cahalane, 1939; Knipe, 1977; Heffelfinger, 2006). The deer prefer areas that receive more than 25 cm mean annual precipitation, and are most abundant in areas receiving more than 38 cm mean annual precipitation (Knipe, 1977). Winters are mild and moist at lower elevations, whereas April to the first of July is usually hot and dry (Anthony, 1976). The remainder of the summer is usually characterized by a rainy season consisting of thunderstorms, followed by light precipitation from mid-September through mid-November. Unlike deer in northern climates, winters are generally not stressful for Coues white-tailed deer; however, summers may be

FIGURE 18.2 Coues white-tailed deer habitat in Arizona. (Photo by T. E. Fulbright. With permission.)

stressful because of elevated temperatures and deficient water (Ockenfels and Brooks, 1994). Periodic droughts are common in the region and severe drought may result in high mortality of Coues white-tailed deer (Anthony, 1976).

Coues white-tailed deer are found primarily in Madrean evergreen woodlands dominated by oaks and in oak–pine woodlands (Knipe, 1977) (Figure 18.2). They prefer grassland and oak–pine forest over pine–oak forest (Galindo-Leal et al., 1993). They also occur in riparian deciduous forests, montane coniferous forests, desert grasslands, upland desert, and interior chaparral where oaks occur (Knipe, 1977). Thermal cover provided by trees is important for Coues white-tailed deer to cope with elevated daytime temperatures during summer (Ockenfels and Brooks, 1994). Diurnal bed sites used during summer in Arizona have at least 60% woody plant canopy cover consisting primarily of Mexican blue oak, juniper, and velvet mesquite.

Interactions with Other Herbivores

Coues white-tailed deer avoided areas where stocking rates of cattle were high in Durango, Mexico (Galindo-Leal et al., 1993). Coues white-tailed deer and cattle diets in Durango, Mexico, overlapped 51% during the wet portion of the year when both species were intensively using forbs (Gallina, 1993). Little diet overlap occurred when forbs were not abundant; cattle ate primarily grass whereas shrub and tree species composed 85% of deer diets averaged across the year. Oaks and madrone were the two most important plant genera in deer diets.

Degree of association between Coues white-tailed deer and desert mule deer varies considerably among mountain ranges; consequently, factors determining when the two species are sympatric are unclear (Anthony and Smith, 1977). Plant species composition of the diets of the two deer species overlap substantially. Diets of Coues white-tailed deer, however, are more species diverse than mule deer diets. Coues white-tailed deer prefer areas with greater than 40% canopy cover of woody vegetation, whereas mule deer prefer more open habitats (Ockenfels et al., 1991).

Predators

Cougars are the major predator of Coues white-tailed deer (Knipe, 1977; Rosas-Rosas et al., 2003). Coyotes are also a major predator, particularly of fawns. Other predators include bobcats, golden eagles, black bears, and dogs (Knipe, 1977).

Food Habits

Coues white-tailed deer switch diet composition among seasons and during wet and dry cycles (Anthony, 1976; Ockenfels et al., 1991). Browse dominates overall diets of the subspecies, but forbs constitute a large portion of the diet when they are available (White, 1961; Urness et al., 1971; Ockenfels et al., 1991). Plants species composition of diets is highly diverse; more than 610 different plant species are consumed by Coues white-tailed deer (Knipe, 1977).

Texas White-tailed Deer

Regional Climate and Habitat Preferences

Texas white-tailed deer occur across a variety of climatic regimes, from areas with less than 40 cm of mean annual precipitation on the western edge of their range to greater than 90 cm on the eastern edge, and from semiarid steppe to humid subtropical climates. They occupy habitats lower in elevation than Carmen Mountains and Coues white-tailed deer. Texas white-tailed deer densities are commonly greatest in bottomlands and ephemeral drainages, a pattern evident in white-tailed deer habitat selection throughout their western range (Darr and Klebenow, 1975; Zwank et al., 1979; Beasom et al., 1982). In the southern Great Plains, white-tailed deer are primarily associated with woodland communities along riparian areas, but may forage in adjacent grasslands and croplands (Choate, 1987; Compton et al., 1988; Volk et al., 2007). They may also use afforested areas, such as shelterbelts, to a limited extent (Menzel, 1984).

Texas white-tailed deer are widespread in a variety of different ecological regions and plant communities to the south and east of the southern Great Plains. These include honey mesquite, juniper, and shin oak communities of the Rolling Plains; the woodlands of the Cross Timbers and Prairies region that includes a variety of oak species, Ashe juniper, honey mesquite, and other woody species; and the Edwards Plateau characterized by live oak and other oak species, Ashe juniper in the eastern portion and redberry juniper in the west and north, and honey mesquite. Further south, Texas white-tailed deer are abundant in the Tamaulipan thornscrub vegetation of southern Texas and in the coastal prairies and marshes ecological region.

Texas white-tailed deer prefer woody plant communities for bedding, security, and thermal cover, but commonly forage in open areas dominated by grasses and forbs during early morning, late evening, and nighttime (Inglis et al., 1986; Rothley, 2001). Openings dominated by herbaceous vegetation intermixed between stands of woody vegetation often serve as core areas for nighttime feeding (Inglis et al., 1986). White-tailed deer in eastern Kansas, for example, selected areas with woody vegetation interspersed with grassland and avoided extensive open grasslands (Volk et al., 2007). Most foraging, however, occurred within the grassland patches rather than in or beside woodland patches. Use of open areas for feeding is probably related to greater abundance of preferred forage, such as forbs, in the openings (Stewart et al., 2000; Volk et al., 2007).

Woody canopy cover requirements have been more thoroughly studied for Texas white-tailed deer than for the other subspecies in this chapter. Preferences of deer for woody plant cover vary seasonally, between sexes, and with the amount of disturbance by humans. During periods when temperatures are not extreme, deer may be catholic in selection of woody canopy cover (Steuter and Wright, 1980). Dense woody cover provides important thermal cover during winter in cold environments and during summer in warm environments (Mysterud and Ostbye, 1999). Importance of thermal cover during summer may be substantial in torrid environments such as southern Texas. In a study conducted in southern Texas during summer, deer density increased with increasing percent woody canopy cover, possibly because deer were seeking dense cover for thermoregulation (Steuter and Wright, 1980). Greatest densities occurred in areas with 60–97% woody plant canopy cover.

Woody plants also provide security or screening cover for white-tailed deer. Security cover is a vertical cover where a deer can hide from predators and humans. Degree of selection for security cover depends on sex, age, predators, and disturbance by humans. Females with fawns in eastern southern Texas select areas with the densest available brush, possibly to avoid predators (Kie and Bowyer, 1999). Males select more open habitats than females do in eastern southern Texas. Conversely, mature males in the more arid western portion of southern Texas select areas with dense security cover and woody canopy cover greater than 85% (Pollock et al., 1994). Fawns in southern Texas select bed sites with mid-to-tall grasses; half of the bed sites found in one study were near a woody plant (Hyde et al., 1987). White-tailed deer in west Texas selected bed sites averaging 54% woody plant cover (Avey et al., 2003). White-tailed deer may avoid open habitats during hunting season and select areas with dense cover, in addition to becoming more nocturnal (Kilgo et al., 1998).

Interactions with Other Herbivores

Mule deer populations overlap Texas white-tailed deer populations in much of the western edge of the range of white-tailed deer from the Trans-Pecos region of Texas to the High Plains. White-tailed deer tend to be more abundant, and mule deer less abundant, in Trans-Pecos Texas when woody plant canopy cover exceeds 50% (Wiggers and Beasom, 1986). Habitats in Trans-Pecos Texas with high densities of white-tailed deer averaged 63% woody plant canopy cover compared to 43% woody plant canopy cover for areas with high densities of mule deer. In the western Edwards Plateau, mule deer use areas with steeper slopes than do white-tailed deer (Avey et al., 2003). Shrub canopy cover used for bed sites in the western Edwards Plateau was greater for white-tailed deer (54%) than for mule deer (38%). Foraging white-tailed deer in the western Edwards Plateau used areas averaging 38% woody plant canopy cover compared to 21% for mule deer. Mule deer primarily use areas above 870 m in elevation that are dominated by juniper, whereas white-tailed deer in the western Edwards Plateau use honey mesquite-dominated areas less than 840 m in elevation (Brunjes et al., 2006). Males of both species avoid areas consisting of small thickets of dense honey mesquite, juniper, or mixed shrubs. Females of both deer species select areas of dense brush, particularly during fawning. Spatial overlap between the two species occurs because certain areas are used by individuals of both deer species.

Mule deer at the Rocky Mountain arsenal in Colorado selected plant communities based primarily on forage availability, whereas white-tailed deer selected plant communities based primarily on availability of cover for security or thermoregulation (Whittaker and Lindzey, 2004). Diet composition of mule deer and white-tailed deer at the Rocky Mountain arsenal was similar, but segregation in use of space between the two species minimized potential for competition.

Plant species composition of the diets of Texas white-tailed deer and cattle have minimal overlap when precipitation is adequate, livestock stocking rates are light to moderate, and vegetation is not dormant (Thill and Martin, 1989; Ortega et al., 1997a). High cattle density, drought, and cold, dry winters result in increased overlap in diet botanical composition. Greater potential for competition for forage between cattle and white-tailed deer exists in vegetation communities preferred by both species. Cattle avoid rocky terrain, areas of dense brush with minimal grass biomass, and sites distant from water resulting in minimal spatial overlap with deer in those areas (Owens et al., 1991; Cooper et al., 2008).

Although spatial distribution of white-tailed deer and cattle may overlap, deer avoid close proximity to cattle. Cattle and white-tailed deer may both heavily use riparian habitats; however, deer avoid these habitats when cattle are present (Compton et al., 1988; Cooper et al., 2008). Texas white-tailed deer avoid high concentrations of cattle and select pastures that are not grazed if available within their home range (McMahan, 1966; Cohen et al., 1989). In research in Texas, deer avoided cattle only when they approached to within 50 m of each other (Cooper et al., 2008). Texas white-tailed deer may avoid sheep more strongly than cattle, possibly because of greater diet overlap with sheep (Darr and Klebenow, 1975).

Predators

White-tailed deer are the preferred food of cougars in the South Texas Plains (Harveson et al., 2000). Coyotes are the primary predator of white-tailed deer in habitats where cougars are absent or rare (Cook et al., 1971; Whittaker and Lindzey, 1999). Most predation by coyotes is on fawns (Cook et al., 1971; Garner et al., 1976).

Food Habits

Plant species and plant part composition of white-tailed deer diets are diverse. Researchers in the Rolling Plains of Texas, for example, reported 239 different plant species in deer diets (Burke, 2003). This diversity may be important for deer to (1) consume and mix plant species that complement one another in nutrient content, enabling them to optimize the nutrient content of their diets in a manner that meets nutritional needs; and (2) buffer effects of secondary compounds on digestion and nutrient assimilation while concomitantly taking advantage of beneficial effects of these compounds such as reducing internal parasites (Provenza et al., 2009). The greater the diversity of plant species in the habitat, the more likely deer can mix nutritious diets and ameliorate effects of toxic plants.

Presence of a variety of different plant species in the habitat facilitates shifting of diet composition seasonally and during droughts. Texas white-tailed deer prefer forbs over browse, grasses, and succulents (Fulbright and Ortega-S., 2006). Forbs that dominate deer diets during spring or high rainfall years may be replaced by more heat or drought-tolerant species during dry years or summer (Dillard et al., 2005). Most forbs are cool season plants that decline in abundance and availability during warmer portions of the year. Browse may often dominate deer diets during warmer parts of the year when forbs are lacking (Arnold and Drawe, 1979; Waid et al., 1984; Richardson et al., 2008).

Browse increases in importance in deer diets during droughts because lack of rainfall results in reduced forb abundance. In the Rolling Plains of Texas, for example, browse declined from 57% of deer diets during drought to 39% of diets during a year with greater rainfall; forbs increased from 18% of diets during drought to 38% of diets during the wetter year (Richardson et al., 2008). Further east in the Cross Timbers and Prairies region of Texas, browse in diets declined from 46% during a dry summer to 29% during a wet summer, whereas forbs increased from 13% of diets during the dry summer to 43% of diets during the wet summer (Dillard et al., 2005). Browse also increases in diets when dry winters and freezing temperatures result in decreased forb availability.

Diet composition of Texas white-tailed deer varies geographically. Forb production and seasonal availability decline along a gradient from humid and subhumid to arid and semiarid climates; consequently, predominance of forbs in annual deer diets tends to decline, and browse tends to increase, along this gradient of decreasing annual precipitation (Fulbright and Ortega-S., 2006).

Diet composition also varies between sexes. Differences in diet composition between sexes may result in part from use of separate habitats during the nonbreeding season (Stewart et al., 2003). The larger body size of males relative to females may enable males to consume diets higher in fiber (Barboza and Bowyer, 2000; Bowyer et al., 2002) (Figure 6.5). Greater nutrient demands of females during gestation and lactation may also influence differences in diet composition between sexes.

Although browse and forbs are often the major items in deer diets, mast may be critically important at times. In southwest Texas, bite-count studies showed that tame deer habituated to their environment consumed 40–75% of their diet as mast, primarily pricklypear fruits, during spring, summer, and autumn (Timmons et al., 2010). High mast consumption by deer apparently enabled them to meet metabolizable energy needs. In addition to pricklypear fruits, live oak acorns and mesquite beans are important in deer diets (Dillard et al., 2005; Richardson et al., 2008).

Population Status

Columbian white-tailed deer were listed as a federally endangered species in 1968. Following listing, the population in Douglas County, Oregon, increased sufficiently that the subspecies was delisted as endangered in Oregon in 2003 (U.S. Fish and Wildlife Service, 2008). Limited, controlled buck hunting began in 2005. The population in Oregon exceeded 5000 animals in 2008 (U.S. Fish and Wildlife Service, 2008). Fawn survival in the Oregon population is low, and the population may be at the carrying capacity of the currently occupied habitat (Ricca et al., 2002). Dispersal of Columbian white-tailed deer from the Oregon population into adjacent areas of their historic range had not occurred during the two decades preceding 2002. Reasons for lack of dispersal are unknown. In Washington, Columbian white-tailed deer remain listed as an endangered species with a population size of about 250 in 2002 (Brookshier, 2004).

The estimated population of northwest white-tailed deer in British Columbia was about 61,000 in 1997 (Shackleton and Cowan, 1999). An estimated 200,000 northwest white-tailed deer are in Idaho (Idaho Fish and Game, 2004).

Populations of Carmen Mountains white-tailed deer are relatively stable (Texas Parks and Wildlife Department, 2007). Estimated population size in Big Bend National Park, Texas, during 1968–1974 ranged from 649 to 1244 individuals (Krausman and Ables, 1981). Coues white-tailed deer declined in several mountain ranges in Arizona during the 25-year period before 1977 (Anthony and Smith, 1977). A possible reason for the decline was competition with mule deer which was intensified by vegetation changes resulting from overgrazing by livestock and fire suppression. Hunting was unlikely to be a cause of the decline of white-tailed deer because mule deer are more susceptible to hunting mortality. Estimated population of Coues white-tailed deer in Arizona in 2009 was 82,000 (Arizona Game and Fish Department, 2009). In Arizona, 1500–7000 white-tailed deer are harvested each year. Texas white-tailed deer populations are increasing and appear to be expanding their range in western Texas (Texas Parks and Wildlife Department, 2007).

Land Use, Human Impacts, and Habitat Management

Columbian and Northwest White-tailed Deer

Land Use and Human Impacts

In pre-Columbian times, fires set by native Americans maintained the oak–savanna woodland preferred by Columbian white-tailed deer (Thilenius, 1968; Smith, 1981). Much of this habitat, however, has been altered by suppression of periodic fire and replaced by cultivated fields and housing developments (Smith, 1981; Ricca et al., 2002). Some human-imposed changes within the past century possibly benefited Columbian white-tailed deer such as planting of improved pastures and creation of edge habitat through clearing of oak woodlands (Smith, 1985). Creation of coniferous forest openings through clear-cutting may also have benefitted deer, along with predator control programs associated with livestock management.

Mature conifers provide a substrate for growth of lichens, an important component of deer winter diets in the northern Rocky Mountains. Conifers also provide important thermal cover and reduce snow accumulation beneath their canopies during winter. Decreasing Douglas fir in overstories, increasing spacing between trees, and reducing conifers in the understory of ponderosa pine communities potentially reduces white-tailed deer populations by reducing lichen litterfall and thermal cover (Baty et al., 1996).

Home Range Sizes

Convex polygon estimates of Columbian white-tailed deer home ranges in Oregon were 21.1 ha for females and 32.8 ha for males (Smith, 1981). Mean home range size of fawns in Oregon was 18 ha; areas where fawns concentrated activities were 2.4 ha (Ricca et al., 2003). In southwestern Washington, mean home range size for Columbian white-tailed deer ranged from 158.5 ha for females to 192.2 ha for males (Gavin et al., 1984).

Habitat-Management Approaches

In riparian areas where woody plants are absent, planting native trees and shrubs such as cottonwood, spruce, alder, salal, ninebark, dogwood, and elderberry provides browse and cover for Columbian white-tailed deer (Suring and Vohs, 1979; Brookshier, 2004). Excluding cattle by fencing selected wooded areas increases use of those areas by deer (Suring and Vohs, 1979).

Managing forests to maintain high rates of lichen litterfall benefits white-tailed deer populations in the northwest (Ward and Marcum, 2005). Creating small clear-cuts to provide foraging areas and maintaining uneven-aged stands in close proximity result in a mix of different forest successional stages and may benefit white-tailed deer (Hanley, 1996; Idaho Fish and Game, 2004). Use of clear-cuts is greater when depth of downed material is less than 0.5 m (Lyon and Jensen, 1980). Optimum deer use of clear-cuts occurs when they are about 24 ha in size.

Basal spraying with herbicides can be used to kill crowns of shrubs such as Rocky Mountain maple and stimulate resprouting to increase availability of browse (Mueggler, 1966). Herbicides can also be used to kill shrubs such as creambush rockspirea and ninebark; however, herbicides should be used with caution because desirable browse species such as redstem ceanothus may be adversely affected.

Carmen Mountains and Coues White-tailed Deer

Land Use and Human Impacts

Much of the research on Carmen Mountains white-tailed deer has been conducted in areas such as Big Bend National Park where human impacts are minimal. In the early 1970s, less than 5 km of roads in white-tailed deer habitat existed in the park and road impacts such as vehicle collisions were minimal (Krausman and Ables, 1981). Outside of the National Park, some of the habitats occupied by Carmen Mountains white-tailed deer are overgrazed by livestock (Krausman et al., 1978; Krausman and Ables, 1981).

Heavy grazing by cattle negatively impacts Coues deer populations, although the two species may be compatible if grazing by cattle is not excessive (Anthony and Smith, 1977; Ockenfels et al., 1991; Galindo-Leal et al., 1993; Gallina, 1993). Presence of cattle at water sources inhibits use by deer (Maghini and Smith, 1990). Roads negatively impact Coues white-tailed deer. In the Santa Rita Mountains in Arizona, deer avoided areas within 400 m of graded dirt roads (Ockenfels et al., 1991).

Home Range Sizes

Core areas of female and male Coues white-tailed deer in Arizona were estimated to average 189 and 447 ha, respectively (Ockenfels et al., 1991). Home ranges of females and males averaged 518 and 1057 ha, respectively.

Habitat-Management Approaches

Areas containing Mexican blue oak and juniper should be protected from fuelwood cutting to maintain thermal cover for Coues white-tailed deer (Ockenfels and Brooks, 1994). Livestock grazing should be managed so that no more than 35% of the total annual production of herbaceous vegetation is utilized (Galt et al., 2000).

Water Needs

Carmen Mountains and Coues white-tailed deer are spatially distributed based on availability of water sources (Krausman et al., 1978; Krausman and Ables, 1981; Rosenstock et al., 1999). In Arizona, Coues white-tailed deer selected areas within 800 m of water sources and avoided areas greater than 1200 m from water (Ockenfels et al., 1991).

Texas White-tailed Deer

Land Use and Human Impacts

Range-Management Practices

Most habitat occupied by Texas white-tailed deer is rangeland—arid and semiarid land managed to support grazing by native and domestic herbivores (Fulbright and Ortega-S., 2006). Range-management practices that may impact white-tailed deer populations and their habitat include brush management, livestock grazing, fencing, and water development.

Woody plants have increased substantially in abundance during the past two centuries in much of the range of Texas white-tailed deer (Van Auken, 2000). This increase in woody plants has probably benefitted white-tailed deer because of their need for woody cover. Livestock producers, however, often prefer reducing woody plant abundance to increase grasses for forage and increase visibility of livestock. A large number of mechanical and chemical methods have been developed to reduce abundance of woody plants (Fulbright and Ortega-S., 2006). Mechanical approaches include entire plant removal using heavy equipment to uproot woody plants, and removal of the aerial portions of woody plants leaving the crown and root system intact. Entire plant removal includes individual plant treatments such as grubbing and excavating, and large-scale plant removal by root plowing. Root plows are large V-shaped blades attached to crawler tractors that sever the roots and crowns of woody plants below the soil surface. Another plant removal method is to pull large, heavy anchor chains between two crawler tractors to uproot woody plants. Treatments in which the aerial portions of woody plants are removed, but the crown and roots remain intact include roller chopping, mowing, heavy discing, and aeration. Woody plants readily resprout following treatments that leave crowns and roots intact, and woody canopy cover similar to pretreatment levels develops in less than a decade following treatment.

A variety of herbicides are used to kill woody plants. These are used with varying levels of success depending on timing of application, method of application, and species of brush (Hamilton et al., 2004). Prescribed burning is also used to reduce canopy cover of woody plants (Fulbright and Ortega-S., 2006). Prescribed burns are conducted under predetermined conditions of wind speed, relative humidity, fuel moisture, and fine fuel load. Fire is effective in killing woody plant species that are not capable of sprouting from crowns, roots, and other subterranean organs such as Ashe juniper and big sagebrush. Most of the woody plants within the range of Texas white-tailed deer, such as honey mesquite, are sprouting species. Aerial portions of sprouting woody plants may be killed by fire, but mortality of the plants is usually low (Owens et al., 2002; Dacy and Fulbright, 2009). Sprouting shrub species typically grow to prefire height within a few years following death of aerial portions from fire.

Large-scale (larger than 259 ha) clearing of woody plants by mechanical methods such as root plowing and chaining reduces white-tailed deer population densities (Davis and Winkler, 1968; Darr and Klebenow, 1975). Population densities in cleared areas are commonly only 33–50% of the population density present in nearby shrubland (Inglis et al., 1986). Important browse species for white-tailed deer are reduced in abundance in plant communities that reestablish following root plowing on certain soils (Fulbright and Beasom, 1987; Ruthven et al., 1993).

Herbicide treatments often leave standing dead woody vegetation that provides screening cover, and deer may return to the treated areas once forbs recover (Tanner et al., 1978; Beasom et al., 1982; Leslie et al., 1996). Most of the research on effects of herbicide treatments on deer densities should be interpreted with caution because treatments were done in strips or on a relatively small scale with woody plant communities surrounding the treated patches. Inferences can be made based on these studies about effects of treatments on deer distribution, but not on populations. Nutritional status of deer does not appear to be impaired by herbicide treatments; in certain cases browse nutritional quality may be temporarily higher posttreatment (Fulbright and Garza, 1991; Soper et al., 1993b).

In areas where Texas white-tailed deer and mule deer populations overlap, increasing woody plant canopy cover favors white-tailed deer over mule deer (Wiggers and Beasom, 1986). Reducing woody plant canopy cover in those areas using brush management may favor mule deer (Avey et al., 2003).

Excessive grazing by livestock is detrimental to white-tailed deer habitat and nutritional status of white-tailed deer (McMahan and Ramsey, 1965; Bryant et al., 1979; Warren and Krysl, 1983). For livestock and deer to be compatible, livestock densities must be sufficiently low such that adequate forage remains to support deer, and to avoid damage to deer habitat (Bryant et al., 1979; Thill, 1984; Gallina, 1993). Potential for negative impacts of livestock grazing on white-tailed deer habitat and nutrition increases with decreasing habitat productivity. Sheep and goats are more competitive with white-tailed deer than are cattle and are more likely than cattle to damage white-tailed deer habitat and reduce deer productivity (McMahan, 1964; McMahan and Ramsey, 1965; Bryant et al., 1979).

Livestock fences are generally constructed of either barbed wire or net wire and are about 1.4 m tall. Movements of white-tailed deer are generally not impeded by these fences; deer are able to jump over or go under them (VerCauteren et al., 2006; Meek et al., 2008). Deer may become entangled in these fences while trying to cross them, and fences may be a source of mortality for dispersing yearling males in particular (McCoy et al., 2005).

Presence of livestock water developments does not appear to increase productivity or population size of Texas white-tailed deer in semiarid environments (Rosenstock et al., 1999). Cattle tend to congregate around water sources, which may reduce use of livestock water developments by white-tailed deer (Prasad and Guthery, 1986; Cooper et al., 2008). Use of the same water sources by deer and cattle has potential for disease transmission. Cattle infected with bovine viral diarrhea virus, for example, may potentially transmit the virus to white-tailed deer fawns if they are using shared resources such as watering facilities (Passler et al., 2009).

Highways and Roads

More than 127,138 km of state-maintained road rights-of-way exists in Texas. Road kill causes mortality in Texas white-tailed deer; however, it is unclear how many are killed each year. In 1992, it was estimated that highway mortality exceeded 19,000 deer/year in Texas (Romin and Bissonette, 1996). Texas Parks and Wildlife Department estimated in 2000 that more than 10,000 vehicle collisions occur each year with deer (Sullivan and Messmer, 2003). In comparison, about 400,000 white-tailed deer are harvested annually by hunters in Texas. In the zone between the Edwards Plateau and South Texas Plains of Texas, half of the relocations of white-tailed deer outfitted with collars containing global position systems were within 100 m of a ranch road (Cooper et al., 2008). Deer were attracted to roads in part because of spreading of corn along roads by hunters during November through January.

Home Range Sizes

Home range size of Texas white-tailed deer decreases with increasing habitat productivity. In an area with average annual precipitation of 51 cm, for example, home ranges were twice the size of those in an area with 93 cm average annual precipitation (Inglis et al., 1986). Home range size also decreases with increasing juxtaposition and interspersion of plant communities, food, water, and cover because deer can meet their needs in a relatively smaller area (Beier and McCullough, 1990). Males generally have larger home ranges than females, possibly because the larger body size of males results in a need for greater space to meet nutrient demands. Older males (at least 7 years) have smaller home ranges than younger males (Hellickson et al., 2008). Home range size also varies seasonally. Home ranges of males were larger during spring than other times of the year in western South Texas (Hellickson et al., 2008). Variation in deer density and availability of females in estrus may influence home range size of males during the rut; search distances for receptive females are shorter in high-density deer populations (Inglis et al., 1986).

Habitat-Management Approaches

Creating Landscape Mosaics

Brush management during much of the twentieth century focused on improving rangeland for livestock with little emphasis on wildlife (Hamilton et al., 2004). Growing economic value of white-tailed deer and other wildlife species during the late twentieth century fostered a paradigm shift from a livestock focus to applying brush management as a wildlife habitat-management tool. Application of brush management in linear strips, blocks, or other patterns to create mosaics of woody plant stands interspersed with herbaceous plant stands is a common approach to improving brush-dominated landscapes for white-tailed deer (Figure 18.3). This pattern is based on the idea of providing areas dominated by herbaceous vegetation for white-tailed deer to forage and providing areas dominated by woody plants for security and thermal cover, and browse when forbs are lacking (Fulbright and Ortega-S., 2006). Brush sculpting is another approach to increasing interspersion of woody plant and herbaceous patches. Brush sculpting involves selective removal of brush to accomplish multiple-use objectives including habitat improvement for wildlife and increased forage for livestock (Fulbright, 1997; Ansley et al., 2003). Brush eventually reestablishes in portions of the landscape regardless of the brush removal approach used (Fulbright and

FIGURE 18.3 Application of brush management in patterns to create mosaics of woody plant stands interspersed with herbaceous plant stands is a common approach to improving brush-dominated landscapes for white-tailed deer. (Photo by T. E. Fulbright. With permission.)

Beasom, 1987; Montemayor et al., 1991; Nolte et al., 1994; Schindler and Fulbright, 2003). Mechanical treatments such as root plowing, discing, and roller chopping may result in a temporary increase in forbs eaten by white-tailed deer (Rollins and Bryant, 1986; Bozzo et al., 1992a; Ruthven et al., 1993; Fulbright, 1999). Sprouts produced from woody plants subjected to treatments that kill only the aerial portions of the plants are more nutritious than mature growth on untreated plants. Sprouts may be heavily used by deer for several weeks following treatment (Fulbright et al., 1991; Reynolds et al., 1992; Stewart et al., 2000; Schindler et al., 2004).

Presence of a mosaic of treated patches with sprouting shrubs and tracts of untreated woody plants results in spatial redistribution of deer. Deer are attracted to the patches containing recently produced woody plant sprouts and increased abundance of forbs. Deer densities commonly increase temporarily in areas where brush management has been conducted in strips or blocks (Rollins et al., 1988; Bozzo et al., 1992b).

Increased densities of a wildlife species in an area following brush management and temporary increases in forage abundance and nutrient content are often interpreted as "habitat improvement." These responses, however, do not necessarily indicate sustained improvement in habitat (Van Horne, 1983). Productivity of wildlife populations is a more reliable indicator of how brush management affects wildlife. Indices of increased productivity include increased population size, increased reproduction, or increased body mass. Documentation that brush management improves white-tailed deer productivity is minimal in the scientific literature. A short-term study (1987–1989) in Oklahoma is one of the few in which a measure of productivity was documented; in the study, treatment with fire and herbicides increased carcass mass of deer compared to untreated controls (Soper et al., 1993a).

Cattle Grazing as a Tool to Increase Forbs

Cattle grazing has been recommended as a tool for increasing herbaceous plant species diversity and forbs, thereby enhancing the supply of forage for white-tailed deer (Harmel and Litton, 1981). However, use of cattle grazing to manipulate white-tailed deer habitat is of questionable value in semiarid and arid environments, in part because effects of grazing on plant species richness and forb abundance depend on habitat productivity (Figure 18.4). In productive habitats dominated by grasses, increasing intensity of cattle grazing increases plant species richness up to a threshold beyond which species richness declines (Olff and Ritchie, 1998; Oba et al., 2001; Herrera and Pellmyr, 2002). Cattle grazing may have little effect on plant species richness, or plant species richness may decline with increasing grazing intensity in less productive habitats or in landscapes that have been overgrazed by livestock.

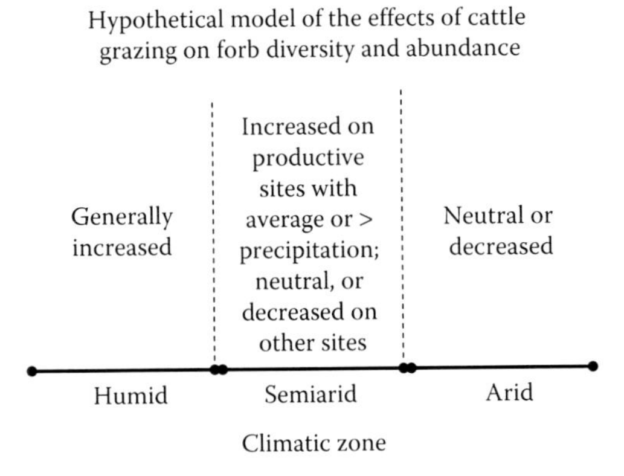

FIGURE 18.4 Hypothetical effects of grazing by cattle on abundance and species diversity of forbs.

Cattle grazing at light or moderate intensities may increase forbs in productive habitats dominated by grasses (Thill and Martin, 1986). Conversely, during winter or dry periods, cattle grazing may reduce deer forage, even at moderate grazing intensities (Thill and Martin, 1986, 1989; Ortega et al., 1997b; Jenks and Leslie, 2003). In addition, grazing by cattle may increase annual forbs, but concomitantly reduce abundance of perennial forbs (Hayes and Holl, 2003; Hickman et al., 2004). Perennial forbs provide a more stable food supply for white-tailed deer than annuals because annuals are only temporarily available. Documentation that forbs important in deer diets increase in response to cattle grazing in semiarid and arid environments appears lacking in the literature (Fulbright and Ortega-S., 2006). Unequivocal documentation that white-tailed deer productivity is greater in habitats grazed by cattle or other livestock species than in ungrazed habitats is also lacking. Researchers in Oklahoma and Arkansas, for example, found that fetuses/doe were greater in habitats not grazed by cattle than in grazed habitats; however, carcass mass of deer was higher in a grazed area during summer (Jenks and Leslie, 2003). More research is needed to clarify relationships between cattle grazing as a habitat-management tool and deer productivity.

Supplemental Feeding and Habitat

Supplemental feeding of white-tailed deer with corn, cottonseed, commercially produced pellets high in protein, and planted food plots is a widespread practice within the range of Texas white-tailed deer. Availability of supplemental feed may result in spatial redistribution of deer and alter foraging pressures on vegetation. Female deer provided supplemental feed reduce the size of their core areas (Cooper et al., 2006). Browsing by deer on native vegetation is greater near feeders and intensity of use declines with distance from feeders (Cooper et al., 2006).

The idea that providing supplemental feed may artificially increase carrying capacity for deer is controversial (Brown and Cooper, 2006). In southern Texas, forbs comprised a similar percentage of deer diets whether deer used food plots or not (Hehman and Fulbright, 1997). Consuming highly nutritious rations relieves time constraints on foraging. Consequently, white-tailed deer provided supplemental feed may spend more time foraging on highly palatable plants than deer not provided supplemental feed (Murden and Risenhoover, 1993). Results of both of these studies suggest that providing artificial sources of feed as rationale to allow deer densities to increase may result in overuse of palatable native forbs.

Providing supplemental feed is not widely accepted in western North America as a management practice outside of the range of Texas white-tailed deer, except during winter emergencies. Objections to supplemental feeding are complex and range from ethical issues surrounding hunting and wildness of deer, to concerns regarding disease transmission (Brown and Cooper, 2006).

Water Needs

Texas white-tailed deer readily use surface water, but there is no documentation that availability of surface water is essential to their survival (Rosenstock et al., 1999). Several studies have documented selection by deer for areas near water. In the western edge of the Edwards Plateau in Texas, female white-tailed deer selected home ranges near permanent water sources but selection of areas near permanent water was less pronounced in males (Brunjes et al., 2006). Both sexes appeared to prefer areas within 0.8 km of water in southern Texas (Hyde, 1987).

Management Issues and Trends

Columbian and Northwest White-tailed Deer

Degradation of riparian areas is the major factor that has reduced populations of Columbian white-tailed deer (Brookshier, 2004). Protection of oak woodlands and remaining riparian areas is critically important (Brookshier, 2004; Lyman, 2006). Restoring deciduous plant communities in cleared areas, particularly to connect riparian corridors, may benefit deer populations.

Columbian white-tailed deer exist in two distinct populations with separate classifications. Presence of a population along the lower Columbia River in Washington and northwest Oregon with endangered status complicates management of the subspecies and the habitats it occupies (D. G. Whittaker, Oregon Department of Game and Fish, personal communication). The Oregon Department of Fish and Wildlife has greater flexibility in capturing and moving animals in the Douglas County population than in the endangered population. Growing interest among hunters in harvesting whitetails in Douglas County from this unique subspecies may also create management pressures.

Developing snow interception models to map areas of winter range is important for management of northwest white-tailed deer (Poole and Mowat, 2005). Open areas used for foraging also need to be mapped. In northeastern Oregon, northwest white-tailed deer are generally not hunted or managed as a separate species from mule deer (D. G. Whittaker, Oregon Department of Game and Fish, personal communication). Other states, such as Idaho, manage white-tailed deer separately from mule deer (Idaho Fish and Game, 2004). Idaho Fish and Game collects samples from deer captured by agency personnel to determine exposure of white-tailed deer to diseases such as chronic wasting disease, tuberculosis, and epizootic hemorrhagic disease, and to determine the relative importance of these diseases to deer within the state.

Carmen Mountains and Coues White-tailed Deer

Climate in the southwestern United States is predicted to become warmer and drier during the twenty-first century (Seager et al., 2007; Archer and Predick, 2008). Assuming that these predictions are correct, replacement of woodland plant communities as desert plant communities expand to higher elevations may result in loss of Carmen Mountains white-tailed deer habitat. While desert plant communities may expand to higher elevations, oak woodland communities preferred by Coues white-tailed deer may also expand to higher elevations, so it is unclear whether or not climate warming will impact the subspecies (J. Heffelfinger, Arizona Game and Fish Department, personal communication). Water availability may also decline as a result of a drier climate, which could affect white-tailed deer distribution and abundance (Archer and Predick, 2008).

Texas White-tailed Deer

Desert plant communities are predicted to expand eastward into the range of Texas white-tailed deer as the climate dries and becomes warmer (Seager et al., 2007; Archer and Predick, 2008). Changes in traditional management approaches may become necessary if the habitat becomes more xerophytic. For example, white-tailed deer management has traditionally been based on density-dependent population dynamics (Chapter 5). Populations in the western portion of the range of Texas white-tailed deer may become more likely to exhibit density-independent population dynamics as the climate becomes drier. Thermal cover may increase in importance to deer for maintaining thermal equilibrium and energy balance as the climate warms. Practices such as supplemental feeding may become more widespread as habitats become less productive, unless changes in the United States economy and in energy costs make the practice less affordable for landowners and hunters.

Research Needs

Columbian and Northwestern White-tailed Deer

Fawn survival is low in Columbian white-tailed deer in Oregon (Ricca et al., 2002). Research is needed to clarify density-dependent and density-independent effects on fawn survival and related factors such as nutritional status of females. Habitat management for the species can be improved by gaining more information on effects of cattle grazing and prescribed fire on deer habitat and diet composition. A final need is additional research to clarify the subspecies status of Columbian white-tailed deer.

Basic population ecology of northwest white-tailed deer including survival, mortality rates, and productivity is not well understood (Idaho Fish and Game, 2004). More information is needed on habitat use, and more reliable methods of estimating population density are needed. A better understanding of the effects of density-dependent processes on partitioning space, habitat, and forage among white-tailed deer, mule deer, elk, and cattle is needed (Stewart et al., 2002; Ward and Marcum, 2005). More research is needed on the relationship between lichen availability during winter and forest stand characteristics, on intensity of use of lichens by deer, and the nutritional value of lichens relative to other winter forages (Hodgman and Bowyer, 1985; Ward and Marcum, 2005).

Sky Island White-tailed Deer: Carmen Mountains and Coues

Information on the population status, population dynamics, and habitat relationships of Carmen Mountains white-tailed deer is limited. Research to develop predictions of potential effects of climate drying and warming on Sky Island white-tailed deer ecology is needed. In particular, better understanding of how climate change may influence competitive interrelationships between white-tailed deer and mule deer is needed.

Texas White-tailed Deer

Wind farm development is rapidly expanding within the range of Texas white-tailed deer. Most of the research on wind farms has focused on avian species (Kuvlesky et al., 2007). Research on the effects of disturbance and extensive road systems resulting from wind farm development on white-tailed deer is lacking.

Brush management, livestock grazing, or both affect most of the range of Texas white-tailed deer. Most of the research on the effects of these land-management practices on white-tailed deer has been relatively short term. Long-term studies of the effects of these practices on white-tailed deer productivity are needed.

Supplemental feeding may result in excessive use of vegetation, particularly if white-tailed deer populations are managed for greater densities (Murden and Risenhoover, 1993; Cooper et al., 2006). More research to determine how supplemental feeding affects foraging ecology, productivity, behavior, and population dynamics of white-tailed deer is needed. The manner in which supplemental feeding and increased deer densities may influence vegetation dynamics and how impacts may vary with habitat productivity are poorly understood.

Nonnative grasses including buffelgrass, Lehman lovegrass, and old-world bluestems are widely planted to provide livestock forage within the geographic range of Texas white-tailed deer. These grasses are highly competitive with native vegetation and can reduce forb production and species diversity (Jackson, 2005; Sands et al., 2009). One result of projected climate changes is a substantial increase in nonnative grasses (Archer and Predick, 2008). Research is needed to determine the impact of planting these grasses on white-tailed deer diets and nutrition.

Research is needed to predict potential changes in white-tailed deer distribution resulting from climate change and subsequent changes in habitat. Potential effects of climate change on white-tailed deer and desert mule deer interrelationships on the western edge of the range of Texas white-tailed deer are unclear. Drying and increasing temperature would be expected to favor mule deer over white-tailed deer, but increased ambient carbon dioxide concentrations may result in increased woody plant cover, which would favor white-tailed deer (Bond and Midgley, 2001).

Summary

Many commonalities in ecology and management exist among the subspecies of white-tailed deer in western North America. Cougars are a major predator of white-tailed deer in western North America; in their absence, coyotes are the major predator. Riparian areas provide critical habitat for white-tailed deer in western North America. Conservation and restoration of these habitats are essential to sustain and increase white-tailed deer populations. Amount of woody plant cover is a major factor separating white-tailed deer and desert mule deer habitat; reducing woody canopy cover below 40% favors mule deer. Although woody plant cover is important, white-tailed deer commonly prefer to forage in patches dominated by herbaceous vegetation.

White-tailed deer are adapted to higher stages of succession in much of western North America. Range-management practices including livestock grazing and brush management alter white-tailed deer habitat, but their effects on white-tailed deer productivity are unclear. Habitat manipulation with mechanical methods, grazing, or other forms of disturbance is commonly done to benefit white-tailed deer; however, documentation of improved deer productivity resulting from these practices is scant.

White-tailed deer are successful across a wide range of environmental conditions in western North America, inhabiting areas ranging from semiarid to greater than 250 cm of annual precipitation. They inhabit a wide range of plant communities, from mesic coniferous forests to xerophytic woodland. The extreme adaptability of white-tailed deer is further demonstrated by their diverse diets.

White-tailed deer are able to shift diet composition in response to drought and overgrazing; this diversity in diet selection is important to their survival in the semiarid regions of western North America. Availability of surface water is critical to white-tailed deer in the arid mountain ranges of the southwest; surface water becomes less important in regions with higher rainfall. Effects of climate change and energy development on white-tailed deer populations are important future issues, particularly for deer in the southern part of western North America.

Acknowledgments

I thank Don Whittaker for reviewing an early draft of this manuscript, and R. Terry Bowyer and Robert G. Anthony for reviewing portions of the manuscript. I also thank Kelley Stewart, Jim Heffelfinger, and John Bissonette for providing helpful information.

REFERENCES

Ansley, R. J., B. A. Kramp, and D. L. Jones. 2003. Converting mesquite thickets to savanna through foliage modification with clopyralid. *Journal of Range Management* 56:72–80.

Anthony, R. G. 1976. Influence of drought on diets and numbers of desert deer. *Journal of Wildlife Management* 40:140–144.

Anthony, R. G. and N. S. Smith. 1977. Ecological relationships between mule deer and white-tailed deer in southeastern Arizona. *Ecological Monographs* 47:255–277.

Archer, S. R. and K. I. Predick. 2008. Climate change and ecosystems of the southwestern United States. *Rangelands* 30:23–28.

Arizona Game and Fish Department. 2009. *Hunt Arizona, 2009 Edition: Survey, Hunt, and Harvest Data for Big and Small Game.* Phoenix, AZ: Arizona Game and Fish Department.

Arnold, L. A. and D. L. Drawe. 1979. Seasonal food habits of white-tailed deer in the south Texas plains. *Journal of Range Management* 32:175–178.

Avey, J. T., W. B. Ballard, M. C. Wallace et al. 2003. Habitat relationships between sympatric mule deer and white-tailed deer in Texas. *Southwestern Naturalist* 48:644–653.

Ballinger, S. W., L. H. Blankenship, J. W. Bickham, and S. M. Carr. 1992. Allozyme and mitochondrial DNA analysis of a hybrid zone between white-tailed deer and mule deer (*Odocoileus*) in west Texas. *Biochemical Genetics* 30:1–11.

Barboza, P. S. and R. T. Bowyer. 2000. Sexual segregation in dimorphic deer: A new gastrocentric hypothesis. *Journal of Mammalogy* 81:473–489.

Baty, G. R., C. L. Marcum, M. J. Thompson, and J. M. Hillis. 1996. Potential effects of ecosystem management on cervids wintering in ponderosa pine habitats. *Intermountain Journal of Sciences* 2:1–7.

Beasom, S. L., J. M. Inglis, and C. J. Scifres. 1982. Vegetation and white-tailed deer responses to herbicide treatment of a mesquite drainage type. *Journal of Range Management* 35:790–794.

Beier, P. and D. R. McCullough. 1990. Factors influencing white-tailed deer activity patterns and habitat use. *Wildlife Monographs* 109:1–51.

Bond, W. J. and G. F. Midgley. 2001. A proposed CO_2-controlled mechanism of woody plant invasion in grasslands and savannas. *Global Change Biology* 6:865–869.

Bowyer, R. T., K. M. Stewart, S. A. Wolfe et al. 2002. Assessing sexual segregation in deer. *Journal of Wildlife Management* 66:536–544.

Bozzo, J. A., S. L. Beasom, and T. E. Fulbright. 1992a. Vegetation responses to 2 brush management practices in south Texas. *Journal of Range Management* 45:170–175.

Bozzo, J. A., S. L. Beasom, and T. E. Fulbright. 1992b. White-tailed deer use of rangeland following browse rejuvenation. *Journal of Range Management* 45:496–499.

Brookshier, J. 2004. Columbia white-tailed deer *Odocoileus virginianus leucurus*. In *Management Recommendations for Washington's Priority Species—Volume V: Mammals*, 1–9. Olympia, WA: Washington Department of Fish and Wildlife.

Brown, R. D. and S. M. Cooper. 2006. The nutritional, ecological, and ethical arguments against baiting and feeding white-tailed deer. *Wildlife Society Bulletin* 34:519–524.

Brunjes, K. J., W. B. Ballard, M. H. Humphrey et al. 2006. Habitat use by sympatric mule and white-tailed deer in Texas. *Journal of Wildlife Management* 70:1351–1359.

Bryant, F. C., M. M. Kothmann, and L. B. Merrill. 1979. Diets of sheep, angora goats, Spanish goats and white-tailed deer under excellent range conditions. *Journal of Range Management* 32:412–417.

Burke, K. M. 2003. Seasonal diets and selectivity of white-tailed deer in the Rolling Plains ecological region. MS thesis, Southwest Texas State University.

Cahalane, V. H. 1939. Mammals of the Chiricahua Mountains, Cochise County, Arizona. *Journal of Mammalogy* 20:418–440.

Carr, S. M., S. W. Ballinger, J. N. Derr, L. H. Blankenship, and J. W. Bickham. 1986. Mitochondrial DNA analysis of hybridization between sympatric white-tailed deer and mule deer in west Texas. *Proceedings of the National Academy of Sciences of the United States of America* 83:9576–9580.

Choate, J. R. 1987. Post-settlement history of mammals in western Kansas. *Southwestern Naturalist* 32:157–168.

Cohen, W. E., D. L. Drawe, F. C. Bryant, and L. C. Bradley. 1989. Observations on white-tailed deer and habitat response to livestock grazing in south Texas. *Journal of Range Management* 42:361–365.

Compton, B. B., R. J. Mackie, and G. L. Dusek. 1988. Factors influencing distribution of white-tailed deer in riparian habitats. *Journal of Wildlife Management* 52:544–548.

Cook, R. S., M. White, D. O. Trainer, and W. C. Glazener. 1971. Mortality of young white-tailed deer fawns in south Texas. *Journal of Wildlife Management* 35:47–56.

Cooley, H. S., H. S. Robinson, R. B. Wielgus, and C. S. Lambert. 2008. Cougar prey selection in a white-tailed deer and mule deer community. *Journal of Wildlife Management* 72:99–106.

Cooper, S. M., M. K. Owens, R. M. Cooper, and T. F. Ginnett. 2006. Effect of supplemental feeding on spatial distribution and browse utilization by white-tailed deer in semi-arid rangeland. *Journal of Arid Environments* 66:716–726.

Cooper, S. M., H. L. Perotto-Baldivieso, M. K. Owens, M. G. Meek, and M. Figueroa-Pagán. 2008. Distribution and interaction of white-tailed deer and cattle in a semi-arid grazing system. *Agriculture, Ecosystems and Environment* 127:85–92.

Cronin, M. A., E. R. Vyse, and D. G. Cameron. 1988. Genetic relationships between mule deer and white-tailed deer in Montana. *Journal of Wildlife Management* 52:320–328.

Dacy, E. C. and T. E. Fulbright. 2009. Survival of sprouting shrubs following summer fire: Effects of morphological and spatial characteristics. *Rangeland Ecology and Management* 62:179–185.

Darr, G. W. and D. A. Klebenow. 1975. Deer, brush control, and livestock on the Texas Rolling Plains. *Journal of Range Management* 28:115–119.

Davis, R. B. and C. K. Winkler. 1968. Brush vs cleared range as deer habitat in southern Texas. *Journal of Wildlife Management* 32:321–329.

Day, G. I. 1980. Characteristics and measurements of captive hybrid deer in Arizona. *Southwestern Naturalist* 25:434–438.

Derr, J. N. 1991. Genetic interactions between white-tailed and mule deer in the southwestern United States. *Journal of Wildlife Management* 55:228–237.

Dillard, J., S. Jester, J. Baccus, R. Simpson, and L. Poor. 2005. *White-tailed Deer Food Habits and Preferences in the Cross Timbers and Prairies Region of Texas*. Austin, TX: Texas Parks and Wildlife Department.

Dusek, G. L., A. K. Wood, C. A. Sime, and J. T. Morgan. 2005. Forage use by white-tailed deer in northwest Montana from an historical perspective. *Intermountain Journal of Sciences* 11:58–65.

Fulbright, T. E. 1997. Designing shrubland landscapes to optimize habitat for white-tailed deer. In *Proceedings of the Brush Sculptors Symposium*, eds. D. Rollins, D. N. Ueckert, and C. G. Brown, 61–67. College Station, TX: Texas Agricultural Extension Service.

Fulbright, T. E. 1999. Response of white-tailed deer foods to discing in a semiarid habitat. *Journal of Range Management* 52:346–350.

Fulbright, T. E. and S. L. Beasom. 1987. Long-term effects of mechanical treatments on white-tailed deer browse. *Wildlife Society Bulletin* 15:560–564.

Fulbright, T. E. and A. Garza, Jr. 1991. Forage yield and white-tailed deer diets following live oak control. *Journal of Range Management* 44:451–455.

Fulbright, T. E. and J. A. Ortega-S. 2006. *White-tailed Deer Habitat: Ecology and Management on Rangelands*. College Station, TX: Texas A&M University Press.

Fulbright, T. E., J. P. Reynolds, S. L. Beasom, and S. Demarais. 1991. Mineral content of guajillo regrowth following roller chopping. *Journal of Range Management* 44:520–522.

Galindo-Leal, C., A. Morales G. and M. Weber R. 1993. Distribution and abundance of Coues deer and cattle in Michilia Biosphere Reserve, Mexico. *Southwestern Naturalist* 38:127–135.

Gallina, S. 1993. White-tailed deer and cattle diets at La Michilia, Durango, Mexico. *Journal of Range Management* 46:487–492.

Galt, D., F. Molinar, J. Navarro, J. Joseph, and J. Holechek. 2000. Grazing capacity and stocking rate. *Rangelands* 22:7–11.

Garner, G. W., J. A. Morrison, and J. C. Lewis. 1976. Mortality of white-tailed deer fawns in the Wichita Mountains, Oklahoma. *Proceedings of the Annual Conference of the Southeastern Association of Fish and Wildlife Agencies* 30:493–506.

Gavin, T. A. and B. May. 1988. Taxonomic status and genetic purity of Columbian white-tailed deer. *Journal of Wildlife Management* 52:1–10.

Gavin, T. A., L. H. Suring, P. A. Vohs, Jr., and E. C. Meslow. 1984. Population characteristics, spatial organization, and natural mortality in the Columbian white-tailed deer. *Wildlife Monographs* 91:1–41.

Gerlach, D., S. Atwater, and J. Schnell, eds. 1994. *Deer*. Mechanicsburg, PA: Stackpole Books.

Hamilton, W. T., A. McGinty, D. N. Ueckert, C. W. Hanselka, and M. R. Lee, eds. 2004. *Brush Management: Past, Present, and Future*. College Station, TX: Texas A&M University Press.

Hanley, T. A. 1996. Potential role of deer (Cervidae) as ecological indicators of forest management. *Forest Ecology and Management* 88:199–204.

Harmel, D. E. and G. W. Litton. 1981. *Deer Management in the Edwards Plateau of Texas*. Austin, TX: Texas Parks and Wildlife Department.

Harveson, L. A., M. E. Tewes, N. J. Silvy, and J. Rutledge. 2000. Prey use by mountain lions in southern Texas. *Southwestern Naturalist* 45:472–476.

Hatter, I., G. Harcombe, L. Stanlake, and A. Bethune. 2000. *White-tailed Deer in British Columbia*. Victoria, BC: British Columbia Ministry of Environment, Lands, and Parks.

Hayes, G. F. and K. D. Holl. 2003. Cattle grazing impacts on annual forbs and vegetation composition of mesic grasslands in California. *Conservation Biology* 17:1694–1702.

Heffelfinger, J. 2006. *Deer of the Southwest*. College Station, TX: Texas A&M University Press.

Hehman, M. W. and T. E. Fulbright. 1997. Use of warm-season food plots by white-tailed deer. *Journal of Wildlife Management* 61:1108–1115.

Hellickson, M. W., T. A. Campbell, K. V. Miller, R. L. Marchinton, and C. A. DeYoung. 2008. Seasonal ranges and site fidelity of adult male white-tailed deer (*Odocoileus virginianus*) in southern Texas. *Southwestern Naturalist* 53:1–8.

Herrera, C. M. and O. Pellmyr, eds. 2002. *Plant–Animal Interactions: An Evolutionary Approach*. Oxford: Blackwell Science.

Hickman, K. R., D. C. Hartnett, R. C. Cochran, and E. O. Clenton. 2004. Grazing management effects on plant species diversity in tallgrass prairie. *Journal of Range Management* 57:58–65.

Hodgman, T. P. and R. T. Bowyer. 1985. Winter use of arboreal lichens, ascomycetes, by white-tailed deer, *Odocoileus virginianus*, in Maine. *Canadian Field-Naturalist* 99:313–316.

Hudson, R. J., D. M. Hebert, and V. C. Brink. 1976. Occupational patterns of wildlife on a major east Kootenay winter–spring range. *Journal of Range Management* 29:38–43.

Hyde, K. J. 1987. Effects of short-duration grazing on white-tailed deer. MS thesis, Texas A&I University.

Hyde, K. J., C. A. DeYoung, and A. Garza. 1987. Bed sites of white-tailed deer fawns in south Texas. *Proceedings of the Annual Conference of the Southeastern Association of Fish and Wildlife Agencies* 41:288–293.

Idaho Fish and Game. 2004. *2005–2014 White-tailed Deer Plan*. Boise, ID: Idaho Department of Fish and Game. http://fishandgame.idaho.gov/cms/wildlife/plans/wtd. Accessed September 2, 2009.

Inglis, J. M., B. A. Brown, C. A. McMahan, and R. E. Hood. 1986. *Deer–Brush Relationships on the Rio Grande Plain, Texas*. College Station, TX: Kleberg Studies in Natural Resources, Texas Agricultural Experiment Station.

Jackson, J. 2005. Is there a relationship between herbaceous species richness and buffel grass (*Cenchrus ciliaris*)? *Austral Ecology* 30:505–517.

Jenkins, K. J. and R. G. Wright. 1988. Resource partitioning and competition among cervids in the northern Rocky Mountains. *Journal of Applied Ecology* 25:11–24.

Jenks, J. A. and D. M. Leslie, Jr. 2003. Effect of domestic cattle on the condition of female White-tailed deer in southern pine–bluestem forests. *Acta Theriologica* 48:131–144.

Kie, J. G. and R. T. Bowyer. 1999. Sexual segregation in White-tailed deer: Density-dependent changes in use of space, habitat selection, and dietary niche. *Journal of Mammalogy* 80:1004–1020.

Kilgo, J. C., R. F. Labisky, and D. E. Fritzen. 1998. Influences of hunting on the behavior of White-tailed deer: Implications for conservation of the Florida panther. *Conservation Biology* 12:1359–1364.

Kingery, J. L., J. C. Mosley, and K. C. Bordwell. 1996. Dietary overlap among cattle and cervids in northern Idaho forests. *Journal of Range Management* 49:8–15.

Knipe, T. 1977. The Arizona White-tailed deer. Arizona Game and Fish Department special Report 6. Phoenix, AZ.

Krausman, P. R. 1978. Forage relationships between two deer species in Big Bend National Park, Texas. *Journal of Wildlife Management* 42:101–107.

Krausman, P. R. and E. D. Ables. 1981. Ecology of the Carmen Mountains White-tailed deer. U.S. National Park Service Scientific Monograph Series 15. Washington, DC.

Krausman, P. R., D. J. Schmidly, and E. D. Ables. 1978. Comments on the taxonomic status, distribution, and habitat of the Carmen Mountains White-tailed deer (*Odocoileus virginianus carminis*) in Trans-Pecos Texas. *Southwestern Naturalist* 23:577–589.

Kunkel, K., D. H. Pletscher, and T. K. Ruth. 2003. Habitat factors affecting hunting success of cougars and wolves in northwestern Montana. In *Proceedings of the Sixth Mountain Lion Workshop*, eds. L. A. Harveson, P. M. Harveson, R. W. Adams, 64–64. San Antonio, TX.

Kunkel, K. E., D. H. Pletscher, D. K. Boyd, R. R. Ream, and M. W. Fairchild. 2004. Factors correlated with foraging behavior of wolves in and near Glacier National Park, Montana. *Journal of Wildlife Management* 68:167–178.

Kuvlesky, W. P., L. A. Brennan, M. L. Morrison, K. K. Boydston, B. M. Ballard, and F. C. Bryant. 2007. Wind energy development and wildlife conservation: Challenges and opportunities. *Journal of Wildlife Management* 71:2487–2498.

Leopold, B. D. and P. R. Krausman. 1986. Diets of 3 predators in Big Bend National Park, Texas. *Journal of Wildlife Management* 50:290–295.

Leslie, D. M., Jr., R. B. Soper, R. L. Lochmiller, and D. M. Engle. 1996. Habitat use by White-tailed deer on Cross Timbers rangeland following brush management. *Journal of Range Management* 49:401–406.

Lingle, S. 2002. Coyote predation and habitat segregation of White-tailed deer and mule deer. *Ecology* 83:2037–2048.

Lingle, S., S. Pellis, and W. Wilson. 2005. Interspecific variation in antipredator behaviour leads to differential vulnerability of mule deer and White-tailed deer fawns early in life. *Journal of Ecology* 74:1140–1149.

Lopez, R. G. 2006. Genetic structuring of Coues White-tailed deer in the southwestern United States. MS thesis, Northern Arizona University.

Lyman, R. L. 2006. Late prehistoric and early historic abundance of Columbian White-tailed deer, Portland Basin, Washington and Oregon, USA. *Journal of Wildlife Management* 70:278–282.

Lyon, L. J. and C. E. Jensen. 1980. Management implications of elk and deer use of clear-cuts in Montana. *Journal of Wildlife Management* 44:352–362.

Maghini, M. T. and N. S. Smith. 1990. Water use and diurnal seasonal ranges of Coues White-tailed deer. In *Managing Wildlife in the Southwest*, eds. P. R. Krausman and N. S. Smith, 21–34. Phoenix, AZ: Arizona Chapter of the Wildlife Society.

Martinka, C. J. 1968. Habitat relationships of White-tailed and mule deer in northern Montana. *Journal of Wildlife Management* 32:558–565.

McClure, M. F., J. A. Bissonette, M. R. Conover, and D. D. Austin. 1997. Range expansion of White-tailed deer (*Odocoileus virginianus*) into urban and agricultural areas of Utah. *Great Basin Naturalist* 57:278–280.

McCoy, J. E., D. G. Hewitt, and F. C. Bryant. 2005. Dispersal by yearling male White-tailed deer and implications for management. *Journal of Wildlife Management* 69:366–376.

McMahan, C. A. 1964. Comparative food habits of deer and three classes of livestock. *Journal of Wildlife Management* 28:798–808.

McMahan, C. A. 1966. Suitability of grazing enclosures for deer and livestock research on the Kerr Wildlife Management Area, Texas. *Journal of Wildlife Management* 30:151–162.

McMahan, C. A. and C. W. Ramsey. 1965. Response of deer and livestock to controlled grazing in central Texas. *Journal of Range Management* 18:1–7.

Meek, M. G., S. M. Cooper, M. K. Owens, R. M. Cooper, and A. L. Wappel. 2008. White-tailed deer distribution in response to patch burning on rangeland. *Journal of Arid Environments* 72:2026–2033.

Menzel, K. E. 1984. Central and southern plains. In *White-tailed Deer: Ecology and Management*, ed. L. K. Halls, 449–456. Harrisburg, PA: Stackpole Books.

Montemayor, E., T. E. Fulbright, L. W. Brothers, B. J. Schat, and D. Cassels. 1991. Long-term effects of rangeland disking on White-tailed deer browse in south Texas. *Journal of Range Management* 44:246–248.

Mueggler, W. F. 1966. Herbicide treatment of browse on a big-game winter range in northern Idaho. *Journal of Wildlife Management* 30:141–151.

Murden, S. B. and K. L. Risenhoover. 1993. Effects of habitat enrichment on patterns of diet selection. *Ecological Applications* 3:497–505.

Mysterud, A. and E. Ostbye. 1999. Cover as a habitat element for temperate ungulates: Effects on habitat selection and demography. *Wildlife Society Bulletin* 27:385–394.

Nolte, K. R., T. M. Gabor, M. W. Hehman, M. A. Asleson, T. E. Fulbright, and J. C. Rutledge. 1994. Long-term effects of brush management on vegetation diversity in ephemeral drainages. *Journal of Range Management* 47:457–459.

Oba, G., O. R. Vetaas, and N. C. Stenseth. 2001. Relationships between biomass and plant species richness in arid-zone grazing lands. *Journal of Applied Ecology* 38:836–845.

Ockenfels, R. A. and D. E. Brooks. 1994. Summer diurnal bed sites of Coues White-tailed deer. *Journal of Wildlife Management* 58:70–75.

Ockenfels, R. A., D. E. Brooks, and C. H. Lewis. 1991. General ecology of Coues White-tailed deer in the Santa Rita mountains. Arizona Game and Fish Department Technical Report 6. Phoenix, AZ.

Olff, H. and M. E. Ritchie. 1998. Effects of herbivores on grassland plant diversity. *Trends in Ecology and Evolution* 13:261–265.

Ortega, I. M., S. Soltero-Gardea, F. C. Bryant, and D. L. Drawe. 1997a. Evaluating grazing strategies for cattle: Deer and cattle food partitioning. *Journal of Range Management* 50:622–630.

Ortega, I. M., S. Soltero-Gardea, D. L. Drawe, and F. C. Bryant. 1997b. Evaluating grazing strategies for cattle: Nutrition of cattle and deer. *Journal of Range Management* 50:631–637.

Owens, M. K., K. L. Launchbaugh, and J. W. Holloway. 1991. Pasture characteristics affecting spatial distribution of utilization by cattle in mixed brush communities. *Journal of Range Management* 44:118–123.
Owens, M. K., J. W. Mackley, and C. J. Carroll. 2002. Vegetation dynamics following seasonal fires in mixed mesquite/acacia savannas. *Journal of Range Management* 55:509–516.
Passler, T., P. H. Walz, S. S. Ditchkoff et al. 2009. Cohabitation of pregnant White-tailed deer and cattle persistently infected with bovine viral diarrhea virus results in persistently infected fawns. *Veterinary Microbiology* 134:362–367.
Pauley, G. R., J. M. Peek, and P. Zager. 1993. Predicting White-tailed deer habitat use in northern Idaho. *Journal of Wildlife Management* 57:904–913.
Pollock, M. T., D. G. Whittaker, S. Demarais, and R. E. Zaiglin. 1994. Vegetation characteristics influencing site selection by male White-tailed deer in Texas. *Journal of Range Management* 47:235–239.
Poole, K. G. and G. Mowat. 2005. Winter habitat relationships of deer and elk in the temperate interior mountains of British Columbia. *Wildlife Society Bulletin* 33:1288–1302.
Prasad, N. L. N. S. and F. S. Guthery. 1986. Wildlife use of livestock water under short duration and continuous grazing. *Wildlife Society Bulletin* 14:450–454.
Provenza, F., J. Villalba, R. Wiedmeier et al. 2009. Value of plant diversity for diet mixing and sequencing in herbivores. *Rangelands* 31:45–49.
Reynolds, J. P., T. E. Fulbright, and S. L. Beasom. 1992. Mechanical rejuvenation to dampen seasonal variation in chemical composition of browse. *Journal of Range Management* 45:589–592.
Ricca, M. A., R. G. Anthony, D. H. Jackson, and S. A. Wolfe. 2002. Survival of Columbian White-tailed deer in western Oregon. *Journal of Wildlife Management* 66:1255–1266.
Ricca, M. A., R. G. Anthony, D. H. Jackson, and S. A. Wolfe. 2003. Spatial use and habitat associations of Columbian White-tailed deer fawns in southwestern Oregon. *Northwest Science* 77:72–80.
Richardson, C., J. Lionberger, and G. Miller. 2008. *White-tailed Deer Management in the Rolling Plains of Texas*. Austin, TX: Texas Parks and Wildlife Department.
Robinson, H. S., R. B. Wielgus, and J. C. Gwilliam. 2002. Cougar predation and population growth of sympatric mule deer and White-tailed deer. *Canadian Journal of Zoology* 80:556–568.
Rollins, D. and F. C. Bryant. 1986. Floral changes following mechanical brush removal in central Texas. *Journal of Range Management* 39:237–240.
Rollins, D., F. C. Bryant, D. D. Waid, and L. C. Bradley. 1988. Deer response to brush management in central Texas. *Wildlife Society Bulletin* 16:277–284.
Romin, L. A. and J. A. Bissonette. 1996. Deer: Vehicle collisions: Status of state monitoring activities and mitigation efforts. *Wildlife Society Bulletin* 24:276–283.
Rosas-Rosas, O. C., R. Valdez, L. C. Bender, and D. Daniel. 2003. Food habits of pumas in northwestern Sonora, Mexico. *Wildlife Society Bulletin* 31:528–535.
Rosenstock, S. S., W. B. Ballard, and J. C. Devos, Jr. 1999. Viewpoint: Benefits and impacts of wildlife water developments. *Journal of Range Management* 52:302–311.
Rothley, K. D. 2001. Manipulative, multi-standard test of a White-tailed deer habitat suitability model. *Journal of Wildlife Management* 65:953–963.
Rue, L. L. 2004. *The Deer of North America*. Guilford, CT: The Lyons Press.
Ruthven, D. C., III, T. E. Fulbright, S. L. Beasom, and E. C. Hellgren. 1993. Long-term effects of root plowing on vegetation in the eastern south Texas plains. *Journal of Range Management* 46:351–354.
Sands, J. P., L. A. Brennan, F. Hernandez et al. 2009. Impacts of buffelgrass (*Pennisetum ciliare*) on a forb community in south Texas. *Invasive Plant Science and Management* 2:130–140.
Schindler, J. R. and T. E. Fulbright. 2003. Roller chopping effects on Tamaulipan scrub community composition. *Journal of Range Management* 56:585–590.
Schindler, J. R., T. E. Fulbright, and T. D. A. Forbes. 2004. Shrub regrowth, antiherbivore defenses, and nutritional value following fire. *Journal of Range Management* 57:178–186.
Seager, R., M. Ting, I. Held et al. 2007. Model projections of an imminent transition to a more arid climate in southwestern North America. *Science* 316:1181–1184.
Shackleton, D. M. and I. M. Cowan. 1999. *Hoofed Mammals of British Columbia*. Vancouver, BC: University of British Columbia Press.
Smith, W. P. 1981. Status and habitat use of Columbian White-tailed deer in Douglas county, Oregon. MS thesis, Oregon State University.

Smith, W. P. 1985. Current geographic distribution and abundance of Columbian White-tailed deer, *Odocoileus virginianus leucurus* (Douglas). *Northwest Science* 59:243–251.

Smith, W. P. 1987. Dispersion and habitat use by sympatric Columbian White-tailed deer and Columbian black-tailed deer. *Journal of Mammalogy* 68:337–347.

Soper, R. B., R. L. Lochmiller, J. D. M. Leslie, and D. M. Engle. 1993a. Condition and diet quality of White-tailed deer in response to vegetation management in central Oklahoma. *Proceedings of the Oklahoma Academy of Science* 73:53–61.

Soper, R. B., R. L. Lochmiller, D. M. Leslie, Jr., and D. M. Engle. 1993b. Nutritional quality of browse after brush management on Cross Timbers rangeland. *Journal of Range Management* 46:399–410.

Squires, J. R. and L. F. Ruggiero. 2007. Winter prey selection of Canada lynx in northwestern Montana. *Journal of Wildlife Management* 71:310–315.

Steuter, A. A. and H. A. Wright. 1980. White-tailed deer densities and brush cover on the Rio Grande plain. *Journal of Range Management* 33:328–331.

Stewart, K. M., R. T. Bowyer, J. G. Kie, N. J. Cimon, and B. K. Johnson. 2002. Temporospatial distributions of elk, mule deer, and cattle: Resource partitioning and competitive displacement. *Journal of Mammalogy* 83:229–244.

Stewart, K. M., T. E. Fulbright, and D. L. Drawe. 2000. White-tailed deer use of clearings relative to forage availability. *Journal of Wildlife Management* 64:733–741.

Stewart, K. M., T. E. Fulbright, D. L. Drawe, and R. T. Bowyer. 2003. Sexual segregation in White-tailed deer: Responses to habitat manipulations. *Wildlife Society Bulletin* 31:1210–1217.

Stubblefield, S. S., R. J. Warren, and B. R. Murphy. 1986. Hybridization of free-ranging White-tailed and mule deer in Texas. *Journal of Wildlife Management* 50:688–690.

Sullivan, T. L. and T. A. Messmer. 2003. Perceptions of deer–vehicle collision management by state wildlife agency and department of transportation administrators. *Wildlife Society Bulletin* 31:163–173.

Suring, L. H. and P. A. Vohs, Jr. 1979. Habitat use by Columbian White-tailed deer. *Journal of Wildlife Management* 43:610–619.

Tanner, G. W., J. M. Inglis, and L. H. Blankenship. 1978. Acute impact of herbicide strip treatment on mixed-brush White-tailed deer habitat on the northern Rio Grande plain. *Journal of Range Management* 31:386–391.

Texas Parks and Wildlife Department. 2007. *White-tailed Deer*. Austin, TX: Texas Parks and Wildlife Department. http://www.tpwd.state.tx.us. Accessed July 15, 2009.

Thilenius, J. F. 1968. The *Quercus garryana* forests of the Willamette Valley, Oregon. *Ecology* 49:1124–1133.

Thilenius, J. F. and K. E. Hungerford. 1967. Browse use by cattle and deer in northern Idaho. *Journal of Wildlife Management* 31:141–145.

Thill, R. E. 1984. Deer and cattle diets on Louisiana pine–hardwood sites. *Journal of Wildlife Management* 48:788–798.

Thill, R. E. and A. Martin, Jr. 1986. Deer and cattle diet overlap on Louisiana pine–bluestem range. *Journal of Wildlife Management* 50:707–713.

Thill, R. E. and A. Martin, Jr. 1989. Deer and cattle diets on heavily grazed pine–bluestem range. *Journal of Wildlife Management* 53:540–548.

Timmons, G. R., D. G. Hewitt, C. A. DeYoung, T. E. Fulbright, and D. A. Draeger. 2010. Does supplemental feed increase selective foraging in a browsing ungulate? *Journal of Wildlife Management* 74:995–1002.

Urness, P. J., W. Green, and R. K. Watkins. 1971. Nutrient intake of deer in Arizona chaparral and desert habitats. *Journal of Wildlife Management* 35:469–475.

U.S. Fish and Wildlife Service. 2008. *Species Fact Sheet: Columbian White-tailed Deer*. U.S. Fish and Wildlife Service, Pacific Region, Oregon Fish and Wildlife Office. http://www.fws.gov/oregonfwo/Species/Data/ColumbianWhiteTailedDeer. Accessed June 11, 2009.

Van Auken, O. W. 2000. Shrub invasions of North American semiarid grasslands. *Annual Review of Ecology and Systematics* 31:197–215.

Van Horne, B. 1983. Density as a misleading indicator of habitat quality. *Journal of Wildlife Management* 47:893–901.

VerCauteren, K. C., M. J. Lavelle, and S. Hygnstrom. 2006. Fences and deer-damage management: A review of designs and efficacy. *Wildlife Society Bulletin* 34:191–200.

Volk, M. D., D. W. Kaufman, and G. A. Kaufman. 2007. Diurnal activity and habitat associations of White-tailed deer in tallgrass prairie of eastern Kansas. *Transactions of the Kansas Academy of Science* 110:145–154.

Waid, D. D., R. J. Warren, and D. Rollins. 1984. Seasonal deer diets in central Texas and their response to brush control. *Southwestern Naturalist* 29:301–307.

Ward, R. L. and C. L. Marcum. 2005. Lichen litterfall consumption by wintering deer and elk in western Montana. *Journal of Wildlife Management* 69:1081–1089.

Warren, R. J. and L. J. Krysl. 1983. White-tailed deer food habits and nutritional status as affected by grazing and deer-harvest management. *Journal of Range Management* 36:104–109.

Warshall, P. 1995. The Madrean sky island archipelago: A planetary overview. In *Proceedings Biodiversity and Management of the Madrean Archipelago: The Sky Islands of Southwestern United States and Northern Mexico*, eds. L. F. Debano, P. F. Ffolliott, A. Ortega-Rubio, G. J. Hamre, and C. B. Edminster, 6–18. Tucson, AZ: U. S. Forest Service, Rocky Mountain Forest and Range Experiment Station.

White, R. W. 1961. Some foods of the White-tailed deer in southern Arizona. *Journal of Wildlife Management* 25:404–409.

Whittaker, D. G. and F. G. Lindzey. 1999. Effect of coyote predation on early fawn survival in sympatric deer species. *Wildlife Society Bulletin* 27:256–262.

Whittaker, D. G. and F. G. Lindzey. 2004. Habitat use patterns of sympatric deer species on Rocky Mountain Arsenal, Colorado. *Wildlife Society Bulletin* 32:1114–1123.

Wiggers, E. P. and S. L. Beasom. 1986. Characterization of sympatric or adjacent habitats of 2 deer species in west Texas. *Journal of Wildlife Management* 50:129–134.

Wright, A. L. and R. G. Kelsey. 1997. Effects of spotted knapweed on a cervid winter–spring range in Idaho. *Journal of Range Management* 50:487–496.

Zwank, P. J., R. D. Sparrowe, W. R. Porath, and O. Torgerson. 1979. Utilization of threatened bottomland habitats by White-tailed deer. *Wildlife Society Bulletin* 7:226–232.

19

Managing White-tailed Deer: Latin America

J. Alfonso Ortega-S., Salvador Mandujano, Jorge G. Villarreal-González, Ma. Isabel Di Mare, Hugo López-Arevalo, Misael Molina, and Martín Correa-Viana

CONTENTS

Latin America includes Mexico and Central and South America, with a great diversity of ecosystems and human cultures. Climate, elevation, soils, topography, vegetation, and wildlife are highly diverse, ranging from arid to tropical ecosystems. Language is one of the few characteristics common throughout Latin America; another commonality is the presence of the magnificent white-tailed deer, also known in Latin America as "venado cola blanca" and more than 30 other names in different indigenous dialects (Gallina et al., 2010). Twenty-four subspecies of white-tailed deer have been reported in Latin America (Chapter 1). The type of ecosystems and habitats occupied by deer in Mexico vary from the Sonoran desert in northwestern Mexico with 100 mm of annual precipitation, to the tropical rain forest in Chiapas with more than 3000 mm/year of rainfall. Morphological characteristics of white-tailed deer in Latin America also vary, from the majestic *O. v. texanus* subspecies with Boone and Crockett scores that can surpass 200 points to the beautiful *O. v. truei* in Costa Rica with antler scores of less than 100 points at maturity. Deer in Latin America have been an important game animal since pre-Columbian cultures; one of the 20 days of the Aztec calendar or sun stone was dedicated to the deer. The importance of deer for food, leather, and other products such as projectile points, bone scrapers, and needles has been reported by Méndez (1984) and more recently by Gallina et al. (2010).

Prominent Ecological Influences

Matorral Shrubs

Northern Mexico encompasses three major ecoregions, the Sonoran desert, the Chihuahuan desert, and the Tamaulipan biotic province. The Sonoran desert represents an area of 223,000 km^2 (World Wildlife Foundation, 2009). Mountains in central Sonora reach 2000 m elevation. Most of the Sonoran desert receives 120–300 mm/year precipitation during two wet seasons, one generally during December through March, and the other July through September. Succulents that store water for later use are common. Daytime temperatures are about 40°C for much of the summer. Hard freezes are uncommon during mild winters. Lower elevation areas are dominated by dense communities of creosote and white bursage (McNab and Avers, 1994), but on slopes and at higher elevations, palo verde, ironwood, saguaros, and other tall cacti are abundant (Turner et al., 1995).

The Chihuahuan desert region in Mexico is located within the Central Plateau, with the Sierra Madre Occidental to the west and the Sierra Madre Oriental to the east. Elevations in the Chihuahuan desert range from 600 to about 1675 m; with hot summers and cold winters, winter frosts are common in all areas and the northern portion regularly receives hard freezes. The Chihuahuan desert is predominantly a shrub desert. Creosote bush is abundant. Tarbush is more scattered but may form extensive stands under proper soil and moisture conditions. Other common plants include mariola, four-winged saltbush, and honey mesquite. Succulents include a variety of small to medium-sized cacti, yuccas, and agaves such as lechuguilla. Herbaceous vegetation includes grasses such as black grama and tobosagrass. Ocotillo, sotol, and barrel cactus are also common (University of Texas at El Paso, 2009).

The Tamaulipan biotic province encompasses 141,500 km^2 and begins in eastern Coahuila at the base of the Sierra Madre Oriental and proceeds east to include the northern part of the state of Tamaulipas. Elevations range from sea level to 300 m at the northern boundary of the ecoregion. Before alteration, mesquite and curly mesquite grass were characteristic plants of this region. Other common shrubs are lotebush and white brush. Portions of this region consist of open mesquite woodlands with an abundant understory of grasses which often contain a layer of taller species such as hooded windmillgrass, and a layer of shorter grasses such as grama. Prickly pear or nopal may be present in dense stands replacing many of the shrubs and grasses (Crosswhite, 1980).

Temperate Forests

Oak and pine forest communities are typical of mountain areas in Mexico and are the predominant vegetation type of temperate semihumid areas. Except for the states of Yucatán and Quintana Roo, temperate forests can be found in all Mexican states from sea level to 3000 m elevation, although more than 95% of this vegetation type is located between 1200 and 2800 m elevation. Oak forest is the dominant vegetation of the Sierra Madre Oriental, and is also common in the Sierra Madre Occidental, Eje Volcánico Transversal, Sierra Madre del Sur, Sierras del Norte de Oaxaca, and the states of Chiapas and Baja California. Oak forest occupies 5% of Mexico and may contain a single or several oak species. Plant families associated with oak forest are: Asteraceae, Poaceae, Fabaceae, Lamiaceae, Euphorbiaceae, Rosaceae, Onagraceae, Apiaceae, Scrophulariaceae, Commelinaceae, Rubiaceae, Pteridaceae, Cyperaceae. Coniferous forests common in temperate areas can be found from sea level to high elevation in semiarid, semihumid, and humid climates. Some species may be found only in special soil conditions. Coniferous forests occupy about 15% of Mexico with pine and pine/oak being most prominent, followed by juniper and fir forests.

Tropical Dry Forests

Many different names are used for vegetation types included under this definition, such as tropical and subtropical dry forests, seasonally dry tropical forests, caatinga, mesotrophic, mesophilous, or mesophytic forest, and semideciduous or deciduous forest (Murphy and Lugo, 1986). The largest areas of seasonally dry tropical forests in South America are found on the Caribbean coasts of Colombia and Venezuela, and north-eastern Brazil (the "caatingas"). Other smaller and more isolated areas occur in dry valleys in the Andes of Bolivia, Perú, Ecuador, and Colombia, coastal Ecuador and northern Perú, the "Mato Grosso de Goiás" in central Brazil, and scattered throughout the Brazilian cerrado biome. In Central America, seasonally dry forests are concentrated along the Pacific coast from Guanacaste in northern Costa Rica, to just north of the Tropic of Cancer in the Mexican state of Sonora. Within all of these areas, seasonally dry tropical forests occur within a complex of vegetation types depending on soil, climatic, and topographic conditions.

Seasonally dry tropical forests occur where rainfall is less than 1600 mm/year, with at least five months receiving less than 100 mm (Bullock et al., 1995). Vegetation is mostly deciduous during the dry season, with the level of deciduousness increasing as rainfall declines, though in the driest forests, evergreen and succulent species are common. Seasonally dry tropical forests have a lower canopy height and lower basal area than tropical rain forests (Murphy and Lugo, 1986), and thorny species are often prominent. Ecological processes are strongly seasonal, and net primary productivity is lower than in rain forests because growth only takes place during the wet season. There is a build up of leaf litter during the dry season because sunlight penetrates to the forest floor and decomposition ceases in the low relative humidity. Flowering and fruiting phenologies are strongly seasonal, and many species flower synchronously during the transition between the dry and wet seasons when trees are still leafless. Conspicuous flowers and wind-dispersed seeds are common, in contrast to rain forests. Leguminosae and Bignoniaceae dominate the woody floras of these forests throughout their range, along with Anacardiaceae, Myrtaceae, Rubiaceae, Sapindaceae, Euphorbiaceae, Flacourtiaceae, and Capparidaceae, which are also represented. Cactaceae are prominent in the understory, particularly at the latitudinal extremes, and are an important element in the diversity of these forests. Seasonally dry tropical forests usually have a closed woody canopy, with a sparse ground flora consisting of few grasses, Bromeliaceae, Compositae, Malvaceae, and Marantaceae. Seasonally dry tropical forests occur on fertile soils with a moderate-to-high pH and nutrient status and low aluminum concentrations. Such soils are favorable for agriculture, which has resulted in enormous destruction of these forests in many areas, a problem exacerbated by the large human populations in many neotropical dry forest life zones (Murphy and Lugo, 1995).

Tropical Evergreen or Rain Forests

This vegetation type is highly productive because moisture and sunlight are not limiting. Tropical evergreen forests are found through much of east and southeast Mexico, from southeast San Luis Potosi and Veracruz, through parts of Hidalgo, Puebla, and Oaxaca, to northern Chiapas and portions of Tabasco where drainage is proper for vegetation growth. Campeche and Quintana Roo have extensive rain forests. About 12.8% of Mexico is covered with rain forest, primarily between sea level and 1500 m elevation. Mean annual temperature never goes below 20°C and rarely exceeds 26°C. Mean annual rainfall ranges from 1500 to over 4000 mm. The evergreen forest is a complex biological community in which evergreen trees 25 m in height predominate. An interesting characteristic of this plant community is the abundance of woody creeping plants from the families Annonaceae, Bignoniaceae, Euphorbiaceae, Fabaceae, Moraceae, Rubiaceae, Sapindaceae, and Sapotaceae, which in many cases are large with abundant foliage and compete with overstory trees. In contrast to the dry forest, tropical evergreen forests are not good habitat for white-tailed deer and densities are low.

South American Savannas

The savannas in South America are located in the Llanos Orientales in Colombia and the Llanos Occidentales in Venezuela. They represent 16% of the tropical savannas in the world. Geographically, the Llanos are part of the Orinoco watershed (Rippstein et al., 2001). The Llanos of Venezuela are vast (235,530 km^2) plains bounded to the north by the Cordillera de La Costa, to the west by the Andes and to the south by the Guayana Shield. The Llanos comprises four subregions: the piedemonte, which is adjacent to the Andes, the high plains or upper llanos, the alluvial overflow plain or lower llanos, and the aeolian plains (Sarmiento, 1983). The savannas experience a dry season (December–April), and the wet season (May–November) during which up to 90% of annual rainfall occurs. Mean annual temperature varies from 26°C to 28°C. Native vegetation in the Llanos is generally herbaceous, although riparian forests are interspersed in savannas. Some savannas contain low growing trees and others are dominated by grasses. Historically, the orinoquia savannas have been used for livestock production; consequently, removal of woody cover and fire has been used to promote open areas dominated by herbaceous plants consumed by cattle (Rippstein et al., 2001).

Andean Páramo

The páramos occur from north Peru to Sierra Nevada de Santa Martha in Colombia, including Ecuador, and extending to the Sierra Nevada de Merida in Venezuela. In Colombia the páramos are patchy at high elevations in the Andes and cover about 2.5% of the country (Sturm and Rangel, 1985). This vegetation type is characterized by two or three vegetation strata where herbaceous vegetation is dominant. Important genera are *Chusquea*, *Calamagrostis*, *Archytophyllum*, *Hypericum*, *Jamesonia*, and *Espeletia*. Species composition and the altitudinal limit of the páramo vary depending on variations in moisture. The Andean páramo is characteristically dominated by rosette shrubs of the genus *Espeletia* and small trees of the genus *Hypericum* (Vareschi, 1992). Isolated patches of trees occur in areas protected from wind, at the base of mountains. The Andean páramo contains small trees of the genera *Chaetolepis* (Melastomataceae), *Hesperomeles* (Rosaceae), *Pernettia* (Ericaceae), *Bacharis*, and *Stevia* (Asteraceae), which form the overstory. Herbs such as *Calamagrostis*, *Agrostis*, *Bromus*, *Poa*, and *Aciachne* (Poaceae), *Castilleja* (Scrophulariaceae), *Lupinus* (Fabaceae), *Hintherhunera*, *Bidens* (Asteraceae), *Luzula* (Juncaceae), *Arenaria* (Caryophyllaceae), *Geranium* (Geraniaceae), and *Acaena* (Rosaceae) compose the understory. Forests of silky phacelia up to 15 m tall are located above 3000 m elevation (Universidad de los Andes, 1999). Shrubs, such as *Hesperomeles* (Rosaceae), *Preopanax* (Araliaceae), *Vallea* (Elaeocarpaceae), *Gramaderia*, *Rapanea* (Myrsinaceae), and *Scalonia* (Saxifragaceae) are found in the medium and inferior layers.

Mexico

Mexico is one of the 12 countries in the world considered mega diverse; with only 1.3% of the world's land area, Mexico hosts 10–15% of the world's flora and fauna. Mexico ranks first in the world in diversity of reptiles (717 species), second in mammals (525 species), fourth in amphibians (295 species), and tenth in angiosperms (25,000 species; SEMARNAP, 1997). This great biological diversity is the result of the interaction of many factors including climate, geology, and topography, as well as the presence of the Gulf of Mexico on the east and the Pacific Ocean on the west (Ceballos and Oliva, 2005; SEMARNAP, 1997). White-tailed deer occupy 98% of Mexico (181,525,233 ha), being found throughout the country except on the Baja California Peninsula.

Morphological Variation

Thirty-eight subspecies of white-tailed deer have been recognized by Smith (1991); 14 subspecies can be found in Mexico (Hall, 1981; Figure 19.1; Table 19.1). Definition of the subspecies concept is controversial from the biological perspective, even when it is frequently used for legislation, conservation, and management purposes (Chapter 1). Several authors have reported information on systematics and distribution of white-tailed deer in Mexico; Mearns (1898), Merriam (1898), Allen (1903), Cowan (1936), Goldman and Kellogg (1940), Armstrong et al. (1972), and Genoways and Jones (1975). Classification of subspecies currently accepted for white-tailed deer is based exclusively on qualitative characteristics such as body and skull size, pelage color, and size and shape of antlers (Kellogg, 1956). Detailed studies of subspecies morphology and genetics, as well as white-tailed deer genetic variability in Mexico, are scarce (Krausman et al., 1978; Logan-López et al., 2006, 2007; O. Chassin, Universidad Michoacana de San Nicolás de Hidalgo, personal communication).

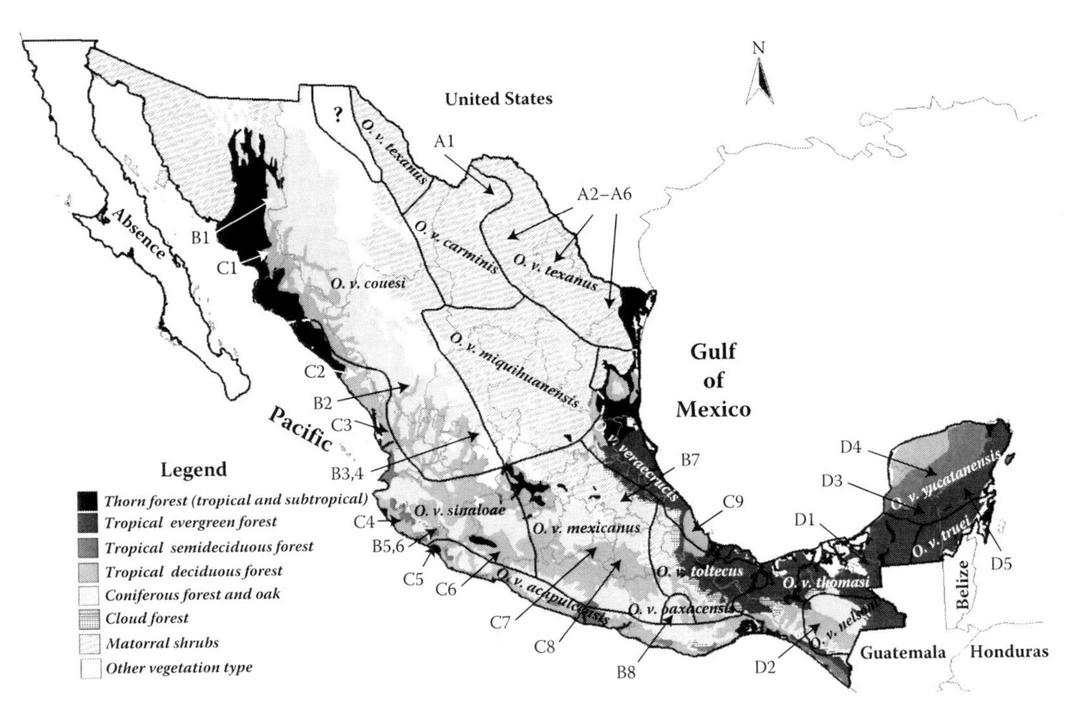

FIGURE 19.1 Range of white-tailed deer subspecies across principal vegetation types in Mexico, with locations of density estimates referenced in Table 19.1 shown using arrows.

TABLE 19.1

Estimates of White-tailed Deer Density (Deer/km²), Subspecies ("?" where Subspecies is not Clear), and Location across the Principal Vegetation Types in Mexico

Key	Deer/km²	Subspecies	State	References
Matorral Shrubs				
A1	4.3	*car*	Coahuila	Villarreal and Rodríguez (1998)
A2	25.0	*tex*	Coahuila	Villarreal and López (1987)
A3	15.9	*tex*	Nuevo León	Villarreal (1995)
A4	8.0	*tex*	Nuevo León	Villarreal (1998)
A5	25.0	*tex*		Villarreal (2009)
A6	8.0	*tex*	Tamaulipas	Villarreal (1999)
Mixed Oak–Pine Forests				
B1	6	*cou*	Sonora	Coronel-Arellano et al. (2009)
B2	10–20	*cou*	Durango	Ezcurra and Gallina (1981) Galindo-Leal et al. (1993)
B3	1.5	*cou*	Aguascalientes	Romo and Gallina (1988)
B4	4–11	*cou*	Aguascalientes	Villalobos (1998), Kobelkowsky-Sosa and Palacio-Núñez (2002)
B5	4.8	*sin*	Jalisco	Valenzuela (1994)
B6	5–10	*sin*	Jalisco	Zavala (1992)
B7	2.1	*mex*	Hidalgo	Sánchez-Rojas et al. (2009)
B8	1–3	*oax?*	Oaxaca	Ortíz-Martínez et al. (2005), Briones-Salas and García-Cruz (2005)
Tropical Dry Forests				
C1	9.3	*cou*	Sonora	Coronel-Arellano et al. (2009)
C2	<5	*sin*	Sinaloa	Ahumada et al. (1998)
C3	<5	*cou*	Nayarit	Ávila-Gómez (2008)
C4	10–14	*sin*	Jalisco	Mandujano and Gallina (1995)
C5	<5	*sin*	Colima	Castillo (1998)
C6	6–10	*sin*	Michoacán	Yañéz (2009)
C7	9–18	*mex*	Morelos	García-Sierra and Monroy (1985), Corona (2003), Hernández et al. (2008)
C8	1–6	*mex*	Puebla	Villarreal-Espino (2006), López-Téllez et al. (2007)
C9	1–2	*ver*	Veracruz	Gallina et al. (2009)
Tropical Semideciduous and Evergreen Forests				
D1	<1	*tho*	Tabasco	Zúñiga-Sánchez et al. (2008)
D2	<1	*nel?*	Chiapas	Naranjo et al. (2004a)
D3	<1	*tho?*	Campeche	Reyna-Hurtado and Tanner (2007)
D4	<5	*yuc*	Yucatán	Segovia and Hernández-Betancourt (2003)
D5	4–8	*yuc*	Quintana Roo	Ávila-Gómez (1996); González-Marín et al. (2008)

The key reference is shown in Figure 19.1.

Northern Mexico hosts four subspecies of white-tailed deer, *O. v. couesi*, *O. v. carminis*, *O. v. texanus*, and *O. v. miquihuanensis*, which occupy areas of 515,052 km² (28.3% of Mexico), 187,028 km² (10.3%), 68,358 km² (3.8%), and 174,142 km² (9.6%), respectively. Morphological differences among these four subspecies are marked. *O. v. texanus* is the largest, with a total length of 1829 mm, height at shoulder of 1048 mm, and a condylobasal skull length of 287.5 mm (Taylor, 1956; Villarreal, 1999). Villarreal (1999) reported an average weight of 60 kg for young males one to two years old, and up to 80 kg for adult males. Twenty-six typical trophies of this subspecies are registered in the Boone and Crockett (B&C) book of records with a B&C score of more than 160, and eight are registered in the nontypical category with more than 185 B&C points. *O. v. carminis* adult males can have a total body length of 1520 mm, height at shoulder of 793 mm, and condylobasal skull length of 246 mm (Taylor, 1956). Antlers of adult males in the area of San Buenaventura, Coahuila can score 120–130 B&C points with eight to 10 antler points (Villarreal, 1999). *O. v. miquihuanensis* are similar in size to *O. v. carminis* but smaller than *O. v. texanus*. Average total length for this subspecies is 1530 mm, height at shoulder is 820 mm, and condylobasal skull length is 247 mm (Taylor, 1956). Antler size of mature *O. v. miquihuanensis* are similar in size to three-year-old *O. v. texanus* males and can reach 90–120 B&C with eight to 10 points.

O. v. couesi adult males can have a total body length of 1530 mm, height at shoulder 890 mm, and condylobasal skull length of 241 mm (Taylor, 1956). Antler size can reach 100–120 B&C points for adult males, with eight to 10 antler points (Medina, 1986). Seventy-eight Mexican typical trophies of this subspecies are registered in the B&C book of records with more than 100 B&C points, and 18 within the nontypical category with more than 155 B&C points.

For management purposes the 14 subspecies of white-tailed deer in Mexico may be classified in three groups or ecoregions: (1) North-eastern, which includes *O. v. texanus*, *O. v. miquihuanensis*, and *O. v. carminis*, inhabiting primarily matorral shrubs; (2) Pacific and Central, which includes *O. v. couesi*, *O. v. mexicanus*, *O. v. sinaloae*, *O. v. oaxacensis*, and *O. v. acapulcensis*, occurring in temperate oak–pine forest and tropical dry forest; and (3) Gulf and southern, which encompasses *O. v. veracrucis*, *O. v. thomasi*, *O. v. toltecus*, *O. v. nelsoni*, *O. v. truei*, and *O. v. yucatanensis*, all associated with tropical rain forest, semideciduous forest, and cloud forest (Mandujano et al., 2010).

White-tailed Deer Diseases

Information on wildlife diseases in Mexico is scarce. Romero-Castañón et al. (2008) surveyed gastrointestinal and external parasites in five wild and three domestic ungulates in Selva Lacandona, Chiapas. Ectoparasites included 14 species of Ixodidae and one hippoboscid. Ten endoparasites were new host records and 15 were new geographic records for Mexico. Morisita's similarity index revealed the greatest similarity in parasite communities between the two deer species and the two peccary species, while the greatest similarity between wild and domestic species was between cattle and white-tailed deer. Montes et al. (1998) studied gastrointestinal parasites and egg and oocyst output in feces of captive white-tailed deer in Yucatán. Seven genera were identified (*Haemonchus* spp., *Cooperia* spp., *Trichuris* spp., *Strongyloides* spp., *Moniezia* spp., *Isospora* spp., and *Eimeria* spp.) representing five orders. The most frequent genera were *Haemonchus*, *Isospora*, and *Eimeria*. The genus *Isospora* was reported for the first time in deer of this region; although, it was not possible to explain the source of this parasite. The frequency and number of eggs and oocysts in feces were variable during the year and increased during the rainy season. There was a positive correlation between relative humidity, environmental temperature, and rainfall with the Coccidia and Strongylida orders. In the central zone of Yucatán, meteorological conditions during the rainy season are favorable for development of gastrointestinal parasitism which increases infection risk for deer.

Cantú et al. (2007) reported for the first time the presence of *Babesia bovis* and *B. bigemina* in white-tailed deer from two ranches in northern Mexico using nested PCR and the sequence of the positive samples. In another study, 4.2% of 427 white-tailed deer blood samples from the states of Coahuila, Nuevo León, and Tamaulipas were positive to *B. bigemina* and 1.7% were positive to *B. bovis* (Cantú et al., 2009). These authors suggested white-tailed deer may serve as a reservoir for *B. bigemina* and *B. bovis*; however, no evidence is available to determine whether white-tailed deer could be a suitable host for *Babesia* spp. to complete their transmission cycle.

Martínez et al. (1999) studied serum samples from 350 white-tailed deer from north-eastern Mexico to assess prevalence of antibody activity against five infectious diseases of ruminants. Prevalence rate was 81% for all serotypes of bluetongue virus (BTV), 72% for epizootic hemorrhagic disease virus (EHDV), 3% for *Borrelia burgdorferi*, 69% for *Anaplasma marginale*, and 0% for *Brucella abortus*, *B. melitensis*, and *B. ovis*. These are diseases that affect domestic ruminants, and deer may act as a reservoir for infection. In addition, if deer are translocated, they may introduce pathogens to formerly disease-free areas. The high seroprevalence of BTV and EHDV cannot be related to the presence of hemorrhagic disease in deer of this region. This is the first report to indicate *B. burgdorferi* infection of deer in Mexico. Despite the high prevalence of *A. marginale* titers, it is uncertain if deer play a role in the epidemiology of cattle anaplasmosis in the region. Apparently, white-tailed deer are unimportant in the epidemiology of brucellosis of both cattle and goats in northeastern Mexico.

Cantú-Martínez et al. (2008) estimated prevalence of antibodies against *Babesia bovis* and *B. bigemina* in sera from 165 white-tailed deer in northeastern Mexico. Seropositivity of deer was 87% for *Babesia* spp., with 53% prevalence for *B. bigemina* and 75% for *B. bovis*. Risk of infection with *B. bovis* was 2.4 times greater for females than for males, but sex was not a risk factor for *B. bigemina* infection. Cantú et al. (2008) reported prevalence of antibodies to *Leptospira* spp., infectious bovine rhinotracheitis, and bovine viral diarrhea virus with values of 5.6%, 41.1%, and 63.5%, respectively. Based on these results, white-tailed deer and cattle may share disease agents when sympatric in northern Mexico.

Population Status

Estimates of white-tailed deer population density vary greatly among regions and vegetation types (Table 19.1, Figure 19.1). Sampling errors, hunting intensity, and differences in sampling and analytical techniques contribute to the variation. However, these factors alone are insufficient to explain the magnitude of variation reported. Ecological factors determining carrying capacity and human impact seem to explain much of the density pattern. Semiarid habitats of northern Mexico, temperate forests of oak–pine, and tropical dry forests support higher densities than tropical semideciduous and rain forests. Protected areas and some management areas may maintain higher deer densities compared with areas where hunting is not closely regulated and heavily disturbed areas. Using rainfall as a predictor to estimate ungulate biomass across neotropical forests, Mandujano and Naranjo (2010) predicted that tropical dry forests could support threefold greater biomass than tropical rain forests, and grassland/savannas support higher ungulate biomass than tropical forests. Therefore, a gradient of white-tailed deer densities may be expected in which matorral shrub could maintain the highest density (>15 deer/km^2), temperate and tropical dry forests medium densities (8–14 deer/km^2), tropical semideciduous forests a lower density (<6 deer/km^2), and tropical rain forests the lowest density (<1 deer/km^2). Field estimations seem to confirm these results. Deer densities in Mexico are higher in matorral shrubs and lower in evergreen forests (Table 19.1).

White-tailed deer diets vary greatly depending on vegetation type and season from the coastal marshes in Suriname with only 14 species consumed to different areas of Venezuela, Colombia, and Mexico with over 100 species including forbs, grasses, shrubs, and trees (Table 19.2). Home range size of white-tailed deer is generally larger for males than females. Home range size for males varies from 26 ha in the tropical dry forest in Mexico to 378 h in similar vegetation type in Colombia (Table 19.3).

Land Use and Human Impacts

Human activities in Mexico have caused vast land use/cover change (LUCC) that threatens white-tailed deer habitat and biodiversity in general. Deforestation rate estimates range from 365,000 to more than 1.5 million ha/year (SARH 1992, 1994). Using LUCC at the national level, Mas et al. (2004) estimated that from 1976 to 2000, more than 20,000 km^2 of temperate forest, 60,000 km^2 of tropical forest, and 45,000 km^2 of matorral shrubs were cleared, representing an annual average rate of LUCC of 0.25%, 0.76%, and 0.33% per year, respectively. Large areas of forest and matorral shrubs were converted to man-made cover types (principally cropland and pastures), a small number of man-made areas were reconverted to forest or other vegetative covers (mainly matorral shrubs), and large areas of both temperate and tropical forests were disturbed and transformed to secondary forests which in some cases are used by white-tailed deer. Areas of LUCC are concentrated primarily in the states of Yucatán, Chiapas, Nuevo León, and Tamaulipas where most deforestation has occurred.

At a local level, LUCC patterns have been documented. For example, in tropical dry forest of western Mexico, forests are replaced by agriculture in flatlands, pastures are established on slopes, and wood is extracted from hill crests. If cultivated fields and pastures are not continuously maintained, thorny vegetation develops within one to three years. If left untouched, this secondary vegetation becomes a low forest dominated by *Acacia* and *Mimosa* spp. persisting for at least 20 years (Burgos and Mass, 2005).

Habitat Management

Traditional Practices

Slash-burn-cultivation agriculture creates a habitat mosaic that in some circumstances may improve habitat quality for white-tailed deer and promote hunting success (Greenberg, 1992; Escamilla et al., 2000).

TABLE 19.2

White-tailed Deer Diets in Different Habitats and Countries in Latin America

Habitat Type and Country	Family	Species	Main Food Items		Main Species	References
			Rainy Season	Dry Season		
Mesquite and chaparral-mixed grass, Nuevo León, Mexico		79	Forbs Grasses	Leaves	*Acacia wrightii, Acacia farnesiana, Castela texana, Celtis pallida, Porlieria angustifolia, Cercidium macrum, Desmanthus virgatus, Eysenhardtia polystachya, Karwinskia humboltiana, Leucaena leucocephala, Pithecellobium ebano, Prosopis glandulosa, Zanthoxylum fagara, Opuntia engelmanni*	Ramírez-Lozano (2004), Villarreal (1999)
Temperate deciduous and mixed forest, La Michilia, Mexico	41	135	Forbs Leaves	Leaves	*Phoradendron bolleanum, Phoradendron villosum, Arctostaphylos pungens, A. longifolia, A. polifolia, Pithecellobium leptophyllum Condalia hookeri, Juniperus durangensis, Quercus* spp., *Arbutus glandulosa*	Gallina et al. (1981)
Temperate deciduous and mixed forest, Aguscalientes, Mexico	17	31	Leaves Forbs	Leaves Forbs	*Arctostaphylos pungens, Quercus rugosa, Castilleja canescens, Viola barroetana, Juniperus depeana*	Clemente-Sánchez (1984)
Tropical dry forest, Chamela, Mexico	20	82	Leaves Vines Forbs	Leaves Vines Grasses Mast	*Spondias purpurea, Rourea glabra* *Acalypha langiana, Coursetia caribaea* *Tephrosia leiocarpa, Cardiospermum halicacabum, Thouinia paucidentata,* *Ayenia micranta*	Arceo et al. (2005)
Tropical dry forest, Mixteca Poblana, Mexico	50	133	Leaves Stems	Mast Leaves	*Acacia pennatula, Acacia subangulata, Leucaena leucocephala, Eysenhardtia polystachya, Pithecellobium dulce, Haematoxylum brasiletto, Herpalyce leceneriana, Mimosa luisiana, Cerdium praecox, Pachyrrisus* sp., *Agave kerchovei, Quercus glaucoides, Waltheria americana, Montanoa* sp.	Villarreal-Espino (2006)
Tropical semideciduous rain forest, Calakmul Mexico		26	Leaves Stems	Mast	*Brosimum alicastrum, Manilkara sapota*	Weber (2006)
Tropical semideciduous rain forest, Yucatán, Mexico			Leaves Stems	Mast Leaves	*Abutilon gaumeri, Bursera simaruba, Caesalpinia violacea, Capsicum* spp., *Colubrina greggii, Eugenia axillaris, Gymnopodium floribundum, Ipomoea batatas, Jacquemontia pentantha, Merremia aegyptia, Metopium brownei, Neomillspaughia emarginata, Pachyrrhizus erosus, Phaseolus* spp., *Quamoclit coccinea, Salvia coccinea, Viguiera den tata, Brosimum alicastrum*	Mandujano and Rico-Gray (1991)

continued

TABLE 19.2 (continued)

White-tailed Deer Diets in Different Habitats and Countries in Latin America

Habitat Type and Country	Family	Species	Main Food Items		Main Species	References
			Rainy Season	Dry Season		
Finca La Emilia, Cañas, Guanacaste, Costa Rica		48			*Desmodium* spp., *Mimosa pudica, Centrosema pubescens, Sida* spp., *Lippia alba, Mesechites* sp. and the fruits of *Pithecellobium saman, Solanum americana* and *Sorghum bicolor*	Espach and Sáenz (1994)
Colombia	45	112	Forbs Leaves	Leaves Grasses	*Achyranthes aspera, Alternanthera pungens, Mesechites trifidus, Maximiliana regia, Ageratina tinifolia, Baccharis nitida, Bidens andicola, Bidens cynapifolia,* cf. *Brickellia diffusa, Erechtites hieraciifolius, Leucanthemum máximum, Taraxacum officinale, Hypericum thuyoides, Commelina erecta, Citrullus lanatus, Cucurbita máxima, Plutarchia guascensis, Chamaesyce hirta, Abarema jupunba, Desmodium adscendens, Pseudosamanea guachapele, Rhynchosia minima, Samanea saman, Trifolium repens, Malvastrum coromandelianum, Byrsonima crassifolia, Morus alba, Myrica parvifolia, Axonopus conpressus, Bothriochloa pertusa, Calamagrostis effusa, Holcus lanatus, Pseudechinolaena polystachya, Melpomene flabelliformis, Talinum triangulare, Borreria alata, Zanthoxylum rhoifolium, Aragoa abietina, Capsicum annuum, Guazuma ulmifolia, Turnera ulmifolia*	Ramos (1995), Mora and Mosquera (2000)
Tropical forests, plains, and cultivated areas, Venezuela	23	125	Forbs Grasses	Leaves Mast	*Copernica tectorum, Acacia macracantha, Caesalpinia coriaria, Guazuma ulmifolia, Guttardia divaricata, Platymiscium pinnatum, Randia hebecarpa, Enterolobium cyclocarpum, Pithecellobium carabobense, P. saman, Copernicia tectorum, Genipa caruto, Spondias mombin*	Granado (1989), Danields (1987)
Páramos, Venezuela			Forbs Grasses	Forbs Grasses	*Aciachne pulvinata, Dantonia secundiflora, Poa pauciflora, Trisetum foliosum, Carex albolutescens, Rumex acetosella, Acaena elongata. A. cilindrostrachya, Hesperomelles pernettyoides, Lachemilla hirta, L. polylepis, Oxalis spiralis, Boletus aedulis*	Molina-Molina and Arias (1998)
Coastal marshes, Suriname	10	14			*Avicenia germinans, Sesuvium portulacastrum*	Branan et al. (1985)

Rainy season includes spring, summer, and autumn in temperate regions, and dry season includes winter.

TABLE 19.3

Home Range Size, Daily Movements, and Habitat Use by White-tailed Deer in Different Regions of Latin America

Habitat Type and Country	Home Range (ha)	Sampling Size	Daily Movements (km)	Daily Activity (h)	Habitat Use and Preference	Reference
Temperate forests Manatlan Biosphere Reserve, Mexico	16.0–17.9 in wet season, and 12.9–15.9 in dry season	Six females		1900–0000 and 0900–1200	Cloud forest, shrub and grazing areas; while pine–oak, agriculture zones, and regenerating areas were used less	González-Pérez (2003)
Tropical dry forests Chamela, Mexico	21.0–34.0 female in wet season and 11.0 female, 26.0 male in dry season	One female One male	1.98–2.58 female in wet season and 1.5 female, 2.5 male in dry season	Night female Day male	Used tropical dry forest and semideciduous tropical forest, but prefers the first	Sánchez-Rojas et al. (1997)
Palo Verde National Park, Bagaces, Guanacaste, Costa Rica	14.8 juvenile 7.7 adult	Juvenile and adult does	1.612 juvenile 1.665 adult		Mainly deciduous forest, but also shrubland, evergreen forest, and a small part of pastures	Rodríguez et al. (1985b)
Finca La Emilia (FLE) in Cañas, Guanacaste, Costa Rica	18.3	Four female fawns, One buck and Three does		Where hunting is permitted, deer usually move during dawn or dusk. In places where there is no hunting or human disturbance, deer move freely at different hours and even may approach humans	Grazing land, forest plantations (*Pithecellobium saman*), cultivated land (sorghum and fruits), riparian vegetation, haparral, jaragua grass (*Hyparrhenia rufa*) fields, forests of guacimo (Guazuma ulmifolia) trees, chaparrals and open habitats with some forest cover	Sáenz (1990, 1994)
Semicautiverio, Colombia	12.0–17.0 female 17.0–27 male	One female One male		0800 and 1800		Camargo-Sanabria (2005), Mateus-Gutierrez (2005)
Tropical dry forest, Colombia	114.0 female 378.0 male	Two female Two male				Gómez-Giraldo (2005)
Savannah, Portuguesa, Los Llanos, Venezuela			4.8	0600 and 0800 h, near and during dusk, and around midnight until 0200 hours	Open savannahs, chaparrals, gallery forest and cultivated areas in both seasons In western llanos: deciduous and gallery forest during wet and rainy seasons, and open areas of savannahs covered with pasture, and grasses during the wet season	Correa-Viana (1994, 2000)
Paramos, Venezuela				0900–1100 h and 1500–1700 h	Used units are the rosette-shrubs of *Espeletia schultzii—Hypericum laricifolium,* the ecotones of the *Pololylepis sericea* and the exotic pine forest	Molina-Molina and Arias (1998), Molina-Molina (2003b)

In such agricultural systems, a short period of cropping is followed by a relatively long period of fallow; secondary vegetation becomes dominant shortly after cultivation. Fallow secondary forests remain productive for years under varying degrees of management and yield multiple benefits for local people. Hunting for wildlife such as white-tailed deer is a common practice in these sites (Mandujano and Rico-Gray, 1991; González-Pérez and Briones-Salas, 2000; Naranjo et al., 2004a; Reyna-Hurtado and Tanner, 2007). Use of secondary forests also relieves pressure on primary forests. Secondary forests serve as alternate sites for cultivation and sources of timber, food, gums, fiber, and other products, as well as for biodiversity conservation. Currently, researchers and extension specialists actively promote improvement of fallowed areas with more ecologically and economically viable species as part of an effort to advance agroforestry, thereby maintaining forest cover while improving the well-being of local populations (Villarreal-Espino et al., 2004; López-Tellez et al., 2008).

Active Habitat Management

White-tailed deer habitat in northern Mexico has been managed for decades to accomplish specific objectives in cattle and wildlife operations. The southern United States and northern Mexico share three main ecoregions: the Sonoran and Chihuahuan deserts, and the Tamaulipan biotic province. Much of the information generated on both sides of the border may be adapted for use on ranches in both countries. Fulbright and Ortega-S. (2006) described white-tailed deer habitat requirements, nutrition, and population and habitat management on rangelands in Oklahoma, Texas, and northern Mexico. Specific information for locations in Mexico including deer habitat requirements, diets, reproduction, and diseases has been published by Villarreal (1999).

Reducing brush canopy cover and density may improve habitat for white-tailed deer by increasing herbaceous vegetation, providing openings for feeding, and increasing nutritional quality, accessibility, and palatability of browse (Drawe et al., 1999). However, the decision to apply brush management is serious because of its expense and long-term implications. Repeated treatments are necessary to maintain short-term benefits and to postpone long-term detriments after brush control has been applied. Lack of long-term planning, including maintenance treatments, often results in habitat degradation. The first decision to make when planning brush management is deciding areas of habitat that must remain undisturbed. Drainage areas with diverse woody plants that are continuous throughout the landscape are valuable as protective cover and movement corridors. Areas that are relatively undisturbed, diverse, and productive without human interference should be left untouched (Fulbright and Ortega-S., 2006).

Fulbright and Ortega-S. (2006) presented their concept of the ideal landscape characteristics for white-tailed deer; each of these attributes should be present within the home range of a deer in the management unit.

- The landscape consists of areas dominated by woody plants interspersed with openings.
- As a general guideline, created clearings should comprise no more than 40% of the landscape in arid and semiarid bioclimatic zones. No clearing should be done if woody plant canopy cover is 60% or less.
- Created clearings should comprise no more than 60% of the landscape in dry subhumid bioclimatic zones. No clearing should be done if woody plant canopy cover is 40% or less.
- Clearings should be 8–16 ha in size, interspersed and juxtaposed within the woodland matrix, and surrounded by brush sufficient to provide screening cover.
- Areas dominated by woody plants should be interconnected to facilitate deer movements from woody cover to clearings and back.
- Preferably, 30–40% of the landscape with highly diverse brush should never be subjected to any type of brush management.
- A portion of the landscape remaining in brush should include dense stands with 85% or greater woody plant canopy cover; these should be well interspersed throughout the management unit to provide optimum habitat for males and hiding habitat for females with fawns.
- Drainage areas lined with woody plants should be continuous throughout the landscape.

- Areas that support stands of prickly pear should be interspersed throughout the landscape, particularly in the South Texas Plains and adjacent areas of northern Mexico.
- Tall, single-stemmed mesquites, oaks, sugar hackberries, and other desirable shade trees should be left intact within cleared areas.
- Brush clearing should be avoided in plant communities and sites that receive high levels of deer use relative to other plant communities.
- Important habitat features should be dispersed across the landscape and juxtaposed to each other.

In comparison with the United States and some parts of northern Mexico where habitat management has been common for several decades, habitat management for white-tailed deer in central and southern Mexico occurs only in a few specific sites, such as tropical dry forests in the Mixteca Poblana and the Sierra-Costa of Michoacán (Mendoza et al., 2006; Villarreal-Espino, 2006). Habitat improvement projects include planting browse species such as ciruelo, prickly pear, leucaena, and huisache, and watering habitat with drip irrigation systems. Watershed management includes reforestation to protect riparian areas, stabilize soil, or improve aquifer recharge, protection of water springs, installation of water troughs to improve water distribution, and establishment of food plots with plants such as red mombin and prickly pear cactus that may serve as water sources during the dry season. Other practices, such as planting small areas with corn or beans, shredding native vegetation to promote the regrowth of browse, and planting valuable browse species, are used to improve forage availability for white-tailed deer (Mendoza et al., 2006).

Habitat Evaluation Procedures

Habitat evaluation procedures (HEP) has been developed for some temperate and tropical sites to determine habitat potential for white-tailed deer (Mandujano, 1994; Delfín-Alfonso and Gallina, 2007). For example, before reintroducing deer in Nuevo San Juan Parangaricutiro, Michoacán, HEP was used to determine sites that best met white-tailed deer habitat requirements (González et al., 2003). At a regional level, HEP and GIS have been used in central Veracruz (5330 km^2) to define sites for deer hunting, conservation, restoration, and reintroduction (Delfín-Alfonso et al., 2009). In the tropical dry forests of Mixteca Poblana, HEP was used to define management units, referred to as Unidad de Manejo para la Conservación de la Fauna (UMAs) (Delfín-Alfonso and Gallina, 2007).

Using a completely different approach to evaluate habitat potential for white-tailed deer in the Bajo Balsas region (700,000 ha), Yáñez (2009) used niche modeling through a MaxEnt algorithm. Using presence and density data at a local level, this author modeled deer distribution and defined optimal areas to sustain high deer densities (>12 deer/km^2), suboptimal areas (up to 12 deer/km^2), and vulnerable sites (<6 deer/km^2). This approach is being applied in other sites, such as the Tehuacán-Cuicatlán Biosphere Reserve.

Management Issues and Trends

History and Legislation of White-tailed Deer Management

In pre-Hispanic Mexico, white-tailed deer were appreciated by all the Mesoamerican cultures; Mexica, Tolteca, Zapoteca, Huasteca, Tarasca, and Maya, which developed in south and central Mexico, and also by the nomadic groups in northern Mexico. Other wildlife species were also appreciated and respected as part of the cultural mythology and mysticism. However, many of the cultural values and respect for wildlife of pre-Hispanic times were lost with European colonization in the sixteenth century. This loss was a result of the introduction into natural ecosystems of domestic animals from Europe. Livestock was highly valued and appreciated for the following 400 years, while wildlife was relegated to a public commodity of no economic value and its use was unrestricted (Retana, 2006). For this reason, many species of large mammals were almost extinct in Mexico by the end of the 1800s.

In Mexico the first rules and regulations for hunting were established in 1870; however, it was not until 1940 when the first federal "Ley Federal de Caza" (federal hunting law) was established. This law declared wildlife as state property and established hunting regulations that prohibited nocturnal hunting with artificial light and harvest of pregnant does and their offspring. However, this law continued allowing commercial wildlife hunting which caused overharvest of many wildlife species including jaguar, ocelot, and black bear. In 1950 a new "Ley Federal de Caza" was enacted in which wildlife continued being state property but commercial hunting was prohibited and only sport hunting was authorized. Hunters were required to be members of a hunting club and a hunting license was mandatory. In 2000, the "Ley General de Vida Silvestre" was enacted which included all aspects related to conservation, management, and sustainable utilization of flora and fauna. In this new law, wildlife was still considered state property but the right to use flora and fauna in a sustainable manner was conferred to landowners. Landowners are required to register their properties as an UMA (Wildlife Management Unit for Conservation and Sustainable Use) and develop a management plan. Sport hunting in Mexico is allowed only on properties registered as UMAs and landowners are the only ones allowed to commercially use these resources; therefore, wildlife now has economical value for the landowner and in this context, interest in wildlife conservation has increased markedly.

Traditional Use

White-tailed deer are appreciated throughout Mexico as a food, both for direct consumption and for commercial sale, and for use in handicrafts and recreation. White-tailed deer have also been part of the religious ceremonies of native cultures (Mandujano and Rico-Gray, 1991; Greenberg, 1992; Montiel-Ortega et al., 1999; González-Pérez and Briones-Salas, 2000; Naranjo et al., 2004a). A preferred deer hunting technique of the Mayas in Yucatán is the *batida* or drive hunt. The person who actually shoots a deer during such a hunt has the privilege of looking for the *Tunich* (a small good-luck stone sometimes found in a deer's stomach), and taking the deer's right leg, head, stomach, and liver. The rest of the carcass is divided among other party members. Deer are shared with the whole village only when hunting is associated with a religious celebration known as a *Ch'u' Chaak*.

As in other regions where slash-burn-cultivation is practiced, forest fallows provide multiple benefits to local residents (Abizaid and Coomes, 2004). Secondary vegetation replenishes soil nutrients and eliminates weeds. Firewood is collected from secondary forests, as are palm fronds, wood, vines, and other materials used in construction of houses, granaries, furniture, and fences. Valuable timber such as tropical cedar may also be extracted from fallows at least 30 years old. Households in Nuevo Becal in the state of Quintana Roo hunt many species in addition to deer in fallowed lands when walking to and from their swiddens, or during special hunting expeditions.

UMA Model

An overview and critique of the UMA model has been published by Weber et al. (2006). According to the "Ley General de Vida Silvestre" the UMA model promotes alternative production schemes that are compatible with ecosystem conservation through the planned and rational use of renewable natural resources. UMAs create additional economic opportunities for the rural sector of the society traditionally dedicated to livestock production, row cropping, or other agricultural activities (SEMARNAP, 1997). The UMA model was well accepted because it provided private and communal landowners with a new perspective concerning benefits of natural resource conservation combined with additional sources of nontraditional income. The UMAs can be classified into two categories: (1) extensive, which involve management of wildlife populations in natural habitats, and (2) intensive, which includes zoos and wildlife farming operations in small, fenced feeding areas (Dirección General de Vida Silvestre, 2007). As of March 2009, 9748 UMAs were registered encompassing nearly 34 million ha (17% of the country; Dirección General de Vida Silvestre, 2010). Most UMAs are located in northern Mexico in the states of Sonora, Chihuahua, Coahuila, Nuevo León, and Tamaulipas (Gallina and Escobedo-Morales, 2009).

The UMA model was created in part because of pressure exerted on SEMARNAT by the Asociación Nacional de Ganaderos Diversificados y Criadores de Fauna (ANGADI; National Diversified Cattleman and Wildlife Breeders Association) and UMAs were implemented first in north-eastern Mexico (Weber et al., 2006). The UMA model has successfully diversified income of landowners in northern Mexico because of landowner experience managing cattle–wildlife operations, the large size of properties which increases the number of animals that can be harvested, and proximity to affluent hunters in the United States. In this way, sport hunting has become an economic activity of great importance (Gallina and Escobedo-Morales, 2009). Villarreal (1999) indicated that about 10,000 bucks of the subspecies *texanus* are harvested annually in 1500 UMAs in north-eastern Mexico; and the value of one *texanus* buck is equal to the value of 20 calves. Given the interest and economic importance of sport hunting in this area, many ranches have reduced cattle stocking rates to provide better habitat for white-tailed deer. From 1992 to 2008, approximately 17,000 female deer were translocated for repopulation purposes in the states of Coahuila, Nuevo León, and Tamaulipas.

Despite these benefits, wildlife conservation results of the UMA model have not been properly evaluated and are unclear (González-Marín et al., 2003), even for UMAs that have been economically successful and promoted conservation of commercially important wildlife (Gallina and Escobedo-Morales, 2009). Part of the problem is that some UMAs have introduced exotic wildlife species to increase hunting opportunities. According to Medellín (2005), 660 UMAs in 16 Mexican states have exotic species; additionally, five states have reported feral populations of some exotic species. Competition between exotic and native species for resources represents a threat to native wildlife populations. Disease transmission from exotics to native wildlife is another potential hazard (Gallina and Escobedo-Morales, 2009).

Intensive UMAs with captive white-tailed deer have increased in recent years. Improving "genetic quality" of white-tailed deer has been one of the main objectives, as well as sale of deer for repopulation programs. In the state of Nuevo León, 39 white-tailed deer-intensive UMAs are registered with Nuevo León Parks and Wildlife Department (Parques y Vida Silvestre de Nuevo León). Intensive UMAs with native white-tailed deer may be valuable for repopulation if translocations are of deer subspecies native to the area. Translocation of white-tailed deer outside their natural distribution jeopardizes integrity of white-tailed deer subspecies, which not only have ecological value but may have economic value as unique trophies for sport hunters.

Implementation of the UMA model in southern Mexico has been more complicated. Most of the land is in *ejidos* (communal land) where poverty, ignorance, dependence on government subsides, and apathy for government initiatives are common (Weber et al., 2006). Additionally, high biological diversity in the tropics offers more opportunities for productive diversification compared to the focus on sport hunting characteristic of northern Mexico. For example, in the states of Campeche and Quintana Roo, hunters may pursue up to 12 bird and mammal species in addition to viewing unique wildlife, such as jaguars, visiting archaeological sites, and participating in other ecotourism activities.

A problem with the UMA model is accurately censusing wildlife populations and evaluating habitat. Population estimation is difficult in the tropics because visibility is poor, and thus accurate information is often not available to determine harvest rates. Other problems include surveying methods used, frequency of surveys, subsistence hunting, and poaching (Weber et al., 2006). Monitoring protocols to evaluate composition and trends of deer populations and habitat quality are necessary to make timely management decisions to maintain natural composition, integrity, and functionality of ecosystems (Rojo-Curiel et al., 2007). More specifically, wildlife scientists can help improve monitoring protocols to ensure viability and sustainable harvest of deer populations. Finally, creation of interdisciplinary groups will improve education, administration, extension, commercialization, and sustainable management which are necessary to offer high-quality sport hunting and nature tourism programs.

Hunting sustainability in tropical forests has been assessed using many methods and models (Bodmer and Robinson, 2004). Naranjo et al. (2004b) evaluated the sustainability of ungulate hunting in the tropical rain forests of La Lacandona, Chiapas. Local people in this area extract 17.5 kg/km^2/year of wild ungulates, making this practice sustainable for brocket deer and collared peccary, but unsustainable for tapir and white-lipped peccary. In tropical dry forest at Chamela, on the Pacific coast of Mexico, carrying capacity for white-tailed deer has been estimated to be 16.5–17.2 deer/km^2, with a potential annual harvest of 2.1 deer/km^2 (Mandujano, 2007).

Status and Conservation Areas

In Mexico, wildlife conservation and management is officially based on two schemes: (1) Natural Protected Areas (Areas Naturales Protegidas; ANP) and (2) Wildlife Management Unit for Conservation (UMA).

Mandujano and González-Zamora (2009) recently evaluated ANP and UMA to assess whether these areas satisfy the minimum critical area (MCA) necessary to support a minimum viable population (MVP) of white-tailed deer. A MCA of 1667–50,000 ha is necessary to support a MVP of 500 deer, whereas 16,670–500,000 ha is needed for long-term viability of 5000 deer, depending on regional deer density. Some types of ANPs such as Biosphere Reserves, Protection Areas of Flora and Fauna, and Protection Areas of Natural Resources satisfy MCA requirements better than other ANP categories such as National Park, Sanctuary, and Monument. In general, nearly all UMAs are smaller than the MCA for a population of 500 deer. Most large ANPs and UMAs occur in northern Mexico, so that viability of deer populations in other parts of the country is uncertain. The proposal of Priority Terrestrial Regions (RTP) could complement the need for reserves in the central part of the country. Mandujano and González-Zamora (2009) suggested a regional network system of conservation reserves and wildlife management units integrating ANP, UMA, and RTP at regional scales through source–sink, metapopulation, and archipelago reserve models. Additionally, some ANPs are research sites where biological information on deer is generated, which is true of La Michilía in Durango, La Sierra de La Laguna in Baja California Sur, La Estación de Biología Chamela and Manantlán in Jalisco, Bajo Balsas in Michoacán, La Sierra de Huautla in Morelos, El Edén in Quintana Roo, Calakmul in Campeche, and La Lacandona in Chiapas (Gallina et al., 2007).

Costa Rica

Geographic Description of Region

Most Costa Rican territory is potential habitat for white-tailed deer. Costa Rica is a hilly country with a high mountain range, reaching nearly 4000 m elevation, crossing the country from northwest to southeast and dividing the country in two slopes, Atlantic and Pacific. Most lowland is near the coasts and a rich soil plateau occurs in the central region where most of the human population has settled. Consequently, Costa Rican geography is dominated by the two coasts and mountains which cover more than half of the territory (Stiles and Skutch, 1998). These elements and northerly winds influence the climate. Currently, most forest lands occur either in protected areas, riparian forest, or in high hill country. Other lands are urban, industrial, or agricultural. Agricultural lands offer open areas with scattered trees, shrub rows, and water sources where deer could thrive.

Morphological Variation

Costa Rican deer weigh 30–50 kg, with bucks being heavier than does. Their height is nearly 1 m (Mora, 2000). Two subspecies have been recognized: *O. v. truei* and *O. v. chiriquensis* (Goodwin, 1946). Genetic analyses remain to be conducted to confirm their status. Local people indicate that bucks most commonly have four, and rarely six, antler points.

Prominent Ecological Influences

Predators

Several vertebrates prey on Costa Rican white-tailed deer at different ages. Coyotes, crocodiles, and large felines such as cougars and jaguars (Goodwin, 1946) may feed on young and adult deer, whereas boas may feed on fawns (Rodríguez and Solís-Rivera, 1988; Vaughan, 1991; Janzen, 1986).

Health

Diseases of white-tailed deer in Costa Rica have scarcely been addressed. López-Pizarro (1985) reported anaplasmosis in white-tailed deer from San Lucas Island and suggested that piroplasmosis could also be expected. Mora (1983) conducted serological and fecal analyses of seven deer from Tilarán to identify parasites and BTV. Deer tested negative for parasites, but five of six samples were seropositives to BTV.

Additional disease and parasite studies are important, because white-tailed deer are known to be hosts to serious pathogens and parasites of cervids and other animals. Primary concerns are *Borellia burgdorferi* and *Parelaphostrongylus tenuis*. *Borellia*, the agent of Lyme disease, is currently under study by a Costa Rican interuniversity research project. *P. tenuis* is a meningeal parasite that is not pathogenic to white-tailed deer, but is known to cause serious disease in other North American cervids. The life cycle of *P. tenuis* includes snails as intermediary hosts. When snails are ingested by other ruminant species, severe neurologic disease may occur as the adult worms burrow through the brain and central nervous system (Carreno et al., 2001). The only other Costa Rican native cervid that could be affected is the brocket deer.

Several parasites have recently been identified in Costa Rican white-tailed deer. Carreno et al. (2001) reported two does from Guanacaste infected by 18 species of parasites, including six species of ticks and nematodes. Ticks found were: *Amblyomma parvin*, *Haemaphysalis juxtakochi*, *Riphicephalus microplus*, *Ixodes affinis*, and *Anocentor nitens*. Although the tick species found are not known to be reservoirs of *Borrelia burgdorferi* and no cases of Lyme disease have been reported in Costa Rica, it is important to conduct surveillance studies because some of the tick species are closely related to Lyme disease vectors (Brooks and Hoberg, 2006). Carreno et al. (2001) also found the hippoboscid fly, the larvae of the nasopharyngeal botfly (*Cephenemyia jellisoni*), and the helminths *Eucyathostomum webbi*, *Gongylonema pulchrum*, *P. tenuis*, and *Paramphistomum liorchis*. *Onchocerca cervipedis* and *Taenia omissa* were encapsulated in the lung parenchyma. They also found *Setaria yehi*, a formerly undescribed species of *Ashworthius*, in a doe from Guanacaste National Park. This new species of nematode not formerly found in cervidae from the Western Hemisphere was later described by Hoberg et al. (2002). Durán (2003) observed a magpie-jay on the head of a doe in Guanacaste. The jay was apparently foraging on ectoparasites, a behavior that may improve health of the deer.

Population Status

Most projects that have estimated population abundance or trends have used direct counts (López-Pizarro, 1985) or track counts (Gómez, 1997; Piedra-Castro and Maffei, 2000; Pacheco-Figueroa, 2003). Pacheco-Figueroa (2003) studied deer at Barra Honda National Park in Guanacaste. She compared relative abundance of deer along human-disturbed and nondisturbed transects and found more deer tracks on the nondisturbed transects. Local inhabitants feel that deer numbers have decreased possibly due to hunting, fires, or deforestation. While most inhabitants (105 of 167) indicated they had seen deer visiting crops, 87% did not consider white-tailed deer a pest. When farmers where interviewed, 59% declared they had previous problems with deer. Gómez (1997) studied density of white-tailed deer in Finca la Emilia in Cañas, Guanacaste, using track counts along linear transects in five habitat types: sugarcane fields, fruit fields, riparian forests, forest, and pastures. He obtained a density of 3.3 deer/km^2. Higher densities were found in sugarcane fields and fruit fields, whereas density was least in pastures.

Land Use and Human Impacts

Irby and Vaughan (1996) analyzed changes in western Costa Rican land cover and land use to determine if changes in technology and land use practices have impacted white-tailed deer habitat. They determined that, despite human population increases, there was not an increase of rural houses and roads in the study area and that closed canopy cover had increased fourfold. They noted a change in rural housing distribution related to power-line distribution. Unexpectedly, these changes seem to have resulted in increased habitat for wildlife and higher human tolerance for wild vertebrates.

Fire Effects and Management

Costa Rica has two seasons determined by precipitation, the dry (December–April) and the wet (May–November). These seasons are more defined in the northern Pacific region where many plant species are deciduous compared to the Caribbean slope, were precipitation is higher and vegetation tends to be evergreen. Every year, there is a constant fight against fires during the dry season, mainly on wildlands on the Pacific slope. These fires are primarily caused by human negligence; however, some are started to clear vegetation before seeding, and some may be used by poachers to force game toward waterholes. These fires may last hours, days, or even weeks, and are a threat to wildlife as well as humans.

Habitat Use and Evaluation

Segura-Lopez (1995) studied presence and habitat use of white-tailed deer in the northwest Pacific dry forest and created a HEP model to predict potential white-tailed deer habitat (Segura-Lopez and Fallas, 1996). The resulting map predicted the most important regions for white-tailed deer persistence and resulted in an important tool for the decision making process. For example, the map was used in an Environmental Impact Assessment of a proposed agricultural irrigation project in Guanacaste. Model results indicated that prime white-tailed deer habitat could be negatively affected due to habitat loss and installation of structures that may impair deer movements or even cause mortality (Segura-Lopez, 1998). Options to mitigate project impacts were proposed.

Few studies on nutritional ecology have been carried out in Latin America; therefore, there is little information on food habits and less on chemical composition of deer food. Several authors provided insight on plant parts and plant species used by Costa Rican deer (Mena, 1978; Janzen, 1983; Mora and Moreira, 1984; López-Pizarro, 1985; McCoy and Vaughan, 1985; and Solís et al., 1986). Later studies focused on deer food habits including forage composition, forage quality, and food preferences, resulting in a preliminary list of deer food species in Costa Rica (Di Mare, 1986; Di Mare-Hering, 1994; Espach and Sáenz, 1994).

A nutritional study of white-tailed deer food from San Lucas Island showed that forages were generally lower quality than temperate zone forages; primarily because of high lignin concentrations. Mineral concentrations were determined for 18 major white-tailed deer food plants. Low phosphorus to calcium ratios (P:Ca ratio) were a problem in most forage plants during both wet and dry seasons. Low concentrations of P are common in high rainfall environments where P is lost by leaching. A proper P:Ca ratio, generally between 1:2 and 2:1, is important for adequate absorption of both minerals, although dietary vitamin D makes this ratio less important because it promotes absorption of both minerals. Deer on San Lucas Island may not obtain sufficient P from forage, although they may be getting P from some other source, such as mineral licks (Rodríguez et al., 1985a) or leaves covered with seabird droppings.

Janzen (1983) indicated that Costa Rican bucks have smaller antlers compared to northern deer, and that a set of antlers with four tines weighed about 25% of a four-tined set of antlers from a deer in Minnesota. He suspected a shortage of Ca-rich food as a possible cause. It remains to be studied if antler size in Costa Rican bucks is mostly genetically determined or if diet plays an important role. If the later, antler size could be managed through better nutrition.

Management Issues and Trends

Importance in Costa Rica

White-tailed deer are important to Costa Ricans for many reasons: as a game species that provides trophy antlers, meat, hides, and a source of protein; as an attractive species to observe (ecotourism); as an ecologically important seed disperser and prey species; as a source of inspiration in art (poetry, paintings, sculptures, songs, etc.); and for religious beliefs. Indigenous people used all parts of white-tailed deer including its bones (Janzen, 1983, 1986; McCoy and Vaughan, 1985; Solís-Rivera, 1994; Vaughan and Rodríguez, 1994; Mora, 2000). The national cultural importance of white-tailed deer was evident when the species was officially declared the symbol of the Guanacaste Province. White-tailed deer have been one of the most important and sought-after game species in Costa Rica (López-Pizarro, 1980, 1985; McCoy and Vaughan, 1985; Solís-Rivera, 1994). On the other hand, there have been concerns regarding the possible role of deer in damaging crops and recently, health concerns have arisen regarding the role of white-tailed deer as a host for disease agents (Carreno et al., 2001; Hoberg et al., 2002).

White-tailed deer were abundant in Costa Rica in the nineteenth century and it has been documented that a good hunter could obtain 35 deer during one hunting trip. Deer meat was so abundant that chronicles indicate rural people would not use the meat, but would hunt deer only for the hides and antlers (Solís-Rivera, 1994). White-tailed deer were an important source of fine leather. Hides were obtained for exportation to the United States, United Kingdom, and France. By 1910, hide exports reached 34,000 kg/year. During the 1940s, 10,000–60,000 deer were harvested annually and their meat was less expensive than beef (Janzen, 1983). Exports decreased drastically after 1910 and ceased by 1950. Furthermore, commercial exports of wild deer leather became illegal in 1961 with passage of Law No. 4551 that made it illegal to sell wildlife, its products, or its remains (Gobierno de Costa Rica, 1961). These concepts remain in current law No. 7317 (Gobierno de Costa Rica, 1992).

White-tailed deer are still considered a game species and every year the government determines the legal quota through publication of a decree for the hunting season where bag limit, seasons, hours of day, and sites are established. Currently, the bag limit is two bucks/year and nonresident foreigners are not allowed to hunt (Gobierno de Costa Rica, 2009). Costa Rican wildlife laws are highly developed; however, monitoring and law enforcement are inadequate because of budget limitations in the wildlife department. Adequate personnel, training, and equipment are also lacking. Collaborative efforts among the Ministry, nongovernmental organizations, and universities may help provide research capability and training to strengthen the wildlife department.

Captive white-tailed deer in Costa Rica are found in zoos, zoocriaderos, and rescue centers. There is no commercial ranching or farming of deer and the law requires that only third-generation captive-raised deer may be used for commercial purposes. There are a small number of zoos (e.g., "Parque Bolívar" in San Jose, "Zoológico La Marina" in Ciudad Quesada, "Africa Mia" in Guanacaste) that maintain small deer populations. Zoocriaderos are a second option to legally maintain deer in captivity, but commercialization is prohibited. White-tailed deer are kept in zoocriaderos, despite costs involved, mainly because people enjoy their presence. There are 20–30 legal zoocriaderos in Costa Rica, each with no more than 20 deer. Rescue centers are the third legal avenue for deer to be held in captivity and are places where animals are brought for temporary care when orphaned or injured. These deer are released to the wild as soon as they are rehabilitated; otherwise, they are relocated in a zoo or other place where people provide permanent care.

Hunting preserves have been discussed (Fernández, 1987), but expense and lack of well organized hunting associations have been a limitation. Sustainable use of Costa Rican white-tailed deer was discussed previously, and management strategies have been proposed initially for San Lucas Island, and later for Costa Rican deer (Solís and Brenes, 1987; Vaughan, 1991, 1994; Solís-Rivera and Vaughan, 1994). The only continuous activity regarding deer use in the country is the hunting season that is set annually by the government (Gobierno de Costa Rica, 1992).

Conservation and Management Efforts

White tailed-deer are one of the most important wildlife and game species in Costa Rica, and research on the species began in the 1980s. However, two events occurred by 1984 that contributed to a better understanding of deer biology. By the 1980s, the need to improve wildlife management in Latin America was identified and a meeting was held with wildlife government officials from the U.S. Fish and Wildlife Service and directors of several Latin American countries to determine where to initiate a program to strengthen wildlife management. Costa Rica was the chosen country. Therefore, the "Regional Program on Wildlife Management for Mesoamerica and the Caribbean" (or PRMVS) began in 1984 in Costa Rica, at the Universidad Nacional, and a graduate program to strengthen wildlife research and train students from the region began. Simultaneously, there was interest in addressing the potential of white-tailed deer in the region, which in turn was a good opportunity to develop a model project. Given the characteristics of white-tailed deer and its potential, the conservation strategies available (Teer, 1987, 1994a,b), and the existing deer habitat in north-western Costa Rica, the potential for a model project on the species was positively evaluated by independent observers, and the project was approved. Hence, the long-term study on "Restoration of white-tailed deer on north-western Costa Rica" began. It was supported by the U.S. World Wildlife Fund (Vaughan, 1994) and conducted in collaboration with the PRMVS, providing a fertile ground for development of research projects, either as student theses or collaborative efforts with the Costa Rican Wildlife Service.

Because of the importance of the species, some research had been conducted previously. However, with the restoration project, knowledge regarding Costa Rica's main game species increased rapidly. Research needs were identified and a series of projects and theses were conducted on subjects such as nutrition, social behavior, reproductive behavior, movements, habitat use, translocation, and repopulation. Research on white-tailed deer biology was conducted at San Lucas Island and Palo Verde Wildlife Refuge and two sites for reintroduction of white-tailed deer were selected: Finca la Emilia in Cañas, Guanacaste, and the town of Cóbano in Puntarenas, in the Nicoya Peninsula. In both cases, reasons for site selection included the presence of favorable conditions for deer survival (habitat, water, vegetation cover) and local community support of the project. Reintroduction projects were carried out in Finca la Emilia and Cóbano with deer from San Lucas Island and from Palo Verde Wildlife Refuge. Follow up studies were conducted to evaluate habitat use, food habits, herd composition, fawn behavior, movements, home ranges, and post-release survival and behavior (Solís, 1986; Rodríguez, 1989; Sáenz, 1990; Rodríguez and Vaughan, 1994; Sáenz and Vaughan, 1998). After the restoration project was completed, other research projects were conducted covering topics such as evaluation of actual and potential deer habitat, habitat use, deer abundance, human perceptions of white-tailed deer, and deer and human conflicts (e.g., Vaughan et al., 1994; Segura-Lopez, 1995; Gómez, 1997; Cabrera-Rentería, 1998; Pacheco-Figueroa, 2003).

Translocation and Reintroduction Efforts

White-tailed deer have the potential to occur across the country. Translocations of deer may be more common than previously thought, because ranchers apparently move deer among properties. Deer, for example, were introduced to San Lucas Island (López-Pizarro, 1985; Di Mare, 1986; Solís, 1986) and Tortuga Island. Reports of a deer introduction on Coco Island date to 1934 when an expedition to the island was made in search of treasure. A later introduction occurred on an island of high endemism, which is now a National Park and where introduced species may have caused severe damage (Montoya-Maquín, 2004). A 1996 decree authorized extirpation of white-tailed deer from this National Park.

As part of the white-tailed deer restoration project, two white-tailed deer reintroductions were conducted in northwestern Costa Rica. An initial reintroduction was performed in Cañas, Guanacaste, where deer from San Lucas Island and Palo Verde Wildlife Refuge were translocated to Finca La Emilia (Sáenz, 1990, 1994; Sáenz et al., 1994). Wild pregnant does, female fawns, and bucks were captured on the mainland and relocated to Finca La Emilia to increase the deer population. Translocated deer had an 80% survival rate and a 100% reproductive rate. Success of the translocation was probably due to availability of good deer habitat, no hunting pressure, and the behavioral plasticity of deer allowing them to adapt to new areas. Next, deer from San Lucas Island were translocated to Cóbano in the Nicoya Peninsula, Puntarenas (Calvopiña, 1990; Hernández, 1993, 1994; Irby and Calvopiña, 1994). The restoration project provided biologists experience in bottle-raising fawns, helped develop knowledge of deer health, nutrition, and behavior, and provided a better understanding of the costs of a reintroduction program. Deer movements and survival were monitored using radio-telemetry, and food habits were recorded. Fawns translocated to Finca la Emilia were cared for by farm personnel. In Cóbano, each fawn was given to a local family and its care became the duty of the housewife. An unexpected result was that this closeness of local people to fawns contributed to a high commitment of Cóbano people to deer conservation and to better knowledge of deer biology. It may also have reduced local deer hunting. Experience gained through these efforts contributed to better dissemination of deer information to the general public (Rodríguez and Solís, 1994a,b), and to a preliminary white-tailed deer management plan (Solís-Rivera and Vaughan, 1994).

Colombia and Venezuela

Potential distribution of white-tailed deer in Colombia includes most of the country, from the savannahs and open forest of Orinoquía and Amazonía, to the Caribe plateau from Córdoba to south of La

Guajira, the Alto Magdalena region, Tolima, Cundinamarca, Huila, and Valle del Dagua, even when records indicate its presence only in the "savannahs del Yari." White-tailed deer can be found up to 4000 m elevation in the Central and Oriental Mountainous region all the way to the Los Andes de Nariño. In Los Nevados National Park, located in the Central Mountain region, white-tailed deer were last reported in the 1960s; however, Mateus (2006) reported sightings recently. In the Oriental mountain region, white-tailed deer are present in Nariño (Laguna de La Cocha), Cundinamarca (Páramo de Sumapaz, Páramo de Chingaza), Boyacá (Sierra Nevada del Cocuy, Iguaque, Páramo de Pisba), north of Santander (Serranía de Los Motilones), as well as in the Sierra Nevada de Santa Marta (Cuervo et al., 1986; Alberico et al., 2000; López-Arévalo and González-Hernández, 2006). White-tailed deer specimens in the main Colombian scientific collections originate from the departamentos of Casanare, Meta, Vichada, Vaupés, Caquetá, Cesar, Santander, Boyacá, Cundinamarca, and Cauca.

In Venezuela, white-tailed deer or "venado caramerudo" are found over most of the country, including the Margarita Island and the northeast part of the Cordillera de Mérida (Brokx, 1984; Molina-Molina, 2000). However, it is in the savannahs (the Llanos region) where the species is most abundant (Figure 1.14). The major part of this range is covered by palmares (palms growing on flooded sites), dry deciduous forest, riparian corridors, gallery forest, and groves of tree or matas with a diverse understory of grass and forbs. Most savannahs are less than 200 m elevation. Because of their irregular micro relief, these savannahs were distinguished by Ramia (1962) as banco (highest elevation of ground), bajío (intermediate elevation), and estero (lower elevation).

Morphological Variation

Five subspecies of white-tailed deer have been recognized in Colombia according to UICN (2001) criteria: *O. v. goudotii*, *O. v. ustus*, and *O. v. curassavicus*; *O. v. apurensis*, which is designated as a subspecies of concern, and *O. v. tropicales* which has been designated as endangered and may be extinct (Hernández-Camacho et al., 1992; López-Arévalo and González-Hernández, 2006).

Crania and mandibles of Venezuelan white-tailed deer were compared by Molina and Molinari (1999) who found the more divergent forms were from Margarita Island and the Mérida Andean region. Margarita Island deer had smaller mandibles, small body size, and some unique cranial and mandible traits. Deer from the Mérida Andean region had a different shaped mandible. The authors concluded that the remaining Venezuelan forms constitute a single group. They also proposed that the Margaritan and Andean forms are distinct species: *Odocoileus margaritae* and *O. lasiotis*, respectively. The remaining forms must be included as *O. cariacou*. More recent analyses (Moscarella et al., 2003; Molinari, 2007) confirms unique taxonomic units on Margarita Island and in the Merida Andean highlands, but does not lend support to these deer being a species other than *O. virginianus* (Chapter 1).

Population Status

White-tailed deer population density in Colombia is not well known. Increasing number of sightings in some areas of páramos at the Chingaza National Park indicate an increasing population during the past 15 years; however, more rigorous estimates are needed. In all, 11–44 deer/km^2 were estimated using direct counts along linear transects in the municipality of Paz de Ariporo, Casanare in the Colombian Orinoquía. More males than females and also more adults than juveniles were observed.

In Mucubají, Venezuela, a 540-ha area of alpine meadows of páramos consisting of tussock grasses, herbaceous shrubs, valleys and slopes, pine forest, and various species of frailejón, the white-tailed deer population was 213–252 deer (39–46 deer/km^2; Correa-Viana, 1994). Molina-Molina (2003a) estimated a population of 240 deer in the same area. The size and status of the white-tailed deer population on Margarita Island is critical. This population could be near extinction. Only deer in captivity (<12) were observed in the locality of Macanao (Molina-Molina, 1996). Geographical isolation, high human population density, illegal hunting, and habitat destruction are the main causes of the deer population decline on the island.

Management Issues and Trends

Traditional Use

Archeological records indicate white-tailed deer were valued for hunting and other uses in Colombia 12,000 years ago. For example, groups of nomadic hunters from the Altiplano Cundinamarques preferred white-tailed deer and curí as food sources (Peña and Pinto, 1996). Nomadic hunters from the Altiplano Cundiboyacense used white-tailed deer as a food source, bones were used to manufacture tools, and hides were used for cold weather protection 9000 years ago (Correal and Van der Hammen, 1977; Rincón, 2003). Presently, white-tailed deer are one of the most popular wildlife species for subsistence and sport hunting. White-tailed deer are also used as pets. Hides and venison are sold in local markets. In rural areas of the Andes, white-tailed deer were used for religious ceremonies not too long ago. No detailed studies exist on current use of white-tailed deer for subsistence, except for recent records from municipalities in the states of Boyacá (Blanco and Zabala, 2005), Cundinamarca, Meta, and Casanare, among others.

White-tailed Deer in Captivity

Presently, Colombian law allows individuals to raise any species of wildlife on farms or hunting preserves as long as the activities have no negative effects on free-ranging wildlife populations, as stated in the Ley 611 of 2000 and the Resolución 1317 of 2000. However, no legally constituted white-tailed deer farms exist; although, proposals have been developed for some regional entities (Corporaciones Autónomas Regionales) to establish white-tailed deer farms (Mateus, 2006). Smaller captive populations occur in zoos, state parks, and small farms. Guzmán-Lenis (2005) analyzed white-tailed deer management experiences in six zoos, two farms, and a private reserve. Professionals responsible for animal welfare were interviewed. More than 100 captive white-tailed deer were reported between 2003 and 2004.

In Venezuela, white-tailed deer are one of the most sought species by peasants, indigenous people, and sport hunters. In the Llanos region, deer meat is mainly consumed as a complementary source of animal protein by farmers, and for subsistence by indigenous people. Hunting white-tailed deer is a recreational experience for sport hunters. In the páramos of the Cordillera de Mérida, white-tailed deer are hunted by Andean farmers for food and to reduce crop depredation (Molina, 2004). Antlers, tails, and skins are used to make ornaments. Antlers and hooves are also considered hunting trophies, and some bones are used for medicinal purposes.

Conservation Status

The main factors affecting populations of free ranging white-tailed deer in Colombia are the same factors that affect other wildlife species: habitat fragmentation, habitat degradation, interaction with domestic animals, and excessive subsistence hunting and poaching (López-Arévalo and Morales-Jiménez, 2004). White-tailed deer are valuable as a big game species in communities where free ranging white-tailed deer are still found. There is an increasing need to develop management and conservation strategies. Presently, white-tailed deer populations have been reported in 10 National Parks within the Protected Areas System, in two wildlife sanctuaries of flora and fauna in the Andes region, and in only one Natural National Park in the Orinoquia and Amazonìa Colombiana. Reintroduction of white-tailed deer into some areas of the national parks system has been suggested as one conservation strategy, along with design of biological corridors joining parks in the Andes region (Matallana-Tobón, 2001).

In Venezuela, white-tailed deer populations are not managed, in a strict sense, through programs directed by government institutions. Nevertheless, wildlife protection laws prohibit hunting of white-tailed deer, especially in national parks, natural refuges, and fauna and biosphere reserves. These laws enable the national government to prepare management plans for wildlife. In the 1980s the Ministerio de Agricultura y Cría and the Ministerio del Ambiente y de los Recursos Naturales Renovables promoted extensive discussion of establishing experimental areas for hunting, hunting reserves, habitat restoration, and other management options for white-tailed deer; however, no actions have been initiated.

Meanwhile, illegal hunting of the species occurs throughout its distribution. Meat of harvested deer is mainly used for subsistence of farmers and indigenous people, and to a lesser extent to satisfy the incipient demand for "carne de monte" in restaurants. In the llanos region there are many private lands known as hatos, haciendas, or fincas where owners follow local regulations, including hunting bans, bag limits, and selective harvest. Some of the most successful examples of deer management are related to ecotourism, especially in the savannahs of Apure (i.e., Hato El Frío, and Hato El Cedral) and Cojedes (i.e., Hato Piñero). In the páramos, neither the condition of National Parks nor the critical condition of remnant deer populations has been sufficient to establish a management program. Molina-Molina (2000) presented a plan to improve management of white-tailed deer, which unfortunately has not yet been given full consideration.

Research Needs

According to Mandujano (2004) and Gallina and Mandujano (2009), research objectives for management and conservation of ungulates in Mexico should be

- To develop management approaches for subspecies of white-tailed deer in tropical forests.
- Basic studies on habitat requirements and use, home ranges, population dynamics, and habitat management are needed for most of the subspecies of white-tailed deer. Only five of 14 white-tailed deer subspecies have been studied to any extent in Mexico (*O. v. texanus*, *O. v. couesi*, *O. v. sinaloae*, *O. v. mexicanus*, and *O. v. yucatanensis*).
- Assess the effectiveness of UMAs for conservation and sustainable use of ungulates. Determine whether different conservation, use, and management practices may be necessary for deer in tropical forests.
- Analyze current harvest and determine sustainable harvest of ungulates by indigenous communities in areas with high and low hunting pressure.
- Study fragmentation, sink–source models, and metapopulations of white-tailed deer from a landscape perspective.
- Research the role of ungulates in altering the structure and composition of vegetation in tropical habitats and the consequences of their absence.
- Studies on taxonomy, genetics, and distribution of white-tailed deer.
- Analyze effects of human population growth, agriculture technology, cattle ranching, and habitat transformation on ungulate populations and their habitat.

In Costa Rica, Colombia, and Venezuela research needs include:

- Identification of white-tailed deer subspecies based on genetics and morphological characteristics; meanwhile, careful consideration must be given to reintroduction of white-tailed deer from low elevation populations to the Andean region.
- Confirm the presence of white-tailed deer in protected areas where management plans report it.
- Evaluate population status in at least one area of each region where the presence of white-tailed deer has been confirmed.
- Determine competitive relationships between white-tailed deer and domestic animals and other wildlife species.
- Evaluate the effect of water availability on movements and feeding habits of deer as well as the spatial and seasonal use of habitat by deer of different ages and sex.
- Determine the incidence of disease and parasites in white-tailed deer and rates of transmission to other wildlife species, livestock, and humans.
- Evaluate social and economic viability of conservation schemes using either free-ranging or captive deer.

REFERENCES

Abizaid, C. and O. T. Coomes. 2004. Land use and forest fallowing dynamics in seasonally dry tropical forests of the southern Yucatan Peninsula, Mexico. *Land Use Policy* 21:71–84.

Ahumada, R., M. Flores-Fuentes, and E. Torres. 1998. Distribución y abundancia del venado cola blanca (*Odocoileus virginianus sinaloae*), en la sierra de Navachiste, Sinaloa. *VI Simposio sobre Venados en México*. Universidad Nacional Autonoma de Mexico, Xalapa, Ver.

Alberico, M., A. Cadena, J. Hernández-Camacho, and Y. Muñoz-Saba. 2000. Mamíferos (Synapsida: Theria) de Colombia. *Biota Colombiana* 1:43–75.

Allen, J. A. 1903. A new deer and a new lynxs from the state of Sinaloa, Mexico. *Bulletin of the American Museum of Natural History* 19:613–615.

Arceo, G., S. Mandujano, S. Gallina, and L. A. Perez-Jimenez. 2005. Diversity of diet of white-tailed deer in a Mexican tropical forest. *Mammalia* 69:159–168.

Armstrong, D. M., J. K. Jones Jr., and E. C. Birney. 1972. Mammals from the Mexican state of Sinaloa. III. Carnivora and Artiodactyla. *Journal of Mammalogy* 52:48–61.

Ávila-Gómez, G. 1996. Estudio de abundancia y distribución del venado en selvas de Quintana Roo. *V Simposio sobre Venados en México*. Universidad Nacional Autonoma de Mexico, Quintana Roo, Mexico.

Ávila-Gómez, G. 2008. El contexto de la caza actual y futura del venado cola blanca coues (*Odocoileus virginianus couesi*) en la Sierra de Nayarit. *XI Simposio Sobre Venados en México*. Universidad Nacional Autonoma de Mexico, México, D.F.

Blanco, L. and A. I. Zabala. 2005. Recopilación del conocimiento local sobre el venado cola blanca (*Odocoileus virginianus*) como base inicial para su conservación en la zona amortiguadora del Parque Nacional Natural Pisba, en los municipios de Socha y Tasco (Boyacá, Colombia). MS thesis, Universidad Pedagógica y Tecnológica de Colombia, Tunja, Colombia.

Bodmer, R. E. and J. G. Robinson. 2004. Evaluating the sustainability of hunting in the Neotropics. In *People in Nature: Wildlife Conservation in South and Central America*, eds. K. M. Silvus, R. E. Bodmer, and J. M. V. Fragoso, 299–323. New York, NY: Columbia University Press.

Branan, W. B., M. C. M. Werkhoven, and R. L. Marchinton. 1985. Food habits of brocket and white-tailed deer in Suriname. *Journal of Wildlife Management* 49:972–976.

Briones-Salas, M. and C. García-Cruz. 2005. Estimación de la densidad del venado cola blanca (*Odocoileus virginianus oaxacensis*) en la sierra norte de Oaxaca. *Revista Mexicana de Mastozoología* 9:141–145.

Brokx, P. A. 1984. South America. In *White-tailed Deer Ecology and Management*, ed. L. K. Halls, 525–546. Harrisburg, PA: Stackpole Books.

Brooks, D. R. and E. P. Hoberg. 2006. Systematics and emerging infectious diseases: From management to solution. *Journal of Parasitology* 92:426–429.

Bullock, S. H., H. A. Mooney, and E. Medina. 1995. *Seasonally Dry Forests*. Cambridge, UK: Cambridge University Press.

Burgos, A. and J. M. Mass. 2005. Vegetation change associated with land-use in tropical dry forest areas of western Mexico. *Agriculture, Ecosystems and Environment* 104:475–481.

Cabrera Rentería, J. A. 1998. Uso de ojos de agua por grandes y medianos mamíferos en el Sector Santa Rosa, Area de Conservación Guanacaste. MS thesis, Universidad Nacional, Heredia, Costa Rica.

Calvopiña, J. E. 1990. Reintroducción del venado cola blanca (*Odocoileus virginianus*) a Cóbano, Puntarenas, Costa Rica. MS Thesis, Universidad Nacional. Costa Rica.

Camargo-Sanabria, A. A. 2005. Evaluación preliminar del área de acción y patrón de actividad del venado cola blanca (*Odocoileus virginianus*), como parte de una alternativa de manejo *ex situ* en un bosque seco tropical, (Cundinamarca, Colombia). MS thesis, Universidad Nacional de Colombia, Bogotá, Colombia.

Cantú, A., J. A. Ortega-S., J. Mosqueda, Z. Garcia-Vasquez, S. E. Henke, and J. E. George. 2007. Immunological and molecular identification of *Babesia bovis* and *Babesia bigemina* in free ranging deer in northern Mexico. *Journal of Wildlife Diseases* 43:504–507.

Cantú, A., J. A. Ortega-S., J. Mosqueda, Z. Garcia-Vasquez, S. E. Henke, and J. E. George. 2008. Prevalence of infectious agents in free-ranging white-tailed deer in northeastern Mexico. *Journal of Wildlife Diseases* 44:504–507.

Cantú, A., J. A. Ortega-S., J. Mosqueda, Z. Garcia-Vasquez, S. E. Henke, and J. E. George. 2009. Epizotiology of *Babesia bovis* and *Babesia bigemina* in free-ranging white-tailed deer in northeastern Mexico. *Journal of Parasitology* 95:536–542.

Cantú-Martínez, M. A., J. A. Salinas-Melendez, J. J. Zarate-Ramos, R. Avalos-Ramirez, A. Martinez-Munoz, and J. C. Segura-Correa. 2008. Prevalence of antibodies against *Babesia bigemina* and *Babesia bovis* in white-tailed deer (*Odocoileus virginianus texanus*) in farms of northeastern Mexico. *Journal of Animal and Veterinary Advances* 7:121–123.

Carreno, R. A., L. A. Durden, D. R. Brooks, A. Abrams, and E. P. Hoberg. 2001. *Parelaphostrongylus tenuis* (Nematoda: Protostrongylidae) and other parasites of white-tailed deer (*Odocoileus virginianus*) in Costa Rica. *Comparative Parasitology* 68:177–184.

Castillo, A. 1998. Estimación poblacional del venado cola blanca en tres localidades de Colima. BS thesis, Benemérita Universidad Autónoma de Puebla, Puebla, Mexico.

Ceballos, G. and G. Oliva. 2005. *Los mamíferos silvestres de México*. México, D.F.: CONABIO y Fondo de Cultura Económica.

Clemente-Sánchez, S. 1984. Utilización de la vegetación nativa en la alimentación del venado cola blanca (*Odocoileus virginianus* Hays) en el estado de Aguascalientes. MS thesis, Colegio Postgraduados Chapingo, Edo. México.

Corona, P. 2003. Bases biológicas para el aprovechamiento del venado cola blanca en el Ejido El Limón de Cuachichinola, Municipio de Tepalcingo, Morelos. MS thesis, Instituto de Ecología A. C., Xalapa, Ver., México.

Coronel-Arellano, H., C. A. López González, and C. N. Moreno-Arzate. 2009. Pueden las variables de paisaje predecir la abundancia de venado cola blanca? El caso del noroeste de México. *Tropical Conservation Science* 2:229–236.

Correa-Viana, M. 1994. Actividad diaria y selección de hábitat por el venado caramerudo, *Odocoileus virginianus*, en Masaguaral, estado Guárico, Venezuela. *Biollania* 10:33–42.

Correa-Viana, M. 2000. Movimientos, actividad y uso del hábitat de venados liberados en la finca El Jaimero, Portuguesa, Venezuela. PhD dissertation, Universidad Central de Venezuela, Maracay.

Correal, U. G. and T. Van der Hammen. 1977. Investigaciones arqueológicas en los abrigos rocosos del Tequendama: 11,000 años de prehistoria en la Sabana de Bogotá. Bogotá, Colombia: Biblioteca Banco Popular.

Cowan, I. McT. 1936. Distribution and variation in deer (genus *Odocoileus*) of the Pacific coastal region of North America. *California Fish and Game* 22:155–246.

Crosswhite, F. S. 1980. Dry country plants of the south Texas plains. *Desert Plants* 2:141–179.

Cuervo, A., J. Hernández, and A. Cadena. 1986. Lista actualizada de los mamíferos de Colombia. Anotaciones sobre su distribución. *Caldasia* 15: 471–501.

Danields, H. 1987. Ecología nutricional del venado caramerudo (*Odocoileus virginianus gymnotis*) en los llanos centrales. PhD dissertation, Universidad Central de Venezuela, Caracas.

Delfín-Alfonso, C. A. and S. Gallina. 2007. Modelo de evaluación del hábitat para el venado cola blanca en un bosque tropical caducifolio en México. In *Escarabajos, Diversidad y Conservación Biológica: Ensayos en Homenaje al Dr. Gonzalo Halffter*, eds. M. Zunino and A. Melic, 193–202. Zaragoza, España: Mografías Tercer Milenio.

Delfín-Alfonso, C. A., S. Gallina, and C. A. López-González. 2009. Evaluación del hábitat del venado cola blanca utilizando modelos espaciales y sus implicaciones para el manejo en el centro de Veracruz, México. *Tropical Conservation Science* 2:215–228.

Dirección General de Vida Silvestre. 2007. *Sistemas de Unidades de Manejo*. Dirección General de Vida Silvestre, Secretaria del Medio Ambiente y Recursos Naturales, México, D.F.

Dirección General de Vida Silvestre. 2010. *Sistema de Unidades de Manejo para la Conservación de la Vida Silvestre (SUMA)*. Secretaria de Medio Ambiente y Recursos Naturales. México, D.F. www.semarnat.gob.mx/gestionambiental/vidasilvestre/ Pages/sumas.aspx. Accessed June 7, 2010.

Di Mare, M. I. 1986. Food habits of an insular neotropical white-tailed deer (*Odocoileus virginianus*) population. MS thesis, Colorado State University.

Di Mare-Hering, M. I. 1994. Hábitos alimentarios del venado cola blanca en la Isla San Lucas, Puntarenas, Costa Rica. In *Ecología y Manejo del Venado Cola Blanca en México y Costa Rica,* eds. C. Vaughan-Dickhaut and M.A. Rodríguez-Ramírez, 73–90. Heredia, Costa Rica: Universidad Nacional.

Drawe, L. E., J. A. Ortega-S., and T. E. Fulbright. 1999. Habitat management for white-tailed deer. Proceedings of a wildlife management workshop, Consorcio Técnico del Noreste de Mexico. Saltillo, Coah., Mexico.

Durán, F. J. 2003. Forrajeo de urraca copetona, *Calocitta formosa* (Aves: Corvidae), sobre venado cola blanca *Odocoileus virginianus* (Mammalia: Cervidae). *Brenesia* 59/60:89–90.

Escamilla, A., M. Sanvicente, M. Sosa, and C. Galindo-Leal. 2000. Habitat mosaic, wildlife availability, and hunting in the tropical forest of Calakmul, Mexico. *Conservation Biology* 14:1592–1601.

Espach, H. and J. C. Sáenz. 1994. Comportamiento de cervatos criados en cautiverio y reintroducidos en la Finca La Emilia, Costa Rica. In *Ecología y Manejo del Venado Cola Blanca en México y Costa Rica*, eds. C. Vaughan and M. A. Rodríguez, 163–180. Heredia, Costa Rica: Universidad Nacional.

Ezcurra, E. and S. Gallina. 1981. Biology and population dynamics of white-tailed deer in northwestern Mexico. In *Deer Biology, Habitat Requirements and Management in Western North America*, eds. P. F. Ffolliott and S. Gallina, 77–108. Mexico, D.F.: Instituto de Ecología, A.C.

Fernández, R. 1987. Condición actual del venado cola blanca (*Odocoileus virginianus*) con base en experiencias de caza. In *Actas del Primer Taller Sobre el Venado Cola Blanca (Odocoileus virginianus) del Pacífico seco*, eds. V. Solís, M. Rodríguez, and C. Vaughan, 55–61. Heredia, Costa Rica: Universidad Nacional.

Fulbright, T. E. and J. A. Ortega-S. 2006. *White-tailed Deer Habitat Ecology and Management on Rangelands*. College Station, TX: Texas A&M University Press.

Galindo-Leal, C., A. Morales, and M. Weber. 1993. Distribution and abundance of Coues deer and cattle in Michilia Biosphere Reserve, Mexico. *Southwestern Naturalist* 38:127–135.

Gallina, S. and L. A. Escobedo-Morales. 2009. Análisis sobre las Unidades de Manejo (UMAs) de ciervo rojo (*Cervus elaphus* Linnaeus, 1758) y wapiti (*Cervus canadensis* Erxleben, 1777) en México: Problemática para la conservación de los ungulados nativos. *Tropical Conservation Science* 2:251–265.

Gallina, S. and S. Mandujano. 2009. Ecology, management and conservation of ungulate species in Mexico. *Tropical Conservation Science* 2:116–127.

Gallina, S., S. Mandujano, J. Bello, H. López-Fernandez, and M. Weber. 2009. White-tailed deer *Odocoileus virginianus* (Zimmermann, 1780). In *Neotropical Cérvidos*, ed. S. González, 101–118. Montevideo, Uruguay: Deer specialist Group, IUCN.

Gallina, G., S. Mandujano, J. Bello, H. López-Fernandez, and M. Weber. 2010. White-tailed deer *Odocoileus virginianus* (Zimmermann, 1780). In *Neotropical Cervidology: Biology and Medicine of Latin American Deer*. eds. J. M. Duarte and S. González, 101–118. IJaboticabal, Brazil: Funep and Gland, Switzerland: IUCN.

Gallina, S., S. Mandujano, and C. Delfín-Alfonso. 2007. Importancia de las Áreas Naturales Protegidas para conservar y generar conocimiento biológico de las especies de venados en México. In *Hacia una Cultura de Conservación de la Diversidad*, eds. G. Halffter, S. Guevara, and A. Melic, 187–196. Zaragoza, España: Sociedad Entomológica Aragonesa.

Gallina, S., E. Maury, and V. Serrano. 1981. Food habits of white-tailed deer. In *Deer Biology, Habitat Requirements and Management in Western North America*, eds. P. F. Ffolliott and S. Gallina, 135–148. Mexico, D.F.: Instituto de Ecología, A.C.

García-Sierra, L. and R. Monroy. 1985. Estimación de la población del venado cola blanca (*Odocoileus virginianus*) en la selva baja caducifolia del sureste del estado de Morelos. *III Simposio sobre Fauna Silvestre*. Universidad Nacional Autonoma de Mexico, México D. F.

Genoways, H. H. and J. K. Jones, Jr. 1975. Annotated checklist of mammals of the Yucatan Peninsula, Mexico. IV. Carnivora, Sirenia, Perissodactyla, Artiodactyla. *Occasional Papers Museum Texas Tech University* 26:1–22.

Gobierno de Costa Rica. 1961. Ley No. 4551. Ley de Conservación de la Fauna Silvestre. Asamblea Legislativa del Gobierno de Costa Rica. San José. Asamblea Legislativa. CR.

Gobierno de Costa Rica. 1992. Ley No. 7317. Ley de Conservación de la Vida Silvestre. Asamblea Legislativa del Gobierno de Costa Rica. La Gaceta No. 235 del 07 diciembre 1992. 51 p. San José. Asamblea Legislativa. CR.

Gobierno de Costa Rica. 2009. DECRETOS. Nº 34967-MINAET. Regulaciones para la Caza Menor y Mayor fuera de las áreas silvestres protegidas y, de la pesca en áreas silvestres protegidas. Decreto presidencial. La Gaceta, 19 de enero del 2009.

Goldman, E. A. and A. R. Kellogg. 1940. Ten new white-tailed deer from North and Middle America. *Proceedings Biological Society Washington* 53:81–90.

Gómez, L. G. 1997. Estimación de la densidad y tendencia poblacional del venado cola blanca (*Odocoileus virginianus*) en una finca ganadero-cañera del bosque seco de Costa Rica. MS thesis, Universidad Nacional, Herdia, Costa Rica.

Gómez-Giraldo, C. 2005. Radio-telemetría aplicad a la reintroducción del venado cola blanca (*Odocoileus virginianus*). MS thesis, Universidad Nacional de Colombia, Medellín, Colombia.

González, A., J. Lobato, A. Velásquez, and A. Torres. 2003. El manejo del venado cola blanca: La experiencia de una comunidad indígena para el manejo y uso sustentable de la vida silvestre. In *Las Enseñanzas de San Juan Investigación Participativa Para el Manejo Integral de Recursos Naturales*, eds. A. Velazquez, A. Torres, and G. Bocco, 531–547. México, D.F.: INE-SEMARNAT.

González-Marín, R., S. Gallina, S. Mandujano, and M. Weber. 2008. Densidad y distribución de ungulados silvestres en la Reserva Ecológica El Edén, Quintana Roo, México. *Acta Zoológica Mexicana (nueva serie)* 24:73–93.

González-Marín, R. M., E. Montes, and J. Santos. 2003. Caracterización de las unidades de manejo para la conservación, manejo y aprovechamiento sustentable de la fauna silvestre en Yucatán, México. *Tropical and Subtropical Agroecosystems* 2:13–21.

González-Pérez, G. E. 2003. Uso del hábitat y área de actividad del venado cola blanca (*Odocoileus virginianus sinaloae* J. Allen) en la Estación Científica Las Joyas, Reserva de la Biosfera de Manantlán, Jalisco. MS thesis, Universidad Autónoma de México. México, D.F.

González-Pérez, G. and M. Briones-Salas. 2000. Venado cola blanca (*Odocoileus virginianus*) en comunidades indígenas de Oaxaca. *Investigación Hoy* 94:20–27.

Goodwin, G. G. 1946. *Odocoileus Rafinesque*. *Bulletin of the American Museum of Natural History* 87:447–449.

Granado, N. 1989. Dieta del venado caramerudo (*Odocoileus virginianus gymnotis*) en El Socorro, estado Guárico. BS thesis, Universidad Central de Venezuela, Caracas, Venezuela.

Greenberg, L. S. Z. 1992. Garden hunting among the Yucatecan Maya: A coevolutionary history of wildlife and culture. *Etnoecológica* 1:23–33.

Guzmán-Lenis, A. 2005. Análisis de las experiencias colombianas de manejo *ex situ* de venado cola blanca (*Odocoileus virginianus*) como aporte a su conservación. BS thesis, Universidad Nacional de Colombia, Bogotá, Colombia.

Hall, E. R. 1981. *The Mammals of North America*. New York, NY: John Wiley and Sons.

Hernández, L. 1993. La extensión y la educación ambiental en el manejo de venado cola blanca (*Odocoileus virginianus*) en Cóbano, provincia de Puntarenas, Costa Rica. *Biocenosis* 10:64–67.

Hernández, L. 1994. La extensión y la educación ambiental en el manejo del venado cola blanca en Cóbano, provincia de Puntarenas, Costa Rica. In *Ecología y Manejo del Venado Cola Blanca en México y Costa Rica*, eds. C. Vaughan and M.A. Rodríguez, 350–368. Heredia, Costa Rica: Universidad Nacional.

Hernández, S. D. A., B. G. T. González , R. J. L. Zaragoza, and D. E. Cortés. 2008. Densidad de población del venado cola blanca mexicano (*Odocoileus virginianus mexicanus*), en la sierra de Huautla, Morelos. *XI Simposio Sobre Venados en México*. Universidad Nacional Autonoma de Mexico, México, D.F.

Hernández-Camacho, J., A. Hurtado, R. Ortiz, and T. Walschburger. 1992. Centros de endemismo. In *La Diversidad Biológica de Ibero América*, ed. G. Halffter, 175–190. México, D.F.: Instituto de Ecología, A.C.

Hoberg, E. R., A. Abrams, R. A. Carreno, and J. R. Lichtenfels. 2002. *Ashworthius patriciapilittae* n. sp (Trichostrongyloidea: Haemonchinae), an abomasal nematode in *Odocoileus virginianus* from Costa Rica, and a new record for species of the genus in the Western Hemisphere. *Journal of Parasitology* 88:1187–1199.

Irby, L. R. and J. Calvopiña. 1994. Uso del hábitat por el venado cola blanca reintroducido en la Península de Nicoya, Costa Rica. In *Ecología y Manejo del Venado Cola Blanca en México y Costa Rica*, eds. C. Vaughan and M. A. Rodríguez, 333–347. Heredia, Costa Rica: Universidad Nacional.

Irby, L. R. and C. Vaughan. 1996. Deer habitat in western Costa Rica: Impacts of changing technology and land use. *Wildlife Society Bulletin* 24:660–666.

Janzen, D. H. 1983. *Odocoileus virginianus* (venado, venado cola blanca, white-tailed deer). In *Costa Rican Natural History*, ed. D. H. Janzen, 481–483. Chicago, IL: University of Chicago Press.

Janzen, D. H. 1986. El venado cola blanca del bosque seco tropical, In *Actas del Primer Taller Nacional Sobre el Venado Cola Blanca (Odocoileus virginianus) del Pacífico Seco, Costa Rica*, eds. V. Solís, M. Rodríguez, and C. Vaughan, 9–13. Heredia, Costa Rica: Universidad Nacional.

Kellogg, R. 1956. What and where are the whitetails?. In *The Deer of North America*, ed. W. P. Taylor, 31–55. Harrisburg, PA: Stackpole Press.

Kobelkowsky-Sosa, R. and J. Palacio-Núñez. 2002. Evaluación del hábitat y estado poblacional del venado cola blanca (*Odocoileus virginianus*, Hays) en ranchos cinegéticos de la Sierra Fría, Aguascalientes. *VIII Simposio sobre Venados en México*. Universidad Nacional Autonoma de Mexico, Universidad Autónoma de Tlaxcala y ANGADI.

Krausman, P. R., D. J. Schmidly, and E. D. Ables. 1978. Comments on the taxonomic status, distribution, and habitat of the Carmen Mountains white-tailed deer (*Odocoileus virginianus carminis*) in Trans-Pecos Texas. *Southwestern Naturalist* 23:577–590.

Logan-López, K., E. Cienfuegos-Rivas, F. C. Sánchez, G. Mendoza, A. M. Sifuentes, and L. A. Taranfo. 2006. Caracterización morfométrica de cuatro subespecies de venado cola blanca (*Odocoileus virginianus*) en la zona noreste de México. *Revista Científica* 1:14–22.

Logan-López, K., E. Cienfuegos-Rivas, A. M. Sifuentes et al. 2007. Patrones de variación genética en cuatro subespecies de venado cola blanca del noreste de México. *Agrociencia* 41:13–21.

López-Arévalo, H. F. and A. González-Hernández. 2006. Venado sabanero. *Odocoileus virginianus*. In *Libro Rojo de Los Mamíferos de Colombia*, eds. J. V. Rodríguez-M, M. Alberico, F. Trujillo, and J. Jorgenson, 114–121. Bogota, Colombia: Conservación Internacional Colombia, Ministerio de Ambiente, Vivienda y Desarrollo Territorial.

López-Arévalo, H. F. and A. L. Morales-Jimenez. 2004. Conservación de los mamíferos colombianos. In *Mamíferos Terrestres y Voladores de Colombia,* eds. A. L. Morales-Jimenez, F. Sánchez, K. Poveda, and A. Cadena, 24–34. Bogota, Colombia: Ramos López Editorial.

López-Pizarro, E. 1980. Aspectos etológicos del venado cola blanca [*Odocoileus virginianus*] de Costa Rica. *Desarrollo Agropecuario e Industrial* 7:113–114.

López-Pizarro, E. 1985. Análisis de población del venado cola blanca (*Odocoileus virginianus*) en la Isla San Lucas, Golfo de Nicoya, Costa Rica. Convenio MAG-UNA. In *Investigaciones Sobre Fauna Silvestre de Costa Rica*, 47–50. San José, Costa Rica: Editorial Universidad Estatal a Distancia.

López-Téllez, M. C., M. E. Hernández, D. Jiménez, and G. Yañez. 2008. Diagnostico del manejo extensivo de venado cola blanca (*Odocoileus virginianus*) en las UMA de la Mixteca Poblana. *XI Simposio Sobre Venados en México*. Facultad de Medicina Veterinaria y Zootecnia. UNAM, México, D.F.

López-Tellez, M. C., S. Mandujano, and G. Yañez. 2007. Densidad poblacional y características del hábitat del venado cola blanca (*Odocoileus virginianus mexicanus*) en un bosque tropical seco de Puebla. *Acta Zoológica Mexicana (nueva serie)* 22:45–56.

Mandujano, S. 1994. Método para evaluar el hábitat del venado cola blanca en un bosque de coníferas. In *Ecología y Manejo del Venado Cola Blanca en México y Costa Rica*, eds. C. Vaughan and M.A. Rodríguez, 283–297. Heredia, Costa Rica: Universidad Nacional. Editorial UNA.

Mandujano, S. 2004. Análisis bibliográfico de los estudios de venados en México. *Acta Zoológica Mexicana (Nueva Serie)* 20:211–251.

Mandujano, S. 2007. Carrying capacity and potential production of ungulates for human use in a Mexican tropical dry forest. *Biotropica* 39:19–524.

Mandujano, S., C. A. Delfin-Alfonso, and S. Gallina. 2010. Comparison of geographic distribution models of white-tailed deer *Odocoileus virginianus* (Zimmermann, 1780) subspecies in Mexico: Biological and management implications. *Therya* 1: 41–68.

Mandujano, S. and S. Gallina. 1995. Comparison of deer censusing methods in a tropical dry forest. *Wildlife Society Bulletin* 23:180–186.

Mandujano, S. and A. González-Zamora. 2009. Evaluation of natural conservation areas and wildlife management units to support minimum viable populations of white-tailed deer in Mexico. *Tropical Conservation Science* 2:237–250.

Mandujano, S. and E. J. Naranjo. 2010. Ungulate biomass across a rainfall gradient: A comparison of Neotropical and Paleotropical forests and local analysis in Mexico. *Journal of Tropical Ecology* 26:13–23.

Mandujano, S. and V. Rico-Gray. 1991. Hunting, use, and knowledge of the biology of the white-tailed deer, *Odocoileus virginianus* (Hays), by the Maya of central Yucatan, Mexico. *Journal of Ethnobiology* 11:175–183.

Martinez, A., A. Salinas, F. Martinez, A. Cantu, and D. K. Miller. 1999. Serosurvey for selected disease agents in white-tailed deer from Mexico. *Journal of Wildlife Diseases* 35:799–803.

Mas, J. F., A. Velázquez, J. Reyes et al. 2004. Assessing land use/cover changes: A nationwide multidate spatial database for Mexico. *International Journal of Applied Earth Observation and Geoinformation* 5:249–261.

Matallana-Tobón, C. L. 2001. Propuestas de corredores biológicos entre el Parque Nacional Natural Chingaza y el Parque Nacional Natural Sumapaz (Cordillera Oriental, Colombia). BS thesis, Pontificia Universidad Javeriana, Bogotá.

Mateus, G. C. 2006. Informe final del estudio biológico que determine la viabilidad de la reintroducción del venado cola blanca, *Odocoileus virginianus*, en el Parque Nacional Natural Los Nevados. Informe final del contrato de prestación de servicios 116/2005 Unidad especial de Parques Nacionales Naturales.

Mateus-Gutiérrez, C. 2005. Evaluación preliminar de la dieta y monitoreo del movimiento del venado cola blanca (Odocoileus virginianus) en semicautiverio en un bosque seco tropical (Cundinamarca, Colombia). BS thesis, Universidad Nacional de Colombia, Bogotá, Colombia.

McCoy, M. and C. Vaughan. 1985. *Resultados Preliminares del Estudio del Venado Cola Blanca (Odocoileus virginianus) en Costa Rica*. Convenio MAG-UNA. Investigaciones sobre fauna silvestre de Costa Rica. Editorial Universidad Estatal a Distancia. San José, Costa Rica.

McNab, W. H. and P. E. Avers (compilers). 1994. Ecological subregions of the United States: Section descriptions. WO-WSA-5. USDA Forest Service, Ecosystem Management, Washington, DC.

Mearns, E. A. 1898. Description of a new deer (*Dorcelaphus texanus*) from Texas and northern Mexico. *Proceedings of the Biological Society of Washington* 12:23–26.

Medellín L., R. A. 2005. Vertebrados superiores exóticos en México: Diversidad, distribución y efectos potenciales. Instituto de Ecología, Universidad Nacional Autónoma de México. Comisión Nacional para el conocimiento y Uso de la Biodiversidad. Proyecto U020. México City, México.

Medina, F., J. A. 1986. Programa de conservación y aprovechamiento cinegético del venado cola blanca en la "Sierra Fría", Aguascalientes. Memorias del Primer Simposio sobre el Venado en México. Universidad Nacional Autónoma de México. México, D.F.

Mena, R. A. 1978. *Fauna y Caza en Costa Rica*. San José, Costa Rica: Litografía e Imprenta LIL, S.A.

Méndez, E. 1984. White-tailed deer populations and habitats of Mexico and Central America. In *White-tailed Deer Ecology and Management*, ed. L. K. Halls, 513–524. Harrisburg, PA: Stackpole Books.

Mendoza, J., A. López, V. Tapia and G. Ramos. 2006. Rescate, conservación, manejo y aprovechamiento sustentable del venado cola blanca, *Odocoileus virginianus acapulcensis*, en la sierra-costa de Michoacán, una experiencia de la UCCOFF A.C. X Simposio sobre Venados en México. Universidad Nacional Autónoma de México, México, D.F.

Merriam, C. H. 1898. The earliest generic name from the North American deer, with descriptions of five new species and subspecies. *Proceedings of the Biological Society of Washington* 12:99–104.

Molina, M. 2004. Conocimiento de la biología del venado de páramo (Mammalia, Cervidae, *Odocoileus*) por los campesinos de los Andes de Mérida. *Boletín Antropológico* 61:269–285.

Molina-Molina, M. 1996. Revisión taxonómica de los *Odocoileus* (Mammalia, Artiodactyla, Cervidae) de Venezuela, con aportes a la historia natural y conservación del venado de páramo. BS thesis, Universidad de Los Andes. Mérida, Venezuela.

Molina-Molina, M. 2000. Bases ecológicas y plan de manejo para la conservación del venado de páramo (*Odocoileus lasiotis* Osgood 1914). MS thesis, Universidad Nacional Experimental de los Llanos Ezequiel Zamora. Guanare, Portuguesa, Venezuela.

Molina-Molina, M. 2003a. Estimaciones de densidad poblacional del venado de páramo *Odocoileus lasiotis* Osgood 1914 (Mammalia, Cervidae) en la Sierra Nevada de Mérida, Venezuela. *Revista de Ecología Latinoamericana* 10:1–10.

Molina-Molina, M. 2003b. Uso de hábitat por el venado de páramo *Odocoileus lasiotis* Osgood, 1914 (Mammalia, Cervidae) en el Parque Nacional Sierra Nevada, Venezuela. *Revista de Ecología Latinoamericana* 10:1–10.

Molina-Molina, M. and Arias, J. H. 1998. Población y uso del hábitat del venado de páramo *Odocoileus lasiotis* (Artiodactyla: Cervidae) en Venezuela. *Revista de Biología Tropical* 46:817–820.

Molina-Molina, M. and J. Molinari. 1999. Taxonomy of Venezuelan white-tailed deer (*Odocoileus*, Cervidae, Mammalia), based on cranial and mandibular traits. *Canadian Journal of Zoology* 77:632–645.

Molinari, J. 2007. Variacion geografica en los venados de cola blanca (Cervidae, *Odocoileus*) de Venezueala, con enfasis en *O. margaritae*, la especie de la Isla de Margarita. *Memoria de la Fundacion La Salle de Ciencias Naturales* 167:29–72.

Montes, R. C., R. I. Rodríguez, and J. F. Torres, 1998. Seguimiento anual de la parasitosis gastrointestinal de venados cola blanca *Odocoileus virginianus* (Artiodactyla: Cervidae) en cautiverio en Yucatán, México. *Revista de Biología Tropical* 46:821–827.

Montiel-Ortega, S., L. M. Arias, and F. Dickinson. 1999. La cacería tradicional en el norte de Yucatán: Una práctica comunitaria. *Revista Geografía Agrícola* 29:43–51.

Montoya-Maquín, J. M. 2004. *Vertebrados Terrestres Alóctonos de la Isla del Coco, San José, Costa Rica.* Costa Rica: Fundación Amigos Isla del Coco/Area de Conservación Isla del Coco.

Mora, L. G. 1983. Estudio preliminar acerca del venado cola blanca en cautiverio. BS thesis, Universidad Nacional, Heredia, Costa Rica.

Mora, J. M. 2000. *Mamíferos Silvestres de Costa Rica.* San José, Costa Rica: Editorial Universidad Estatal a Distancia.

Mora, J. M. and I. Moreira. 1984. *Mamíferos de Costa Rica.* San José, Costa Rica: Editorial Universidad Estatal a Distancia.

Mora, C. A. and S. E. Mosquera. 2000. Estudio preliminar del comportamiento alimenticio del venado cola blanca (*Odocoileus virginianus goudotii*) en el ecosistema de subpáramo y páramo del Parque Nacional Natural Chingaza en Cundinamarca-Meta, Colombia. BS thesis, Universidad Nacional de Colombia, Bogotá, Colombia.

Moscarella, R. A., M. Aguilera, and A. A. Escalante. 2003. Phylogeography, population structure, and implications for conservation of white-tailed deer (*Odocoileus virginianus*) in Venezuela. *Journal of Mammalogy* 84:1300–1315.

Murphy, P. and A. E. Lugo. 1995. Dry forest of Central America and the Caribbean. In *Seasonally Dry Tropical Forest*, eds. S. H. Bullock, H. A. Mooney, and E. Medina, 146–194. Cambridge: Cambridge University Press.

Murphy, P. G. and A. E. Lugo. 1986. Ecology of tropical dry forests. *Annual Review of Ecology and Systematics* 17:67–88.

Naranjo, E. J., J. E. Bolaños, M. M. Guerra, and R. E. Bodmer. 2004a. Hunting sustainability of ungulate populations in the Lacandon forest, México. In *People in Nature: Wildlife Conservation in South and Central America*, eds. K. M. Silvus, R. E. Bodmer, and J. M. Fragoso, 324–343. New York, NY: Columbia University Press.

Naranjo, E. J., M. M. Guerra, R. E. Bodmer, and J. E. Bolaños. 2004b. Subsistence hunting by three ethnic groups of the Lacandon forest, México. *Journal of Ethnobiology* 24:233–253.

Ortíz-Martínez, T., S. Gallina, M. Briones-Salas and G. González-Pérez. 2005. Densidad poblacional y caracterización del hábitat del venado cola blanca (*Odocoileus virginianus oaxacensis*, Goldman y Kellog, 1940) en un bosque templado de la sierra norte de Oaxaca, México. *Acta Zoológica Mexicana* (n.s.) 21: 65–78.

Pacheco-Figueroa, C. J. 2003. Situación actual del venado cola blanca (*Odocoileus virginianus*) y caracterización socieconómica de los vecinos del Parque Nacional Barra Honda, Costa Rica. MS thesis, Universidad Nacional, Heredia, Costa Rica.

Peña, G. and M. Pinto. 1996. *Mamíferos más comunes en sitios precerámicos de la sabana de Bogotá. Guía ilustrada para arqueólogos.* Santa Fe de Bogota, Colombia: Academia Colombiana de Ciencias Exactas, Físicas y Naturales.

Piedra-Castro, L. and L. Maffei. 2000. Efecto de las actividades humanas sobre la diversidad de mamíferos terrestres en un gradiente altitudinal. *Revista de Biología Tropical* 48:263–264.

Ramia, M. 1962. *Sabanas Llaneras.* Ministerio de Agricultura y Cría. Caracas, Venezuela.

Ramírez-Lozano, R. G. 2004. *Nutrición del Venado Cola Blanca.* Monterrey, Mexico: Unión Ganadera Regional de Nuevo León. Fundación Produce, Nuevo, A. C.

Ramos, D. 1995. Determinación de la dieta y utilización del hábitat del venado cola blanca en el PNN Chingaza, cordillera oriental, Colombia. BS thesis, Pontificia Universidad Javeriana, Bogotá, Colombia.

Retana, O. G. 2006. *Fauna Silvestre de México: Aspectos Históricos de su Gestión y Conservación.* Mexico, D.F.: Fondo de Cultura Económico, México, D.F.

Reyna-Hurtado, R. and G. W. Tanner. 2007. Ungulate relative abundance in hunted and no-hunted sites in Calakmul forest (Southern Mexico). *Biodiversity & Conservation* 16:743–756.

Rincón, L. 2003. La fauna arqueológica del sitio San Carlos, Municipio de Funza, Sabana de Bogotá. Tomo 1 y 2. BS thesis, Universidad Nacional de Colombia, sede Bogotá.

Rippstein, G., E. Amézquita, G. Escobar, and C. Grollier. 2001. Condiciones naturales de la sabana. In *Agroecología y Biodiversidad de las Sabanas en los Llanos Orientales de Colombia*, eds. G. Rippstein, G. Escobar, and F. Motta, 1–22. Cali, Colombia: Centro Internacional de Agricultura tropical.

Rodríguez, M. A. 1989. Tamaño y composición de los grupos sociales del hato de venado cola blanca (*Odocoileus virginianus*) de la Isla San Lucas, Costa Rica. BS thesis, Universidad Nacional, Heredia, Costa Rica.
Rodríguez, M. A. and V. Solís. 1994a. Ciclo de vida del venado cola blanca en la Isla San Lucas, Costa Rica. In *Ecología y Manejo del Venado Cola Blanca en México y Costa Rica*, eds. C. Vaughan and M. A. Rodríguez, 63–71. Heredia, Costa Rica: Universidad Nacional.
Rodríguez, M. A. and V. Solís. 1994b. Ciclo de astas del venado cola blanca en el Refugio Nacional de Vida Silvestre Palo Verde, Guanacaste, Costa Rica. In *Ecología y Manejo del Venado Cola Blanca en México y Costa Rica*, eds. C. Vaughan and M. A. Rodríguez, 103–110. Heredia, Costa Rica: Universidad Nacional.
Rodríguez, M. A. and V. Solís-Rivera. 1988. Venado cola blanca: Los primeros seis meses de vida. Heredia. UNA/MIRENEM. CR. Serie Venado Cola Blanca. No. 2.
Rodríguez, M. A. and C. Vaughan. 1994. Tamaño y composición de los grupos sociales del venado cola blanca en la Isla San Lucas, Costa Rica. In *Ecología y Manejo del Venado Cola Blanca en México y Costa Rica*, eds. C. Vaughan and M. A. Rodríguez, 131–162. Heredia, Costa Rica: Universidad Nacional.
Rodríguez, M. A., C. Vaughan, and M. McCoy. 1985a. Composición elemental de algunos afloramientos minerales utilizados por los ungulados silvestres en Palo Verde. San José, Costa Rica: Editorial Universidad Estatal a Distancia.
Rodríguez, M. A., C. Vaughan, V. Villalobos, and M. McCoy. 1985b. Notas sobre los movimientos del venado cola blanca (*Odocoileus virginianus Rafinesque*) en un bosque tropical seco de Costa Rica. San José, Costa Rica: Editorial Universidad Estatal a Distancia.
Rojo-Curiel, A., J. L. Cruz, G. Solano, and R. Hernández. 2007. Plan de manejo tipo de venado cola blanca en zonas templadas y tropicales en México. DGVS, SEMARNAT, México. D.F.
Romero-Castañón, S., B. G. Ferguson, D. Guiris et al. 2008. Comparative parasitology of wild and domestic ungulates in the Selva Lacandona, Chiapas, Mexico. *Comparative Parasitology* 75:115–126.
Romo, M. and S. Gallina. 1988. Estudio de la población del venado cola blanca (*Odocoileus virginianus*) en la Sierra San Blas de Pabellón del estado de Aguascalientes. II Simposio sobre el Venado en México. Universidad Nacional Autonoma de Mexico, México, D. F.
Sáenz, J. 1990. Ecología de dos grupos de venados cola blanca (*Odocoileus virginianus*) liberados en un nuevo hábitat. BS thesis, Universidad Nacional, Heredia, Costa Rica.
Sáenz, J. and C. Vaughan. 1998. Ambito de hogar y utilización de hábitat de dos grupos de venados cola blanca *Odocoileus virginianus* (Artiodactyla: Cervidae) reubicados en un ambiente tropical. *Revista de Biología Tropical* 46:1185–1197.
Sáenz, J. C. 1994. Reintroducción del venado cola blanca en el noroeste de Costa Rica. In *Ecología y Manejo del Venado Cola Blanca en México y Costa Rica*, eds. C. Vaughan and M. A. Rodríguez, 383–416. Heredia, Costa Rica: Universidad Nacional.
Sáenz, J. C., M. A. Rodríguez, and H. Espach. 1994. Captura y traslado de venados cola blanca para reintroducción. In *Ecología y Manejo del Venado Cola Blanca en México y Costa Rica*, eds. C. Vaughan and M. A. Rodríguez, 417–428. Heredia, Costa Rica: Universidad Nacional.
Sánchez-Rojas, G., C. Aguilar-Miguel, and E. Hernández-Cid. 2009. Estudio poblacional y uso de hábitat por el venado cola blanca (*Odocoileus virginianus*) en un bosque templado de la Sierra de Pachuca, Hidalgo, México. *Tropical Conservation Science* 2:204–214.
Sánchez-Rojas, G., S. Gallina, and S. Mandujano. 1997. Áreas de actividad y uso del hábitat de dos venados cola blanca (*Odocoileus virginianus*) en un bosque tropical de la costa Jalisco, México. *Acta Zoológica Mexicana (nueva serie)* 72:39–54.
SARH. 1992. Inventario forestal nacional de gran visión. Subsecretaría Forestal y de Fauna Silvestre, Secretaría de Agricultura y Recursos Hidráulicos (SARH), México.
SARH. 1994. Inventario nacional forestal periódico. Memoria Nacional. Subsecretaria Forestal y de Fauna Silvestre, Secretaria de Agricultura y Recursos Hidráulicos (SARH), México.
Sarmiento, G. 1983. The savannahs of tropical América. In *Tropical Savannahs*, ed. F. Bourliere, 245–288. Amsterdam: Elsevier.
Segovia, A. and S. Hernández-Betancourt. 2003. La cacería de subsistencia en Tzucacab, Yucatán, México. *Tropical and Subtropical Agroecosystems* 2:49.
Segura-Lopez, W. 1995. Uso de sensores remotos y sistemas de información geográfica en la evaluación del habitat potencial del venado cola blanca (*Odocoileus virginianus*), Bagaces, Guanacaste, Costa Rica. MS thesis, Universidad Nacional, Heredia, Costa Rica.

Segura-López, W. 1998. Application of the HEP methodology and use of GIS to identify priority sites for the management of white-tailed deer. In *GIS Methodologies for Developing Conservation Strategies: Tropical Forest Recovery and Wildlife Management in Costa Rica*, eds. B. G. Savitsky and T. E. Lacher, Jr., 127–137. New York, NY: Columbia University Press.

Segura-Lopez, W. and J. Fallas. 1996. Uso de sistemas de información geográfica para analizar impactos de actividades humanas sobre el hábitat potencial del venado cola blanca (*Odocoileus virginianus*), Guanacaste, Costa Rica. *Revista Geográfica de América Central* 32/33:123–129.

SEMARNAP. 1997. Secretaría de medio ambiente, recursos naturales y pesca. Programa de Conservación de la Vida Silvestre y Diversificación Productiva en el Sector Rural (México 1997–2000). Instituto Nacional de Ecología de la SEMARNAP. México, D.F.

Smith, W. P. 1991. *Odocoileus virginianus*. *Mammalian Species* 388:1–13.

Solís, V. 1986. Ecology and behavior of white-tailed deer fawns on San Lucas Island, Costa Rica. MS thesis, University of Kansas.

Solís, V. and S. Brenes. 1987. El uso sostenido del venado cola blanca en Costa Rica: Un recurso natural renovable. *Biocenosis*. 3:63–65.

Solís, V., M. Rodríguez, and C. Vaughan. 1986. Actas del primer taller nacional sobre el venado cola blanca (Odocoileus virginianus) del Pacífico Seco, Costa Rica. Heredia, Costa Rica: Universidad Nacional.

Solís-Rivera, V. 1994. Uso tradicional y conservación del venado cola blanca en Costa Rica. In *Ecología y Manejo del Venado Cola Blanca en México y Costa Rica*, eds. C. Vaughan and M. A. Rodríguez, 351–357. Heredia, Costa Rica: Universidad Nacional.

Solís-Rivera, V. and C. Vaughan. 1994. Plan de manejo del venado cola blanca en la Isla San Lucas. In *Ecología y Manejo del Venado Cola Blanca en México y Costa Rica*, eds. C. Vaughan and M. A. Rodríguez, 429–444. Heredia, Costa Rica: Universidad Nacional.

Stiles, F. G. and A. F. Skutch. 1998. *Aves de Costa Rica*. San José, Costa Rica: Editorial Instituto Nacional de Biodiversidad.

Sturm, H. and O. Rangel. 1985. *Ecología de los Páramos Andinos: Una Visión Preliminar Integrada*. Bogotá, Colombia: Instituto de Ciencias Naturales, Museo de Historia Natural.

Taylor, W. P. 1956. *The Deer of North America*. Harrisburg, PA: Stackpole Company.

Teer, J. 1987. Aspectos socioeconómicos del venado cola blanca en Texas. In *Actas del Primer Taller Sobre el Venado Cola Blanca (Odocoileus virginianus) del Pacífico Seco*, eds. V. Solís, M. Rodríguez, and C. Vaughan, 19–22. Heredia, Costa Rica: Universidad Nacional.

Teer, J. 1994a. El venado cola blanca: Historia y principios de manejo. In *Ecología y Manejo del Venado Cola Blanca en México y Costa Rica*, eds. C. Vaughan and M. A. Rodríguez, 33–47. Heredia, Costa Rica: Universidad Nacional.

Teer, J. 1994b. Prefacio. In *Ecología y Manejo del Venado Cola Blanca en México y Costa Rica*, eds. C. Vaughan and M. A. Rodríguez, 17–20. Heredia, Costa Rica: Universidad Nacional.

Turner, R. M., J. E. Bowers, and T. L. Burgess. 1995. *Sonoran Desert Plants: An Ecological Atlas*. Tucson, AZ: University of Arizona Press.

UICN. 2001. *Categorias y Criterios de la Lista Roja de la UICN*. Version 3.1. Oxford, Reino Unido: Information Press.

Universidad de los Andes. 1999. Evaluación ecológica rápida del Parque Nacional Sierra Nevada. Mérida. Unidad de Prestacion de Servicios y Proyectos Forestales, Geograficos, Agropecuarios y Ambientales - Technical Report. Merida, Venezuela.

University of Texas at El Paso. 2009. Chihuahuan Desert Home Page. http://museum.utep.edu/chih/chihdes.htm. Accessed July 7, 2009.

Valenzuela, D. 1994. Estimación de la densidad y distribución de la población del venado cola blanca en el bosque La Primavera, Jalisco, México. In *Ecología y Manejo del Venado Cola Blanca en México y Costa Rica*, eds. C. Vaughan and M. A. Rodríguez, 247–262. Heredia, Costa Rica: Universidad Nacional.

Vareschi, W. 1992. *Ecología de la Vegetación Tropical*. Caracas, Venezuela: Sociedad Venezolana de Ciencias Naturales.

Vaughan, C. 1991. White-tailed deer management in Costa Rica. In *Neotropical Wildlife Use and Conservation*, eds. J. G. Robinson and K. H. Redford, 288–299. Chicago, IL: University of Chicago Press.

Vaughan, C., 1994. Prólogo. In *Ecología y Manejo del Venado Cola Blanca en México y Costa Rica*, eds. C. Vaughan and M. A. Rodríguez, 21–23. Heredia, Costa Rica: Universidad Nacional.

Vaughan, C., K. Fisher, M. A. Rodríguez, P. Johns, and M. Smith. 1994. Variabilidad genética en la población de venado cola Blanca en la Isla San Lucas, Costa Rica, y las implicaciones para su manejo. In *Ecología y Manejo del Venado Cola Blanca en México y Costa Rica*, eds. C. Vaughan and M. A. Rodríguez, 91–103. Heredia, Costa Rica: Universidad Nacional.

Vaughan, C. and M. A. Rodríguez. 1994. Introducción. Conservación del venado cola blanca en el Neotrópico. In *Ecología y Manejo del Venado Cola Blanca en México y Costa Rica*, eds. C. Vaughan and M. A. Rodríguez, 25–30. Heredia, Costa Rica: Universidad Nacional.

Villalobos, V. 1998. El venado cola blanca en la Sierra Fría, Aguascalientes. Cuaderno de Trabajo. *Agricultura y Recursos Naturales* 89:1–33.

Villarreal, J. 1999. *Venado Cola Blanca: Manejo y Aprovechamiento Cinegético*. Monterrey, N. L., México: Unión Ganadera Regional de Nuevo León.

Villarreal, J. 2009. Propone el CEFFSNL regionalización cinegética de venados cola blanca mexicanos para libro récords internacionales del SCI. *Boletín Cuenca Palo Blanco* 4:3–7.

Villarreal G., J. G. 1995. Estimación de las poblaciones silvestres de venado cola blanca Texano *Odocoileus virginianus texanus* del noreste de México. *Memorias del XIII Simposio sobre Fauna Silvestre*. Universidad Nacional Autonoma de México. México, D.F.

Villarreal G., J. G. 1998. Evaluación de la transferencia con fines de repoblación de venado cola blanca Texano *Odocoileus virginianus texanus* en un matorral xerófilo de Cerralvo, Nuevo León. *Memorias del VI Simposio sobre Venados de México*. Universidad Nacional Autonoma de México. México, D.F.

Villarreal G., J. G., and G. López, F. 1987. Historia del desarrollo de un rancho cinegético. Revista DUMAC, Volumen IX, Número 5. Monterrey, Nuevo León, México.

Villarreal G., J. G., and J. G. Rodríguez, S. 1998. Estimación de la densidad y composición de la población de venado cola blanca "del Carmen" *Odocoileus virginianus carminis* en un predio de las Serranías de San Buenaventura, Coahuila. *Memorias del XVI Simposio sobre Fauna Silvestre*. Universidad Nacional Autonoma de México. México, D.F.

Villarreal-Espino, O. 2006. El venado cola blanca en la mixteca Poblana: Conceptos y métodos para su conservación y manejo. Puebla, Pue., Mexico: Fundación Produce Puebla A. C.

Villarreal-Espino Barros, O. A., R. V. Guevara Viera, R. Reséndiz Martínez, J. C. Castillo Correo, and F. J. Tomé Torres. 2004. Diversificación productiva por medio del venado cola blanca (*Odocoileus virginianus veraecrucis*), en el Campo Experimental "Las Margaritas", México. *Deer Specialist Group Newsletter* 19.

Weber, M. 2006. Ecología y conservación de tres especies simpátricas de venados en Calakmul, Campeche. *X Simposio sobre Venados en México*. Universidad Nacional Autonoma de México. México, D.F.

Weber, M., G. García-Marmolejo, and R. Reyna-Hurtado. 2006. The tragedy of the commons: Wildlife management units (UMAs) in southeastern Mexico. *Wildlife Society Buletin* 34:1480–1488.

World Wildlife Foundation. 2009. http://www.worldwildlife.org/wildworld/profilesterrestrial/na/na1310_full.html. Accessed 7 July 2009.

Yañez, C. 2009. Distribución y densidad poblacional del venado cola blanca en el Bajo Balsas, Michoacán. MS thesis, Instituto de Ecología A. C., Xalapa, Ver. México.

Zavala, G. 1992. Estimación poblacional del venado cola blanca (*Odocoileus virginianus*) en la Estación Científica Las Joyas, Reserva de la Biosfera Sierra de Manantlán, Jalisco. BS thesis, Universidad de Guadalajara. Guadalajara, Mexico.

Zúñiga-Sánchez, J. S., F. M. Contreras-Moreno, and J. Bello. 2008. Densidad poblacional y preferencia de hábitat del venado cola blanca (*Odocoileus virginianus thomasi*), en el Municipio de Tenosique, Tabasco. XI Simposio Sobre Venados en México. Universidad Nacional Autonoma de Mexico. México, D.F.

20

Managing White-tailed Deer: Exurban, Suburban, and Urban Environments

Jacob L. Bowman

CONTENTS

Introduction

Overabundant deer populations have recently gained national attention in exurban, suburban, and urban landscapes (DeNicola et al., 2000) (Figure 20.1). Development patterns have provided a habitat mosaic

FIGURE 20.1 A deer standing near a clear browse line at Manassas National Battlefield, a suburban park. (Photo by E. Tymkiw. With permission.)

ideal for white-tailed deer because undeveloped land interspersed among developed areas provides deer refuges. Deer management in these landscapes provides new challenges for wildlife agencies, which have typically focused on wildlife management in rural landscapes. These human dominated landscapes provide not only new challenges from a biological standpoint, but also from a human dimension perspective. Attitudes in these landscapes are typically highly polarized (Porter, 1997; Swihart and DeNicola, 1997) so deer management in these landscapes has focused on the political and sociological aspects (Green et al., 1997; Kilpatrick et al., 1997; Messmer et al., 1997; Stout et al., 1997). Porter (1997) suggested that a greater emphasis on deer ecology was necessary to not only manage these deer populations but also to gain public approval. Management techniques to reduce deer–human conflicts in these landscapes are often based on limited ecological knowledge or knowledge gained from the rural landscape; because a paucity of information is available for white-tailed deer ecology in developed landscapes, additional research is needed in these areas.

Description of the Landscape

In 1950, 1% of the area of the United States was suburban or urban and the number had doubled by 2000 (Brown et al., 2005). In contrast, exurbia occupied only 5% in 1950 but increased fivefold to 25% in 2000 (Brown et al., 2005). Although these landscapes represented only 27% of the area of the United States in 2000, 81% of the human population lived in exurban, suburban, and urban landscapes (Brown et al., 2005). The high human density in these landscapes provides unique challenges for deer managers, who are more familiar with deer management in rural settings. Developed landscapes are a patchwork of suitable and unsuitable habitat with the degree of suitability decreasing and habitat patches becoming more isolated along a gradient from rural through exurban, suburban, and urban landscapes. These human-dominated landscapes are characterized by developed areas surrounding natural areas, which are typically in some form of land preservation. Many natural areas remaining in these landscapes were devoid of deer until the 1970s or later, so residents adjacent to these areas may only have recent experience with white-tailed

FIGURE 20.2 A common deer–human interaction is browsing on landscaping plants like these heavily browsed arborvitae. (Photo by D. Tallamy. With permission.)

deer and little or no experience with deer harvest. Additionally, because of the lack of hunter harvest, deer populations in these areas expanded quickly and are considered overabundant in many places (Figure 20.2). The primary challenges in these landscapes are that the density of human settlement often precludes the use of normal hunting seasons to control deer abundance and human support for harvesting deer has been low.

Exurban

The exurban landscape is defined by having about 6–25 homes/km^2 and includes development on the periphery of cities and rural developments interspersed with agriculture (Hansen et al., 2005). Exurbia sits on the gradient from the suburban landscape to the rural landscape (Theobald, 2001; Brown et al., 2005; Hansen et al., 2005; Radeloff et al., 2005). An important aspect of exburia from a wildlife conservation perspective is that habitat patches remain connected unlike the disjunct spatial distribution of deer habitat in urban and suburban landscapes (Odell and Knight, 2001). Instead of deer habitat being isolated in these landscapes, it often surrounds developed areas or the development is at such low densities that

FIGURE 20.3 Example of an exurban landscape with low-density development within deer habitat.

much of the area is deer habitat (Figure 20.3). Exurban landscapes will provide new challenges to wildlife managers because the density of homes may reduce or preclude the use of harvest from controlling white-tailed deer abundance (Harden et al., 2005; Storm et al., 2007a).

Suburban

The suburban landscape is defined by having 25–250 homes/km^2 (Brown et al., 2005; Hansen et al., 2005). This landscape is characterized by subdivisions with houses on lots varying in size from 0.1 to 2 ha and these subdivisions typically include protected areas of deer habitat often along riparian corridors (Odell and Knight, 2001). Although habitat patches are not as connected as in exurban landscapes, connectivity is sufficient to allow relatively high use of the landscape by deer (Figure 20.4). In many areas, deer use protected areas as refuges and forage into suburban developments, which often provide high-quality foraging opportunities without risk of predation or harvest. Human experience with deer in these landscapes is highly variable but is generally characterized by a lack of knowledge of basic deer ecology and management techniques. Most residents have little or no experience with harvest and have been opposed to harvest as a management option until relatively recently. Use of traditional harvest in these landscapes is less likely than in rural areas because of the density of human development but harvest is typically still feasible on large public land areas and remnant agricultural lands.

Urban

The urban landscape is defined by having more than 250 homes/km^2 (Brown et al., 2005; Hansen et al., 2005) (Figure 20.5). This landscape is characterized by high-density development with deer habitat limited to protected areas (e.g., parks) or larger industrial complexes with woodlots or other undeveloped habitat patches. Habitat patches have little connectivity and deer populations are largely sedentary in these patches. Human experience with deer in these landscapes is typically very low and few people have any knowledge of basic deer ecology and management techniques. Most residents have no experience with harvest and are opposed to harvest as a management option. Density of human development typically precludes use of traditional harvest in these landscapes but controlled hunts are effectively used in many areas.

FIGURE 20.4 Example of a suburban landscape with subdivision interspersed among deer habitat.

FIGURE 20.5 Example of an urban landscape with high-density development and little deer habitat.

Ecology

Spatial Ecology

Home Range

Home range size is dictated by suitable habitat availability (Rhoads et al., 2010). In exurban, suburban, and urban landscapes the extent and distribution of development and the degree of habitat fragmentation likely affect home range size. Annual home range sizes in suburban and urban landscapes varied from 43 ha in Connecticut to 62 ha in Illinois (Kilpatrick and Sphor, 2000a; Etter et al., 2002). These estimates are nearly a quarter of the size (192 ha) reported in an exurban landscape in Maryland (Rhoads et al., 2010). Seasonal home range size in exurban, suburban, and urban landscapes followed the same trend documented in rural landscapes with home ranges during fawning being the smallest and increasing in size into autumn (Kilpatrick and Sphor, 2000b; Etter et al., 2002; Storm et al., 2007b; Rhoads et al., 2010). The degree of variation among seasonal home ranges in suburban and urban landscapes has varied from none (Kilpatrick and Sphor, 2000b) to extensive (Etter et al., 2002; Grund et al., 2002), whereas both studies in exurban landscapes documented extensive seasonal variation (Storm et al., 2007b; Rhoads et al., 2010). Although Kilpatrick et al. (2001) suggested a positive relationship between home range size and deer density, home range sizes may be more constrained by fragmentation and/or habitat quality. Annual home range size tends to decrease with increasing levels of development. Although home range size varies across different landscapes, home range fidelity was high in all landscapes (Grund et al., 2002; Rhoads et al., 2010). This high degree of home range fidelity supports the idea that localized management can reduce deer–human conflicts on the local scale (McNulty et al., 1997).

Habitat Use

Research on spatial and temporal use of exurban, suburban, and urban landscapes has been more consistent than that on home range size. A general shift toward use of residential areas occurs in exurban and suburban landscapes in contrast to avoidance of these areas in more rural landscapes. Female deer in

Minnesota used park shrubland more than its availability from spring through autumn, whereas females used residential wooded areas more than its availability during winter (Grund et al., 2002). Similarly, a shift in core use areas toward residential areas during winter in Connecticut was hypothesized to result from greater food availability in residential areas (Kilpatrick and Spohr, 2000b). Storm et al. (2007b) documented deer avoiding areas with human dwellings during the fawning season but increased use of these areas during winter. Temporal and spatial dynamics of deer movements could affect success of management activities because deer use of residential areas may reduce the effectiveness of lethal control measures on adjacent undeveloped areas (Kilpatrick and Spohr, 2000a).

Movement Rates and Patterns

White-tailed deer are typically considered crepuscular and nocturnal. Grund (1998) documented seasonal variation in movement patterns in a suburban landscape in which most movement occurred at dawn during winter, at dawn and dusk during spring, and during the middle of the night in summer and autumn. In contrast, Rhoads et al. (2010) documented a consistent pattern for movement rates in an exurban landscape among seasons. Deer always had the greatest movement rates at dusk with the next greatest rates during night and at dawn, whereas deer moved the least during the day (Rhoads et al., 2010). Human activity may alter deer movement patterns. Deer in an exurban landscape shifted their movements to be more nocturnal on Friday to Sunday when human activity was greatest (J. L. Bowman, unpublished data). Additional research is needed on movement patterns to aid in developing techniques for controlling overabundant populations.

Dispersal

A paucity of data exists for dispersal in exurban, suburban, and urban landscapes. In a study of factors that affect dispersal, Long et al. (2005) documented that deer density did not influence dispersal rate or dispersal distance. Forest cover did not influence dispersal rate but did negatively influence dispersal distance (Long et al., 2005). Because the amount of forest cover decreases on a gradient from exurban, through suburban, to urban, dispersal distances would be expected to increase with increasing development; however, dispersal distances were less than 9 km for all but one deer (34 km) in a suburban/urban landscape in Illinois (Etter et al., 2002), similar to dispersal distances in forested environments (Long et al., 2005). Additional research on dispersal is required in these human-dominated landscapes to elucidate the trend in dispersal distances related to development. Biologists need to understand how quickly and from what distance recolonization (i.e., dispersal) of habitat patches can occur to better aid in development of localized management regimes to reduce deer–human conflicts.

Population Ecology

Population Abundance/Density

Deer are generally considered overabundant in most suburban and urban areas but abundance is more variable in exurban landscapes. Hansen and Beringer (1997) recommended a deer density of 15 deer/km^2 to provide for viewing opportunities while balancing the needs of the ecosystem. Tymkiw (2010) recommended a deer density less than 20 deer/km^2 to conserve avian abundance and richness. Deer densities in suburban and urban landscapes are typically two to five times (35–91 deer/km^2) this recommended density (Kilpatrick et al., 2001; Bowman, 2006). Ebersole et al. (2007) estimated 48 deer/km^2 in an exurban area in Maryland. Deer densities in exurbia appear to be highly variable and a function of harvest pressure.

The sex and age ratio of the deer population in human dominated landscapes is more related to harvest pressure than human development. In an unhunted suburban and urban landscape, the adult population was composed of 66% females and 33% males (DeNicola et al., 2008). In an exurban landscape subjected to harvest, the adult population was 84% female and 16% male (Ebersole et al., 2007). The unhunted population was composed of 40% fawns, whereas fawns accounted for 45% of the hunted population (Ebersole et al., 2007; DeNicola et al., 2008).

Reproductive Ecology

The reproductive ecology of deer in exurban, suburban, and urban landscapes should be similar to rural landscapes, after accounting for the effect of high deer density on production. Ebersole et al. (2007), in an exurban landscape, estimated 1.76, 1.44, 0.06 fetuses/female for adults, yearlings, and fawns, respectively, values similar to those for more rural landscapes. A negative relationship of fetuses/female and population abundance has been observed in Maryland, but the reduction in fetuses was limited to yearlings and fawns (J. L. Bowman, unpublished data). DeNicola et al. (2008), in a suburban landscape, documented 1 fawn/doe at recruitment age, whereas Ebersole et al. (2007) documented a slightly lower value (0.88 fawns/doe). Additional research is needed to understand better the relationship between fawn survival and deer abundance.

Survival/Cause-Specific Mortality

Adult

Adult survival rates in exurban, suburban, and urban landscape are often greater than rural landscapes because of the lack of or reduced hunter harvest. Etter et al. (2002), in a suburban/urban landscape, estimated 82% and 83% adult survival for males and females, respectively, and similar survival rates among seasons. Vehicle collision accounted for 90–100% of the documented mortalities, whereas harvest accounted for less than 5% of documented mortalities (Etter et al., 2002). Storm et al. (2007b), in an exurban landscape, estimated an annual survival of 87% for females. In contrast to Etter et al. (2002), vehicle mortality accounted for only 15%, whereas harvest was much more important and accounted for 85% of mortalities (Storm et al., 2007b). Ebersole et al. (2007), in another exurban area, documented a much lower annual survival rate (66%) but harvest mortality was much more important. The proportion of mortality causes (14% vehicle and 86% harvest) was similar to Storm et al. (2007b) but mortality rates were triple those of Storm et al. (2007b). In the Florida Keys, adult survival rates for males and females ranged from 41% to 84% and 71% to 89%, respectively (Lopez et al., 2003). Vehicle mortality accounted for 74% of all mortalities (Lopez et al., 2003). The gradient of increasing vehicle mortality as human development increases is not surprising but is of great concern given that average deer–vehicle collisions costs are greater than $3000/accident (Bissonette et al., 2008) and human injuries are not uncommon. The decrease in harvest mortality as human development increases is a function of safety concerns related to harvest. To more effectively manage deer in these landscapes, additional research is needed to understand better the survival rates and causes of mortality.

Fawn

A paucity of information exists for fawn survival in exurban, suburban, and urban landscapes. Etter et al. (2002) documented an 83% ($n = 6$) survival rate to six weeks of age. The only documented mortality was from a domestic dog (Etter et al., 2002). Prevalence of dogs in these landscapes could be an important cause of mortality compared to rural landscapes. Saalfeld and Ditchkoff (2007) estimated a 33% survival rate to eight weeks of age in an exurban landscape. Predation by coyotes or suspected predation accounted for 62% of the mortalities and emaciation accounted for 25% (Saalfeld and Ditchkoff, 2007). The survival rate observed by Etter et al. (2002) in a suburban/urban landscape was well above the average rate for rural environments, whereas the survival rate observed by Saalfeld and Ditchkoff (2007) in an exurban landscape was well below the average for rural environments. With so few studies conducted in these landscapes, the true average and range of survival rates for these landscapes is unknown. Additional research is needed to provide better estimates of fawn survival in these landscapes.

Evaluation of Deer Management Options

Deer–human conflicts can be alleviated by altering deer behavior (i.e., repellents and fencing) or by reducing the deer population. Public opinion typically favors nonlethal methods such as altering deer behavior or fertility control. Techniques that alter deer behavior are typically only effective at low or moderate deer densities and do not address the actual problem of an overabundant deer herd. Population reduction has been documented to reduce deer–human conflicts (DeNicola et al., 2000; Henderson et al., 2000) and the most common deer management techniques discussed are lethal control (e.g., controlled hunts or sharpshooting), translocation, and fertility control (DeNicola et al., 1997; Coffey and Johnston, 1997; McShea and Rappole, 1997b; Messmer et al., 1997, Muller et al., 1997; Stout et al., 1997; DeNicola et al., 2000; Warren, 2000).

Human Dimension

The human dimension of deer management is particularly important in exurban, suburban, and urban landscapes. For many years, deer management was based on deer ecology and hunter attitudes (i.e., desires), with attitudes of the general public generally ambivalent and having little impact on deer management. Early efforts to control deer populations in these human dominated landscapes were often met with resistance from some residents who were unfamiliar with deer management techniques. An effort by animal rights organizations in the 1980s and 1990s was effective at spreading propaganda and delaying deer management efforts. These organizations have found less sympathy from the public as information about the detrimental effects of deer became more publicized and the public gained more personal experience with deer (e.g., human conflicts or overabundant deer). Acceptance capacity is affected by experience with deer, such that individuals with more negative experiences (i.e., deer–vehicle collisions) are more likely to support lethal control (Lischka et al., 2008). Support for lethal control also increases as experience with controlled hunts increases (Norman and Bowman, 2003).

Understanding attitudes toward deer and the origin of those attitudes is important for an effective deer management program in exurban, suburban, and urban landscapes. One major problem with public opinion surveys is misunderstanding or misrepresenting the results of these surveys. Storm et al. (2007a) documented 82% of landowners experienced damage to vegetation plantings but only 11% considered the damage intolerable. Based on Storm et al. (2007a) some might suggest that deer are a major problem, whereas others could say deer are not a problem because only 11% of respondents had intolerable damage. Surveys of attitudes should be carefully designed and conducted by professionals with experience in survey research.

Another issue with using public opinion for deer management is that these opinions are often not based on facts. Stout et al. (1997) documented that most respondents supported using trapping and transferring deer or contraception to control deer numbers in a suburban New York community. They determined that these attitudes were rooted in an ethical resistance to killing deer (Stout et al., 1997). These results are common throughout suburban and urban areas where residents have little familiarity with harvest as a management tool. However, use of controlled hunts and an increase in television programs on hunting (e.g., ESPN) have made hunting more acceptable (Norman and Bowman, 2003).

A successful technique used to evaluate and incorporate the human dimension into deer management decisions is the citizen task force or stakeholder group. Selecting individuals must be done carefully so that all viewpoints and user groups are represented. Including individuals that are well known and respected in the community will increase community support and "buy-in," and efforts should be made to avoid stacking the committee with too many people of one viewpoint. Once the group is selected, meetings should be carefully scheduled to ensure maximum attendance and meetings should be cancelled if low attendance is expected. Wildlife professionals should determine what advice they would like from their stakeholder group before selecting their group. Generally, determining a population objective and methods to achieve this objective are the common goals of these groups. Because of the diverse opinion of stakeholders in these groups, unanimous consent is not expected, rather a consensus is the ultimate goal (Curtis and Hauber, 1997). Use of citizen task forces is a powerful tool to involve stakeholders and gain consensus for deer management.

Lethal Methods

Regulated Hunting

Many options are often presented for managing deer abundance in exurban, suburban, and urban landscapes but regulated hunting is the most common and effective technique (DeNicola et al., 2000). Using the free labor of hunters to achieve specified management goals makes hunting the most cost-effective method of population control (Ishmael and Rongstad, 1984). Although hunting is cost-effective, land ownership patterns complicate implementation (Messmer et al., 1997). Implementing harvest regimes across the landscape typically is not feasible because of the varying opinions of landowners, which typically limits harvest to undeveloped areas. Implementation can be particularly problematic within suburban and urban landscapes. Another limitation is that developed areas often contain small pockets of deer habitat that function as deer refuges and reduce the efficacy of harvest on adjacent undeveloped areas (Rhoads, 2006). The rigidity of state regulations is another important limitation. Often state wildlife agencies aggressively control allocation of antlerless permits or have overly restrictive bag limits and/or season lengths while complaining that hunting is ineffective (Brown et al., 2000). Even with these limitations, hunting remains the only management technique that can be used across the entire landscape to manage deer abundance.

Biologists can manipulate the harvest regime to alter deer abundance to more closely match population goals. Development of a sound population abundance objective based on consensus is important to the success of any management action. Hansen and Beringer (1997) and Bowman (2006) suggested an abundance of 15 deer/km^2 as a management goal to balance the desires of the public and concerns about deer damage. Tracking population abundance is more critical in these landscapes than in more rural landscapes because of greater public scrutiny of harvest as a management tool. Managers should develop a long-term monitoring plan, which includes population abundance and reports of human–deer conflicts. Although monitoring population abundance is important, it is less important than understanding how harvest is affecting prevalence of human–deer conflicts. Demonstrating a reduction in human–deer conflicts is often important to retain consensus for harvest as a management tool.

One of the major objections to hunting is human safety. The general public often views hunting as unsafe for the nonhunter. Because hunting accidents are rare (less than 7 per 100,000 hunters) and very rarely involve nonhunters, hunting is a safe method to reduce deer abundance (Smith et al., 2005). The threat of hunters accidentally injuring nonhunters is likely overstated. In contrast, deer are a greater threat to public safety. Deer annually injure about 29,000 people and cause at least 200 fatalities (Conover et al., 1995, Chapter 13). Managers should educate the public that deer are likely a greater threat to human safety than hunters.

Traditional Harvest

The simplest harvest regime to implement in exurban, suburban, and urban landscapes is traditional harvest, which follows the state established guidelines for seasons and bag limits. Although Brown et al. (2000) raised numerous issues with management of deer populations using traditional harvest in New York, McDonald et al. (2007) documented that public hunting can be used to reduce and maintain a deer population in Massachusetts. Traditional harvest is probably more effective in exurban landscapes when compared to suburban and urban because of increased hunter access and fewer developed areas to act as refuges.

Use of traditional harvest in these landscapes has several limitations. The primary concern is one of human health (i.e., accidental shooting of a nonhunter). Although official statistics are lacking, the occurrence of hunter shooting nonhunters is almost nonexistent. Hunter proximity to development is one of the most regulated aspects in these landscapes (see safety zones below). Weapon type is a function of the safety concerns and regulations are often complicated. State wildlife agencies normally have general regulations about weapon type (i.e., use of rifles versus shotguns) but state and local law enforcement agencies may implement very restrictive regulations (i.e., prohibition of any type of weapon, including archery equipment), which preclude harvest as a management tool. Hunter access can also be problematic. Only 19% of landowners in an exurban Illinois landscape permitted hunters on their property (Storm et al., 2007a). Landowners that do not allow hunting often unknowingly provide a refuge for deer from harvest, which exasperates population abundance issues.

Traditional harvest is probably most effective on large tracts of public lands and agriculture areas in exurban landscapes. Many state agencies have liberal bag limits and season length in counties containing greater amounts of development to encourage hunter harvest. Additionally, public acceptance of hunting as a management tool increases as the public becomes more experienced with harvest and understands that it is a safe, effective method to reduce deer abundance (Norman and Bowman, 2003). As the amount of development increases on the landscape, the traditional harvest regime will be less effective and desirable and will likely be replaced by more structured but effective harvest regimes (i.e., controlled hunts).

Controlled Hunts

In many situations, traditional harvest regimes are not appropriate because of safety concerns. In these situations, controlled deer hunts are used to manage deer populations (Kilpatrick et al., 1997; McDonald et al., 1998; Kilpatrick and Walter 1999; Ebersole et al., 2007). During controlled hunts, the number and locations of hunters are strictly regulated to provide a safe environment for hunters and the general public (Kilpatrick et al., 1997; McDonald et al., 1998; Kilpatrick and Walter, 1999). Specific management objectives are achieved by regulating the number and sex of harvested animals (McDonald et al., 1998; Ebersole et al., 2007). Controlled hunts must be careful designed to be effective because they are so rigidly structured. Most of the structuring is to ensure safety of the public and participating hunters. Primary considerations are training, weapon type, hunter location, hunt timing, hunt length, and hunter density (Hansen and Beringer, 1997).

Controlled hunts generally begin with an application from a hunter. An application fee should be used to off-set the cost of the application process and administration costs associated with hunt establishment. The cost can vary from $14–$117/deer depending on the amount of law enforcement involvement (Kilpatrick and Walter, 1999; Doerr et al., 2001). The approach for selecting hunters varies from a standard lottery to preference given to hunters who assist with hunt preparation. Some states require hunters to pass a shooting proficiency test and/or a hunt briefing (Kilpatrick et al., 2007). Kilpatrick and Walter (1999) documented that shooting proficiency tests can improve hunter success and public acceptance because these tests reduce wounding loss.

The type of weapon permitted is also related to safety. Hansen and Beringer (1997) recommend archery on areas less than 450 ha, muzzleloaders on areas greater than 1000 ha, and centerfire rifles on areas over 1500 ha. The structure of the hunt is probably more important than the size of the area for determining weapon type. Hunting from elevated stands and requiring hunters to stay at designated locations can improve safety when using firearms. During some controlled hunts, the direction in which a hunter can fire is tightly regulated. Another concern is the sound of a firearm discharging which may alarm local residents. Notifying adjacent landowners and local law enforcement of the dates and times of the hunt along with providing contact information where questions or concerns can be directed will reduce public complaints. Additionally, the areas are typically closed to other users during the hunt with the exception of archery hunts.

Hunter location can be used to insure safety of the general public and hunt participants. Details about how a hunter is assigned a location vary widely by hunt location and weapon type. The most relaxed system allows hunters to choose their location anywhere within the hunt area, whereas the most rigid structure specifies that hunters must stay within 5 m of an assigned location for the duration of the hunt. Use of archery equipment is more associated with relaxed assignment of location. The best approach for management of archery hunters is to designate units based on roads and streams within the area; then a specified number of hunters are assigned to each unit but the exact location within the unit is not specified. The use of firearms often necessitates more rigid controls on hunter location. A major problem with rigid assignment of hunter location is that the hunt staff selecting the sites may select them for reasons other than maximizing harvest (e.g., safety or convenience).

Timing of the hunt can be critical to maximize hunter participation. Controlled hunts conducted when other deer hunting is not permitted increases hunter participation (Hansen and Beringer, 1997; Kilpatrick et al., 1997). Hunts conducted later in the season (December and January) tend to have greater antlerless harvest because hunters are less selective than earlier in the season (Hansen and Beringer, 1997). Late season hunts often have the advantage of conflicting with fewer other user groups because few people use these areas during winter. Another approach to prevent conflict with other user groups is to conduct hunts early in the week (i.e., Monday and Tuesday), but the disadvantage of early week hunts is decreased hunter participation.

Length of the hunt can be critical for maximizing hunter success. Hunts should be short duration (one–two days) followed by a rest period (no hunting) and then additional hunts if necessary (Rhoads, 2006). Hunts longer than two days often have severely reduced hunter success after the second day (Hansen and

Beringer, 1997). Reduction in hunter success stems from most hunters being highly successful (Hansen and Beringer, 1997) or greater deer use of refuge areas, which reduces hunt efficacy (Rhoads, 2006).

Hunter density is important for maximizing safety, hunter success, and hunter satisfaction. Hansen and Beringer (1997) recommended 14 hunters/km^2 for archery hunts and 12/km^2 for firearm hunts. Areas with a greater proportion of forest can support greater hunter densities. Managers can use the careful placement of hunters to allow for increased hunter density. Additionally, when hunters are dropped off at their hunt location, it increases safety and permits increased hunter densities.

Each controlled hunt location is unique and requires special considerations. Rhoads (2006) documented the impact of refuges and suggested that managers should carefully consider stand site placement to limit refuges. Some areas use limited drives through refuge areas to push deer past hunters; however, caution should be used with this approach because it could increase wounding loss (Ebersole et al., 2007). Hunter quality is also a concern. In many areas, better quality hunters do not enjoy the rigid structure of controlled hunts and opt not to participate. The loss of these quality hunters may reduce hunt efficacy. One method used to increase hunter participation is the use of special tags, which do not count against a hunters seasonal bag limit (Hansen and Beringer, 1997).

Damage Permits

Crop damage can be a major problem for agricultural producers in exurban landscapes (Chapter 13). Hunters are the most cost-effective way to reduce crop damage (Conover, 2001). Therefore, crop damage permits are often issued by state agencies to assist agricultural producers to reduce deer damage to crops. The permits range from highly restrictive to very liberal. At the highly restrictive end of the perspective, permits may only be used by the producer during the normal hunting season, whereas the very liberal perspective allows deer to be harvested at any time of the year including night shooting by anyone designated by the landowner. The effectiveness of crop damage permits varies widely and is primarily a function of landowner effort. Generally, these programs are more effective sociologically than biologically (Horton and Craven, 1997). Agricultural producers feel better because they have options, but typically few landowners avail themselves of the opportunity to use these permits (Horton and Craven, 1997). These programs should be used sparingly as public attitudes toward removing deer out-of-season are often not positive, particularly when females are caring for fawns. Most crop damage situations originate because of a lack of hunter access and effort during the hunting season. Producers complain about damage but are unwilling to allow hunters access to alleviate the situation or hold hunters accountable for a lack of effort.

Safety Zone Issues

A safety zone is an area around an inhabited building where it is illegal to hunt. These areas are intended to provide protection to landowners but may create refuges for deer, which inhibits deer management. Most states outside of the northeastern United States do not have safety zones even in highly developed areas (e.g., Virginia and North Carolina) and depend on local laws to dictate safe regulations.

The weapon used by hunters should dictate the size of the safety zone. In the Northeast, safety zones vary from 91 to 183 m for firearms and 46–183 m for archery equipment. Variation in the size of safety zones among states and the fact that safety zone size is not based on weapon type suggest the arbitrary nature of these regulations. Only Pennsylvania and Rhode Island have reduced safety zone sizes for archery equipment. Additionally, requiring hunting from elevated stands would force hunters to fire projectiles downward reducing the probability that a projectile could travel a long distance, which would permit reduced safety zone sizes.

A major problem with safety zones is the assumptions that these regulations make hunting safer for the nonhunting public. The term "safety zone" is a misnomer because unsafe behavior is not prevented by these regulations, which is the primary cause of hunting accidents. State agencies should focus on promoting safe behavior which is more important to public safety than these regulations.

An unexpected consequence of safe zones is the creation of refuges (Harden et al., 2005). The 183 m safety zone regulation in northern Delaware prevents harvest of deer on 61% of deer habitat (J. L. Bowman, unpublished data) (Figure 20.6). The lack of access to such a large amount of deer habitat will certainly preclude effective management of the population. As development increases in the Northeast, the issue of safety zones will become more important.

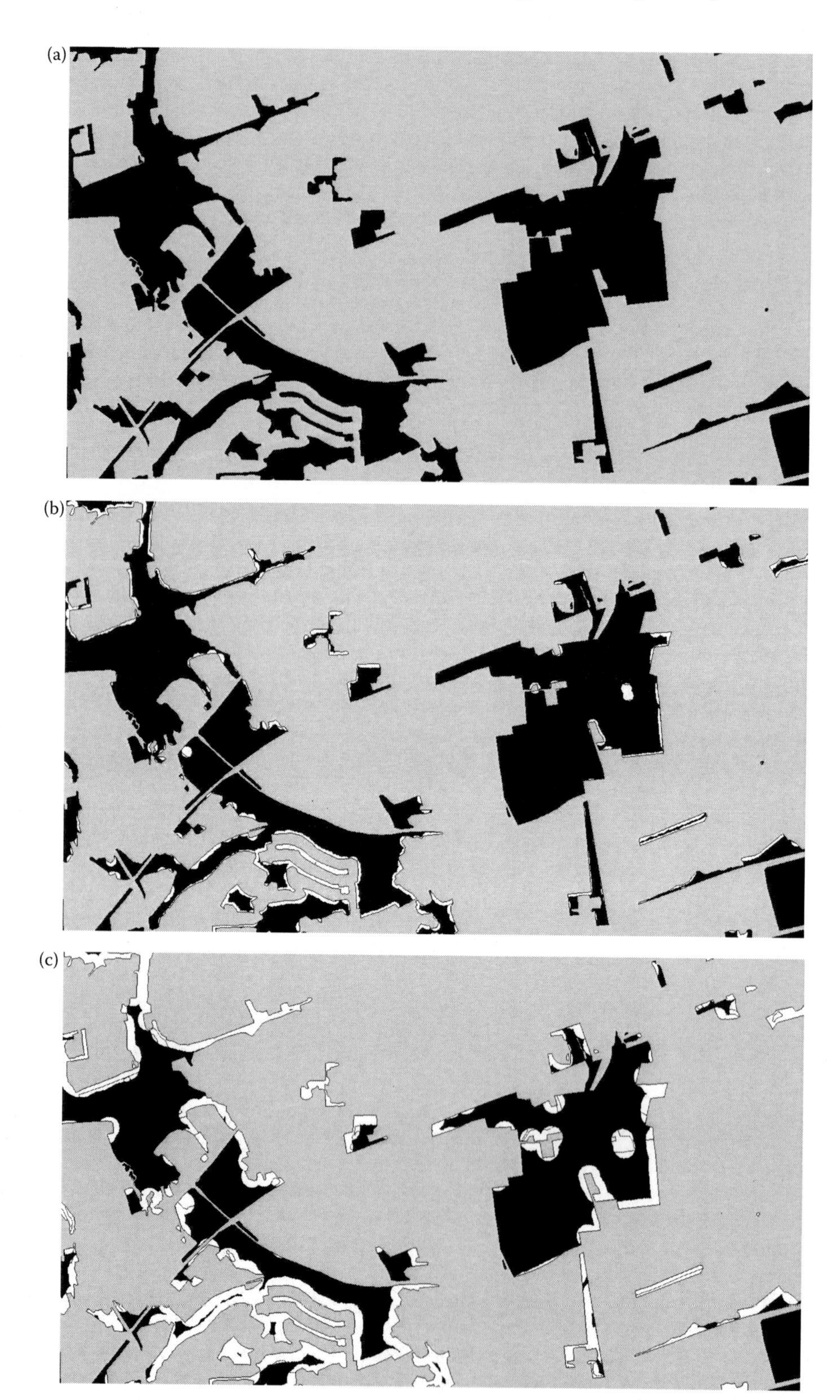

FIGURE 20.6 Total deer habitat (black), nonhabitat (gray), and deer habitat in which hunting is allowed (a) without a safety zone, (b) within a 46 m safety zone (white), (c) within a 91 m safety zone (white), and (d) within a 183 m safety zone (white) in northern New Castle County, Delaware, 2008.

FIGURE 20.6 *continued*

Baiting

Baiting is controversial because it is considered unethical by some and others express disease concerns. Recently, studies on bovine tuberculosis in Michigan suggested that banning baiting because of concerns about disease transmission are valid (Miller et al., 2003; Palmer and Whipple, 2006). In areas without serious diseases like bovine tuberculosis or chronic wasting disease the benefits of baiting to increase deer harvest may outweigh the disadvantages. Kilpatrick et al. (2010) documented that archery hunters using bait killed more deer than those not using bait. Because many landowners do not permit access to the property for hunting (Storm et al., 2007a), baiting may be one method to attract those deer to adjacent properties where they could be harvested. Deer will shift their core use areas to overlap a bait site if the site is in the deer's home range (Kilpatrick and Stober, 2002). Although no deer used a bait site outside its home range in Kilpatrick and Stober's (2002) study, all deer with bait sites in their home range used a bait site. Because baiting did not affect home range or core use area size nor did deer find bait sites outside their home range (Kilpatrick and Stober, 2002), baiting will only be effective on small spatial scales.

Sharpshooters

Sharpshooters have been used in many areas to reduce deer abundance. DeNicola and Williams (2008) documented a 49–78% reduction in deer–vehicle collisions when deer abundance was reduced by 54–76% using sharpshooting. Because sharpshooters are highly trained, safety concerns of citizens and local governments are often alleviated, permitting removal of deer from areas in close proximity to human development (DeNicola et al., 1997; Doerr et al., 2001). Additionally, culled deer are typically donated to a local food bank, which increases the acceptability of these programs. Sharpshooting is best suited for corporate campuses, government facilities, and city parks when traditional hunting or controlled hunts are not feasible. Sharpshooting also has promise for reducing densities in parks and other areas where hunters have been unable to achieve success. Sharpshooting permits rapid removal of a large number of deer in a short period of time, which can be difficult to achieve with hunter harvest in human-dominated landscapes. In these situations, sharpshooters would be used to reduce the population and hunters would be used to maintain that population in the future. The greatest disadvantage of sharpshooter operations is the cost, which ranges from $90–260/deer and is dependent upon the daily removal rate (DeNicola et al., 1997). In areas with high deer densities, removal rates can exceed 50 deer/day. Although this technique can effectively reduce the population, additional operations will continuously be needed to maintain the population at desired levels.

Trap and Euthanasia

Trapping and then euthanizing deer is rarely considered because of cost and the fact that animals are killed. Schwartz et al. (1997) recommended a gunshot as the most efficient and humane method of euthanasia. In areas where discharge of firearms is not legal or socially acceptable, they recommended using succinylcholine chloride to immobilize the deer, followed by euthanasia with a bolt gun (Schwartz et al., 1997). Use of this technique should still allow for donation of the deer to a food bank. Because the deer must be trapped, the time expenditure will likely be much greater than for sharpshooting. Although this technique may have appeal to some, it will cost more than sharpshooting and it is not any safer than other lethal control techniques.

Reintroduce Predators

White-tailed deer have many natural predators, few of which remain in exurban, suburban, and urban landscapes. Predators most likely to have an impact on deer populations are black bears, coyotes, gray wolves, and mountain lions. The larger predators (i.e., bear, wolves, and mountain lions) need very large home ranges and would be unlikely to survive in suburban and urban landscapes where suitable habitat is highly fragmented. The possibility of human conflicts with these larger predators also makes reintroduction unlikely (Bowman et al., 2001, 2004). Even if larger predators were reintroduced, deer evolved in the presence of these predators and are unlikely to be severely impacted by their reintroduction. Although coyotes do inhabit these human dominated landscapes, coyotes have not been documented controlling deer abundance.

Nonlethal

Allow Nature to Take Its Course

Allowing nature to take its course is a commonly suggested solution to deer overabundance. Deer evolved in the presence of many predators including humans. Their high reproductive rate allows them to persist even with high predation rates. Without population control, deer will increase until they reach carrying capacity or irrupt past carrying capacity. Deer in such a population will be in poor health and subject to severe population fluctuations as are often noted for irruptive populations. Habitat will be severely degraded prior to any population crash and will take many years to recover.

Overabundant deer populations affect various trophic levels of the ecosystems they inhabit (deCalesta, 1997; McShea et al., 1997; Waller and Alverson, 1997; Augustine and McNaughton, 1998) (Chapter 12) because deer are keystone herbivores (Waller and Alverson, 1997). Deer browsing can alter plant communities (Horsley and Marquis, 1983; McShea et al., 1997; Waller and Alverson, 1997; Augustine and McNaughton, 1998; Anderson et al., 2001) and inhibit forest regeneration (Harlow and Downing, 1970; Tilghman, 1989). Changes to plant communities then affect other species like birds, which decline at greater deer densities (deCalesta, 1994; McShea and Rappole, 1997a, 2000; Pietz and Gransfors, 2000). Allowing nature to take its course would be devastating for the deer population as well as the entire ecosystem.

Fencing

In exurban, suburban, and urban landscapes, fencing is an effective method to exclude deer from an area and reduce deer–human conflicts (Figure 20.7). Fencing creates a barrier to deer movement, which prevents damage in the protected area. The effectiveness of the barrier is dependent on the type of fence and the desire of the deer to enter the protected area (VerCauteren et al., 2006). Regulations in some areas preclude the use of fencing or limit the types of fence permitted. Fencing only prevents damage to the protected area but avoids the greater issue of deer overabundance. Excluding deer in one area will force them into adjacent areas, often in greater concentrations, causing more intensive deer–human conflicts.

FIGURE 20.7 Fencing is used to protect many types of plants like saplings. (Photo by D. Tallamy. With permission.)

The cost of fencing ranges from less than \$2/m to greater than \$20/m (Hygnstrom and Craven, 1988; Rosenberry et al., 2001; VerCauteren et al., 2006). Some common fence types are plastic or wire mesh, woven-wire mesh, slanted wire-mesh, barbed-wire, and electric. The most effective fences are the wire mesh and woven wire and they are typically the most expensive (VerCauteren et al., 2006). The least expensive options are barbed wire and electric fences, but these are typically not as effective (VerCauteren et al., 2006). Hygnstrom and Craven (1988) demonstrated that a single-strand electric fence could be used to protect corn. Electric fences are probably not an option in many suburban and urban areas because of liability concerns. The height of the fence is dependent upon the size of the area protected. Areas less than 5 m^2 can be protected with fencing as short as 1 m. Smaller gardens (up to 30 × 30 m) can be protected with fencing as short as 2 m. In areas where visual aesthetics is a concern, black fence material and poles should be used to reduce fence visibility. All fencing requires regular maintenance to be effective and to increase the life of the fence.

Repellents

Repellents are a commonly suggested technique for reducing deer–human conflicts. As their names suggest, repellents attempt to repel deer from feeding on treated foliage. Two general categories of repellents are available: odor and taste repellents. Odor repellents are applied near the area of desired protection or directly to the plants. These repellents have an unpleasant odor that repels deer (Abbott et al., 1990). Taste repellents require the animal to bite the protected foliage to discover the unpleasant taste (Belant et al., 1997). Many repellents work by using both properties.

Repellents are most effective when applied prior to deer browsing. Regular application is necessary for most repellents as rainfall and plant growth reduces repellent effectiveness. Many researchers have documented variable success with repellents, likely relating to a difference in deer density among studies (Conover, 1984, 1987; Sullivan et al., 1985; Hygnstrom and Craven, 1988; Byers et al., 1989; Milunas et al., 1994). No repellent prevents 100% of the damage but repellents requiring more frequent applications are most effective (Ward and Williams, 2010).

As deer abundance increases, the effectiveness of repellents decreases. At high deer densities, deer will browse on plants treated with repellents because resources are limited. Repellents are not a substitute for sound deer management. Although repellents may reduce deer–human conflicts at low-to-moderate densities, the ecological impact of overabundant deer is not affected without population reduction.

Supplemental Feeding

Supplemental feeding is a commonly suggested alternative to lethal control. Most research has focused on reducing mortality during winter (Ozoga and Verme, 1982; Baker and Hobbs, 1985; Doenier et al., 1997). Ozoga and Verme (1982) documented increased physical condition and reproductive condition of supplementally fed deer in Michigan; however, increased reproductive potential of the population will inhibit attempts to reduce deer abundance. Additionally, concentrating deer in small areas, as is typical with supplemental feeding, increases the probability of disease transmission and deer–human conflicts. Some pathogens can subsist on feed for more than a week, which suggests that supplemental feeding should be avoided because of the disease concerns (Miller et al., 2003; Palmer and Whipple, 2006). The cost of most long-term or large-scale supplemental feeding programs is prohibitive.

Trap and Transfer

The public typically supports translocations as a tool to reduce deer populations. Although translocations are very expensive (Beringer et al., 2002), they have been used occasionally when lethal control was deemed unacceptable. Success of these translocation efforts has been highly variable and most designed studies have used primarily females (Diehl, 1988; Jones and Witham, 1990; Jones et al., 1997; Beringer et al., 2002). Survival and home range establishment at the release site of translocated deer are the two primary measures of success. Survival probability is a function of the movement rates and patterns of translocated deer. Most large mammal species have been documented moving long distances to travel back to their original capture location after being translocated (Rogers, 1988; Nelson, 1994).

Several studies have documented below normal survival of translocated deer (Diehl, 1988; Jones and Witham, 1990; Jones et al., 1997; Cromwell et al., 1999). Vehicular collisions followed by hunting were the primary mortality factors for translocated deer in Illinois (Jones and Witham, 1990). Conducting translocations after the hunting season should reduce the importance of hunting as a mortality factor (Jones and Witham, 1990). Even when translocated deer survive, most deer do not remain at the release site. Only 40% of translocated deer survived and remained on the release area after one year in South Texas (Foley et al., 2008). In South Carolina, only 50% of translocated deer remained on the release site (Cromwell et al., 1999).

Most deer translocated more than 30 km typically settle within 30 km of their release site (Rogers, 1988). The homing ability of deer appears to decrease with increasing distance from the release site and most deer translocated more than 100 km do not return to their capture location (Rogers, 1988). Although deer may not return to their capture site, they often disperse from their release location toward their capture location. Therefore, to maximize translocation success, deer should be released at least 100 km from their capture location, after the hunting season, and on an area of sufficient size to allow deer to disperse homeward without coming in contact with large quantities of vehicular traffic. In most areas, locating a suitable release sites is difficult because few areas remain unpopulated with deer. Therefore, translocations are generally not considered a viable management tool for controlling deer numbers. Additionally, because deer must be captured and moved, the cost of this technique (greater than $1000/deer) far exceeds most other options.

Fertility Control

A commonly suggested option for reducing deer abundance is fertility control (Warren, 2000). Currently, fertility control is experimental in free-ranging populations of deer, so wildlife researchers are attempting to determine if fertility control can be a practical management alternative (Walter et al., 2002; Rutberg et al., 2004; Gionfriddo et al., 2009). Fertility control is a term used to describe methods that inhibit reproduction (Warren, 2000). The primary approaches used to inhibit reproduction are contraception, contragestation, and sterilization. Contraception is the most commonly suggested approach and it attempts to block conception by stopping ovulation or fertilization (Warren, 2000). Contragestation is less well known. The approach attempts to block pregnancy after conception occurs (e.g., abortion; Warren, 2000). The least common of the three approaches is sterilization, which is permanent infertility

(Warren, 2000). The intent of fertility control is to reduce the reproductive output of the population so that it is less than or equal to the mortality rate.

Most research on fertility control has focused on females because it is more likely to be effective on females. Because males can breed multiple females (Chapter 10), missing a small number of males would render fertility control ineffective. Four methods have been researched for fertility control: surgical sterilization, synthetic steroid hormones, immunocontraceptive vaccines, and abortion-inducing hormones. Although surgical sterilization is permanent, it has limited applicability for free-ranging deer because it requires capture and field surgery on each deer, making it cost prohibitive and logistically impractical (Warren, 2000; Merrill et al., 2003). Abortion-inducing hormones work by terminating pregnancy (Warren, 2000). Some ethical concerns have been raised because this method involves terminating pregnancy. This method requires that the animals be treated while they are pregnant, which necessitates a large proportion of the population being treated each year. Synthetic steroid hormones are used to prevent ovulation or conception (Warren, 2000). Although an orally administered delivery system has been tested, it requires a daily dose, which in not feasible in free-ranging deer (Warren, 2000). Subcutaneous implants effective for one–two years have been used, but questions remain about the secondary impacts on nontarget wildlife and humans (Warren, 2000). Immunocontraceptive vaccines have the greatest promise. These vaccines inhibit conception in a mode similar to an animal's immune response to diseases (Warren, 2000, Chapter 2). They stimulate the animal's immune system to form antibodies, which attack protein hormones or reproductive cells. This immune response effectively blocks fertilization of the ova (Warren, 2000). These vaccines must be injected via syringe or dart-gun and require annual boosters (Warren, 2000).

Efforts to develop an effective fertility control agent in free-ranging deer have focused on immunocontraception. The main barriers to development of an effective agent are: (1) a vaccine with long-term efficacy (greater than two years), (2) being able to treat an appropriate proportion of the population, and (3) cost, which is a function of the first two. In March 2010, the first commercially available control agent (GonaCon) was registered by the Environmental Protection Agency for use on free-ranging deer. GonaCon was 88% and 47% effective the first and second years after treatment, respectively, in preventing the production of fawns (Gionfriddo et al., 2009). Although these contraception rates seem high, retreatment will still be required on an annual basis to ensure contraception of a large proportion of the population. The annual survival rate of adult females is typically high, so a large proportion (70–90%) must be treated for these methods to be effective (Warren, 2000). Rudolph et al. (2000) suggested that the effort required for treating a population is a function of population abundance, encounter rates, abundance of breeding females, and efficacy of the fertility control agent. Rutberg et al. (2004) were able to treat 70–90% of the deer population on a 233 ha government campus of the National Institute of Stands and Technology (NIST) in Gaithersburg, Maryland. Even with such a great treatment rate, they observed at least 0.2 fawns/does. This reproductive rate would allow does to more than replace themselves before dying, preventing a long-term population decline and likely producing an increasing population. Another barrier to treating large proportions of the population could be high immigration rates, which would swamp the effect of the control agent (Rudolph et al., 2000). Cost is the final barrier. Walter et al. (2002) estimated treatment would cost $1128/treated deer. The true cost is likely highly variable and dependent upon efficacy of the fertility control agent and encounter rates of females in the population. On areas where females are easily located (i.e., NIST) (Rutberg et al., 2004), treatment could cost a quarter of Walter et al.'s (2002) estimate.

Unfortunately, a great deal of confusion surrounds the status of fertility control agents. The lack of public understanding regarding availability and practicality of fertility control has caused unnecessary delays in implementation of effective management programs, because fertility control is perceived as the ideal solution. Although research has suggested that fertility control will require greater effort than lethal control to implement (Nielsen et al., 1997), fertility control of some form is typically a popular option for population control in suburban landscapes (Muller et al., 1997; Stout et al., 1997; Warren, 2000). Although fertility control at the individual level is possible, the debate continues on whether fertility control will work at the population level in free-ranging populations of white-tailed deer (Rudolph et al., 2000). One of the greatest problems with wildlife research on fertility is the reluctance of many wildlife professionals to accept that it could be an option for management (Warren, 1995).

With advances in treatments and delivery systems, fertility control will likely become an option for small isolated government facilities or public parks but it is unlikely to ever be effective on large free-ranging populations. Research continues on the efficacy of these agents, but currently contraception is not a viable option for management of free-ranging deer (DeNicola et al., 2000; Warren, 2000; Rutberg et al., 2004).

Acknowledgments

I appreciate reviews of the manuscript by David Kalb, Melissa Miller, and Liz Tymkiw. I acknowledge the Department of Entomology and Wildlife Ecology at the University of Delaware for providing support during the drafting of this chapter.

REFERENCES

Abbott, D. H., D. A. Baines, C. G. Faulkes, D. C. Jennens, P. C. Y. K. Ning, and A. J. Tomlinson. 1990. A natural deer repellent: Chemistry and behaviour. In *Chemical Signals in Vertebrates 5*, eds. D. W. MacDonald, D. Müller-Schwarze, and S. E. Natynczuk, 599–609. Oxford: Oxford Press.

Anderson, R. C., E. A. Corbett, M. R. Anderson, G. A. Corbett, and T. A. Kelley. 2001. High white-tailed deer density has negative impact on tallgrass prairie forbs. *Journal of the Torrey Botanical Society* 128:381–392.

Augustine, D. J. and S. J. McNaughton. 1998. Ungulate effects on the functional species composition of plant communities: Herbivore selectivity and plant tolerance. *Journal of Wildlife Management* 62:1165–1183.

Baker, D. L. and N. T. Hobbs. 1985. Emergency feeding of mule deer during winter: Tests of a supplemental ration. *Journal of Wildlife Management* 49:934–942.

Belant, J. L., S. K. Ickes, L. A. Tyson, and T. W. Seasmans. 1997. Comparison of four particulate substances as wildlife feeding repellents. *Crop Protection* 16:439–447.

Beringer, J., L. P. Hansen, J. A. Demand, J. Sartwell, M. Wallendorf, and R. Mange. 2002. Efficacy of translocation to control urban deer in Missouri: Costs, efficiency, and outcome. *Wildlife Society Bulletin* 30:767–774.

Bissonette, J. A., C. A. Kassar, and L. J. Cook. 2008. Assessment of costs associated with deer–vehicle collisions: Human death and injury, vehicle damage, and deer loss. *Human–Wildlife Conflicts* 21:17–27.

Bowman, J. L. 2006. *An Evaluation of Deer Management in Delaware*. Dover, DE: Delaware Division of Fish and Wildlife.

Bowman, J. L., B. D. Leopold, F. J. Vilella, D. A. Gill, and H. A. Jacobson. 2001. Attitudes of landowners toward black bears compared between areas of high and low bear populations. *Ursus* 12:153–160.

Bowman, J. L., B. D. Leopold, F. J. Vilella, and H. A. Jacobson. 2004. A spatially explicit model, derived from demographic variables, to predict attitudes toward black bear restoration. *Journal of Wildlife Management* 68:223–232.

Brown, D. G., K. M. Johnson, T. R. Loveland, and D. M. Theobald. 2005. Rural land-use trends in the conterminous United States, 1950–2000. *Journal of Wildlife Management* 15:1851–1863.

Brown, T. L., D. J. Decker, S. J. Riley et al. 2000. The future of hunting as a mechanism to control white-tailed deer populations. *Wildlife Society Bulletin* 28:797–807.

Byers, R. E., D. H. Carbaugh, and C. N. Presley. 1989. Screening odor and taste repellents for control of white-tailed deer browse to apples or apple shoots. *Journal of Environmental Horticulture* 8:185–189.

Coffey, M. A. and G. H. Johnston. 1997. A planning process for managing white-tailed deer in protected areas: Integrated pest management. *Wildlife Society Bulletin* 25:443–439.

Conover, M. R. 1984. Effectiveness of repellents in reducing deer damage in nurseries. *Wildlife Society Bulletin* 12:399–404.

Conover, M. R. 1987. Comparison of two repellents for reducing deer damage to Japanese yews during winter. *Wildlife Society Bulletin* 15:265–268.

Conover, M. R. 2001. Effect of hunting and trapping on wildlife damage. *Wildlife Society Bulletin* 29:521–532.

Conover, M. R., W. C. Pitt, K. K. Kessler, T. J. DuBow, and W. A. Sanborn. 1995. Review of human injuries, illnesses, and economic losses caused by wildlife in the United States. *Wildlife Society Bulletin* 23:407–414.

Cromwell, J. A., R. J. Warren, and D. W. Henderson. 1999. Live-capture and small-scale relocation of urban deer on Hilton Head Island, South Carolina. *Wildlife Society Bulletin* 27:1025–1031.

Curtis, P. D. and J. R. Hauber. 1997. Public involvement in deer management decisions: Consensus versus consent. *Wildlife Society Bulletin* 25:399–403

deCalesta, D. S. 1994. Effect of white-tailed deer on songbirds within managed forests in Pennsylvania. *Journal of Wildlife Management* 58:711–718.

deCalesta, D. S. 1997. Deer and ecosystem management. In *The Science of Overabundance: Deer Ecology and Population Management*, eds. W. J. McShea, H. B. Underwood, and J. H. Rappole, 267–279. Washington, DC: Smithsonian Institute Press.

DeNicola, A. J., D. R. Etter, and T. Almendinger. 2008. Demographics of non-hunted white-tailed deer populations in suburban areas. *Human–Wildlife Conflicts* 2:102–109.

DeNicola, A. J., K. C. VerCauteren, P. D. Curtis, and S. E. Hygnstrom. 2000. *Managing White-tailed Deer in Suburban Environments: A Technical Guide*. Ithaca, NY: Cornell Cooperative Extension, Cornell University.

DeNicola, A. J., S. J. Weber, C. A. Bridges, and J. L. Stokes. 1997. Nontraditional techniques for management of overabundant deer populations. *Wildlife Society Bulletin* 25:496–499.

DeNicola, A. J. and S. C. Williams. 2008. Sharpshooting suburban white-tailed deer reduces deer–vehicle collisions. *Human–Wildlife Conflicts* 2:28–33.

Diehl, S. R. 1988. The translocation of urban white-tailed deer. In *Translocation of Wild Animals*, eds. L. Nielsen and R. D. Brown, 239–249. Milwaukee, WI: Humane Society.

Doenier, P. B., G. D. DelGiudice, and M. R. Riggs. 1997. Effects of winter supplemental feeding on browse consumption by white-tailed deer. *Wildlife Society Bulletin* 25:235–243.

Doerr, M. L., J. B. McAninch, and E. P. Wiggers. 2001. Comparison of 4 methods to reduce white-tailed deer abundance in an urban community. *Wildlife Society Bulletin* 29:1105–1113.

Ebersole, R., J. L. Bowman, and B. Eyler. 2007. Efficacy of an exurban controlled hunt. *Proceedings of the Annual Conference of Southeastern Association of Fish and Wildlife Agencies* 61:68–75.

Etter, D. R., K. M. Hollis, T. R. Van Deelen et al. 2002. Survival and movements of white-tailed deer in suburban Chicago, Illinois. *Journal of Wildlife Management* 66:500–510.

Foley, A. M., B. Pierce, D. G. Hewitt et al. 2008. Survival and movements of translocated white-tailed deer in south Texas. *Proceedings of the Annual Conference of the Southeast Association of Fish and Wildlife Agencies* 62:25–30.

Gionfriddo, J. P., J. D. Eisemann, K. J. Sullivan et al. 2009. Field test of a single-injection gonadotrophin-releasing hormone immunocontraceptive vaccine in female white-tailed deer. *Wildlife Research* 36:177–184.

Green, D., G. R. Askins, and P. D. West. 1997. Public opinion: Obstacle or aid to sound deer management? *Wildlife Society Bulletin* 25:367–370.

Grund, M. D. 1998. Movement patterns and habitat use of an urban white-tailed deer population in Bloomington, Minnesota. MS thesis, University of Missouri.

Grund, M. D., J. B. McAninch, E. P. Wiggers. 2002. Seasonal movement patterns and habitat use of female white-tailed deer associated with an urban park. *Journal of Wildlife Management* 66:123–130.

Hansen, A. J., R. L. Knight, J. M. Marzluff et al. 2005. Effects of exurban development on biodiversity: Patterns, mechanisms, and research needs. *Ecological Applications* 15:1893–1905.

Hansen, L. and J. Beringer. 1997. Managed hunts to control white-tailed deer populations on urban public areas in Missouri. *Wildlife Society Bulletin* 25: 484–487.

Harden, C. D., A. Woolf, and J. Roseberry. 2005. Influence of exurban development on hunting opportunity, hunter distribution, and harvest efficiency of white-tailed deer. *Wildlife Society Bulletin* 33:233–242.

Harlow, R. F. and R. L. Downing. 1970. Deer browsing and hardwood regeneration in the southern Appalachians. *Journal of Forestry* 68:298–300.

Henderson, D. W., R. J. Warren, D. H. Newman, J. M. Bowker, J. S. Cromwell, and J. J. Jackson. 2000. Human perceptions before and after a 50% reduction in an urban deer herd's density. *Wildlife Society Bulletin* 28:911–918.

Horsley, N. W. and D. A. Marquis. 1983. Interference by weeds and deer with Allegheny hardwood reproduction. *Canadian Journal of Forest Resources* 13:61–69.

Horton, R. R. and S. R. Craven. 1997. Perceptions of shooting-permit use for deer damage abatement in Wisconsin. *Wildlife Society Bulletin* 25:330–336.

Hygnstrom, S. E. and S. R. Craven. 1988. Electric fences and commercial repellents for reducing deer damage in cornfields. *Wildlife Society Bulletin* 16:291–296.

Ishmael, W. E. and O. J. Rongstad. 1984. Economics of an urban deer-removal program. *Wildlife Society Bulletin* 12:394–398.

Jones, J. M. and J. H. Witham. 1990. Post-translocation survival and movements of metropolitan white-tailed deer. *Wildlife Society Bulletin* 18:434–441.

Jones, M. L., N. E. Mathews, and W. F. Palmer. 1997. Influence of social organization on dispersal and survival of translocated female white-tailed deer. *Wildlife Society Bulletin* 25:272–278.

Kilpatrick, H. J., A. M. Labonte, and J. S. Barclay. 2007. Factors affecting bowhunter access in suburban areas. *Journal of Wildlife Management* 71:2102–2105.

Kilpatrick, H. J., A. M. Labonte, and J. S. Barclay. 2010. Use of bait to increase archery deer harvest in an urban-suburban landscape. *Journal of Wildlife Management* 74:714–718.

Kilpatrick, H. J. and S. M. Spohr. 2000a. Movements of female white-tailed deer in a suburban landscape: A management perspective. *Wildlife Society Bulletin* 28:1038–1045.

Kilpatrick, H. J. and S. M. Spohr. 2000b. Spatial and temporal use of a suburban landscape by female white-tailed deer. *Wildlife Society Bulletin* 28:1023–1029.

Kilpatrick, H. J., S. M. Spohr, and G. G. Chasko. 1997. A controlled deer hunt on a state-owned coastal reserve in Connecticut: Controversies, strategies, and results. *Wildlife Society Bulletin* 25:451–456.

Kilpatrick, H. J., S. M. Sphor, and K. K. Lima. 2001. Effects of population reduction on home ranges of female white-tailed deer at high densities. *Canadian Journal of Zoology* 79:949–954.

Kilpatrick, H. J. and W. A. Stober. 2002. Effects of temporary bait sites on movements of suburban white-tailed deer. *Wildlife Society Bulletin* 30:760–766.

Kilpatrick, H. J. and W. D. Walter. 1999. A controlled archery deer hunt in a residential community: Cost, effectiveness, and deer recovery rates. *Wildlife Society Bulletin* 27:115–123.

Lischka, S. A., S. J. Riley, and B. A. Rudolph. 2008. Effects of impact perception on acceptance capacity for white-tailed deer. *Journal of Wildlife Management* 72:502–509.

Long, E. S., D. R. Diefenbach, C. S. Rosenberry, B. D. Wallingford, and M. D. Grund. 2005. Forest cover influences dispersal distance of white-tailed deer. *Journal of Mammalogy* 86:623–629.

Lopez, R. R., M. E. P. Viera, N. J. Silvy, P. A. Frank, S. W. Whisenant, and D. A. Jones. 2003. Survival, mortality, and life expectancy of Florida Key deer. *Journal of Wildlife Management* 67:34–45.

McDonald, J. E., Jr., D. E. Clarks, and W. A. Woytek. 2007. Reduction and maintenance of a white-tailed deer herd in central Massachusetts. *Journal of Wildlife Management* 71:1585–1593.

McDonald, J. E., M. R. Ellingwood, and G. M. Vecellio. 1998. *Case Studies in Controlled Deer Hunting*. Concord, NH: New Hampshire Fish and Game Department.

McNulty, S. A., W. F. Porter, N. E. Matthews, and J. A. Hill. 1997. Localized management for reducing white-tailed deer populations. *Wildlife Society Bulletin* 25:265–271.

McShea, W. J. and J. H. Rappole. 1997a. Herbivores and the ecology of forest understory birds. In *The Science of Overabundance: Deer Ecology and Population Management*, eds. W. J. McShea, H. B. Underwood, and J. H. Rappole, 298–309. Washington, DC: Smithsonian Institute Press.

McShea, W. J. and J. H. Rappole. 1997b. The science and politics of managing deer within a protected area. *Wildlife Society Bulletin* 25:443–446.

McShea, W. J. and J. H. Rappole. 2000. Managing the abundance and diversity of breeding bird populations through manipulation of deer populations. *Conservation Biology* 14:1161–1170.

McShea, W. J., H. B. Underwood, and J. H. Rappole. 1997. *The Science of Overabundance: Deer Ecology and Population Management*. Washington, DC: Smithsonian Institute Press.

Merrill, J. A., E. G. Cooch, and P. D. Curtis. 2003. Time to reduction: Factors influencing management efficacy in sterilizing overabundant white-tailed deer. *Journal of Wildlife Management* 67:267–279.

Messmer, T. A., S. M. George, and L. Cornicelli. 1997. Legal considerations regarding lethal and nonlethal approaches to managing urban deer. *Wildlife Society Bulletin* 25:424–429.

Miller, R., J. B. Kaneene, S. D. Fitzgerald, and S. M. Schmitt. 2003. Evaluation of the influence of supplemental feeding of white-tailed deer (*Odocoileus virginianus*) on the prevalence of bovine tuberculosis in the Michigan wild deer population. *Journal of Wildlife Diseases* 39:84–95.

Milunas, M. C., A. F. Rhoads, and J. R. Mason. 1994. Effectiveness of odour repellents for protecting ornamental shrubs from browsing by white-tailed deer. *Crop Protection* 13:393–397.

Muller, L. I., R. J. Warren, and D. L. Evans. 1997. Theory and practice of immunocontraception in wild mammals. *Wildlife Society Bulletin* 25:504–514.

Nelson, M. E. 1994. Migration bearing and distance memory by translocated white-tailed deer, *Odocoileus virginianus*. *Canadian Field Naturalist* 108:74–76.

Nielsen, C. K., W. F. Porter, and H. B. Underwood. 1997. An adaptive management approach to controlling suburban deer. *Wildlife Society Bulletin* 25:470–477.

Norman, P. C. and J. L. Bowman. 2003. Attitudes of residents toward a managed deer hunting program in the vicinity of suburban parks. Paper presented at the 26th Annual Meeting of the Southeast Deer Study Group, Chattanooga, TN.

Odell, E. A. and R. L. Knight. 2001. Songbird and medium-sized mammal communities associated with exurban development in Pitkin County, Colorado. *Conservation Biology* 15:1143–1150.

Ozoga, J. J. and L. J. Verme. 1982. Physical and reproductive characteristics of a supplementally fed white-tailed deer herd. *Journal of Wildlife Management* 46:281–301.

Palmer, M. V. and D. L. Whipple. 2006. Survival of *Mycobacterium bovis* on feedstuffs commonly used as supplemental feed for white-tailed deer (*Odocoileus virginianus*). *Journal of Wildlife Diseases* 42:853–858.

Pietz, P. J. and D. A. Gransfors. 2000. White-tailed deer (*Odocoileus virginianus*) predation on grassland songbird nestlings. *American Midland Naturalist* 144:419–422.

Porter, W. F. 1997. Ignorance, arrogance, and the process of managing overabundant deer. *Wildlife Society Bulletin* 25:408–412.

Radeloff, V. C., R. B. Hammer, S. I. Stewart, J. S. Fried, S. S. Holcomb, and J. F. McKeefry. 2005. The wildland-urban interface in the United States. *Ecological Applications* 15:799–805.

Rhoads, C. L. 2006. Spatial ecology and response to a controlled hunt of female white-tailed deer in an exurban park. MS thesis, University of Delaware.

Rhoads, C. L., J. L. Bowman, and B. Eyler. 2010. Home range and movement rates of female exurban white-tailed deer. *Journal of Wildlife Management* 74:987–994.

Rogers, L. L. 1988. Homing tendencies of large mammals: A review. *In Translocation of Wild Animals*, eds. L. Nielsen and R. D. Brown, 76–92. Milwaukee, WI: Humane Society.

Rosenberry, C. S., L. I. Muller, and M. C. Conner. 2001. Movable, deer-proof fencing. *Wildlife Society Bulletin* 29:754–757.

Rudolph, B. A., W. F. Porter, and H. B. Underwood. 2000. Evaluating immunocontraception for managing suburban white-tailed deer in Irondequoit, New York. *Journal of Wildlife Management* 64:463–473.

Rutberg, A. T., R. E. Naugle, L. A. Thiele, and I. K. M. Liu. 2004. Effects of immunocontraception on a suburban population of white-tailed deer *Odocoilieus virginianus*. *Biological Conservation* 116:243–250.

Saalfeld, S. T. and S. S. Ditchkoff. 2007. Survival of neonatal white-tailed deer in an exurban population. *Journal of Wildlife Management* 71:940–944.

Schwartz, J. A., R. J. Warren, D. W. Henderson, D. A. Osborn, and D. J. Kessler. 1997. Captive and field tests of a method for immobilization and euthanasia of urban deer. *Wildlife Society Bulletin* 25:532–541.

Smith, J. L., G. C. Wood, and E. J. Lengerich. 2005. Hunting-related shooting incidents in Pennsylvania, 1987–1999. *Journal of Trauma-Injury Infection and Critical Care* 58:582–590.

Storm, D. J., C. K. Nielsen, E. M. Schauber, and A. Woolf. 2007a. Deer–human conflict and hunter access in an exurban landscape. *Human–Wildlife Conflicts* 1:53–59.

Storm, D. J., C. K. Nielsen, E. M. Schauber, and A. Woolf. 2007b. Space use and survival of white-tailed deer in an exurban landscape. *Journal of Wildlife Management* 71:1170–1176.

Stout, R. J., B. A. Knuth, and P. D. Curtis. 1997. Preferences of suburban landowners for deer management techniques: A step towards better communication. *Wildlife Society Bulletin* 25:348–359.

Sullivan, T. P., L. O. Nordstrom, and D. S. Sullivan. 1985. Use of predator odors as repellents to reduce feeding damage by herbivores II. *Journal of Chemical Ecology* 11:921–935.

Swihart, R. K. and A. J. DeNicola. 1997. Public involvement, science, management, and the overabundance of deer: Can we avoid a hostage crisis? *Wildlife Society Bulletin* 25:382–387.

Theobald, D. M. 2001. Land-use dynamics beyond the American urban fringe. *Geographical Review* 91:544–564.

Tilghman, N. G. 1989. Impacts of white-tailed deer on forest regeneration in northwestern Pennsylvania. *Journal of Wildlife Management* 53:524–532.

Tymkiw, E. L. 2010. The effect of white-tailed deer density on breeding songbirds in Delaware. MS thesis, University of Delaware.

VerCauteren, K. C., M. J. Lavelle, and S. Hygnstrom. 2006. Fences and deer-damage management: A review of designs and efficacy. *Wildlife Society Bulletin* 34:191–200.

Waller, D. M. and W. S. Alverson. 1997. The white-tailed deer: A keystone herbivore. *Wildlife Society Bulletin* 25:217–226.

Walter, W. D., P. J. Perkins, A. T. Rutberg, and H. J. Kilpatrick. 2002. Evaluation of immunocontraception in a free-ranging suburban white-tailed deer herd. *Wildlife Society Bulletin* 30:186–192.

Ward, J. S. and S. C. Williams. 2010. Effectiveness of deer repellents in Connecticut. *Human–Wildlife Interactions* 41:56–66.

Warren, R. J. 1995. Should wildlife biologists be involved in wildlife contraception research and management? *Wildlife Society Bulletin* 23:441–444.

Warren, R. J. 2000. *Fertility Control in Urban Deer: Questions and Answers*. Gainesville, FL: American Archery Council Field Publication FP-1.

Part IV

The Future

21

The Future of White-tailed Deer Management

Brian P. Murphy

CONTENTS

Introduction

The restoration and recovery of white-tailed deer is a global wildlife management success with few parallels. Throughout most of the twentieth century, wildlife agencies catered to a narrow group of constituents—hunters, anglers, and trappers. For decades, deer hunter numbers and agency deer management programs increased as deer populations expanded. Traditional management strategies maximized buck harvests and resulted in high hunter success and satisfaction. By the 1980s, signs of change began to emerge on many fronts. Deer populations were overabundant in many areas, deer herd sex ratios were highly unbalanced favoring females, license sales were declining, hunter expectations were changing, and wildlife agency responsibilities were expanding.

By the year 2000, more than 30 million whitetails existed in North America, and it had become the most abundant and economically important big-game animal in the world (U.S. Fish and Wildlife Service, 2007; Quality Deer Management Association, 2008; Adams et al., 2009, 2010). However, in some areas, the whitetail's recovery was too successful resulting in significant ecological, social, and economic impacts (Conover et al., 1995; Latham et al., 2005; Chapters 12 and 13). The future of white-tailed deer management will require development of new and innovative solutions to meet the needs of an increasingly diverse stakeholder group. The purpose of this chapter is to examine current trends and challenges and discuss how these may impact the future of white-tailed deer hunting and management.

Deer Management Trends

Quality Deer Management

One of the most significant trends in deer hunting and management has been the shift from traditional deer management to quality deer management (QDM) (Green and Stowe, 2000). The goal of QDM is to balance deer herds within existing habitat constraints while improving adult sex ratios and buck age structures (Miller and Marchinton, 1995). The philosophy originated in Texas in the late 1960s, but was slow to spread to other states until formation of the Quality Deer Management Association (QDMA) in 1988 (Brothers and Ray, 1975; Miller and Marchinton, 1995). Today, QDM practices are common among hunters and an increasing number of wildlife agencies. For example, by 2008, 22 states in the United States had implemented some form of antler restriction to reduce the number of young bucks in the harvest (Adams et al., 2009) (Figure 11.2). Additionally, an increasing number of hunters voluntarily restrict buck harvest beyond legal requirements. Collier and Krementz (2006) reported that 60% of hunting clubs registered in the Arkansas Deer Camp Program used buck harvest restrictions in excess of state regulations (at least three points on one antler at time of study). Collectively, these changes have significantly impacted the age structure of the buck harvest (Figure 21.1). This is evidenced by the percentage of 1.5-year-old bucks in the United States' buck harvest, which declined from 62% in 1989 to 41% in 2009 (Adams et al., 2010).

Increasing support for QDM also has positively impacted antlerless deer harvests. For example, 1999 was the first year hunters in the United States harvested more antlerless than antlered deer (Quality Deer Management Association, 2001). In 2005, about 65% of states recorded higher harvests for antlerless deer than antlered bucks (Figure 11.5). This trend was most noticeable in the Midwest where herd productivity is high. From 1999 to 2005, antlerless harvests in Illinois, Indiana, Iowa, Kentucky, Ohio, and Missouri increased an average of 78% (Adams et al., 2009). These changes resulted from an increased willingness among hunters to harvest antlerless deer and liberalized agency regulations governing antlerless deer harvest. Without question, QDM has created a new paradigm with significant positive implications for the future of deer management.

FIGURE 21.1 With growing participation in Quality Deer Management among hunters, landowners, and wildlife agencies, quality bucks, like this 3.5-year-old taken on a QDM cooperative in Georgia, have become increasingly common in recent years. (Photo by L. Thomas, Jr., Quality Deer Management Association. With permission.)

Habitat Enhancement

Growth in popularity of QDM has fueled many associated trends, especially on private lands. One such trend is the use of food plots to increase nutrition available to deer and to aid in harvest (Murphy, 2006). Though no precise estimate exists, the area currently planted in wildlife food plots is higher today than ever before. This is supported by an 86% increase in the number of food plot seed distributors listed in a major outdoor catalog from 2003 to 2008 (Adams et al., 2009). Food plots have become more sophisticated with complex mixtures of annuals and perennials to maximize palatability, production, and seasonal availability (Kammermeyer et al., 2006; Harper, 2008). Food plot intensity also has increased, especially on small properties. Today it is common for 5% or more of a property to be planted in year-round food plots for whitetails, though as little as 1.5% has been shown to impact deer herd harvest and health (Kammermeyer and Moser, 1990).

Another increasingly common management practice involves managing natural vegetation to enhance habitat. This is especially true on millions of hectares of intensively managed pine plantations in the southern United States owned or managed by the forest industry, most of which are leased to hunters. Edwards et al. (2004) reported four-fold increases in leaf biomass and digestible protein in mature pine plantations following treatment with herbicide, prescribed fire, and fertilization. This compared favorably to a warm-season planting of cowpeas (a commonly planted summer annual) in both production of digestible protein and cost. Once relatively uncommon, other practices such as establishment of native warm season grasses, timber stand improvement, and creation of wildlife orchards are growing in popularity among deer hunters and managers.

While numerous habitat management practices can improve deer quality and quantity, their impacts on wildlife conservation, species biodiversity, and land stewardship are arguably more significant (Figure 21.2). Thousands of hunters and landowners across North America, many driven by the desire to produce and harvest mature whitetail bucks, have become active deer and habitat managers. This has increased habitat quality and diversity on millions of hectares throughout the whitetail's range. Dougherty (2006) used the term "Leopold Landscapes" to describe these intensively managed and biologically diverse properties. Ricks (2010) documented benefits of perennial clover food plots in eastern hardwood forests to numerous small mammals and songbirds. Miller (2010) documented 77 bird, 23 mammal, 36 reptile and amphibian, and over 300 plant species on a QDM-managed wildlife cooperative in Mississippi that utilized various prescribed fire and herbicide treatments to enhance wildlife habitat and timber production.

Habitat management practices which increase early successional habitat, including young forested habitat, are important given recent declines in the percentages of such habitat in United States' forests. For example, as of 2008, only 18% of forestland was in early succession or a young forest stage, down from

FIGURE 21.2 Interest in improving the quality of deer available for harvest has resulted in increased habitat quality and diversity on millions of hectares throughout North America to the benefit of game and nongame species alike. (Photo by L. Thomas, Jr., Quality Deer Management Association. With permission.)

22% in the past two decades (U.S.D.A. Forest Service, 2008). Declines were most notable in Arkansas, New York, Michigan, Pennsylvania, and Tennessee, which maintain only 11–19% in early succession or young forested habitat (U.S.D.A. Forest Service, 2008). Given the importance of these states to the future of deer hunting, increasing the number of Leopold Landscapes in these states offers significant potential for wildlife management and conservation.

Deer Management Cooperatives

Another significant trend is increasing participation in wildlife management cooperatives. While early cooperatives date to the mid-1950s in Texas, cooperatives did not become common in Texas until the 1980s (Texas Parks and Wildlife Department, 2004). By 2004, more than 160 cooperatives encompassing 0.8 million hectares existed in the Lone Star state (Texas Parks and Wildlife Department, 2004). By the early 1990s, growing interest in deer management fueled establishment of "QDM cooperatives" throughout the whitetail's range (Murphy et al., 2001). QDM cooperatives are groups of landowners and hunters working together to improve the quality of the deer herd and hunting experiences on their collective acreage (Murphy et al., 2001). Most cooperatives are voluntary and vary in size, number of participants, and structure according to site-specific goals and constraints. A major benefit of QDM cooperatives is that they enable small properties to gain the management advantages of a larger landowner. Thus, it is not surprising that cooperatives have become increasingly popular in areas like the Midwest where the average size of hunting properties often is less than 50 ha. Another benefit is the ability to pool harvest and observation data enabling better deer hunting and management. Miller (2010) reported increased doe harvest, mature buck harvest, body weights, lactation rates, and hunter satisfaction following establishment of a QDM cooperative in Mississippi.

Given growing interest in QDM and benefits of a landscape-level approach to deer management, continued growth of cooperatives is likely. For example, a 2008 survey of QDMA members revealed that 34% were involved in cooperatives. Of those cooperatives, 80% were 1000 ha in size or smaller and 90% involved 10 or fewer participating landowners. Among QDMA members not participating in cooperatives, the two most common reasons cited were lack of trust that their neighbors would adhere to QDM guidelines and lack of interest among neighbors. These findings stress the importance of communication and trust to the success of future deer management efforts.

Private Ownership of Recreational Hunting Lands

Increasing ownership of recreational lands among hunters is another significant trend. From 1991 to 2006, the number of sportsmen in the United States owning land for hunting increased 56% from 853,000 to 1.3 million (U.S. Fish and Wildlife Service, 1993, 2007). This trend appears to be accelerating. From 1991 to 2001, land ownership among hunters in the United States increased 24%; whereas, it increased 26% from 2001 to 2006 (U.S. Fish and Wildlife Service, 1993, 2002, 2007). As of 2006, 1.3 million hunters in the United States owned 54 million hectares, or about 40 hectares each. It is notable that increases in land ownership from 1991 to 2006 were accompanied by an 11% decline in the number of hunters leasing land (U.S. Fish and Wildlife Service, 1993, 2007). In fact, by 2001, the number of hunters owning land exceeded the number leasing land. These data suggest a rapidly changing dynamic regarding deer hunting and management on private lands.

Broad societal trends in demographics, economics, and a desire for a high-quality and intimate hunting experience have fueled this trend. The "baby boomer" generation is the largest and most affluent in United States history and many have the time and interest in owning land for hunting. Not surprisingly, land ownership among hunters is positively correlated with age, education, and avidity, though not strongly with income (Leonard, 2004).

Historically, rural land values were determined primarily by capitalized values of expected earnings from timber, livestock, or agricultural crops. Today, values often are influenced by the land's ability to provide quality recreation and therefore may be undervalued by traditional appraisal methods (Henderson and Moore, 2005). Previous surveys have confirmed that recreation is a significant motivation for farmland purchases by non-farmers (Gilliland et al., 2003; Henderson and Akers, 2009). Despite numerous examples where deer hunting has influenced land values at time of sale, research in this area remains limited. Nevertheless, the importance of hunting to land values is increasing and likely to continue doing so in the future.

Challenges

Hunter Participation

One of the most significant challenges facing the future of deer hunting and management is the decline in hunter participation. In most areas, deer–hunter numbers peaked in the 1970s or 1980s and declined gradually since (U.S. Fish and Wildlife Service, 2007, 2010). In 2006, only 5%, or approximately 12.5 million Americans 16 years and older, hunted (U.S. Fish and Wildlife Service, 2007). Of those, approximately 10.7 million (86%) hunted big game and 10.1 million (94%) of those hunted deer. These figures stress the importance of deer and deer hunters to the future of hunting in North America.

According to the National Shooting Sports Foundation, 35 states lost hunters from 1996 to 2006 (National Shooting Sports Foundation, 2008). While this report included all hunters, the decline was most noticeable in key deer states such as Pennsylvania and Michigan where license sales dropped 6% and 8%, respectively, during this period. These states each lost more hunters than existed in Connecticut, Delaware, Hawaii, Massachusetts, and Rhode Island combined (Adams et al., 2009).

Despite the downward trend in overall hunter numbers, the number of big-game hunters has remained more stable than other hunter groups with only a 5% decline from 1996 to 2006 (U.S. Fish and Wildlife Service, 2007). Of the 15 states in which deer hunter numbers increased from 1996 to 2006, many including Oklahoma, North Dakota, and South Dakota likely benefited from increased hunting opportunity as whitetail populations expanded along the "western front" of their geographic range (Adams et al., 2009, 2010). Thus, it appears hunter losses in the East, at least to some extent, were offset by increases in the West. This trend is likely to slow over time as whitetails occupy all available habitat. Additionally, the average age of big-game hunters in the United States was 41.5 years in 2005 and is increasing (National Shooting Sports Foundation, 2005). If this trend continues, further declines in hunter numbers will occur in the future as more cease hunting due to age or health.

Reasons for declining hunter participation are numerous and complex, but generally result from an increasingly urban society with little connection to nature (Responsive Management, 2005, 2006a). According to the U.S. Census Bureau (1995, 2000), the percentage of the United States population that lived in a rural area declined from 36% in 1950 to 21% in 2000. This trend is important because hunting is positively correlated with living in a rural area (Responsive Management, 2006b). An analysis of the 2006 National Survey of Fishing, Hunting and Wildlife-Associated Recreation revealed that hunter recruitment among urban residents was down sharply compared to non-urban residents (Leonard, 2007). Therefore, not only is the United States becoming more urban, those living in urban areas are becoming less likely to hunt.

This profound detachment from nature, coined "urbanism" by Marchinton (1997, pp. 21–23) and "nature deficit disorder" by Louv (2008), presents a formidable challenge for the future of hunting. In his 1997 address to the Southeast Deer Study Group meeting in Charleston, South Carolina, Marchinton stated, "Urbanism results from the plastic environment and the profound ignorance of the natural world by those living in it. Today's urbanites are three or four generations removed from rural life and have few, if any, direct functional ties to or a spiritual bonding with the natural world." Clearly, significant barriers exist to increasing hunter numbers in a society becoming disconnected from and disinterested in nature.

The decline in young hunter participation is especially concerning. From 1996 to 2006, the number of hunters 16–24-years old declined from 2.1 million to 1.5 million (U.S. Fish and Wildlife Service, 1997, 2007). The decline was most noticeable in states with restrictive regulations or laws governing youth hunter participation (Southwick Associates, 2005). In 2004, the National Shooting Sports Foundation, U.S. Sportsmen's Alliance, and the National Wild Turkey Federation launched the Families Afield Initiative which encouraged state agencies to eliminate hunter age restrictions and ease hunter education requirements. As of 2009, Families Afield legislation had been passed in 28 states with over 283,000 apprentice licenses sold (National Shooting Sports Foundation, 2010). Many states also have implemented special hunting seasons for youth. Another noteworthy initiative is the National Archery in the Schools Program launched in Kentucky in 2003. This program teaches target archery to students in grades 4 to 12. By 2010, the program had expanded to more than 5500 schools throughout the United States and Canada and reached more than 1.2 million students annually (T. Bennett, National Archery in the Schools Program, personal communication). While not a hunter recruitment program, surveys have revealed that more than 50% of participants expressed interest in hunting after taking the course. These statistics compare favorably with hunting participation data for 6- to 15-year-olds which declined from around 2 million in 1980 to 1.7 million in 1995, but increased to nearly 1.8 million by 2005 (U.S. Fish and Wildlife Service, 2007).

While deer hunting remains a male-dominated activity, the number of female hunters has increased in recent decades. During the 1980s, the percentage of female hunters more than doubled while the percentage of male hunters declined 16% (Duda et al., 1995). According to the U.S. Fish and Wildlife Service, there were approximately 1.2 million female hunters 16-years old and older in 2006 (U.S. Fish and Wildlife Service, 2007). Additional research indicated that the number of female hunters, as a group, is not as stable as the number of male hunters (Duda, 2001). Whereas most male hunters hunt every year, this is not true for female hunters. One notable exception is the number of female big-game hunters, especially female bowhunters, which remained stable or increased from 1996 to 2001 (U.S. Fish and Wildlife Service, 1997, 2002; National Shooting Sports Foundation, 2003). Additionally, females hunt predominately for utilitarian and familial reasons. Research suggests that female hunters are twice as likely as male hunters to hunt for meat (47% female, 22% male), nearly three times more likely to hunt to be with friends and family (27% female, 11% male), less than half as likely to hunt for the sport and recreation (20% female, 45% male), and a third less likely to hunt to be close to nature (7% female, 22% male; Responsive Management 1995).

Current trends in youth and female participation in deer hunting, when combined with the numerous hunter-recruitment programs currently being undertaken by wildlife agencies and hunting/conservation organizations, provide guarded optimism for the future of hunting in North America.

Public Support for Hunting

Maintaining public support for hunting in the future is another major challenge. Given the likelihood hunters will represent a lower percentage of citizens in the future, demonstrating the environmental, economic, and societal benefits of deer hunting will be essential. Hunters provide a free ecological service by controlling deer populations that could otherwise negatively impact wildlife, forest, and human health. Deer hunters also are the backbone of the United States hunting economy given that nearly four times more hunters hunt deer than the next closest species (Leonard, 2004; Adams et al., 2009). In 2006, deer hunters spent $12.4 billion on their favorite pastime, or 52% of total expenditures for all game species (National Shooting Sports Foundation, 2008). Societal benefits of deer hunting are widespread and numerous, ranging from providing millions of meals of organic meat for human consumption, to reducing the number of deer–vehicle collisions, to supporting rural economies throughout North America. The effectiveness of conveying these messages to the non-hunting public will directly affect support for hunting in the future.

Recent research revealed that 76% of Americans approve of hunting while only 16% oppose (Responsive Management, 2006a). However, the degree of public support varies considerably by species hunted and motivation. In 2006, deer hunting enjoyed the highest degree of public support (78%) among all species examined (Responsive Management, 2006a). Overall, support was higher for species such as deer, turkey, and waterfowl than for predators such as bears and mountain lions.

A 2006 survey revealed that public support for hunting was highest (85%) when undertaken for meat consumption (Responsive Management, 2006a). Moreover, support was higher when conducted for animal population control (83%) or wildlife management (81%) than for personal motivations such as hunting for sport (53%) or to harvest a trophy (28%).

However, use of the terms "sport" and "trophy" likely bias survey results due to misconceptions regarding their true meanings. Regardless, these findings suggest that public support of hunting in the future will be influenced more by the benefits to wildlife populations and habitat than to hunters. Thus, careful consideration regarding the potential impact of a particular hunting activity or terminology on the overall level of public support for hunting is imperative.

Hunter Access

A major challenge facing the future of deer hunting is ensuring access to lands on which to hunt. This is especially true for private lands where most big-game hunting occurs. In 2006, 80% of big-game hunters hunted private lands compared to 35% who hunted public lands (some hunted both private and public lands; U.S. Fish and Wildlife Service, 2007). This disparity likely is greater for white-tailed deer than for western big-game species given the decreased availability of public land in the eastern United States. For example, the percentage of public land ranges from 58% to 80% in the Mountain and Pacific regions, respectively, compared to 6–16% in other regions (Leonard, 2004). Therefore, while only 16% of big-game hunters in the United States hunted only on public lands in 2006 (U.S. Fish and Wildlife Service, 2007), the percentage likely is lower for those hunting white-tailed deer. Addressing this difference is paramount to devising effective and equitable access solutions.

In some western states such as Idaho, Kansas, Utah, and Wyoming, state wildlife agencies compensate landowners for providing free public access for hunting (Responsive Management and National Shooting Sports Foundation, 2010). While these programs have opened millions of hectares of private land to hunting, similar programs remain uncommon in the eastern United States where most whitetail hunting occurs. Likely causes include smaller average property size, increased private ownership of hunting lands, and higher hunter densities.

Landowner compensation by hunters through hunting leases is a common access strategy, especially in the southern United States. Hunting leases can be controversial and often are cited by hunters as a barrier to access (Responsive Management and National Shooting Sports Foundation, 2008). However, as of 2006, only 6.9% of hunters in the United States leased land for hunting (U.S. Fish and Wildlife Service, 2007). Moreover, the number of hunters leasing land in the United States declined 14% from 982,000 in 2001 to 860,000 in 2006, compared to 2% growth from 1991 to 2001 (U.S. Fish and Wildlife Service, 1993, 2002, 2007; Leonard, 2004). This trend was reflected in the area leased, which declined 4% from 91.2 million hectares in 2001 to 87.8 million hectares in 2006. Possible reasons for the decline include increasing land ownership among hunters, loss of hunting lands to alternative land uses, and the divestiture of millions of hectares of timberlands in the southern United States by the forest products industry during the late 1990s and early 2000s.

Hunter involvement in leases varies considerably by region and is inversely correlated to availability of public land. For example, less than 5% of hunters lease land for hunting in the Pacific and Mountain regions (western half of United States) compared to nearly 25% in the West South Central region (Arkansas, Louisiana, Oklahoma, and Texas) (Leonard, 2004). Lease participation also varies by preferred hunting method. Archery and muzzleloader hunters are more than twice as likely to lease land for deer hunting than rifle and pistol hunters (15.1% and 7.2%, respectively) (Leonard, 2004).

Hunting lease rates vary from $2 to 185/ha, with most in the southern United States ranging from $7–25/ha compared to $20–77/ha in Illinois (Harper et al., 1999; Henderson and Moore, 2005; Eberle and Wallace, 2008). Lease rates are influenced by numerous variables including property size and location, habitat quality, deer density, buck quality, availability of other game, proximity to metro areas, and land values (Baen, 1997; Henderson and Moore, 2005; Eberle and Wallace, 2008).

When viewed in the context of hunter access, leasing is a double-edged sword. Leases provide financial incentives for landowners to open lands to hunting and security for participating hunters. They also provide opportunities for hunters to participate in management programs designed to improve deer quality and hunting experiences. Therefore it is not surprising that hunting effort is positively correlated with leasing. Those hunting deer more than 25 days annually are nearly three times as likely (17.4%) to lease land than those hunting 6–12 days (6.2%), and more than five times as likely as those hunting five days or less (3.2%) (Leonard, 2004). Wildlife and habitat management on leased lands can lead to habitat retention and improvement, increased wildlife biodiversity, and enhanced hunter satisfaction (Miller, 2010). Leases also can result in reduced property damage by deer and improved property security, especially for absentee landowners. However, leases can displace local hunters and provide hunting opportunity to fewer hunters. In general, leasing is positively correlated with a hunter's age, income, education, and population density of the hunter's residence (Leonard, 2004). Consequently, leases may present barriers for hunters with lower levels of education or income and those residing in rural areas.

Gaining access to private lands that are not leased or otherwise hunted is another significant challenge. This is especially true for lands owned by individuals rather than corporations, as hunters have reported experiencing greater difficulties securing access to these lands (Responsive Management and National Shooting Sports Foundation, 2008). Many landowners cite poor hunter behavior, safety, and liability as reasons they do not allow hunting. When combined with declining awareness of and participation in hunting, it is not surprising that an increasing number of properties feature "posted" or "no hunting" signs on trees, fences, and gates (Figure 14.19). This trend has significant implications for the future of deer management.

Wildlife agencies and hunters must establish landowner education and outreach programs that emphasize safety and promote ethical hunting behavior to improve access for deer hunting and management. Establishment of more comprehensive landowner liability laws would facilitate these efforts (Responsive Management, 2009). While many states have statutes that shield landowners from civil liability from hunting-related activities, most do not apply if the landowner receives compensation for hunting (e.g., lease or commercial hunting operation) and none cover legal defense costs (Harper et al., 1999; Wright et al., 2002). Thus, even in states with strong liability laws, an increasing number of landowners and hunters are purchasing specialized insurance that provides comprehensive liability coverage for hunting-related activities. Given current trends, it is likely that social, economic, and legal barriers will make accessing private land for hunting more difficult and costly in the future.

Overabundance

With white-tailed deer populations at record densities in many areas and hunter numbers in decline, managing overabundant deer populations will become increasingly challenging. Admittedly, the term "overabundant" is value-laden and relative. Even when whitetail populations are maintained within biological carrying capacity, they can exceed social carrying capacity and be considered overabundant by some segments of society (McCabe and McCabe, 1997). Therefore, in this chapter, overabundance is used in a broad sense to refer to deer populations exceeding either biological or social carrying capacity.

Managing overabundant urban deer populations is an increasing area of responsibility for many wildlife agencies, especially because the human population in the United States is expected to surpass 400 million by 2050 (U.S. Census Bureau, 2004) (Figure 21.3). Conover et al. (1995) estimated the number

FIGURE 21.3 Overabundant urban whitetail populations will present an increasing management challenge for state and provincial wildlife agencies in the future as human populations increase and hunting becomes more restricted in urban areas. (Photo by L. Thomas, Jr., Quality Deer Management Association. With permission.)

of deer–vehicle collisions at more than 1 million annually during the early 1990s. The estimated annual impacts of these collisions include 155–211 human fatalities, 13,713–29,000 human injuries, and more than $1 billion in property damage (Conover et al., 1995; Williams and Wells, 2005). Abundant whitetail populations contributed to a 25-fold increase in Lyme disease cases between 1982 and 2000 (DeNicola et al., 2000). Whitetails also cause significant agricultural, timber and household damage, conservatively estimated at more than $2 billion annually (Chapter 13).

Many wildlife agencies have established urban deer management plans incorporating a wide range of control measures, from extended seasons in urban areas, to controlled hunts with primitive weapons and use of private or governmental wildlife control specialists (DeNicola et al., 2000). Other techniques, such as immunocontraception, sterilization, repellants, and relocation, though costly and generally ineffective, likely will continue to be considered in specific situations given public interest in non-lethal control options.

State wildlife agencies have responded to burgeoning deer populations in rural areas by expanding hunting seasons and bag limits for antlerless deer, such that antlerless deer harvest is essentially unlimited in some areas. When combined with hunter education, most efforts have resulted in increased antlerless harvests, though some appear to be reaching a point of diminishing returns. Even with few regulatory constraints on harvest, hunters typically become self-limiting with respect to harvesting antlerless deer (Responsive Management, 2004). Some agencies have resorted to programs such as earn-a-buck, which require hunters to harvest one or more antlerless deer prior to being eligible to harvest a buck. As of 2008, at least 11 states used this strategy in all or a portion of their states. Though successful in some areas, surveys have revealed strong opposition to this strategy among hunters, often exceeding 50% (Responsive Management, 2004, 2005).

Another barrier to increasing antlerless deer harvests concerns use of harvested deer. Despite numerous state-sponsored "Hunters for the Hungry" programs and private venison distribution programs such as Farmers and Hunters Feeding the Hungry (FHFH), insufficient funding and infrastructure have precluded mainstream hunter participation. Experience has revealed that while many hunters are willing to harvest additional deer, few are willing to pay processing and distribution costs. Many states have experimented with measures to increase funding for their venison-distribution programs. In 1999, Maryland became the first state to offer a voluntary $1 contribution with hunting license sales to support venison-distribution programs in their state. This effort proved only marginally effective and was abandoned in 2002 with the passage of Senate Bill 599, which earmarked $100,000 annually for venison distribution programs in Maryland. In 2008, Ohio implemented an annual $100,000 subsidy to support venison distribution programs within the state. In 2010, Indiana implemented a voluntary $1 contribution with hunting license purchases. In many other states, efforts to implement either a voluntary or dedicated funding source for venison distribution programs have been met with opposition by state wildlife agencies, governmental officials, and even hunters concerned over agency funding or license fee increases (R. Wilson, FHFH, personal communication). Therefore, without significant changes in funding and infrastructure, existing venison distribution programs are unlikely to measurably impact whitetail overabundance, except in localized situations.

It is worth noting that the primary source of non-governmental funding for FHFH programs, which existed in 30 states and one Canadian province in 2010, came from the non-hunting sector through community grants and private contributions (R. Wilson, FHFH, personal communication). Thus, broader recognition by the non-hunting public of the societal benefits of feeding the nation's hungry and controlling overabundant deer populations offers a new and largely untapped funding source for these programs in the future.

Diseases

Diseases are another potential threat to the future of white-tailed deer. In general, two categories of disease threats pose the greatest risk: (1) those with potential to directly or indirectly reduce or eliminate whitetail populations across their geographic range, and (2) those with significant human health implications that could erode hunter participation or public support for hunting. Diseases of white-tailed deer have been the subject of extensive research for more than 50 years as evidenced by the formation of the Southeastern Cooperative Deer Disease Study (later renamed the Southeastern Cooperative Wildlife Disease Study) in 1957. However, diseases such as chronic wasting disease (CWD) and bovine tuberculosis (BT) have heightened awareness regarding the potential impact of diseases on the future of the species. While encouraging that CWD and BT are not widespread at this time, their cost to state and provincial wildlife agencies remains significant. Disease surveillance efforts that once involved testing a handful of suspect animals annually now involve systematic monitoring and considerable costs (Adams et al., 2010).

One of the greatest threats to the future of deer hunting and management is the discovery of a disease, pathogen, or toxin in wild deer with significant human health implications. The impacts of such a discovery are difficult to comprehend because public support for hunting is largely predicated on hunters consuming the animals they harvest (Responsive Management and National Shooting Sports Foundation, 2008). The initial public reaction to the discovery of CWD in Wisconsin provides a good example. Given the similarities of CWD to mad cow disease, fears of widespread human infection and possible death quickly spread throughout Wisconsin and all of North America, often fueled by erroneous media reports. However, as of 2010, there remained no evidence of CWD transmissibility to humans (Chronic Wasting Disease Alliance, 2010a).

Another public health scare occurred in 2008 following reports of elevated lead levels in hunter-harvested deer and processed venison in North Dakota (Hunt et al., 2009). This resulted in thousands of pounds of venison intended for the less fortunate being discarded. A subsequent study by the Centers for Disease Control revealed that while mean lead levels were higher in hunters than in non-hunters, they were not above the nationwide average (Iqbal et al., 2009). These examples stress the importance of basing management decisions on science and fact.

Predators

During the past century, most whitetail populations, especially those in the eastern United States, thrived in a relative absence of large predators. However, during the past 20 years predator populations have increased dramatically in size and distribution. This is especially true for coyotes, which are relative newcomers to the eastern United States and portions of Canada. Initially, many wildlife managers considered coyotes an insignificant source of mortality at the population level. However, recent research has challenged this view. Kilgo et al. (2009) reported that 44 of 60 (73%) fawns monitored in their South Carolina study died prior to being recruited into the fall hunting population. Of all fawn deaths for which the source of mortality could be determined, coyote predation accounted for 64–84% during the three-year study. VanGilder et al. (2009) reported 193–256% increase in fawn survival following intensive predator removal on an 800-ha study site in Alabama.

While coyotes are the primary fawn predator throughout much of the whitetail's range, black bears, wolves, and bobcats also can play significant roles. A Minnesota study attributed 49% and 51% of predator-related fawn mortality to black bears and wolves, respectively (Kunkel and Mech, 1994). A subsequent Minnesota study by Carstensen et al. (2009) reported that predation accounted for 86% of fawn mortality, with black bears causing 57% and 38% of all mortalities during years one and two of the study, respectively, and bobcats causing 50% during year two. Wolves accounted for only 2% of predator-related fawn mortalities in this study. A Pennsylvania study by Vreeland et al. (2004) attributed 33% of fawn predation events to black bears and 37% to coyotes.

While impacts of predators on fawn survival are increasingly well documented, predator impacts at the population level remain poorly understood. However, a growing body of anecdotal evidence suggests predators can significantly reduce or limit recovery of some deer herds. Predator impacts likely are influenced by a complex mix of factors including deer density, predator density, availability of alternative prey, habitat quality, fawn recruitment, and intensity of deer harvest (Chapter 8). In areas with high deer densities or inadequate removals through hunting or other mortality sources, predators will be an asset. In areas with low deer densities, marginal habitat, or high hunter harvest, predation may prove problematic. Thus, predators will present both challenges and opportunities for future deer managers.

Privatization

Of all challenges facing the future of deer hunting and management, few are more complex and contentious than privatization. While high-fencing and captive deer breeding are widely considered practices which privatize the resource, effects of management practices such as supplemental feeding, baiting, and mineral supplementation are less clear and widely debated among hunters and wildlife professionals alike.

A primary tenet of the North American Model of Wildlife Conservation is that wildlife belongs to all people (Geist, 2006). This principle is problematic at best as it relates to white-tailed deer hunting and management on private lands. The legal framework for land ownership in North America provides de facto ownership of wildlife and broad authority to restrict access (Benson, 1992). A more relevant discussion is the extent to which society should allow these de facto rights to become incorporated in common law.

The large and growing number of captive cervid operations (deer farms, deer breeding facilities, and hunting enclosures) throughout North America is an emerging concern to state and provincial wildlife agencies (Figure 21.4). Anderson et al. (2007) estimated the number of captive cervid operations in the United States at 7828 in 2007. Five states were home to more than 500 facilities including Texas (1006), Pennsylvania (1000), Minnesota (722), Ohio (666), and Wisconsin (611). Only two states, South Carolina and Utah, were reported to have no captive cervid operations. However, due to variations in regulations, additional cervid operations exist, including several in South Carolina (C. Ruth, South Carolina Department of Natural Resources, personal communication). The total economic impact of this industry in 2007 was about $3 billion and it supported more than 29,000 jobs (Anderson et al., 2007). In Canada, as of 2002, only three of 13 provinces (Labrador/Newfoundland, the Nunavut, and Northwest Territories) had no captive cervid operations (Van Putten et al., 2002). The remaining 10 were home to approximately 2600 facilities.

As with other niche industries, the value of breeder bucks, breeder does, fawns, and semen straws is influenced by supply and demand. However, the sustainability of this industry appears likely given the widespread and growing demand for bucks with large antlers. This is evidenced by the success of the 1st Annual All American Whitetail Deer Auction held by the Texas Deer Association in 2009. This event attracted 138 deer breeders from eight states (Alabama, Iowa, Ohio, Oklahoma, Pennsylvania, Missouri, Texas, and Wisconsin; Texas Deer Association, 2009) and reportedly grossed more than $1.5 million.

The growth of the deer breeding industry has heightened concerns regarding its potential impact on wild deer herds and the future of hunting. One area of concern is the impact of selective breeding on antler size and configuration. Today, it is not uncommon for 1.5-year-old bucks bred in captivity to

FIGURE 21.4 The large and growing number of deer farms, deer breeding facilities, and hunting enclosures is an emerging concern given their implications with privatization, disease transmission, and public support for hunting. (Photo by L. Thomas, Jr., Quality Deer Management Association. With permission.)

have antlers with gross Boone and Crockett scores exceeding 200 in., with scores for some adult bucks exceeding 400 in. (Texas Deer Association, 2009). While state and provincial regulations prevent or tightly govern captive-bred bucks being released into the wild, and fair chase guidelines from groups such as the Boone and Crockett Club prevent bucks killed inside enclosures from being entered into the record books, the potential for abuse remains a valid concern.

Another area of concern is the potential for disease introduction to wild deer herds, especially given the number of documented escapes. For example, from 2006 to 2009, 595 escapes were reported in Wisconsin, including 266 whitetails, 78 elk, and 251 exotics (P. Dunn, Wisconsin Department of Natural Resources, personal communication). Given the presence of CWD in both wild and captive herds in Wisconsin, wildlife managers are understandably concerned. In 2004, Michigan, another CWD-positive state, documented 456 previously unreported escapes. Thus, it is undeniable that captive cervid facilities pose tangible threats to wild whitetail populations. However, regulatory oversight of these facilities varies considerably. According to the CWD Alliance, state or provincial departments of agriculture have jurisdiction over captive cervids in 10 states and four Canadian provinces (Chronic Wasting Disease Alliance, 2010b). State or provincial wildlife agencies have authority in nine states and five provinces. In the remaining 31 states and one province, authority is shared between agencies. In many cases, regulatory authority is a complex mix of lobbied and statutory law, which creates confusion, loopholes, and limits management efforts. Given that state and provincial wildlife agencies, not departments of agriculture, are entrusted to ensure the health and sustainability of wild deer populations, the current regulatory approach presents valid concerns for the future of deer management.

Growing interest among hunters in improving the size, health, distribution, and harvest of white-tailed deer has increased involvement in supplemental feeding, baiting, and mineral supplementation and placed pressure on state agencies to legalize these practices where not currently permitted. As of 2009 in the continental United States, supplemental feeding was legal in 29 states, partially permitted in eight states, prohibited in eight states, and unknown in three states. Baiting was legal in 15 states, partially permitted in seven states, prohibited in 24 states, and unknown in two states. In most states, minerals are governed under existing baiting or supplemental feeding regulations.

The nutritional, ecological, and ethical arguments against supplemental feeding and baiting are well documented (Brown and Cooper, 2006). The potential role of these practices in transmission of diseases such as CWD and BT has led some state agencies to prohibit or greatly reduce these practices, especially in or near known disease areas, though hunter resistance has occasionally hampered agency efforts (Rudolph et al., 2006; Van Deelen et al., 2006). Despite potential implications for disease transmission, privatization, and decreased public acceptance of hunting, continued growth of these practices is likely.

No state has been at the epicenter of the privatization discussion and debate more than Texas. A state well known for strong landowner rights and minimal governmental regulation, Texas landowners have fundamentally similar legal rights for both deer and livestock. For example, under the state's Trap, Transport, and Transplant (TTT) program, landowners can capture wild deer and relocate them to another property or place them in breeding pens before releasing them into the wild (Texas Parks and Wildlife Department, 2010a). Landowners and managers also have the greatest management flexibility on private lands under the state's Managed Lands Deer Permit (MLDP) program. This program features three potential levels of involvement ranging from increased antlerless harvest opportunities at level one to the ability to harvest antlerless deer and antlered bucks by any lawful means during an extended hunting season framework at level three (Texas Parks and Wildlife, 2010b). All MLDP programs require written wildlife management plans and increasing requirements for population data, harvest data, and habitat management. While Texas is clearly unique, broader societal trends toward privatization likely will increase demand for and participation in these practices throughout the whitetail's range in the future.

Agency Funding

A looming challenge for the future is how to ensure adequate funding for state and provincial wildlife agencies. For nearly a century, hunters, trappers, and firearm owners have funded wildlife conservation and management through license sales and Pittman–Robertson funds. However, some contend that iron-triangle relationships exist among wildlife agencies, consumptive users (e.g., hunters and anglers), and policy makers that exclude other stakeholders from the decision-making process (Gill, 2004). While a gross oversimplification, it is undeniable that sportsmen have been the most influential special interest group. However, given declining license sales and increasing agency responsibilities, the traditional user-pay approach appears unsustainable. Simply extending this approach to other narrow groups (e.g., birdwatchers and campers) fails to recognize the economic limitations of this strategy (Gill, 2004).

From a philosophical perspective, some argue the user-pay approach is inconsistent with the fundamental premise of the Public Trust Doctrine that wildlife is owned by no one and held in trust by governments for public benefit (Jacobson et al., 2010). Gill (1996) suggested that the current user-pay approach blurs the line between public and special interest and erodes scientific credibility and public trust. McCabe and McCabe (1997) argued that use of Pittman–Robertson funds for game species was not only justified given hunters' contributions, but necessary as an incentive to expand the funding base for all wildlife. It is not debatable, however, that hunters have been the physical and financial engine behind the world's most successful wildlife conservation program to the benefit of both game and nongame wildlife.

Jacobson et al. (2010) argued the need for establishment of a broad-based and reliable nonvoluntary contribution mechanism. In theory, a centralized funding mechanism would encourage all citizens to provide input on wildlife conservation and management issues. The closest such systems exist in Missouri, Arkansas, and Virginia where state wildlife agencies receive a portion of sales tax revenues (Organ and Fritzell, 2000). In Missouri, hunters have benefited from a private-lands program that would not have been possible without the additional funding (L. Hanson, Missouri Department of Natural Resources, personal communication). Thus, despite concerns that hunters will become marginalized under a broad-based funding mechanism, this approach offers a viable alternative to the traditional user-pay system.

Hunter Support within the Wildlife Profession

A growing challenge is how to maintain hunter support within the wildlife profession as it evolves to meet the needs of non-consumptive users. Brown et al. (2006) documented declines in hunting participation among members of The Wildlife Society (TWS) from 1994 to 2004. In 2004, participation in big-game hunting was highest among state agency personnel (79%), followed by TWS members and federal agency employees at 50% each. In all three groups, consumptive use of wildlife was greater among wildlife professionals with 10 or more years of experience (Brown et al., 2006). Thus, if these trends continue, hunting participation among wildlife professionals will decline, which could erode the level of support for hunting within the profession. According to McCabe and McCabe (1997, p. 21), "It is no less irresponsible for wildlifers to dismiss hunting as a practical and practicable management method … any more than it is to suppose that all regulated hunting is biologically defensible."

University programs and student attitudes toward consumptive use of wildlife also are changing as evidenced by a survey of 14 representative fish and wildlife programs at major universities in the United States (Organ and Fritzell, 2000). In this study, senior wildlife faculty reported that hunting participation among students declined 10–60% over the past 20 years while students with anti-hunting attitudes increased 30–50%. Changes in course content most commonly included an increased focus on conservation biology and human dimensions and a decreased emphasis on species harvest management. While these changes likely are a reflection of broad societal changes, increasing agency mandates, and changing employment opportunities within the wildlife profession, they pose potential future barriers to consumptive use of wildlife.

The Future Outlook

The future of deer management will be shaped by an increasingly complex mix of biological, social, political, and economic forces. These forces will determine the public's acceptance of hunting, the role of hunters in wildlife management, and agency policies and priorities. Heberlein (1991, p. 528) stated, "Our current and highly successful system of scientific, rational, and government-centered wildlife management is something like a heavily loaded Great Lakes freighter, constructed in the latter half of the nineteenth century with its course set in the first decades of the twentieth." What course will whitetail hunting and management take in the twenty-first century? The answer will be determined by the effectiveness with which hunters, wildlife agencies, and the public establish ecologically responsible and adaptive solutions to future challenges. This process will impact stakeholders in unique ways, with some likely outcomes for each group discussed below.

Hunters

Despite recent trends, the rate of deer-hunter decline is slowing and likely will stabilize in response to increasing youth and female participation and numerous hunter recruitment, retention, and land-access initiatives. Thus, fears regarding the demise of deer hunters in North America appear unfounded, at least for the foreseeable future. Deer hunters will certainly become more diverse. Increasing participation among females, Hispanics, and other non-traditional groups will alter the profile of the average deer hunter. One example of a non-traditional group entering the hunting arena is locavores, or those who procure their food from local sources in or near their communities. This growing movement has generated interest in hunting among urban residents seeking sources of hormone- and antibiotic-free meat from animals raised in humane conditions. Importantly, most locavores have no prior experience with or involvement in hunting. If sustainable, this movement offers a large and previously untapped source for future deer hunters. The impact of this group on deer hunting and management could be substantial given increasing urban deer populations.

Future deer hunters will be better educated and more engaged as noted by Miller and Marchinton (1995, p. 2), "The deer hunting tradition has changed in recent years and will continue to do so. In many regions, hunters are no longer satisfied with the status quo. They are better educated and more sophisticated than ever before." Deer hunters also will become more specialized, organized, and involved in the decision-making process (Ditton et al., 1992; Adams et al., 2009; Figure 21.5). However, given declining hunter participation within the wildlife profession and the likelihood of continued agency shift toward non-consumptive wildlife uses, hunter influence on wildlife management will be challenged. The possibility of waning influence has led to proliferation of non-governmental organizations (NGOs) and increased collaboration on a wide range of hunting and conservation issues (Adams et al., 2000). Current examples include the American Wildlife Conservation Partners and Theodore Roosevelt Conservation Partners, which represent dozens of hunting and conservation organizations and millions of sportsmen.

FIGURE 21.5 Future deer hunters will be better educated, more engaged, more organized, and more involved in the decision-making process than those in the past. (Photo by L. Thomas, Jr., Quality Deer Management Association. With permission.)

Hunters also have become more politically engaged as evidenced by the formation of groups such as the National Assembly of Sportsmen's Caucuses and the Congressional Sportsman's Foundation. Individual deer advocacy organizations, such as the Quality Deer Management Association, also will play increasingly significant roles in deer hunting and management policy. The degree to which hunters and conservation organizations can present a large and unified voice will largely determine their impact on agency deer management programs.

Involvement in intensive deer management on private lands will increase, which will continue fueling associated trends including private ownership of hunting lands, food plots, habitat management, supplemental feeding, and mineral supplementation. When combined with increasing habitat fragmentation, the average hunting property in the future will be smaller, but more intensively managed. Smaller property ownership will increase the number of QDM cooperatives and, possibly, the number of high-fenced properties. Continued expansion of the captive cervid industry also is likely unless profitability substantially declines or diseases provide the necessary social and political support to restrict industry growth.

Given current trends, is North America heading down the "European pathway" where only the wealthy or well-connected will have access to deer hunt in the future? Current data suggest that participation in hunting among those with limited incomes will decline. For example, from 1995 to 2005, the hunting initiation rate of children was more than four times lower in households with incomes less than $25,000 than those with incomes more than $40,000 (Leonard, 2007). Therefore, hunting will almost assuredly become a pastime of the middle and upper classes in the future. Similarly, increasing costs associated with land ownership and management suggest access to private land will become more restricted and expensive, a trend expected for any finite resource. However, land ownership in North America is not restricted to the socially elite as it is in many other regions of the world. In 2006, approximately 10% of sportsmen hunted on lands they owned (U.S. Fish and Wildlife Service, 2007). Research by Leonard (2004) revealed that the average land ownership rate among deer hunters was above 6% for all income levels. Thus, land ownership will remain a viable option for many deer hunters in the foreseeable future.

An important safeguard regarding the democracy of hunting in the future is the Public Trust Doctrine which requires the government to preserve and maintain certain resources for the public's reasonable use (Geist, 2006). This common law extends to recreational activities, including hunting, on public lands. While this will help ensure continued access to public lands, it does little to ensure the quality of hunting experiences on those lands. Thus, the quality of deer hunting on public lands remains a legitimate area of concern for the future, especially given the increasing influence of special interest groups on agency priorities and practices on such lands.

Trends associated with increasing land ownership and habitat management among hunters provide genuine optimism for the future of wildlife conservation and biodiversity. This is especially true in the eastern United States where less public land exists. As of 2006, hunters owned more than 54 million hectares in the United States (U.S. Fish and Wildlife Service, 2007). A 2008 QDMA survey revealed that its members owned more than 6 million hectares in the United States. Often guided by dreams of quality deer and quality hunting experiences, properties throughout North America are being transformed into Leopold Landscapes. Therefore, not only should deer hunters be thanked for their role in restoring whitetail populations and providing the lion's share of funding for wildlife conservation and management, they should be credited with providing a brighter future for most North American wildlife.

Agencies

State and provincial wildlife agencies will experience many changes in the future as license revenues decline and responsibilities increase. There will be fewer wildlife biologists with hunting experience and a decreased emphasis on deer management programs and services. Hunter education and outreach efforts will decline, except where they assist in conflict resolution (e.g., urban, agricultural, or disease areas). Reduced agency resources for deer management and hunter recruitment will increase opportunities for private wildlife consultants, university extension programs, and NGOs. Enforcement of deer hunting regulations also will decline with decreasing funding and expanding wildlife officer responsibilities.

Agencies will increasingly view deer hunters as one of many "customers" they serve rather than their core constituency. New licenses and fees will emerge, especially for non-consumptive users. Agencies will collect detailed demographic data on each group and use segmented marketing strategies to attract and retain customers. They will use traditional business models to provide internal and political justification and to guide management decisions. In essence, wildlife agencies will resemble private enterprise in the twenty-first century. Accordingly, many functions currently provided by wildlife agencies will be provided by the private sector in the future (Figure 21.6). Current examples include wildlife consulting and urban deer management in the United States and hunter education in Ontario, Canada.

Agencies will become more centralized, automated, and Web-based. Hunter surveys and telecheck systems will replace mandatory deer check stations. Traditional hunting license agents will be replaced by centralized Web- and telephone-based systems. Physical copies of hunting regulations, and perhaps hunting licenses, will give way to digital or electronic versions. When combined with inevitable license fee increases, deer hunting will be more expensive and technology-based in the future. This will exacerbate participation barriers for the poor and uneducated.

Hunter demand for older bucks will lead to expansion of antler restriction programs (e.g., antler point restrictions) on private and public lands. This will stimulate development of more biologically sound techniques to protect young bucks like those currently in Mississippi (minimum antler spread or beam length) and portions of Texas (slot limit). Deer managers on private land will have increased flexibility and decreased agency oversight. Increasing hunter demand for QDM opportunities on public lands will present significant agency challenges, especially given staff and budget constraints. Several states have implemented antler restrictions or quota hunts on public lands, though such opportunities remain limited in most regions. Agencies also will be challenged with devising new population estimation models as buck harvests become increasingly biased toward older bucks.

Efforts to control overabundant herds will result in further season, bag limit, and weapons expansions, especially for antlerless deer. However, once these efforts are no longer possible, deer management will become compromised, especially on private lands. Consequently, where hunters and landowners desire abundant deer populations on private lands, such populations are likely to persist. However, in forested and agricultural areas where landowners desire deer densities lower than those that can be achieved through traditional hunting and crop depredation programs, new approaches likely will emerge.

Efforts to increase flexibility for landowners and agricultural producers to control deer herds on private lands have occurred in several states. Examples include a "shoot and let lay" proposal in Delaware (House Bill 387, 2006) and one in Nebraska to allow harvest of deer at night without seasons or bag limits (Legislative Bill 836, 2010). These initiatives were met with stiff resistance from the wildlife and sporting communities and were unsuccessful, though such demands are likely to continue.

While contrary to the foundation of the North American Model of Wildlife Conservation, a commercial industry for wild-harvested white-tailed deer remains possible, if not likely, given current trends.

FIGURE 21.6 Given current trends, many services currently provided by wildlife agencies likely will be provided by the private sector in the future like providing detailed, site-specific guidance to landowners and hunters. (Photo by L. Thomas, Jr., Quality Deer Management Association. With permission.)

Such programs are already common for native cervids in Europe and introduced cervids in New Zealand. Agencies will increasingly embrace private deer control companies in urban areas where hunting is not feasible. However, given agency resistance to commercialization of wildlife, dedicated funding for state and private venison distribution programs is likely, especially where such programs help achieve adequate deer harvests in urban and agricultural areas.

Agencies also will face increasing challenges with diseases and predators. Monitoring and management of diseases such as CWD and BT will continue to drain critical agency resources. Other disease issues likely will emerge, some with the potential to impact human health or the livestock industry. Depending on the situation, predators will become either an ally or an impediment to effective deer management. In the long term, predators likely will play an increasingly important role in balancing deer populations with broader conservation objectives. Clearly, the role of wildlife agencies will be very different in the twenty-first century.

Public

Given the high degree of public support for deer hunting, animal rights organizations are unlikely to significantly impact deer management efforts in the near term. This is especially true given the high value Americans place on individual rights and freedoms to pursue cultural traditions (Johnson, 1997). However, support for hunting is likely to decline as fewer citizens hunt or know someone who does. A potential safeguard exists if hunters and wildlife professionals can effectively demonstrate the positive role of hunting in wildlife conservation. To achieve such recognition, the public will require increasing levels of hunter education, proficiency, and accountability, especially in or near urban areas. Basic hunter education courses will give way to more comprehensive programs like those in Europe. Increased firearms restrictions could exacerbate these requirements. This forecast, while in conflict with the current trend to reduce hunter education requirements through Families Afield legislation, is based on a longer term view. Regardless, the public must recognize the vital services hunters provide society or be willing to pay for those services.

One thing is certain; deer management will look very different in the future. As society becomes more urbanized, diverse, and detached from nature, its expectations of wildlife agencies and hunters regarding deer management will fuel many changes. Demands on agencies to resolve deer–human conflicts in agricultural and urban areas will increase. Agency involvement in deer management on private lands will diminish as landowners and hunters are afforded increased flexibility within a broad regulatory framework. Deer hunters will become one of many customers agencies serve, with cost–benefit models guiding expenditures for hunter recruitment and retention, agency services, and management intensity on public lands. Hunters will become more educated, affluent, involved in intensive deer management activities, and engaged in advocacy and politics.

These predictions are based on current trends and therefore not meant to suggest a concrete path for the future. According to Alan Kay, inventor of the windows-based platform that revolutionized computer technology, "The best way to predict the future is to invent it." The challenge is determining what we are trying to invent, what tools will be necessary, and if our invention has the functional and operational capacity to provide a sustainable future for North America's most hunted and most economically important big-game species.

REFERENCES

Adams, C. E., N. Wilkins, and J. L. Cooke. 2000. A place to hunt: Organizational change in recreational hunting, using Texas as a case study. *Wildlife Society Bulletin* 28:788–796.

Adams, K., J. Hamilton, and M. Ross. 2009. *QDMA's Whitetail Report 2009*. Bogart, GA: Quality Deer Management Association.

Adams, K., J. Hamilton, and M. Ross. 2010. *QDMA's Whitetail Report 2010*. Bogart, GA: Quality Deer Management Association.

Anderson, D. P., B. J. Frosch, and J. L. Outlaw. 2007. *Economic Impact of the United States Cervid Farming Industry*. College Station, TX: Texas A&M University Agricultural and Food Policy Center.

Baen, J. S. 1997. The growing importance and value implications of recreational hunting leases to agricultural land investors. *Journal of Real Estate Research* 14:399–414.

Benson, D. E. 1992. Commercialization of wildlife: A value-added incentive for conservation. In *The Biology of Deer,* ed. R. D. Brown, 539–553. New York, NY: Springer-Verlag.

Brothers, A. and M. E. Ray, Jr. 1975. *Producing Quality Whitetails*. Laredo, TX: Fiesta Publishing Company.

Brown, R. D. and S. M. Cooper. 2006. The nutritional, ecological, and ethical arguments against baiting and feeding white-tailed deer. *Wildlife Society Bulletin* 34:519–524.

Brown, T. L., N. A. Connelly, and D. J. Decker. 2006. *Participation in and Orientation of Wildlife Professionals Toward Consumptive Wildlife Use: A Resurvey,* 06–01. Ithaca, NY: Cornell University Human Dimensions Research Unit.

Carstensen, M., G. D. Delgiudice, B. A. Sampson, and D. W. Kuehn. 2009. Survival, birth characteristics, and cause-specific mortality of white-tailed deer neonates. *Journal of Wildlife Management* 73:175–183.

Chronic Wasting Disease Alliance. 2010a. *Learn About CWD:FAQ*. http://www.cwd-info.org/index.php/fuse-action/about.faqDetail/ID/44f5d3ac2048816 aacedd9d28fb50f8. Accessed March 30, 2010.

Chronic Wasting Disease Alliance. 2010b. *Chronic Wasting Disease and Cervidae Regulations in North America*. http://www.cwd-info.org/index.php/fuseaction/policy.regulationsMap. Accessed April 1, 2010.

Collier, B. A. and D. G. Krementz. 2006. White-tailed deer management practices on private lands in Arkansas. *Wildlife Society Bulletin* 34:307–313.

Conover, M. R., W. C. Pitt, K. K. Kessler, T. J. DuBow, and W. A. Sanborn. 1995. Review of human injuries, illness, and economic losses caused by wildlife in the United States. *Wildlife Society Bulletin* 23:407–414.

DeNicola, A. J., K. C. VerCauteren, P. D. Curtis, and S. E. Hygnstrom. 2000. *Managing White-tailed Deer in Suburban Environments: A Technical Guide*. Ithaca, NY: Cornell University Media and Technology Services Resource Center.

Ditton, R. B., D. K. Loomis, and S. Choi. 1992. Recreation specialization: Re-conceptualization from a social world's perspective. *Journal of Leisure Research* 24:33–51.

Dougherty, C. 2006. Leopold landscapes. In *Quality Food Plots: Your Guide to Better Deer and Better Deer Hunting*, eds. K. Kammermeyer, K. Miller, and L. Thomas, 204–209. Bogart, GA: Quality Deer Management Association.

Duda, M. D. 2001. The hunting mind: Women and hunting. *North American Hunter* November:35–37.

Duda, M. D., S. J. Bissell, and K. C. Young. 1995. *Factors Related to Hunting and Fishing Participation in the United States: Phase V: Final report*. Harrisonburg, PA: Responsive Management.

Eberle, P. R. and R. Wallace. 2008. Recreational leases as means to increase landowner income. *Journal of the ASFMRA* 71:28–36.

Edwards, S. L., S. Demarais, B. Watkins, and B. K. Strickland. 2004. White-tailed deer forage production in managed and unmanaged pine stands and summer food plots in Mississippi. *Wildlife Society Bulletin* 32:739–745.

Geist, V. 2006. The North American model of wildlife conservation. In *Gaining Ground: In Pursuit of Ecological Sustainability*, ed. D. M. Lavine, 285–293. Guelph, Ontario: International Fund for Animal Welfare and Limerick-University of Limerick.

Gill, R. B. 1996. The wildlife professional subculture: The case study of the crazy aunt. *Human Dimensions of Wildlife* 1:60–69.

Gill, R. B. 2004. Challenges of change: Natural resource management professionals engage their future. In *Society and Natural Resources: A Summary of Knowledge,* eds. M. J. Manfredo, J. J. Vaske, B. L. Bruyere, D. R. Field, and P. Brown, 35–46. Jefferson City, MO: Modern Litho.

Gilliland, C. E., J. Robertson, and H. Cover. 2003. *Texas Rural Land Prices, 2003*. College Station, TX: Texas A&M University.

Green, D. and J. P. Stowe, Jr. 2000. Quality deer management: Ethical and social issues. *Human Dimensions of Wildlife* 5:62–71.

Harper, C. A. 2008. *A Guide to Successful Wildlife Food Plots: Blending Science and Common Sense*. Knoxville, TN: University of Tennessee Institute of Agriculture.

Harper, C. A., C. E. Dixon, P. M. Jakus, and D. A. Barefield. 1999. *Earning Additional Income Through Hunt Leases on Private Land*. Knoxville, TN: Agricultural Extension Service, University of Tennessee, PB1627.

Heberlein, T. A. 1991. Changing attitudes and funding for wildlife—Preserving the sport hunter. *Wildlife Society Bulletin* 19:528–534.

Henderson, J. and M. Akers. 2009. Cropland values strengthen. *Survey of 10th District Agricultural Credit Conditions* fourth quarter 2009.

Henderson, J. and S. Moore. 2005. *The Impact of Wildlife Recreation on Farmland Values.* Kansas City, MO: The Federal Reserve Bank of Kansas City Economic Research Department, RWP 05–10.

Hunt, W. G., R. T. Watson, J. L. Oaks et al. 2009. Lead bullet fragments in venison from rifle-killed deer: Potential for human dietary exposure. *PLos ONE* 4:1–6.

Iqbal, S., W. Blumenthal, C. Kennedy et al. 2009. Hunting with lead: Association between blood lead levels and wild game consumption. *Environmental Research* 109:952–959.

Jacobson, C. A., J. F. Organ, D. J. Decker, G. R. Batcheller, and L. Carpenter. 2010. A conservation institution for the 21st century: Implications for state wildlife agencies. *Journal of Wildlife Management* 74:203–209.

Johnson, P. 1997. *A History of the American People*. New York, NY: Harper Collins.

Kammermeyer, K., K. V. Miller, and L. Thomas. 2006. *Quality Food Plots: Your Guide to Better Deer and Better Deer Hunting*. Bogart, GA: Quality Deer Management Association.

Kammermeyer, K. E. and E. B. Moser. 1990. The effect of food plots, roads, and other variables on deer harvest in Northeastern Georgia. *Proceedings of the Annual Conference of the Southeast Association of Fish and Wildlife Agencies* 44:364–373.

Kilgo, J. C., H. S. Ray, M. J. Goode, M. A. Vukovich, and C. Ruth. 2009. Impact of coyotes on fawn survival in South Carolina. Paper presented at the 32nd annual meeting of the Southeast Deer Study Group, Roanoke, VA.

Kunkel, K. E. and L. D. Mech. 1994. Wolf and bear predation on white-tailed deer fawns in Minnesota. *Canadian Journal of Zoology* 72:1557–1565.

Latham, R. E., J. Beyea, M. Benner et al. 2005. *Managing White-tailed Deer in Forest Habitat from an Ecosystem Perspective: Pennsylvania Case Study.* Report by the Deer Management Forum for Audubon Pennsylvania and Pennsylvania Habitat Alliance. Harrisburg, PA.

Leonard, J. 2004. *Deer Hunting in the United States: An Analysis of Hunter Demographics and Behavior.* Arlington, VA: U.S. Fish and Wildlife Service.

Leonard, J. 2007. *Fishing and Hunting Recruitment and Retention in the U.S. from 1990 to 2005.* Arlington, VA: U.S. Fish and Wildlife Service.

Louv, R. 2008. *Last Child in the Woods*. Chapel Hill, NC: Algonquin Books of Chapel Hill.

Marchinton, R. L. 1997. Obstacles to the future of deer management. *Quality Whitetails* 4:21–23.

McCabe, T. R. and R. E. McCabe. 1997. Recounting whitetails past. In *The Science of Overabundance: Deer Ecology and Population Management*, eds. W. J. McShea, H. B. Underwood, and J. H. Rappole, 11–26. Washington, DC: Smithsonian Institution Press.

Miller, D. A. 2010. Lessons from the piney woods: Blending deer management and commercial forestry to benefit biodiversity. *The Wildlife Professional* 4:34–37.

Miller, K. V. and R. L. Marchinton, eds. 1995. *Quality Whitetails: The Why and How of Quality Deer Management*. Mechanicsburg, PA: Stackpole Books.

Murphy, B. 2006. Introduction to quality food plots. In *Quality Food Plots: Your Guide to Better Deer and Better Deer Hunting*, eds. K. Kammermeyer, K. Miller, and L. Thomas, 4–11. Bogart, GA: Quality Deer Management Association.

Murphy, B., D. Stewart, S. Demarais et al. 2001. *Developing Successful Quality Deer Management Cooperatives*. Bogart, GA: Quality Deer Management Association.

National Shooting Sports Foundation. 2003. *Women's Participation in the Shooting Sports.* Newtown, CT: National Shooting Sports Foundation Information Services Report.

National Shooting Sports Foundation. 2005. A profile of today's hunter: Social and economic highlights. *Industry Intelligence Reports* 1:1–16.

National Shooting Sports Foundation. 2008. *Industry Reference Guide*. Newtown, CT: National Shooting Sports Foundation.

National Shooting Sports Foundation. 2010. http://www.familiesafield.org/Successes.cfm. Accessed March 30, 2010.

Organ, J. F. and E. K. Fritzell. 2000. Trends in consumptive recreation and the wildlife profession. *Wildlife Society Bulletin* 28:780–787.

Quality Deer Management Association. 2001. *White-tailed Deer Density Map—1999.* Bogart, GA: Quality Deer Management Association.

Quality Deer Management Association. 2008. *White-tailed Deer Density Map—2005*. Bogart, GA: Quality Deer Management Association.

Responsive Management. 1995. *Factors Related to Hunting and Fishing Participation in the United States: Final Report*. Harrisonburg, VA: Responsive Management.

Responsive Management. 2004. *Opinions and Attitudes of Georgia Residents, Hunters, and Landowners Toward Deer Management in Georgia*. Harrisonburg, VA: Responsive Management.

Responsive Management. 2005. *Opinions of the General Population, Hunters, and Farmers Regarding Deer Management in Delaware*. Harrisonburg, VA: Responsive Management.

Responsive Management. 2006a. Sportsmen's attitudes. *Unpublished Survey about Various Hunting and Fishing Issues*. Harrisonburg, VA: Responsive Management.

Responsive Management. 2006b. *The Public's Attitudes Toward and Participation in the Shooting Sports*. Harrisonburg, VA: Responsive Management.

Responsive Management. 2009. *Issues Related to Hunting Access in the United States: Wisconsin Results*. Harrisonburg, VA: Responsive Management.

Responsive Management and National Shooting Sports Foundation. 2008. *The Future of Hunting and the Shooting Sports: Research-Based Recruitment and Retention Strategies*. Harrisonburg, VA: Responsive Management.

Responsive Management and National Shooting Sports Foundation. 2010. *Issues Related to Hunting Access in the United States*. Harrisonburg, VA: Responsive Management.

Ricks, W. E. 2010. Response of non-game species to white-tailed deer food plots in Appalachian hardwood forests. MS thesis, University of Georgia.

Rudolph, B. A., S. J. Riley, G. J. Hickling, B. J. Frawley, M. S. Garner, and S. R. Winterstein. 2006. Regulating hunter baiting for white-tailed deer in Michigan: Biological and social considerations. *Wildlife Society Bulletin* 34:314–321.

Southwick Associates. 2005. *Families Afield: Revised Youth Hunter Report*. Fernandina Beach, FL: Southwick Associates.

Texas Deer Association. 2009. *The All American Invitational Whitetail Deer Auction*. San Antonio, TX: Texas Deer Association.

Texas Parks and Wildlife Department. 2004. *A Guide for Wildlife Management Associations and Co-Ops*. Austin, TX: Texas Parks and Wildlife Department.

Texas Parks and Wildlife Department. 2010a. http://www.tpwd.state.tx.us/publications/pwdforms/media/pwd_1135_w7000_trap_transport_transplant_permit_app.pdf. Accessed March 28, 2010.

Texas Parks and Wildlife Department. 2010b. http://www.tpwd.state.tx.us/publications/pwdforms/media/pwd_1047_w7000_mldp_program_information_white_tailed_deer.pdf. Accessed March 28, 2010.

U.S. Census Bureau. 1995. Table 1. Urban and rural population: 1900–1990. http:/www.census.gov/population/censusdata/urpop0090.txt. Accessed May 13, 2010.

U.S. Census Bureau. 2000. Census 2000 summary file 1, matrix p1. http://factfinder.census.gov/servlet/GCTTable?_bm=y&-geo_id=01000US&-_box_head_nbr=GCT-P1&-ds_name=DEC_2000_SF1_U&-redoLog=false&-mt_name=DEC_2000_SF1_U_GCTP1_US1&-format=US-1. Accessed May 13, 2010.

U.S.D.A. Forest Service. 2008. *Inventory and Analysis Mapmaker 3.0 Program*. http://fia.fs.fed.us/tools-data/other/default.asp. Accessed December 8, 2008.

U.S. Fish and Wildlife Service. 1993. *1991 National Survey of Fishing, Hunting, and Wildlife-Associated Recreation*. Washington, DC: U.S. Fish and Wildlife Service.

U.S. Fish and Wildlife Service. 1997. *1996 National Survey of Fishing, Hunting, and Wildlife-Associated Recreation*. Washington, DC: U.S. Fish and Wildlife Service.

U.S. Fish and Wildlife Service. 2002. *2001 National Survey of Fishing, Hunting, and Wildlife-Associated Recreation*. Washington, DC: U.S. Fish and Wildlife Service.

U.S. Fish and Wildlife Service. 2007. *2006 National Survey of Fishing, Hunting, and Wildlife-Associated Recreation*. Washington, DC: U.S. Fish and Wildlife Service.

U.S. Fish and Wildlife Service. 2010. U.S. Fish and Wildlife Service national hunting license report. http://wsfrprograms.fws.gov/Subpages/LicenseInfo/ HuntingLicCertHistory.pdf. Accessed March 30, 2010.

VanGilder, C. L., G. R. Woods, and K. V. Miller. 2009. Effects of an intensive predator removal on white-tailed deer recruitment in northeastern Alabama. *Proceedings of the Annual Conference of the Southeastern Association of Fish and Wildlife Agencies* 63:11–16.

Van Deelen, T. R., B. Dhuey, K. R. McCaffery and R. E. Rolley. 2006. Relative effects of baiting and supplemental antlerless seasons on Wisconsin's 2003 deer harvest. *Wildlife Society Bulletin* 34:322–328.

Van Putten, M., J. R. Clark, P. Kavits, and G. Chavarria. 2002. *A Guide to Reporting on Chronic Wasting Disease*. Reston, VA: National Wildlife Federation.

Vreeland, J. K., D. R. Diefenbach, and B. D. Wallingford. 2004. Survival rates, mortality causes, and habitats of Pennsylvania white-tailed deer fawns. *Wildlife Society Bulletin* 32:542–553.

Williams, A. F. and J. K. Wells. 2005. Characteristics of vehicle–animal crashes in which vehicle occupants are killed. *Traffic Injury Prevention* 6:56–59.

Wright, B. A., R. A. Kaiser, and S. Nicholls. 2002. Rural landowner liability for recreational injuries: Myths, perceptions and realities. *Journal of Soil and Water Conservation* 57:183–191.

Appendix: Scientific Names of Plants and Animals Referenced in the Text

Animals

Common Name	Scientific Name
Alpaca	*Lama pacos*
Antelope, blackbuck	*Antilope cervicapra*
Antelope, roan	*Hippotragus equinus*
Aoudad	*Ammotragus lervia*
Badger	*Taxidea taxus*
Badger, European	*Meles meles*
Bear, black	*Ursus americanus*
Bear, grizzly	*Ursus arctos*
Beaver	*Castor canadensis*
Bison	*Bison bison*
Bison, American	*Bison bison*
Blackbuck	*Antilope cervicapra*
Boa	*Boa* sp.
Boar, wild	*Sus scrofa*
Bobcat	*Lynx rufus*
Buffalo, African	*Syncerus caffer*
Buffalo, Asiatic water	*Bubalis bubalis*
Camel, dromedary	*Camelus dromedarius*
Caribou	*Rangifer tarandus*
Cattle	*Bos* spp.
Cattle, domestic	*Bos taurus*
Chamois	*Rupicapra rupicapra*
Chevrotain	*Moschiola meminna, Hyemoschus aquaticus*
Chipmunk, eastern	*Tamias striatus*
Chital	*Axis axis*
Cougar	*Puma concolor*
Coyote	*Canis latrans*
Crocodile	*Crocodylus* spp.
Deer, axis	*Axis axis*
Deer, black-tailed	*Odocoileus hemionus columbianus, O. h. sitkensis*
Deer, brocket	*Mazama* spp.
Deer, Chinese water	*Hydropotes inermis*
Deer, Columbian black-tailed	*Odocoileus hemionus columbianus*
Deer, Columbian white-tailed	*Odocoileus virginianus leucurus*
Deer, desert mule	*Odocoileus hemionus crooki*
Deer, fallow	*Dama dama*
Deer, Key	*Odocoileus virginianus clavium*

continued

(continued)

Common Name	Scientific Name
Deer, marsh	*Blastocerus dichotomus*
Deer, mouse	*Tragulus* spp.
Deer, mule	*Odocoileus hemionus*
Deer, musk	*Moschus* spp.
Deer, pampas	*Ozotoceros bezoarticus*
Deer, Peré David's	*Elaphurus davidianus*
Deer, red	*Cervus elaphus*
Deer, red brocket	*Mazama americana, M. rufina, M. temama*
Deer, roe	*Capreolus capreolus*
Deer, sika	*Cervus nippon*
Deer, tufted	*Elaphodus cephalophus*
Deer, white-tailed	*Odocoileus virginianus*
Dog	*Canis lupus familiaris*
Dove, mourning	*Zenaida macroura*
Eagle, golden	*Aquila chrysaetos*
Elephant, African	*Loxodonta africana*
Elephant, Asian	*Elephas maximus*
Elk	*Cervus elaphus*
Elk, North American	*Cervus elaphus*
Elk, Roosevelt	*Cervus elaphus roosevelti*
Fox, gray	*Urocyon cinereoargenteus*
Fox, swift	*Vulpes velox*
Foxes	*Vulpes* spp., *Urocyon* sp.,
Goat, domestic	*Capra hircus*
Goat, mountain	*Oreamnos americanus*
Ground sloth, Rusconi's	*Eremotherium rusconii*
Hare, arctic	*Lepus arcticus*
Hares	*Lepus* spp.
Hartebeest	*Alcelaphus buselaphus*
Hippopotamus	*Hippopotamus amphibius*
Hog, feral	*Sus scrofa*
Horses	*Equus* spp.
Huemul	*Hippocamelus bisulcus*
Jackrabbit, black-tailed	*Lepus californicus*
Jaguar	*Panthera onca*
Kudu, greater	*Tragelaphus strepsiceros*
Lion, mountain	*Puma concolor*
Llamas	*Lama glama*
Lynx	*Lynx canadensis*
Magpie-jay	*Calocitta formosa*
Mammoth, Columbian	*Mammuthus columbi*
Mammoth, woolly	*Mammuthus primigenius*
Mastodon, American	*Mammut americanum*
Moose	*Alces alces*
Mouflon	*Ovis orientalis orientalis*
Mouse, deer	*Peromyscus maniculatus*
Mouse, white-footed	*Peromyscus leucopus*
Muntjac	*Muntiacus* spp.
Muskoxen	*Ovibos moschatus*

(continued)

Common Name	Scientific Name
Nilgai	*Boselaphus tragocamelus*
Ocelot	*Leopardus pardalis*
Opossum	*Didelphis virginiana*
Oryx, Arabian	*Oryx leucoryx*
Otter, river	*Lontra canadensis*
Peccary, collared	*Pecari tajacu*
Peccary, white-lipped	*Tayassu pecari*
Pigeon, passenger	*Ectopistes migratoriu*
Possum, brush-tailed	*Trichosurus vulpecula*
Pronghorn	*Antilocapra americana*
Pudu	*Pudu* spp.
Rabbit, cottontail	*Sylvilagus* spp.
Rabbit, swamp	*Sylvilagus aquaticus*
Raccoon	*Procyon lotor*
Reindeer	*Rangifer tarandus*
Rhinos	Family Rhinocerotidae
Saiga	*Saiga* spp.
Salamander, red-backed	*Plethodon cinereus*
Sambar	*Rusa unicolor*
Sheep	*Ovis* spp.
Sheep, Barbaro	*Ovis aries*
Sheep, bighorn	*Ovis canadensis*
Sheep, desert bighorn	*Ovis canadensis nelsoni*
Sheep, domestic	*Ovis aries*
Shrew, short-tailed	*Blarina brevicauda*
Skunk, striped	*Mephitis mephitis*
Snake, garter	*Thamnophis sirtalis*
Squirrel	Family *Sciuridae*
Squirrel, fox	*Sciurus niger*
Squirrel, gray	*Sciurus carolinensis*
Swine, feral	*Sus scrofa*
Tapirs	*Tapirus* spp.
Taruka	*Hippocamelus antisensis*
Turkey, wild	*Meleagris gallopavo*
Vole, red-backed	*Clethrionomys gapperi*
Wildebeest	*Connochaetes* spp.
Wildebeest	*Connochaetes gnou*
Wolf	*Canis lupus*
Wolf, gray	*Canis lupus*
Wolf, Mexican	*Canis lupus baileyi*
Wolf, red	*Canis lupus*
Wolf, timber	*Canis lupus*
Woodchuck	*Marmota monax*
Yak	*Bos grunniens*
Zebra	*Equus zebra*

Plants

Common Name	Scientific Name
Alder	*Alnus* spp.
Alder, red	*Alnus rubra*
Alfalfa	*Medicago sativa*
Arborvitae	*Thuja occidentalis*
Ash	*Fraxinus* spp.
Ash, American mountain	*Sorbus americana*
Ash, green	*Fraxinus pennsylvanica*
Aspen	*Populus* spp.
Aspen, quaking	*Populus tremuloides*
Beadlily, queencup	*Clintonia uniflora*
Beech	*Fagus grandifolia*
Beech, American	*Fagus grandifolia*
Birch	*Betula* spp.
Birch, paper	*Betula papyrifera*
Birch, yellow	*Betula alleghaniensis*
Blackberry, Allegheny	*Rubus allegheniensis*
Blackhaw	*Viburnum prunifolium*
Blueberry	*Vaccinium* spp.
Bluestem, broomsedge	*Andropogon viginicus*
Boxelder	*Acer negundo*
Buffaloberry, silver	*Shepherdia argentea*
Buffelgrass	*Pennisetum ciliare*
Bursage, white	*Ambrosia dumosa*
Cactus, barrel	*Ferocactus* spp.
Ceanothus, redstem	*Ceanothus sanguineus*
Cedar	*Juniperus* spp.
Cedar, eastern red	*Juniperus virginiana*
Cedar, northern white	*Thuja occidentalis*
Cedar, tropical	*Cedrela fissilis*
Cedar, western red	*Thuja plicata*
Cherry, black	*Prunus serotina*
Choke cherry	*Prunus virginiana*
Chokecherry, common	*Prunus virginiana*
Ciruelo	*Spondias purpurea*
Cloudberry	*Rubus chamaemorus*
Clover	*Trifolium* spp.
Corn	*Zea mays*
Cotton	*Gossypium* spp.
Cotton	*Gossypium hirsutum*
Cottonwood	*Populus deltoides*
Cottonwood	*Populus* spp.
Cottonwood, eastern	*Populus deltoides*
Cowpeas	*Vigna* spp.
Creosote	*Larrea tridentata*
Cypress	*Cupressus* spp.

(continued)

Common Name	Scientific Name
Dogwood	*Cornus* spp.
Dogwood, alternate leaf	*Cornus alternifolia*
Elderberry	*Sambucus* spp.
Elm	*Ulmus* spp.
Elm, American	*Ulmus americana*
Farkleberry	*Vaccinium arboreum*
Fern, hay-scented	*Dennstaedia punctilobula*
Fir	*Abies* spp.
Fir, balsam	*Abies balsamea*
Fir, Douglas	*Pseudotsuga menziesii*
Fir, grand	*Abies grandis*
Fireweed	*Epilobium angustifolium*
Foxtail	*Alopecurus* spp.
Frailejón	*Espeletia* spp.
Ginseng, American	*Panax quinquefolius*
Grama	*Boutelous* spp.
Grama, black	*Boutelous eriopoda*
Grape	*Vitis* spp.
Grape, Oregon	*Mahonia aquifolium*
Guajillo	*Acacia berlandieri*
Gum	*Liquidambar* spp., *Nyssa* spp., *Eucalyptus* spp.
Hackberry	*Celtis* spp.
Hazelnut	*Corylus* spp.
Hemlock, eastern	*Tsuga canadensis*
Hemlock, western	*Tsuga heterophylla*
Hickory	*Carya* spp.
Holly, deciduous	*Ilex decidua*
Honeylocust	*Gleditsia triacanthos*
Huisache	*Acacia farnesiana*
Ironwood	*Eusideroxylon* spp.
Juniper	*Juniperus* spp.
Juniper, Ashe	*Juniperus ashei*
Juniper, redberry	*Juniperus pinchotii*
Juniper, Rocky Mountain	*Juniperus scopulorum*
Knapweed, spotted	*Centaurea stoebe*
Lablab	*Lablab* spp.
Lambsquarter	*Chenopodium album*
Lechuguilla	*Agave lechuguilla*
Leucaena	*Leucaena* spp.
Lotebush	*Ziziphus obtusifolia*
Lovegrass, Lehman	*Eragrostis lehmanniana*
Madrone	*Arbutus* spp.
Mahogany, mountain	*Cercocarpus* spp.
Maple	*Acer* spp.
Maple, mountain	*Acer spicatum*
Maple, Rocky Mountain	*Acer glabrum*
Maple, red	*Acer rubrum*
Maple, sugar	*Acer saccharum*
Mariola	*Parthenium incanum*

continued

(continued)

Common Name	Scientific Name
Mesquite grass, curly	*Hilaria* spp.
Mesquite, honey	*Prosopis glandulosa*
Mesquite, velvet	*Prosopis velutina*
Mombin, red	*Spondias purpurea*
Mustard, garlic	*Alliara petiolata*
Ninebark	*Physocarpus inalvaceus*
Ninebark	*Physocarpus* spp.
Oak	*Quercus* spp.
Oak, bur	*Quercus macrocarpa*
Oak, live	*Quercus virginiana*
Oak, Mexican blue	*Quercus oblongifolia*
Oak, northern red	*Quercus rubra*
Oak, scrub	*Quercus berberidifolia*
Oak, shin	*Quercus havardii*
Oak, Shumard	*Quercus shumardii*
Oak, swamp chestnut	*Quercus michauxii*
Oak, water	*Quercus nigra*
Oak, white	*Quercus alba*
Oak, willow	*Quercus phellos*
Oat	*Avena* spp.
Ocotillo	*Fouquieria splendens*
Palo verde	*Parkinsonia* spp.
Pecan	*Carya illinoinensis*
Persimmon, common	*Diospyros virginiana*
Persimmon, Texas	*Diospyros texana*
Pine	*Pinus* spp.
Pine, jack	*Pinus banksiana*
Pine, loblolly	*Pinus taeda*
Pine, lodgepole	*Pinus contorta*
Pine, longleaf	*Pinus palustris*
Pine, pinyon	*Pinus edulis, P. monophylla*
Pine, ponderosa	*Pinus ponderosa*
Pine, sand	*Pinus clausa*
Pine, short leaf	*Pinus echinata*
Pine, slash	*Pinus elliottii*
Pricklypear	*Opuntia* spp.
Red alder	*Alnus rubra*
Redbud	*Cercis canadensis*
Rockspirea, creambush	*Holodiscus dumosus*
Rose	*Rosa* spp.
Rye	*Secale cereale*
Sagebrush, big	*Artemisia tridentata*
Saguaro	*Carnegiea gigantea*
Salal	*Gaultheria shallon*
Saltbush, fourwing	*Atriplex canescens*
Serviceberry	*Amelanchier* spp.
Serviceberry, western	*Amelanchier alnifolia*
Snowberry	*Symphoricarpus* spp.
Snowberry western	*Symphoricarpos occidentalis*
Snowbrush	*Breynia disticha*

(continued)

Common Name	Scientific Name
Soapberry, western	*Sapindus saponaria drummondii*
Sorghum	*Sorghum bicolor*
Sotol	*Dasylirion* spp.
Soybean	*Glycine max*
Spicebush	*Lindera* spp.
Spruce	*Picea* spp.
Spruce, black	*Picea mariana*
Spruce, sitka	*Picea sitchensis*
Spruce, white	*Picea glauca*
Stiltgrass, Japanese	*Microstegium vimineum*
Strawberry bush	*Euonymus kiautschovicus, E. obovatus*
Sumac	*Rhus* spp.
Sunflower	*Helianthus annuus*
Switchgrass	*Panicum virgatum*
Tarbush	*Holocarpha virgata*
Tobosagrass	*Pleuraphis mutica*
Trillium, red	*Trillium erectum*
Trillium, white	*Trillium grandiflorum*
Trillium, white-flowered	*Trillium grandiflorum*
Vetch	*Vicia* spp.
Wheat	*Triticum* spp.
Wheat, spring	*Triticum durum*
Wheat, winter	*Triticum aestivum*
White brush	*Aloysia gratissima*
Willow	*Salix* spp.
Willow, desert	*Chilopsis linearis*
Windmill grass, hooded	*Chloris cucullata*
Yaupon	*Ilex vomitoria*
Yew, Canada	*Taxus canadensis*

Index

A

B

E

F

M

P

Q

R

T

U

V

W